FOURTH EDITION

TECHNOLOGY OF MACHINE TOOLS

STEVE F. KRAR
J. WILLIAM OSWALD

GLENCOE

Macmillan/McGraw-Hill

New York, New York
Columbus, Ohio
Mission Hills, California
Peoria, Illinois

Production *York Production Services*
Cover Photograph *Gill C. Kenny, The Image Bank, Inc.*
Cover Design *Ed Smith Design*

Library of Congress Cataloging-in-Publication Data

Krar, Stephen F.
Technology of machine tools / S.F. Krar, J.W. Oswald.—4th ed. p. cm.
Includes index.
ISBN 0-07-035563-0
1. Machine-tools 2. Machine-shop practice. I. Oswald, John William. II. Title.
TJ1185.K668 1990
621.9'02—dc20 89-12112
CIP

TECHNOLOGY OF MACHINE TOOLS, FOURTH EDITION

Imprint 1993

Send all inquiries to:
GLENCOE DIVISION
Macmillan/McGraw-Hill
936 Eastwind Drive
Westerville, Ohio 43081

5 6 7 8 9 10 11 12 13 14 15 VH 00 99 98 97 96 95 94 93

ISBN 0-07-035563-0

CONTENTS

PREFACE *ix*
About the Authors *x*
Acknowledgments *xi*

SECTION 1 Introduction to Machine Tools **1**

UNIT **1** History of Machines **2**

SECTION 2 Machine Trade Opportunities **11**

UNIT **2** Careers in the Metalworking Industry **12**
3 Getting the Job **17**

SECTION 3 Safety **21**

UNIT **4** Safety in the Machine Shop **22**

SECTION 4 Job Planning **27**

UNIT **5** Engineering Drawings **28**
6 Machining Procedures for Various Workpieces **32**

SECTION 5 Measurement **39**

UNIT **7** Basic Measurement **42**
8 Squares and Surface Plates **47**
9 Micrometers **51**
10 Vernier Calipers **59**
11 Inside-, Depth-, and Height-Measuring Instruments **63**
12 Gage Blocks **72**
13 Angular Measurement **77**
14 Gages **82**
15 Comparison Measurement **88**
16 The Coordinate Measuring System **97**

17 Measuring with Light Waves 101

18 Surface Finish Measurement 105

SECTION 6 Layout Tools and Procedures 111

UNIT 19 Basic Layout Materials, Tools, and Accessories 112

20 Basic or Semiprecision Layout 119

21 Precision Layout 122

SECTION 7 Hand Tools and Bench Work 129

UNIT 22 Holding, Striking, and Assembling Tools 130

23 Hand-Type Cutting Tools 135

24 Thread-Cutting Tools and Procedures 141

25 Finishing Processes—Reaming, Broaching, and Lapping 146

26 Bearings 152

SECTION 8 Metal Cutting Technology 157

UNIT 27 Physics of Metal Cutting 158

28 Machinability of Metals 163

29 Cutting Tools 168

30 Operating Conditions and Tool Life 178

31 Carbide Cutting Tools 181

32 Diamond, Ceramic and Cermet Cutting Tools 196

33 Polycrystalline Cutting Tools 207

34 Cutting Fluids—Types and Applications 214

SECTION 9 Metal-Cutting Saws 225

UNIT 35 Types of Metal Saws 226

36 Contour Bandsaw Parts and Accessories 230

37 Contour Bandsaw Operations 236

SECTION 10 Drilling Machines 247

UNIT 38 Drill Presses 248

39 Drilling Machine Accessories 252

40 Twist Drills 259

41 Cutting Speeds and Feeds 268

42 Drilling Holes 271

43 Reaming 278

44 Drill Press Operations 284

SECTION 11 The Lathe 293

UNIT 45 Engine Lathe Parts 297

46 Lathe Accessories 301

47 Cutting Speeds, Feeds, and Depth of Cut 311

48 Lathe Safety 316

49 Mounting, Removing, and Aligning Lathe Centers 318

50 Grinding Lathe Cutting Tools 321

51 Facing Between Centers 324

52 Machining Between Centers 328

53 Knurling, Grooving, and Form Turning 335

54 Tapers and Taper Turning 341

55 Threads and Thread Cutting 352

56 Steady Rests, Follower Rests, Mandrels 371

57 Machining in a Chuck 376

58 Drilling, Boring, Reaming, and Tapping 386

SECTION 12 Milling Machines 391

UNIT 59 Milling Machines and Accessories 392

60 Milling Cutters 400

61 Cutting Speeds, Feeds, and Depth of Cut 407

62 Milling Machine Setups 415

63 Milling Operations 421

64 The Indexing or Dividing Head 426

65 Gears 435

66 Gear Cutting 441

67 Helical Milling 448

68 Cam, Rack, Worm, And Clutch Milling 465

69 The Vertical Milling Machine—Construction and Operation 473

70 Special Milling Operations 483

SECTION 13 The Jig Borer and Jig Grinder 491

UNIT 71 The Jig Borer 492

72 The Jig-Boring Holes 498

73 The Jig Grinder 509

SECTION 14 Computer-Age Machining 519

UNIT 74 The Computer 520

75 Numerical Control 523

76 Computer-Aided Design 539

77 Chucking and Turning Centers 543

78 Machining Centers 548

79 Robotics 554

80 Manufacturing Systems 560

81 Factories of the Future 564

SECTION 15 Grinding 569

UNIT 82 Types of Abrasives 570

83 Surface Grinders and Accessories 587

84 Surface-Grinding Operations 598

85 Cylindrical Grinders 608

86 Universal Cutter and Tool Grinder 618

SECTION 16 Metallurgy 631

UNIT 87 Manufacture and Properties of Steel 632

88 Heat Treatment of Steel 643

89 Testing of Metals and Nonferrous Metals 659

SECTION 17 Hydraulics 671

UNIT 90 Hydraulic Circuits and Components 672

SECTION **18** Special Processes **683**

UNIT **91** Electro-Chemical Machining and Electrolytic Grinding **684**

92 Electrical Discharge Machining **690**

93 Forming Processes **700**

94 The Laser **710**

Index 735

PREFACE

No single invention since the Industrial Revolution has left such an impact on society as the computer. During the last two or three decades, basic computers have been applied to machine tools to program and control various machine operations. Computers have steadily improved until there are now highly sophisticated units capable of controlling the operation of a single machine, a group of machines, or soon even a complete manufacturing plant, therefore a section titled "Computer-Age Machining" is included in this edition. New machine tools and processes have been developed to reduce the large production gap between North American and overseas workers. In order for these new machine tools to reach their full potential, new cutting tools are being continually developed to produce accurate parts more quickly and at competitive prices. With this in mind, the authors have included and expanded machining processes such as Flexible Manufacturing Systems and added new cutting tools and materials such as Polycrystalline Cubic Boron Nitride, Polycrystalline Diamond, and SG Ceramic Aluminum Oxide.

This book is based on the authors' many years of practical experience as skilled workers in the trade and as specialists in teaching. To keep abreast of rapid technological change, the authors have researched the latest technical information available, and have visited industries which are leaders in their field. Many sections of this book were reviewed by key personnel in various manufacturing firms, so that the most accurate and up-to-date information is presented. The authors appreciated the opportunity to incorporate into the text the suggestions and recommendations made by these people.

The fourth edition of *Technology of Machine Tools* is presented in unit form with each unit introduced with a set of objectives followed by related theory and operational sequence. Each operation is explained in a step-by-step procedure which students can readily follow. Advanced operations are introduced by problems, followed by step-by-step solutions and matching procedures. Review questions at the end of each unit can be used for review or for homework assignments to prepare students for subsequent operations. To make this text easily understood, each unit contains many new illustrations and photographs. Color has been used throughout to emphasize important points and to make the illustrations more meaningful.

Throughout the text, dual dimensioning (U.S. Customary: inch; SI: metric) is used. The need for a working knowledge of the metric system cannot be overstressed, because most of the countries in the world are already on or in the process of converting to the metric system. Since those involved in the machine shop trade will have to be familiar with both the metric and inch systems during the changeover period, wherever possible, metric tools and information are included. All inch dimensions or quantities in this book have been generally rounded to the nearest metric equivalent, which is shown in brackets following each dimension or quantity.

The purpose of this text is to assist instructors in giving students basic training in the operation of machine tools, and in helping students understand the latest machining processes and developments. The material is organized so that instructors may easily select those topics that are most suitable for class projects, or that suit the individual differences of students.

A technician in the machine shop trade should be neat, develop sound work habits, and have a good knowledge of mathematics, print reading, and computers. To keep up to date on technological changes, technicians must continue to expand their knowledge by reading specialized texts, trade literature, and magazine articles in this field.

Steve F. Krar
J. William Oswald

About the Authors

STEVE F. KRAR majored in Machine Shop Practice and spent fifteen years in the trade, first as a machinist and finally as a tool and diemaker. After this period he entered Teachers' College and graduated from the University of Toronto with a Specialist's Certificate in Machine Shop Practice. During his nineteen years of teaching, Mr. Krar was active in Vocational and Technical education and served on the executive of many educational organizations. For ten years he was on the summer staff of the University of Toronto, involved in teacher training programs. Active in machine tool associations, Steve Krar has been a senior member of the Society of Manufacturing Engineers for over thirty years. He is a co-author of the McGraw-Hill text *Machine Tool Operations.* He is also co-author of the following McGraw-Hill Ryerson Ltd. publications: *Machine Shop Training, Machine Shop Operations,* and the overhead transparency kits: *Machine Tools, Measurement and Layout, Threads and Testing Equipment,* and *Cutting Tools.*

J. WILLIAM OSWALD served an apprenticeship in machine shop, and after sixteen years in the trade attended Teachers' College at the University of Toronto. After graduation, he received a Specialist's Certificate in Machine Shop Practice and taught machine shop work for twenty-five years. During this time he attended several up-grading courses in the operation of the latest machine shop and testing equipment. For several years Mr. Oswald served on the teacher-training staffs at the University of Toronto and the University of Western Ontario. He had also worked with various technical educational committees and organizations. He is a co-author of the McGraw-Hill text *Machine Tool Operations.* Mr. Oswald is also a co-author of *Machine Shop Operations,* and the overhead transparency kits: *Machine Tools, Measurement and Layout, Threads and Testing Equipment,* and *Cutting Tools.*

Acknowledgments

The authors wish to express their sincere thanks and appreciation to Alice H. Krar for her untiring devotion in reading, typing, and checking the manuscript for this text. Without her supreme efforts, this text could not have been produced.

Special thanks are due to Mr. H. Bacsu, Westlane Secondary School, Niagara Falls; Don Matthews, formerly of San Joaquin Delta College, Stockton, California; and to all the teachers who offered suggestions which we were happy to include.

Our deep thanks go to the following firms which reviewed sections of the manuscript and offered suggestions which were incorporated to make this text as accurate and up to date as possible: American Superior Electric Co., Bendix Corporation, FAG Bearing Co. Ltd., Federal Products Corporation, General Electric Co., Norton Co., and Moore Special Tool Co. Cincinnati Milacron Inc. was most helpful in offering advice and in reviewing several sections on milling and special processes.

We are grateful to the following firms who have assisted in the preparation of this text by supplying illustrations and technical information.

Allen Bradley Co.
Allen, Chas. G. & Co.
American Chain and Cable Co. Inc., Wilson Instrument Division
"American Machinist"
American Superior Electric Co. Ltd.
Ametek Testing Equipment
Armstrong Bros. Tool Co.
Ash Precision Equipment Inc.
Atlas Press Co., Clausing Division
Avco-Bay State Abrasive Company
Bausch & Lomb Inc.
Bendix Corporation, Automation Group
Bethlehem Steel Corporation
Bickle Co. H. W.
Boston Gear Works
Bridgeport Machines Division of Textron Inc.
Brown & Sharpe Manufacturing Co.
Buffalo Forge Co.
Butterfield Division, Union Twist Drill Co.
Canadian Blower and Forge Co. Ltd.
Canadian Tap and Die Co. Ltd.
Carborundum Company
Cincinnati Lathe and Tool Co.
Cincinnati Milacron Inc.
Cincinnati Shaper Co.
Clausing Corporation
Cleveland Tapping Machine Co.
Cleveland Twist Drill (Canada) Ltd.
Colchester Lathe & Tool Co.
Coleman Engineering Co. Inc.
Concentric Tool Corp.
Covel Manufacturing Co.
Delmar Publishers Inc.
Delta File Works
DeVlieg Machine Co.
Dillon, W: C. and Co. Inc.
DoALL Company
Elliott Machine Tools
Emco Maier Corp.
Enco Manufacturing Co.
Everett Industries Inc.
Ex-Cell-O Corp.
Explosive Fabricators Division, Tyco Corp.
FAG Bearing Co. Ltd.
Federal Products Corp.
Firth-Brown Tools (Canada) Ltd.
Frontier Equipment Ltd.
General Electric Co. Ltd.
General Motors of Canada
Greenfield Tap and Die Co.
Hones, Charles H. Inc.
Inland Steel Co.
Jacobs Manufacturing Co.
Jones and Lamson Division of Waterbury Farrel
Kaiser Steel Corp.
Kostel Enterprises Ltd.
Kalamazoo Machine Tool Co.
KTS Industries
LeBlond, R. K. Machine Tool Co.
Lionite Abrasives Ltd.
Lodge & Shipley

Mahr Gage Co. Inc.
Mitutoyo
Mobil Oil Corporation
Moore Special Tool Co.
Morse Twist Drill and Machine Co.
MTI Corporation
National Broach & Machine Division, Lear Siegler Inc.
National Twist Drill & Tool Co.
Neill, James & Co. (Sheffield) Ltd.
Nicholson File Co. of Canada Ltd.
Norton Company of Canada Ltd.
Powder Metallurgy Parts Manufacturers' Association
Pratt & Whitney Co. Inc., Machine Tool Division
Precision Diamond Tool Co.
Retor Developments Ltd.
Rockford Machine Tool Co.
Rockwell International
Shore Instrument & Mfg. Co. Inc.
Slocomb, J. T. Co.
South Bend Lathe Inc.
Standard-Modern Tool Co. Ltd.
Standard Tool Co.
Stanley Tools Division, Stanley Works
Starrett, L. S. Co.
Sun Oil Co.
Taft-Peirce Manufacturing Co.
Taper Micrometer Corp.
Thompson Grinder Co.
Union Carbide Corp., Linde Division
United States Steel Corporation
Volstro Manufacturing Co. Inc.
Watts Bros. Tool Works
Weldon Tool Co.
Wells Manufacturing Corp.
Whitman & Barnes
Wickman A. C. Ltd.
Wilkie Brothers Foundation
Williams, A. R., Machinery Co. Ltd.
Williams, J. H. & Co.
Wilton Machinery Co.
Woodworth, W. J., and J. D. Woodworth

SECTION 1

Introduction to Machine Tools

The progress of humanity throughout the ages has been governed by the types of tools available. Ever since primitive people used rocks as hammers or as weapons to kill animals for food, tools have governed our standard of living. The use of fire to extract metals from ore led to the development of newer and better tools. The harnessing of water led to the development of hydropower, which greatly improved humanity's well-being.

With the industrial revolution in the mid-18th century, early machine tools were developed and were continually improved. The development of machine tools and related technologies advanced rapidly during and immediately after World Wars I and II. Since World War II, processes such as numerical control, electro-machining, computer-aided design (CAD), computer-aided manufacturing (CAM), and flexible manufacturing systems (FMS) greatly altered manufacturing methods.

Today we are living in a society greatly affected by the development of the computer. Computers affect the growing and sale of food, manufacturing processes, and even entertainment. Although the computer influences our everyday lives, it is still important that you as a student or apprentice be able to perform basic operations on standard machine tools. This knowledge will provide the necessary background for a person seeking a career in the machine tool trade.

UNIT 1

History of Machines

The high standard of living we enjoy today did not just happen. It has been the result of the development of highly efficient machine tools over the past several decades. Processed foods, automobiles, telephones, televisions, refrigerators, clothing, books, and practically everything else we use are produced by machinery.

OBJECTIVES

After completing this unit you will be familiar with:

1. The development of tools throughout history
2. The standard types of machine tools used in shops
3. The newly developed space-age machines and processes

The history of machine tools began during the stone age (over 5000 years ago), when the only tools were hand tools made of wood, animal bones, or stone (Fig. 1-1).

Between 2000 and 2500 B.C., stone spears and axes were replaced with copper and bronze implements and power supplied by humans was in a few cases replaced with animal power. It was during this bronze age that human beings first enjoyed "power-operated" tools.

Around 1000 B.C., the iron age dawned, and most bronze tools were replaced with more durable iron implements. After smiths learned to harden and temper iron, its use became widespread. Tools and weapons were greatly improved, and animals were domesticated to provide power for some of these tools, such as the plow. During the iron age, all commodities required by humans, such as housing and shipbuilding materials, wagons, and furniture, were handmade by the skilled craftsmen of that era.

About 300 years ago, the iron age became the machine age. In the 17th century, people began exploring new sources of energy. Water power began to replace human and animal power. With this new power came improved machines and, as production increased, more products became available. Machines continued to be improved, and the boring machine made it possible for James Watt to produce the first steam engine in 1776, beginning the industrial revolution. The steam engine made it possible to provide power to any area where it was needed. With quickening speed, machines were improved and new ones invented. Newly designed pumps reclaimed thousands of acres of the Netherlands from the sea. Mills and plants which had depended on water power were converted to steam power to produce flour, cloth, and lumber more efficiently. Steam engines replaced sails and steel replaced wood in the shipbuilding industry. Railways sprang up, unifying countries, and steamboats connected the continents. Steam-driven tractors and improved farm machinery lightened the farmer's task. As machines improved, further sources of power were developed. Generators were made to produce electricity and diesel and gasoline engines were developed.

With further sources of energy available, industry grew and new and better machines were built. Progress continued slowly during the first part of the 20th century except for spurts during the two world wars. The Second World War sparked an urgent need for new and better machines, which resulted in more efficient production (Fig. 1-2).

Since the 1950s, progress has been rapid and we are now in the space age. Calculators, computers, robots, and automated machines and plants are commonplace. The atom has been harnessed and nuclear power is used to produce electricity and to drive ships. We have traveled to the moon and outer space, all because of fantastic technological developments. Machines can mass produce parts to millionths of an inch accuracy. The fields of measurement, machining, and metallurgy have become very sophisticated. All these factors have produced a high standard of living for us. All

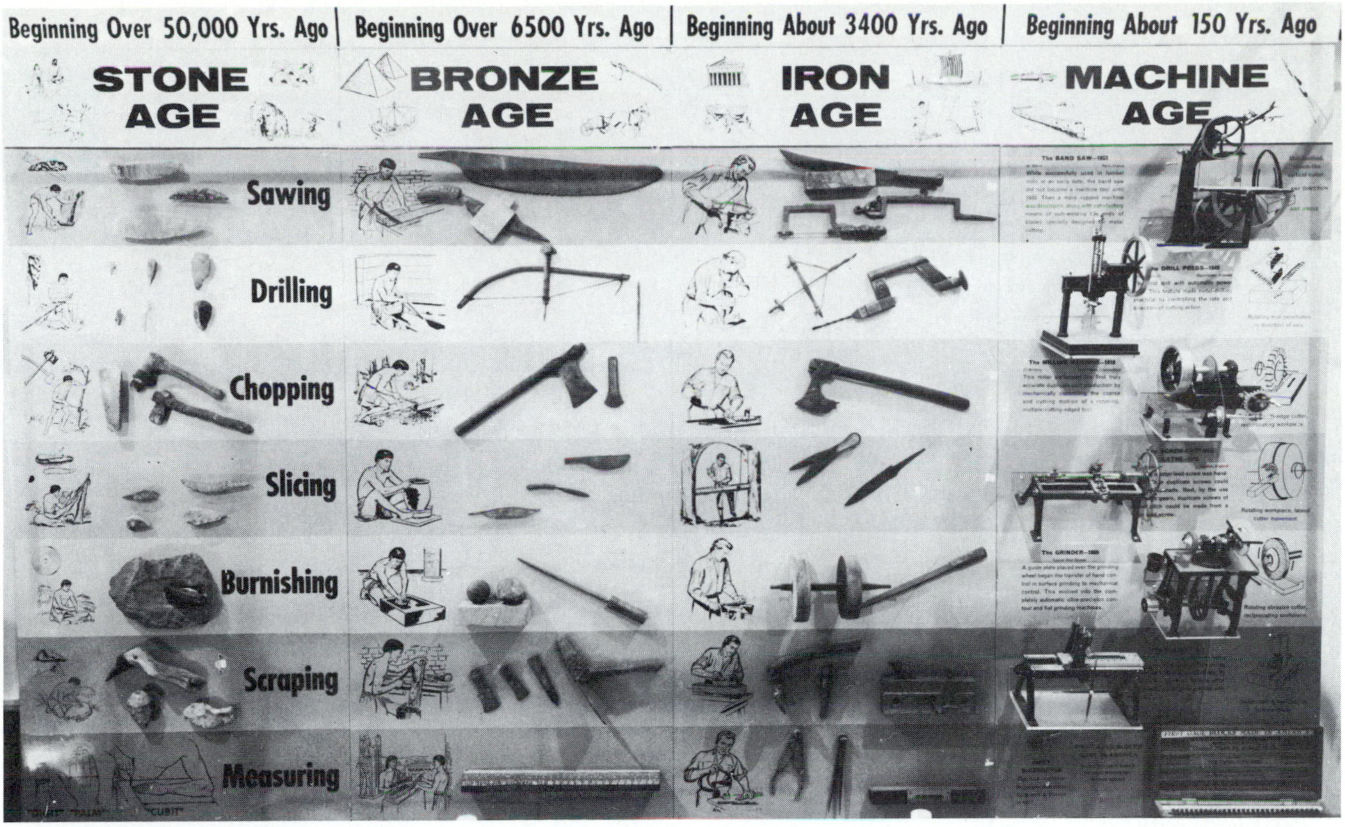
Beginning Over 50,000 Yrs. Ago
Beginning Over 6500 Yrs. Ago
Beginning About 3400 Yrs. Ago
Beginning About 150 Yrs. Ago
STONE AGE
BRONZE AGE
IRON AGE
MACHINE AGE
Sawing
Drilling
Chopping
Slicing
Burnishing
Scraping
Measuring

Figure 1-1 *The development of hand tools over the years. (Courtesy DoAll Company.)*

Figure 1-2 *New machine tools were developed during the mid-20th century. (Courtesy DoAll Company.)*

of us, regardless of our occupation or status, are dependent on machines and/or their products (Fig. 1-3).

Through constant improvement, modern machine tools have become more accurate and efficient. Improved production and accuracy have been made possible through the application of hydraulics, pneumatics, fluidics, and electronic devices such as numerical control to basic machine tools.

COMMON MACHINE TOOLS

Machine tools are generally power-driven metal-cutting or -forming machines used to shape metals by:

- The removal of chips
- Pressing, drawing, or shearing
- Controlled electrical machining processes

Any machine tool generally has the capability of:

- Holding and supporting the workpiece
- Holding and supporting a cutting tool
- Imparting a suitable movement (rotating or reciprocating) to the cutting tool or the work
- Feeding the cutting tool or the work so that the desired cutting action and accuracy will be achieved

The machine tool industry is divided into several different categories, such as the general machine shop, the tool-

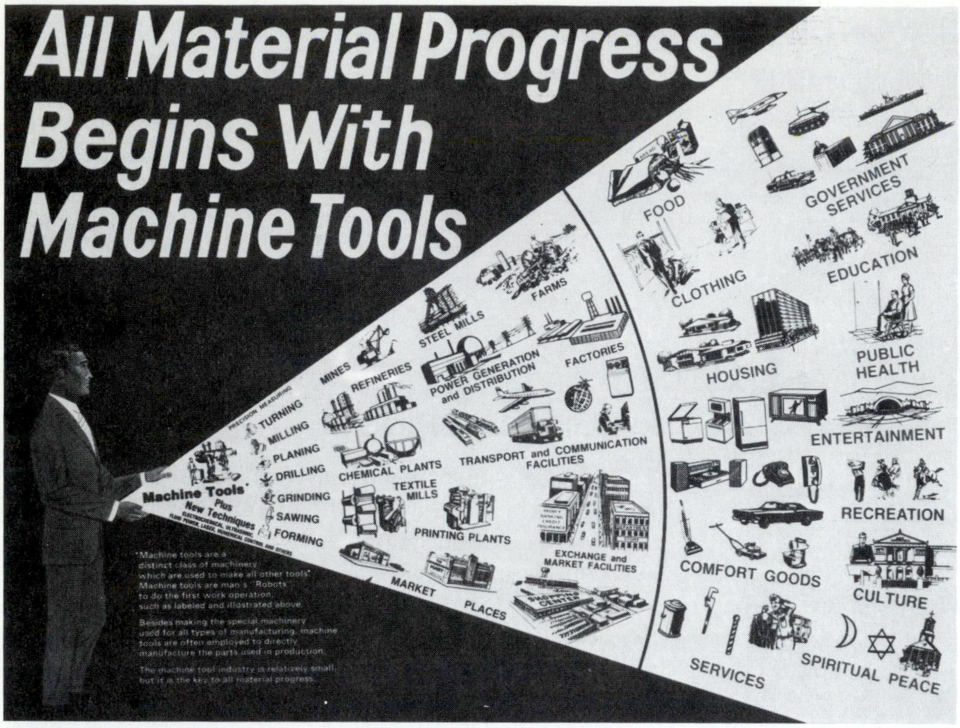
All Material Progress
Begins With
Machine Tools
Machine Tools
Plus
New Techniques
TURNING
MILLING
PLANING
DRILLING
GRINDING
SAWING
FORMING
MINES
REFINERIES
STEEL MILLS
FARMS
POWER GENERATION and DISTRIBUTION
FACTORIES
CHEMICAL PLANTS
TRANSPORT and COMMUNICATION FACILITIES
TEXTILE MILLS
PRINTING PLANTS
EXCHANGE and MARKET FACILITIES
MARKET
PLACES
FOOD
CLOTHING
GOVERNMENT SERVICES
EDUCATION
HOUSING
PUBLIC HEALTH
ENTERTAINMENT
RECREATION
COMFORT GOODS
CULTURE
SERVICES
SPIRITUAL PEACE

Figure 1-3 *Machine tools produce tools and machines for manufacturing all types of products. (Courtesy DoAll Company.)*

room, and the production shop. The machine tools found in the metal trade fall into three broad categories:

1. *Chip-producing machines,* which form metal to size and shape by cutting away the unwanted sections. These machine tools generally alter the shape of steel products by casting, forging, or rolling in a steel mill.
2. *Non-chip-producing machines,* which form metal to size and shape by pressing, drawing, punching, or shearing. These machine tools generally alter the shape of sheet steel products and also produce parts which need little or no machining by compressing granular or powdered metallic materials.
3. *New-generation machines,* which were developed to perform operations would be very difficult, if not impossible, to perform on chip- or non-chip-producing machines. Electro-discharge, electro-chemical and laser machines, for example, use either electrical or chemical energy to form metal to size and shape.

A general machine shop contains a number of standard machine tools that are basic to the production of a variety of metal components. Operations such as turning, boring, threading, drilling, reaming, sawing, milling, filing, and grinding are most commonly performed in a machine shop. Machines such as the drill press, engine lathe, power saw, shaper, milling machine, and grinder are usually considered the *basic machine tools* in a machine shop (Fig. 1-4).

STANDARD MACHINE TOOLS

DRILL PRESS

The drill press or drilling machine (Fig. 1-5), probably the first mechanical device developed prehistorically, is used primarily to produce round holes. Drill presses range from the simple hobby type to the more complex automatic and numerical control machines used for production purposes. The function of a drill press is to grip and revolve the cutting tool (generally a twist drill) so that a hole can be produced in a piece of metal or other material. Operations such as drilling, reaming, spot facing, countersinking, counterboring, and tapping are commonly performed on a drill press.

ENGINE LATHE

The engine lathe (Fig. 1-6) is used to produce round work. The workpiece, held by a work-holding device mounted on the lathe spindle, is revolved against a cutting tool, which produces a cylindrical form. Straight turning, tapering, facing, drilling, boring, reaming, and thread cutting are some of the common operations performed on a lathe.

Figure 1-4 *Common machine tools found in a machine shop. (Courtesy DoAll Company.)*

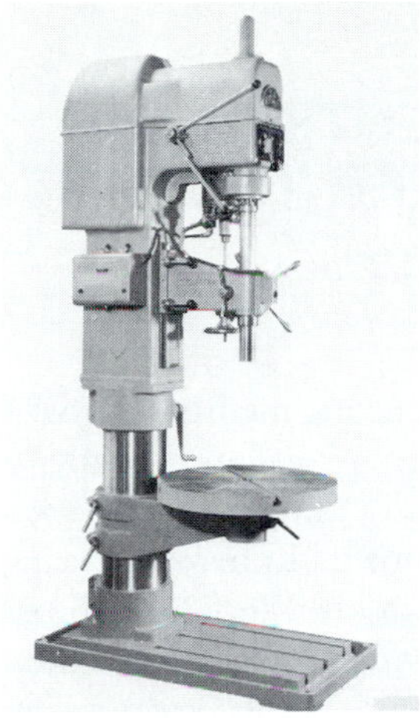

Figure 1-5 *A standard upright drill press.*

Monarch

Figure 1-6 *An engine lathe is used to produce round work. (Courtesy Monarch Machine Tool Company.)*

METAL SAW

The metal-cutting saws are used to cut metal to the proper length and shape. There are two main types of metal-cutting saws: the bandsaw (horizontal and vertical) and the reciprocating cutoff saw. On the vertical bandsaw (Fig. 1-7) the workpiece is held on the table and brought into contact with the continuous-cutting saw blade. It can be used to cut work to length and shape. The horizontal bandsaw and the reciprocating saw are used to cut work to length only. The material is held in a vise and the saw blade is brought into contact with the work.

MILLING MACHINE

The horizontal milling machine (Fig. 1-8) and the vertical milling machine are two of the most useful and versatile machine tools. Both machines use one or more rotating milling cutters having single or multiple cutting edges. The workpiece, which may be held in a vise, fixture, accessory, or fastened to the table, is fed into the revolving cutter. Equipped with proper accessories, milling machines are capable of performing a wide variety of operations such as

DoALL

Figure 1-7 *A contour-cutting bandsaw. (Courtesy DoAll Company.)*

drilling, reaming, boring, counterboring, and spot facing, and of producing flat and contour surfaces, grooves, gear teeth, and helical forms.

GRINDER

Grinders use an abrasive cutting tool to bring a workpiece to an accurate size and produce a high surface finish. In the grinding process, the surface of the work is brought into contact with the revolving grinding wheel. The most common types of grinders are the surface, cylindrical, cutter and tool, and bench or pedestal.

Surface grinders (Fig. 1-9) are used to produce flat, angular, or contoured surfaces on a workpiece.
Cylindrical grinders are used to produce internal and external diameters, which may be straight, tapered, or contoured.
Cutter and tool grinders are generally used to sharpen milling machine cutters.
Bench or pedestal grinders are used for offhand grinding and the sharpening of cutting tools such as chisels, punches, drills, and lathe and shaper tools.

SPECIAL MACHINE TOOLS

Special machine tools are designed to perform all the operations necessary to produce a single component. Special-purpose machine tools include gear-generating machines; centerless, cam, and thread grinders; turret lathes; and automatic screw machines.

SPACE-AGE MACHINES AND PROCESSES

The introduction of *electro-discharge machining, electrochemical machining, electrolytic grinding,* and *laser machining* has made it possible to machine new space-age materials and to produce shapes which were difficult or often impossible to produce by other methods.

Numerical control and *computer-numerically controlled* (CNC) machines, more recent additions to the array of machine tools, have greatly increased production and improved the quality of finished products. Consistent accuracy over many hundreds of parts is one of the features of these machines.

The numerical control principle has also been applied to *robots,* which are now capable of handling materials and changing machine tool accessories as easily and probably more efficiently than a person can (Fig. 1-10). *Robotics* has become the fastest-growing phase of the manufacturing industry.

Since its development, the *laser* has been applied to several areas of manufacturing. Lasers are now used increasingly for cutting and welding all types of metals—even those that have been impossible to cut or weld by other methods. Laser beams can pierce diamonds and any other known material and are also used in extremely accurate measuring and surveying devices, and as sensing devices.

With the introduction of numerous special machines and special cutting tools, production has increased tremendously over that attained with standard machine methods. Many products are produced automatically by a continuous flow of finished parts from these special machines. Product control and high production rates allow us to enjoy the pleasure and convenience of automobiles, power lawn mowers, automatic washers, stoves, and scores of other modern products. Without the basic machine tools required for mass production and automation, the costs of many luxuries that we now enjoy would be prohibitive.

MAJOR DEVELOPMENTS IN METALWORKING OVER THE PAST HALF CENTURY

Prior to the 20th century, manufacturing methods changed very slowly. Mass production had not developed nearly to the stage that we know now. It wasn't until the early 1930s that new and outstanding developments in manufacturing began to affect manufacturing processes. Since then, progress has been so rapid that now some of the newer developments astound most of us. It is because of this progress over the past fifty years or so that we in North America enjoy one of the highest standards of living in the world.

Manufacturing prior to 1932 was done on standard types of machine tools with little or no automation. Engine lathes, turret lathes, drill presses, shapers, planers, and horizontal milling machines were the common machine tools of the day. Most of the cutting tools were made of carbon steel or early grades of high-speed steel, which were not very efficient by today's standards. Production was slow and much of the work was finished by hand. This resulted in high costs of the items produced in relation to wages paid to the workers.

In the early 1930s, machine tool manufacturers took advantage of the lull in production and sales caused by the Great Depression to upgrade their machines by improving flexibility and controls. Thus began the trend leading to the machines of the present.

According to the Society of Manufacturing Engineers and the Machine Tool Builders Association, the following chronology lists the most important developments in metalworking during the past half century.

CINCINNATI

Figure 1-8 *A horizontal milling machine. (Courtesy Cincinnati Milacron Inc.)*

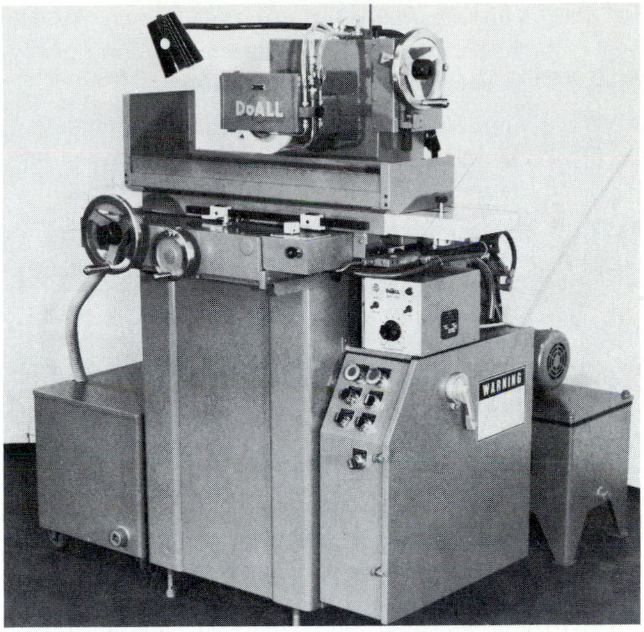
DoALL
WARNING

Figure 1-9 *A surface grinder is used to grind flat surfaces. (Courtesy DoAll Company.)*

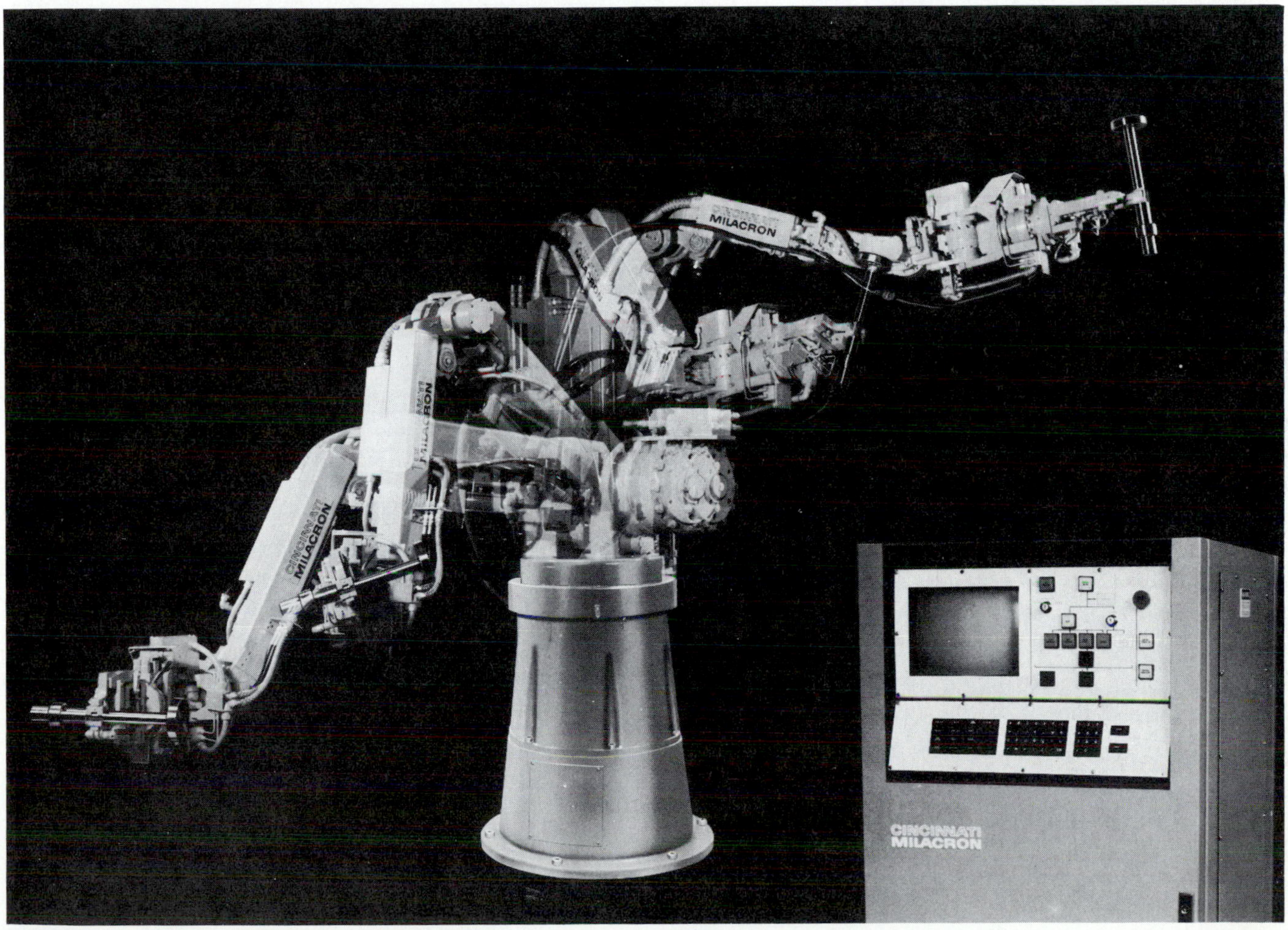
MILACRON
CINCINNATI MILACRON
CINCINNATI MILACRON

Figure 1-10 *Robots are finding ever-increasing applications in industry. (Courtesy Cincinnati Milacron Inc.)*

1932: *The electric gauge* was designed to provide greater accuracy and reduce inspection time by 75 percent over previous methods.

1933: *Crankshaft grinding* enabled the grinding of crankshafts in one setup, compared to the several setups and operations previously required.

1934: *The profilograph* was capable of magnifying surface inaccuracies of millionths of an inch and made contour forms easier to inspect.

1935: *Contour bandsawing* provided a fast and economical method of cutting metal to size and shape.

1936: *Induction heat treating* greatly reduced the amount of time required to produce a hard outer surface on a workpiece.

1937: *Automatic surface-broach grinding* automatically located the correct tooth spacing for surface broaches.

1938: *The Bridgeport mill* provided more versatility than the horizontal milling machine and increased metal-removal rates.

1939: *Aircraft components press* was designed to form duralumin parts for all-metal airplanes.

1940: *The jig grinder* provided the means of accurately locating holes in hardened workpieces.

1941: *First plant for producing electric alloy steel* provided a more accurate control of heat in order to produce specialty steels.

1942: *Heliarc welding* was a new process developed for welding magnesium on a production basis.

1943: *Air gaging* provided a means of gaging parts more quickly and accurately than was previously possible.

1944: *The 60,000 r/min grinder motor* was developed to provide small grinding wheels (1/8 in. diameter or less) with sufficient speed for efficient grinding.

1945: *Man-au-trol control* was the first hydraulic–electric control system introduced for automatic machine control.

1946: *ENIAC digital computer* was the first all-electronic general-purpose computer introduced and would eventually help with design problems.

1947: *Automatic size control* provided a means of automatically boring, honing, and gauging the size of engine block cylinders.

1948: *Cardamatic milling* automatically controlled by punched cards the cycle of a milling machine.

1949: *Ultrasonic inspection* provided a nondestructive method of testing materials by means of extremely high frequency sound waves.

1950: *Electronic hardness testing* was a quick and accurate type of test based on the magnetism retention of a part against a standard.

1951: *Method X electrical discharge machining* provided a means of removing metal from the workpiece by means of a spark of high density and short duration.

1952: *Numerical control* introduced a system attached to a milling machine whereby the table and cutting-tool movements were controlled by punched tape.

1953: *Project tinkertoy* was a system developed to automatically manufacture and assemble electronic circuit elements.

1954: *Indexible insert tooling* introduced a throwaway type of carbide cutting-tool insert, which could be turned over and used on both sides. This eliminated the need for expensive cutting-tool maintenance.

1955: *The numericord system* was the first completely automatic control for machines, provided by means of electronic control and magnetic tape.

1956: *The gear-honing process* provided a method used after heat treating to remove nicks and burrs from a gear and form it to its correct specifications.

1957: *Borazon* was an extremely hard abrasive developed from cubic boron nitride as an inexpensive substitute for diamond in certain applications.

1958: *The machining center* introduced a computer-controlled machine with a tape-controlled tool-changer capable of performing milling, drilling, tapping, and boring on a workpiece as large as an 18-in. cube.

1959: *The APT (automatically programmed tool) programming language* was a 107-word computer language used by programmers to write programs using data from engineering drawings.

1960: *Ultra-high-speed machining* was based on the principle that at extremely high cutting speeds (2500 sf/min and higher) the tool temperature and horsepower required to machine a workpiece drop. Speeds of 18,000 sf/min were used and speeds of up to 360,000 sf/min were planned.

1961: *The industrial robot* provided a single-armed device that can manipulate parts or tools through a sequence of operations or motions, as controlled by computer programs.

1962: *Computer-controlled steelmaking* introduced a system in which every steelmaking variable, from order and raw-material requirements to the finished product, is computer controlled.

1963: *The ADAPT programming language* provided a program compatible with APT language, used only about one-half of APT's vocabulary words, and was designed for use with small computers to control machine operations.

1964: *DAC-1—Design augmented by computers* was a computer system that allowed the computer to read drawings from paper or film and to generate new drawings by use of the keyboard and a lightpen.

1965: *System/360* introduced a large mainframe computer capable of responding in billionths of a second and which became the standard in industry for the next decade.

1966: *The single-layer metal-bonded diamond grinding wheel* was a diamond-impregnated grinding wheel, contoured to the profile of the workpiece, and reduced the grinding time required for certain parts from 10 hours to 10 minutes.

1967: *Computer numerical control* provided a computer control system that combined the functions of separate tape preparation equipment, numerical control, and program and part verification into one unit.

1968: *Direct numerical control* allowed the operation of machines directly from the mainframe computer without the use of tapes.

1969: *The programmable controller* was a smaller, single-purpose computer that could control as many as 64 machines using APT-created programs.

1970: *System GEMINI* provided a system whereby a supervisory computer and a distribution computer control several machines in the total manufacture of a part. This system was the forerunner of the automatic factory.

1971: *Robotic sensory capabilities* permitted a robot to "feel" for objects by means of a sensor applied to the robot's gripping fingers or vacuum cup.

1972: *Hummingbird press* was a 60-ton automated press with speeds of up to 1600 strokes per minute and a feed rate of 400 ft/min.

1973: *Robotic vision* was a robot system that utilized a television camera and image processing equipment to permit the robot to "see" and prevent the arm from bumping other parts as it travels to the desired location.

1974: *Remote machine diagnostics* allowed diagnosis of CNC machine problems in a plant by a computer in the manufacturer's head office by tying both computers into the telephone system.

1975: *Super CIMFORM grinding wheel* was a long-life vitrified aluminum oxide grinding wheel, developed for high production, and which cut grinding costs by 25 percent.

1976: *CAM-I automated process planning,* when a part is required, allows the computer to determine the "family" the part belongs to, calls up the drawing, makes any necessary modifications, and then directs the production of the part in the shop.

1977: *Distributed plant management systems* allowed a DNC computer system in a plant to be controlled and programmed by a remote computer that may not even be located in the same plant.

1978: *Automated programmable assembly systems* were designed to increase production by the use of several programmed robots to assemble component parts into a unit.

1979: *YMS-50 flexible manufacturing systems* linked standard NC machine modules with a parts-handling device and provided total computer control of the system.

1980: *The variable mission automatic toolhead changer* stores and installs cutting tools, as programmed, on as many as 18 multiple spindle heads.

1981: *Grinding center* provides computer-controlled grinding that can be programmed for as many as 48 different grinds on a workpiece.

1982–1988: *Continuing manufacturing developments include:*

- Polycrystalline superabrasive cutting tools
- Computer-integrated manufacturing (CIM)
- Group technology (GT)
- Just-in-time (JIT) manufacturing
- Adaptive controls (AC)
- Manufacturing automation protocol (MAP)
- Artificial intelligence (AI)

REVIEW QUESTIONS

1. Briefly trace the development of tools from the stone age to the industrial revolution.
2. Why are machine tools so important to our society?

3. How have improved production and accuracy been achieved with basic machine tools?
4. Name three categories of machine tools used in metalworking.
5. List five operations that can be performed on each of the following:
   a. drill press
   b. lathe
   c. milling machine
6. Name four types of grinders found in a machine shop.
7. What is the importance of the electro-machining processes?
8. What effect has numerical and computer control had on manufacturing?
9. State two applications of robots.
10. What is the importance of lasers in modern industry?

SECTION

2

Machine Trade Opportunities

Almost all products used by people, whether in farming, mining, manufacturing, construction, transportation, communication, or the professions, are dependent on machine tools for their manufacture. Constant improvements to and efficient use of machine tools affect the standard of living of any nation. Only through machine tools have we been able to enjoy the automobile, airplane, television, home furnishings, appliances, and many other products on which we rely in our daily lives. Through constant improvement, modern machine tools have become more accurate and efficient. Improved production and accuracy have been made possible through the application of hydraulics and electronic devices such as numerical control, computer numerical control, direct numerical control, and lasers to basic machine tools.

UNIT 2

Careers in the Metalworking Industry

The advancing technology, new ideas, new products, and special processes and manufacturing techniques are creating new and more specialized jobs. To advance in the machine trade, a person must keep up-to-date with modern technology. A young person leaving school may be employed in an average of five jobs in his or her lifetime, three of which do not even exist today. Industry is always on the lookout for bright young people who are conscientious and do not hesitate to assume responsibility. To be successful, do the job to the best of your ability and never be satisfied with inferior workmanship. Always try to produce quality products, at a reasonable price, in order to compete with foreign products, which are of such serious concern to North American industry.

OBJECTIVES

After completing this unit you will:

1. Know the various types of jobs available in the metalworking industry
2. Know the type of work each job entails

Many different careers are available in the metalworking industry. Choosing the right one depends on the skill, initiative, and qualifications of an individual. Metalworking offers exciting opportunities to any ambitious young person who is willing to accept the challenge of working to close tolerances and producing intricate parts. To be successful in this trade, the individual must also possess characteristics such as care of self, orderliness, accuracy, confidence, and safe work habits.

APPRENTICESHIP TRAINING

One of the best ways to learn a skilled trade is through an apprenticeship program. An apprentice (Fig. 2-1) is a person who is employed to learn a trade under the guidance of skilled tradespeople. The apprenticeship program is set up in conjunction with and under the supervision of the company, the Department of Labor Federal Bureau of Apprenticeship, and the trade union. It is usually about two- to four-years' duration and includes on-the-job training and related theory or classroom work. This period of time may be reduced by the completion of approved courses or because of previous experience in the trade.

To qualify for an apprenticeship, the individual should have completed a high school program or its equivalent. Mechanical ability with a good standing in mathematics, science, English writing skills, and mechanical drawing is desirable. Apprentices earn as they learn; the wage scale increases periodically during the training program.

Upon completion of an apprenticeship program, a certificate is granted, which qualifies a person to apply for journeyman status in the trade. Further opportunities in the trade are limited only by the person's initiative and interest. It is quite possible for an apprentice eventually to become an engineer, tool designer, supervisor, or shop owner.

MACHINE OPERATOR

Machine tool operators are classified as semiskilled tradespeople (Fig. 2-2). They are usually rated and paid according to their job classification, skill, and knowledge. The

Figure 2-1 *An apprentice learns the trade under the guidance of a skilled tradesperson. (Courtesy DoAll Company.)*

CINCINNATI

Figure 2-2 *A machine operator generally operates only one type of machine. (Courtesy Cincinnati Milacron Inc.)*

class A operator possesses more skill and knowledge than class B and C operators. For example, a class A operator should be able to operate the machine and:

- Make necessary machine setups
- Adjust cutting tools
- Calculate cutting speeds and feeds
- Read and understand drawings
- Read and use precision measuring tools

With the continued advancement in computer-controlled machines and programmable robots, gradually operators' jobs will be minimized. However, some machine tool operators may qualify with advanced technology courses and remain employed as operators on computer numerical control (CNC) turning centers, machining centers, and CNC robots.

MAINTENANCE MACHINIST

The maintenance machinist needs a combination of mechanical, rigging, and carpentry skills. The time for apprenticeship varies but usually ranges from two to four years and includes the necessary related theory for the job. During the apprenticeship the trainee works with qualified tradespeople. A maintenance machinist may be required to:

- Move and/or install machinery including production lines
- Read drawings and calculate sizes, fits, and tolerances of machine parts
- Repair machines by replacing and fitting new parts
- Dismantle and install equipment

To become a maintenance machinist apprentice, a student should have good technical training and be a high school graduate. A general knowledge of electricity, carpentry, sheet metal, and the machine tool trade is helpful.

The job outlook is good. Most large industries are expected to employ a work force of maintenance machinists to maintain and install machinery and production lines for the foreseeable future.

MACHINIST

Machinists (Fig. 2-3) are skilled workers who can efficiently operate all standard machine tools. Machinists must be able to read drawings and use precision measuring instruments and hand tools. They must have acquired enough knowledge and developed sound judgment to perform any bench, layout, or machine tool operation. In addition, they should be capable of making mathematical calculations required for setting up and machining any part. Machinists should have a thorough knowledge of metallurgy and heat treating. They should also have a basic understanding of welding, hydraulics, electricity, and pneumatics, and be familiar with computer technology.

Figure 2-3 *A machinist is skilled in the operation of all machines. (Courtesy Cincinnati Lathe & Tool Company.)*

Figure 2-4 *This production driller produces many identical parts.*

TYPES OF MACHINE SHOPS

A machinist may qualify to work in a variety of shops. The three most common are maintenance, production, and jobbing shops.

A *maintenance shop* is usually connected with a manufacturing plant, lumber mill, or a foundry. A maintenance machinist generally makes and replaces parts for all types of setup and cutting tools and production machinery. The machinist must be able to operate all machine tools and be familiar with bench operations such as layout, fitting, and assembly.

A *production shop* may be connected with a large factory or plant which makes many types of identical machined parts, such as pulleys, shafts, bushings, motors, and sheet-metal pieces. A person working in a production shop generally operates one type of machine tool and often produces identical parts (Fig. 2-4).

A *jobbing shop* is generally equipped with a variety of standard machine tools and perhaps a few production machines, such as a turret lathe and punch and shear presses. A jobbing shop may be required to do a variety of tasks, usually under contract to other companies. This work may involve the production of jigs, fixtures, dies, molds, tools, or short runs of special parts. A person working in a jobbing shop generally is a qualified machinist, toolmaker, or mold maker and would be required to operate all types of machine tools and measuring equipment.

TOOL AND DIEMAKER

A *tool and diemaker* is a highly skilled craftsman who must be able to make different types of dies, molds, cutting tools, jigs, and fixtures. These tools may be used in the mass production of metal, plastic, or other parts. For example, to make a die to produce a 90° bracket in a punch press, the tool and diemaker must be able to select, machine, and heat treat the steel for the die components. He or she should also know what production method will be used to produce the part since this information will help to produce a better die. For a mold used to produce a plastic handle in an injection molding machine, the tool and diemaker must know the type of plastic used, finish required, and the process used in production.

To qualify as a tool and diemaker, a person should serve an apprenticeship, have above-average mechanical ability, and be able to operate all standard machine tools. This work also requires a broad knowledge of shop mathematics, print reading, machine drafting, principles of design, machining operations, metallurgy, heat treating, computers, and space-age machining processes.

TECHNICIAN

A *technician* is a person who works at a level between the professional engineer and the machinist. The technician may assist the engineer in making cost estimates of products, preparing technical reports on plant operation, or program a numerical control machine.

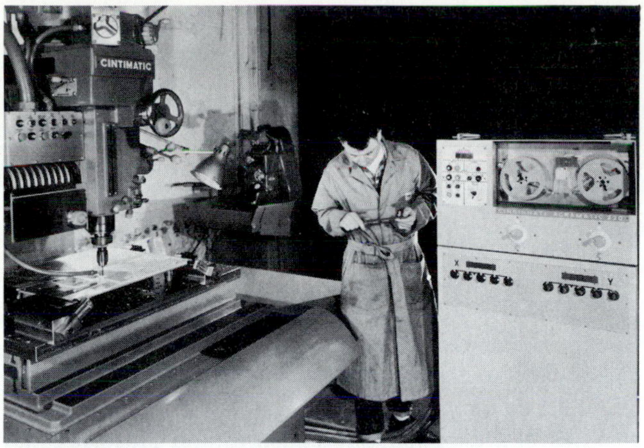
CINTIMATIC

Figure 2-5 *A technician is often required to check the setup and operation of a machine program. (Courtesy Cincinnati Milacron Inc.)*

A technician should have completed high school and have at least two additional years of education at a community college, technical institute, or university. A technician must also possess a good knowledge of drafting, mathematics, and technical writing.

Opportunities for technicians are becoming more plentiful because of the development of machine tools such as numerical control, turning centers, and electro-machining processes. Technicians are usually trained in only one area of technology, such as electrical, manufacturing, machine tool, or metallurgy. Some technicians may need knowledge beyond that of their specialty field. For example, a machine tool technician (Fig. 2-5) should have a knowledge of industrial machines and manufacturing processes in order to know the best method of manufacturing a product. However, it is not necessary for a technician to perform as a skilled machinist. A technician may qualify as a technologist after at least one year of on-the-job training under a technologist or engineer.

TECHNOLOGIST

A *technologist* works at a level between a graduate engineer and a technician. Most technologists are three- or four-year graduates from a community or technical college. Their studies generally include physics, advanced mathematics, chemistry, engineering graphics, computer programming, business organization, and management.

An *engineering technologist* may do many jobs normally performed by an engineer, such as design studies, production planning, laboratory experiments, and supervision of technicians. This permits the engineer to work in other important areas. Technologists are often employed to serve in a middle management position within a large company. A technologist may be employed in many areas, such as quality and cost control, production control, labor relations, training, and product analysis. A technologist may qualify as an engineer by obtaining further education at the university level and by passing a qualifying examination.

INSPECTORS (QUALITY CONTROL)

An *inspector* (Fig. 2-6) checks and examines machined parts to determine whether they meet the specifications on the drawing. If the parts are not within the limits shown on the drawing, they will not fit together and function properly when assembled. This task is very important since parts made in one country may be assembled in another or be interchanged with worn or broken parts.

An inspector should have a technical or vocational education and be familiar with measuring tools and inspection processes. On-the-job training may take from several weeks to several years, depending on the job, the items to be inspected, and the technical knowledge required. An inspector may need varying degrees of skill depending on the size, cost, type, and tolerances required on the finished workpiece. A good inspector should be able to:

- Understand and read mechanical drawings
- Make basic mathematical calculations
- Use micrometers, gages, comparators, and precision measuring instruments

SHEFFIELD

Figure 2-6 *An inspector checks the dimensions of finished part and products. (Courtesy The Bendix Corporation.)*

INSTRUMENT MAKERS

Instrument makers are highly skilled tool and diemakers who work directly with scientists and engineers. Instrument makers should be able to operate all precision machine tools in order to make measuring instruments, gages, and special machines for testing purposes. Generally, instrument makers must have more training than a machinist or a tool and diemaker. They must work to closer tolerances and limits than a machinist. Also, they usually must perform all the work on the instrument or gage being produced.

Instrument makers generally serve four or five years of apprenticeship. They may also learn the craft by transferring from a toolmaker or machinist trade and further training on the job.

With new processes in manufacturing, the demand for instrument makers will remain fairly close to the demand for skilled tradespeople. Instrument makers are employed by research centers, scientific laboratories, manufacturers of gages and measuring tools, and government and standards organizations.

PROFESSIONS

Many areas are open to the engineering graduate. Teaching is one of the most satisfying and challenging professions. Graduation from a college, including teacher training course work, is required. On-the-job industrial experience is not always a prerequisite, but it will prove helpful in teaching. However, some states and/or some schools require on-the-job experience to qualify for technical and vocational certification. With industrial experience a person may teach industrial arts or in certain subject areas in a technical school, vocational school, or community college.

Engineers in industry are responsible for designing and developing new products and production methods and for redesigning and improving existing products. Most engineers specialize in a specific discipline of engineering, such as:

- Civil
- Metallurgical
- Aerospace
- Mechanical
- Electrical
- Electronics

A bachelor's degree in engineering is usually required to enter the profession. However, in some branches of engineering, such as tool and manufacturing, the individual progresses through a program of practical experience and obtains a certificate after passing a qualifying examination. Because of the variety of engineering jobs available, many women, as well as men, are entering the profession.

NUMERICAL CONTROL PROGRAMMERS

Tool programmers for numerical control machines must be thoroughly familiar with all machining techniques. A programmer should be familiar with a skilled machine operator's job and know all the sequences and procedures for producing a machined part. In order to program a machine to produce a part, an operator must be able to:

- Read working drawings
- Select the best tools for the machining operation
- Calculate speeds and feeds for different materials and types of cutting tools
- Estimate production costs and simulate machine tool processes

Many vocational schools, technical schools, and universities offer CNC and NC (numerical control) programming courses. Computer-assisted machining (CAM) is perhaps one of the fastest-growing specialized areas in the machining industry. Learning how to program a specialized turning center is only the beginning of the challenge that industry will be offering in the future. A good education is essential, including a background in electronics, mathematics, and manufacturing processes.

REVIEW QUESTIONS

1. Define an apprentice.
2. Name three desirable qualities a person should have for apprenticeship training.
3. Explain the difference between a machinist and a machine operator.
4. Briefly explain the difference between a jobbing and a production shop.
5. Define a tool and diemaker.
6. How can a person become a tool and diemaker?
7. Explain the difference between a technician and a technologist.
8. Briefly define the duties of an inspector and an instrument maker.
9. What qualifications must a person have in order to become a technical teacher?
10. List four areas in industry that require an engineer's qualifications.

UNIT 3

Getting the Job

After graduation or leaving school, your most important task probably is that of finding a full-time job. *Choosing the right job is very important.*

After consulting with the school guidance counselor, state employment service, and any other agency which may be helpful, start looking into the job which appeals to you. In most areas aptitude tests are available to help you decide on the career you may wish to explore.

OBJECTIVES

After completing this unit, you will be able to:
1. Prepare a comprehensive resumé
2. Arrange for a job interview.
3. Prepare for and follow up on the job interview.

ASSESS YOUR ABILITIES

In order to help determine where your interests lie, ask yourself:

- What type of work do I like?
- What type of work do I not like?
- What jobs have I done with some success?
- What skills have I acquired at school?
- What have I done in my part-time work that has been outstanding?
- Do I enjoy taking apart and repairing appliances and items which are not working?

EXPLORE YOUR INTERESTS

After having narrowed the field, then:

- Read any books on the chosen subject(s) (Fig. 3-1).
- Talk to people who do the type of work that you are considering. Ask them about the job and job opportunities.
- Talk further with your school guidance counselor.
- Consult state or federal employment services.

When you have gathered enough information to discuss reasonably the type of work you are interested in, check the help-wanted ads in the newspapers.

Figure 3-1 *Read as much as possible about a job that might appeal to you.*

MAKING A RÉSUMÉ

When you have decided on the job you wish to apply for, prepare a résumé listing the following facts in a logical order.

- Your full name.

- Address, including zip code and telephone number, including area code.
- Social security number.
- Father's and mother's names and address(es).
- Education—schools attended and grade or level completed.
- Other special training that may be helpful; for example, first-aid courses.
- Special interests and hobbies.
- Sports in which you are active or interested.
- Any organizations to which you belong or are active in.
- Previous employment. List the names of firms and places where you have worked and in reverse chronological order include the following information:

  Dates of employment.
  Type of work and equipment operated (if any).
  Supervisor's name.
  Salary.
  Reason for leaving. (Be honest because this information is usually available if the prospective employer chooses to phone your former employer.)

- A list of at least three persons who may be contacted for character references. Include addresses and telephone numbers. Be sure to ask permission to use their names *before* listing them on an application.

Many state employment services and government agencies have pamphlets that can be used as a guide to prepare résumés. Also available is information on how to get and keep a job, which can be especially valuable to a person looking for that first job.

FACTS ABOUT INTERVIEWS

ARRANGING AN INTERVIEW

After completing your résumé, submit it with a covering letter to the personnel manager of the company you are interested in. Be sure to include a request for an interview. In many cases you may phone the company and make an appointment for an interview. In this case, leave the résumé with the person who interviews you.

QUESTIONS COMMONLY ASKED DURING AN INTERVIEW

Prior to the interview, consider the following questions, which are often asked by employers, and the probable reason for their asking them. You should be prepared with satisfactory answers.

Question: Why would you like to work here?
Reason: To see if you have gathered any information about the company before the interview.

Question: What were your best subjects at school?
Reason: This will reveal some of your interests and abilities.

Question: What sports or activities did you participate in when attending school?
Reason: To find out your interests and abilities outside of school. Also to see if you can work as part of a team.

Question: What type of job do you hope to have five years from now?
Reason: To assess your ambition and initiative.

Question: At what salary would you expect to start?
Reason: To see if you are familiar with the "going rates" and to see how you assess your abilities.

Question: What do you have to offer for the job?
Reason: To give you a chance to outline your abilities.

Question: What type of books or plays are you interested in?
Reason: To assess your interests and often your environment.

Question: How did you get along with your previous employer?
Reason: Your answer may reveal whether you are a complainer and a person who is hard to get along with.

Question: Why are you applying for this job?
Reason: To see if you have checked into this particular job and aren't just looking for any job.

THE INTERVIEW

In preparing for the interview you should consider the following:

- Be sure of the address, the office, and the time.
- Know the name and position of the person that you are to see. This information can be obtained by a phone call to the company prior to the interview.
- Be neatly dressed and groomed. Remember, a neat applicant usually commands more attention (Fig. 3-2).
- Be on time.
- Display confidence when you introduce yourself to the interviewer.
- During the interview be honest. Emphasize your good qualities and abilities, but don't bluff.
- Know enough about the company to enter into a discussion with the interviewer.

AFTER THE INTERVIEW

- Thank the interviewer and ask when you may expect to hear from him or her (Fig. 3-3).

Figure 3-2 *Be at ease while you are being interviewed.*

Figure 3-3 *Always thank the interviewer—it may pay off!*

- If you are offered a job, accept it (or reject it) as soon as possible. Never leave the prospective employer "hanging," awaiting your decision. If you aren't interested, explain why.
- The next day, send the interviewer a short letter expressing your appreciation for the valuable time he or she took for the interview.
- If you don't hear from the company in a reasonable time (7 to 10 days) call and ask for the person who interviewed you. Ask if they have made a decision yet and, if not, when you might expect one.
- If you don't get the first job, apply to other companies. Don't stop looking.
- Try to learn from each interview, which will eventually help you to get the right job.
- After several unsuccessful interviews, you might seek professional counseling from the Department of Labor, a vocational school, or a community college.

POINTS TO REMEMBER

- The job will not find you. You must find the job.
- Know the type of work that you want and don't offer to take just any job.
- Look for the type of work that you feel will be interesting. You will be much more successful if you like your work.

REVIEW QUESTIONS

1. List four things that you should consider when attempting to assess your abilities.
2. Name four ways to obtain further information about a trade or job.
3. Assume that you are applying for a job; prepare a personal résumé that you would submit to an employer.
4. Name three methods of arranging an interview.
5. List four important points that you must consider in preparing for an interview.
6. Name four important points that you should follow after an interview.

SECTION

3

Safety

There has never been a more meaningful saying than "Safety Is Everyone's Business." To become a skilled craftsman, it is very important for you to learn to work safely, taking into consideration not only your own safety but the safety of your fellow workers.

In general, everyone has a tendency to be careless about safety at times. We take chances every day by not wearing seat belts, walking under ladders, cluttering the work area, and doing many other careless and unsafe things. It seems that people tend to feel that accidents always happen to others. However, be sure to remember that a moment of carelessness can result in an accident which can affect you for the rest of your life. A loss of eyesight due to not wearing safety glasses or the loss of a limb due to loose clothing caught in a machine can seriously affect or end your career in the machine tool trade. *THINK SAFE, WORK SAFE, and BE SAFE.*

UNIT

4

Safety in the Machine Shop

All hand and machine tools can be dangerous if used improperly or carelessly. Working safely is one of the first things a student or apprentice should learn because the safe way is usually the correct and most efficient way. A person learning to operate machine tools *must first* learn the safety regulations and precautions for each tool or machine. Far too many accidents are caused by careless work habits. It is easier and much more sensible to develop safe work habits than to suffer the consequences of an accident. *SAFETY IS EVERYONE'S BUSINESS AND RESPONSIBILITY.*

OBJECTIVES

After completing this unit you will be able to:

1. Recognize safe and unsafe work practices in a shop
2. Identify and correct hazards in the shop area
3. Perform your job in a manner that is safe to you and other workers

SAFETY ON THE JOB

The safety programs initiated by accident prevention associations, safety councils, government agencies, and industrial firms are constantly attempting to reduce the number of industrial accidents. *Nevertheless,* each year accidents which could have been avoided result not only in millions of dollars' worth of lost time and production but also in a great deal of pain, many lasting physical handicaps, or even the death of workers. Modern machine tools are equipped with safety features, but it is still the operator's responsibility to use these machines wisely and safely.

Accidents don't just happen; they are caused. The cause of an accident can usually be traced to carelessness on someone's part. Accidents can be avoided, and a person learning the machine tool trade must first develop safe work habits.

A safe worker should:

- Be neat, tidy, and safely dressed for the job he or she is performing
- Develop a responsibility for personal safety and the safety of fellow workers
- Think safely and work safely at all times

SAFETY IN THE SHOP

Safety in a machine shop may be divided into two broad categories:

- Those practices which will prevent injury to workers.
- Those practices which will prevent damage to machines and equipment. Too often damaged equipment results in personal injuries.

In considering these categories, we must consider personal grooming, proper housekeeping (including machine maintenance), safe work practices, and fire prevention.

PERSONAL GROOMING

The following rules should be observed when working in a machine shop.

1. Always wear approved safety glasses in any area of the machine shop. Most plants now insist that all employees and visitors wear safety glasses or some other eye protection device when entering a shop area. Several types of eye protection devices are available for use in the machine shop:

a. The most common are *plain safety glasses* with side shields (Fig. 4-1A). These glasses offer sufficient eye protection when an operator is operating any machine or performing any bench or assembly operation. The lenses are made of shatterproof glass, and the side shields protect the sides of the eyes from flying particles.

b. *Plastic safety goggles* (Fig. 4-1B) are generally worn by anyone wearing prescription eyeglasses. These goggles are soft, flexible plastic and fit closely around the upper cheeks and forehead. Unfortunately, they have a tendency to fog up in warm temperatures.

c. *Face shields* (Fig. 4-1C) may also be used by those wearing prescription glasses. The plastic shield gives full face protection and permits air circulation between the face and the shield, thus preventing fogging up in most situations. These shields, as well as approved protective clothing and gloves, *must* be worn when an operator is heating and quenching materials during heat-treating operations or when there is any danger of hot flying particles. In industry, some companies provide their employees with prescription safety glasses, which eliminate the need for protective goggles or shields.

NOTE: Never think that because you are wearing glasses your eyes are safe. If the lenses are not made of approved safety shatterproof glass, serious eye injury can occur.

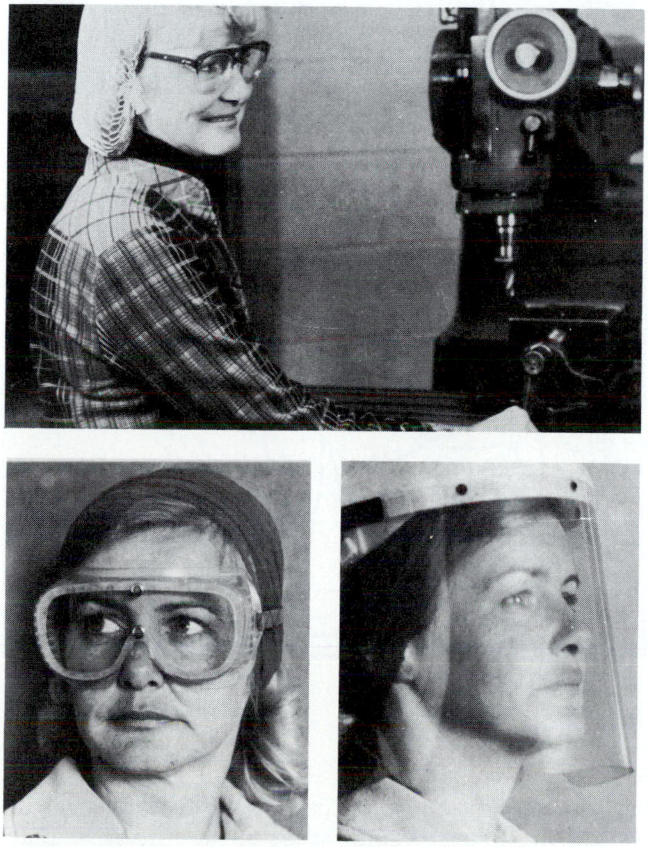

Figure 4-1 *Types of safety glasses: (A) plain; (B) plastic goggles; (C) face shields.*

2. Never wear loose clothing when operating any machine (Fig. 4-2).

a. Always roll up your sleeves or wear short sleeves.

b. Clothing should be made of hard, smooth material that will not catch easily in a machine. Loose-fitting sweaters should not be worn for this reason.

c. Remove or tuck in a necktie before starting a machine. If you want to wear a tie, make it a bow tie.

d. When wearing a shop apron, *always tie it at the back* and never in front of you so that the apron strings will not get caught in rotating parts (Fig. 4-3).

Figure 4-2 *Loose clothing can easily be caught in moving parts of machinery.*

Figure 4-3 *Tie aprons behind your back to keep the ties from being caught in machinery. (Courtesy Whitman & Barnes.)*

3. Remove wrist watches, rings, and bracelets; these can get caught in the machine, causing painful and often serious injury (Fig. 4-4).
4. Never wear gloves when operating a machine.
5. Long hair must be protected by a hair net or approved protective shop cap (Fig. 4-5). One of the most common accidents on a drill press is caused by long, unprotected hair getting caught in a revolving drill.
6. Canvas shoes or open-toed sandals must *never* be worn in a machine shop because they offer no protection to the feet against sharp chips or falling objects. In industry, most companies make it mandatory for employees to wear safety shoes.

HOUSEKEEPING

The operator should remember that *good housekeeping will never interfere with safety or efficiency;* therefore, the following points should be observed.

1. *Always stop the machine before you attempt to clean it.*
2. Always keep the machine and hand tools clean. Oily surfaces can be dangerous. Metal chips left on the table surface may interfere with the safe clamping of a workpiece.
3. Always use a brush and not a cloth to remove any chips.
4. Oily surfaces should be cleaned with a cloth.
5. Do not place tools and materials on the machine table—use a bench near the machine.
6. Keep the floor free from oil and grease (Fig. 4-6).
7. Sweep up the metal chips on the floor frequently. They become embedded in the soles of shoes and can cause dangerous slippage if a person walks on a terrazzo or concrete floor. Use a scraper, mounted on the floor near the door, to remove these chips before leaving the shop (Fig. 4-7).
8. Never place tools or materials on the floor close to a machine where they will interfere with the operator's ability to move safely around the machine (Fig. 4-8).
9. Return bar stock to the storage rack after cutting off the required length (Fig. 4-9).
10. Never use compressed air to remove chips from a machine. Not only is it a dangerous practice because of flying metal chips, but small chips and dirt can become wedged between machine parts and cause undue wear.

SAFE WORK PRACTICES

1. *Do not operate any machine before understanding its mechanism and knowing how to stop it quickly.* Knowing how to stop a machine quickly can prevent a serious injury.
2. Before operating any machine, be sure that the safety devices are in place and in working order. Remember, safety devices are for the operator's protection and should not be removed.

SPINDLE R.P.M.
LOW
HIGH
SLOW RANGE
FAST RANGE

Figure 4-4 *Wearing rings and watches can cause serious injury.*

Figure 4-5 *Long hair must be protected by a hair net or approved shop cap.*

Figure 4-6 *Grease and oil on the floor can cause dangerous falls.*

3. Always disconnect the power and lock it off at the switch box when making repairs to any machine (Fig. 4-10). Place a sign on the machine noting that it is out of order.

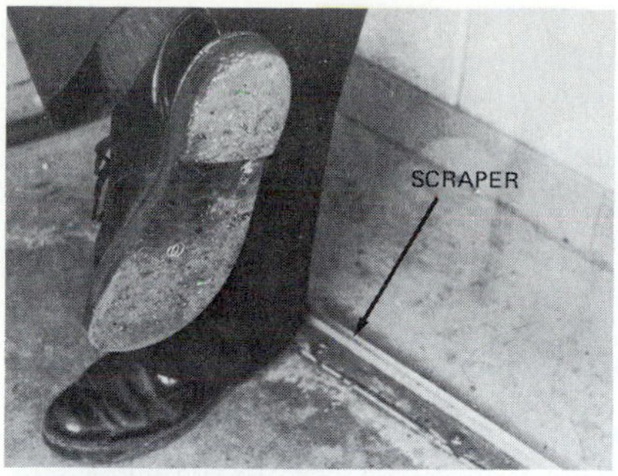
SCRAPER

Figure 4-7 *Remove chips from the soles of your shoes before leaving the shop.*

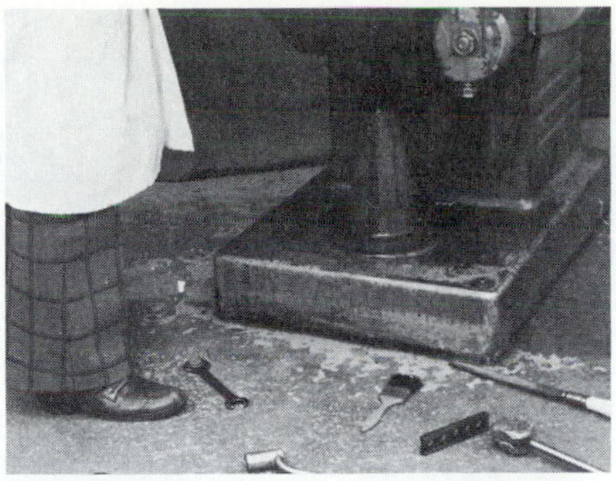

Figure 4-8 *Poor housekeeping can lead to accidents.*

Figure 4-9 *Store stock safely in a material stock rack.*

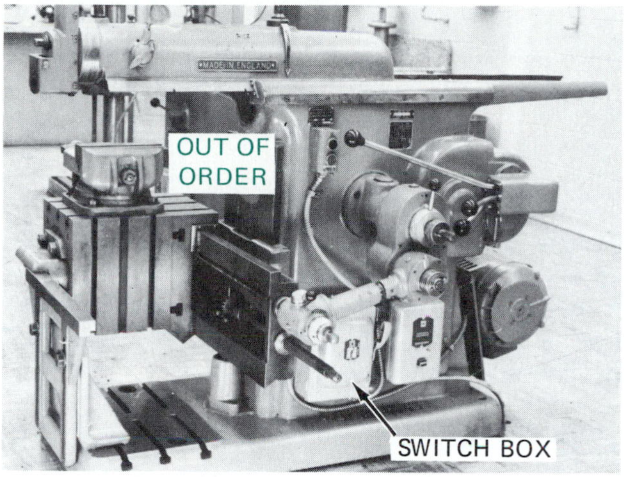
MADE IN ENGLAND
OUT OF ORDER
SWITCH BOX

Figure 4-10 *Power switches must be locked off before you repair or adjust a machine.*

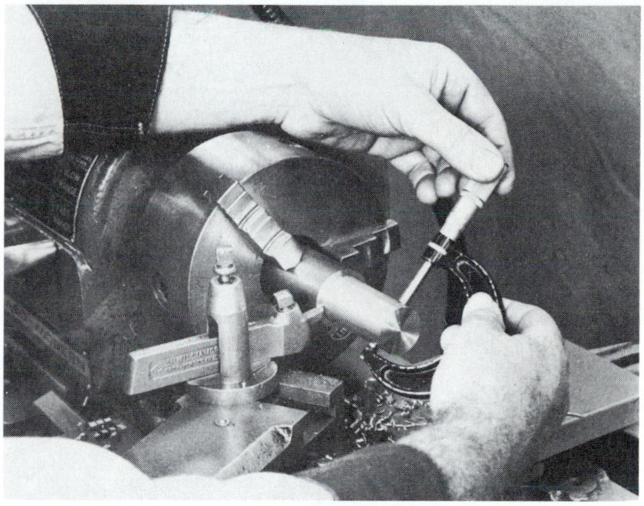

Figure 4-11 *The machine must be stopped before you measure a workpiece.*

4. Always be sure that the cutting tool and the workpiece are properly mounted before starting the machine.
5. *Keep hands away from moving parts.* It is dangerous practice to "feel" the surface of revolving work or to stop a machine by hand.
6. *Always stop a machine before measuring, cleaning, or making any adjustments.* It is dangerous to do any type of work around moving parts of a machine (Fig. 4-11).
7. *Never use a rag near the moving parts of a machine.* The rag may be drawn into the machine, along with the hand that is holding it.
8. *Never have more than one person operate a machine at the same time.* Not knowing what the other person would or would not do has caused many accidents.
9. *Get first aid immediately for any injury, no matter how small.* Report the injury and be sure that the smallest cut is treated to prevent the chance of a serious infection.
10. Before you handle any workpiece, remove all burrs and sharp edges with a file.
11. Do not attempt to lift heavy or odd-shaped objects which are difficult to handle on your own.
12. For heavy objects, follow safe lifting practices:
   a. Assume a squatting position with your knees bent and back straight.
   b. Grasp the workpiece firmly.
   c. Lift the object by straightening your legs and keeping your back straight (Fig. 4-12). This procedure uses the leg muscles and prevents injury to the back.
13. Be sure the work is clamped securely in the vise or to a machine table.
14. Whenever work is clamped, be sure the bolts are placed closer to the workpiece than to the clamping blocks.
15. Never start a machine until you are sure that the cutting tool and machine parts will clear the workpiece (Fig. 4-13).
16. Use the proper wrench for the job, and replace nuts with worn corners.
17. It is safer to pull on a wrench than to push on it.

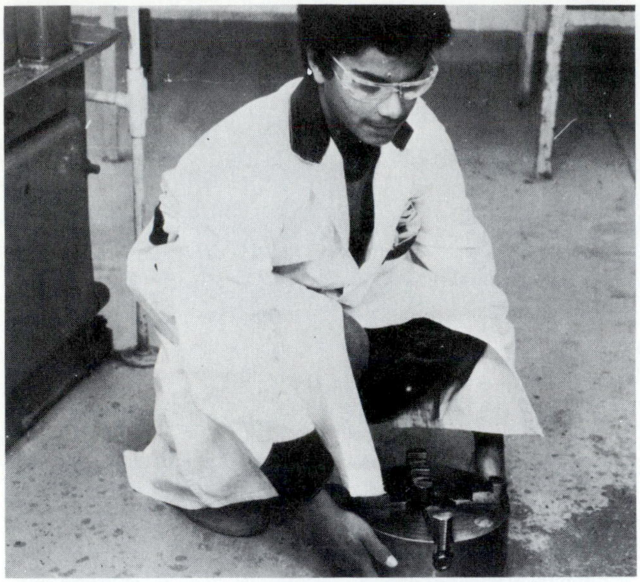

Figure 4-12 *Follow recommended lifting procedures to avoid back injury.*

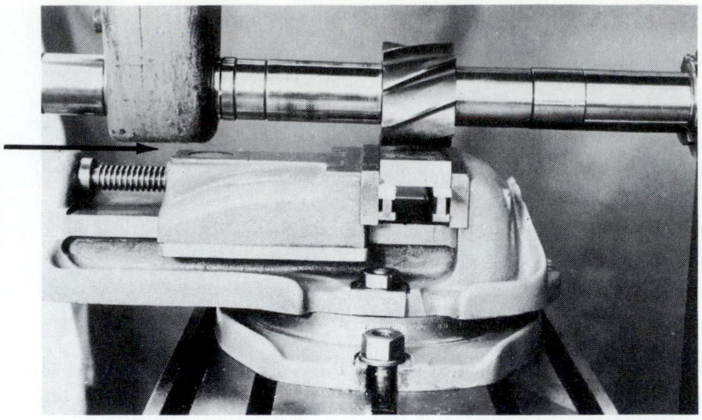

Figure 4-13 *Make sure that the cutting tool and machine parts will clear the workpiece.*

FIRE PREVENTION

1. *Always* dispose of oily rags in proper metal containers.
2. Be sure of the proper procedure before lighting a gas furnace.
3. Know the location and the operation of every fire extinguisher in the shop.
4. Know the location of the nearest fire exit from the building.
5. Know the location of the nearest fire-alarm box and its operating procedure.
6. When using a welding or cutting torch, be sure to direct the sparks away from any combustible material.

REVIEW QUESTIONS

1. What must be learned before operating a machine tool for the first time?
2. List three qualities of a safe worker.

PERSONAL GROOMING

3. Name three types of eye protection which may be found in a shop.
4. State four precautions which must be observed in regard to clothing worn in a shop.
5. Why should gloves not be worn when operating a machine?
6. How must long hair be protected?

HOUSEKEEPING

7. Why must a cloth not be used to remove chips?
8. Why should shoe soles be scraped before leaving the shop?
9. State two reasons why compressed air should not be used for cleaning machines.

SAFE WORK PRACTICES

10. State three precautions to observe before operating any machine.
11. Describe the procedure to follow for lifting a heavy object.
12. What should you do immediately after receiving any injury?

FIRE PREVENTION

13. What three fire prevention factors should everyone become familiar with before starting to work in a machine shop?

SECTION 4

Job Planning

Machine shop work consists of machining a variety of parts (round, flat, contour) and either assembling them into a unit or using them separately to perform some operation. It is important that the sequence of operations be carefully planned in order to produce a part quickly and accurately. Improper planning or following a wrong sequence of operations often results in spoiled work.

UNIT 5

Engineering Drawings

Engineering drawing is the common language by which draftspersons, tool designers, and engineers indicate to the machinist and toolmaker the physical requirements of a part. Drawings are made up of a variety of lines, which represent surfaces, edges, and contours of a workpiece. By adding symbols, dimension lines and sizes, and word notes, the draftsperson can indicate the exact specifications of each individual part. The student must become familiar with this language so that he or she can interpret lines and symbols in order to produce the parts quickly and accurately.

A complete product is usually shown on an *assembly drawing* by the drafter. Each part or component of the product is then shown on a *detailed drawing,* which is reproduced as copies called *prints.* The prints are used by the machinist or toolmaker to produce the individual parts which eventually will make up the complete product. Some of the more common lines and symbols will be reviewed briefly. However, it must be assumed that the student already has a basic knowledge of print reading or that a good book on this subject is available for review purposes.

OBJECTIVES

After completing this unit you will be able to:

1. Comprehend the meaning of the various lines used on engineering drawings
2. Recognize the various symbols used to convey information
3. Read and understand engineering drawings or prints

TYPES OF DRAWINGS AND LINES

In order to describe the shape of noncylindrical parts accurately on a drawing or print, the draftsperson uses the *orthographic view* or *projection method.* The orthographic view shows the part from three sides: the front, the top, and the right-hand side (Fig. 5-1). These three views enable the draftsperson to describe a part or object so completely that the machinist knows exactly what is required.

Cylindrical parts are generally shown on prints in two views: the front and the right side (Fig. 5-2). However, if a part contains many details, it may be necessary to use the top, bottom, or left-side views to describe the part accurately to the machinist.

In many cases, complicated interior forms are difficult to describe in the usual manner by a draftsperson. Whenever this occurs, a *sectional view,* which is obtained by making an imaginary cut through an object, is presented. This imaginary cut can be made in a straight line in any direction to best expose the interior contour or form of a part (Fig. 5-3).

A wide variety of *standard lines* are used in engineering drawings in order for the designer to indicate to the machinist exactly what is required. Thick, thin, broken, wavy, and section lines are used on shop or engineering drawings. See Table 5-1 for examples, including the description and purpose of some of the more common lines used on shop drawings.

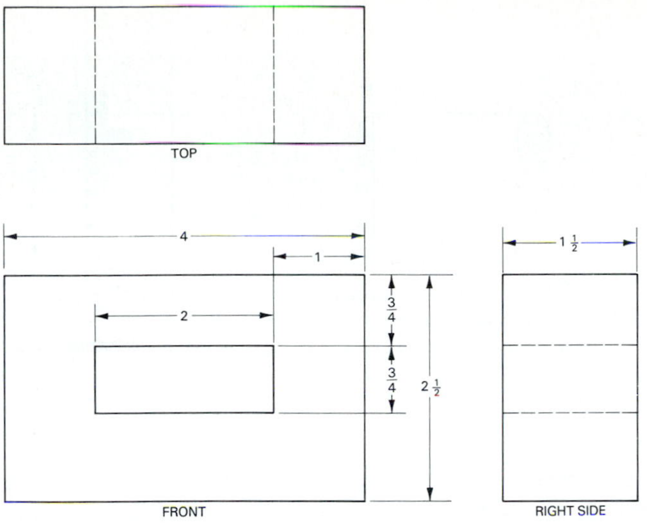

TOP
4
1
2
3/4
3/4
2 1/2
1 1/2
FRONT
RIGHT SIDE

Figure 5-1 *Three views of orthographic projection make it easier to describe the details of a part.*

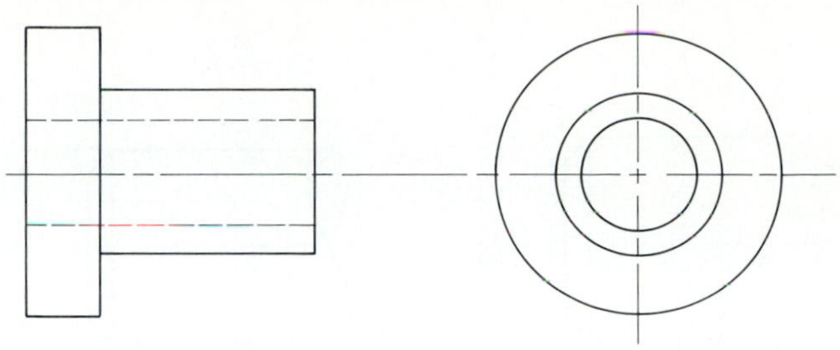

Figure 5-2 *Cylindrical parts are generally shown in two views.*

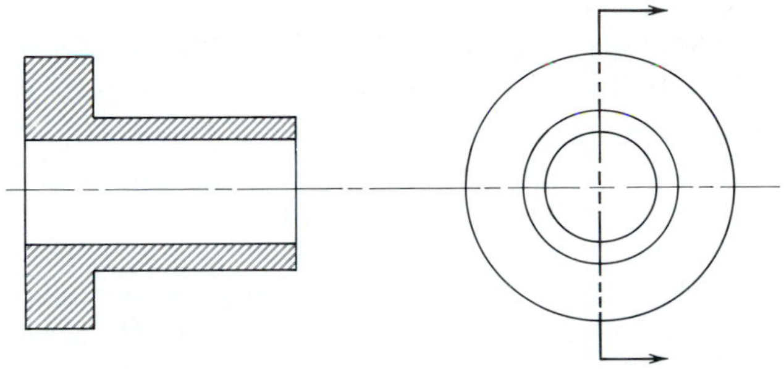

Figure 5-3 *Section views are used to show complicated interior forms.*

DRAFTING TERMS AND SYMBOLS

Common drafting terms and symbols are used on shop and engineering drawings in order for the designer to describe each part accurately. If it were not for the universal use of terms, symbols, and abbreviations, the designer would have

Table 5-1 Common Lines Used on Shop Drawings

Example	Name	Description	Use
a	Object lines	Thick black lines approximately 1/32 in. wide (the width may vary to suit drawing size).	Indicate the visible form or edges of an object.
b	Hidden lines	Medium-weight black lines of 1/8-in. long dashes and 1/16-in. spaces.	Indicate the hidden contours of an object.
c	Center lines	Thin lines with alternating long lines and short dashes. Long lines from 1/2 to 3 in. long. Short dashes 1/16 to 1/8 in. long, spaces 1/16 in. long.	Indicate the centers of holes, cylindrical objects, and other sections.
1 1/2 d	Dimension lines	Thin black lines with an arrowhead at each end and a space in the center for a dimension.	Indicate the dimensions of an object.
e	Cutting-plane lines	Thick black lines make up a series of one long line and two short dashes. Arrowheads show the line of sight from where the section is taken.	Show the imagined section cut.
f	Cross-section lines	Fine evenly spaced parallel lines at 45°. Line spacing is in proportion to the part size.	Show the surfaces exposed when a section is cut.

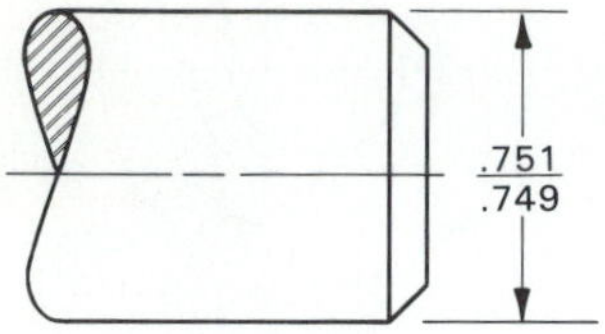

.751
.749

Figure 5-4 *Limits show the largest and smallest size of a part.*

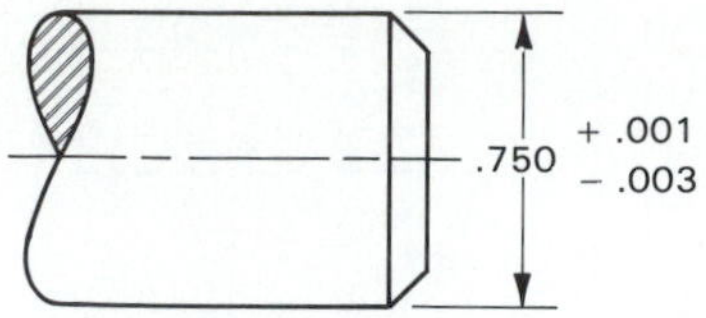

.750
+ .001
− .003

Figure 5-5 *Tolerance is the permissible variation of a specified size.*

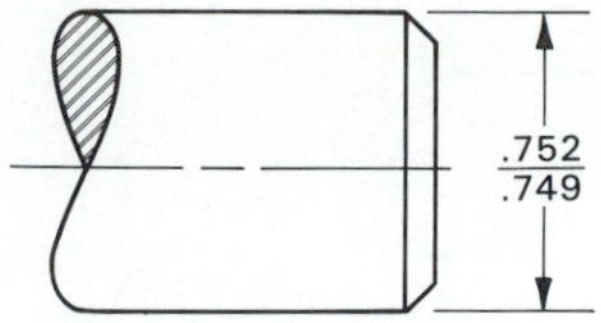

.752
.749

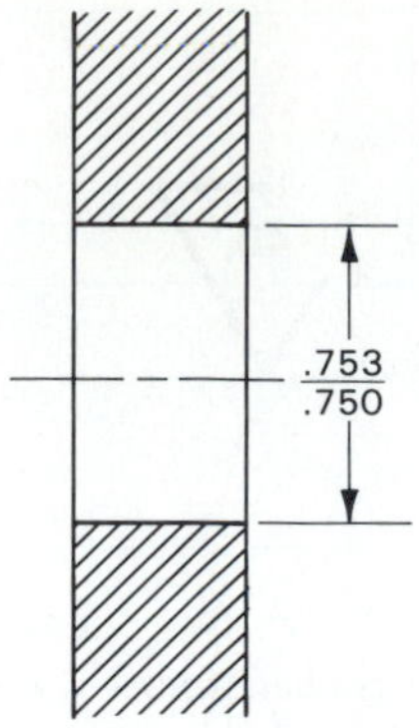

.753
.750

Figure 5-6 *Allowance is the intentional difference in the sizes of mating parts.*

to make extensive notes to describe exactly what is required. These notes not only would be cumbersome but could be misunderstood and therefore result in costly errors. Some of the common drafting terms and symbols are explained in the following paragraphs and examples.

Limits (Fig. 5-4) are the largest and the smallest permissible dimensions of a part (the maximum and minimum dimensions). On a shop drawing both sizes would be given.

EXAMPLE

0.751 largest dimension
0.749 smallest dimension

Tolerance (Fig. 5-5) is the permissible variation of the size of a part. On a drawing the basic dimension plus or minus the variation allowed is given.

EXAMPLE

$$0.750 \begin{array}{l} +0.001 \\ -0.003 \end{array}$$

The tolerance in this case would be 0.004 (the difference between +0.001 oversize or −0.003 undersize).

Allowance (Fig. 5-6) is the intentional difference in the sizes of mating parts, such as the diameter of a shaft and the size of the hole. On a shop drawing both the shaft and the hole would be indicated with maximum and minimum sizes in order to produce the best fit.

Fit is the range of tightness between two mating parts. There are two general classes of fits:

1. *Clearance fits,* whereby a part may revolve or move in relation to a mating part
2. *Interference fits,* whereby two parts are forced together to act as a single piece

Scale size is used on most shop or engineering drawings because it would be impossible to draw parts to exact size; some drawings would be too large, and others would be too small. The scale size of a drawing is generally found in the title block and indicates the scale to which the drawing has been made, which is a representative measurement.

Scale	Definition
1:1	Drawing is made to the actual size of part, or full scale
1:2	Drawing is made to one-half the actual size of the part
2:1	Drawing is made to twice the actual size of the part

SYMBOLS

Some of the symbols and abbreviations used on shop drawings indicate the surface finish, type of material, roughness symbols, and common machine shop terms and operations.

Surface Symbols (SEE Pg 106)

Surface finish is the deviation from the nominal surface caused by the machining operation. Surface finish includes roughness, waviness, lay, and flaws, and is measured by a surface finish indicator in microinches (μin.).

The surface finish mark, used in many cases, indicates which surface of the part must be finished. The number in the $\sqrt{}$ indicates the quality of finish required on the surface (Fig. 5-7). In this case the *roughness height* or the measurement of the finely spaced irregularities caused by the cutting tool cannot exceed 40 μin.

If the surface of a part must be finished to exact specifications, each part of the specification is indicated on the symbol (Fig. 5-8) as follows:

40	Surface finish (ROUGHNESS HEIGHT) in microinches
0.002	Waviness height in thousandths of an inch
0.001	Roughness width in thousandths of an inch
⊥	Machining marks run perpendicular to the boundary of the surface indicated

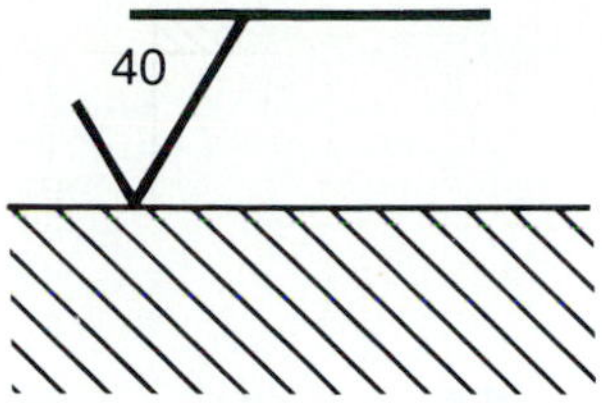
40

Figure 5-7 *Surface finish symbols indicate the type and finish of the surface.*

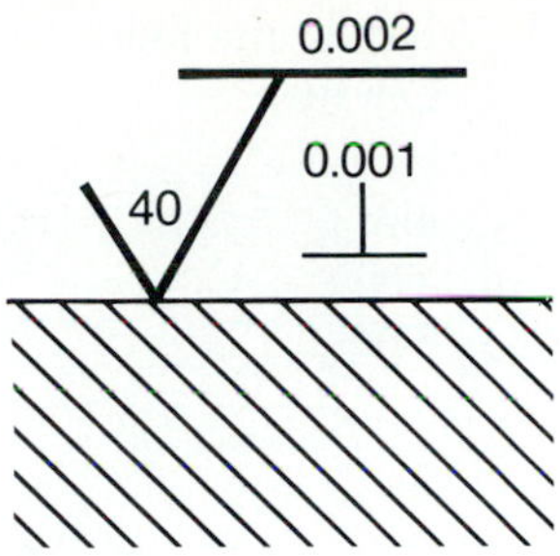
0.002
0.001
40

Figure 5-8 *Surface finish specifications.*

Material Symbols

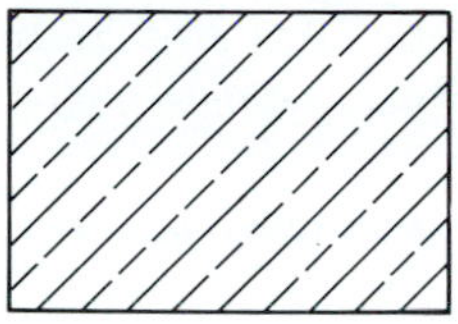

Represents copper, brass, bronze, etc.

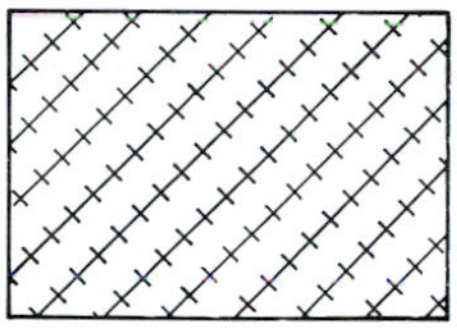

Represents aluminum, magnesium, and their alloys.

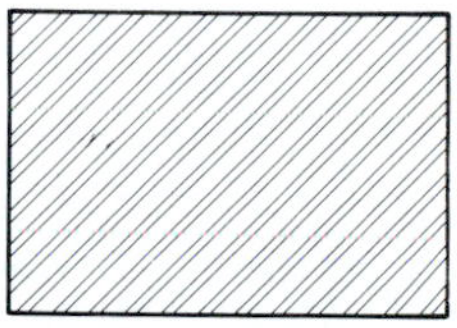

Represents steel and wrought iron.

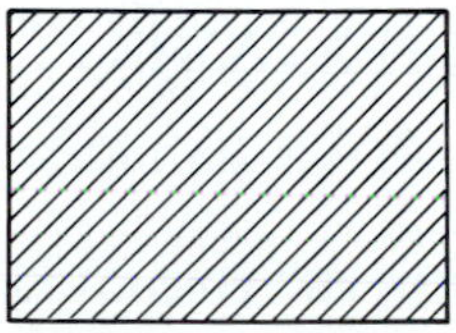

Represents cast iron and malleable iron.

Figure 5-9 *Symbols used to indicate types of material.*

The following symbols indicate the direction of the lay (marks produced by machining operations on work surfaces).

=	Parallel to the boundary line of the surface indicated by the symbol
X	Angular in both directions on the surface indicated by the symbol
M	Multidirectional
C	Approximately circular to the center of the surface indicated by the symbol
R	Approximately radial to the center of the surface indicated by the symbol

Figure 5-9 shows the drafting symbols used to indicate some of the most common materials used in a machine shop.

Common Machine Shop Abbreviations	
CBORE	Counterbore
CSK	Countersink
DIA	Diameter
∅	Diameter
HDN	Harden
L	Lead
LH	Left hand
mm	Millimeter
NC	National coarse
NF	National fine
P	Pitch
R	Radius
Rc	Rockwell hardness test
RH	Right hand
THD	Thread or threads
TIR	Total indicated run-out
TPI	Threads per inch
UNC	Unified national coarse
UNF	Unified national form

REVIEW QUESTIONS

1. How can a drafter indicate the exact specifications required for a part?
2. What is the purpose of:
   a. An assembly drawing?
   b. A detailed drawing?
3. What is the purpose of an orthographic view?
4. Why are section views shown?
5. What lines are used to show:
   a. The form of a part?
   b. The centers of holes, objects, or sections?
   c. The exposed surfaces of where a section is cut?
6. Define:
   a. Limits

b. Tolerance
c. Allowance
7. How is half-scale indicated on an engineering drawing?
8. Define each part of the surface-finish symbol

$$\overset{0.003}{\sqrt[60]{0.002}}$$

9. What do the following abbreviations mean?
a. CBORE
b. HDN
c. mm
d. THD
e. TIR

UNIT 6

Machining Procedures for Various Workpieces

Planning the procedures for machining any part so that it can be produced accurately and quickly is very important. Many parts have been spoiled because the incorrect sequence has been followed in the machining process. Although it would be impossible to list the exact sequence of operations that would apply to every type and shape of workpiece, some general rules should be followed to machine a part accurately and in the shortest time possible.

OBJECTIVES

After completing this unit you should be able to:
1. Plan the sequence of operations and machine round work mounted between lathe centers
2. Plan the sequence of operations and machine round work mounted in a lathe chuck
3. Plan the sequence of operations and machine flat workpieces

MACHINING PROCEDURES FOR ROUND WORK

Most of the work produced in a machine shop is round and is turned to size on a lathe. In industry much of the round work is held in a chuck. A larger percentage of work in school shops is machined between centers because of the need to reset work more often. In either case it is important to follow the correct machining sequence of operations in order to prevent spoiling work, which often happens when incorrect procedures are followed.

GENERAL RULES FOR ROUND WORK

1. Rough-turn all diameters to within 1/32 inch (in.) [0.79 millimeter (mm)] of the size required.
   - Machine the largest diameter first and progress to the smallest.
   - If the small diameters are rough-turned first, it is quite possible that the work would bend when the large diameters are machined.
2. Rough-turn all steps and shoulders to within 1/32 in. (0.79 mm) of the length required (Fig. 6-1).
   - Be sure to measure all lengths from one end of the workpiece.

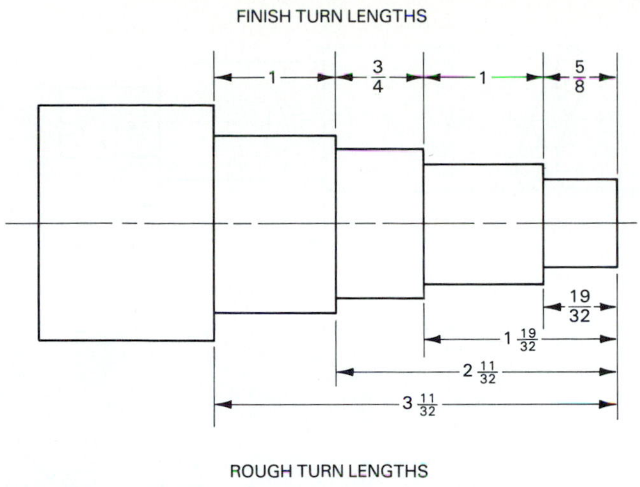
FINISH TURN LENGTHS
1
3/4
1
5/8
19/32
1 19/32
2 11/32
3 11/32
ROUGH TURN LENGTHS

Figure 6-1 *A sample part, showing rough- and finish-turned lengths.*

- If all measurements are not taken from the end of the workpiece, the length of each step would be 1/32 in. (0.79 mm) shorter than required. If four steps were required, the length of the fourth step would be 1/8 in. (3.17 mm) shorter than required (4 × 1/32 in., or 4 × 0.79 mm) and would leave too much material for the finishing operation.

3. If any special operations such as knurling or grooving are required, they should be done next.
4. Cool the workpiece before starting the finishing operations.
   - Metal expands from the friction caused by the machining process, and any measurements taken while work is hot are incorrect.
   - When the workpiece is too cold, the diameters of round work will be smaller than required.
5. Finish-turn all diameters and lengths.
   - Finish the largest diameters first and work down to the smallest diameter.
   - Finish the shoulder of one step to the correct length and then cut the diameter to size.

WORKPIECES REQUIRING CENTER HOLES

There are times when it is necessary to machine the entire length of a round workpiece. When this is required, usually on shorter workpieces, center holes are drilled in each end. The workpiece shown in Fig. 6-2 is a typical part which can be machined between the centers on a lathe.

Machining Sequence

1. Cut off a piece of steel 1/8 in. (3 mm) longer and 1/8 in. (3 mm) larger in diameter than required.
   - In this case the diameter of the steel cut off would be 1 5/8 in. (41 mm) and its length would be 16 1/8 in. (409 mm).
2. Hold the workpiece in a three-jaw chuck, face one end square, and then drill the center hole.
3. Face the other end to length and then drill the center hole.
4. Mount the workpiece between the centers on a lathe.
5. Rough-turn the largest diameter to within 1/32 in. (0.79 mm) finish size or 1 17/32 in. (39 mm).

NOTE: The purpose of the rough cut is to remove excess metal as quickly as possible.

6. Finish-turn the diameter to be knurled.

NOTE: The purpose of the finish cut is to cut work to the required size and produce a good surface finish.

7. Knurl the 1 1/2-in. (38-mm) diameter.
8. Machine the 45° chamfer on the end.
9. Reverse the work in the lathe, being sure to protect the knurl from the lathe dog with a piece of soft metal.
10. Rough-turn the 1 1/4-in. (31-mm) diameter (Fig. 6-2) to 1 9/32 in. (32 mm) (Fig. 6-3).
    - Be sure to leave the length of this section 1/8 in. (3 mm) short [12 7/8 in. (327 mm) from the end] to allow for finishing the 1/8-in. radius.
11. Rough-turn the 1 1/8-in. (28-mm) diameter (Fig. 6-3) to 1 5/32 in. (29 mm).

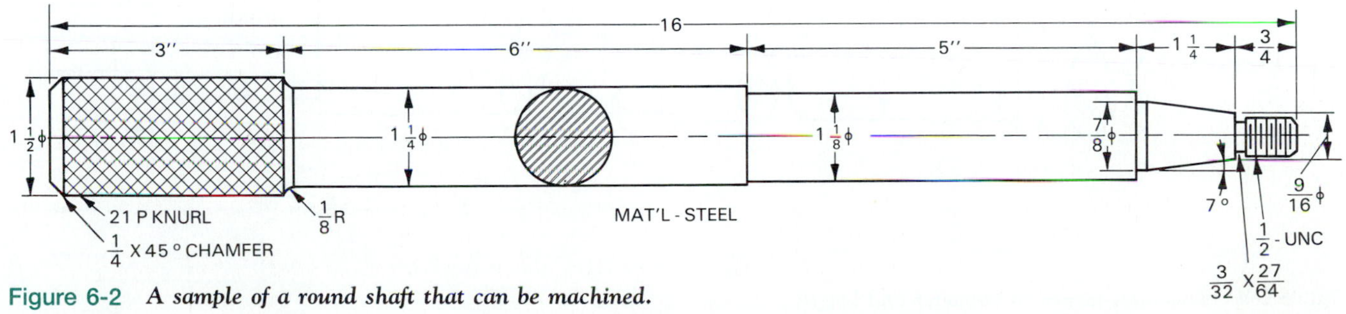
16
3″
6″
5″
1 1/4
3/4
1 1/2 ϕ
1 1/4 ϕ
1 1/8 ϕ
7/8 ϕ
7°
9/16 ϕ
21 P KNURL
1/4 X 45° CHAMFER
1/8 R
MAT'L - STEEL
1/2 - UNC
3/32 X 27/64

Figure 6-2 *A sample of a round shaft that can be machined.*

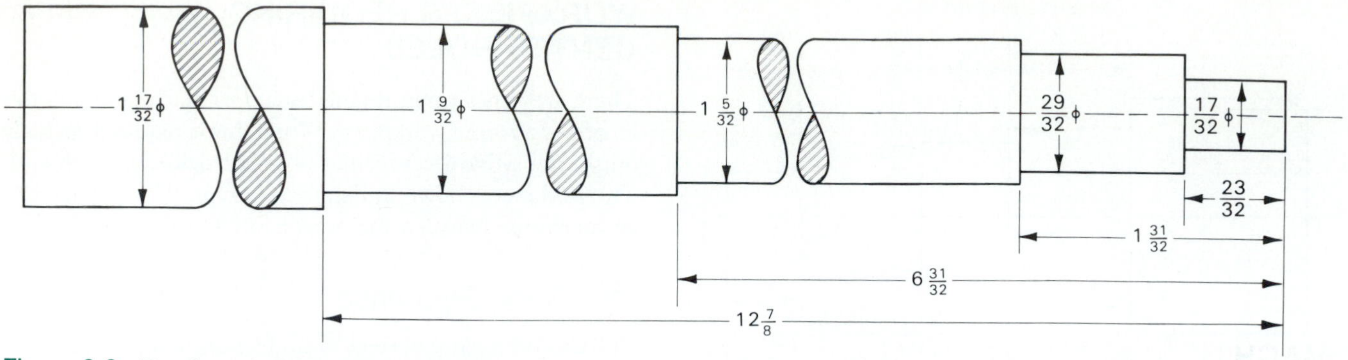
1 17/32 ϕ
1 9/32 ϕ
1 5/32 ϕ
29/32 ϕ
17/32 ϕ
23/32
1 31/32
6 31/32
12 7/8

Figure 6-3 *Rough-turned diameters and lengths of a shaft.*

- Leave the length of this section 1/32 in. (0.79 mm) short [6 31/32 in. (177 mm) from the end] to allow for finishing the shoulder.

12. Rough-turn the 7/8-in. (22-mm) diameter to 29/32 in. (23 mm).
    - Leave the length of this section 1/32 in. (0.79 mm) short [1 31/32 in. (50 mm) from the end] to allow for finishing the shoulder.
13. Rough-turn the 1/2-in. (13-mm) diameter to 17/32 in. (13.48 mm).
    - Machine the length of this section to 23/32 in. (18 mm).
14. Cool the work to room temperature before starting the finishing operations.
15. Finish-turn the 1 1/4-in. (32-mm) diameter to 12 7/8 in. (327 mm) from the end.
16. Mount a 1/8 in. (3-mm)-radius tool and finish the corner to the correct length (Fig. 6-4).
17. Finish-turn the 1 1/8-in. (28-mm) diameter to 7 in. (177 mm) from the end.
18. Finish-turn the 7/8-in. (22-mm) diameter to 2 in. (50 mm) from the end.
19. Set the compound rest to 7° and machine the taper to size.
20. Finish-turn the 1/2-in. (13-mm) diameter to 3/4 in. (19 mm) from the end.
21. With a cut-off tool cut the groove at the end of the 1/2-in. (13-mm) diameter (Fig. 6-5).
22. Chamfer the end of the section to be threaded.
23. Set the lathe for threading and cut the thread to size.

WORKPIECES HELD IN A CHUCK

The procedure for machining the external surfaces of round workpieces held in a chuck (three-jaw, four-jaw, collet, etc.) is basically the same as for machining work held between lathe centers. However, if both external and internal surfaces must be machined on work held in a chuck, the sequence of some operations is changed.

Whenever work is held in a chuck for machining, it is very important that the workpiece be held short for rigidity and to prevent accidents. Never let work extend more than *three times its diameter* beyond the chuck jaws unless it is supported by some means, such as a steady rest or center.

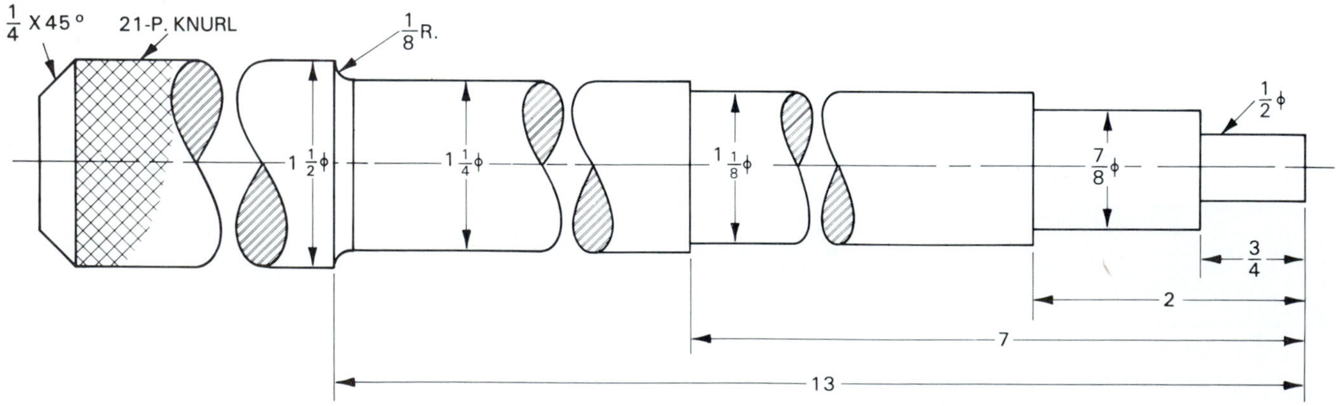
1/4 X 45°
21-P. KNURL
1/8 R.
1 1/2 ϕ
1 1/4 ϕ
1 1/8 ϕ
7/8 ϕ
1/2 ϕ
3/4
2
7
13

Figure 6-4 *The shaft turned to diameter and length.*

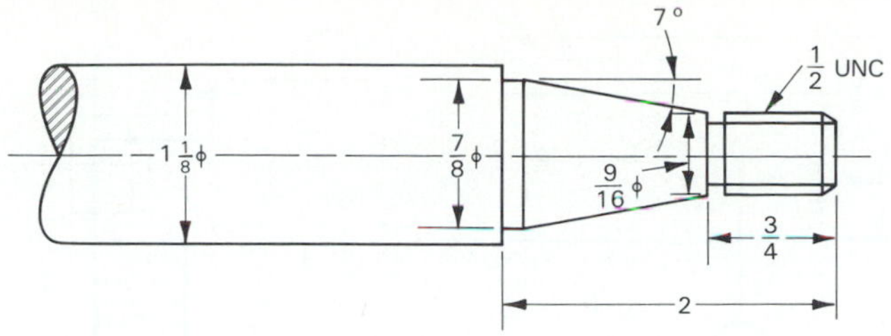

7°
1/2 UNC
1 1/8 ϕ
7/8 ϕ
9/16 ϕ
3/4
2

Figure 6-5 *Special operations completed on a shaft.*

MACHINING EXTERNAL AND INTERNAL DIAMETERS IN A CHUCK

To machine the part shown in Fig. 6-6, the following sequence of operations is suggested.

1. Cut off a piece of steel 1/8 in. (3 mm) larger in diameter and 1/2 in. (13 mm) longer than required.
   - In this case the rough diameter would be 2 1/8 in. (54 mm).
   - The length would be 3 7/8 in. (98 mm); the extra length is to allow the piece to be gripped in the chuck.
2. Mount and center the workpiece in a four-jaw chuck, gripping only 5/16 to 3/8 in. (8 to 9.5 mm) of the material in the chuck jaws.
   - A three-jaw chuck would not hold this size workpiece securely enough for the internal and external machining operations.
3. Face the end of the work square.
   - Remove only the minimum amount of material required to square the end.
4. Rough-turn the three external diameters, starting with the largest and progressing to the smallest to within 1/32 in. (0.79 mm) of size and length.
5. Mount a drill chuck in the tailstock spindle and center-drill the work.
6. Drill a 1/2-in. (13-mm)-diameter hole through the work.
7. Mount a 15/16-in. (24-mm) drill in the tailstock and drill through the work.
8. Mount a boring bar in the toolpost and bore the 1-in. (25-mm) ream hole to 0.968 in. (24.58 mm) in diameter.

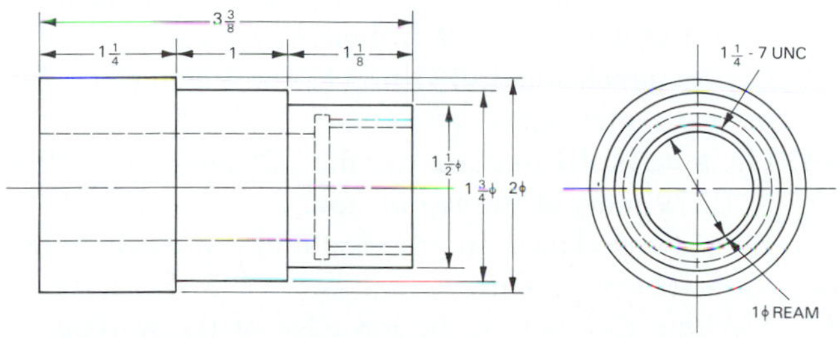

3 3/8
1 1/4
1
1 1/8
1 1/2ϕ
1 3/4ϕ
2ϕ
1 1/4 - 7 UNC
1ϕ REAM

Figure 6-6 *A round part requiring internal and external machining.*

9. Bore the 1 1/4 in.-7 UNC threaded section to the tap-drill size, which is 1.107 in. (28 mm).

$$\text{Tap Drill Size} = \text{TDS} = D - P$$

where D = diameter
P = pitch

10. Cut the groove at the end of the section to be threaded to length and a little deeper than the major diameter of the thread.
11. Mount a threading tool in the boring bar and cut the 1 1/4 in.-7 UNC thread to size.
12. Mount a 1-in. (25-mm) reamer in the tailstock and cut the hole to size.
13. Finish-turn the external diameters to size and length starting with the largest and working down to the smallest.
14. Reverse the workpiece in the chuck and protect the finish diameter with a piece of soft metal between it and the chuck jaws.
15. Face the end surface to the proper length.

MACHINING FLAT WORKPIECES

Since there are so many variations in size and shape of flat workpieces, it is difficult to give specific machining rules for each. Some general rules are listed, but they may have to be modified to suit particular workpieces.

1. Select and cut off the material a little larger than the size required.
2. Machine all surfaces to size in a milling machine in a proper sequence of surfaces.
3. Lay out the physical contours of the part, such as angles, steps, radii, and so on.
4. Lightly prick-punch the layout lines which indicate the surfaces to be cut.
5. Remove large sections of the workpiece on a contour bandsaw.
6. Machine all forms such as steps, angles, radii, and grooves.
7. Lay out all hole locations and with dividers scribe the reference circle.
8. Drill all holes and tap those which require threads.
9. Ream holes.
10. Surface-grind any surfaces which require it.

OPERATIONS SEQUENCE FOR A SAMPLE FLAT PART

The part shown in Fig. 6-7 is used only as an example to illustrate a sequence of operations that should be followed when machining similar parts. These are not meant to be hard-and-fast rules but only guides.

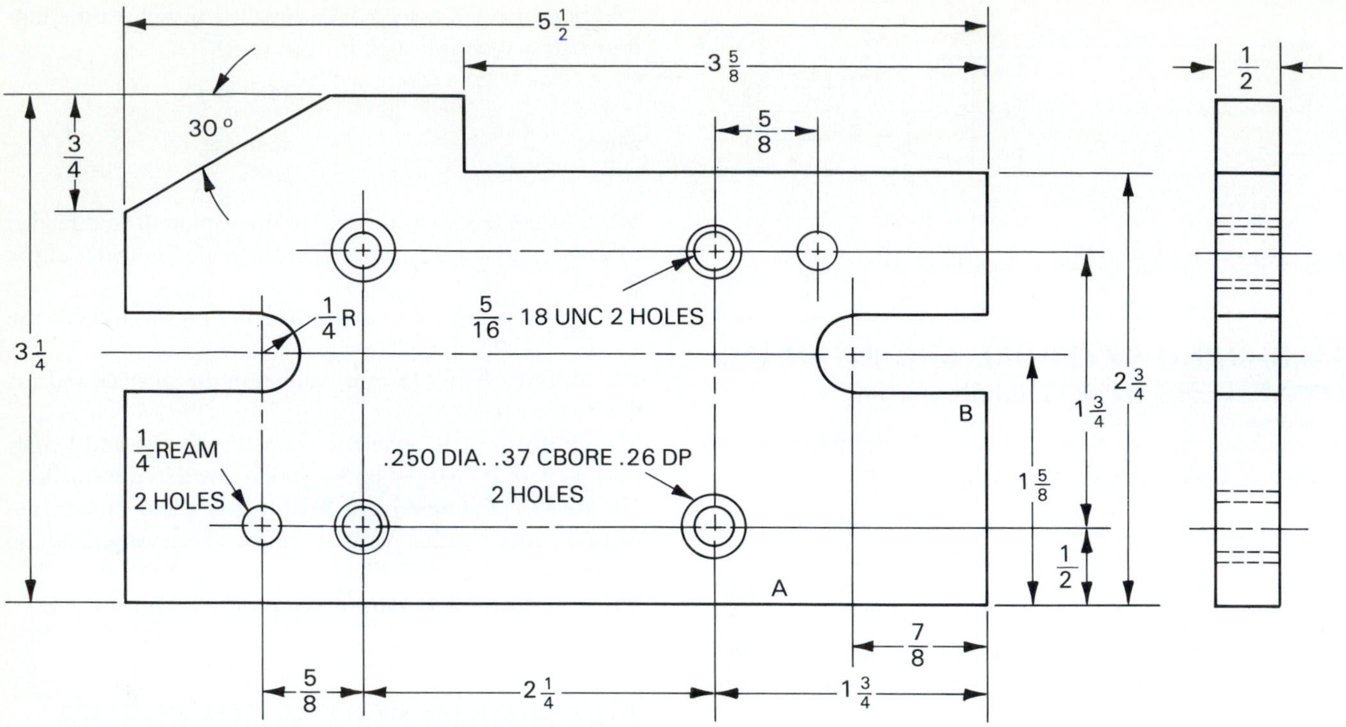

5 1/2
3 5/8
5/8
1/2
30°
3/4
3 1/4
1/4 R
5/16 - 18 UNC 2 HOLES
2 3/4
1 3/4
B
1/4 REAM
2 HOLES
.250 DIA. .37 CBORE .26 DP
2 HOLES
1 5/8
1/2
A
7/8
5/8
2 1/4
1 3/4

Figure 6-7 *A typical flat part that must be laid out and machined.*

The sequence of operations suggested for the sample part shown in Fig. 6-7 is different from those suggested for machining a block square and parallel as outlined in Unit 69 because:

1. The part is relatively thin and has a large surface area.
2. Since at least ⅛ in. (3 mm) of work should be above the vise jaws, it would be difficult to use a round bar between the work and movable jaw for machining the large flat surfaces.
3. A small inaccuracy (out-of-squareness) on the narrow edge would create a greater error when the large surface was machined.

Procedure

1. Cut off a piece of steel ⅝ in. (16 mm) × 3⅜ in. (86 mm) × 5⅝ in. (143 mm) long.
2. In a milling machine, finish one of the larger surfaces (face) first.
3. Turn the workpiece over and machine the other face to ½ in. (13 mm) thick.
4. Machine one edge square with the face.
5. Machine an adjacent edge square (at 90°) with the first edge.
6. Place the longest finished edge (A) down in the machine vise and cut the opposite edge to 3¼ in. (83 mm) wide.

NOTE: Leave 0.010 in. (0.25 mm) on each surface to be ground.

7. Place the narrower finished edge (B) down in the machine vise and cut the opposite edge to 5½ in. (140 mm) long.
8. With edge A as a reference surface, lay out all the horizontal dimensions with an adjustable square, surface gage, or height gage (Fig. 6-8).
9. With edge B as a reference surface, lay out all the vertical dimensions with an adjustable square, surface gage, or height gage (Fig. 6-9).
10. Use a bevel protractor to lay out the 30° angle on the upper right-hand edge.
11. With a divider set to ¼ in. (6 mm), draw the arcs for the two center slots.
12. With a sharp prick punch, lightly mark all the surfaces to be cut and the centers of all hole locations.
13. Center-punch and drill ½-in. (13-mm) diameter holes for the two center slots.
14. On a vertical bandsaw, cut the 30° angle to within 1/32 in. (0.79 mm) of the layout line.
15. Place the workpiece in a vertical mill and machine the two ½-in. (13-mm) slots.
16. Machine the step on the top edge of the workpiece.
17. Set the work to 30° in the machine vise and finish the 30° angle.

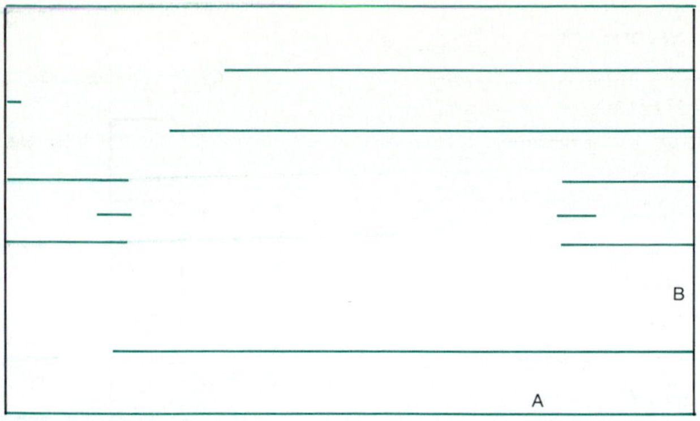
B
A

Figure 6-8 *Lay out all horizontal lines using edge A as the reference surface.*

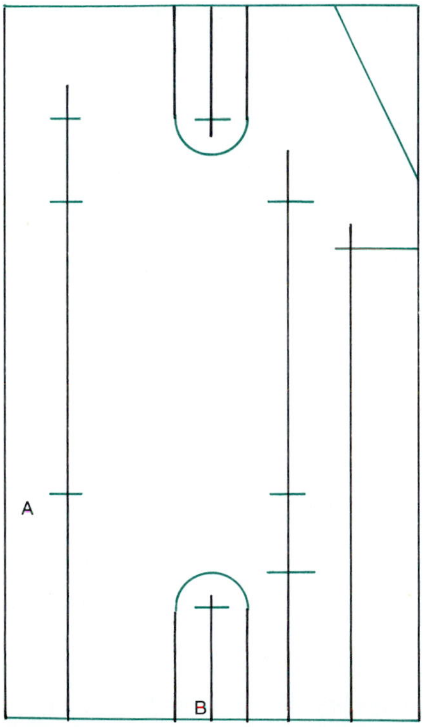
A
B

Figure 6-9 *Lay out all vertical lines using edge B as the reference surface.*

18. Prick-punch the hole locations, scribe reference circles, and then center-punch all hole centers.
19. Center-drill all hole locations.
20. Drill and counterbore the holes for the 1/4 in.-20 NC screws.
21. Tap drill the 5/16 in.-18 thread holes. (F drill or 6.5 mm)
22. Drill the 1/4-in. (6 mm) ream holes to 15/64 in. (5.5 mm).
23. Countersink all holes to be tapped slightly larger than their finished size.
24. Ream the 1/4-in. (6 mm) holes to size.
25. Tap the 5/16 in.-18 UNC holes.

REVIEW QUESTIONS

1. Why is it not advisable to rough-turn small diameters first?
2. To what size should work be rough-turned?
3. Why must all measurements be taken from one end of a workpiece?
4. Why is it important that the workpiece be cooled before finish-turning?
5. Why can a workpiece be bent during the knurling operation?
6. Why is the work cut off longer than required when machining a chuck?
7. How far should a 6-in.-long workpiece, 1 in. in diameter, extend beyond the chuck jaws?
8. How deep should the groove be cut for the internal section to be threaded?
9. How is a finish diameter protected from the chuck jaws?
10. How much material should be left on a flat surface for grinding?
11. When using a bandsaw to remove excess material, how close to the layout line should the cut be made?
12. When machining flat surfaces, what surface should be machined first?

SECTION

5

Measurement

The world has depended on some form of measurement system since the beginning of civilization. The Egyptians, for example, used a unit of length called the cubit, a unit equal to the length of the forearm from the middle finger to the elbow. James Watt, on the other hand, improved his steam engine by maintaining its tolerances to the thickness of a thin shilling, an English coin. The days of such crude measurements, however, are gone. Today we live in a demanding world where products must be built to very precise tolerances. These same products may start out as components built by several subindustries and then be used by other industries in the manufacture of final consumer products. From start to finish a product may be utilized by several industries located in widely separate places, often nations away, and then be sold at still other locations. *Interchangeable manufacture,* world trade, and the need for high precision have all contributed to the need for a *highly accurate international* measurement system. Consequently, in 1960 the International System of Units (SI) was developed to satisfy this need.

Presently, there are two major systems of measurement used in the world. The inch system, often called the English system of measurement, is still widely used in the United States and Canada. However, since over 90 percent of the world's population uses some form of metric measurement, the need for universal adoption of metric measurement is evident.

METRIC (DECIMAL) SYSTEMS

On December 8, 1975, the U.S. Senate passed Metric Bill S100 "to facilitate and encourage the substitution of metric measurement units for customary measurement units in education, trade, commerce and all other sections of the economy of the United States. . . ." On January 16, 1970, the Canadian government adopted SI for implementation throughout Canada by 1980.

Although both the United States and Canada are now reasonably committed to conversion to the metric system as rapidly as possible, it is likely to be some years before all machine tools and measuring devices are redesigned or converted. The change to the metric system in the machine shop will be gradual because of the long life expectancy of the costly machine tools and measuring equipment involved. It is probable, therefore, that people involved in the machine tool trade will have to be familiar with both the metric and the inch systems during the long changeover period.

THE CHANGEOVER PERIOD

Although precision tool manufacturers are producing most measuring tools in metric sizes, the use of these tools is not widespread because of the reluctance of industry to make the costly changeover. Consequently, many present-day machinists will probably have to be familiar with both the inch and the metric systems of measurement.

Since the changeover will be gradual, it is safe to assume that students graduating from schools for the next several years must be taught a dual system of measurement (inch and metric) until the majority of industries have changed to the metric system. With this in mind, we have used dual measurements throughout this book so that the student will be able to work effectively in both systems.

To address these problems, we adopted the following policy for this book to enable you to work effectively in both systems now, while permitting an easy transition to full metric use as new materials and tools become available.

1. Where general measurements or references to quantity are not related specifically to metric standards, tools, or products, inch units are given with the metric equivalent in parentheses.
2. Where the student may be exposed to equipment designed to both metric and inch standards, separate information is given on both types of equipment in *exact* dimensions.
3. Where only inch standards, tools, or products exist at the present time, inch measurements are given with a metric conversion, to two decimal places, provided in parentheses.

SYMBOLS FOR USE WITH SI

Here is a list of some common SI quantities, names, and symbols that you are likely to encounter in your work in the machine shop.

Quantity	Name	Symbol
length*	meter	m
volume*	liter	ℓ and l
mass*	gram	g
time	minute	min
	second	s
force	newton	N
pressure, stress*	pascal	Pa
temperature	degree Celsius	°C
area*	square meter	m^2
velocity (speed)	meters per minute and meters per second	m/min and m/s
angles	degrees	°
	minutes	′
	seconds	″
electric potential	volt	V
electric current	ampere	A
frequency	hertz	Hz
electric capacitance	farad	F

Here is a list of prefixes often used with the quantities indicated by * above.

Prefix	Meaning	Multiplier	Symbol
micro	one-millionth	0.000 001	μ
milli	one-thousandth	0.001	m
centi	one-hundredth	0.01	c
deci	one-tenth	0.1	d
deka	ten	10	da
hecto	one hundred	100	h
kilo	one thousand	1 000	k
mega	one million	1 000 000	M

UNIT

7

Basic Measurement

Basic measurement can be termed as those measurements taken by use of a rule or any other nonprecision measuring tool, whether it be on the inch or metric standard.

OBJECTIVES

After completing this unit you should be able to:
1. Identify several types of steel rules
2. Measure round and flat work to 1/64-in. accuracy with a rule
3. Measure with spring calipers and a rule

INCH SYSTEM

The unit of length in the inch system is the inch, which may be divided into fractional or decimal fraction divisions. The fractional system is based on the binary system, or base 2. The binary fractions commonly used in this system are 1/2, 1/4, 1/8, 1/16, 1/32, and 1/64. The decimal-fraction system has base 10, so any number may be written as a product of ten and/or a fraction of ten.

Value	Fraction	Decimal
one-tenth	1/10	0.1
one-hundredth	1/100	0.01
one-thousandth	1/1000	0.001
one ten-thousandth	1/10,000	0.0001
one hundred-thousandth	1/100,000	0.00001
one millionth	1/1,000,000	0.000001

METRIC SYSTEM

Linear metric dimensions are expressed in multiples and submultiples of the meter. In the machine tool trade, the millimeter is used to express most metric dimensions. Fractions of the millimeter are expressed in decimals.

A brief comparison of common inch and metric equivalents shows:

1 yd = 36 in.
1 m = 39.37 in.
1000 m = 1 km
1 km = 0.621 mi
1 mi = 1.609 km

Table 7-1 shows inch–metric comparisons for common inch–metric system measurements.

NOTE: In machine shop metric measurements, most dimensions will be given in millimeters (mm). Very large dimensions would be given in meters (m) and millimeters (mm). For metric–inch conversion tables and decimal equivalents, see the Appendix of Tables at the end of this book.

FRACTIONAL MEASUREMENT

Fractional dimensions, often called *scale dimensions,* can be measured with such instruments as rulers or calipers. The steel rules used in machine shop work are graduated either in binary-fractional divisions of 1, 1/2, 1/4, 1/8, 1/16, 1/32, and 1/64 of an inch (Fig. 7-1) or in decimal fractional divisions, decimeters, centimeters, millimeters, and half-millimeters (Fig. 7-2). Divisions of 1/64 of an inch or

Table 7-1

Inch Size	Metric Size			
	Millimeter (mm)	Centimeter (cm)	Decimeter (dm)	Meter (m)
1 in.	25.4	2.54	0.254	0.0254
1 ft	304.8	30.48	3.048	0.3048
1 yd	914.4	91.44	9.144	0.9144

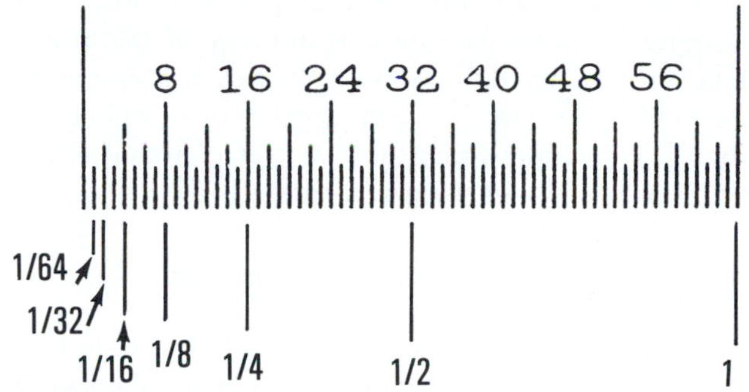

8 16 24 32 40 48 56
1/64
1/32
1/16
1/8
1/4
1/2
1

Figure 7-1 *Fractional divisions of an inch.*

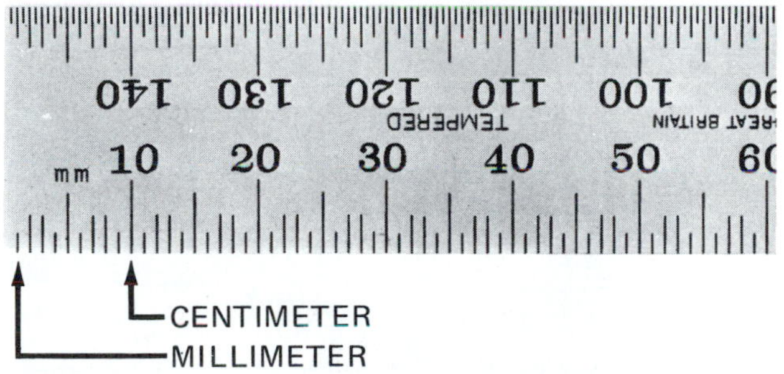

mm 10 20 30 40 50 60
CENTIMETER
MILLIMETER

Figure 7-2 *Metric rules are usually graduated in millimeters and half-millimeters. (Courtesy Kostel Enterprises Limited.)*

0.50 mm are about as fine as can be seen on a rule without the use of a magnifying glass. Precision measuring instruments such as micrometers and verniers are required when metric-drawn prints show any dimensions of less than 0.50 mm or when inch-drawn prints show any dimensions in decimals.

STEEL RULES

METRIC STEEL RULES

Metric steel rules (Fig. 7-3), usually graduated in millimeter and half-millimeter, are used for making linear metric measurements which do not require great accuracy. A wide variety of metric rules are available in lengths from 15 cm to 1 m. The 15-cm rule shown in Fig. 7-3 is 2.4 mm, or about 3⁄32 in., shorter than a standard 6-in. rule.

FRACTIONAL STEEL RULES

The common binary fractions found on inch steel rules are 1⁄64, 1⁄32, 1⁄16, and 1⁄8 of an inch. Several varieties of inch steel rules may be used in machine shop work, such as *spring-tempered, flexible, narrow,* and *hook.* Lengths range from 1 to 72 in. Again, these rules are used for measurements which do not require great accuracy. Spring-tempered quick-reading 6-in. rules (Fig. 7-4) with No. 4 graduations are the most frequently used inch rules in machine shop work. These rules have four separate scales, two on each side. The front is graduated in eighths and sixteenths and the back is graduated in thirty-seconds and sixty-fourths of an inch. Every fourth line is numbered to make reading in thirty-seconds and sixty-fourths easier and quicker.

Hook rules (Fig. 7-5) are used to make accurate measurements from a shoulder, step, or edge of a workpiece.

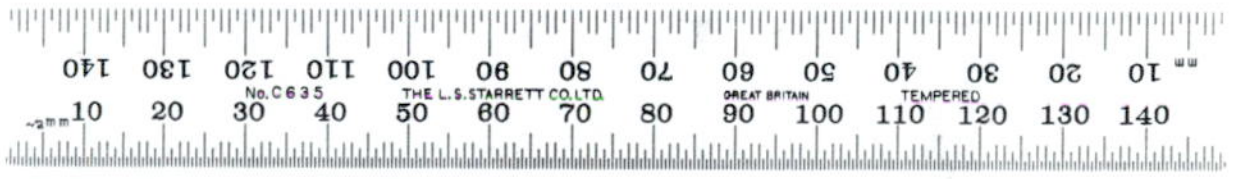

10 20 30 40 50 60 70 80 90 100 110 120 130 140
THE L.S.STARRETT CO.LTD.

Figure 7-3 *A 15-cm metric rule.*

1 2 3 4 5
No.C604RE
L.S.STARRETT CO. LTD.
TEMPERED

Figure 7-4 *A spring-tempered (quick-reading) 6-in. rule.*

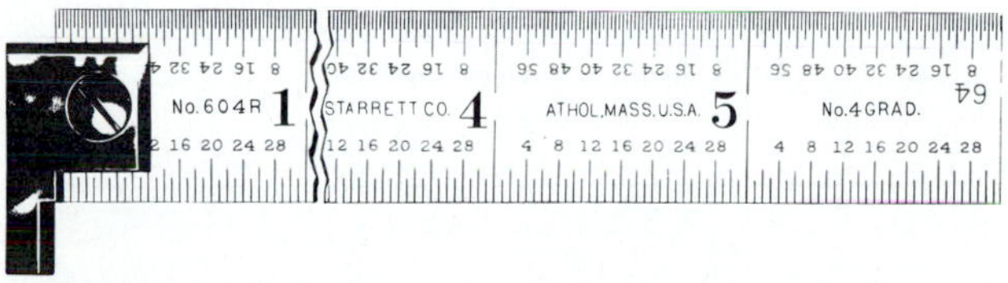

No.604R 1
STARRETT CO. 4
ATHOL,MASS.U.S.A. 5
No.4 GRAD.

Figure 7-5 *A hook rule is used to make accurate measurements from an edge or shoulder. (Courtesy The L. S. Starrett Company.)*

They may also be used to measure flanges, circular pieces, and for setting inside calipers to a dimension.

Short-length rules (Fig. 7-6) are useful in measuring small openings and hard-to-reach locations where an ordinary rule cannot be used. Five small rules come to a set; they range between 1/4 and 1 in. in length and can be interchanged in the holder.

Decimal rules (Fig. 7-7) are most often used when it is necessary to make linear measurements smaller than 1/64 in. Since linear dimensions are sometimes specified on drawings in decimals, these rules are very useful to the machinist. The most common graduations found on decimal rules are 0.100 (1/10 of an inch), 0.050 (1/20 of an inch), 0.020 (1/50 of an inch), and 0.010 (1/100 of an inch). A 6-in. decimal rule is shown in Fig. 7-8.

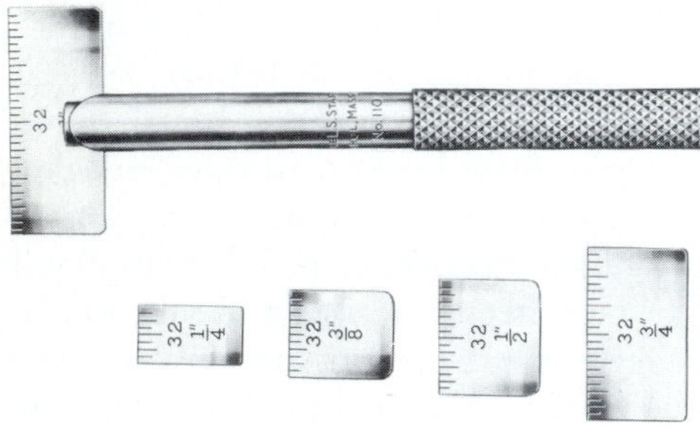

Figure 7-6 *Short-length rules are used for measuring small openings. (Courtesy The L. S. Starrett Company.)*

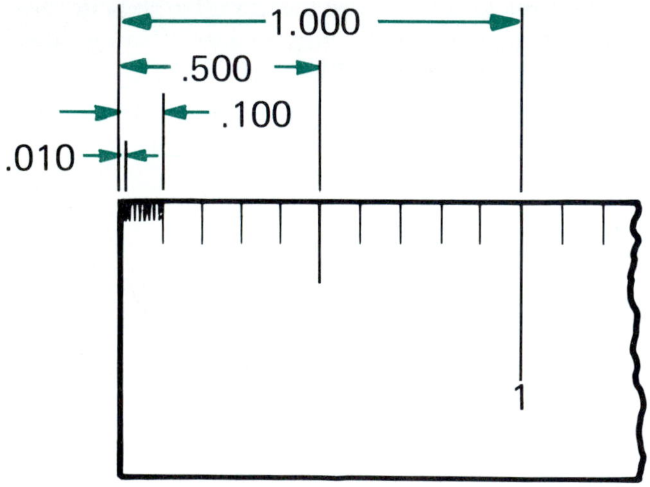
1.000
.500
.100
.010
1

Figure 7-7 *Decimal graduations on a rule provide an accurate and simple form of measurement.*

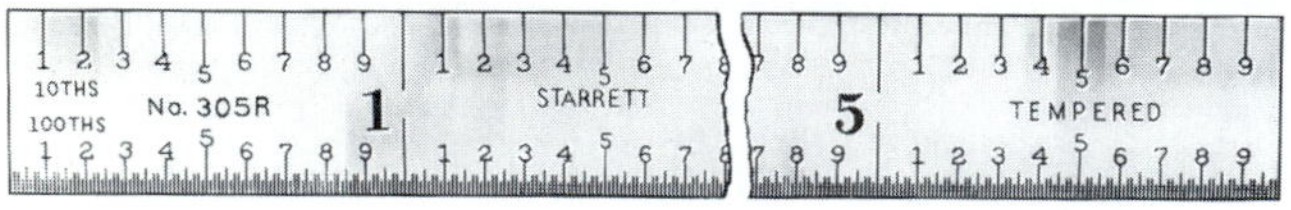
10THS
100THS
No. 305R
STARRETT
TEMPERED

Figure 7-8 *Common graduations found on a 6-in. decimal rule. (Courtesy The L. S. Starrett Company.)*

MEASURING LENGTHS

With a reasonable amount of care, fairly accurate measurements can be made using steel rules. Whenever possible, butt the end of a rule against a shoulder or step (Fig. 7-9) to ensure an accurate measurement.

Through constant use, the end of a steel rule becomes worn. Measurements taken from the end, therefore, are often inaccurate. Fairly accurate measurements of flat work can be made by placing the 1-in. or 1-cm graduation line on the edge of the work, taking the measurement, and subtracting 1 in. or 1 cm from the reading (Fig. 7-10). When measuring flat work, be sure that the edge of the rule is parallel to the edge of the work. If the rule is placed at an angle to the edge (Fig. 7-11), the measurement will not be accurate. When measuring the diameter of round stock, start from the 1-in. or 1-cm graduation line.

THE RULE AS A STRAIGHTEDGE

The edges of a steel rule are ground flat. The rule may therefore be used as a straightedge to test the flatness of workpieces. The edge of a rule should be placed on the

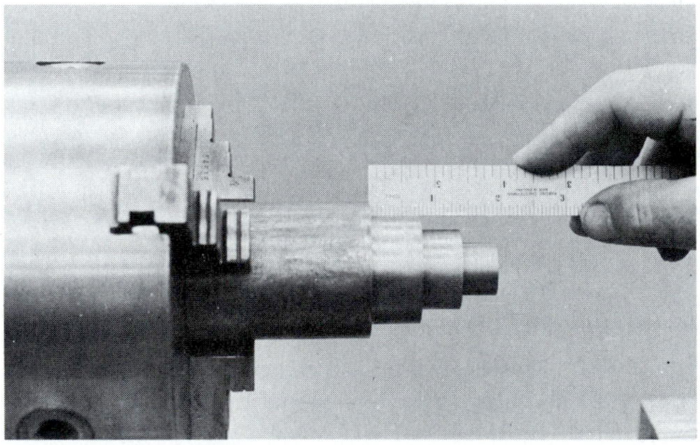

Figure 7-9 *Butting a rule against a shoulder.*

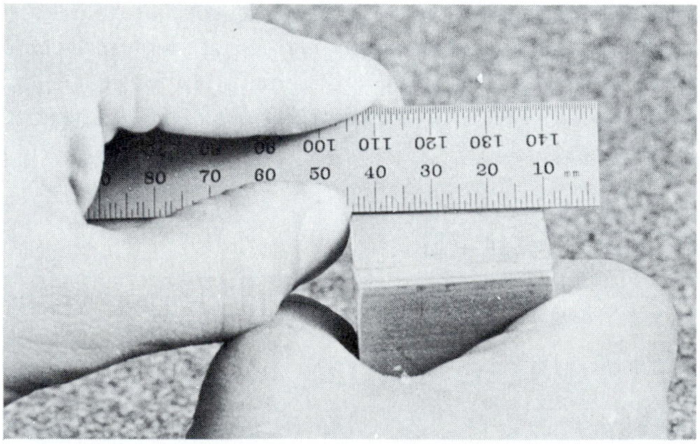

Figure 7-10 *Measuring with a rule starting at the 1-cm line. (Courtesy Kostel Enterprises Limited.)*

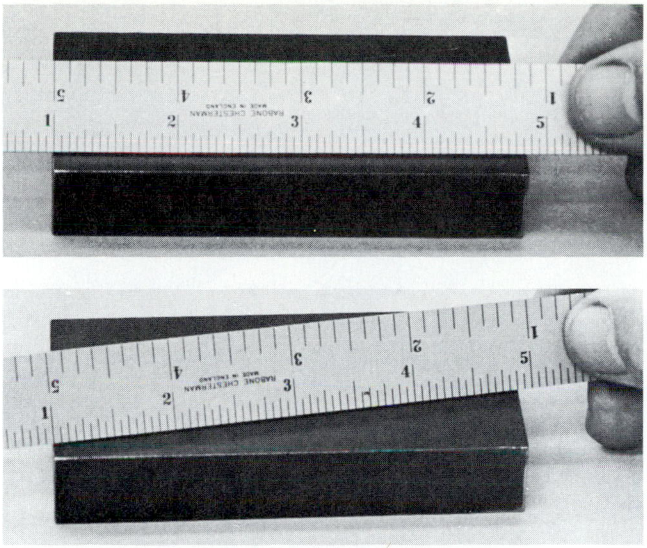

Figure 7-11 *The rule must be held parallel to the edge of the workpiece; otherwise, the measurement will not be correct.*

work surface, which is then held up to the light. Inaccuracies as small as a few thousandths of an inch or 0.05 mm can easily be seen by this method.

OUTSIDE CALIPERS

Outside calipers are tools used to measure the outside surface of either round or flat work. They are made in several styles, such as *spring joint* and *firm joint calipers.* The spring joint caliper consists of two curved legs, a spring, and an adjusting nut. The outside spring joint caliper is most commonly used because it can easily be adjusted to size. The caliper itself cannot be read directly and therefore must be set to a steel rule or a standard-size gage. Therefore, calipers should not be used when an accuracy of less than 0.007 to 0.008 in. (0.18 to 0.20 mm) is required.

SETTING AN OUTSIDE CALIPER TO A RULE

When calipers are set to a rule, it is important that the end of the rule be in good condition and not worn or damaged. Proceed as follows:

1. Place the rule in the left hand with the forefinger extending slightly beyond the end of the rule (Fig. 7-12).
2. Hold the caliper in the right hand and place one leg of the caliper over the end of the rule, supporting it with the left forefinger.

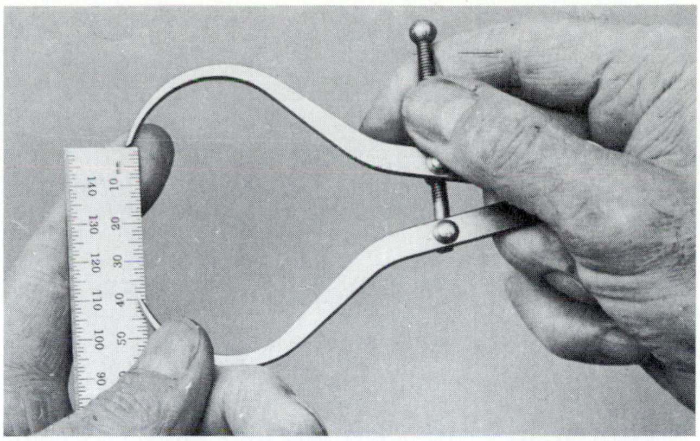

Figure 7-12 *Setting an outside caliper to size with a rule.*

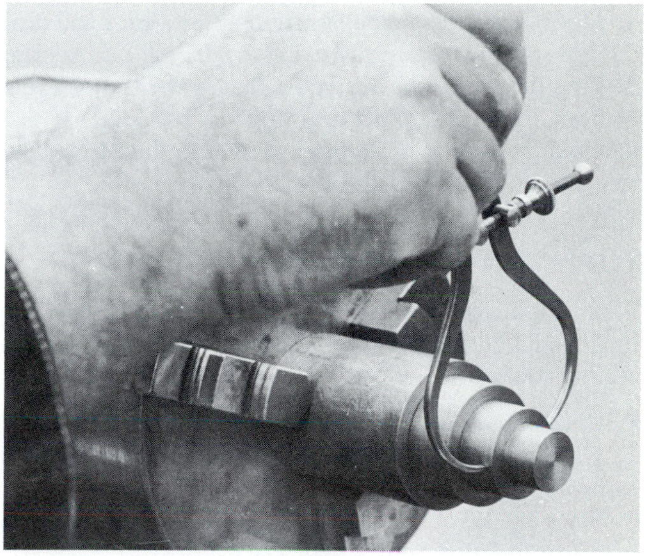

Figure 7-13 *Checking a diameter with an outside caliper.*

3. Hold both legs of the caliper parallel to the edge of the rule and turn the adjusting nut until the end of the lower caliper leg splits the desired graduation on the rule.

MEASURING WITH OUTSIDE CALIPERS

Never attempt to measure work while it is revolving or otherwise moving. Not only is it a dangerous practice, which could result in an accident, but any measurements taken will not be accurate. Proceed as follows:

1. Place the caliper on the work with both legs at right angles to the centerline of the work (Fig. 7-13).
2. Hold the top of the caliper (spring) lightly between the thumb and forefinger.
3. The diameter is correct when the caliper will just slide over the work by its own weight.

NOTE: *Never* force the caliper over the diameter.

INSIDE CALIPERS

Inside calipers are used to measure the diameter of holes or the width of keyways and slots. They are made in several styles, such as *spring joint* and *firm joint calipers.*

MEASURING AN INSIDE DIAMETER

Fairly accurate measurements of holes and slots may be made using an inside caliper and a rule. Proceed as follows:

1. Place one leg of the caliper near the botttom edge of the hole (Fig. 7-14).
2. Hold the caliper leg in this position with a finger.
3. Keep the caliper legs vertical or parallel to the hole.
4. Move the top leg in the direction of the arrows and turn the adjusting nut until a slight drag is felt on the caliper leg.
5. Find the size of the setting by placing the end of a rule and one leg of the caliper against a flat surface.
6. Hold the legs of the caliper parallel to the edge of the rule and note the reading on the rule.

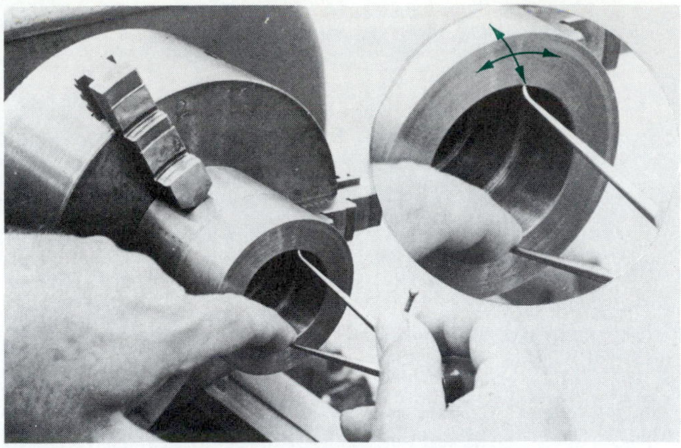

Figure 7-14 *Adjusting an inside caliper to the size of a hole.*

TRANSFERRING MEASUREMENTS

When an accurate measurement is required, the caliper setting should be checked with an outside micrometer. Proceed as follows:

1. Check the accuracy of the micrometer (Unit 9).
2. Hold the micrometer in the right hand so that you can easily adjust it with the thumb and forefinger (Fig. 7-15).
3. Place one leg of the caliper on the micrometer anvil and hold it in position with a finger.
4. Rock the top leg of the caliper in the direction of the arrows.
5. Adjust the micrometer thimble until *only a slight drag* is felt as the caliper leg passes over the measuring face.

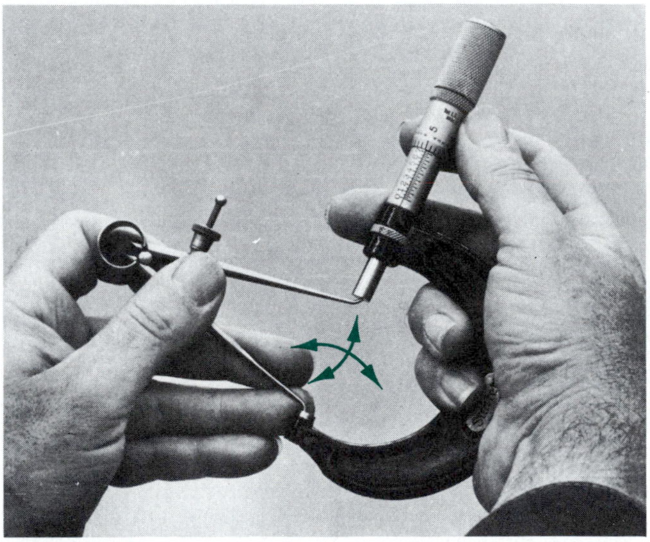

Figure 7-15 *An inside caliper setting being checked with a micrometer.*

REVIEW QUESTIONS

1. Name two systems of measurement presently used in North America.
2. What is the common unit of length in the SI system?
3. How are metric rules usually graduated?
4. Name four types of steel rules used in machine shop work.
5. Describe a rule with No. 4 graduations.
6. State the purpose of:
   a. Hook rules b. Decimal rules
7. Name and describe two types of outside calipers.
8. What is the procedure for setting an outside caliper to a size?
9. Explain how you would know when the work is the same size as the caliper setting.
10. Explain the procedure for setting an inside caliper to the size of a hole.
11. In point form, list the procedure for checking an inside caliper setting with a micrometer.

UNIT 8

Squares and Surface Plates

The square is a very important tool used by the machinist for layout, inspection, and setup purposes. Squares are manufactured to various degrees of accuracy, ranging from semiprecision to precision squares. Precision squares are hardened and accurately ground.

OBJECTIVES

After completing this unit you will be able to identify and explain the uses of:

1. The machinist's combination square
2. Three types of solid and adjustable squares
3. Two types of surface plates

MACHINIST'S COMBINATION SQUARE

The combination square is a basic tool used by the machinist for quickly checking 90 and 45° angles. It is part of a combination set (Fig. 8-1) which includes the square head, the center head, the bevel protractor, and a graduated grooved rule to which the various heads may be attached. In addition to its use for laying out and checking angles, the combination square may also be used as a depth gage (Fig. 8-2) or to measure the length of work to reasonable accuracy. Further uses of the combination set will be discussed in layout work (Unit 19).

PRECISION SQUARES

Precision squares are used chiefly for inspection and setup purposes. They are hardened and accurately ground and must be handled carefully to preserve their accuracy. A great variety of squares are manufactured for specific purposes. The squares are all variations of either the *solid square* or the *adjustable square.*

BEVELED-EDGE SQUARES

The better-quality standard squares used in inspection have a beveled-edge blade, which is hardened and ground. The beveled edge allows the blade to make line contact with the work, thereby permitting a more accurate check. Two methods of using a beveled-edge square for checking purposes are illustrated in Figs. 8-3 and 8-4. In Fig. 8-3, if the work is square (90°), both pieces of paper will be tight between the square and the work. In Fig. 8-4 the light is shut out only where the blade makes line contact with the surface of the work. Light shows through where the blade of the square does not make contact with the surface being checked.

TOOLMAKER'S SURFACE PLATE SQUARE

The *toolmaker's surface plate square* (Fig. 8-5) provides a convenient method of checking work for squareness on a surface plate. Since it is of one-piece construction, there is little chance of any inaccuracy developing, as is the case with a blade and beam square.

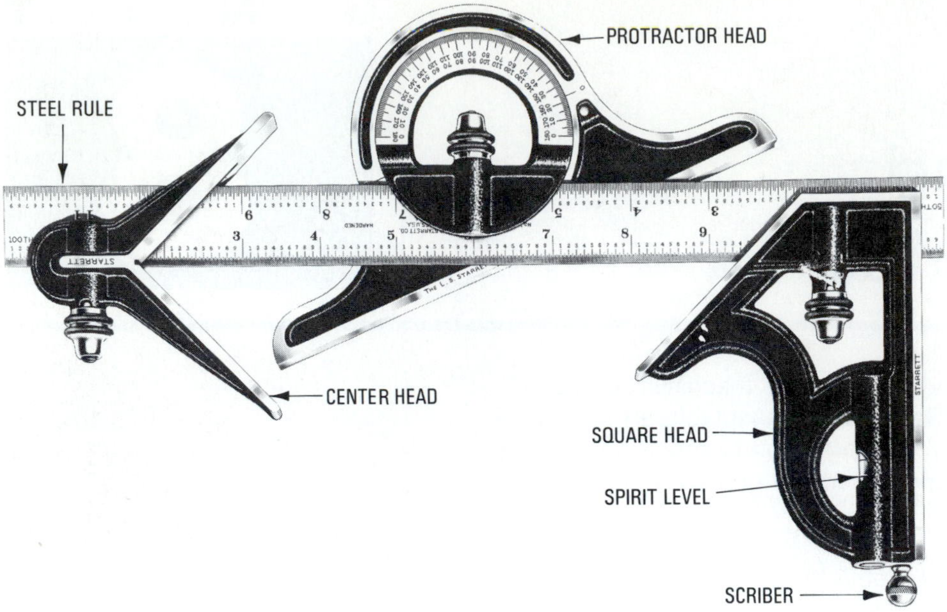
PROTRACTOR HEAD
STEEL RULE
CENTER HEAD
SQUARE HEAD
SPIRIT LEVEL
SCRIBER

Figure 8-1 *The combination set may be used for laying out and checking work.*

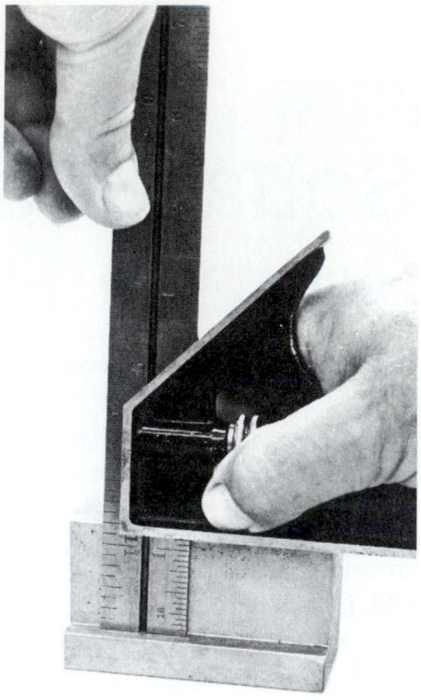

Figure 8-2 *The combination (adjustable) square being used as a depth gage.*

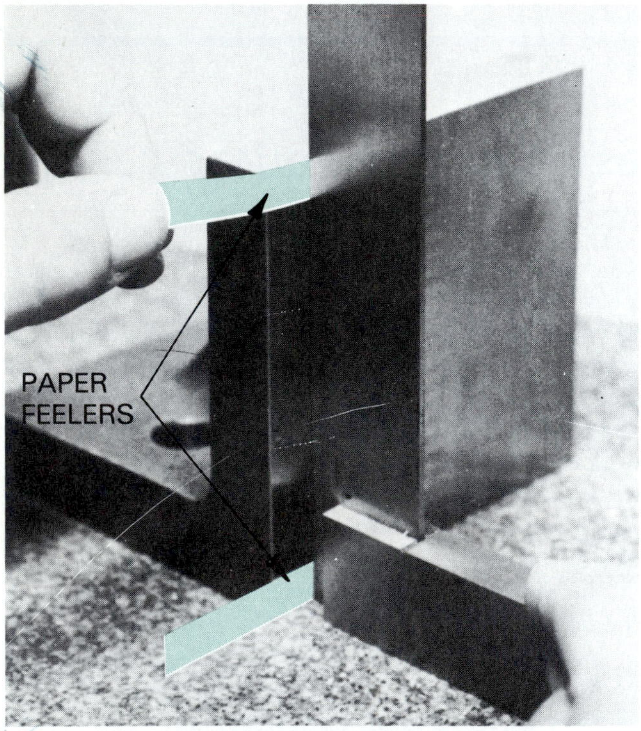
PAPER
FEELERS

Figure 8-3 *Using paper between the blade of the square and the workpiece to check for squareness.*

Figure 8-4 *Light shines through where the blade of the square does not make line contact with the surface.*

CYLINDRICAL SQUARES

Cylindrical squares are commonly used as master squares against which other squares are checked. The square consists of a thick-walled alloy steel cylinder, which has been hardened, ground, and lapped. The outside diameter is a nearly true cylinder, and the ends are ground and lapped square with the axis. The ends are recessed and notched to decrease the inaccuracy from dust and to reduce friction. When a cylindrical square is used, it must be set carefully on a clean surface plate and rotated slightly to force particles of dust and dirt into the end notches; the square can then make the proper contact with surface plate. Cylindrical squares provide perfect line contact with the part being checked.

Another type of cylindrical square is the *direct-reading cylindrical square* (Fig. 8-6), which indicates directly the amount that the part is out of square. One end of the cylinder is lapped square with the axis, while the other end is

Figure 8-5 *A toolmaker's surface plate square.*

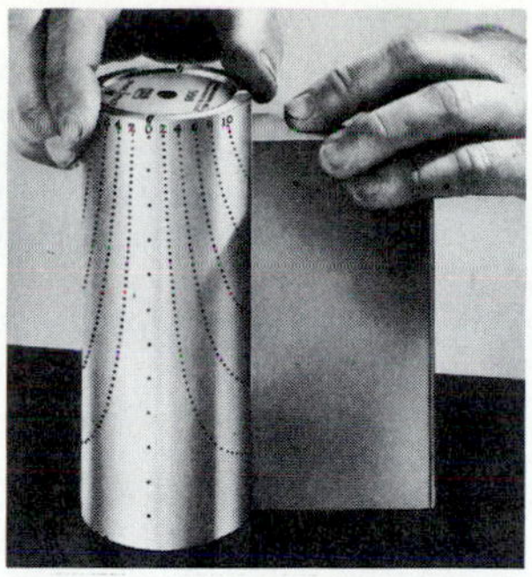

Figure 8-6 *A direct-reading cylindrical square being used to check a part for squareness.*

ground and lapped slightly out of square. The circumference is etched with several series of dots which form elliptical curved lines. Each curve is numbered at the top to indicate the amount, in ten-thousandths of an inch (0.0001), that the workpiece is out of square over the length of the square. Absolute squareness, or *zero deviation,* is indicated by an etched, vertical dotted line on the square.

When used, the square is carefully placed in contact with the work and turned until no light is seen between it and the part being inspected. The uppermost curved line in contact with the work is noted and followed to the top where the number shows the amount the work is out of square. This square may also be used as a conventional cylindrical square if the opposite end, which is ground and lapped square with the axis, is used.

ADJUSTABLE SQUARES

The *adjustable square,* while not providing the accuracy of a good solid square, is used by the toolmaker where it would be impossible to use a fixed square.

A *diemaker's square* (Fig. 8-7) is used to check the clearance angle on dies. The blade is adjusted to the angle of the workpiece by means of a blade-adjusting screw. This angular setting must then be checked with a protractor. Another form of diemaker's square is the *direct-reading* type, which indicates the angle at which the blade is set (Fig. 8-8).

ADJUSTABLE MICROMETER SQUARE

The *adjustable micrometer square* (Fig. 8-9) may be used to check a part for squareness accurately. When a piece of work is being checked and light shows between the blade

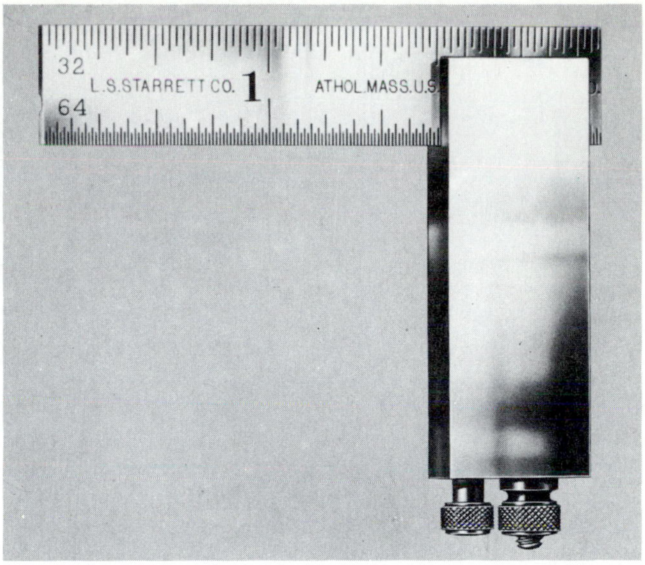
32
L.S.STARRETT CO.
1
ATHOL MASS.U.S.
64

Figure 8-7 *A diemaker's square is useful for checking die clearance.*

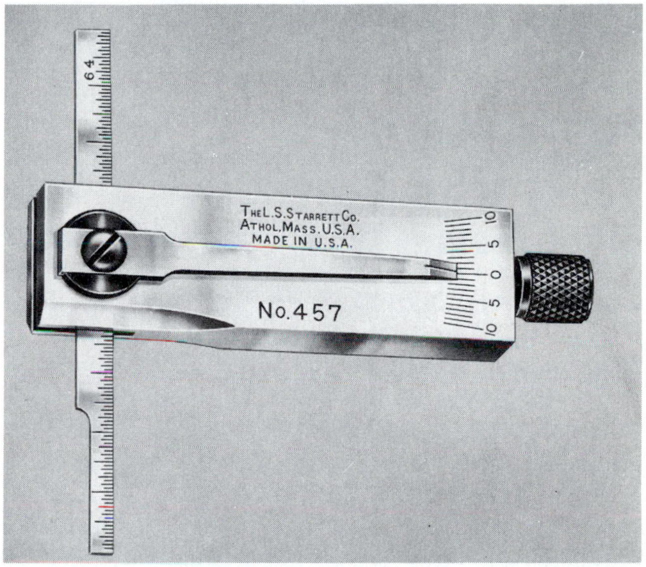
64
THE L.S.STARRETT CO.
ATHOL,MASS.U.S.A.
MADE IN U.S.A.
No.457

Figure 8-8 *The direct-reading diemaker's square indicates the angle at which the blade is set.*

and the work, turn the micrometer head until the full length of the blade, which may be tilted, touches the work. The amount the part is out of square may be read from the micrometer head. When the micrometer head is set at zero, the blade is perfectly square with the beam.

STRAIGHTEDGES

A *straightedge* is used to check surfaces for flatness and to act as a guide for scribing long, straight lines in layout

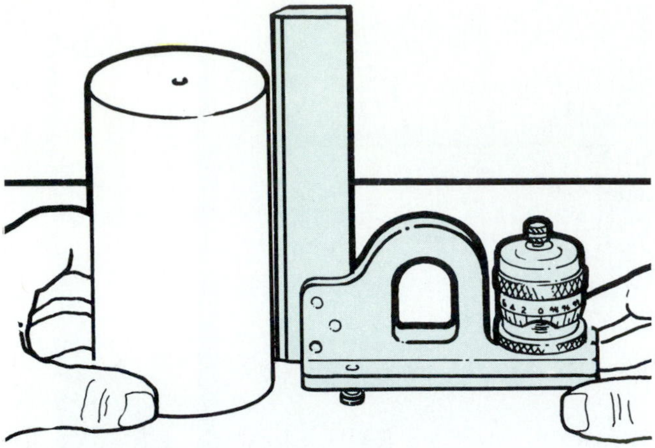

Figure 8-9 *The amount that a part is out of square can be read on an adjustable micrometer square. (Courtesy Ash Precision Equipment Inc.)*

work. Straightedges are generally rectangular bars of hardened and accurately ground steel, having both edges flat and parallel. They are supplied with either plain or beveled edges. Long straightedges are generally made of cast iron with ribbed construction.

SURFACE PLATES

A *surface plate* is a rigid block of granite or cast iron, the flat surface of which is used as a reference plane for layout, setup, and inspection work. Surface plates generally have a three-point suspension to prevent rocking when mounted on an uneven surface.

Cast-iron plates are well ribbed and supported to resist deflection under heavy loads. They are made of close-grained cast iron, which has high strength and good wear-resistance qualities. After a cast-iron surface plate has been machined, its surface must be scraped by hand to a flat plane. This operation is long and tedious; therefore, the cost of these plates is high.

Granite surface plates (Fig. 8-10) have many advantages over cast-iron plates and are replacing them in many shops. They may be manufactured from gray, pink, or black granite and are obtainable in several degrees of accuracy. Extremely flat finishes are produced by lapping.

The advantages of granite plates are:

1. They are not appreciably affected by temperature change.
2. Granite will not burr, as does cast iron; therefore the accuracy is not impaired.
3. They are nonmagnetic.
4. They are rustproof.
5. Abrasives will not embed themselves as easily in the surface; thus they may be used near grinding machines.

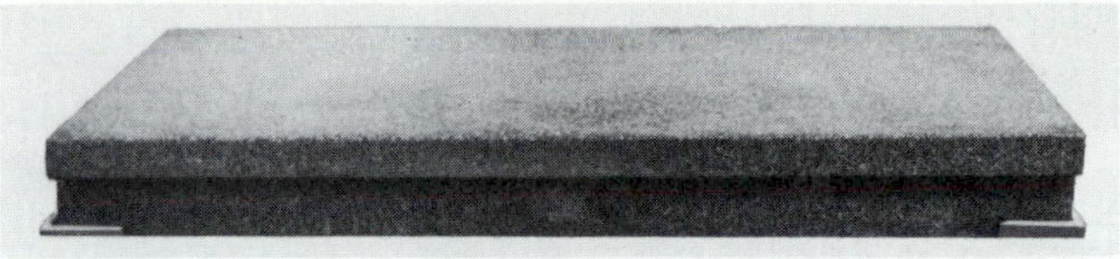

Figure 8-10 *Granite surface plates are not affected by changes in humidity and temperature.*

CARE OF SURFACE PLATES

1. Keep surface plates clean at all times, and wipe them with a dry cloth before using.
2. Clean them occasionally with solvent or surface-plate cleaner to remove any film.
3. Protect them with a wooden cover when they are not in use.
4. Use parallels whenever possible to prevent damage to plates by rough parts or castings.
5. Remove burrs from the workpiece before placing it on the plate.
6. Slide heavy parts onto the plate rather than place them directly on the plate, since a part might fall and damage the plate.
7. Remove all burrs from cast-iron plates by honing.
8. When they are not in regular use, cover cast-iron plates with a thin film of oil to prevent rusting.
9. Center punching or prick punching layout lines should not be done on a surface plate since these plates will not withstand impact forces.

REVIEW QUESTIONS

PRECISION SQUARES

1. Name two types of solid squares and state the advantage of each.
2. Why are beveled-edge squares used in inspection work?
3. What procedure should be followed when using a cylindrical square?
4. State the purpose of a diemaker's square.
5. How can the angle of the workpiece be determined using each type of diemaker's square?

SURFACE PLATES

6. What is the purpose of a surface plate?
7. Name three types of granite used in making surface plates.
8. State five advantages of granite over cast-iron surface plates.
9. List eight ways of caring for surface plates.

PRECISION MEASURING TOOLS—UNITS 9–18

Although fixed gages are convenient for checking the upper and lower limits of external and internal dimensions, they do not measure the actual size of the part. The machinist must use some form of precision measuring instrument to obtain this desired size. Precision measuring tools may be divided into five categories, namely, tools used for outside measurement, inside measurement, depth measurement, thread measurement, and height measurement.

UNIT 9

Micrometers

The *micrometer caliper,* usually called the *micrometer,* is the most commonly used measuring instrument when accuracy is required. The *standard inch micrometer,* shown in a cutaway view in Fig. 9-1, measures accurately to 0.001 in. Since many phases of modern manufacturing require greater accuracy, the *vernier micrometer,* capable of even finer measurements, is being used to an increasing extent.

The standard metric micrometer (Fig. 9-5) measures in hundredths of a millimeter, whereas the vernier metric micrometer measures up to 0.002 mm.

The only difference in construction and reading between the standard inch and the vernier micrometer is the addition of the vernier scale on the sleeve above the index or centerline.

OBJECTIVES

After completing this unit you should be able to:

1. Identify the most common types of outside micrometers and their uses
2. Measure the size of a variety of objects to within 0.001-in. accuracy
3. Read vernier micrometers to 0.0001-in. accuracy
4. Measure the size of a variety of objects to within 0.01-mm accuracy

PRINCIPLE OF THE INCH MICROMETER

In order to understand the principle of the inch micrometer, the student should be familiar with two important thread terms:

- *Pitch,* which is the distance from a point on one thread to a corresponding point on the next thread. For inch threads pitch is expressed as $1/N$ (number of threads). For metric threads it is expressed in millimeters.
- *Lead,* which is the distance a screw thread advances axially in one complete revolution or turn.

Since there are 40 threads per inch on the micrometer, the pitch is 1/40 (0.025) in. Therefore, one complete revolution of the spindle will either increase or decrease the distance between the measuring faces by 1/40 (0.025) in. The 1-in. distance marked on the micrometer sleeve is divided into 40 equal divisions, each of which equals 1/40 (0.025) in.

If the micrometer is closed until the measuring faces just touch, the zero line on the thimble should line up with the index line on the sleeve (barrel). If the thimble is revolved counterclockwise one complete revolution, one line will appear on the sleeve. Each line on the sleeve indicates 0.025 in. Thus, if three lines were showing on the sleeve (or barrel), the micrometer would have opened 3 × 0.025, or 0.075 in.

Every *fourth* line on the sleeve is longer than the others and is numbered to permit easy reading. Each numbered line indicates a distance of 0.100 in. For example, #4 showing on the sleeve indicates a distance between the measuring faces of 4 × 0.100, or 0.400 in.

The thimble has 25 equal divisions about its circumference. Since one turn moves the thimble 0.025 in., one division would represent 1/25 of 0.025 or 0.001. Therefore, each line on the thimble represents 0.001 inch.

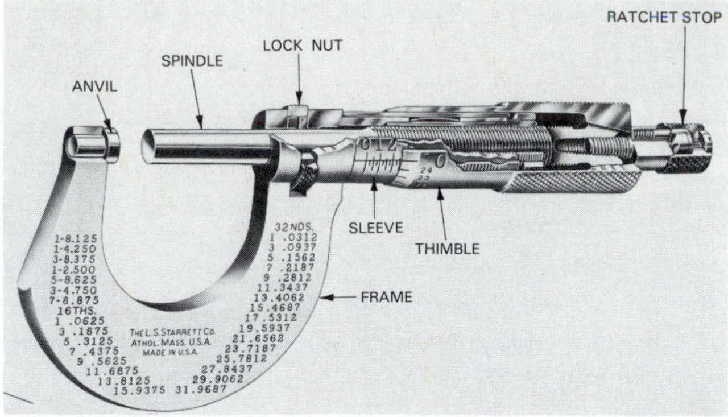
RATCHET STOP
LOCK NUT
SPINDLE
ANVIL
SLEEVE
THIMBLE
FRAME

Figure 9-1 *Cutaway view of a standard micrometer with a ratchet stop.*

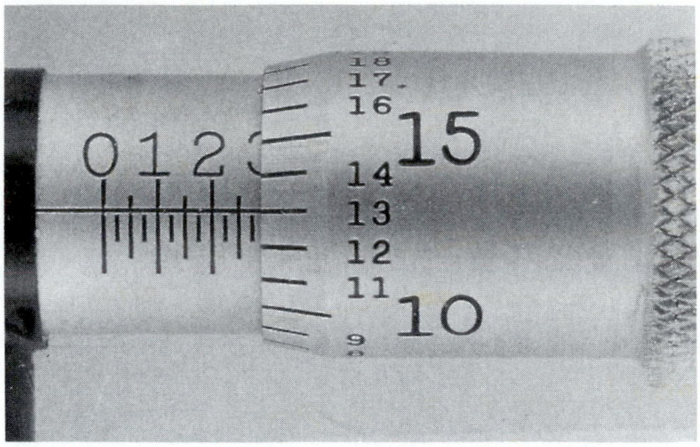

Figure 9-2 *An inch micrometer reading of 0.288 in. (Courtesy Kostel Enterprises Limited.)*

TO READ A STANDARD INCH MICROMETER

1. Note the last number showing on the sleeve. Multiply that number by 0.100.
2. Note the number of small lines visible to the right of the last number shown. Multiply that number by 0.025.
3. Add the number of divisions on the thimble from zero to the line that coincides with the index line on the sleeve.

In Fig. 9-2,

■ #2 is shown on the sleeve	2 × 0.100 = 0.200
■ 3 lines are visible past the number	3 × 0.025 = 0.075
■ #13 line on thimble coincides with the index line	13 × 0.001 = 0.013
Total reading	0.288 in.

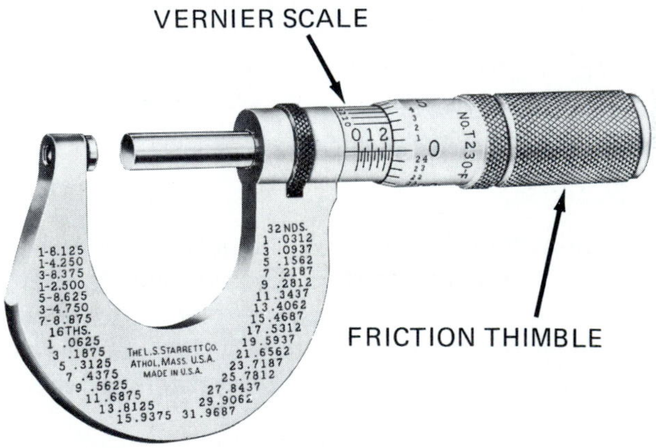
VERNIER SCALE
FRICTION THIMBLE

Figure 9-3 *An inch vernier micrometer caliper with a friction thimble.*

VERNIER MICROMETER

The *inch vernier micrometer* (Fig. 9-3) has, in addition to the graduations found on a standard micrometer, a *vernier scale* on the sleeve. This vernier scale consists of 10 divisions that run *parallel to and above* the index line. These 10 divisions on the sleeve occupy the same distance as 9 divisions (0.009) on the thimble. One division on the vernier scale, therefore, represents 1/10 × 0.009, or 0.0009 in. Since one graduation on the thimble represents 0.001 or 0.0010 in., the difference between one thimble division and one vernier scale division represents 0.0010 − 0.0009, or 0.0001. Therefore, each division on the vernier scale has a value of 0.0001 in.

TO READ A VERNIER MICROMETER

1. Read the vernier micrometer as you would a standard micrometer.
2. Note the line on the vernier scale that coincides with a line on the thimble. This line will indicate the number of ten-thousandths that must be added to the above reading.

Refer to Fig. 9-4. The reading of the vernier micrometer is as follows:

■ #2 is shown on the sleeve	2 × 0.100	= 0.200
■ 1 line is visible past the number	1 × 0.025	= 0.025
■ #11 line on the thimble is just past the index line	11 × 0.001	= 0.011
■ In Fig. 9-4 #3 line on the vernier scale coincides with a line on the thimble	3 × 0.0001	= 0.0003
Total reading		0.2363 in.

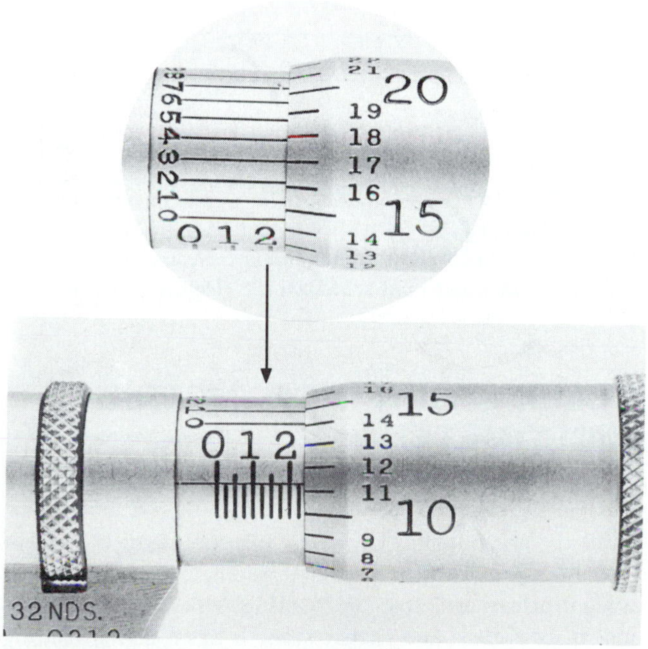
20
19
18
17
16
15
14
0 1 2
15
14
13
12
11
10
9
8
0 1 2
32 NDS.

Figure 9-4 *An inch vernier micrometer reading of 0.2363 in. (Courtesy Kostel Enterprises Limited.)*

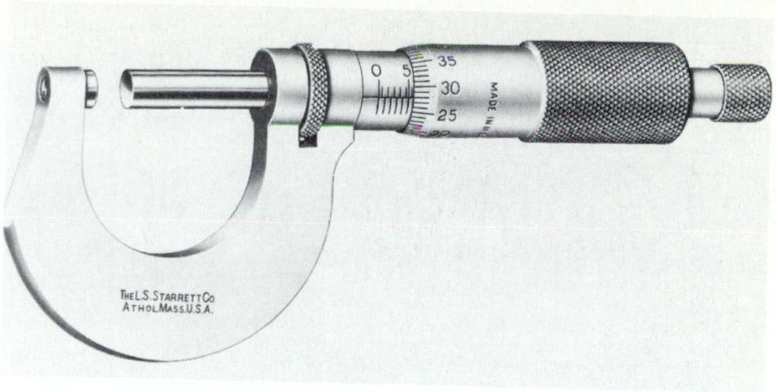
0 5
35
30
25
MADE IN
THE L.S.STARRETT CO.
ATHOL.MASS.U.S.A.

Figure 9-5 *A metric micrometer measures in hundredths of a millimeter.*

METRIC MICROMETER

The *metric micrometer* (Fig. 9-5) is similar to the inch micrometer with two exceptions: the pitch of the spindle screw and the graduations on the sleeve and thimble. The pitch of the screw is 0.5 mm; therefore, a complete revolution of the thimble increases or decreases the distance between the measuring faces 0.5 mm. Above the index line on the sleeve, the graduations are in millimeters (from 0 to 25) with every fifth line being numbered. Below the index line, each millimeter is subdivided into two equal parts of 0.5 mm, which corresponds to the pitch of the thread. It is apparent, therefore, that two turns of the thimble will be required to move the spindle 1 mm.

The circumference of the thimble is divided into 50 equal divisions, with every fifth line being numbered. Since one revolution of the thimble advances the spindle 0.5 mm, each graduation on the thimble equals $\frac{1}{50} \times 0.5$ mm = 0.01 mm.

TO READ A METRIC MICROMETER

1. Note the number of the last main division showing *above the line* to the left of the thimble. Multiply that number by 1 mm.
2. If there is a half-millimeter line showing *below the index line,* between the whole millimeter and the thimble, then add 0.5 mm.
3. Add the number of the line on the thimble that coincides with the index line.

In Fig. 9-6 there are

■ 17 lines above the index line	17 × 1	= 17.00
■ 1 line below the index line	1 × 0.50 =	0.50
■ 11 lines on the thimble	11 × 0.01 =	0.11
Total reading		17.61 mm

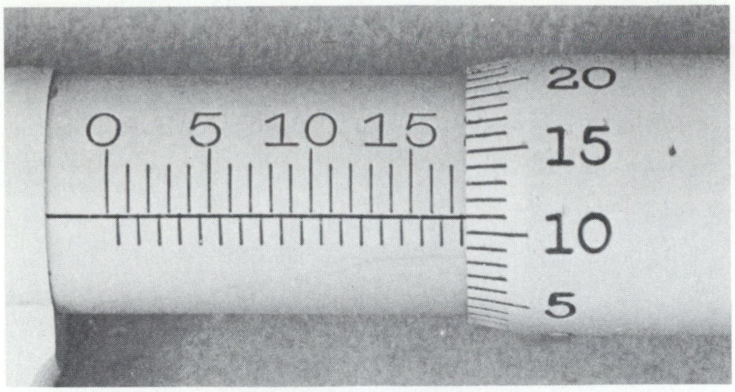

0 5 10 15
20
15
10
5

Figure 9-6 *A metric micrometer reading of 17.61 mm. (Courtesy Kostel Enterprises Limited.)*

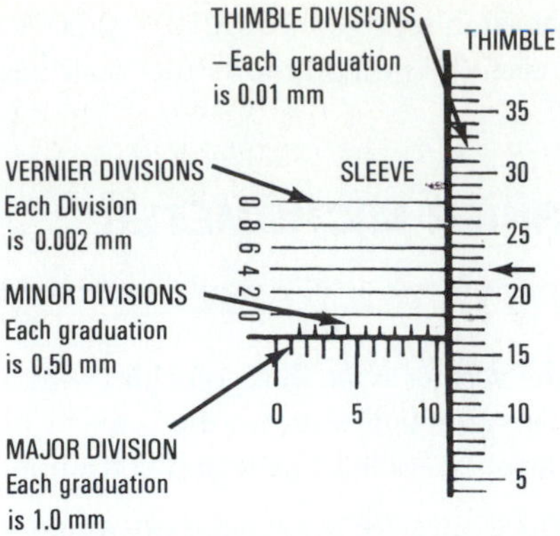

THIMBLE DIVISIONS
–Each graduation
is 0.01 mm
THIMBLE
VERNIER DIVISIONS
Each Division
is 0.002 mm
SLEEVE
MINOR DIVISIONS
Each graduation
is 0.50 mm
MAJOR DIVISION
Each graduation
is 1.0 mm
0 2 4 6 8 0
0 5 10
35
30
25
20
15
10
5

Figure 9-7 *A metric vernier micrometer reading of 10.164 mm.*

METRIC VERNIER MICROMETER

The *metric vernier micrometer,* in addition to the graduations found on the standard micrometer, has five vernier divisions on the barrel, each representing 0.002 mm. In the vernier micrometer reading illustrated in Fig. 9-7, each major division (below the index line) has a value of 1 mm. Each minor division (above the index line) has a value of 0.50 mm. There are 50 divisions around the thimble, each having a value of 0.01 mm.

TO READ A METRIC VERNIER MICROMETER

1. Read the micrometer as you would a standard metric micrometer.
2. Note the line on the vernier scale that coincides with a line on the thimble. This line will indicate the number of two-thousandths of a millimeter that must be added to the above reading.

Refer to Fig. 9-7. The reading of the metric vernier micrometer would be

Major divisions (below the index line)	10 × 1.00 = 10.00
Minor divisions (above the index line)	0 × 0.50 = 0.00
Thimble divisions	16 × 0.01 = 0.16
The second vernier division coincides with a thimble line	2 × 0.002 = 0.004
Total reading	10.164 mm

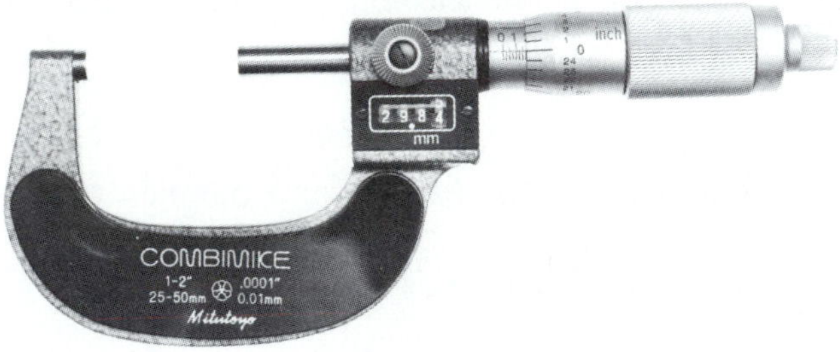

COMBIMIKE
1-2" .0001"
25-50mm 0.01mm
Mitutoyo
mm

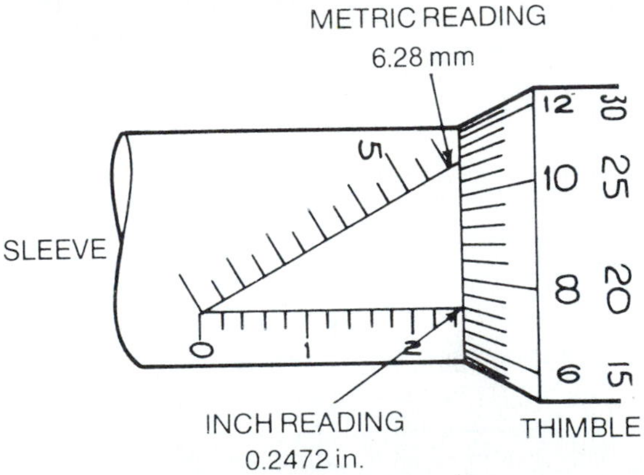

METRIC READING
6.28 mm
SLEEVE
INCH READING
0.2472 in.
THIMBLE

Figure 9-8 *(A) A combination inch–metric micrometer with a digital reading for one system. (Courtesy Mitutoyo/MTI Canada Ltd.) (B) A combination inch–metric micrometer that has dual scales on the sleeve and the barrel.*

COMBINATION INCH–METRIC MICROMETER

With the gradual shift to metric measurement, a dual dimension system will be needed for some time. The combination inch–metric micrometer (Fig. 9-8A) will give readings in both inch and metric sizes. It has a digital reading for one system and a standard barrel and thimble reading for the other.

Another type of inch–metric micrometer (Fig. 9-8B) has dual scales on the sleeve and the barrel. A horizontal scale on the sleeve (usually black) and the left-hand scale on the barrel represent inch readings. The angular scale on the sleeve (usually red) and the right-hand scale on the barrel represent metric readings.

MICROMETER ADJUSTMENTS

Proper care and use of a micrometer is necessary to preserve its accuracy and keep adjustments to a minimum. Minor adjustments to micrometers can easily be made; however, it is extremely important that all parts of the micrometer be kept free from dust and foreign matter during any adjustment.

TO REMOVE PLAY IN THE MICROMETER THREADS

To remove play (looseness) in the spindle threads due to wear:

1. Back off the thimble, as shown in Fig. 9-9.
2. Insert the C-spanner into the slot or hole of the adjusting nut.
3. Turn the adjusting nut clockwise until play between the threads has been eliminated.

NOTE: After the micrometer has been adjusted, the spindle should advance freely while the ratchet stop or friction thimble is being turned.

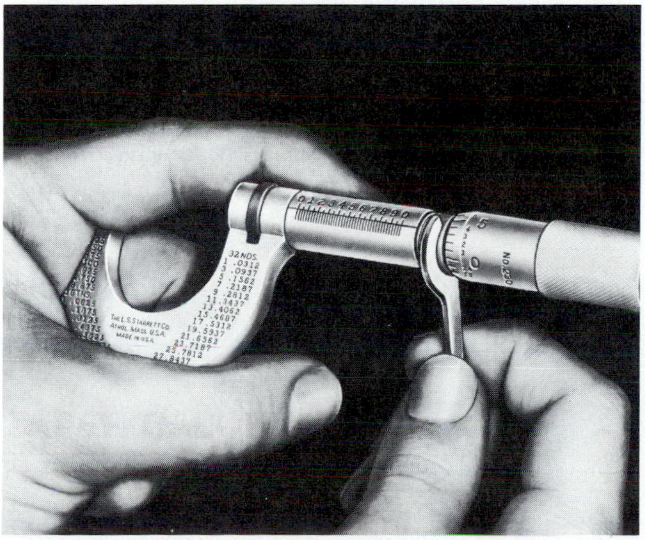

Figure 9-9 *Removing the play in micrometer spindle screw threads. (Courtesy The L. S. Starrett Company.)*

TESTING THE ACCURACY OF MICROMETERS

The accuracy of a micrometer should be tested periodically to ensure that the work produced is the size required. Always make sure that both measuring faces are clean before checking a micrometer for accuracy.

To test a 1 in. or 25 mm micrometer, first clean the measuring faces and then turn the thimble using the friction thimble or ratchet stop until the measuring faces contact each other. If the zero line on the thimble coincides with the center (index) line on the sleeve, the micrometer is accurate. Micrometers can also be checked for accuracy by measuring a gage block or other known standard.

The reading of the micrometer must be the same as the gage block or standard. Any micrometer which is not accurate should be adjusted by a qualified person.

TO ADJUST THE ACCURACY OF A MICROMETER

Should the accuracy of a micrometer require adjustment:

1. Clean the measuring faces and inspect them for damage.
2. Close the measuring faces carefully by turning the ratchet stop or friction thimble.
3. Insert the C-spanner into the hole or slot provided in the sleeve (Fig. 9-10).
4. Carefully turn the *sleeve* until the index line on the sleeve coincides with the zero line on the thimble.
5. Recheck the accuracy of the micrometer by opening the

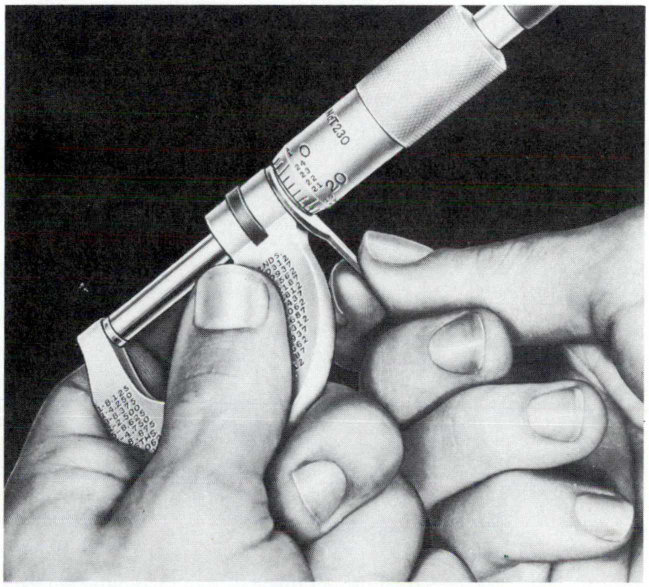

Figure 9-10 *Resetting the accuracy of a micrometer. (Courtesy The L. S. Starrett Company.)*

micrometer and then closing the measuring faces by turning the ratchet stop or friction thimble.

SPECIAL PURPOSE MICROMETERS

Although the design of most micrometers is fairly standard, certain refinements may be added to the basic design, if desired. Items such as the lock ring, ratchet, friction thimble, carbide measuring faces, and anvil extensions increase the accuracy and range of these instruments. Some of the more common types of micrometers used in the machine tool industry are shown in Figures 9-11–9-17.

The *direct-reading micrometer,* Fig. 9-11, has graduations on the thimble and barrel as in a standard micrometer in addition to a digital readout built into the frame. The exact micrometer reading at any point within its range is shown in the readout. Some micrometers combine both the standard inch reading on the thimble and barrel with a millimeter reading.

The *large-frame micrometer,* Fig. 9-12, is made for easier, faster precision measuring of large outside diameters (up to 60 in.). The frame is made of special steel to give it extreme rigidity and the lightest possible weight. Interchangeable anvils give each micrometer a range of 6 in.

The *Mul-T-Anvil micrometer,* Fig. 9-13, comes equipped with round and flat anvils, which are interchangeable. The round (rod) anvil is used to measure the wall thickness of tubing and cylinders and for measuring from a hole to an edge. The flat anvil is used to measure the distance from the inside of slots and grooves to an edge.

The *indicating micrometer,* Fig. 9-14, uses an indicating dial and a movable anvil to permit accurate measurements to ten-thousandths of an inch (0.002 mm). This micrometer may be used as a comparator by setting it to a particular size with gage blocks or a standard and locking

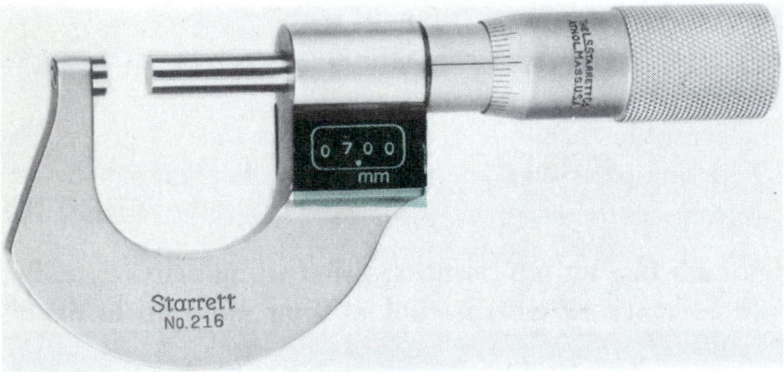
0 7 0 0
mm
Starrett
NO.216

Figure 9-11 *A direct-reading micrometer has graduations like those in a standard micrometer and a digital readout built into the frame. (Courtesy The L. S. Starrett Company.)*

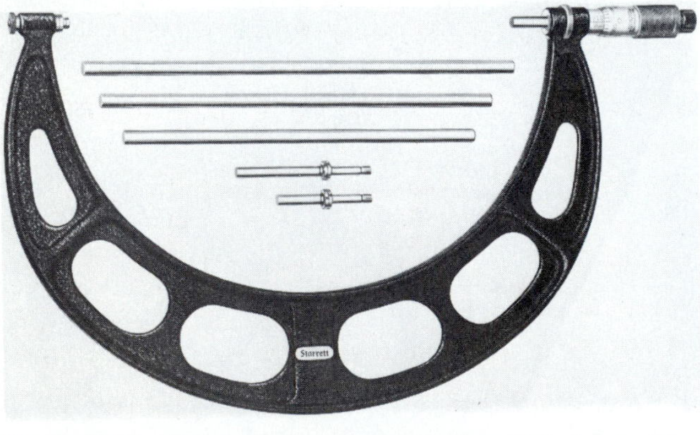

Figure 9-12 *A large-frame micrometer has interchangeable anvils that increase the range of the micrometer. (Courtesy The L. S. Starrett Company.)*

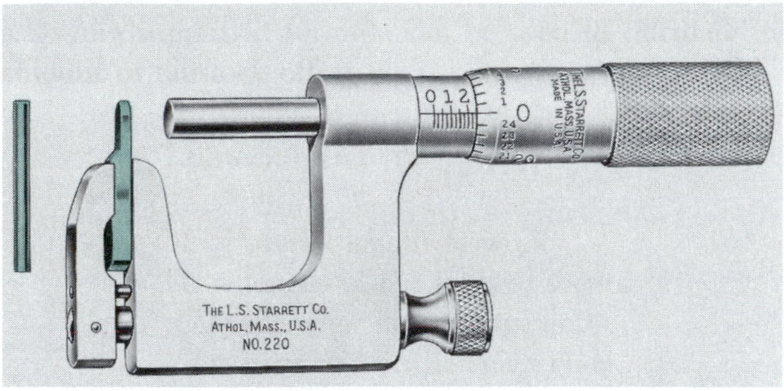
THE L.S. STARRETT CO.
ATHOL, MASS., U.S.A.
NO.220

Figure 9-13 *The Mult-T-Anvil micrometer is used for measuring tubing and distances from a slot to an edge. (Courtesy The L. S. Starrett Company.)*

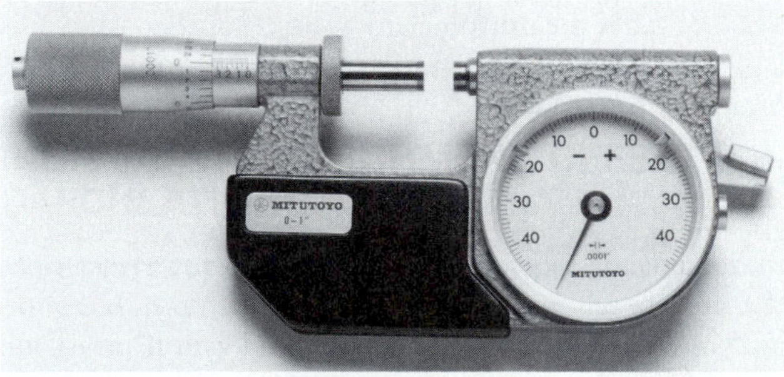
MITUTOYO

Figure 9-14 *An indicating micrometer may be used as a comparator to check parts to ten-thousandths of an inch (0.002 mm). (Courtesy Mitutoyo: MTI Ltd.)*

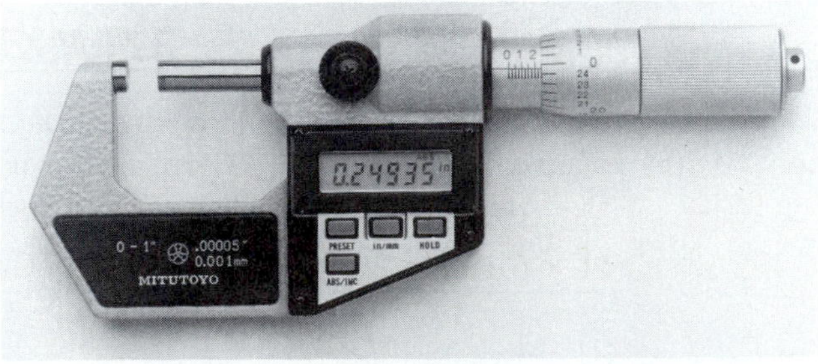
0.24935
PRESET
HOLD
0 - 1"
.00005"
0.001mm
MITUTOYO

Figure 9-15 *The Digi-Matic micrometer provides a digital display of readings accurate to 50 millionths of an inch. (Courtesy Mitutoyo: MTI Ltd.)*

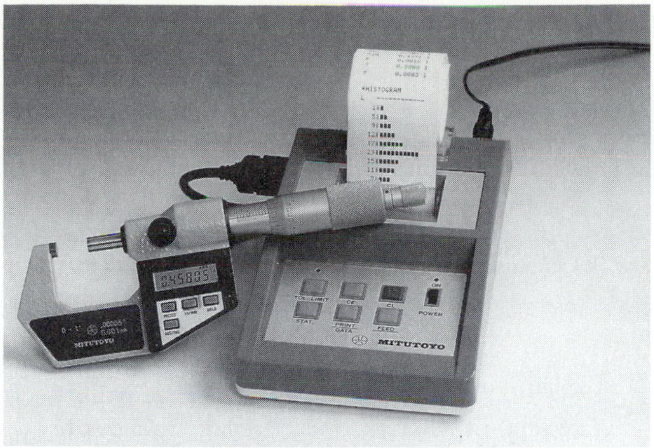
MITUTOYO

Figure 9-16 *The Digi-Matic micrometer with statistical process control is a miniature data processor. (Courtesy Mitutoyo: MTI Ltd.)*

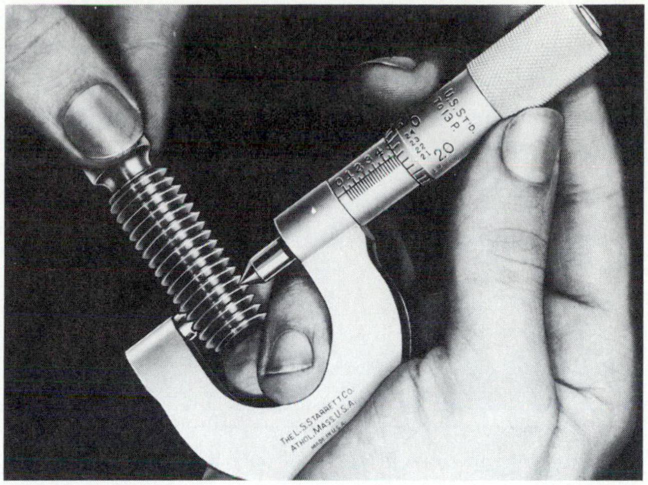

Figure 9-17 *A screw thread micrometer measures the pitch diameter of a thread. (Courtesy The L. S. Starrett Company.)*

the spindle. The tolerance arms are then set to the required limits and each workpiece can be compared with the micrometer setting.

The *Digi-Matic micrometer,* Fig. 9-15, is used as a hand gage for inspecting small parts. It is accurate up to 50 millionths of an inch (0.00127 mm) and displays readings in both inch or metric sizes.

The *Digi-Matic micrometer with statistical process control,* Fig. 9-16, provides a stand-alone inspection system that can be used on the production floor. This unit can be interfaced with a personal or host computer, providing valuable statistics on production quality.

SCREW THREAD MICROMETERS

Sharp-V, American National, Unified, and International Organization for Standardization (ISO) threads may be measured with reasonable accuracy with a screw thread micrometer. This type of micrometer has a pointed spindle and a double-V swivel anvil, which are shaped to contact the pitch diameter of the thread being measured (Fig. 9-17). The micrometer reading indicates the pitch diameter of the thread, which is equal to the outside diameter less the depth of one thread.

Each thread micrometer is limited to measuring a certain range of threads; this range is stamped on the micrometer frame. One-inch thread micrometers are manufactured in four ranges to cover the following range of threads per inch (TPI):

8 to 13 TPI
14 to 20 TPI
22 to 30 TPI
32 to 40 TPI

Metric-thread micrometers are available in sizes from 0 to 25 mm, 25 to 50 mm, 50 to 75 mm, and 75 to 100 mm. A set of 12 anvil and spindle inserts is available for thread pitches from 0.4 to 6 mm.

To check the accuracy of a thread micrometer, carefully bring the measuring faces into light contact; the micrometer reading for this setting should be zero.

When measuring threads, the micrometer gives a slightly distorted reading because of the helix angle of the thread. In order to overcome this inaccuracy, set the thread micrometer to a thread plug gage or to a thread which must be duplicated.

TO MEASURE WITH A THREAD MICROMETER

1. Select the correct thread micrometer to suit the number of threads per inch of the workpiece, or pitch in mm for an ISO thread.
2. Thoroughly clean the measuring surfaces.
3. Check the micrometer for accuracy by bringing the measuring faces together; the reading should be zero.
4. Clean the thread to be measured.
5. Set the micrometer to the required thread plug gage and note the reading.
6. Fit the swivel anvil onto the threaded workpiece.
7. Adjust the spindle until the point just bears against the opposite side of the thread.
8. Carefully roll the micrometer over the thread to get the proper "feel."
9. Note the readings and compare them to the micrometer reading of the thread plug gage.

Threads may also be checked by the *screw thread comparator micrometer,* which has two conical measuring surfaces. Since it does not measure the pitch important to set this instrument to a thread plug gage before measuring a threaded workpiece. The screw micrometer is used for quick comparison of threads, as well as for checking small grooves and recesses where regular micrometers cannot be used.

When thread micrometers or comparators are not available, threads may be accurately checked by the *three-wire method,* which we discuss fully in Unit 49.

REVIEW QUESTIONS

MICROMETERS

1. How many threads per inch are there on a standard inch micrometer?
2. What is the value of:
   a. Each line on the sleeve?
   b. Each numbered line on the sleeve?
   c. Each line on the thimble?
3. Read the following standard micrometer settings.

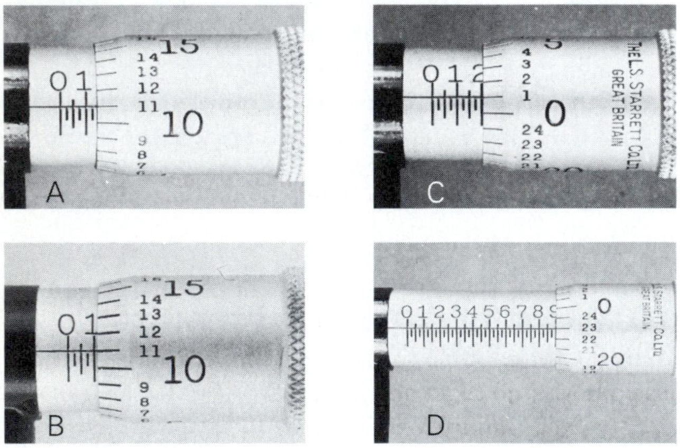
A
C
B
D

4. Describe briefly the principle of the vernier micrometer.
5. Describe the procedure for reading a vernier micrometer.
6. Read the following vernier micrometer settings.

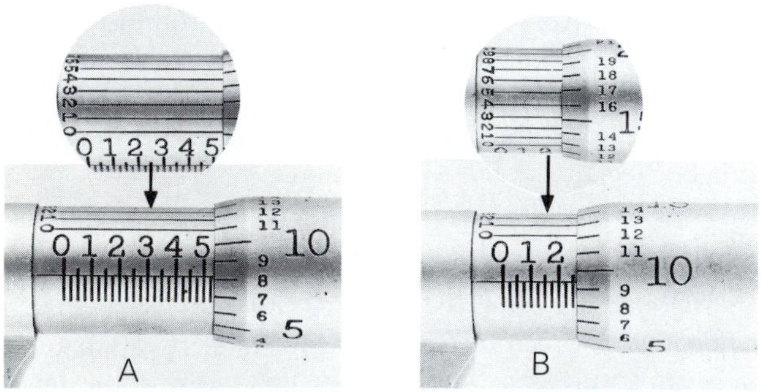
A
B

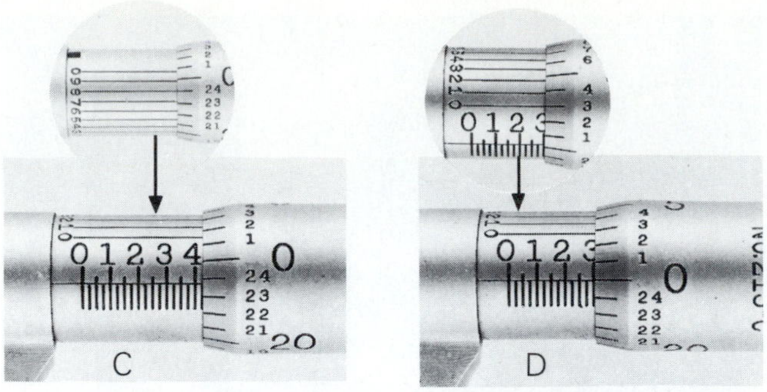
C
D

Explain how to adjust a micrometer:
   a. To remove play in the spindle threads
   b. For accuracy

METRIC MICROMETERS

8. What are the basic differences between a metric and an inch micrometer?
9. What is the value of one division on:
   a. The sleeve above the index line?
   b. The sleeve below the index line?
   c. The thimble?
10. Read the following metric micrometer settings.

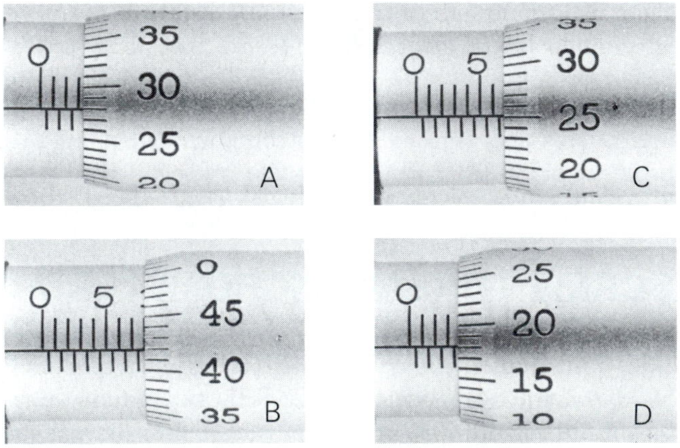
A
C
B
D

INDICATING MICROMETER

11. State two uses for an indicating micrometer.
12. What is the purpose of the relieving button on the indicating micrometer?

SCREW THREAD MICROMETERS

13. Describe the construction of the contact points of a screw thread micrometer.
14. What dimension of the thread is indicated on a screw thread micrometer reading?
15. List the four ranges covered by screw thread micrometers.
16. How may threads be measured *accurately* with a screw thread micrometer?

UNIT 10

Vernier Calipers

Vernier calipers are precision measuring tools used to make accurate measurements to within 0.001 in. for inch verniers or to 0.02 mm for metric verniers. The bar and the movable jaw may be graduated on both sides or both edges. One side is used to take outside measurements; the other, to take inside measurements. Vernier calipers are available in inch and metric graduations; however, some types have both inch and metric graduations on the same caliper.

OBJECTIVES

After completing this unit you should be able to:

1. Measure workpieces to within an accuracy of 0.001 in. with a 25-division inch vernier caliper
2. Measure workpieces to within an accuracy of 0.001 in. using a 50-division inch vernier caliper
3. Measure workpieces to within an accuracy of 0.02 mm using a metric vernier caliper

VERNIER CALIPERS

Vernier calipers (Fig. 10-1) are precision tools used to make accurate measurements to within 0.001 in. or 0.02 mm, depending on whether they are inch or metric vernier calipers.

PARTS OF THE VERNIER CALIPER

The vernier caliper, regardless of the standard of measurement used, consists of an L-shaped frame and a movable jaw.

The L-shaped frame consists of a *bar,* which shows the *main scale graduations,* and the *fixed jaw.* The *movable jaw,* which slides along the bar, contains the *vernier scale.* Adjustments for size are made by means of an *adjusting nut.* Readings may be locked into place by means of the *clamp screws.*

Most bars are graduated on both sides or on both edges, one for outside measurements and the other for inside measurements. The outer tips of the jaws are cut away to form nibs, which permit inside measurements to be taken. Inch vernier calipers are manufactured with both 25- and 50-division vernier scales. The 50-division scale is much easier to read than the 25-division scale. Metric vernier calipers graduated in millimeters are also available.

Some vernier calipers are provided with two small indentations, or points, on the bar and a movable jaw, which may be used to set dividers accurately to a specific dimension or radius.

The bar of the vernier caliper with the 25-division vernier scale on the movable jaw is graduated exactly the same as a micrometer. Each inch is divided into 40 equal divisions, each having a value of 0.025 in. Every fourth line, representing 1/10 or 0.100, is numbered. The vernier scale on the movable jaw has 25 equal divisions, each representing 0.001. The 25 divisions on the vernier scale, which are 0.600 in. in length, are equal to 24 divisions on the bar. The difference between *one* division on the bar and one vernier division equals 0.025 − 0.024, or 0.001 in. Therefore, only one line of the vernier scale will line up exactly with a line on the bar at any one setting.

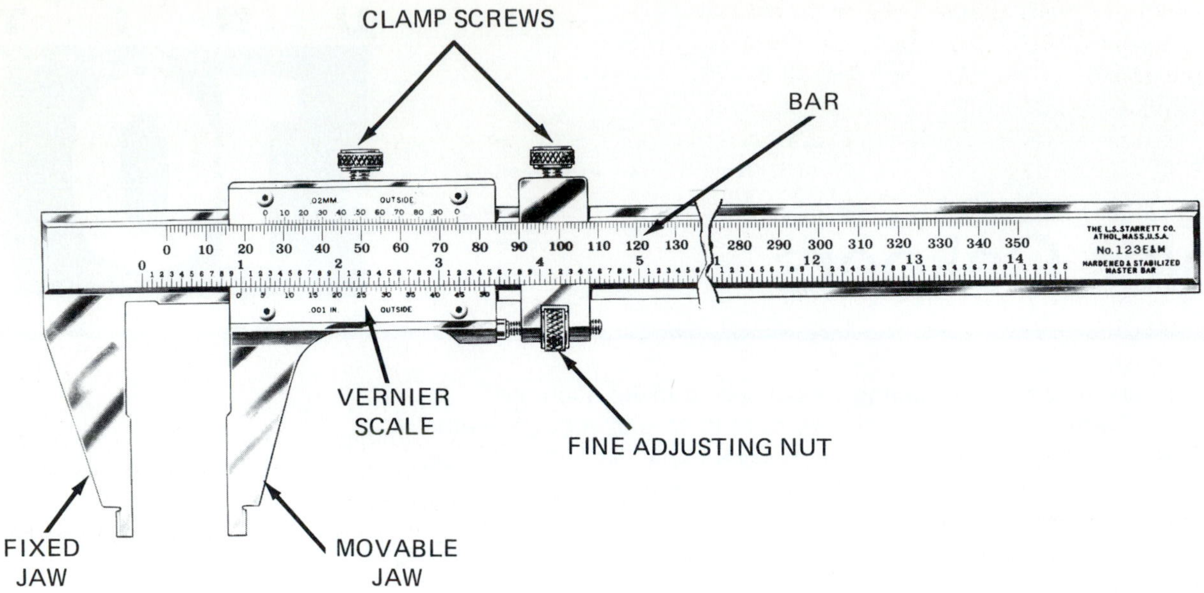
CLAMP SCREWS
BAR
VERNIER
SCALE
FINE ADJUSTING NUT
FIXED
JAW
MOVABLE
JAW
Figure 10-1 *Parts of a vernier caliper.*

MEASURING A WORKPIECE WITH A 25-DIVISION INCH VERNIER CALIPER

1. Remove all burrs from the workpiece and clean the surface to be measured.
2. Open the jaws enough to pass over the work.
3. Close the jaws against the work and lock the right-hand clamp screw.
4. Turn the adjusting screw until the jaws *just touch* the work surface. Be sure that the jaws are in place by attempting to move the bar slightly sideways and vertically while turning the adjusting nut.
5. Lock the clamp screw on the movable jaw.
6. Read the measurement shown in Fig. 10-2 as follows:

■ The large #1 on the bar		= 1.000
■ The small #4 past the #1	4 × 0.100	= 0.400
■ One line is visible past the #4	1 × 0.025	= 0.025
■ The 11th line of the vernier scale coincides with a line on the bar	11 × 0.001	= 0.011
Total reading		1.436 in.

THE 50-DIVISION INCH VERNIER CALIPER

Because 25-division vernier calipers are often difficult to read, many vernier calipers are now manufactured with 50 divisions (equal to 49 on the main scale) on the vernier

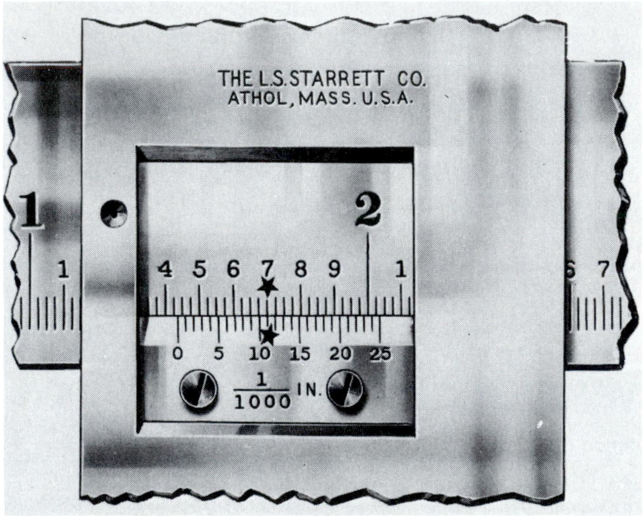
THE L.S.STARRETT CO.
ATHOL, MASS. U.S.A.
1000 IN.
Figure 10-2 *A 25-division inch vernier caliper reading of 1.436 in. (Courtesy The L. S. Starrett Company.)*

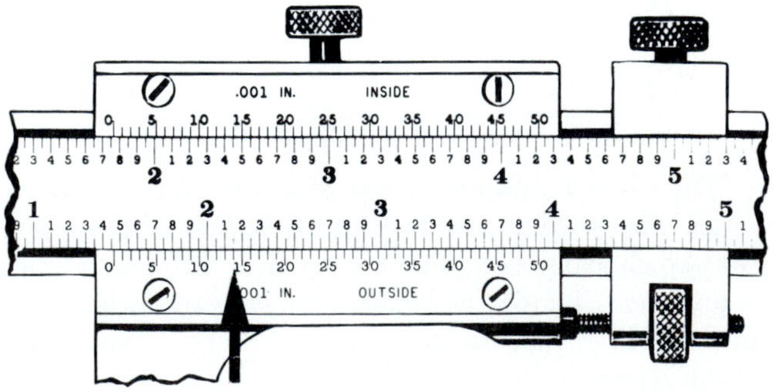
.001 IN.
INSIDE
OUTSIDE
Figure 10-3 *A 50-division inch vernier caliper reading of 1.464 in.*

scale of the movable jaw. Each of these scales on the bar and movable jaw is equal to 2.450 in. in length. Each division on the bar then equals 2.450 divided by 49 divisions, or 0.050 in. in length. Each division on the vernier scale would equal 2.450 divided by 50 divisions, or 0.049 in. in length. The *difference in length* between one main scale division and one vernier division equals 0.050 − 0.049, or 0.001 in.

Each line on the main scale of a 50-division vernier caliper has a value of 0.050 in. Each line on the vernier scale has a value of 0.001 in. In Fig. 10-3,

■ The large #1 on the bar		= 1.000
■ The small #4 past the #1	4 × 0.100	= 0.400
■ 1 line is visible past #4	1 × 0.050	= 0.050
■ The 14th line on the vernier scale coincides with a line on the bar	14 × 0.001	= 0.014
Total reading		1.464 in.

THE METRIC VERNIER CALIPER

Vernier calipers are also made with metric readings and many have both metric and inch graduations on the same instrument (Fig. 10-4). The parts of metric vernier calipers are the same as those of the inch vernier.

The *main scale* is graduated in millimeters and every main division is numbered. Each numbered division has a value of 10 mm, for example, #1 represents 10 mm, #2 represents 20 mm, etc. There are 50 graduations on the sliding or *vernier scale* with every fifth one being numbered. These 50 graduations occupy the same space as 49 graduations on the main scale (49 mm). Therefore,

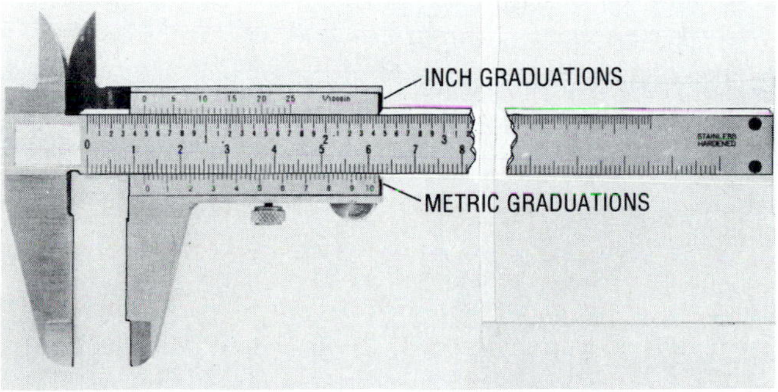
INCH GRADUATIONS
METRIC GRADUATIONS

Figure 10-4 *A vernier caliper with both inch and metric readings.*

$$1 \text{ vernier division} = \frac{49}{50} = 0.98 \text{ mm}$$

The difference between 1 main scale division and 1 vernier scale division is

$$1 - 0.98 = 0.02 \text{ mm}$$

TO READ A METRIC VERNIER CALIPER

1. The last numbered division on the bar to the left of the vernier scale represents the number of millimeters multiplied by 10.
2. Note how many full graduations are showing between this numbered division and the zero on the vernier scale. Multiply this number by 1 mm.
3. Find the line on the vernier scale which coincides with a line on the bar. Multiply this number by 0.02 mm.

In Fig. 10-5,

■ The large #4 graduation on the bar	4 × 10	= 40
■ Three full lines past the #4 graduation	3 × 1	= 3
■ The 9th line on the vernier scale coincides with a line on the bar	9 × 0.02	= 0.18
Total reading		43.18 mm

DIRECT-READING DIAL CALIPER

Because it is easier to read, the *direct-reading dial caliper* is gradually replacing the standard vernier caliper. Dial calipers are manufactured in inch and/or metric standards and are available with digital readout. A dial indicator, the hand of which is attached to a pinion, is mounted on the sliding jaw. For the metric dial caliper (Fig. 10-6) one revolution of the hand represents 2 mm of travel; one revolution on

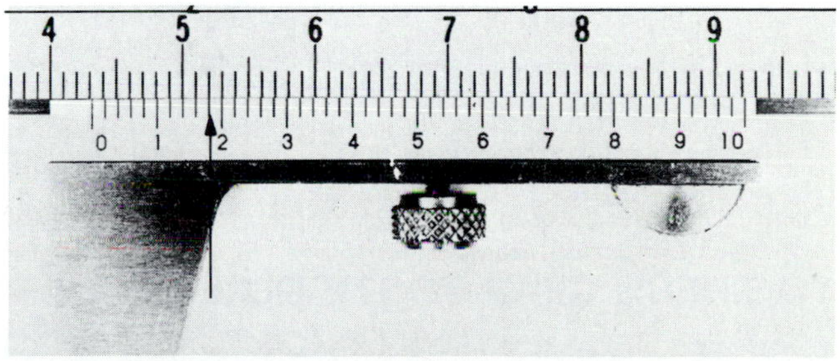
4 5 6 7 8 9
0 1 2 3 4 5 6 7 8 9 10

Figure 10-5 *A metric vernier caliper reading of 43.18 mm.*

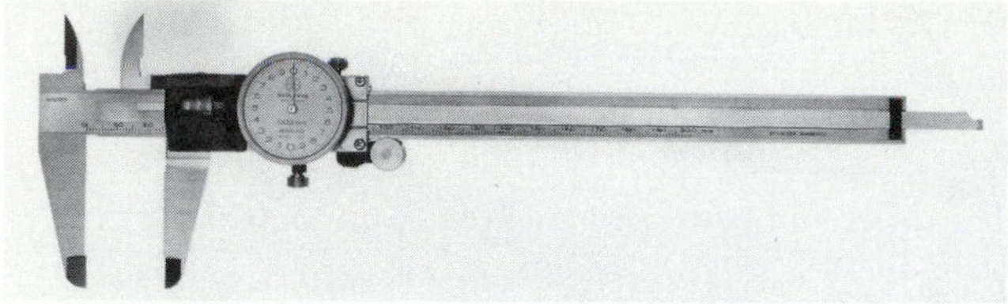

Figure 10-6 *A dial caliper with digital readout provides a quick and accurate method of measurement.*

the inch caliper may represent 0.100 or 0.200 in. of travel, depending on the manufacturer. Most direct-reading calipers have a narrow sliding blade attached to the sliding jaw (and dial). This narrow blade permits the dial caliper to be used as an efficient and accurate depth gage. (See Unit 11.)

GEAR-TOOTH VERNIER CALIPER

The *gear-tooth vernier caliper* (Fig. 10-7) consists basically of a vernier caliper and a vernier depth gage mounted at right angles to each other. This caliper is used to measure the thickness of gear teeth at the pitch-circle line. It may be used also to measure the width of Acme threads at the pitch line. Inch instruments are manufactured in two ranges: 20-diametral pitch to 2-diametral pitch and 10-diametral pitch to 1-diametral pitch, measuring in thousandths of an inch.

Metric gear-tooth vernier calipers are graduated in steps of 0.02 mm and come in the 1.25- to 12.00-mm range and the 2.5- to 25.00-mm range.

CORRECTED ADDENDUM

Measuring gear teeth requires setting the gear-tooth vernier caliper to the *corrected addendum,* which is a point slightly lower than the *true addendum* of the gear. If the vernier depth slide is set to the true addendum, the caliper will measure the sides of the tooth at a point above the pitch circle and not give a true measurement (Fig. 10-8A and B). Most handbooks contain tables giving the corrected addendum of gear teeth. If tables are not available, the following formula may be used to calculate the corrected addendum.

$$\mathrm{CA} = A + \left[\tfrac{1}{2}\mathrm{PD}\left(1 - \cos\frac{90^\circ}{N}\right)\right]$$

where CA = corrected addendum
A = addendum
PD = pitch diameter
N = number of teeth in the gear

TO CHECK GEAR-TOOTH DIMENSIONS

1. Set the vertical slide to the *corrected addendum* of the gear.

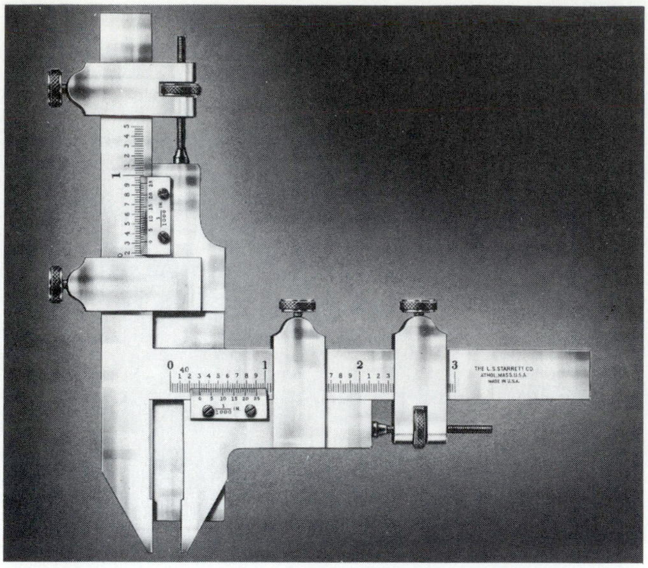

Figure 10-7 *A gear-tooth vernier caliper.*

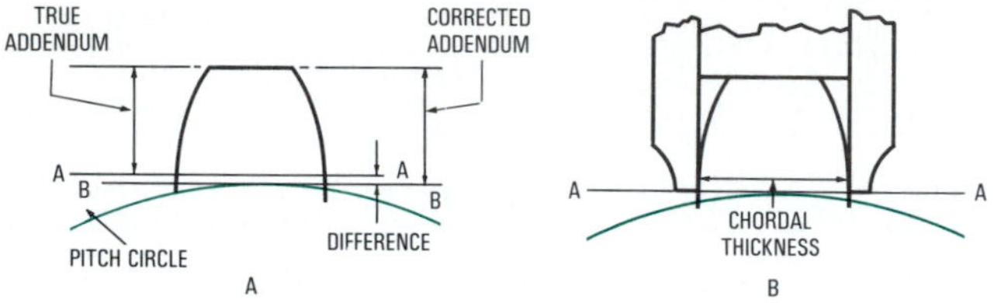
TRUE ADDENDUM
CORRECTED ADDENDUM
A
B
A
B
PITCH CIRCLE
DIFFERENCE
A
A
A
CHORDAL THICKNESS
B

Figure 10-8 *Gear-tooth measurement showing the (A) true and (B) corrected addendum.*

2. Place the slide on top of the gear tooth to be measured.
3. Using the adjusting nut, set the horizontal slide to the thickness of the tooth.
4. Note the reading on the horizontal scale and check it with the required thickness.

REVIEW QUESTIONS

VERNIER CALIPER

1. Describe the principle of:
   a. The 25-division vernier
   b. The 50-division vernier
2. Describe the procedure for reading a vernier caliper.
3. What are the following vernier caliper readings?

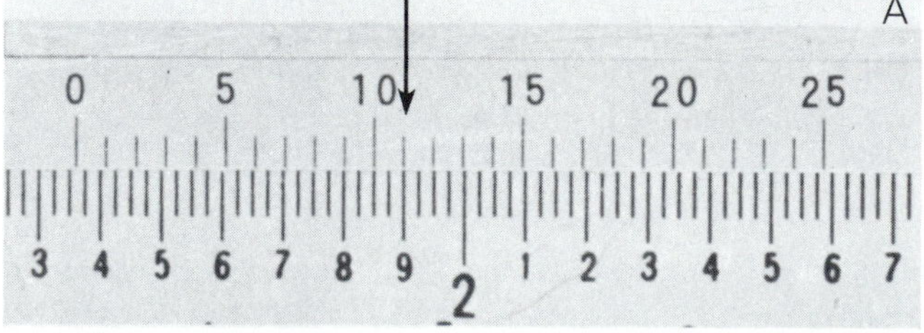
A
0 5 10 15 20 25
3 4 5 6 7 8 9 2 1 2 3 4 5 6 7

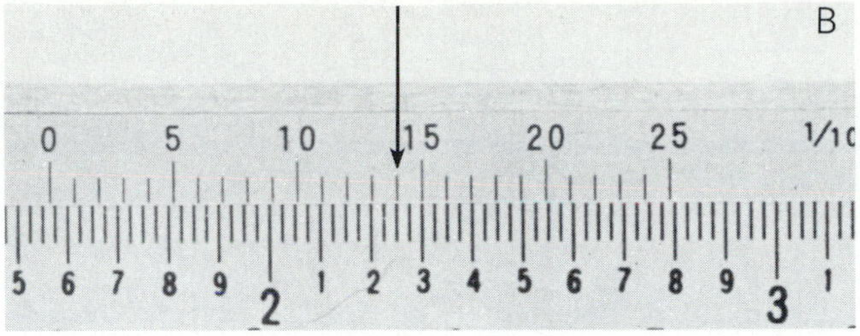
B
1/10

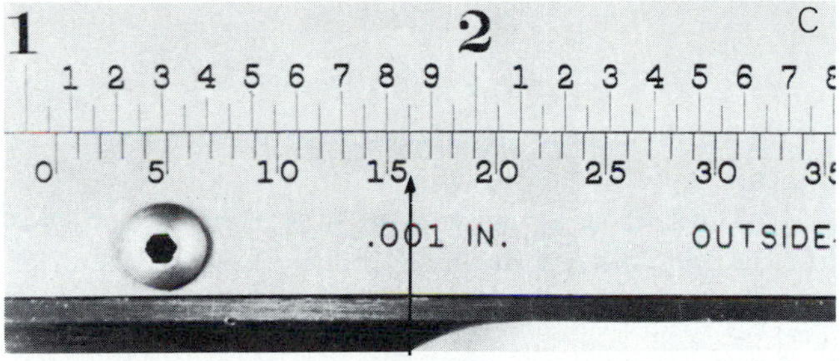
C
.001 IN.
OUTSIDE

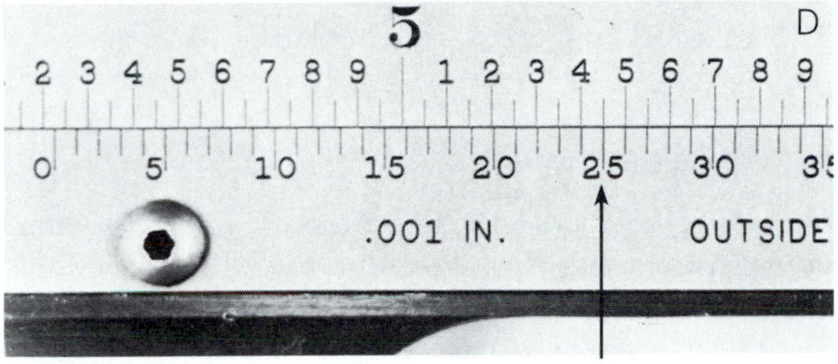
D
.001 IN.
OUTSIDE

METRIC VERNIER CALIPER

4. Describe the principle of the metric vernier caliper.
5. Read the following metric caliper settings.

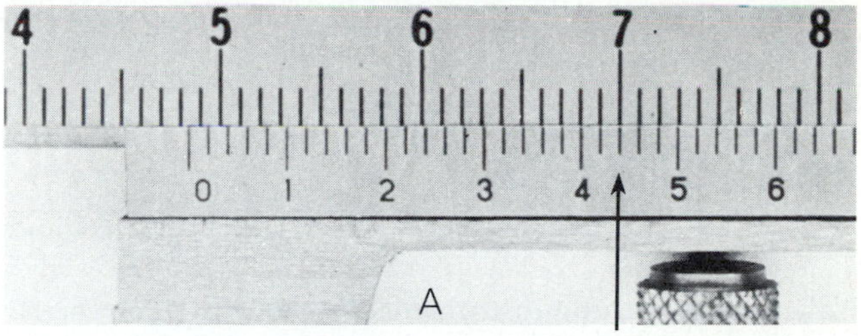
A

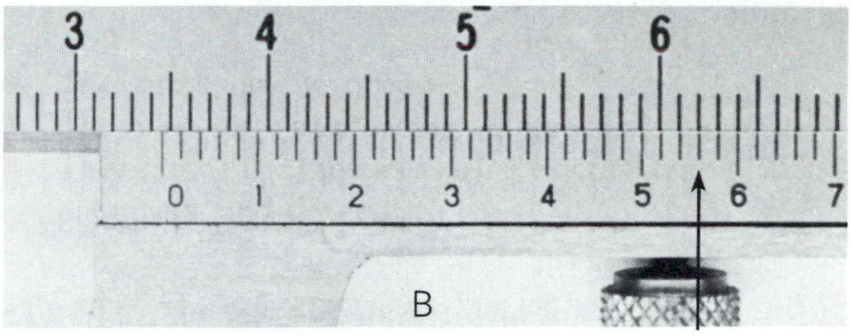
B

GEAR-TOOTH VERNIER CALIPER

6. A 16-diametral pitch gear has 54 teeth.
   a. Calculate the corrected addendum.
   b. Explain how to measure the gear tooth.

UNIT 11

Inside-, Depth-, and Height-Measuring Instruments

Because of the great variety of measurements required in machine shop work, many types of measuring tools are available. These permit the machinist to measure not only outside sizes but also inside diameters, depths, and heights. Direct-reading instruments are most commonly used and generally the most accurate; however, because of the shape or size of the part, transfer-type instruments may be required.

OBJECTIVES

After completing this unit you should be able to:

1. Measure hole diameters to within 0.001-in. (0.02-mm) accuracy using inside micrometers and inside micrometer calipers
2. Measure depths of slots and grooves to an accuracy of 0.001 in. (0.02 mm)

3. Measure heights to an accuracy of 0.001 in. (0.02 mm) using a vernier height gage.

INSIDE-MEASURING INSTRUMENTS

All inside-measuring instruments fall into two categories: direct-reading and transfer-type.

With *direct-reading instruments* the size of the hole can be read on the instrument being used to measure the hole. The most common direct-reading instruments are the inside micrometer, the Intrimik, and the vernier caliper.

Transfer-type instruments are set to the diameter of the hole and then this size is transferred to an outside micrometer to determine the actual size. The most common transfer-type instruments are inside calipers, small hole gages, and telescopic gages.

DIRECT-READING INSTRUMENTS

Inside Micrometer Calipers

The inside micrometer caliper (Fig. 11-1) is designed for measuring holes, slots, and grooves, from 0.200 to 2.000 in. in size for the inch-designed instruments, or 5 to 50 mm in size for metric instruments. The nibs or ends of the jaws are hardened and ground to a small radius to permit accurate measurement. A locking nut provided with this micrometer can be used to lock it at any desired size.

The inside micrometer caliper is based on the same principle as a standard micrometer, except that the barrel readings on some calipers are reversed (as shown in Fig. 11-1). Extreme care must be taken in reading this type of instrument. Other inside micrometer calipers have the readings on the spindle and are read in the same manner as a standard outside micrometer. Inside micrometer calipers are special-purpose tools and are not used in mass production measurement.

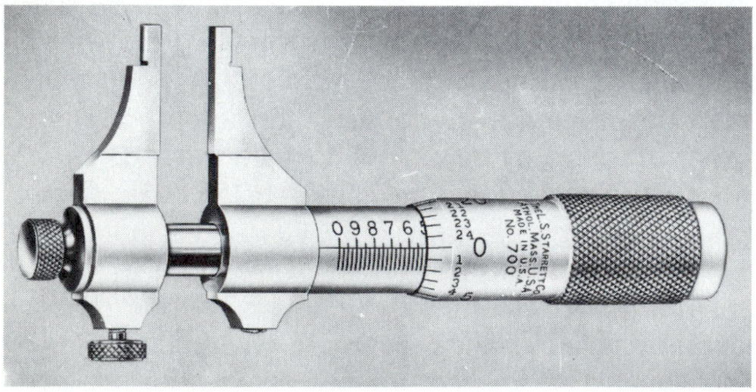
0 9 8 7 6
0
No. 700
Figure 11-1 *An inside micrometer caliper. (Courtesy The L. S. Starrett Company.)*

To Use an Inside Micrometer Caliper

1. Adjust the jaws to slightly less than the diameter to be measured.
2. Hold the fixed jaw against one side of the hole and adjust the movable jaw until the proper "feel" is obtained.

NOTE: Move the movable jaw back and forth to ensure that the measurement taken is across the true diameter.

3. Set the lock nut, remove the instrument, and check the reading.

Inside Micrometers

For internal measurements larger than 1½ in. or 40 mm, inside micrometers (Fig. 11-2) are used. The inside micrometer set consists of a micrometer head, having a range ½ or 1 in.; several extension rods of different lengths, which may be inserted into the head; and a ½-in. spacing collar. These sets cover a range from 1½ to over 100 in., or from 40 to 1000 mm for metric tools. Sets that are used for the larger ranges generally have hollow tubes rather than rods for greater rigidity.

The inside micrometer is read in the same manner as the standard micrometer. Since there is no locking nut on the inside micrometer, the thimble nut is adjusted to a tighter fit on the spindle thread to prevent a change in the setting while it is being removed from the hole.

To Measure with an Inside Micrometer

1. Measure the size of the hole with a rule.
2. Insert the correct micrometer extension rod after having carefully cleaned the shoulders of the rod and the micrometer head.
3. Align the zero marks on the rod and micrometer head.
4. Hold the rod firmly against the micrometer head and tighten the knurled set screw.
5. Adjust the micrometer to slightly less than the diameter to be measured.
6. Hold the head in a fixed position and adjust the micrometer to the hole size while moving the rod end in the direction of the arrows (Fig. 11-3).

NOTE: When a micrometer has been properly adjusted to size, there should be a slight drag when the rod end is moved past the centerline of the hole.

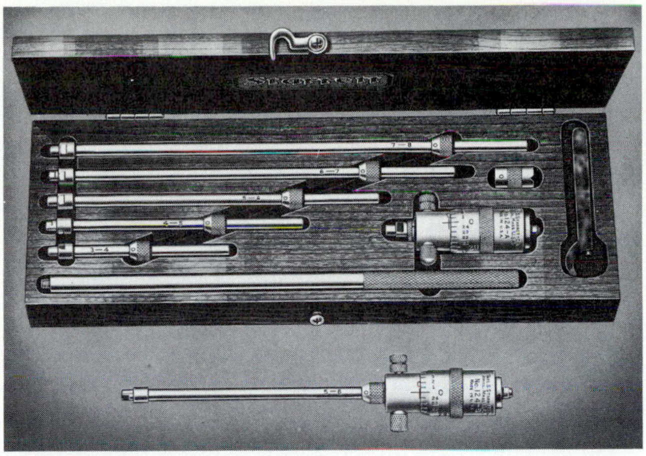

Figure 11-2 *An inside micrometer set can measure a large range of sizes.*

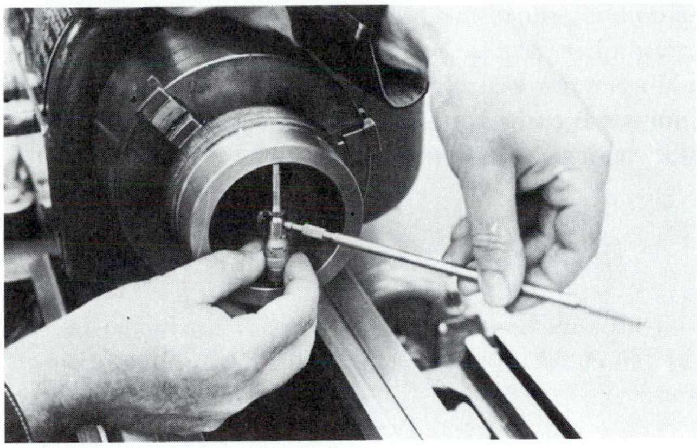

Figure 11-3 *Using an inside micrometer to measure the size of a hole.*

7. Carefully remove the micrometer and note the reading.
8. To this reading, add the length of the extension rod and collar.

Intrimik

A difficulty encountered in measuring hole sizes with instruments employing only two measuring faces is that of properly measuring the diameter and not a chord of the circle. An instrument which eliminates this problem is the *Intrimik* (Fig. 11-4).

The Intrimik consists of a head with three contact points spaced 120° apart; this head is attached to a micrometer-type body. The contact points are forced out to contact the inside of the hole by means of a tapered or conical plug attached to the micrometer spindle (Fig. 11-5). The construction of a head with three contact points permits the Intrimik to be self-centering and self-aligning. It is more accurate than other methods because it provides a direct reading, eliminating the necessity of transferring measurements to determine hole size as with telescope or small hole gages.

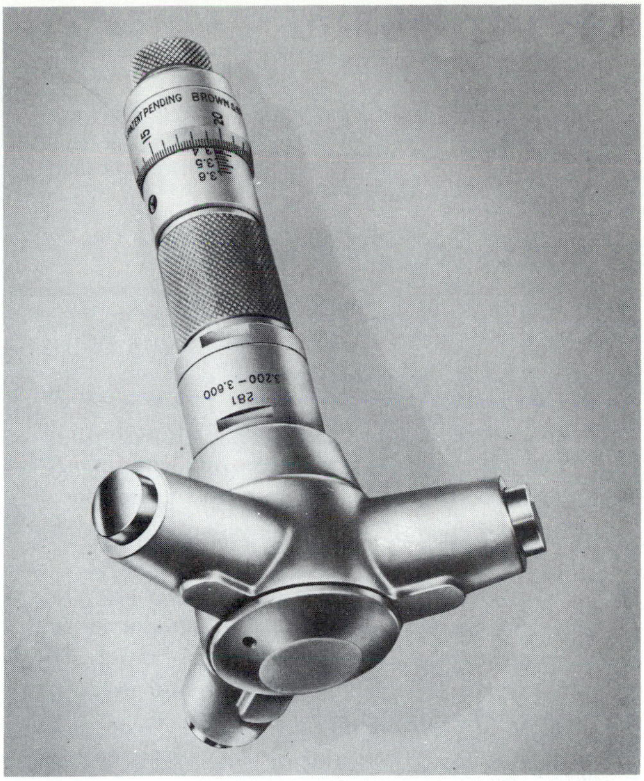

Figure 11-4 *The Intrimik, which has three contact points, measures holes accurately. (Courtesy Brown & Sharpe Co.)*

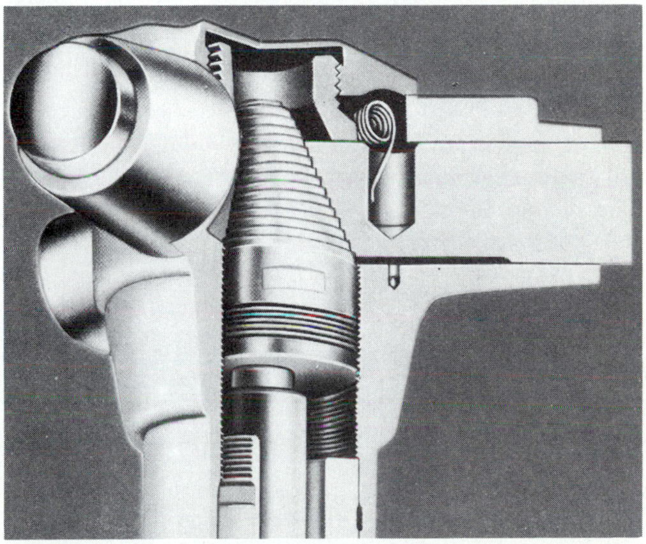

Figure 11-5 *Construction of the Intrimik head. (Courtesy Brown & Sharpe Co.)*

The range of these instruments is from 0.275 to 12.000 in., and the accuracy varies between 0.0001 and 0.0005 in., depending on the head used. Metric Intrimiks have a range from 6 to 300 mm, with graduations in 0.001 mm. The accuracy of the Intrimik should be checked periodically with a setting ring or master ring gage.

TRANSFER TYPE INSTRUMENTS

In *transfer measurement* the size of an object is taken with an instrument which is not capable of giving a direct reading. The size is then determined by measuring the setting of the instrument with a direct-reading instrument or gage of a known size.

Small Hole Gages

Small hole gages are available in sets of four, covering a range from ⅛ to ½ in. (3 to 13 mm). They are manufactured in two types (Fig. 11-6A and B).

The small hole gages shown in Fig. 11-6A have a small, round end, or ball, and are used for measuring holes, slots, grooves, and recesses which are too small for inside calipers or telescope gages. Those shown in Fig. 11-6B have a flat bottom and are used for similar purposes. The flat bottom permits the measurement of shallow slots, recesses, and holes impossible to gage with the rounded type.

Both types are of similar construction and are adjusted to size by turning the knurled knob on the top. This draws up a tapered plunger, causing the two halves of the ball to open up and contact the hole.

To Use a Small Hole Gage

Small hole gages require extreme care in setting since it is easy to get an incorrect setting when checking the diameter of a hole. Proceed as follows:

1. Measure the hole to be checked with a rule.
2. Select the proper small hole gage.
3. Clean the hole and the gage.
4. Adjust the gage until it is slightly smaller than the hole and insert it into the hole.
5. Adjust the gage until it can be felt just touching the sides of the hole or slot.
6. Swing the handle back and forth, and adjust the knurled end until the proper "feel" is obtained across the widest dimension of the ball.
7. Remove the gage and check the size with an outside micrometer.

NOTE: It is important to obtain the same "feel" when transferring the measurement and when adjusting the gage to the hole.

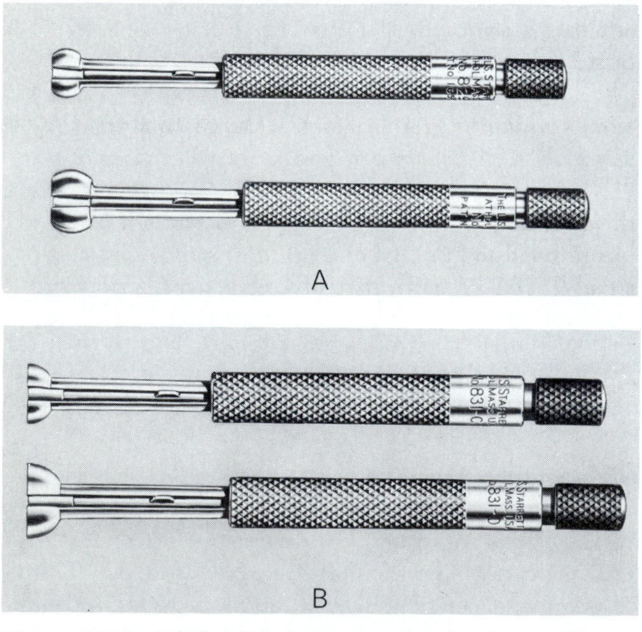
A
B

Figure 11-6 *(A) Small hole gages with a hardened-ball end; (B) small hole gage with a flat-bottom end. (Courtesy The L. S. Starrett Company.)*

Telescope Gages

Telescope gages (Fig. 11-7) are used to obtain the size of holes, slots, and recesses from 5⁄16 to 6 in. (8 to 152 mm). They are T-shaped instruments, each consisting of a pair of telescoping tubes or plungers connected to a handle. The plungers are spring-loaded to force them apart. The knurled knob on the end of the handle locks the plungers into position when it is turned in a clockwise direction.

NOTE: In some sets, only one plunger moves.

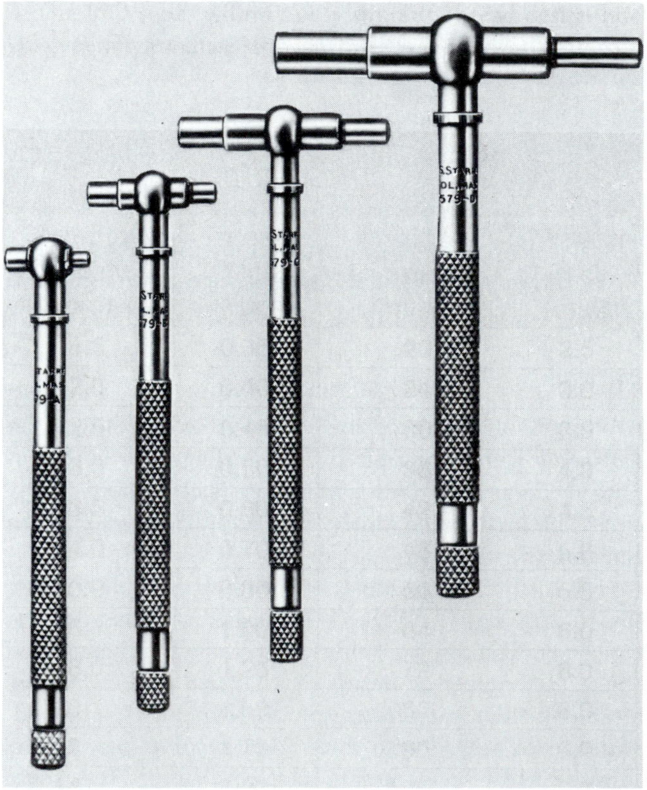

Figure 11-7 *A set of telescope gages. (Courtesy The L. S. Starrett Company.)*

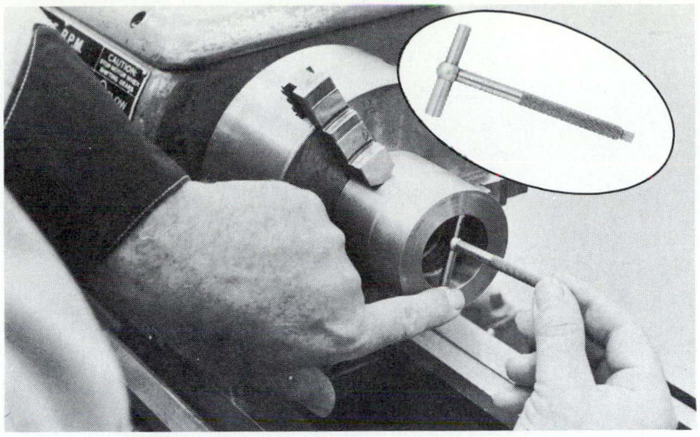

Figure 11-8 *Setting a telescope gage to the hole diameter.*

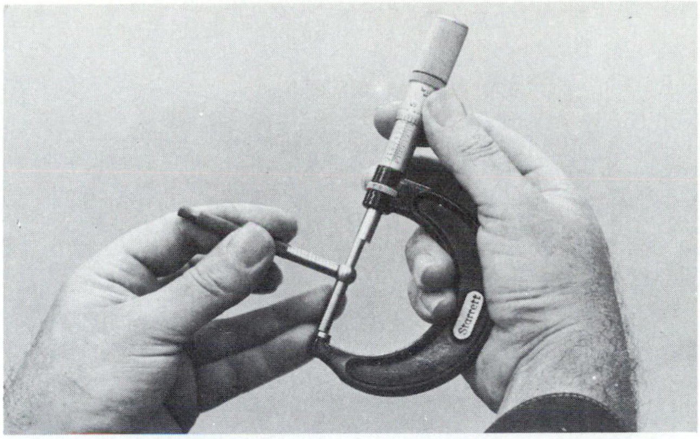

Figure 11-9 *Measuring the telescope gage setting with a micrometer.*

To Measure Using a Telescope Gage

1. Measure the hole size and select the proper gage.
2. Clean the gage and the hole.
3. Depress the plungers until they are slightly smaller than the hole diameter and lightly tighten the knurled knob.
4. Insert it into the hole and, with the handle tilted upward slightly, loosen the knurled knob to release the plungers.
5. *Lightly* snug up the knurled knob.
6. Hold the bottom leg of the telescope gage in position with one hand.
7. Move the handle downward through the center while slightly moving the top leg from side to side (Fig. 11-8).
8. Tighten the knurled knob to lock the plungers in position.
9. Recheck the "feel" on the gage by testing it in the hole again.
10. Check the gage size with outside micrometers, maintaining the same "feel" as in the hole (Fig. 11-9).

Dial Bore Gages

A quick and accurate method of checking hole diameters and bores for size, out-of-round, taper, bell-mouth, hourglass, or barrel shapes, is by means of the *dial bore gage* (Fig. 11-10).

Gaging is accomplished by three spring-loaded centralizing plungers in the head, one of which actuates the dial indicator, graduated in ten-thousandths of an inch, or in 0.01-mm graduations for metric tools.

These instruments are available in six sizes to cover a range from 3 to 12 in. or from 75 to 300 mm. Each instrument is supplied with extensions to increase its range. The dial bore gage must be set to size with a master gage; the hole size is then compared to the gage setting. Should the hole size vary, it is not necessary to adjust the gage as long as the size remains within the range of the gage.

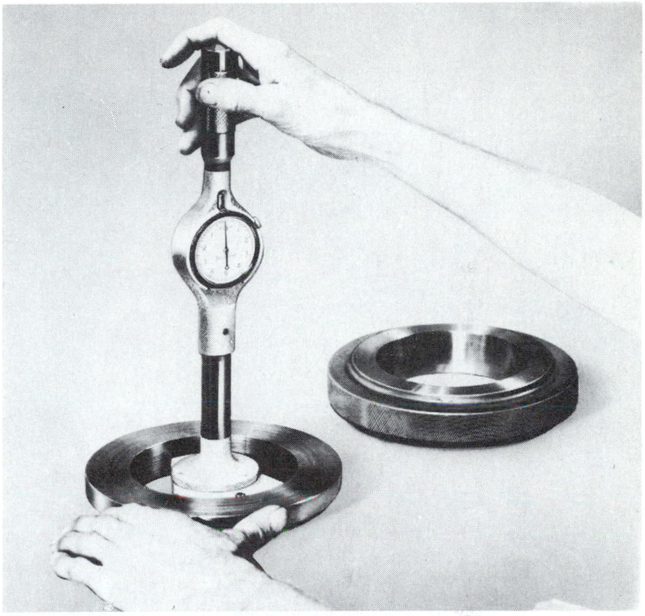

Figure 11-10 *Dial bore gages provide a quick and accurate method of measuring hole diameters.*

DEPTH MEASUREMENT

Although rules and various attachments can be used for measuring depth, the depth micrometer and the depth vernier are most commonly used where accuracy is required.

MICROMETER DEPTH GAGE

Micrometer depth gages are used for measuring the depth of blind holes, slots, recesses, and projections. Each gage consists of a flat base attached to a micrometer sleeve. An extension rod of the required length fits through the sleeve and protrudes through the base (Fig. 11-11). This rod is held in position by a threaded cap on the top of the thimble.

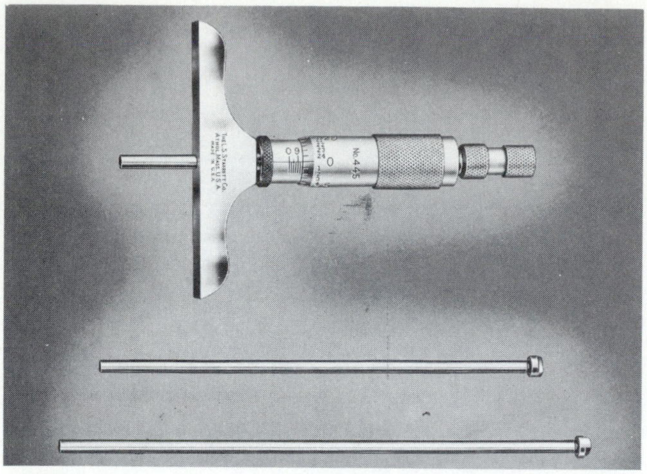
No. 445

Figure 11-11 *A micrometer depth gage and extension rods. (Courtesy The L. S. Starrett Company.)*

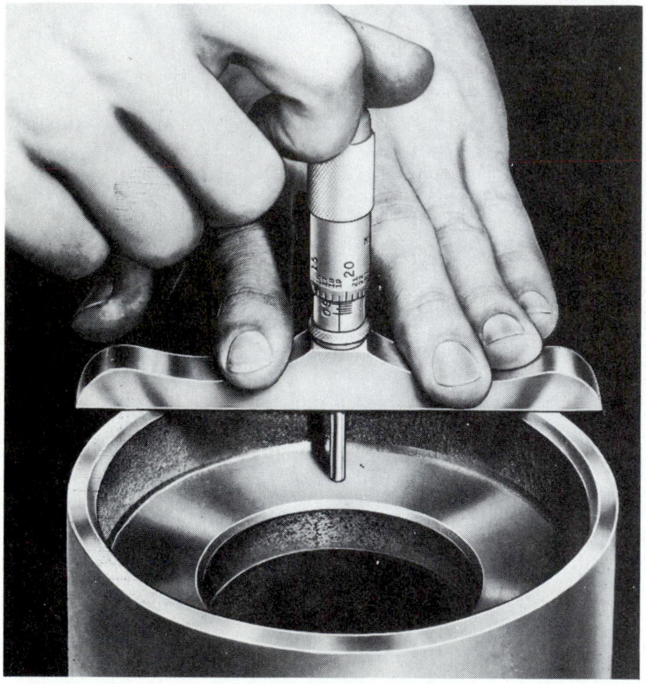

Figure 11-12 *Measuring the depth of a shoulder.*

Micrometer extension rods are available in various lengths, providing a range of up to 9 in. or 225 mm for metric tools. The micrometer screw has a range of ½ or 1 in., or up to 25 mm for metric tools. Depth micrometers are available with both round or flat rods, which are *not interchangeable* with other depth micrometers. The accuracy of these micrometers is controlled by a nut on the end of each extension rod which can be adjusted if necessary.

To Measure with a Micrometer Depth Gage

1. Remove burrs from the edge of the hole and the face of the workpiece.

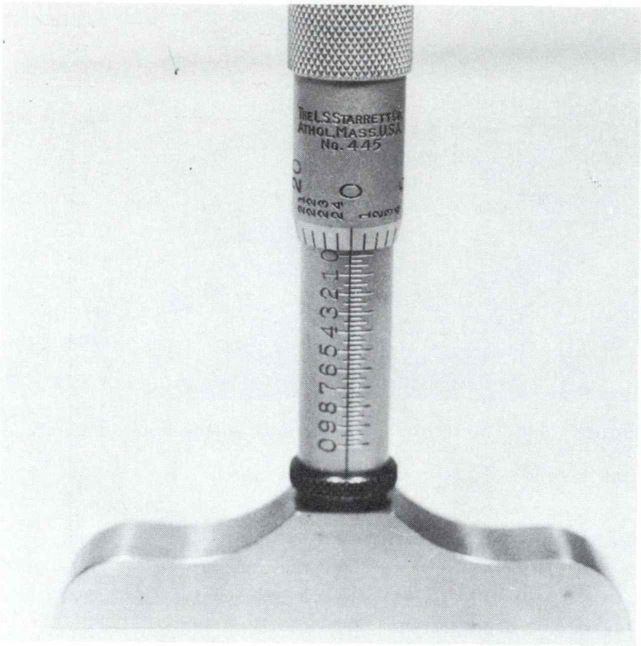
No. 445

Figure 11-13 *Graduations on a depth micrometer are reversed from those on an outside micrometer.*

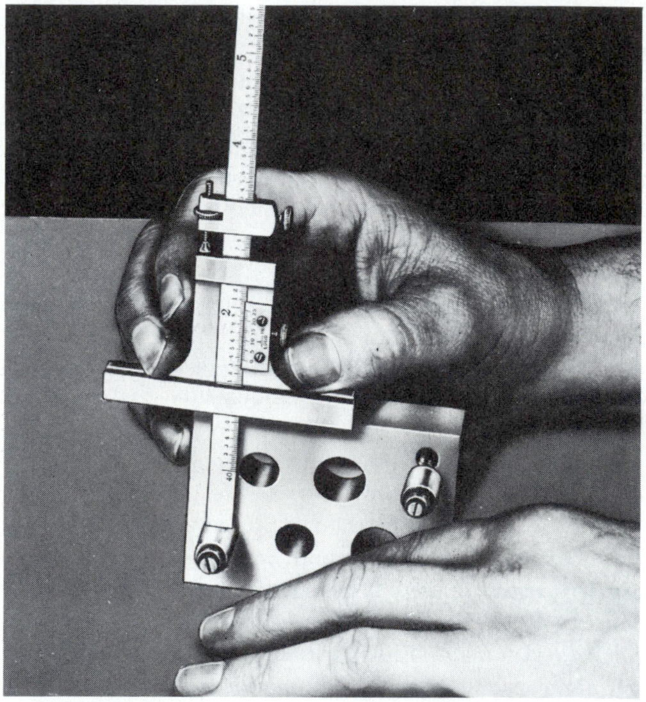

Figure 11-14 *Checking the position of toolmaker's buttons using a vernier depth gage.*

2. Clean the work surface and the base of the micrometer.
3. Hold the micrometer base firmly against the surface of the work (Fig. 11-12).
4. Rotate the thimble lightly with the tip of one finger in a clockwise direction until the bottom of the extension rod touches the bottom of the hole or recess.

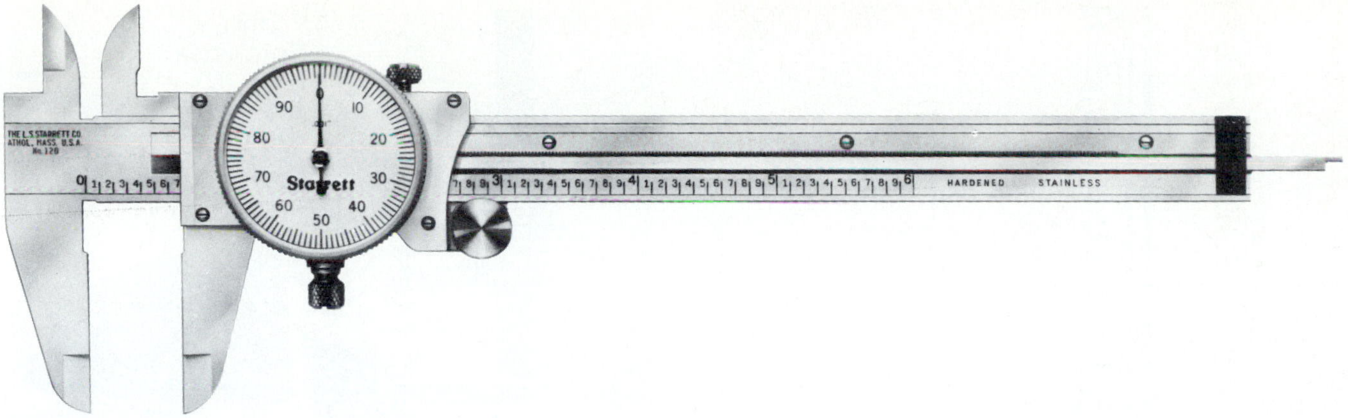
Starrett
HARDENED
STAINLESS

Figure 11-15 *A dial caliper with a depth gage blade. (Courtesy The L. S. Starrett Company.)*

5. Recheck the micrometer setting a few times to make sure that not too much pressure was applied in the setting.
6. Carefully note the reading.

NOTE: *The numbers on the thimble and the sleeve are the reverse of those on a standard micrometer* (Fig. 11-13).

VERNIER DEPTH GAGES

The depths of holes, slots, and recesses may also be measured by a vernier depth gage. This instrument is read in the same manner as a standard vernier caliper. Figure 11-14 illustrates how the toolmaker's button may be set up with this instrument.

Depth measurements may also be made with certain types of vernier or dial calipers which are provided with a thin sliding blade or depth gage attached to the movable jaw (Fig. 11-15). The blade protrudes from the end of the bar opposite the sliding jaw. The caliper is placed vertically over the depth to be measured, and the end of the bar is held against the shoulder while the blade is inserted into the hole to be measured. Depth readings are identical to standard vernier readings.

HEIGHT MEASUREMENT

Accurate height measurement is very important in layout and inspection work. With the proper attachments, the vernier height gage is a very useful and versatile tool for these purposes. Where extreme accuracy is required, gage blocks or a precision height gage may be used.

VERNIER HEIGHT GAGE

The vernier height gage is a precision instrument used in toolrooms and inspection departments on layout and jig and fixture work to measure and mark off distances accurately. These instruments are available in a variety of sizes—from 12 to 72 in. or from 300 to 1000 mm—and can be accurately set at any height to within 0.001 in. or 0.02 mm, respectively. Basically, a vernier height gage is a vernier caliper with a hardened, ground, and lapped base instead of a fixed jaw and is always used with a surface plate or an accurate flat surface. The sliding jaw assembly can be raised or lowered to any position along the beam. Fine adjustments are made by means of an adjusting nut. The vernier height gage is read in the same manner as a vernier caliper.

The vernier height gage is very well suited to accurate layout work and may be used for this purpose if a scriber is mounted on the movable jaw (Fig. 11-16). The scriber height may be set either by means of the vernier scale or by setting the scriber to the top of a gage block buildup of the desired height.

The *offset scriber* (Fig. 11-17) is a vernier height gage attachment, which permits setting heights from the face of the surface plate. When using this attachment, it is not necessary to consider the height of the base or the width of the scriber and clamp.

A *depth gage attachment* may be fastened to the movable jaw, permitting the measurement of height differences which may be difficult to measure by other methods.

Another important use for the vernier height gage is in inspection work. A dial indicator may be fastened to the movable jaw of the height gage (Fig. 11-18), and distances between holes or surfaces can be checked to within an accuracy of 0.001 in. (0.02 mm) on the vernier scale. If greater accuracy (0.0001 in. or less) is required, the indicator may be used in conjunction with gage blocks.

In Fig. 11-18, the height gage is being used to check the location of reamed holes in relation to the edges of the plate and to each other.

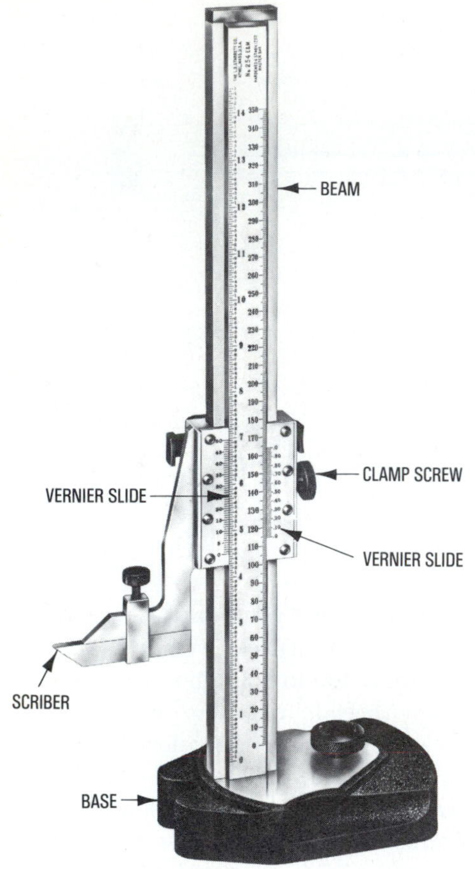
BEAM
CLAMP SCREW
VERNIER SLIDE
VERNIER SLIDE
SCRIBER
BASE

Figure 11-16 *A vernier height gage is used for layout and inspection work.*

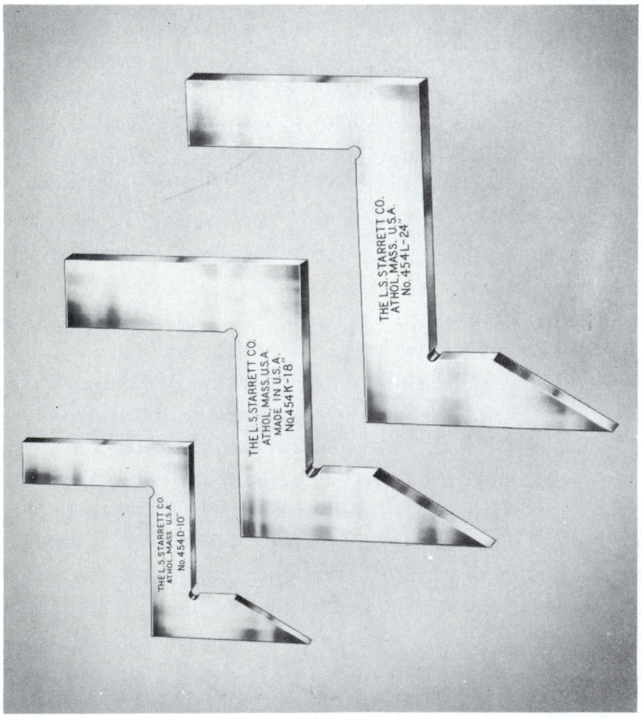
THE L.S.STARRETT CO.
ATHOL MASS. U.S.A.
No. 454L-24"
THE L.S.STARRETT CO.
ATHOL MASS. U.S.A.
MADE IN U.S.A.
No.454K-18"
THE L.S.STARRETT CO.
ATHOL MASS. U.S.A.
No. 454D-10"

Figure 11-17 *Offset scribers are used with a vernier height gage for accurate layout work.*

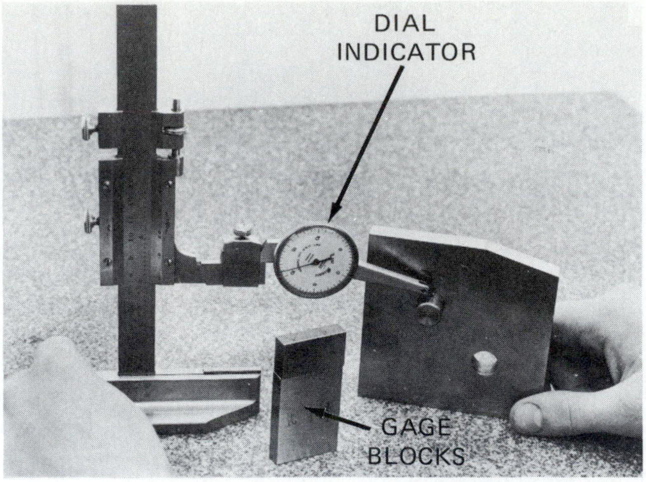
DIAL
INDICATOR
GAGE
BLOCKS

Figure 11-18 *Using a height gage and dial indicator to check a height.*

To Measure with a Vernier Height Gage and Dial Indicator

1. Thoroughly clean the surface plate, height gage base, and work surface.
2. Place a finished edge of the work on the surface plate and clamp it against an angle plate if necessary.
3. Insert a snug-fitting plug into the hole to be checked with about ½ in. (13 mm) projecting beyond the work.
4. Mount the dial indicator on the movable jaw of the height gage.
5. Adjust the movable jaw until the indicator almost touches the surface plate. (SEE Pg 90, FIG. 15-4B)
6. Lock the upper slide of the height gage and use the adjusting nut to ~~move~~ LOWER the indicator until the dial registers about one-quarter turn. (i.e. DEFINITLY TOUCHES PLATE)
7. Set the indicator dial to zero.
8. Record the reading of the vernier height gage.
9. Adjust the vernier height gage until the indicator registers zero on the top of the plug. Record this vernier height gage reading.
10. From this reading, subtract the initial reading plus half the diameter of the plug. This will indicate the distance from the surface plate to the center of the hole.
11. Check other hole heights using the same procedure.

To Measure Heights Using Gage Blocks

When hole locations must be accurate to 0.0005 in. (0.010 mm) or less, a gage block buildup is made for the proper dimension from the surface plate to the top of a plug fitted in the hole. Proceed as follows:

1. Prepare the required gage block buildup (the center of the hole height plus one-half the hole diameter). (See Unit 12.)

2. Mount a suitable dial indicator on a surface or vernier height gage.
3. Set the dial indicator to register zero on the top of the gage blocks.
4. Move the indicator over the top of the plug. The difference between the gage block buildup and the top of the plug will register on the dial indicator (Fig. 11-18).

PRECISION HEIGHT GAGE

The precision height gage (Fig. 11-19) provides a quick and accurate method of setting any height within the range of the instrument, eliminating the need for calculating and assembling specified gage blocks for comparative measurements. A surface plate is used as the reference surface.

The precision height gage is made from a hardened and ground round steel bar, with ground and lapped measuring steps or disks spaced exactly at 1-in. (25-mm) intervals. The measuring bar or column is raised or lowered by turning the large micrometer thimble, which is graduated by steps of 0.0001 in., or 0.002 mm if the instrument is metric. The column may be raised or lowered a full inch or 25 mm, permitting any reading from zero to the range of the instrument in increments of 0.0001 in. or 0.002 mm, respectively. Some models have a vernier scale below the micrometer thimble; readings in increments of 0.000 010 in. or 0.000 25 mm are then possible. It is important that the accuracy of precision height gages be checked periodically with a master set of gage blocks.

Figure 11-19 *Checking hole locations with a precision height gage and a height transfer gage. (Courtesy ExCello Corporation.)*

Height gages are available in models of 6, 12, 24, and 36 in., and their range may be increased by the use of riser blocks under the base. Metric gages range from 300 to 600 mm.

To Use a Precision Height Gage

1. Clean the surface plate and the feet of the height gage.
2. Clean the bottom of the work to be checked and place it on the surface plate, using parallels and an angle plate if required.
3. Insert plugs into the holes to be checked.
4. Mount a dial indicator on the movable jaw of a vernier height gage.
5. Adjust the height gage until the dial indicator registers approximately 0.015 in. (0.40 mm) across the top of the plug.
6. Turn the dial of the indicator to zero.
7. Move the dial indicator over the nearest disk of the precision height gage and raise the column by turning the micrometer until the dial indicator reads zero.
8. Check the micrometer reading. This reading will indicate the distance from the surface plate to the top of the plug.
9. Subtract half the diameter of the plug from this reading.

NOTE: If the work is set on parallels, this height must be subtracted from the precision height gage reading.

REVIEW QUESTIONS

INSIDE MICROMETER CALIPERS

1. a. On what type of inside micrometer calipers are the readings reversed to an outside micrometer?
   b. What type are read the same as an outside micrometer?
2. What precautions must be observed in taking a measurement with an inside micrometer caliper?

INSIDE MICROMETERS

3. What construction feature compensates for a lock nut on inside micrometers?
4. What precautions must be taken when:
   a. Assembling the inside micrometer and extension rod?
   b. Using the inside micrometer?
5. What is the correct "feel" with an inside micrometer?

SMALL HOLE GAGES

6. Name two types of small hole gages and state the purpose of each.
7. What precaution must be observed when using a small hole gage to obtain a dimension?

TELESCOPE GAGES

8. List the steps required to measure a hole with a telescope gage.

DIAL BORE GAGES

9. What hole defects may be conveniently measured with a dial bore gage?

INTRIMIK

10. Why is an Intrimik particularly suited to measure hole sizes?
11. Why is an Intrimik more accurate than a telescope gage for measuring hole size?

MICROMETER DEPTH GAGE

12. How is the accuracy of a micrometer depth gage adjusted?
13. How must the workpiece be prepared prior to measuring the depth of a hole or slot with a micrometer depth gage?
14. Explain the procedure for measuring a depth with a depth micrometer.
15. How does the reading of a depth micrometer differ from that of a standard outside micrometer?

VERNIER HEIGHT GAGE

16. State the two main applications for the vernier height gage.
17. What accessories are required for a vernier height gage to check *accurately* the height of a workpiece?

PRECISION HEIGHT GAGE

18. What are the advantages of using a precision height gage in lieu of a gage block buildup?
19. What dimension(s) must be subtracted from the reading so that the correct reading for the height of a hole being checked will be obtained?

UNIT 12

Gage Blocks

Interchangeable manufacture requires an accurate standard of measurement in order to function efficiently. Gage blocks, the acceptable standard of accuracy, have provided industry with a means of maintaining sizes to specific standards or tolerances. This feature has led to high rates of production and has made interchangeable manufacture possible.

OBJECTIVES

After completing this unit you will be able to:
1. Explain the use and application of gage blocks
2. Calculate inch and metric gage block buildups
3. Prepare and wring gage blocks together

GAGE BLOCK MANUFACTURE

Gage blocks are rectangular blocks of hardened and ground alloy steel which have been stabilized through alternate cycles of extreme heat and cold until the crystalline structure of the metal is left without strain. The two measuring surfaces are lapped and polished to an optically flat surface and to a specific size accurate within a range of 2 to 8 millionths of an inch (50 to 200 millionths of a millimeter). The size of each block is stamped on one of its surfaces. Chrome-plated gage blocks are also available and when long wear is desirable, carbide blocks are used. Great care is exercised in their manufacture; the final calibration is made under ideal conditions where the temperature is maintained at 68°F (20°C). Therefore any gage block is accurate to size *only* when measured at the standard temperature of 68°F (20°C).

Ceramic gage blocks, made of zirconia one of the most durable materials on earth, have recently been introduced. They have ten times the abrasion resistance of common steel blocks, need no special maintenance, and have a thermal expansion coefficient very close to steel. Some of the main features of zirconia ceramic gage blocks are:

- corrosion resistant
- no detrimental effects as a result of handling
- superior abrasion resistance
- thermal expansion coefficient close to steel
- resistance to impact
- free from burrs
- wring together tightly

USES

Industry has found gage blocks to be invaluable tools. Because of their extreme accuracy, they are used for the following purposes.

1. To check the dimensional accuracy of fixed gages to determine the extent of wear, growth, or shrinkage
2. To calibrate adjustable gages, such as micrometers and vernier calipers, imparting accuracy to these instruments
3. To set comparators, dial indicators, and height gages to exact dimensions
4. To set sine bars and sine plates when extreme accuracy is required in angular setups
5. For precision layout with the use of attachments
6. To make machine tool setups
7. To measure and inspect the accuracy of finished parts

GAGE BLOCK SETS

INCH STANDARD GAGE BLOCKS

Gage blocks are manufactured in sets that vary from a few blocks to as many as 115. The most commonly used is the 83-piece set (Fig. 12-1) from which it is possible to make over 120,000 different measurements, ranging from one hundred-thousandths of an inch to over 25 inches. The blocks that comprise an 83-piece set are listed in Table 12-1.

Two *wear* blocks are supplied with an 83-piece set. Some manufacturers supply 0.050-in. wear blocks; others provide

Table 12-1 Sizes in an 83-Piece Set of Inch Standard Gage Blocks

First: 0.0001-in. Series—9 Blocks									
0.1001	0.1002	0.1003	0.1004	0.1005	0.1006	0.1007	0.1008	0.1009	
Second: 0.001-in. Series—49 Blocks									
0.101	0.102	0.103	0.104	0.105	0.106	0.107	0.108	0.109	
0.110	0.111	0.112	0.113	0.114	0.115	0.116	0.117	0.118	
0.119	0.120	0.121	0.122	0.123	0.124	0.125	0.126	0.127	
0.128	0.129	0.130	0.131	0.132	0.133	0.134	0.135	0.136	
0.137	0.138	0.139	0.140	0.141	0.142	0.143	0.144	0.145	
0.146	0.147	0.148	0.149						
Third: 0.050-in. Series—19 Blocks									
0.050	0.100	0.150	0.200	0.250	0.300	0.350	0.400	0.450	0.500
0.550	0.600	0.650	0.700	0.750	0.800	0.850	0.900	0.950	
Fourth: 1.000-in. Series—4 Blocks									
1.000	2.000	3.000	4.000						
Two 0.050-in. Wear Blocks									

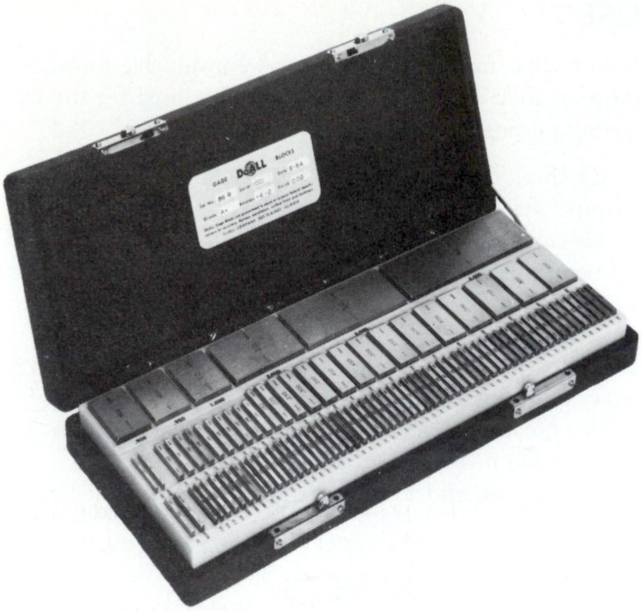

Figure 12-1 *An 83-piece set of gage blocks.*

0.100-in. wear blocks. They should be used at each end of a combination, especially if the blocks will be in contact with hard surfaces or abrasives. In this way, the wear which occurs during use will be on the two wear blocks only, rather than on many blocks, prolonging the useful life and accuracy of the set. During use, it is considered good practice always to expose the same face of the wear block to the work surface. A good habit to acquire is always to have the words "Wear Block" appear on the outside of the combination. In this way, all the wear will be on one surface and the wringing quality of the other surface will be preserved.

METRIC GAGE BLOCKS

Metric gage blocks are supplied by most manufacturers in sets of 47, 88, and 113 blocks. The most common is the 88-piece set (Table 12-2). Each of these sets contains a pair of 2-mm wear blocks. These blocks are used at each end of the buildup to prolong the accuracy of the other blocks in the set.

ACCURACY

Gage blocks in inch and metric standards are manufactured to three common degrees of accuracy, depending on the purpose for which they are used.

1. The *Class AA* set, commonly called a *laboratory* or *master* set, is accurate to ±0.000 002 in. in the inch standard; the metric set, to ±0.000 05 mm. These gage blocks are used in temperature-controlled laboratories as references to compare or check the accuracy of working gages.
2. The *Class A* set is used for inspection purposes and is accurate to ±0.000 004 in. in the inch standard; the metric set, to +0.000 15 mm and −0.000 05 mm.
3. The *Class B* set, commonly called the *working* set, is accurate to ±0.000 008 in. in the inch standard; the met-

Table 12-2 Sizes in an 88-Piece Set of Metric Gage Blocks

0.001-mm Series—9 Blocks								
1.001	1.002	1.003	1.004	1.005	1.006	1.007	1.008	1.009
0.01-mm Series—49 Blocks								
1.01	1.02	1.03	1.04	1.05	1.06	1.07	1.08	1.09
1.10	1.11	1.12	1.13	1.14	1.15	1.16	1.17	1.18
1.19	1.20	1.21	1.22	1.23	1.24	1.25	1.26	1.27
1.28	1.29	1.30	1.31	1.32	1.33	1.34	1.35	1.36
1.37	1.38	1.39	1.40	1.41	1.42	1.43	1.44	1.45
1.46	1.47	1.48	1.49					
0.5-mm Series—1 Block								
0.5								
0.5-mm Series—18 Blocks								
1	1.5	2	2.5	3	3.5	4	4.5	5
5.5	6	6.5	7	7.5	8	8.5	9	9.5
10-mm Series—9 Blocks								
10	20	30	40	50	60	70	80	90
Two 2-mm Wear Blocks								

ric set, to +0.000 25 mm and −0.000 15 mm. These blocks are used in the shop for machine tool setups, layout work, and measurement.

THE EFFECT OF TEMPERATURE

While the effect of temperature on ordinary measuring instruments is negligible, changes in temperature are important when precision gage blocks are handled. Gage blocks have been calibrated at 68°F (20°C), but human body temperature is about 98°F (37°C). A 1°F (0.5°C) rise in temperature will cause a 4-in. (100-mm) stack of gage blocks to expand approximately 0.000 025 in. (0.0006 mm); therefore, these blocks should be handled as little as possible. The following suggestions are offered to eliminate as much temperature-change error as possible.

1. Handle gage blocks only when they must be moved.
2. Hold them by hand for as little time as possible.
3. Hold them between the tips of the fingers so that the area of contact is small, or use insulated tweezers.
4. Have the work and gage blocks at the same temperature. If a temperature-controlled room is not available, both the work and gage blocks may be placed in kerosene until both are at the same temperature.
5. Where extreme accuracy is necessary, use insulating gloves and tweezers to prevent temperature change during handling.

GAGE BLOCK BUILDUPS

Gage blocks are manufactured to great accuracy; they adhere to each other so well when wrung together properly that they can withstand a 200-pound (lb) [890-newton (N)] pull. Many theories have been advanced to explain this adhesion. Scientists have felt that it may be atmospheric pressure, molecular attraction, the extremely flat surfaces of the blocks, or a minute film of oil which gives the blocks this quality. Possibly a combination of any of these factors could be responsible.

When the blocks required to make up a dimension are being calculated, the following procedure should be followed to save time, reduce the chance of error, and use as few blocks as possible. For example, if a measurement of 1.6428 in. is required, proceed as follows:

Step	Procedure	
1	Write the dimension required on the paper	1.6428
2	Deduct the size of two wear blocks: 2 × .050 in.	0.1000
	Remainder	1.5428
3	Use a block which will eliminate the right-hand digit	0.1008
	Remainder	1.4420
4	Use a block which will eliminate the right-hand digit and at the same time bring the digit to the left of it to a zero (0) or a five (5)	0.142
	Remainder	1.300
5	Continue to eliminate the digits from the right to the left until the dimension required is attained	0.300
	Remainder	1.000
6	Use a 1.000-in. block	1.000
	Remainder	0.000

To eliminate the possibility of subtraction error while making a buildup, use two columns for this calculation. As the following example illustrates, the gage blocks are subtracted from the original dimension in the left-hand column, and the right-hand column is used as a check column. For example, to build up a dimension of 3.8716 in., proceed as follows:

	Procedure Column	Check Column
1. Dimension required	3.8716 in.	
2. Two wear blocks (2 × 0.050 in.)	0.100	0.100
	3.7716	
3. Use 0.1006 in.	0.1006	0.1006
	3.6710	
4. Use 0.121 in.	0.121	0.121
	3.550	
5. Use 0.550 in.	0.550	0.550
	3.000	
6. Use 3.000 in.	3.000	3.000
	0.000 in.	3.8716 in.

When using metric blocks for buildup of 57.150 mm, proceed as follows:

Step	Procedure	
1	Write the dimension required on the paper	57.150
2	Deduct the size of 2 wear blocks (2 × 2 mm)	4.000
	Remainder	53.150

3	Use a block which will eliminate the right-hand digit	1.050
	Remainder	52.100
4	Use a block which will eliminate the right-hand digit	1.100
	Remainder	51.000
5	Use a 1-mm block	1.000
	Remainder	50.000
6	Use a 50-mm block	50.000
	Remainder	0.000

For a metric buildup of 27.781 mm, proceed as follows:

	Procedure Column	Check Column
1. Dimension required	27.781 mm	
2. Two wear blocks (2 × 2 mm)	4.000	4.000
	23.781	
3. Use 1.001 mm	1.001	1.001
	22.780	
4. Use 1.080 mm	1.080	1.080
	21.700	
5. Use 1.700 mm	1.700	1.700
	20.000	
6. Use 20.000 mm	20.000	20.000
	0.000 mm	27.781 mm

To Wring Blocks Together

When wringing blocks together, take care not to damage them. The correct sequence of movement to wring blocks together, illustrated in Fig. 12-2, is as follows:

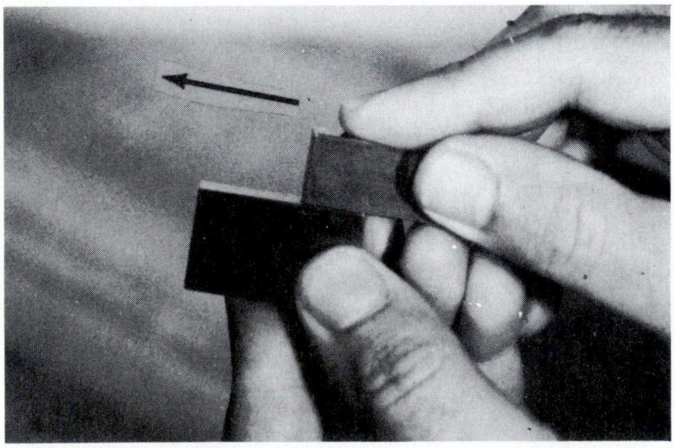

Figure 12-2 *Procedure for wringing gage blocks.*

1. Clean the blocks with a clean, soft cloth.
2. Wipe each of the contacting surfaces on the clean palm of the hand or on the wrist. This procedure removes any dust particles left by the cloth and also applies a light film of oil.
3. Place the end of one block over the end of another block as shown in Figure 12-2.
4. While applying pressure on the two blocks, slide one block over the other.

NOTE: If the blocks do not adhere to each other, it is generally because the blocks have not been thoroughly cleaned.

Care of Gage Blocks

1. Gage blocks should always be protected from dust and dirt by being kept in a closed case when not in use.
2. Gages should not be handled unnecessarily since they absorb heat from the hand. Should this occur, the gage blocks must be permitted to return to room temperature before use.
3. Fingering of lapped surfaces should be avoided to prevent tarnishing and rusting.
4. Care should be taken not to drop gage blocks or scratch their lapped surfaces.
5. Immediately after use, each block should be cleaned, oiled, and replaced in the storage case.
6. Before gage blocks are wrung together, their faces must be free from oil and dust.
7. Gage blocks should never be left wrung together for any length of time. The slight moisture between the blocks can cause rusting, which will permanently damage the blocks.

REVIEW QUESTIONS

1. How are gage blocks stabilized, and why is this necessary?
2. State five general uses for gage blocks.
3. For what purpose are wear blocks used?
4. How should wear blocks always be assembled into a buildup?
5. State the difference between a master set and a working set of gage blocks.
6. What precautions are necessary when handling gage blocks to minimize the effect of heat on the blocks?
7. List five precautions necessary for the proper care of gage blocks.
8. Calculate the gage blocks required for the following buildups. (Use the check column for accuracy.)
   a. 2.1743 in. b. 6.2937 in. c. 7.8923 in. d. 32.079 mm e. 74.213 mm f. 89.694 mm

UNIT

13

Angular Measurement

Precise angular measurement and setups constitute an important phase of machine shop work. The most commonly used tools for accurately laying out and measuring angles are the universal bevel protractor, sine bar, and sine plate.

OBJECTIVES

After completing this unit you will be able to:

1. Make angular measurements to an accuracy of 5′ (minutes) of a degree using a universal bevel protractor
2. Make angular measurements to less than 5′ of a degree using a sine bar, gage blocks, and a dial indicator

THE UNIVERSAL BEVEL PROTRACTOR

The *universal bevel protractor* (Fig. 13-1) is a precision instrument capable of measuring angles to within 5′ (0.083°). It consists of a *base* to which a *vernier scale* is attached. A *protractor dial,* graduated in degrees, with every tenth degree being numbered, is mounted on the circular section of the base. A *sliding blade* is fitted into this dial; it may be extended in either direction and set at any angle to the base. The blade and the dial are rotated as a unit. Fine adjustments are obtained with a small knurled-headed pinion, which when turned, engages with a gear attached to the blade mount. The protractor dial may be locked in any desired position by means of the *dial clamp nut.*

The vernier protractor shown in Fig. 13-2 is being used to measure an obtuse angle, or an angle greater than 90° but less than 180°. An *acute-angle attachment* is fastened to the vernier protractor to measure angles of less than 90° (Fig. 13-3).

The vernier protractor dial, or main scale, is divided into two arcs of 180°. Each arc is divided into two quadrants of 90° and has graduations from 0 to 90° to the left and right of the zero (0) line, with every tenth degree being numbered.

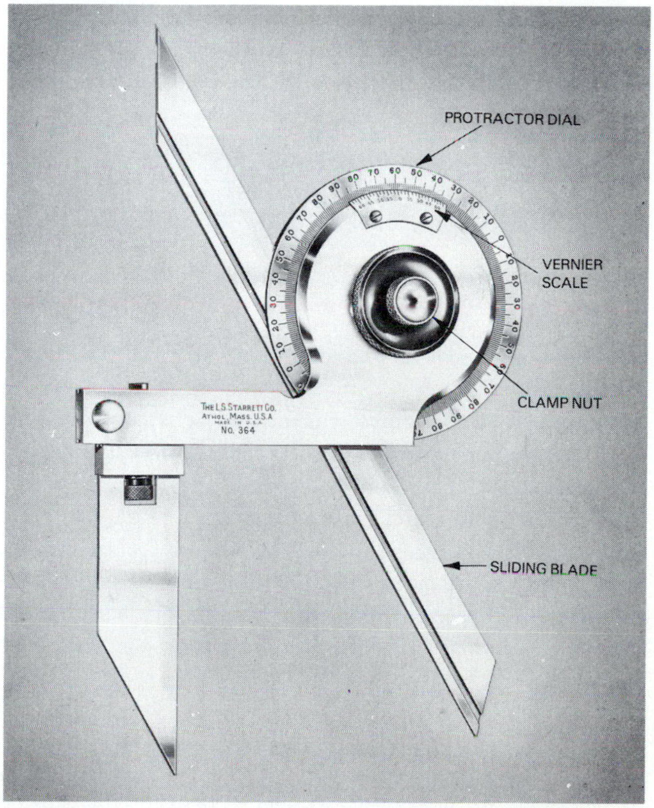

PROTRACTOR DIAL
VERNIER SCALE
CLAMP NUT
SLIDING BLADE
THE L.S. STARRETT CO.
ATHOL, MASS. U.S.A.
NO. 364

Figure 13-1 *The universal bevel protractor can measure angles accurately.*

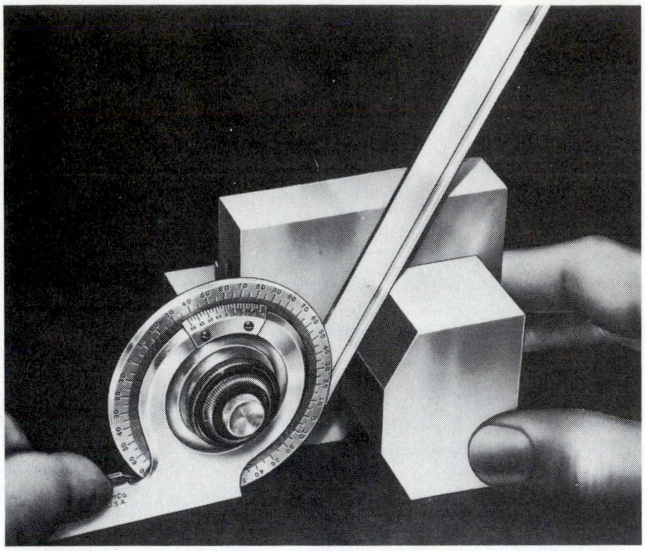

Figure 13-2 *Measuring an obtuse angle using a universal bevel protractor.*

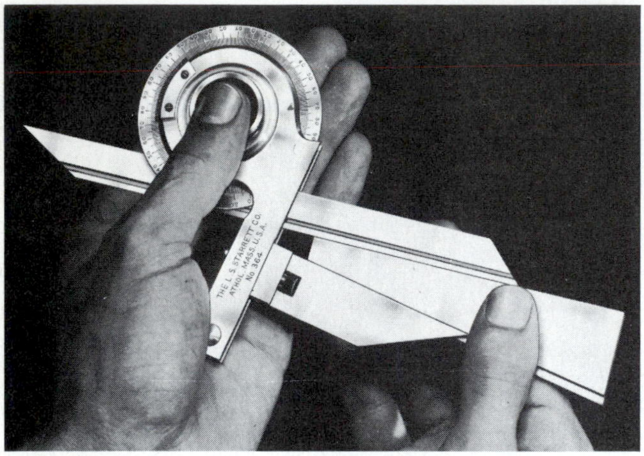

Figure 13-3 *Measuring an acute angle.*

The vernier scale is divided into 12 spaces on each side of the 0 line, which occupy the same space as 23° on the protractor dial. By simple calculation, it is easy to prove that one vernier space is 5′, or less than two graduations on the main scale. If zero on the vernier scale coincides with a line on the main scale, the reading will be in degrees only. However, if any other line on the vernier scale coincides with a line on the main scale, the number of vernier graduations beyond the zero should be multiplied by 5′ and added to the number of full degrees indicated on the protractor dial.

TO READ A VERNIER PROTRACTOR

1. Note the number of whole degrees between the zero on the main scale and the zero on the vernier scale.
2. Proceeding in the *same direction* beyond the zero on the vernier scale, note which vernier line coincides with a main scale line.
3. Multiply this number by 5′ and add it to the number of degrees on the protractor dial.

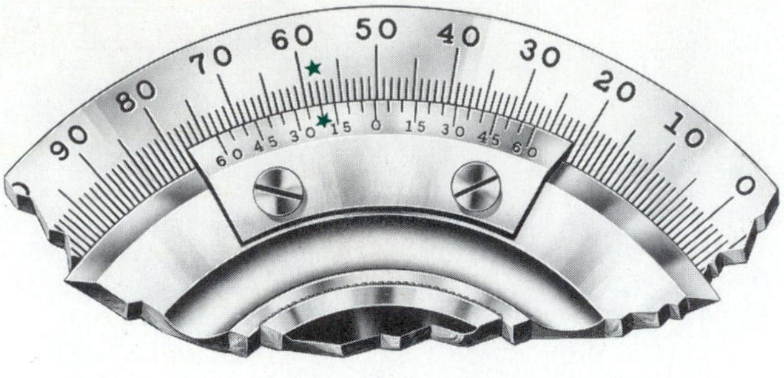
90 80 70 60 50 40 30 20 10 0
60 45 30 15 0 15 30 45 60

Figure 13-4 *A vernier protractor reading of 50°20′.*

In Fig. 13-4, the angular reading is calculated as follows: The number of degrees indicated on the main scale is 50 plus. The fourth line on the vernier scale *to the left* of the zero coincides with a line on the main scale; therefore, the reading is

Number of full degrees	= 50°
Value of vernier scale (4 × 5′)	= 20′
Reading	= 50°20′

NOTE: A double check of the reading would locate the vernier scale line on the other side of zero, which coincides with a protractor scale line. This line should always equal the complement of 60′. In Fig. 13-4, the 40′ line to the right of the zero coincides with a line on the protractor scale. This reading, when added to the 20′ on the left of the scale, is equal to 60′, or 1°.

THE SINE BAR

A *sine bar* (Fig. 13-5) is used when the accuracy of an angle must be checked to less than 5′ or work must be located to a given angle within close limits. The sine bar consists of a steel bar with two cylinders of equal diameter fastened near the ends. The centers of these cylinders are on a line exactly at 90° to the edge of the bar. The distance between the centers of these lapped cylinders is usually 5 or 10 in. on inch sine bars; 125 or 250 mm on metric sine bars. Sine bars are generally made of stabilized tool steel, hardened, ground, and lapped to extreme accuracy. They are used on surface plates, and any desired angle can be set by raising one end of the bar to a predetermined height with gage blocks.

Sine bars are generally made 5 in. or multiples of 5 in. in length. That is, the lapped cylinders are 5 in. ±0.0002 or 10 in. ±0.000 25 between centers. The face of the sine

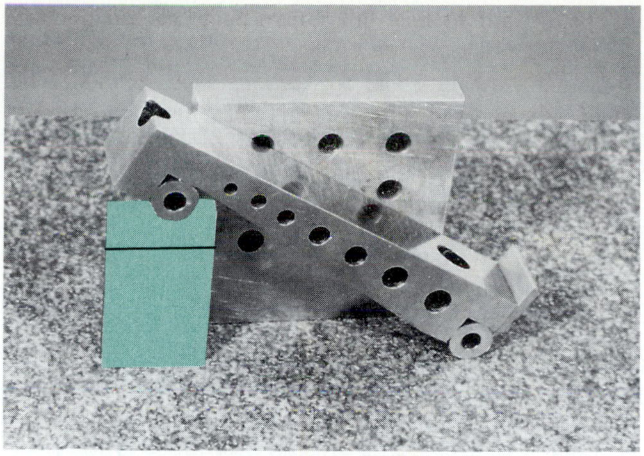

Figure 13-5 *A 5-in. sine bar with gage block buildup is used to set up work to an angle.*

bar is accurate to within 0.000 05 in. in 5 in. In theory, the sine bar is the hypotenuse of a right-angle triangle. The gage block buildup forms the side opposite, and the face of the surface plate forms the side adjacent in the triangle.

Using trigonometry, it is possible to calculate the side opposite, or gage block buildup, for any angle between 0 and 90° as follows:

$$\text{Sine of the angle} = \frac{\text{side opposite}}{\text{hypotenuse}}$$
$$= \frac{\text{gage block buildup}}{\text{length of sine bar}}$$

When using a 5-in. sine bar, this would become:

$$\text{Sine of the angle} = \frac{\text{buildup}}{5}$$

Therefore, by transposition, the gage block buildup for any required angle with a 5-in. bar is

$$\text{Buildup} = 5 \times \text{sine of the required angle}$$

EXAMPLE

Calculate the gage block buildup required to set a 5-in. sine bar to an angle of 30°.

$$\begin{aligned}\text{Buildup} &= 5 \sin 30^\circ \\ &= 5(0.5000) \\ &= 2.5000 \text{ in.}\end{aligned}$$

NOTE: This formula is applied only to angles up to 60°.

When an angle greater than 60° is to be checked, it is better to set up the work using the complement of the angle (Fig. 13-6A). The angle plate is then turned 90° to produce the correct angle (Fig. 13-6B). The reason is that when the sine bar is in a near-horizontal position, a small change in the height of the buildup will produce a smaller change in the angle than when the sine bar is in the near-vertical position. This change in gage block height may be shown by calculating the buildups required both for 75° and for the complementary angle of 15°.

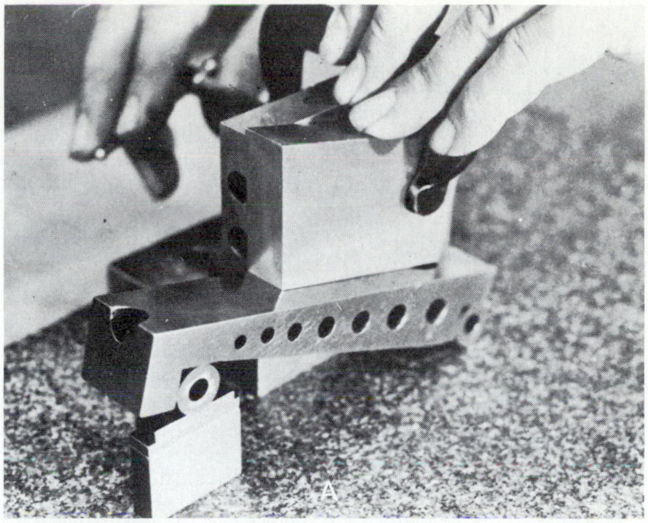

Figure 13-6 *Setting up for an angle greater than 60°. (A) Set the sine bar to the complement of the angle; (B) turn the angle plate 90° on its side. (Courtesy Kostel Enterprises Limited.)*

Buildup required for:

$$\begin{aligned}75^\circ 1' &= 5 \sin 75^\circ 1' \ (0.9660) \\ &= 4.8300 \text{ in.} \\ 75^\circ &= 5 \sin 75^\circ \ (0.96592) \\ &= 4.82960 \text{ in.}\end{aligned}$$

Difference in the buildup for 1'
$$= 0.00040 \text{ in.}$$

Buildup required for:

$$\begin{aligned}15^\circ 1' &= 5 \sin 15^\circ 1' \ (0.25910) \\ &= 1.29550 \text{ in.}\end{aligned}$$

$$15° = 5 \sin 15° \; (0.25882) = 1.29410 \text{ in.}$$

Difference in buildup for 1′ = 0.00140 in.

This example shows that exactly 3.5 times the buildup is required to produce a change of 1′ at 15°, than is required for 1′ at 75°. Therefore, a small inaccuracy in setup would result in a smaller error at a smaller angle than it would at a larger one. If the complementary angles of 80 and 10° are used, this ratio increases to over 5:1.

When small angles are to be checked, it is sometimes impossible to get a buildup small enough to place under one end of the sine bar. In such situations, it will be necessary to place gage blocks under both rolls of the sine bar, having a net difference in measurement equal to the required buildup. For example, the buildup required for 2° is 0.1745 in. Since it is impossible to make this buildup, it is necessary to place the buildup for 1.1745 in. under one roll and a 1.000-in. block under the other roll, giving a net difference of 0.1745 in.

Before the sine bar is used to check a taper, it is necessary to calculate the angle of the taper so that the proper gage block buildup may be made. Figure 13-7A and B illustrates how this is done.

In the right-angled triangle *ABD*:

$$\text{Tan } \frac{a}{2} = \frac{\frac{1}{2}}{12} = 0.04166 = 2°23'10''$$

$$\therefore a = 4°46'20''$$

From this solution, the following formula for solving the included angle, when the taper per foot (*tpf*) is known, is derived.

$$\text{Tan } \frac{a}{2} = \frac{tpf}{24}$$

NOTE: When calculating the angle of a taper *do not* use the formula tan $a = tpf/12$ since triangle *ABC* is not a right-angle triangle.

By transposition, if the included angle is given, the *tpf* may be calculated as follows:

$$tpf = \tan \frac{1}{2} a \times 24$$

If the *tpf* is not known, the angle may be calculated as shown in Fig. 13-7B. That is,

$$AC = \frac{1.750 - 1.000}{2} = \frac{0.750}{2} = 0.375$$

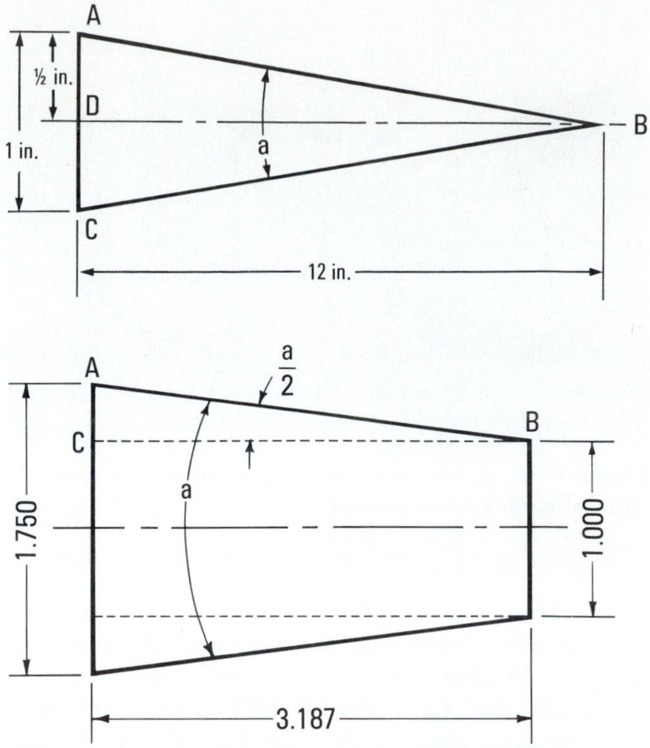

A
½ in.
D
B
1 in.
a
C
12 in.
a
2
A
B
C
a
1.750
1.000
3.187

Figure 13-7 *The angle formed by a taper of 1 in./ft.*

$$\text{Tan } \frac{a}{2} = \frac{0.375}{3.187} = 0.11766 = 6°42'22''$$

$$\therefore a = 13°24'44''$$

To check the accuracy of this taper using a 5-in. sine bar, we calculate the buildup, as follows:

$$\text{Buildup} = 5 \sin 13°24'44'' \; (0.23196) = 1.1598 \text{ in.}$$

Metric tapers are expressed as a ratio of 1 mm per unit of length; for example, a taper having a ratio of 1:20 would taper 1 mm in diameter in 20 mm of length. (See Unit 54.)

Tapers can be checked conveniently and accurately with a *taper micrometer.* This measuring instrument measures work to sine bar accuracy while the work is still in the machine. (See "Checking a Taper" in Unit 54.)

The *sine plate* (Fig. 13-8) is based on the same principle as the sine bar and is similar in construction except that it is wider. Sine bars are up to 1 in. in width, but sine plates are generally more than 2 in. wide. They have several tapped holes in the surface which permit the work to be clamped to the surface of the sine plate. An end stop on a sine plate prevents the workpiece from moving during machining.

Sine plates may be hinged to a base (Fig. 13-9) and are often called *sine tables.* Both types are supplied in 5- and

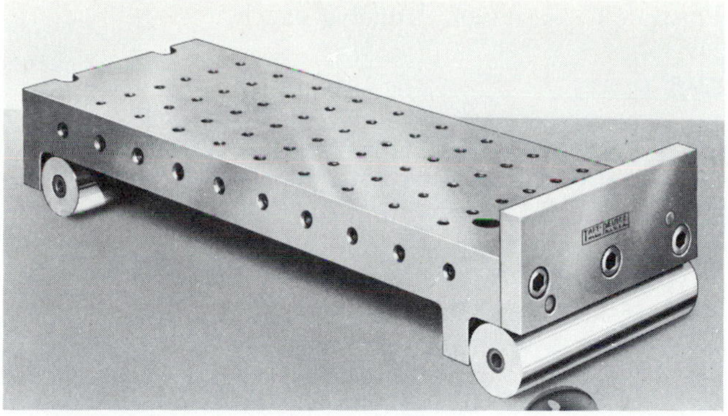

Figure 13-8 *A workpiece may be clamped to a sine plate. (Courtesy The Taft-Pierce Manufacturing Company.)*

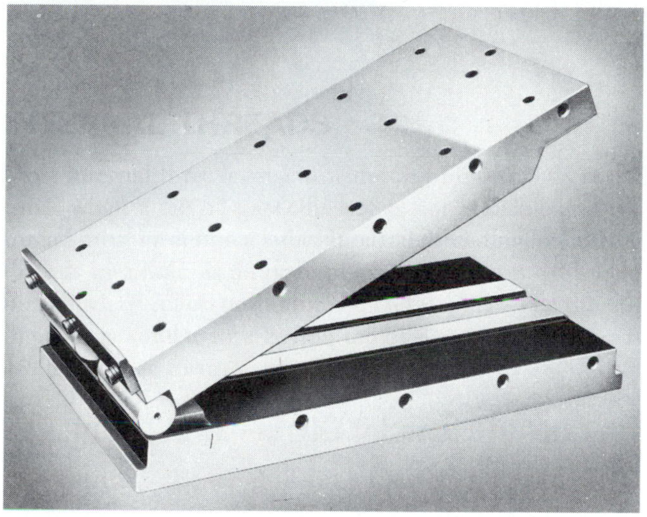

Figure 13-9 *A hinged sine plate may be clamped to the machine table.*

10-in. lengths and have a step or groove of 0.100 or 0.200 in. deep ground in the base to permit the buildup for small angles to be placed under the free roll.

THE COMPOUND SINE PLATE OR TABLE

The *compound sine plate* (Fig. 13-10) consists of one sine plate superimposed on another sine plate. The lower plate is hinged to a base and may be tilted to any angle from 0 to 60° by placing gage blocks under the free roll or cylinder. The upper base is hinged to the lower base so that its cylinder and hinge are at right angles to those of the lower plate. The upper plate may also be tilted to any angle up to 60°. This feature permits the setting of compound angles (angles in two directions).

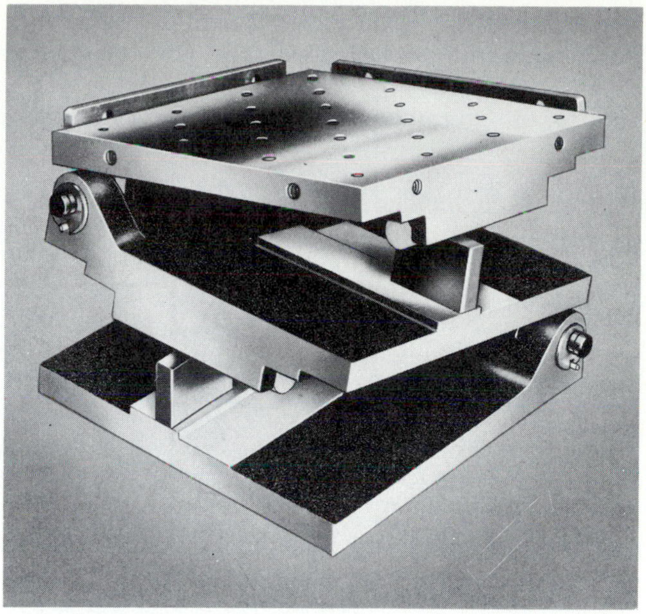

Figure 13-10 *A compound sine plate allows the setting of angles in two directions.*

Compound sine plates have a side and an end plate on the top table to facilitate setting up work square with the table edge and to prevent movement of the work during machining. A step or groove of 0.100 or 0.200 in. deep ground in the base and in the lower table permits the setup for small angles, as the gage block buildup may be placed in the groove.

Since most angles are machined in fixtures, sine plates are not used until the finishing operation, which is generally grinding. To facilitate the holding of parts, both simple and compound sine plates are available with built-in magnetic chucks.

REVIEW QUESTIONS

UNIVERSAL BEVEL PROTRACTOR

1. Name the parts of a universal bevel protractor and state the purpose of each.
2. Describe the principle of the vernier protractor.
3. Sketch a vernier protractor reading of:
   a. 34°20′   b. 17°45′

SINE BAR

4. Describe the construction and principle of a sine bar.
5. What are the accuracies of the 5-in. and the 10-in. sine bars?
6. Calculate the gage block buildup for the following angles using a 5-in. sine bar:
   a. 7°40′   b. 25°50′   c. 40°10′

7. What procedure should be followed to check an angle of 72° using a sine bar and gage blocks? Why is this procedure recommended?

8. In calculating the angle of a taper, why is the formula

$$\text{Tan}\,\frac{1}{2}a = \frac{tpf}{24} \text{ used, rather than}$$

$$\text{Tan}\,a = \frac{tpf}{12}?$$

Illustrate by means of a suitable sketch.

SINE PLATE

9. Describe a sine plate and state its purpose.
10. What is the advantage of a hinged sine plate?
11. What is the purpose of a compound sine plate?

UNIT 14

Gages

Although modern production processes have reached a high degree of precision, producing a part to an *exact* size would be far too costly. Consequently, industrial production methods, including interchangeable manufacture, permit certain variations from the exact dimensions specified while ensuring that the component part will fit into the unit at the time of assembly.

To determine the sizes of the various parts, an inspector generally uses some type of gage. Gages used in industry vary from the simpler type of fixed gage to the sophisticated electronic and laser devices used to measure extremely fine variations.

To ensure that the part will meet the specifications, certain basic terms are used, and these terms are usually added to the drawing of the part. These terms apply to all forms of measurement and inspection and should be understood by both the machinist and the inspector.

OBJECTIVES

After completing this unit you will be able to:
1. Recognize and describe the uses of three types of plug and ring gages
2. Check the accuracy of a part with a plug or ring gage
3. Check the accuracy of a part using a snap gage

BASIC TERMS

The following basic terms are used to express the exact size of a part and allowable variations from that size. An example is provided in Table 14-1.

Basic dimension is the exact size of a part from which all limiting variations are made.

Limits are the maximum and minimum dimensions of a part (the high and low dimensions).

Tolerance is the permissible variation of a part. Tolerance is often shown on the drawing by the basic dimension

Table 14-1 An Example of Limits and Tolerances

Nominal size	3 in.	
Basic dimension	3.000	(Decimal equivalent of nominal size)
Basic dimension and amount of bilateral tolerance permitted	3.000	±0.002
Limits	3.002	(Largest size permitted)
	2.998	(Smallest size permitted)
Tolerance	0.004	(Difference between minimum and maximum limits)

plus or minus the amount of variation allowed. If a part on a drawing was dimensioned to 3.000 in. ±0.002 (76 mm ±0.05), the tolerance would be 0.004 in. (0.10 mm).

If the tolerance is in one direction only, that is, plus *or* minus, it is said to be *unilateral tolerance*. However, if the tolerance is both plus *and* minus, it is referred to as *bilateral tolerance*.

Allowance is the intentional difference in the dimensions of mating parts. For example, it is the difference between the maximum diameter of the shaft and the minimum diameter of the mating bore.

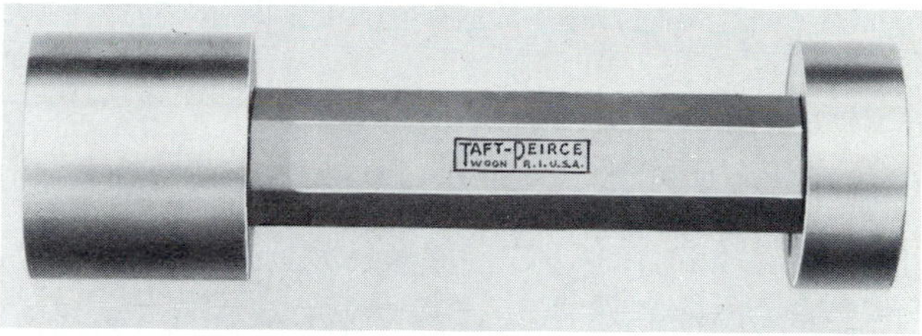

TAFT-PEIRCE
WOON R.I. U.S.A.

Figure 14-1 *A cylindrical plug gage is used to check hole sizes. (Courtesy Taft-Peirce Manufacturing Company.)*

FIXED GAGES

Fixed gages are used for inspection purposes because they provide a quick means of checking a specific dimension. These gages must be easy to use and accurately finished to the required tolerance. They are generally finished to one-tenth the tolerance they are designed to control. For example, if the tolerance of a piece being checked is to be maintained at 0.001 in. (0.02 mm), the gage must be finished to within 0.0001 in. (0.002 mm) of the required size.

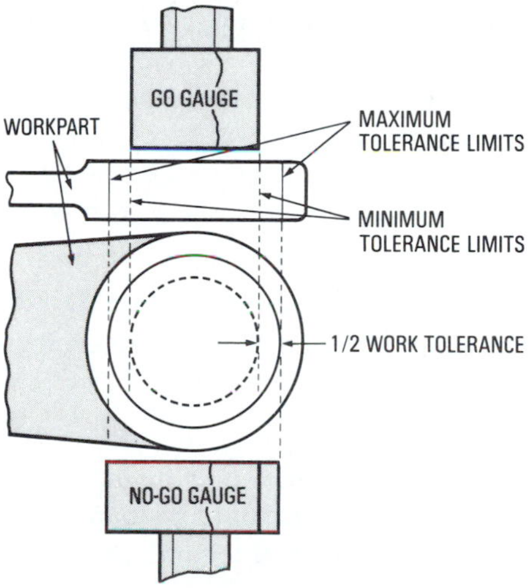

GO GAUGE
WORKPART
MAXIMUM
TOLERANCE LIMITS
MINIMUM
TOLERANCE LIMITS
1/2 WORK TOLERANCE
NO-GO GAUGE

Figure 14-2 *The "go" end of the gage checks the minimum tolerance limit, while the "no-go" end checks the maximum tolerance. (Courtesy The Bendix Corporation.)*

CYLINDRICAL PLUG GAGES

Plain *cylindrical plug gages* (Fig. 14-1) are used for checking the inside diameter of a straight hole and are generally of the "go" and "no-go" variety. This type of gage consists of a handle and plug on each end ground and/or lapped to a specific size. The smaller-diameter plug, or the "go" gage, checks the lower limit of the hole. The larger-diameter plug, or the "no-go" gage, checks the upper limit of the hole (Fig. 14-2). For instance, if a hole size is to be maintained at 1.000 in. ±0.0005 (25.4 mm ±0.012) the "go" end of the gage would be designed to fit into a hole 0.9995 in. (25.38 mm) in diameter. The larger end ("no-go") would not fit into any hole smaller than 1.0005 in. (25.41 mm) in diameter.

The dimensions of these gages are usually stamped on the handle at each end adjacent to the plug gage. The "go" end is made longer than the "no-go" end for easy identification. Sometimes a groove is cut on the handle near the "no-go" end to distinguish it from the "go" end.

Due to the wear caused by the constant use of plug gages, many of them are made with carbide tips, which greatly increases gage life.

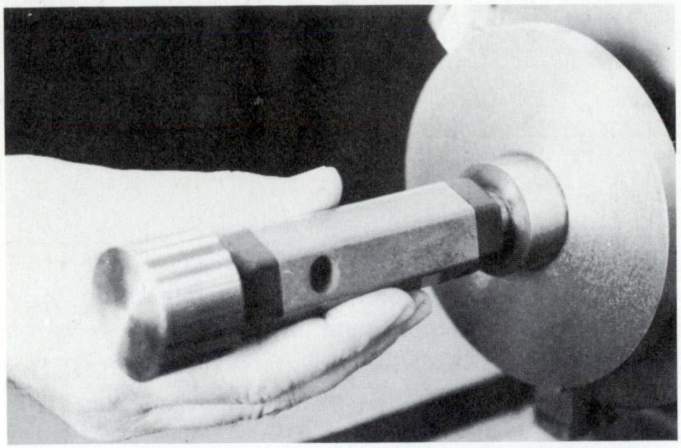

Figure 14-3 *Checking a hole size with a plug gage.*

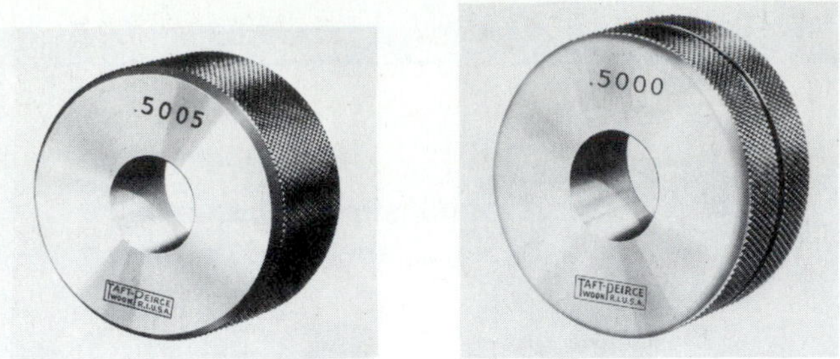
.5005
.5000

Figure 14-4 *Plain ring gages are used to check the diameter of round workpieces. (Courtesy Taft-Peirce Manufacturing Company.)*

TO USE A CYLINDRICAL PLUG GAGE

1. Select a plug gage of the correct size and tolerance for the hole being checked.
2. Clean both ends of the gage and the hole in the workpiece with a clean, dry cloth.
3. Check the gage (both ends) and the workpiece for nicks and burrs.
4. Wipe both ends of the gage with an oily cloth to deposit a thin film of oil on the surfaces.
5. Start the "go" end *squarely* into the hole (Fig. 14-3). If the hole is within the limits, the gage will enter easily. *Do not force or turn it.* The plug should enter the hole for its full length, and there should be no excessive play between the plug and the part.

NOTE: If the gage enters the hole only part way, the hole is tapered. Excessive play or looseness in one direction indicates that the hole is elliptical (out of round).

6. After the hole has been checked with the "go" end, it should be checked with the "no-go" end. This end should not begin to enter the hole. An entry of more than 1/16 in. (1.5 mm) indicates an oversize, bell-mouth, or tapered hole.

PLAIN RING GAGES

Plain ring gages, used to check the outside diameter of pieces, are ground and lapped internally to the desired size. The size is stamped on the side of the gage. The outside diameter is knurled, and the "no-go" end is identified by an annular groove on the knurled surface (Fig. 14-4). The precautions and procedures for using a ring gage are similar to those outlined for a plug gage and should be followed carefully.

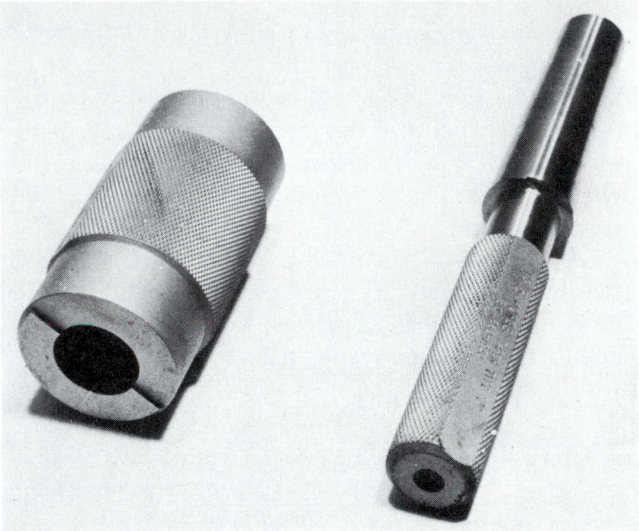

Figure 14-5 *Taper ring and plug gages.*

TAPER PLUG GAGES

Taper plug gages (Fig. 14-5), made with standard or special tapers, are used to check the size of the hole and taper accuracy. Some of these gages have "go" and "no-go" rings scribed on them. If the gage fits into the hole between these two rings, the hole is within the required tolerance. Other taper plug gages have steps ground on the large end to indicate the limits. The rings or steps measure hole-size limits only. A wobble between the plug gage and the hole is evidence of an incorrect taper.

TO CHECK AN INTERNAL TAPER USING A TAPER PLUG GAGE

1. Select the proper taper gage for the hole being checked.
2. Wipe the gage and the hole with a clean, dry cloth.
3. Check both the gage and hole for nicks and burrs.
4. Apply a *thin* coating of Prussian blue to the surface of the plug gage.

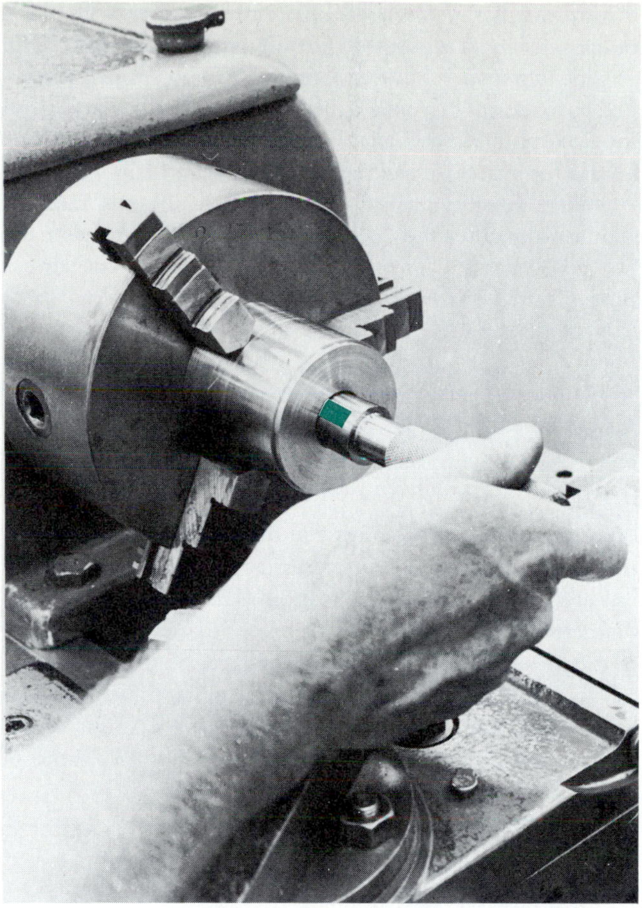

Figure 14-6 *Checking a tapered hole with a taper plug gage.*

5. Insert the plug gage into the hole as far as it will go (Fig. 14-6).
6. Maintaining light end pressure on the plug gage, rotate it *counterclockwise* for approximately one-quarter turn.
7. Check the diameter of the hole. A proper size is indicated when the edge of the workpiece lies between the limit steps or lines on the gage.

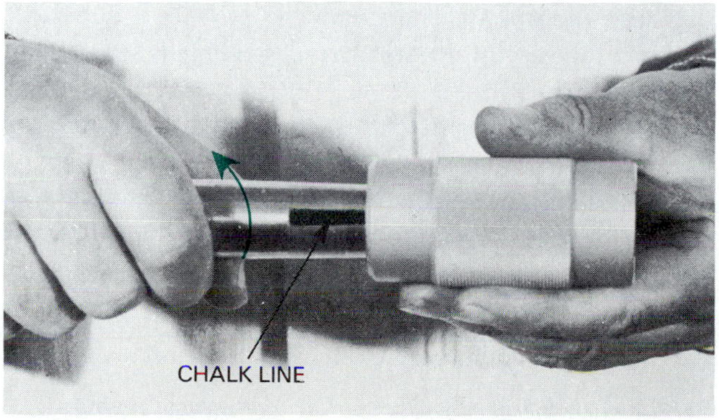
CHALK LINE

Figure 14-7 *Checking the accuracy of a taper using chalk lines.*

8. Check the taper of the hole by attempting to move the gage radially in the hole. Any error in the taper will be indicated by play at either end between the hole and the gage. Movement or play at the large end indicates excessive taper; movement at the small end indicates insufficient taper.
9. Remove the gage from the hole to see if the bluing has rubbed off evenly along the length of the gage, a result which would indicate a proper fit. A poor fit is evident if the bluing has been rubbed off more at one end than the other.

TAPER RING GAGES

Taper ring gages (Fig. 14-5) are used to check both the accuracy and the outside diameter of the taper. Ring gages often have scribed lines or a step ground on the small end to indicate the "go" and "no-go" dimensions.

For a taper ring gage, the precautions and procedures are similar to those outlined for a taper plug gage. However, when work which has not been ground or polished is being checked, three equally spaced chalk lines around the circumference and extending for the full length of the tapered section may be used to indicate the accuracy of the taper (Fig. 14-7). If the work has been ground or polished, it is advisable to use three thin lines of Prussian blue.

CARE OF PLUG AND RING GAGES

Gage life is dependent on the following factors.

1. Materials from which the gage is made
2. Material of the part being checked
3. Class of fit required
4. Proper care of the gage

In order to preserve the accuracy and life of gages:

1. Store gages in divided wooden trays to protect them from being nicked or burred.
2. Check them frequently for size and accuracy.
3. Correctly align gages with the workpiece to prevent binding.
4. Do not force or twist a plain plug or ring gage. Forcing or twisting will cause excessive wear.
5. Clean the gage and workpiece thoroughly before checking the part.
6. Use a light film of oil on the gage to help prevent binding.
7. Make provision for air to escape when gaging blind holes with a plug gage.
8. Have gages and work at room temperature to ensure accuracy and prevent damage to the gage.
9. Never use an inspection gage as a working gage.

THREAD PLUG GAGES

Internal threads are checked with *thread plug gages* (Fig. 14-8) of the "go" and "no-go" variety and are based on the same principle as cylindrical plug gages.

When a thread plug gage is used, the "go" end, which is the longer end, should be turned in flush to the bottom of the hole. The "no-go" end should just start into the hole and become quite snug before the third thread enters.

Since thread plug gages are quite expensive, certain precautions should be observed in their use:

1. Thread plug gages have a chip groove cut along the thread to clear loose chips. Do not depend on this feature to remove burrs or loose chips. To prolong the life of the gage, it is advisable to remove burrs and loose chips (wherever possible) by means of an old tap.
2. Before using the thread plug gage, apply a little oil to its surface.
3. Never force the gage.

THREAD RING GAGES

The most popular gage of this type is the *adjustable thread ring gage.* It is used to check the accuracy of an external thread and has a threaded hole in the center with three radial slots and a setscrew to permit small adjustments. The outside diameter is knurled, and the "no-go" gage is identified by an annular groove cut into the knurled surface. Both the "go" and "no-go" gages are generally assembled in one holder for ease of checking the part (Fig. 14-9).

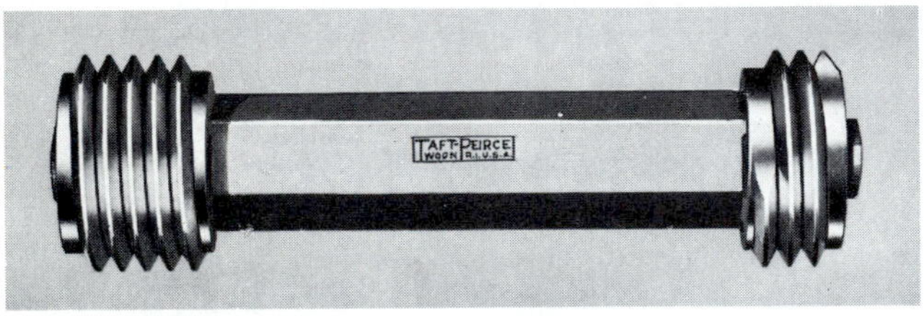

Figure 14-8 *Thread plug gages are used to check the size and accuracy of an internal thread. (Courtesy Taft-Peirce Manufacturing Company.)*

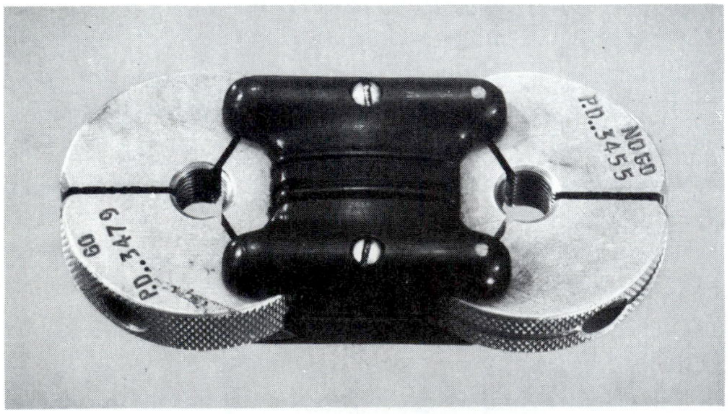
NO GO
P.D..3455
GO
P.D..3479

Figure 14-9 *"Go" and "no-go" gage thread ring gages in a holder. (Courtesy The Bendix Corporation.)*

When these gages are used, the thread being checked should fully enter the "go" gage, but should not enter the "no-go" gage by more than 1½ turns. Before checking a thread, remove any dirt, grit, or burrs. A little oil will help to prolong the life of the gage.

SNAP GAGES

Snap gages, one of the most common types of comparative measuring instruments, are faster to use than micrometers but are limited in their application. They are used to check diameters within certain limits by comparing the part size to the preset dimension of the snap gage. Snap gages generally have a C-shaped frame with adjustable gaging anvils or rolls, which are set to the "go" and "no-go" limits of the part. These gages are supplied in several styles, some of which are shown in Fig. 14-10.

TO USE A SNAP GAGE TO CHECK A DIMENSION

Proper use of a snap gage is required to prevent springing the gage and marring the work surface. Proceed as follows:

1. Thoroughly clean the anvils of the gage.
2. Set the "go" and "no-go" anvils to the required limits, using gage blocks or some other standard.
3. Lock the anvils in position and recheck the accuracy of the settings.

NOTE: If a dial-indicator snap gage is used, set the bezel (outer ring) of the indicator to read 0 and lock it into position.

4. Clean the surface of the work.
5. Hold the gage in the right hand, keeping it square with the work.
6. With the left hand, hold the lower anvil in position on the workpiece.
7. Push the gage over the work surface with a rolling motion. Only light hand pressure should be used to pass the "go" pins.

NOTE: Do not force the gage; if the work is the correct size, the gage should pass easily over the work.

8. Advance the gage until the "no-go" anvils or rolls contact the work. If the gage stops at this point, the work is within the limits.

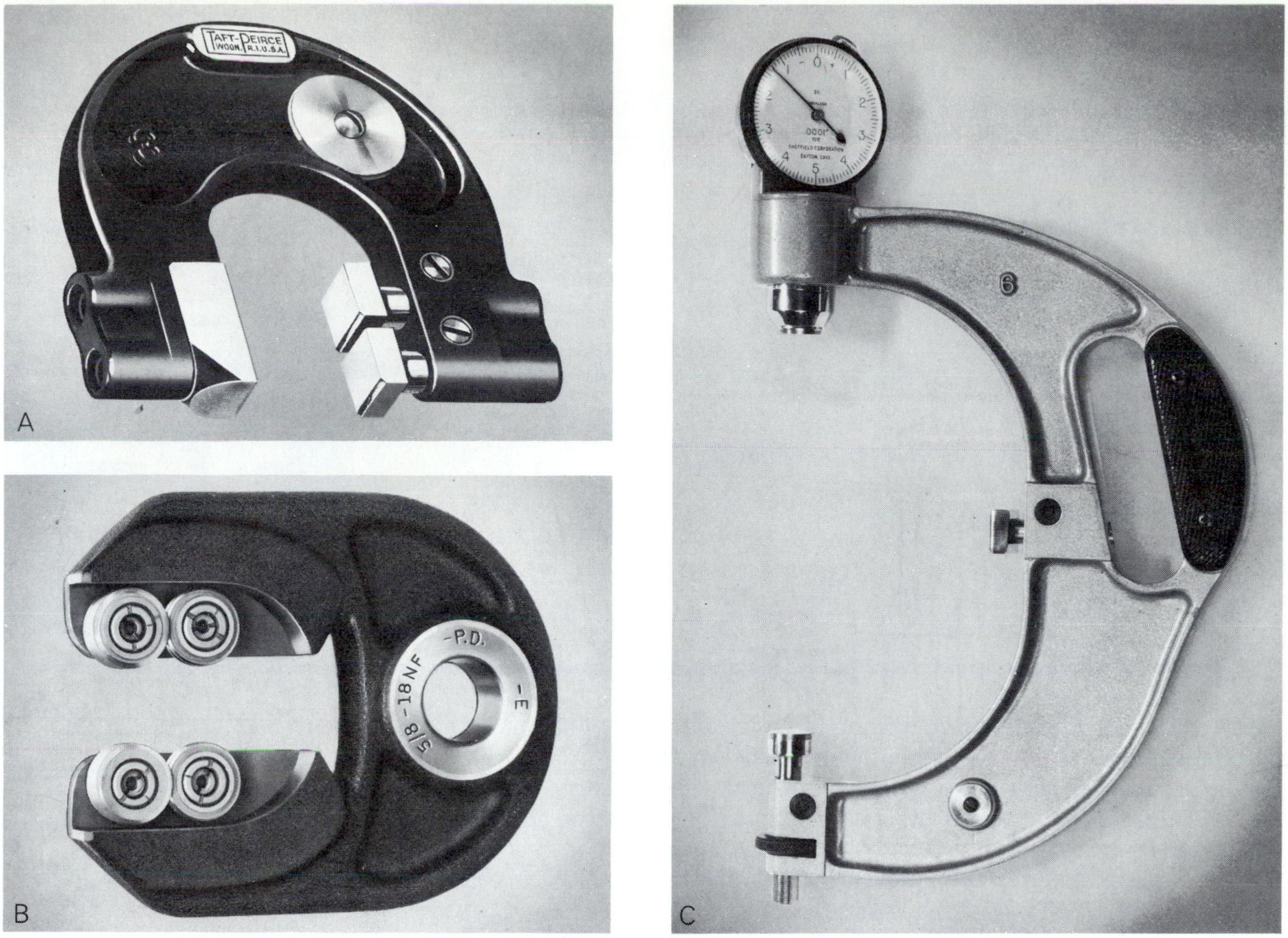
TAFT-PEIRCE
WOON. R.I. U.S.A.
A
5/8-18NF
-P.D.
-E
B
C

Figure 14-10 *Various types of snap gages: (A) adjustable snap gage (Courtesy Taft-Peirce Manufacturing Company); (B) adjustable roll snap gage (Courtesy The Bendix Corporation); (C) dial indicator snap gage (Courtesy Taft-Peirce Manufacturing Company).*

REVIEW QUESTIONS

1. For each of the following dimensions, indicate the basic diameter, upper limit, lower limit, and tolerance.
   a. 1.750 + 0.002 in.
       − 0.000 in.
   b. 0.625 + 0.0015 in.
       − 0.0000 in.
   c. 12.50 ± 0.02 mm
   d. 20.500 + 0.000 mm
       − 0.015 mm
   e. 0.500 ± 0.005 mm

2. State whether the tolerances for each size in Question 1 are unilateral or bilateral.

FIXED GAGES

3. What purpose do fixed gages serve in industry?
4. To what tolerance are fixed gages finished?
5. If a hole size is to be maintained at 1.750 in. ±0.002 (44 mm ±0.05), what would be the sizes of the "go" and "no-go" gages?

CYLINDRICAL PLUG GAGES

6. How are the "go" and "no-go" ends of a cylindrical plug gage identified?
7. What precautions must be observed when using a cylindrical plug gage?

PLAIN RING GAGES

8. How is a "no-go" ring gage distinguished from a "go" ring gage?

TAPER PLUG AND RING GAGES

9. How may the limits of a taper plug gage be indicated on the gage?

10. List the precautions to observe when checking with a taper plug or ring gage.
11. When may chalk be used to check an external taper and when should bluing be used?
12. Why should a taper plug or ring gage be rotated no more than one-quarter of a turn when checking a taper?

THREAD PLUG AND RING GAGES

13. What parts of a thread are checked with a "go" gage? A "no-go" gage?
14. List three precautions to observe when using a thread plug gage.
15. How should an external thread of the proper size fit into the thread ring gage?
16. Describe a snap gage.
17. What advantage does the dial indicator snap gage have over an adjustable snap gage?
18. List the precautions necessary when checking a workpiece with a snap gage.

UNIT 15

Comparison Measurement

Manufacturing processes have now become so precise that component parts are often made in several different places and then shipped to a central location for final assembly. In order for this process of *interchangeable manufacture* to be economical, there must be some assurance that these parts will fit on assembly. The components are therefore made to within certain limits, and further inspection or *quality control* ensures that only properly sized parts will be used.

Much of this inspection is done rapidly, accurately, and economically by a process called *comparison measurement.* It consists of comparing the measurement of the part to a known standard or master of the exact dimension required. Basically, *comparators* are gages that incorporate some means of amplification to compare the part size to a set standard, usually gage blocks.

Mechanical, optical, and mechanical–optical comparators, and air, electrical, and electronic gages are all utilized for comparison measurement.

OBJECTIVES

After completing this unit you will be able to:
1. Explain the principle of comparison measurement
2. Identify four types of comparators and describe their use
3. Measure to within 0.0005-in. (0.01-mm) accuracy with a dial indicator, mechanical and optical comparator, or air and electronic gages

COMPARATORS

A comparator may be classified as any instrument which is used to compare the size of a workpiece to a known standard. The simplest form of comparator is a dial indicator mounted on a surface gage. All comparators are provided with some means of amplification by which variations from the basic dimensions can be easily noted.

DIAL INDICATORS

Dial indicators are used to compare sizes and measurements to a known standard and to check the alignment of machine tools, fixtures, and workpieces prior to machining. Many types of dial indicators operate on a gear and rack principle (Fig. 15-1). A *rack* cut on the *plunger* or *spindle* is in mesh with a *pinion,* which in turn is connected to a *gear train.* Any movement of the spindle is then magnified and transmitted to a *hand* or *pointer* over a *graduated dial.* Inch-designed dials may be graduated in thousandths of an inch or less. The dial, attached to a *bezel,* may be adjusted and locked in any position.

During use, the contact point on the end of the spindle bears against the work and is held in constant engagement with the work surface by the rack spring. A hair spring is attached to the gear that meshes with the center pinion. This flat spiral spring takes up the backlash from the gear train and prevents any lost motion from affecting the accuracy of the gage.

Dial indicators are generally of two types: the continuous-reading dial indicator and the dial test indicator.

The *continuous-reading dial indicator* (Fig. 15-2), numbered clockwise for 360°, is available as a regular-range and a long-range indicator. The regular-range dial indicator has only about 2½ revolutions of travel. It is generally used for comparison measurement and setup purposes. The long-range dial indicator (Fig. 15-2) is often used to indicate table travel or cutting tool movement on machine tools. It has a second, smaller hand that indicates the number of revolutions that the large hand has traveled.

Dial test indicators (Fig. 15-1) may have a balanced-type dial, that is, one which reads both to the right and left from 0 and indicates a plus or minus value. Indicators of this type have a total spindle travel of only 2½ revolutions. These instruments may be equipped with tolerance pointers to indicate the permissible variation of the part being measured.

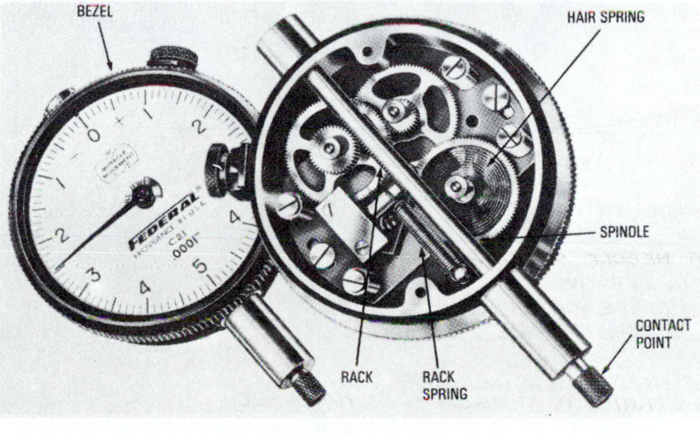
BEZEL
HAIR SPRING
SPINDLE
CONTACT POINT
RACK
RACK SPRING

Figure 15-1 *A balanced-type dial test indicator, showing the internal mechanism. (Courtesy Federal Products Corporation.)*

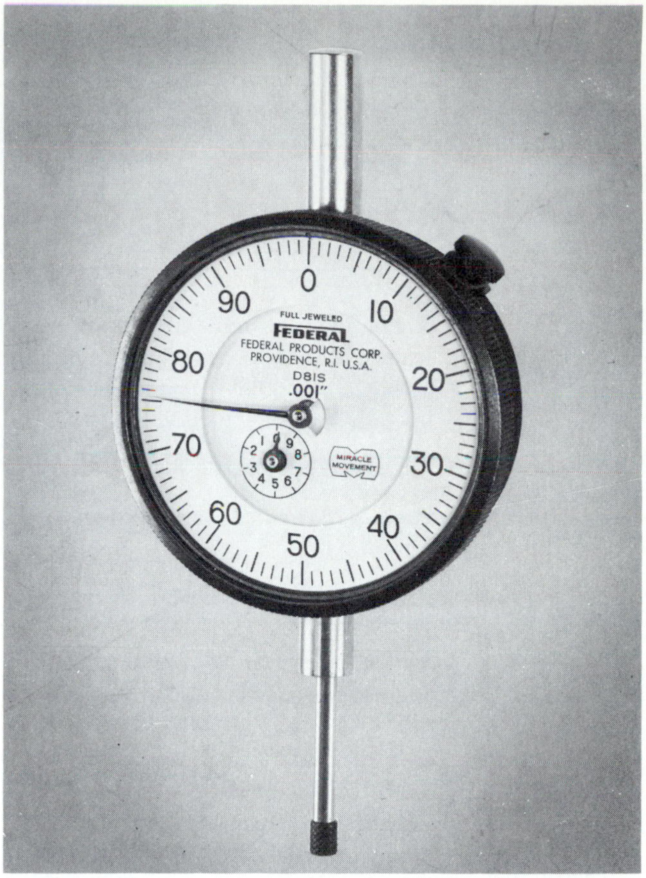
FULL JEWELED
FEDERAL
FEDERAL PRODUCTS CORP.
PROVIDENCE, R.I. U.S.A.
D8IS
.001"
MIRACLE MOVEMENT

Figure 15-2 *A continuous-reading dial indicator. (Courtesy Federal Products Corporation.)*

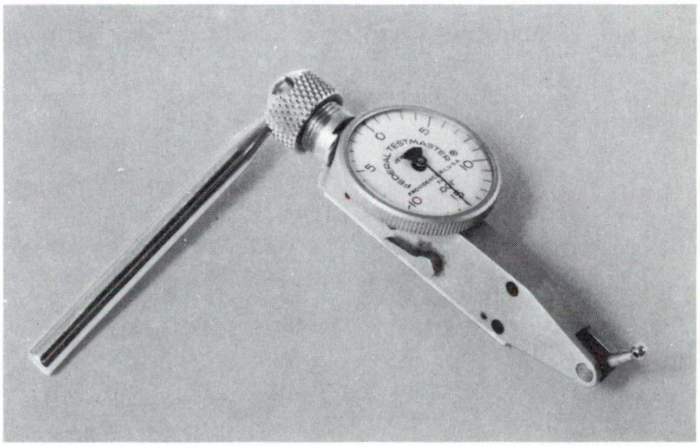

Figure 15-3 *The universal dial test indicator.*

Perpendicular dial test indicators, or back plunger indicators, have the spindle at right angles (90°) to the dial. They are used extensively in setting up lathe work and for machine table alignment.

The *universal* dial test indicator (Fig. 15-3) has a contact point that may be set at several positions through a

Figure 15-4 *(A) A universal dial test indicator being used to center a workpiece with the machine spindle; (B) checking measurements with a dial height gage. (Courtesy Federal Products Corporation.)*

180° arc. This type of indicator may be conveniently used to check internal and external surfaces. Figure 15-4A and B illustrates typical applications of this type of indicator.

Metric dial indicators (Fig. 15-5) are available in both the balanced and continuous-reading types. The type used for inspection purposes is usually graduated in 0.002 mm and has a range of 0.5 mm. The regular indicators are usually graduated in 0.01 mm and have a range up to 25 mm.

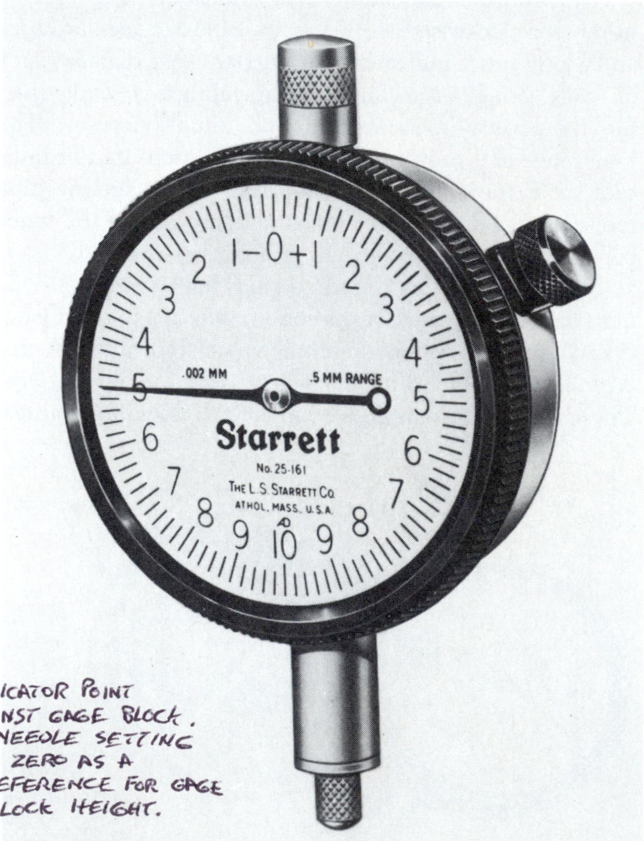
.002 MM
.5 MM RANGE
Starrett
No. 25-161
The L. S. Starrett Co.
Athol, Mass. U.S.A.

Figure 15-5 *A metric dial indicator with a balanced dial.*

TO MEASURE WITH A DIAL TEST INDICATOR AND HEIGHT GAGE

1. Clean the face of the surface plate and the vernier height gage.
2. Mount the dial test indicator on the movable jaw of the height gage (Fig. 15-4B).
3. Lower the movable jaw until the indicator point just touches the top of a gage block resting on the surface plate. SEE Pg 70 FIG. 11-18
4. Tighten the upper locking screw on the vernier and loosen the lower locking screw.
5. Carefully turn the adjusting nut until the indicator needle registers approximately one-quarter turn.
6. Turn the bezel to set the indicator to zero. THIS PUTS INDICATOR POINT FIRMLY AGAINST GAGE BLOCK. THEN ADJUST NEEDLE SETTING TO ZERO AS A REFERENCE FOR GAGE BLOCK HEIGHT.
7. Note the reading on the vernier and record it on a piece of paper.
8. Raise the indicator to the height of the first hole to be measured.
9. Adjust the vernier until the indicator reads zero. A ZERO NEEDLE READING OCCURS WHEN VERNIER IS AT HEIGHT TO BE MEASURED.
10. Note the vernier reading again and record it.
11. Subtract the first reading from the second and add the height of the gage block.

12. Proceed in this manner to record the location of all other holes.

MECHANICAL AND ELECTRONIC COMPARATORS

The *mechanical comparator* consists of a base, a column, and a gaging head. Various mechanical comparators operate on different principles. Some are based on the gear and rack principle used in some dial indicators; others use a system of levers similar to the universal dial indicator.

Mechanical comparators are gradually being replaced by electronic indicators and comparators. The *Digi-Matic indicator,* Fig. 15-6, is capable of readings in increments of 0.0001 in. or 0.001 mm. It can be interfaced with a data recorder, minicomputer, or host computer to provide statistical data based on inspection results.

The *electronic comparator* (Fig. 15-7), a highly accurate form of comparator, uses the Wheatstone bridge circuit to transform minute changes in spindle movement into a relatively large needle movement on the gage. This degree of magnification is controlled by a selector on the front of the *amplifier.* The widely spaced graduations represent zero values from 0.0001 to 0.000 01 in. (0.002 mm to 0.0002 mm), depending on the scale selected.

When it is not necessary to know the exact dimension of a part, but only whether it falls within the required limits, a signal light attachment may be installed on the gage. When a workpiece is tested, an amber light indicates that it is within the prescribed limits, a red light indicates that the workpiece is too small, and a blue light indicates that it is too large.

Electronic units may also be used as height gages by mounting a rectangular gage head on a height gage stand (Fig. 15-8). This method is particularly suited to the checking of soft, highly polished surfaces because of the light gaging pressure required.

All types of comparators are used in inspection to check the size of a part against a master gage. The variation between the part and the master is shown on a scale as a plus or minus quantity.

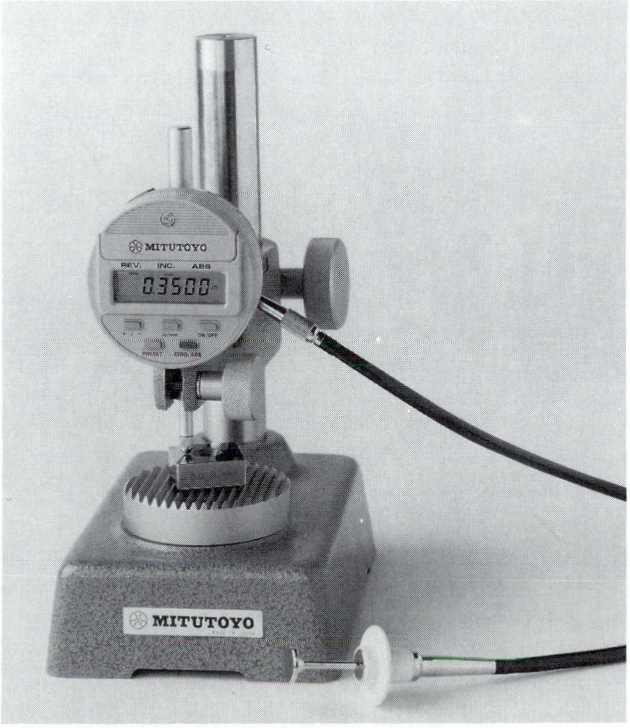
MITUTOYO
MITUTOYO

Figure 15-6 *The Digi-Matic indicator can make readings in increments of 0.0001 in. or 0.001 mm. (Courtesy Mitutoyo: MTI Ltd.)*

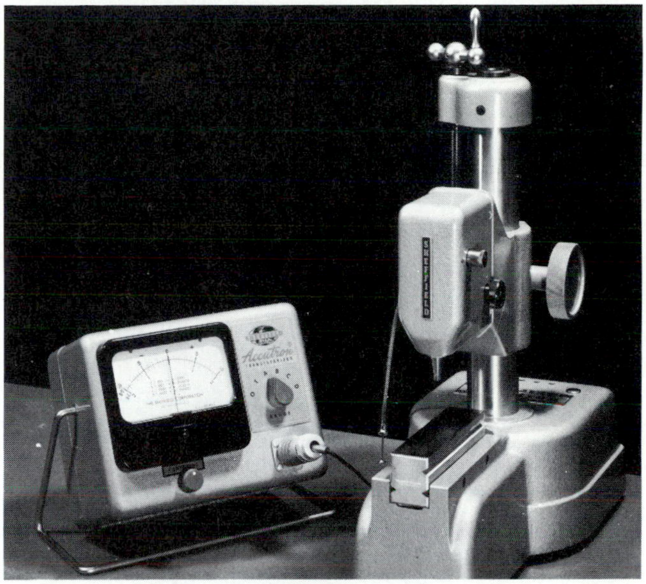

Figure 15-7 *The value of the scale divisions on an electronic comparator can be readily changed to suit the accuracy required.*

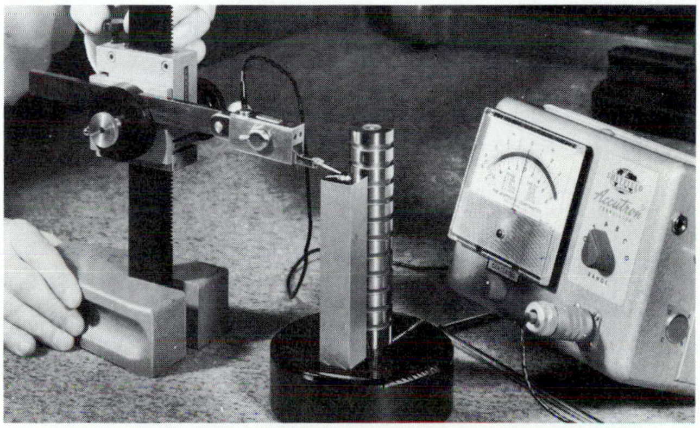

Figure 15-8 *The electronic gage with a rectangular head is used to check heights accurately. (Courtesy The Bendix Corporation.)*

TO MEASURE WITH A MECHANICAL COMPARATOR

1. Clean the anvil and master gage of the required size.
2. Place the master on the anvil.
3. Carefully lower the gaging head until the stylus touches the master and indicates a movement of the needle.
4. Lock the gaging head to the column.
5. Adjust the needle to zero using the fine adjustment knob, and set the limit pointers on the face.
6. Recheck the setting by removing the master and replacing it.
7. Set the tolerance pointers to the upper and lower tolerances of the part being checked.
8. Substitute the work being gaged for the master and note the reading. If the reading is to the right of zero, the work is too large; if to the left, it is too small. If the needle stops between the tolerance pointers, the work is within the permitted limits.

OPTICAL COMPARATORS

An *optical comparator* or *shadowgraph* (Fig. 15-9) projects an enlarged shadow onto a screen where it may be compared to lines or to a master form which indicates the limits

Figure 15-9 *Intricate forms can be checked easily with an optical comparator.*

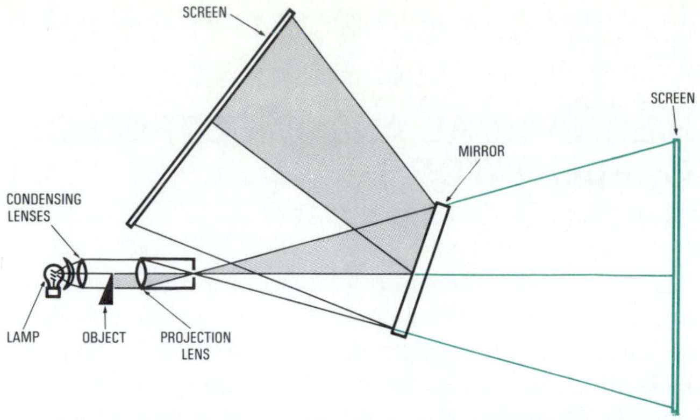
SCREEN
SCREEN
MIRROR
CONDENSING LENSES
LAMP
OBJECT
PROJECTION LENS

Figure 15-10 *The principle of the optical comparator.*

of the dimensions or the contour of the part being checked. The optical comparator is a fast, accurate means of measuring or comparing the workpiece with a master. It is often used when the workpiece is difficult to check by other methods. Optical comparators are particularly suited to checking extremely small or odd-shaped parts, which would be difficult to inspect without the use of expensive gages.

Optical comparators are available in bench and floor models, which are identical in principle and operation (Fig. 15-10). Light from a *lamp* passes through a *condenser lens* and is projected against the workpiece. The shadow caused by the workpiece is transmitted through a *projecting lens* system, which magnifies the image and casts it onto a *mirror.* The image is then reflected to the *viewing screen* and is further magnified in this process.

The extent of the image magnification depends on the lens used. Interchangeable lenses for optical comparators are available in the following magnifications: 5×, 10×, 31.25×, 50×, 62.5×, 90×, 100×, and 125×.

A comparator chart or master form mounted on the viewing screen is used to compare the accuracy of the enlarged image of the workpiece being inspected. Charts are usually made of translucent material, such as cellulose acetate or frosted glass. Many different charts are available for special jobs, but the most commonly used are linear-measuring, radius, and angular charts. A vernier protractor screen is also available for checking angles. *Since charts are available in several magnifications, care must be taken to use a chart of the same magnification as the lens mounted on the comparator.*

Many accessories are available for the comparator, increasing the versatility of the machine. Some of the most common are *tilting work centers,* which permit the workpiece to be tilted to the required helix angle for checking threads; a *micrometer work stage,* which permits quick and accurate measuring of dimensions in both directions; and *gage blocks, measuring rods,* and *dial indicators* used on comparators for checking measurement. The surface of the workpiece may be checked by a *surface illuminator,* which lights up the face of the workpiece adjacent to the projecting lens system and permits this image to be projected onto the screen.

TO CHECK THE ANGLE OF A 60° THREAD USING AN OPTICAL COMPARATOR (FIG. 15-11)

1. Mount the correct lens in the comparator.
2. Mount the tilting centers on the micrometer cross-slide stage.
3. Set the tilting centers to the helix angle of the thread.
4. Set the workpiece between centers.
5. Mount the vernier protractor chart and align it horizontally on the screen.
6. Turn on the light switch.
7. Focus the lens so that a clear image appears on the screen.
8. Move the micrometer cross-slide stage until the thread image is centralized on the screen.
9. Revolve the vernier protractor chart to show a reading of 30°.
10. Adjust the cross-slides until the image coincides with the protractor line.
11. Check the other side of the thread in the same manner.

NOTE: If the thread angle is not correct or square with the centerline, adjust the vernier protractor chart to measure the angle of the thread image.

Other dimensions of the thread, such as depth, diameters, and width of flats, may be measured with micrometer-measuring stages or devices such as rods, gage blocks, and indicators.

MECHANICAL–OPTICAL COMPARATORS

The *mechanical–optical comparator* (Fig. 15-12), or the *reed*-type comparator, combines a reed mechanism with a light beam to cast a shadow on a magnified scale to indicate the dimensional variation of the part. It consists of a base, a column, and a gaging head which contains the reed mechanisms and light source.

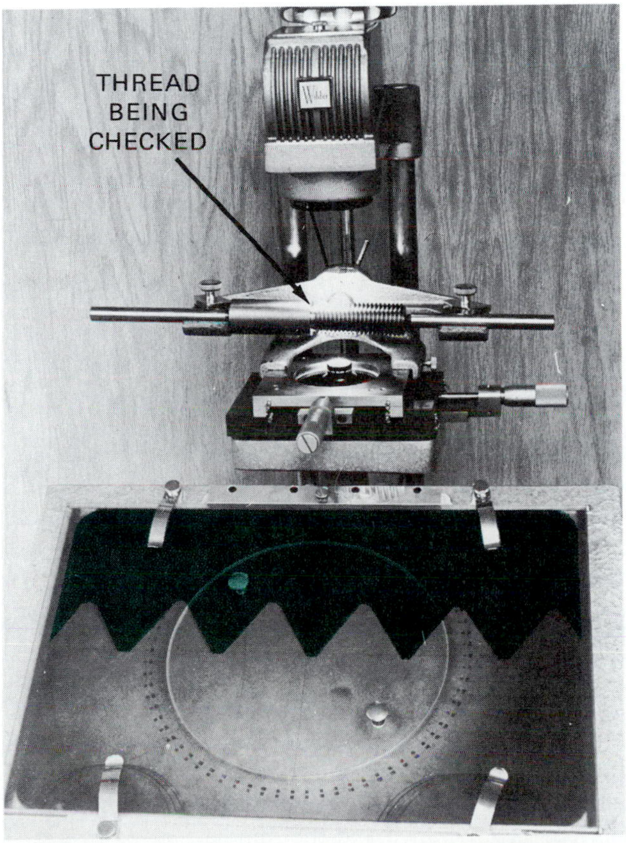
THREAD
BEING
CHECKED

Figure 15-11 *Checking a thread form on an optical comparator.*

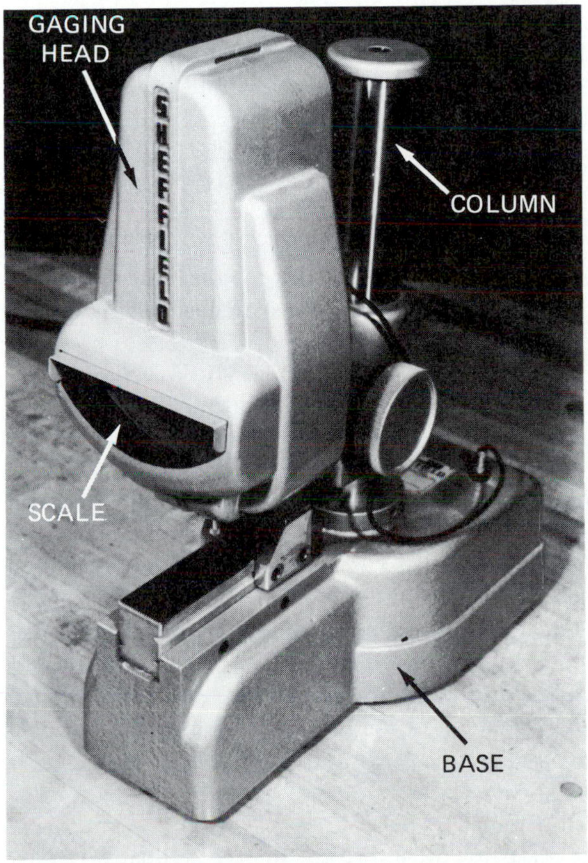
GAGING
HEAD
SHEFFIELD
COLUMN
SCALE
BASE

Figure 15-12 *A reed-type comparator uses both the mechanical and optical principles of measurement. (Courtesy The Bendix Corporation.)*

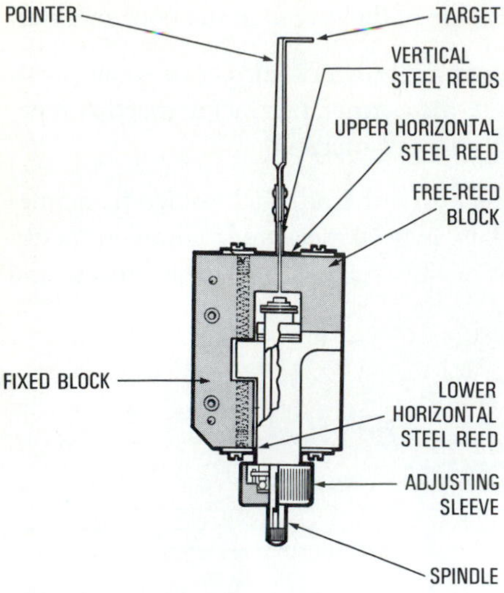
POINTER
TARGET
VERTICAL
STEEL REEDS
UPPER HORIZONTAL
STEEL REED
FREE-REED
BLOCK
FIXED BLOCK
LOWER
HORIZONTAL
STEEL REED
ADJUSTING
SLEEVE
SPINDLE

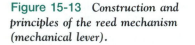

Figure 15-13 *Construction and principles of the reed mechanism (mechanical lever).*

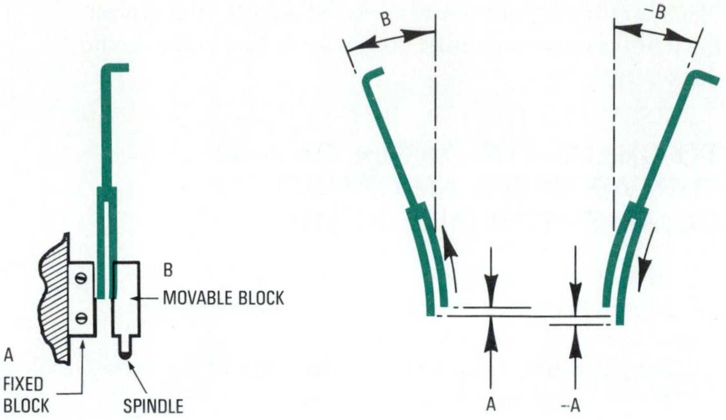
B
MOVABLE BLOCK
A
FIXED
BLOCK
SPINDLE
B
-B
A
-A

Figure 15-14 *(A) Cutaway section showing the blocks and reed mechanism; (B) the upper part of the reed deflects a greater distance than the lower part.*

THE REED MECHANISM

Figure 15-13 illustrates the principle of the reed mechanism. A fixed steel block A and a movable block B have two pieces of spring steel, or reeds, attached to them (Fig. 15-14A). The upper ends of the reeds are joined and connected to a pointer. Since block A is fixed, any movement of the spindle attached to block B will move this block up or down, causing the pointer to move a much greater distance to the right or left (Fig. 15-14B).

A light beam passing through an aperture illuminates the scale (Fig. 15-15). The pointer, with a *target* attached, is located below the aperture so that a movement of block B will cause the target to interrupt the light beam, casting a highly amplified shadow on the scale. The movement of the pointer and target is obviously greater than the movement of the spindle. Also, the shadow cast on the scale will be larger than the movement of the target. Therefore, the measurement on the scale will be much greater than the movement of the spindle.

To illustrate the total magnification of this instrument, suppose that the ratio of target movement to the stylus is 25:1, and the ratio to the light beam lever is 20:1. By combining these two movements, the shadow cast would then be 25×20 or 500 times as large as the movement of the stylus. This large amplification would permit extremely accurate gaging to be performed on this type of instrument. Reed-type comparators are manufactured with magnifications from 500:1 to 20 000:1.

The scales for these instruments are graduated in plus and minus with zero being in the center of the scale. The value for each graduation is marked on the scale of every machine.

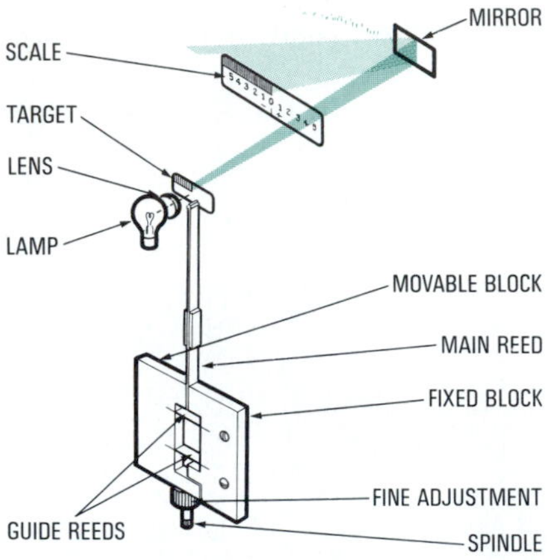
MIRROR
SCALE
TARGET
LENS
LAMP
MOVABLE BLOCK
MAIN REED
FIXED BLOCK
FINE ADJUSTMENT
GUIDE REEDS
SPINDLE

Figure 15-15 *The light beam or optical lever amplifies target movement.*

To Measure with a Reed Comparator

1. Raise the gaging head above the required height, and clean the anvil and master thoroughly.
2. Place the master gage or gage block buildup on the anvil.
3. Carefully lower the gaging head until the end of the spindle *just touches the master.*

NOTE: A shadow will begin to appear on the left side of the scale.

4. Clamp the gaging head to the column.
5. Turn the adjusting sleeve until the shadow coincides with the zero on the scale.
6. Remove the gage blocks and carefully slide the workpiece between the anvil and the spindle.
7. Note the reading. If the shadow is to the right of zero, the part is oversize; if to the left, it is undersize.

AIR GAGES OR PNEUMATIC COMPARATORS

Air gaging, a form of comparison measurement, is used to compare workpiece dimensions with those of a master gage by means of air pressure or flow. Air gages are of two types: the *flow* or *column type* (Fig. 15-16), which indicates air velocity, and the *pressure type* (Fig. 15-18), which indicates air pressure in the system.

COLUMN-TYPE AIR GAGE

After air has been passed through a filter and a regulator, it is supplied to the gage at about 10 psi (69 kPa) (Fig. 15-17). The air flows through a transparent tapered tube in which a float is suspended by this air flow. The top of the tube is connected to the gaging head by a plastic tube.

The air flowing through the gage exhausts through the passages in the gaging head into the clearance between the head and the workpiece. The rate of flow is proportional to the clearance indicated by the position of the float in the column. The gage is set to a master, and the float is then positioned by means of an adjusting knob. The upper and lower limits for the workpiece are then set. If the hole in the workpiece is larger than the hole size of the master, more air will flow through the gaging head, and the float will rise higher in the tube. Conversely, if the hole is smaller than the master, the float will fall in the tube. Amplification from 1000:1 to 40 000:1 may be obtained with this type of gage. Snap-, ring-, and plug-type gaging heads may be fitted to this type of gaging device.

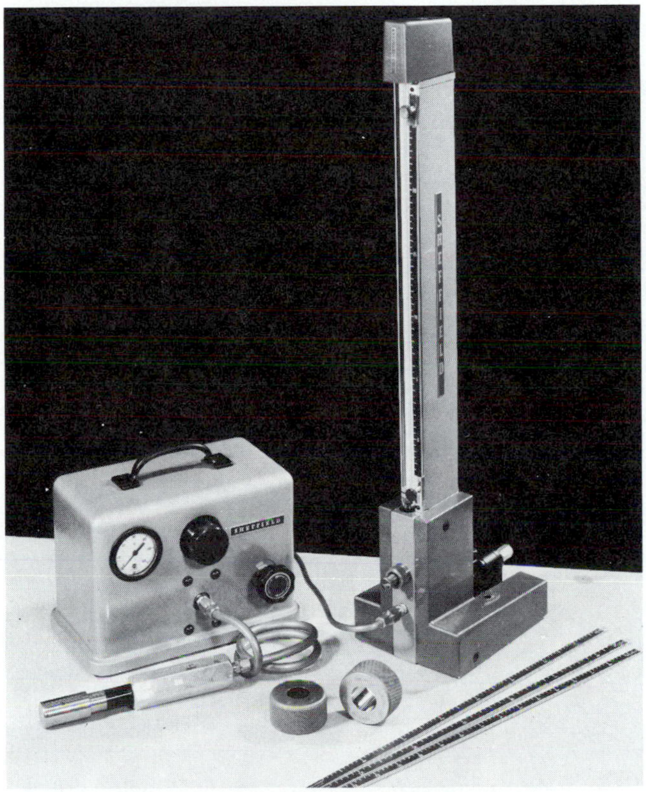

Figure 15-16 *A flow- or column-type air gage.*

PRESSURE-TYPE AIR GAGE

In the pressure-type air gage (Fig. 15-18), air passes through a filter and regulator and is then divided into two channels (Fig. 15-19). The air in the *reference channel* escapes to atmosphere through a zero-setting valve. The air in the *measuring channel* escapes to atmosphere through the gage head jets. The two channels are connected by an extremely accurate differential pressure meter.

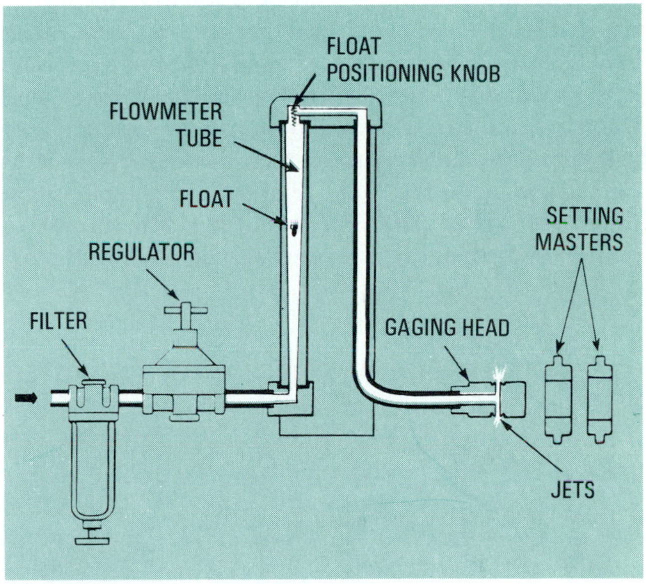
FLOAT
POSITIONING KNOB
FLOWMETER
TUBE
FLOAT
SETTING
MASTERS
REGULATOR
FILTER
GAGING HEAD
JETS

Figure 15-17 *Principle of the column-type air gage.*

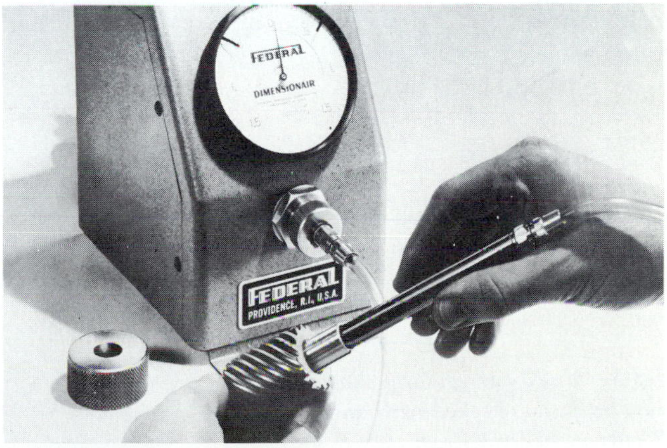
FEDERAL
DIMENSIONAIR
FEDERAL
PROVIDENCE, R.I., U.S.A.

Figure 15-18 *Gaging a hole using a pressure-type air gage.*

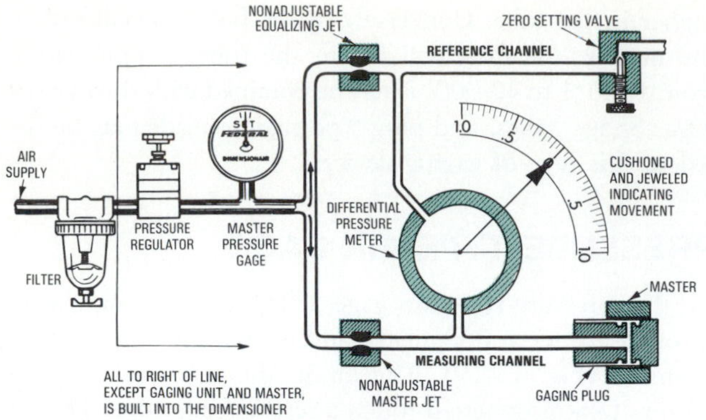
NONADJUSTABLE
EQUALIZING JET
ZERO SETTING VALVE
REFERENCE CHANNEL
AIR
SUPPLY
CUSHIONED
AND JEWELED
INDICATING
MOVEMENT
DIFFERENTIAL
PRESSURE
METER
PRESSURE
REGULATOR
MASTER
PRESSURE
GAGE
FILTER
MASTER
MEASURING CHANNEL
ALL TO RIGHT OF LINE,
EXCEPT GAGING UNIT AND MASTER,
IS BUILT INTO THE DIMENSIONER
NONADJUSTABLE
MASTER JET
GAGING PLUG

Figure 15-19 *Operation of the pressure-type air gaging system.*

The master is placed over the gaging spindle and the zero-setting valve is adjusted until the gage needle indicates zero. Any deviation in the workpiece size from the master size changes the reading. If the workpiece is too large, more air will escape through the gaging plug; therefore, pressure in the measuring channel will be less and the dial gage hand will move counterclockwise, indicating how much the part is oversize. A diameter smaller than the master gage indicates a reading on the right size of the dial. Amplification from 2500:1 to 20 000:1 may be obtained with this type of gage. Pressure-type air gages may also be fitted with plug, ring, or snap gaging heads for a wide variety of measuring jobs.

Air gages are widely used since they have several advantages over other types of comparators:

1. Holes may be checked for taper, out-of-roundness, concentricity, and irregularity more easily than with mechanical gages (Fig. 15-20).
2. The gage does not touch the workpiece; therefore, there is little chance of marring the finish.
3. Gaging heads last longer than fixed gages because wear is reduced between the head and the workpiece.
4. Less skill is required to use this type of gaging equipment than other types.
5. Gages may be used at a machine or bench.
6. More than one diameter may be checked at the same time.

REVIEW QUESTIONS

COMPARISON MEASUREMENT

1. Describe:
   a. Quality control
   b. Comparison measurement

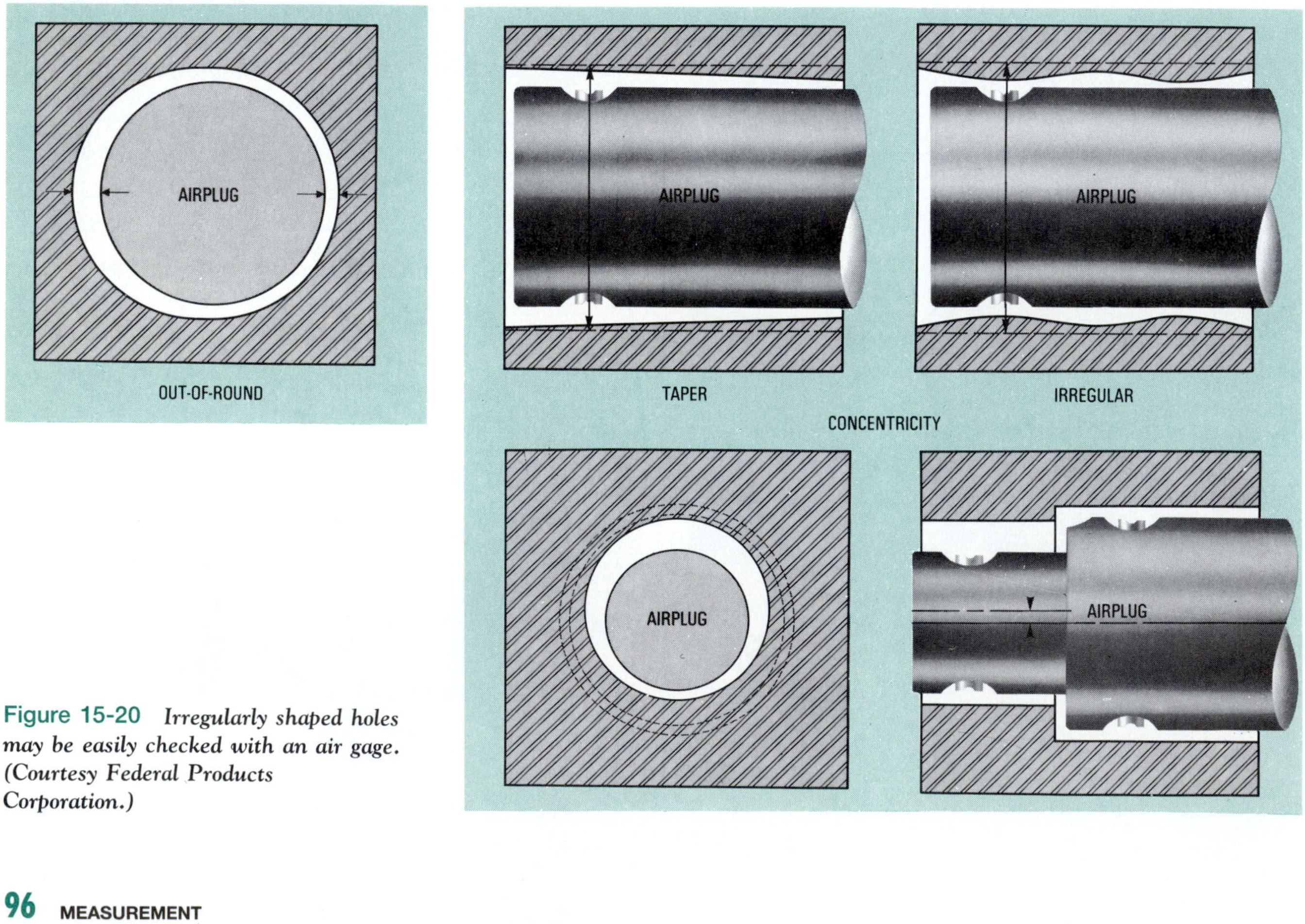
AIRPLUG
OUT-OF-ROUND
AIRPLUG
TAPER
AIRPLUG
IRREGULAR
CONCENTRICITY
AIRPLUG
AIRPLUG

Figure 15-20 *Irregularly shaped holes may be easily checked with an air gage. (Courtesy Federal Products Corporation.)*

DIAL INDICATORS

2. What is the difference between a regular-range dial indicator and a long-range indicator?
3. Compare a perpendicular dial indicator with a dial test indicator.
4. How are metric dial indicators usually graduated?

COMPARATORS

5. Define a comparator.
6. List three principles used in mechanical comparators.
7. Why is high amplification necessary in any comparison measurement process?
8. Describe the procedure for measuring a workpiece with a reed-type comparator.

OPTICAL COMPARATORS

9. List the advantages of an optical comparator.
10. Describe the principle of an optical comparator. Illustrate by means of a suitable sketch.
11. What precautions are necessary when charts are used on an optical comparator?

AIR GAGES

12. Describe the principle of the column-type air gage. Illustrate by means of a suitable sketch.
13. Describe and neatly illustrate the principle of the pressure-type air gage.
14. List six advantages of air gages.

ELECTRONIC COMPARATORS

15. What circuit is employed in electronic comparators?
16. Describe the operation of the signal light attachment for electrical or electronic gages.

UNIT 16

The Coordinate Measuring System

Coordinate measurement is a method used to speed up layout, machining, and inspection by having all measurements dimensioned from three rectangular coordinates (surfaces of the workpiece): *x*, *y*, and *z*. Coordinate measuring systems are capable of measuring to 0.000 050 in. or 0.001 mm. This high degree of accuracy is achieved by the use of a series of dark and light bands called the *moiré fringe pattern.* This system has been applied extensively to machine tools such as lathes, milling machines, and jig borers (Fig. 16-1A and B) to speed up the measurement process, thus reducing overall machining time. Coordinate measuring machines (Fig. 16-1C), widely used in the field of measurement and inspection, provide a quick and accurate means of checking machined parts prior to assembly.

Various units may be mounted on a machine tool to give readings for the *x* (length), *y* (width), and *z* (depth) axes. A unit providing direct readout in degrees, minutes, and seconds is available for angular positioning and measurement.

OBJECTIVES

After completing this unit you will know:

1. The purpose of and how to apply the coordinate measuring system
2. The main components and operation of the measuring unit
3. The advantages of the coordinate measuring system

+ 12.664
− 7.5870
A
14.2637
7.8191
3.5243
X
Y
Z
B
C

Figure 16-1 *(A) A lathe equipped with a coordinate measuring system; (B) the digital readout shows the exact location of the cutting tool or workpiece; (C) the coordinate measuring system being used for inspection purposes. (Courtesy The Bendix Corporation.)*

PARTS OF THE MEASURING UNIT

The *measuring unit* (Fig. 16-2) has three basic components: *a machined spar with a calibrated grating* (A), *a reading head* (B), and *a counter with a digital readout display* (C). The main element in this system is an accurately ruled grating (Fig. 16-2A) of the desired length of travel. The face of this grating for systems having 0.0001-in. (0.002-mm) resolution, is etched for its entire length with lines spaced 0.001 in. (0.02 mm) apart. For machines capable of greater accuracy, the grating spar is etched with 2500 lines per inch, so that the digital readout resolution is 0.000 050 in. (0.001 mm). Metric readings are obtained by pushing a button on the side of the box.

A *transparent index grating,* having the same graduations as the grating spar, is mounted in the reading head (Fig. 16-2B). The index grating is positioned so that its lines are at a slight angle to the lines on the spar. The reading head is mounted so that the index grating is positioned just 0.002 in. (0.05 mm) above the main grating. Mounted in the reading head are a small light, a collimating lens, and four photoelectric cells (Fig. 16-3).

NOTE: This electro-optical system "counts" the moiré fringes, and produces the high resolution measurement accuracy of the Sheffield Cordax Measuring System.

PRINCIPLE OF THE MOIRÉ FRINGE

To illustrate the moiré fringe pattern, we can draw a series of equally spaced lines on two pieces of plastic sheeting (Fig. 16-4). One sheet should then be placed over the other at an angle, as in Fig. 16-4. Where the lines cross, dark bands appear. If the top sheet is moved to the right, the position of these bands will shift down, and the pattern will appear as a series of dark and light bands moving vertically on the sheet. Thus, any longitudinal movement will produce a vertical movement of the light bands. This principle can also be illustrated by placing one comb on another at a small angle and moving one across the face of the other.

OPERATION OF THE MEASURING UNIT

Since the etched lines on the reading-head grating are at an angle to the lines on the main spar, a series of bands would

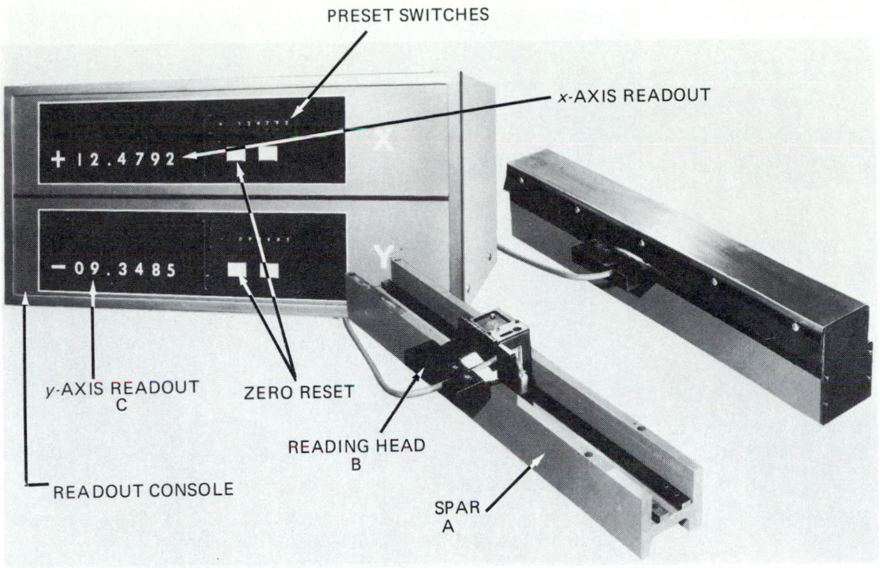

PRESET SWITCHES
x-AXIS READOUT
+12.4792
−09.3485
y-AXIS READOUT
C
ZERO RESET
READING HEAD
B
READOUT CONSOLE
SPAR
A

Figure 16-2 *Components of a coordinate measuring unit.*

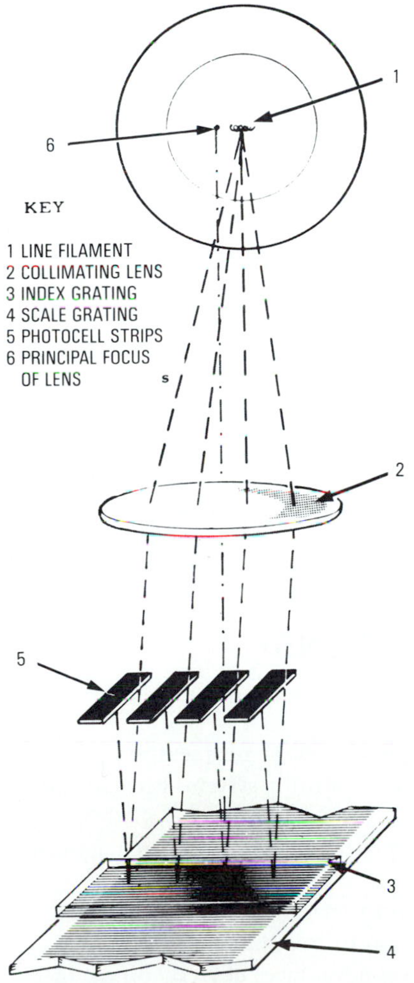

1
6
KEY
1 LINE FILAMENT
2 COLLIMATING LENS
3 INDEX GRATING
4 SCALE GRATING
5 PHOTOCELL STRIPS
6 PRINCIPAL FOCUS OF LENS
2
5
3
4

Figure 16-3 *Principle of the coordinate measuring unit.*

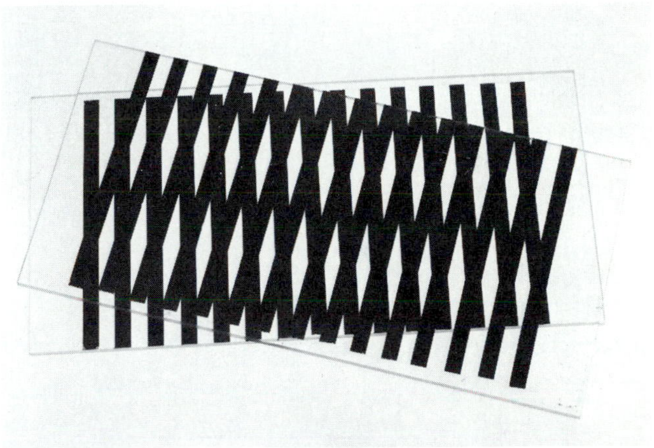

Figure 16-4 *Principle of the moiré fringe pattern. (Courtesy The Bendix Corporation.)*

appear when viewed from directly above. Let's consider one band only, to explain the operation of the system.

Light from the lamp (Fig. 16-3) passes through the collimating lens and is converted into a parallel beam of light. This beam strikes the fringe pattern and is reflected back up to one of four photoelectric cells, which convert the fringe pattern into an electrical signal. As the reading head is moved longitudinally, the band moves vertically on the face of the spar and will be picked up by the next photoelectric cell creating another signal (Fig. 16-5). These signals from the photoelectric cells (output signals) are transmitted to the digital readout display box where they indicate accurately the head travel at any point. A signal from the next photoelectric cell will increase or decrease the reading on the digital readout box by 0.0001 in. (0.002 mm), depending on the direction of the movement.

NOTE: The fringe pattern (Fig. 16-4) shifts laterally and continuously across the grating path.

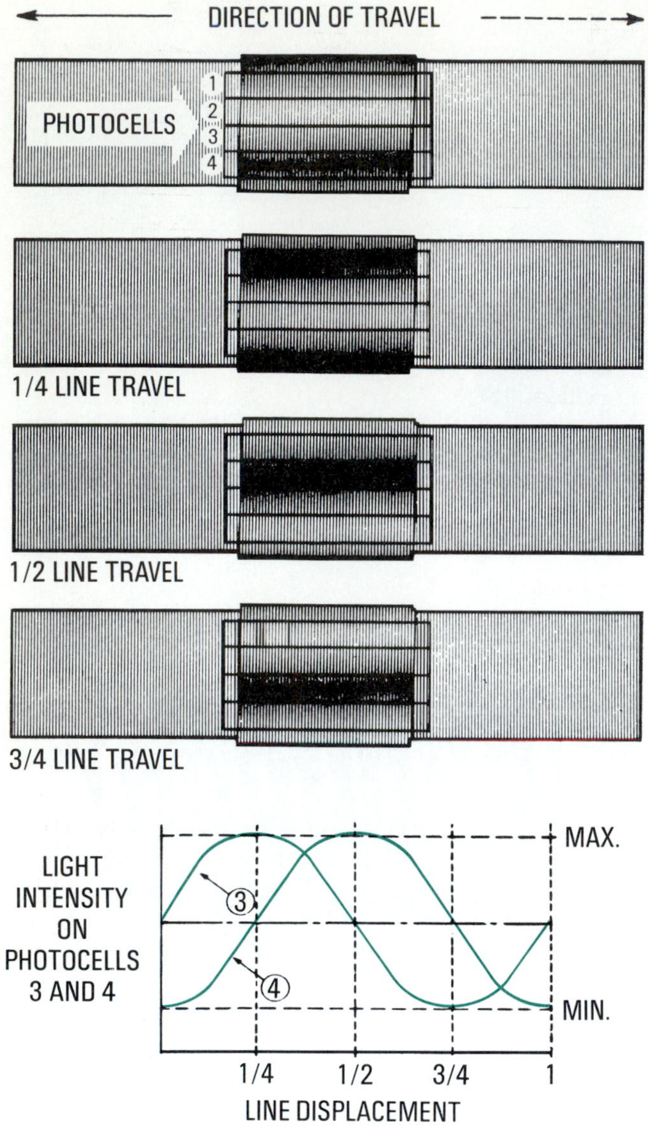
DIRECTION OF TRAVEL
PHOTOCELLS
1
2
3
4
1/4 LINE TRAVEL
1/2 LINE TRAVEL
3/4 LINE TRAVEL
LIGHT INTENSITY ON PHOTOCELLS 3 AND 4
MAX.
MIN.
3
4
1/4
1/2
3/4
1
LINE DISPLACEMENT

Figure 16-5 *Lateral movement of the fringe pattern.*

ADVANTAGES OF THE COORDINATE MEASURING SYSTEM

1. Readout boxes provide clear, visible numbers, which eliminates the possibility of misreading a dial gage.
2. It provides a constant readout of the tool (or table) position.
3. The reading indicates the exact position of the tool (or table) and is not affected by machine or lead screw wear.
4. This system eliminates the need for gage blocks and measuring rods on jig borers and vertical milling machines.
5. The need for operator calculations and the inherent possibility of errors are eliminated.
6. Machine setup time is greatly reduced.
7. Production is increased since the workpiece need only be checked for one size. For example, when a workpiece having several diameters is to be machined in a lathe, it is necessary to measure the first diameter only. Once the machine has been set to this size, all other diameters will be accurate.
8. The need for operator skill is reduced.
9. Scrap and rework are virtually eliminated.
10. Most machine tools can be fitted with this system.

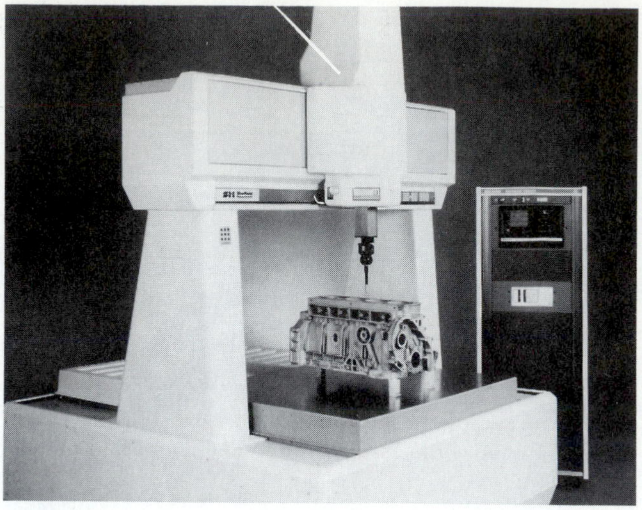

Figure 16-6 *Coordinate measuring machines can be programmed to automatically inspect parts. (Courtesy Sheffield Measurement Division.)*

Coordinate measuring machines, Fig. 16-6, were developed to speed up the process of measuring parts produced on numerical control (NC) machine tools. What NC is to manufacturing, the coordinate measuring machine (CMM) is to inspection. It can measure almost any shape with extreme accuracy and without the use of special gages. Coordinate measuring machines, which are computer controlled, have eliminated the problems of operator error; the need for long, complex, and inefficient conventional measuring systems; and the low productivity common to previous inspection methods. These measuring machines can be installed on production lines to automate inspection, minimize operator error, and provide uniform part quality.

REVIEW QUESTIONS

1. What is the principle of coordinate measurement?
2. Why have coordinate measuring systems been so widely accepted by industry?
3. State two different applications of coordinate measuring systems.
4. Name the three main parts of the measuring unit.
5. Describe the operation of this measuring system.
6. State seven important advantages of a coordinate measuring system.

UNIT

17

Measuring with Light Waves

Two of the most precise measuring methods are those using optical flats and the laser. Although based on different principles, both use a source of monochromatic light to produce highly accurate measurements.

OBJECTIVES

After completing this unit you will be able to:

1. Check pieces for size, flatness, and parallelism, using optical flats
2. Describe the operation of a laser interferometer
3. Explain the application of lasers to measurement

MEASURING WITH OPTICAL FLATS

One of the most accurate and reliable means of measuring is with light waves. *Optical flats* (Fig. 17-1), used with a monochromatic light, are utilized to check work for flatness (Fig. 17-4), parallelism (Fig. 17-5), and size (Fig. 17-6).

Optical flats are disks of clear fused quartz, lapped to within a few millionths of an inch of flatness. They are generally used with a helium light source which produces a greenish-yellow light, having a wavelength of 23.1323 μin. (0.587 56 μm).

The optical flat, a perfectly flat, transparent disk, is placed on the surface of the work to be checked. The functioning surface of the optical flat is the surface adjacent to the workpiece. It is transparent and capable of reflecting light; therefore, all light waves that strike this surface are split into two parts (Fig. 17-2). One part is reflected back by the lower surface of the flat. The other part passes through this surface and is reflected by the upper surface of the work. Whenever the reflected split portions of two light waves cross each other, or *interfere,* they become visible and produce dark interference bands or fringe lines. This happens whenever the distance between the lower surface of the flat and the upper surface of the workpiece is *only one-half of a wavelength* or multiples thereof (Fig. 17-2).

Since the wavelength of helium light is 23.1323 μin. (0.587 56 μm, or millionths of a meter), each half-wavelength will represent 11.6 μin. (0.293 μm). Each dark band then represents a progression of 11.6 μin. (0.293 μm) above the point of contact between the workpiece and the optical flat. Therefore, when a height is checked, the number of bands between two points on a

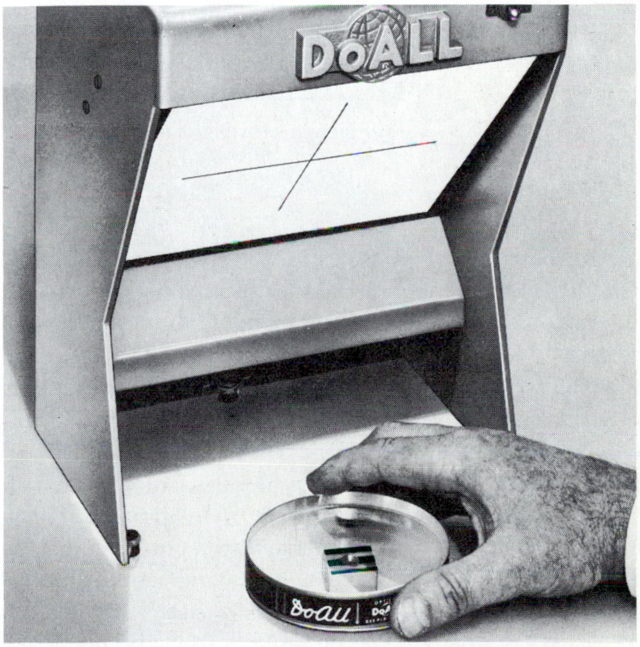
DoALL
DoAll

Figure 17-1 *Checking a gage block using an optical flat and a helium light source. (Courtesy DoAll Company.)*

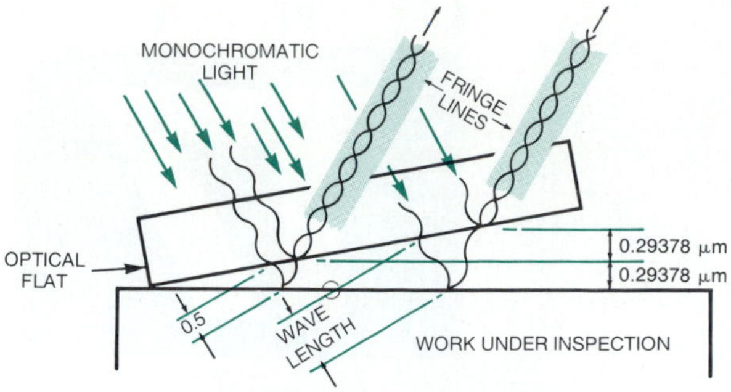

MONOCHROMATIC LIGHT
FRINGE LINES
OPTICAL FLAT
0.29378 μm
0.29378 μm
0.5
WAVE LENGTH
WORK UNDER INSPECTION

Figure 17-2 *Principle of the optical flat.*

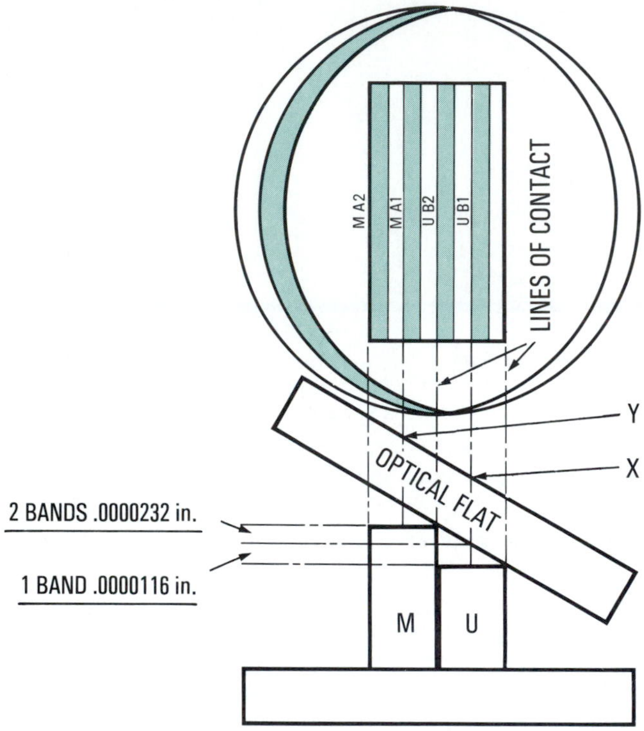

M A2
M A1
U B2
U B1
LINES OF CONTACT
Y
X
OPTICAL FLAT
2 BANDS .0000232 in.
1 BAND .0000116 in.
M
U

Figure 17-3 *Checking the height of a block with a master block.*

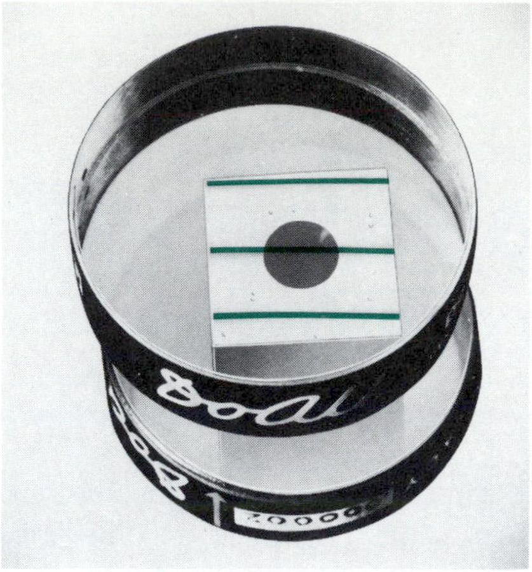

Figure 17-4 *Checking flatness with an optical flat. (Courtesy DoAll Company.)*

Figure 17-5 *Checking parallelism with an optical flat. (Courtesy DoAll Company.)*

surface multiplied by 0.000 011 6 (0.293 μm) will indicate the height difference between two surfaces.

For comparison measurements, the difference in height between a master block and the workpiece can be determined as shown in Fig. 17-3. This illustrates the method used for checking accurately the height of an unknown surface by comparing it with a gage block of a known height. It is necessary to first know which block is larger before the unknown block can be measured. To determine which block is larger, apply finger pressure to points *X* and *Y*. If this pressure at *X* does not change the band pattern and the pressure at *Y* causes the bands to separate, the master block (*M*) is larger. If the opposite is true, the unknown block (*U*) is larger. In Fig. 17-3, two bands appear on the low block; therefore, the unknown block is two bands, or 2 × 11.6 = 23.2 μin. (2 × 0.293 = 0.586 μm) less than the master.

HOW TO INTERPRET THE BANDS

Referring to Fig. 17-4, since the lines show only a slight curve, the block would be convex or very slightly higher in the center. The block is 2 × 11.6 μin., or 0.000 023 2 in., out of flat.

In Fig. 17-5, since two bands appear on the master block, the workpiece is 2 × 11.6 or 23.2 μin. smaller or larger than the master. If pressure on the master causes the spacing of the bands to widen, the part is smaller. Note the curve in the band, which shows that the workpiece is not exactly parallel. It is about one-half band, or 5.8 μin., out of parallel.

In Fig. 17-6, the master block is on the left. Note the three bands on the master block and the six bands on the

Figure 17-6 *Checking size with an optical flat. (Courtesy DoAll Company.)*

block of unknown size on the right. The unknown block then has three more bands than the master. (The block with more bands is always smaller.) Also note that the lines on the unknown block are straight and evenly spaced, which indicates that it is parallel over its length. Since the lines slope down toward the line of contact, the left side of the unknown block is lower by one-half band, or 5.8 μin. (11.6 ÷ 2).

MEASUREMENT WITH LASERS

Besides its many other uses in medicine and industry, the laser provides one of the most accurate means of measurement. The laser-light measuring device, called an *interferometer* (Fig. 17-7), measures changes in position (alignment) by means of light-wave interference.

As Fig. 17-7 shows, the laser beam is split into two parts by the *beam splitter*. One of these beams (green) is transmitted through the beam splitter to a *motion-sensitive mirror* back to the beam splitter at point X. From this point, where the two beams rejoin, the recombined beams are transmitted to the *detector*.

If there is no movement, both portions of the beam stay in the same phase and the light reaching the detector will remain constant (Fig. 17-8). If there is any movement at the sensitive mirror, the beam (black) reflected from that mirror will be altered and will fluctuate in and out of phase with the other (green) beam (Fig. 17-9). When this beam (black) is out-of-phase, it then cancels the other beam (green) causing the light to fluctuate. This fluctuating light pattern registers on the beam detector where the number of fluctuations are computed relative to the laser wavelength and the precise movement is displayed on a readout box.

Because of the very thin, straight beam produced by a laser interferometer, these devices are used for very precise

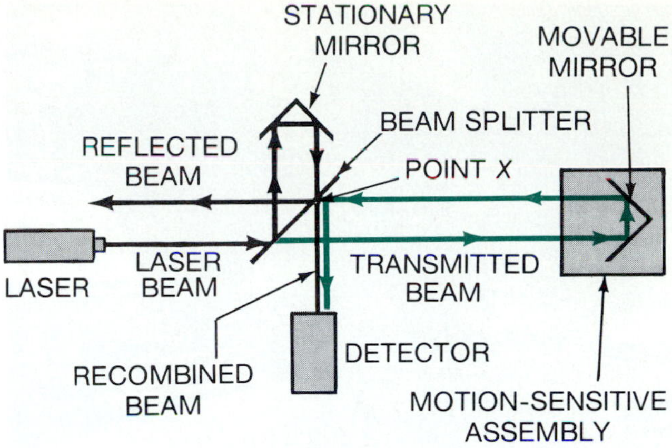
STATIONARY MIRROR
MOVABLE MIRROR
BEAM SPLITTER
REFLECTED BEAM
POINT X
LASER
LASER BEAM
TRANSMITTED BEAM
DETECTOR
RECOMBINED BEAM
MOTION-SENSITIVE ASSEMBLY

Figure 17-7 *Principle of the laser interferometer.*

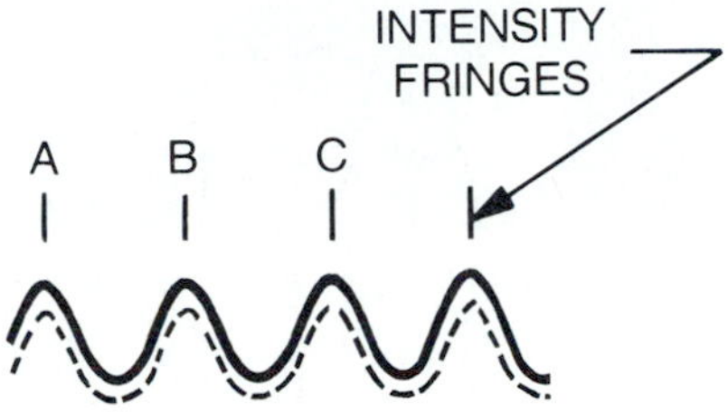
INTENSITY FRINGES
A
B
C

Figure 17-8 *The laser beam's wave components are in phase.*

A a B b C c

Figure 17-9 *The laser beam's wave components are out of phase.*

linear measurement and alignment in the production of large machines. They are also used to calibrate precision machines and measuring devices. Laser devices may also be used to check machine setups (Fig. 17-10). A laser beam is projected against the work and measurements are made by the beam and displayed on a digital readout panel.

Because of their very thin, straight beam characteristics, lasers are used extensively in construction and surveying. They may be used to indicate the exact location for positioning girders on tall buildings or for establishing directional lines for a tunnel being constructed under a river. They are also widely used for establishing grades for sewer and drainage systems.

Lasers have become an established part of the space program. Laser beams are used to indicate how far the spaceship is from a given planet. They are also used for military purposes for missile range finding, guidance, and tracking. Operation of the laser and its other industrial applications will be covered in Unit 94.

CINCINNATI
MILACRON
WARNING

Figure 17-10 *A laser interferometer being used to check the alignment of machine parts.*

THE LASERMIKE

The *LaserMike* (Fig. 17-11) is an optical micrometer that is simple in principle, yet highly accurate in operation. The heart of the instrument is a helium–neon laser beam that is projected in a straight line with virtually no diffusion.

The beam is directed to mirrors mounted on the shaft of a precision electric motor. During rotation, these mirrors "scan" the laser beam through an optical lens, which aligns the beams in parallel and projects them toward a receiving lens. When an object is placed in the center of the laser beam, it creates a shadow segment in the scan path, which is deflected by the photo cell enabling the unit to determine the edges of the object. A high-frequency, crystal clock times the interval between edges and converts time to linear dimensions.

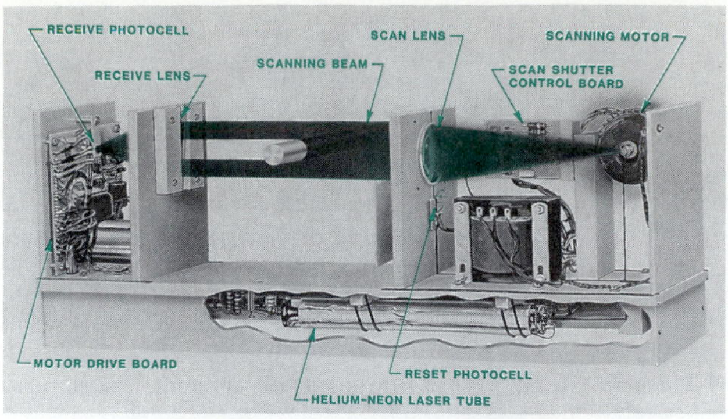
RECEIVE PHOTOCELL
SCAN LENS
SCANNING MOTOR
SCANNING BEAM
RECEIVE LENS
SCAN SHUTTER CONTROL BOARD
MOTOR DRIVE BOARD
RESET PHOTOCELL
HELIUM-NEON LASER TUBE

Figure 17-11 *The LaserMike, an optical micrometer, makes use of a helium-neon laser beam. (Courtesy LaserMike Div. Techmet Co.)*

Without even touching the product, LaserMike provides instant readouts to accuracies never before possible with other conventional measurement techniques. Moreover, measurement speeds permit accurate measurements while the product is in motion as well as at rest.

REVIEW QUESTIONS

OPTICAL FLATS

1. Describe an optical flat.
2. What light source is used with optical flats and what is its wavelength?
3. Describe and illustrate in detail the principle of an optical flat.
4. When measuring the height of a block using a gage block and an optical flat, how is it possible to determine the higher block?

LASERS

5. Name five measurement applications for an interferometer.
6. Name the four main parts of an interferometer.
7. Briefly describe the operation of an interferometer.

UNIT 18

Surface Finish Measurement

Modern technology has demanded improved surface finishes to ensure proper functioning and long life of machine parts. Pistons, bearings, and gears depend to a great extent on a good surface finish for proper functioning and therefore require little or no break-in period. Finer finishes often require additional operations, such as lapping or honing. These higher finishes are not always required on a part and only result in higher production costs. To prevent overfinishing a part, the desired finish is indicated on the shop drawing. This information specifying the degree of finish is conveyed to the machinist by a system of symbols devised by the American Standards Association (ASA). These symbols provide a standard system of determining and indicating surface finish. The inch unit of surface finish measurement is the *microinch* (μin.). The metric unit for surface finish is the *micrometer* (μm).

OBJECTIVES

After completing this unit you will be able to:

1. Interpret the surface finish symbols that appear on a drawing
2. Use a surface finish indicator to measure the surface finish of a part

The most common instrument used to measure finish is the *surface indicator* (Fig. 18-1). This device consists of a *tracer head* and an *amplifier*. The tracer head houses a diamond stylus, having a point radius of 0.0005 in. (0.013 mm), which bears against the surface of the work. It may be moved along the work surface by hand or it may be motor-driven. Any movement of the stylus caused by surface irregularities is converted into electrical fluctuations by the tracer head. These signals are magnified by the amplifier and registered on the meter by an indicator hand or needle. The reading shown on the meter indicates the *average* height of surface roughness or the departure of this surface from the reference (center) line.

Readings may be in either *arithmetic average roughness height* (*Ra*) or *root mean square* (*rms*). A highly magnified cross section of a workpiece would appear as shown in Fig. 18-2, with "hills and valleys" above and below the centerline. To calculate the surface finish without a surface indicator, the height of these deviations must be measured and recorded as shown. The Ra or rms could then be calculated as in Fig. 18-2. The rms is considered the better method of determining surface roughness since it emphasizes extreme deviations.

For accurate determination of the surface finish, the indicator must first be calibrated by setting it to a precision reference surface on a test block calibrated to ASA standards. The symbols used to identify surface finishes and characteristics are shown in Fig. 18-3.

SURFACE FINISH DEFINITIONS

Surface deviations—any departures from the nominal surface in the form of waviness, roughness, flaws, lay, and profile.

Waviness—surface irregularities that deviate from the mean surface in the form of waves; they may be caused by vibrations in the machine or work and are generally widely spaced.

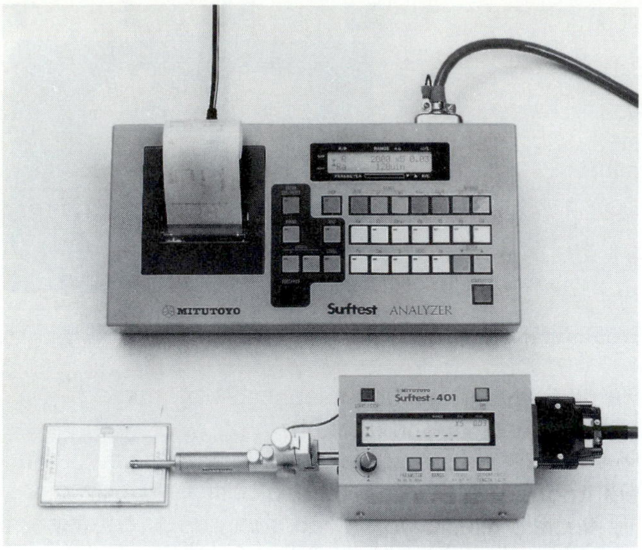
MITUTOYO
Surftest ANALYZER
Surftest-401

Figure 18-1 *A surface indicator can accurately check the roughness of a surface. (Courtesy Mitutoyo: MTI Ltd.)*

a = 3	a^2 = 9
b = 19	b^2 = 361
c = 22	c^2 = 484
d = 15	d^2 = 225
e = 30	e^2 = 900
f = 19	f^2 = 361
g = 27	g^2 = 729
h = 19	h^2 = 361
i = 30	i^2 = 900
j = 12	j^2 = 144
k = 22	k^2 = 484
l = 14	l^2 = 196
m = 5	m^2 = 25
Totals 237	5179

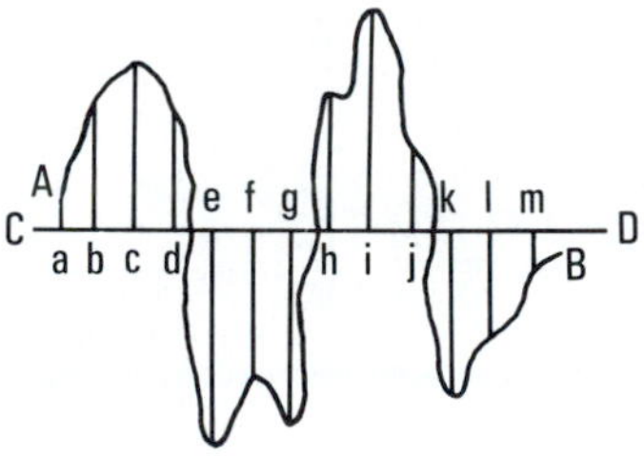
A
C
D
B
a b c d
e f g
h i j
k l m

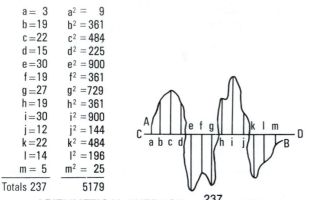

$$\text{ARITHMETICAL AVERAGE} = \frac{237}{13} = 18.2\ \mu\text{ in.}$$

$$\text{ROOT MEAN SQUARE AVERAGE} = \sqrt{\frac{5179}{13}} = 19.9\ \mu\text{ in.}$$

Figure 18-2 *Calculation of arithmetic average roughness height (Ra) and root mean square (rms).*

Waviness height—the peak-to-valley distance in inches or millimeters.

Waviness width—the distance between successive waviness peaks or valleys in inches or millimeters.

Roughness—relatively finely spaced irregularities superimposed on the waviness pattern and caused by the cutting tool or the abrasive grain action and the machine feed. These irregularities are much narrower than the waviness pattern.

Roughness height—the Ra deviation measured normal to the centerline in microinches or micrometers. NOMINAL SURFACE

Roughness width—the distance between successive roughness peaks parallel to the nominal surface in inches or millimeters.

Roughness width cutoff—the greatest spacing of repetitive surface irregularities to be included in the measurement of roughness height. It must always be greater than the roughness width. Standard values are 0.003, 0.010, 0.030, 0.100, 0.300 and 1 in. (0.075, 0.25, 0.76, 2.54, 7.62, and 25.4 mm).

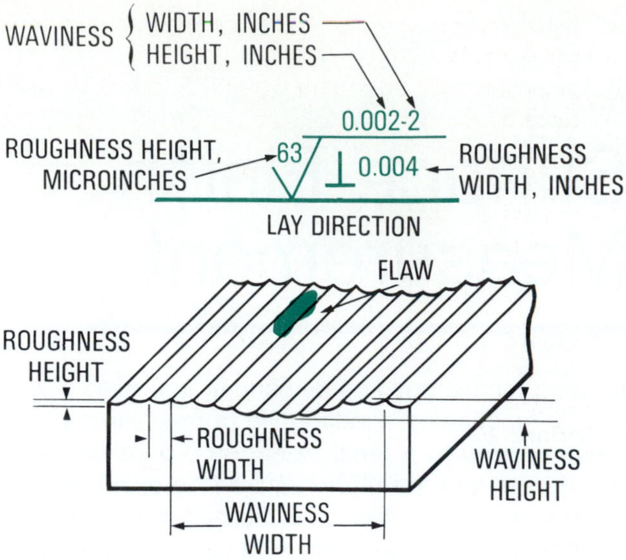
WAVINESS
WIDTH, INCHES
HEIGHT, INCHES
0.002-2
ROUGHNESS HEIGHT, MICROINCHES
63
0.004
ROUGHNESS WIDTH, INCHES
LAY DIRECTION
FLAW
ROUGHNESS HEIGHT
ROUGHNESS WIDTH
WAVINESS HEIGHT
WAVINESS WIDTH

Figure 18-3 *Surface characteristics and symbols.*

Flaws—irregularities such as scratches, holes, cracks, ridges, or hollows that do not follow a regular pattern as in the case of waviness and roughness.

Lay—the direction of the predominant surface pattern caused by the machining process.

Profile—the contour of a specified section through a surface.

Microinch or micrometer—the unit of measurement used to measure the surface finish. The microinch is equal to 0.000 001 in.; the micrometer, to 0.000 001 m.

The following symbols indicate the direction of the lay (Fig. 18-4).

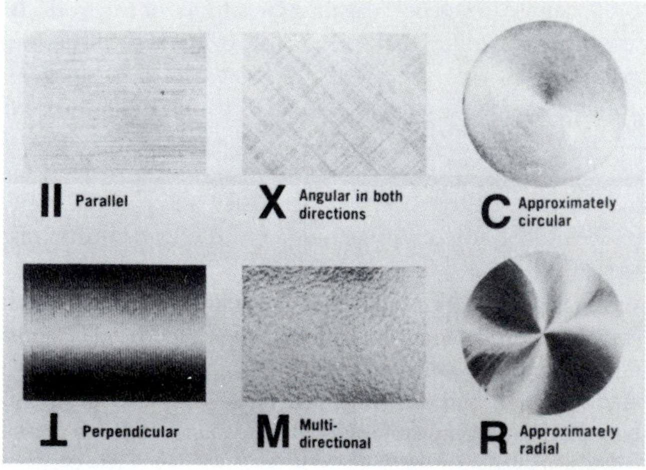
Parallel
X Angular in both directions
C Approximately circular
Perpendicular
M Multi-directional
R Approximately radial

Figure 18-4 *Surface symbols used to designate the direction of the lay. (Courtesy GAR Electroforming Limited.)*

‖ parallel to the boundary line of the surface indicated by the symbol

⊥ perpendicular to the boundary line of the surface indicated by the symbol

X angular in both directions on the surface indicated by the symbol

M multidirectional

C approximately circular to the center of the surface indicated by the symbol

R approximately radial in relation to the center of the surface indicated by the symbol

Average Surface Roughness Produced by Standard Machining Processes

	Microinches	Micrometers
Turning	100–250	2.5–6.3
Drilling	100–200	2.5–5.1
Reaming	50–150	1.3–3.8
Grinding	20–100	0.5–2.5
Honing	5–20	0.13–0.5
Lapping	1–10	0.025–0.254

TO MEASURE SURFACE FINISH WITH A SURFACE INDICATOR

1. Turn the switch on and allow the instrument to warm up for approximately 3 min.
2. Check the machine calibration by moving the stylus over the 125-μin. (3.2-μm) test block at approximately ⅛ in./s (3 mm/s).
3. If necessary, adjust the calibration control so that the instrument registers the same as the test block.
4. Unless otherwise specified, use the 0.030-in. (0.76-mm) cutoff range for surface roughness of 30 μin. (0.76 μm) or more. For surfaces of less than 30 μin. (0.76 μm), use the 0.010-in. (0.25-mm) cutoff range.

NOTE: When measuring a surface with an unknown roughness, set the range switch at a high setting to avoid damaging the instrument. After an initial test, the range switch may be turned to a finer setting for an accurate surface reading.

5. Thoroughly clean the surface to be measured to ensure accurate readings and reduce wear on the rider cap protecting the stylus.
6. With a smooth, steady movement of the stylus, trace the work surface at approximately ⅛ in./s (3 mm/s).
7. Note the reading from the meter scale.

A more elaborate device for measuring surface finish is the *surface analyzer.* It utilizes a recording device to reproduce the surface irregularities on a graduated chart, providing an ink-line record.

Although the surface indicator is the most common, other methods may be used to measure surface finish with reasonable accuracy during machining processes, including:

1. *Comparison blocks,* which are used for comparing the finish on the workpiece with the calibrated finish on a test block using the fingernail test.
2. *Commercial sets of standard finished specimens,* which have up to 25 different surface finish samples. They consist of blocks or plates having surfaces varying from the smoothest to the roughest likely to be required (Fig. 18-5).

These specimens are used to check the finish of the machined part against the sample finish to determine approxi-

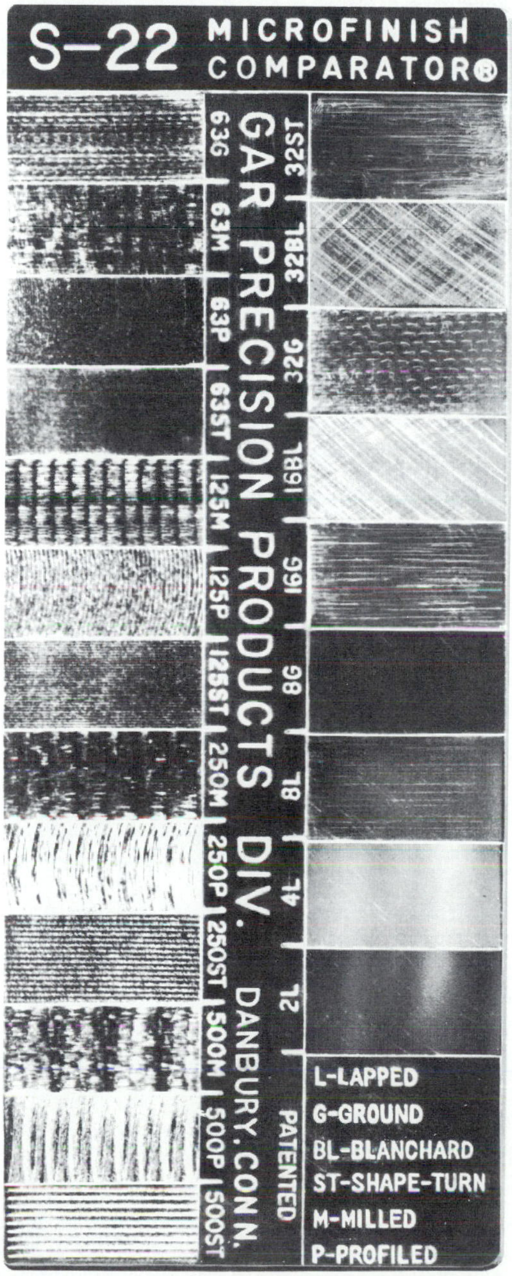
S-22
MICROFINISH
COMPARATOR®
GAR PRECISION PRODUCTS DIV. DANBURY, CONN.
PATENTED
32ST
32BL
32G
16BL
16G
8G
8L
4L
2L
63G
63M
63P
63ST
125M
125P
125ST
250M
250P
250ST
500M
500P
500ST
L-LAPPED
G-GROUND
BL-BLANCHARD
ST-SHAPE-TURN
M-MILLED
P-PROFILED

Figure 18-5 *A visual surface roughness comparator gage. (Courtesy GAR Electroforming Limited.)*

Table 18-1 Surface Finishes Obtained by Various Machining Operations*

Tool	Operation	Material	Speed	Feed	Tool	Analyzer Setting: Cutoff	Analyzer Setting: Range	Surface Finish, rms
Cutoff saw	Sawing	2½-in. diameter aluminum	320 ft/min (97.5 m/min)	—	10-pitch saw	0.030 in. (0.76 mm)	1000	300–400
Shaper	Shaping a flat surface	Machine steel	100 ft/min (30.5 m/min)	0.005 in. (0.13 mm)	3/64 in. radius HSS	0.030 in. (0.76 mm)	300	225–250
Vertical milling machine	Fly cutting (flat surface)	Machine steel	820 r/min	0.015 in. (0.38 mm)	1/16 in. radius Stellite	0.030 in. (0.76 mm)	300	125–150
Horizontal milling machine	Slab milling	Cast aluminum	225 r/min	2½ in./min (63.5 mm/min)	Slab cutter, 4 in. diameter HSS	0.030 in. (0.76 mm)	100	40–50
Lathe	Turning	2½-in. diameter aluminum	500 r/min	0.010 in. (0.25 mm)	3/64 in. radius HSS	0.030 in. (0.76 mm)	300	100–200
	Turning	2½-in. diameter aluminum	500 r/min	0.007 in. (0.18 mm)	5/64 in. radius HSS	0.030 in. (0.76 mm)	100	50–60
	Facing	2-in. diameter aluminum	600 r/min	0.010 in. (0.25 mm)	1/32 in. radius HSS	0.030 in. (0.76 mm)	300	200–225
	Facing	2-in. diameter aluminum	800 r/min	0.005 in. (0.13 mm)	1/32 in. radius HSS	0.030 in. (0.76 mm)	100	30–40
	Filing	¾-in. diameter machine steel	1200 r/min	—	10-in. lathe file	0.010 in. (0.25 mm)	100	50–60
	Polishing	¾-in. diameter machine steel	1200 r/min	—	#120 abrasive cloth	0.010 in. (0.25 mm)	30	13–15
	Machine reaming	Aluminum	500 r/min	—	machine reamer HSS, ¾ in. diameter	0.030 in. (0.76 mm)	100	25–32
Surface grinder	Grinding a flat surface	Machine steel	—	0.030 in. (0.76 mm)	60-grit grinding wheel	0.003 in. (0.076 mm)	10	7–9
Cutter and tool grinder	Cylindrical grinding	1-in. machine steel	—	Hand (slow)	46-grit grinding wheel	0.010 in. (0.25 mm)	30	12–15
Lapping	Flat lapping	⅞ in. × 5½ in. tool steel (hardened)	—	Hand	600-grit abrasive	0.010 in. (0.25 mm)	10	1–2
	Cylindrical lapping	½-in. diameter tool steel (hardened)	—	Hand	600-grit abrasive	0.010 in. (0.25 mm)	10	1–2

*Metric figures given represent soft conversions.

mately the finish produced on the part. It is often difficult to determine the finish visually. In such cases, the surfaces may be compared by moving the tip of your fingernail over the two surfaces.

Table 18-1 shows the results obtained on pieces of round and flat metal by various machining operations. A model B-1 110 Brush surface analyzer was used to obtain the readings. The speeds, feeds, and tool radii shown are those recommended for a high-speed steel toolbit.

REVIEW QUESTIONS

1. Explain why present-day standards for surface finish are very important to industry.

2. Define the following surface finish terms: microinch, lay, flaw, roughness, waviness, root mean square (rms), and arithmetic average roughness height (Ra).

3. Explain each symbol and number (inches) as it applies to surface finish:

0.002 – 2
√50 = 0.020

4. Explain what the following lay symbols represent: =, ⊥, X, M, C, and R.

5. Describe briefly the principle and operation of a surface indicator and a surface analyzer in measuring surface finish.

SECTION 6

Layout Tools and Procedures

Laying out is the process of scribing or marking center points, circles, arcs, or straight lines on metal to indicate the shape of the object, the amount of metal to be removed during the machining process, and the position of the holes to be drilled. The layout helps the machinist determine the amount of material to be removed, although the size for rough and finish cuts must be checked by actual measurement.

All layouts should be made from a *baseline* or finished surface to ensure an accurate layout, correct dimensions, and proper location of holes. The importance of proper layout cannot be over-emphasized. The accuracy of the finished product depends greatly on the accuracy of the layout. The drawing does not show how much material must be removed from each surface of the casting or workpiece; it merely shows which surfaces are to be finished.

The layout for holes, whether in cored or solid material, is as important as the layout for other dimensions of the workpiece.

Layouts may be of two types—*basic* or *semiprecision* and *precision*. A semiprecision layout may involve the use of basic measuring and layout tools, such as a rule and surface gage. It is generally not as accurate as a precision layout, which requires the use of more accurate equipment such as the vernier height gage. If the part does not have to be precise, time should not be spent in making a precision layout. Therefore, keep the layout as simple as workpiece requirements permit.

UNIT 19

Basic Layout Materials, Tools, and Accessories

The accuracy of a layout is very important to the accuracy of the finished product. If the layout is not correct, the workpiece will not be usable. A student should therefore realize that good layout entails the proper and careful use of all layout tools.

OBJECTIVES

After completing this unit you will be able to:

1. Prepare a work surface for layout
2. Use and care for various types of surface plates
3. Identify and use the main basic layout tools and accessories

LAYOUT SOLUTIONS

The surface of the metal is usually coated with a layout solution to make layout lines visible. Several types of layout solutions are available. Regardless of the type used, the surface should be clean and free of grease.

The most commonly used layout solution is *layout dye*, or *bluing* (Fig. 19-1). This quick-drying solution, when coated lightly on the surface of any metal, will produce a background for sharp, clear-cut lines. Layout dye may be applied with a cloth, brush, or dauber, or sprayed on the work surface.

STEEL
INK
Figure 19-1 *The work surface should be coated with layout dye before starting a layout.*

A copper-colored surface can be produced if the clean surface of a steel workpiece is coated with a copper sulfate ($CuSO_4$) solution to which a few drops of sulfuric acid have been added. When this solution is used, the surface of the workpiece must be absolutely clean and free from grease and finger marks.

NOTE: Copper sulfate should be used on ferrous metal only. It is particularly useful where hot chips are produced, which could affect other layout materials and blur the layout lines.

A mixture of vermilion powder and shellac is often used for aluminum since some layout compounds corrode aluminum. Alcohol should be used to thin this solution or to remove it from the workpiece.

The surfaces of castings and hot-rolled steel are often prepared for layout by merely chalking the surface. A mixture of lime and alcohol, which readily clings to the rough surface of the castings, is often used for this purpose.

LAYOUT TABLES AND SURFACE PLATES

Layout work may be performed on a layout table (Fig. 19-2) or on a surface plate (Fig. 19-3) made of granite or cast iron. Granite layout tables and plates are considered better than cast-iron ones because they:

- Do not become burred
- Do not rust
- Are not affected by temperature change
- Do not have internal stresses and therefore will not warp or distort
- Are nonmagnetic
- Can be used for checking near grinding machines since abrasive particles will not embed in the surface
- Are cheaper than a similar sized cast-iron plate

Granite plates are available in three colors: black, pink, and gray. Black granite plates are considered superior because they are harder, denser, less porous, and therefore less liable to absorb moisture.

Figure 19-2 *A granite layout table provides an accurate reference surface for layout work.*

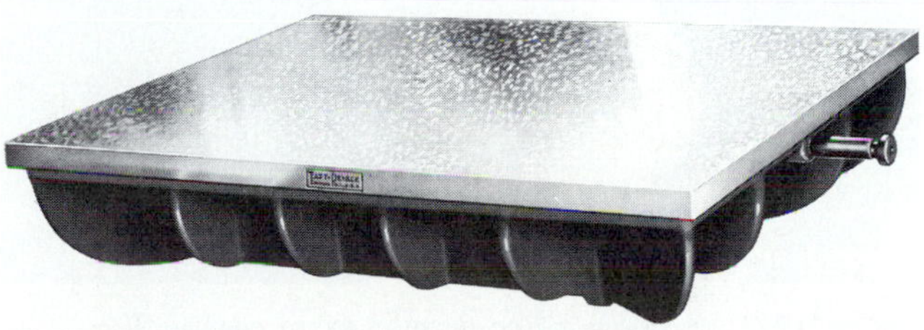

Figure 19-3 *A cast-iron surface plate.*

CARE OF SURFACE PLATES AND LAYOUT TABLES

Although surface plates are rugged, their accuracy can be easily destroyed. The following precautions should be taken with regard to surface plates.

- Keep the working surface clean.
- Cover the plate or table when it is not in use.
- Carefully place the work on the surface plate—do not drop it onto the plate.
- Use parallels under the workpiece whenever possible.
- Never hammer or punch any layout on a surface plate.
- Remove burrs from cast-iron plates and always protect their surfaces with a thin film of oil and a cover when they are not in use.

SCRIBERS

The *scriber* (Fig. 19-4) has a hardened steel point, or points, and may be used in conjunction with a combination square, a rule, or a straightedge to draw straight lines. On some scribers, one end is bent at an angle to allow marking lines in hard-to-reach places. To be accurate, any layout requires fine lines; therefore, the scriber point must always be sharp. The points of scribers, hermaphrodite calipers, dividers, and trammels should be honed frequently on a fine oilstone (Fig. 19-5) to maintain their sharpness. When extremely fine lines are required, knife-edge scribers should be used.

DIVIDERS AND TRAMMELS

Dividers are used for scribing arcs and circles on a layout and for transferring measurements. The spring divider (Fig. 19-6), the most common type, is available in sizes from 3 to 12 in. Larger circles and arcs may be scribed with a trammel (Fig. 19-7).

A *trammel* consists of a *beam* on which *two sliding* or *adjustable heads* with *scriber points* are mounted. Some trammels may have an *adjusting screw* for fine adjustment. Rods or beams of different lengths can be used to increase the capacity of the trammel. When a circle is laid out from

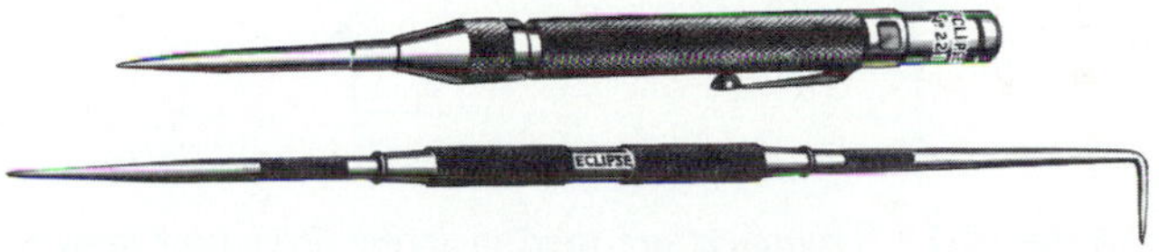

Figure 19-4 *A pocket and double-end scriber.*

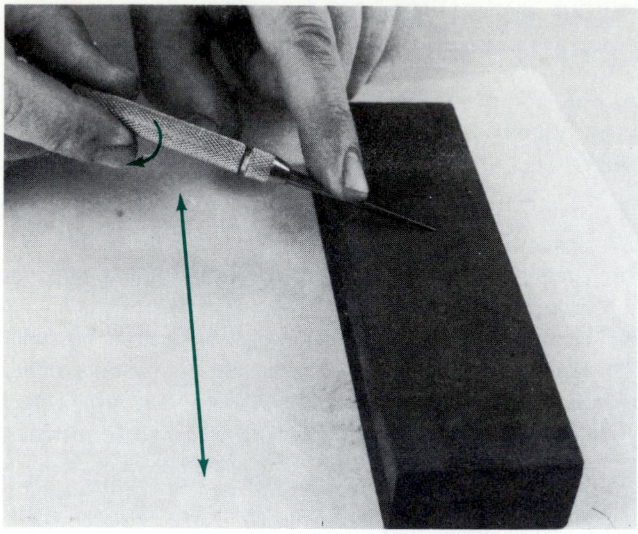

Figure 19-5 *Sharpening a scriber on an oilstone.*

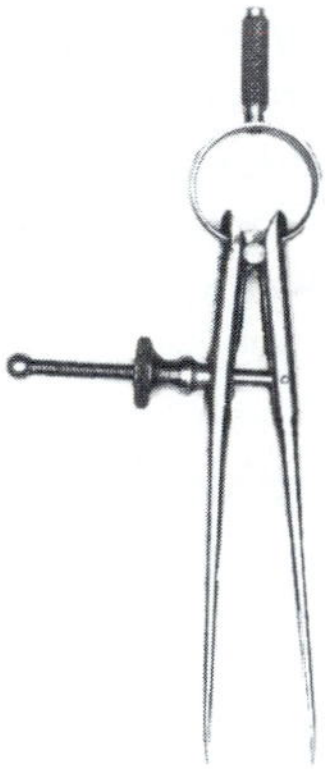

Figure 19-6 *A spring divider can be used to scribe arcs and circles.*

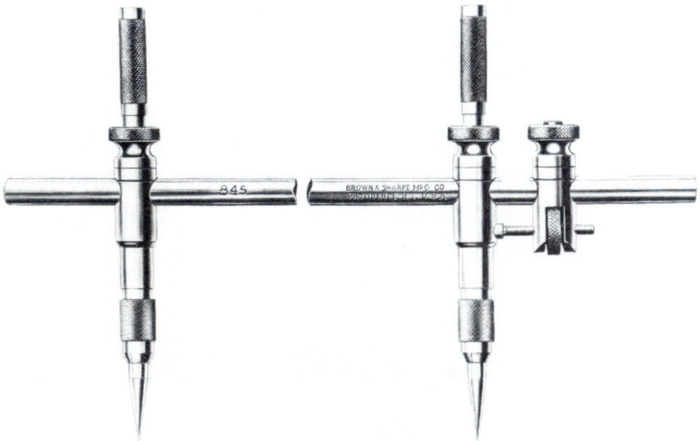
845

Figure 19-7 *Trammels are used to scribe large arcs and circles.*

a hole, a ball attachment may be substituted for one of the scriber points.

HERMAPHRODITE CALIPERS

The *hermaphrodite caliper* (Fig. 19-8) has one curved leg and one straight leg, which contains a scriber point. It is used for laying out lines parallel to an edge (Fig. 19-8) or for locating the center of round or irregular-shaped stock (Fig. 19-9).

When setting the hermaphrodite caliper to a size, place the bent leg on the end of the rule and adjust the other leg until the scriber point is at the desired graduation. When scribing parallel lines with this tool, be sure always to hold the scriber at 90° to the edge of the workpiece.

SQUARES

Squares are used to lay out lines at right angles (90°) to a machined edge, to test the accuracy of surfaces which must be square (90° to each other), and to set up work for machining.

Adjustable squares are used for general-purpose work. The *solid square* (Fig. 19-10), made up of two parts, the beam and the blade, is used where greater accuracy is required. Extremely accurate solid squares called *master squares* are used to check the accuracy of other squares.

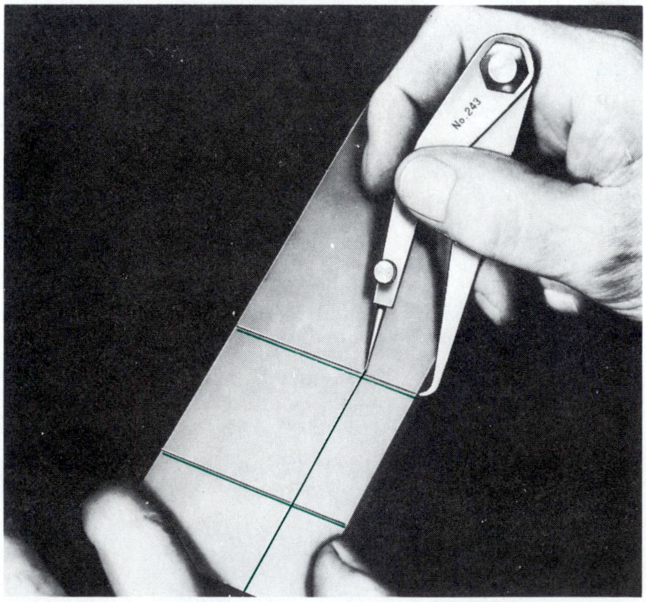
No. 243

Figure 19-8 *Scribing a line parallel to an edge with a hermaphrodite caliper.*

THE COMBINATION SET

The *combination set* (Fig. 19-11), used extensively in layout work, consists of a steel rule, square head, bevel protractor, and center head.

The *steel rule* may be fitted to the other three parts of the combination set for various layout, setup, or checking operations.

The *square head* and the rule or *combination square* may be used to lay out lines parallel to an edge (Fig. 19-12). It is also used to lay out angles at 45 and 90° to an edge (Fig. 19-13). The square head may be moved along the rule to any position. The square head is also used to check 45 and 90° angles and measure depths.

When mounted on a rule, the *bevel protractor* is used to lay out and check various angles. The protractor can be adjusted to any angle from 0 to 180°. The accuracy of this protractor is $\pm 0.5°$ (30′). A universal bevel protractor may be used if an accuracy of 5′ is required.

The *center head* forms a center square when mounted on a rule. It may be used for locating the centers on the ends of round, square, and octagonal stock.

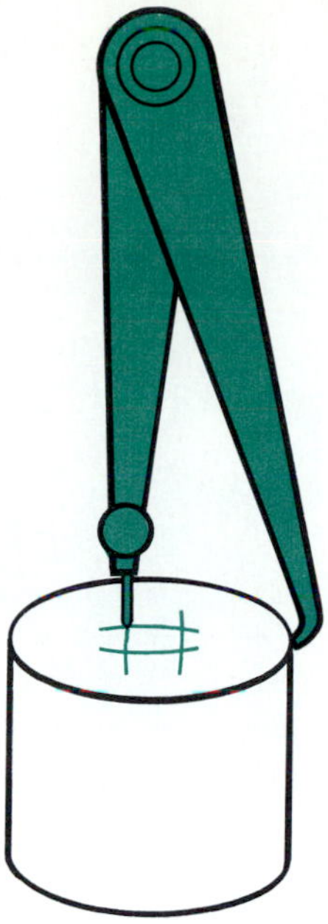

Figure 19-9 *Locating the center of a round workpiece with a hermaphrodite caliper.*

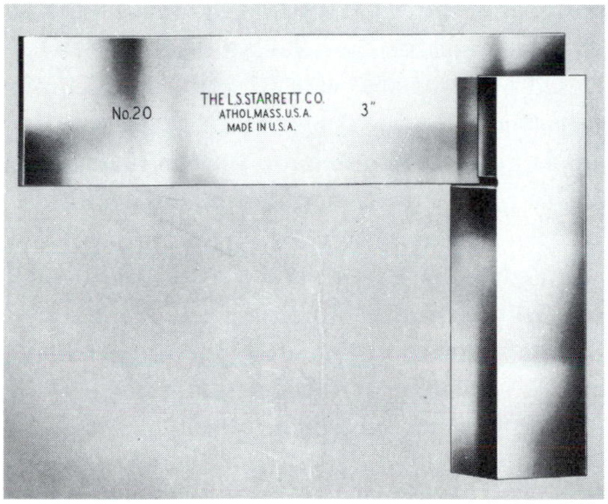
No.20
THE L.S.STARRETT CO.
ATHOL.MASS.U.S.A.
MADE IN U.S.A.
3"

Figure 19-10 *The solid square is used for inspection purposes.*

SURFACE GAGE

The *surface gage* (Fig. 19-14) is used with a surface plate or any flat surface to scribe layout lines on a workpiece. It consists of a *base, spindle,* and *scriber.*

The surface gage may be set to the required dimension

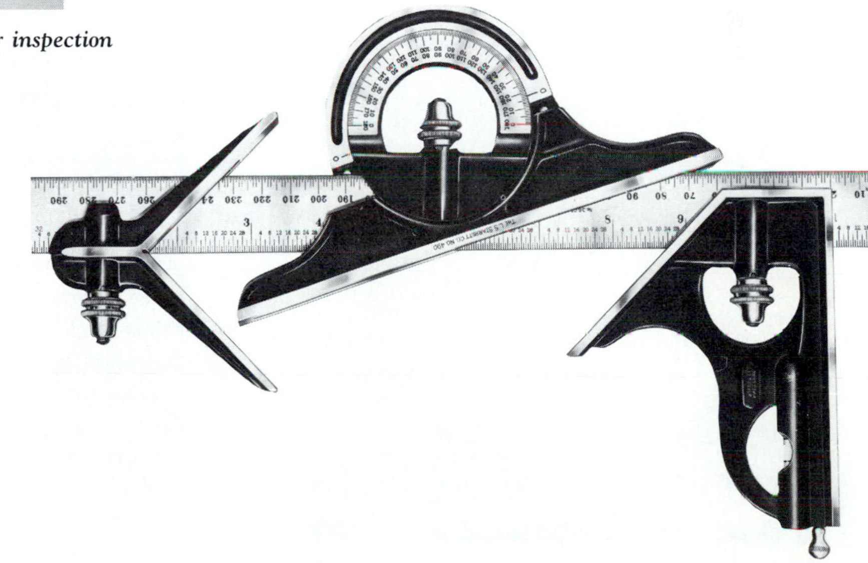

Figure 19-11 *The combination set is used for laying out and checking work.*

Figure 19-12 *Scribing a line parallel to an edge.*

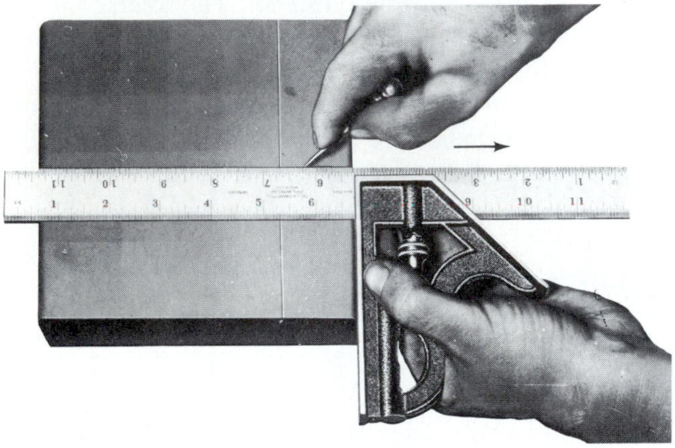

Figure 19-13 *Scribing a line at 90° to an edge.*

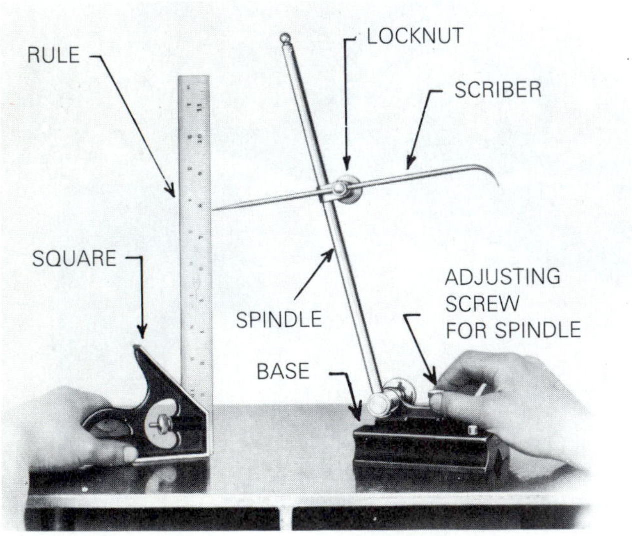
RULE
LOCKNUT
SCRIBER
SQUARE
SPINDLE
ADJUSTING
SCREW
FOR SPINDLE
BASE

Figure 19-14 *Setting a surface gage to a dimension with a combination square.*

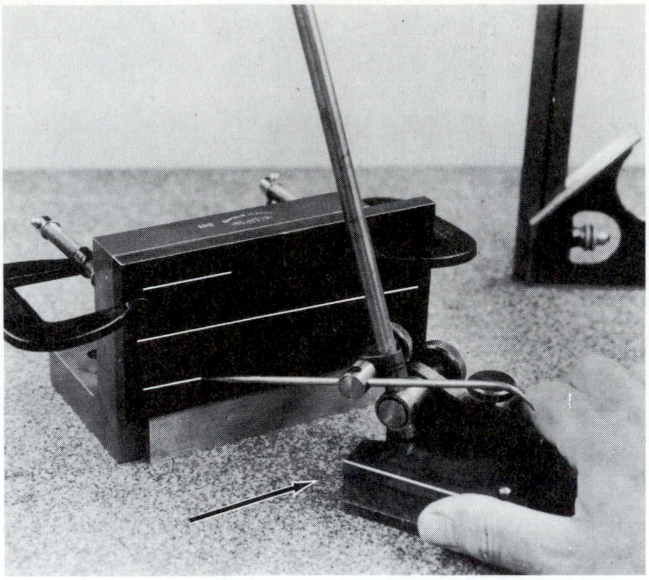

Figure 19-15 *Using a surface gage to lay out lines parallel to the top of a surface plate.*

by using a combination square (Fig. 19-14). This method is usually accurate enough for most layout work. A surface gage may be used on a surface plate to scribe parallel lines on a workpiece (Fig. 19-15). When the workpiece is fastened to an angle plate, horizontal and vertical lines can be laid out in one setup by merely rotating the angle plate 90° on the surface plate and scribing the line(s).

Most surface gages have two pins in the base, which when pushed down, are used to guide the surface gage along the edge of the workpiece or surface plate. Some surface gages have a V-groove machined in the base, which allows them to be used on cylindrical work.

LAYOUT OR PRICK PUNCHES AND CENTER PUNCHES

The *layout* or *prick punch* and the *center punch* (Fig. 19-16) differ only in the angle of the point. The prick punch is ground to an angle of 30 to 60° and is used to mark permanently the location of layout lines. The narrower angle of this punch makes a smaller and neater indentation in the metal surface.

The center punch is ground to an angle of 90° and is used to mark the location of the centers of holes. The wider indentation permits easier and more accurate starting of a drill point.

After the layout lines have been scribed on the workpiece, they should be permanently marked by means of layout or prick-punch marks. This step will ensure that the layout line location will still be visible should the line be

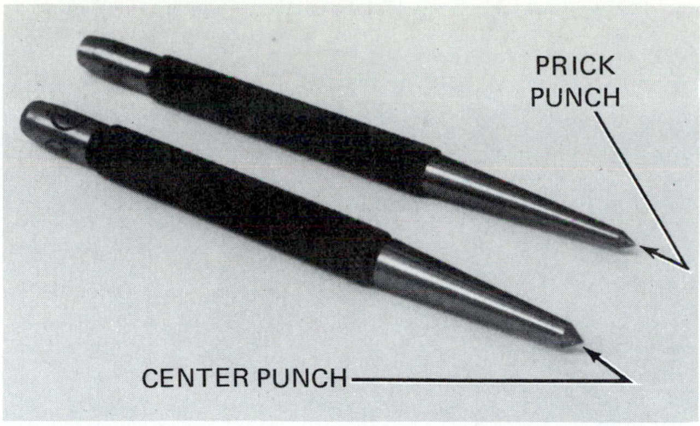
PRICK
PUNCH
CENTER PUNCH

Figure 19-16 *A prick punch and a center punch are used in layout work.*

rubbed off through handling. The intersection of the centerlines of a circle should be carefully prick-punched and then enlarged with a center punch.

NOTE: Extreme care should be taken when the intersections of layout lines are being prick-punched. Regardless of how accurate the layout is, it is impossible to mark locations with a prick punch to closer than 0.003 to 0.004 in. (0.07 to 0.10 mm).

Uniform layout punch marks may be obtained with an automatic center punch (Fig. 19-17). This punch contains a striking block in the body, which is released by downward pressure; the block then strikes the punch proper, causing it to indent the metal. The impression size can be changed by adjusting the tension on the screw cap on the upper end of the tool. This type of punch produces marks uniform in size, which improves the appearance of the workpiece. Automatic center punches may also have a spacing attachment to provide uniform spacing of layout punch marks.

LAYOUT ACCESSORIES

In addition to regular layout tools, certain accessories are helpful in layout work. When lines are required on the face

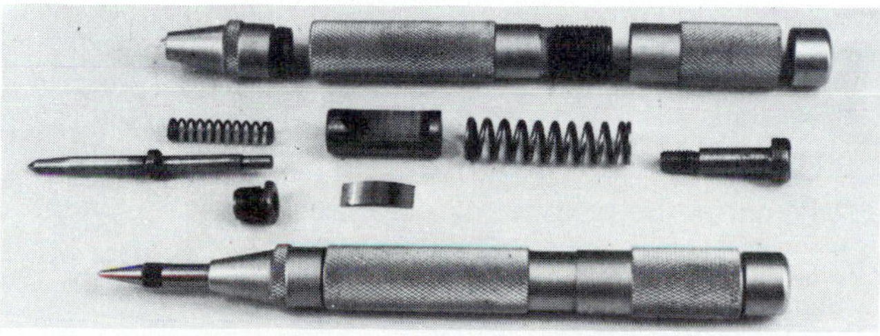

Figure 19-17 *An automatic center punch produces uniform indentations on a layout.*

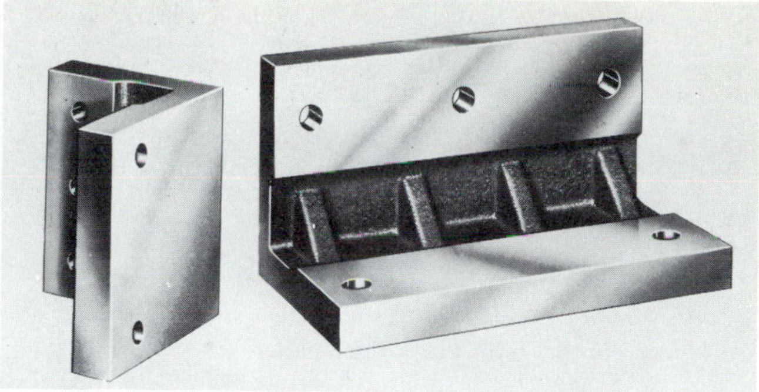

Figure 19-18 *Angle plates have sides 90° to each other.*

of a plate, it is customary to clamp the work to an *angle plate* (Fig. 19-18) with a *toolmaker's clamp*. This will hold the work in a vertical plane so that the layout lines will be accurately positioned. Since an accurate angle plate will have all adjacent surfaces at 90° to each other, it is possible to scribe intersecting 90° lines accurately. This is achieved by scribing all the horizontal lines on the workpiece, then turning the angle plate on its side and scribing the intersecting lines (Fig. 19-19A and B).

Parallels (Fig. 19-20) may be used when it is necessary to raise the workpiece to a desired height and to maintain the work surface parallel to the top of the surface plate.

V-blocks are used to hold round work for layout and for inspection. They may be used singly or in pairs. Some blocks are so constructed that they may be rotated 90° on

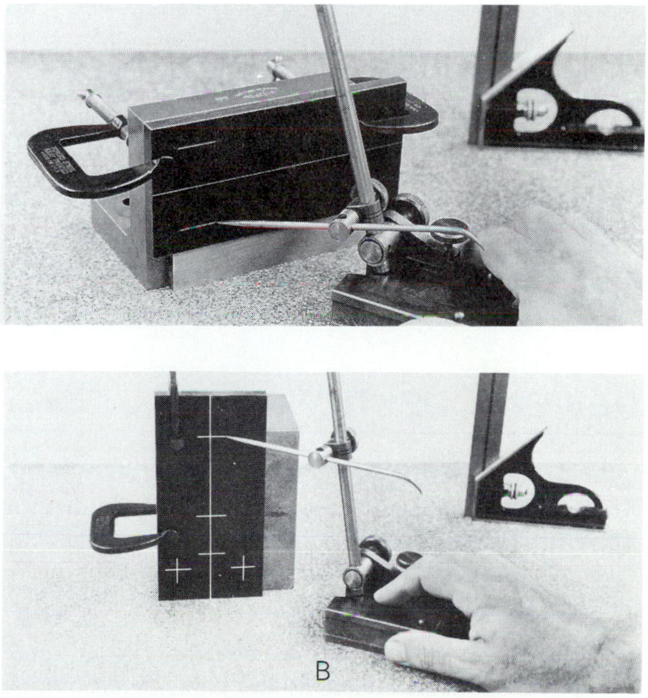
B

Figure 19-19 *(A) Scribing horizontal lines with a surface gage; (B) the angle plate turned on its side to scribe vertical lines. (Courtesy Kostel Enterprises Limited.)*

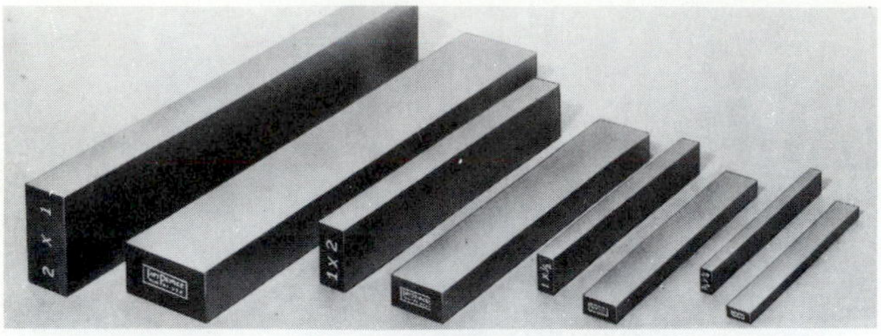

Figure 19-20 *Parallels keep the bottom surface of the workpiece parallel with the surface plate.*

Figure 19-21 *Lines may be scribed conveniently at 90° with a special V-block.*

their sides without the work having to be removed. This feature permits laying out lines at 90° on a shaft without changing the position of the work in the V-block (Fig. 19-21).

Keyseat rules are used to lay out keyseats on shafts or to draw lines parallel to the axis of the shaft. A solid keyseat rule resembles two straightedges machined at 90° to each other (Fig. 19-22A). *Keyseat clamps,* when attached to a rule or straightedge, will convert it to a keyseat rule (Fig. 19-22B).

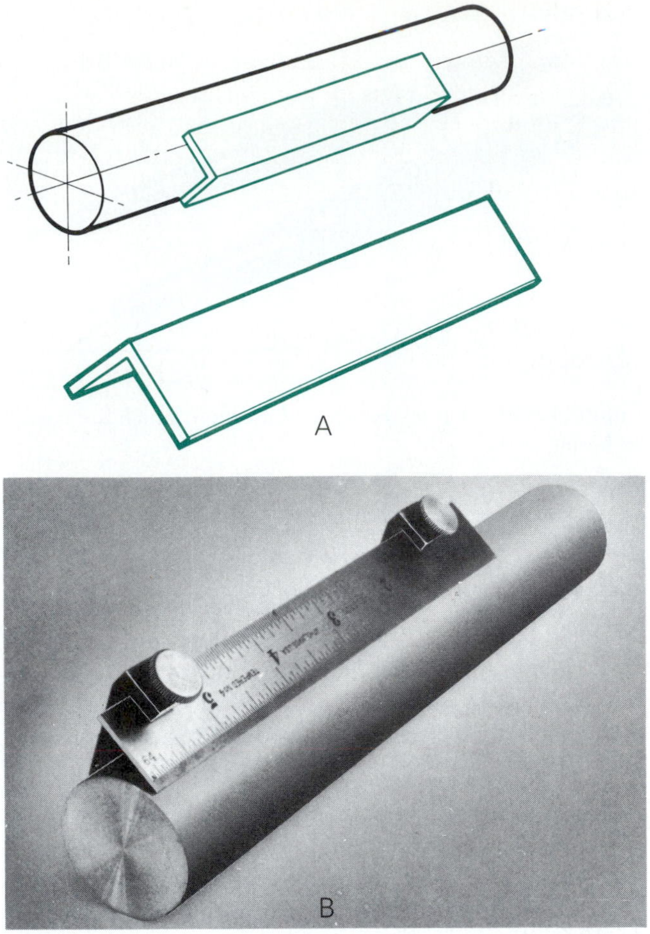
A
B

Figure 19-22 *Keyseat rules are used to scribe lines parallel to the axis of a cylinder. (A) Solid keyseat rule; (B) keyseat rule and a clamp.*

REVIEW QUESTIONS

1. State two reasons why a layout is necessary.
2. Why should a layout be as simple as possible?

LAYOUT MATERIALS

3. What is the purpose of a layout solution?
4. Name four layout solutions and state one application of each.
5. State two methods of preparing the surface of a casting prior to laying out the workpiece.

LAYOUT TABLES AND SURFACE PLATES

6. State five reasons why granite surface plates are considered better than cast-iron ones.
7. List five precautions to be observed in the care of surface plates.

SCRIBERS

8. Why must the point of a scriber always be sharp?

DIVIDERS AND TRAMMELS

9. For what purpose are dividers used when laying out a workpiece?
10. What is the purpose of a trammel?
11. How can a circle be laid out concentric with a hole?

HERMAPHRODITE CALIPERS

12. List two uses for hermaphrodite calipers.

SQUARES

13. Name two types of squares used in layout work.
14. List four main parts of a combination set.
15. State three uses for the combination square.
16. What is the accuracy of the bevel protractor?
17. For what purpose is the center head used?

SURFACE GAGE

18. Name the three main parts of the surface gage.
19. What is the purpose of the two pins in the base of the surface gage?

LAYOUT OR PRICK PUNCHES AND CENTER PUNCHES

20. How may the layout be made "permanent"?
21. What is the purpose of:
    a. A prick punch?
    b. A center punch?
22. To what angle is each ground?

LAYOUT ACCESSORIES

23. List four layout accessories and state the purpose of each.

UNIT 20

Basic or Semiprecision Layout

Basic or semiprecision layout requires the use of the basic tools described in Unit 19. Remember that the layout should be kept as simple as possible in order to save time and reduce the chance of error. Since the accuracy of the layout will affect the accuracy of the finished workpiece, care should be exercised in the layout procedure.

Although the layout required will naturally not be the same for each workpiece, certain procedures should be followed in any layout. The jobs described in this unit are intended to acquaint the reader with basic layout procedures.

OBJECTIVES

After completing this unit you will be able to:
1. Lay out a workpiece to an accuracy of ± 0.007 in.
2. Lay out straight lines using the combination square and surface gage
3. Lay out hole centers, arcs, and circles

TO LAY OUT HOLE LOCATIONS, SLOTS, AND RADII

1. Study Fig. 20-1 and select the proper stock.
2. Cut off the stock, allowing enough material to square the ends if required.
3. Remove all burrs.
4. Clean the surface thoroughly and apply layout dye.
5. Place a suitable angle plate on a surface plate.

NOTE: Clean both plates (Fig. 20-2).

6. Clamp the work to the angle plate with a finished edge (A) of the part against the surface plate or on a parallel. Leave one end of the angle protruding beyond the workpiece.

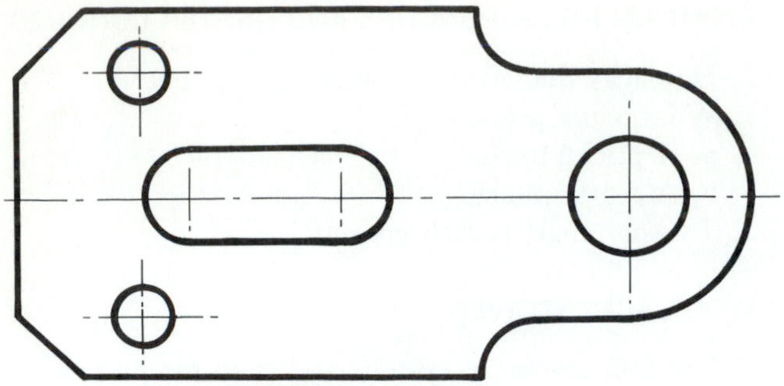

Figure 20-1 *A layout exercise.*

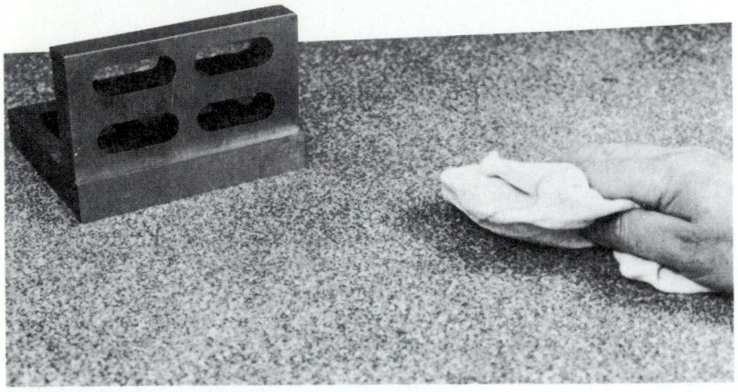

Figure 20-2 *Place a suitable angle plate on the clean surface.*

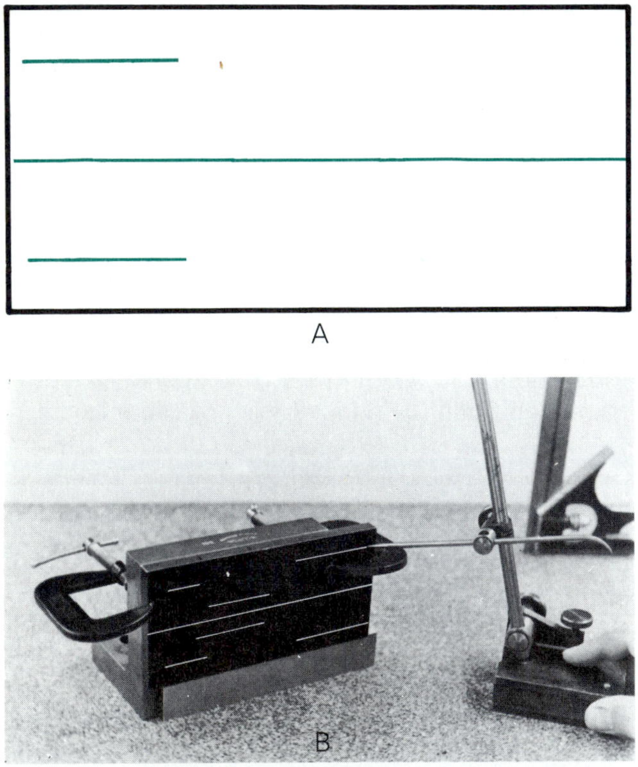

Figure 20-3 *(A) Centerlines are scribed parallel to the base; (B) using a surface gage to scribe parallel lines.*

7. With the surface gage set to the proper height, scribe a centerline for the full length of the workpiece (Fig. 20-3A).
8. Using the centerline as a reference, set the surface gage for each horizontal line and scribe the center lines for all hole and radii locations (Fig. 20-3B).
9. With the work still clamped to the angle plate, turn the angle plate 90° with edge B down and scribe the baseline at the bottom of the workpiece (Fig. 20-4A).
10. Using the baseline as a reference line, locate and scribe the other centerlines for each hole or arc (Fig. 20-4B).

NOTE: All measurements for any location must be taken from the baseline or finished edge.

11. Locate the starting points for the angular layout (Fig. 20-4A).
12. Remove the workpiece from the angle plate.
13. Carefully prick-punch the center of all hole or radii locations.
14. Using a divider set as required, scribe all circles and arcs (Fig. 20-5A).
15. Scribe any lines required to connect the arcs or circles (Fig. 20-6).
16. Draw in the angular lines.

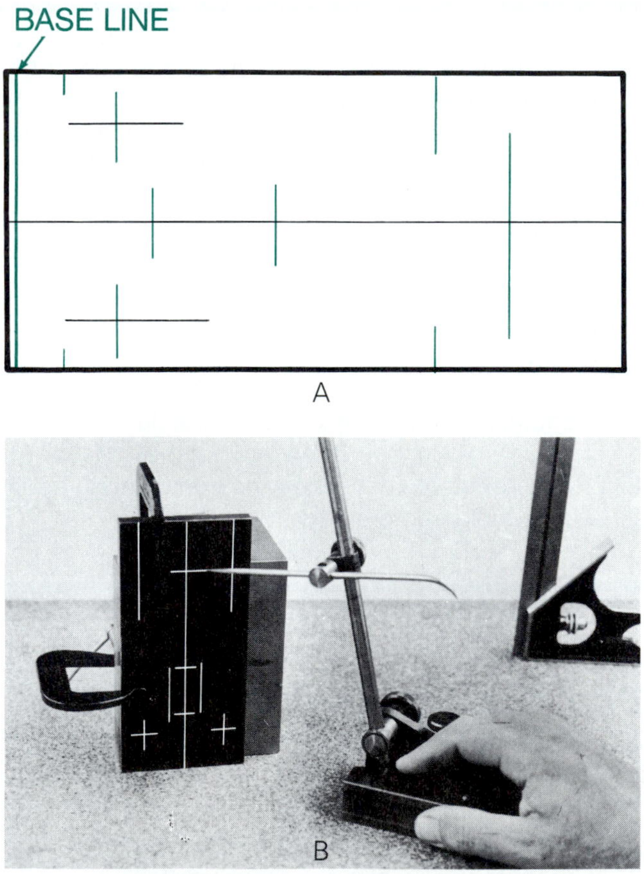
BASE LINE
A
B

Figure 20-4 *(A) Centerlines scribed. (B) The angle plate turned to 90° to scribe vertical lines.*

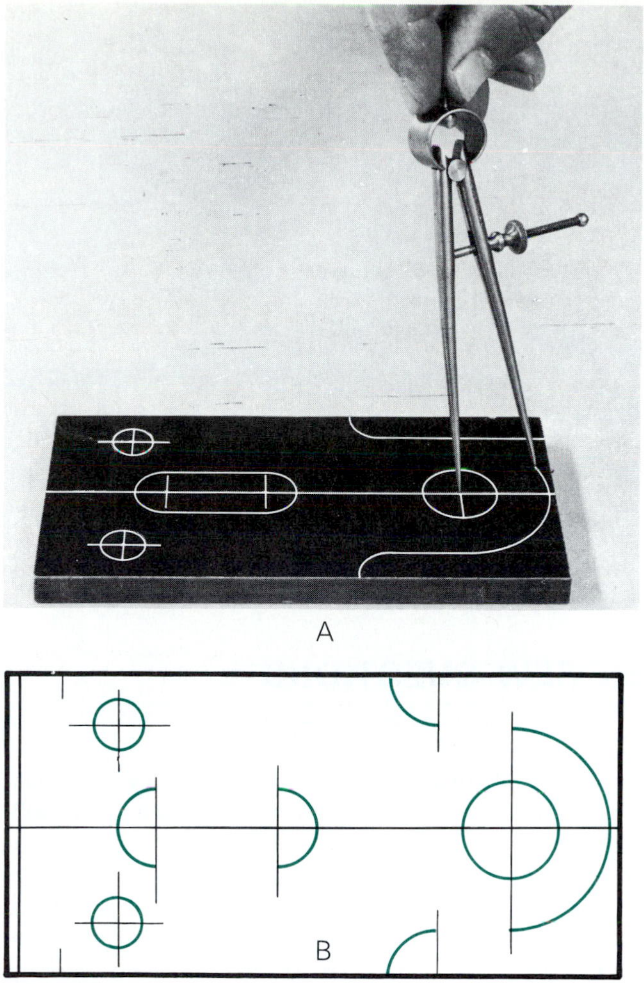
A
B

Figure 20-5 *(A) Arcs and circles are scribed with a divider; (B) arcs and circles are scribed.*

TO LAY OUT A CASTING HAVING A CORED HOLE

When a casting which requires a hole in it is molded in a foundry, a core is used to produce the rough hole, which may have to be machined later. Often the core shifts out of place and the hole is cast off center, as shown in Fig. 20-7. If the hole must be machined concentric with the outside of the casting, it may be necessary to lay out the location of the hole. Proceed as follows:

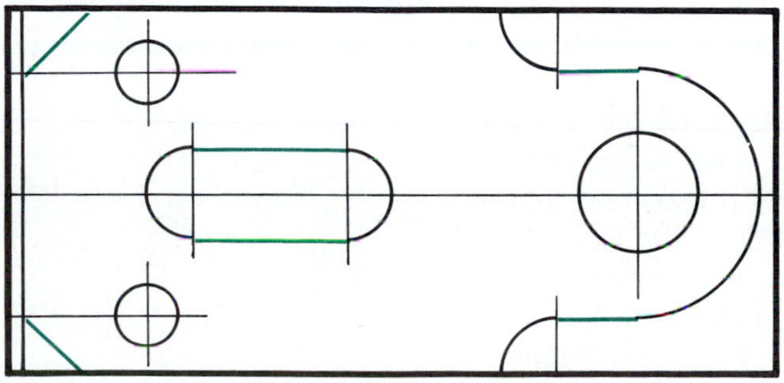

Figure 20-6 *Arcs and circles are connected.*

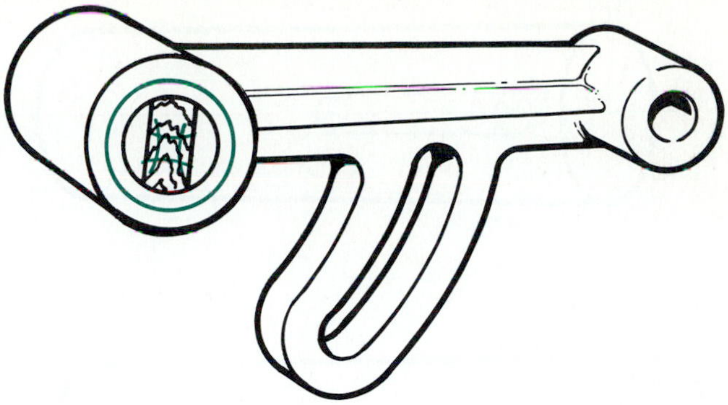

Figure 20-7 *Method of centering a hole location on a casting.*

1. Grind the scale off the surface to be laid out.
2. Tap a tightly fitting wooden piece into the cast hole (Fig. 20-7).
3. Coat the surface to be laid out and the wooden piece with a solution of slaked lime and alcohol or layout dye.
4. With the hermaphrodite calipers scribe four arcs as shown, using the outside diameter of the shoulder as the reference surface.
5. Using the intersection of these arcs as a center, scribe a circle of the required diameter on the casting. The hole should be concentric with the outside of the casting (Fig. 20-7).
6. Prick-punch the layout line at about eight equidistant points around the layout circle.

TO LAY OUT A KEYSEAT IN A SHAFT

A keyseat is a recessed groove cut in a shaft and into which a key is fitted to prevent a mating part, such as a pulley or gear, from turning on the shaft. Laying out a keyseat requires much care, particularly if the mating parts must maintain a certain relative position. Proceed as follows:

1. Apply layout dye to the end of the shaft and to the area where the keyseat is to be laid out.
2. Mount the workpiece in a V-block.
3. Set the surface gage scriber to the center of the shaft.
4. Scribe a line across the end and continue it along the shaft to the keyseat location (Fig. 20-8).
5. Rotate the workpiece in the V-block and mark the length and the position of the keyseat on the shaft.

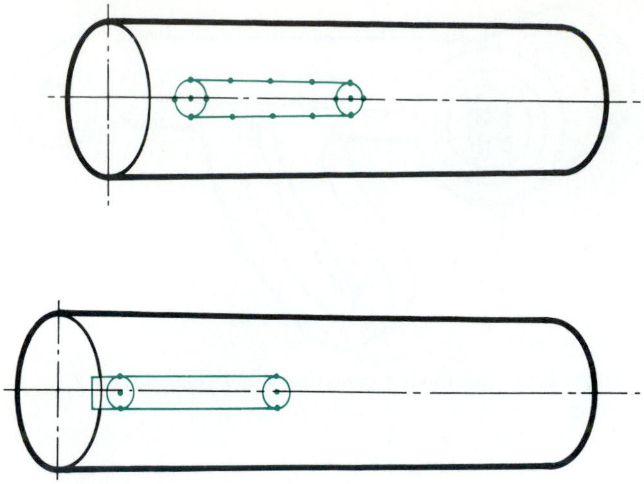

Figure 20-8 *Keyseat layouts.*

6. Set the dividers to half the width of the keyseat, and scribe a circle at each end of the layout (Fig. 20-8).
7. Using a keyseat rule and scriber, connect the circles with a line on either side of the centerline and tangent to the circles.

NOTE: If a keyseat rule is not available, the circles may be connected using the surface gage.

8. Prick-punch the layout of the keyseat, and center-punch the centers of the circles.
9. If it is necessary to drill holes at the end of the keyseat, set up the shaft by aligning the end layout line in a vertical position with a square (Fig. 20-9).

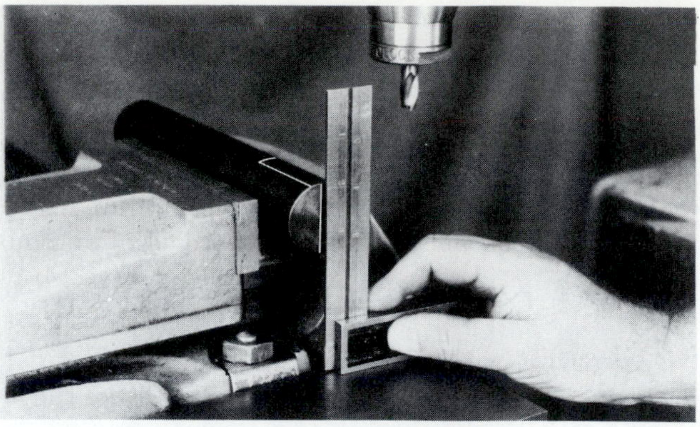

Figure 20-9 *Aligning the keyseat layout in a vise.*

REVIEW QUESTIONS

1. Outline the main steps that should be followed in making a layout consisting of straight lines, slots, and radii.
2. Name three common tools used to make basic or semiprecision layouts.
3. Describe how to lay out a cored hole concentric with the outside shoulder of a casting.
4. List the main steps for laying out a keyseat on a shaft.

UNIT 21

Precision Layout

The accuracy of the finished workpiece is generally determined by the accuracy of the layout; therefore, great care must be used when laying out. In order to make a precision layout, a person must be able to read and understand drawings, select and use the proper layout tools for the job, and accurately transfer measurements from the drawing to the workpiece. After completing a layout, check all layout work against the sizes on the working drawings to make sure that the layout is accurate. When the layout lines must be accurate to within 0.001 in. (0.02 mm), a *vernier height gage* may be used (Fig. 21-1).

When hole locations and dimension lines are made on a layout, they are generally made from two machined edges called *reference surfaces* using *x* and *y* coordinates. Any layout composed of holes, angles, and lines may be calculated using trigonometry to determine the coordinate measurements. Once the coordinates have been determined, they can be used for setting up the workpiece and accurately positioning the holes for machining. Another method of calculating hole locations is by means of the Woodworth Coordinate Factors and Angles tables in the Appendix of Tables in this book.

OBJECTIVES

After completing this unit you will be able to:
1. Make a precision layout using the vernier height gage
2. Use the Woodworth Coordinate Factors and Angles tables to calculate equidistant hole locations
3. Make precision layouts using the sine bar and gage blocks

THE VERNIER HEIGHT GAGE

The *vernier height gage* may be used to measure or mark off vertical distances to ±0.001 in. (0.02 mm) accuracy. The main parts of the vernier height gage (Fig. 21-2) are the *base, beam, vernier slide,* and *scriber,* which is attached to the vernier slide when making layouts. Other accessories, such as a dial indicator or a depth gage attachment, may be added to the slide for measurement and inspection work. The graduations on the beam and the vernier slide are the same as those on a vernier caliper and the readings are made in the same manner as with a vernier caliper.

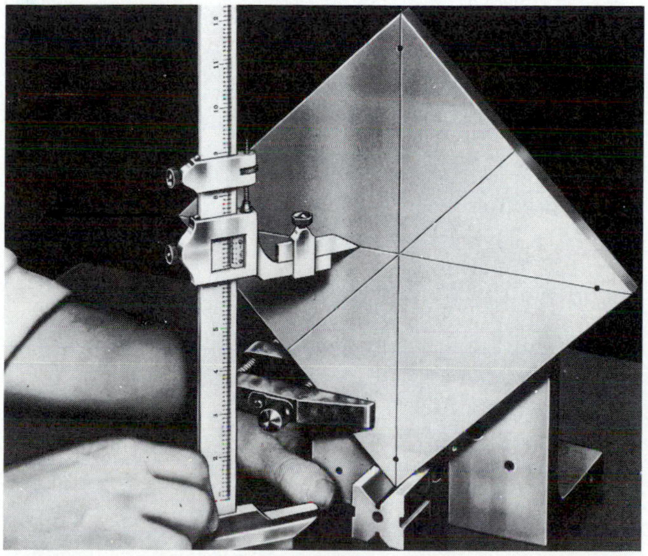

Figure 21-1 *A vernier height gage is used when an accurate layout is required.*

TO MAKE A PRECISION LAYOUT USING A VERNIER HEIGHT GAGE

It is required to lay out the position of five equally spaced holes on a 5-in.-diameter circle located in the center of a 7-in.-*square* steel plate (See Fig. 21-3). Proceed as follows:

1. Refer to the drawing of the required workpiece (Fig. 21-3).
2. Remove all burrs from the workpiece.
3. Apply layout dye to the surface and mount it on an angle plate.

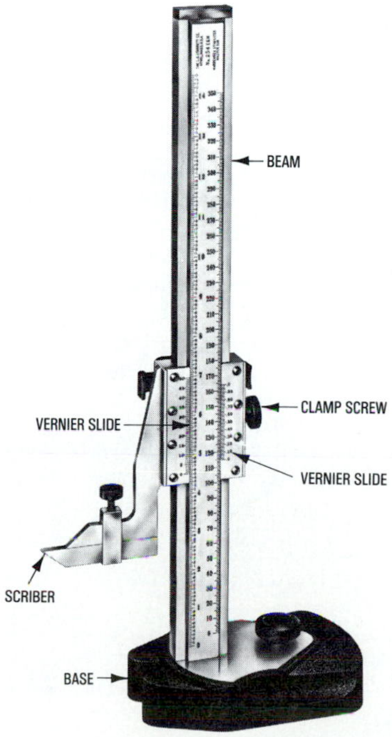
BEAM
CLAMP SCREW
VERNIER SLIDE
VERNIER SLIDE
SCRIBER
BASE

Figure 21-2 *The main parts of a vernier height gage.*

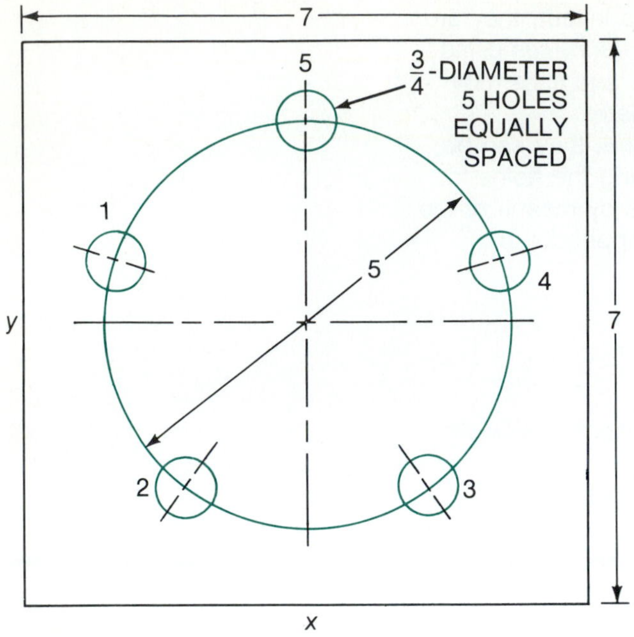

7
5
3/4-DIAMETER
5 HOLES
EQUALLY
SPACED
1
5
4
y
7
2
3
x

Figure 21-3 *Layout of five equally spaced holes on a circle.*

4. Clean the surface of the layout table, the angle plate, and the base of the height gage.

5. Mount an offset scriber (Fig. 21-4) on the vernier slide and clamp it in position.

6. Move the vernier slide and scriber down until the scriber touches the top of the surface plate.

7. Check the reading on the vernier scale. The zero mark on the vernier should align exactly with the zero graduation on the beam. If zero on the vernier does not coincide with zero on the beam, recheck the assembly of the scriber and the vernier slide.

8. Refer to the Coordinate Factors and Angles tables in the Appendix for the coordinates for five equidistant holes.

9. Calculate the location of all five holes as follows:

Hole 1

Horizontal distance from left-hand edge Y
= (diameter of circle × factor for A) + 1.000
= (5 × 0.024472) + 1.000
= 0.12236 + 1.000
= 1.122

Vertical distance from upper edge X
= (diameter × factor for B) + 1.000
= (5 × 0.345492) + 1.000
= 1.72746 + 1.000
= 2.727

Hole 2

Distance from edge Y
= (5 × 0.206107) + 1.000
= 1.030535 + 1.000
= 2.031

Distance from edge X
= (5 × 0.904508) + 1.000
= 4.52254 + 1.000
= 5.523

Hole 3

Distance from edge Y
= (5 × 0.793893) + 1.000
= 3.969465 + 1.000
= 4.969

Distance from edge X
= (5 × 0.904508) + 1.000
= 4.52254 + 1.000
= 5.523 (same as for hole 2)

Hole 4

Distance from edge Y
= (5 × 0.975528) + 1.000
= 4.87764 + 1.000
= 5.878

Distance from edge X
= (5 × 0.345492) + 1.000
= 1.72746 + 1.000
= 2.727 (same as for hole 1)

Hole 5

Distance from edge Y
= (5 × 0.5000) + 1.000
= 2.5000 + 1.000
= 3.500

Distance from edge X
= (5 × 0.000) + 1.000
= 0.000 + 1.000
= 1.000

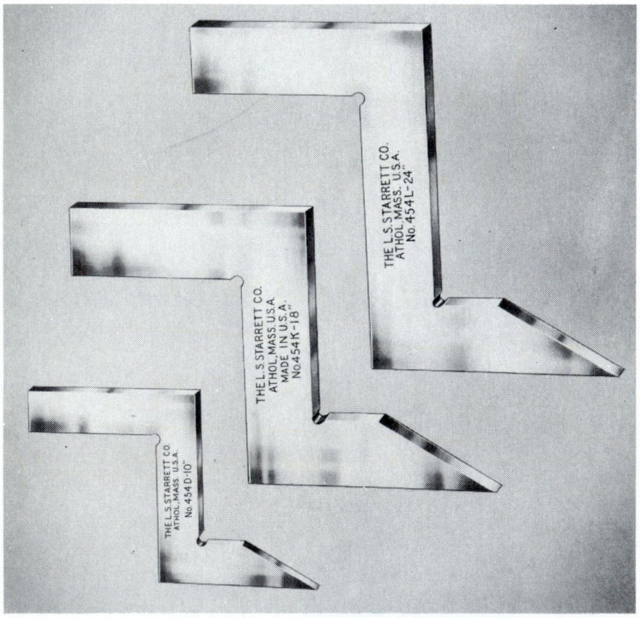

Figure 21-4 *Offset scribers.*

10. Place edge Y on the layout table surface.
11. Set the height gage to 1.122 and scribe a line by *drawing* the scriber across the face of the workpiece at the location for hole 1.
12. Set the height gage to each of the following settings. After each setting is made, scribe the line for the appropriate hole location (Fig. 21-5).

Hole 2—2.031
Hole 3—4.969
Hole 4—5.878
Hole 5—3.500

13. Rotate the angle plate and the work 90° and place edge X on the layout table.
14. Set the vernier height gage to 2.727.
15. Scribe the intersecting lines at the centers of holes 1 and 4.
16. Set the height gage to 5.523.
17. Scribe the intersecting lines for the centers of holes 2 and 3.
18. Set the height gage to 1.000 and scribe the intersecting line for hole 5 (Fig. 21-6).
19. Remove the workpiece from the angle plate and place it on the bench with the layout surface up.
20. Using a sharp prick punch and a magnifying glass, carefully mark the centers of the holes at the intersecting lines.
21. Set the dividers to ⅜ in. and scribe the five ¾-in. circles.
22. Carefully prick-punch the circumference of each circle at four equidistant points to ensure the permanency of the layout (Fig. 21-7).

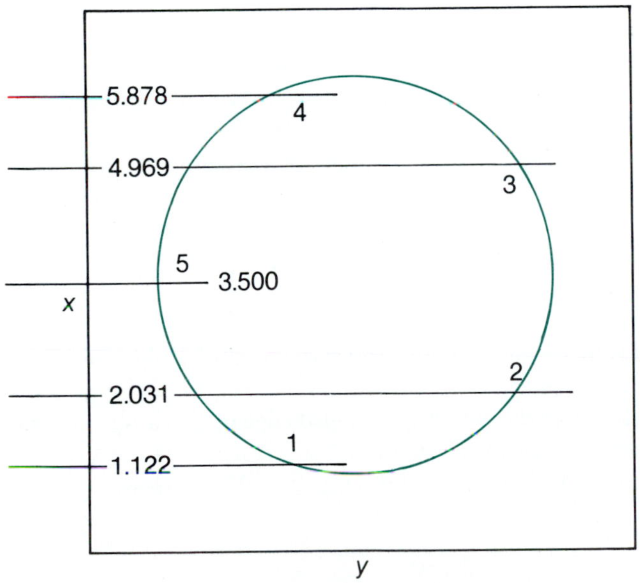
5.878
4
4.969
3
5
3.500
x
2.031
2
1
1.122
y

Figure 21-5 *The horizontal positions of five holes.*

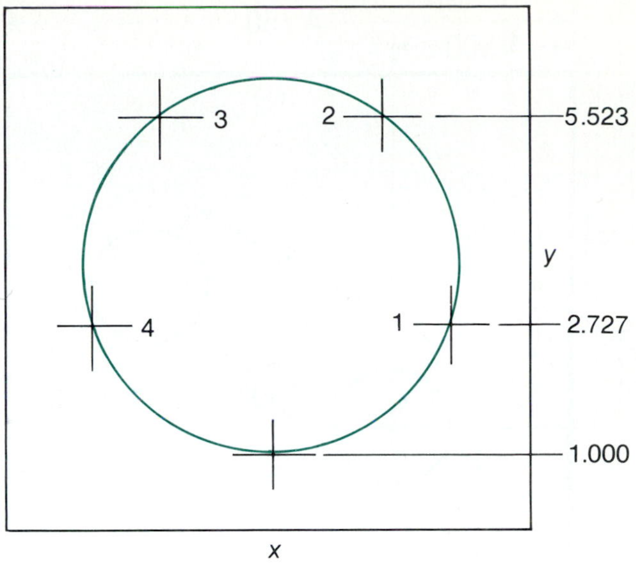
3
2
5.523
y
4
1
2.727
1.000
x

Figure 21-6 *The vertical distance of each hole.*

TO MAKE A PRECISION LAYOUT USING A SINE BAR, GAGE BLOCKS, AND A VERNIER HEIGHT GAGE

If a more accurate layout is required for hole positions and angles (Fig. 21-8), it may be done by using a sine bar, gage blocks, and height gage to accurately establish the hole loca-

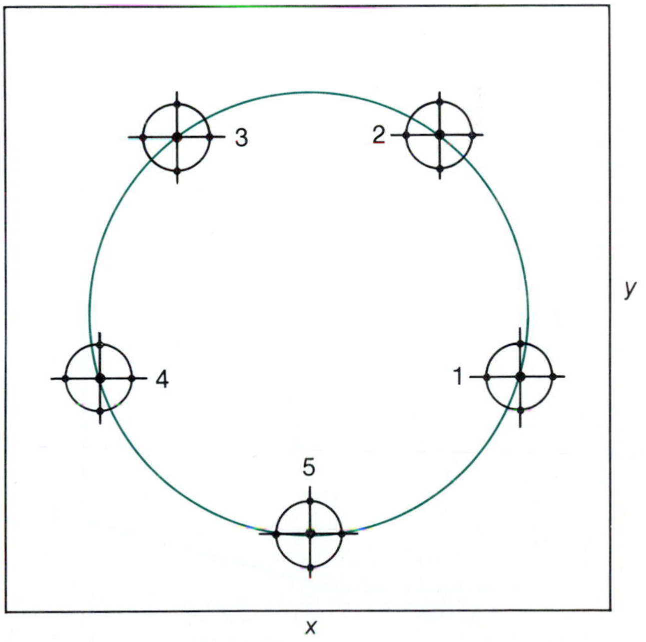
3
2
y
4
1
5
x

Figure 21-7 *Prick punch marks on the circles ensure the permanency of the layout.*

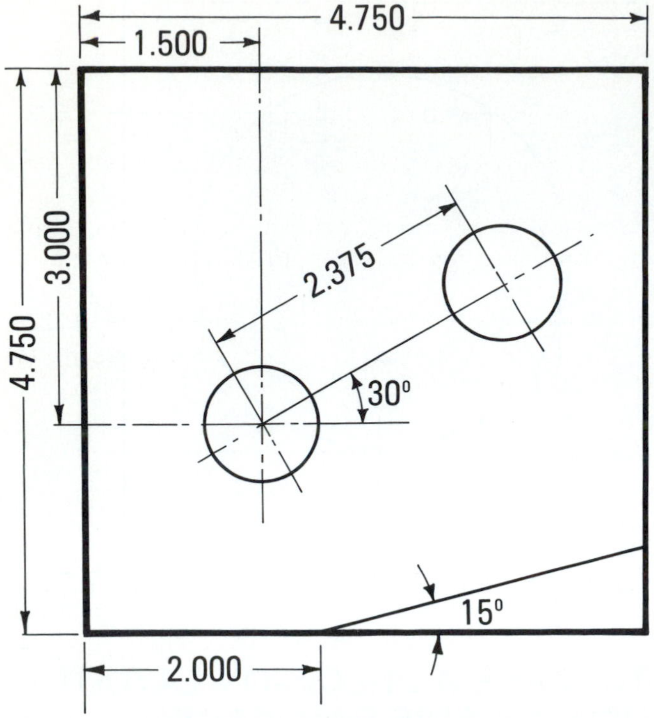

4.750
1.500
3.000
4.750
2.375
30°
15°
2.000

Figure 21-8 *An accurate layout for precision work.*

tions and their *X* and *Y* axes. The use of coordinates is preferred to locate hole centers since the work can be set up on a jig borer or vertical mill using these same coordinate measurements to position the holes for machining. This type of layout example is shown in Fig. 21-9. Proceed as follows:

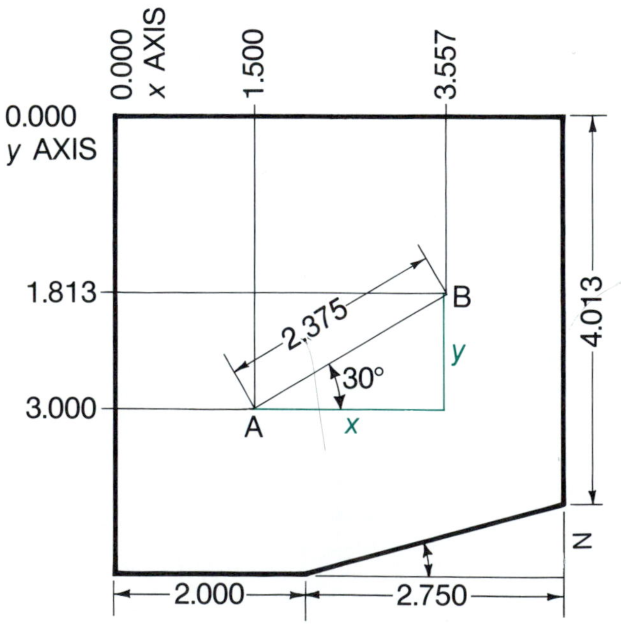

0.000
x AXIS
1.500
3.557
0.000
y AXIS
1.813
3.000
B
2.375
y
30°
A
x
4.013
N
2.000
2.750

Figure 21-9 *The position of hole B is calculated using trigonometry.*

1. Check the drawing for the required dimensions (Fig. 21-8).
2. Machine and grind a plate square to size 4.750 in. × 4.750 in.
3. Clean the surface of the workpiece and coat it with layout dye.
4. Set the work on edge on a surface plate and clamp it to an angle plate.
5. Using a vernier height gage, scribe the centerlines for hole A (Fig. 21-9).
6. Calculate the position of hole B as follows:

Length of side X

$$\frac{X}{2.375} = \cos 30°$$

$$X = 2.375 \cos 30° = 2.375 \times 0.86603 = 2.0568 \text{ in.}$$

Length of side Y

$$\frac{Y}{2.375} = \sin 30°$$

$$Y = 2.375 \sin 30° = 2.375 \times 0.5000 = 1.1875 \text{ in.}$$

The position of the center of hole B is then 2.0568 in. to the right of the centerline for hole A and 1.1875 in. above the centerline for hole A. It is now possible to position the hole using the coordinate method:

The vertical center line for hole B would then be located 1.500 + 2.0568 = 3.5568 in. along the X axis.
The horizontal centerline for hole B would be located 3.000 − 1.1875 = 1.8125 in. along the Y axis.

7. Using a vernier height gage, mark off the centerline for hole B.
8. Remove the workpiece from the angle plate.
9. Carefully prick-punch the intersection of the centerlines of these holes using a sharp punch and a magnifying glass.
10. To lay out the 15° angle at the corner of the plate, calculate the buildup for 15° using a 5-in. sine bar.

$$\text{Buildup} = 5 \sin 15° = 5 \times 0.25882 = 1.2941 \text{ in.}$$

11. Place the buildup under one end of a sine bar (on a surface plate).
12. Place the workpiece, angled edge up, on a sine bar and clamp to an angle plate.
13. Calculate length *DF* (Fig. 21-10) as follows:

$$\frac{DF}{2.750} = \tan 15° = 0.26795$$

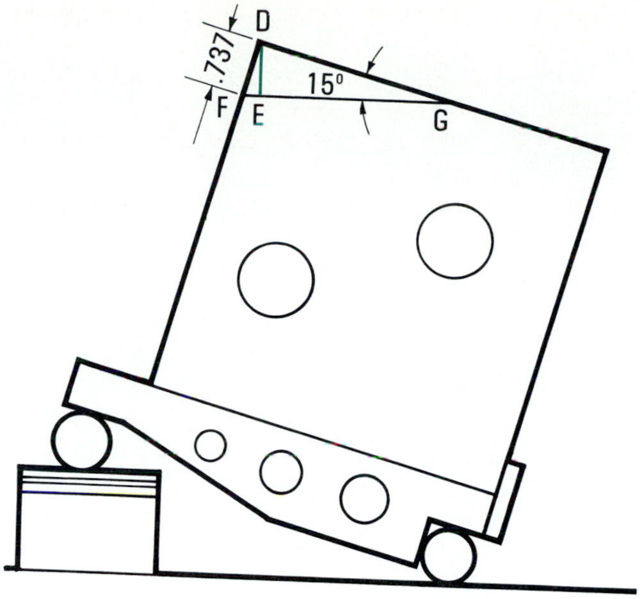
D
.737
15°
F
E
G

Figure 21-10 *To locate line* EG *accurately, distance* DE *must be known.*

$$DF = 0.26795 \times 2.750$$
$$= 0.7368 \text{ in.}$$

14. Calculate length *DE* (Fig. 21-10). In order to scribe the line accurately at 15° as required in Fig. 21-8, it is necessary first to calculate the length of line *DE* since this is the vertical distance below *D* that the line must be scribed. By previous calculations, it was determined that *DF* = 0.737 in. In the triangle *DEF* the angle *FDE* is 15°.

$$\therefore \frac{DE}{DF} = \cos 15^\circ$$

$$DE = \cos 15^\circ \times DF$$
$$= 0.96592 \times 0.737$$
$$= 0.7118$$
$$= 0.712 \text{ in.}$$

15. Set the scriber on the vernier height gage to the uppermost corner of the plate (point *D*).
16. Lower the scriber 0.712 in. and scribe line *GF*. This will locate point *G* at a position 2.000 in. from the side of the plate as required in Fig. 21-8.
17. Remove the workpiece from the angle plate.
18. Lightly prick-punch the layout line using a magnifying glass and a sharp prick punch.

REVIEW QUESTIONS

1. Name three requirements that a machinist must meet in order to make a precision layout.
2. List four main parts of a vernier height gage.
3. How can the vertical and horizontal distances of equally spaced holes on a circle be calculated from the edges of a workpiece?
4. Calculate the vertical and horizontal distances for three equally spaced holes on a 4-in.-diameter circle located in the center of a 6-in.-square plate.
5. How can the hole centers at intersecting lines be accurately marked?
6. How can accurate angular lines be laid out?
7. Calculate the gage block buildup for setting a sine bar to an 18° angle.

SECTION

7

Hand Tools and Bench Work

The machine tool trade may be divided into two categories: hand tool and machine tool operations.

Although this era is looked upon as the machine age, the importance of hand tool operations or bench work should not be overlooked. Bench work includes the operations of laying out, fitting, and assembling. These operations may involve sawing, chipping, filing, polishing, scraping, reaming, and threading. A good machinist should be capable of using all hand tools skillfully. Effective selection and use of these tools is possible only with continued practice.

UNIT 22

Holding, Striking, and Assembling Tools

Hand tools may be divided into two classes: *noncutting* and *cutting.* Noncutting tools include vises, hammers, screwdrivers, wrenches, and pliers, which are used basically for holding, assembling, or dismantling parts.

OBJECTIVES

After completing this unit you will be able to:
1. Select various tools used for holding, assembling, or dismantling workpieces
2. Properly use these tools for holding, assembling, and dismantling workpieces

THE BENCH VISE

The *machinist's,* or *bench, vise* (Fig. 22-1) is used to hold small work securely for sawing, chipping, filing, polishing, drilling, reaming, and tapping operations.

Vises are mounted close to the edge of the bench; they permit long work to be held in a vertical position. Vises may be made of cast iron or cast steel. Vise size is determined by the width of the jaws.

A machinist's vise may be of the solid-base or swivel-base type. The swivel-base vise (Fig. 22-1) differs from the solid-base vise in having a swivel plate attached to the bottom of the vise. This plate allows the vise to be swung into any circular position. To grip finished work or soft material, use jaw caps made of brass, aluminum, or copper to protect the work surface from being marred or damaged.

HAMMERS

Many different types of hammers are used by the machinist, the most common being the *ball-peen hammer* (Fig. 22-2). The larger striking surface is called the *face,* and the smaller, rounded end is the *peen.* Ball-peen hammers are made in a variety of sizes, with head masses ranging from approximately 2 ounces (oz) to 3 pounds (lb) (55 to 1400 g). The smaller sizes are used for layout work and the larger ones for general work. The peen is generally used in riveting or peening operations.

Soft-faced hammers (Fig. 22-3A) have heads made of plastic, rawhide, copper, or lead. These heads are fastened

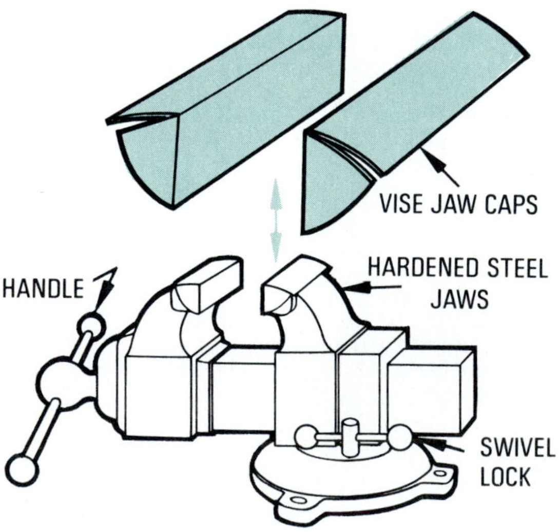
VISE JAW CAPS
HANDLE
HARDENED STEEL JAWS
SWIVEL LOCK

Figure 22-1 *A swivel-base bench vise can be rotated to any position.*

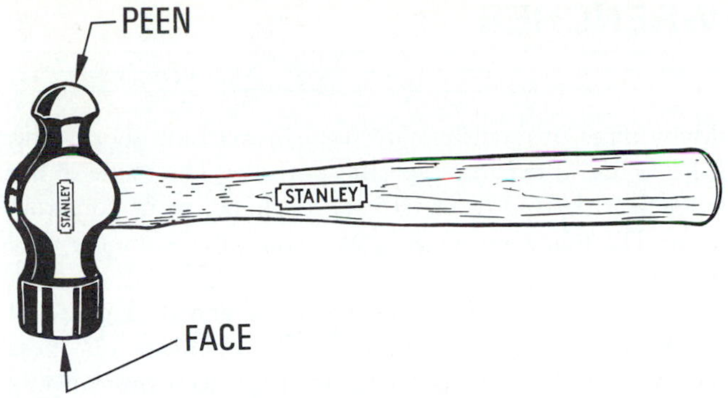
PEEN
STANLEY
STANLEY
FACE

Figure 22-2 *A ball-peen hammer. (Courtesy Stanley Tools Division, Stanley Works.)*

to a steel body and can be replaced when worn. Soft-faced hammers are used in assembling or dismantling parts so the finished surface of the work will not be marred. Lead hammers are often used to seat the workpiece properly on parallels when setting up work in a vise for milling or shaping operations. Plastic hammer heads, which have been filled with lead or steel shot, are gradually replacing the lead hammer since they do not lose their shape and last much longer than the lead hammer heads.

When using a hammer, always grasp it at the end of the handle to provide better balance and greater striking force. This grip also tends to keep the hammer face flat on the work and reduces the chance of damage to the workface.

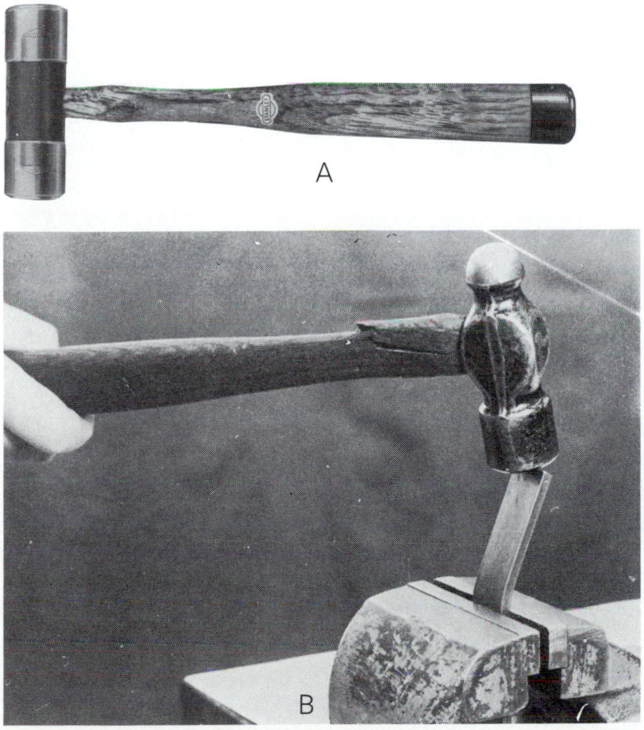
A
B

Figure 22-3 *(A) A soft-faced hammer; (B) cracked hammer handles are dangerous to use.*

The following safety precautions should always be observed when using a hammer.

1. Be sure that the handle is solid and not cracked (Fig. 22-3B).
2. See that the head is tight on the handle and secured with a proper wedge to keep the handle expanded in the head.
3. Never use a hammer with a greasy handle or when your hands are greasy.
4. Never strike two hammer faces together. The faces have been hardened and a metal chip may fly off, causing an injury.

SCREWDRIVERS

Screwdrivers are manufactured in a variety of shapes, types, and sizes. The two most common types used in a machine shop are the standard or flat blade (Fig. 22-4) and the Phillips screwdriver (Fig. 22-5). Both types are manufactured in various sizes and styles, such as standard shank, stubby shank, and offset (Fig. 22-6).

Phillips screwdrivers have a +-shaped tip for use with Phillips-type recessed screw heads (Fig. 22-7). These screwdrivers are manufactured in four sizes: #1, #2, #3, and #4, to suit the various-sized recesses in the heads of fasteners. Care must be taken to use the proper size screwdriver. Too small a screwdriver will damage both the tip and the

STANDARD

Figure 22-4 *A standard screwdriver.*

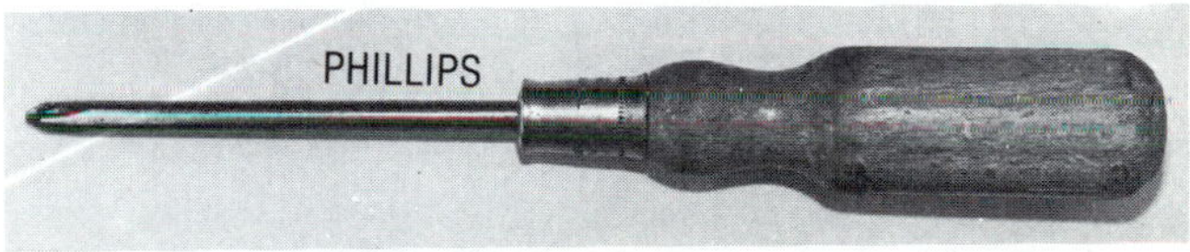
PHILLIPS

Figure 22-5 *A Phillips screwdriver.*

STANLEY NO. 669 MADE IN U.S.A.

Figure 22-6 *An offset screwdriver is often used in confined spaces.*

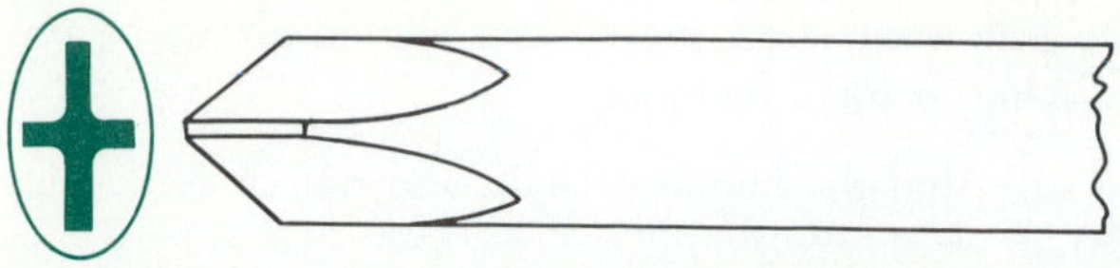

Figure 22-7 *The tip of a Phillips screwdriver fits into the socket in the screw.*

recess in the screw head. The screwdriver should be held firmly in the recess and square with the screw.

Blades for smaller standard screwdrivers are generally made of round stock, and blades for larger ones are often square, so that a wrench may be applied for leverage.

CARE OF A SCREWDRIVER

1. Choose the correct size of screwdriver for the job. If too small a screwdriver is used, both the screw slot and the tip of the screwdriver may become damaged.
2. Do not use the screwdriver as a pry, chisel, or wedge.
3. When the tip of a standard screwdriver becomes worn or broken, it should be redressed to shape (Fig. 22-8).

REGRINDING A STANDARD SCREWDRIVER BLADE

When regrinding a screwdriver tip, make the sides of the blade slightly concave by holding the side of the blade tangential to the periphery of the grinding wheel (Fig. 22-8). Grind an equal amount off each side of the blade. This shape will enable the blade to maintain a better grip in the slot. Be sure to retain the original taper, width, and thickness of the tip and grind the end square with the centerline of the blade.

NOTE: When grinding, remove a minimum amount of metal so as not to grind past the hardened zone in the tip. Quench the tip frequently in cold water so as not to draw the temper from the blade.

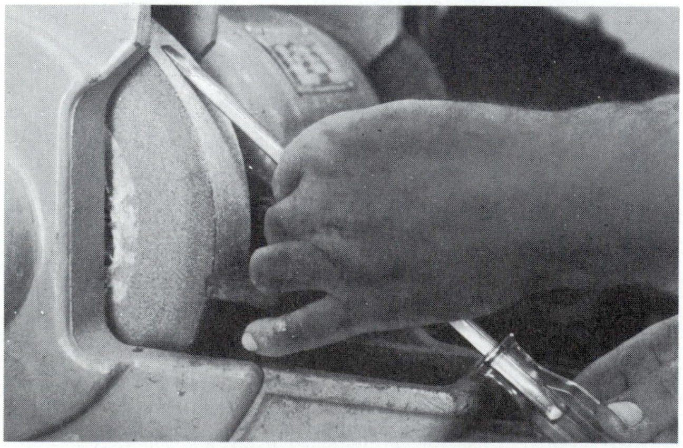

Figure 22-8 *Regrinding a standard screwdriver blade.*

WRENCHES

Many types of wrenches are used in machine shop work, each being suited for a specific purpose. The name of the wrench is derived from its use, its shape, or its construction. The following types of wrenches are commonly used in a machine shop.

Open-end wrenches may be single-ended (Fig. 22-9) or double-ended (Fig. 22-10). The openings on these wrenches are usually offset at a 15° angle to permit turning the nut or bolt head in limited spaces by "flopping" the wrench.

Double-ended wrenches usually have a different-sized opening at each end to accommodate two different sizes of bolt heads or nuts. These wrenches are available in both inch and metric sizes.

Box-end 12-point wrenches (Fig. 22-11) completely surround the nut and are useful in close quarters where only a small rotation of the nut can be obtained at one time. The box end has 12 precisely cut notches around the inside face, which fit closely over the points on the outside of the nut. Because this wrench cannot slip when the proper size is used, it is preferred over most other styles of wrenches. These wrenches usually have a different size at each end and are available in inch and metric sizes.

Socket wrenches (Fig. 22-12) are similar to box wrenches in that they are usually made with 12 points and surround the nut. These sockets are also available in inch and metric sizes. Several types of drives, including ratchet and torque-wrench handles, are available for the various sockets. When nuts or bolts must be tightened to within certain limits to prevent warping, socket wrenches are used in conjunction with a torque-wrench handle.

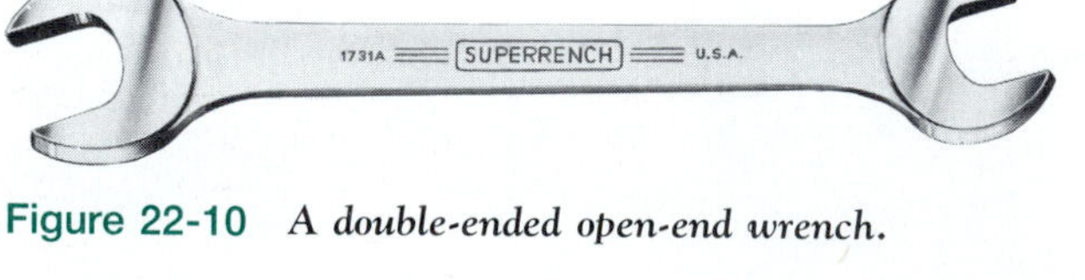
1704
SUPERRENCH
U.S.A.

Figure 22-9 *A single-ended open-end wrench.*

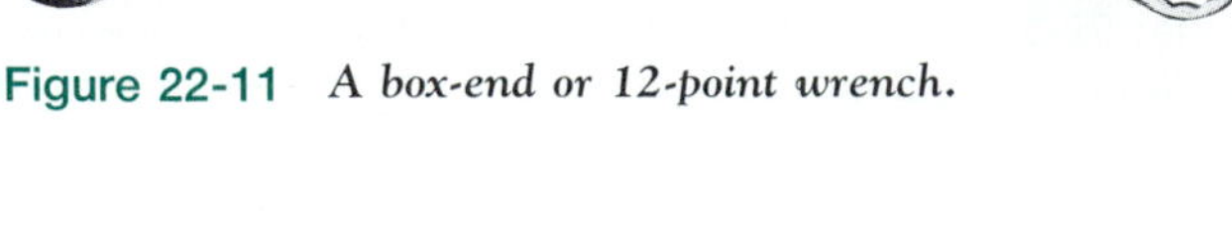
1731A
SUPERRENCH
U.S.A.

Figure 22-10 *A double-ended open-end wrench.*

Figure 22-11 *A box-end or 12-point wrench.*

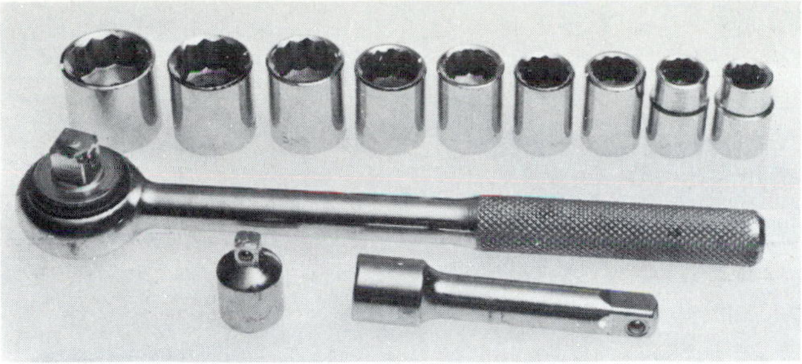

Figure 22-12 *A set of socket wrenches.*

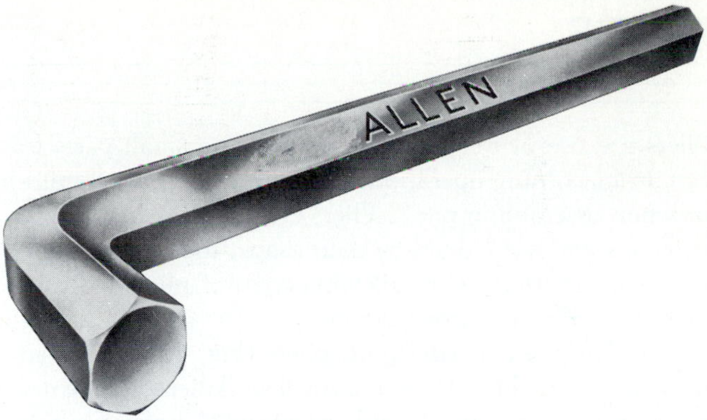
ALLEN

Figure 22-14 *An Allen wrench.*

Adjustable wrenches (Fig. 22-13) may be adjusted to within a certain range to fit several sizes of nuts or bolt heads. This wrench is particularly useful for odd-size nuts or when another wrench of the proper size is not available. Unfortunately, this type of wrench can slip when not properly adjusted to the flats of the nut. This may result in injury to the operator and damage to the corners of the nut.

When using an adjustable wrench, it should be tightened securely to the faces of the nut and the turning force applied in the direction indicated in Fig. 22-13.

Allen setscrew wrenches (Fig. 22-14), commonly called *Allen wrenches,* are hexagonal and fit into the recesses of socket head setscrews. They are made of tool steel and are available in sets to fit a wide variety of screw sizes. The indicated size of the wrench is the distance across the flats of the wrench. Usually this distance is one-half the outside diameter of the Allen setscrew in which it is used. These wrenches are available in both inch and metric sizes.

Pin spanner wrenches are specialized wrenches generally supplied by the machine tool manufacturer for use on specific machines. They are supplied in various types. Fixed-face spanners and adjustable-face spanners (Fig. 22-15A) are positioned in two holes on the face of a special nut or threaded fitting on a machine.

A *hook-pin spanner* (Fig. 22-15B) is used on the circumference of a round nut, the pin of the spanner fitting into a hole in the periphery of the nut.

HINTS ON USING WRENCHES

1. Always select a wrench which fits the nut or bolt properly. A wrench that is too large may slip off the nut and possibly cause an accident.
2. Whenever possible, *pull* rather than push on a wrench in order to avoid injury if the wrench should slip.
3. Always be sure that the nut is fully seated in the wrench jaw.
4. Use a wrench in the same plane as the nut or bolt head.
5. When tightening or loosening a nut, give it a sharp quick jerk, which is more effective than a steady pull.
6. Put a drop of oil on the threads when assembling a bolt and nut to ensure easier removal later.

Figure 22-13 *The correct method of using an adjustable wrench.*

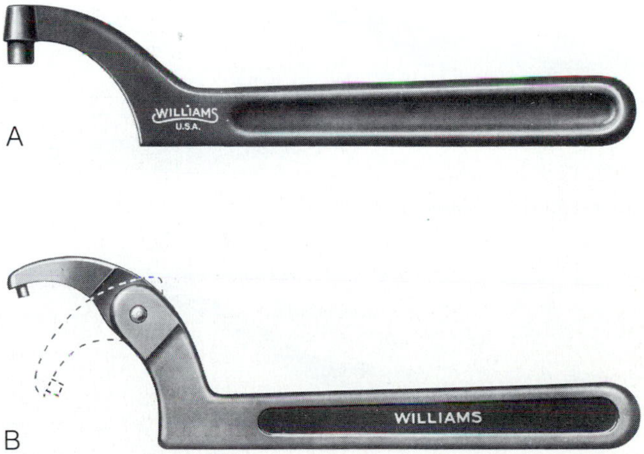
A
WILLIAMS
U.S.A.
B
WILLIAMS

Figure 22-15 *(A) A pin spanner wrench; (B) a hook-pin spanner wrench.*

PLIERS

Pliers are useful for gripping and holding small parts for certain machining operations (such as drilling small holes) or when assembling parts. Pliers are made in many types and sizes and are named by their shape, their function, or their construction. The following types of pliers are commonly used in a machine shop.

Combination, or *slip-joint, pliers* (Fig. 22-16) are adjustable, to grip both large and small workpieces. They may be used to grip certain work when small holes must be drilled or for bending or twisting light, thin materials.

Side-cutting pliers (Fig. 22-17) are used mainly for cutting, gripping, and bending of small-diameter (1/8 in. or less) rods or wire.

Needle-nose pliers (Fig. 22-18) are available in both straight- and bent-nose types. They are useful for holding very small parts, for positioning them in hard-to-get-at places, and for bending or forming wire.

Diagonal cutters (Fig. 22-19) are used solely for cutting wire and small pieces of soft metal.

Vise-grip pliers (Fig. 22-20) provide extremely high gripping power because of the adjustable lever action. The screw in the handle allows adjustment to various sizes. This type of plier is available in several different styles, such as standard jaws, needle jaws, and C-clamp jaws.

Figure 22-16 *Slip-joint or combination pliers.*

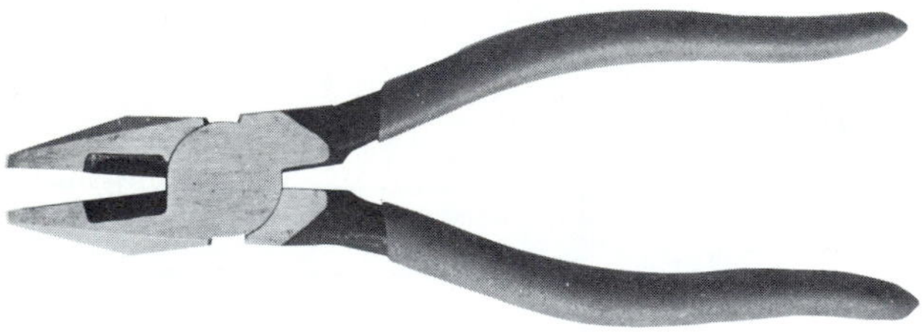

Figure 22-17 *Side-cutting pliers.*

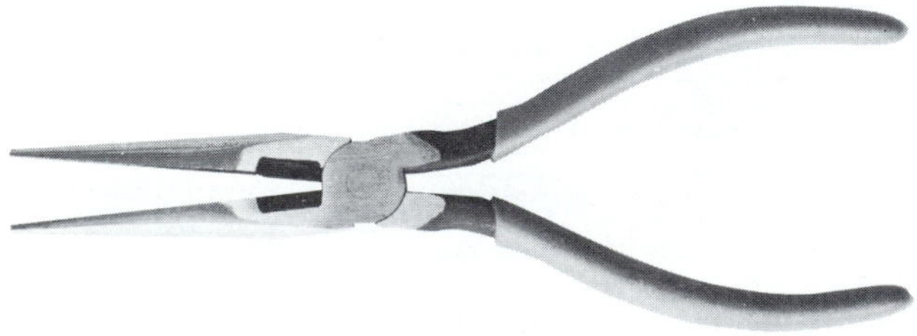

Figure 22-18 *Needle-nose pliers.*

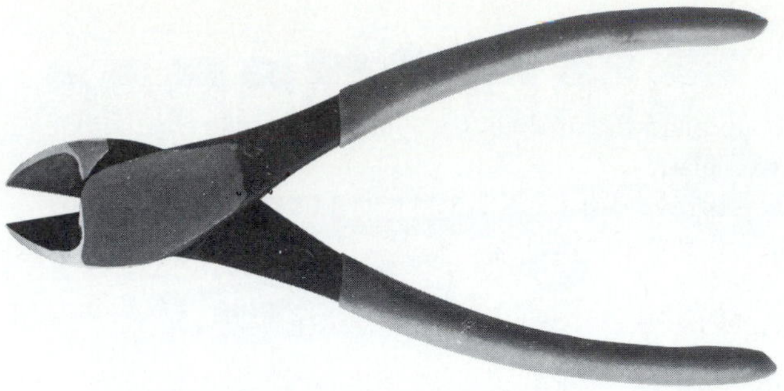

Figure 22-19 *Diagonal cutters.*

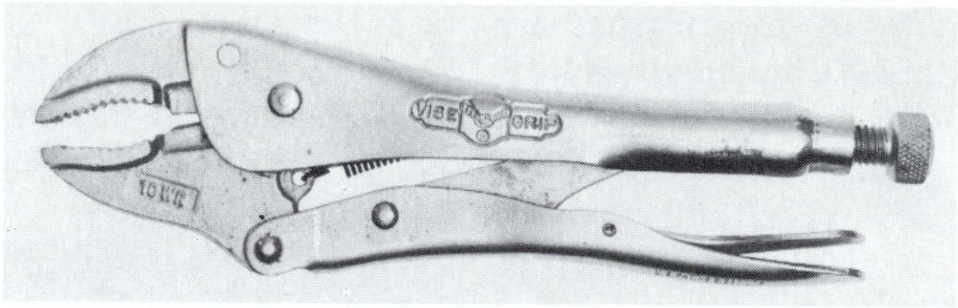

Figure 22-20 *Vise-grip pliers.*

HINTS ON USING PLIERS

The following points should be observed if pliers are to give the proper service.

1. Never use a plier instead of a wrench.
2. Never attempt to cut large-diameter or heat-treated material with pliers. This may cause the jaws to distort or the handle to break.
3. Always keep pliers clean and lubricated.

REVIEW QUESTIONS

THE BENCH VISE

1. What is the advantage of the swivel-base vise over the solid-base vise?
2. How may finished work be held in a vise without the surface being marred or damaged?

HAMMERS

3. Describe the most common hammer used by a machinist.
4. For what purpose are soft-faced hammers used?
5. State three safety rules which should be observed when using a hammer.

SCREWDRIVERS

6. List three important ways to take care of a screwdriver.
7. Explain the procedure for regrinding the tip of a screwdriver blade.
8. List two precautions that should be observed in using a Phillips screwdriver.

WRENCHES

9. Why are open-end wrenches offset about 15° to the handle?
10. Why is a properly sized box wrench preferred to other types of wrenches?
11. What advantage does a socket wrench have over a box wrench?
12. What precaution should be observed when using an adjustable wrench?
13. What will happen if excess pressure is applied to an adjustable wrench or pressure applied on the wrong jaw?
14. What is the cross-sectional shape of an Allen wrench and for what purpose is an Allen wrench used?
15. Where is a hook-pin spanner used?
16. State four hints which you feel are useful when using any wrench.

PLIERS

17. Name four types of pliers and state one use for each.
18. What advantage do vise grips have over other types of pliers?

UNIT 23

Hand-Type Cutting Tools

Although most metal cutting can be done more easily, quickly, and accurately on a machine, it is often necessary to perform certain metal-cutting operations at a bench or on a job. Such operations include sawing, filing, scraping, reaming, and tapping. It is therefore important that the prospective machinist knows how to use hand-type cutting tools properly.

OBJECTIVES

After completing this unit you will be able to:

1. Select and use the proper hacksaw blade for sawing a variety of materials
2. Select and use a variety of files to perform various filing operations
3. Identify and know the purpose of rotary files, ground burrs, and scrapers

SAWING, FILING, AND SCRAPING

Hacksaws, files, and scrapers are very common tools in the machine shop and often the most incorrectly used and abused. The proper use of these tools will not come immediately. It is only through practice that the student or apprentice will become proficient in their use.

THE HAND HACKSAW

The *pistol-grip hand hacksaw* (Fig. 23-1) is composed of three main parts: the *frame*, the *handle*, and the *blade*. The frame can be either solid or adjustable. The solid frame is more rigid and will accommodate blades of only one specific length. The adjustable frame is more commonly used

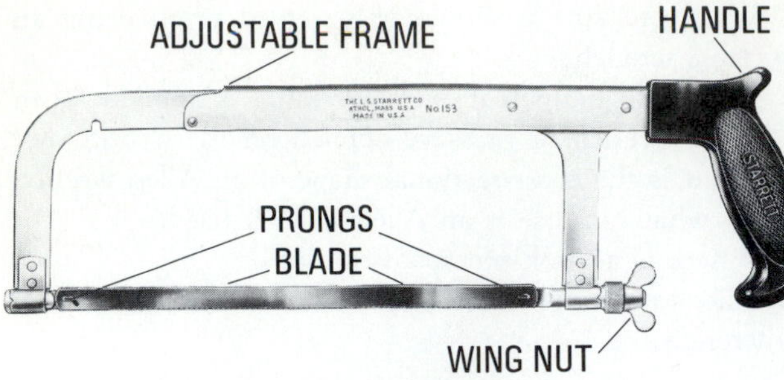
ADJUSTABLE FRAME
HANDLE
PRONGS
BLADE
WING NUT

Figure 23-1 *Parts of a hand hacksaw.*

and will take blades which range from 10 to 12 in. (250 to 300 mm) long. A wing nut at the back of the frame provides adjustment for blade tensioning.

Hacksaw blades are made of high-speed molybdenum or tungsten-alloy steel that has been hardened and tempered. There are two types: the solid, or all-hard, blade and the flexible blade. Solid blades are hardened throughout and are very brittle. They break easily if not used properly. Only the teeth of the flexible blade are hardened, while the back of the blade is soft and flexible. Although this type of blade will stand more abuse than the all-hard blade, it will not last as long in general use.

Solid blades are usually used on brass, tool steel, cast iron, and larger sections of mild steel since they do not run out of line when pressure is applied. Flexible blades may be used on channel iron, tubing, copper, and aluminum since they do not break as easily on material with thin cross sections.

Blades are manufactured in various pitches (number of teeth per inch), such as 14, 18, 24, and 32. The pitch is the most important factor to consider when selecting the proper blade for a job. An 18-tooth blade (18 teeth per inch) is recommended for general use. When selecting a blade, choose as coarse a blade as possible in order to provide plenty of chip clearance and to cut through the work as quickly as possible. The blade selected should have at least *two teeth in contact with the work* at all times. This will prevent the work from jamming between the teeth and stripping the teeth from the blade. Figure 23-2 provides a guide for proper blade selection.

TO USE THE HAND HACKSAW

1. Check to make sure that the blade is of the proper pitch for the job and that the teeth point *away* from the handle.
2. Adjust the blade tension so that the blade cannot flex or bend.
3. Mount the stock in the vise so that the cut will be about ¼ in. (6 mm) from the vise jaws.
4. Grasp the hacksaw as shown in Fig. 23-3. Assume a comfortable stance, standing erect with the left foot slightly ahead of the right foot.
5. Start the saw cut just outside and parallel to a previously scribed line.

NOTE: File a V-shaped nick at the starting point to help start the saw blade at the right spot.

6. After the cut has started, apply pressure only on the forward stroke. Use about 50 strokes per minute.
7. When cutting thin material, hold the saw at an angle to

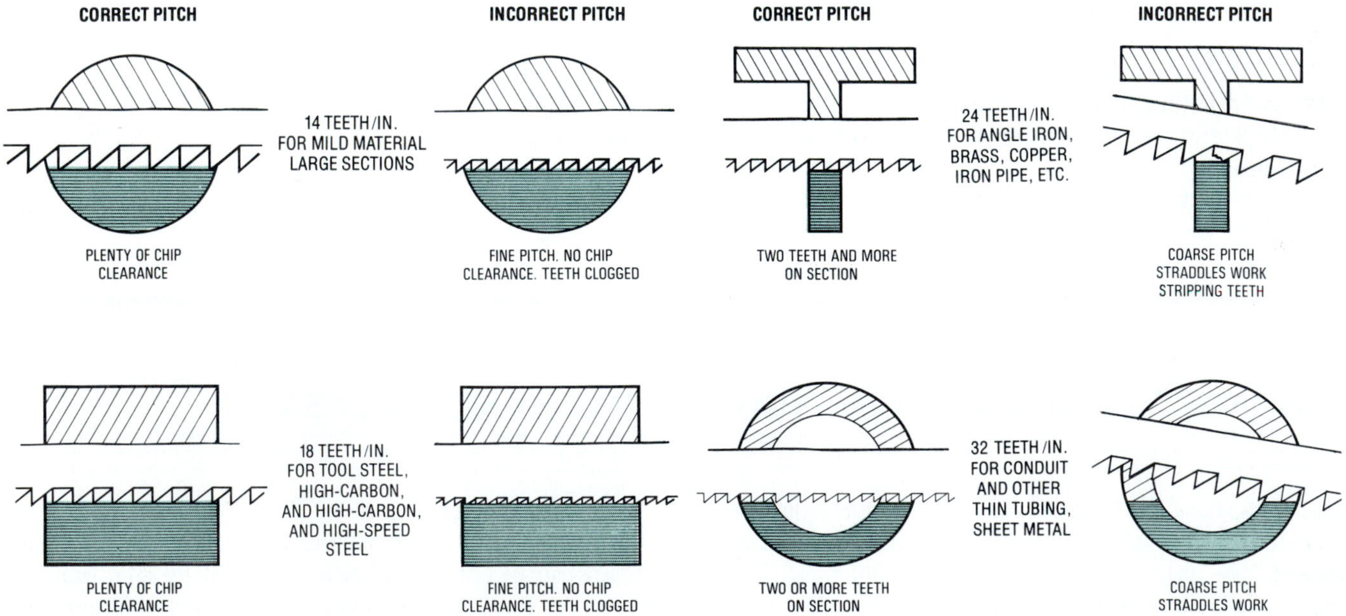
CORRECT PITCH
INCORRECT PITCH
CORRECT PITCH
INCORRECT PITCH
14 TEETH/IN.
FOR MILD MATERIAL
LARGE SECTIONS
24 TEETH/IN.
FOR ANGLE IRON,
BRASS, COPPER,
IRON PIPE, ETC.
PLENTY OF CHIP
CLEARANCE
FINE PITCH. NO CHIP
CLEARANCE. TEETH CLOGGED
TWO TEETH AND MORE
ON SECTION
COARSE PITCH
STRADDLES WORK
STRIPPING TEETH
18 TEETH/IN.
FOR TOOL STEEL,
HIGH-CARBON,
AND HIGH-CARBON,
AND HIGH-SPEED
STEEL
32 TEETH/IN.
FOR CONDUIT
AND OTHER
THIN TUBING,
SHEET METAL
PLENTY OF CHIP
CLEARANCE
FINE PITCH. NO CHIP
CLEARANCE. TEETH CLOGGED
TWO OR MORE TEETH
ON SECTION
COARSE PITCH
STRADDLES WORK

Figure 23-2 *Selection of the proper blade pitch is very important.*

WOOD PIECES
THIN METAL
Figure 23-3 *The correct method of holding a hacksaw.*

have at least two teeth in contact with the work at all times. Thin work is often clamped between two pieces of wood, and the cut is made through all pieces (Fig. 23-3).

8. When nearing the end of the cut, slow down to control the saw as it breaks through the material.

NOTE: If a saw blade breaks or becomes dull in a partly finished cut, replace the blade and rotate the work one-half turn so that the old cut is at the bottom. A new blade will bind in an old cut and the "set" of the new teeth will be ruined quickly.

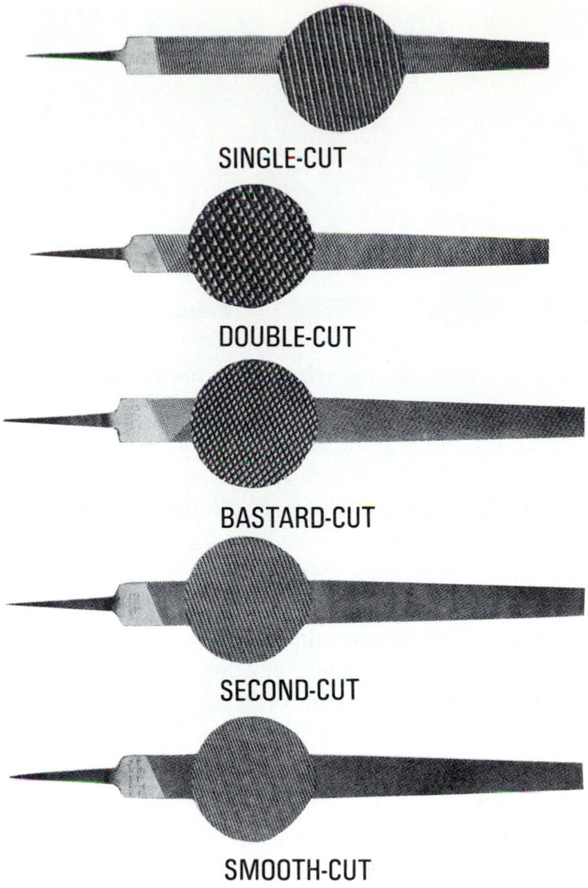
SINGLE-CUT
DOUBLE-CUT
BASTARD-CUT
SECOND-CUT
SMOOTH-CUT
Figure 23-5 *File classification.*

FILES

A file is a hand cutting tool made of high-carbon steel, having a series of teeth cut on its body by parallel chisel cuts. The parts of a file are shown in Fig. 23-4. Files are used to remove surplus metal and to produce finished surfaces. Files are manufactured in a variety of types and shapes, each for a specific purpose. They may be divided into two classes: single cut and double cut (Fig. 23-5).

Single-cut files have a single row of parallel teeth running diagonally across the face. They include *mill, long-angle lathe,* and *saw files.* Single-cut files are used when a smooth finish is desired or when hard materials are to be finished.

Double-cut files have two intersecting rows of teeth. The first row is usually coarser and is called the *overcut.* The other row is called the *upcut.* These intersecting rows produce hundreds of cutting teeth, which provide for fast removal of metal and easy clearing of chips.

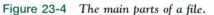

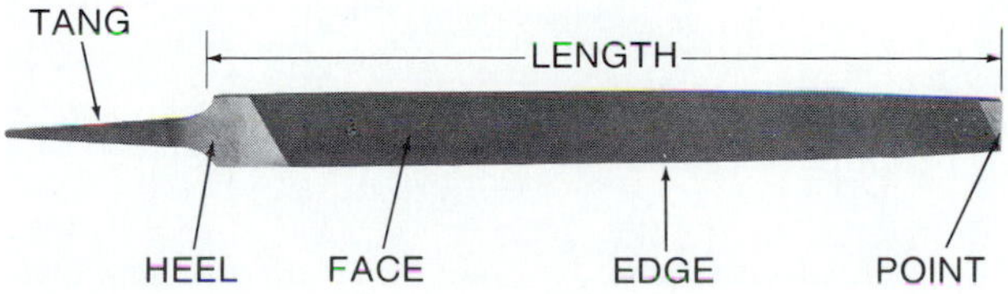
TANG
LENGTH
HEEL
FACE
EDGE
POINT
Figure 23-4 *The main parts of a file.*

DEGREES OF COARSENESS

Both single- and double-cut files are manufactured in various degrees of coarseness, such as *rough, coarse, bastard, second-cut, smooth,* and *dead smooth.* Those most commonly used by the machinist are the bastard, second cut, and smooth (Fig 23-5).

MACHINIST FILES

The types of files most commonly used by machinists are the *flat, hand, round, half-round, square, pillar, three-quarter* (triangular), *warding,* and *knife* (Fig. 23-6).

Care of Files

Because files are relatively inexpensive hand tools, they are often abused. Proper care, selection, and use are most important if good results are to be obtained with files. The following points should be observed in the care of files.

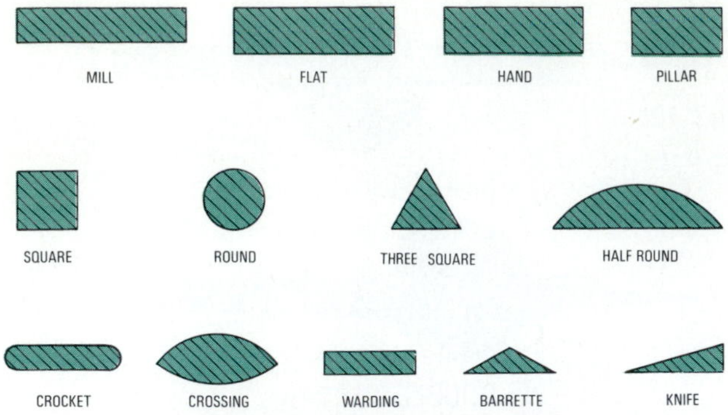
MILL
FLAT
HAND
PILLAR
SQUARE
ROUND
THREE SQUARE
HALF ROUND
CROCKET
CROSSING
WARDING
BARRETTE
KNIFE
Cross-sectional views of machinists' files.

1. Do not store files where they will rub together. Hang or store them separately.
2. Never use a file as a pry or a hammer. Since the file is hard, it snaps easily, often causing small pieces to fly, which may result in a serious eye injury.
3. Do not knock a file on a vise or other metallic object to clean it. Always use a file card or brush for this purpose (Fig. 23-7).
4. Apply pressure only on the forward stroke when filing. Pressure on the return stroke will dull the file.
5. Do not press too hard on a new file. Too much pressure tends to break off the cutting edges and shorten the life of the file.
6. Too much pressure also results in "pinning" (*small particles being wedged between the teeth*), which causes scratches on the surface of the work. Keep the file clean. A piece of brass, copper, or wood pushed through the teeth will remove the "pins." Applying chalk to the face of the file will lessen the tendency for the file to become clogged (pinned).

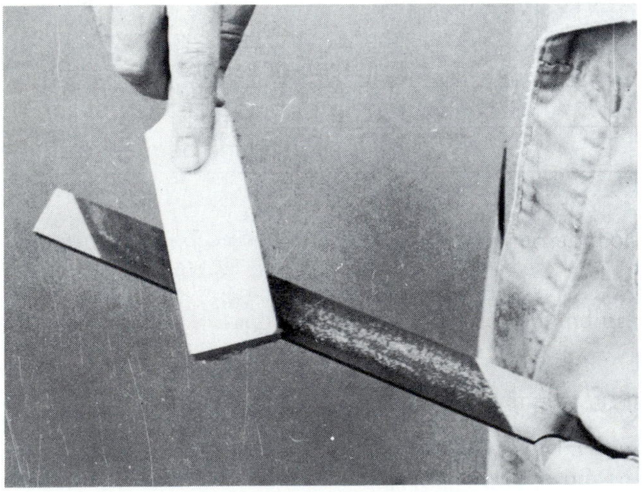
Figure 23-7 *Cleaning a file with a file brush.*

FILING PRACTICE

Filing is an important hand operation and one that can be mastered only through patience and practice. The following points should be observed when cross filing.

1. *Never* use a file without a handle. Ignoring this rule is a dangerous practice. Serious hand injury may result should the file slip.
2. Fasten the work to be filed, at about elbow height, in a vise.
3. To produce a flat surface, hold the right hand, right forearm, and left hand in a horizontal plane (Fig. 23-8). Push the file across the work face in a straight line and do not rock the file.
4. Apply pressure only on the forward stroke.
5. Never rub the fingers or hand across a surface being filed. Grease or oil from the hand causes the file to slide over instead of cutting the work. Oil will also clog the file.
6. Keep the file clean by using a file card frequently.

For rough filing, use a double-cut file and cross the stroke at regular intervals to help keep the surface flat and straight (Fig. 23-9). When finishing, use a single-cut file and take shorter strokes to keep the file flat.

Test the work for flatness occasionally by laying the edge of a steel rule across its surface. Use a steel square to test the squareness of one surface to another.

DRAW FILING

Draw filing is used to produce a smooth, flat surface on the workpiece. This method of filing removes file marks and scratches left by cross filing.

NOTE: When draw filing, hold the file as shown in Fig. 23-10 and move the file back and forth along the length of the work.

POLISHING

After a surface has been filed, it may be finished with abrasive cloth to remove small scratches left by the file. This

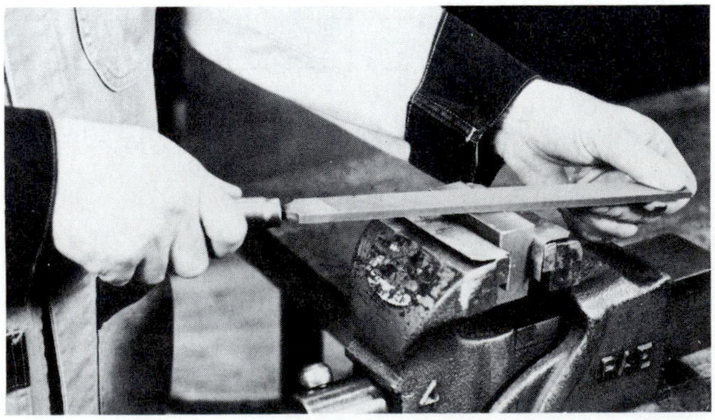
Figure 23-8 *Always hold a file level when filing.*

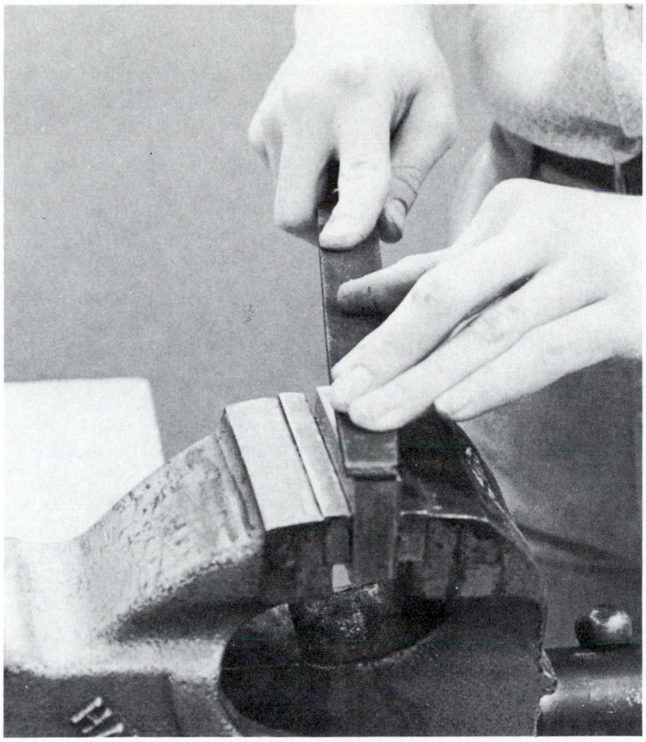

Figure 23-9 *Cross-filing will show any high spots on the workpiece surface.*

may be done with a piece of abrasive cloth held under the file, which is moved back and forth along the work.

SPECIAL FILES

Long-angle lathe files are used for filing on a lathe because they provide a better shearing action than mill files. The long angle of the teeth tends to clean the file, helps eliminate chatter, and reduces the possibility of tearing the metal.

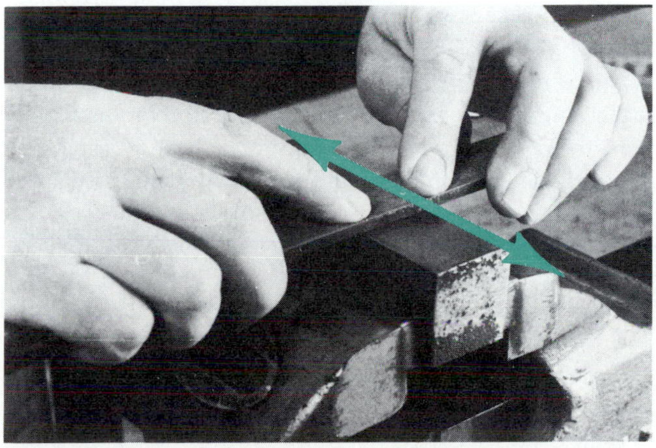

Figure 23-10 *Draw filing is used to produce a flat, smooth surface.*

Aluminum files are designed for soft, ductile metals, such as aluminum and white metal, because regular files tend to clog quickly when used on this type of material. The modified tooth construction on aluminum files tends to reduce clogging. The upcut tooth is deep, and the overcut is fine. This produces small scallops on the upcut which breaks up the chips and permits them to clear more easily.

Brass files have a small upcut angle and a fine, long-angle overcut which produces small, easily cleared chips. The almost straight upcut prevents grooving the surface of the work.

Shear tooth files combine a long angle and a single-cut coarse tooth for filing materials such as brass, aluminum, copper, plastics, and hard rubber.

PRECISION FILES

Precision files include swiss pattern, needle, and riffler files. *Swiss pattern* and *needle files* (Fig. 23-11) are small files having very fine tooth cuts and round integral handles. They are made in several shapes and are generally used in tool and die shops for finishing delicate and intricate pieces. *Die sinker rifflers* are curved up at the ends to permit filing the bottom surface of a die cavity.

ROTARY FILES AND BURRS

The increased use of portable electric and pneumatic power tools has developed the widespread application of rotary files and burrs. The wide range of shapes and sizes available makes these tools particularly suitable for metal pattern making and die sinking.

Rotary file teeth (Fig. 23-12) are cut and form broken lines in contrast to the unbroken flutes of the ground burr (Fig. 23-13). The teeth of the rotary file tend to dissipate the heat of friction, a feature which makes this tool particularly useful for work on tough die steels, forgings, and scaly surfaces.

Ground burrs (Fig. 23-13) may be made of high-speed steel or carbide. The flutes of a burr are generally machine ground to a master burr to ensure uniformity of tooth shape and size. Ground high-speed burrs are used more efficiently on nonferrous metals, such as aluminum, brass, bronze, and magnesium, since they have better chip clearance than rotary files.

Carbide burrs may be used on hard or soft materials with good results and will last up to 100 times longer than a high-speed steel burr.

USING HIGH-SPEED STEEL ROTARY FILES AND BURRS

For best results, rotary files or burrs should be used in the following manner.

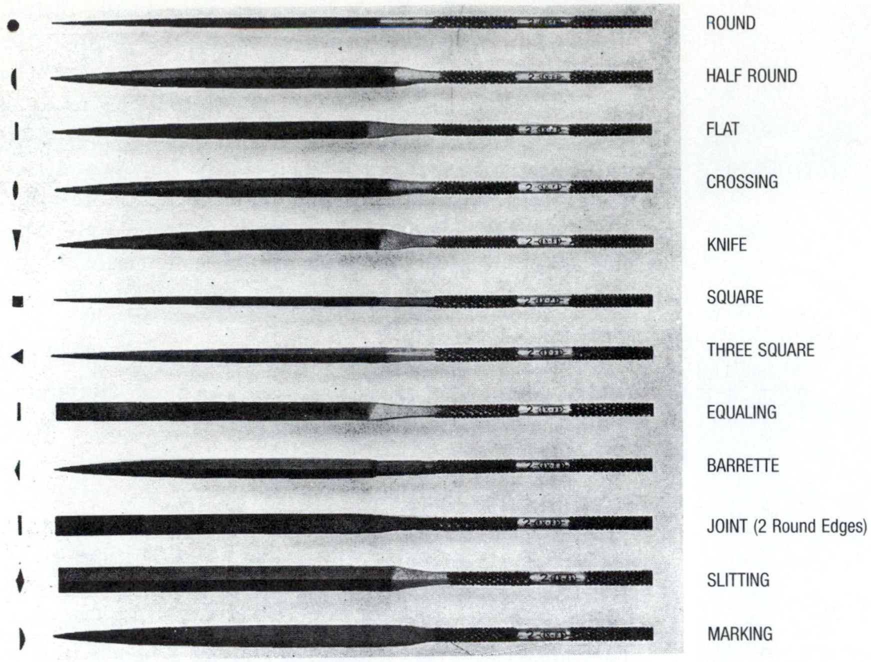
ROUND
HALF ROUND
FLAT
CROSSING
KNIFE
SQUARE
THREE SQUARE
EQUALING
BARRETTE
JOINT (2 Round Edges)
SLITTING
MARKING

Figure 23-11 *Needle files are used for intricate work. (Courtesy Nicholson File Co. of Canada Ltd.)*

1. Move the file or burr at an even rate to produce a smooth surface. An uneven rate of pressure produces surfaces with ridges and hollows.
2. Use the proper speed for the burr diameter or file as recommended by the manufacturer.
3. Use only sharp burrs or files.
4. For more accurate control of the burr or file, grip the grinder as close as possible to its end.
5. Medium-cut burrs and files generally provide satisfactory metal removal and finish for most jobs. If greater stock removal is required, use a coarse burr or file. For an extra-smooth finish, use a fine burr or file.

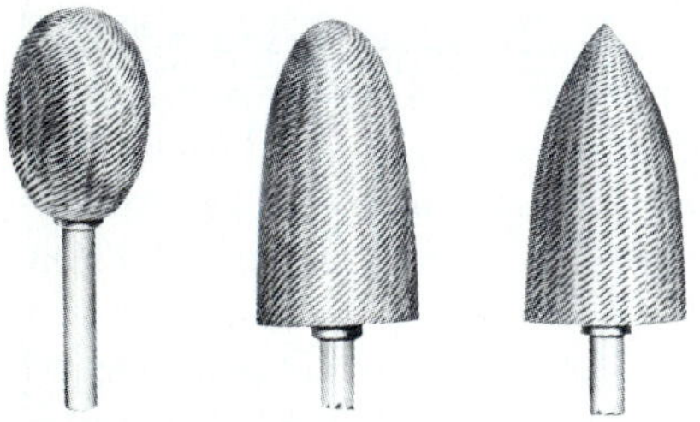

Figure 23-12 *The broken-line teeth of rotary files tend to dissipate heat quickly. (Courtesy Nicholson File Co. of Canada Ltd.)*

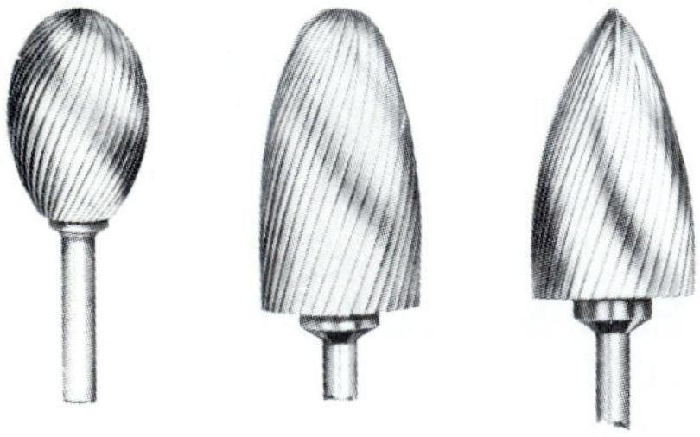

Figure 23-13 *Ground burrs are used on nonferrous metals. (Courtesy Nicholson File Co. of Canada Ltd.)*

SCRAPERS

When a truer surface is required than can be produced by machining, the surface may be finished by scraping. However, this is a long and tedious process. Most bearing surfaces (flat and curved) are now finished by grinding, honing, or broaching.

Scraping is a process of removing small amounts of metal from specific areas to produce an accurate bearing surface. It is used to produce flat surfaces or in fitting brass and babbitt bearings to shafts.

Scrapers are made in various shapes, depending on the surface to be scraped (Fig. 23-14). They are generally made of high-grade tool steel, hardened and tempered. Carbide-tipped scrapers are very popular because they maintain the cutting edge longer than other types.

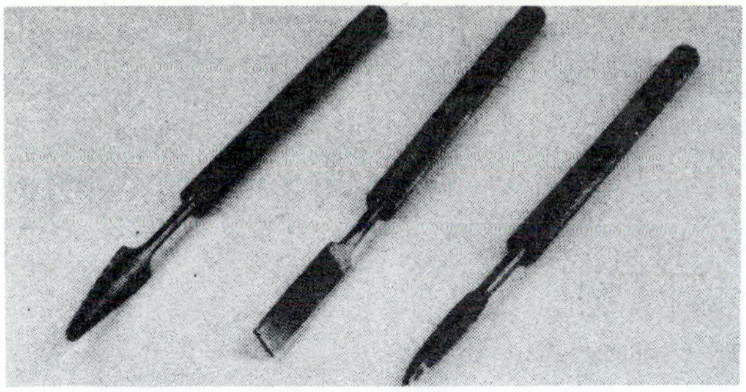

Figure 23-14 *A set of hand scrapers.*

REVIEW QUESTIONS

THE HAND HACKSAW

1. Compare the flexible blade and the solid, or all-hard, hacksaw blade.
2. What pitch hacksaw blade should be selected to cut:
   a. Tool steel?
   b. Thin-wall tubing?
   c. Angle iron and copper?
3. What procedure is recommended if a saw blade breaks or becomes dull in a partially finished cut?

FILES

4. Describe and state the purpose of:
   a. Single-cut files
   b. Double-cut files
5. Name the most commonly used degrees of coarseness in which files are manufactured.
6. List four important aspects of file care.
7. How can pinning of a file be kept to a minimum?
8. Describe and state the purpose of:
   a. Long-angle lathe files
   b. Aluminum files
   c. Shear tooth files
9. Describe and state the purpose of:
   a. Swiss pattern files
   b. Die sinker rifflers
10. Compare rotary files and ground burrs.
11. List three important considerations in the use of rotary files or ground burrs.
12. Describe the correct procedure for filing a flat surface.

UNIT 24

Thread-Cutting Tools and Procedures

Threads may be cut internally using a tap and externally using a die. The proper selection and use of these threading tools is an important part of machine shop work.

OBJECTIVES

After completing this unit you will be able to:

1. Calculate the tap drill size for inch and metric taps
2. Cut internal threads using a variety of taps
3. Know the methods used to remove broken taps from a hole
4. Cut external threads using a variety of dies

HAND TAPS

Taps are cutting tools used to cut internal threads. They are made from high-quality tool steel, hardened and ground. Two, three, or four flutes are cut lengthwise across the threads to form cutting edges, provide room for the chips, and admit cutting fluid to lubricate the tap. The end of the shank is square so that a tap wrench (Fig. 24-1A, B) can be used to turn the tap into a hole. For inch taps, the major diameter, number of threads per inch, and type of thread are usually found stamped on the shank of a tap. For example, ½ in.—13 UNC represents:

½ in. = major diameter of the tap
13 = number of threads per inch
UNC = Unified National Coarse (a type of thread)

Hand taps are usually made in sets of three: *taper, plug,* and *bottoming taps* (Fig. 24-2A).

A *taper tap* is tapered from the end approximately six threads and is used to start a thread easily. It can be used for tapping a hole which goes *through* the work, as well as for starting a *blind* hole (one that does not go all the way through).

A *plug tap* is tapered for approximately three threads. Sometimes the plug tap is the only tap used to thread a hole going through a workpiece.

A *bottoming tap* is not tapered but chamfered at the end for one thread. It is used for threading to the bottom of a blind hole. When tapping a blind hole, first use the taper tap, then the plug tap, and complete the hole with a bottoming tap.

Taps may also be identified by an annular ring or rings cut around the shank of the tap. One ring around the shank indicates that it is a taper tap, two rings indicate a plug tap, and three rings indicate a bottoming tap (Fig. 24-2B).

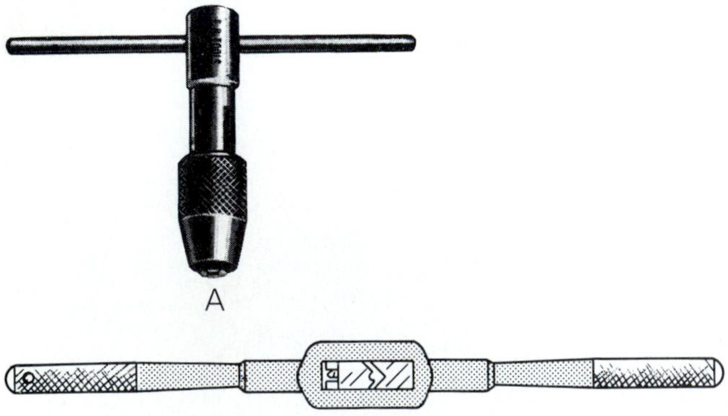
A
B

Figure 24-1 *(A) A T-handle tap wrench; (B) a double-ended adjustable tap wrench.*

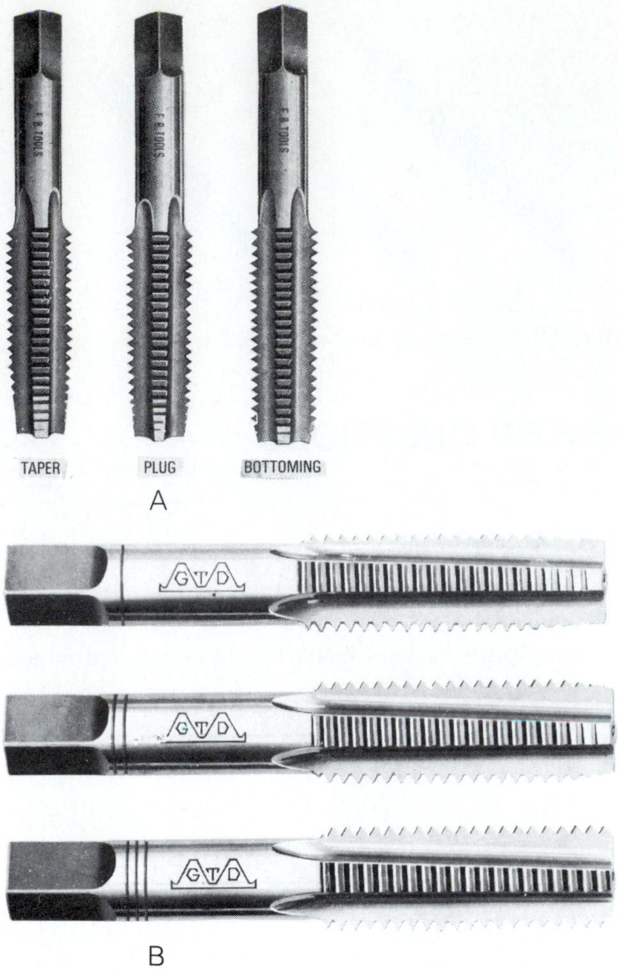
TAPER
PLUG
BOTTOMING
A
B

Figure 24-2 *(A) A set of hand taps; (B) taps may be identified by an annular ring on the shank.*

TAP DRILL SIZE

Before a tap is used, the hole must be drilled to the correct tap drill size (Fig. 24-3). This is the drill size that would leave the proper amount of material in the hole for a tap to cut a thread. The tap drill is always smaller than the tap and leaves enough material in the hole for the tap to produce 75 percent of a full thread.

When a chart is not available, the tap-drill size for any American, National, or Unified thread can be found easily by applying this simple formula: SEE PP 353-359 FOR DETAILS ON THREADS

$$TDS = D - \frac{1}{N}$$

where TDS = tap drill size
D = major diameter of tap
N = number of threads per inch

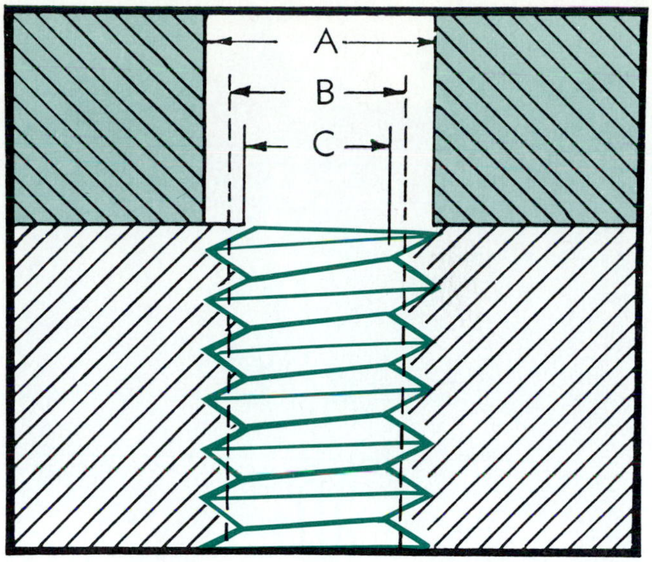
A
B
C

Figure 24-3 *Cross section of a tapped hole:* A = *Body size;* B = *tap drill size;* C = *minor diameter.*

EXAMPLE

Find the tap drill size for a 7/8 in.—9 NC tap.

$$\begin{aligned} TDS &= 7/8 - 1/9 \\ &= 0.875 - 0.111 \\ &= 0.764 \text{ in.} \end{aligned}$$

The nearest drill size to 0.764 in. is 0.765 in. (49/64). Therefore, 49/64 in. is the tap drill size for a 7/8 in.—9 NC tap.

METRIC TAPS

Although there are several thread forms and standards in the metric thread system, the International Standards Organization (ISO) has adopted a standard metric thread, which will be used in the United States, Canada, and many other countries throughout the world. This new series will have only 25 thread sizes, ranging from 1.6 to 100 mm diameter. See Table 6 in the Appendix for the size and pitch of the threads in this series. Also see Unit 55 for the thread form and dimensions of the ISO metric thread.

Like inch taps, metric taps are available in sets of three: *taper, plug,* and *bottoming taps.* They are identified by the letter M, followed by the nominal diameter of the thread in millimeters ~~times~~ MINUS the pitch in millimeters. Thus, a tap with the markings M 4—0.7 would indicate:

M—a metric thread
4—the nominal diameter of the thread in millimeters
0.7—the pitch of the thread in millimeters

TAP DRILL SIZES FOR METRIC TAPS

The tap drill size for metric taps is calculated in the same manner as for U.S. Standard threads.

$$TDS = \text{major diameter (mm)} - \text{pitch (mm)}$$

EXAMPLE

Find the tap drill size for a 22 − 2.5 mm thread.

$$\begin{aligned} TDS &= 22 - 2.5 \\ &= 19.5 \text{ mm} \end{aligned}$$

TAPPING A HOLE

Tapping is the operation of cutting an internal thread using a tap and tap wrench. Because taps are hard and brittle, they are easily broken. *Extreme* care must be used when tapping a hole to prevent breakage. A broken tap in a hole is difficult to remove and often results in scrapping the work.

To Tap a Hole by Hand

1. Select the correct taps and tap wrench for the job.
2. Apply a suitable cutting fluid to the tap.

NOTE: No cutting fluid is required for tapping brass or cast iron.

3. Place the tap in the hole as vertically as possible, press downward on the wrench, applying equal pressure on both handles, and turn clockwise (for right-hand thread) for about two turns.
4. Remove the tap wrench and check the tap for squareness.

NOTE: Check at two positions at 90° to each other (Fig. 24-4).

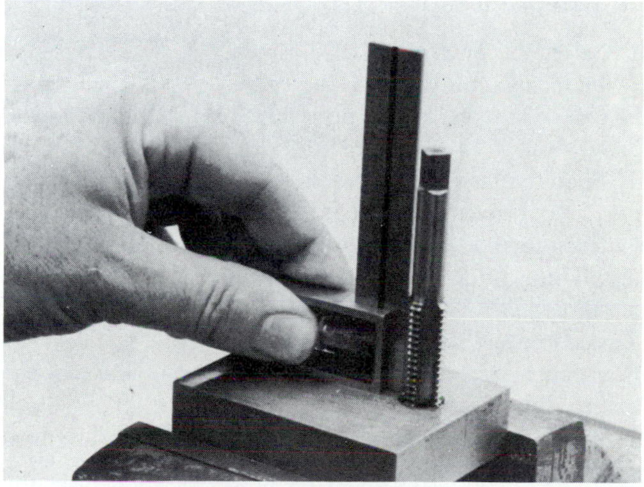

Figure 24-4 *Checking a tap for squareness while holding the workpiece in a vise.*

5. If the tap has not entered squarely, remove it from the hole and restart it by applying pressure in the direction from which the tap leans. *Be careful* not to exert too much pressure in the straightening process.
6. When a tap has been properly started, feed it into the hole by turning the tap wrench.
7. Turn the tap clockwise one-quarter turn, and then turn it backward about one-half turn to break the chip. Turning must be done with a steady motion to prevent the tap from breaking.

NOTE: When tapping blind holes, use all three taps in order: taper, plug, and then the bottoming tap. Before using the bottoming tap, remove all the chips from the hole and be careful not to hit the bottom of the hole with the tap.

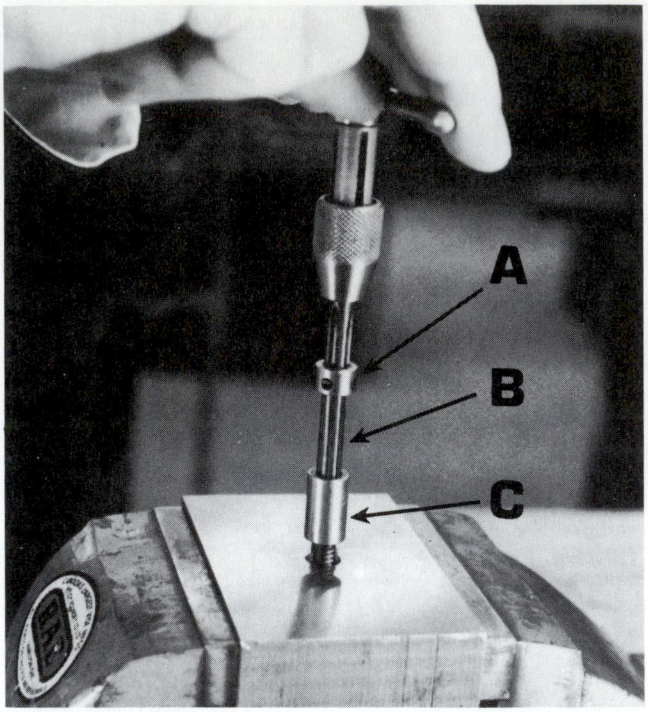

A
B
C

Figure 24-5 *A tap extractor being used to remove a broken tap.*

REMOVING BROKEN TAPS

If extreme care is not used when cutting a thread, particularly in a blind hole, the tap may break in the hole and considerable work will be required to remove it. In some cases it may not be possible to remove it and another piece of work must be started.

Several methods may be used to remove a broken tap; some may be successful, others may not.

Tap Extractor

The tap extractor is a tool (Fig. 24-5) that has four fingers that slip into the flutes of a broken tap. It is adjustable in order to support the fingers close to the broken tap, even when the broken end is below the surface of the work. A wrench is fitted to the extractor and turned counterclockwise to remove a right-hand tap. Tap extractors are made to fit all sizes of taps.

To Remove a Broken Tap Using a Tap Extractor

1. Select the proper extractor for the tap to be removed.
2. Slide *collar A,* to which the fingers are attached, down *body B* so that the fingers project well below the end of the body.
3. Slide the fingers into the flutes of the broken tap, making sure they go down into the hole as far as possible.
4. Slide the body down until it rests on top of the broken tap. This will give the maximum support to the fingers.
5. Slide *collar C* down until it rests on top of the work. This also provides support for the fingers.
6. Apply a wrench to the square section on the top of the body.
7. Turn the wrench *gently* in a counterclockwise direction.

NOTE: Do not force the extractor as this will damage the fingers. It may be necessary to turn the wrench back and forth carefully to free the tap sufficiently to back it out.

Drilling

If the broken tap is made of carbon steel, it may be possible to drill it out. Proceed as follows:

1. Heat the broken tap to a bright red color and allow it to cool *slowly.*
2. Center-punch the tap as close to the center as possible.
3. Using a drill considerably smaller than the distance between opposite flutes, proceed *carefully* to drill a hole through the broken tap.
4. Enlarge this hole to remove as much of the metal between the flutes as possible.
5. Collapse the remaining part with a punch and remove the pieces.

Acid Method

If the broken tap is made of high-speed steel and cannot be removed with a tap extractor, it is sometimes possible to remove it by the acid method. Proceed as follows:

1. Dilute one part nitric acid with five parts water.
2. Inject this mixture into the hole. The acid will act on the steel and loosen the tap.
3. Remove the tap with an extractor or a pair of pliers.
4. Wash the remaining acid from the thread with water so that the acid will not continue to act on the threads.

Tap Disintegrators

Taps may sometimes be removed successfully by a tap disintegrator, which may be held in the spindle of a drill press. The disintegrator uses the *electrical discharge principle* to cut its way through the tap, using a hollow brass tube as an electrode. Taps may also be removed using the same method on any electrical discharge machine (see Unit 92).

THREADING DIES

Threading dies are used to cut external threads on round work. The most common threading dies are the solid, adjustable split, and the adjustable and removable screw plate die.

The *solid die* (Fig. 24-6) is used for chasing or recutting damaged threads and may be driven by a suitable wrench. It is not adjustable.

The *adjustable split die* (Fig. 24-7) has an adjusting screw which permits an adjustment over or under the standard depth of thread. This type of die fits into a die stock (Fig. 24-8).

The *adjustable screw plate die* (Fig. 24-9) is probably a more efficient die since it provides for greater adjustment than the split die. Two die halves are held securely in a collet by means of a threaded plate, which also acts as a guide when threading. The plate, when tightened into the collet, forces the die halves with tapered sides into the tapered slot of the collet. Adjustment is provided by means of two adjusting screws which bear against each die half. The threaded section at the bottom of each die half is tapered to provide for easy starting of the die. Note that the upper side of each die half is stamped with the manufacturer's name, and the lower side of each is stamped with the same serial number. Care should be taken in assembling the die to make sure that both serial numbers are facing down. *Never* use two die halves with different serial numbers.

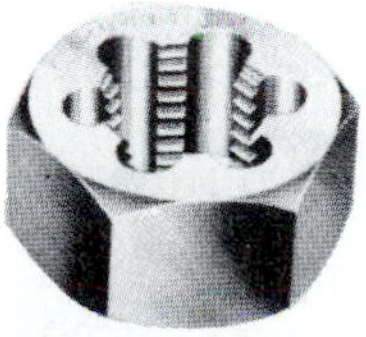

Figure 24-6 *A solid die nut.*

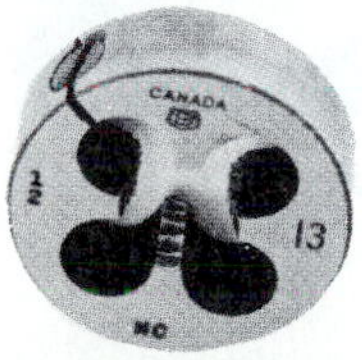
CANADA
13
NC

Figure 24-7 *An adjustable, round split die.*

BAY STATE

Figure 24-8 *A die stock is used to turn the die onto the workpiece.*

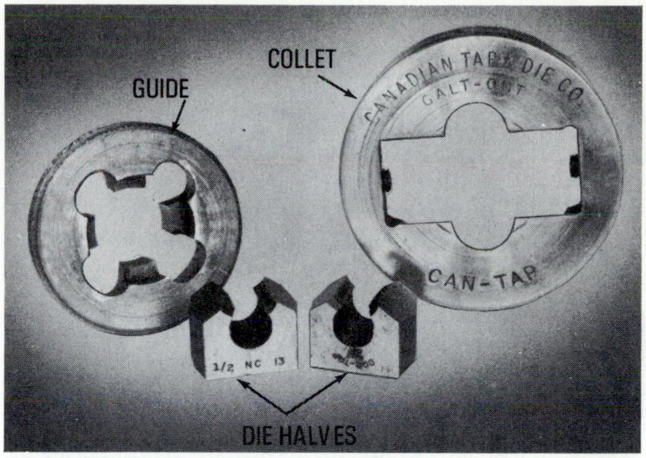
COLLET
GUIDE
CAN-TAP
DIE HALVES

Figure 24-9 *An adjustable screw plate die.*

TO THREAD WITH A HAND DIE

1. Chamfer the end of the workpiece with a file or on a grinder.
2. Fasten the work securely in a vise.
3. Select the proper die and die stock.
4. Lubricate the tapered end of the die with a suitable cutting lubricant.
5. Place the tapered end of the die squarely on the workpiece (Fig. 24-10).
6. Press down on the die stock handles and turn clockwise several turns.
7. Check the die to see that it has started squarely with the work.
8. If it is not square, remove the die from the work and restart it squarely.
9. Turn the die forward one turn and then reverse it approximately one-half turn to break the chip.
10. During the threading process apply cutting fluid frequently.

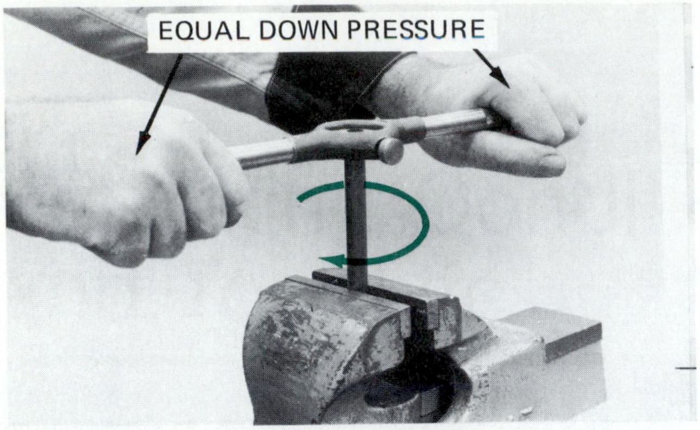
EQUAL DOWN PRESSURE
Figure 24-10 *Start the tapered end of the die on the work.*

CAUTION: When cutting a long thread, keep the arms and hands clear of the sharp threads coming through the die.

If the thread must be cut to a shoulder, remove the die and restart it with the tapered side of the die facing up. Complete the thread, being careful not to hit the shoulder, otherwise the work may be bent and the die broken.

REVIEW QUESTIONS

HAND TAPS

1. Name, describe, and state the purpose of the three taps in a set.
2. Define tap drill size.
3. Use the formula and calculate the tap drill size for:
   a. ½ in.—13 NC tap b. M 42—4.5 mm tap
4. Why should care be used when a hole is being tapped?
5. Explain the procedure for correcting a tap which has not started squarely.
6. Briefly outline the procedure for tapping a hole using the tap and reamer aligner.
7. Briefly explain the method of removing a broken tap using a tap extractor.

THREADING DIES

8. For what purpose are threading dies used?
9. State the purpose of the adjustable split die and the solid die.
10. Explain the procedure for starting a die on the work.
11. What procedure should be followed when it is necessary to cut a thread to a shoulder?

UNIT 25

Finishing Processes—Reaming, Broaching, and Lapping

Hand cutting tools are generally used to remove only small amounts of metal and are designed to do specific operations.

Reamers, available in a wide variety of types and sizes, are used to bring a hole to size and produce a good finish.

Broaches, when used in a machine shop, are generally used with an arbor press to produce special shapes in the workpiece. The broach, which is a multitooth cutting tool of the exact shape and size desired, is forced through a hole in the workpiece to reproduce its shape in the metal.

Lapping is a process whereby very fine abrasive powder, embedded in a proper tool, is used to remove minute amounts of material from a surface.

OBJECTIVES

After completing this unit you will be able to:

1. Identify and explain the purpose of several types of hand reamers
2. Ream a hole accurately with a hand reamer
3. Cut a keyway in a workpiece using a broach and arbor press
4. Lap a hole or an external diameter of a workpiece to size and finish

HAND REAMERS

A hand reamer is a tool used to finish drilled holes accurately and provide a good finish. Reaming is generally performed by machine, but there are times when a hand reamer must be used to finish a hole. Hand reamers, when used properly, will produce holes accurate to size, shape, and finish.

TYPES OF HAND REAMERS

The *solid hand reamer* (Fig. 25-1) may be made of carbon steel or high-speed steel. These straight reamers are available in inch sizes from $\frac{1}{8}$ to $1\frac{1}{2}$ in. in diameter and in metric sizes from 1 to 26 mm in diameter. For easy starting, the cutting end of the reamer is ground to a slight taper for a distance equal to the diameter of the reamer. Solid reamers are not adjustable and may have straight or helical flutes. Straight-fluted reamers should not be used on work with a keyway or any other interruption, since chatter and poor finish will result. Since hand reamers are designed to remove only small amounts of metal, no more than 0.005 in. or 0.12 mm, should be left for reaming, depending on the diameter of the hole. A square on the end of the shank provides the means of driving the reamer with a tap wrench.

The *expansion hand reamer* (Fig. 25-2) is designed to permit an adjustment of approximately 0.006 in. (0.15 mm) above the nominal diameter. The reamer is made hollow and has slots along the length of the cutting section. A tapered threaded plug fitted into the end of the reamer provides for limited expansion. If the reamer is expanded too much, it will break easily. For *inch expansion hand reamers,* the limit of adjustment is 0.006 in. over the nominal size on reamers up to $\frac{1}{2}$ in. and about 0.015 in. on reamers over $\frac{1}{2}$ in. *Metric expansion hand reamers* are available in sizes from 4 to 25 mm. The maximum amount of expansion on these reamers is 1 percent over the nominal size. For example, a 10-mm-diameter reamer can be expanded to 10.01 mm (10 + 1 percent). The cutting end of the reamer is ground to a slight taper for easy starting.

The *adjustable hand reamer* (Fig. 25-3) has tapered slots along the entire length of the body. The inner edges of the cutting blades have a corresponding taper so that the blades remain parallel for any setting. The blades are adjusted to size by upper and lower adjusting nuts.

The blades on inch adjustable hand reamers have an adjustment range of $\frac{1}{32}$ in. on the smaller reamers to almost $\frac{5}{16}$ in. on the larger ones. They are manufactured in sizes $\frac{1}{4}$ to 3 in. in diameter. *Metric adjustable hand reamers* are available in sizes from #000 (adjustable from 6.4 to 7.2 mm) to #16 (adjustable from 80 to 95 mm).

Taper reamers are made to standard tapers and are used to finish tapered holes accurately and smoothly. They may be made with either spiral or straight teeth. Because of its shearing action and its tendency to reduce chatter, the spiral-fluted reamer is superior to the straight one. A *roughing reamer* (Fig. 25-4), with nicks ground at intervals along the teeth, is used for more rapid removal of surplus metal. These nicks or grooves break up the chips into smaller sections; they prevent the tooth from cutting and overloading along its entire length. When a roughing

Figure 25-1 *A solid hand reamer. (Courtesy Cleveland Twist Drill Company.)*

Figure 25-2 *An expansion hand reamer.*

Figure 25-3 *An adjustable hand reamer.*

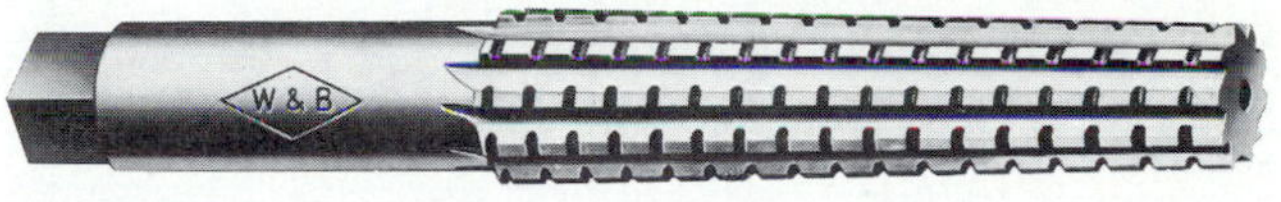
W&B

Figure 25-4 *A roughing taper reamer. (Courtesy Whitman and Barnes.)*

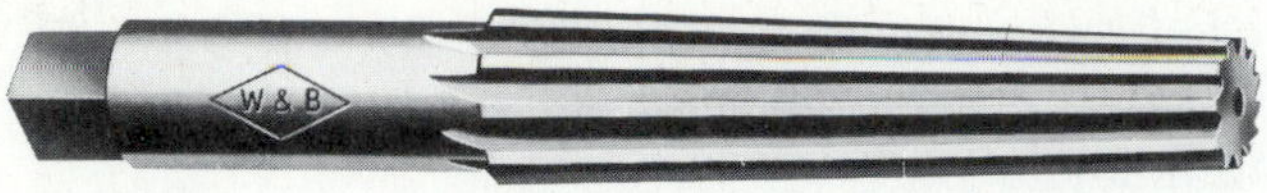
W & B

Figure 25-5 *A finishing taper reamer. (Courtesy Whitman and Barnes.)*

reamer is not available, an old taper reamer is often used before finishing the hole with a finishing reamer.

The *finishing taper reamer* (Fig. 25-5) is used after the roughing reamer to finish the hole smoothly and to size. This reamer, which has either straight or left-hand spiral flutes, is designed to remove only a small amount of metal [about 0.010 in. (0.25 mm)] from the hole. Since taper reamers do not clear themselves readily, they should be removed frequently from the hole and the chips cleared from their flutes.

REAMING PRECAUTIONS

1. Never turn a reamer backward (counterclockwise) as it will dull the cutting teeth.
2. Use a cutting lubricant where required.
3. Always use a helical-fluted reamer in a hole that has a keyway or oil groove cut in it.
4. Never attempt to remove too much material with a hand reamer; about 0.010 in. (0.25 mm) is the maximum.
5. Frequently clear a taper reamer (and the hole) of chips.

TO REAM A HOLE WITH A STRAIGHT HAND REAMER

1. Check the size of the drilled hole. It should be between 0.004 and 0.005 in. (0.10 and 0.12 mm) smaller than the finished hole size.
2. Place the end of the reamer in the hole and place the tap wrench on the square end of the reamer.
3. Rotate the reamer clockwise to allow it to align itself with the hole (Fig. 25-6).
4. Check the reamer for squareness with the work by testing it with a square at several points on the circumference.
5. Brush cutting fluid over the end of the reamer if required.
6. Rotate the reamer slowly in a clockwise direction and apply downward pressure. Feed should be fairly rapid and steady to prevent the reamer from chattering.

NOTE: The rate of feed should be about one-quarter the diameter of the reamer for each turn.

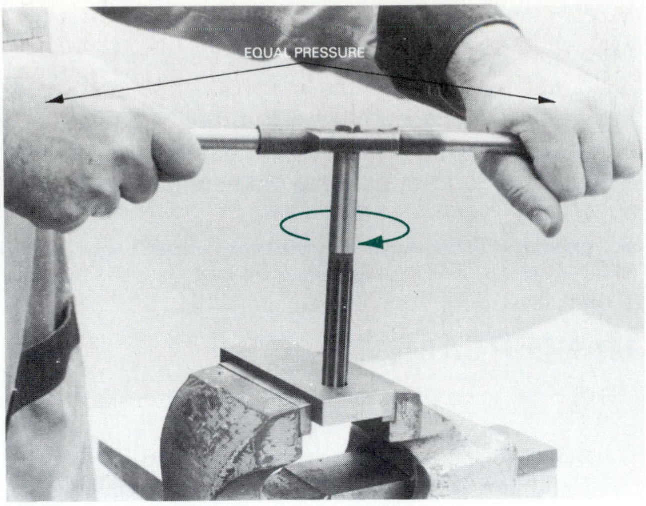
EQUAL PRESSURE

Figure 25-6 *Turn the reamer clockwise when starting it in a hole.*

BROACHING

Broaching is a process in which a special tapered multi-toothed cutter is forced through an opening or along the outside of a piece of work to enlarge or change the shape of the hole or to form the outside to a desired shape.

Broaching was first used for producing internal shapes, such as keyways, splines, and other odd internal shapes (Fig. 25-7). Its application has been extended to exterior surfaces, such as the flat face on automotive engine blocks and cylinder heads. Most broaching is now performed on special machines which either pull or push the broach through or along the material. Hand broaches are used in the machine shop for operations such as keyway cutting.

The cutting action of a broach is performed by a series of successive teeth, each protruding about 0.003 in. (0.07 mm) farther than the preceding tooth (Fig. 25-8). The last three teeth are generally of the same depth and provide the finish cut.

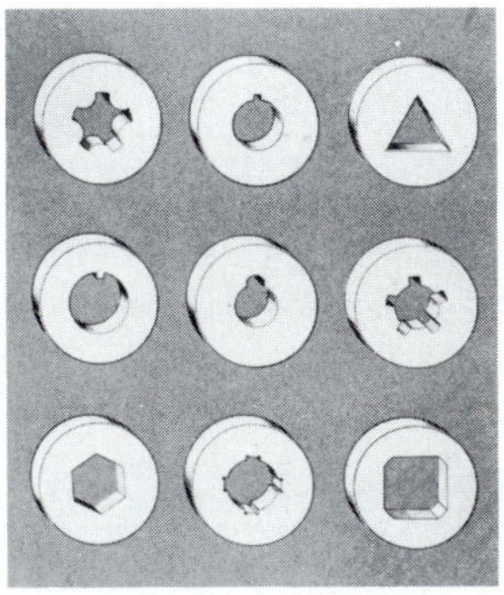

Figure 25-7 *Examples of internal broaching.*

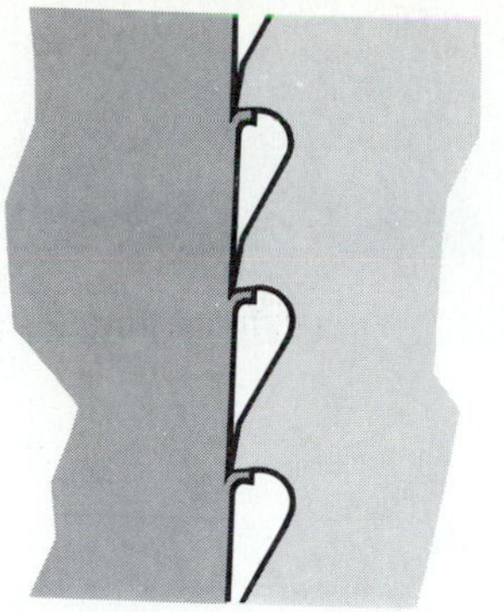

Figure 25-8 *The cutting action of a broach.*

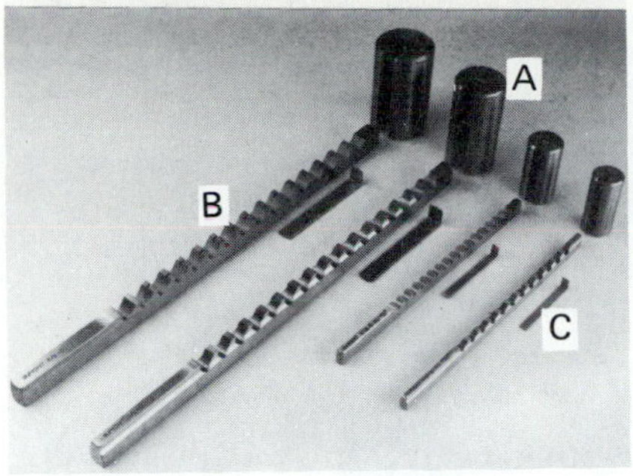
A
B
C

Figure 25-10 *A broach set for cutting internal keyways.*

Broaching has many advantages and an extremely wide range of applications:

1. Machining almost any irregular shape is possible, providing it is parallel to the broach axis.
2. It is rapid; the entire machining process is usually completed in one pass.
3. Roughing and finishing cuts are generally combined in the same operation.
4. A variety of forms, either internal or external, may be cut simultaneously and the entire width of a surface may be machined in one pass, thus eliminating the need for a machining operation.

CUTTING A KEYWAY WITH A BROACH

Keyways may be cut by hand in the machine shop quickly and accurately by means of a broach set and an arbor press (Fig. 25-9). A broach set (Fig. 25-10) covers a wide range of keyways and is a particularly useful piece of equipment when many keyways must be cut. The equipment necessary to cut a keyway is a bushing (A) to suit the hole size in the workpiece, a broach (B) the size of the keyway to be cut, and shims (C) to increase the depth of the cut of the broach.

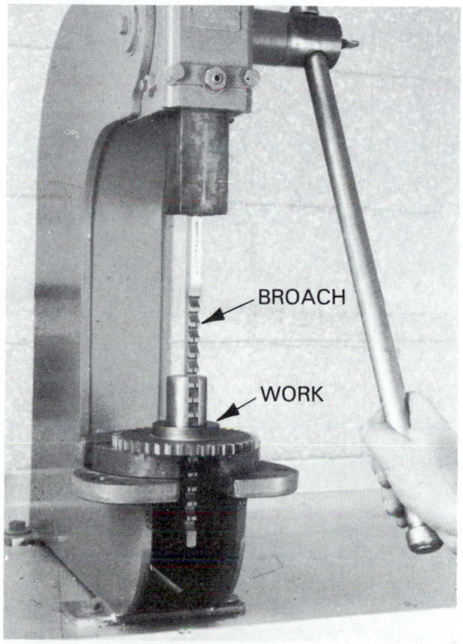
BROACH
WORK

Figure 25-9 *Using an arbor press to cut a keyway with a broach.*

Proceed as follows:

1. Determine the keyway size required for the size of the workpiece.
2. Select the proper broach, bushing, and shims.
3. Place the workpiece on the arbor press. Use an opening on the base smaller than the opening in the workpiece so that the bushing will be properly supported.
4. Insert the bushing and the broach into the opening. Apply cutting fluid if the workpiece is made of steel.
5. Check the broach to be sure that it has started squarely in the hole.
6. Press the broach through the workpiece, maintaining constant pressure on the arbor-press handle.
7. Remove the broach, insert one shim, and press the broach through the hole.
8. Insert the second shim, if required, and press the broach through again. This will cut the keyway to the proper depth (Fig. 25-11).
9. Remove the bushing, broach, and shims.

LAPPING

Lapping is an abrading process used to remove minute amounts of metal from a surface which must be flat, accurate to size, and smooth. Lapping may be performed for any of the following reasons:

1. To increase the wear life of a part
2. To improve accuracy and surface finish
3. To improve surface flatness
4. To provide better seals and eliminate the need for gaskets.

SHIMS

Figure 25-11 *Two shims are used for making the final pass with a broach.*

Lapping may be performed by hand or machine, depending on the nature of the job. Lapping is intended to remove only about 0.0005 in. (0.01 mm) of material. Lapping by hand is a long, tedious process and should be avoided unless absolutely necessary.

LAPPING ABRASIVES

Both natural and artificial abrasives are used for lapping. Flour of emery and fine powders made of silicon carbide or aluminum oxide are used extensively. Abrasives used for rough lapping should be no coarser than 150 grit; fine powders used for finishing run up to about 600 grit. For fine work, diamond dust, generally in paste form, is used.

TYPES OF LAPS

Laps may be used to finish flat surfaces, holes, or the outside of cylinders. In each case, the lap material must be *softer* than the workpiece.

Flat Laps

Laps for producing flat surfaces are made from close-grained cast iron. For the roughing operation or "blocking down," the lapping plate should be scored with narrow grooves about ½ in. (13 mm) apart, both lengthwise and crosswise or diagonally to form a square or diamond pattern (Fig. 25-12A). Finish lapping is done on a smooth cast-iron plate (Fig. 25-12B).

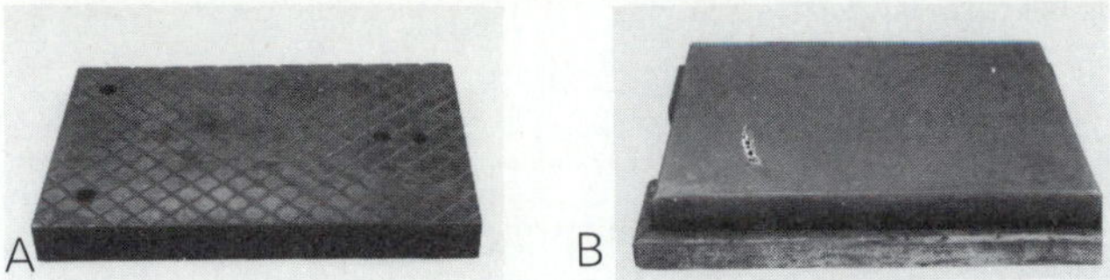
A
B

Figure 25-12 *(A) A roughing lapping plate; (B) a finishing lapping plate.*

Charging the Flat Lapping Plate

Spread a thin coating of abrasive powder over the surface of the plate and press the particles into the surface of the lap with a hardened steel block or roll. Rub as little as possible. When the entire surface appears to be charged, clean the surface with varsol and examine it for bright spots. If any bright spots appear, recharge the lap and continue until the entire surface assumes a gray appearance after it has been cleaned.

Lapping a Flat Surface

If work is to be roughed down, oil should be used on the roughing plate as a lubricant. As the work is rubbed over the lap, the abrasive powder will be washed from the grooves and act between the surface of the work and the lap. If the work has been surface-ground, rough lapping or "blocking down" is not required.

Proceed as follows:

1. Place a little varsol on a finish-lapping plate which has been properly charged.
2. Place the work on top of the plate and gently push it back and forth over the full surface of the lap using an irregular movement. *Do not stay in one spot.*
3. Continue this movement with a light pressure until the desired surface finish is obtained.

Precautions to Be Observed

1. Do not stay in one area; cover the full surface of the lap.
2. Never add a fresh supply of loose abrasive. If required, recharge the lap.
3. Never press too hard on the work because the lap will become stripped in places.
4. Always keep the lap moist.

Internal Laps

Holes may be accurately finished to size and smoothness by lapping. Internal laps may be made of brass, copper, or lead and may be of three types.

The *lead lap* (Fig. 25-13A) is made by pouring lead around a tapered mandrel which has a groove along its length. The lap is turned to a running fit into the hole and is then sometimes slit on the outside to trap the loose abrasive during the lapping operation. Adjust by lightly tapping

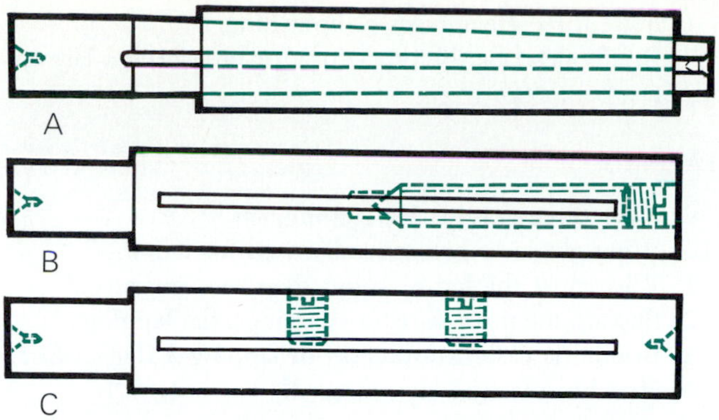
A
B
C

Figure 25-13 *Various types of internal laps: (A) lead lap; (B) copper lap; (C) adjustable lap.*

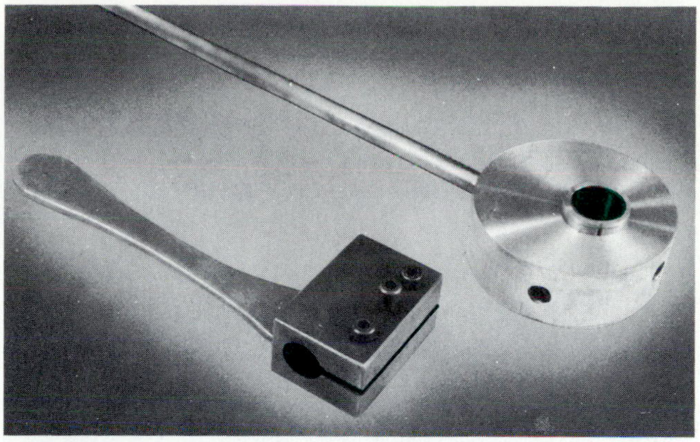

Figure 25-14 *External laps.*

the large end of the mandrel on a soft block. This will cause the lead sleeve to move along the mandrel and expand.

The *internal lap* (Fig. 25-13B) may be made of copper, brass, or cast iron. A threaded-taper plug fits into the end of the lap which is slit for almost its entire length. The lap diameter may be adjusted by the threaded-taper plug.

The *adjustable lap* (Fig. 25-13C) may be made from copper or brass. The lap is split for almost its full length, but both ends remain solid. Slight adjustment is provided by means of two setscrews in the center section of the lap.

Charging and Using an Internal Lap

Before charging, the lap should be a running fit in the hole. Proceed as follows:

1. Sprinkle some lapping powder evenly on a flat plate.
2. Roll the lap over the powder, applying sufficient pressure to embed the abrasive into the surface of the lap.
3. Remove any excess powder.
4. Mount a lathe dog on the end of the lap.
5. Fit the workpiece over the end of the lap.

NOTE: The lap should now be a wringing fit in the hole of the work and about 2.5 times the length of the workpiece.

6. Place some oil or varsol on the lap.
7. Mount the lap and the work between lathe centers.
8. Set the machine to run at a slow speed, 150 to 200 r/min for a 1-in. (25-mm) diameter.
9. Hold the work securely and start the machine.
10. Run the work back and forth along the entire length of the lap.
11. Remove the work and rinse it in varsol to remove the abrasive and to bring it to room temperature.
12. Gage the hole for size.

NOTE: Always keep the lap moist and never add loose abrasive to the lap. Loose abrasive will cause the work to become bell-mouthed at the ends. If more abrasive is necessary, recharge the lap and adjust as required.

External Laps

External laps are used to finish the outside of cylindrical workpieces. They may be of several forms (Fig. 25-14); however, the basic design is the same. External laps may be made of cast iron or may have a split brass bushing mounted inside by means of a setscrew. There must be some provision for adjusting the lap.

Charging and Using an External Lap

1. Mount the workpiece in a three-jaw chuck on the lathe or drill press.
2. Adjust the lap until it is a running fit on the workpiece.
3. Grip the end of the lap in a vise.
4. Sprinkle abrasive powder in the hole.
5. With a hardened steel pin, roll the abrasive evenly around the inside surface of the lap.
6. Remove any excess lapping powder.
7. Place the lap on the workpiece. It should now be a wringing fit.
8. Set the machine to run at a slow speed [150 to 200 r/min for a 1-in.-diameter (25-mm) workpiece].
9. Add some varsol to the workpiece and the lap.
10. Hold the lap securely and start the machine.
11. Move the lap back and forth along the work.

NOTE: Always keep the lap moist.

12. In order to gage the work, remove the lap and clean the workpiece with varsol.

REVIEW QUESTIONS

HAND REAMERS

1. What is the purpose of a hand reamer?
2. Describe and state the purpose of:

a. The solid hand reamer
b. The expansion hand reamer
c. The taper finishing reamer

3. How much metal should be removed with a hand reamer?
4. List four important precautions to be observed while reaming.

BROACHING

5. Define the term *broaching.*
6. Describe the cutting action of a broach.
7. State three advantages of broaching.
8. Briefly describe the procedure for broaching a keyway on an arbor press.

LAPPING

9. State three reasons for lapping.
10. What abrasives are generally used for lapping?
11. Why must the lap be softer than the workpiece?
12. Explain the procedure for charging a flat lapping plate.
13. Briefly describe the process of lapping a flat surface.
14. How are internal laps charged?

UNIT 26

Bearings

Bearings contribute to the smooth operation of rotating parts of machinery and motors. They are used to support and position a shaft and to reduce the friction created by the rotating part, particularly when under load. They must also be capable of absorbing and transmitting loads at the required speeds and temperatures. Bearings may be divided into two general classifications: *plain* and *rolling.*

OBJECTIVES

After completing this unit you will be able to:

1. Identify and explain the purpose of the different types of bearings
2. Install a ball or roller bearing so that it operates properly

PLAIN-TYPE BEARINGS

Plain-type bearings operate on the oil-film principle; that is, when rotating, the shaft is actually supported on a thin film of oil. Plain bearings are generally used in machines or components running at slower speeds. The bearing material is usually of a different type from the shaft, although great success has been claimed with hardened steel shafts rotating in hardened steel bearings at relatively high speeds. Plain bearings may be *solid, split,* or *thrust* bearings.

Solid bearings (Fig. 26-1) are of the sleeve type and often found in electric motors. They may be made of bronze, sintered bronze, or even cast iron in slowly rotating equipment. Straight sleeve bearings have no provision for adjustment and often present problems when not kept

Figure 26-1 *A plain-type sleeve bearing has no adjustment.*

properly lubricated. Sleeve bearings on some machines have a slight taper on the outside to provide for wear adjustment. Bearings of sintered bronze are made by the powder metallurgy process and often contain graphite between the bronze particles, which aids in the lubrication of the bearing. The porosity created by the powder metallurgy process also provides a reservoir for oil. This tends to create a much longer lasting bearing than the plain bronze bushing. Sleeve bearings are usually made to standard sizes and are easily replaceable.

Split bearings are often used on larger machines, which operate at slower speeds. They may be made of bronze, bronze with babbitt, or have a babbitt metal lining. Adjustment is usually provided for by means of laminated shims between the upper and lower bearing halves. Oil grooves chipped or machined in the bearing provide a channel for lubrication.

Thrust bearings are used to take the longitudinal thrust of a shaft. They may be flat, kidney-shaped, babbitt-faced pieces called *shoes.* They bear against a collar or collars on a shaft, which rotates in oil. The bearings operate on the oil-wedge principle whereby the oil is drawn up by the rotating shaft and forms a wedge between the bearing and the collar of the shaft. These bearings may be used on large equipment where there is considerable end thrust, such as on a boat propeller shaft. They prevent damage to the equipment and maintain the position of the rotating part.

ROLLING-TYPE (ANTIFRICTION) BEARINGS

Rolling bearings (ball and roller) are used in preference to plain bearings for several reasons:

1. They have a lower coefficient of friction, especially at start-up.
2. They are compact in design.
3. They have high running and dimensional accuracy.
4. They do not wear as much as plain-type bearings.
5. They are easily replaced because of standardized sizes.

The following rolling bearing types are commonly available.

Radial Bearings	
Ball Bearings	**Roller Bearings**
Single-row deep-groove	Cylindrical
Double-row deep-groove	Double-row spherical
Single-row angular contact	Single-row spherical
Double-row angular contact	Tapered
Self-aligning	Needle
Thrust Bearings	
Ball	Spherical roller
Cylindrical roller	Taper roller

TYPES OF RADIAL BEARINGS

Ball bearings (Fig. 26-2) are widely used because of their low coefficient of friction and their suitability for high speeds. They are capable of absorbing medium to high radial and thrust loads. Ball bearings may be of single-row design for light to medium loads and of double-row design for heavy loads. There is no way to adjust this type of bearing.

Single-row angular contact ball bearings (Fig. 26-3A) are designed for high radial and high thrust loads in one direction. *Double-row angular contact ball bearings* (Fig. 26-3B) are designed for high radial and high thrust loads in both directions. Both types are suitable for high speeds. When mounting single-row angular contact ball bearings, make sure that the thrust is applied to the proper side of the bearing.

TYPES OF ROLLER BEARINGS

When it is necessary to support heavy loads under medium to high speeds, roller bearings are used because their basic dynamic load rating is higher than that of ball bearings of the same dimensions. The higher basic dynamic load rating is achieved by the greater area of contact between the rolling elements and the bearing raceways.

Figure 26-2 *Single- and double-row ball bearings.*

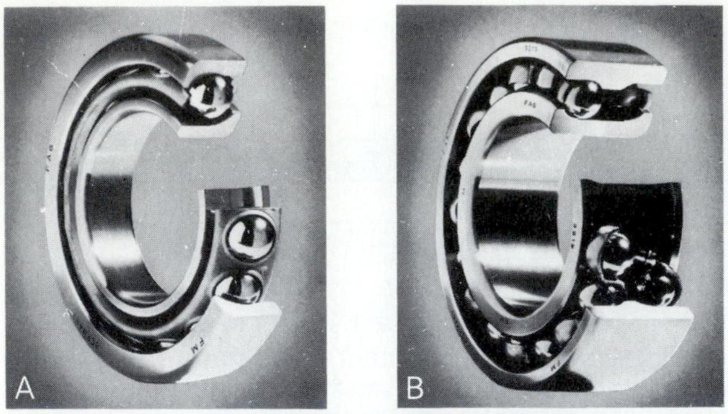

Figure 26-3 *Angular contact bearings provide for a high radial and axial thrust. (A) Single-row bearings; (B) double-row bearings.*

Cylindrical roller bearings (Fig. 26-4A) may be single or double row. They are designed for heavy radial loads at medium to high speeds.

Double-row self-aligning roller bearings (Fig. 26-4B) consist of two rows of rollers, either spherical or barrel-shaped, revolving in a spherical outer race. They are capable of absorbing very high radial loads and moderate thrust loads in both directions. The self-aligning feature of these bearings permits shaft deflection up to 0.5° without impairing the basic dynamic load rating.

Tapered roller bearings (Fig. 26-4C) are manufactured as single-row, double-row, or four-row bearings. They are capable of absorbing high radial and high thrust loads at moderate speeds. Provision for adjustment makes these bearings versatile. They are used extensively in machine tool manufacture.

Needle bearings are long cylindrical rollers of small diameter. They are used where a longer bearing surface is required and where limited installation space is available. Needle bearings are designed for radial loads and moderate speeds.

TYPES OF THRUST BEARINGS

Thrust bearings consist of a set of balls or rollers held in a retainer ring between two races or washers. They are designed for heavy thrust loads and, in some instances, for combined high thrust and medium radial load.

Ball thrust bearings (Fig. 26-5A) are designed for medium to high thrust loads at moderate speeds. No radial load may be applied.

There are two types of *roller thrust bearings:*

1. *Cylindrical roller thrust bearings* (Fig. 26-5B) are designed for high thrust load and low speed. No radial load can be applied.
2. *Spherical roller thrust bearings* are capable of absorbing very high thrust loads and moderate radial loads at low speeds. The track in the housing washer is spherical, which permits a shaft misalignment of 3°.

BEARING INSTALLATION

When a bearing must be replaced, it is advisable to select a duplicate bearing for replacement. If another make of bearing is chosen, the maker's catalogue should be consulted so that a bearing of the same specifications is used. Once the replacement bearing has been obtained, the following precautions should be observed in the installation of rolling bearings.

1. Check the shaft and housing tolerances.
2. Make sure tolerances are within the range recommended by the bearing supplier.
3. Clean the installation area and mating parts.
4. Do not unwrap a bearing until it is required for installation.
5. Do not expose a bearing to dust or dirt.

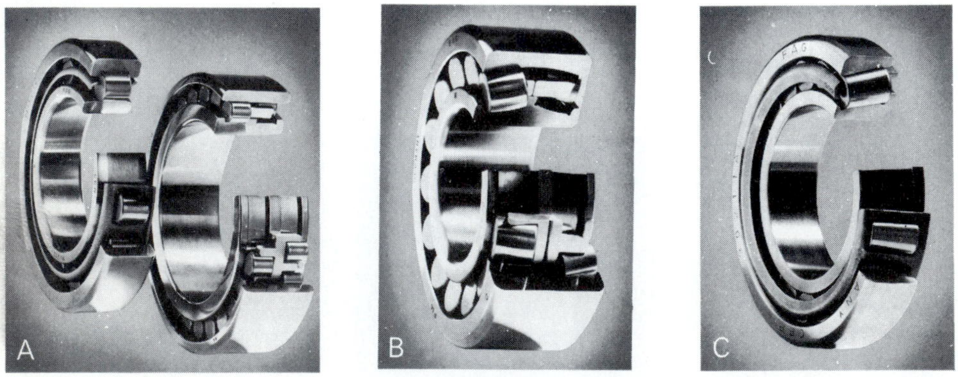

Figure 26-4 *Types of roller bearings: (A) single- and double-row cylindrical; (B) double-row self-aligning; (C) taper. (Courtesy Fisher Bearing Manufacturing Limited.)*

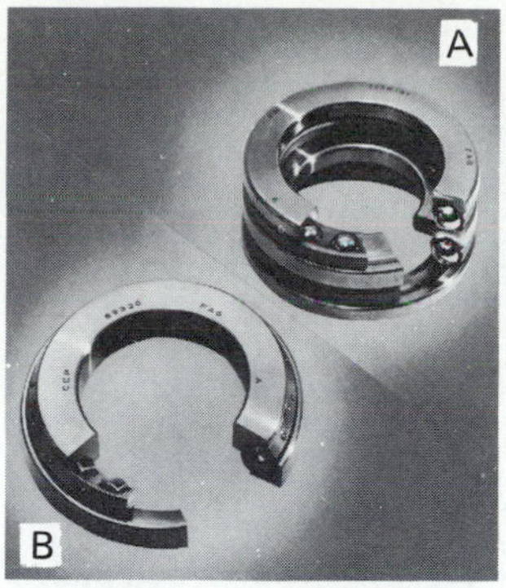
A
B

Figure 26-5 *Thrust bearings: (A) ball; (B) roller.*

6. Do not wash a new bearing as this will remove the protective film.
7. Under no condition mount the bearing by exerting force over or through the rolling elements.

NOTE: The outer ring is generally a hand push fit into the housing while the inner ring has a light to heavy interference fit on the shaft (depending on the application).

LUBRICATION OF BALL AND ROLLER BEARINGS

Bearings operating under moderate speed and temperature conditions are generally lubricated with grease, since grease can easily be retained in the bearings and housings. It also tends to create a seal to keep out dirt and foreign matter. Oil lubrication is generally used on bearings operating at high speeds and temperatures and also in such mechanisms as closed gear trains.

Double-sealed or shielded bearings are lubricated by the manufacturer and, unless special provisions have been made, do not have to be relubricated.

PRECAUTIONS IN HANDLING BEARING LUBRICANTS

Because of the danger of small particles of dirt or grit entering the bearings, grease or oil should be stored in covered containers. Grease fittings and oil caps should be wiped clean before lubricant is added. Overlubrication of ball and roller bearings must be avoided since it may result in high operating temperatures, rapid deterioration of the lubricant, and premature bearing failure.

REVIEW QUESTIONS

1. Explain the difference between a plain-type and an antifriction-type bearing.
2. Why are antifriction bearings preferred over plain-type bearings?
3. Name three types of ball bearings and state the purpose of each.
4. Describe and state the purpose of any three types of roller bearings.
5. List three important precautions which should be observed when bearings are being installed.

SECTION

8

Metal-Cutting Technology

Industry recognizes that in order to operate economically, metals used in the manufacture of its products must be machined efficiently. To cut metals efficiently requires not only a knowledge of the metal to be cut, but also how the cutting-tool material and its shape will perform under various machining conditions. Cutting-tool angles, rakes, and clearances have assumed increasing importance in metal cutting. Many new cutting-tool materials have been introduced in the last few decades, which of necessity, have led to improved machine construction, higher cutting speeds, and increased productivity.

Many materials are machined most efficiently with cutting fluids; others are not. With the appearance of new and varying alloys, new cutting fluids are constantly being developed. All these factors combine to make metal-cutting theory a challenging and constantly researched field in the machine tool industry.

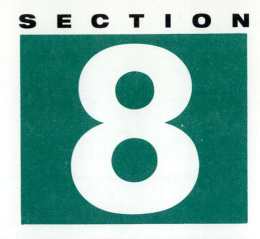

UNIT

27

Physics of Metal Cutting

Human beings have been using tools to cut metal for hundreds of years without really understanding how the metal was cut or what was occurring where the cutting tool met the metal. For many years, it was felt that the metal ahead of the cutting tool split in a manner similar to the way that wood splits in front of an ax (Fig. 27-1). This, according to the original theory, accounted for the wear which occurred on the cutting tool face some distance away from the cutting edge. An early cutting-fluid advertisement illustrated this same theory by showing the metal splitting in front of a cutting tool during a lathe-turning operation (Fig. 27-2).

OBJECTIVES

After completing this unit you will be able to:

1. Define the various terms which apply to metal cutting
2. Explain the flow patterns of metal as it is being cut
3. Recognize the three types of chips produced from various metals

NEED FOR METAL-CUTTING RESEARCH

The manufacture of dimensionally accurate, closely fitting parts is essential to interchangeable manufacture. The accuracy and wearability of mating surfaces is directly proportional to the surface finish produced on the part. Every year in the United States alone, over 15 million tons (13.6 million tonnes) of metal are cut into chips at a cost of over $10 billion. To reduce the cost of machining, prolong the life of cutting tools, and maintain high surface finishes, it was essential for research to be done in the area of metal cutting. Since the Second World War, a great deal of research has been conducted in areas such as the theory of metal cutting, measurement of cutting forces and temperatures, machinability of metals, machining economics, and the theory of cutting-fluid action. This research has found that the metal of a workpiece, instead of rupturing or breaking a little before the cutting tool, is compressed and then flows up the face of the cutting tool. New cutting tools, speeds

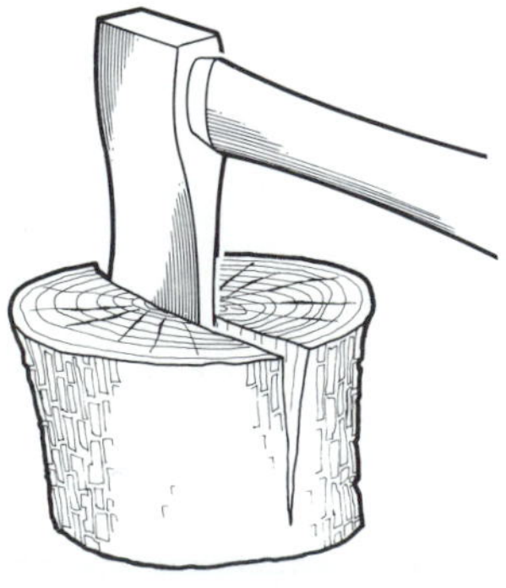

Figure 27-1 *The simile of an ax splitting wood was often used incorrectly to illustrate the action of a cutting tool.*

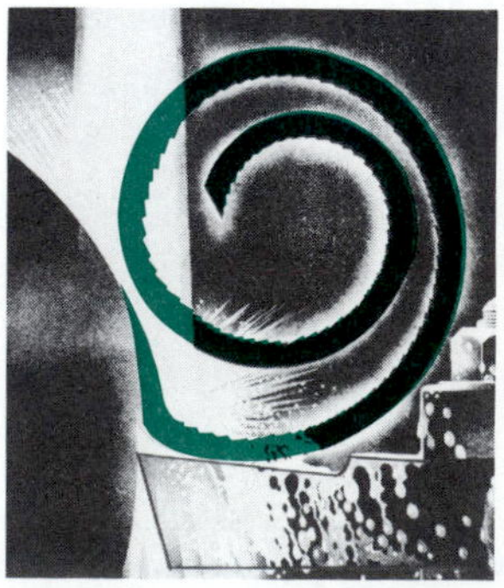

Figure 27-2 *A false conception of metal cutting.*

and feeds, cutting-tool angles and clearances, and cutting fluids have been developed as a result of this research. These developments have greatly assisted in the economical machining of metals; however, much work remains to be done before all the factors affecting surface finish, tool life, and machine output can be controlled.

METAL-CUTTING TERMINOLOGY

A number of terms resulted from the research conducted on metal cutting and it may be wise to clearly define these terms.

A *built-up edge* is a layer of compressed metal from the material being cut, which adheres to and piles up on the face of the cutting tool edge during a machining operation (Fig. 27-3).

The *chip–tool interface* is that portion of the face of the cutting tool upon which the chip slides as it is cut from the metal (Fig. 27-3).

Crystal elongation is the distortion of the crystal structure of the work material that occurs during a machining operation (Fig. 27-4).

The *deformed zone* is the area in which the work material is deformed during cutting.

Plastic deformation is the deformation of the work material that occurs in the shear zone during a cutting action (Fig. 27-4).

Plastic flow is the flow of metal that occurs on the shear plane, which extends from the cutting-tool edge *A* to the corner between the chip and the work surface *B* in Fig. 27-4.

A *rupture* is the tear that occurs when brittle materials, such as cast iron, are cut and the chip breaks away from the work surface. This generally occurs when discontinuous or segmented chips are produced.

The *shear angle* or *plane* is the angle of the area of material where plastic deformation occurs (Fig. 27-3).

The *shear zone* is the area where plastic deformation of the metal occurs. It is along a plane *AB* from the cutting edge of the tool to the original work surface (Fig. 27-4).

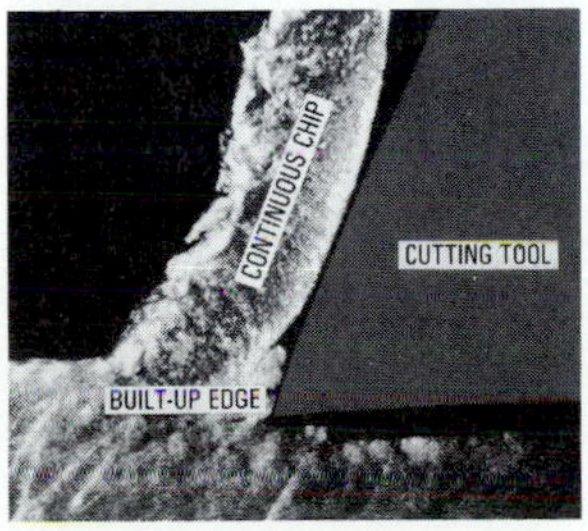
CONTINUOUS CHIP
CUTTING TOOL
BUILT-UP EDGE

Figure 27-3 *Chip–tool interface. (Courtesy Cincinnati Milacron Inc.)*

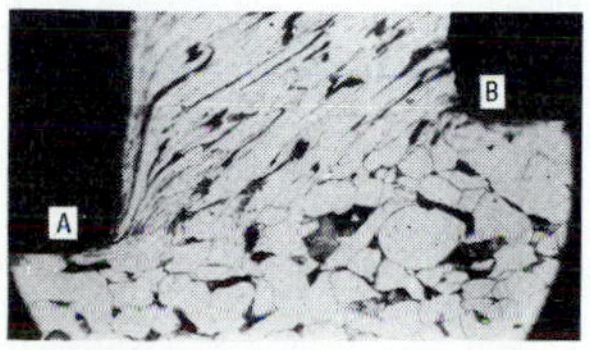
B
A

Figure 27-4 *photomicrograph of a chip, showing crystal elongation and plastic deformation.*

PLASTIC FLOW OF METAL

In order to understand more fully what occurs in the metal while it is being deformed in the shear zone, researchers made many tests on various types of materials. They used flat punches on ductile material to study the stress pattern, direction of material flow, and the distortion created in the metal. To observe what occurs when pressure is applied to one spot, they used blocks of photoelastic materials, such as celluloid and Bakelite. Researchers also used polarized light to observe stress lines created when pressure is exerted on the punch. Using a suitable analyzer, they saw a series of colored bands known as *isochromatics.* They used three different types of punches (flat, narrow-faced, and knife-edge) on photoelastic material to create the various stresses observed.

FLAT PUNCH

When a flat punch is forced into a block of photoelastic material, the lines of constant maximum shear stress appear, indicating the distribution of the stress. In Fig. 27-5, the shape of these stress lines, or isochromatics, appears as a family of curves almost passing through the corners of the flat punch. The greatest concentration of stress lines occurs at each corner of the punch, and larger circular stress lines appear farther away from the punch. The spacing of the isochromatics is relatively wide.

NARROW-FACED PUNCH

When a narrow-faced punch is forced into a block of photoelastic material, the stress lines are still concentrated at the punch corners and where the punch meets the top

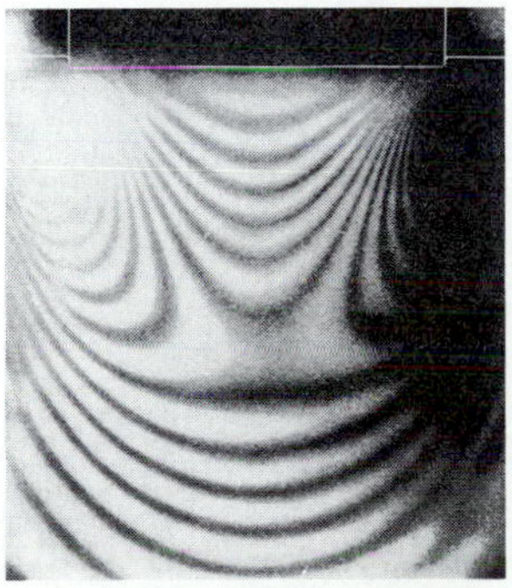

Figure 27-5 *The stress distribution created by a flat punch in photoelastic material. (Courtesy Cincinnati Milacron Inc.)*

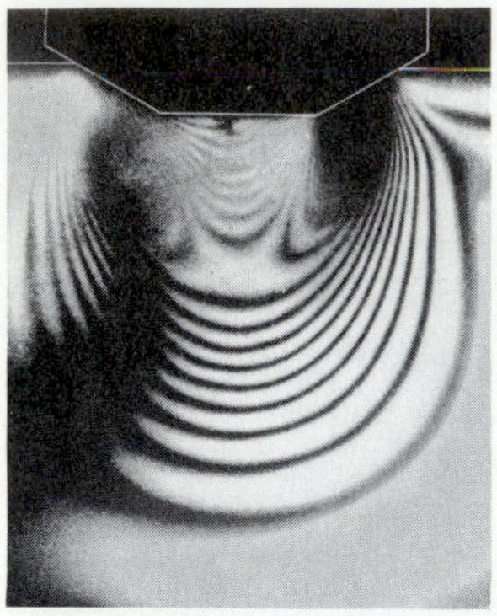

Figure 27-6 *The stress distribution created by a narrow-faced punch. (Courtesy Cincinnati Milacron Inc.)*

surface of the work. As can be noted in Fig. 27-6, the isochromatics are spaced closer than with the flat punch.

KNIFE-EDGE PUNCH

When a knife-edge punch is forced into the block of photoelastic material (Fig. 27-7), the isochromatics become a series of circles tangent to the two faces of the punch. In this case, the flow of material occurs upward from the point toward the free area along the faces of the punch.

When a cutting tool engages a workpiece, this flow of material takes place, and the compressed material escapes up the tool face. As the tool advances, opposition to the upward flow of material creates stresses in the material ahead of the tool due to the friction of the chip flow up the tool face. These stresses are somewhat relieved by the plastic flow or rupture of the material along a plane leading from the cutting edge of the tool to the surface of the unmachined metal. From Figs. 27-5 to 27-7, it can be concluded that internal stresses are created during metal-cutting operations and:

1. Because of the forces exerted by the cutting tool, compression occurs in the work material.
2. As the cutting tool or work moves forward during a cut, the stress lines concentrate at the cutting-tool edge and radiate from there in the material (Fig. 27-7).
3. This concentration of stresses causes the chip to shear from the material and flow along the chip–tool interface.

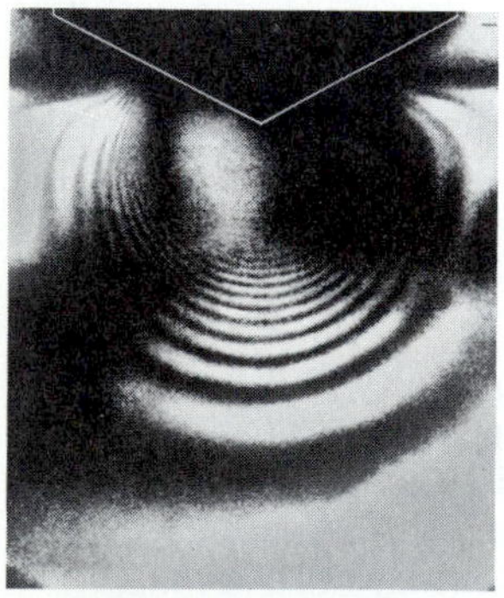

Figure 27-7 *The stress lines created by a knife-edge punch. (Courtesy Cincinnati Milacron Inc.)*

4. By either plastic flow or rupture, the metal tries to flow along the chip–tool interface. Since most metals are ductile to some degree, a plastic flow generally occurs.

The plastic flow or rupture which occurs as the metal flows along the chip–tool interface determines the type of chip produced. When brittle materials such as cast iron are being cut, the metal has a tendency to rupture and produce discontinuous or segmented chips. When relatively ductile metals are being cut, a plastic flow occurs, and continuous or flow-type chips are produced.

CHIP TYPES

Machining operations performed on lathes, milling machines, or similar machine tools produce chips of three basic types: discontinuous chip (Fig. 27-8), continuous chip (Fig. 27-9), and continuous chip with a built-up edge (Fig. 27-10).

TYPE 1—DISCONTINUOUS (SEGMENTED) CHIP

Discontinuous or segmented chips (Fig. 27-8) are produced when brittle metals such as cast iron and hard bronze are cut and even when some ductile metals are cut under poor cutting conditions. As the point of the cutting tool contacts the metal (Fig. 27-11A), some compression occurs, as can be noted in Fig. 27-11B and C, and the chip begins flowing along the chip–tool interface. As more stress is applied to brittle metal by the cutting action, the metal compresses until it reaches a point where rupture occurs (Fig. 27-11D), and the chip separates from the unmachined portion (Fig. 27-11E). This cycle is repeated indefinitely during the cutting operation with the rupture of each segment occurring on the shear angle or plane. Generally, as a result of these

Figure 27-8 *A discontinuous chip. (Courtesy Cincinnati Milacron Inc.)*

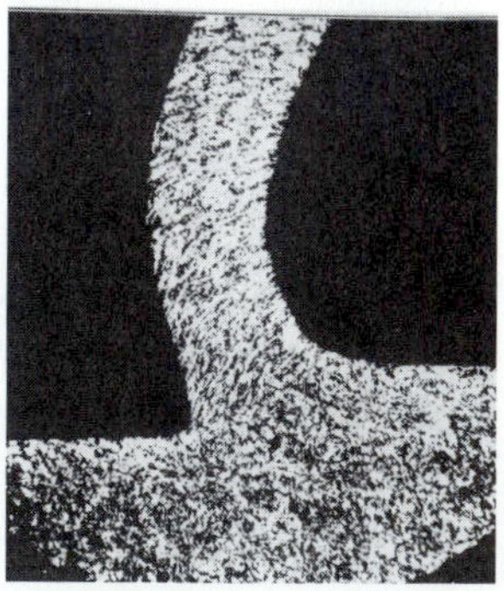

Figure 27-9 *A continuous chip. (Courtesy Cincinnati Milacron Inc.)*

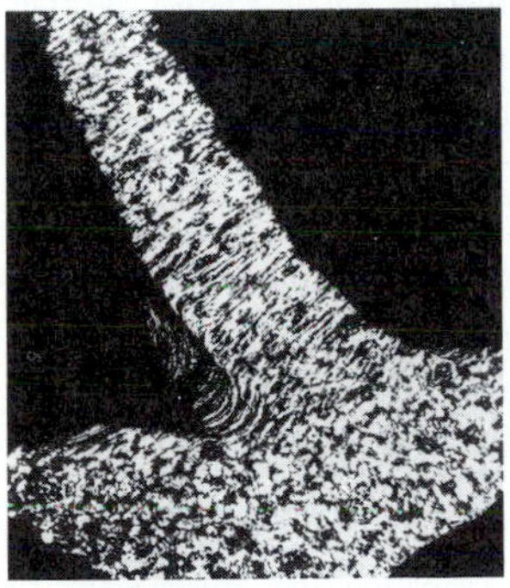

Figure 27-10 *A continuous chip with a built-up edge. (Courtesy Cincinnati Milacron Inc.)*

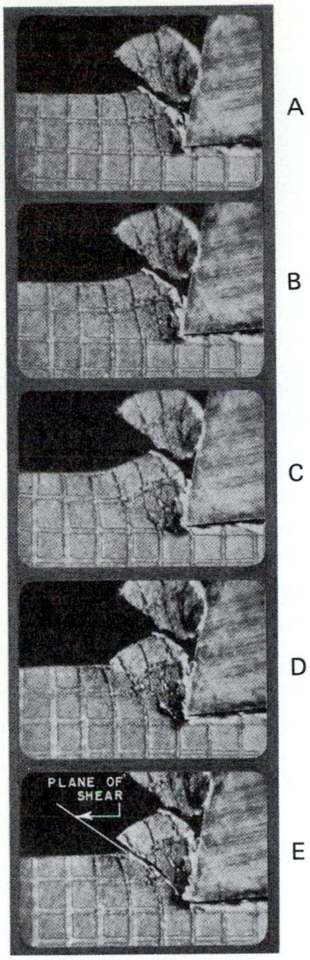
A
B
C
D
PLANE OF SHEAR
E

Figure 27-11 *Formation of a discontinuous chip.*

successive ruptures, a poor surface is produced on the workpiece.

Machine vibration or tool chatter sometimes causes discontinuous chips to be produced when ductile metal is cut.

The following conditions favor the production of a Type 1 discontinuous chip:

1. Brittle work material
2. Small rake angle on the cutting tool
3. Large chip thickness (coarse feed)
4. Low cutting speed
5. Excessive machine chatter.

TYPE 2—CONTINUOUS CHIP

The Type 2 chip is a continuous ribbon produced when the flow of metal next to the tool face is not greatly retarded by a built-up edge or friction at the chip–tool interface. The continuous ribbon chip is considered ideal for efficient cutting action because it results in better surface finishes.

When ductile materials are cut, plastic flow in the metal takes place by the deformed metal sliding on a great number of crystallographic slip planes. As is the case with the Type 1 chip, fractures or ruptures do not occur because of the ductile nature of the metal.

In Fig. 27-9 it can be seen that the crystal structure of the ductile metal is elongated when it is compressed by the action of the cutting tool and as the chip separates from the metal. The process of chip formation occurs in a single plane extending from the cutting tool to the unmachined work surface. The area where plastic deformation of the crystal structure and shear occurs is called the *shear zone* (illustrated in Fig. 27-4 by the line *AB*). The angle on which the chip separates from the metal is called the *shear plane* or *shear angle.*

The mechanics of chip formation can best be understood with the aid of the schematic diagram in Fig. 27-12. As the cutting action progresses, the metal immediately ahead of the cutting tool is compressed with a resultant deformation (elongation) of the crystal structure. This elongation does not take place in the direction of shear. As this process of compression and elongation continues, the material above the cutting edge is forced along the chip–tool interface and away from the work.

Machine steel generally forms a continuous (unbroken) chip with little or no built-up edge when machined with a cemented-carbide cutting tool or a high-speed steel toolbit and cutting fluid (Fig. 27-13). In order to reduce the amount of resistance occurring as the compressed chip slides along the chip–tool interface, a suitable rake angle is ground on the tool, and cutting fluid is used during the cutting operation. These features allow the compressed chip to flow relatively freely along the chip–tool interface. A shiny layer on the back of a continuous-type chip indi-

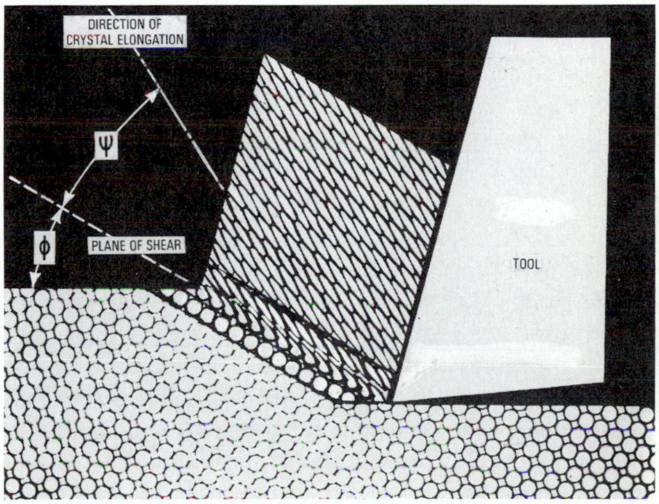
DIRECTION OF CRYSTAL ELONGATION
Ψ
Φ
PLANE OF SHEAR
TOOL

Figure 27-12 *A schematic diagram showing formation of a continuous chip and deformation of the crystal structure. (Courtesy Cincinnati Milacron Inc.)*

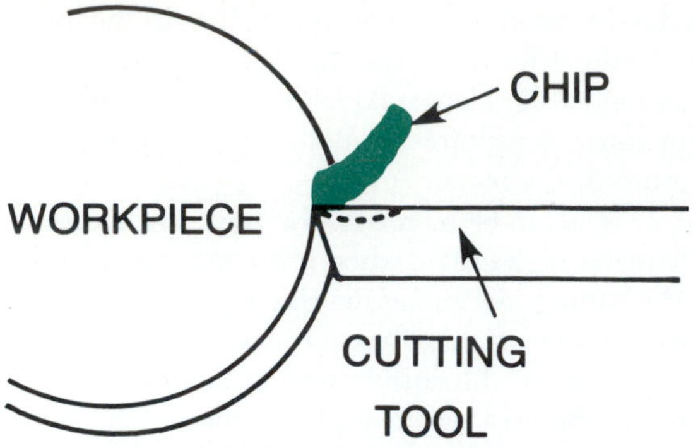
CHIP
WORKPIECE
CUTTING
TOOL

Figure 27-13 *A continuous chip is generally formed when a carbide or high-speed toolbit is used to machine steel.*

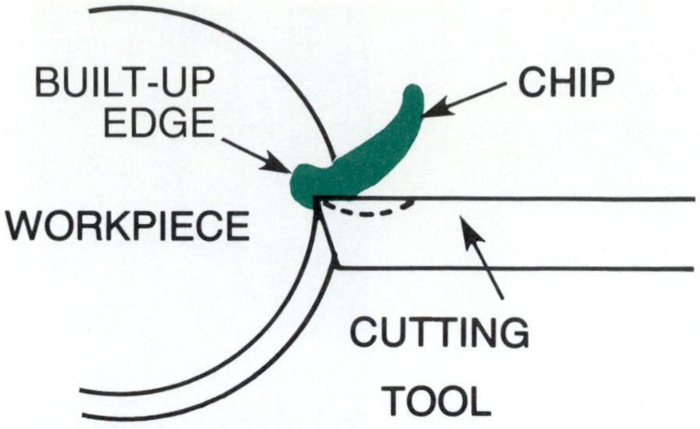
BUILT-UP
EDGE
CHIP
WORKPIECE
CUTTING
TOOL

Figure 27-14 *A built-up edge is formed when workpiece fragments become welded to the tool face.*

cates ideal cutting conditions with little resistance to chip flow.

The conditions favorable to producing a Type 2 chip are:

1. Ductile work material
2. Small chip thickness (relatively fine feeds)
3. Sharp cutting-tool edge
4. A large rake angle on the cutting tool
5. High cutting speeds
6. Cutting tool and work kept cool by use of cutting fluids
7. A minimum of resistance to chip flow by:
   a. A high polish on the cutting-tool face
   b. Use of cutting fluids to prevent the formation of a built-up edge
   c. Use of cutting-tool materials, such as cemented carbides, which have a low coefficient of friction
   d. Free-machining materials (those alloyed with elements such as lead, phosphor, and sulphur).

TYPE 3—CONTINUOUS CHIP WITH A BUILT-UP EDGE

Low-carbon machine steel and many high-carbon alloyed steels, when cut at a low cutting speed with a high-speed steel cutting tool and without the use of cutting fluids, generally produce a continuous-type chip with a built-up edge (Fig. 27-10).

The metal ahead of the cutting tool is compressed and forms a chip, which begins to flow along the chip–tool interface (Fig. 27-14). As a result of the high temperature, the high pressure, and the high frictional resistance against the flow of the chip along the chip–tool interface, small particles of metal begin adhering to the edge of the cutting tool while the chip shears away. As the cutting process continues, more particles adhere to the cutting tool; a larger buildup results, which affects the cutting action. The built-up edge increases in size and becomes more unstable; eventually a point is reached where fragments are torn off. Portions of the fragments which break off stick to both the chip and the workpiece (Fig. 27-15). The buildup and breakdown of the built-up edge occur rapidly during cutting action and cover the machined surface with a multitude of built-up fragments usually identified by a rough, grainy surface. These fragments adhere to and score the machined surface; the result is a poor surface finish.

The continuous chip with the built-up edge, as well as being the main cause of surface roughness, also shortens the cutting-tool life. When a cutting tool starts to dull, it creates a rubbing or compressing action on the workpiece, which generally produces work-hardened surfaces. This type of chip affects cutting-tool life in two ways:

1. The fragments of the built-up edge abrade the tool flank as they escape with the workpiece and chip.

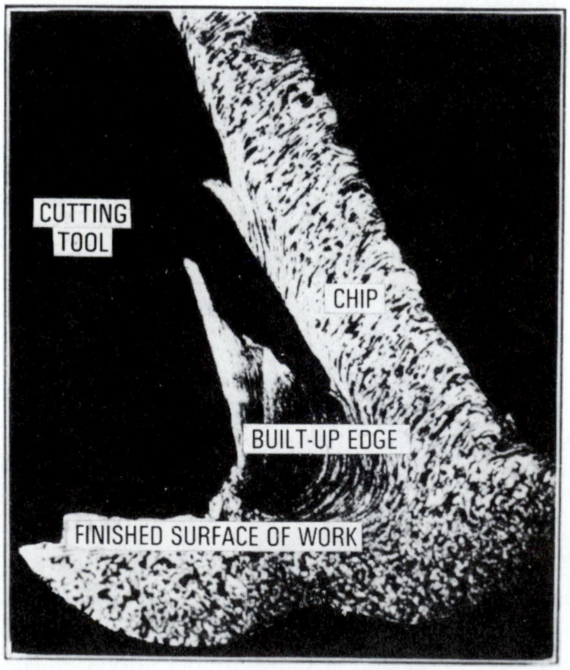
CUTTING
TOOL
CHIP
BUILT-UP EDGE
FINISHED SURFACE OF WORK

Figure 27-15 *A Type 3 continuous chip with a built-up edge being formed. (Courtesy Cincinnati Milacron Inc.)*

2. A cratering effect is caused a short distance back from the cutting edge where the chip contacts the tool face. As this cratering continues, it eventually extends closer to the cutting edge until fracture or breakdown occurs.

REVIEW QUESTIONS

1. Explain the original theory of what occurred during a metal-cutting operation.
2. Why was extensive research into metal cutting carried out?
3. Explain the new theory of the metal-cutting operation.

METAL-CUTTING TERMINOLOGY

4. Define the following metal-cutting terms:
   a. Built-up edge
   b. Chip–tool interface
   c. Plastic deformation
   d. Shear angle or plane

PLASTIC FLOW OF METAL

5. Why was research conducted to determine the plastic flow in metal?
6. Briefly describe what occurs when the following are forced into a block of photoelastic material:
   a. A flat punch
   b. A narrow-faced punch
   c. A knife-edge punch
7. Describe the two fundamental processes involved in metal cutting.

CHIP FORMATION

8. Describe briefly how each of the following chip types is produced:
   a. Discontinuous
   b. Continuous
   c. Continuous with a built-up edge
9. Which is the most desirable type of chip? Give reasons for your answer.
10. What conditions must be present in order to produce the Type 2 chip?
11. Explain how a built-up edge is formed and state its effect on cutting-tool life.

UNIT 28

Machinability of Metals

Machinability describes the ease or difficulty with which a metal can be machined. Such factors as cutting-tool life, surface finish produced, and power required must be considered. Machinability has been measured by the length of cutting-tool life in minutes or by the rate of stock removal in relation to the cutting speed employed, that is, depth of cut. For finish cuts, machinability refers to the life of the cutting tool and the ease with which a good surface finish is produced.

OBJECTIVES

After completing this unit you will be able to:

1. Explain the factors that affect the machinability of metals
2. Describe the effect of positive and negative rake angles on a cutting tool
3. Assess the effects of temperature and cutting fluids on the surface finish produced

GRAIN STRUCTURE

The machinability of a metal is affected by its microstructure and will vary if the metal has been annealed. The ductility and shear strength of a metal can be modified greatly by operations such as annealing, normalizing, and stress relieving. Certain chemical and physical modifications of steel will improve their machinability. Free-machining steels have generally been modified in the following manner by:

1. The addition of sulfur
2. The addition of lead
3. The addition of sodium sulfite
4. Cold working, which modifies the ductility

By making these (free-machining) modifications to the steel, three main machining characteristics become evident:

1. Tool life is increased.
2. A better surface finish is produced.
3. Lower power consumption is required for machining.

LOW-CARBON (MACHINE) STEEL

The microstructure of *low-carbon steel* may have large areas of ferrite (iron) interspersed with small areas of pearlite (Fig. 28-1A and B). Ferrite is soft, with high ductility and low strength, whereas pearlite, a combination of ferrite (iron) and iron carbide, has low ductility and high strength. When the amount of ferrite in steel is greater than pearlite—or the ferrite is arranged in alternate layers with pearlite (Fig. 28-1C and D)—the amount of power required to remove material increases and the surface finish produced is poor. Figure 28-2A and B illustrates a more desirable microstructure in steel because the pearlite is well distributed, and the material is therefore better for machining purposes.

HIGH-CARBON (TOOL) STEEL

A greater amount of pearlite is present in *high-carbon (tool) steel* because of the higher carbon content. The greater the amount of pearlite (low ductility and high strength) present in the steel, the more difficult it becomes to machine the steel efficiently. It is therefore desirable to anneal these steels to alter their microstructures and, as a result, improve their machining qualities.

ALLOY STEEL

Alloy steels are combinations of two or more metals. These steels generally are slightly more difficult to machine than low- or high-carbon steels. In order to improve their machining qualities, combinations of sulfur and lead or sulfur and manganese in proper proportions are sometimes added to alloy steels. A combination of normalizing and annealing is also used with some types of alloy steels to create desirable machining characteristics. The machining of *stainless steel,* generally difficult because of its work-hardening qualities, can be greatly eased by the addition of selenium.

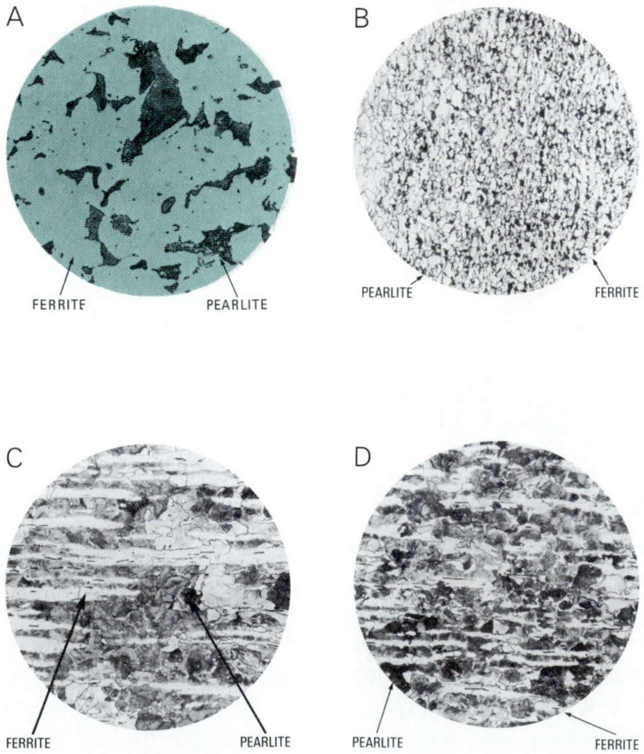
A
FERRITE
PEARLITE
B
PEARLITE
FERRITE
C
FERRITE
PEARLITE
D
PEARLITE
FERRITE

Figure 28-1 *Photomicrographs indicating undesirable steel microstructures. (Courtesy Cincinnati Milacron Inc.)*

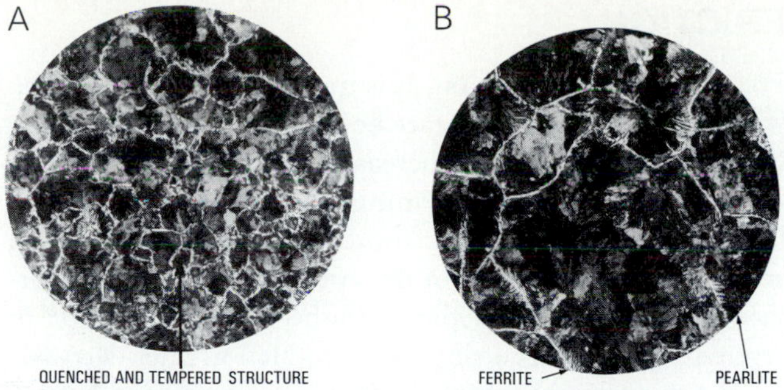
A
B
QUENCHED AND TEMPERED STRUCTURE
FERRITE
PEARLITE

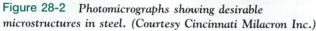

Figure 28-2 *Photomicrographs showing desirable microstructures in steel. (Courtesy Cincinnati Milacron Inc.)*

CAST IRON

Cast iron, consisting generally of ferrite, iron carbide, and free carbon, forms an important group of materials used by industry. The microstructure of cast iron can be controlled by the addition of alloys, the method of casting, the rate of cooling, and by heat treating. *White cast iron* (Fig. 28-3A), cooled rapidly after casting, is usually hard and brittle because of the formation of hard iron carbide. *Gray cast iron* (Fig. 28-3B) is cooled gradually; its structure is composed of compound pearlite, a mixture of fine ferrite and iron carbide, and flakes of graphite. Because of the gradual cooling, it is softer and therefore easier to machine.

Iron carbide and the presence of sand on the outer surface of the casting generally make cast iron a little difficult to machine. Through annealing, the microstructure is altered. The iron carbide is broken down into graphitic carbon and ferrite; the cast iron is thus made easier to machine. The addition of silicon, sulfur, and manganese gives cast iron different qualities and improves its machinability.

ALUMINUM

Pure *aluminum* is generally more difficult to machine than most aluminum alloys. It produces long, stringy chips and is much harder on the cutting tool because of its abrasive nature.

Most alloys of aluminum can be cut at high speeds, yielding a good surface finish and long tool life. Hardened and tempered alloys are generally easier to machine than annealed alloys and produce a better surface finish. Alloys containing silicon are more difficult to machine since the chips tear, rather than shear, from the work, thus producing a poorer surface finish. Cutting fluid is generally used when heavy cuts and feeds are used for machining aluminum or its alloys.

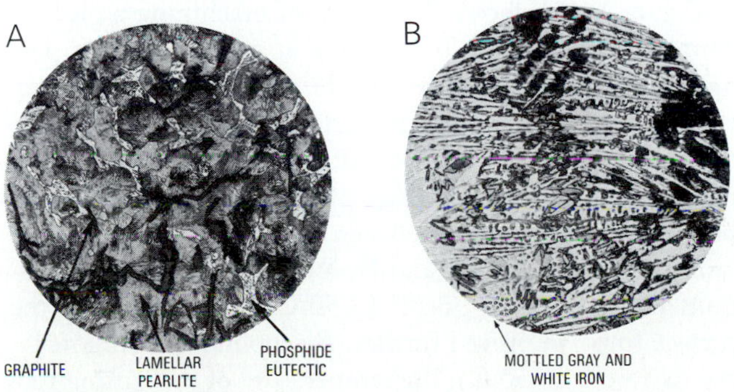
A
B
GRAPHITE
LAMELLAR PEARLITE
PHOSPHIDE EUTECTIC
MOTTLED GRAY AND WHITE IRON

Figure 28-3 *(A) The microstructure of white cast iron; (B) the microstructure of gray cast iron. (Courtesy Cincinnati Milacron Inc.)*

COPPER

Copper is a heavy, soft, reddish-colored metal refined from copper ore (copper sulfide). It has high electrical and thermal conductivity, good corrosion resistance, strength, and is easily welded, brazed, or soldered. It is very ductile and is easily drawn into wire and tubing. Since copper work hardens readily, it must be heated at about 1200°F (648.8°C) and quenched in water to anneal.

Because of its softness, copper does not machine well. The long chips produced in drilling and tapping tend to clog the flutes of the cutting tool and must be cleared frequently. Sawing and milling operations require cutters with good chip clearance. Coolant should be used to minimize heat and aid the cutting action.

Copper-Base Alloys

Brass, an alloy of copper and zinc, has good corrosion resistance and is easily formed, machined, and cast. There are several forms of brass. *Alpha brasses* containing up to 36 percent zinc are suitable for cold working. *Alpha + beta brasses* containing 54 to 62 percent copper are used in hot working of this alloy. Small amounts of tin or antimony are added to alpha brasses to minimize the pitting effect of salt water on this alloy. Brass alloys are used for water and gasoline line fittings, tubing, tanks, radiator cores, and rivets.

Bronze, the term which originally referred to an alloy of copper and tin, has now been extended to include all alloys except copper–zinc alloys, which contain up to 12 percent of the principal alloying element.

Phosphor-bronze contain about 90 percent copper, 10 percent tin, and a very small amount of phosphorus, which acts as a hardener. This metal has high strength, toughness, and corrosion resistance and is used for lock washers, cotter pins, springs, and clutch discs.

Silicon-bronze (a copper–silicon alloy) contains less than 5 percent silicon and is the strongest of the work-hardenable copper alloys. It has the mechanical properties of machine steel and the corrosion resistance of copper. It is used for tanks, pressure vessels, and hydraulic pressure lines.

Aluminum-bronze (a copper–aluminum alloy) contains

between 4 and 11 percent aluminum. Other elements such as iron, nickel, manganese, and silicon are added to aluminum bronzes. Iron (up to 5 percent) increases the strength and refines the grain. The addition of nickel (up to 5 percent) has effects similar to those of iron. Silicon (up to 2 percent) improves machinability. Manganese promotes soundness in casting. Aluminum-bronzes have good corrosion resistance and strength and are used for condenser tubes, pressure vessels, nuts, and bolts.

Beryllium-bronze (copper and beryllium), containing up to about 2 percent beryllium, is easily formed in the annealed condition. It has a high tensile strength and fatigue strength in the hardened condition. Beryllium-bronze is used for surgical instruments, bolts, nuts, and screws.

THE EFFECTS OF TEMPERATURE AND FRICTION

In the process of cutting metals, heat is created by:

1. The plastic deformation occurring in the metal during the process of forming a chip
2. The friction created by the chips sliding along the cutting tool face

The cutting temperature varies with each type of metal and increases with the cutting speed employed and the rate of metal removal. The *greatest heat* is generated when ductile material of high tensile strength, such as steel, is being cut. The *lowest heat* is generated when soft material of low tensile strength, such as aluminum, is being cut. The maximum temperature attained during the cutting action will affect cutting-tool life, quality of the surface finish, rate of production, and accuracy of the workpiece.

At times, the temperature of metal immediately ahead of the cutting tool comes close to the melting temperature of the metal being cut. This great heat affects the life of the cutting tool. High-speed steel cutting tools cannot withstand the same high temperatures that cemented-carbide tools can without the cutting edges breaking down.

High-speed steel cutting tools are capable of cutting metal even when the cutting tools turn red from the cutting action. This property is known as *red hardness* and occurs at temperatures above 900°F (482°C). However, when the temperature exceeds 1000°F (538°C), the edge of the cutting tool will begin to break down.

Cemented-carbide cutting tools can be used efficiently at temperatures up to 1600°F (871°C). The carbide tools are harder than high-speed steel tools, have greater wear resistance, and perform extremely well under red-hardness operating conditions. Therefore, much higher cutting speeds can be used with carbide tools than with high-speed tools.

FRICTION

For efficient cutting action, it is important that the friction between the chip and tool face be kept as low as possible. As the coefficient of friction increases, there is a greater possibility of a built-up edge forming on the cutting edge. The larger the built-up edge, the more friction is created, which results in the breakdown of the cutting edge and poor surface finish. Every time the machine must be stopped to regrind or replace a cutting tool, production rates decrease.

The temperature created by friction also affects the accuracy of the machined part. Even though the workpiece does not reach the same temperature as the cutting-tool point, it is still high enough to cause the metal to expand. If a part that has been heated by the cutting action is machined to size, the part will be smaller than required when it cools to room temperature. A good supply of cutting fluid will help reduce friction at the chip–tool interface and also help maintain efficient cutting temperatures.

SURFACE FINISH

Many factors affect the surface finish produced by a machining operation, the most common being the feed rate, the nose radius of the tool, the cutting speed, the rigidity of the machining operation, and the temperature generated during the machining process.

If a high temperature is created during the cutting action, there is a marked tendency for a rough surface finish to result. The reason for this is that at high temperatures metal particles tend to adhere to the cutting tool and form a built-up edge. A direct relationship between the temperature of the workpiece and the quality of the surface finish is illustrated in Fig. 28-4.

Figure 28-4A shows the results of machining a piece of aluminum without cutting fluid at 200°F (93°C). The rough surface finish indicates the presence of a built-up edge on the cutting tool. The same piece of aluminum was machined under the same conditions, but at a room temperature of 75°F (24°C) (Fig. 28-4B). A considerable improvement can be noted between the surface finishes of the samples in Figs. 28-4A and B. When the piece of aluminum was cooled to −60°F (−50°C) and machined, the surface finish improved further. By cooling the work material to −60°F (−50°C), the temperature of the cutting-tool edge was reduced considerably and resulted in a much better surface finish (Fig. 28-4C) than that produced at 200°F (93°C).

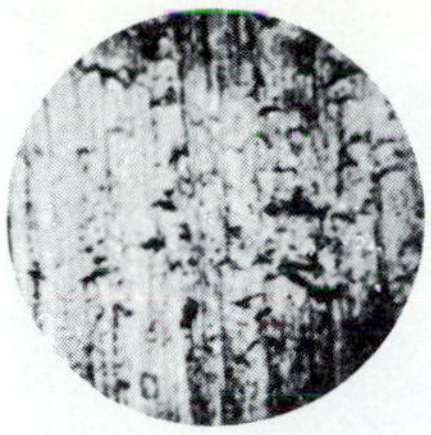

A

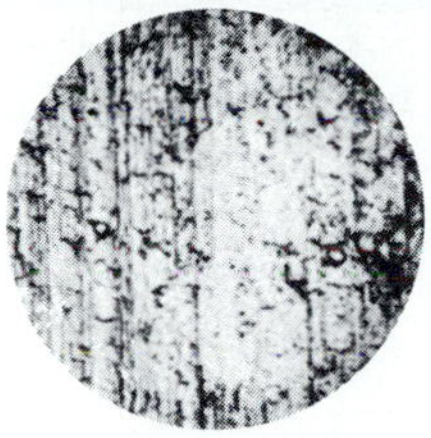

B

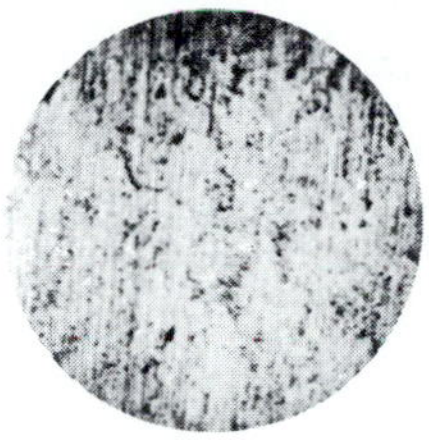

C

Figure 28-4 *(A) Aluminum machined at 200°F (93°C); (B) aluminum machined at 75°F (24°C); (C) aluminum machined after cooling to −60°F (−50°C).*

EFFECTS OF CUTTING FLUIDS

Cutting fluids are important to most machining operations in that they make it possible to cut metals at higher rates of speed. They perform three important functions.

1. They reduce the temperature of the cutting action.
2. They reduce the friction of the chips sliding along the tool face.
3. They decrease tool wear and increase tool life.

There are three types of cutting fluids: cutting oils, emulsifiable (soluble oils), and chemical (synthetic) cutting fluids. Some cutting fluids form a nonmetallic film on the metal surface, which prevents the chip from sticking to the cutting edge. This prevents a built-up edge from forming and as a result, a better surface finish is produced. The surface finish of most metals can be improved considerably by the use of the proper cutting fluids.

Cutting fluids are generally used for machining steel, alloy steel, brass, and bronze with high-speed steel cutting tools. As a rule, cutting fluids are not generally used with cemented-carbide tools unless a great quantity of cutting fluid can be applied to ensure uniform temperatures to prevent the carbide inserts from cracking. Cast iron, aluminum, and magnesium alloys are generally machined dry; however, cutting fluids have been used with good results in some cases.

REVIEW QUESTIONS

MACHINABILITY OF METALS

1. Define *machinability*.
2. What factors affect the machinability of a metal?
3. Compare the microstructure of low-carbon and high-carbon steels with respect to their machinability.
4. How can the machining qualities of alloy steels be improved?
5. Why is pure aluminum more difficult to machine than most aluminum alloys?
6. What can be done to improve the machining of aluminum and its alloys?

EFFECTS OF TEMPERATURE AND FRICTION

7. Name two methods by which heat is created during machining.
8. How does high temperature affect a machining operation?
9. Why is it important that friction between the chip and tool be kept to a minimum?

SURFACE FINISH

10. What common factors determine surface finish?
11. Why does high temperature affect the surface finish produced?

EFFECTS OF CUTTING FLUIDS

12. List four ways in which cutting fluids assist the machining of metals.
13. What precaution should be taken when cutting fluids are used with carbide tools?

UNIT 29

Cutting Tools

One of the most important components in the machining process is the cutting tool, the performance of which will determine the efficiency of the operation. Consequently, much thought should be given not only to the selection of the cutting-tool material but also to the cutting-tool angles required to machine a workpiece material properly.

There are basically two types of cutting tools (excluding abrasives): single- and multi-point tools. Since both types must have rake and clearance angles ground or formed on them in order to cut, the nomenclature of cutting-tool points will apply to both types. The lathe tool, which is the most common single-point tool, will be discussed in greater detail. The principles involved in this type of cutting tool will then be related to multi-point cutting tools for ease of understanding.

OBJECTIVES

After completing this unit you will be able to:

1. Use the nomenclature of a cutting-tool point
2. Explain the purpose of each type of rake and clearance angle
3. Identify the applications of various types of cutting-tool materials
4. Describe the cutting action of different types of machines

CUTTING-TOOL MATERIALS

Lathe toolbits are generally made of five materials: high-speed steel, cast alloys (such as stellite), cemented carbides, ceramics, and cermets. More exotic cutting tool materials, such as polycrystalline cubic boron nitride (PCBN), commonly called Borazon, and polycrystalline diamond (PCD), are finding wide use in the metal-working industry because of the increased productivity they offer. Borazon is used to machine hardened alloy steels and tough superalloys. Polycrystalline diamond cutting tools are used to machine nonferrous and nonmetallic materials requiring close tolerances and a high surface finish. The properties possessed by each of these materials are different and the application of each depends on the material being machined and the condition of the machine.

Lathe toolbits should possess the following properties.

1. They should be hard.
2. They should be wear-resistant.
3. They should be capable of maintaining a *red hardness* during the machining operation. Red hardness is the ability of a cutting tool material to maintain a sharp cutting edge even when it turns red, due to the high heat produced at the work–tool interface during the cutting operation.
4. They should be able to withstand shock during the cutting operation.
5. They should be shaped so that the edge can penetrate the work. The shape will be determined by the cutting-tool material, the material being cut, and the angle of keenness.

HIGH-SPEED STEEL TOOLBITS

Probably the toolbit most commonly used in schools for lathe operations is the high-speed steel toolbit. High-speed steels may contain combinations of tungsten, chromium, vanadium, molybdenum, and cobalt. They are capable of taking heavy cuts, withstanding shock, and maintaining a sharp cutting edge under red heat.

High-speed steel toolbits are generally of two types: molybdenum-base (Group M) and tungsten-base (Group T). The most widely used tungsten-base toolbit is known as T_1, which is sometimes called 18-4-1 since it contains about 18 percent tungsten, 4 percent chromium, and 1 percent vanadium.

A general-purpose molybdenum-base high-speed steel toolbit is known as M_1 or 8-2-1. This alloy contains about 8 percent molybdenum, 2 percent tungsten, 1 percent vanadium, and 4 percent chromium.

These two types are general-purpose tools; if more red hardness is desired, a tool containing more cobalt should be selected. Since there are many different grades of high-speed steel toolbits, one should refer to the manufacturer's recommendations for a toolbit for a specific job. Table 29-1 indicates the properties imparted to a high-speed steel toolbit by the various alloying elements.

CAST ALLOY TOOLBITS

Cast alloy (stellite) toolbits usually contain 25 to 35 percent chromium, 4 to 25 percent tungsten, and 1 to 3 percent carbon; the remainder is cobalt. These toolbits have high hardness, high resistance to wear, and excellent red-hardness qualities. Since they are cast, they are weaker and more brittle than high-speed steel toolbits. Stellite toolbits are capable of high speeds and feeds on deep uninterrupted cuts. They may be operated at about two to two-and-a-half times the speed of a high-speed steel toolbit.

NOTE: When grinding stellite toolbits, apply only light pressure and do not quench the toolbit in water.

CEMENTED-CARBIDE TOOLBITS

Cemented-carbide toolbits (Fig. 29-1) are capable of cutting speeds three to four times those of high-speed steel toolbits. They have low toughness but high hardness and excellent red-hardness qualities.

Cemented carbide consists of tungsten carbide sintered in a cobalt matrix. Sometimes other materials such as titanium or tantalum may be added before sintering to give the material the desired properties.

Straight tungsten carbide toolbits are used to machine cast iron and nonferrous materials. Since they crater easily and wear rapidly, they are not suitable for machining steel. Crater-resistant carbides, which are used for machining steel, are made by adding titanium and/or tantalum carbides to the tungsten carbide and cobalt.

Figure 29-1 *A variety of cemented carbide tool inserts.*

Different grades of carbides are manufactured for different work requirements. Those used for heavy roughing cuts will contain more cobalt than those used for finishing cuts, which are more brittle and have greater wear resistance at higher finishing speeds.

COATED CARBIDE TOOLBITS

Coated carbide cutting tools are made by depositing a very thin layer of wear-resistant material, such as titanium nitride, titanium carbide, or aluminum oxide (ceramic) on the cutting edge of the tool. This fused layer increases lubricity and thus improves the wear resistance of the cutting edge by 200–500 percent and will lower the breakage resistance by up to 20 percent while providing longer life and increased cutting speeds.

Titanium-coated inserts offer greater wear resistance at speeds below 500 sf/min., whereas ceramic coated tips are best suited for higher cutting speeds. Both types of insert may be used for cutting steels, cast irons, and nonferrous materials.

CERAMIC TOOLBITS

A *ceramic* is a heat-resistant material produced without a metallic bonding agent such as cobalt. Aluminum oxide is the most popular material used to make ceramic cutting tools. Titanium oxide or titanium carbide may be used as an additive, depending on the cutting tool application.

Ceramic tools (Fig. 29-2) permit higher cutting speeds, increased tool life, and better surface finish than do carbide tools. However, they are much weaker than carbide or coated carbide tools and must be used in shock-free or low-shock situations.

CERMET TOOLBITS

A *cermet* is a cutting-tool insert composed of ceramics and metal. Most cermets are made from aluminum oxide, titanium carbide, and zirconium oxide compacted and com-

Figure 29-2 *A variety of ceramic cutting tool inserts.*

Table 29-1 The Effect of Alloying Elements on Steel

Effect	Element											
	Carbon	Chromium	Cobalt	Lead	Manganese	Molybdenum	Nickel	Phosphorus	Silicon	Sulfur	Tungsten	Vanadium
Increases tensile strength	X	X			X	X	X					
Increases hardness	X	X										
Increases wear resistance	X	X			X		X				X	
Increases hardenability	X	X			X	X	X					X
Increases ductility					X							
Increases elastic limit		X				X						
Increases rust resistance		X					X					
Increases abrasion resistance		X			X							
Increases toughness		X				X	X					X
Increases shock resistance		X					X					X
Increases fatigue resistance												X
Decreases ductility	X	X										
Decreases toughness			X									
Raises critical temperature		X	X								X	
Lowers critical temperature					X		X					
Causes hot shortness										X		
Causes cold shortness								X				
Imparts red hardness			X			X					X	
Imparts fine grain structure					X							X
Reduces deformation					X		X					
Acts as deoxidizer					X				X			
Acts as desulphurizer					X							
Imparts oil hardening properties		X			X	X	X					
Imparts air hardening properties					X	X						
Eliminates blow holes								X				
Creates soundness in casting									X			
Facilitates rolling and forging					X				X			
Improves machinability				X						X		

pressed under intense heat. The advantages of cermet toolbits are:

- They exceed the equivalent tool life of coated and uncoated carbides.
- They can be used for machining at high temperatures.
- They produce an improved surface finish, which eliminates the need for grinding and provides greater dimensional control.
- They may be used to machine steels up to 66 Rc hardness.

DIAMOND TOOLBITS

Diamond tools are used mainly to machine nonferrous metals and abrasive nonmetallics. Single-crystal natural diamonds have high-wear but low shock-resistant factors. The new type of diamond tooling (polycrystalline diamonds) consists of tiny synthetic diamonds fused together and bonded to a suitable carbide substrate. Polycrystalline cutting tools offer greater wear and shock resistance and greatly increased cutting speeds. Polycrystalline diamond tools offer improved surface finish, better part-size control, up to 100 times greater tool life than carbide tools, and increased productivity.

CUBIC BORON NITRIDE TOOLBITS

Cubic boron nitride (Borazon) is next to diamond on the hardness scale. These cutting tools are made by bonding a layer of polycrystalline cubic boron nitride to a cemented-carbide substrate, which provides good shock resistance. They offer exceptionally high wear resistance and edge life and may be used to machine high-temperature alloys and hardened ferrous alloys.

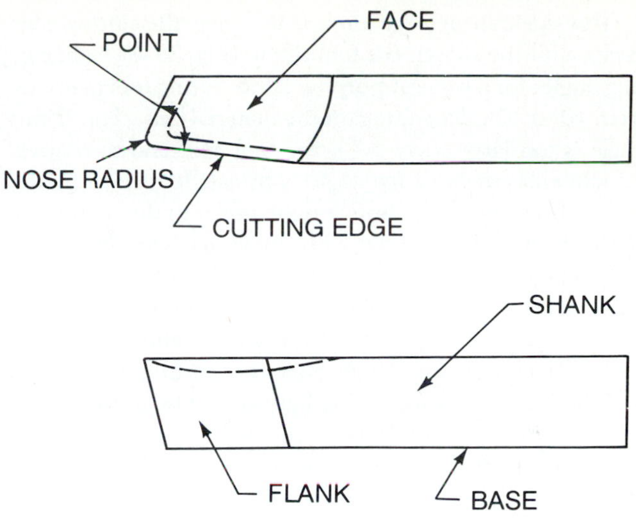
FACE
POINT
NOSE RADIUS
CUTTING EDGE
SHANK
FLANK
BASE

Figure 29-3 *Angles and clearances for lathe cutting tools.*

CUTTING-TOOL NOMENCLATURE

Cutting tools used on a lathe are generally single-pointed, and although the shape of the tool is changed for various applications, the same nomenclature applies to all cutting tools (Fig. 29-3).

The *base* is the bottom surface of the tool shank.

The *cutting edge* is the leading edge of the toolbit that does the cutting.

The *face* is the surface against which the chip bears as it is separated from the work.

The *flank* is the surface of the tool adjacent to and below the cutting edge.

The *nose* is the tip of the cutting tool formed by the junction of the cutting edge and the front face.

The *nose radius* is the radius to which the nose is ground. The size of the radius will affect the finish. For rough turning, a small nose radius [about 1/64 in. (0.38 mm)] is used. A larger radius [about 1/16 to 1/8 in. (1.5 to 3 mm)] is used for finish cuts.

The *point* is the end of the tool that has been ground for cutting purposes.

The *shank* is the body of the toolbit or the part held in the toolholder.

LATHE TOOLBIT ANGLES AND CLEARANCES

Proper toolbit performance depends on the clearance and rake angles, which must be ground on the toolbit. Although these angles vary for different materials, the nomenclature is the same for all toolbits (see Table 29-2).

Table 29-2 Recommended Angles for Single-Point Carbide Tools

Material	Side Relief	End Relief	Side Rake	Back Rake	Angle of Keenness
Aluminum	12	8	15	35	27
Brass	10	8	5 to −4	0	6 to 15
Bronze	10	8	5 to −4	0	6 to 15
Cast iron	10	8	12	5	22
Copper	12	10	20	16	32
Machine steel	10 to 12	8	12 to 18	8 to 15	22 to 30
Tool steel	10	8	12	8	22
Stainless steel	10	8	15 to 20	8	18

The *side cutting edge angle* is the angle the cutting edge forms with the side of the tool shank (Fig. 29-4). Side cutting angles for a general-purpose lathe cutting tool may vary from 10 to 20°, depending on the material being cut. If this angle is too large (over 30°), the tool will tend to chatter.

The *end cutting edge angle* is the angle formed by the end cutting edge and a line at right angles to the center line of the toolbit (Fig. 29-4). This angle may vary from 5 to 30°, depending on the type of cut and finish desired. An angle of 5 to 15° is satisfactory for roughing cuts; angles between 15 and 30° are used for general-purpose turning tools. The larger angle permits the cutting tool to be swivelled to the left for taking light cuts close to the dog or chuck, or when turning to a shoulder.

The *side relief (clearance) angle* is the angle ground on the flank of the tool below the cutting edge (Figs. 29-4 and 5). This angle is generally 6 to 10°. The side clearance on a toolbit permits the cutting tool to advance lengthwise into the rotating work and prevents the flank from rubbing against the workpiece.

The *end relief (clearance) angle* is the angle ground below the nose of the toolbit, which permits the cutting tool to be fed into the work. It is generally 10 to 15° for general-purpose tools (Figs. 29-4 and 5). This angle must be measured when the toolbit is held in the toolholder. The end relief angle varies with the hardness and type of material and the type of cut being made. The end relief angle is smaller for harder materials, providing support under the cutting edge.

The *side rake angle* is the angle at which the face is ground away from the cutting edge. For general-purpose toolbits, the side rake is generally 14° (Figs. 29-4 and 5). Side rake creates a keener cutting edge and allows the chips to flow away quickly. For softer materials, the side rake angle is generally increased. Side rake may be either positive or negative, depending on the material being cut.

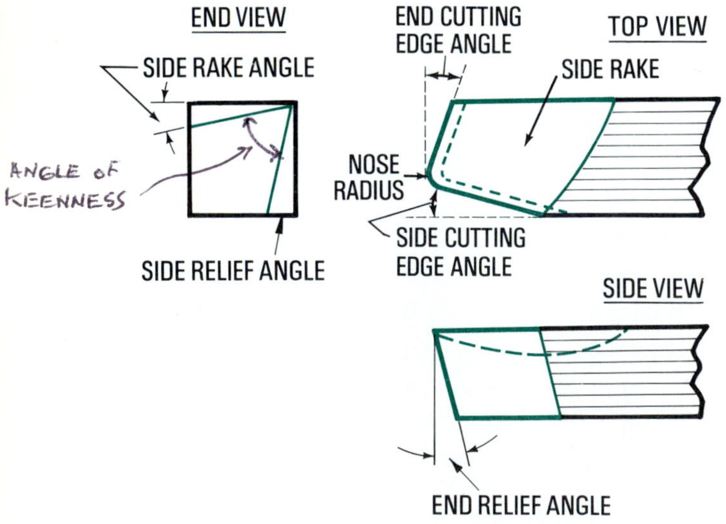

END VIEW
SIDE RAKE ANGLE
ANGLE OF KEENNESS
SIDE RELIEF ANGLE
END CUTTING EDGE ANGLE
TOP VIEW
SIDE RAKE
NOSE RADIUS
SIDE CUTTING EDGE ANGLE
SIDE VIEW
END RELIEF ANGLE

Figure 29-4 *Nomenclature of a general-purpose lathe toolbit.*

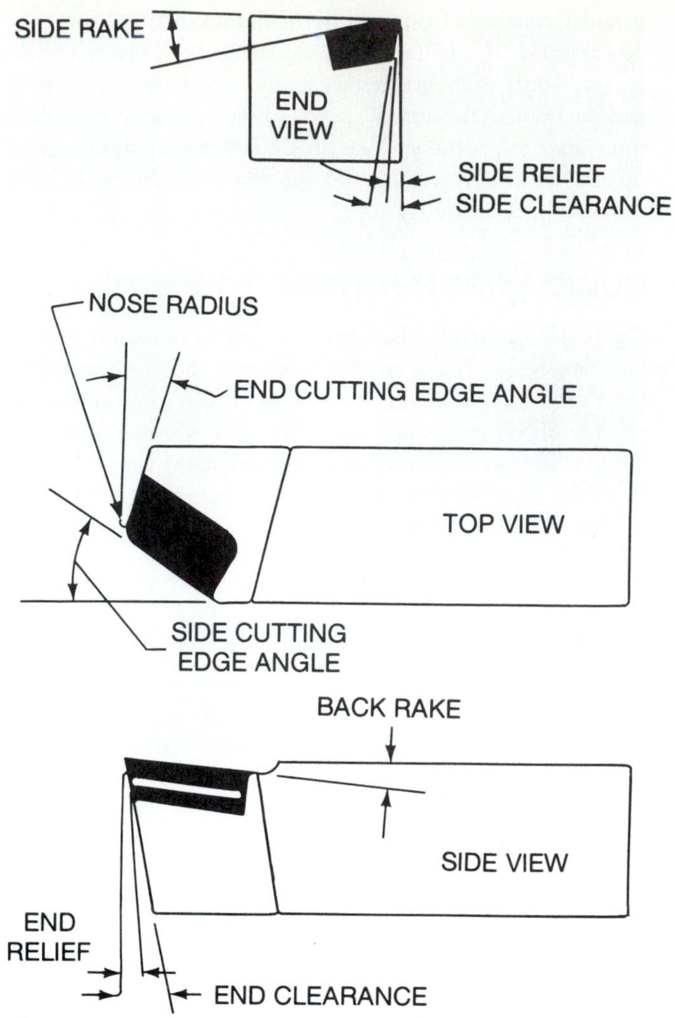

SIDE RAKE
END VIEW
SIDE RELIEF
SIDE CLEARANCE
NOSE RADIUS
END CUTTING EDGE ANGLE
TOP VIEW
SIDE CUTTING EDGE ANGLE
BACK RAKE
SIDE VIEW
END RELIEF
END CLEARANCE

Figure 29-5 *Lathe cutting tool angles and clearances.*

The *angle of keenness* is the included angle produced by grinding side rake and side clearance on a toolbit (Fig. 29-4). This angle may be altered, depending on the type of material being machined, and will be greater (closer to 90°) for harder materials.

The *back (top) rake angle* is the backward slope of the tool face away from the nose. The back rake angle is generally about 20° and is provided for in the toolholder (Fig. 29-5). Back rake permits the chips to flow away from the point of the cutting tool. Two types of back or top rake angles are provided on cutting tools and are always found on the top of the toolbit:

- *Positive rake* (Fig. 29-6A), where the point of the cutting tool and the cutting edge contact metal first and the chip moves *down the face* of the toolbit.
- *Negative rake* (Fig. 29-6B), where the face of the cutting tool contacts the metal first and the chip is forced *up the face* of the toolbit.

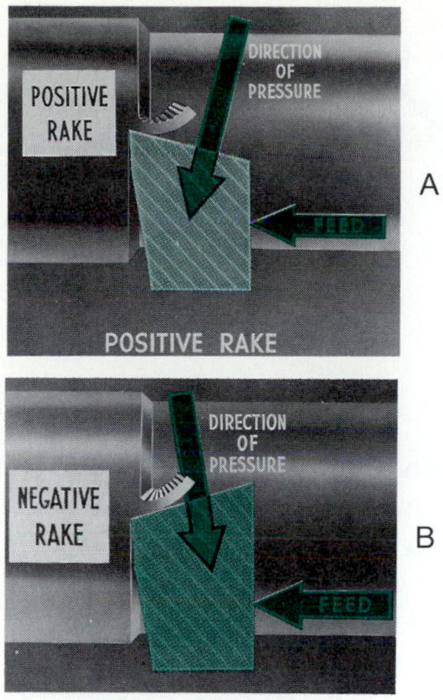
POSITIVE RAKE
DIRECTION OF PRESSURE
FEED
POSITIVE RAKE
A
NEGATIVE RAKE
DIRECTION OF PRESSURE
FEED
B

Figure 29-6 *(A) Large rake angle ground on a cutting tool. (B) Negative rake angle ground on a cutting tool.*

Each type of rake angle serves a specific purpose. The type used depends upon the machining operation being performed and the characteristics of the work material. Rake angles can be ground on cutting tools or, in the case of cutting-tool inserts, they can be held in suitable holders, which provide the rake angle desired.

POSITIVE RAKE ANGLE

A positive rake angle (Fig. 29-6A) is considered best for the efficient removal of metal. It creates a large shear angle at the shear zone, reduces friction and heat, and allows the chip to flow freely along the chip–tool interface. Positive-rake-angle cutting tools are generally used for continuous cuts on ductile materials which are not too hard or abrasive. Even though positive-rake-angle tools remove metal efficiently, they are not recommended for all work materials or cutting applications. The following factors must be considered when the type and the amount of rake angle for a cutting tool are being determined.

1. The hardness of the metal to be cut
2. The type of cutting operation (continuous or interrupted)
3. The material and shape of the cutting tool
4. The strength of the cutting edge

NEGATIVE RAKE ANGLE

A negative rake angle (Fig. 29-6B) is used for interrupted cuts and when the metal is tough or abrasive. A negative-rake-angle on the tool creates a small shear angle and a long shear zone; therefore more friction and heat are created. Although the increase in heat may seem to be a disadvantage, it is desirable when tough metals are machined with carbide cutting tools. Face-milling cutters with carbide tool inserts are a good example of the use of negative rake for interrupted and high-speed cutting.

The advantages of negative rake on cutting tools are:

- The shock from the work meeting the cutting tool is on the tool's face, not its point or edge, which prolongs the life of the tool.
- The hard outer scale on the metal does not come into contact with the cutting edge.
- Surfaces with interrupted cuts can be readily machined.
- Higher cutting speeds can be utilized.

The shape of a chip can be altered in a number of ways to improve the cutting action and reduce the amount of power required. A continuous *straight* ribbon chip on a lathe can be changed to a continuous curled ribbon by:

1. Changing the *angle of the keenness* (the included angle produced by grinding the side rake and side clearance) on a toolbit (Fig. 29-4).
2. Grinding a chip breaker behind the cutting edge of the toolbit.

A helix angle on a milling cutter affects the cutting performance by providing a shearing action when the chip is removed.

CUTTING-TOOL SHAPE

The shape of the cutting tool is very important to the efficient removal of metal. Every time a machine must be stopped to recondition or replace a worn cutting tool, production rates decrease. The life of a cutting tool is generally reported as:

1. The number of minutes that the tool has been cutting
2. The length of material cut
3. The number of cubic inches or cubic centimeters (cm^3) of material removed
4. In the case of drills, the number of inches or millimeters of hole depth drilled

To prolong cutting-tool life, reduce the friction between the chip and the tool as much as possible. This can be accomplished by providing the cutting tool with a suitable rake angle and by highly polishing the cutting-tool face with a honing stone. The polished cutting face reduces the friction on the chip-tool interface, reduces the size of the built-up edge, and generally results in better surface finish. The rake angle on cutting tools allows chips to flow away freely and reduces friction and the amount of power required for the machining operation.

The rake angle of the cutting tool also affects the shear angle or plane of the metal, which in turn determines the area of plastic deformation. If a *large rake angle* is ground on the cutting tool, a large shear angle is created in the metal during the cutting action (Fig. 29-6A). The results of a large shear angle are:

1. A thin chip is produced.
2. The shear zone is relatively short.
3. Less heat is created in the shear zone.
4. Good surface finish is produced.
5. Less power is required for the machining operation.

A small or a negative rake angle on the cutting tool (Fig. 29-6B) creates a small shear angle in the metal during the cutting process with the following results:

1. A thick chip is produced.
2. The shear zone is long.
3. More heat is produced.
4. The surface finish is not quite as good as with large rake angle cutting tools.
5. More horsepower is required for the machining operation.

TOOL LIFE

Tool life, or the number of parts produced by a cutting-tool edge before regrinding is required, is a very important cost factor in manufacturing a part or product. Consequently, cutting tools must be reground at the first sign of dullness. If a tool is used beyond this point, it will break down rapidly and much more tool material will have to be removed when regrinding, thus shortening the tool's life.

In order to detect the time when a cutting tool should be changed, most modern machines are equipped with indicators that show the horsepower used during the machining operation. When a tool becomes dull, more horsepower is required for the operation, which will show on the indicator. When this occurs, the tool should be reconditioned immediately.

The wear or abrasion of the cutting tool will determine its life. Three types of wear are generally associated with cutting tools: *flank wear, nose wear,* and *crater wear* (Fig. 29-7).

Flank wear occurs on the side of the cutting edge as a result of friction between the side of the cutting-tool edge and the metal being machined. Too much flank wear increases friction and makes more power necessary for machining. When the flank wear is 0.015 to 0.030 in. (0.38 to 0.76 mm) long, the tool should be reground.

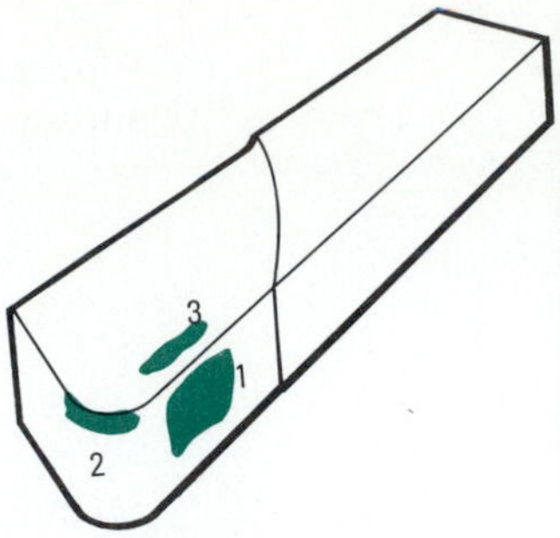
3
1
2

Figure 29-7 *The tool wear areas of a cutting tool. (1) flank wear; (2) nose wear; (3) crater wear.*

Nose wear occurs on the nose or point of the cutting tool as a result of friction between the nose and the metal being machined. Wear on the nose of the cutting tool affects the quality of the surface finish on the workpiece.

Crater wear occurs a slight distance away from the cutting edge as a result of the chips sliding along the chip–tool interface due to a built-up edge on the cutting tool. Too much crater wear eventually breaks down the cutting edge.

The following factors affect the life of a cutting tool:

1. The type of material being cut
2. The microstructure of the material
3. The hardness of the material
4. The type of surface on the metal (smooth or scaly)
5. The material of the cutting tool
6. The profile of the cutting tool
7. The type of machining operation being performed
8. Speed, feed, and depth of cut

PRINCIPLES OF MACHINING

TURNING

A high proportion of the work machined in a shop is turned on a lathe. The workpiece is held securely in a chuck or between lathe centers. A turning tool, mounted in a holder and set to a given depth of cut, is fed parallel to the axis of the work to reduce the diameter of the workpiece (Fig. 29-8).

As the workpiece revolves and the cutting tool is fed along the axis, material is separated by the edge of the cutting tool. A chip forms and slides along the cutting tool's upper surface created by the side rake. The *angle of keenness* (Fig. 29-4), which permits the tool to remove the metal as the tool is fed along the workpiece, is formed by the side relief angle clearance and the side rake ground on the toolbit. Let's assume that we are cutting a piece of machine steel; if the rake and relief clearance angles are correct and the proper speed and feed are used, a continuous chip should be formed. If the angle of keenness is too small, the edge of the tool will break down too soon. If the angle is too great, that is, little or no side rake, the metal will not be removed as effectively and greater torque will be required to remove it. In either case, a built-up edge and rough surface finish will result.

PLANING

The cutting tool used in the shaper or planer is basically the same shape as the lathe tool for machining similar materials. It should have the proper rake and clearance angles ground on it to machine the workpiece efficiently. The cutting action of the planer is illustrated in Fig. 29-9. The workpiece is moved back and forth under the cutting tool, which is fed sideways a set amount at the end of each table reversal.

PLAIN MILLING

A milling cutter is a multitooth tool having several equally spaced cutting edges (teeth) around its periphery. Each tooth may be considered to be a single-point cutting tool and must have proper rake and clearance angles to cut effectively. Figure 29-10 shows the chip formation produced by a helical milling cutter.

The workpiece, held in a vise or fastened to the table, is fed into the horizontal revolving cutter by hand or by automatic table feed. As the work is fed into the cutter, each tooth makes successive cuts, which produce a smooth flat or profiled surface, depending on the shape of the cutter used (Fig. 29-11). The nomenclature of a plain milling cutter is shown in Fig. 29-12.

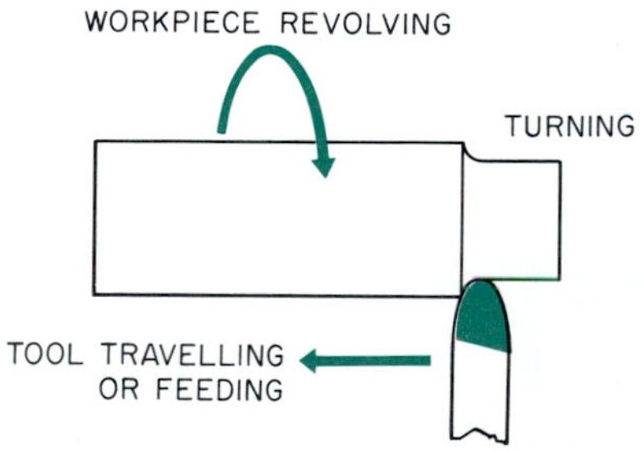
WORKPIECE REVOLVING
TURNING
TOOL TRAVELLING
OR FEEDING
Figure 29-8 *The cutting action of a lathe or turning center.*

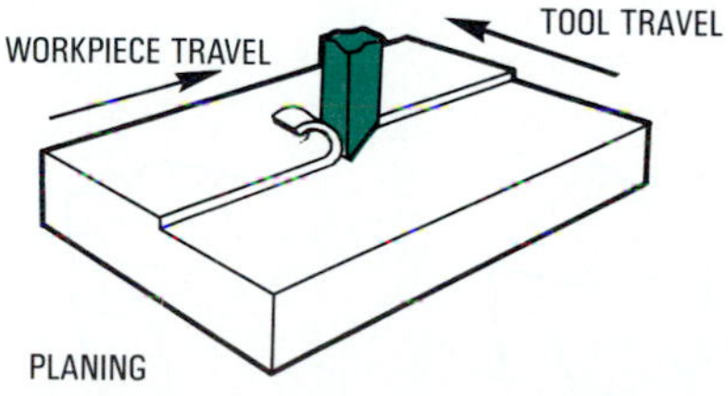
WORKPIECE TRAVEL
TOOL TRAVEL
PLANING
Figure 29-9 *The cutting action of a planer.*

Figure 29-10 *The type of chip produced by a helical milling cutter. (Courtesy Cincinnati Milacron Inc.)*

END AND FACE MILLING

End mills (Fig. 29-13) are multitooth cutters held vertically in a vertical milling machine spindle or attachment. They are used primarily for cutting slots or grooves, whereas shell end mills and face mills are used primarily for producing flat surfaces. The workpiece is held in a vise or clamped to the table and is fed into the revolving cutter by hand or automatically. When end milling, the cutting is done by the periphery of the teeth. The nomenclature of an end mill is shown in Fig. 29-14.

The inserted blade face mill consists of a body that holds

Figure 29-11 *As the workpiece is fed into the revolving cutter, each tooth removes material from the work. (Courtesy Cincinnati Milacron Inc.)*

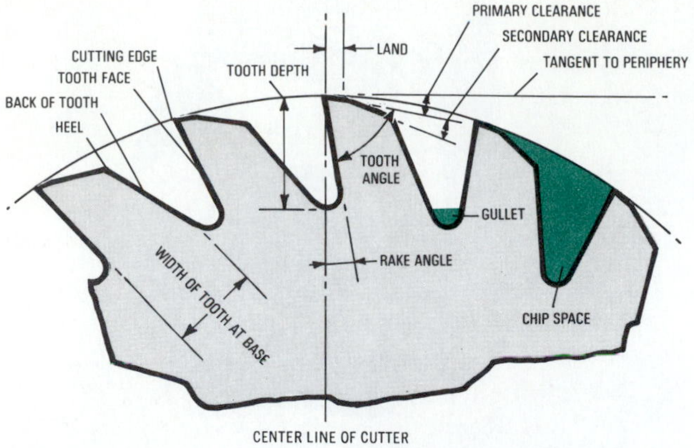
PRIMARY CLEARANCE
SECONDARY CLEARANCE
LAND
TANGENT TO PERIPHERY
CUTTING EDGE
TOOTH FACE
TOOTH DEPTH
BACK OF TOOTH
HEEL
TOOTH ANGLE
GULLET
RAKE ANGLE
WIDTH OF TOOTH AT BASE
CHIP SPACE
CENTER LINE OF CUTTER

Figure 29-12 *The nomenclature of a plain milling cutter.*

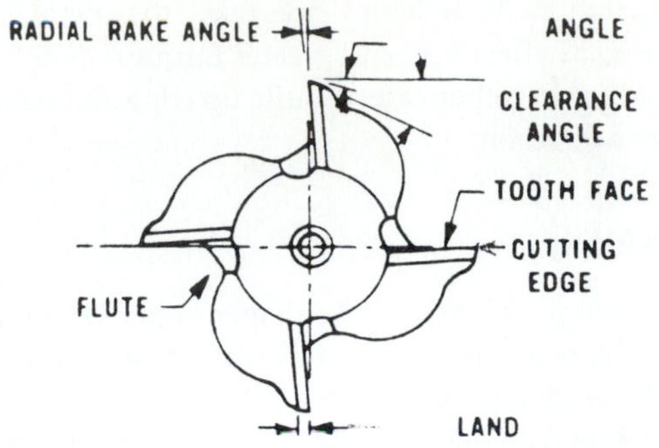
RADIAL RAKE ANGLE
ANGLE
CLEARANCE ANGLE
TOOTH FACE
CUTTING EDGE
FLUTE
LAND

several equally spaced inserts at the required rake angle. The lower edge of each insert has a relief or clearance angle ground on it. Since the cutting action occurs at the lower corner of the insert, generally each corner is chamfered to give it strength and to prevent the corners from breaking off (Fig. 29-15).

DRILLING

The drill is a multi-edge cutting tool that cuts on the point. The drill's cutting edges or lips are provided with lip clearance (Fig. 29-16) to permit the point to penetrate the workpiece as the drill revolves. It may be fed into the work manually or automatically. The rake angle is provided by helical-shaped flutes that slope away from the cutting edge. As in other cutting tools, the included angle between the rake angle and the clearance angle is known as the angle of keenness. Cutting-point angles for a standard drill are shown in Fig. 29-16; chip formation is shown in Fig. 29-17.

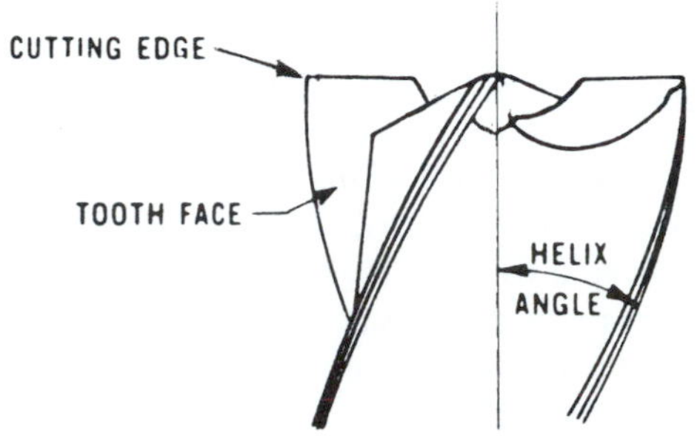
CUTTING EDGE
TOOTH FACE
HELIX ANGLE

Figure 29-14 *The nomenclature of an end mill.*

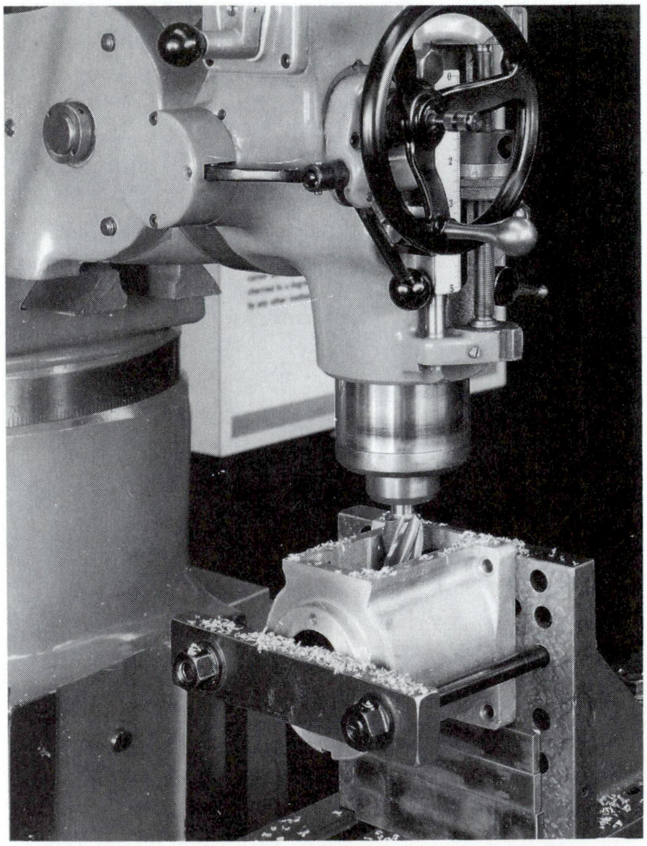

Figure 29-13 *End mills generally cut on the periphery.*

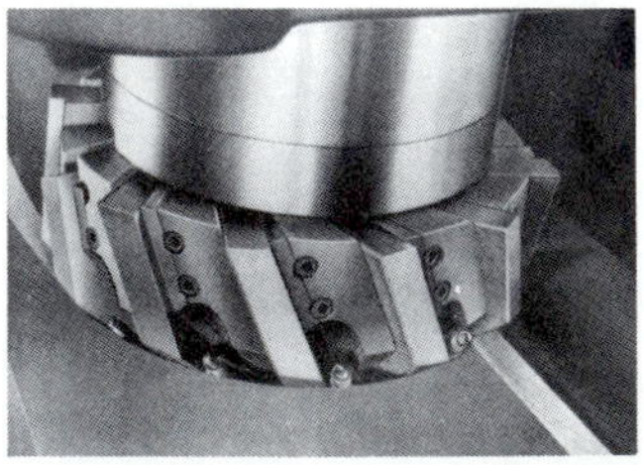

Figure 29-15 *The corner of each tooth on face milling cutters is generally chamfered to provide strength and prevent the teeth from chipping.*

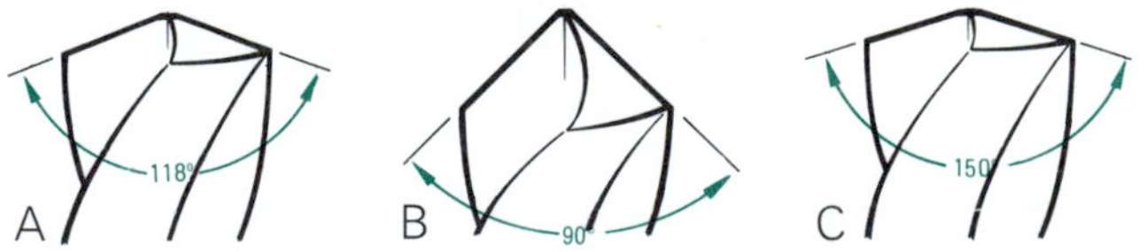
118°
A
B
90°
150°
C

Figure 29-16 *The characteristics of a drill point.*

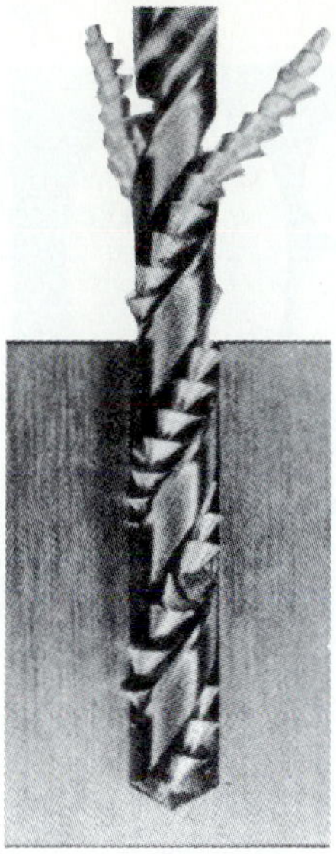

Figure 29-17 *The chip formation of a drill.*

REVIEW QUESTIONS

CUTTING-TOOL MATERIALS

1. What properties should a cutting tool possess?
2. What elements are found in high-speed steel toolbits?
3. State the precaution which should be taken when grinding stellite toolbits.
4. List three qualities of a carbide toolbit.
5. For what purpose are straight tungsten carbide toolbits used?
6. What two substances may be added to tungsten carbide to make it crater resistant?
7. List four advantages of coated carbide toolbits over conventional carbide toolbits.
8. State the uses of:
   a. Titanium coated inserts
   b. Ceramic coated inserts
9. List three advantages of ceramic toolbits.
10. For what application should ceramic toolbits not be used?
11. Describe a polycrystalline diamond toolbit.
12. What are the main applications of polycrystalline diamond tools?
13. List six advantages of polycrystalline diamond toolbits.
14. How are polycrystalline cubic boron nitride toolbits made?
15. State two advantages of polycrystalline cubic boron nitride toolbits.
16. List two applications of polycrystalline boron nitride toolbits.

CUTTING-TOOL NOMENCLATURE

17. Make neat sketches of a single-point cutting tool and label the following parts:
   a. face — d. point
   b. cutting edge — e. flank
   c. nose — f. shank

LATHE TOOLBIT ANGLES AND CLEARANCES

18. State the purpose of the following:
   a. side cutting edge angle — d. side rake angle
   b. side relief angle — e. back rake
   c. end relief angle — f. angle of keenness
19. Draw sketches to illustrate the angles in Question 18.

CUTTING-TOOL SHAPE

20. Name two methods that can be used to improve the life of a cutting tool.
21. What results can be expected from grinding a large rake angle on a cutting tool?
22. How does negative rake on a cutting tool affect the cutting process?

TOOL LIFE

23. Define *flank wear, nose wear,* and *crater wear.*
24. List six important factors that affect the life of a cutting tool.
25. Name the two types of cutting tools.

CUTTING IN A LATHE

26. What will the effect be if the angle of keenness is:
   a. Too small? — b. Too large?
27. List two results of Question 26a and b.

PLAIN MILLING

28. Describe a plain milling cutter.
29. Describe the cutting action of a milling machine.
30. What is another name for the tooth angle in Fig. 29-12?

END AND FACE MILLING

31. What is the purpose of:
   a. End milling? — b. Face milling?
32. How is the rake angle achieved on the blades of an inserted-tooth face mill?
33. Why are inserted-tooth cutters chamfered on the corners?

DRILLING

34. Why is lip clearance ground on a drill?
35. What two surfaces provide the angle of keenness for the drill?

UNIT

30

Operating Conditions and Tool Life

The productivity gap between North American and overseas workers and intense global competition is forcing many companies to renew their commitment to product quality, while at the same time reduce manufacturing costs. During the past decade or so, marked gains in productivity have resulted from high-technology automation, numerical control machine tools, flexible manufacturing systems, and other innovations. These new more rigid machine tools and systems with higher horsepower are much more productive than the machines they have replaced. However, realizing their full potential depends on the use of reliable *high-efficiency cutting tools* to repeatedly produce accurate parts at prices that make them competitive with overseas production.

OBJECTIVES

After completing this unit you will be able to:

1. Describe the effect of cutting conditions on cutting-tool life
2. Explain the effect of cutting conditions on metal-removal rates
3. State the advantages of new cutting-tool materials
4. Calculate the economic performance and cost analysis for a machining operation

OPERATING CONDITIONS

There has been a continual search for new cutting-tool materials in order to increase productivity. High-speed steels were gradually replaced by cast alloys, carbides, ceramics, and cermets. Superabrasive high-efficiency cutting tools such as polycrystalline diamonds (PCD) and polycrystalline cubic boron nitride (PCBN) are now finding wide use in the metalworking industry. The productivity of these new high-efficiency cutting tools far surpasses that of other cutting tools for certain applications.

In order to produce parts efficiently, optimum operating conditions must be achieved during production. Three operating variables—cutting speed, feed rate, and depth of cut—influence metal-removal rate and tool life (Fig. 30-1). This unit will explain how to achieve these optimum conditions, which will result in maximum productivity and minimum cost per piece produced.

DEPTH OF CUT, FEED RATE, AND CUTTING SPEED

The *metal-removal rate* (MRR) is the rate at which metal is removed from an unfinished part and is measured in cubic inches or cubic centimeters per minute. Whenever any one of the three variables (cutting speed, feed, and depth of cut) is changed, MRR will change accordingly. For example, if the cutting speed or the depth of cut is increased by 25 percent, MRR will increase by 25 percent, but the life of the cutting tool will be reduced. However, a change in each of the variables will affect cutting-tool life differently. This can be proven by setting up a test piece on a lathe.

EFFECTS OF CHANGING OPERATING CONDITIONS

Let's assume that a test piece is being machined. The lathe is set to the proper r/min for the material being cut. The

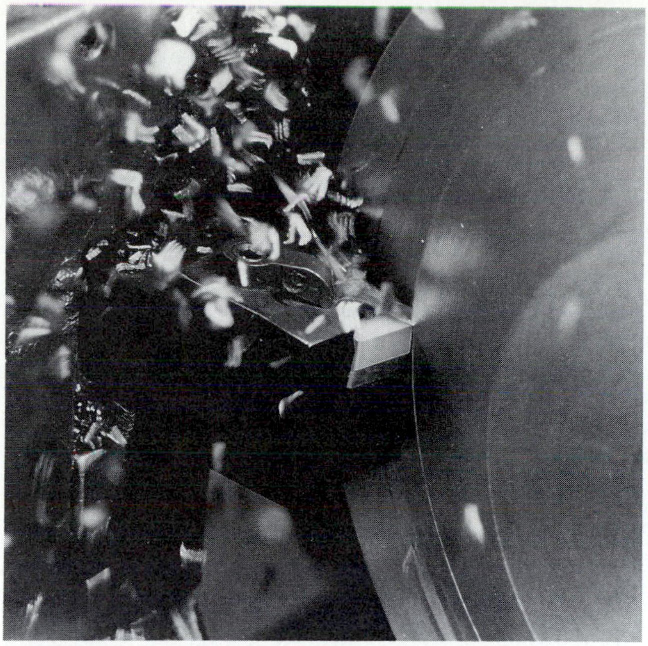

Figure 30-1 *The best operating conditions result in the best metal removal rates and an increase in productivity. (Courtesy GE Superabrasives.)*

feed rate has been selected and the depth of cut has been set at ten times the rate of feed, which is generally accepted as the minimum depth. After the test cut has been made and the tool life established, increase each variable by 50 percent and note the effects on tool life (Fig. 30-2), the results are approximately as follows:

- Increasing the depth of cut by 50 percent reduces tool life by 15 percent.
- Increasing the feed rate by 50 percent reduces tool life by 60 percent.
- Increasing the cutting speed by 50 percent reduces tool life by 90 percent.

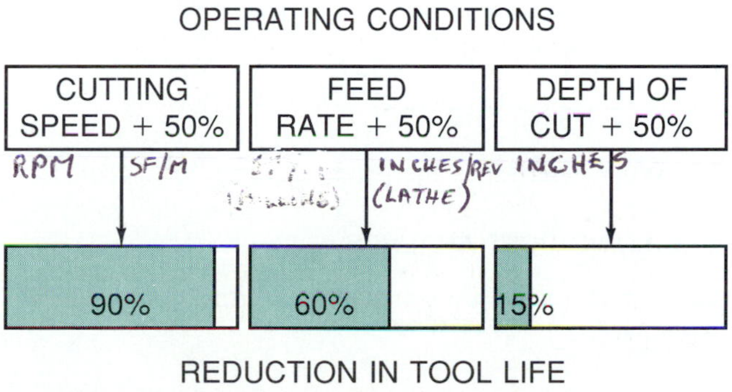
OPERATING CONDITIONS
CUTTING SPEED + 50%
FEED RATE + 50%
DEPTH OF CUT + 50%
90%
60%
15%
REDUCTION IN TOOL LIFE

Figure 30-2 *Conditions which affect single-point machining operations.*

Based on Fig. 30-2, it can be assumed that:

- Changes in the depth of cut have the least effect on tool life.
- Changes in the feed rate have a greater effect than depth-of-cut changes on tool life.
- Changes in the cutting speed of any material have a greater effect than either depth-of-cut or feed-rate changes on tool life.

GENERAL OPERATING CONDITION RULES

From the previous test, it is evident that the selection of the proper cutting speed is the most critical factor to consider when establishing the optimum or ideal operating conditions. If the cutting speed is too low, fewer parts will be produced. At very low speeds, a built-up edge may occur on the toolbit, requiring tool changes. If the cutting speed is too high, the tool will break down quickly, requiring frequent tool changes. Thus,

The optimum cutting speed for any job should balance the metal-removal rate and cutting-tool life.

When considering the best feed rate and depth of cut, always choose the heaviest depth of cut and feed rate possible, because they will reduce tool life much less than too high a cutting speed. Thus,

The optimum feed rate should balance the metal-removal rate and cutting-tool life.

In summary, the maximum production rate is achieved by a combination of cutting speed, feed, and depth of cut that attains the highest output for which the sum of machining time and tool-change time is a minimum. Factors that affect production rate include:

1. Inadequate horsepower, which limits the metal-removal rate.
2. Surface finish requirements, which may limit the feed rate.
3. Machine rigidity, which may not be sufficient to withstand cutting forces, feed rate, and depth of cut.
4. Rigidity of the part being machined, which may limit the depth of cut.

ECONOMIC PERFORMANCE

When the cost of any machining operation is analyzed, many factors must be considered in order to arrive at a true cost. The most important factor affecting the metal-removal rate is the type of cutting tool used. The *cost of a conventional cutting tool* may be very low, but the *cost of using the tool* may be very high. On the other hand, the *cost of a superabrasive cutting tool* such as polycrystalline diamond or cubic boron nitride may be relatively high, but the *cost of using the tool* may be low enough to justify its use.

In order to arrive at a true picture of whether a cutting tool will be economical or not, two factors must be considered for the total machining cost equation: the cost of using the cutting tool, and the price of the cutting tool. Let us examine what factors must be considered to arrive at the total machining costs per part produced, Fig. 30-3.

Cost of Using the Tool

1. The ability of a cutting tool to remove stock will determine the production rate, and also the amount of labor and investment required to produce parts.
2. Any tool's ability to remove stock is governed by the number of times that a tool must be reconditioned or replaced in order to produce accurate work and a good surface finish.
3. The rate that a cutting tool wears will influence how often a worn tool must be removed from a machine and replaced.
4. The cutting tool must be reconditioned and stored in inventory which affects the total machining cost.

MACHINING COST ANALYSIS

Many factors must be considered in order to analyze the true cost of any machining operation. One of the most important elements which governs the rate at which metal is removed is the choice of the cutting tool used. Even though one cutting tool may be *more expensive* than another tool, it can be *more economical* in the long run if it removes material faster and at less tool cost per part.

In order to obtain the lowest possible costs in any machining operation, tool engineers and methods personnel must look at the total operation and carefully examine each element that can affect the cost picture. The many variables which affect a machining operation are listed in Fig. 30-3. All of these factors play a part in determining how a cutting tool performs during a machining operation.

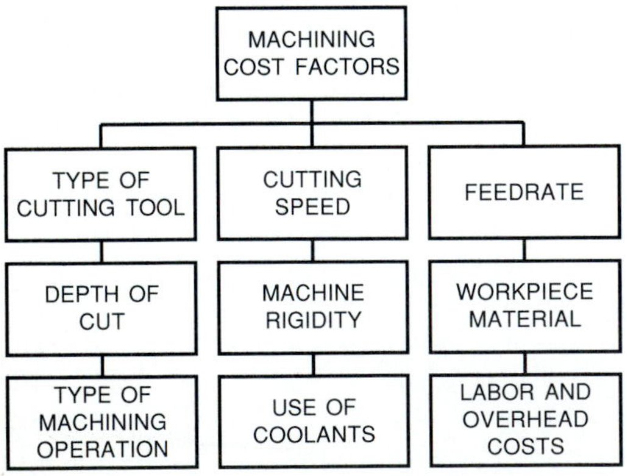
MACHINING COST FACTORS
TYPE OF CUTTING TOOL
CUTTING SPEED
FEEDRATE
DEPTH OF CUT
MACHINE RIGIDITY
WORKPIECE MATERIAL
TYPE OF MACHINING OPERATION
USE OF COOLANTS
LABOR AND OVERHEAD COSTS

Figure 30-3 *Factors which affect the cost of machining a part.*

REVIEW QUESTIONS

OPERATING CONDITIONS

1. What are the three factors that influence cutting-tool life?

DEPTH OF CUT, FEED RATE, AND CUTTING SPEED

2. How is metal removal rate measured?

EFFECTS OF CHANGING OPERATING CONDITIONS

3. What is the generally accepted minimum depth of cut?
4. Which factor increase causes the greatest reduction in cutting-tool life?

GENERAL OPERATING CONDITION RULES

5. What may occur if too slow a cutting speed is used?
6. What is the optimum or ideal cutting speed for any job?
7. List four factors that affect production rates.

ECONOMIC PERFORMANCE

8. What is the most important factor affecting metal-removal rate?
9. What two factors must be considered when arriving at the total cost of a part?

COST OF USING THE TOOL

10. What determines the production rate of a toolbit?

MACHINING COST ANALYSIS

11. List six of the most important cost factors in machining a part.

UNIT 31

Carbide Cutting Tools

Carbide was first used for cutting tools in Germany during World War I as a substitute for diamonds. During the 1930s, various additives were discovered which generally improved the quality and performance of carbide tools were discovered. Since that time, various types of cemented (sintered) carbides have been developed to suit different materials and machining operations. Cemented carbides are similar to steel in appearance, but they are so hard that the diamond is almost the only material which will scratch them. They have been accepted by industry because they have good wear resistance and can operate efficiently at cutting speeds ranging from 150 to 1200 sf/min (46 to 366 m/min). Carbide tools can machine metals at speeds that cause the cutting edge to become red hot without losing its hardness or sharpness.

OBJECTIVES

After completing this unit you will be able to:

1. Identify and state the purpose of the two main types of carbide grades
2. Select the proper grade of carbide for various workpiece materials
3. Select the proper speeds and feeds for carbide tools

MANUFACTURE OF CEMENTED CARBIDES

Cemented carbides are products of the powder metallurgy process; they consist primarily of minute particles of tungsten and carbon powders cemented together under heat by a metal of lower melting point, usually cobalt. Powdered metals such as tantalum, titanium, and niobium are also used in the manufacture of cemented carbides to provide cutting tools with various characteristics. The entire operation of producing cemented-carbide products is illustrated in Fig. 31-1.

BLENDING

Five types of powders are used in the manufacture of cemented-carbide tools: tungsten carbide, titanium carbide, tantalum carbide, niobium carbide, and cobalt. One or any combination of these carbide powders and cobalt (the binder) are blended in different proportions, depending on the grade of carbide desired. This powder is mixed in alcohol; the mixing process takes anywhere from 24 to 190 h. After the powder and alcohol have been thoroughly mixed, the alcohol is drained off, and paraffin is added to simplify the pressing operation.

COMPACTION

After the powders have been thoroughly mixed, they must be molded to shape and size. Five different methods may be used to compact the powder to shape (Fig. 31-2): the extrusion process, the hot press, the isostatic press, the ingot press, or the pill press. The green (pressed) compacts are soft and must be presintered to dissolve the paraffin and lightly bond the particles so that they may be handled easily.

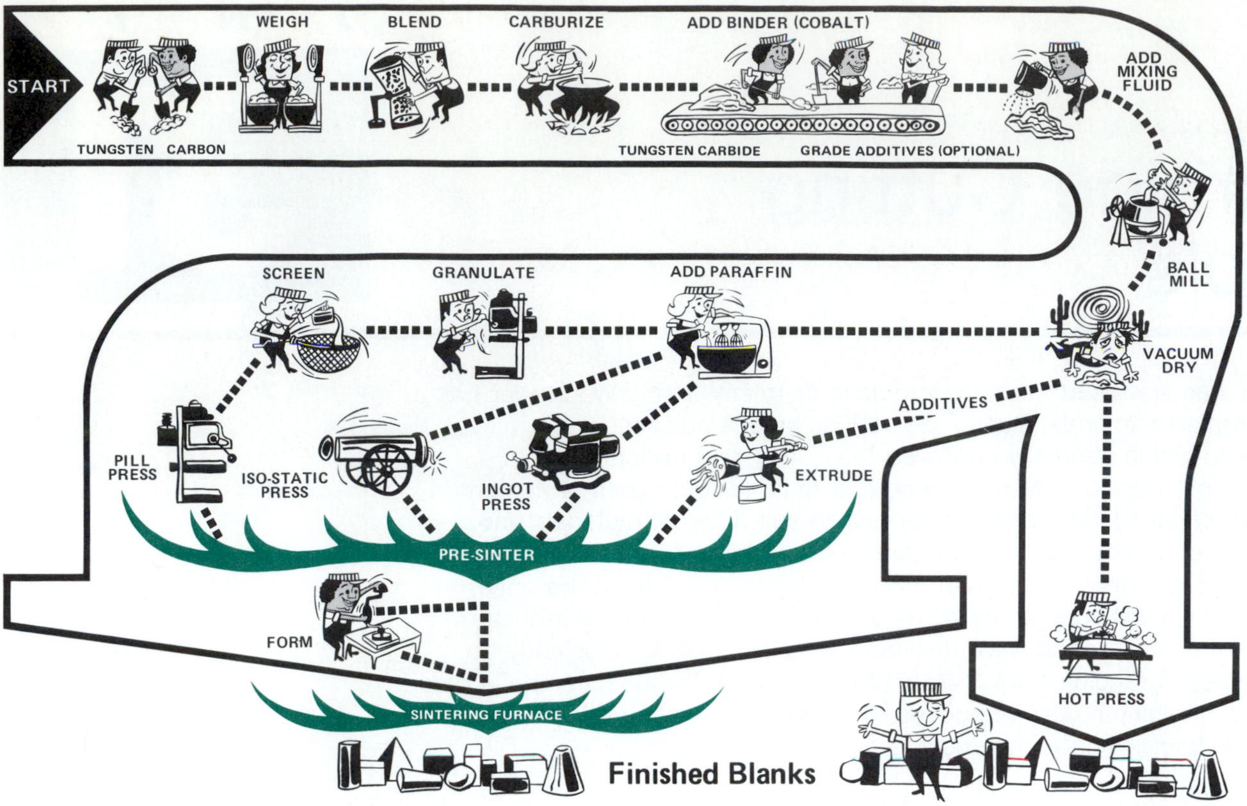
START
TUNGSTEN CARBON
WEIGH
BLEND
CARBURIZE
ADD BINDER (COBALT)
TUNGSTEN CARBIDE
GRADE ADDITIVES (OPTIONAL)
ADD MIXING FLUID
BALL MILL
VACUUM DRY
ADD PARAFFIN
GRANULATE
SCREEN
ADDITIVES
PILL PRESS
ISO-STATIC PRESS
INGOT PRESS
EXTRUDE
PRE-SINTER
FORM
HOT PRESS
SINTERING FURNACE
Finished Blanks

Figure 31-1 *The process of manufacturing cemented-carbide products. (Courtesy General Electric Co. Ltd.)*

PRESINTERING

The green compacts are heated to about 1500°F (815°C) in a furnace under a protective atmosphere of hydrogen. After this operation, carbide blanks have the consistency of chalk and may be machined to the required shape and approximately 40 percent oversize to allow for the shrinkage which occurs during final sintering.

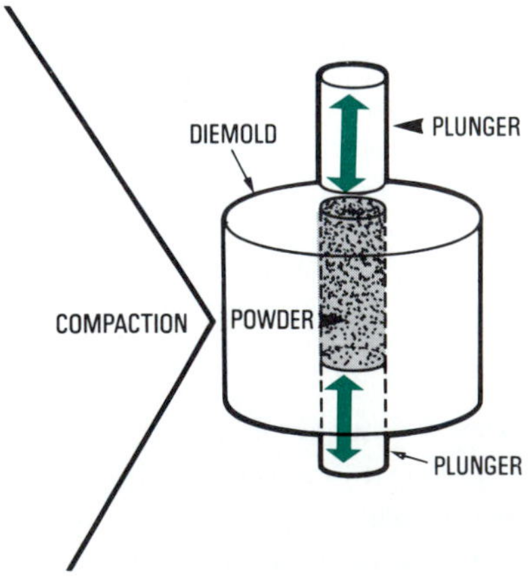
DIEMOLD
PLUNGER
COMPACTION
POWDER
PLUNGER

Figure 31-2 *Compacting carbide powders to shape and size. (Courtesy General Electric Co. Ltd.)*

SINTERING

Sintering, the last step in the process, converts the presintered machined blanks into cemented carbide. Sintering is carried out in either a hydrogen atmosphere or a vacuum, depending on the grade of carbide manufactured, at temperatures between 2550 and 2730°F (1400 and 1500°C). During the sintering operation, the binder (cobalt) unites and cements the carbide powders into a dense structure of extremely hard carbide crystals.

CEMENTED-CARBIDE APPLICATIONS

Because of the extreme hardness and good wear-resistance properties of cemented carbide, it has been used extensively in the manufacture of metal-cutting tools. Drills, reamers, milling cutters, and shaper and lathe cutting tools are only a few examples of uses for cemented carbides. These tools may have cemented-carbide tips, either brazed or held mechanically, on the cutting edge.

Cemented carbides were first used successfully in machining operations as lathe cutting tools. The majority of cemented-carbide tools in use are single-point cutting tools used on machines such as lathes and milling machines.

TYPES OF CARBIDE LATHE CUTTING TOOLS

Cemented-carbide lathe cutting tools are available in two types: the brazed-tip type and the throwaway insert type.

Brazed-Tip Carbide Tools

Cemented-carbide tips can be brazed to steel shanks and are available in a wide variety of styles and sizes (Fig. 31-3). Brazed-tip carbide tools are rigid and are generally used for production turning purposes.

Throwaway Inserts

Cemented-carbide indexible throwaway inserts (Fig. 31-4) are made in a wide variety of shapes, such as triangular, square, diamond, and round. These inserts are held mechanically in a special holder (Fig. 31-5). When one cutting edge becomes dull, it may be quickly indexed or turned in the holder and a new cutting edge will be presented. A triangular insert has three cutting edges on the top surface and three on the bottom, for a total of six cutting edges. When all six cutting edges are dull, the carbide insert is discarded and replaced with a new insert.

Cemented-carbide throwaway inserts are becoming more popular than the brazed-tip carbide tools because:

1. Less time is required to change to a new cutting edge.
2. The amount of machine downtime is reduced considerably and production is thus increased.

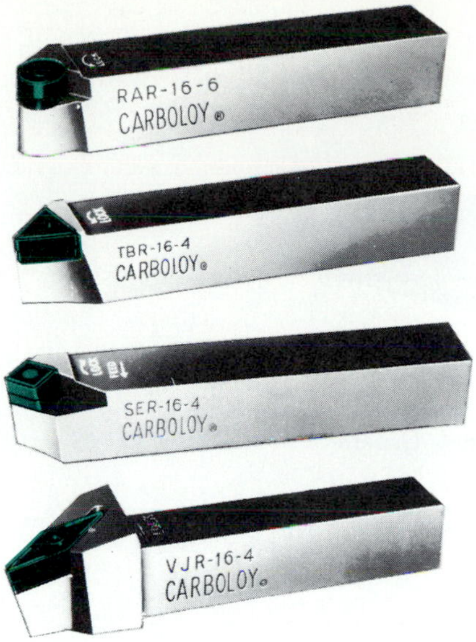
RAR-16-6
CARBOLOY®
TBR-16-4
CARBOLOY®
SER-16-4
CARBOLOY®
VJR-16-4
CARBOLOY®

Figure 31-4 *A variety of indexible throwaway cemented-carbide inserts. (Courtesy General Electric Co. Ltd.)*

3. The time normally spent in regrinding a tool is eliminated.
4. Faster speeds and feeds can be used with throwaway inserts.
5. The cost of diamond wheels, required for grinding carbide tools, is eliminated.
6. Throwaway inserts are cheaper than brazed-tip tools.

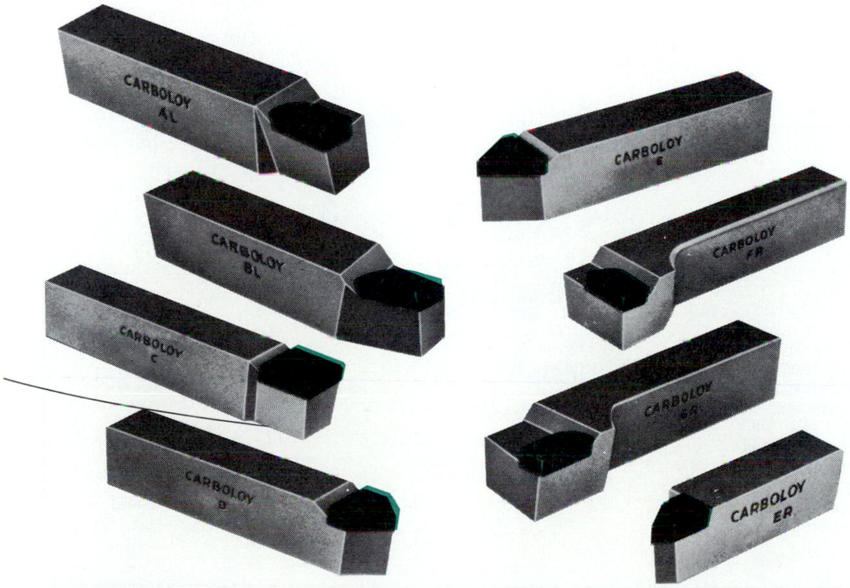
CARBOLOY
4L
CARBOLOY
BL
CARBOLOY
C
CARBOLOY
CARBOLOY
CARBOLOY
FR
CARBOLOY
CARBOLOY
ER

Figure 31-3 *A variety of brazed carbide tools. (Courtesy General Electric Co. Ltd.)*

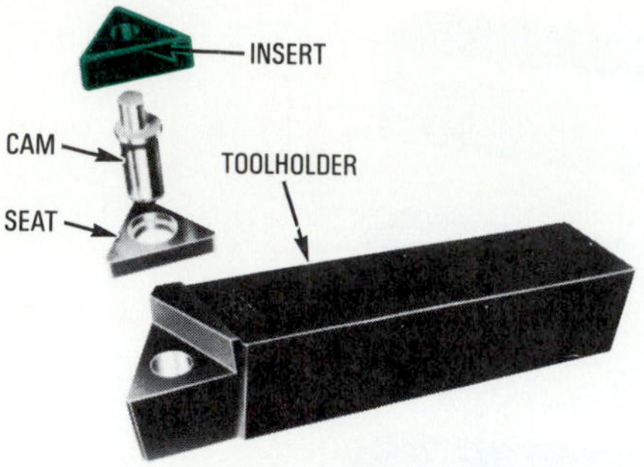
INSERT
CAM
TOOLHOLDER
SEAT

Figure 31-5 *A toolholder for indexible throwaway cemented-carbide inserts. (Courtesy General Electric Co. Ltd.)*

CEMENTED-CARBIDE INSERT IDENTIFICATION

The American Standards Association has developed a system by which throwaway inserts can be identified quickly and accurately. This system has been generally adopted by manufacturers of cemented-carbide inserts (Table 31-1).

GRADES OF CEMENTED CARBIDES

There are two main groups of carbides from which various grades can be selected: the straight tungsten carbide grades

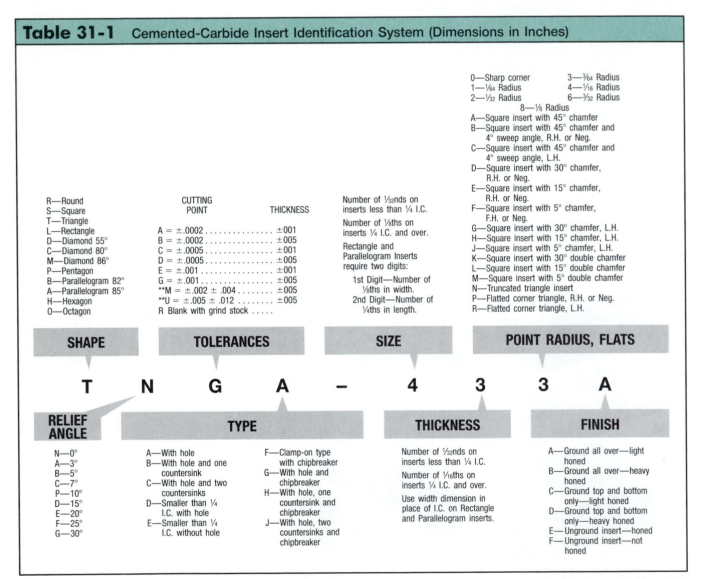

Table 31-1 Cemented-Carbide Insert Identification System (Dimensions in Inches)

Example: **T N G A – 4 3 3 A**

SHAPE (T)

R—Round
S—Square
T—Triangle
L—Rectangle
D—Diamond 55°
C—Diamond 80°
M—Diamond 86°
P—Pentagon
B—Parallelogram 82°
A—Parallelogram 85°
H—Hexagon
O—Octagon

RELIEF ANGLE (N)

N—0°
A—3°
B—5°
C—7°
P—10°
D—15°
E—20°
F—25°
G—30°

TOLERANCES (G)

	CUTTING POINT	THICKNESS
A	±.0002	±.001
B	±.0002	±.005
C	±.0005	±.001
D	±.0005	±.005
E	±.001	±.001
G	±.001	±.005
**M	±.002 ± .004	±.005
**U	±.005 ± .012	±.005
R	Blank with grind stock	

TYPE (A)

A—With hole
B—With hole and one countersink
C—With hole and two countersinks
D—Smaller than 1/4 I.C. with hole
E—Smaller than 1/4 I.C. without hole
F—Clamp-on type with chipbreaker
G—With hole and chipbreaker
H—With hole, one countersink and chipbreaker
J—With hole, two countersinks and chipbreaker

SIZE (4)

Number of 1/32nds on inserts less than 1/4 I.C.

Number of 1/8ths on inserts 1/4 I.C. and over.

Rectangle and Parallelogram Inserts require two digits:
1st Digit—Number of 1/8ths in width.
2nd Digit—Number of 1/4ths in length.

THICKNESS (3)

Number of 1/32nds on inserts less than 1/4 I.C.

Number of 1/16ths on inserts 1/4 I.C. and over.

Use width dimension in place of I.C. on Rectangle and Parallelogram inserts.

POINT RADIUS, FLATS (3)

0—Sharp corner
1—1/64 Radius
2—1/32 Radius
3—3/64 Radius
4—1/16 Radius
6—3/32 Radius
8—1/8 Radius
A—Square insert with 45° chamfer
B—Square insert with 45° chamfer and 4° sweep angle, R.H. or Neg.
C—Square insert with 45° chamfer and 4° sweep angle, L.H.
D—Square insert with 30° chamfer, R.H. or Neg.
E—Square insert with 15° chamfer, R.H. or Neg.
F—Square insert with 5° chamfer, F.H. or Neg.
G—Square insert with 30° chamfer, L.H.
H—Square insert with 15° chamfer, L.H.
J—Square insert with 5° chamfer, L.H.
K—Square insert with 30° double chamfer
L—Square insert with 15° double chamfer
M—Square insert with 5° double chamfer
N—Truncated triangle insert
P—Flatted corner triangle, R.H. or Neg.
R—Flatted corner triangle, L.H.

FINISH (A)

A—Ground all over—light honed
B—Ground all over—heavy honed
C—Ground top and bottom only—light honed
D—Ground top and bottom only—heavy honed
E—Unground insert—honed
F—Unground insert—not honed

*Shall be used only when required.
**Exact tolerance is determined by size of insert.

Source: General Electric Co. Ltd.

and the crater-resistant grades containing titanium and/or tantalum carbide.

The *straight tungsten carbide grades,* containing only tungsten carbide and cobalt, are the strongest and most wear resistant. Generally they are used for machining cast iron and nonmetals. Tungsten carbide grades are not very satisfactory for steel because of their tendency to crater and rapid tool failure.

The size of the tungsten carbide particles and the percentage of cobalt used determine the qualities of tungsten carbide tools:

1. The finer the grain particles, the lower the tool toughness
2. The finer the grain particles, the higher the tool hardness
3. The higher the hardness, the greater the wear resistance
4. The lower the cobalt content, the lower the tool toughness
5. The lower the cobalt content, the higher the hardness

For maximum tool life, always select a tungsten carbide grade with the lowest cobalt content and the finest grain size possible which gives satisfactory performance without breakage.

Crater-resistant grades contain titanium carbide and tantalum carbide in addition to the basic components of tungsten carbide and cobalt. These grades are used for machining most steels.

The addition of tantalum carbide and/or titanium carbide provide tools with various characteristics:

1. The addition of titanium carbide provides resistance to tool cratering. The higher the titanium content, the greater is the resistance to cratering.
2. As the titanium carbide content is increased, the toughness of the tool is decreased.
3. As the titanium carbide content is increased, the abrasive wear resistance at the cutting edge is lowered.
4. Tantalum carbide additions have effects similar to tungsten carbide on the resistance to cratering and strength.
5. Tantalum carbide gives good crater resistance without affecting the abrasive wear resistance.
6. The addition of tantalum carbide increases the tool's resistance to deformation.

Because of the wide variety of cemented-carbide compositions available and the rapid development of new, improved types, no standard grade classification system has been developed. Follow the manufacturer's recommendations as to the type and grade of carbide for each specific application. A few general rules will assist in the selection of the proper cemented-carbide grade:

1. Always use a grade with the lowest cobalt content and the finest grain size (strong enough to eliminate breakage).
2. To combat abrasive wear only, use straight tungsten carbide grades.
3. To combat cratering, seizing, welding, and galling, use titanium carbide grades.
4. For crater and abrasive wear resistance, use tantalum carbide grades.
5. For heavy cuts in steel, when heat and pressure might deform the cutting edge, use tantalum carbide grades.

COATED CARBIDE INSERTS

A relatively recent development in the search for better tooling has been the coating of carbide cutting tools with titanium nitride. Coated carbide inserts give longer tool life, greater productivity, and freer-flowing chips. The coating acts as a permanent lubricant, greatly reducing cutting forces, heat generation, and wear. This permits higher speeds to be used during the machining process, particularly when a good surface finish is required. The lubricity and antiweld characteristics of the coating greatly reduce the amount of heat and stress generated when making a cut. The coating also greatly improves the tool life.

Several different materials and processes are used in coating carbide inserts, including: titanium nitride, titanium carbide, aluminum oxide (ceramic), and combinations of these materials. In one process, a surface coating of titanium carbide is formed by a reaction between titanium tetrachloride, methane, and hydrogen. In a furnace having a temperature of about 1830°F (1000°C) a layer of titanium carbide 0.0002 in. (0.005 mm) is bonded to the surface of the cemented-carbide core. This process imparts a surface hardness of almost three times the surface hardness of the standard (uncoated) carbide insert and increases tool life by up to five times that of an uncoated carbide cutter.

In another process, the inserts are triple coated. Strong, wear-resistant titanium carbide forms the innermost layer. This is followed by a thick layer of aluminum oxide, which provides toughness, shock resistance, and chemical stability at high temperatures. A third, very thin layer composed of titanium nitride is applied over the aluminum oxide. This provides a lower coefficient of friction and reduces the tendency to form a built-up edge.

TOOL GEOMETRY

The geometry of cutting tools refers to the various angles and clearances machined or ground on the tool faces. Although the terms and definitions relating to single-point cutting tools vary greatly, the ones adopted by the American Society of Mechanical Engineers (ASME) and currently in general use are illustrated in Fig. 31-6.

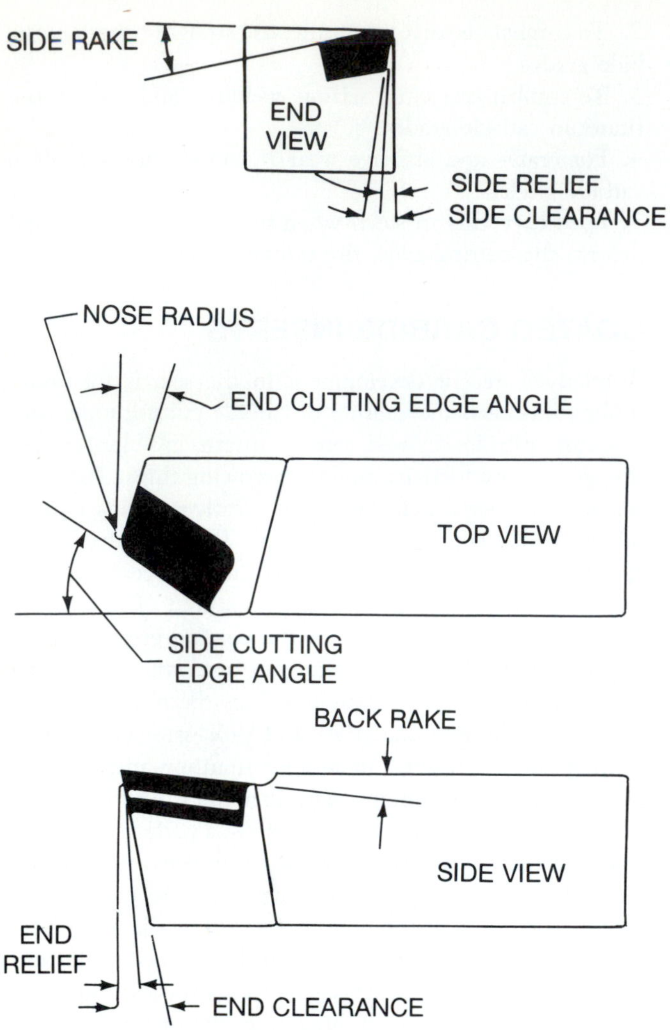
SIDE RAKE
END VIEW
SIDE RELIEF
SIDE CLEARANCE
NOSE RADIUS
END CUTTING EDGE ANGLE
TOP VIEW
SIDE CUTTING EDGE ANGLE
BACK RAKE
SIDE VIEW
END RELIEF
END CLEARANCE

Figure 31-6 *Single-point tool nomenclature.*

CUTTING-TOOL TERMS

Front, or End, Relief (Clearance)

The front relief, or clearance (Fig. 31-6) allows the end of the cutting tool to enter the work. This clearance should be just enough to prevent the tool from rubbing. Too much front clearance will reduce the support under the point and cause rapid tool failure.

Side Relief (Clearance)

The side relief, or clearance (Fig. 31-6) permits the side of the tool to advance into the work. Too little side clearance will prevent the tool from cutting, and excessive heat will be generated by the rubbing action. Too much side clearance will weaken the cutting edge and cause it to chip.

Side Cutting Edge Angle

The angle of the side of the cutting edge that meets the work may be either positive or negative. A negative side cutting angle (Fig. 31-6) is preferred because it protects the point of the tool at both the start and the end of a cut; this is especially useful on work that has a hard abrasive scale.

Nose Radius

The nose radius strengthens the finishing point of the tool and improves the surface finish on the work. The nose radius of most cutting tools should be approximately twice the amount of feed per revolution. Too large a nose radius may cause chatter; too small a radius weakens the point of the cutting tool.

To obtain maximum efficiency with carbide tools, keep the nose radius as small as possible. Use Table 31-2 to find the proper nose radius for the depth of cut and feed being used.

Side Rake

The side rake angle should be as large as possible, without weakening the cutting edge, to allow the chips to escape readily (Fig. 31-7A). The amount of side rake will be determined by the type and grade of the cutting tool, the type of material being cut, and the feed per revolution. The included angle formed by the side rake and side clearance is called the angle of keenness. This angle will vary, depending on the material being cut. For difficult-to-machine metals, it may be advisable to use a small side rake angle or at times even negative side rake (Fig. 31-7B). Figure 31-8 shows suggested side rakes for a variety of materials.

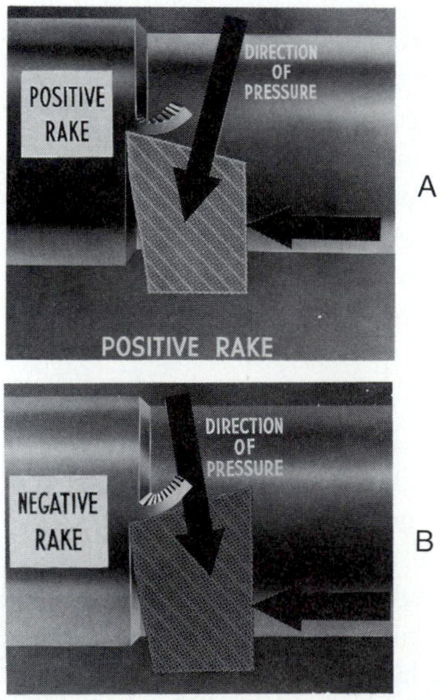
POSITIVE RAKE
DIRECTION OF PRESSURE
POSITIVE RAKE
A
NEGATIVE RAKE
DIRECTION OF PRESSURE
B

Figure 31-7 *(A) Positive rake side angle on a cutting tool; (B) negative rake side angle on a cutting tool. (Courtesy A. C. Wickman Limited.)*

Table 31-2 Nose Radius Nomograph*

Feed per Rev., in.	Nose Radius, in.	Depth of Cut, in.

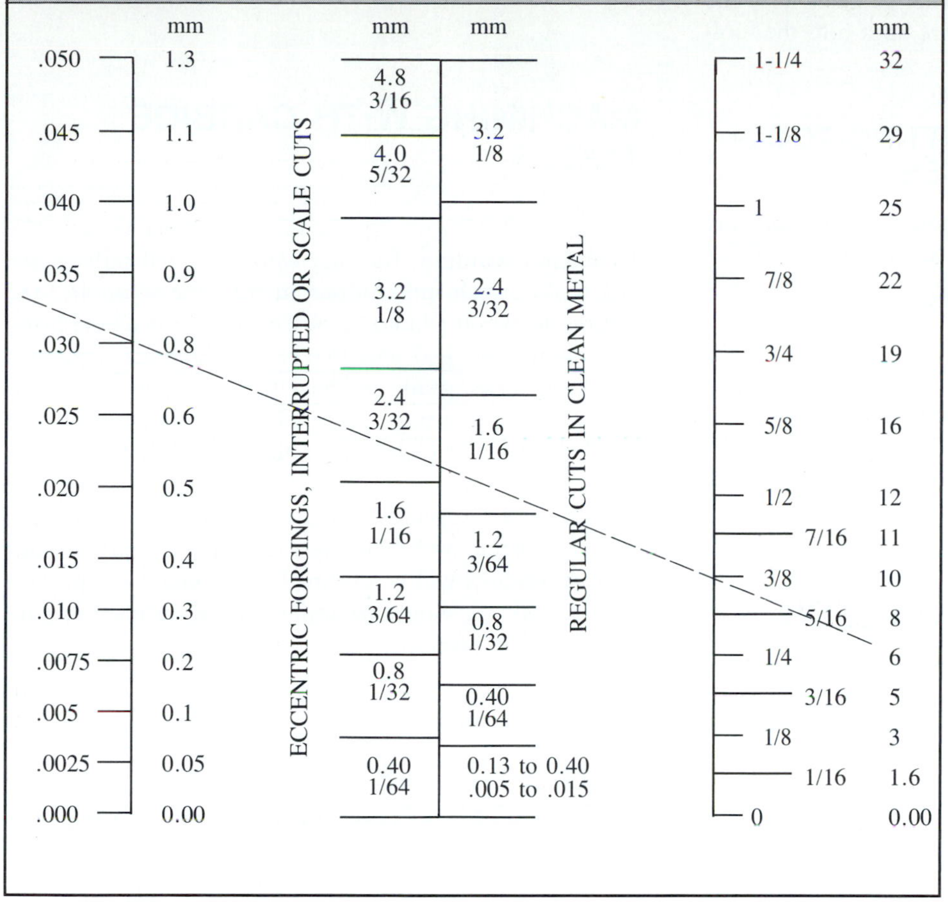

mm
.050 1.3
.045 1.1
.040 1.0
.035 0.9
.030 0.8
.025 0.6
.020 0.5
.015 0.4
.010 0.3
.0075 0.2
.005 0.1
.0025 0.05
.000 0.00
ECCENTRIC FORGINGS, INTERRUPTED OR SCALE CUTS
mm
4.8 3/16
4.0 5/32
3.2 1/8
2.4 3/32
1.6 1/16
1.2 3/64
0.8 1/32
0.40 1/64
mm
3.2 1/8
2.4 3/32
1.6 1/16
1.2 3/64
0.8 1/32
0.40 1/64
0.13 to 0.40 .005 to .015
REGULAR CUTS IN CLEAN METAL
mm
1-1/4 32
1-1/8 29
1 25
7/8 22
3/4 19
5/8 16
1/2 12
7/16 11
3/8 10
5/16 8
1/4 6
3/16 5
1/8 3
1/16 1.6
0 0.00

*To obtain a maximum efficiency with carbide tools, the nose radius should be kept small. Use chart as a guide. Use straightedge to join feed with depth of cut. Use nose radius where line crosses.
Source: General Electric Co. Ltd.

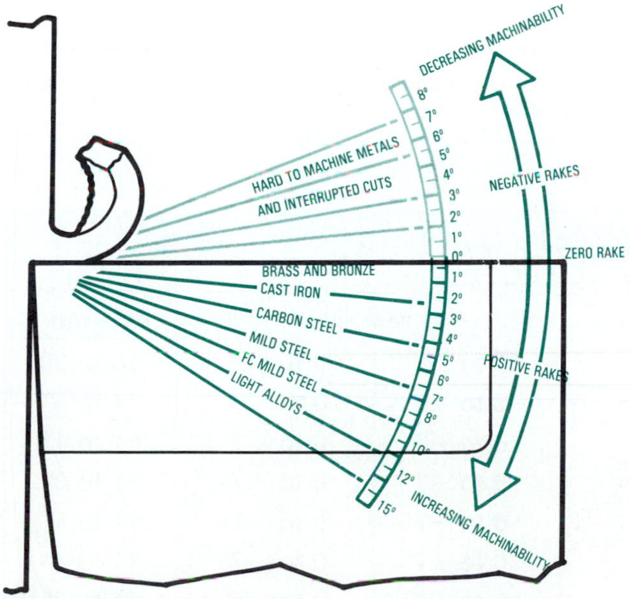

DECREASING MACHINABILITY
HARD TO MACHINE METALS
AND INTERRUPTED CUTS
NEGATIVE RAKES
ZERO RAKE
BRASS AND BRONZE
CAST IRON
CARBON STEEL
MILD STEEL
FC MILD STEEL
LIGHT ALLOYS
POSITIVE RAKES
INCREASING MACHINABILITY

Figure 31-8 *Recommended side rake angles for various workpiece materials. (Courtesy A. C. Wickman Limited.)*

Back Rake

The back rake angle is the angle formed between the top face of the tool and the top of the tool shank. It may be positive, negative, or neutral. When a tool has *negative back rake,* the top face of the tool slopes upward away from the point (Fig. 31-9A). Negative back rake protects the tool point from cutting pressure and also from the abrasive action of hard materials and scale. When a tool has *positive back rake* (Fig. 31-9B), the top face of the tool slopes down-

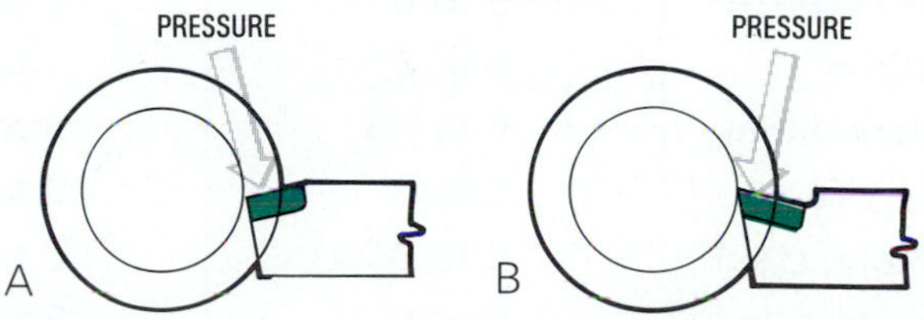

PRESSURE
PRESSURE
A
B

Figure 31-9 *(A) Negative back rake; (B) positive back rake.*

ward away from the point. This allows the chips to flow away freely from the cutting edge.

NOTE: Because throwaway inserts are generally flat, the required side and back rake angles are built into the toolholder by the manufacturer.

CEMENTED-CARBIDE CUTTING-TOOL ANGLES AND CLEARANCES

The angles and clearances of single-point carbide tools vary greatly and generally depend on three factors:

1. The hardness of the cutting tool
2. The workpiece material
3. The type of cutting operation

Table 31-3 lists recommended cutting-tool angles and clearances for a variety of materials. These may have to be altered slightly to suit various conditions encountered while machining.

CUTTING SPEEDS AND FEEDS

Many variables influence the speeds, feeds, and depth of cut which should be used with cemented-carbide tools. Some of the most important factors are:

1. The type and hardness of the work material
2. The grade and shape of the cutting tool
3. The rigidity of the cutting tool
4. The rigidity of the work and machine
5. The power rating of the machine

Table 31-4 gives the recommended cutting speeds and feeds for single-point carbide tools. These should be used as a guide and may have to be altered slightly to suit the machining operation.

Table 31-5 illustrates the nomograph method of determining the cutting speed in feet or meters per minute when the Brinell hardness of the steel is known.

MACHINING WITH CARBIDE TOOLS

To obtain maximum efficiency with cemented-carbide cutting tools, certain precautions in machine setup and the cutting operation should be observed. The machine being used should be rigid and free from vibrations, equipped with heat-treated gears, and have sufficient power to maintain a constant cutting speed. The work and cutting tool should be held as rigidly as possible to avoid chatter or keep it to a minimum.

Single-point carbide cutting tools are more commonly used on a lathe, and therefore the setups and precautions for this machine will be outlined. The same basic precautions and setups should be applied when carbide tools are used on other machines.

SUGGESTIONS FOR USING CEMENTED-CARBIDE CUTTING TOOLS

WORK SETUP

1. Work mounted in a chuck or other work-holding device must be held firmly enough to prevent it from slipping or chattering.

Table 31-3 Recommended Angles for Single-Point Carbide Tools

Material	End Relief (Front Clearance)	Side Relief (Side Clearance)	Side Rake	Back Rake	Angle of Keenness
Aluminum	6 to 10°	6 to 10°	10 to 20°	0 to 10°	16 to 30°
Brass, bronze	6 to 8°	6 to 8°	+8 to −5°	0 to −5°	14 to 3°
Cast iron	5 to 8°	5 to 8°	+6 to −7°	0 to −7°	11 to 1°
Machine steel	5 to 10°	5 to 10°	+6 to −7°	0 to −7°	11 to 3°
Tool steel	5 to 8°	5 to 8°	+6 to −7°	0 to −7°	11 to 1°
Stainless steel	5 to 8°	5 to 8°	+6 to −7°	0 to −7°	11 to 1°
Titanium alloys	5 to 8°	5 to 8°	+6 to −5°	0 to −5°	11 to 3°

Note: Use the lower range of these figures for hard-to-machine metals and interrupted cuts.

Table 31-4 Recommended Cutting Speeds and Feeds for Single-Point Carbide Tools

Material	Depth of Cut		Feed per Revolution		Cutting Speed	
	in.	mm	in.	mm	ft/min	m/min
Aluminum	0.005–0.015	0.15–0.40	0.002–0.005	0.05–0.15	700–1000	215–305
	0.020–0.090	0.50–2.30	0.005–0.015	0.15–0.40	450–700	135–215
	0.100–0.200	2.55–5.10	0.015–0.030	0.40–0.75	300–450	90–135
	0.300–0.700	7.60–17.80	0.030–0.090	0.75–2.30	100–200	30–60
Brass, bronze	0.005–0.015	0.15–0.40	0.002–0.005	0.05–0.15	700–800	215–245
	0.020–0.090	0.50–2.30	0.005–0.015	0.15–0.40	600–700	185–215
	0.100–0.200	2.55–5.10	0.015–0.030	0.40–0.75	500–600	150–185
	0.300–0.700	7.60–17.80	0.030–0.090	0.75–2.30	200–400	60–120
Cast iron (medium)	0.005–0.015	0.15–0.40	0.002–0.005	0.05–0.15	350–450	105–135
	0.020–0.090	0.50–2.30	0.005–0.015	0.15–0.40	250–350	75–105
	0.100–0.200	2.55–5.10	0.015–0.030	0.40–0.75	200–250	60–75
	0.300–0.700	7.60–17.80	0.030–0.090	0.75–2.30	75–150	25–45
Machine steel	0.005–0.015	0.15–0.40	0.002–0.005	0.05–0.15	700–1000	215–305
	0.020–0.090	0.50–2.30	0.005–0.015	0.15–0.40	550–700	170–215
	0.100–0.200	2.55–5.10	0.015–0.030	0.40–0.75	400–550	120–170
	0.300–0.700	7.60–17.80	0.030–0.090	0.75–2.30	150–300	45–90
Tool steel	0.005–0.015	0.15–0.40	0.002–0.005	0.05–0.15	500–750	150–230
	0.020–0.090	0.50–2.30	0.005–0.015	0.15–0.40	400–500	120–150
	0.100–0.200	2.55–5.10	0.015–0.030	0.40–0.75	300–400	90–120
	0.300–0.700	7.60–17.80	0.030–0.090	0.75–2.30	100–300	30–90
Stainless steel	0.005–0.015	0.15–0.40	0.002–0.005	0.05–0.15	375–500	115–150
	0.020–0.090	0.50–2.30	0.005–0.015	0.15–0.40	300–375	90–115
	0.100–0.200	2.55–5.10	0.015–0.030	0.40–0.75	250–300	75–90
	0.300–0.700	7.60–17.80	0.030–0.090	0.75–2.30	75–175	25–55
Titanium alloys	0.005–0.015	0.15–0.40	0.002–0.005	0.05–0.15	300–400	90–120
	0.020–0.090	0.50–2.30	0.005–0.015	0.15–0.40	200–300	60–90
	0.100–0.200	2.55–5.10	0.015–0.030	0.40–0.75	175–200	55–60
	0.300–0.700	7.60–17.80	0.030–0.090	0.75–2.30	50–125	15–40

Note: Millimeter and inch speeds are approximate equivalents.
Source: General Electric Co. Ltd.

2. A revolving center should be used in the tailstock for turning work between centers.
3. The tailstock spindle should be extended only a minimum distance and locked securely to ensure rigidity.
4. The tailstock should be clamped firmly to the lathe bed to prevent loosening.

TOOL SELECTION

1. Use a cutting tool with the proper rake and clearances for the material being cut.
2. Hone the cutting edge for good performance and longer tool life.
3. If the work permits, use a side cutting edge angle (Fig. 31-10A) large enough that the tool can be eased into the work. This helps to protect the nose of the cutting tool (weakest part) from shock and wear as it enters or leaves the work.
4. Use the largest nose radius (Fig. 31-10B) which operating conditions will permit. Too large a nose radius causes chattering; too small a nose radius causes the point to break down quickly. Use Table 31-2 to determine the correct nose radius to use for the amount of feed and depth of cut being used.

TOOL SETUP

1. Carbide tools preferably should be held in a sturdy, turret-type holder (Fig. 31-11). The amount of tool overhang should be just enough for chip clearance.

Table 31-5 Carbide Cutting Speed Nomograph

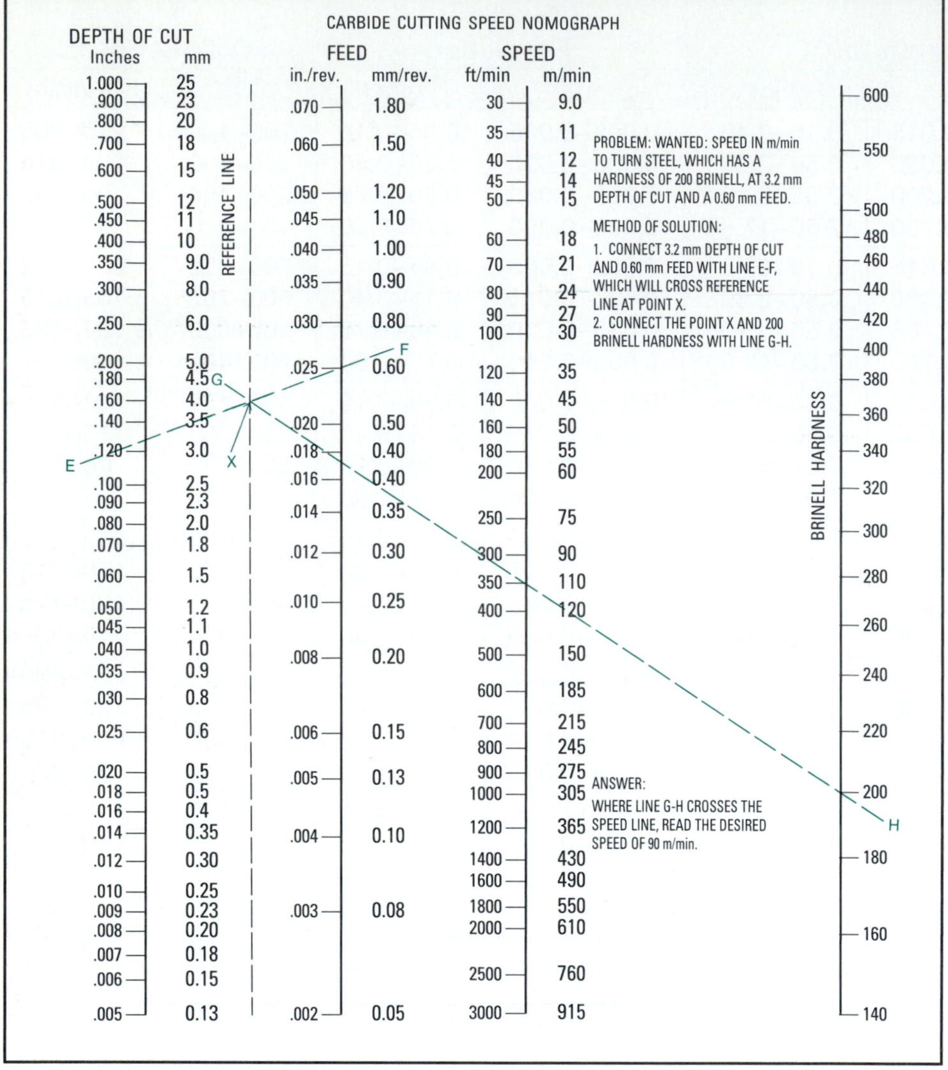
CARBIDE CUTTING SPEED NOMOGRAPH
DEPTH OF CUT
Inches mm
1.000 25
.900 23
.800 20
.700 18
.600 15
.500 12
.450 11
.400 10
.350 9.0
.300 8.0
.250 6.0
.200 5.0
.180 4.5
.160 4.0
.140 3.5
.120 3.0
.100 2.5
.090 2.3
.080 2.0
.070 1.8
.060 1.5
.050 1.2
.045 1.1
.040 1.0
.035 0.9
.030 0.8
.025 0.6
.020 0.5
.018 0.5
.016 0.4
.014 0.35
.012 0.30
.010 0.25
.009 0.23
.008 0.20
.007 0.18
.006 0.15
.005 0.13
REFERENCE LINE
FEED
in./rev. mm/rev.
.070 1.80
.060 1.50
.050 1.20
.045 1.10
.040 1.00
.035 0.90
.030 0.80
.025 0.60
.020 0.50
.018 0.40
.016 0.40
.014 0.35
.012 0.30
.010 0.25
.008 0.20
.006 0.15
.005 0.13
.004 0.10
.003 0.08
.002 0.05
SPEED
ft/min m/min
30 9.0
35 11
40 12
45 14
50 15
60 18
70 21
80 24
90 27
100 30
120 35
140 45
160 50
180 55
200 60
250 75
300 90
350 110
400 120
500 150
600 185
700 215
800 245
900 275
1000 305
1200 365
1400 430
1600 490
1800 550
2000 610
2500 760
3000 915
E
F
G
H
X
PROBLEM: WANTED: SPEED IN m/min TO TURN STEEL, WHICH HAS A HARDNESS OF 200 BRINELL, AT 3.2 mm DEPTH OF CUT AND A 0.60 mm FEED.
METHOD OF SOLUTION:
1. CONNECT 3.2 mm DEPTH OF CUT AND 0.60 mm FEED WITH LINE E-F, WHICH WILL CROSS REFERENCE LINE AT POINT X.
2. CONNECT THE POINT X AND 200 BRINELL HARDNESS WITH LINE G-H.
ANSWER:
WHERE LINE G-H CROSSES THE SPEED LINE, READ THE DESIRED SPEED OF 90 m/min.
BRINELL HARDNESS
600
550
500
480
460
440
420
400
380
360
340
320
300
280
260
240
220
200
180
160
140

Note: Inch and millimeter speeds are approximate equivalents.
Source: General Electric Co. Ltd.

2. The cutting tool should be set exactly on center; when it is above or below center, the tool angles and clearances change in relation to the job and result in poor cutting action.
3. Carbide tools are designed to operate while the bottom of the tool shank is in a *horizontal position* (Fig. 31-12).
4. If a rocker-type toolpost is being used (Fig. 31-13):
   a. Remove the rocker.
   b. Invert the rocker base.
   c. Shim the tool to the correct height.
   d. Use a special carbide toolholder (having no rake) when machining with carbide toolbits (Fig. 31-14).
5. When setting up a carbide tool, always keep it from touching the work and machine parts to avoid damaging the tool point.

MACHINE SETUP

1. Always make sure that the machine has an adequate power rating for the machining operation and that there is no slippage in the clutch and belts.
2. Set the correct speed for the material being cut and the operation being performed.
   a. Too high a speed will cause rapid tool failure; too low a speed will result in inefficient cutting action and poor production rates.
   b. Use Table 31-4 or 31-5 to calculate the correct speed for the type of material being machined.
3. Set the machine at a feed which will give good metal-removal rates and still provide the surface finish desired.

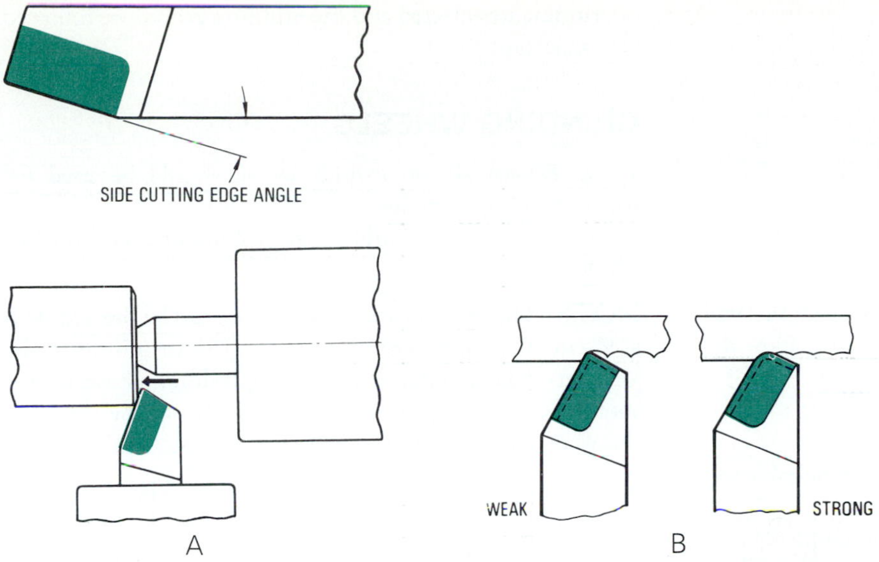
SIDE CUTTING EDGE ANGLE
WEAK
STRONG
A
B
Figure 31-10 *(A) A side cutting edge angle protects the point of the cutting tool; (B) a large nose radius strengthens the tool point and produces a better surface finish. (Courtesy General Electric Co. Ltd.)*

a. Too light a feed causes rubbing, which may result in hardening of the material being cut.
b. Too coarse a feed slows down the machine, creates excessive heat, and results in premature tool failure.

CUTTING OPERATION

1. Never bring the tool point against work that is stationary; this will damage the cutting edge.
2. Always use the heaviest depth of cut possible for the machine and size of cutting tool.

Figure 31-12 *The heavy-duty toolpost permits rigid setting of the cutting tool for heavy cuts.*

Figure 31-11 *Turret-type toolholders hold carbide tools securely.*

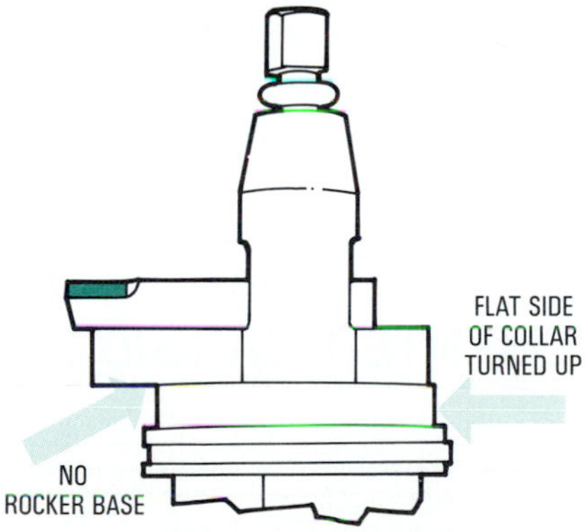
FLAT SIDE
OF COLLAR
TURNED UP
NO
ROCKER BASE
Figure 31-13 *To hold carbide cutting tools in a rocker-type toolpost, remove the rocker, invert the rocker base, and shim the tool to the correct height.*

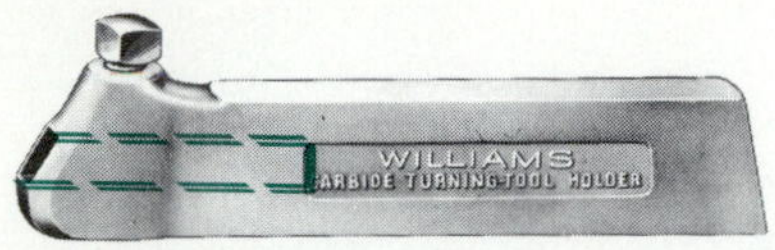
WILLIAMS
CARBIDE TURNING TOOL HOLDER

Figure 31-14 *The hole in a carbide toolholder is straight and not slanted as in standard toolholders. (Courtesy J. H. Williams & Co.)*

3. Never stop a machine while the feed is engaged; this will break the cutting edge. Always stop the feed and allow the tool to clear itself before stopping the machine.
4. Never continue to use a dull cutting tool.
5. A dull cutting tool may be recognized by:
   a. Work being produced oversize and with a glazed finish
   b. A rough and ragged finish
   c. A change in the shape or color of the chips
6. Apply cutting fluid only if:
   a. It can be applied under pressure
   b. It can be directed at the point of cutting and kept there at all times

TOOL SELECTION AND APPLICATION GUIDE

To obtain the best results from cemented-carbide tools, the proper type and style of insert must be used, and the cutting tool must be set up and used properly. The points illustrated in Table 31-6 should be followed as closely as possible in order to obtain the most efficient metal-removal rates and produce the most cost-effective machining operation. Other factors that can affect the optimum life of carbide tools are:

1. The horsepower available on the machine tool
2. The rigidity of the machine tool and toolholders
3. The shape of the workpiece and the setup
4. The speed and feed rates used for the machining operation

GRINDING CEMENTED-CARBIDE TOOLS

The efficiency of a cutting tool determines to a large extent the efficiency of the machine tool on which it is being used. A tool that has been improperly ground cannot perform well, and the cutting edge will soon break down. In order to grind cemented-carbide cutting tools successfully, the type of grinding wheel used and the grinding procedure followed are important.

GRINDING WHEELS

1. An 80-grit silicon carbide wheel should be used for rough grinding carbides.
2. A 100-grit silicon carbide wheel should be used for finish grinding carbides.

NOTE: The silicon carbide wheels should be dressed with a 1/16 in. (1.5 mm) crown (Fig. 31-15) to minimize the amount of heat generated during grinding. These wheels should be used to grind only the carbide, not the steel tool shank of a brazed carbide toolbit.

3. Aluminum oxide grinding wheels should be used if it is necessary to grind the steel shank of a brazed carbide tool.
4. Diamond grinding wheels (100 grit) are excellent for finish grinding of carbides for general work. Where high finishes on the tool and work are desired, a 220-grit diamond wheel is recommended.

TYPE OF GRINDER

1. A heavy-duty grinder should be used for grinding carbides because the cutting pressures required to remove carbide are from 5 to 10 times as great as those for grinding high-speed steel tools.
2. The grinder should be equipped with an adjustable table and a protractor (Fig. 31-16) so that the necessary tool angles and clearances may be ground accurately.

TOOL GRINDING

1. Regrind the cutting tool to the angles and clearances recommended by the manufacturer.
2. Use silicon carbide wheels for rough grinding. Diamond wheels should be used when high surface finishes are required.
3. When grinding, move the carbide tool back and forth over the grinding wheel face to keep the amount of heat generated to a minimum.
4. *Never quench carbide tools* which become hot during grinding; allow them to cool gradually. The shock of

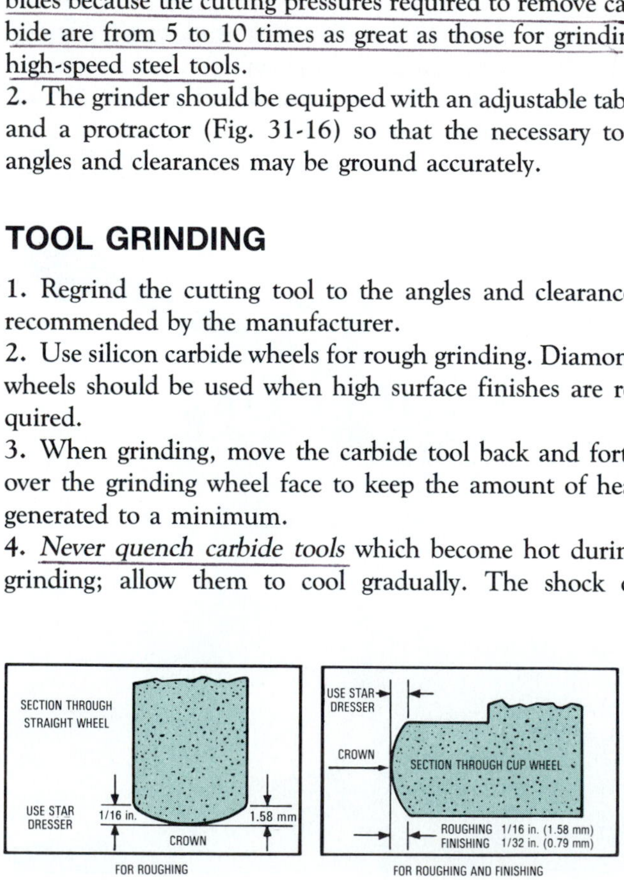
SECTION THROUGH STRAIGHT WHEEL
USE STAR DRESSER
1/16 in.
1.58 mm
CROWN
FOR ROUGHING
USE STAR DRESSER
CROWN
SECTION THROUGH CUP WHEEL
ROUGHING 1/16 in. (1.58 mm)
FINISHING 1/32 in. (0.79 mm)
FOR ROUGHING AND FINISHING

Figure 31-15 *A crown dressed on a silicon carbide wheel reduces heat when carbide tools are being ground.*

Table 31-6 Tooling Selection and Application Guide

Follow these fundamental techniques for more efficient metal removal.

1. Always select standard products whenever the operating conditions allow.
BENEFITS:
- Less expensive
- Proven design
- Availability
- Interchangeable

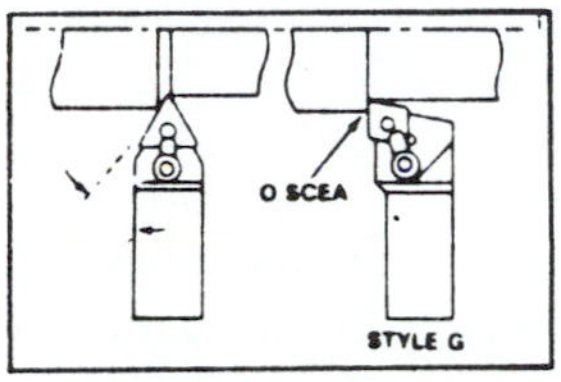
0 SCEA
STYLE G

2. Always select the largest side cutting (or end cutting) edge angle the workpiece will allow.
BENEFITS:
- Thin chips
- Dissipates heat
- Protects nose radius
- Reduces insert notching

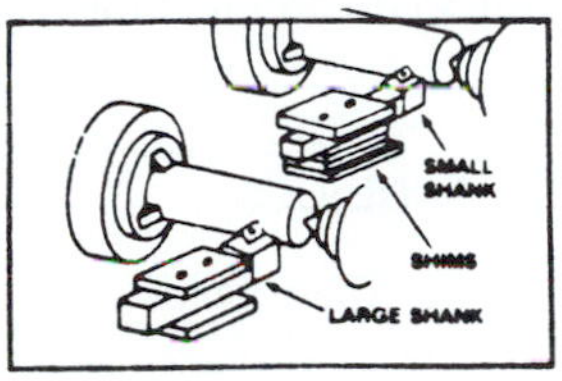
SMALL SHANK
SHIMS
LARGE SHANK

3. Always select the largest toolholder shank the machine tool will allow.
BENEFITS:
- Minimize deflection
- Reduces overhang ratio

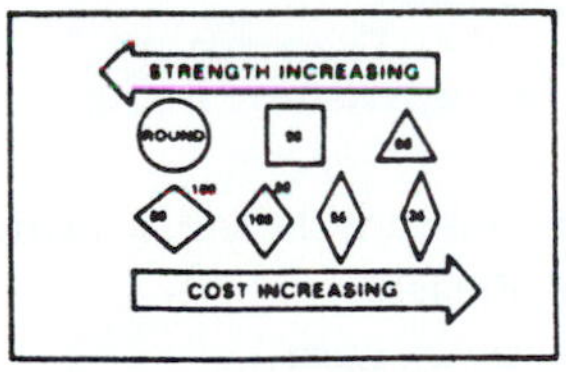
STRENGTH INCREASING
ROUND
COST INCREASING

4. Always select the strongest insert shape the workpiece will allow.
BENEFITS:
- Productivity
- Optimum grade selection
- Lower cost/edge

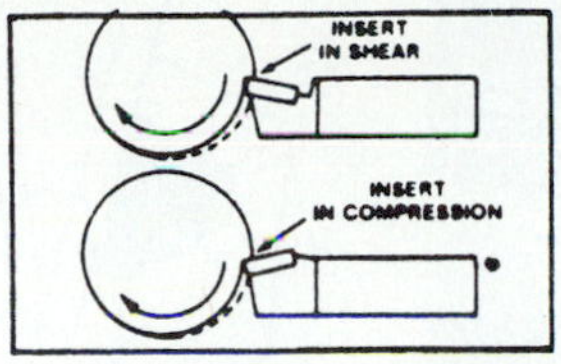
INSERT IN SHEAR
INSERT IN COMPRESSION

5. Always select negative rake geometry whenever the workpiece or the machine tool will allow.
BENEFITS:
- Double the cutting edges
- Greater strength
- Dissipates heat

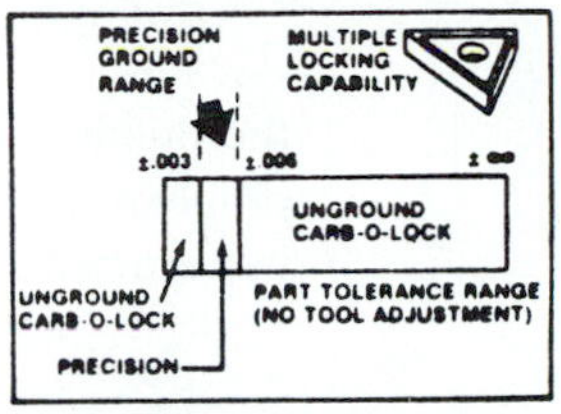
PRECISION GROUND RANGE
MULTIPLE LOCKING CAPABILITY
UNGROUND CARB-O-LOCK
UNGROUND CARB-O-LOCK
PART TOLERANCE RANGE (NO TOOL ADJUSTMENT)
PRECISION

6. Always select a Carb-O-Lock® insert whenever possible.
BENEFITS:
- Less expensive
- May be multiple locked
- Strong unground cutting edge
- Chip control groove
- Covers most workpiece tolerances

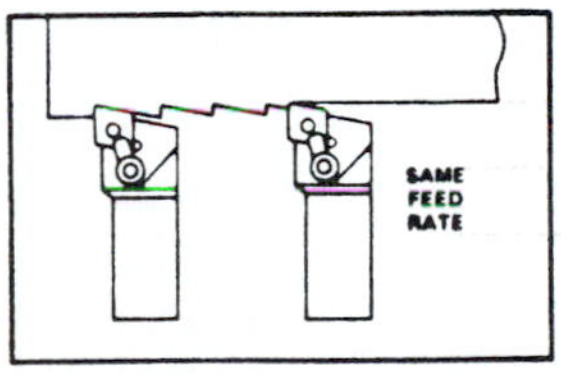
SAME FEED RATE

7. Always select the largest insert nose radius that either the workpiece or the operating conditions will allow.
BENEFITS:
- Improves finish
- Thins chips
- Dissipates heat
- Greater strength

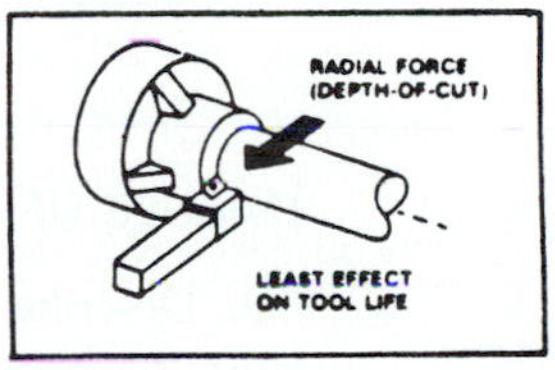
RADIAL FORCE (DEPTH-OF-CUT)
LEAST EFFECT ON TOOL LIFE

8. Always select the largest depth-of-cut that either the workpiece or the machine tool will allow.
BENEFITS:
- Greater productivity
- Negligible effect on tool life

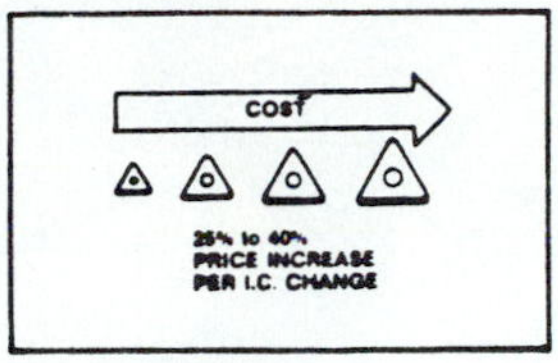
COST
25% to 40% PRICE INCREASE PER I.C. CHANGE

9. Always select the smallest insert size the operating conditions will allow.
BENEFITS:
- Less expensive

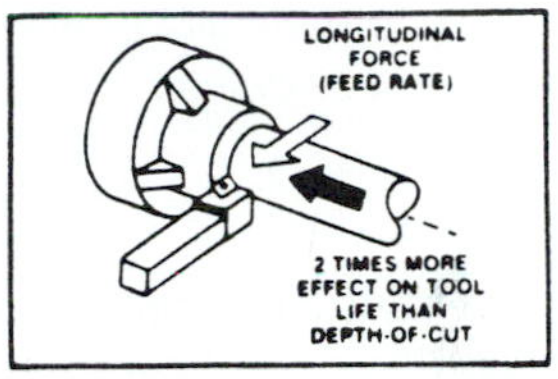
LONGITUDINAL FORCE (FEED RATE)
2 TIMES MORE EFFECT ON TOOL LIFE THAN DEPTH-OF-CUT

10. Always select the highest feed rate that either the workpiece or the machine tool will allow.
BENEFITS:
- Greater productivity
- Minimal effect on tool life

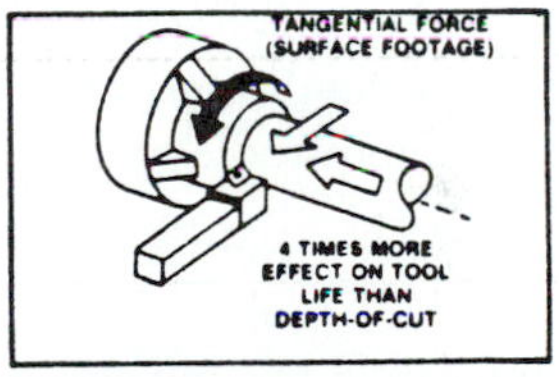
TANGENTIAL FORCE (SURFACE FOOTAGE)
4 TIMES MORE EFFECT ON TOOL LIFE THAN DEPTH-OF-CUT

11. Always select speed within the HI-E surface footage range.
BENEFITS:
- Minimum cost
- Maximum production

The correct selection of tooling and operating conditions is the first step toward lowering tool costs and increasing productivity.
(Courtesy Carboloy Inc.)

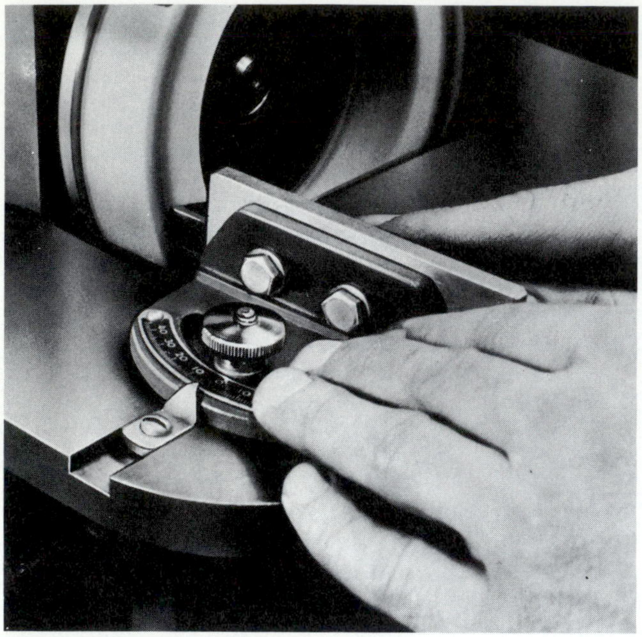

Figure 31-16 *A carbide tool grinder equipped with an adjustable table and protractor.*

quenching carbide tools creates heat checks and results in rapid tool failure.

HONING

After carbide tools have been ground, it is important that the cutting edge be honed. The purpose of honing is to remove the fine, ragged edge left by the grinding wheel. This fine, nicked edge is fragile and will break down quickly under machining conditions. Honing often means the difference between tool success or failure.

1. A 320-grit silicon carbide or diamond hone is recommended for carbide tools..
2. On carbides used for cutting steel, a 45° chamfer (Fig. 31-17) 0.002 to 0.004 in. (0.05 to 0.10 mm) wide should be honed on the cutting edge.
3. Carbide tools used for aluminum, magnesium, and plastics should not be chamfered with a hone. The fine, nicked edge should be honed to produce a sharp, keen cutting edge.

CEMENTED-CARBIDE TOOL PROBLEMS

When problems occur during machining with carbide cutting tools, consult Table 31-7 for their possible causes and remedies. It is wise to change only one thing at a time until

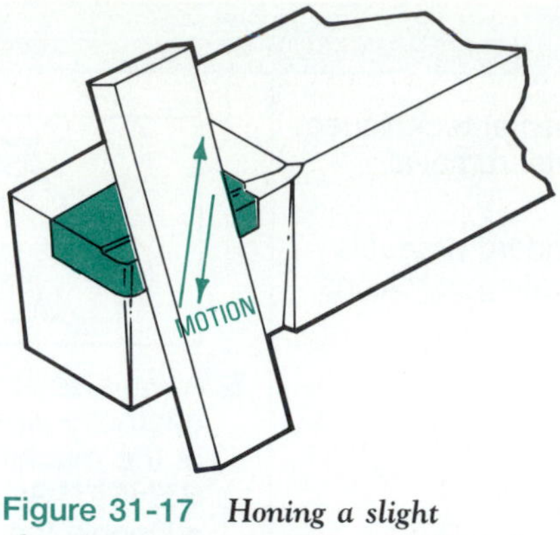
MOTION

Figure 31-17 *Honing a slight chamfer on the edge of a carbide tool will produce a stronger cutting edge.*

the problem is corrected. In this way the real cause of the problem can be determined and steps taken to guard against its recurrence.

REVIEW QUESTIONS

CARBIDE CUTTING TOOLS

1. State four reasons why cemented-carbide tools have been widely accepted by industry.

MANUFACTURE OF CEMENTED CARBIDES

2. Name five powders used in the manufacture of cemented carbides.
3. What purpose does cobalt serve?
4. Describe the process of compaction.
5. What is the purpose of presintering?
6. Describe the sintering process.

CEMENTED-CARBIDE APPLICATIONS

7. Why is cemented carbide used extensively in the manufacture of cutting tools?
8. Name two types of cemented-carbide lathe cutting tools, and state the advantages of each.
9. Identify the following throwaway insert:

SNG—321—A

GRADES OF CEMENTED CARBIDES

10. Describe straight tungsten carbide tools, and state the purpose for which they are used.
11. Explain how the size of the tungsten carbide particles and the percentage of cobalt affect the grade of carbide.
12. How does the addition of titanium carbide affect the cutting tool?

Table 31-7 Carbide Tool Problems, Causes, and Remedies

Problem	Cause	Remedy
Brazing failure and cracks	Improper braze material Too much heat Improper cooling Tip not wiped on Contact surfaces dirty	Use the sandwich braze. Silver solder at 1400°F (760°C). Cool slowly. Slide tip back and forth and tap gently. Thoroughly clean both surfaces.
Built-up edge (on the tool)	Cutting speed too slow Insufficient rake angle Wrong grade of carbide	Increase the speed. Increase rake angle. Change to titanium and/or tantalum grade.
Chipped and broken cutting edge	Incorrect tool geometry Improper carbide grade Cutting edge not honed Improper speed, feed or depth of cut Too much tool overhang Lack of rigidity in setup or machine Machine stopped during cut	Increase side cutting edge angle. Decrease end cutting edge angle. Use a negative rake. Increase the nose radius. Decrease the relief angles. Use a tougher grade. Hone cutting edge at 45°. Change one or all as required. Reduce overhang or use a larger shank. Locate and correct. Disengage feed before stopping the machine.
Cratering	Improper grade Speed too high Wrong tool geometry	Select harder or crater-resistant grade. Reduce the speed. Increase side rake.
Grinding cracks	Improper wheel Wheel glazed Tool quenched while hot Grinding carbide and steel shank with the same wheel	Select proper wheel. Dress often to a $\frac{1}{16}$-in. crown. *Do not quench;* allow tool to cool slowly. Relieve the steel shank with an aluminum oxide wheel first.
Tool wear (excessive)	Improper carbide grade Cutting speed too high Feed too light Too little relief angle	Select a wear-resistant grade. Reduce the speed. Increase the feed rate. Increase the relief angle.

13. What properties does the addition of tantalum carbide provide?
14. List four important rules to observe for selecting a cemented-carbide grade.

COATED CARBIDE INSERTS

15. List three advantages that coated carbide inserts have over standard carbide inserts.
16. What material is deposited on the carbide substrate when a single coating is applied?
17. Describe the process of triple coating.

TOOL GEOMETRY

18. Define and state the purpose of:
    a. front or end relief
    b. Side relief
    c. Side rake
    d. Angle of keenness
19. a. What is the purpose of the side cutting edge angle?
    b. How is the angle of keenness formed?
20. State the purpose of the nose radius on a cutting tool. How large should it be?
21. Name two types of back rake and state the purpose of each.
22. What three factors determine the angles and clearances of single-point carbide tools?

CUTTING SPEEDS AND FEEDS

23. Name four important factors that influence the speed, feed, and depth of cut for carbide tools.
24. Use Table 31-5 to determine what cutting speed should be used for:
    a. Taking a $\frac{3}{32}$-in. depth of cut on 260 Brinell steel with a 0.015 in. feed

b. A cut 1.50 mm deep on 300 Brinell steel using a 0.20-mm feed

MACHINING WITH CARBIDE TOOLS

25. List two important precautions that should be observed when work is being set up on a lathe.
26. Why should the cutting edge of a carbide tool be honed?
27. What occurs if the nose radius on the cutting tool is
   a. Too large? b. Too small?
28. Using Table 31-2, determine the nose radius for a carbide tool taking:
   a. A ¼-in.-deep cut at .020-in. feed
   b. A 3.2-mm depth of cut using a 0.40-mm feed
29. How should carbide tools be set up for machining?
30. Explain how carbide tools should be set up in a rocker-type toolpost.
31. What precautions should be taken when setting up a machine for cutting with carbide tools?
32. Discuss the effects of:
   a. Too light a feed b. Too coarse a feed
33. Why should a machine never be stopped while the tool is engaged in a cut?
34. Explain how a dull cutting tool may be recognized.

GRINDING CEMENTED-CARBIDE TOOLS

35. What type of grinding wheels are used to grind carbide tools?
36. How should silicon carbide wheels be dressed? Explain why.
37. List three important precautions that should be observed in the grinding of carbide tools.
38. Why is it important not to quench carbide tools?
39. What is the purpose of honing carbide tools?
40. How should carbide tools used for steel be honed?

CEMENTED-CARBIDE TOOL PROBLEMS

41. List the factors that would cause the following problems:
   a. Cratering
   b. Grinding cracks
   c. Chipped or broken cutting edge

UNIT 32

Diamond, Ceramic, and Cermet Cutting Tools

Since the development of carbide cutting tools, industry has continued to research and develop better cutting tools that are capable of operating at greater speeds, feeds, and depths of cut. Although no tool has been found that will perform all jobs perfectly, great strides have been made in the development of cutting tools.

OBJECTIVES

After completing this unit you will be able to:

1. Explain the purpose and application of diamond cutting tools
2. State the uses of two types of ceramic tools
3. Describe the types and application of cermet tools

DIAMOND CUTTING TOOLS

Since diamond is the hardest known material, it would be natural to assume that it could be used to cut other materials effectively. Two types of diamonds are used in industry: *natural* (or mined) and manufactured. Natural or mined diamonds were once widely used for machining nonmetallic and nonferrous materials, but are being replaced by manufactured diamonds, which are superior in performance in most cases. These diamonds are used to machine hard-to-finish materials and produce excellent surface finishes.

MANUFACTURED DIAMONDS

Diamond, the hardest substance known, was used primarily in machine shop work for truing and dressing grinding wheels. Because of the high cost of natural diamonds, industry began to look for cheaper, more reliable sources. In 1954, the General Electric Company, after four years of research, produced manufactured diamonds in its laboratory. In 1957, GE, after more research and testing, began the commercial production of these diamonds.

Many forms of carbon were used in experiments to manufacture diamonds. After much experimentation with various materials, the first success came when carbon and iron sulfide in a granite tube closed with tantalum disks were subjected to a pressure of approximately 1.5 million psi (10,342,500 KPa) and temperatures of between 2550 and 4260°F (1400 and 2350°C) in the "Belt" furnace (Fig. 32-1). Various diamond configurations are produced by using other metal catalysts, such as chromium, manganese, tantalum, cobalt, nickel, or platinum, in place of iron. The temperatures used must be high enough to melt the metal saturated with carbon and start the diamond growth.

Types of Manufactured Diamonds

Because the temperature, pressure, and catalyst-solvent can be varied, it is possible to produce diamonds of the size, shape, and crystal structure best suited to a particular need.

Type RVG Diamond

This manufactured diamond is an elongated, friable crystal with rough edges (Fig. 32-2A). The letters RVG indicate that this type may be used with a resinoid or vitrified bond for grinding ultrahard materials such as tungsten carbide, silicon carbide, and space-age alloys. The RVG diamond may be used for wet and dry grinding.

Type MBG-II Diamond

This tough, blocky shaped crystal is not as friable as the RVG type and is used in metal-bonded grinding (MBG) wheels (Fig. 32-2B). It is used for grinding cemented carbides, sapphires, and ceramics, as well as in electrolytic grinding.

Type MBS Diamond

This is a blocky, extremely tough crystal with a smooth, regular surface which is not very friable (Fig. 32-2C). It is used in metal-bonded saws (MBS) to cut concrete, marble, tile, granite, stone, and masonry. The diamonds may be

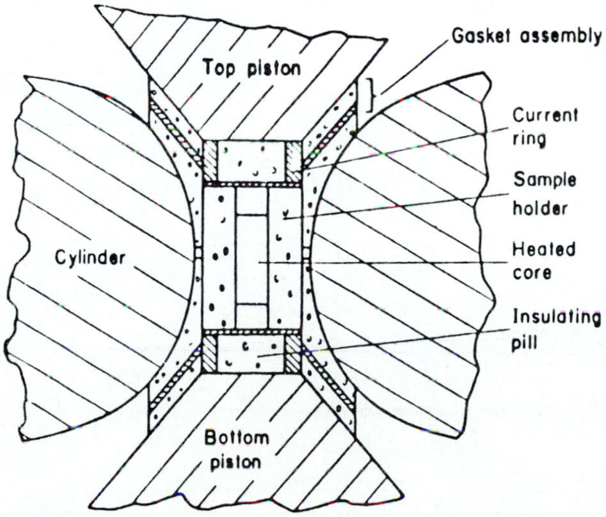

Gasket assembly
Top piston
Current ring
Sample holder
Heated core
Insulating pill
Cylinder
Bottom piston

Figure 32-1 *Diamonds are manufactured in a high-temperature and high-pressure belt-type apparatus. (Courtesy GE Superabrasives.)*

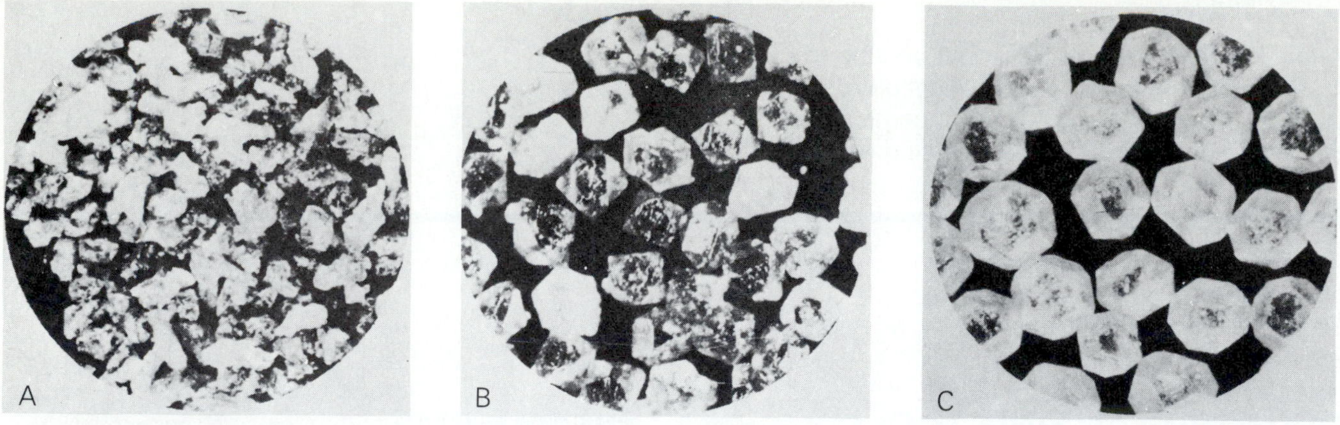
A
B
C

Figure 32-2 *(A) Type RVG diamond is used to grind ultrahard materials. (B) Type MBG-11 is a tough diamond crystal used in metal-bonded grinding wheels. (C) Type MBS is a very tough, large crystal used in metal-bonded saws. (Courtesy GE Superabrasives.)*

coated with nickel or copper to provide a better holding surface in the bond and to prolong the life of the wheel.

ADVANTAGES OF DIAMOND CUTTING TOOLS

Diamond-tipped cutting tools (Fig. 32-3) are used to machine nonferrous and nonmetallic materials which require a high surface finish and extremely close tolerances. They are used primarily as finishing tools because they are brittle and do not resist shock or cutting pressure as well as carbide cutting tools. Single-point, diamond-tipped cutting tools are available in various shapes for turning, boring, grooving, and special forming.

The main advantages of diamond-tipped cutting tools are:

Figure 32-3 *Diamond-tipped cutting tools are used to make shallow cuts at very high speeds.*

1. They can be operated at high cutting speeds, and production can be increased to 10 to 15 times that of other cutting tools.
2. Surface finishes of 5 μin. (0.127 μm) or less can be obtained easily. In many cases, the necessity of finishing operations on the workpiece is eliminated.
3. They are very hard and resist abrasion. Much longer runs are therefore possible on abrasive materials.
4. Closer tolerance work can be produced with diamond-tipped cutting tools.
5. Minute cuts, as low as 0.0005 in. (0.012 mm) deep, can be taken from the inside or outside diameter.
6. Metallic particles do not build up (weld) on the cutting edge.

USE OF DIAMOND CUTTING TOOLS

Some of the most successful applications of diamond cutting tools have been in the turning of metallic (nonferrous) and nonmetallic materials. The most common of these materials are listed.

Metallic Materials

1. *Light metals,* such as aluminum, duraluminum, and magnesium alloys
2. *Soft metals,* such as copper, brass, and zinc alloys
3. *Bearing metals,* such as bronze and babbitt
4. *Precious metals,* such as silver, gold, and platinum

For finishing cuts in these metallic materials, diamond tools can be expected to increase production from 10 to 15 times that possible with any other cutting tool.

Nonmetallic Materials

Some of the most common materials machined with diamond cutting tools are hard and soft rubber, all types of plastics, carbon, graphite, and ceramics. In some cases, diamond tools will increase production 20 to 50 times that of carbide tools.

CUTTING SPEEDS AND FEEDS

In general, diamond-tipped cutting tools operate most efficiently with shallow cuts at high cutting speeds and fine feeds. They are not recommended for materials in which the temperature of the chip or the heat generated at the chip–tool interface exceeds 1400°F (760°C). High cutting speeds, fine feeds, and shallow cuts are used. The heat generated is entirely dissipated in the chip as it leaves the cutting edge. Low cutting speeds and heavy cuts create more heat and damage the diamond tip.

There is an ideal cutting speed for each type of material–machine combination. The minimum cutting speed for diamond tools should be 250 to 300 sf/min (76 to 91 m/min). The conditions of the machine will determine the maximum cutting speed for each job. Cutting speeds as high as 10,000 sf/min (3048 m/min) have been used for some applications. Table 32-1 lists cutting speed, feed, and depth of cut ranges for various material groups.

HINTS ON THE USE OF DIAMOND TOOLS

Diamond cutting tools will perform more efficiently and have longer life if the following hints and precautions are observed.

1. Diamond-tipped points should be designed with maximum included point angle and radius for added strength (Fig. 32-4).
2. Diamond tools should always be handled with care, especially when setups are being made. The cutting edges

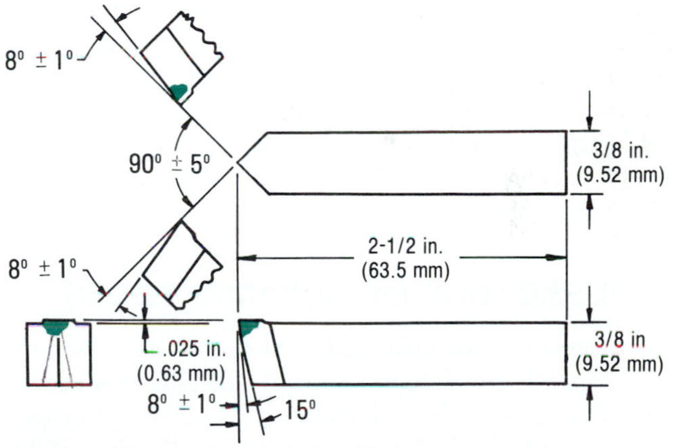

8° ± 1°
90° ± 5°
8° ± 1°
3/8 in.
(9.52 mm)
2-1/2 in.
(63.5 mm)
.025 in.
(0.63 mm)
8° ± 1°
15°
3/8 in
(9.52 mm)

Figure 32-4 *Diamond-tipped cutting tool angles and clearances.*

Table 32-1 Diamond Cutting-Tool Data

Material	Cutting Speed		Feed (per rev.)		Depth of Cut	
	ft/min	m/min	in.	mm	in.	mm
Metallic (nonferrous)	250–10,000	75–3050	0.0008–0.004	0.002–0.10	0.0005–0.024	0.001–0.60
Nonmetallic	250–3300	75–1005	0.0008–0.024	0.002–0.60	0.0008–0.060	0.002–1.50

should *never* be checked with a micrometer or bumped with a height gage.

3. Diamond tools should always be stored in separate containers, with rubber protectors over the tips, so that they will not be fractured by coming into contact with other tools.

4. The machine tool should be as free of vibration as possible. Any vibration can result in tool failure.

5. A very rigid setup, with the diamond tip set exactly on center, should be used.

6. The work should be roughed out with a carbide tool. This step will establish an even work surface and bring the work close to size.

7. Diamond tools should always be fed into the work while the work is revolving. Never stop a machine during a cut.

8. Interrupted cuts, especially on hard metals, will shorten tool life.

CERAMIC CUTTING TOOLS

The first ceramic (cemented-oxide) cutting-tool inserts were put on the market in 1956; they were the result of many years of research. At first these ceramic inserts were inconsistent. Weak and unsatisfactory results were obtained because of lack of knowledge and improper use. Since then, the strength of ceramic cutting tools has nearly doubled, their uniformity and quality have been greatly improved, and they are now widely accepted by industry. Ceramic cutting tools are being used successfully in the machining of hard ferrous materials and cast iron. As a result, lower costs, increased productivity, and better results are being gained. In some operations, ceramic toolbits can be operated at three to four times the speed of carbide toolbits.

MANUFACTURE OF CERAMIC TOOLS

Most ceramic or cemented-oxide cutting tools are manufactured primarily from aluminum oxide. Bauxite (a hydrated alumina form of aluminum oxide) is chemically processed and converted into a denser, crystalline form called *alpha alumina.* Fine grains (micron size) are obtained from the precipitation of the alumina or from the precipitation of the decomposed alumina compound.

Ceramic tool inserts are produced by either *cold* or *hot pressing.* In cold pressing, the fine alumina powder is compressed into the required form and then sintered in a furnace at 2912° to 3092°F (1600° to 1700°C). Hot pressing combines forming and sintering, with pressure and heat being applied simultaneously. Certain amounts of titanium oxide or magnesium oxide are added for certain types of ceramics to aid in the sintering process and to retard growth. After the inserts have been formed, they are finished with diamond-impregnated grinding wheels.

Further research has led to the development of stronger ceramic cutting tools. Aluminum oxide (Al_2O_3) and zirconium oxide (ZrO_2) are mixed in powder form, cold-pressed into the required shape, and sintered. This white ceramic insert has proved successful in machining high-tensile-strength materials (up to 42 Rockwell C hardness) at speeds of up to 2000 sf/min. (609 m/min.) However, the most common applications of ceramic inserts are in the general machining of steel, where there are no heavy, interrupted cuts and where negative rakes can be used. This type of toolbit has the highest hot-hardness strength of any cutting-tool material and gives excellent surface finish. No coolant is required with ceramic tools since most of the heat goes into the chip and not into the workpiece. Ceramic inserts should be used with toolholders which have a fixed or adjustable chipbreaker.

The most common ceramic cutting tool is the *throwaway insert* (Fig. 32-5), which is fastened in a mechanical holder. Throwaway inserts are available in many styles, such as triangular, square, rectangular, and round. These inserts are indexible; when a cutting edge becomes dull, a sharp edge can be obtained by indexing (turning) the insert in the holder.

Cemented ceramic tools (Fig. 32-6) are the most economical, especially if the tool shape must be altered from the standard shape. The ceramic insert is bonded to a steel shank with an epoxy glue. This method of holding the ceramic inserts almost eliminates the strains caused by clamping inserts in mechanical holders.

CERAMIC TOOL APPLICATIONS

Ceramic tools were intended to supplement rather than replace carbide tools. They are extremely valuable for specific applications and must be carefully selected and used. Ceramic tools can be used to replace carbide tools which wear rapidly in use but should never be used to replace carbide tools which are breaking.

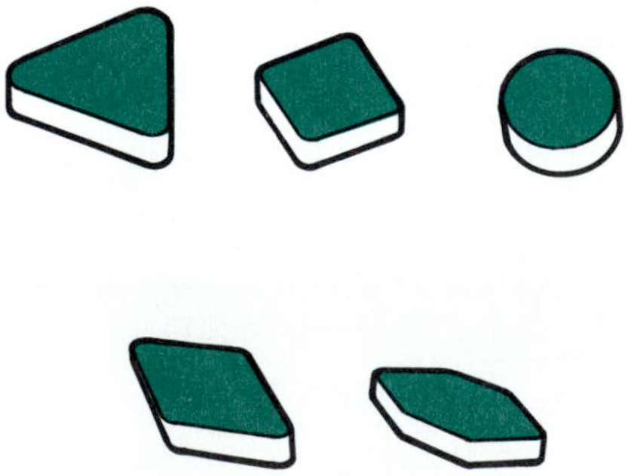

Figure 32-5 *Throwaway ceramic inserts are available in a variety of sizes.*

Figure 32-6 *Ceramic tips may be cemented to the toolshank.*

Ceramics are being used successfully for:

1. High-speed, single-point turning, boring, and facing operations, with continuous cutting action
2. Finishing operations on ferrous and nonferrous materials
3. Light, interrupted finishing cuts on steel or cast iron; heavy, interrupted cuts on cast iron only if there is adequate rigidity in the machine and tool
4. Machining castings when other tools break down because of the abrasive action of sand, inclusions, or hard scale
5. Cutting hard steels up to a hardness of Rockwell C 66, which previously could be machined only by grinding
6. Any operation in which size and finish of the part must be controlled and in which previous tools have not proved satisfactory

FACTORS AFFECTING CERAMIC TOOL PERFORMANCE

As a result of an extensive research and testing program, several factors have been found which significantly affect the performance of ceramic tools. The following factors must be considered for optimum results to be derived from ceramic cutting tools.

1. Accurate and rigid machine tools are essential when ceramic tools are used. Machines with loose bearings, inaccurate spindles, slipping clutches, or any imbalance will result in the ceramic tools chipping and in premature failure.
2. The machine tool must be equipped with ample power and be capable of maintaining the high speeds necessary for ceramics.
3. Tool mounting and toolholder rigidity are as important as machine rigidity. An intermediate plate between the toolholder clamp and ceramic insert should be used to distribute clamping pressure.
4. The overhang of the toolholder should be kept to a minimum: no more than 1½ times the shank thickness.
5. Negative rake inserts give the best results because less force is applied directly to the ceramic tip.
6. A large nose radius and a large side cutting edge angle on the ceramic insert reduces its tendency to chip.
7. Cutting fluids are generally not required because ceramics remain cool during the machining operation. If cutting fluids are required, a continuous and copious flow should be used to prevent thermal shock to the tool.
8. As the cutting speed or the hardness of the workpiece increases, the ratio of feed to the depth of cut must be checked. Always make a deeper cut with a light feed rather than a light cut with a heavy feed. Most ceramic tools are capable of cuts as deep as one-half the width of the cutting surface on the insert.
9. Toolholders with fixed or adjustable chipbreakers are best for use with ceramic inserts. Adjust the chipbreakers to produce a figure 6 or 9 curled chip.

ADVANTAGES AND DISADVANTAGES OF CERAMIC TOOLS

Advantages

Ceramic cutting tools, when properly mounted in suitable holders and used on accurate, rigid machines, offer the following advantages.

1. Machining time is reduced because of the higher cutting speeds possible. Speeds ranging from 50 to 200 percent higher than those used for carbide tools are common.
2. High stock removal rates and increased productivity result because heavy depths of cut can be made at high surface speeds.
3. A ceramic tool used under proper conditions lasts from 3 to 10 times longer than a plain carbide tool and will exceed the life of coated carbide tools.
4. Ceramic tools retain their strength and hardness at high machining temperatures [in excess of 2000°F (1093°C)].
5. More accurate size control of the workpiece is possible because of the greater wear resistance of ceramic tools.
6. Ceramic cutting tools withstand the abrasion of sand and of inclusions found in castings.
7. A better surface finish than is possible with other types of cutting tools is produced.
8. Heat-treated materials as hard as Rockwell C 66 can be readily machined.

Disadvantages

Although ceramic cutting tools have many advantages, they have the following limitations.

1. Ceramic tools are brittle and therefore tend to chip easily.
2. They are satisfactory for interrupted cuts only under ideal conditions.
3. The initial cost of ceramics is higher than that of carbides.
4. A more rigid machine than is necessary for other cutting tools is required.
5. Considerably more power and higher cutting speeds are required for ceramics to cut efficiently.

CERAMIC TOOL GEOMETRY

The geometry of ceramic tools depends on five main factors:

1. The material to be machined
2. The operation being performed
3. The condition of the machine
4. The rigidity of the work setup
5. The rigidity of the toolholding device

Although the final geometry of a ceramic cutting tool depends on these specific factors, some general considerations should also be mentioned.

Rake Angles

Because ceramic tools are brittle, negative rake angles are generally preferred. A negative rake angle allows the shock of the cutting force to be absorbed behind the tip, thus protecting the cutting edge. Negative rake angles from 2 to 30° are used for machining ferrous and nonferrous metals. Positive rake angles are used for machining nonmetals such as rubber, graphite, and carbon (Table 32-2).

Side Clearance

A side clearance angle is desirable for ceramic cutting tools whenever possible. The side clearance angle must not be too great, otherwise the cutting edge will be weakened and tend to chip.

Front Clearance

The front clearance angle should be only large enough to prevent the tool from rubbing on the workpiece. If this angle is too great, the ceramic tool becomes susceptible to chipping.

End Cutting Edge Angle

This end cutting edge angle governs the strength of the tool and the area of contact between the work and the end of the cutting tool. If properly designed, it will remove the crests resulting from feed lines (Fig. 32-7) and improve the surface finish.

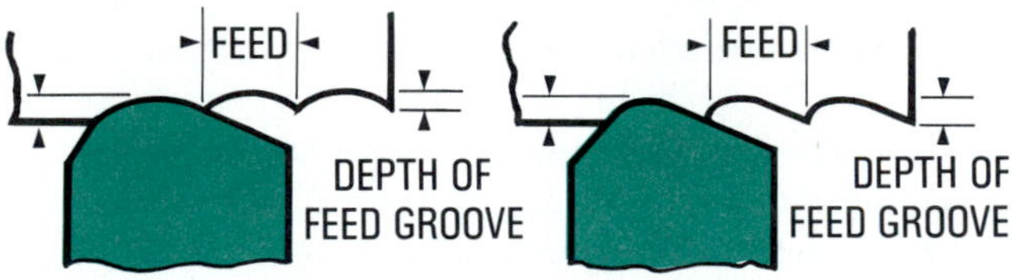
FEED
DEPTH OF FEED GROOVE
FEED
DEPTH OF FEED GROOVE

Figure 32-7 *The relationship of the nose radius and end cutting edge angle to the surface finish produced.*

Table 32-2 Suggested Rake and Relief Angles for Ceramic Tools

Workpiece Material	Rake Angles (degrees)	Relief Angles (degrees)
Carbon and alloy steels: Annealed and heat-treated	Neg. 2 to 7	2 to 7
Cast iron: Hard or chilled, Gray or ductile	0 to Neg. 7	2 to 7
Nonferrous: Hard or soft	0 to Neg. 7	2 to 7
Nonmetallics: Wood, paper, green ceramics, fiber, asbestos, rubber, carbon, graphite	0 to Pos. 10	6 to 18

*For uninterrupted turning with tool point on centerline of workpiece.
Source: Courtesy The Carborundum Company.

Nose Radius

The nose radius (Fig. 32-7) has two important functions: to strengthen the weakest part of the tool and to improve the surface finish of the workpiece. It should be as large as possible without chatter or vibration resulting.

Cutting Edge Chamfer

A small chamfer, or radius, on the cutting edge is recommended for ceramic tools, especially on heavy cuts and hard materials. This strengthens and protects the cutting edge. A 0.002- to 0.008-in. (0.05- to 0.20-mm) radius, or chamfer, is recommended for machining steel. For heavy roughing cuts and hard materials, a 0.030 to 0.060 in. (0.76 to 1.52 mm) chamber gives satisfactory results.

CUTTING SPEEDS

When ceramics are used in machining, the highest cutting speed possible, considering the machine tool's limitations, which gives reasonable tool life should be used. Less heat is generated when machining is done with ceramics than with any other type of cutting tool, because there is a lower coefficient of friction between the chip, work, and tool surface. Since most of the heat generated escapes with the chip, the cutting speed can be from 2 to 10 times higher than with other cutting tools. Table 32-3 lists the recommended speeds for various materials under ideal conditions. Should lower speeds be necessary to suit various machine or setup conditions, ceramic cutting tools will still perform well.

Table 32-3 Recommended Cutting Speeds for Ceramic Cutting Tools

Workpiece Material	Material Condition or Type	Roughing Cut		Finishing Cut		Recommended Tool Geometry (Type of Rake Angle)	Recommended Coolant
		Depth >0.062 in. Feed 0.015–0.030 in.	Depth >1.60 mm Feed 0.40–0.75 mm	Depth <0.062 in. Feed 0.010 in.	Depth <1.60 mm Feed 0.25 mm		
Carbon and tool steels	Annealed	300–1500	90–455	600–2000	185–610	Neg.	None
	Heat-treated	300–1000	90–305	500–1200	150–365	Neg.	
	Scale	300–800	90–245			Neg. honed edge	
Alloy steels	Annealed	300–800	90–245	400–1400	120–425	Neg.	None
	Heat-treated	300–800	90–245	300–1000	90–305	Neg. honed edge	
	Scale	300–600	90–185			Neg. honed edge	
High-speed steel	Annealed	100–800	30–245	100–1000	30–305	Neg.	None
	Heat-treated	100–600	30–185	100–600	30–385	Neg. honed edge	
	Scale	100–600	30–185			Neg. honed edge	
Stainless steel	300 series	300–1000	90–305	400–1200	120–365	Pos. and neg.	Sulfur base oil
	400 series	300–1000	90–305	400–1200	120–365	Neg.	
Cast iron	Gray iron	200–800	60–245	200–2000	60–610	Pos. and neg.	None
	Pearlitic	200–800	60–245	200–2000	60–610	Neg.	
	Ductile	200–600	60–185	200–1400	60–427	Neg.	
	Chilled	100–600	30–185	200–1400	60–427	Neg. honed edge	
Copper and alloys	Pure	400–800	120–245	600–1400	185–425	Pos. and neg.	Mist coolant
	Brass	400–800	120–245	600–1200	185–365	Pos. and neg.	Mist coolant
	Bronze	150–800	45–245	150–1000	45–305	Pos. and neg.	Mist coolant
Aluminum alloys*		400–2000	120–610	600–3000	185–915	Pos.	None
Magnesium alloys		800–10,000	245–3050	800–10,000	245–3050	Pos.	None
Non-metallics	Green ceramics	300–600	90–185	500–1000	150–305	Pos.	None
	Rubber	300–1000	90–305	400–1200	120–365	Pos.	None
	Carbon	400–1000	120–305	600–2000	185–610	Pos.	None
Plastics		300–1000	90–305	400–1200	120–365	Pos.	None

*Alumina-based cutting rods have a tendency to develop a built-up cutting edge on certain aluminum alloys.
Source: Courtesy A. C. Wickman Limited.

CERAMIC TOOL PROBLEMS

Ceramic tools remove metal faster than most types of cutting tools and therefore should be selected carefully for the type of material and the operation. Some points to consider when selecting ceramic tools are:

1. The tool should be large enough for the job. It cannot be too large, but can easily be too small.
2. The style (tool geometry) should be right for the type of operation and material. A tool designed for one job is not necessarily right for another.

Some of the more common problems which occur with ceramic cutting tools and their possible causes are listed in Table 32-4.

GRINDING CERAMIC TOOLS

It is not recommended that ceramic tools be ground; however, with the proper care, these tools may be resharpened successfully. Resinoid-bonded, diamond-impregnated wheels are recommended for grinding ceramic tools. A coarse-grit wheel should be used for rough grinding, and at least a 220-grit wheel should be used for finish grinding. Because ceramic tools are extremely notch-sensitive, all surfaces should be ground as smoothly as possible to avoid notches or grinding lines at the cutting edge. The cutting edge should be honed or lapped after grinding to remove any notches left by grinding and to avoid the wedging action that would occur if material were to enter these notches.

Table 32-4 Ceramic Tool Problems and Possible Causes

Problem	Possible Causes	Problem	Possible Causes
Chatter	1. Tool not on center 2. Insufficient end relief and/or clearance 3. Too much rake angle 4. Too much overhang or tool too small 5. Nose radius too large 6. Feed too heavy 7. Lack of rigidity in the tool or machine 8. Insufficient power or slipping clutch	Torn finish	1. Lack of rigidity 2. Dull tool 3. Speed too slow 4. Chipbreaker too narrow or too deep 5. Improper grinding
		Wear	1. Speed too high or feed too light 2. Nose radius too large 3. Improper grinding
Chipping	1. Lack of rigidity 2. Saw-toothed or too keen a cutting edge 3. Chipbreaker too narrow or too deep 4. Chatter 5. Scale or inclusions in the workpiece 6. Improper grinding 7. Too much end relief 8. Defective toolholder	Cracking or breaking	1. Insert surfaces not flat 2. Insert not seated solidly 3. Stopping workpiece while tool is engaged 4. Worn or chipped cutting edges 5. Feed too heavy 6. Improperly applied coolant 7. Too much rake angle or end relief 8. Too much overhang or tool too small 9. Lack of rigidity in setup 10. Speed too slow 11. Too much variation in cut for size of tool 12. Chatter 13. Grinding cracks
Cratering	1. Chipbreaker set too close to the edge 2. Nose radius too large 3. Side cutting edge angle too great		

CERMET CUTTING TOOLS

As a result of continued research aimed at improving the strength of ceramic cutting tools, cermet tools were developed in about 1960. Cermets are toolbits made of various ceramic and metallic combinations.

TYPES OF CERMET TOOLS

There are two types of cermet tools: those composed of titanium carbide (TiC) based materials and those containing titanium nitride (TiN) based materials.

Titanium carbide (TiC) *cermets* have a nickel and molybdenum binder and are produced by cold pressing and sintering in a vacuum. They are used extensively for finishing cast irons and steels that require high speeds and light to moderate feeds.

Recently titanium nitride has been added to titanium carbide to produce titanium carbide–titanium nitride (TiC–TiN) cermets. Other materials such as molybdenum carbide, vanadium carbide, zirconium carbide and others may be added, depending on the application.

Because of their high productivity, cermets are considered a cost-effective replacement for coated and uncoated carbide and ceramic toolbits (Fig. 32-8). However, cermets are not recommended for use with hardened ferrous metals (over 45 Rc) or nonferrous metals.

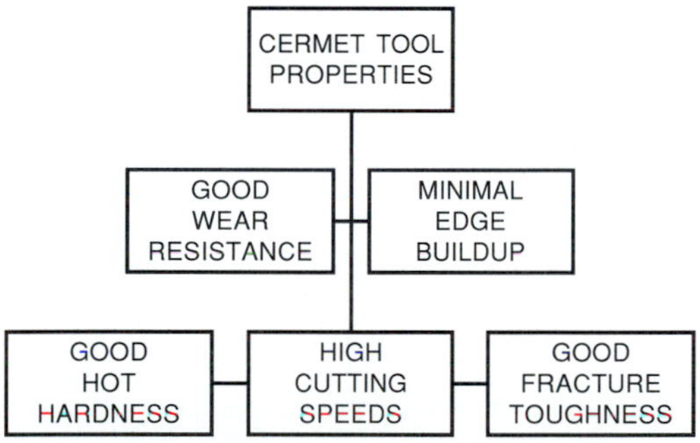
CERMET TOOL PROPERTIES
GOOD WEAR RESISTANCE
MINIMAL EDGE BUILDUP
GOOD HOT HARDNESS
HIGH CUTTING SPEEDS
GOOD FRACTURE TOUGHNESS

Figure 32-8 *The characteristics of cermet tools make them very cost effective.*

CHARACTERISTICS OF CERMET TOOLS

The main characteristics of cermet tools are:

- They have great wear resistance and permit higher cutting speeds than do carbide tools.
- Edge buildup and cratering are minimal, which increases tool life.
- They possess high hot-hardness qualities greater than carbide tools but less than ceramic tools.
- They have a lower thermal conductivity than carbide tools because most of the heat goes into the chip, and they can therefore operate at higher cutting speeds.
- Fracture toughness is greater than for ceramic tools but less than for carbide tools.

ADVANTAGES OF CERMET TOOLS

Cermet tools have the following advantages.

1. The surface finish is better than that produced with carbides under the same conditions, which often eliminates the need for finish grinding.
2. High wear resistance permits close tolerances for extended periods, ensuring accuracy of size for larger batches of parts.
3. Cutting speeds can be higher than with carbides for the same tool life.
4. When operated at the same cutting speed as carbide tools, cermet tool life is longer.
5. The cost per insert is less than that of coated carbide inserts and equal to that of plain carbide inserts.

USE OF CERMET TOOLS

Titanium carbide cermets are the hardest cermets and are used to fill the gap between tough tungsten carbide inserts and the hard, brittle ceramic tools. They are used mainly for machining steels and cast irons where high speeds and moderate feeds may be used (Table 32-5).

Titanium carbide–titanium nitride inserts are used for semifinish and finish machining of harder cast irons and

Table 32-5 Table 32-5 Suggested Cutting Conditions for Cermet Cutting Tools

Material	Hardness (Brinell)	Cutting Speed	Feed	Depth of Cut
Cast irons	100–250	200–1200 sf/min. (60–366 m/min.)	0.002–0.016 in. (0.05–0.40 mm)	0.187–0.250 in. (4.74–6.35 mm)
Steel, carbon	100–250	160–1200 sf/min. (48–366 m/min.)	0.002–0.016 in. (0.05–0.40 mm)	0.200–0.300 in. (5.08–7.62 mm)
Steels, alloys and stainless	250–400	150–1000 sf/min. (46–305 m/min.)	0.002–0.016 in. (0.05–0.40 mm)	0.187–0.300 in. (4.74–7.62 mm)

steels (less than 45 Rc) such as alloy steel, stainless steel, armor plate, and powder metallurgy parts.

REVIEW QUESTIONS

DIAMOND CUTTING TOOLS

1. Name two types of diamonds used in industry.
2. List four main advantages of diamond cutting tools.
3. Explain where diamond cutting tools may be used successfully.

CUTTING SPEEDS AND FEEDS

4. What speed, feed, and depth of cut should be used for diamond tools? Explain why.
5. Why should the diamond tool be as rigid as possible and the machine free from vibration?
6. List five important precautions to be observed when using diamond cutting tools.

MANUFACTURE OF CERAMIC TOOLS

7. Briefly explain how cemented-oxide cutting tools are manufactured.
8. What is the difference between ceramic and cermet cutting tools?

CERAMIC TOOL APPLICATIONS

9. Name four important applications for ceramic tools.
10. List four main factors which affect the performance of ceramic tools.

ADVANTAGES AND DISADVANTAGES OF CERAMIC TOOLS

11. Name five advantages of ceramic tools.
12. What are the disadvantages of using ceramic cutting tools?

CERAMIC TOOL GEOMETRY

13. List five factors that determine the geometry of ceramic tools.
14. Fully explain the rake angle of ceramic tools.
15. Define *cutting edge chamfer* and state its purpose.

CUTTING SPEEDS

16. Explain why high cutting speeds can be used with ceramic tools.
17. What two points should be kept in mind when ceramic tools are being selected?
18. List the factors that may cause the following problems.
    a. Chipping
    b. Chatter
    c. Wear

GRINDING CERAMIC TOOLS

19. Explain how ceramic tools should be ground.
20. What is the purpose of honing?

CERMET CUTTING TOOLS

21. What is a cermet toolbit?
22. Name the two basic types of cermets.
23. What is a TiC–TiN toolbit?
24. State four characteristics of a cermet cutting tool.
25. List four advantages of cermet cutting tools.
26. State the conditions under which titanium carbide toolbits may be used.
27. What are the applications of TiC–TiN toolbits?

UNIT

33

Polycrystalline Cutting Tools

The metalworking industry has a relatively short history of developing large-scale cutting tools. It began about the time that Henry Bessemer invented the commercial method of making steel. Before that time, steel blades were produced and forged by hand processes that were kept secret and passed on through a father-and-son tradition. Carbon-steel cutting tools were gradually replaced by the introduction of high-speed steel cutting tools in the late 1800's. While successful and still used today, high-speed steel tools, in turn, gave way to the more productive cemented-carbide and coated carbide cutting tools for most production and many toolroom cutting operations.

A new cutting-tool technology was born in about 1954, when the General Electric Company produced manufactured diamond. Later, a polycrystalline layer composed of thousands of small diamond or cubic boron nitride abrasive particles was fused to a cemented-carbide substrate (base) to produce a cutting tool having a long-wearing, superior cutting edge. Because of its excellent abrasion resistance, it soon became a highly efficient cutting tool, which rewrote all production records for machining abrasive nonmetallic, nonferrous materials.

OBJECTIVES

After completing this unit you will be able to:

1. Explain the manufacture and properties of polycrystalline tools
2. Select the proper type and size of polycrystalline cutting tools
3. Set up a tool and machine for cutting with polycrystalline tools

MANUFACTURE OF POLYCRYSTALLINE CUTTING TOOLS

There are two distinct types of polycrystalline cutting tools: polycrystalline cubic boron nitride and polycrystalline diamond. Each type is used to machine a certain class of materials. The manufacture of polycrystalline cutting-tool blanks involves basically the same process for both types of tool.

The manufacture of polycrystalline cutting-tool blanks (Fig. 33-1) was a major step in the production of new and more efficient superabrasive cutting tools. A layer of polycrystalline diamond or cubic boron nitride—approximately 0.020 in. (0.5 mm) thick—is fused on a cemented-carbide substrate or base by a high-temperature [3090 to 3275°F (1700 to 1800°C)], high-pressure process [about 1 million psi (6,895,000 kPa)]. The substrate is composed of tiny grains of tungsten carbide cemented tightly together by a cobalt metal binder. Under the high-heat, high-pressure conditions, the cobalt liquifies, flows up and sweeps around the diamond or cubic boron nitride abrasive particles, and

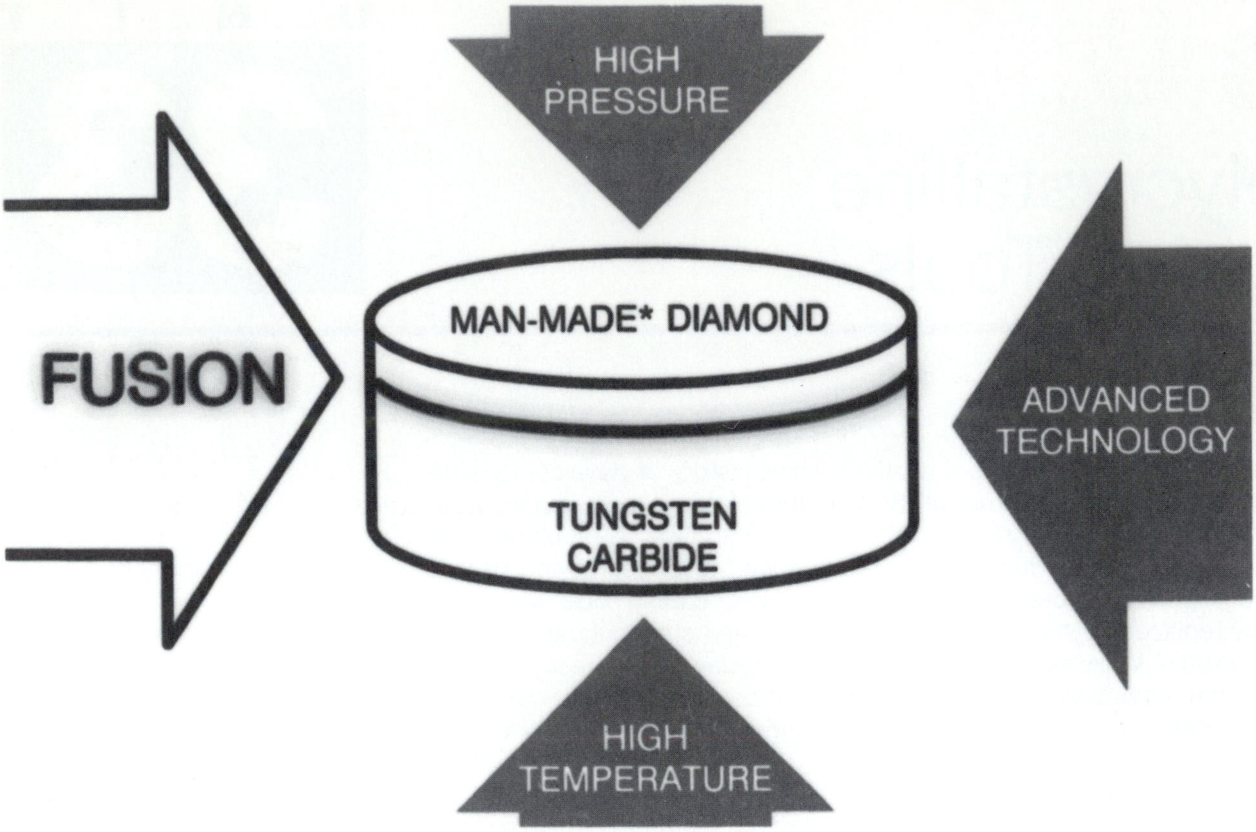
HIGH
PRESSURE
FUSION
MAN-MADE* DIAMOND
TUNGSTEN
CARBIDE
ADVANCED
TECHNOLOGY
HIGH
TEMPERATURE

Figure 33-1 *Polycrystalline tool blanks consist of a layer of polycrystalline diamond or cubic boron nitride abrasive fused to a cemented carbide base. (Courtesy GE Superabrasives.)*

serves as a catalyst that promotes intergrowth (fusing the abrasive particles). This process forms what is known as a *polycrystalline mass.*

POLYCRYSTALLINE CUBIC BORON NITRIDE (PCBN) TOOLS

The polycrystalline structure of cubic boron nitride (CBN) features nondirectional, consistent properties that resist chipping and cracking and provide uniform hardness and abrasion resistance in all directions. Therefore, it was inevitable that experiments would be conducted to see if these qualities could be built into turning and milling cutting-tool blanks and inserts. These experiments were successful, making possible the application of superabrasive cutting tools to machining operations that previously were very difficult, if not impossible, to perform by conventional methods. Polycrystalline cubic boron nitride (PCBN) blanks and inserts (Fig. 33-2) can be operated at higher cutting speeds, take deeper cuts, and machine hardened steels and high-temperature alloys (Rc 35 and harder) such as Iconel, Rene, Waspaloy, Stellite, and Colmonoy.

PROPERTIES OF POLYCRYSTALLINE CUBIC BORON NITRIDE

Cubic boron nitride is a synthetic material that is not found in nature. PCBN cutting tools contain the four main properties that cutting tools must have to cut extremely hard or abrasive materials at high metal-removal rates: hardness, abrasion resistance, compressive strength, and thermal conductivity (Fig. 33-3).

Hardness

Figure 33-4 indicates the Knoop hardness of the common abrasives. Cubic boron nitride, next to diamond in hardness, is about twice as hard as silicon carbide and aluminum oxide. PCBN cutting tools have good impact resistance, high strength, and high hardness in all directions because of the random orientation of the tiny CBN crystals.

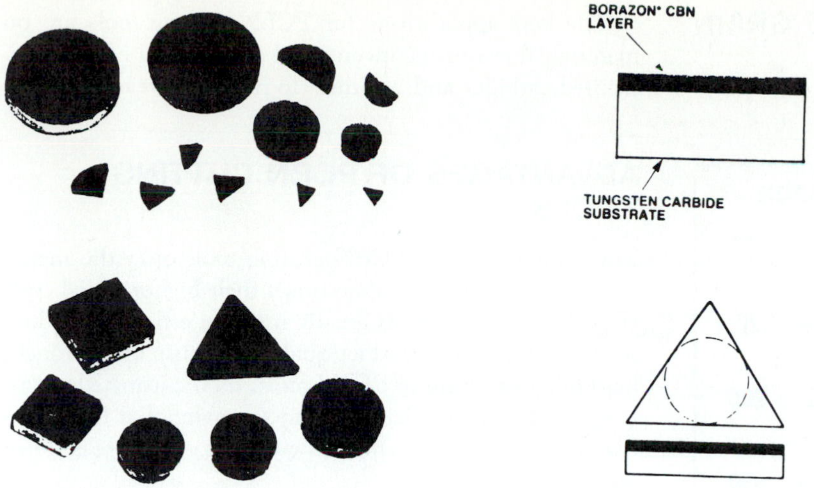

BORAZON* CBN LAYER
TUNGSTEN CARBIDE SUBSTRATE

Figure 33-2 *The types and sizes of PCBN tool blank shapes and sizes available. (Courtesy GE Superabrasives.)*

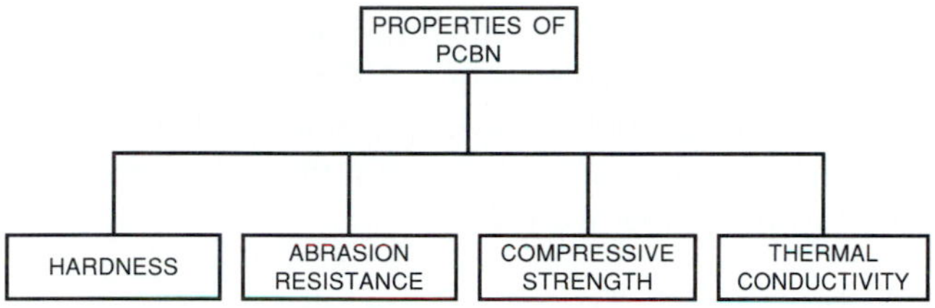

PROPERTIES OF PCBN
HARDNESS
ABRASION RESISTANCE
COMPRESSIVE STRENGTH
THERMAL CONDUCTIVITY

Figure 33-3 *The main properties of polycrystalline cubic boron nitride cutting tools.*

Abrasion Resistance

Figure 33-5 shows the relative abrasion resistance of CBN in relation to conventional abrasives and diamond. Compared to conventional cutting tools, PCBN tools maintain their sharp cutting edges much longer, thereby increasing productivity while at the same time producing dimensionally accurate parts.

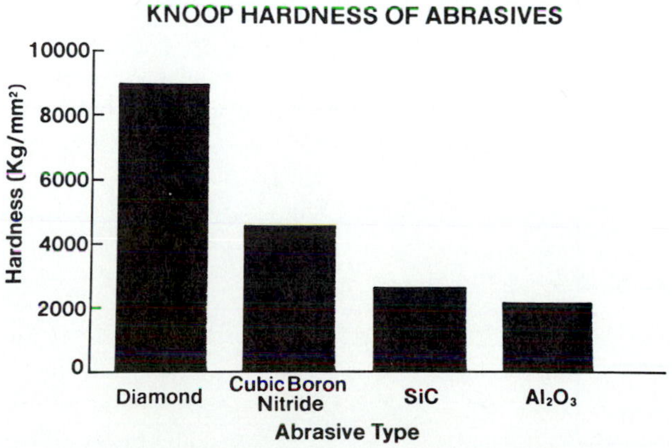

KNOOP HARDNESS OF ABRASIVES
Hardness (Kg/mm²)
10000
8000
6000
4000
2000
0
Diamond
Cubic Boron Nitride
SiC
Al₂O₃
Abrasive Type

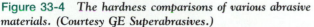

Figure 33-4 *The hardness comparisons of various abrasive materials. (Courtesy GE Superabrasives.)*

Compressive Strength

Compressive strength is defined as the maximum stress in compression that a material will take before it ruptures or breaks. The high compressive strength of CBN crystals give PCBN tools excellent qualities for withstanding the forces created during high metal-removal rates and the shock of severe interrupted cuts.

Thermal Conductivity

Because they have excellent thermal conductivity, PCBN cutting tools allow greater heat dissipation or transfer, especially when they are used to cut hard, abrasive, tough materials at high material-removal rates. The high cutting temperatures created at the cutting-tool–workpiece interface would weaken or soften conventional cutting-tool materials.

TYPES OF METAL CUT

Polycrystalline cubic boron nitride cutting tools are used on lathes and turning centers to machine round surfaces and on milling machines and machining centers to machine flat surfaces. These cutting tools have been used successfully for straight turning, facing, boring, grooving, profiling, and milling operations.

The four general types of metals that best lend themselves to the use of PCBN cutting tools and have proven to be cost effective are:

1. *Hardened ferrous metals* of greater than 45 Rc hardness, including:

- Hardened steels, such as 4340, 8620, M-2, and T-15
- Cast irons, such as chilled iron, Ni-Hard, etc.

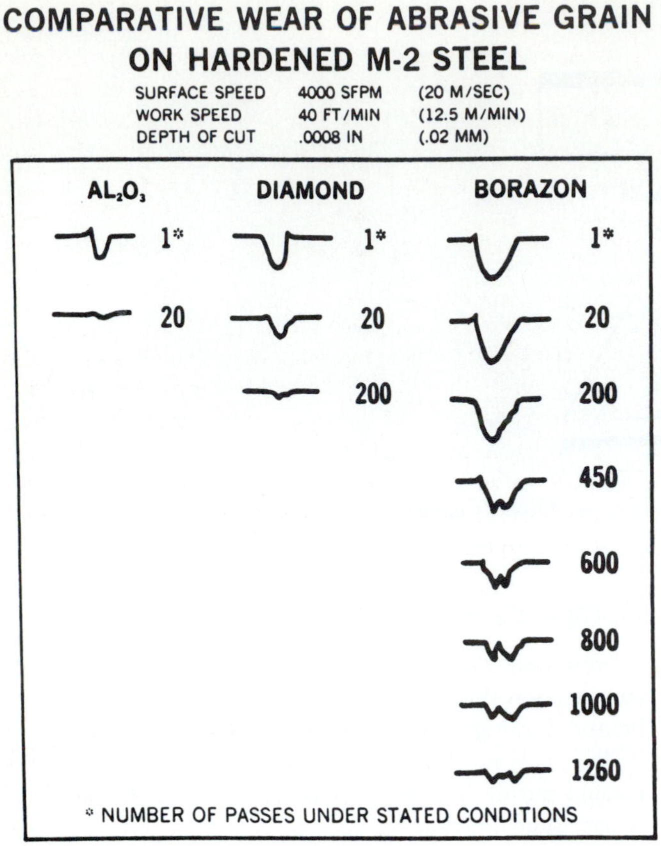
COMPARATIVE WEAR OF ABRASIVE GRAIN ON HARDENED M-2 STEEL
SURFACE SPEED 4000 SFPM (20 M/SEC)
WORK SPEED 40 FT/MIN (12.5 M/MIN)
DEPTH OF CUT .0008 IN (.02 MM)
AL2O3
DIAMOND
BORAZON
1*
20
1*
20
200
1*
20
200
450
600
800
1000
1260
* NUMBER OF PASSES UNDER STATED CONDITIONS

Figure 33-5 *The comparative abrasion or wear resistance of CBN (Borazon) in relation to other abrasives. (Courtesy GE Superabrasives.)*

2. *Abrasive ferrous metals,* such as cast irons ranging from 180 to 240 Brinnel hardness, including:
   - Pearlistic gray cast iron, Ni-Resist, etc.
3. *Heat-resistant alloys,* which most frequently are high-cobalt ferrous alloys used as flame-applied hard surfacing.
4. *Superalloys,* such as high-nickel alloys used in the aerospace industry for jet engine parts.

The best applications for PCBN cutting tools are on materials that cause conventional cutting tool edges of cemented carbides and ceramics to break down too quickly.

ADVANTAGES OF PCBN CUTTING TOOLS

The advantages that PCBN cutting tools offer the metalworking industry more than offset their higher initial costs (Fig. 33-6). These tools greatly improve efficiency, reduce scrap, and increase product quality. Because of the tough, hard microstructure of PCBN tools, their cutting edges last longer. They also efficiently remove material at high rates that would cause conventional cutting tools to break down rapidly.

High Material-Removal Rates

Because PCBN cutting tools are so hard and resist abrasion so well, cutting speeds in the range of 250 to 900 ft/min (274 m/min) and feed rates of 0.010 to 0.020 in. (0.25 to 0.50 mm) are possible. These rates result in higher material-removal rates (as much as three times that of carbide tools) with less tool wear.

Cutting Hard, Tough Materials

Polycrystalline cubic boron nitride tools are capable of efficiently machining all ferrous materials with a Rockwell C hardness of 45 and above. They are also used to machine cobalt-base and nickel-base high-temperature alloys with a Rockwell C hardness of 35 and above.

High-Quality Products

Because the cutting edges of PCBN cutting tools wear very slowly, they produce high-quality parts faster and at a lower cost per piece than do conventional cutting tools. The need to inspect the parts produced is greatly reduced, as is the

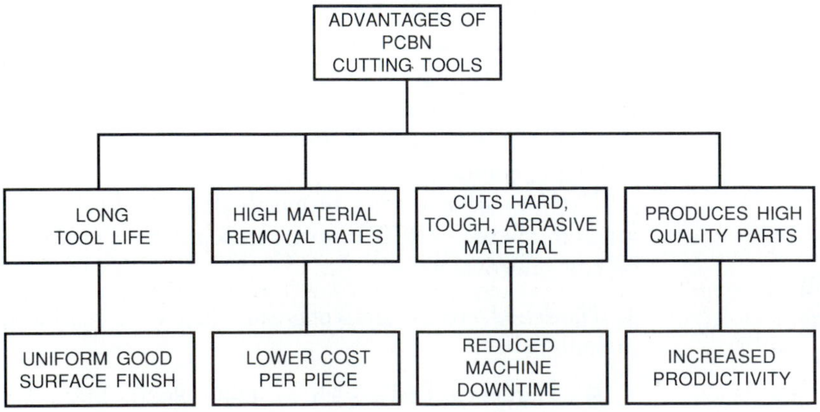
ADVANTAGES OF PCBN CUTTING TOOLS
LONG TOOL LIFE
HIGH MATERIAL REMOVAL RATES
CUTS HARD, TOUGH, ABRASIVE MATERIAL
PRODUCES HIGH QUALITY PARTS
UNIFORM GOOD SURFACE FINISH
LOWER COST PER PIECE
REDUCED MACHINE DOWNTIME
INCREASED PRODUCTIVITY

Figure 33-6 *The main advantages of polycrystalline cubic boron nitride cutting tools.*

adjustment of the machine tool to compensate for cutting-tool wear or maintenance.

Uniform Surface Finish

Surface finishes in the range of 20 to 30 μin. are possible during roughing operations with PCBN cutting tools. On finishing operations, surface finishes in single-digit microinches are possible.

Lower Cost Per Piece

Polycrystalline cubic boron nitride cutting tools stay sharp and cut efficiently throughout long production runs. This results in consistently smoother surface finishes, better control over workpiece shape and size, and fewer cutting-tool changes.

Reduced Machine Downtime

Since PCBN cutting tools stay sharp much longer than carbide or ceramic cutting tools, less time is required to index, change, or recondition cutting tools.

Increased Productivity

All the advantages that PCBN cutting tools offer, such as increased speeds and feeds, long tool life, longer production runs, consistent part quality, and savings in labor costs, combine to increase overall production rates and decrease the manufacturing cost per piece.

POLYCRYSTALLINE DIAMOND (PCD) TOOLS

A new cutting tool technology occurred about 1954 when General Electric produced synthetic manufactured diamond. Later, a polycrystalline diamond (PCD) layer, approximately 0.020 in. (0.5 mm) thick, was fused to a cemented carbide substrate (base) to produce a cutting tool which had a long wearing, superior cutting edge. Because of its excellent abrasion resistance, it soon became a highly efficient cutting tool which increased production when machining abrasive nonmetallic, nonferrous materials.

Although PCD and PCBN tools have many similar properties, each has specific applications. The main difference is that diamond tools are not suitable for machining steels and other ferrous materials. Since diamonds are composed purely of carbon and steel is a "carbon seeking" metal, it has been found that at high cutting speeds, where much heat is produced, the steel develops an affinity for the carbon. In effect, the carbon molecules are attracted to the steel, which causes the edge of the diamond tool to break down rapidly.

TYPES AND SIZES OF PCD TOOLS

Catalyst-bonded PCD, which is the variety most widely used, is available in three microstructure series for various machining applications. The basic difference between the three series is the size of the diamond particle which was used to manufacture the polycrystalline blank. Although manufacturers use different trade names to identify each series, we will discuss them as coarse, medium-fine, and fine.

- *Coarse crystal PCD blanks* (Fig. 33-7A) are designed to cut a wide variety of abrasive nonferrous, nonmetallic materials. They are highly recommended for machining cast aluminum alloys, especially those containing more than 12 percent aluminum. Coarser grade PCD tools are generally more durable than the other types.
- *Medium-fine PCD blanks* (Fig. 33-7B) are composed of fine to medium-fine crystals with a greater size distribution than coarse PCD blanks. These tools are used for machining highly abrasive nonferrous and nonmetallic materials where short cutting edges are acceptable.
- *Fine PCD blanks* (Fig. 33-7C) are manufactured from fine diamond crystals that are fairly uniform in size. The fine grain structure allows production of tools having extremely sharp cutting edges and high finishes on the rake and flank sides. These tools are recommended for applications requiring very fine surface finishes and long cutting edges.

The diamond microstructure plays a major part in determining the characteristics of a PCD tool blank, its applications, wear resistance, and cutting tool life. Remember, a coarse crystal PCD tool should be used when durability is a factor, while a fine crystal PCD tool should be used when high surface finishes are required. PCD cutting tools are generally available as tipped inserts, full-faced inserts, and brazed shank tools (Fig. 33-8).

PROPERTIES OF PCD TOOLS

Polycrystalline diamond cutting tools, which consist of a layer of diamond fused to a cemented tungsten carbide substrate, have properties that make them superabrasive cutting tools. The composite materials found in the *cemented tungsten carbide substrate* (base) provide mechanical properties. This is due to its relatively high thermal conductivity and a relatively low coefficient of thermal expansion. The substrate also provides excellent mechanical support for the polycrystalline diamond layer and imparts a toughness to the finished PCD tool.

The main properties of the *diamond layer* of PCD tools are hardness, abrasion resistance, compressive strength, and thermal conductivity. Diamond, the hardest material

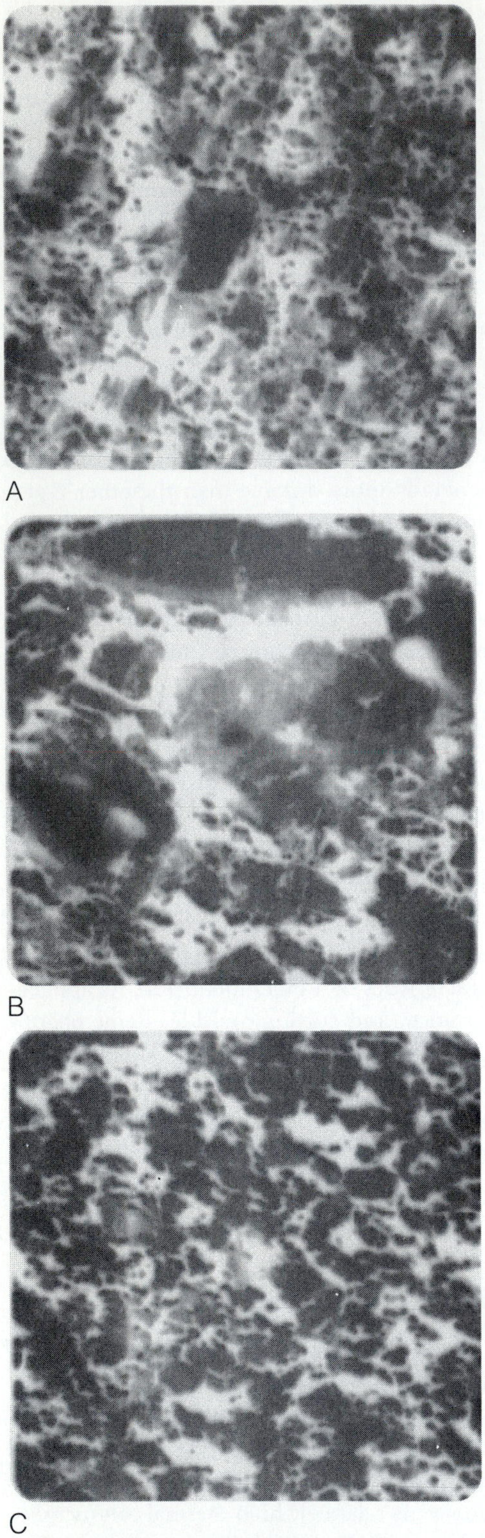
A
B
C

Figure 33-7 *The difference in diamond particle size of the three series of PCD tool blanks. (A) Coarse crystals; (B) medium-fine crystals; (C) fine crystals. (Courtesy GE Superabrasives.)*

known, is about 7000 on the Knoop hardness scale. The hardness and abrasion resistance of the polycrystalline dia-

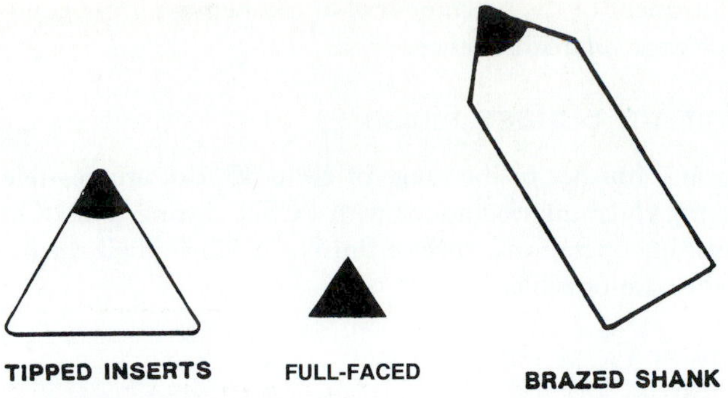
TIPPED INSERTS
FULL-FACED
BRAZED SHANK

Figure 33-8 *The common types of polycrystalline diamond (PCD) cutting tools. (A) Tipped inserts; (B) full-faced inserts; (C) brazed shank. (Courtesy GE Superabrasives.)*

mond layer is due to its inner structure. Unlike that of a single-crystal natural diamond, PCD crystal arrangement is uniform in all directions. There are no hard or soft planes, or weak-bound planes that can lead to severe cracking. Also, no special orientation is required to optimize cutting during machining operations.

The *compressive strength* of the diamond layer is the highest of any cutting tool. This is due to its dense structure which enables PCD cutting tools to withstand the forces created during high metal-removal rates and the shock of interrupted cuts. The *thermal conductivity* of the PCD layer, the highest of any cutting tool, is almost 60 percent higher than that of polycrystalline cubic boron nitride. This allows greater dissipation or transfer of the heat created at the chip–tool interface, especially when tough, abrasive materials are being cut at high metal-removal rates.

ADVANTAGES OF PCD TOOLS

The advantages that PCD cutting tools offer industry more than offset their higher initial cost. Primarily used to machine nonferrous and nonmetallic materials, PCD tools are capable of greatly improving efficiency, reducing scrap, and increasing product quality. Some of the main advantages of polycrystalline diamond cutting tools are:

1. Long tool life
2. Cuts tough, abrasive material
3. High-quality parts
4. Fine surface finishes
5. Reduced machine downtime
6. Increased productivity

TYPES OF MATERIAL CUT

Polycrystalline diamond cutting tools are used to machine nonferrous or nonmetallic materials, primarily where the workpiece is abrasive. These materials are generally considered difficult to machine because of their abrasive character.

Table 33-1 PCD Cutting-Tool Applications

Nonferrous Metals	Nonmetallic Materials	Composites
Aluminum alloys	Alumina, fired	Asbestos
Babbitt alloys	Bakelite	Fiberglass epoxy
Brass alloys	Beryllia	Filled carbons
Bronze alloys	Ceramics	Filled nylon
Copper alloys	Epoxy	Filled phenolic
Lead alloys	Glass	Filled PVC
Manganese alloys	Graphite	Filled silica
Silver, platinum	Macor	Filled Teflon
Tungsten carbides	Rubber, hard	Wood, manufactured
Zinc alloys	Various Plastics	

The largest group of nonferrous metals consists of materials that typically are soft but have hard particles dispersed in them, such as silicon suspended in silicon aluminum or glass fibers in plastic. It is the hard abrasive particle that destroys the cutting edge of conventional tools. Diamond is harder than the abrasive particle and tends to shear the hard particle rather than pushing it out of the way or dulling the cutting edge. PCD blank tools often reach a wear life of 100 times that of cemented carbide tools in such an abrasive machining application.

A growing new category of nonmetallic materials are the ceramics and composites. These materials are both hard and abrasive, and rely on the hardness of the diamond to overcome their own abrasive character. PCD cutting tools are capable of cutting the hard abrasive inclusions in these materials cleanly without the rapid dulling of the cutting edge.

The materials most successfully machined with PCD tools fall into five general categories: silicon–aluminum alloys, copper alloys, tungsten carbide, advanced composites ceramics, and wood composites. Table 33-1 lists the more common materials that can be cost-effectively machined with PCD tools.

REVIEW QUESTIONS

MANUFACTURE OF POLYCRYSTALLINE TOOLS

1. Describe a polycrystalline cutting tool.
2. How are polycrystalline cutting tools manufactured?

PCBN TOOLS

3. What advantages do PCBN cutting tools offer industry?

PROPERTIES OF PCBN

4. Name the four properties that PCBN cutting tools possess.
5. Why is abrasion resistance important to PCBN cutting tools?

TYPES OF METAL CUT

6. List the four general types of metals for which PCBN cutting tools have proven to be cost-effective.

ADVANTAGES OF PCBN TOOLS

7. Why do the cutting edges of PCBN tools last longer and remove material at higher rates than do conventional cutting tools?
8. Name four important advantages of PCBN cutting tools.
9. Explain why increased productivity is possible with PCBN cutting tools.

POLYCRYSTALLINE DIAMOND TOOLS

10. Explain the term *carbon-seeking* and how it affects PCD cutting tools.

TYPES AND SIZES OF PCD TOOLS

11. What type of PCD blank is recommended for machining highly abrasive nonferrous and nonmetallic materials?
12. List the three types of PCD cutting tools.

PROPERTIES OF PCD TOOLS

13. Name the four main properties of the diamond layer of PCD tools.
14. What is the importance of the high compressive strength of the diamond layer of PCD tools?

ADVANTAGES OF PCD TOOLS

15. List four important advantages of PCD cutting tools.

TYPES OF MATERIALS CUT

16. What five general categories of materials can be successfully machined with PCD tools?

UNIT 34

Cutting Fluids—Types and Applications

Cutting fluids are essential in most metal-cutting operations. During a machining process, considerable heat and friction are created by the plastic deformation of metal occurring in the shear zone when the chip slides along the chip–tool interface. This heat and friction cause metal to adhere to the tool's cutting edge, causing the tool to break down; the result is a poor finish and inaccurate work.

The use of cutting fluids is not new, some types having been used for hundreds of years. Centuries ago it was discovered that water dripped onto a grindstone kept the stone from glazing and produced a better surface finish on the part being ground. The major problem was that the water caused the ground part to rust. About 100 years ago, machinists found that wiping tallow on parts before machining helped to produce smoother and more accurate parts. Tallow, the forerunner of cutting fluids, lubricated but did not cool. Lard oils, developed later, lubricated well and had some cooling properties but became rancid quickly. In the early 20th century, soaps were added to water to improve cutting, prevent rust, and provide some lubrication.

In 1936, the development of soluble oils was a great improvement over the previously used cutting oils. These milky-white emulsions combined the high cooling ability of water with the lubricity of petroleum oil. Although economical, they failed to control rust and tended to become rancid.

Chemical cutting fluids were introduced in about 1944. Containing relatively little oil, they depended on chemical agents for lubrication and friction reduction. Chemical emulsions mix easily with water and reduce as well as remove the heat created during machining. Chemical cutting fluids are rapidly increasing in popularity because they provide good rust resistance, do not become rancid quickly, and have good cooling and lubricating qualities.

During a machining process, considerable heat and friction are created. The correct selection and application of cutting fluids can prevent these results by effectively cooling the work and reducing the friction. Cutting fluids cool and lubricate the tool and workpiece. Their use can result in the following economic advantages.

1. *Reduction of tool costs.* Cutting fluids reduce tool wear. Tools last longer, and less time is spent resharpening and resetting them.
2. *Increased speed of production.* Because cutting fluids help reduce heat and friction, higher cutting speeds can be used for machining operations.
3. *Reduction of labor costs.* As cutting tools last longer and require less regrinding when cutting fluids are used, there is less downtime, reducing the cost per part.
4. *Reduction of power costs.* Since friction is reduced by a cutting fluid, less power is required for machining operations and a corresponding saving in power costs is possible.

OBJECTIVES

After completing this unit you will be able to:
1. State the importance and function of cutting fluids
2. Identify three types of cutting fluids and state the purpose of each
3. Apply cutting fluids efficiently for a variety of machining operations

HEAT GENERATED DURING MACHINING

The heat generated at the chip–tool interface must find its way into one of three places: the workpiece, the cutting tool, or the chips (Fig. 34-1). If the *workpiece* receives too much heat, its size will change and a taper will automatically occur as the workpiece expands with the heat. Too much heat will also cause thermal damage to the surface of the workpiece. If the *cutting tool* receives too much heat, the cutting edge will break down rapidly, reducing tool life. The ideal cutting tool is one that can transfer the heat quickly from the cutting zone to some form of cooling system.

Ideally, most of the heat is taken off in the *chips,* which act as a disposable heat sink or reservoir. This heat transfer is indicated by the change in chip color as the heat causes the chips to oxidize. If too little material is removed to form a chip of sufficient mass, as happens when very light feeds and depth of cuts are used, the heat cannot be absorbed by the small chip. It is then forced into the workpiece and the cutting tool.

Cutting fluids assist machining operations by taking away or dissipating the heat created at the chip–tool interface. Fig. 34-1 shows how the heat is dissipated during a typical machining operation. The proper use of some form of cutting fluid or coolant system can dissipate at least 50 percent of the heat created during machining.

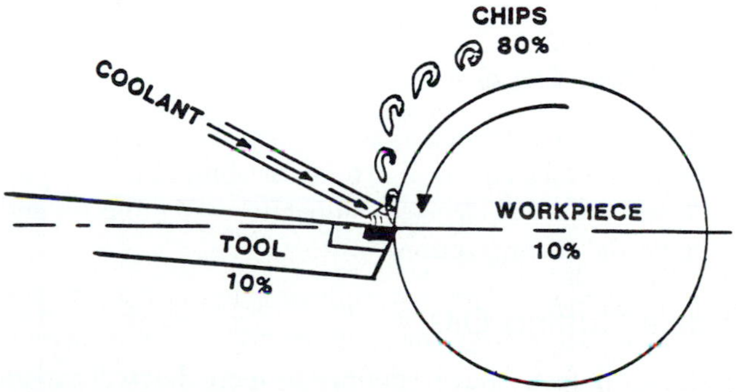
CHIPS
80%
COOLANT
WORKPIECE
10%
TOOL
10%

Figure 34-1 *Most of the heat generated during an ideal machining operation is carried away by the chips. (Courtesy GE Superabrasives.)*

CHARACTERISTICS OF A GOOD CUTTING FLUID

For a cutting fluid to function effectively, it should possess the following desirable characteristics.

1. *Good cooling capacity*—to reduce the cutting temperature, increase tool life and production, and improve dimensional accuracy.
2. *Good lubricating qualities*—to prevent metal from adhering to the cutting edge, and forming a built-up edge, resulting in a poor surface finish and shorter tool life.
3. *Rust resistance*—to prevent stain, rust, or corrosion to the workpiece or machine.
4. *Stability (long life)*—to hold up both in storage and in use.
5. *Resistance to rancidity*—to not become rancid easily.
6. *Nontoxic*—to not cause skin irritation to the operator.
7. *Transparent*—to allow the operator to see the work clearly during machining.
8. *Relatively low viscosity*—to permit the chips and dirt to settle quickly.
9. *Nonflammable*—to avoid burning easily and should preferably be noncombustible. In addition, it should not smoke excessively, form gummy deposits which may cause machine slides to become sticky, or clog the circulating system.

TYPES OF CUTTING FLUIDS

The need for a cutting fluid which possesses as many desirable characteristics as possible has resulted in the development of many different types. The most commonly used cutting fluids are either aqueous- (water-) based solutions or cutting oils. Cutting fluids fall into three categories: cutting oils, emulsifiable oils, and chemical (synthetic) cutting fluids.

CUTTING OILS

Cutting oils are classified as active or inactive. These terms relate to the oil's chemical activity or ability to react with the metal surface at elevated temperatures to protect it and improve the cutting action.

Active Cutting Oils

Active cutting oils may be defined as those that will darken a copper strip immersed in them for 3 h at a temperature of 212°F (100°C). These oils are generally used when steel is being machined and may be either dark or transparent. Dark oils usually contain more sulfur than the transparent types and are considered better for heavy-duty jobs.

Active cutting oils fall into three general categories:

1. *Sulfurized mineral oils* contain from 0.5 to 0.8 percent sulfur. They are generally light-colored, transparent, and have good cooling, lubricating, and antiweld properties. They are useful for the cutting of low-carbon steels and tough, ductile metals. Sulfurized mineral oils stain copper and its alloys and, therefore, are not recommended for use with these metals.
2. *Sulfochlorinated mineral oils* contain up to 3 percent sulfur and 1 percent chlorine. These oils prevent excessive built-up edges from forming and prolong the life of the cutting tool. Sulfochlorinated mineral oils are more effective than sulfurized mineral oils in cutting tough low-carbon and chrome–nickel alloy steels. They are extremely valuable in cutting threads in soft, draggy steel.
3. *Sulfochlorinated fatty oil blends* contain more sulfur than the other types and are effective cutting fluids for heavy-duty machining.

Inactive Cutting Oils

Inactive cutting oils may be defined as those that will not darken a copper strip immersed in them for 3 h at 212°F (100°C). The sulfur contained in an active oil is the natural sulfur of the oil and has no chemical value in the cutting fluid's function during machining. These fluids are termed inactive because the sulfur is so firmly attached to the oil that very little is released to react with the work surface during the cutting action.

Inactive cutting oils fall into four general categories:

1. *Straight mineral oils,* because of their low viscosity, have faster wetting and penetrating factors. They are used for the machining of nonferrous metals, such as aluminum, brass, and magnesium, where lubricating and cooling properties are not essential. Straight mineral oils are also recommended for use in cutting leaded (free-machining) metals and tapping and threading white metal.
2. *Fatty oils,* such as lard and sperm oil, once widely used, find limited applications as cutting fluids today. They are generally used for severe cutting operations on tough nonferrous metals where a sulfurized oil might cause discoloration.
3. *Fatty and mineral oil blends* are combinations of fatty and mineral oils, resulting in better wetting and penetrating qualities than straight mineral oils. These qualities result in better surface finishes on both ferrous and nonferrous metals.
4. *Sulfurized fatty-mineral oil blends* are made by sulfur being combined with fatty oils and then mixed with certain mineral oils. Oils of this type provide excellent antiweld properties and lubricity when cutting pressures are high and tool vibration excessive. Most sulfurized fatty–mineral oil

blends can be used when nonferrous metals are cut to produce high surface finishes. They may also be used on machines when ferrous and nonferrous metals are machined at the same time.

EMULSIFIABLE (SOLUBLE) OILS

An effective cutting fluid should possess high heat conductivity; neither mineral nor fatty oils are very effective as coolants. Water is the best cooling medium known; however, used as a cutting fluid, water alone would cause rust and have little lubricating value. By adding a certain percentage of soluble oil to water, it is possible to add rust resistance and lubricating qualities to water's excellent cooling capabilities.

Emulsifiable, or soluble, oils are mineral oils containing a soaplike material (emulsifier) which makes them soluble in water and causes them to adhere to the workpiece during machining. These emulsifiers break the oil into minute particles and keep them separated in the water for a long period of time. Emulsifiable, or soluble, oils are supplied in concentrated form. From *1 to 5 parts* of concentrate are added to *100 parts* of water. Lean mixtures are used for light machining operations and when cooling is essential. Denser mixtures are used when lubrication and rust prevention are essential.

Soluble oils, because of their good cooling and lubricating qualities, are used when machining is done at high cutting speeds, at low cutting pressures, and when considerable heat is generated.

Three types of emulsifiable, or soluble, oils are manufactured for use under various machining conditions:

1. *Emulsifiable mineral oils* are mineral oils to which various compounds have been added to make the oil soluble in water. These oils are low in cost, provide good cooling and lubrication qualities, and are widely used for general cutting applications.
2. *Superfatted emulsifiable oils* are emulsifiable mineral oils to which some fatty oil has been added. These mixtures provide better lubrication qualities and, therefore, are used for tougher machining operations. Often these soluble oils are used when aluminum is machined.
3. *Extreme-pressure emulsifiable oils* contain sulfur, chlorine, and phosphorus, as well as fatty oils, to provide the added lubrication qualities required for tough machining operations. Extreme-pressure oils are generally mixed with water at the rate of *1 part* oil to *20 parts* water.

CHEMICAL CUTTING FLUIDS

Chemical cutting fluids, sometimes called *synthetic fluids,* have been widely accepted since they were first introduced in about 1945. They are stable, preformed emulsions which contain very little oil and mix easily with water. Chemical cutting fluids depend on chemical agents for lubrication and friction reduction. Some types of chemical cutting fluids contain *extreme-pressure* (EP) lubricants, which react with freshly machined metal under the heat and pressure of a cut to form a solid lubricant. Fluids containing EP lubricants reduce both the *heat of friction* between the chip and tool face and the *heat caused by plastic deformation* of the metal.

The chemical agents found in most synthetic fluids include:

1. *Amines* and *nitrites* for rust prevention
2. *Nitrates* for nitrite stabilization
3. *Phosphates* and *borates* for water softening
4. *Soaps* and *wetting agents* for lubrication
5. *Phosphorus, chlorine,* and *sulfur compounds* for chemical lubrication
6. *Glycols* to act as blending agents
7. *Germicides* to control bacteria growth

As a result of the chemical agents which are added to the cooling qualities of water synthetic fluids provide the following advantages.

1. Good rust control
2. Resistance to rancidity for long periods of time
3. Reduction of the amount of heat generated during cutting
4. Excellent cooling qualities
5. Longer durability than cutting or soluble oils
6. Nonflammable, nonsmoking
7. Nontoxic
8. Easy separation from the work and chips, which makes them clean to work with
9. Quick setting of grit and fine chips so they are not recirculated in the cooling system
10. No clogging of the machine cooling system due to detergent action of the fluid

Three types of chemical cutting fluids are manufactured:

1. *True solution fluids* contain mostly rust inhibitors and are used primarily to prevent rust and provide rapid heat removal in grinding operations. These are generally clear solutions (sometimes a dye is added to color the water) and are mixed *1 part* solution to *50 to 250 parts* water, depending on the application. Some true solution fluids have a tendency to form hard, crystalline deposits when the water evaporates. These deposits may interfere with the operation of chucks, slides, and moving parts.
2. *Wetting-agent types* contain agents to improve the wetting action of water, providing more uniform heat dissipation and antirust action. They also contain mild lubricants, water softeners, and antifoaming agents. Wetting-type chemical cutting fluids are versatile; they have excellent lubricating qualities and provide rapid heat dissipation.

They can be used when machining is done with either high-speed steel or carbide cutting tools.

3. *Wetting-agent types with EP lubricants* are similar to wetting-agent types, but have chlorine, sulfur, or phosphorus additives to provide EP or boundary lubrication effects. They are used for tough machining jobs with either high-speed steel or carbide cutting tools.

CAUTION: Although chemical cutting fluids have been widely accepted and used for many types of metal-cutting operations, certain precautions should be observed regarding their use. Chemical cutting fluids are generally used on ferrous metals; however, many aluminum alloys can be machined successfully with them. Most chemical cutting fluids are not recommended for use on alloys of magnesium, zinc, cadmium, or lead. Certain types of paint (generally poor quality) may be affected by some chemical cutting fluids, which may mar the machine's appearance and allow paint to get into the coolant and clog the system. Before changing to any type of cutting fluid, it is wise to contact suppliers for the right cutting fluid for the machining operation and the metal being cut.

FUNCTIONS OF A CUTTING FLUID

The prime functions of a cutting fluid are to provide both cooling and lubrication. In addition, good cutting fluids prolong cutting-tool life, resist rancidity, and provide rust control.

COOLING

Laboratory tests have proved that the heat produced during machining has a definite bearing on cutting-tool wear. Reducing cutting-tool temperature is important to tool life. Even a small reduction in temperature will greatly extend the life of a cutting tool. For example, if tool temperature were reduced only 50°F (28°C) from 950 to 900°F (510 to 482°C), cutting-tool life would be increased by five times, from 19.5 to 99 min.

There are two sources of heat generated during a cutting action:

1. Plastic deformation of the metal, which occurs immediately ahead of the cutting tool, accounts for approximately two-thirds to three-quarters of the heat generated.
2. Friction resulting from the chip sliding along the cutting-tool face also produces heat.

As previously mentioned, water is the most effective agent for reducing the heat generated during machining. Since water alone causes rusting, soluble oils or chemicals which prevent rust and provide other essential qualities are added to make it a good cutting fluid.

An abundant supply of cutting fluid should be applied to the machining area, at a very low pressure. This will ensure that the machining area will be well covered and that little splashing will occur. The flow of the cutting fluid will help to wash the chips away from the machining area.

LUBRICATING

The lubricating function of a cutting fluid is as important as its cooling function. Heat is generated by the plastic deformation of the metal and by the friction between the chip and tool face. The plastic deformation of metal occurs along the shear plane (Fig. 34-2). Any way of shortening the length of the shear plane would result in a reduction in amount of heat generated.

The only known method of shortening the length of the shear plane for any given shape of a cutting tool and work material is to reduce the friction between the chip and tool face. Figure 34-3 shows a chip sliding along the face of the cutting tool. The enlarged illustration shows the irregularities on the tool face which create areas of friction and tend to cause a built-up edge to form. Note also that because of this friction there is a long shear plane and a small shear angle. The most heat is created at the cutting edge when there is a small shear angle and a long shear plane.

Figure 34-4 illustrates the same depth of cut as in Fig. 34-3, but shows cutting fluid being used to reduce the friction at the chip–tool interface. As soon as the friction is reduced, the shear plane becomes shorter, and the area where plastic deformation occurs is correspondingly

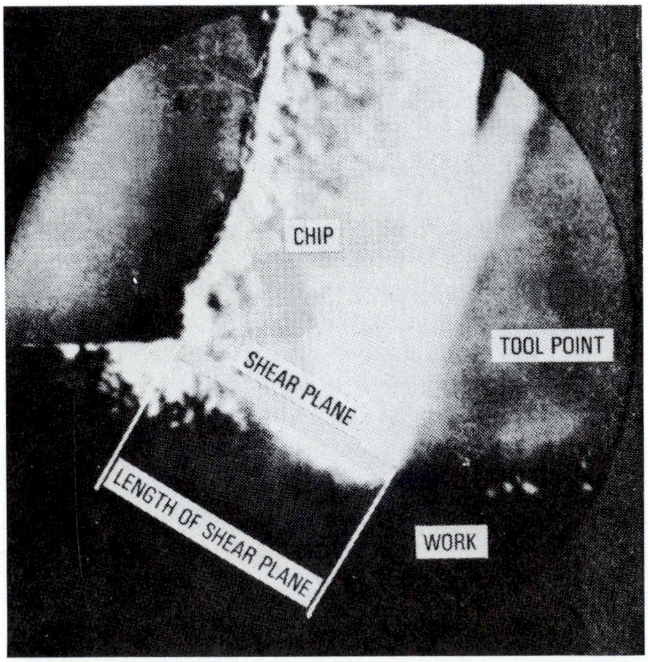

CHIP
TOOL POINT
SHEAR PLANE
LENGTH OF SHEAR PLANE
WORK

Figure 34-2 *During the cutting process the metal is deformed along the shear plane, producing heat. (Courtesy Cincinnati Milacron Inc.)*

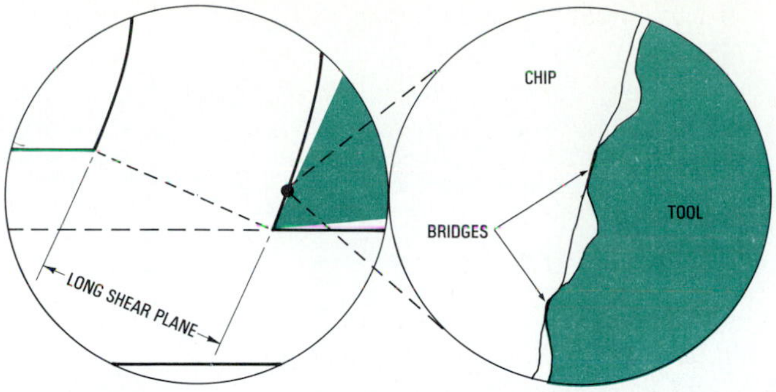
CHIP
BRIDGES
TOOL
LONG SHEAR PLANE

Figure 34-3 *A long shear plane results in great heat at the shear zone. (Courtesy Cincinnati Milacron Inc.)*

smaller. Therefore, by reduction of the friction at the chip–tool interface, both sources of heat (plastic deformation and friction at the chip–tool interface) can be reduced.

The effective life of a cutting tool can be greatly lengthened if the friction and the resultant heat generated are reduced. When steel is machined, the temperature and pressure at the chip–tool interface may reach 1000°F (538°C) and 200,000 psi (1,379,000 kPa), respectively. Under such conditions, some oils and other liquids tend to vaporize or be squeezed from between the chip and tool. Extreme-pressure lubricants reduce the amount of heat-producing friction. The EP chemicals of synthetic fluids combine chemically with the sheared metal of the chip to form solid compounds. These solid compounds or lubricants can withstand high pressure and temperature and allow the chip to slide up the tool face easily even under these conditions.

CUTTING-TOOL LIFE

Heat and friction are the prime causes of cutting-tool breakdown. Decreasing the amount of heat and friction created during a machining operation can greatly increase the life of a cutting tool. Laboratory tests have proved that if the temperature at the chip–tool interface is reduced by as little as 50°F (28°C), the life of the cutting tool increases fivefold. As a result, when cutting fluids are used, faster speeds and feeds can be used; increased production and a reduction in the cost per piece result.

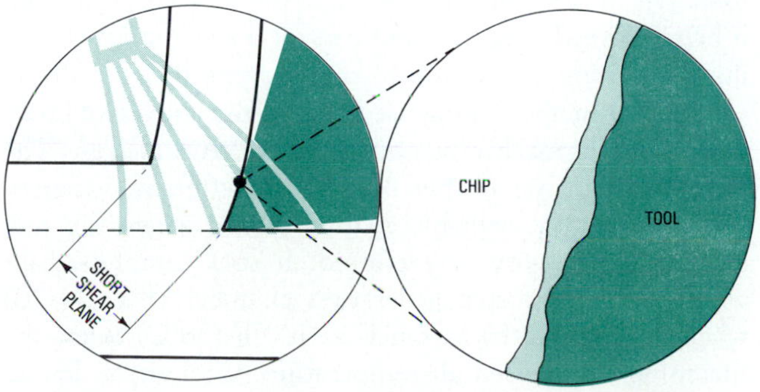
CHIP
TOOL
SHORT SHEAR PLANE

Figure 34-4 *Cutting fluid reduces friction and produces a shorter shear plane. (Courtesy Cincinnati Milacron Inc.)*

During a cut, pieces of metal tend to weld themselves to the tool face, causing a built-up edge to form (Fig. 34-5). If the built-up edge becomes large and flat along the tool face, the effective rake angle of the cutting tool is decreased and more power is required to cut the metal. The built-up edge keeps breaking off and reforming; the result is a poor surface finish, excessive flank wear, and cratering of the tool face. Almost all the roughness of a machined surface is caused by tiny fragments of metal which have been left behind by the built-up edge.

The use of an effective cutting fluid affects the action of a cutting tool in the following ways.

1. It lowers the heat created by the plastic deformation of the metal, thereby increasing cutting-tool life.
2. Friction at the chip–tool interface is decreased, reducing the resultant heat.
3. Less power is required for machining because of the reduced friction.
4. It prevents a built-up edge from forming, resulting in longer tool life.
5. The surface finish of the work is greatly improved.

RUST CONTROL

Cutting fluids used on machine tools should prevent rust from forming; otherwise machine parts and workpieces will be damaged. Cutting oil prevents rust from forming but does not cool as effectively as water. Water is the best and most economical coolant but causes parts to rust unless rust inhibitors are added.

Rust is oxidized iron, or iron that has reacted chemically with oxygen, water, and minerals in the water. Water alone on a piece of steel or iron acts as a medium for the electro-chemical process to start causing corrosion or rust.

Chemical cutting fluids contain rust inhibitors, which prevent the electro-chemical process of rusting. Some types of cutting fluids form a *polar film* on metals, which prevents rusting. This polar film (Fig. 34-6) consists of negatively charged, long, thin molecules, which are attracted and firmly bond themselves to the metal. This invisible film, only molecules thick, is sufficient to prevent the electro-chemical action of rusting. Other types of cutting fluids contain rust inhibitors which form an insulating blanket known as a *passivating film* on the metal surface. These inhibitors combine chemically with the metal and form a nonporous, protective coating which prevents rust.

RANCIDITY CONTROL

When lard oil was the only cutting fluid used, after a few days it would start to spoil and give off an offensive odor. This rancidity is caused by bacteria and other microscopic

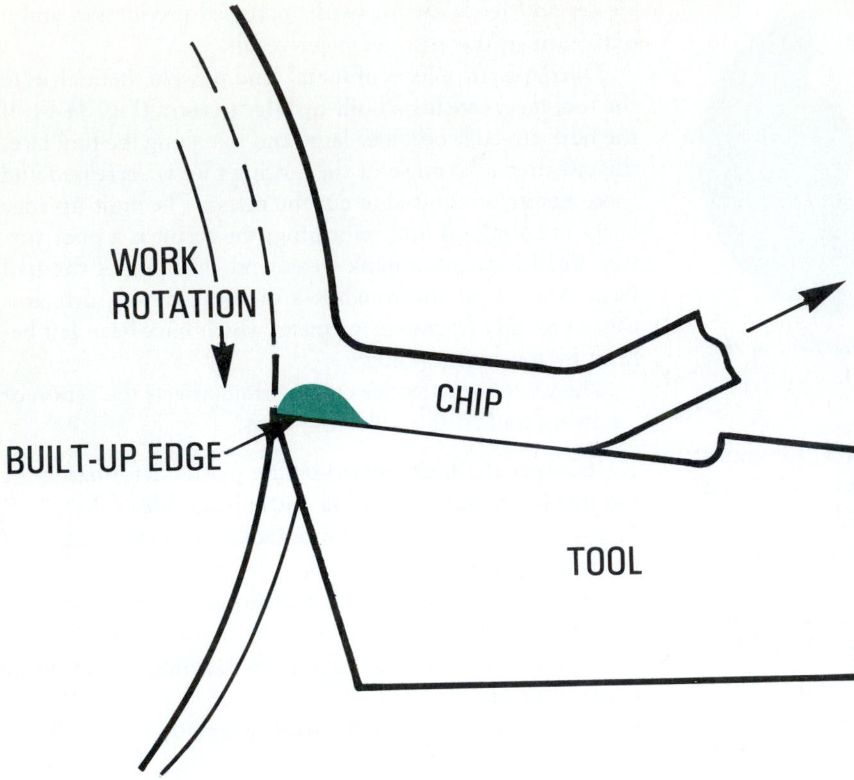
WORK
ROTATION
CHIP
BUILT-UP EDGE
TOOL

Figure 34-5 *A built-up edge is caused by chip fragments being pressure-welded to the cutting-tool face.*

organisms, growing and multiplying almost anywhere and eventually causing bad odors to form. Today, any cutting fluid that has an offensive smell is termed *rancid.*

Most cutting fluids contain bactericides which control the growth of bacteria and make the fluids more resistant to rancidity. The bactericide, which is added to the fluid by the manufacturer, must be strong enough to control the growth of bacteria, but weak enough not to be harmful to the skin of the operator.

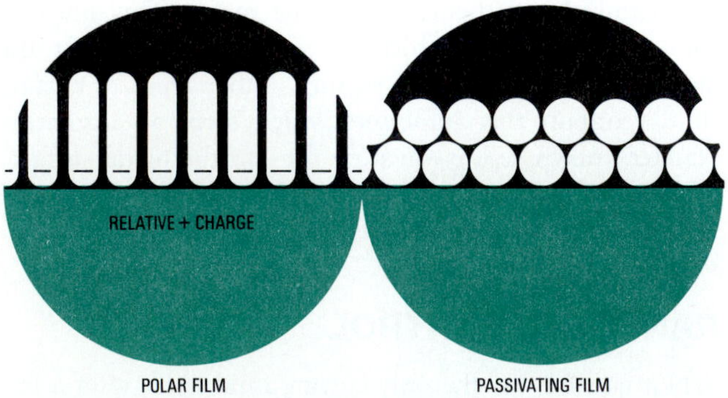
RELATIVE + CHARGE
POLAR FILM
PASSIVATING FILM

Figure 34-6 *Polar and passivating films on metal surfaces prevent rust. (Courtesy Cincinnati Milacron Inc.)*

APPLICATION OF CUTTING FLUIDS

Cutting-tool life and machining operations are greatly influenced by the way that the cutting fluid is applied. It should be supplied in a copious stream under low pressure so that the work and cutting tool are well covered. The rule of thumb is that the inside diameter of the supply nozzle should be about three-quarters the width of the cutting tool. The fluid should be directed to the area where the chip is being formed to reduce and control the heat created during the cutting action and to prolong tool life.

Another method being used to cool the chip–tool interface in various machining operations is refrigerated air. The *refrigerated air system* (Fig. 34-7A) is an effective, inexpensive, and readily available cooling system where dry machining is necessary or preferred. It uses compressed air which enters a vortex generation chamber (Fig. 34-7B) where it is cooled by as much as 100°F (38°C) below the incoming compressed air temperature. Cool air, as low as −40°F (−40°C), can be directed to cool the chip–tool interface and blow the chips away.

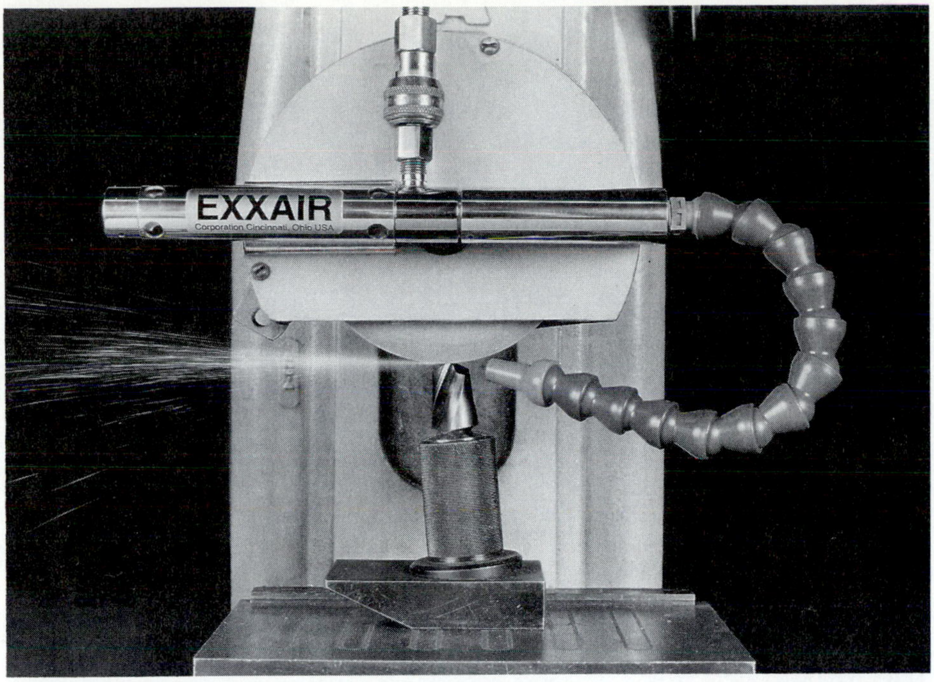
EXXAIR
Corporation Cincinnati, Ohio USA

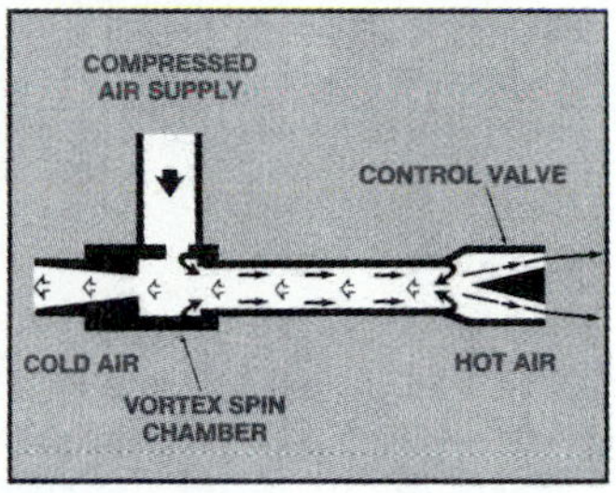
COMPRESSED AIR SUPPLY
CONTROL VALVE
COLD AIR
HOT AIR
VORTEX SPIN CHAMBER

Figure 34-7 *(A) A refrigerated air system being used for cooling purposes during surface grinding; (B) the vortex tube converts compressed air to refrigerated air. (Courtesy Exxair Corp.)*

LATHE-TYPE OPERATIONS

On horizontal-type turning and boring machines, cutting fluid should be applied to that portion of the cutting tool which is producing the chip. For general turning and facing operations, cutting fluid should be supplied directly over the cutting tool, close to the zone of chip formation (Fig. 34-8). In heavy-duty turning and facing operations, cutting fluid should be supplied by two nozzles, one directly above and the other directly below the cutting tool (Fig. 34-9).

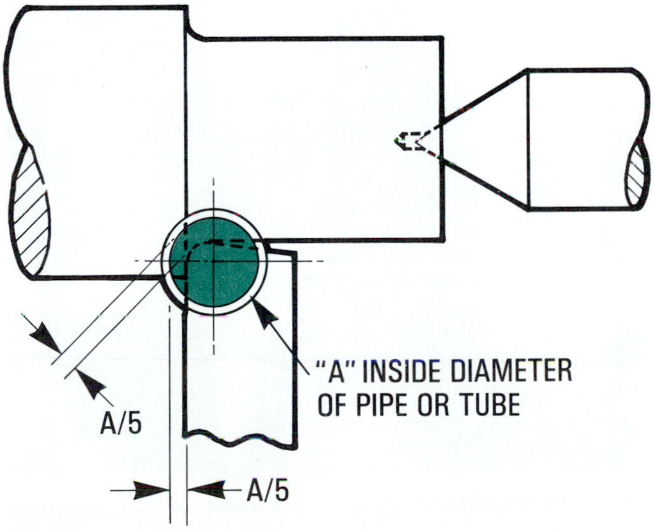
"A" INSIDE DIAMETER
OF PIPE OR TUBE
A/5
A/5

Figure 34-8 *Cutting fluid supplied by one nozzle for general facing and turning operations. (Courtesy Cincinnati Milacron Inc.)*

DRILLING AND REAMING

The most effective method of applying cutting fluids for these operations is to use "oil-feed" drills and hollow-shank reamers. Tools of this type transmit the cutting fluid directly to the cutting edges and at the same time flush the chips out of the hole (Fig. 34-10). When conventional

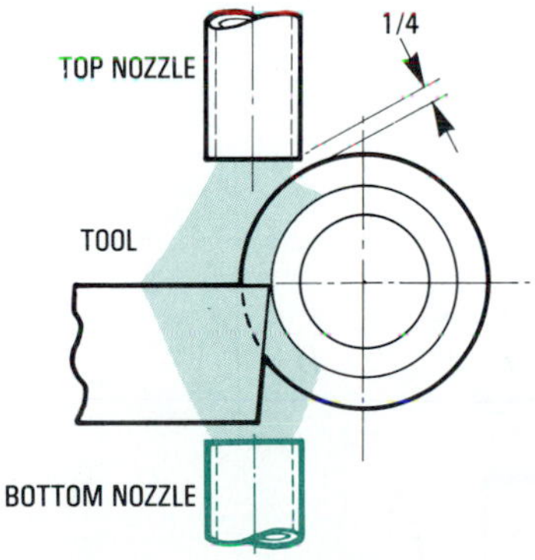
1/4
TOP NOZZLE
TOOL
BOTTOM NOZZLE

Figure 34-9 *Top and bottom nozzles are used to supply cutting fluid for heavy-duty turning operations. (Courtesy Cincinnati Milacron Inc.)*

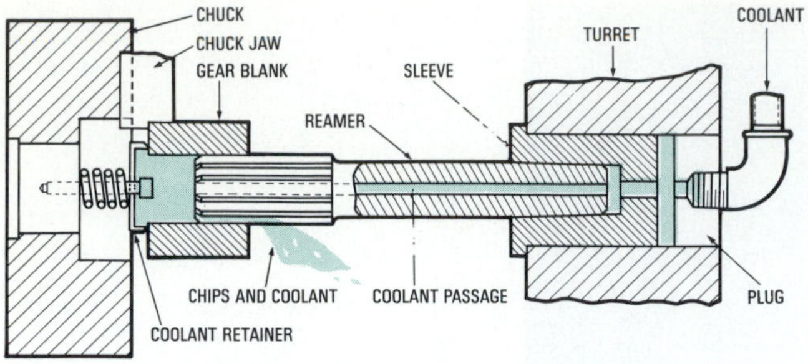
CHUCK
CHUCK JAW
GEAR BLANK
TURRET
COOLANT
SLEEVE
REAMER
CHIPS AND COOLANT
COOLANT PASSAGE
PLUG
COOLANT RETAINER
Figure 34-10 *Cutting fluid is often supplied through a hole in the center of a reamer. (Courtesy Cincinnati Milacron Inc.)*

drills and reamers are used, an abundant supply of fluid should be applied to the cutting edges.

MILLING

In *slab milling,* cutting fluid should be directed to both sides of the cutter by fan-shaped nozzles approximately three-quarters the width of the cutter (Fig. 34-11).

For *face milling,* a ring-type distributor (Fig. 34-12) is recommended in order to flood the cutter completely. Keeping each tooth of the cutter immersed in cutting fluid at all times can increase cutter life almost 100 percent.

GRINDING

Cutting fluid is very important in a grinding operation; it cools the work and keeps the grinding wheel clean. Cutting fluid should be applied in large quantities and under very little pressure.

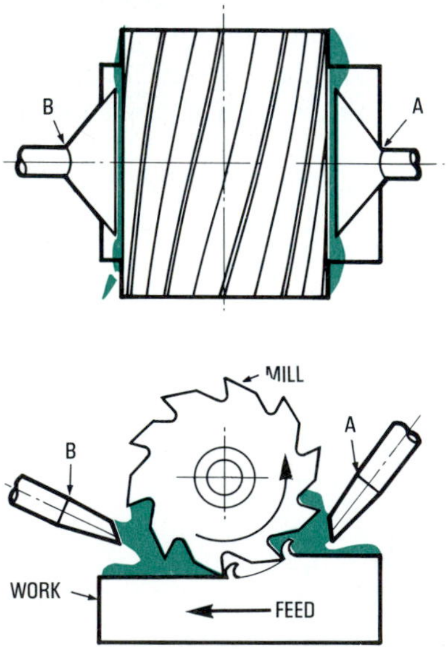
B
A
MILL
B
A
WORK
FEED
Figure 34-11 *Cutting fluid being supplied to both sides of the cutter during slab milling. (Courtesy Cincinnati Milacron Inc.)*

Figure 34-12 *Cutting fluid being applied by a ring-type distributor during a face-milling operation. (Courtesy Cincinnati Milacron Inc.)*

1. *Surface Grinding.* Three methods may be used to apply cutting fluid for surface grinding operations:

a. The *flood method* is the one most commonly used. A steady flow of cutting fluid is applied through a nozzle. Because of the reciprocating table action, most surface-grinding operations can be greatly improved if the fluid is supplied through two nozzles, as in Fig. 34-11.

b. In the *through-the-wheel* method the coolant is fed to a special wheel flange and forced to the periphery of the wheel and to the area of contact by centrifugal force.

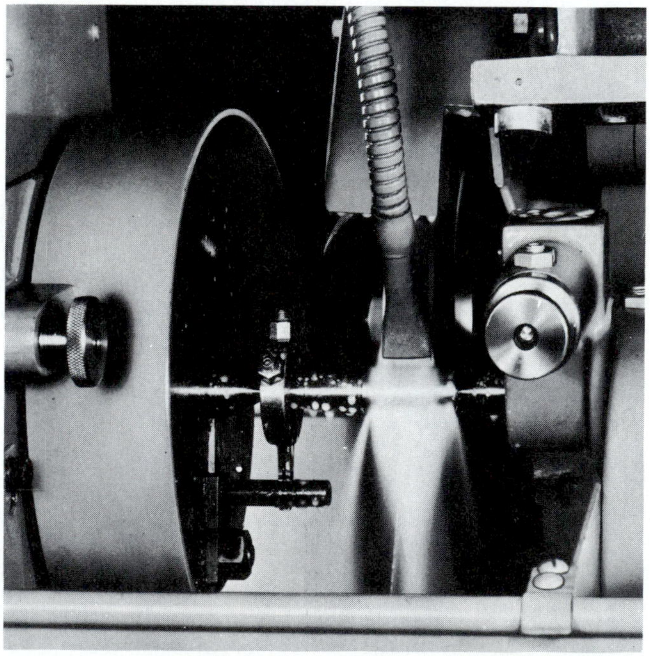
Figure 34-13 *Cutting fluid being applied to the contact zone during cylindrical grinding. (Courtesy Cincinnati Milacron Inc.)*

c. The *spray-mist system* is one of the most effective cooling systems. It uses the atomizer principle, where compressed air passing through a T connection syphons a small amount of coolant from a reservoir and discharges it at the chip–tool interface. Cooling results from the action of the compressed air and evaporation of the mist vapor. The compressed air also blows steel chips away from the cutting tool area, permitting the machine operator to clearly see the operation.

2. *Cylindrical Grinding.* For cylindrical grinding operations, it is important that the entire contact area between the wheel and work be flooded with a steady stream of clean, cool cutting fluid. A fan-shaped nozzle (Fig. 34-13) somewhat wider than the wheel should be used to direct the cutting fluid.

3. *Internal Grinding.* During internal grinding, the cutting fluid must flush the chips and the abrasive wheel particles out of the hole being ground. Because internal grinding practices call for using as large a wheel as possible, it is sometimes difficult to get enough cutting fluid into the

Table 34-1 Recommended Cutting Fluids for Various Materials

Material	Drilling	Reaming	Threading	Turning	Milling
Aluminum	Soluble oil Kerosene Kerosene and lard oil	Soluble oil Kerosene Mineral oil	Soluble oil Kerosene and lard oil	Soluble oil	Soluble oil Lard oil Mineral oil Dry
Brass	Dry Soluble oil Kerosene and lard oil	Dry Soluble oil	Soluble oil Lard oil	Soluble oil	Dry Soluble oil
Bronze	Dry Soluble oil Lard oil	Dry Soluble oil Lard oil	Soluble oil Lard oil	Soluble oil	Dry Soluble oil Lard oil
Cast iron	Dry Air jet Soluble oil	Dry Soluble oil Mineral lard oil	Dry Sulfurized oil Mineral lard oil	Dry Soluble oil	Dry Soluble oil
Copper	Dry Soluble oil Mineral lard oil Kerosene	Soluble oil Lard oil	Soluble oil Lard oil	Soluble oil	Dry Soluble oil
Malleable iron	Dry Soda water	Dry Soda water	Lard oil Soda water	Soluble oil	Dry Soda water
Monel metal	Soluble oil Lard oil	Soluble oil Lard oil	Lard oil	Soluble oil	Soluble oil
Steel alloys	Soluble oil Sulfurized oil Mineral lard oil	Soluble oil Sulfurized oil Mineral lard oil	Sulfurized oil Lard oil	Soluble oil	Soluble oil Mineral lard oil
Steel, machine	Soluble oil Sulfurized oil Mineral lard oil	Soluble oil Mineral lard oil	Soluble oil Mineral lard oil	Soluble oil	Soluble oil Mineral lard oil
Steel, tool	Soluble oil Sulfurized oil Mineral lard oil	Soluble oil Sulfurized oil Lard oil	Sulfurized oil Lard oil	Soluble oil	Soluble oil Lard oil

Note: Chemical cutting fluids can be used successfully for most of the above cutting operations. These concentrates are diluted with water in proportions ranging from 1 part cutting fluid to 15 and as high as 100 parts of water, depending upon the metal being cut and the type of machining operation. When using chemical cutting fluids, it is wise to follow the manufacturer's recommendations for use and mixture.
Source: Courtesy Cincinnati Milacron Incorporated.

hole. A compromise must be made between grinding wheel size and the amount of fluid entering the hole. As much fluid as possible should be applied during internal grinding operations.

REVIEW QUESTIONS

1. What is the function of a modern cutting fluid?
2. Briefly trace the development of cutting fluids.
3. Explain the causes of heat and friction during a machining process.
4. Name four economical advantages of applying correct cutting fluids.

CHARACTERISTICS OF A GOOD CUTTING FLUID

5. List six *important* characteristics that a good cutting fluid should possess.

TYPES OF CUTTING FLUIDS

6. Name three categories into which cutting fluids fall.
7. Describe active and inactive cutting oils.
8. What type of cutting oil should be used for:
   a. Tough, ductile metals?
   b. Heavy-duty machining?
   c. Nonferrous metals?
   d. Threading white metal?
9. Describe the composition of an emulsifiable oil and state its advantages.
10. State the purpose of:
    a. Emulsifiable mineral oil
    b. EP emulsifiable oil
11. Discuss and state six important advantages of chemical cutting fluids.
12. State the purpose of:
    a. True solution fluids
    b. Wetting-agent types with EP lubricants

FUNCTIONS OF A CUTTING FLUID

13. Name five functions of a cutting fluid.
14. Discuss the importance of lubricating and cooling as applied to cutting fluids.
15. Explain how a cutting fluid can change the length of the shear plane.
16. For what purpose are EP lubricants used?
17. What occurs at the cutting–tool face during a cut?
18. What is the main cause of surface roughness?
19. How does the application of cutting fluid affect the cutting tool?
20. Why is the control of rust important?
21. Describe a *polar film,* a *passivating film.*
22. Define *rancidity* and state the purpose of bactericides.

APPLICATION OF CUTTING FLUIDS

23. What is the general recommendation for applying cutting fluids?
24. How should cutting fluid be applied for lathe operations?
25. State two methods of applying cutting fluid for drilling or reaming.
26. Explain how cutting fluid should be applied for:
    a. Slab milling
    b. Face milling
    c. Cylindrical grinding
27. Describe three methods of applying cutting fluid for surface-grinding operations.
28. Why is it sometimes difficult to apply cutting fluid during internal grinding operations?

SECTION

9

Metal-Cutting Saws

Archaeological discoveries show that development of the first crude saw closely followed the origin of the stone ax and knife. The sharp edges of stones were serrated or toothed. This instrument cut by scraping away particles of the object being cut. A great improvement in the quality of saws followed the appearance of copper, bronze, and ferrous metals. With today's steels and hardening methods, many different types of saw blades are available for hand hacksaws and machine power saws.

UNIT

35

Types of Metal Saws

Metal-cutting saws are available in a wide variety of models to suit various cutting-off operations and materials.

OBJECTIVES

After completing this unit you will be able to:

1. Name five types of cutting-off machines and state the advantage of each
2. Select the proper blade to use for cutting various cross sections
3. Install a saw band on a horizontal bandsaw
4. Use a bandsaw to cut off work to an accurate length

METHODS OF CUTTING OFF MATERIAL

Five of the most common methods of cutting off material are *hacksawing, bandsawing, abrasive cutting, cold sawing,* and *friction sawing.* A brief description of each method and its advantages follows.

The *power hacksaw,* which is a reciprocating type of saw, is usually permanently mounted to the floor. The saw frame and blade travel back and forth, with pressure being applied automatically only on the forward stroke. The power hacksaw finds limited use in machine shop work since the saw cuts only on the forward stroke, resulting in considerable wasted motion.

The *horizontal bandsaw* (Fig. 35-1) has a flexible, belt-like "endless" or "one-way" blade, which cuts continuously in one direction. The thin continuous blade travels over the rims of two pulley wheels and passes through roller guide brackets which support the blade and keep it running true. Horizontal bandsaws are available in a wide variety of types and sizes and are becoming increasingly popular because of their high production and versatility.

The *abrasive cutoff saw* (Fig. 35-2) cuts material by means of a thin abrasive wheel revolving at high speed. This type of saw is especially well suited for cutting most metals and materials such as glass and ceramics. It can cut material to close tolerances, and hardened metal does not have to be annealed in order to be cut. Abrasive cutoff can be performed under dry conditions or with a suitable cutting fluid. The use of cutting fluid keeps the work and saw cooler and produces a better surface finish.

The *cold circular cutoff saw* (Fig. 35-3) uses a circular

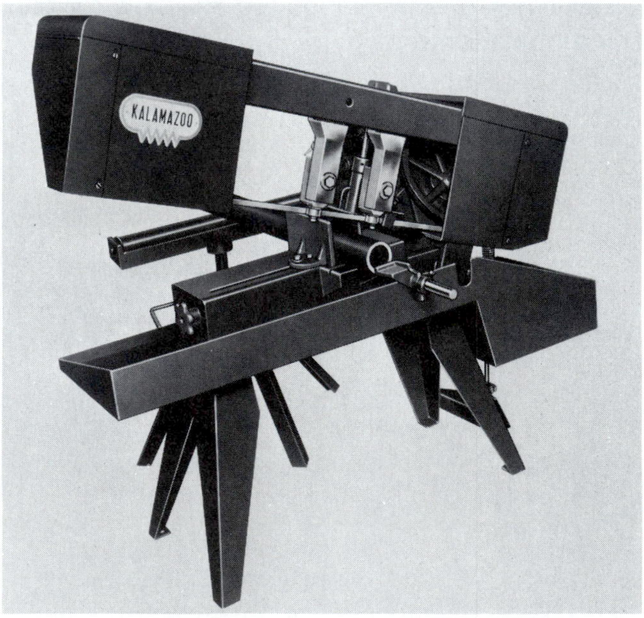
KALAMAZOO

Figure 35-1 *A horizontal bandsaw cuts continuously in one direction.*

Figure 35-2 *An abrasive cutoff saw will cut hardened metals, glass, and ceramics. (Courtesy Everett Industries, Incorporated.)*

blade similar to the one used on a wood-cutting table saw. The saw blade is generally made of chrome–vanadium steel, but carbide-tipped blades are used for some applications. Cold circular saws produce very accurate cuts and are especially suited for cutting aluminum, brass, copper, machine steel, and stainless steel.

Figure 35-3 *A cold circular saw is used for cutting soft or unhardened metals. (Courtesy Everett Industries, Incorporated.)*

Friction sawing is a burning process by which a saw band, with or without saw teeth, is run at high speeds [10 000 to 25 000 sf/min (3048 to 7620 m/min)] to burn or melt its way through the metal. Friction sawing cannot be used on solid metals because of the amount of heat generated; however, it is excellent for cutting structural and honeycombed parts of machine or stainless steel.

HORIZONTAL BANDSAW PARTS

The horizontal bandsaw is the most popular machine used to cut off work. The main operative parts of this saw (Fig. 35-4) are as follows:

- The *saw frame,* hinged at the motor end, has two pulley wheels mounted on it over which the continuous blade passes.
- The *step pulleys* at the motor end are used to vary the speed of the continuous blade to suit the type of material being cut.
- The *roller guide brackets* provide rigidity for a section of the blade and can be adjusted to accommodate various widths of material. These brackets should be adjusted to just clear the width of the work being cut.
- The *blade-tension handle* is used to adjust the tension on the saw blade. The blade should be adjusted to prevent it from wandering or twisting.
- The *vise,* mounted on the table, can be adjusted to hold

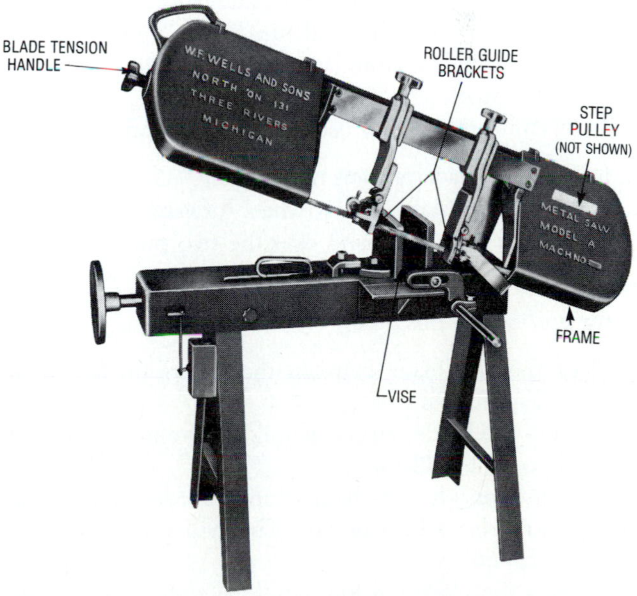
BLADE TENSION HANDLE
W.F. WELLS AND SONS
NORTH ON 131
THREE RIVERS
MICHIGAN
ROLLER GUIDE BRACKETS
STEP PULLEY (NOT SHOWN)
METAL SAW
MODEL A
FRAME
VISE

Figure 35-4 *The main parts of a horizontal bandsaw.*

various sizes of workpieces. It can also be swiveled for making angular cuts on the end of a piece of material.

SAW BLADES

High-speed tungsten and high-speed molybdenum steel are commonly used in the manufacture of saw blades, and for the power hacksaw they are usually hardened completely. Flexible blades used on bandsaws have only the saw teeth hardened.

Saw blades are manufactured in various degrees of coarseness, ranging from 4 to 14 pitch. When cutting large sections, use a coarse or 4-pitch blade, which provides the greatest chip clearance and helps to increase tooth penetration. For cutting tool steel and thin material, a 14-pitch blade is recommended. A 10-pitch blade is recommended for general-purpose sawing. *Metric saw blades* are available in similar sizes but in teeth per 25 mm of length rather than teeth per inch. Therefore, the pitch of a blade having 10 teeth per 25 mm would be 10 ÷ 25 mm or 0.4 mm. Always select a saw blade as coarse as possible, but make sure that *two teeth* of the blade will be in contact with the work at all times. If fewer than two teeth are in contact with the work, the work can be caught in the tooth space (gullet) which would cause the teeth of the blade to strip or break.

INSTALLING A BLADE

When replacing a blade, always make sure that the teeth are pointing in the direction of saw travel or toward the motor end of the machine. The blade tension should be adjusted to prevent the blade from twisting or wandering during a cut. If it is necessary to replace a blade before a cut is finished, rotate the work one-half turn in the vise. This will prevent the new blade from jamming or breaking in the cut made by the worn saw.

To install a saw blade, use the following procedure.

1. Loosen the blade-tension handle (Fig. 35-5).
2. Move the adjustable pulley wheel forward slightly.
3. Mount the new saw band over the two pulleys.

NOTE: *Be sure that the saw teeth are pointing toward the motor end of the machine.*

4. Place the saw blade between the rollers of the guide brackets (Fig. 35-5).
5. Tighten the blade-tension handle only enough to hold the blade on the pulleys.
6. Start and quickly stop the machine in order to make the saw blade revolve a turn or two. This will seat (track) the blade on the pulley.
7. Tighten the blade-tension handle as tightly as possible with *one hand.*

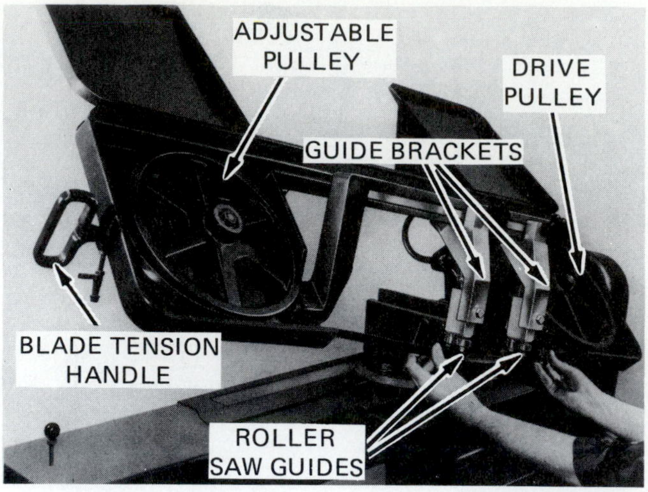
ADJUSTABLE PULLEY
DRIVE PULLEY
GUIDE BRACKETS
BLADE TENSION HANDLE
ROLLER SAW GUIDES

Figure 35-5 *Installing a new blade on a horizontal bandsaw.*

SAWING

For the most efficient sawing, it is important that the correct type and pitch of saw blade be selected and run at the proper speed for the material being cut. Use finer tooth blades when cutting thin cross sections and extra-hard materials. Coarser tooth blades should be used for thick cross sections and material which is soft and stringy. The blade speed should suit the type and thickness of the material being cut. Too fast a blade speed or excessive feeding pressure will dull the saw teeth quickly and cause an inaccurate cut.

To saw work to length, use the following procedure.

1. Check the solid vise jaw with a square to make sure it is at right angles to the saw blade.
2. Place the material in the vise, supporting long pieces with a floor stand (Fig. 35-6).
3. Lower the saw blade until it just clears the work. Keep it in this position by engaging the ratchet lever or by closing the hydraulic valve.
4. Adjust the roller guide brackets until they *just clear* both sides of the material to be cut (Fig. 35-5).
5. Hold a steel rule against the edge of the saw blade and move the material until the correct length is obtained.
6. Always allow 1/16 in. (1.5 mm) for each 1 in. (25 mm) of thickness longer than required to compensate for any *saw run-out* (slightly angular cut caused by hard spots in steel or a dull saw blade).
7. Tighten the vise and recheck the length from the blade to the end of the material to make sure that the work has not moved.
8. Raise the saw frame slightly, release the ratchet lever or open the hydraulic valve, and then start the machine.
9. Lower the blade slowly until it just touches the work.

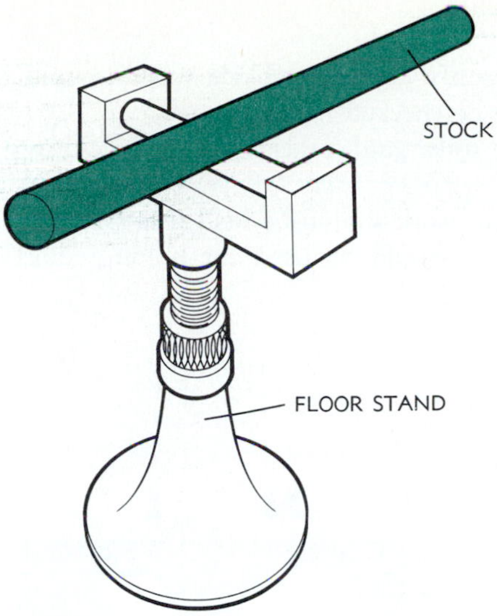
STOCK
FLOOR STAND

Figure 35-6 *A floor stand is used to support long pieces while they are being cut.*

10. When the cut has been completed, the machine will shut off automatically.

SAWING HINTS

1. Never attempt to mount, measure, or remove work unless the saw is stopped.

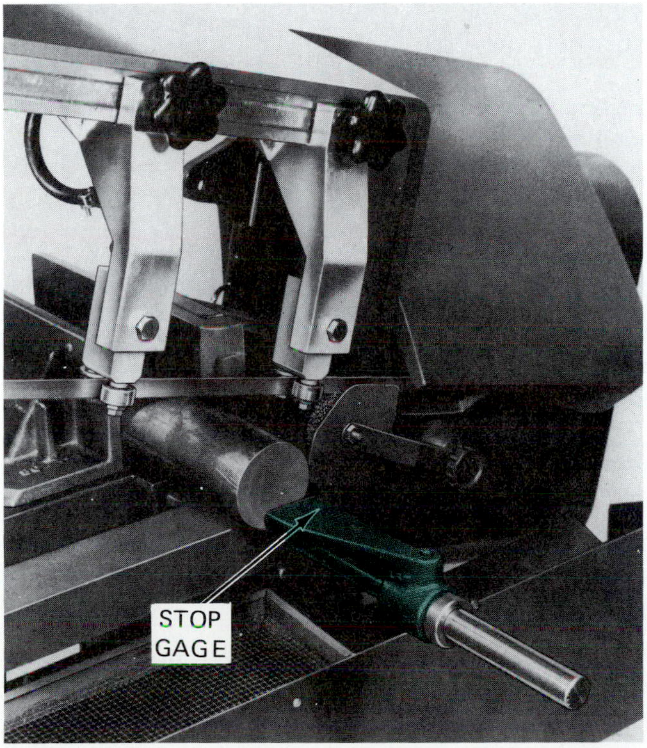
STOP
GAGE

Figure 35-7 *A stop gage is used when many pieces of the same length must be cut.*

SPACER
BLOCK

Figure 35-8 *A spacer block should be used to clamp a short workpiece in a vise.*

2. Guard long material at both ends to prevent anyone from coming in contact with it.
3. Use cutting fluid whenever possible to help prolong the life of the saw blade.
4. When sawing thin pieces, hold the material flat in the vise to prevent the saw teeth from stripping. Thin material can also be placed between two pieces of wood or soft materials when sawing.
5. Use caution when applying extra force to the saw frame, as this generally causes work to be cut out of square.
6. When several pieces of the same length are required, set the stop gage which is supplied with most cutoff saws (Fig. 35-7).
7. When holding short work in a vise, be sure to place a short piece of the same thickness in the opposite end of the vise. This will prevent the vise from twisting when it is tightened (Fig. 35-8).

REVIEW QUESTIONS

METHODS OF CUTTING OFF MATERIAL

1. Name and describe the cutting action of five methods of cutting off material.

HORIZONTAL BANDSAW PARTS

2. What is the purpose of the roller guide brackets?
3. How tight should the blade-tension handle be adjusted?

SAW BLADES

4. Of what material are saw blades made?
5. What pitch blade is recommended for:
 a. Large sections?
 b. Thin sections?
 c. General-purpose sawing?
6. Why should two teeth of a saw blade be in contact with the work at all times?
7. In what direction should the teeth of a saw blade point?
8. How should the workpiece be set for sawing partially cut work with a new saw blade?

SAWING

9. What may happen if too fast a blade speed is used?
10. How can a vise be checked for squareness?
11. How should the roller guide brackets be set for cutting work?
12. Explain how thin work should be held in a vise.
13. What precaution should be used for holding short work in a vise?

UNIT 36

Contour Bandsaw Parts and Accessories

The vertical bandsaw is the latest basic machine tool to be developed. Since its development in the early 1930s, it has been widely accepted by industry as a fast and economical method of cutting metal and other materials.

Band cutoff saws differ from power hacksaws in that they have a continuous cutting action on the workpiece, whereas the latter cuts only on the forward stroke and on a limited section of the blade.

The contour-cutting bandsaw (vertical bandsaw) offers several features not found on other metal-cutting machines. These advantages are illustrated in Fig. 36-1.

OBJECTIVES

After completing this unit you will be able to:
1. Name the main operative parts of a contour-cutting bandsaw and state the purpose of each
2. Select the proper tooth form and set for any cutting application
3. Calculate the length of a saw band for a two-pulley machine

CONTOUR BANDSAW PARTS

The construction of contour-band machines differs from most other machines in that the band machine is generally fabricated from steel rather than being cast. The contour bandsaw consists of three basic parts: *base, column,* and *head* (Fig. 36-2).

BASE

The base of the contour bandsaw supports the column and houses the drive assembly which provides a drive for the saw blade.

- The *lower pulley,* which supports and drives the saw band, is driven by a variable-speed pulley which can be

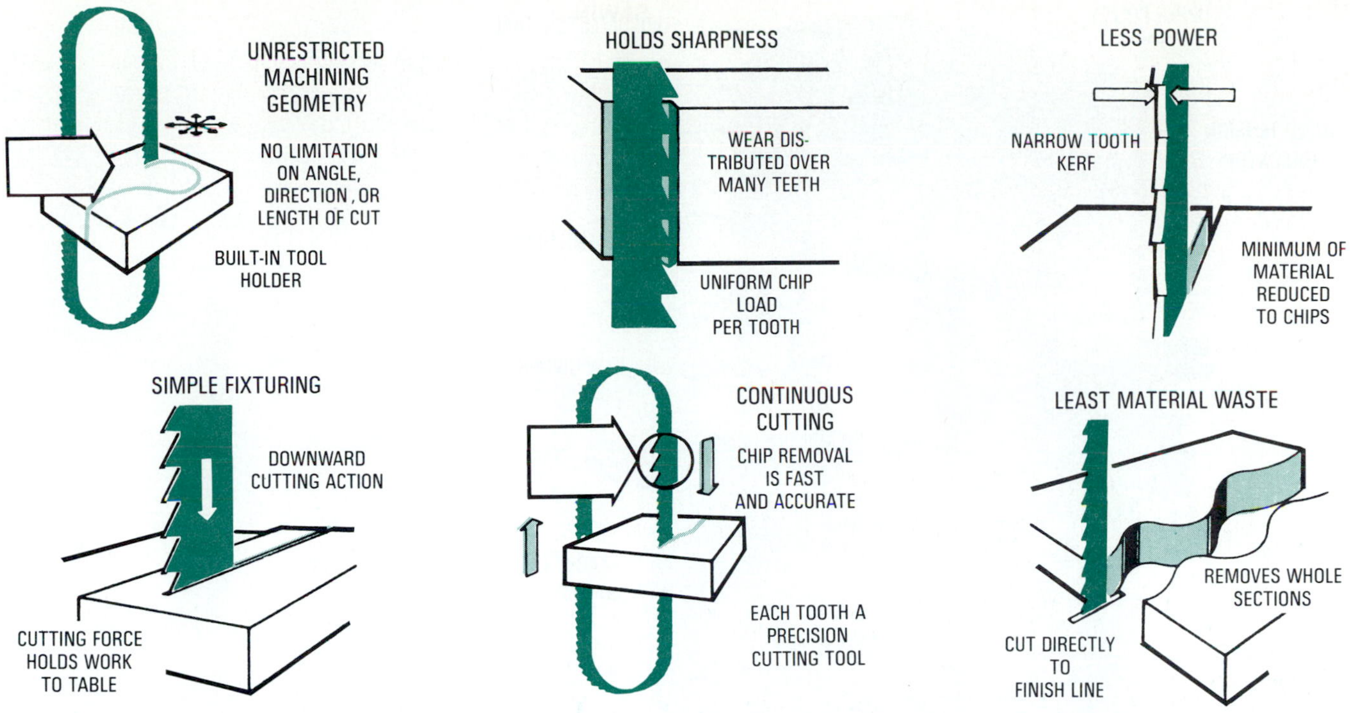
UNRESTRICTED MACHINING GEOMETRY
NO LIMITATION ON ANGLE, DIRECTION, OR LENGTH OF CUT
BUILT-IN TOOL HOLDER
HOLDS SHARPNESS
WEAR DISTRIBUTED OVER MANY TEETH
UNIFORM CHIP LOAD PER TOOTH
LESS POWER
NARROW TOOTH KERF
MINIMUM OF MATERIAL REDUCED TO CHIPS
SIMPLE FIXTURING
DOWNWARD CUTTING ACTION
CUTTING FORCE HOLDS WORK TO TABLE
CONTINUOUS CUTTING
CHIP REMOVAL IS FAST AND ACCURATE
EACH TOOTH A PRECISION CUTTING TOOL
LEAST MATERIAL WASTE
REMOVES WHOLE SECTIONS
CUT DIRECTLY TO FINISH LINE

Figure 36-1 *Advantages of a contour bandsaw. (Courtesy DoAll Company.)*

adjusted to various speeds by the *variable-speed handwheel.*

- The *table* is attached to the base by means of a *trunnion* (Fig. 36-3). It can be tilted 10° to the left and 45° to the right for making angular cuts by turning the *table tilt handwheel.*
- The *lower saw guide* is attached to the trunnion, and it supports the blade to keep it from twisting.
- A removable *filler plate slide* and *center plate* are mounted in the table.

COLUMN

The column supports the *head,* the *left-hand blade guard, welding unit,* and the *variable-speed handwheel.*

- The *variable-speed handwheel* is used to regulate the speed of the bandsaw blade.
- The *blade-tension indicator* and the *speed indicator* are also located in the column.
- The *welding unit* is used to weld, anneal, and grind the saw blade.

HEAD

The parts found in the head of a saw are generally used to guide or support the saw band.

- The *upper saw pulley* supports the saw band, which is adjusted by the tension and tracking controls.
- The *upper saw guide,* attached to the saw guidepost, supports and guides the saw blade to keep it from twisting. It can be adjusted vertically to accommodate various sizes of work.
- The *saw guard* and the *air nozzle* for keeping the area being cut free from chips are also found in the head.

Band machines may be of two types. The machine used in many toolrooms has two band carrier wheels; the larger capacity saws have three carrier wheels (Fig. 36-3). On the larger capacity saws, both upper wheels may be tilted so that the band will track properly. When the blade in this type of machine becomes too short, it need not be discarded. It may be shortened to fit over the upper and lower band carrier wheels (Fig. 36-3); the capacity of the machine is reduced, but this adjustment permits the economical use of the saw blade.

BANDSAW APPLICATIONS

Many operations can be performed faster and easier on the contour bandsaw than on any other machine. In addition to saving time, material is also saved because large sections of a workpiece can be removed as a solid instead of being reduced to chips as on conventional machines. Some of the more common operations performed on a bandsaw are shown in Fig. 36-4A to F.

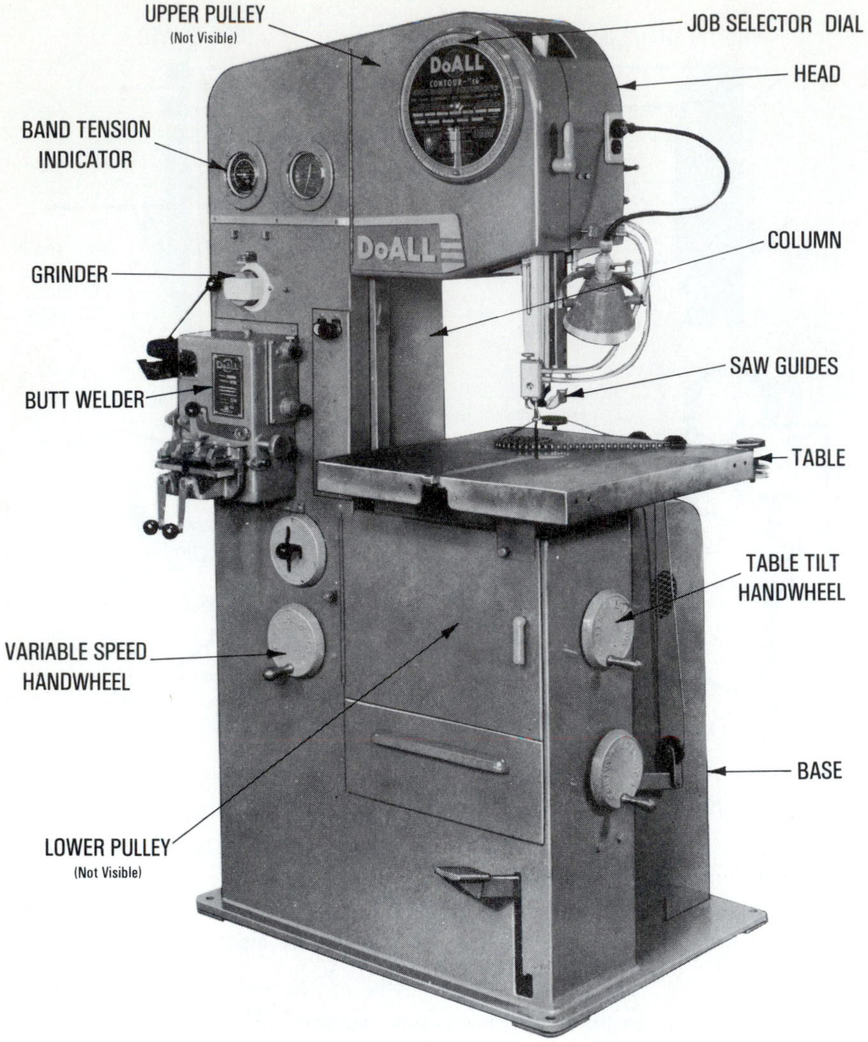
UPPER PULLEY
(Not Visible)
JOB SELECTOR DIAL
HEAD
BAND TENSION
INDICATOR
GRINDER
COLUMN
BUTT WELDER
SAW GUIDES
TABLE
TABLE TILT
HANDWHEEL
VARIABLE SPEED
HANDWHEEL
BASE
LOWER PULLEY
(Not Visible)
DoALL

Figure 36-2 *A contour bandsaw provides an economical means of cutting metals to shape. (Courtesy DoAll Company.)*

TRUNNION

Figure 36-3 *A three-carrier-wheel contour bandsaw with table removed. (Courtesy DoAll Company.)*

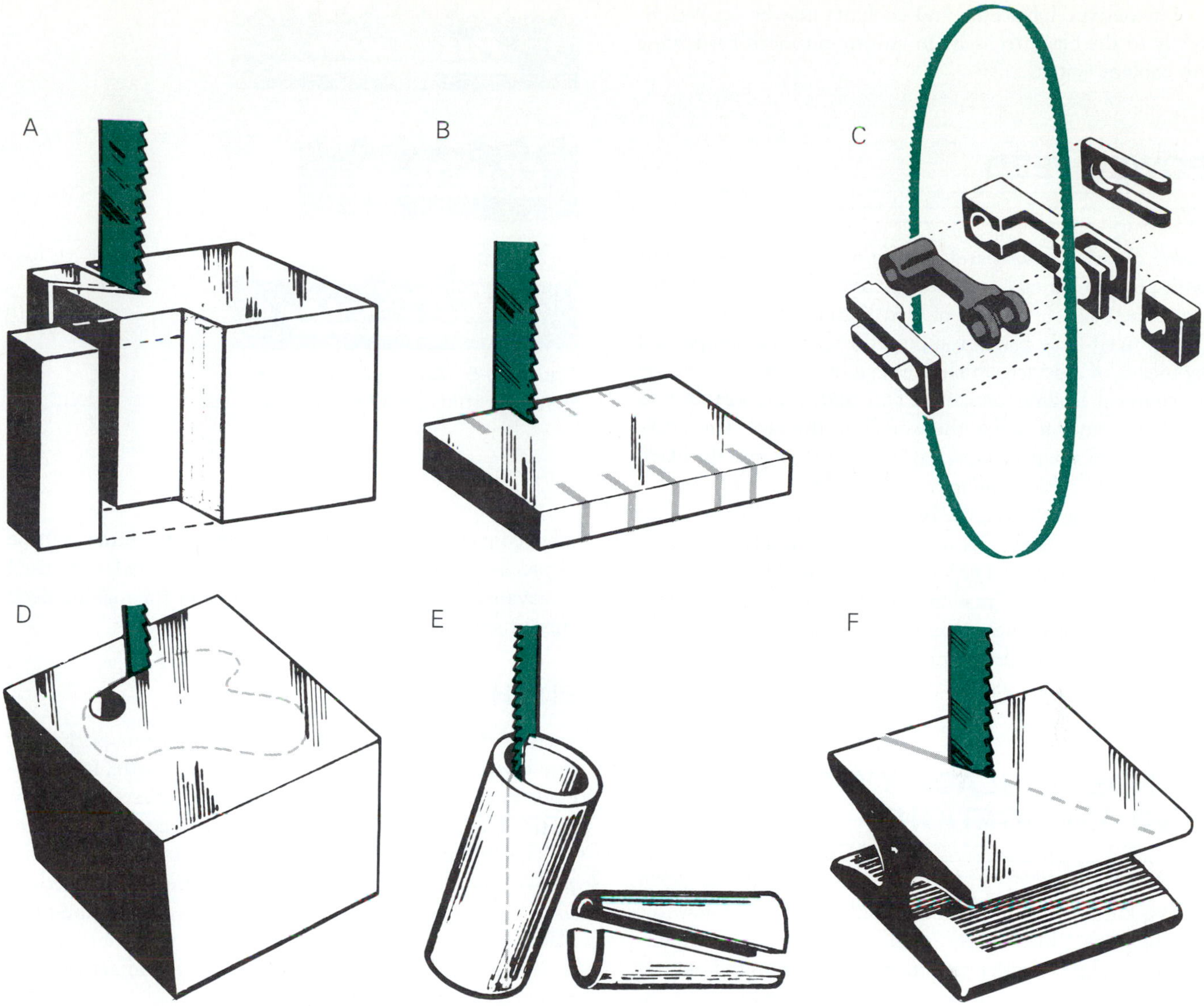

A
B
C
D
E
F

Figure 36-4 *(A) Notching; (B) slotting; (C) three-dimensional shaping; (D) radius cutting; (E) splitting; (F) angular cutting. (Courtesy DoAll Company.)*

- *Notching* (Fig. 36-4A): Sections of metal can be removed in one piece rather than in chips.
- *Slotting* (Fig. 36-4B): This operation can be done quickly and accurately without expensive fixtures.
- *Three-dimensional shaping* (Fig. 36-4C): Complicated shapes may be cut; simply follow the layout lines.
- *Radius cutting* (Fig. 36-4D): Internal or external contours may be cut easily. Internal sections are generally removed in one piece, as shown.
- *Splitting* (Fig. 36-4E): This can be accomplished quickly with a minimum waste of material.
- *Angular cutting* (Fig. 36-4F): The work may be clamped at any angle and fed through the saw. The table may be tilted for compound angles.

COOLANTS

Some machines, particularly the power-feed models, have a cooling system which circulates and discharges coolant against the faces of the blade and work.

In fixed-table machines where coolant is required, a mist coolant system is generally employed. The mist system uses air to atomize the coolant and direct it onto the faces of the blade and the work. This method is very efficient and is recommended for the high-speed machining of nonferrous metals such as aluminum and magnesium alloys. Tough, hard-to-machine alloys can also be cut successfully using the mist coolant system.

Grease-type lubricants and coolants may be applied directly to the blade to assist in cutting on machines having no coolant system.

POWER FEED

Some of the heavier band cutting machines are equipped with power-feed tables. The work and the table are fed toward the blade by means of a hydraulic system.

On fixed-table machines, power feeding is accomplished by means of a device which uses gravity to provide a steady mechanical feeding pressure. This allows the operator to use both hands to guide the work into the saw. The workpiece is held against a work jaw and is forced into the blade by means of cables, pulleys, and weights.

The force applied to the blade can be varied up to about 80 lb (356 N). However, for regular sawing a feeding force of about 30 to 40 lb (133 to 178 N) should be used.

Greater feed force may be used for sawing straight lines than for cutting contours. In order to determine the feed to use for any particular job or operation, consult the job selector dial on the bandsaw.

BANDSAW BLADE TYPES AND APPLICATIONS

Three kinds of blades are commonly used in bandsawing: carbon-alloy, high-speed steel, and tungsten-carbide–tipped blades. In order to obtain the best results from any bandsaw, it is necessary to select the proper blade for the job. Consideration must be given to the kind of saw-blade material and tooth form, pitch set, width, and gage needed for the material being cut.

TOOTH FORMS

Carbon and high-speed steel blades are available in three types of tooth forms (Fig. 36-5). The precision or regular tooth is the most generally used type of tooth. It has a 0° rake angle and about a 30° back clearance angle. It is used when a fine finish and an accurate cut are required.

The claw or hook tooth has a positive rake on the cutting face and slightly less back clearance than the precision or buttress blade. It has the same general application as the buttress-tooth form. It is faster cutting and longer lasting than the buttress tooth but will not produce as smooth a finish.

The buttress or skip tooth is similar to the precision tooth; however, the teeth are spaced farther apart to provide more chip clearance. The tooth angles are the same as those of precision teeth. Buttress, or skip-tooth, blades are used to advantage on thick work sections and for making deep cuts in soft material.

Precision®
Claw Tooth®
Buttress®
Figure 36-5 *Bandsaw blade tooth forms. (Courtesy DoAll Company.)*

PITCH

Each of the tooth forms is available in various pitches, or numbers of teeth per standard reference length. Inch sawblade pitch is determined by the number of teeth per inch; metric blade pitch by the number of teeth per 25 mm.

The thickness of the material to be cut determines the pitch of the blade to be used. When cutting thick materials, use a coarse-pitch blade; thin materials require a fine-pitch blade. When selecting the proper pitch, remember that at least two teeth of the saw blade must be in contact with the material being cut at all times.

SET

The set of a blade is the amount that the teeth are offset on either side of the center to produce clearance for the back of the band or blade. The three common set patterns (Fig. 36-6) are:

- *Raker set,* which has one tooth offset to the right, one offset to the left, and the third tooth is straight. This is the most common pattern and is used for most sawing applications.
- *Wave set,* which has a group of teeth offset to the right and the next group to the left, a pattern that produces a wavelike appearance. Wave-set blades are generally used when the cross section of the workpiece changes, such as on structural steel sections or on pipe.
- *Straight set,* which has one tooth offset to the right and the next to the left. It is used for cutting light nonferrous castings, thin sheet metal, tubing, and Bakelite.

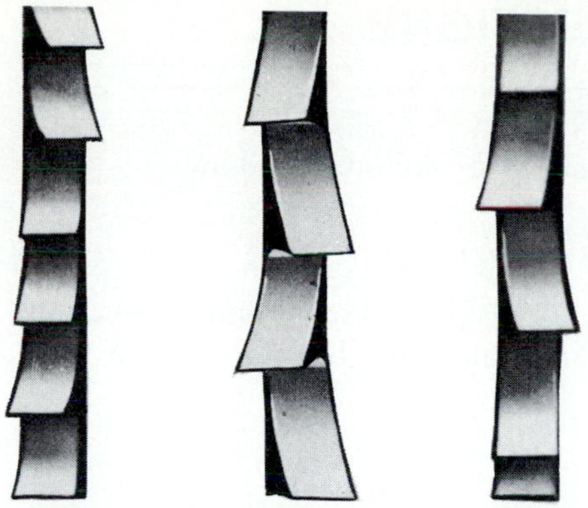
WAVE
STRAIGHT
RAKER

Figure 36-6 *Common set patterns: (A) wave; (B) straight; (C) raker. (Courtesy DoAll Company.)*

WIDTH

When making straight, accurate cuts, select a wide blade. Narrow blades are used to cut small radii. Radius charts, showing the proper width of blade to use for contour sawing, are generally found on all bandsaws. When selecting a blade for contour cutting, choose the widest blade which can cut the smallest radius on the workpiece.

GAGE

The gage is the thickness of the saw blade and has been standardized according to blade width. Blades up to ½ in. (13 mm) wide are 0.025 in. (0.64 mm) thick, ⅝-in. (16-mm) and ¾-in. (19-mm) blades are 0.032 in. (0.81 mm) thick, and 1-in.-wide (25-mm) blades are 0.035 in. (0.89 mm) thick. Since thick blades are stronger than thin blades, the thickest blade possible should be used for sawing tough material.

JOB REQUIREMENTS

The bandsaw operator should be familiar with the various types of blades and be able to select the one which will do the job to the specified finish and accuracy at the lowest cost. Table 36-1 provides a guide for efficient cutting.

BLADE LENGTH

Metal-cutting saw bands are usually packaged in coils about 100 to 150 ft (30 to 45 m) in length. The length required to fit each machine is cut from the coil and its two ends are then welded together to form a continuous band.

To calculate the length required for a two-pulley bandsaw, take twice the center distance (CD) between each pulley and add to it the circumference of one pulley (PC). This is the total length of the saw band (Fig. 36-7).

EXAMPLE

Calculate the length of a saw blade for a bandsaw which has:

a. Two 24-in.-diameter pulleys and a center-to-center distance of 48 in.

b. Two 600-mm-diameter pulleys and a center-to-center distance of 1200 mm.

Table 36-1 Job Requirement Chart

To Increase	Try One or More of the Following								
	Faster Tool Velocity (more teeth per min)	Slower Tool Velocity (fewer teeth per min)	Finer Pitch Band Tool (more teeth and smaller gullets)	Coarser Pitch Band Tool (fewer teeth and larger gullets)	Slower Feeding Rate (decrease chip load)	Faster Feeding Rate (increase chip load)	Medium Feeding Rate	Claw Tooth (positive rake angle)	Precision and Buttress (0° rake angle)
Cutting rate	✓			✓		✓		✓	
Tool life		✓	✓				✓	✓	
Finish	✓		✓		✓				✓
Accuracy	✓				✓				

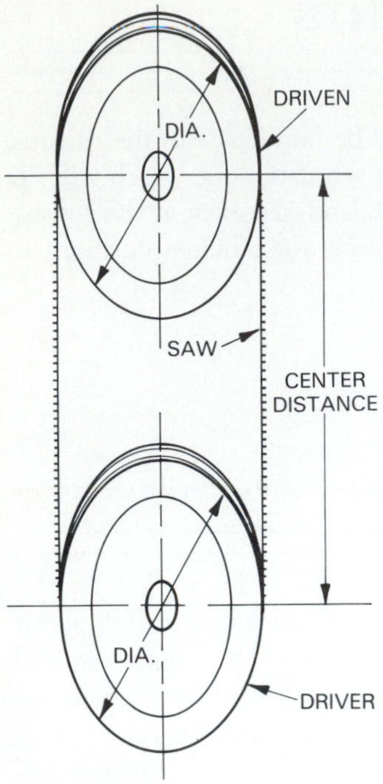
DRIVEN
DIA.
SAW
CENTER DISTANCE
DIA.
DRIVER

Figure 36-7 *Dimensions required to calculate the length of a saw blade.*

SOLUTION

a. Blade length = 2(CD) + PC
= 2(48) + (24 × 3.1416)
= 96 + 75.4
= 171.4 in.

b. Blade length = 2(CD) + PC
= 2(1200) + (600 × 3.1416)
= 2400 + 1885
= 4285 mm

REVIEW QUESTIONS

1. List six advantages of the contour bandsaw.

BANDSAW PARTS

2. How is the blade speed adjusted on a contour bandsaw?
3. What supports and guides the blade to keep it from twisting?

SAW TYPES

4. For what purpose are the following types of saw blades used?
   a. Precision
   b. Claw
   c. Buttress
5. What general rule applies when selecting the pitch of a saw blade?
6. Describe the following blade sets.
   a. Raker
   b. Wave
   c. Straight

JOB REQUIREMENTS

7. How can the following sawing factors be increased?
   a. Tool life
   b. Accuracy

BLADE LENGTH

8. Calculate the length of saw band required for the following contour bandsaws:
   a. Two 30-in.-diameter pulleys with a center-to-center distance of 50 in.
   b. Two 750-mm-diameter pulleys with a center-to-center distance of 1250 mm.

UNIT 37

Contour Bandsaw Operations

The contour bandsaw provides a machinist with the ability to cut material close to the form required quickly while at the same time removing large sections which can be used for other jobs. The versatility of a bandsaw can be increased by using various attachments and cutting tools. Operations such as sawing, filing, polishing, grinding, and friction and high-speed sawing are all possible on a bandsaw with the proper attachments and cutting tools.

OBJECTIVES

After completing this unit you will be able to:

1. Set up the machine and saw external sections to within 1/32 in. (0.8 mm) of layout lines
2. Saw internal sections to within 1/32 in. (0.8 mm) of layout lines
3. Set up a contour bandsaw to file to a layout
4. Know the purpose of special cutting tools used on a contour bandsaw

SAWING EXTERNAL SECTIONS

With the proper machine setups and attachments, a wide variety of operations can be performed on a contour bandsaw. The most common operation is that of sawing external sections. In order to perform an operation quickly and accurately, it is important that a person be able to select, weld, and mount the correct saw to suit the size and type of work material.

TO MOUNT A SAW BAND

1. Select and mount the proper saw inserts in the upper and lower saw guides using the proper gage for the thickness of the blade being used (Fig. 37-1). Allow 0.001- to 0.002-in. (0.02- to 0.50-mm) clearance to ensure that the blade will not bind.
2. Lower the upper saw band carrier wheel to ensure that the blade will slide over the wheels when installed.
3. Mount the blade over the upper and lower wheels with the teeth pointing down toward the table.
4. Adjust the upper wheel until some tension is registered on the blade-tension gage.
5. Set the gearshift lever in the neutral position and turn the upper saw band wheel by hand to make sure that the saw band rides on the center of the crown.

   If the saw is not tracking properly, the upper wheel must be tilted until the saw band rides on the center of the crown. When the blade is tracking properly, it should be very close to, but not touching, the back-up bearings when the saw is not cutting.
6. Re-engage the gearshift lever and close the doors on the upper and lower carrier housings.
7. Replace the filler plate in the table.
8. Lower the upper saw guide as close as possible to the work in order to ensure a straight and accurate cut.
9. Start the machine and adjust the saw band to the proper tension. This is indicated on the band-tension gage and chart.

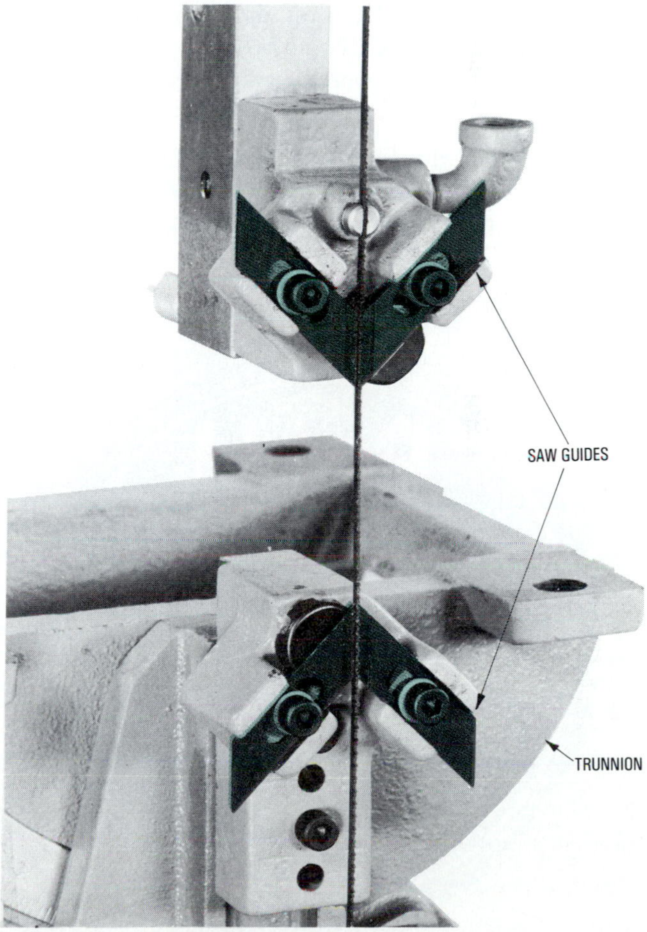

SAW GUIDES
TRUNNION

Figure 37-1 *Upper and lower saw guides of bandsaw with table removed. Note table support, or trunnion. (Courtesy DoAll Company.)*

TO CUT AN EXTERNAL SECTION

1. Study the print and check the layout on the workpiece for accuracy.
2. Check the job selector (Fig. 37-2) and determine the proper blade to use for the job. Consider the material, its thickness, type of cut (straight or curved) required, and the finish desired.

NOTE: When making curved cuts, use as wide a blade as possible.

3. Mount the proper saw blade and guides on the machine.
4. Place the work on the table and lower the upper saw guide until it just clears the work by ¼ in. (6 mm). Clamp the guide in place.

NOTE: It is good practice to drill a hole at every point where a sharp turn must be made to allow the workpiece to be turned easily.

5. Start the machine and make sure that the band is tracking properly.
6. Consult the job selector dial and set the proper speed.
7. Place the material against the work-holding jaw or against a block of wood (Fig. 37-3).
8. Carefully bring the workpiece up to the blade and start the cut.

NOTE: When the workpiece is to be finished by some other machining operation, the cut should be made about $\frac{1}{32}$ in. (0.8 mm) outside the layout line.

DoALL
CONTOUR MACHINE
DoALL COMPANY DES PLAINES, ILLINOIS, U.S.A.
KNIFE
FILE
ABRASIVE
DIAMOND
LINE GRIND
SPIRAL
SCALLOP
FRICTION
PRECISION
BUTTRESS
CLAW TOOTH
OPERATION OF JOB SELECTOR DIAL
SAW BAND RADIUS GUIDE
COOLANT DATA
COOLANT
F-FLOOD M-MIST N-NONE
POWER FEED
WORK THICKNESS
BAND PITCH
BAND VELOCITY
FEET/MINUTE
CONVENTIONAL SAWING
HIGH SPEED SAWING
BAND FILING
BAND POLISHING
LINE GRINDING
DIAMOND SAWING

Figure 37-2 *The job selector dial indicates the correct saw speed for the material being cut. (Courtesy DoAll Company.)*

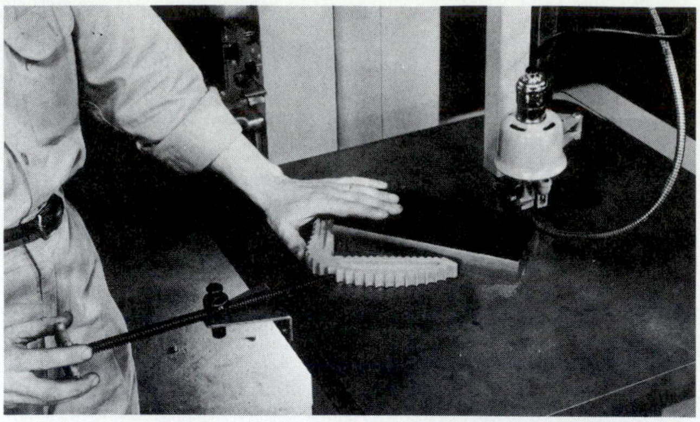

Figure 37-3 *The work-holding jaw is used to feed the workpiece into the saw. (Courtesy DoAll Company.)*

9. Carefully feed the workpiece into the saw. *Do not use too much force. Keep your fingers clear of the moving blade.*
10. Saw to the layout lines.

ACCURACY AND FINISH

As with any other machine, the accuracy and finish produced on a bandsaw depend on the operator's ability to set up and operate the machine properly. Table 37-1 indicates some of the most common sawing problems encountered and their possible causes.

SAWING INTERNAL SECTIONS

The bandsaw is well suited for removing internal sections of a workpiece. A starting hole must be drilled through the section to be removed to allow the saw blade to be inserted and welded. It is also good practice to drill a hole at every point where a sharp turn must be made to allow the workpiece to be turned easily. The saw cut should be made close to the layout line, leaving enough material for the finishing operation.

The butt welder (Fig. 37-4) adds greatly to the versatility of the bandsaw. It permits convenient welding of the blade for the removal of internal sections. Blades can be cut from coil stock and welded into a continuous band; broken blades may be welded and used again. Welders on vertical bandsaws are resistance-type welders, which fuse the ends of the blade when the correct current is applied and the blade is held properly.

In addition to knowing how to select the proper blade for the job, the operator must also be able to weld the blade.

Table 37-1 Sawing Problems and Possible Causes

Problem	Possible Cause: Too Heavy Feed	Too Light Feed	Improper Blade Tracking	Improper Blade Tension	Saw Guides Too Far Apart	Incorrect Saw Speed	Pitch Too Coarse	Pitch Too Fine	Wrong Type of Blade	Blade Dull On One Side	Blade Dull On Both Sides	Machine Too Light
Blade wanders	X		X	X	X					X		
Blade not cutting		X		X	X	X		X	X		X	
Blade dulls quickly	X					X			X			
Poor finish	X			X	X	X	X		X	X	X	X
Severe diagonal waviness	X		X	X	X	X						X
Saw teeth chipping	X						X		X			
Saw teeth clogging						X		X	X			

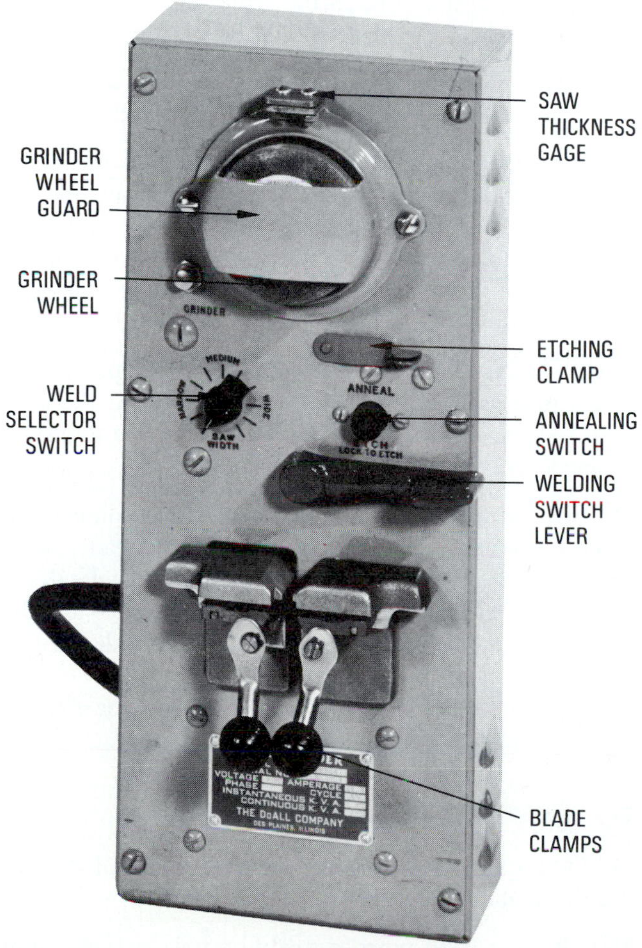
GRINDER WHEEL GUARD
GRINDER WHEEL
WELD SELECTOR SWITCH
SAW THICKNESS GAGE
ETCHING CLAMP
ANNEALING SWITCH
WELDING SWITCH LEVER
BLADE CLAMPS
THE DoALL COMPANY

Figure 37-4 *Parts of a butt welder. (Courtesy DoAll Company.)*

Improper welds will cause the blade to break and require rewelding, which is costly and time-consuming. Unless the weld is as strong as the band itself, it is not a good weld. The butted ends of the welded blade must not overlap in width, set, or pitch of the teeth.

TO WELD A BANDSAW BLADE

1. Select the proper blade for the job by checking the chart on the machine.
2. Determine the length of the blade required.

   The blade length for a two-wheel machine is determined by adding twice the center-to-center distance of the wheels plus the circumference of one wheel. If the upper wheel is extended the full distance, deduct 1 in. (25 mm) to allow for the stretch of the blade.
3. Place the blade in the cutoff shear, and cut it to the required length.

   Make sure that the blade is held straight and against the blade-squaring bar on the shear.
4. If the ends of the blade are straight but not square, hold them firmly with the teeth reversed, as in Fig. 37-5, and grind both ends in one operation.

   If the ends are not square after grinding, it will not matter. When the blade ends are placed together for welding, they will match perfectly.
5. In order to get the proper tooth spacing after the blade has been welded, it is necessary to grind off some teeth at the ends of the blade to the depth of the gullet prior to welding.

   This is necessary since the welding operations on a DoAll welder consume about ¼ in. (6 mm) of the blade length. Other types of welders may consume varying blade lengths during welding. The number of teeth that are ground off will depend on the pitch of

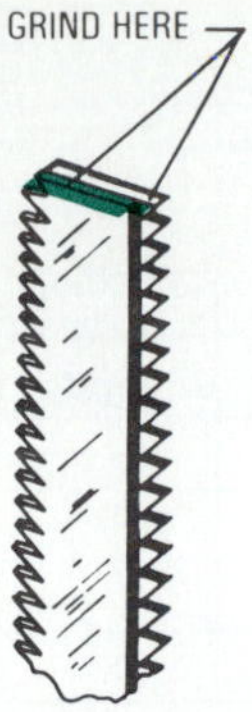
GRIND HERE

Figure 37-5 *Method of grinding blade ends prior to welding. (Courtesy DoAll Company.)*

the blade. Figure 37-6 illustrates the amount to grind off each type of blade from 4 to 10 pitch.

6. Clean the welder jaws and position the inserts for the blade size and pitch.

7. Adjust the jaw pressure by turning the selector handle for the width of the blade being welded.

8. Clamp the blade as shown in Fig. 37-7. Make sure that the blade is against the aligning surface on the back of the jaws and centered between the two jaws.

If the ends of the blades do not butt against each other for the full width, they should be removed and reground.

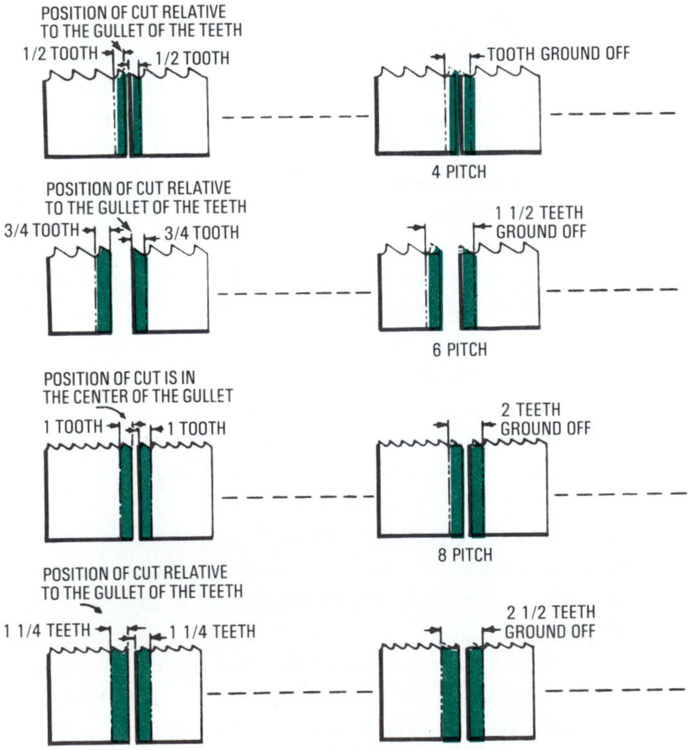
POSITION OF CUT RELATIVE TO THE GULLET OF THE TEETH
1/2 TOOTH
1/2 TOOTH
TOOTH GROUND OFF
4 PITCH
POSITION OF CUT RELATIVE TO THE GULLET OF THE TEETH
3/4 TOOTH
3/4 TOOTH
1 1/2 TEETH GROUND OFF
6 PITCH
POSITION OF CUT IS IN THE CENTER OF THE GULLET
1 TOOTH
1 TOOTH
2 TEETH GROUND OFF
8 PITCH
POSITION OF CUT RELATIVE TO THE GULLET OF THE TEETH
1 1/4 TEETH
1 1/4 TEETH
2 1/2 TEETH GROUND OFF

Figure 37-6 *Amounts to grind off various blades to achieve proper tooth spacing. (Courtesy DoAll Company.)*

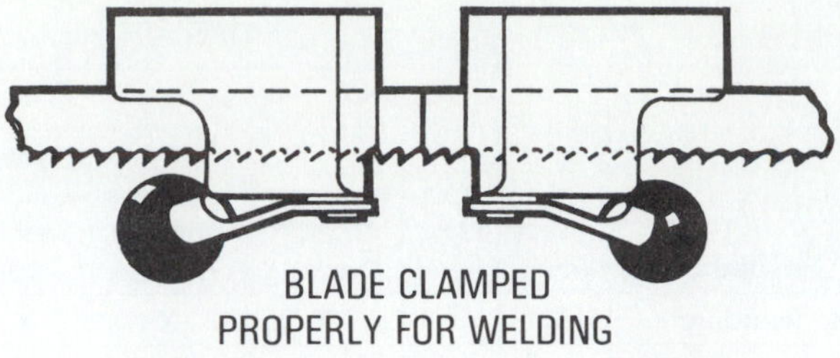
BLADE CLAMPED PROPERLY FOR WELDING

Figure 37-7 *Blade clamped properly for welding.*

9. Depress the welding switch or lever and hold it until the weld has cooled.

NOTE: Stand to one side and wear safety glasses to avoid injury from the welding flash.

10. Release the movable jaw clamp and then release the welding lever.

11. Remove the blade and check the weld for the following points.

a. The flash material should be uniform on both sides.
b. The spacing of the teeth should be uniform.
c. The weld should be in the center of the gullet.
d. The back of the blade should be straight.

NOTE: If the blade does not meet all of these requirements, it should be broken, prepared again, and rewelded.

12. Move the reset lever to the anneal position and clamp the blade, with the weld in the center of the jaws and the teeth to the rear and clearing the jaws.

13. Set the selector switch at the proper setting for annealing the blade. Push in the anneal switch button and jog it intermittently until the band reaches a dull red color.

CAUTION: Do not permit the blade to become too hot at this time as it will air-harden on cooling. As the weld starts to cool, jog the anneal switch occasionally to permit the blade to cool slowly.

14. Remove the blade and grind off the welding flash. Grind the weld to the same thickness as the blade, being careful not to grind the teeth. Continually check the thickness of the blade in the thickness gage located on the welder. When properly ground, it should just slide through the thickness gage (Fig. 37-8).

15. After the blade has been ground to the proper thickness, it is advisable to anneal it to a blue color.

TO SAW INTERNAL SECTIONS

1. Drill a hole slightly larger than the width of the saw blade near the edge of the section to be cut.

NOTE: It is good practice to drill a hole at every point where a sharp turn must be made to allow the workpiece to be turned easily.

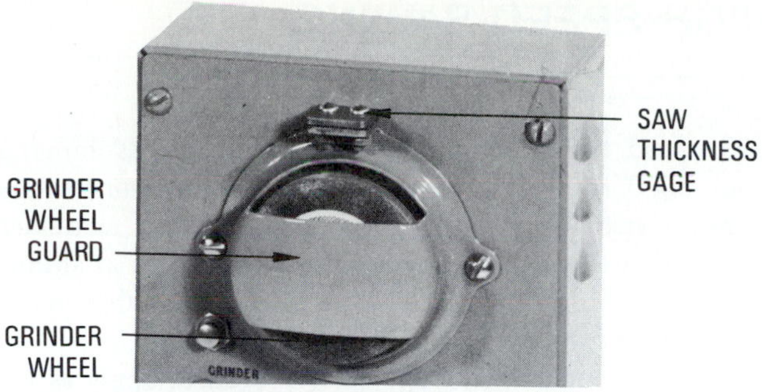
SAW
THICKNESS
GAGE
GRINDER
WHEEL
GUARD
GRINDER
WHEEL

Figure 37-8 *A thickness gage is used to check the thickness of the blade at the weld.*

2. Cut the saw blade and thread it through one of the drilled holes in the workpiece (Fig. 37-9).
3. Weld the blade and then grind off the weld bead to fit the saw thickness gage.
4. Anneal the weld section to remove the brittleness and prevent the blade from breaking.
5. Mount the saw band on the upper and lower pulleys and apply the proper tension for the size of the blade.
6. Insert the table filler plate.
7. Set the machine to the proper speed for the type and thickness of material being cut.
8. Cut out the internal section staying within 1/32 in. (0.8 mm) of the layout line.
9. Remove the saw band from the pulleys.
10. Cut the blade at the weld point on the cutoff shear.
11. Remove the workpiece and the blade.

DRILLED HOLE

Figure 37-9 *Welding a blade for sawing internal sections.*

TRIMMING AND BLANKING DIES

It is possible to use this internal cutting technique to make short-run trimming and blanking dies. By this process, the internal section, or slug, becomes the punch, while the external material forms the die (Fig. 37-10). When doing work of this type, the table must be tilted to provide the proper clearance for the die.

FRICTION SAWING

Friction sawing (Fig. 37-11) is the fastest means of sawing ferrous metals up to 1 in. (25 mm) in thickness. In this process, the metal is fed into a bandsaw which is traveling at a high velocity [up to 15,000 sf/min (4572 m/min)]. The tremendous heat generated by friction brings the metal immediately ahead of the saw teeth to a plastic state, and the teeth easily remove the softened metal. Since the thermal conductivity of steel is very low, the depth to which the metal is softened is only about 0.002 in. (0.50 mm).

The temperature of the saw blade remains quite low, since each tooth is only momentarily in contact with the metal and has time to cool as it travels around the carrier wheels before contacting the metal again. This method of sawing leaves a small burr which is easily removed from the workpiece.

Friction sawing is used on hardened ferrous alloys, armor plate, and parts having a thin wall or section which would be damaged by other sawing methods. Friction saw-

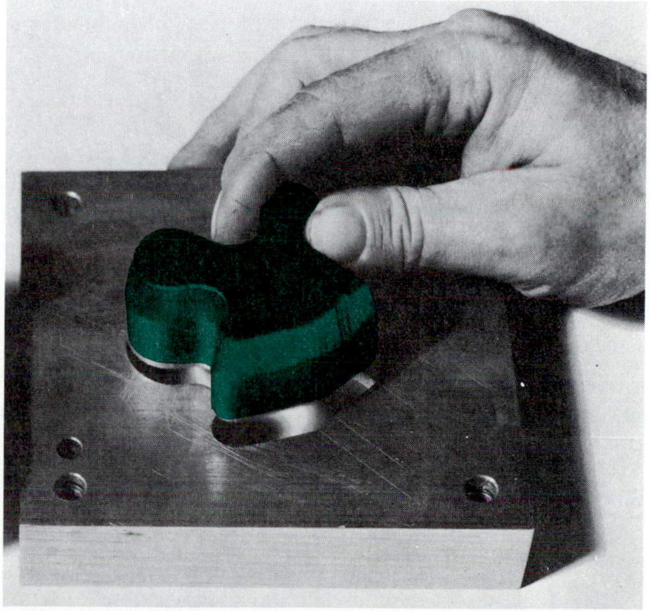

Figure 37-10 *The internal section can be used to make a die punch. (Courtesy DoAll Company.)*

Figure 37-11 *Friction sawing uses a high saw-band speed. (Courtesy DoAll Company.)*

ing is particularly suited to the cutting of stainless-steel alloys, which are difficult to cut because they work-harden so quickly. It is not suitable for sawing most cast irons because grains break off before the metals soften. Aluminum and brass cannot be cut by friction sawing because the high thermal conductivity of the metal does not let the heat concentrate just ahead of the blade. These metals also melt almost immediately after softening and will weld to the blade and clog the teeth. Most thermoplastics react in the same manner as the nonferrous metals.

Friction-sawing machines resemble the standard vertical band machines but are of heavier construction in the frame, bearings, spindles, and guides, to withstand the vibrations created by the high speeds required. The saw band is almost completely covered to protect the operator from the shower of sparks produced by friction sawing.

Friction-sawing bands are made of standard carbon-alloy steel but are thicker than standard blades to provide greater strength. They may be obtained in widths of ½, ¾, and 1 in. (13, 19, and 25 mm) and in 10 and 14 pitch with raker set only.

The teeth on a standard saw blade traveling at the high velocity required for friction sawing would dull very quickly; therefore, the teeth on friction saws are not sharpened. Since dull teeth create more friction than sharp teeth, they are more efficient for friction sawing.

The procedure for setting up a machine for friction sawing is basically the same as for conventional sawing.

HIGH-SPEED SAWING

High-speed bandsawing is performed at speeds ranging from 2000 to 6000 sf/min (609 to 1827 m/min). It is merely a standard sawing procedure performed at higher than standard speeds on nonferrous metals, such as aluminum, brass, bronze, magnesium, and zinc, and other materials, such as wood, plastic, and rubber.

The same machine setups and procedures apply as for conventional sawing. In high-speed sawing, the chips must be removed rapidly; consequently, buttress or claw-tooth blades are the most efficient for this operation.

BAND FILING

When a better finish that that produced by conventional sawing is required on the edge of the workpiece, it may be produced by means of a *band file.*

The band file consists of a steel band onto which are riveted a number of short, interlocking file segments (Fig. 37-12). The ends of the band are locked together to form a continuous loop. Band files may be obtained in flat, oval, and half-round cross sections, in bastard and medium cuts, and in widths of ¼, ⅜, and ½ in. (6, 9.5, and 13 mm). Special file guides and a file-adaptor table-filler plate are used for band filing.

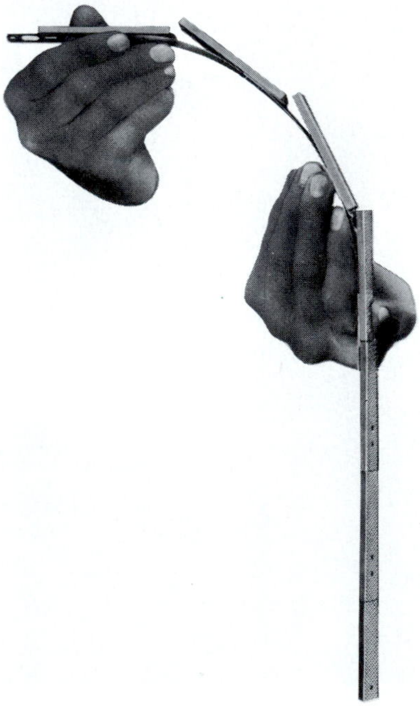

Figure 37-12 *Segments of a band file.*

TO SET UP FOR BAND FILING

1. Select the proper band file for the job by consulting the job selector dial. Consideration must be given to the material being filed and the shape, size, and cut of the file.
2. Set the gearshift lever into neutral position.
3. Remove the saw guides and filler plate.
4. Mount the proper file guide and backup support.
5. Lock the ends of the file blade together.
6. Mount the file band, with the teeth pointing in the proper direction, on the bandsaw carrier wheels (Fig. 37-13).

NOTE: On some makes of machines, or when filing internal sections, it may be necessary to mount the file band and then join the ends.

7. Lightly add tension to the band.
8. Check the alignment and tracking of the file band.
9. Lower the upper guidepost to the proper work thickness. The distance should not exceed 2 in. (50 mm) for a 1/4-in. (6-mm) file band, and 4 in. (100 mm) for a 3/8- or 1/2-in. (9.5- or 13-mm) band.
10. Mount the proper table-filler plate.
11. Set the gearshift lever into low gear and start the machine.
12. Adjust the band file to the proper tension.
13. Adjust the machine to the proper speed for the material being filed.

Figure 37-13
Mounting a file band on a contour bandsaw.

TO FILE ON A CONTOUR BANDSAW

1. Consult the job selector (Fig. 37-2) and set the machine to the proper speed. The best filing speeds are between 50 and 100 ft/min (15 and 30 m/min).
2. Apply light work pressure to the file band; It not only gives a better finish but prevents the teeth from becoming clogged.
3. Keep moving the work back and forth against the file to prevent filing grooves in the work.
4. Use a file card to keep the file clean. Loaded files cause bumpy filing and scratches in the work.

NOTE: Be sure to stop the machine before attempting to clean the file.

ADDITIONAL BAND TOOLS

Although the saw blade and band file are the most commonly used, several other band tools make this machine particularly versatile.

- *Knife-edge blades* (Fig. 37-14) are available with knife, wavy, and scalloped edges, and are used for cutting soft, fibrous materials such as cloth, cardboard, cork, and rubber. Scalloped-edge blades are particularly suited for cutting thin corrugated aluminum. Special roller guides must be used with knife-edge blades.
- *Spiral-edge blades* (Fig. 37-15) are round and have a continuous helical cutting edge around the circumference. This provides a cutting edge of 360° and permits the machining of intricate contours and patterns (Fig. 37-16) without the workpiece having to be turned. Spiral-edge blades require special guides.

Spiral-edge blades are made in two types: the spring-tempered blade, which is used for plastics and wood, and the all-hard blade, which is used for light metals. These

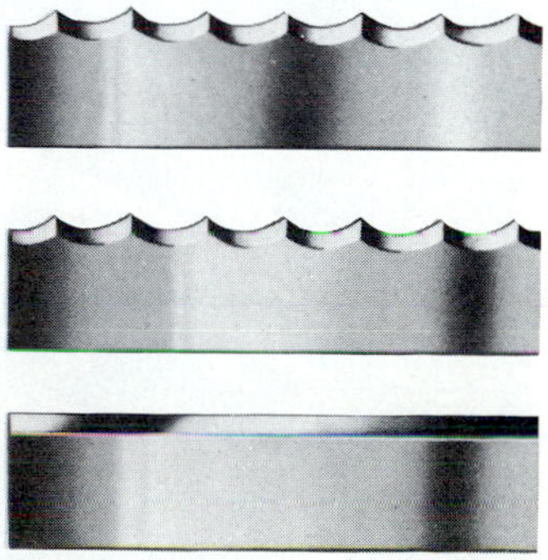

Knife-edge

Figure 37-14
Knife-edge blades.

Figure 37-15 *Spiral-edge blades. (Courtesy DoAll Company.)*

blades are manufactured in diameters of 0.020, 0.040, 0.050, and 0.074 in. (0.50, 1.00, 1.3, and 1.9 mm). Special guides (Fig. 37-16) are used on the machine with this blade. When spiral blades are welded, sheet copper is used to protect the cutting edges from the welder jaws.

Line-grinding bands (Fig. 37-17) have an abrasive (either aluminum oxide or silicon carbide) bonded to the thin edge of the steel band. These bands are used to cut hardened steel alloys and other materials such as brick, marble, and glass, which could not be cut by bandsawing.

This type of machining requires a high speed—3000 to 5000 sf/min (914 to 1524 m/min)—and the use of coolant because of the heat generated. A diamond dressing stick is used to dress line-grinding bands.

Diamond-edge blades (Fig. 37-18) are used to cut superhard space-age materials, as well as ceramics, glass, silicon, and granite. This type of blade has diamond particles fused to the edges of the saw teeth. These blades operate at about 3000 sf/min (914 m/min) and generally require coolant for most efficient operation. Although diamond-edge bands are very expensive, they will outlast 200 steel blades when cutting asbestos-cement pipe.

Polishing bands are used to remove burrs and provide a good finish to surfaces which have been sawed or filed. They may also be used for sharpening carbide toolbits. A polishing band is a continuous loop of 1-in.-wide (25-mm) abrasive cloth manufactured to a specific length to fit the machine. They are available in several grain sizes in both aluminum-oxide and silicon-carbide abrasive.

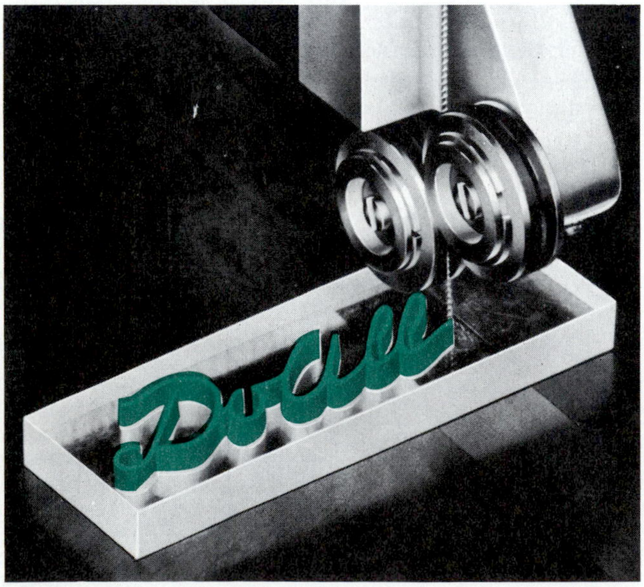

Figure 37-16 *Intricate forms may be cut with spiral-edge blades. Note the special guides. (Courtesy DoAll Company.)*

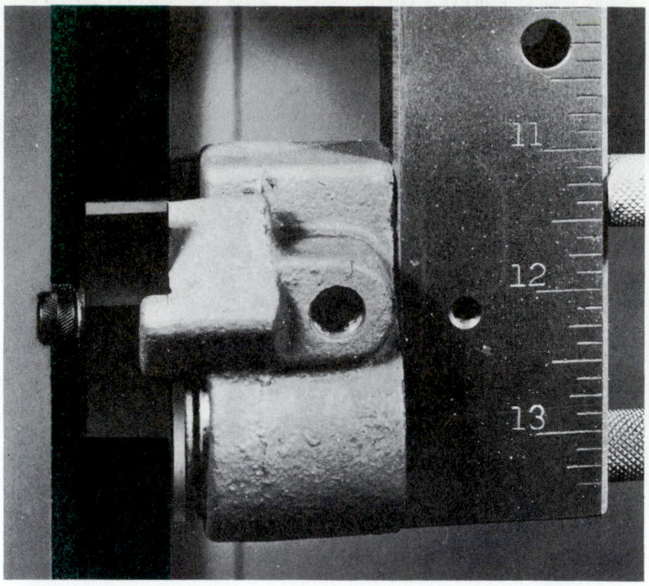

Figure 37-17 *Line-grinding bands are used to cut hardened metals. (Courtesy DoAll Company.)*

The polishing band is mounted in the same manner as a saw band. The special polishing guide uses the same backup support as the file bands. A special polishing-band center plate is used during band polishing. Most polishing bands are marked with an arrow on the back to indicate the direction of travel.

Electro-band machining (Fig. 37-19) is the latest development in band machining and is used to machine such materials as thin-wall tubing, stainless steel, aluminum, and titanium honeycombing.

By this process, a low-voltage, high-amperage current is

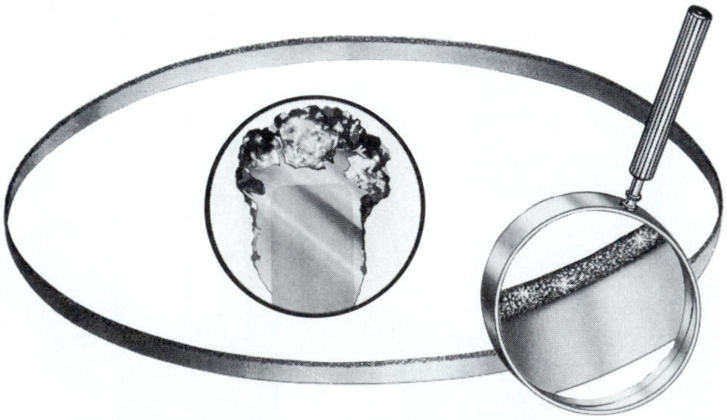

Figure 37-18 *Diamond-edge blades are used to cut superhard materials. (Courtesy DoAll Company.)*

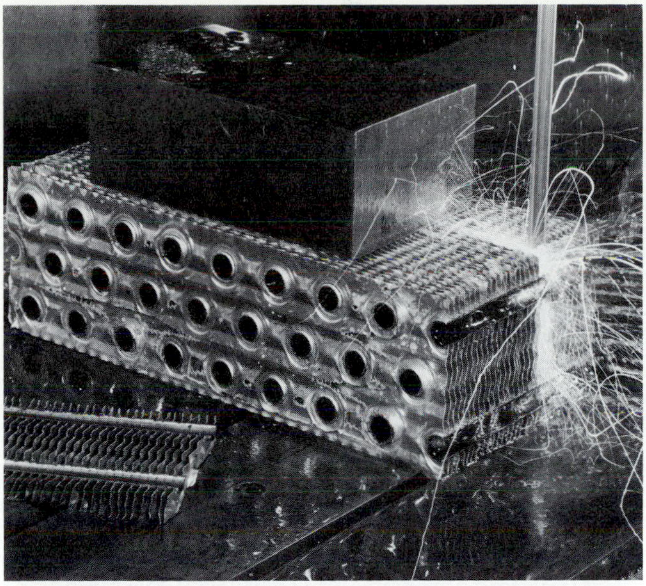

Figure 37-19 *Cutting a heat exchanger by electro-band machining. (Courtesy DoAll Company.)*

fed into the saw blade. The workpiece is connected to the opposite pole of the circuit. When the work comes close to the fast-moving band [6000 sf/min (1827 m/min)], a continuous electric spark passes from the knife edge of the saw to the work. This arc acts on the material and disintegrates it. The blade does not touch the work. Coolant is flooded onto the cutting area to prevent damage to the material by the heat created. Power feed must be used in this type of machining operation.

BANDSAW ATTACHMENTS

Several standard attachments can be obtained which will increase the scope of the band machine. Some of the most common are:

- The *work-holding jaw* (Fig. 37-20) is a device used by the operator to hold and guide the work into the saw. Since it is usually connected to the weight-type power feed, the operator merely steers the work with the work-holding jaw and does not have to apply any feed force.
- The *disk-cutting attachment* (Fig. 37-21) permits the cutting of accurate circles from approximately 2.5 to 30 in. (64 to 760 mm) in diameter.
- The *cutoff and mitering attachment* is used to support the work when square or angular cuts are being made.
- The *ripping fence* provides a means for cutting long sections of flat bar stock or plate into narrow parallel sections.

When special work-holding devices are required, they are generally made in the shop to suit the specific job. These devices are usually attached to the machine table and are called *fixtures*.

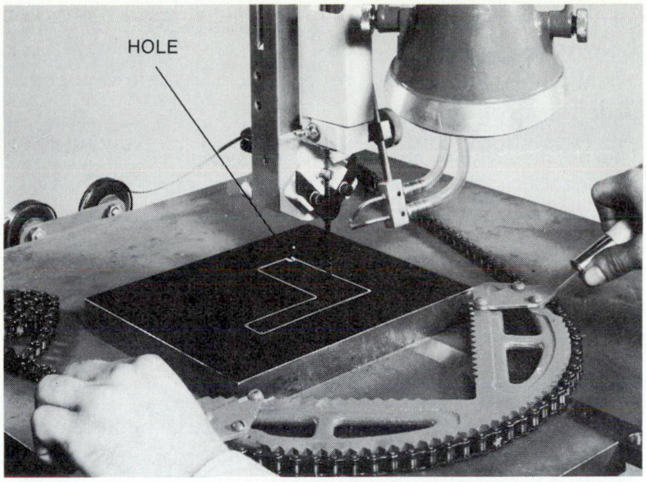
HOLE

Figure 37-20 *A work-holding jaw is used to guide a workpiece into a saw blade. (Courtesy DoAll Company.)*

REVIEW QUESTIONS

1. Name five operations which can be performed on a contour bandsaw.

SAWING EXTERNAL SECTIONS

2. Describe how to make the blade track properly.

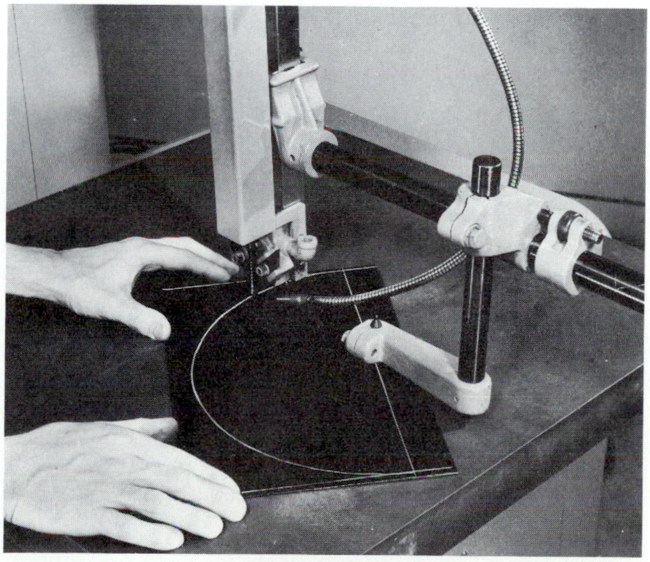

Figure 37-21 *The disk-cutting attachment is used to cut circular shapes. (Courtesy DoAll Company.)*

3. Why must the upper saw guide be close to the top of the work?
4. What factors should be considered when selecting a blade for a contour cut?
5. How should the work be fed into a revolving saw?

ACCURACY AND FINISH

6. How can the following bandsawing problems be corrected?
   a. Blade wander b. Poor finish

SAWING INTERNAL SECTIONS

7. How is the length of a blade calculated when the upper carrier wheel is extended the full distance?
8. Make a sketch to show how the ends of the bandsaw are positioned for grinding prior to welding.
9. In order to obtain the correct tooth spacing, how many teeth should be ground off each end of a 10-pitch blade? A 14-pitch blade?
10. What are the characteristics of a good weld on a bandsaw blade?
11. List the main steps required to remove an internal section from a piece of work.

FRICTION AND HIGH-SPEED SAWING

12. Describe the principle of friction sawing.
13. Why is friction sawing particularly suited to cutting stainless steel?
14. How does high-speed sawing differ from friction sawing?

BAND FILING

15. List the main steps required to set up the machine for band filing.
16. What is the recommended filing speed?
17. How can the band file be kept clean?

ADDITIONAL BAND TOOLS

18. Name five materials that can be cut satisfactorily with knife-edge bands.
19. Name two types of spiral-edge bands and state two uses for each.
20. Describe a line-grinding band and state its purpose.
21. Name five materials which may be cut with a diamond-edge band.
22. Describe the principle of electro-band machining.

BANDSAW ATTACHMENTS

23. Describe and state the purpose of:
   a. A work-holding jaw
   b. A disk-cutting attachment
   c. A mitering attachment

SECTION

10

Drilling Machines

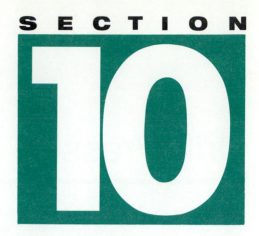

Probably one of the first mechanical devices developed prehistorically was a drill to bore holes in various materials. A bow string was wrapped around an arrow and then rapidly sawed back and forth. This process not only produced fire but it also wore a hole in the wood. The principle of a rotating tool making a hole in various materials is the one on which all drill presses operate. From this basic principle evolved the drill press—one of the most common and useful machines in industry for producing, forming, and finishing holes. Modern drilling machines are made in several types and sizes, ranging from sensitive hand-fed drill presses to the sophisticated tape-controlled production machine of the computer age.

UNIT 38

Drill Presses

The *drilling machine* or *drill press* is essential in any metal-working shop. Fundamentally, a drilling machine consists of a spindle (which turns the drill and which can be advanced into the work, either automatically or by hand) and a work table (which holds the workpiece rigidly in position as the hole is drilled). A drilling machine is used primarily to produce holes in metal; however, operations such as tapping, reaming, counterboring, countersinking, boring, and spot-facing can also be performed.

OBJECTIVES

After completing this unit you will be able to:

1. Identify six standard operations which may be performed on a drill press
2. Identify six types of drill presses and their purposes
3. Name and state the purpose of the main parts of an upright and a radial drill

STANDARD OPERATIONS

Drilling machines may be used for performing a variety of operations besides drilling a round hole. A few of the more standard operations, cutting tools, and work setups will be briefly discussed.

- *Drilling* (Fig. 38-1A) may be defined as the operation of producing a hole by removing metal from a solid mass using a cutting tool called a *twist drill.*
- *Countersinking* (Fig. 38-1B) is the operation of producing a tapered or cone-shaped enlargement to the end of a hole.
- *Reaming* (Fig. 38-1C) is the operation of sizing and producing a smooth round hole from a previously drilled or bored hole with the use of a cutting tool having several cutting edges.
- *Boring* (Fig. 38-1D) is the operation of truing and enlarging a hole by means of a single-point cutting tool, which is usually held in a boring bar.
- *Spot-facing* (Fig. 38-1E) is the operation of smoothing and squaring the surface around a hole to provide a seat for the head of a cap screw or a nut. A boring bar, with a pilot section on the end to fit into the existing hole, is generally fitted with a double-edged cutting tool. The pilot on the bar provides rigidity for the cutting tool and keeps it concentric with the hole. For the spot-facing operation, the work being machined should be securely clamped and the machine set to approximately one-quarter of the drilling speed.
- *Tapping* (Fig. 38-1F) is the operation of cutting internal threads in a hole with a cutting tool called a tap. Special machine or gun taps are used with a tapping attachment when this operation is performed by power in a machine.
- *Counterboring* (Fig. 38-1G) is the operation of enlarging the top of a previously drilled hole to a given depth to provide a square shoulder for the head of a bolt or capscrew.

PRINCIPAL TYPES OF DRILLING MACHINES

A wide variety of drill presses are available, ranging from the simple sensitive drill to highly complex automatic and numerically controlled machines. The size of a drill press

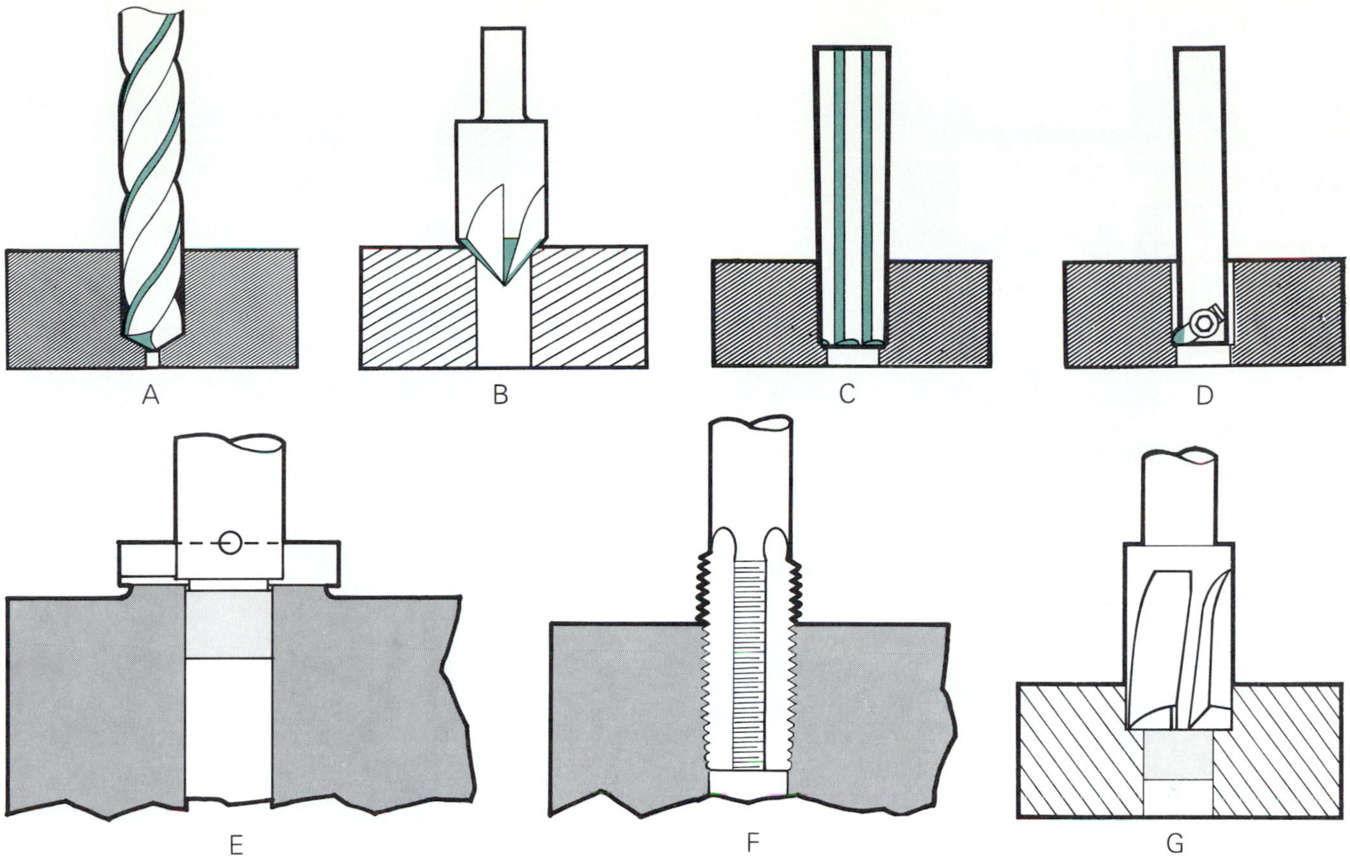
A
B
C
D
E
F
G

Figure 38-1 *(A) Drilling produces a straight hole; (B) countersinking produces a cone-shaped hole; (C) reaming is used to finish a hole; (D) boring is used to true and enlarge a hole; (E) spot-facing produces a square surface; (F) tapping produces internal threads; (G) counterboring produces square shoulders in a hole.*

may be designated in different ways by different companies. Some companies state the size as the distance from the center of the spindle to the column of the machine. Others specify the size by the diameter of the largest circular piece that can be drilled in the center.

SENSITIVE DRILL PRESSES

The simplest type of drilling machine is the sensitive drill press (Fig. 38-2). This type of machine has only a hand feed mechanism, which enables the operator to "feel" how the drill is cutting and to control the downfeed pressure accordingly. Sensitive drill presses are generally light, high-speed machines and are manufactured in bench and floor models.

Sensitive Drill Press Parts

Although drill presses are manufactured in a wide variety of types and sizes, all drilling machines contain certain basic parts. The main parts on the bench and floor models are the *base, column, table,* and *drilling head* (Fig. 38-2). The floor model is larger and has a longer column than the bench type.

Base

The base, usually made of cast iron, provides stability for the machine and also rigid mounting for the column. The base is usually provided with holes so that it may be bolted to a table or bench. The slots or ribs in the base allow the work-holding device or the workpiece to be fastened to the base.

Column

The column is an accurate cylindrical post which fits into the base. The table, which is fitted to the column, may be adjusted to any point between the base and head. The drill press head is mounted near the top of the column.

Table

The table, either round or rectangular in shape, is used to support the workpiece to be machined. The table, whose

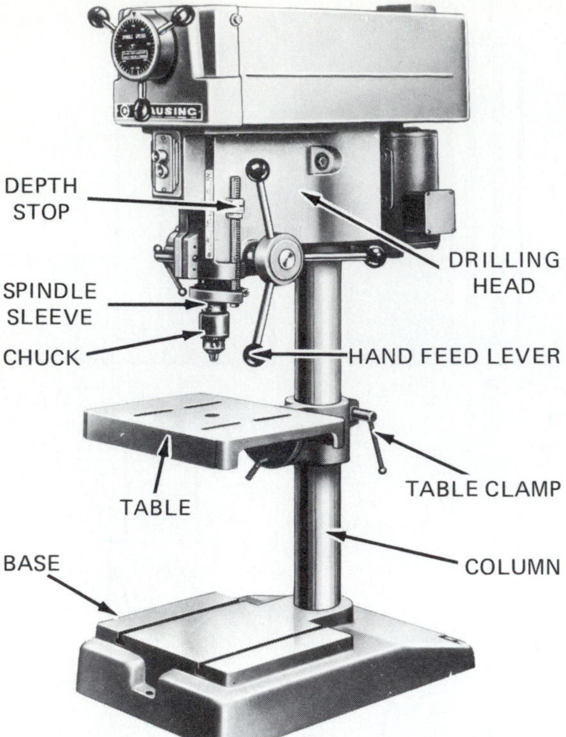

DEPTH STOP
SPINDLE SLEEVE
CHUCK
DRILLING HEAD
HAND FEED LEVER
TABLE CLAMP
TABLE
BASE
COLUMN

Figure 38-2 *A bench-type sensitive drill press. (Courtesy Clausing Corp.)*

surface is at 90° to the column, may be raised, lowered, and swiveled around the column. On some models it is possible to tilt the table in either direction for drilling holes on an angle. Slots are provided in most tables to allow jigs, fixtures, or large workpieces to be clamped directly to the table.

Drilling Head

The head, mounted close to the top of the column, contains the mechanism used to revolve the cutting tool and advance it into the workpiece. The *spindle,* which is a round shaft that holds and drives the cutting tool, is housed in the *spindle sleeve* or *quill.* The spindle sleeve does not revolve but slides up and down inside the head to provide a downfeed for the cutting tool. The end of the spindle may have a tapered hole to hold taper shank tools, or may be threaded or tapered for attaching a *drill chuck* (Fig. 38-2).

The *hand feed lever* is used to control the vertical movement of the spindle sleeve and the cutting tool. A *depth stop* attached to the spindle sleeve can be set to control the depth that a cutting tool enters the workpiece.

UPRIGHT DRILLING MACHINE

The standard *upright drilling machine* (Fig. 38-3) is similar to the sensitive-type drill except that it is larger and heavier.

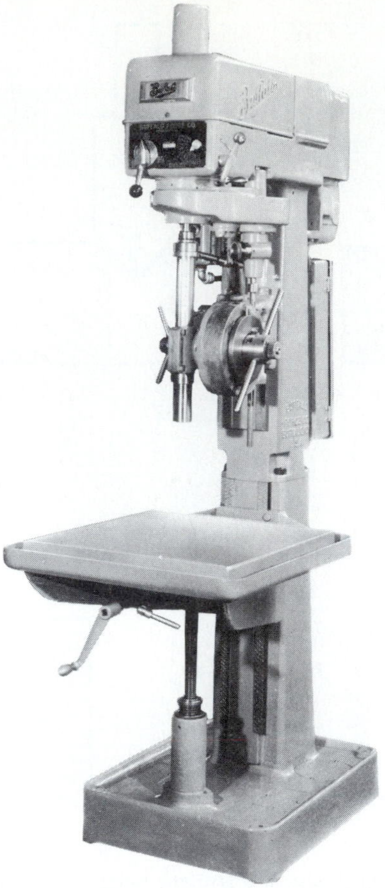

Figure 38-3 *A standard upright drilling machine with a square, or production-type, table.*

The basic differences are:

1. It is equipped with a gearbox to provide a greater variety of speeds.
2. The spindle may be advanced by three methods.
   a. Manually with a hand lever
   b. Manually with a handwheel on most models
   c. Automatically by the feed mechanism
3. The table may be raised or lowered by means of a table-raising mechanism.
4. Some models are equipped with a reservoir in the base for the storage of coolant.

Gang drills (Fig. 38-4) are drilling machines equipped with more than one work or drilling head mounted on a single table. They are used when several operations must be performed on a single job; for example, a drill, reamer, and tap may be mounted on successive spindles so that the work may be advanced quickly from one operation to the next.

For high-speed production work, a number of spindles may be mounted on a single head. This *multispindle head* (Fig. 38-5) may incorporate 20 or more spindles on a single head driven by a drilling machine spindle. Several heads

Figure 38-4 *A gang drilling machine. (Courtesy Buffalo Forge Co.)*

equipped with multispindle attachments may be combined and controlled automatically to drill as many as 100 holes in a single operation. This type of automated drilling is used, for example, by the automotive industry for the drilling of engine blocks.

RADIAL DRILLING MACHINE

The *radial drilling machine* (Fig. 38-6), sometimes called a radial-arm drill, has been developed primarily for the handling of larger workpieces than is possible on upright machines. The advantages of this machine over the upright drill are:

1. Larger and heavier work may be machined.
2. The drilling head may be easily raised or lowered to accommodate various heights of work.
3. The drilling head may be moved rapidly to any desired loction, while the workpiece remains clamped in one position; this feature permits greater production.
4. The machine has more power; thus, larger cutting tools can be used.

Figure 38-5 *A multispindle drilling head.*

5. On universal models, the head may be swiveled so that holes can be drilled on an angle.

Radial Drilling Machine Parts

Base

The base is made of heavy, box-type ribbed cast iron or of welded steel. The base, or pedestal, is used to bolt the machine to the floor and also to provide a coolant reservoir. Large work may be clamped directly to the base for drilling purposes. For convenience in drilling smaller work, a *table* may be bolted to the base.

Column

The column is an upright cylindrical member fitted to the base which supports the radial arm at right angles.

Radial Arm

The arm is attached to the column and may be raised and lowered by means of a *power-driven elevating screw*. The arm may also be swung about the column and may be clamped in any desired position. It also supports the drive motor and drilling head.

Drilling Head

The drilling head is mounted on the arm and may be moved along the length of the arm by means of a *traverse handwheel*. The head may be clamped at any position along the arm. The head houses the change gears and controls for

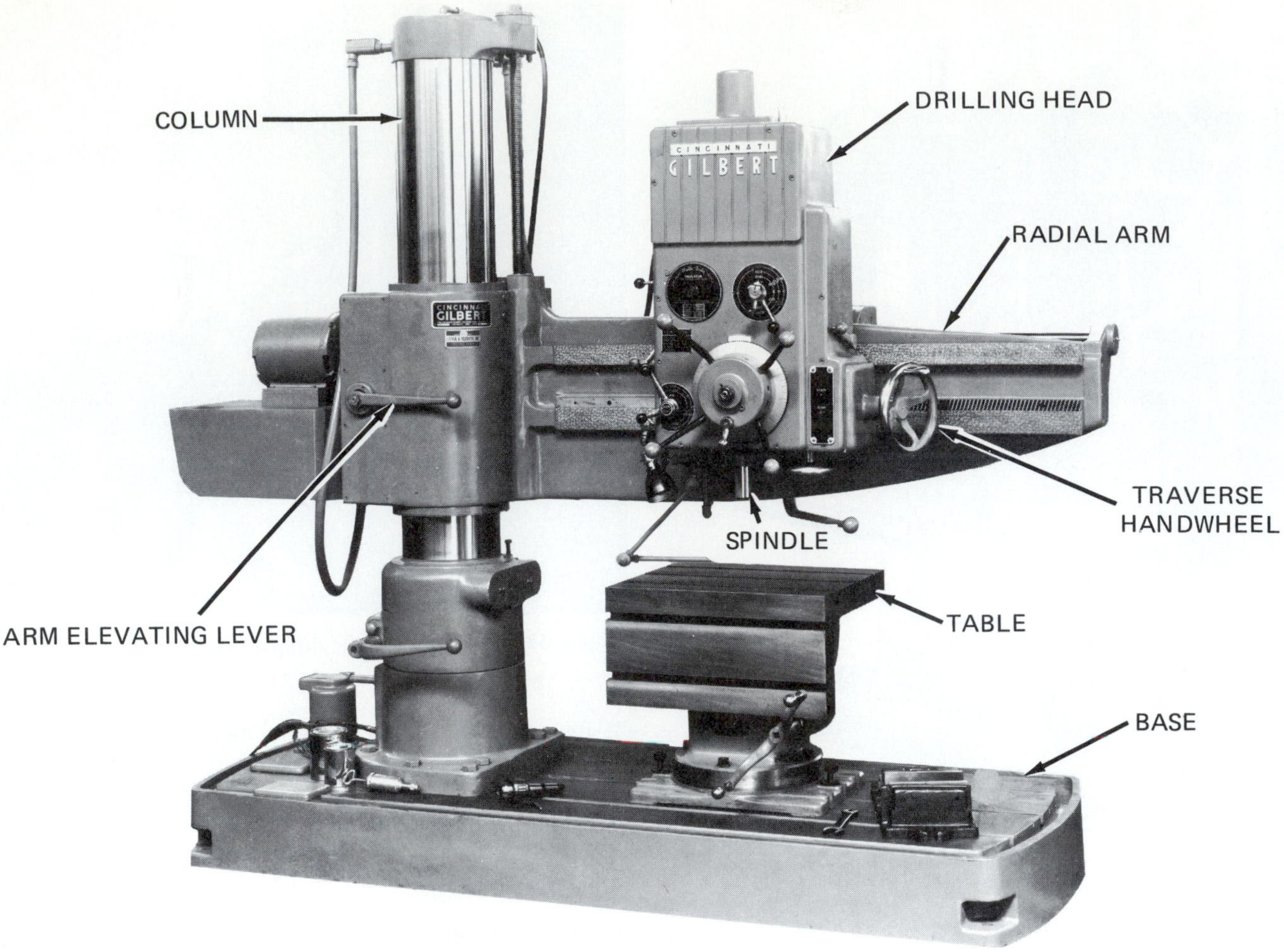
COLUMN
CINCINNATI
GILBERT
DRILLING HEAD
RADIAL ARM
TRAVERSE
HANDWHEEL
SPINDLE
TABLE
ARM ELEVATING LEVER
BASE

Figure 38-6 *A radial drilling machine permits large parts to be drilled. (Courtesy Cincinnati Gilbert.)*

the spindle speeds and feeds. The drill spindle may be raised or lowered manually by means of the spindle-feed handles. When the spindle-feed handles are brought together, automatic feed is provided to the drill spindle.

NUMERICAL CONTROL DRILLING MACHINE

This machine (Fig. 38-7) is a relatively new advance in production drilling. The spindle and table movements are controlled electronically by means of a punched tape or directly from a computer. A tape reader, in the case of punched tape, decodes the information and passes it on to the machine tool. Here the table (and work) is positioned as required and the cutting tool needed to perform the operation is indexed.

After the speed and feed of the cutting tool have been set automatically, the machine starts and the drill or cutting tool enters the workpiece. When it has cut to the proper depth, the cutting tools will retract and another operation such as tapping or reaming may be performed. When the work cycle is complete, another workpiece is positioned under the spindle and the cycle repeated automatically with such precision that the hole positions on every workpiece will be accurate to within ±0.001 in. (0.02 mm).

REVIEW QUESTIONS

STANDARD OPERATIONS

1. Define *drilling, boring,* and *reaming.*
2. What is the difference between spot-facing and counterboring?

TYPES OF DRILLING MACHINES

3. State two methods by which the size of a drill press may be determined.

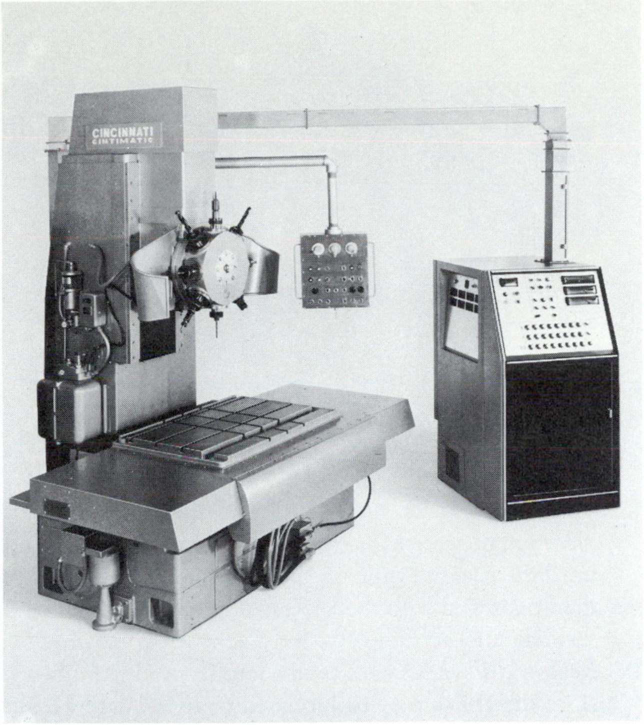
CINCINNATI

Figure 38-7 *A numerical control drilling machine performs automatic drilling cycles by means of punched tape. (Courtesy Cincinnati Milacron Inc.)*

4. Compare a sensitive drill press with an upright drilling machine.
5. Describe a gang drill and state the purpose for which it is used.
6. Describe and state the purpose of the following parts of a sensitive drill press.
   a. Base
   b. Column
   c. Drilling head
   d. Table
   e. Depth stop
7. What are the advantages of a radial drilling machine?
8. Compare the construction of the radial drilling machine to that of a standard upright drill press.
9. Briefly describe the principle of a numerically controlled drilling machine.

UNIT 39

Drilling Machine Accessories

The versatility of the drill press is greatly increased by the various accessories which are available. Drill press accessories fall into two categories:

1. *Tool-holding devices,* which are used to hold or drive the cutting tool
2. *Work-holding devices,* which are used to clamp or hold the workpiece

OBJECTIVES

After completing this unit you will be able to:
1. Identify and use three types of drill-holding devices
2. Identify and use work-holding devices for drilling
3. Set up and clamp work properly for drilling

TOOL-HOLDING DEVICES

The drill press spindle provides a means of holding and driving the cutting tool. It may have a tapered hole to accommodate taper shank tools or its end may be tapered or threaded for mounting a drill chuck. Although there is a variety of tool-holding devices and accessories, the most common found in a machine shop are drill chucks, drill sleeves, and drill sockets.

DRILL CHUCKS

Drill chucks are the most common devices used on a drill press for holding straight-shank cutting tools. Most drill chucks contain three jaws that move simultaneously when the outer sleeve is turned, or on some types of chucks when the outer collar is raised. The three jaws hold the straight shank of a cutting tool securely and cause it to run accurately. There are generally two common types of drill chucks: the key type and the keyless type.

CHUCKS

Straight-shank drills are held in a drill chuck. This chuck may be mounted on the drill press spindle by means of a taper (Fig. 39-1A) or a thread (Fig. 39-1B). Chucks used in larger drill presses are usually held in the spindle by means of a self-holding taper.

There are several types of drill chucks, all of which when properly used will retain their accuracy for years.

- *Key-type drill chucks* (Fig. 39-2) are the most common. They have three jaws that move in or out simultaneously when the outer sleeve is turned. The drill is placed in the chuck and the outer sleeve turned by hand until the jaws are snug on the drill shank. The sleeve is then tightened with the key causing the drill to be held securely and accurately.
- *Keyless drill chucks* are used more in production work since the chuck may be loosened or tightened by hand without a key.
- The *precision keyless chuck* (Fig. 39-3) is designed to hold smaller drills accurately. The drill is changed by turning the outer knurled sleeve.
- The *Jacobs impact keyless chuck* (Fig. 39-4) will hold small or large drills (within the range of the chuck) securely and accurately by means of Rubber-Flex collets. The drill is gripped or released quickly and easily by means of a built-in impact device in the chuck.

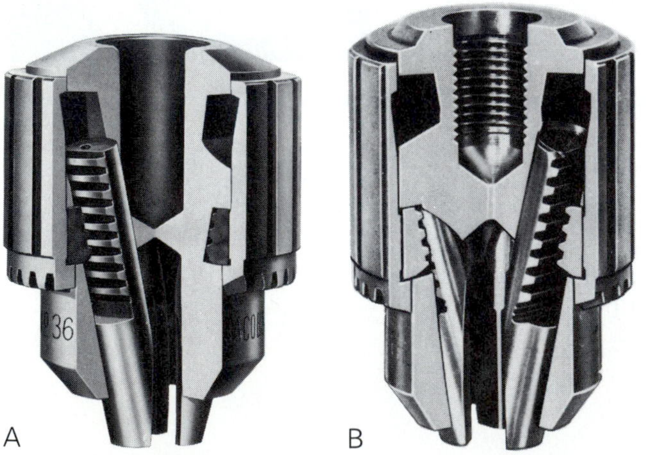

A
B

Figure 39-1 *Methods of mounting drill chucks: (A) Taper-mounted. (Courtesy The Cleveland Twist Drill Company.) (B) Thread-mounted. (Courtesy The Jacobs Manufacturing Company.)*

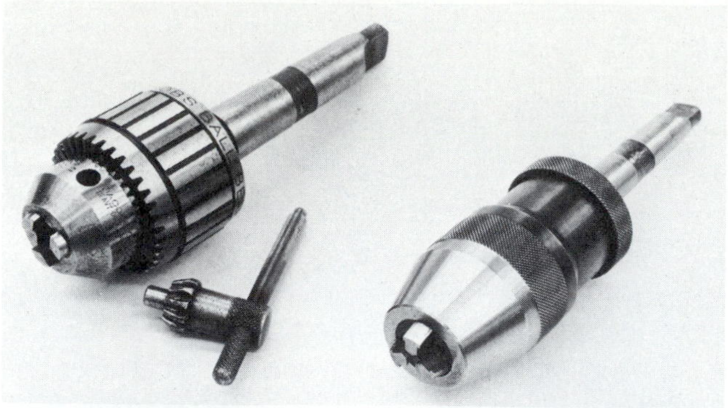

Figure 39-2 *A key-type drill chuck.*

Figure 39-3 *A keyless drill chuck.*

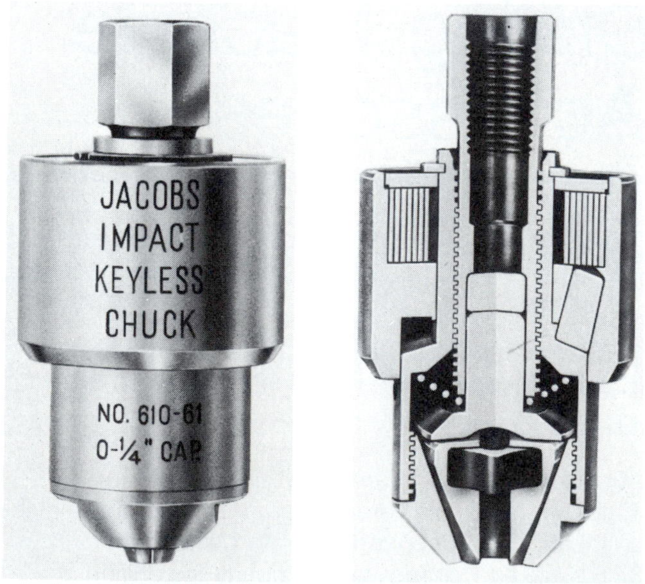

JACOBS
IMPACT
KEYLESS
CHUCK
NO. 610-61
0-¼" CAP

Figure 39-4 *A Jacobs impact keyless chuck. (Courtesy Jacobs Manufacturing Company.)*

A
B

Figure 39-5 *(A) A drill sleeve; (B) a drill socket. (Courtesy The Cleveland Twist Drill Company.)*

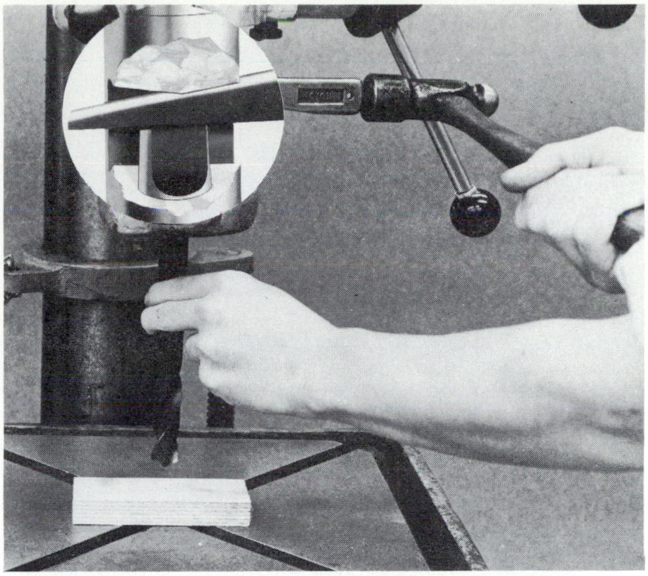

Figure 39-6 *Removing a tapered-shank drill with a drill drift. Note the board that prevents damage to the table if the drill drops. (Courtesy Kostel Enterprises Limited.)*

DRILL SLEEVES AND SOCKETS

The size of the tapered hole in the drill press spindle is generally in proportion to the size of the machine: the larger the machine, the larger the spindle hole. The size of the tapered shank on cutting tools is also manufactured in proportion to the size of the tool. *Drill sleeves* (Fig. 39-5A) are used to adapt the cutting tool shank to the machine spindle if the taper on the cutting tool is smaller than the tapered hole in the spindle.

A *drill socket* (Fig. 39-5B) is used when the hole in the spindle of the drill press is too small for the taper shank of the drill. The drill is first mounted in the socket and then the socket is inserted into the drill press spindle. Drill sockets may also be used as extension sockets to provide extra length.

A flat wedge-shaped tool called a *drill drift* is used to remove tapered-shank drills or accessories from the drill press spindle. When using a drill drift (Fig. 39-6) always place the rounded edge up so that this edge will bear against the round slot in the spindle. A hammer is used to tap the drill drift and loosen the tapered drill shank in the spindle. A board or a piece of masonite should be used to protect the table in case the drill drops when it is being removed.

WORK-HOLDING DEVICES

All workpieces must be fastened securely before cutting operations are performed on a drilling machine. If the work moves or springs during drilling, the drill usually breaks. Serious accidents can be caused by work becoming loose and spinning around during a drilling operation. Some of the commonly used work-holding devices used on drill presses are:

- A *drill vise* (Fig. 39-7) may be used to hold round, rectangular, square, and odd-shaped pieces for any operation that can be performed on a drill press. It is a good practice to clamp or bolt the vise to the drill table when drilling holes over 3/8 in. (9.5 mm) in diameter or to provide a table stop to prevent the vise from swinging during the drilling operation.
- An *angle vise* (Fig. 39-8) has an angular adjustment on its base to allow the operator to drill holes at an angle without tilting the drill press table.
- A *contour vise* (Fig. 39-9) has special movable jaws consisting of several free-moving interlocking segments which automatically adjust to the shape of odd-shaped workpieces when the vise is tightened. These vises are valuable when one operation must be performed on many similar odd-shaped workpieces.
- *V-blocks* (Fig. 39-10A), made out of cast iron or hard-

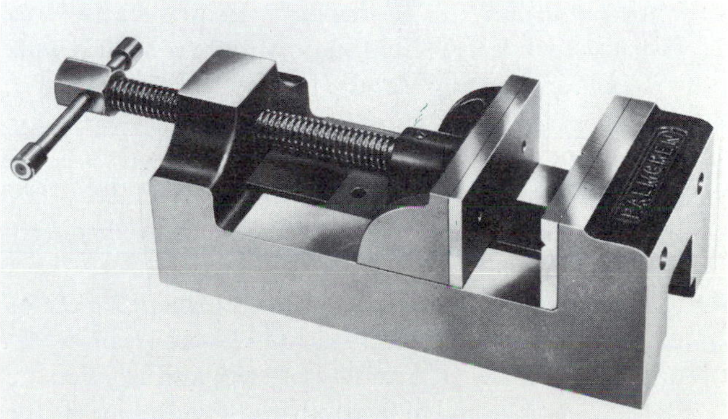
PALMGREN

Figure 39-7 *A drill vise should be clamped to the table when larger holes are being drilled. (Courtesy Rockwell Tool Division Company.)*

Figure 39-8 *An angle vise permits holes to be drilled at an angle in workpieces.*

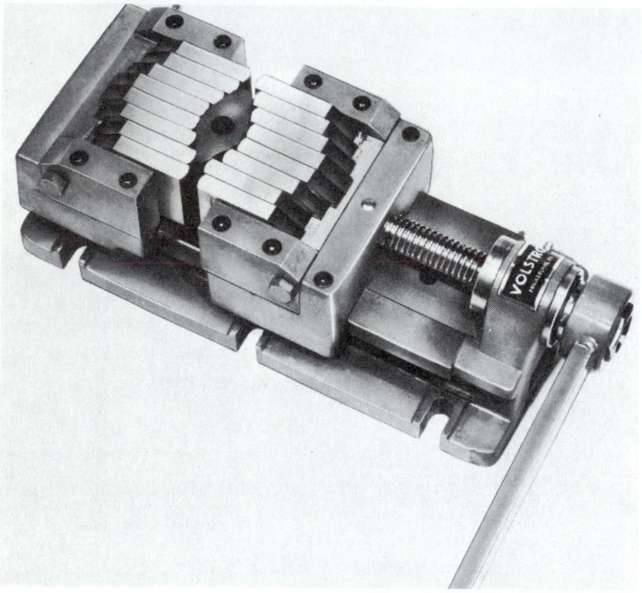

Figure 39-9 *A contour vise automatically adjusts to the shape of the workpiece. (Courtesy Volstro Manufacturing Company.)*

ened steel, are generally used in pairs to support round work for drilling. A U-shaped strap may be used to fasten the work in a V-block, or work may be held with a T-bolt and a strap clamp (Fig. 39-10B).

- *Step blocks* (Fig. 39-11) are used to provide support for the outer end of the strap clamps when work is being fastened for drilling machine operations. They are made in various sizes and steps to accommodate different work heights.
- The *angle plate* (Fig. 39-12A) is an L-shaped piece of cast iron or hardened steel machined to an accurate 90° angle. It is made in a variety of sizes and has slots or holes (clearance and tapped) which provide a means for fastening work for drilling. The angle plate may be bolted or clamped to the table (Fig. 39-12B).
- *Drill jigs* (Fig. 39-13) are used in production for drilling holes in a large number of identical parts. They eliminate the need of laying out a hole location, avoid incorrectly located holes, and allow holes to be drilled quickly and accurately.

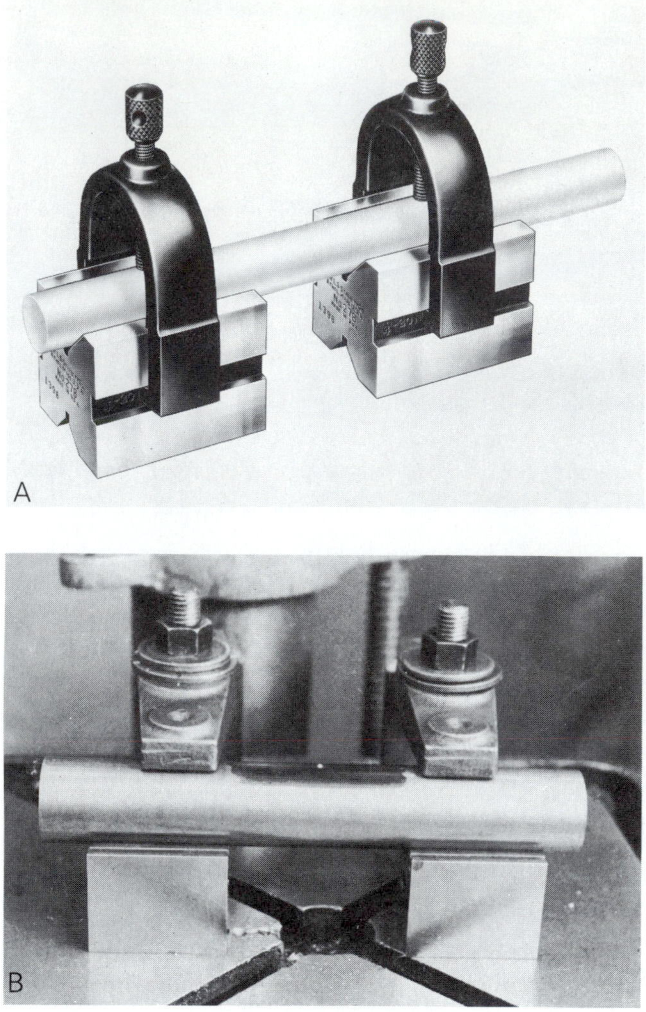
A
B

Figure 39-10 *(A) V-blocks are used to hold round workpieces for drilling. (B) A workpiece clamped in V-blocks. (Courtesy The L. S. Starrett Company.)*

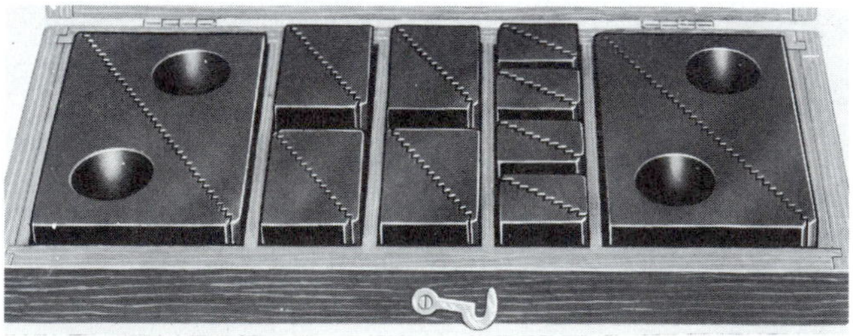

Figure 39-11 *Step blocks support the end of the clamp used to hold the workpiece. (Courtesy Northwestern Tools, Incorporated.)*

- *Clamps* or *straps* (Fig. 39-14), used to fasten work to

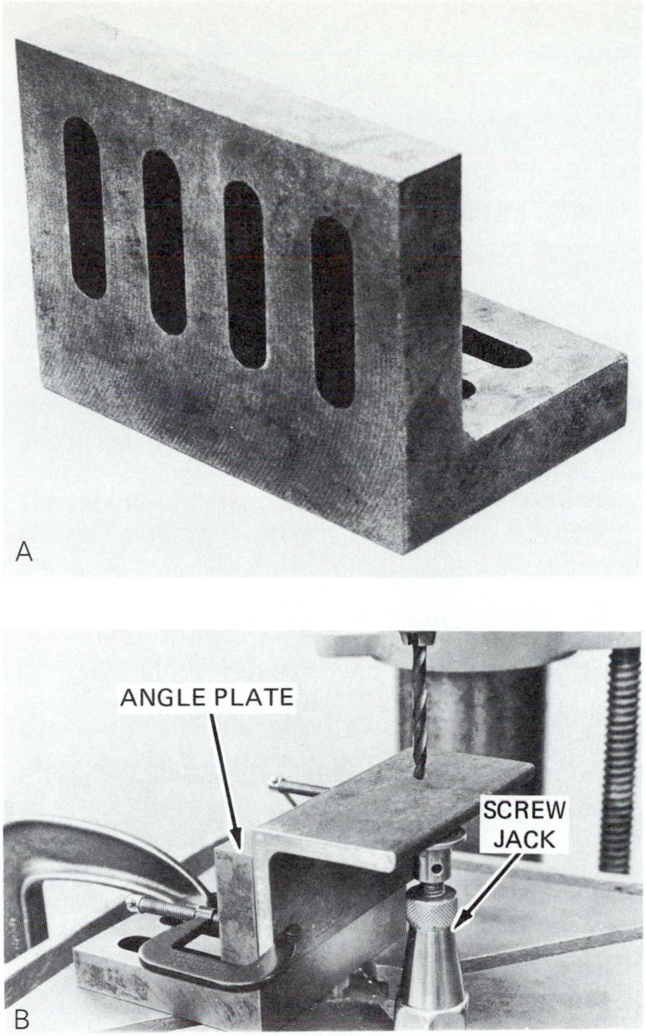
A
ANGLE PLATE
SCREW
JACK
B

Figure 39-12 *(A) Workpieces may be clamped to an angle plate for drilling; (B) a workpiece clamped to an angle plate and supported by a screw jack. (Courtesy Kostel Enterprises Limited.)*

Figure 39-13 *Identical parts are drilled quickly and accurately in a drill jig.*

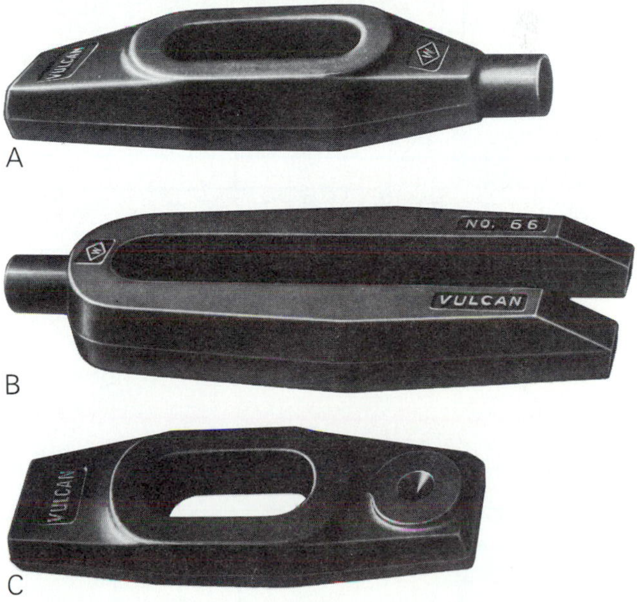
VULCAN
A
NO. 66
VULCAN
B
VULCAN
C

Figure 39-14 *Types of clamps used to fasten work on a drill press: (A) finger clamp; (B) U-clamp; (C) straight clamp. (Courtesy J. W. Williams & Company.)*

the drill table or an angle plate for drilling, are made in various sizes. They are usually supported at the end by a step block and bolted to the table by a T-bolt that fits into the table T-slot. It is a good practice to place the T-bolt in the clamp or strap as close to the work as possible so that pressure will be exerted on the workpiece. Modifications of these clamps are the *double-finger* and *gooseneck* clamps.

CLAMPING STRESSES

Whenever work is clamped for any machining operation, stresses are created. It is important that these clamping stresses should not be great enough to cause springing or distortion of the workpiece.

When work is to be held for drilling, reaming, or any machining operation, it is important that the workpiece be held securely. The clamps, bolts, and step blocks should be properly located and the work clamped firmly enough to prevent movement, yet not enough to cause the workpiece to spring or distort. It is important that the clamping pressures be applied to the work, not to the packing or step block.

Figure 39-15A illustrates the correct clamping procedure, the main pressure being applied to the workpiece.

NOTE: The step block is slightly higher than the workpiece and the bolt is close to the work.

Fig. 39-15B shows a piece of work incorrectly clamped. The bolt is located close to the step block, which is slightly lower than the workpiece. The main clamping pressure with this type of setup is applied to the step block, not the workpiece.

CLAMPING HINTS

The following suggestions are made for clamping work so that good clamping pressure will be obtained and work distortion avoided.

1. Always place the bolt as close as possible to the workpiece (Fig. 39-16).

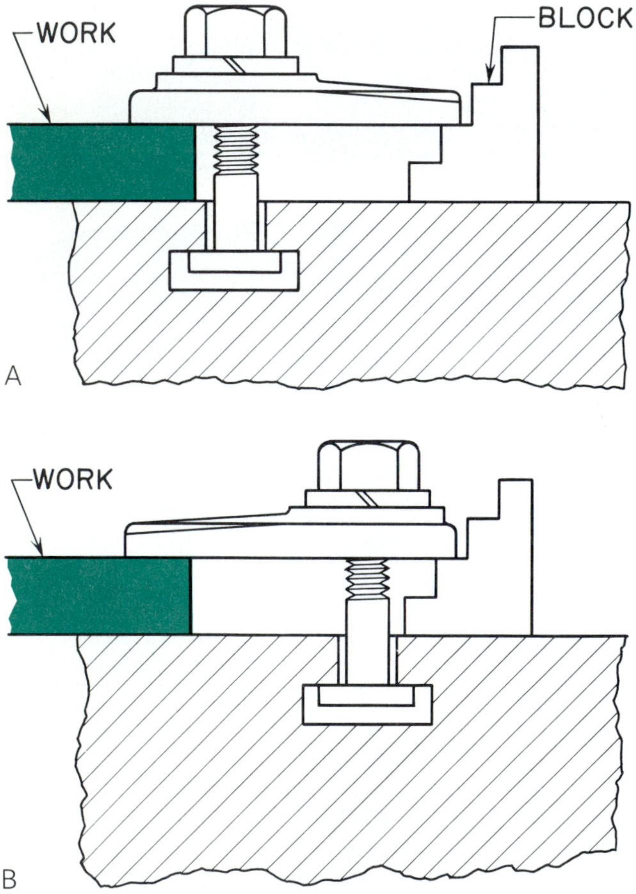
WORK
BLOCK
A
WORK
B

Figure 39-15 *(A) Work is correctly clamped when the bolt is close to the work; (B) work is incorrectly clamped because the clamping pressure is on the step block.*

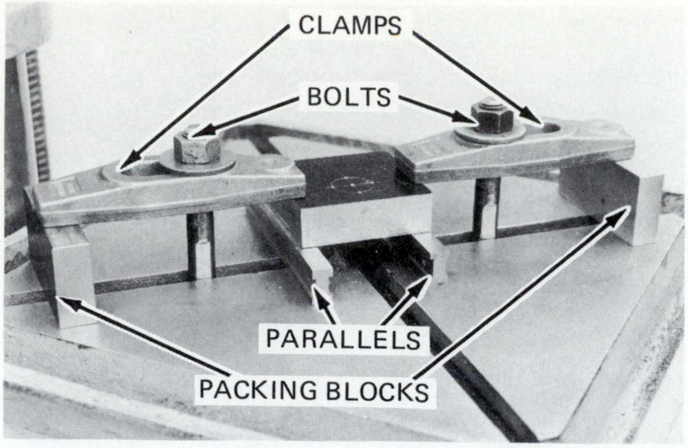
CLAMPS
BOLTS
PARALLELS
PACKING BLOCKS

Figure 39-16 *The bolts and the packing blocks correctly placed for maximum clamping pressure.*

2. Have the packing or step block slightly higher than the work surface being clamped.
3. Insert a piece of paper between the machine table and the workpiece to prevent the work from shifting during the machining process.
4. Place a metal shim between the clamp and workpiece to spread the clamping force over a wider area.
5. It is wise to use a sub-base or liner under a rough casting to prevent damage to the machine table.
6. Parts which do not lie flat on a machine table should be shimmed to prevent the work from rocking. This will prevent distortion when the work is clamped.

REVIEW QUESTIONS

TOOL-HOLDING DEVICES

1. What is the purpose of a drill chuck?
2. How may chucks be secured to the spindle in:
 a. Small drill presses? b. Large drill presses?
3. Name three types of drill chucks.
4. What is the purpose of a:
 a. Drill sleeve? b. Drill socket?
5. How is a tapered shank drill removed from a drill press spindle?

WORK-HOLDING DEVICES

6. Name three types of drill vises and state the purpose of each.
7. What is the purpose of:
 a. V-blocks? b. Step blocks?
8. Describe an angle plate and state its purpose.
9. What is the advantage of a drill jig?

CLAMPING STRESSES

10. Why is it important that work be clamped properly for any machining operation?

11. Explain the procedure for clamping a workpiece properly.

12. List four important clamping hints.

UNIT 40

Twist Drills

Twist drills are end-cutting tools used to produce holes in most types of material. On standard drills, two helical grooves, or flutes, are cut lengthwise around the body of the drill. They provide cutting edges and space for the cuttings to escape during the drilling process. Since drills are among the most efficient cutting tools, it is necessary to know the main parts, how to sharpen the cutting edges, and how to calculate the correct speeds and feeds for drilling various metals in order to use them most efficiently and prolong their life.

OBJECTIVES

After completing this unit you will be able to:

1. Identify the parts of a twist drill
2. Identify four systems of drill sizes and know where each is used
3. Grind the proper angles and clearances on a twist drill

TWIST DRILL PARTS

Most twist drills used in machine shop work today are made of high-speed steel. High-speed steel drills have replaced carbon-steel drills since they can be operated at double the cutting speed and the cutting edge lasts longer. High-speed steel drills are always stamped with the letters "H.S." or "H.S.S." Since the introduction of *carbide-tipped drills,* speeds for production drilling have increased up to 300 percent over high-speed steel drills. Carbide drills have made it possible to drill certain materials that would not be possible with high-speed steels.

A drill (Fig. 40-1) may be divided into three main parts: *shank, body,* and *point.*

SHANK

Generally drills up to ½ in. or 13 mm in diameter have straight shanks, while those over this diameter usually have tapered shanks. *Straight-shank* drills (Fig. 40-2A) are held in a drill chuck; *tapered-shank* drills (Fig. 40-2B) fit into the internal taper of the drill press spindle. A tang is provided on the end of tapered-shank drills to prevent the drill from slipping while it is cutting and to allow the drill to be removed from the spindle or socket without the shank being damaged.

BODY

The *body* is the portion of the drill between the shank and the point. It consists of a number of parts important to the efficiency of the cutting action.

1. The *flutes* are two or more helical grooves cut around the body of the drill. They form the cutting edges, admit cutting fluid, and allow the chips to escape from the hole.
2. The *margin* is the narrow, raised section on the body of the drill. It is immediately next to the flutes and extends along the entire length of the flutes. Its purpose is to provide a full size to the drill body and cutting edges.

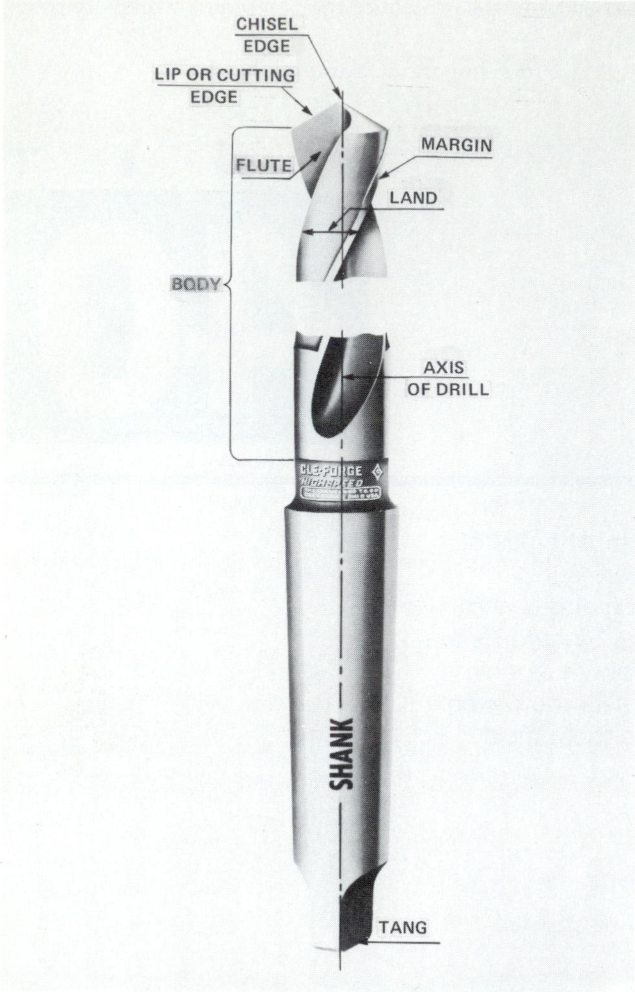
CHISEL
EDGE
LIP OR CUTTING
EDGE
MARGIN
FLUTE
LAND
BODY
AXIS
OF DRILL
SHANK
TANG

Figure 40-1 *The main parts of a twist drill. (Courtesy The Cleveland Twist Drill Company.)*

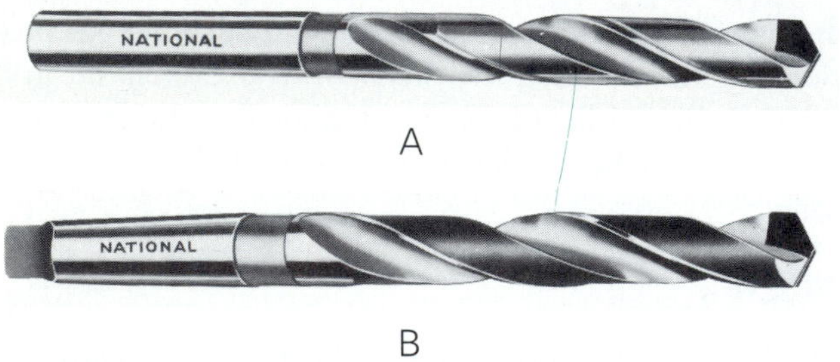
NATIONAL
A
NATIONAL
B

Figure 40-2 *Types of drill shanks: (A) straight; (B) tapered. (Courtesy National Twist Drill & Tool Company.)*

3. The *body clearance* is the undercut portion of the body between the margin and the flutes. It is made smaller to reduce friction between the drill and the hole during the drilling operation.
4. The *web* (Fig. 40-3) is the thin partition in the center of

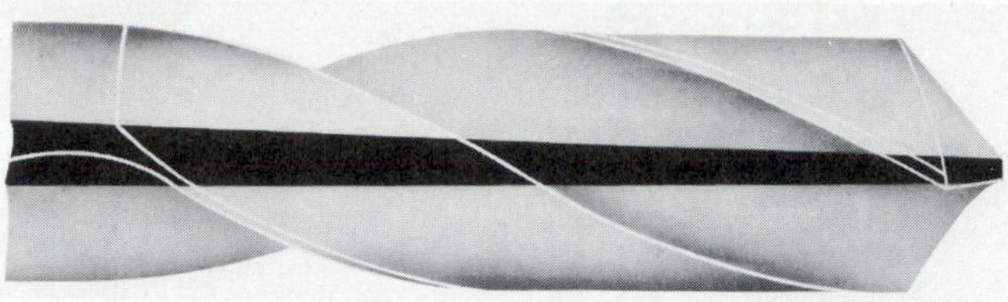

Figure 40-3 *The web is the tapered metal column that separates the flutes. (Courtesy The Cleveland Twist Drill Company.)*

the drill which extends the full length of the flutes. This part forms the chisel edge at the cutting end of the drill (Fig. 40-4). The web gradually increases in thickness toward the shank to give the drill strength.

POINT

The *point* of a twist drill (Fig. 40-4) consists of the chisel edge, lips, lip clearance, and heel. The *chisel edge* is the chisel-shaped portion of the drill point. The *lips* (cutting edges) are formed by the intersection of the flutes. The lips must be of equal length and have the same angle so that the drill will run true and will not cut a hole larger than the size of the drill.

The *lip clearance* is the relief ground on the point of the drill extending from the cutting lips back to the *heel* (Fig. 40-5). The average lip clearance is from 8 to 12°, depending on the hardness or softness of the material to be drilled.

DRILL POINT CHARACTERISTICS

Efficient drilling of the wide variety of materials used by industry requires a great variety of drill points. The most important factors determining the size of the drilled hole are the characteristics of the drill point.

A drill is generally considered a roughing tool capable of

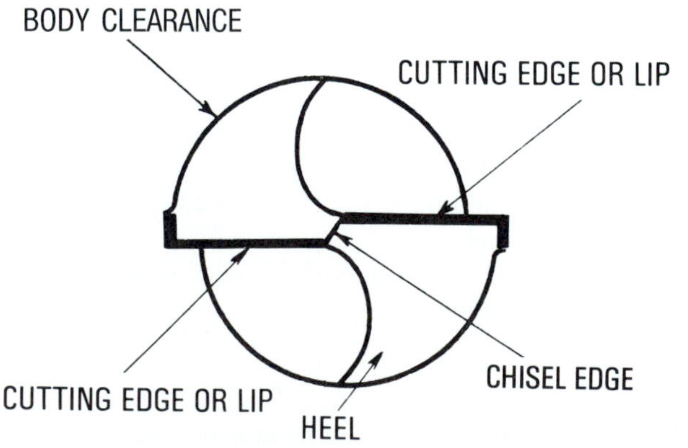
BODY CLEARANCE
CUTTING EDGE OR LIP
CHISEL EDGE
CUTTING EDGE OR LIP
HEEL

Figure 40-4 *The point of a twist drill.*

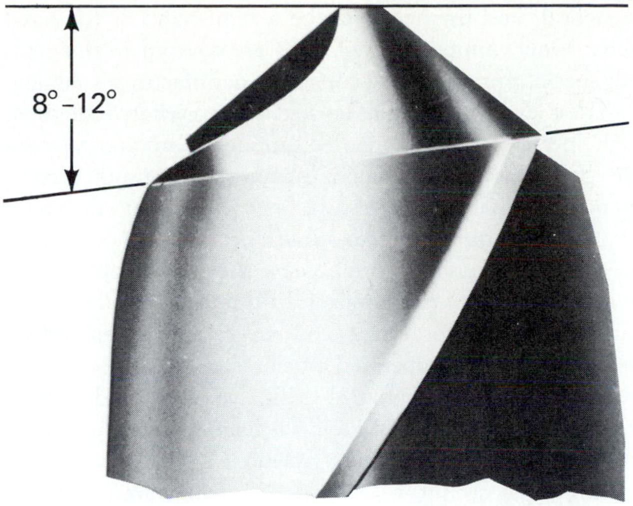
8°-12°

Figure 40-5 *The lip clearance angle of the cutting edge should be 8 to 12°.*

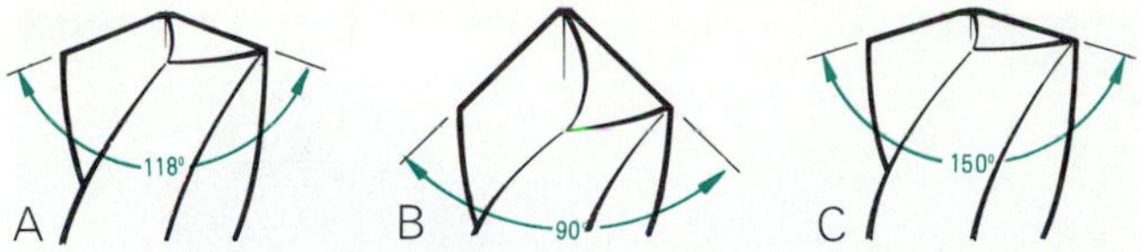
A
118°
B
90°
C
150°

Figure 40-6 *(A) The drill point angle of 118° is suitable for most general work; (B) a drill point angle of 60 to 90° is used for soft material; (C) a drill point angle of 135 to 150° is best for hard materials.*

removing metal quickly. It is not expected to finish a hole to the accuracy possible with a reamer. However, a drill can often be made to cut more accurately and efficiently by proper drill point grinding. The use of various point angles and lip clearances, in conjunction with the thinning of the drill web, will:

1. Control the size, quality, and straightness of the drilled hole
2. Control the size, shape, and formation of the chip
3. Control the chip flow up the flutes
4. Increase the strength of the drill's cutting edges
5. Reduce the rate of wear at the cutting edges
6. Reduce the amount of drilling pressure required
7. Control the amount of burr produced during drilling
8. Reduce the amount of heat generated
9. Permit the use of various speeds and feeds for more efficient drilling

DRILL POINT ANGLES AND CLEARANCES

Drill point angles and clearances are varied to suit the wide variety of material which must be drilled. Three general drill points are commonly used to drill various materials; however, there may be variations of these to suit various drilling conditions.

The *conventional point* (*118°*) shown in Fig. 40-6A is the most commonly used drill point and gives satisfactory results for most general-purpose drilling. The 118° point angle should be ground with 8 to 12° lip clearance for best results. Too much lip clearance weakens the cutting edge and causes the drill to chip and break easily. Too little lip clearance results in the use of heavy drilling pressure; this pressure causes the cutting edges to wear quickly because of the excessive heat generated and also places undue strain on the drill and equipment.

The *long angle point* (*60 to 90°*) shown in Fig. 40-6B is commonly used on low helix drills for the drilling of nonferrous metals, soft cast irons, plastics, fibers, and wood. The lip clearance on long angle point drills is generally from 12 to 15°. On standard drills, a flat may be ground on the face of the lips to prevent the drill from drawing itself into the soft material.

The *flat angle point* (*135 to 150°*) shown in Fig. 40-6C is generally used to drill hard and tough materials. The lip clearance on flat angle point drills is generally only 6 to 8° to provide as much support as possible for the cutting edges. The shorter cutting edge tends to reduce the friction and heat generated during drilling.

SYSTEMS OF DRILL SIZES

Drill sizes are designated under four systems: fractional, number, letter, and millimeter (metric) sizes.

- The *fractional* size drills range from $\frac{1}{64}$ to 4 in. varying in steps of $\frac{1}{64}$ in. from one size to the next.
- The *number* size drills range from #1, measuring 0.228 in., to #97, which measures 0.0059 in.
- The *letter* size drills range from A to Z. Letter-A drill is the smallest in the set (0.234 in.) and Z is the largest (0.413 in.).
- *Millimeter* (*metric*) *drills* are produced in a wide variety of sizes. Miniature metric drills range from 0.04 to 0.09 mm in steps of 0.01 mm. Straight-shank standard metric drills are available in sizes from 0.5 to 20 mm. Taper-shank metric drills are manufactured in sizes from 8 up to 80 mm.

Drill sizes may be checked by using a drill gage (Fig. 40-7). These gages are available in fractional, letter, number, and millimeter sizes. The size of a drill may also be checked by measuring the drill, *over the margins,* with a micrometer (Fig. 40-8).

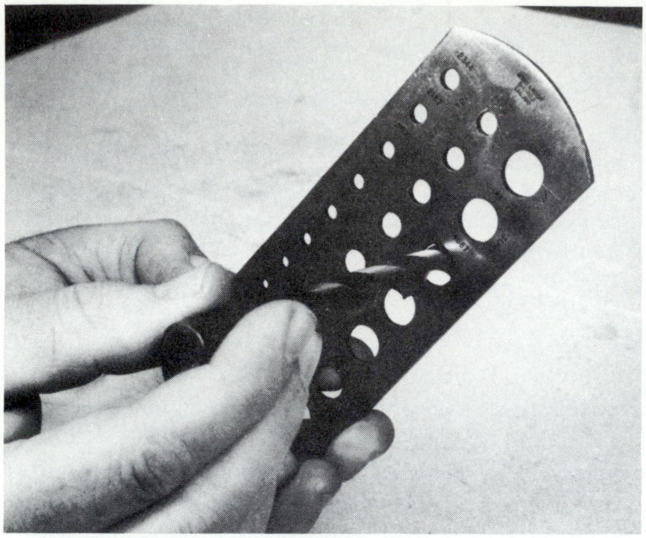

Figure 40-7 *Checking a drill for size using a drill gage.*

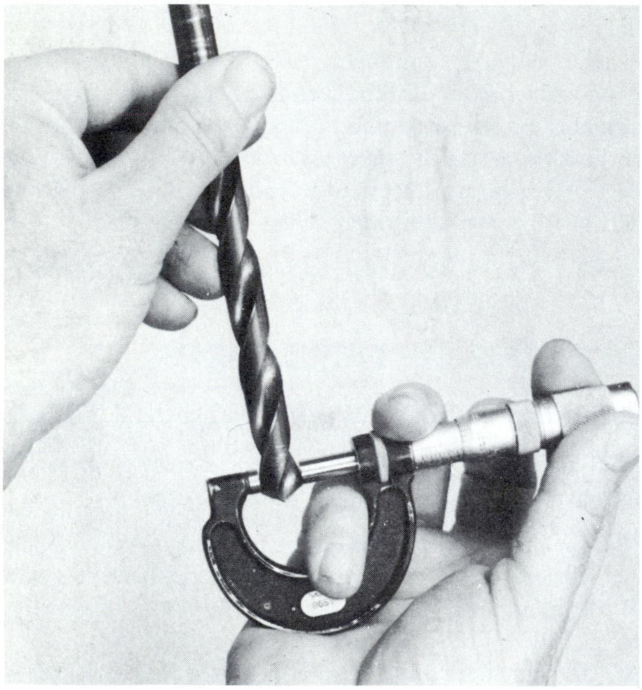

Figure 40-8 *Checking the size of a twist drill using a micrometer.*

TYPES OF DRILLS

A variety of twist-drill styles are manufactured to suit specific drilling operations, types and sizes of material, high production rates, and special applications. The design of drills may vary in the number and width of the flutes, the amount of helix or rake angle of the flutes, or the shape of the land or margin. In addition, the flutes may be straight or helical, and the helix may be a right-hand or left-hand helix. Only commonly used drills are covered in this text. For special-purpose drills, consult a manufacturer's catalog.

Twist drills are manufactured from carbon tool steel, high-speed steel, and cemented carbides. *Carbon-steel drills* are generally used in hobby shops and are not recommended for machine shop work since the cutting edges tend to wear down quickly. *High-speed steel drills* are commonly used in machine shop work because they can be operated at twice the speed of carbon-steel drills and the cutting edges can withstand more heat and wear. *Cemented-carbide drills,* which can be operated much faster (up to three times faster) than high-speed steel drills, are used to drill hard materials. Cemented-carbide drills have found wide use in industry because they can be operated at high speeds, the cutting edges do not wear rapidly, and they are capable of withstanding higher heat.

The most commonly used drill is the *general-purpose drill,* which has two helical flutes (Fig. 40-2). This drill is designed to perform well on a wide variety of materials, equipment, and job conditions. The general-purpose drill can be made to suit different conditions and materials by varying the point angle and the speeds and feeds used. Straight shank drills are commonly known as *general-purpose jobbers length drills.*

The *low-helix drill* was developed primarily to drill brass and thin materials. This type of drill is used to drill shallow holes in some aluminum and magnesium alloys. Because of its design, the low helix drill can remove the large volume of chips formed by high rates of penetration when it is used on machines such as turret lathes and screw machines.

High helix drills (Fig. 40-9) are designed for drilling deep holes in aluminum, copper, die-cast material, and other metals where the chips have a tendency to jam in a hole. The high helix angle (35 to 45°) and the wider flutes of these drills assist in clearing chips from the hole.

A *core drill* (Fig. 40-10), designed with three or four flutes, is used primarily to enlarge cored, drilled, or punched holes. This drill has advantages over the two-fluted drills in productivity and finish. In some cases, a core drill may be used in place of a reamer for finishing a hole. Core drills are produced in sizes from ¼ to 3 in. (6 to 76 mm) in diameter.

Figure 40-9 *A high helix drill. (Courtesy The Cleveland Twist Drill Company.)*

Figure 40-10 *A core drill. (Courtesy The Cleveland Twist Drill Company.)*

Figure 40-11 *An oil hole drill.*

Figure 40-12 *A straight-fluted drill.*

Oil hole drills (Fig. 40-11) have one or two oil holes running from the shank to the cutting point through which compressed air, oil, or cutting fluid can be forced when deep holes are being drilled. These drills are generally used on turret lathes and screw machines. The cutting fluid flowing through the oil holes cools the drill's cutting edges and flushes the chips out of the hole.

Straight-fluted drills (Fig. 40-12) are recommended for drilling operations on soft materials such as brass, bronze, copper, and various types of plastic. The straight flute prevents the drill from drawing itself into the material (digging in) while cutting.

If a straight-fluted drill is not available, a conventional drill can be modified by grinding a small flat approximately 1⁄16 in. (1.5 mm) wide on the face of both cutting edges of the drill.

NOTE: The flat must be ground parallel to the axis of the drill.

Deep hole or *gun drills* (Fig. 40-13) are used for producing holes from approximately 3⁄8 to 3 in. (9.5 to 76 mm) in diameter and as deep as 20 ft (6 m). The most common gun drill consists of a round, tubular stem, on the end of which is fastened a flat, two-fluted drilling insert. Cutting fluid is forced through the center of the stem to flush the chips from the hole. When the drilling insert becomes dull, it can be replaced quickly by loosening one screw, which holds it to the tubular stem.

Spade drills are similar to gun drills in that the cutting end is a flat blade with two cutting lips. Spade drills are usually clamped in a holder (Fig. 40-14) and are easily replaced or sharpened. Spade drills are available in a wide range of sizes from very small microdrills to drills up to 12 in. in diameter. Some of the smaller spade drills have replaceable carbide inserts.

A rather unique drill is the *hard-steel drill* (Fig. 40-15) used for drilling hardened steel. These drills are made from a heat-resistant alloy. As the drill is brought into contact with the workpiece, the fluted, triangular point softens the metal by friction and then removes the softened metal ahead of it, in chip form.

Step drills (Fig. 40-16) are used to drill and countersink or drill and counterbore different sizes of holes in one operation. The drill may have two or more diameters ground on it. Each size or step may be separated by a square or angular shoulder, depending on the purpose of the hole. For example, holes to be tapped should have a slight chamfer at the top of the hole to ease the starting of the tap, protect the thread, and leave the tapped hole free of burrs caused by the tap.

A *saw-type hole cutter* (Fig. 40-17) is a cylindrical diameter cutter with a twist drill in the center to provide a guide for the cutting teeth on the hole cutter. This type of cutter is made in various diameters and is generally used for drilling holes in thin materials. It is especially valuable for drilling holes in pipe and sheet metal, as little burr is produced and the cutter does not have a tendency to jam as it is breaking through.

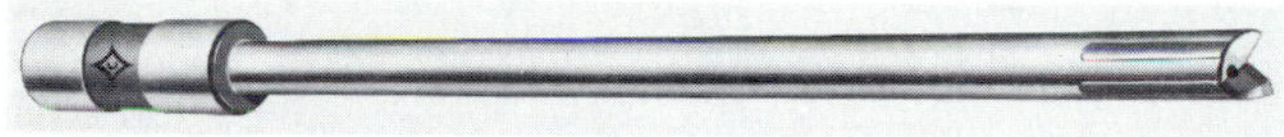

Figure 40-13 *Deep hole, or gun, drill. (Courtesy The Cleveland Twist Drill Company.)*

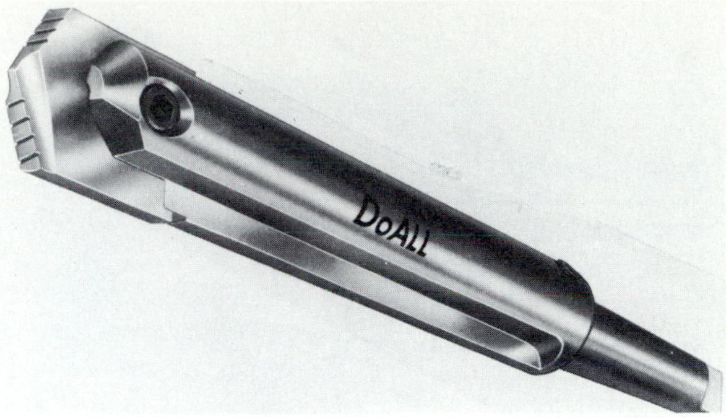
DoALL

Figure 40-14 *A spade drill. (Courtesy DoAll Company.)*

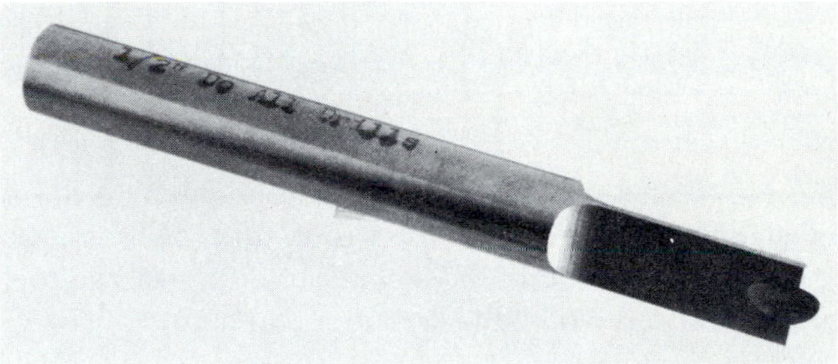

Figure 40-15 *A hard-steel drill. (Courtesy DoAll Company.)*

Figure 40-16 *A step drill. (Courtesy National Twist Drill & Tool Company.)*

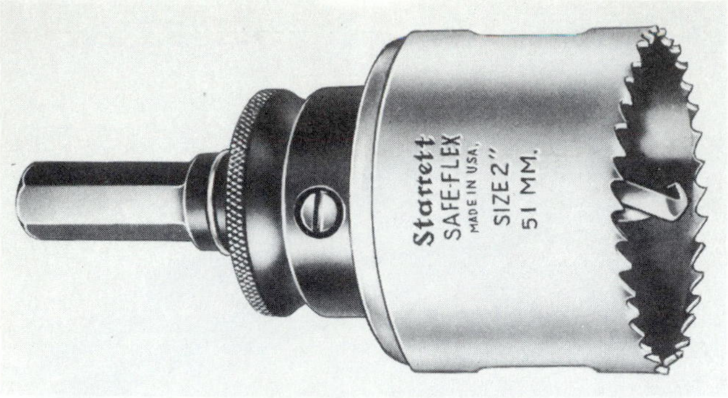
Starrett
SAFE-FLEX
MADE IN USA
SIZE 2"
51 MM.

Figure 40-17 *A saw-type hole cutter. (Courtesy L. S. Starrett Company.)*

DRILLING FACTS AND PROBLEMS

The most common drill problems encountered are illustrated in Fig. 40-18A and B. It is wise to study these various problems to ensure that the amount of drill breakage, regrinding, and downtime is kept to a minimum.

DRILL GRINDING

The cutting efficiency of a drill is determined by the characteristics and condition of the point of the drill. Most new drills are provided with a general-purpose point (118° point angle and an 8 to 12° lip clearance). As a drill is used, the cutting edges may wear and become chipped or the drill may break. Drills are generally resharpened by hand. However, small drill point grinders or drill-sharpening attachments are inexpensive, readily available, and provide more consistent quality than hand grinding.

To ensure that a drill will perform properly, examine the drill point carefully before mounting the drill in the drill press. A properly ground drill should have the following characteristics:

- The length of both cutting lips should be the same.
  Lips of unequal length will force the drill point off center, causing one lip to do more cutting than the other and produce an oversize hole (Fig. 40-19A).
- The angle of both lips should be the same.
  If the angles are unequal, the drill will cut an oversize hole because one lip will do more cutting than the other (Fig. 40-19B).
- The lips should be free from nicks or wear.
- There should be no sign of wear on the margin.

If the drill does not meet all of these requirements, it should be resharpened. If the drill is not resharpened, it will give poor service, produce inaccurate holes, and may break due to excessive drilling strain.

While a drill is being used, there will be signs to indicate that the drill is not cutting properly and should be resharpened. If the drill is not sharpened at the first sign of dullness, it will require extra power to force the slightly dulled

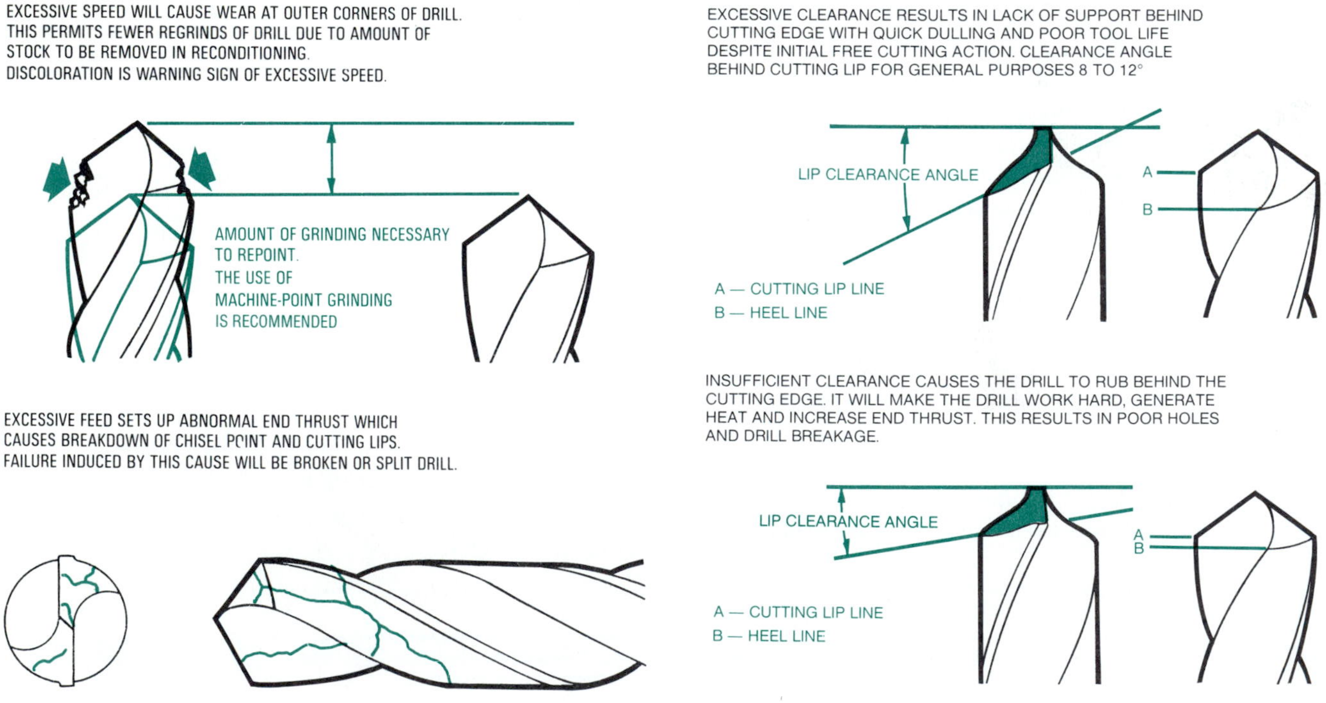
EXCESSIVE SPEED WILL CAUSE WEAR AT OUTER CORNERS OF DRILL. THIS PERMITS FEWER REGRINDS OF DRILL DUE TO AMOUNT OF STOCK TO BE REMOVED IN RECONDITIONING. DISCOLORATION IS WARNING SIGN OF EXCESSIVE SPEED.
AMOUNT OF GRINDING NECESSARY TO REPOINT. THE USE OF MACHINE-POINT GRINDING IS RECOMMENDED
EXCESSIVE FEED SETS UP ABNORMAL END THRUST WHICH CAUSES BREAKDOWN OF CHISEL POINT AND CUTTING LIPS. FAILURE INDUCED BY THIS CAUSE WILL BE BROKEN OR SPLIT DRILL.
EXCESSIVE CLEARANCE RESULTS IN LACK OF SUPPORT BEHIND CUTTING EDGE WITH QUICK DULLING AND POOR TOOL LIFE DESPITE INITIAL FREE CUTTING ACTION. CLEARANCE ANGLE BEHIND CUTTING LIP FOR GENERAL PURPOSES 8 TO 12°
LIP CLEARANCE ANGLE
A
B
A — CUTTING LIP LINE
B — HEEL LINE
INSUFFICIENT CLEARANCE CAUSES THE DRILL TO RUB BEHIND THE CUTTING EDGE. IT WILL MAKE THE DRILL WORK HARD, GENERATE HEAT AND INCREASE END THRUST. THIS RESULTS IN POOR HOLES AND DRILL BREAKAGE.
LIP CLEARANCE ANGLE
A
B
A — CUTTING LIP LINE
B — HEEL LINE

Figure 40-18A *Facts about drills and drilling. (Courtesy Greenfield Tap & Die Company.)*

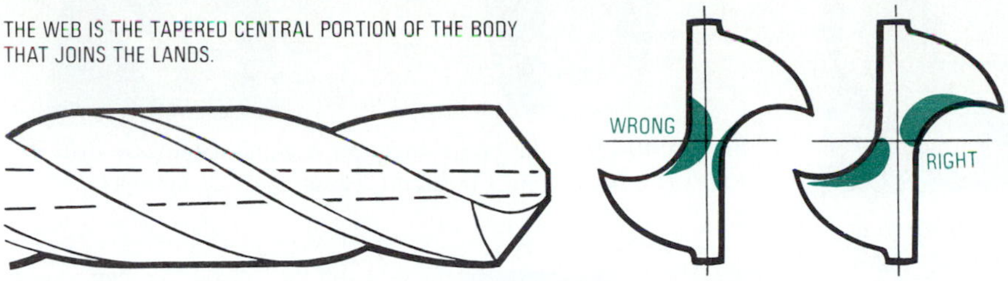
IMPROPER WEB THINNING IS THE RESULT OF TAKING MORE STOCK FROM ONE CUTTING EDGE THAN FROM THE OTHER, THEREBY DESTROYING THE CONCENTRICITY OF THE WEB AND OUTSIDE DIAMETER.
THE WEB IS THE TAPERED CENTRAL PORTION OF THE BODY THAT JOINS THE LANDS.
WRONG
RIGHT

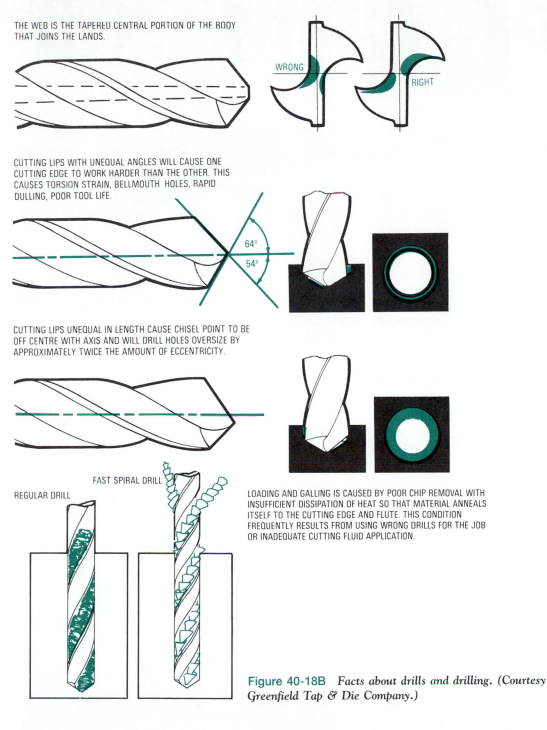

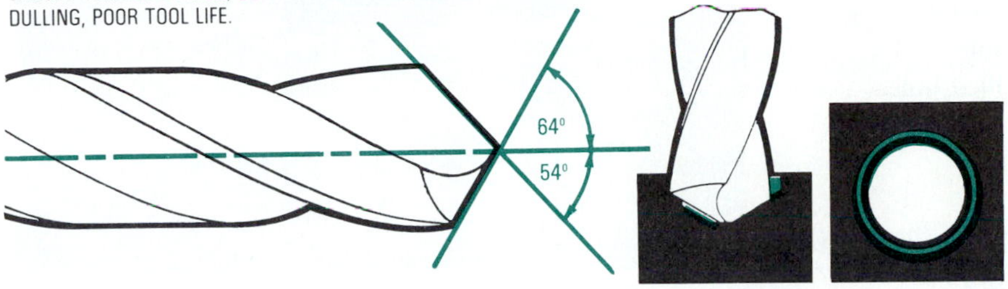
CUTTING LIPS WITH UNEQUAL ANGLES WILL CAUSE ONE CUTTING EDGE TO WORK HARDER THAN THE OTHER. THIS CAUSES TORSION STRAIN, BELLMOUTH HOLES, RAPID DULLING, POOR TOOL LIFE.
64°
54°

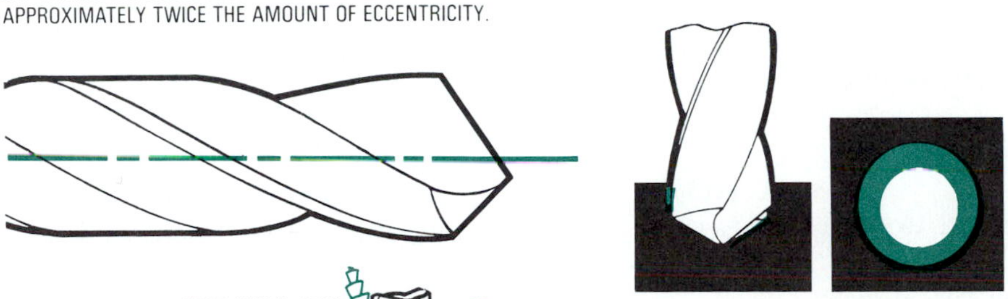
CUTTING LIPS UNEQUAL IN LENGTH CAUSE CHISEL POINT TO BE OFF CENTRE WITH AXIS AND WILL DRILL HOLES OVERSIZE BY APPROXIMATELY TWICE THE AMOUNT OF ECCENTRICITY.

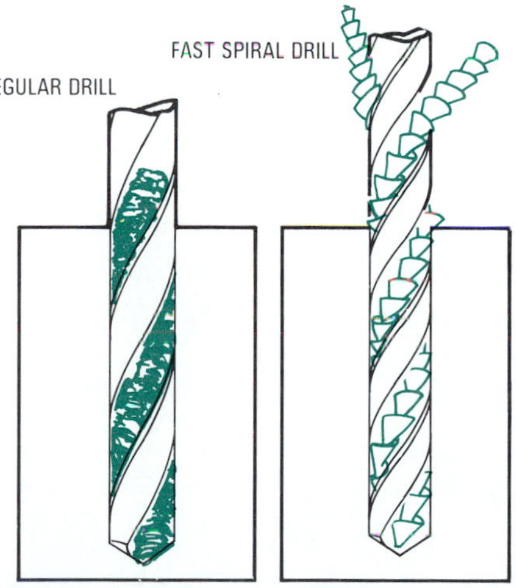
REGULAR DRILL
FAST SPIRAL DRILL
LOADING AND GALLING IS CAUSED BY POOR CHIP REMOVAL WITH INSUFFICIENT DISSIPATION OF HEAT SO THAT MATERIAL ANNEALS ITSELF TO THE CUTTING EDGE AND FLUTE. THIS CONDITION FREQUENTLY RESULTS FROM USING WRONG DRILLS FOR THE JOB OR INADEQUATE CUTTING FLUID APPLICATION.

Figure 40-18B *Facts about drills and drilling. (Courtesy Greenfield Tap & Die Company.)*

drill into the work. This causes more heat to be generated at the cutting lips and results in a faster rate of wear. When any of the following conditions arise while a drill is in use, it should be examined and reground.

- The color and shape of the chips change.
- More drilling pressure is required to force the drill into the work.
- The drill turns blue due to the excessive heat generated while drilling.
- The top of the hole is out of round.
- A poor finish is produced in the hole.

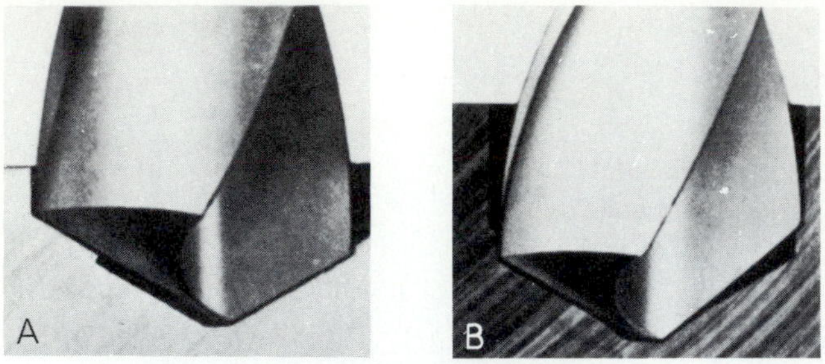
A
B

Figure 40-19 *(A) An incorrect point with lips of unequal lengths; (B) lips with unequal angles produce oversize holes. (Courtesy The Cleveland Twist Drill Company.)*

- The drill chatters when it contacts the metal.
- The drill squeals and may jam in the hole.
- An excessive burr is left around the drilled hole.

CAUSES OF DRILL FAILURE

Drills should not be allowed to become so dull that they cannot cut. Overdulling of any metal-cutting tool generally results in poor production rates, inaccurate work, and the shortening of the tool life. Premature dulling of a drill may be caused by any one of a number of factors.

- The drill speed may be too high for the hardness of the material being cut.
- The feed may be too heavy and overload the cutting lips.
- The feed may be too light and cause the lips to scrape rather than cut.
- There may be hard spots or scale on the work surface.
- The work or drill may not be supported properly, resulting in springing and chatter.
- The drill point is incorrect for the material being drilled.
- The finish on the lips is poor.

To Grind a Drill

A general-purpose drill has an included point angle of 118° and a lip clearance of from 8 to 12° (Fig. 40-20A and B).

1. Be sure to wear approved safety glasses.
2. Check the grinding wheel and dress it, if necessary, to sharpen and/or straighten the wheel face.
3. Adjust the grinder tool rest so that it is within 1/16 in. (1.5 mm) of the wheel face.
4. Examine the drill point and the margins for wear. If there is any wear on the margins, it will be necessary to grind the point of the drill back until all margin wear has been removed.
5. Hold the drill near the point with one hand, and with the other hand hold the shank of the drill slightly lower than the point (Fig. 40-21).

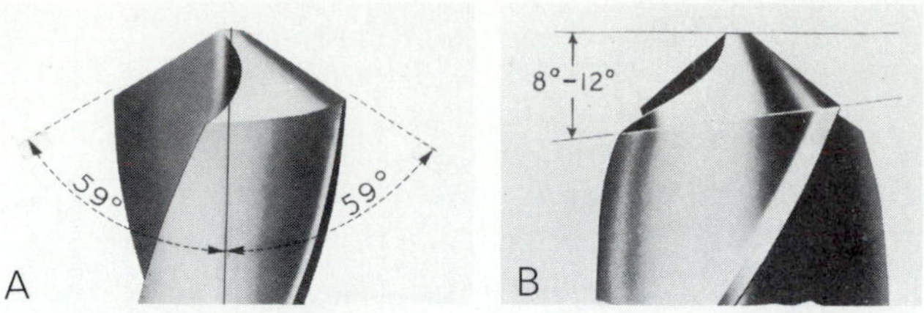
59°
59°
A
8°–12°
B

Figure 40-20 *The point angle for a general-purpose drill is 118°. (Courtesy The Cleveland Twist Drill Company.)*

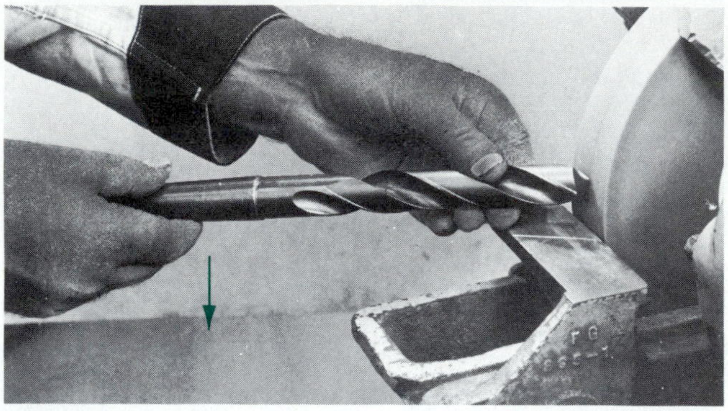

Figure 40-21 *To provide lip clearance, lower the shank of the drill before grinding.*

6. Move the drill so that it is approximately 59° to the face of the grinding wheel (Fig. 40-22).

NOTE: A line scribed on the toolrest at 59° to the wheelface will assist in holding the drill at the proper angle.

7. Hold the lip or cutting edge of the drill parallel to the grinder toolrest.
8. Bring the lip of the drill against the grinding wheel and slowly lower the drill shank. *Do not twist the drill.*
9. Remove the drill from the wheel without moving the position of the body or hands, rotate the drill one-half turn, and grind the other cutting edge.

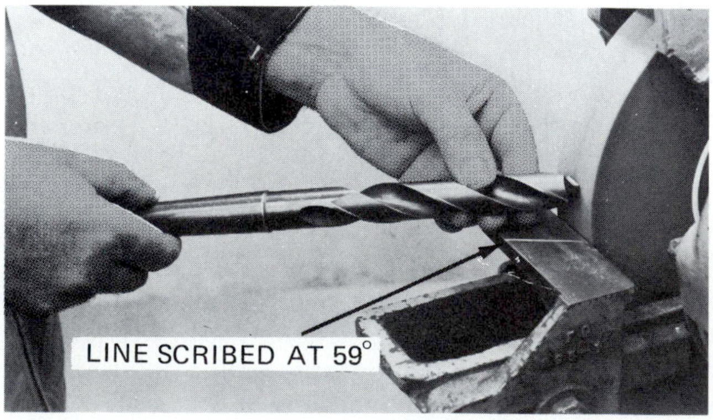
LINE SCRIBED AT 59°

Figure 40-22 *Hold the drill at 59° to the face of the grinding wheel.*

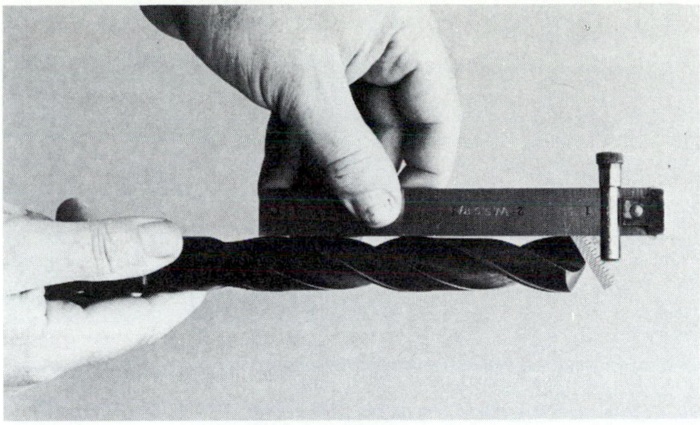

Figure 40-23 *Check the drill point angle with a drill point gage.*

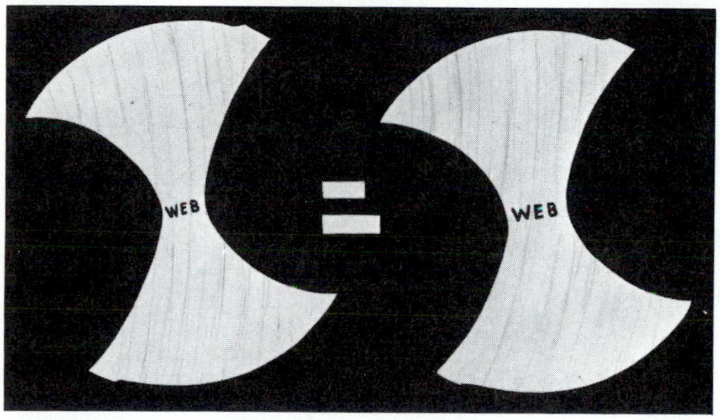
WEB
WEB

Figure 40-24 *Note the difference in web thickness between the drill cross section near the shank (right) and the point (left). (Courtesy The Cleveland Twist Drill Company.)*

10. Check the angle of the drill point and length of the lips with a drill point gage (Fig. 40-23).
11. Repeat operations 6 to 10 until the cutting edges are sharp and the lands are free from wear and nicks.

WEB THINNING

Most drills are manufactured with webs that gradually increase in thickness toward the shank to give the drill strength. As the drill becomes shorter, the web becomes thicker (Fig. 40-24) and more pressure is required in order to cut. This increase in pressure results in more heat, which shortens the drill life. To reduce the amount of drilling pressure and resultant heat, the web of a drill is generally thinned. Webs can be thinned on a special web-thinning grinder, on a tool and cutter grinder, or freehand on a conventional grinder. It is important when thinning a web to grind equal amounts off each edge; otherwise the drill point will be off center (Fig. 40-19A).

REVIEW QUESTIONS

TWIST DRILLS

1. Name three materials used to manufacture drills and state the advantage of any two.
2. Define the *body, web,* and *point* of a twist drill.
3. State the purpose or purposes of each of the following parts.
   a. Tang
   b. Flutes
   c. Margin
   d. Body clearance

DRILL POINT CHARACTERISTICS

4. Why are various drill points and clearances used for drilling operations?
5. Describe and state the purpose of the following drill points.
   a. Conventional
   b. Long angle
   c. Flat angle
6. Why is it necessary to thin the web of a drill?

DRILL SIZES

7. List four systems of drill sizes and give the range of each.
8. State the purpose of the:
   a. High helix drill
   b. Core drill
   c. Oil hole drill
   d. Straight-fluted drill
   e. Gun drill
   f. Hard-steel drill

DRILLING FACTS AND PROBLEMS

9. What problems generally result from the use of excessive speed and excessive feed?
10. Discuss excessive lip clearance and insufficient lip clearance.
11. What is the effect of:
   a. Drills with unequal angles on the cutting lips?
   b. Drills with cutting lips of unequal length?

DRILL GRINDING

12. What are the characteristics of a properly ground drill?
13. List the main steps in grinding a general-purpose drill.

UNIT 41

Cutting Speeds and Feeds

The most important factors which the operator must consider when selecting the proper speeds and feeds are the diameter and material of the cutting tool and the type of material being cut. These factors will determine what speeds and feeds should be used and therefore will affect the amount of time it takes to perform the operation. Production time will be wasted if the speed and/or feed are set too low, while the cutting tool will show premature wear if the speed and/or feed are set too high. The ideal speed and feed for any job is when a combination of the best production rate and the best tool life is attained.

OBJECTIVES

After completing this unit you will be able to:

1. Calculate the revolutions per minute (r/min) for inch and metric size drills
2. Select the feed to be used for various operations
3. Calculate the revolutions per minute for the reaming operation

CUTTING SPEED

The speed of a twist drill is generally referred to as *cutting speed, surface speed,* or *peripheral speed.* It is the distance that a point on the circumference of a drill will travel in 1 min.

A wide range of drills and drill sizes is used to cut various metals; an equally wide range of speeds is required for the drill to cut efficiently. For every job, there is the problem of choosing the drill speed which will result in the best production rates and the least amount of downtime for regrinding the drill. Recommended cutting speeds for drilling various types of materials are shown in Table 41-1. The most economical drilling speed depends upon many variables, such as:

1. The type and hardness of the material
2. The diameter and material of the drill
3. The depth of the hole
4. The type and condition of the drill press
5. The efficiency of the cutting fluid employed
6. The accuracy and quality of the hole required
7. The rigidity of the work setup

Although all these factors are important in the selection of economical drilling speeds, the type of work material and the diameter of the drill are the most important.

When reference is made to the speed at which a drill should revolve, the cutting speed of the material in *surface feet per minute* (sf/min) or *meters per minute* (m/min) is implied unless otherwise stated. The number of revolutions of the drill necessary to attain the proper cutting speed for the metal being machined is called the *revolutions per minute* (r/min). A small drill operating at the same r/min as a larger drill will travel fewer feet per minute; it naturally would cut more efficiently at a higher number of r/min.

REVOLUTIONS PER MINUTE

To determine the correct number of r/min of a drill press spindle for a given size drill, the following should be known.

1. The type of material to be drilled

Table 41-1 Speeds for High-Speed Steel Drills

		Steel Casting		Tool Steel		Cast Iron		Machine Steel		Brass and Aluminum	
Drill size		Cutting Speeds in Feet per Minute or Meters per Minute									
in.	mm	40 ft/min	12 m/min	60 ft/min	18 m/min	80 ft/min	24 m/min	100 ft/min	30 m/min	200 ft/min	60 m/min
1/16	2	2445	1910	3665	2865	4890	3820	6110	4775	12 225	9550
1/8	3	1220	1275	1835	1910	2445	2545	3055	3185	6110	6365
3/16	4	815	955	1220	1430	1630	1910	2035	2385	4075	4775
1/4	5	610	765	915	1145	1220	1530	1530	1910	3055	3820
5/16	6	490	635	735	955	980	1275	1220	1590	2445	3180
3/8	7	405	545	610	820	815	1090	1020	1365	2035	2730
7/16	8	350	475	525	715	700	955	875	1195	1745	2390
1/2	9	305	425	460	635	610	850	765	1060	1530	2120
5/8	10	245	350	365	520	490	695	610	870	1220	1735
3/4	15	205	255	305	380	405	510	510	635	1020	1275
7/8	20	175	190	260	285	350	380	435	475	875	955
1	25	155	150	230	230	305	305	380	380	765	765

Note: There is no direct relationship between the metric and inch drill sizes.

2. The recommended cutting speed of the material
3. The type of material from which the drill is made

Formula (Inch)

$$\text{r/min} = \frac{\text{CS (feet per minute)} \times 12}{\pi D \text{ (drill circumference in inches)}}$$

where CS = the recommended cutting speed in *feet per minute* for the material being drilled.
D = the diameter of the drill being used.

NOTE: The CS would vary, depending on the material from which the drill is made.

Since not all machines can be set to the exact calculated speed, π (3.1416) is divided into 12 to arrive at a simplified formula, which is accurate enough for most drilling operations.

$$\text{r/min} = \frac{\text{CS} \times 4}{D}$$

EXAMPLE

Calculate the r/min required to drill a ½-in. hole in cast iron (CS 80) with a high-speed steel drill.

$$\text{r/min} = \frac{\text{CS} \times 4}{D}$$

$$= \frac{80 \times 4}{1/2}$$

$$= \frac{320}{1/2}$$

$$= 640$$

Formula (Metric)

$$\text{r/min} = \frac{\text{CS (m)}}{\pi D \text{ (mm)}}$$

It is necessary to convert the meters in the numerator to millimeters so that both parts of the equation are in the same unit. To accomplish this, multiply the CS in meters per minute by 1000 to bring it to millimeters per minute.

$$\text{r/min} = \frac{\text{CS} \times 1000}{\pi D}$$

Not all machines have a variable-speed drive and therefore cannot be set to the exact calculated speed. By dividing π (3.1416) into 1000, a simplified formula is derived which is accurate enough for most drilling operations.

$$\text{r/min} = \frac{\text{CS} \times 320}{D}$$

EXAMPLE

Calculate the r/min required to drill a 15-mm hole in tool steel (CS 18) using a high-speed steel drill.

$$\begin{aligned} \text{r/min} &= \frac{\text{CS} \times 320}{D} \\ &= \frac{18 \times 320}{15} \\ &= \frac{5760}{15} \\ &= 384 \end{aligned}$$

FEED

Feed is the distance that a drill advances into the work for each revolution. Drill feeds may be expressed in decimals, fractions of an inch, or millimeters. Since the feed rate is a determining factor in the rate of production and the life of the drill, it should be carefully chosen for each job. The rate of feed is generally governed by:

1. The diameter of the drill
2. The material of the workpiece
3. The condition of the drilling machine

A general rule of thumb is that the feed rate increases as the drill size increases. For example, a ¼-in. (6-mm) drill should have a feed of only 0.002 to 0.004 in. (0.05 to 0.10 mm), while a 1-in. (25-mm) drill will have a feed of 0.010 to 0.025 in. (0.25 to 0.63 mm) per revolution. Too coarse a feed may chip the cutting edges or break the drill. Too light a feed causes a chattering or scraping noise which quickly dulls the cutting edges of the drill.

The drill feeds listed in Table 41-2 are recommended for general-purpose work. When drilling alloy or hard steels, use a somewhat slower feed. Softer metals such as aluminum, brass, or cast iron can usually be drilled with a faster feed. Whenever steel chips coming from the hole turn blue, it is wise to stop the machine and examine the drill. Blue chips indicate too much heat at the cutting edge. This heat is caused by either a dull cutting edge or too high a speed.

CUTTING FLUIDS

Drilling work at the recommended cutting speeds and feeds causes considerable heat to be generated at the drill point. This heat must be dissipated as quickly as possible; otherwise, it will cause a drill to dull rapidly.

The purpose of a *cutting fluid* is to provide both cooling and lubrication. For a liquid to be most effective in dissipating heat, it must be able to absorb heat rapidly, have a good resistance to evaporation, and have a high thermal conductivity. Unfortunately, oil has poor cooling qualities. Water is the best coolant; however, it is rarely used by itself because it promotes rust and has no lubricating value. Basically, a good cutting fluid should:

1. Cool the workpiece and tool
2. Reduce friction
3. Improve the cutting action
4. Protect the work against rusting
5. Provide antiweld properties
6. Wash away the chips

See Unit 34, Table 34-1, for the recommended cutting fluids for various metals.

REVIEW QUESTIONS

DRILLING SPEEDS

1. Why is it important that a drill be operated at the correct speed?
2. Explain the difference between cutting speed and r/min.
3. What factors determine the most economical drilling speed?

Table 41-2 Drill Feeds

Drill Size		Feed per Revolution	
in.	mm	in.	mm
⅛ and smaller	3 and smaller	0.001 to 0.002	0.02 to 0.05
⅛ to ¼	3 to 6	0.002 to 0.004	0.05 to 0.10
¼ to ½	6 to 13	0.004 to 0.007	0.10 to 0.18
½ to 1	13 to 25	0.007 to 0.015	0.18 to 0.38
1 to 1½	25 to 38	0.015 to 0.025	0.38 to 0.63

4. Calculate the r/min which would be required to drill the following holes using a high speed drill:
   a. A 3/8-in.-diameter hole in tool steel
   b. A 1-in.-diameter hole in aluminum
   c. A 9-mm hole in a steel casting
   d. A 20-mm hole in cast iron

CUTTING FLUIDS

5. What is the purpose of a cutting fluid?
6. Name four important qualities that a cutting fluid should have.

UNIT 42

Drilling Holes

Early humans used an arrow wrapped with a bow string to drill holes in bone and wood (Fig. 42-1). Although the technique for drilling holes is much different today, the principle remains the same—pressing a rotating cutting tool into the workpiece. Some of the important factors in drilling are securely fastening the workpiece, observing proper safety precautions, and setting the proper speeds and feeds. The pressure applied to the feed handle and the appearance of the chips are good indicators as to how effective the drilling operation is being performed.

OBJECTIVES

After completing this unit you will be able to:
1. Measure the size of inch and metric drills
2. Drill the correct size center holes in workpieces
3. Drill small and large holes to an accurate location

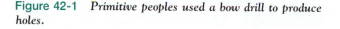

Figure 42-1 *Primitive peoples used a bow drill to produce holes.*

DRILL PRESS SAFETY

Before starting with any drill press operation, it is good practice to observe some basic safety precautions. These precautions will not only ensure your safety, but also will prevent damage to the machine, cutting tool, and the workpiece.

1. *Do not operate any machine before understanding its mechanism and knowing how to stop it quickly. This can prevent serious injury.*
2. Always wear approved safety glasses to protect your eyes.
3. Never attempt to hold the work by hand; a table stop or clamp should be used to prevent the work from spinning (Fig. 42-2).
4. Never set speeds or adjust the work unless the machine is stopped.
5. Keep your head well back from revolving parts of a drill press to prevent your hair from being caught.
6. As the drill begins to break through the work, ease up on the drilling pressure and allow the drill to break through gradually.
7. Always remove burrs from a drilled hole with a file or deburring tool.
8. *Never* leave a chuck key in the drill chuck.
9. *Never* attempt to grab work which may have caught in the drill. *Stop the machine first.*
10. Always keep the floor around a drill press clean and free of tools, chips, and oil (Fig. 42-3). They can cause serious accidents.

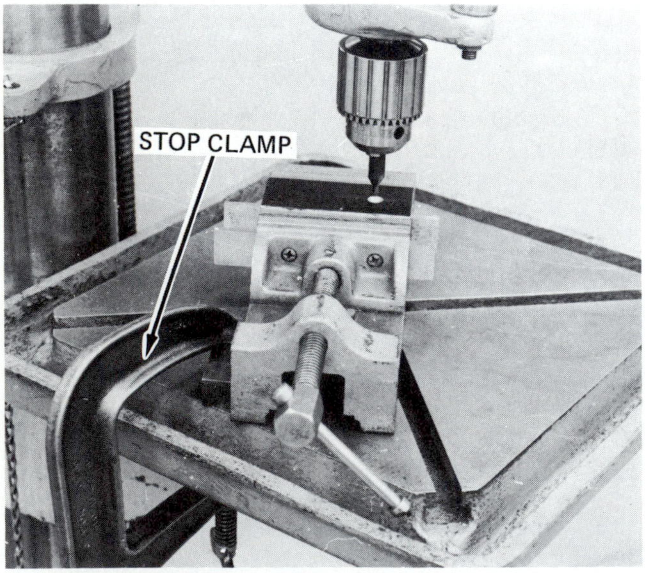
STOP CLAMP

Figure 42-2 *A clamp or table stop should be fastened to the left-hand side of the table.*

Figure 42-3 *Poor housekeeping can lead to accidents.*

DRILLING HINTS

The following hints should help prevent many problems which could affect the accuracy of the hole and the efficiency of the drilling operation.

1. Treat cutting tools with care; they can be damaged through careless use, handling, and storage.
2. Always examine the condition of the drill point before use, and if necessary, resharpen it. *Do not use dull tools.*
3. Make sure that the drill point angle is correct for the type of material to be drilled.
4. Set the correct revolutions per minute (r/min) for the size of the drill and the workpiece material. Too high speed quickly dulls a drill, and too low speed causes a small drill to break.
5. Set up the work so that the drill will not cut into the table, parallels, or drill vise as it breaks through the workpiece.
6. The work should always be clamped securely for the drilling operation. For small-diameter holes, a clamp or stop fastened to the left-hand side of the table will prevent the work from spinning (Fig. 42-2).
7. The end of the workpiece farthest from the hole should be placed on the left-hand side of the table so that if the work catches, it will not swing toward the operator.
8. Always clean a tapered drill shank, sleeve, and the machine spindle before inserting a drill.
9. Use the shortest drill length possible and/or hold it short in the chuck to prevent breakage.
10. It is a good practice to start each hole with a center drill. The small point of the center drill will pick up a center-punch mark accurately; the center-drilled hole will provide a guide for the drill to follow.
11. Thin workpieces, such as sheet metal, should be clamped to a hardwood block for drilling. This prevents the work from catching and also steadies the drill point as it breaks through the workpiece.
12. The chips from each flute should be the same shape; if

the chip turns blue during drilling, check the drill point condition and the speed of the drill press.

13. A drill squeak usually indicates a dull drill. Stop the machine and examine the condition of the drill point; regrind the drill if necessary.

14. When increased pressure must be applied during a drilling operation, the reason is generally a dull drill or a chip caught in the hole between the drill and the work; correct these conditions before proceeding.

MEASURING THE SIZE OF A DRILL

In order to produce a hole to size, the correct size of drill must be used to drill the hole. It is good practice to always check a drill for size before using it to drill a hole. Drills may be checked for size with a drill gage (Fig. 42-4A) or with a micrometer (Fig. 42-4B). Although the size of most drills is stamped on the drill shank, the micrometer is still the most accurate way of measuring the exact size of a drill. When checking a drill for size with a micrometer, always be sure to take the measurement across the margin of the drill.

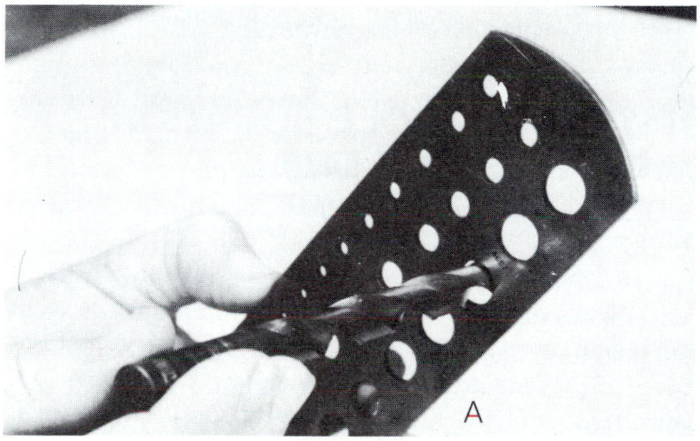
A

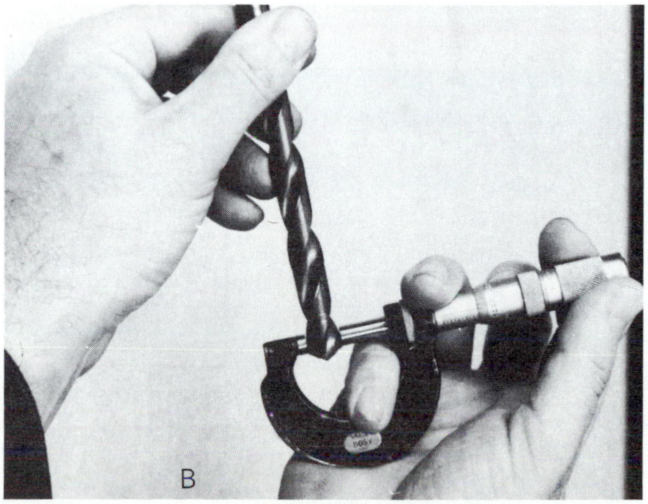
B

Figure 42-4 *(A) Checking a drill for size using a drill gage; (B) checking the size of a twist drill using a micrometer. (Courtesy Kostel Enterprises Limited.)*

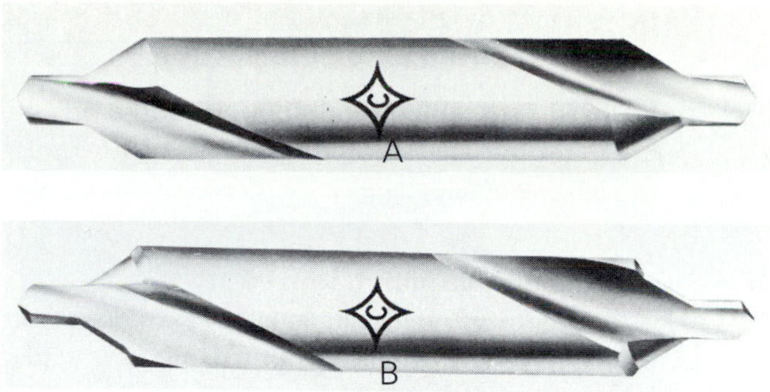
A
B

Figure 42-5 *Two types of center drills: (A) regular type; (B) bell type. (Courtesy Cleveland Twist Drill Company.)*

LATHE CENTER HOLES

Work to be turned between the centers on a lathe must have a hole drilled in each end so that the work may be supported by the lathe centers. A *combination drill and countersink* (Fig. 42-5), more commonly called a *center drill,* is used for this operation.

To ensure an adequate bearing surface for the work on the lathe center, center holes must be drilled to the correct size and depth (Fig. 42-6A).

A center hole which is too shallow is illustrated in Fig. 42-6B. This results in poor support for the work and possible damage to both the lathe center and the work.

Figure 42-6C shows a center hole which has been drilled too deep. The taper on the lathe center cannot contact the taper of the center hole; the result is poor support for the work.

DRILLING LATHE CENTER HOLES

1. Select the proper size of center drill to suit the diameter of the work (see Table 42-1).
2. Fasten the center drill in the drill chuck, having it extend beyond the chuck only about ½ in. (13 mm).
3. Place the work to be center drilled in the drill vise as shown in Fig. 42-7.
4. Set the drill press at the proper speed and start the machine.
5. Locate the center-punch mark in the work directly below the center drill point.
6. Carefully feed the center drill into the center-punch mark in the work for about $^1/_{16}$ in. (1.5 mm).
7. Raise the center drill, apply a few drops of cutting fluid, and continue drilling.
8. Frequently remove the drill from the hole to apply cutting fluid, remove the chips, and measure the diameter of the top of the center hole.
9. Continue drilling until the top of the hole is the proper size.

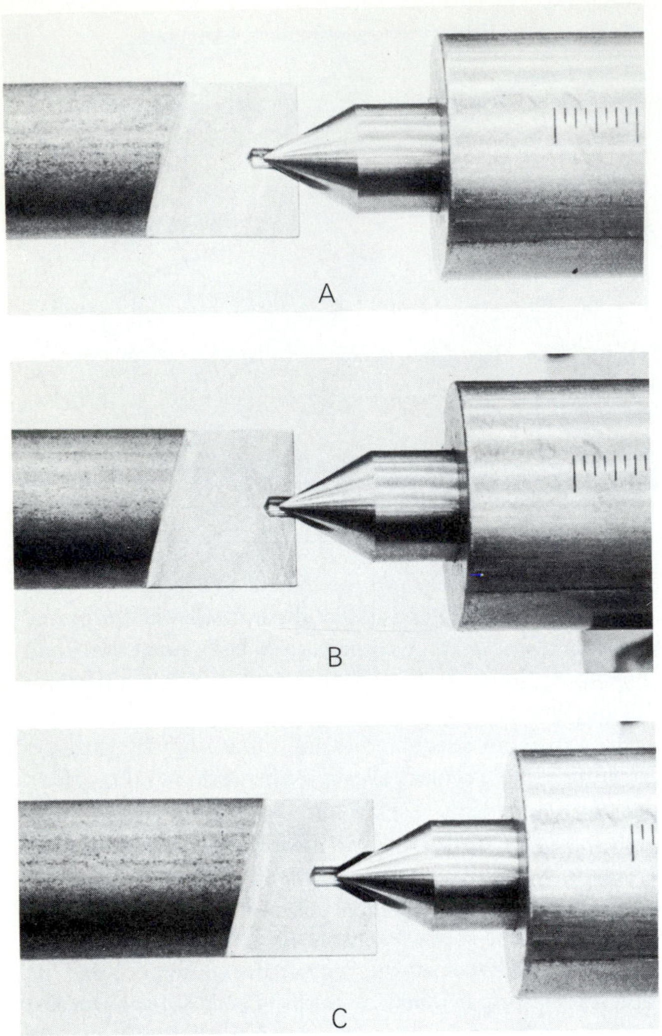
A
B
C

Figure 42-6 *(A) A center hole drilled to the proper depth; (B) a center hole drilled too shallow; (C) a center hole drilled too deep.*

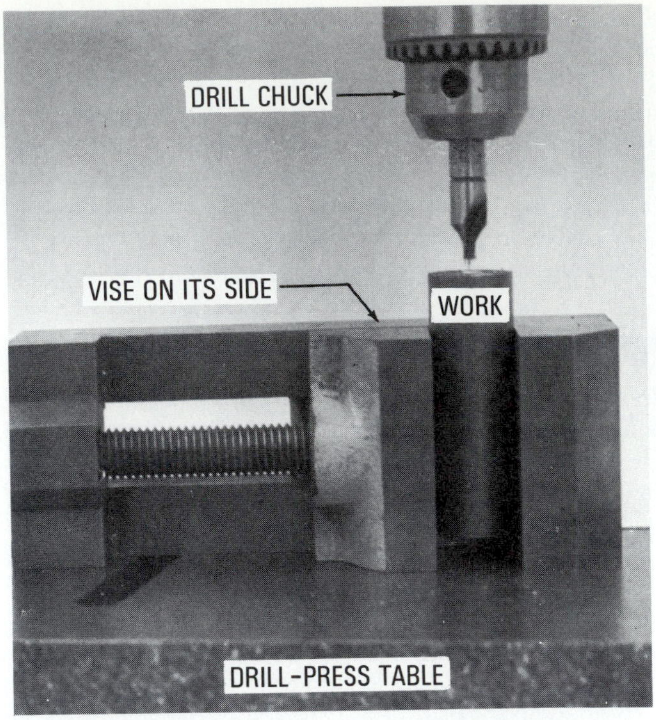
DRILL CHUCK
VISE ON ITS SIDE
WORK
DRILL-PRESS TABLE

Figure 42-7 *Work set up for center-hole drilling.*

SPOTTING A HOLE LOCATION WITH A CENTER DRILL

The chisel edge at the end of the web on most drills is generally wider than the center-punch mark on the work and therefore it is difficult to start a drill at the exact location. To prevent a drill from wandering off center, it is considered good practice to first spot every center-punch mark with a center drill. The small point on the center drill

Table 42-1 Center Drill Sizes

Size		Work Diameter		Diameter of Countersink	Drill Point Diameter	Body Size
Regular Type	Bell Type	in.	mm	in.	in.	in.
1	11	3/16–5/16	3–8	3/32	3/64	1/8
2	12	3/8–1/2	9.5–12.5	9/64	5/64	3/16
3	13	5/8–3/4	15–20	3/16	7/64	1/4
4	14	1–1½	25–40	15/64	1/8	5/16
5	15	2–3	50–75	21/64	3/16	7/16
6	16	3–4	75–100	3/8	7/32	1/2
7	17	4–5	100–125	15/32	1/4	5/8
8	18	6 and over	150 and over	9/16	5/16	3/4

SPOTTED
HOLE
CENTER DRILL

Figure 42-8 *Spotting a hole location with a center drill.*

will accurately follow the center-punch mark and provide a guide for the larger drill which will be used.

1. Mount a small-size center drill in the drill chuck.
2. Mount the work in a vise or set it on the drill press table. *Do not clamp the work or the vise.*
3. Set the drill press speed to about 1500 r/min.
4. Bring the point of the center drill into the center-punch mark and allow the work to center itself with the drill point.
5. Continue drilling until about one-third of the tapered section of the center drill has entered the work (Fig. 42-8).
6. Spot all the holes which are to be drilled.

DRILLING WORK HELD IN A VISE

The most common method of holding small workpieces is by means of a vise, which may be held by hand against a table stop or clamped to the table. When drilling holes larger than ½ in. (13 mm) in diameter, the vise should be clamped to the table.

1. Spot the hole location with a center drill.
2. Mount the correct size drill in the drill chuck.
3. Set the drill press to the proper speed for the size of drill and the type of material to be drilled.
4. Fasten a clamp or stop on the left side of the table (Fig. 42-2).
5. Mount the work on parallels in a drill vise and tighten it securely.
6. With the vise against the table stop, locate the spotted hole under the center of the drill.
7. Start the drill press spindle and begin to drill the hole.
   a. For holes up to ½ in. (13 mm) in diameter, hold the vise against the table *or* stop by hand (Fig. 42-2).
   b. For holes over ½ in. (13 mm) in diameter:
      - Lightly clamp the vise to the table with a clamp.
      - Drill until the full drill point is into the work.
      - With the drill revolving, keep the drill point in the work and tighten the clamp holding the vise securely (Fig. 42-9).
8. Raise the drill occasionally and apply cutting fluid during the drilling operation.
9. Ease up on the drilling pressure as the drill starts to break through the workpiece.

DRILLING TO AN ACCURATE LAYOUT

If a hole must be drilled to an exact location, the position of the hole must be accurately laid out as shown in Fig. 42-10A. During the drilling operation it may be necessary to draw the drill point over so that it is concentric with the layout (Fig. 42-10B).

1. Clean and coat the surface of the work with layout dye.
2. Locate the position of the hole from *two machined edges* of the workpiece and scribe the lines as shown in Fig. 42-10A.
3. Lightly prick-punch where the two lines intersect.
4. Check the accuracy of the punch mark with a magnifying glass and correct if necessary.
5. With a pair of dividers, scribe a circle to indicate the diameter of the hole required (Fig. 42-10B).

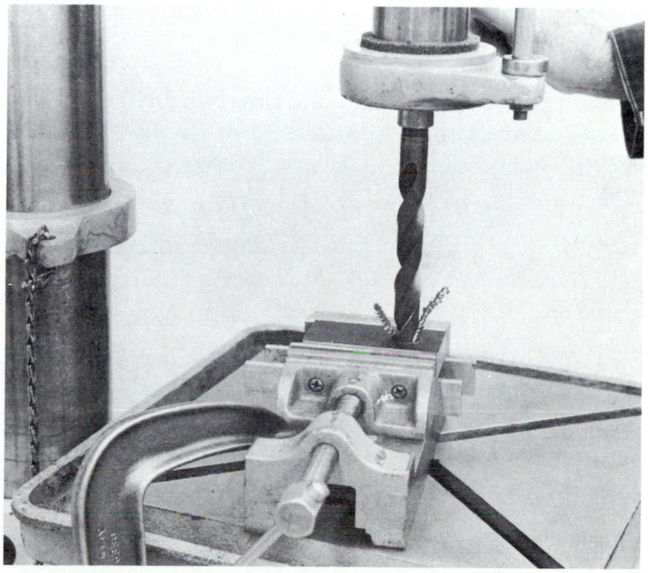

Figure 42-9 *Tighten the table clamp while the drill is revolving in the hole. (Courtesy Kostel Enterprises Limited.)*

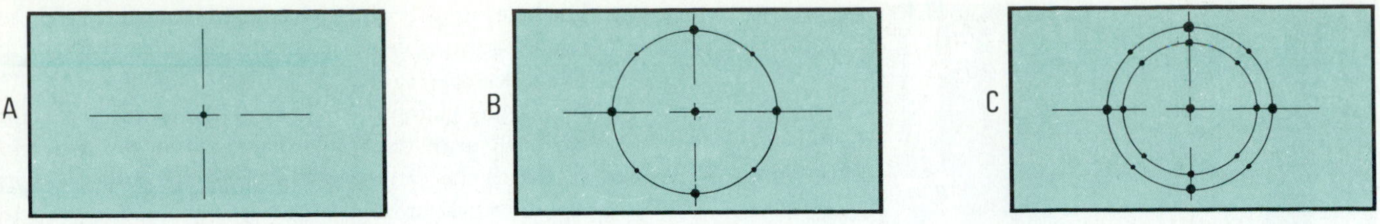
A
B
C

Figure 42-10 *Layout for a hole to be drilled.*

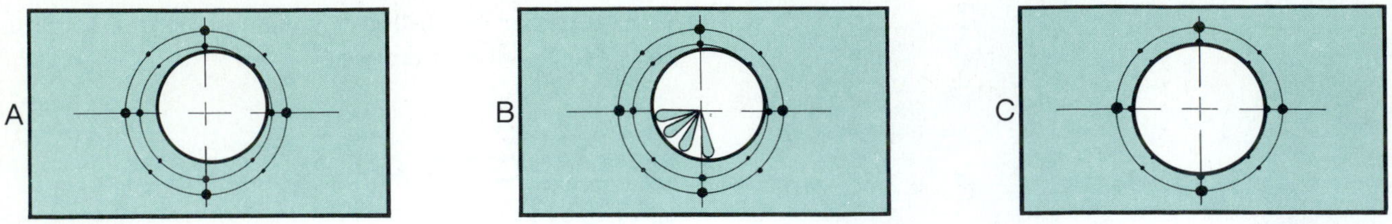
A
B
C

Figure 42-11 *Drawing the drill point to the layout.*

6. Scribe a test circle 1⁄16 in. (1.5 mm) smaller than the hole size.
7. Punch four witness marks on circles up to 3⁄4 in. (19 mm) in diameter, and eight witness marks on larger circles (Fig. 42-10C).
8. Deepen the center of the hole location with a center punch to provide a larger indentation for the drill to follow.
9. Center drill the work to *just beyond the depth of the drill point.*
10. Mount the proper size drill in the machine and drill a hole to a depth equal to one-half to two-thirds of the drill diameter.
11. Examine the drill indentation; it should be concentric with the inner proof circle (Fig. 42-11A).
12. If the spotting is off center, cut shallow V-grooves with a cape or diamond-point chisel on the side toward which the drill must be moved (Fig. 42-11B).
13. Start the drill in the spotted and grooved hole. The drill will be drawn toward the direction of the grooves.
14. Continue cutting grooves into the spotted hole until the drill point is drawn to the center of the scribed circles as shown in Fig. 42-11C.

NOTE: *The drill point must be drawn to the center of the scribed circle before the drill has cut or spotted to the drill's diameter.*

15. Continue to drill the hole to the desired depth.

DRILLING LARGE HOLES

As drills increase in size, the thickness of the web also increases to give the drill added strength. The thicker the web, the thicker will be the point or chisel edge of the drill. As the chisel edge becomes larger, poorer cutting action results and more pressure must be applied for the drilling operation. A thick web will not follow the center-punch mark accurately and the hole may not be drilled in the proper location. Two methods are generally employed to overcome the poor cutting action of a thick web on large drills.

1. The web is thinned.
2. A lead, or pilot, hole is drilled.

The usual procedure for drilling large holes is that first a pilot, or lead, hole (Fig. 42-12), the diameter of which is slightly larger than the thickness of the web, is drilled. *Care must be taken to drill the pilot hole on center.* The pilot hole is then followed with a larger drill. This method may also be used to drill average size holes when the drill press is small and does not have sufficient power to drive the drill through the solid metal.

Never drill a pilot hole any bigger than necessary; otherwise, the larger drill may:

1. Cause chattering
2. Drill the hole out-of-round
3. Spoil the top (mouth) of the hole

The following drilling procedure is recommended.

1. Check the print and select the proper drill for the hole required.
2. Measure the thickness of the web at the point. Select a pilot drill with a diameter slightly larger than the web thickness (Fig. 42-13).
3. Mount the workpiece on the table.
4. Adjust the height and position of the table so that the drill chuck can be removed and the larger drill placed in the spindle without having to lower the table after the pilot hole is drilled. Lock the table securely in this position.

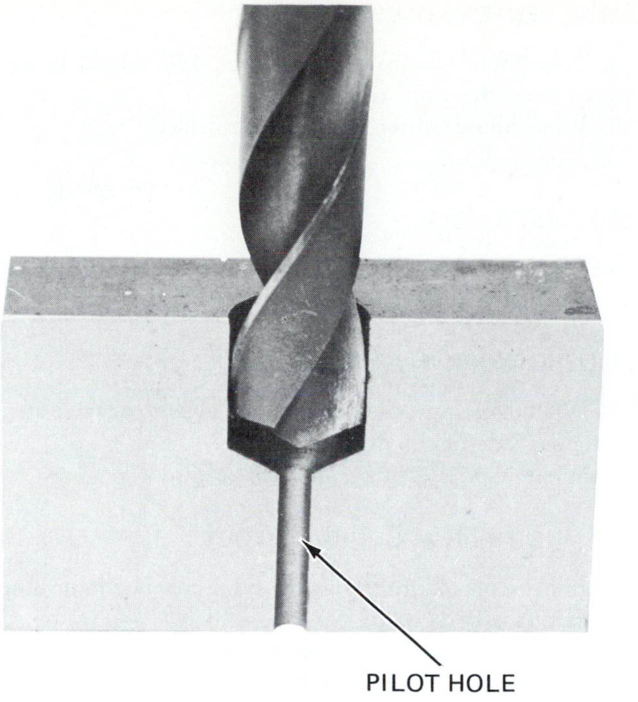
PILOT HOLE

Figure 42-12 *Drilling a pilot, or lead, hole helps the larger drill cut easily and accurately.*

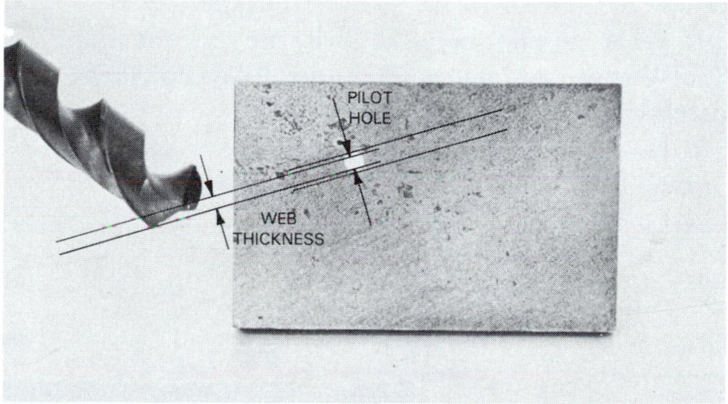
PILOT HOLE
WEB THICKNESS

Figure 42-13 *The size of the pilot drill should be slightly larger than the thickness of the drill web.*

5. Place a center drill in the drill chuck, set the proper spindle speed, and accurately drill a center hole.

NOTE: The center drill should be used first since it is short, rigid, and more likely to follow the center-punch mark.

6. Using the proper size pilot drill and correct spindle speed, drill the pilot hole to the required depth (Fig. 42-12). The work may be lightly clamped at this time.
7. Shut off the machine, leaving the pilot drill in the hole.
8. Clamp the work securely to the table.
9. Raise the drill spindle and remove the drill and drill chuck.
10. Clean the taper shank of the drill and the drill press spindle hole. Remove any burrs on the drill shank with an oilstone.
11. Mount the large drill in the spindle.
12. Set the proper spindle speed and feed and drill the hole to the required depth. If hand feed is used, the feed pressure should be eased as the drill breaks through the work.

DRILLING ROUND WORK IN A V-BLOCK

V-blocks may be used to hold round work for drilling. Round work is seated in the accurately machined V-groove. Small diameters may be held in place with a U-shaped clamp, and larger diameters are fastened with strap clamps.

1. Select a V-block to suit the diameter of round work to be drilled. If the work is long, use a pair of V-blocks.
2. Mount the work in the V-block and then rotate it until the center-punch mark is in the center of the workpiece. With a rule and square check that the distance from both sides is equal (Fig. 42-14).
3. Tighten the U-clamp securely on the work in the V-block *or* hold the work and V-block in a vise as shown in Fig. 42-15.
4. Spot the hole location with a center drill.
5. Mount the proper drill size and set the machine to the correct speed.
6. Drill the hole, being sure that the drill does not hit the V-block or vise when it breaks through the work.

PUNCH MARK

Figure 42-14 *Using a square and a rule to align the center punch mark on a round workpiece. (Courtesy Kostel Enterprises Limited.)*

Figure 42-15 *Turn the workpiece until the line on the end of the work is in line with the blade of the square. (Courtesy Kostel Enterprises Limited.)*

REVIEW QUESTIONS

SAFETY

1. Select three of the most important safety suggestions and explain why they should be observed.

DRILLING HINTS

2. What result might be expected from:
   a. Too high drill speed? b. Too low drill speed?
3. Why is it good practice to start each hole with a center drill?

LATHE CENTER HOLES

4. State three reasons why a center drill should be removed from the work frequently.
5. Why should center holes not be drilled:
   a. Too shallow? b. Too deep?

SPOTTING A HOLE

6. What is the purpose of spotting a hole before drilling?
7. How deep should each hole be spotted?

DRILLING WORK IN A VISE

8. What is the purpose of fastening a clamp or table stop to the left side of the drill press table?
9. When should drilling pressure be eased?

DRILLING TO AN ACCURATE LAYOUT

10. List the procedure for laying out a hole before drilling to a layout.
11. How deep should the hole be drilled before the drill point indentation is examined?
12. Explain how a drill may be drawn over to a layout.

DRILLING LARGE HOLES

13. List three disadvantages of a thick web found on large drills.
14. What are pilot holes and why are they necessary?
15. Why should pilot holes not be drilled any larger than necessary?
16. How high should the drill press table be adjusted when drilling large holes?

UNIT 43

Reaming

Each component in a product must be made to exact standards in order for that product to function properly. Since it is impossible to produce holes which are round, smooth, and accurate to size by drilling, the reaming operation is very important. Reamers are used to enlarge and finish a hole previously formed by drilling or boring. Speed, feed, and reaming allowances are the three main factors which will affect the accuracy and finish of the hole and the life of the reamer.

OBJECTIVES

After completing this unit you will be able to:

1. Identify and state the purpose of hand reamers and machine reamers
2. Explain the advantages of carbide-tipped reamers
3. Calculate the reaming allowance required for each reamer
4. Ream a hole by hand in a drill press
5. Machine ream a hole

REAMERS

A reamer is a rotary cutting tool with several straight or helical cutting edges along its body. It is used to accurately size and smooth a hole which has been previously drilled or bored. Some reamers are operated by hand (hand reamers), while others may be used under power in any type of machine tool (machine reamers).

REAMER PARTS

Reamers generally consist of three main parts: *shank, body,* and *angle of chamfer* (Fig. 43-1).

The *shank,* which may be straight or tapered, is used to drive the reamer. The shank of machine reamers may be round or tapered, while hand reamers have a square on the end to accommodate a tap wrench.

The *body* of a reamer contains several straight or helical grooves, or flutes, and lands (the portion between the flutes). A margin (the top of each tooth) runs from the *angle of chamfer* to the end of the flute. The *body clearance angle* is a relief or clearance behind the margin which reduces friction while the reamer is cutting. The *rake angle* is the angle formed by the face of the tooth when a line is drawn from a point on the front marginal edge through the center of the reamer (Fig. 43-1). If there is no angle on the face of the tooth, the reamer is said to have *radial land.*

The *angle of chamfer* is the part of the reamer which actually does the cutting. It is ground on the end of each tooth and there is clearance behind each chamfered cutting edge. On rose reamers, the angle of chamfer is ground on the end only and the cutting action occurs at this point. On fluted reamers, each tooth is relieved and most of the cutting is done by the reamer teeth.

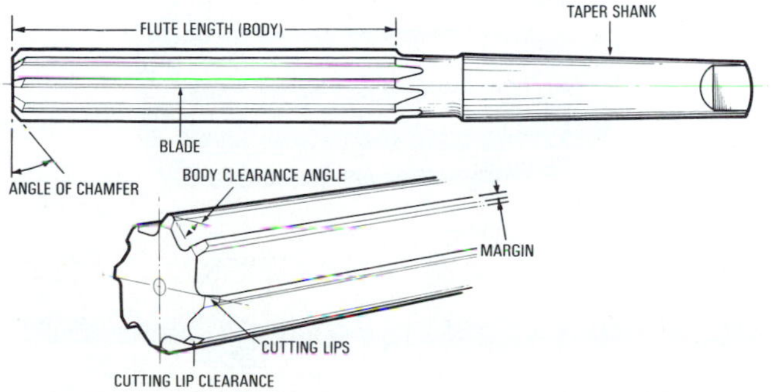
FLUTE LENGTH (BODY)
TAPER SHANK
BLADE
ANGLE OF CHAMFER
BODY CLEARANCE ANGLE
MARGIN
CUTTING LIPS
CUTTING LIP CLEARANCE

Figure 43-1 *The main parts of a reamer.*

TYPES OF REAMERS

Reamers are available in a variety of designs and sizes; however, they all fall into two general classifications: *hand* and *machine* reamers.

Hand Reamers

Hand reamers (Fig. 43-2) are finishing tools used when a hole must be finished to a high degree of accuracy and finish. Holes to be hand reamed should be bored to within 0.003 to 0.005 in. (0.07 to 0.12 mm) of the finish size. *Never* attempt to ream more than 0.005 in (0.12 mm) with a hand reamer.

A square on the shank end allows a wrench to be used for turning the reamer into the hole. The teeth on the end of the reamer are tapered slightly for a distance equal to the reamer diameter so that it can enter the hole to be reamed.

A hand reamer should never be used under mechanical power and should never be turned backwards. When using a hand reamer, keep it true and straight with the hole. The dead center in a lathe or a stub center in a drill press will

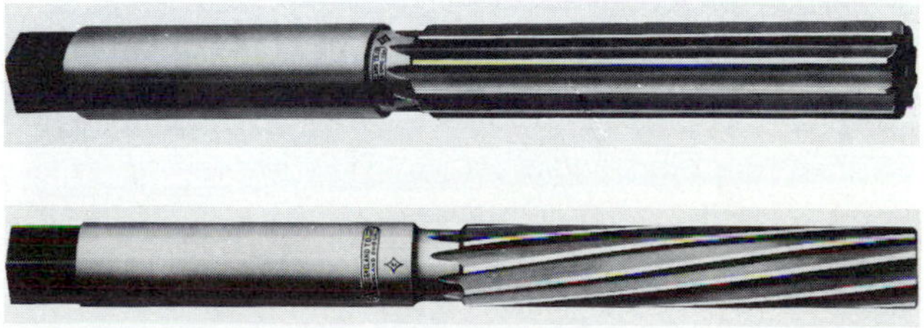

Figure 43-2 *Straight and helical-fluted hand reamers. (Courtesy The Cleveland Twist Drill Company.)*

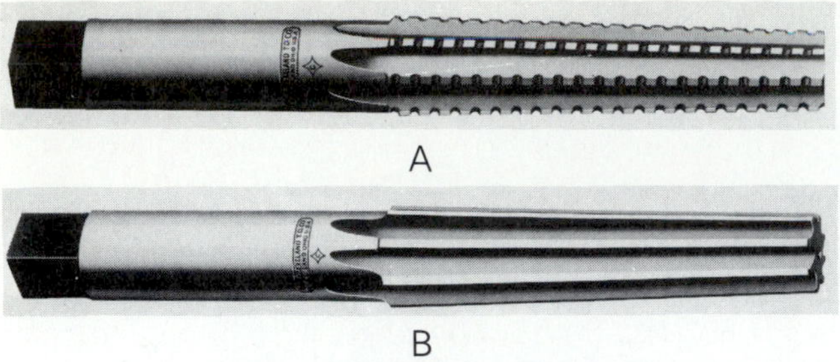
A
B

Figure 43-3 *Taper hand reamers: (A) roughing; (B) finishing. (Courtesy The Cleveland Twist Drill Company.)*

help keep the reamer aligned during the hand-reamer operation.

Taper hand reamers (Fig. 43-3), both roughing and finishing, are available for all standard size tapers. Because chips do not fall out readily, a taper reamer should be removed from the hole and the flutes cleaned frequently.

Machine Reamers

Machine reamers may be used in any machine tool for both roughing and finishing a hole. They are also called *chucking reamers* because of the method used to hold them for the reaming operation. Machine reamers are available in a wide variety of types and styles. Only some of the more common types will be discussed.

Rose reamers (Fig. 43-4A) can be purchased with straight or tapered shanks and with straight or helical flutes. The teeth on the end have a 45° chamfer which is backed off to produce the cutting edge. The lands are nearly as wide as the flutes and are not backed off. Rose reamers cut on the end only and can be used to remove material quickly and bring the hole fairly close to the size required. Rose reamers are usually made 0.003 to 0.005 in. (0.07 to 0.12 mm) under the normal size.

Fluted reamers (Fig. 43-4B) have more teeth than rose reamers for a comparable diameter. The lands are relieved for the entire length, and fluted reamers therefore cut along the side as well as at the chamfer on the end. These reamers are considered finishing tools and are used to bring a hole to size.

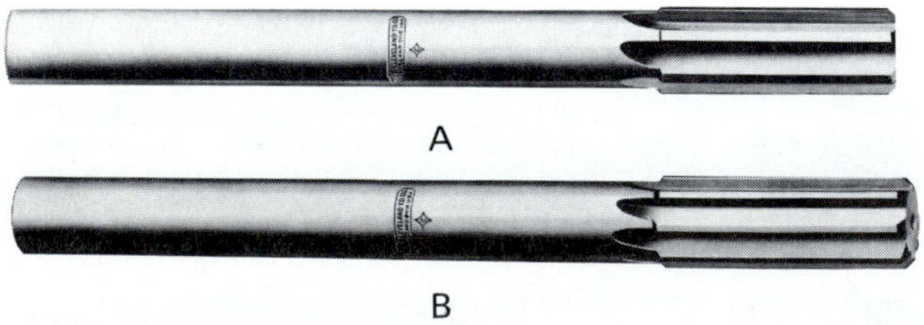
A
B

Figure 43-4 *(A) A rose reamer cuts on the end angle only; (B) the fluted reamer has more teeth than the rose reamer and cuts on the sides and end. (Courtesy The Cleveland Twist Drill Company.)*

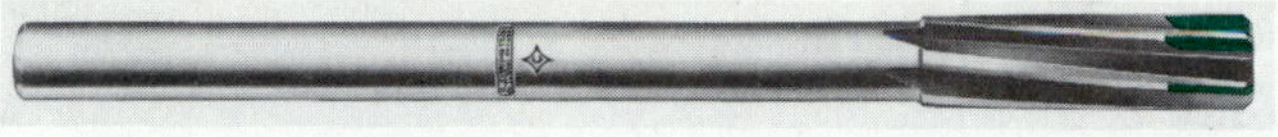

Figure 43-5 *A carbide-tipped reamer.*

Carbide-tipped reamers (Fig. 43-5) were developed to meet the ever-increasing demand for high production rates. They are similar to rose or fluted reamers, except that carbide tips have been brazed to their cutting edges. Because of the hardness of the carbide tips, these reamers resist abrasion and maintain sharp cutting edges even at high temperatures. Carbide-tipped reamers outlast high-speed steel reamers, especially on castings where hard scale or sand is a problem. Because carbide-tipped reamers can be run at higher speeds and still maintain their size, they are used extensively for long production runs.

Shell reamers (Fig. 43-6) are reamer heads mounted on a driving arbor. The shank of the driving arbor may be straight or tapered, depending on the size and type of shell reamer used. Two slots in the end of this reamer fit into lugs on the driving arbor. Sometimes a locking screw in the end of the arbor holds the shell reamer in place. The advantages of shell reamers are:

1. They are economical for larger holes.
2. Various head sizes can be easily interchanged on one arbor.
3. When a reamer becomes worn, it may be thrown away, and the driving arbor can be used with other reamers.

Adjustable reamers (Fig. 43-7) have inserted blades which can be adjusted approximately at 0.015 in. (0.38 mm) over or under the nominal reamer size. The threaded body has a series of tapered grooves cut lengthwise into which blades are fitted. Adjusting nuts on either end can be used to increase or decrease the diameter of the reamer. Hand- or machine-adjustable reamers can be readily sharpened and are available with either high-speed steel or carbide inserts.

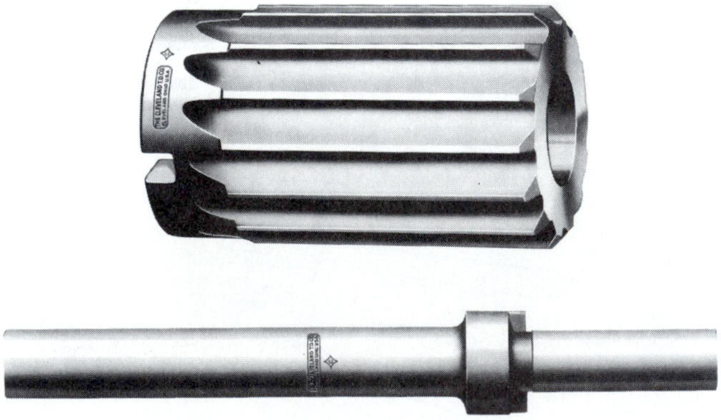

Figure 43-6 *Shell reamers are economical for reaming large holes. (Courtesy The Cleveland Twist Drill Company.)*

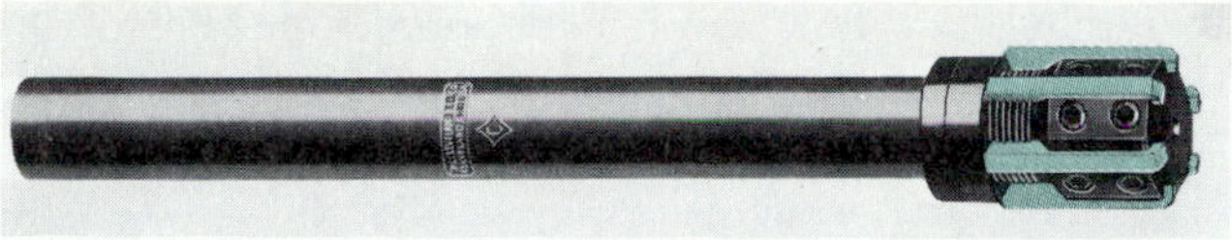

Figure 43-7 *An adjustable reamer with inserted blades. (Courtesy The Cleveland Twist Drill Company.)*

Figure 43-8 *An expansion reamer can be expanded slightly. (Courtesy The Cleveland Twist Drill Company.)*

Expansion reamers (Fig. 43-8) are similar to adjustable reamers; however, the amount they can be expanded is very limited. The body of this reamer is slotted, and a tapered, threaded plug is fitted into the end. Turning this plug will allow a 1-in. (25-mm) reamer to expand up to 0.005 in. (0.12 mm). Expansion reamers are not meant to be oversize reamers, but to give longer life to finishing reamers.

Emergency reamers, drills whose corners (at the lip and land) have been slightly rounded and honed, may be used with fairly good results if a reamer of a particular size is not available. First drill the hole as close as possible to the required size. Then run the reaming drill at a fairly high speed and feed it into the hole slowly.

REAMER CARE

The accuracy and surface finish of a hole, as well as the life of a reamer, depend greatly on the care a reamer receives. It is wise to remember that a reamer is a finishing tool and should be handled carefully.

1. Never turn a reamer backward; this will ruin the cutting edges.
2. Always store reamers in separate containers to prevent the cutting edges from being nicked or burred. Plastic or cardboard tubes make excellent reamer containers.
3. Never roll or drop reamers on metal surfaces, such as bench tops, machines, and plates.
4. When not in use, a reamer should be oiled, especially on the cutting edges, to prevent rusting.
5. A fine, free-cutting grinding wheel should be used for resharpening reamers. Burring of the cutting edges destroys the life of the reamer, while a rough cutting edge produces a rough hole and the reamer dulls quickly.

REAMING ALLOWANCES

The amount of material left in a hole for the reaming operation depends on a number of factors. If a hole has been punched, rough-drilled, or bored, it requires more metal for reaming than a hole which has already been reamed with a roughing reamer. The type of machining operation prior to reaming must be considered as well as the material being reamed.

General rules for the amount of material which should be left in a hole for machine reaming are:

1. For holes up to ½ in. diameter, allow 1/64 in. for reaming.
2. For holes over ½ in. diameter, allow 1/32 in. for reaming.

NOTE: Never leave more than 0.005 in. in a hole for hand reamers up to ½ in. diameter. On larger holes, a proportional allowance should be left to make a good finish possible.

For metric size reamers allow 0.10 mm for holes up to 12 mm in diameter. For holes over 12 mm, allow 0.20 to 0.78 mm for reaming. See Table 43-1 for the recommended allowances for various size holes.

REAMING SPEEDS AND FEEDS

SPEEDS

The selection of the most efficient speed for machine reaming depends on the following factors.

1. The type of material being reamed
2. The rigidity of the setup
3. The tolerance and finish required in the hole

Generally, reaming speeds should be from one-half to two-thirds the speed used for drilling the same material.

Higher reaming speeds can be used when the setup is

Table 43-1 Recommended Stock Allowances for Reaming

Hole Size		Allowance	
in.	mm	in.	mm
¼	6.35	0.010	0.25
½	12.7	0.015	0.38
¾	19.05	0.018	0.45
1	25.4	0.020	0.50
1¼	31.75	0.022	0.55
1½	38.1	0.025	0.63
2	50.8	0.030	0.76
3	76.2	0.045	1.14

rigid; slower speeds should be used when the setup is less rigid. A hole requiring close tolerances and a fine finish should be reamed at slower speeds. The use of coolants improves the surface finish and allows higher speeds to be used.

Reamers do not work well when they chatter; the speed selected should always be low enough to eliminate chatter.

Table 43-2 gives the recommended reaming speeds for high-speed steel reamers. Carbide reamers may be operated at higher speeds.

FEEDS

The feed used for reaming is usually two to three times greater than that used for drilling. The feed rate will vary with the material being reamed; however, it should be approximately 0.001 to 0.004 in. (0.02 to 0.10 mm) per flute per revolution. Feeds which are too low generally result in glazing, excessive reamer wear, and, sometimes, chatter. Too high a feed tends to reduce the hole accuracy and sometimes results in poor surface finish. Generally, feeds should be the highest possible which will still produce the hole accuracy and finish required.

An exception to these feed rates occurs when tapered holes are being reamed. As tapered reamers cut along their entire length, a light feed is necessary. The reamer should be removed occasionally and the flutes cleaned.

REAMING HINTS

1. Examine a reamer and remove all burrs from the cutting edges with a hone so that good surface finishes will be produced.
2. Cutting fluid should be used in the reaming operation to improve the hole finish and prolong the life of the reamer.
3. Helical-fluted reamers should always be used when long holes and those with keyways or oil grooves are reamed.
4. Straight-fluted reamers are generally used when extreme accuracy is required.
5. To obtain hole accuracy and good surface finish, use a roughing reamer first and then a finishing reamer. An old reamer which is slightly undersize may be used as a roughing reamer.
6. *Never, under any circumstances,* turn a reamer backward.
7. *Never* attempt to start a reamer on an uneven surface; the reamer will go toward the point of least resistance and will not produce a straight, round hole.
8. If chatter occurs, stop the machine, reduce the speed, and increase the feed. To overcome the chatter marks, it may be necessary to restart the reamer slowly by pulling the drill press belt by hand.
9. To avoid chatter, select a reamer with an incremental cut (unequally spaced teeth).
10. When hand reaming in a drill press, always use a stub center in the drill press spindle to keep the reamer aligned.

Table 43-2 Recommended Reaming Speeds for High-Speed Steel Reamers

Material	Speed	
	ft/min	m/min
Aluminum	130–200	39–60
Brass	130–180	39–55
Bronze	50–100	15–30
Cast iron	50–80	15–24
Machine steel	50–70	15–21
Steel alloys	30–40	9–12
Stainless steel	40–50	12–15
Magnesium	170–270	52–82

REAMING A STRAIGHT HOLE

Two types of reamers are used in machine shop work—hand reamers and machine reamers. Hand reamers have a square on one end and are used to remove no more than 0.005 in. (0.12 mm) from a hole. Machine reamers may have straight shanks which are held and driven by a drill chuck, or tapered shank which fit directly into the drill press spindle. They are generally used to remove from 1/64 in. (0.40 mm) to 1/32 in. (0.80 mm) of metal from a hole, depending upon the hole diameter.

HAND REAMING A STRAIGHT HOLE

1. Mount the work on parallels in a vise and clamp it securely to the table (Fig. 43-9).
2. Drill the hole to the proper size leaving an allowance for the hand reamer to be used.

NOTE: Reaming allowance should be no more than 0.005 in. (0.12 mm) for a 1-in. (25-mm) diameter reamer.

3. Do not move the location of the work or the table; remove the drill and mount a stub center in the drill chuck (Fig. 43-10).
4. Start the end of the reamer in the drilled hole.
5. Fasten a tap wrench on the reamer.

Figure 43-9 *A workpiece clamped to the table correctly.*

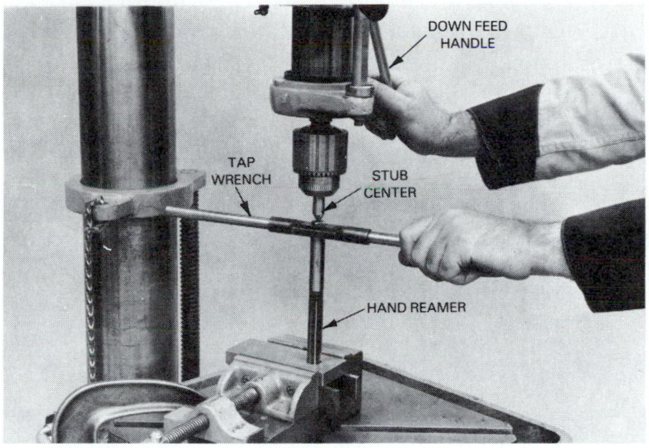
DOWN FEED
HANDLE
TAP
WRENCH
STUB
CENTER
HAND REAMER

Figure 43-10 *Using a stub center to keep the reamer aligned.*

6. Engage the stub center in the center hole on the end of the reamer.
7. With the downfeed lever, apply slight pressure while turning the reamer clockwise by hand.
8. Apply cutting fluid and ream the hole.
9. When removing the reamer, turn it clockwise, never counterclockwise.

MACHINE REAMING A STRAIGHT HOLE

A hole which must be finished to size should be reamed immediately after it has been drilled and while the hole is still aligned with the drill press spindle. This will ensure that the reamer follows the same location as the drill.

1. Mount the work on parallels in a vise and fasten it securely to the table.
2. Select the proper size drill for the reaming allowance required, and drill the hole.

NOTE: *Do not move the work or drill press table at this time.*

3. Mount the proper reamer in the drill press.

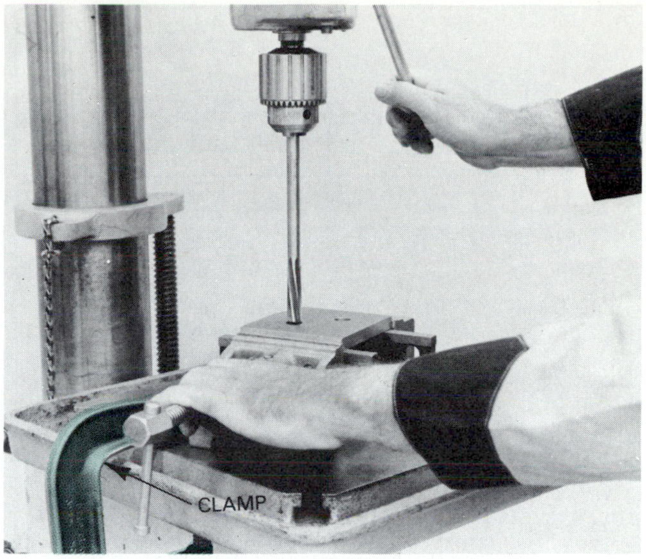
CLAMP

Figure 43-11 *Reaming a hole in the drill press.*

4. Adjust the spindle speed to suit the reamer and the work material.
5. Start the drill press and carefully lower the spindle until the chamfer on the reamer starts to cut (Fig. 43-11).
6. Apply cutting fluid and feed the reamer by applying enough pressure to keep the reamer cutting.
7. Remove the reamer from the hole by raising the downfeed handle.
8. Shut off the machine and remove the burr from the edge of the hole.

REAMING A TAPERED HOLE

When reaming a tapered hole in a workpiece, step-drill the hole prior to reaming. The recommended drills to be used can be determined by referring to Fig. 43-12.

1. Mount the work on parallels in a vise or on the drill table and fasten it securely.
2. Align the hole center under the drill point.
3. Drill a hole $\frac{1}{64}$ in. (0.40 mm) smaller than the smallest diameter of the tapered hole (Fig. 43-12).
4. Obtain a drill $\frac{1}{64}$ in. (0.40 mm) smaller than the size of the finished hole measured at a point *B* (one-half the length of the tapered section).
5. Drill the hole $\frac{1}{16}$ in. (1.5 mm) less than the *C* dimension (Fig. 43-12).
6. Mount a roughing tapered reamer in the drill spindle.
7. Adjust the spindle speed to about one-half the speed used to ream a straight hole.
8. Rough-ream the hole to about 0.005 in. (0.12 mm) undersize while applying cutting fluid.

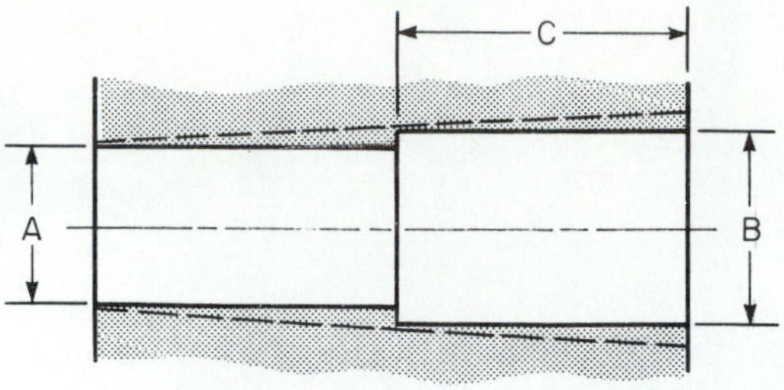
C
A
B

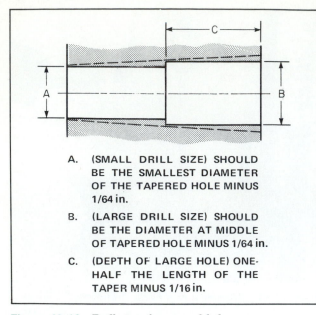

A. (SMALL DRILL SIZE) SHOULD BE THE SMALLEST DIAMETER OF THE TAPERED HOLE MINUS 1/64 in.

B. (LARGE DRILL SIZE) SHOULD BE THE DIAMETER AT MIDDLE OF TAPERED HOLE MINUS 1/64 in.

C. (DEPTH OF LARGE HOLE) ONE-HALF THE LENGTH OF THE TAPER MINUS 1/16 in.

Figure 43-12 *Drill sizes for tapered holes.*

9. Mount the finish reamer, apply cutting fluid, and finish-ream the hole to size.

REVIEW QUESTIONS

REAMERS

1. What is the purpose of a reamer?
2. Define the following reamer parts:
   a. Body
   b. Angle of chamfer
   c. Body clearance angle
   d. Radial land
3. How may a hand reamer be recognized, and for what purpose is it used?
4. Compare a rose reamer and a fluted reamer.
5. State the advantages of carbide-tipped reamers.
6. Describe briefly the following types of reamers and state their purpose.
   a. Shell
   b. Adjustable

REAMER CARE

7. List three important points which should be observed in the care of reamers.

REAMING ALLOWANCES

8. What is the general rule for the amount of material left in a hole for machine reaming?
9. How much should be left in the following holes for machine reaming?
   a. 1¼ in.
   b. 19.05 mm

REAMING HINTS

10. List seven of the most important reaming hints.

REAMING A STRAIGHT HOLE AND A TAPERED HOLE

11. How much material should be left in a hole for hand reaming?
12. How can the reamer be kept in alignment when hand reaming on a drill press?
13. Why should the work or drill press table not be moved before a hole is reamed?
14. Make a neat sketch to show the procedure for drilling in preparation for reaming a tapered hole.

UNIT 44

Drill Press Operations

The drill press is a versatile machine tool and can be used to perform a variety of operations other than the drilling of holes. The variety of cutting and finishing tools available allow operations such as tapping, counterboring, countersinking, boring, spot-facing, and buffing to be performed on a drill press.

OBJECTIVES

After completing this unit you should be able to:

1. Counterbore and countersink holes
2. Select and use the proper tap to thread a hole in a drill press
3. Use three methods to transfer hole locations
4. Finish metal surfaces by buffing

COUNTERBORING

Counterboring is the operation of enlarging the end of a hole which has been drilled previously. A hole is generally counterbored to a depth slightly greater than the head of the bolt, cap screw, or pin which it is to accommodate.

Counterbores (Fig. 44-1) are supplied in a variety of styles, each having a pilot in the end to keep the tool in line with the hole being counterbored. Some counterbores are available with interchangeable pilots to suit a variety of hole sizes.

TO COUNTERBORE A HOLE

1. Set up and fasten the work securely.
2. Drill the proper size of hole in the workpiece to suit the body of the pin or screw.
3. Mount the correct size of counterbore in the drill press (Fig. 44-2).
4. Set the drill press speed to approximately one-quarter that used for drilling.

Figure 44-1 *A set of counterbores.*

Figure 44-2 *A counterbore is used to enlarge the end of a hole.*

5. Bring the counterbore close to the work and see that the pilot turns freely in the drilled hole.
6. Start the machine, apply cutting fluid, and counterbore to the required depth.

COUNTERSINKING

Countersinking is the process of enlarging the top end of a hole to the shape of a cone to accommodate the conical-shaped heads of fasteners so that the head will be flush with or below the surface of the part. These cutting tools, called *countersinks* (Fig. 44-3), are available with various included angles, such as 60, 82, 90, 100, 110, and 120°.

An 82° countersink is used to enlarge the top of a hole so that it will accommodate a flat-head machine screw (Fig. 44-4). The hole is countersunk until the head of the machine screw is flush with or slightly below the top of the work surface (Fig. 44-4). All holes that are to be threaded should be countersunk slightly larger than the tap diameter to protect the start of the thread.

Figure 44-3 *Countersinking produces a tapered hole to fit a flat-head machine screw. (Courtesy Kostel Enterprises Limited.)*

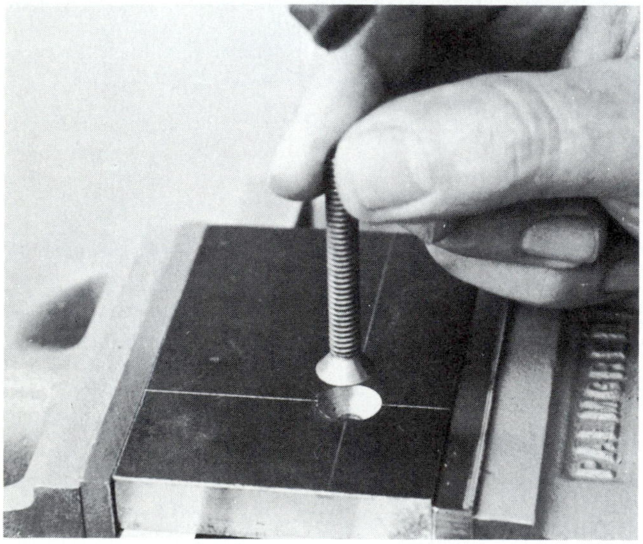

Figure 44-4 *Countersink until the top of the hole is slightly larger than the diameter of the screw head.*

The speed recommended for countersinking is approximately one-quarter of the drilling speed.

TO COUNTERSINK A HOLE FOR A MACHINE SCREW

1. Mount an 82° countersink in the drill chuck.
2. Adjust the spindle speed to about one-half that used for drilling.
3. Place the workpiece on the drill table.
4. With the spindle stopped, lower the countersink into the hole. Clamp the work if necessary. If a pilot-type countersink is used, the pilot should be a slip fit in the drilled hole. The pilot will center the cutting tool and the work.
5. Raise the countersink slightly, start the machine, and feed the countersink by hand until the proper depth is reached. The diameter may be checked by placing an inverted screw in the countersunk hole (Fig. 44-4).
6. If several holes are to be countersunk, set the depth stop so that all of the holes will be the same depth. The gage should be set when the spindle is stationary and the countersink is in the hole.
7. Countersink all of the holes to the depth set on the gage.

TAPPING

Tapping in a drill press may be performed either by hand or under power with the use of a tapping attachment. The advantage of using a drill press for tapping a hole is that the tap can be started squarely and maintained that way through the entire length of the hole.

Machine tapping involves the use of a tapping attachment mounted in the drill press spindle. The tapping operation should be done immediately after the drilling operation to obtain the best accuracy and to avoid duplication of the setup. This sequence is especially important when tapping by hand.

The most common taps used for tapping holes in a drill press are hand taps and machine taps. Hand taps (Fig. 44-5) are available in sets containing the taper, plug, and bottoming taps. When using hand taps in a drill press, it is important that the tap be guided by holding it in a drill chuck or supporting it with a stub center and turning it by hand.

Machine taps are designed to withstand the torque re-

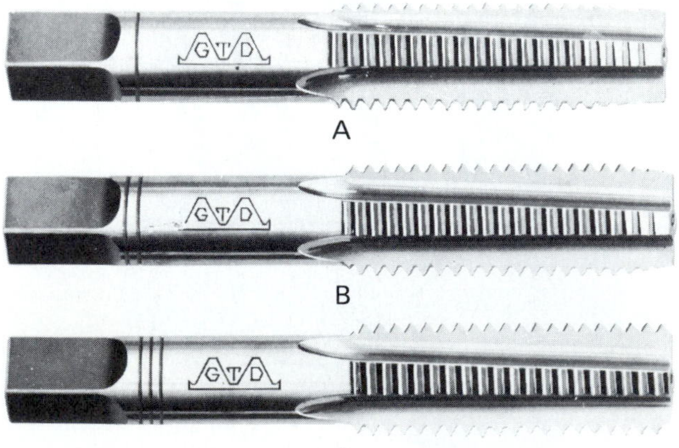
A
B
C

Figure 44-5 *A set of hand taps: (A) taper; (B) plug; (C) bottoming. (Courtesy Greenfield Tap and Die Co.)*

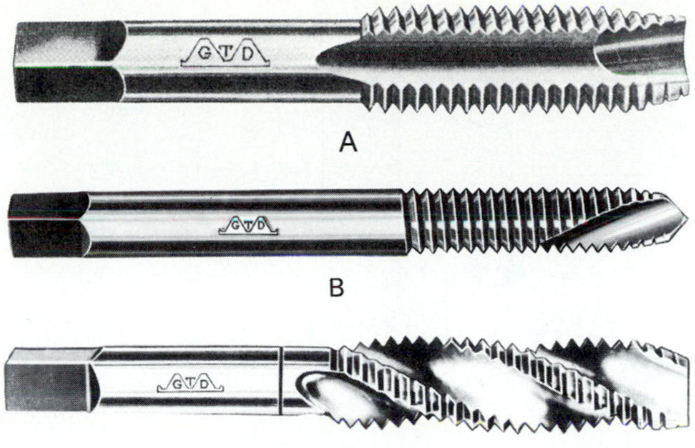
A
B
C

Figure 44-6 *Types of machine taps: (A) gun; (B) stub flute; (C) spiral flute. (Courtesy Greenfield Tap and Die Co.)*

quired to thread the hole and to clear the chips out of the hole quickly. The most common machine taps are the gun taps, stub-flute taps, and spiral-fluted taps (Fig. 44-6).

The *fluteless tap* (Fig 44-7) is actually a forming tool used to produce internal threads in ductile materials such as copper, brass, aluminum, and leaded steels.

TO TAP A HOLE BY HAND IN A DRILL PRESS

1. Mount the work on parallels with the center-punch mark on the work in line with the spindle, and clamp the work securely to the drill press table.
2. Adjust the drill press table height so that the drill may be removed after the hole has been drilled without moving the table or work.
3. Center drill the hole location.
4. Drill the hole to the correct *tap drill size* for the tap to be used.

NOTE: The work or table *must not* be moved after the drilling.

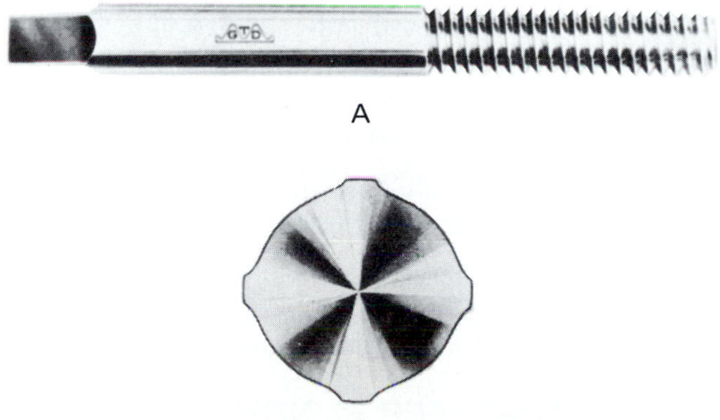
A
B

Figure 44-7 *(A) A fluteless tap; (B) lobes of the tap. (Courtesy Greenfield Tap and Die Co.)*

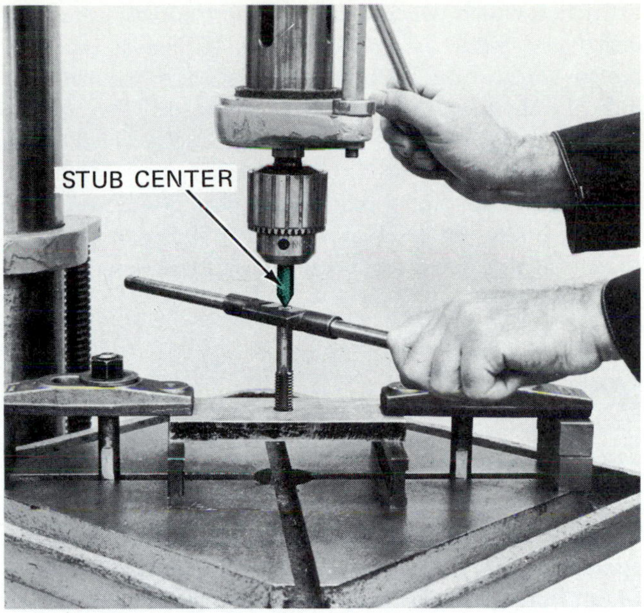
STUB CENTER

Figure 44-8 *Guilding a tap into the workpiece using a stub center held in a drill chuck.*

5. Mount a stub center in the drill chuck (Fig. 44-8),

OR

Remove the drill chuck and mount a special center in the drill press spindle.
6. Fasten a suitable tap wrench on the end of the tap.
7. Place the tap in the drilled hole, and lower the drill press spindle until the center fits into the center hole in the tap shank.
8. Turn the tap wrench clockwise to start the tap into the hole, and at the same time keep the center in light contact with the tap.
9. Continue to tap the hole in the usual manner; keep the tap aligned by applying light pressure on the drill press downfeed lever.

A *tapping attachment* (Fig. 44-9) may be mounted in a drill press spindle to rotate the tap by power. It has a built-

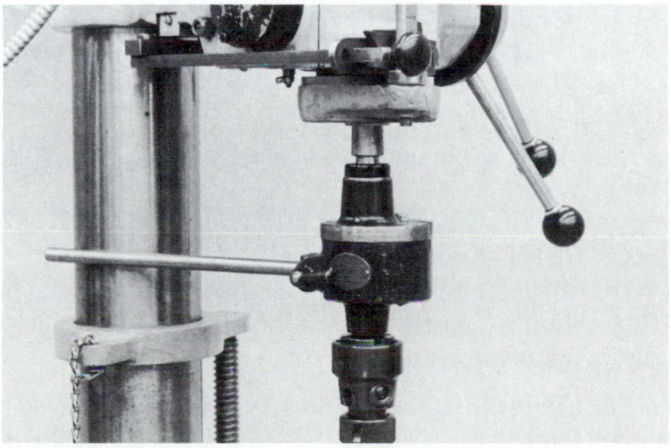

Figure 44-9 *A tapping attachment held in a drill press.*

in friction clutch which drives the tap clockwise when the drill press spindle is fed downward. If there is excessive pressure against the tap because it is stuck or jammed in a hole, the clutch will slip before the tap breaks. The tapping attachment has a reversing mechanism, engaged by the drill press spindle being raised, to back a tap out of the hole.

Two- or three-fluted machine or gun taps are used for tapping under power because of their ability to clear the chips. Tapping speed for most materials ranges from 60 to 100 r/min.

BORING

Boring is the operation of enlarging a drilled or cored hole by means of a single- or double-edged cutting tool held in a boring bar. The hole produced should be concentric, parallel, and perpendicular to the work surface.

Although boring is generally not done on a drill press, sometimes, because of the nature of the workpiece, it may be necessary to use this machine for a boring operation.

Many large drill presses have a hole in the center of the table into which a bushing can be inserted. A pilot on the end of the boring bar fits into the table bushing and keeps the boring bar rigid and aligned during the machining operation.

TO BORE IN A DRILL PRESS

1. Mount as large a boring bar as possible into the drill press spindle.
2. Swing the drill press table to allow the boring bar pilot to enter the table bushing hole.
3. Lock the drill press table in position.
4. Mount the work on parallels to allow the chips to clear the hole and also to prevent cutting into the table (Fig. 44-10).
5. Centralize the hole to be machined to the boring bar, and clamp the work securely in position.
6. Set the drill press to the correct r/min and feed for the material being cut. Roughing feeds may be as high as 0.015 in. (0.38 mm), while finishing feeds are approximately 0.001 to 0.005 in. (0.02 to 0.12 mm).
7. Set the toolbit and make a light trial cut approximately 1/4 in. (6 mm) deep.
8. Stop the machine and measure the size of the hole with inside calipers or telescoping gages.
9. Reset the cutting tool for the depth of cut desired.
10. Continue to bore the hole to the required size by setting the cutting tool for each cut.

NOTE: In order to produce an accurate hole with a good surface finish, keep the drill press spindle sleeve adjusted snugly, use a sharp cutting tool, and apply cutting fluid.

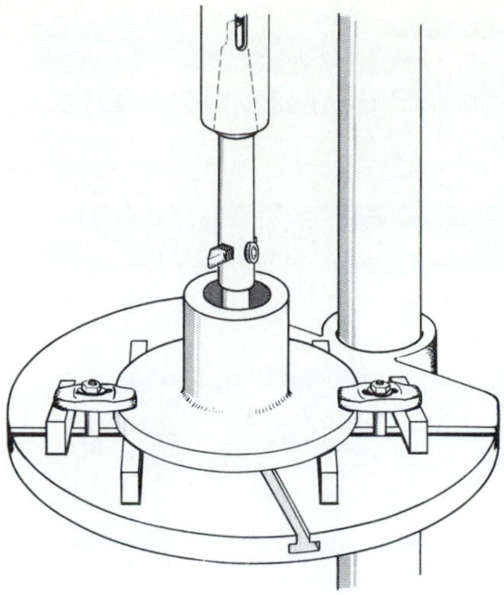

Figure 44-10 *Workpiece and drill press setup for boring a hole.*

TRANSFERRING HOLE LOCATIONS

During the construction of dies, jigs, fixtures, and machine parts, it is often necessary to transfer the location of holes accurately from one part to another. Three common methods of transferring hole locations are:

1. Spotting with a twist drill
2. Using transfer punches
3. Using transfer screws

Regardless of which method is used, the holes in the existing part are used as a master or guide to transfer the hole locations to another part.

TO SPOT WITH A TWIST DRILL

1. Remove the burrs from the mating surfaces on both parts.
2. Align both parts accurately and clamp them together.
3. Mount a drill, the same diameter as the hole to be transferred, in the drill press spindle.
4. Start the drill into the hole of the guide part and spot-drill the second part (Fig 44-11).

NOTE: Never spot-drill deeper than the diameter of the drill.

5. Spot-drill all the holes that are to be transferred.
6. Remove the original part.
7. Drill the spotted holes to the required diameter.

TO USE TRANSFER PUNCHES

1. Remove the burrs from the mating surfaces on both parts.

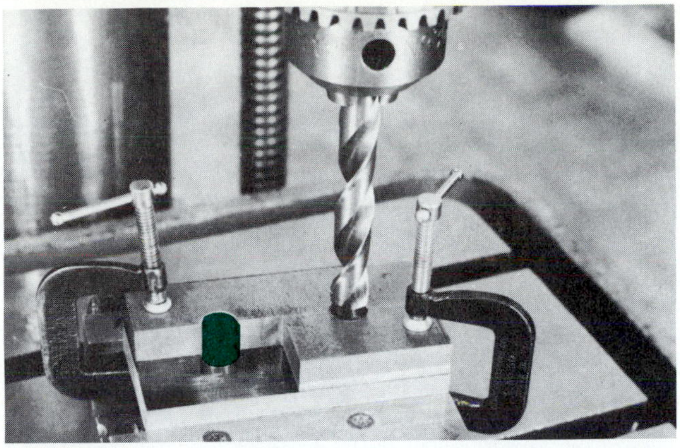

Figure 44-11 *Transferring hole locations by spotting with a drill.*

2. Align both parts accurately and clamp them together.
3. Secure a transfer punch (Fig. 44-12) of the same diameter as the hole to be transferred.
4. Place the punch in the hole and *lightly* strike it with a hammer to mark the hole location.
5. Use the correct size of transfer punch on all the holes that are to be transferred.
6. Remove the original part.
7. Use a divider to lay out proof circles for the holes to be drilled.
8. With a center punch, deepen the existing transfer-punch marks.
9. Use the method outlined in "Drilling to a Layout" (Unit 42) to drill the holes to location accurately.

TO USE TRANSFER SCREWS

Many times it is necessary to transfer the location of threaded holes. This may be easily accomplished by the use of transfer screws (Fig. 44-13) which have been hardened and sharpened to a point. Two flats are ground on the point

Figure 44-12 *Transferring hole locations using a transfer punch.*

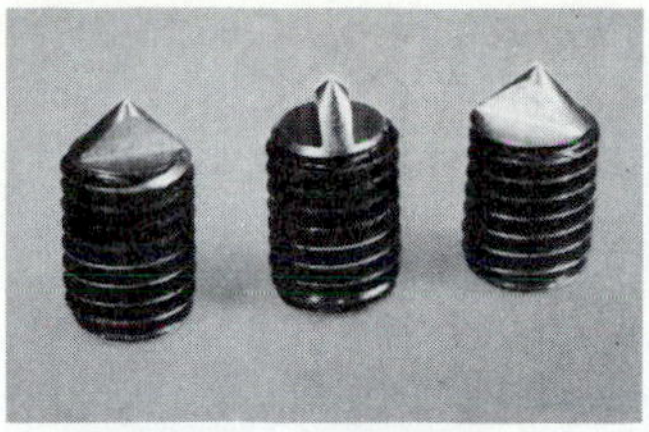

Figure 44-13 *Transfer screws used to transfer threaded hole locations.*

to allow the screws to be threaded into a hole with a small wrench or a pair of needle-nose pliers.

1. Remove all burrs from the mating surfaces.
2. Thread transfer screws into the holes to be transferred, allowing the points to extend beyond the work surface approximately 1/32 in. (0.80 mm).
3. Align both parts accurately and then sharply strike one part with a hammer.
4. Remove the original part, and deepen the marks left by the transfer screws with a center punch.
5. Drill all holes to the required size.

BUFFING

Buffing is the operation of producing a highly polished finish on a round or flat surface. To remove the machine marks on the surface of a workpiece, first use coarse and then fine abrasive cloth. This surface should then be brought into contact with a buffing wheel which has been coated with jeweler's rouge or polishing compound.

BUFFING THE SURFACE OF A WORKPIECE

1. Polish the surface with fine abrasive cloth until all scratches and machine marks are removed (Fig. 44-14).
2. Place a buffing wheel on an arbor and tighten it in a drill chuck.
3. Set the drill spindle to its highest speed.
4. Start the machine and hold the polishing compound against the edge of the revolving wheel until the periphery is coated.

NOTE: Do not apply too much compound as the excess compound will be thrown off the buffing wheel and wasted.

5. Grip the work firmly in both hands and press it against the edge of the buffing wheel (Fig. 44-15).

CAUTION: Do not hold the workpiece with a cloth as it may get caught in the wheel and cause injury to the operator or damage the workpiece, or both.

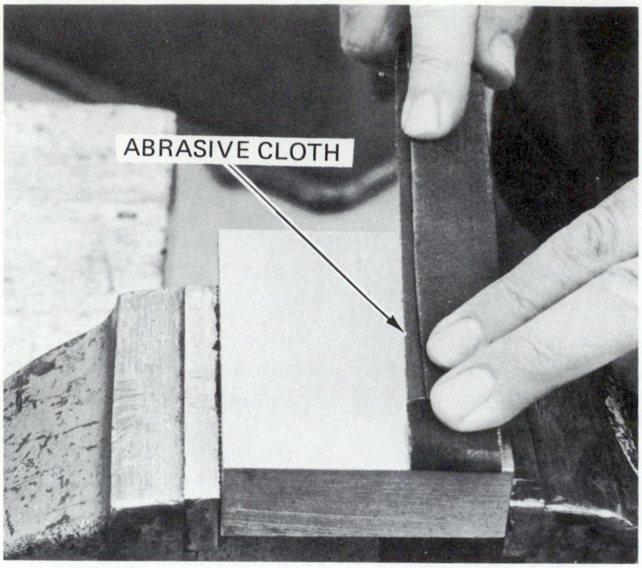
ABRASIVE CLOTH

Figure 44-14 *A workpiece being polished with fine abrasive cloth.*

Figure 44-15 *Buffing improves surface appearance. (Courtesy Kostel Enterprises Limited.)*

6. Slowly turn the surface to be polished while keeping it in contact with the rotating buffing wheel.
7. Continue buffing the work surface until a glossy finish is produced.
8. With a clean, dry, soft cloth remove any excess compound from the surface of the workpiece.

DRILL JIGS

A drill jig is used whenever it is necessary to drill holes to an exact location in a large number of identical parts. Drill jigs are used to save layout time, to avoid incorrectly located holes, and to produce holes accurately and economically. The advantages of using a drill jig are:

1. Since it is not necessary to lay out the hole locations, layout time is eliminated.
2. Each part is quickly and accurately aligned.
3. The part is held in position by a clamping mechanism.
4. The drill jig bushings provide a guide for the drill.
5. The hole locations in each part will be exactly the same; therefore the parts produced are interchangeable.
6. Unskilled labor can be used.

A drill jig (Fig 44-16) is designed so that the part to be drilled may be fastened into it and drilled immediately. Hardened drill jig bushings, used to guide and keep the drill positioned, are located in the drill jig wherever holes must be drilled. When two or more different sizes of holes are to be drilled in the same part, it is preferable to have a gang or multispindle drill press set up. A different size of drill is mounted in each spindle, and the drill jig is passed from one spindle to the next for each hole.

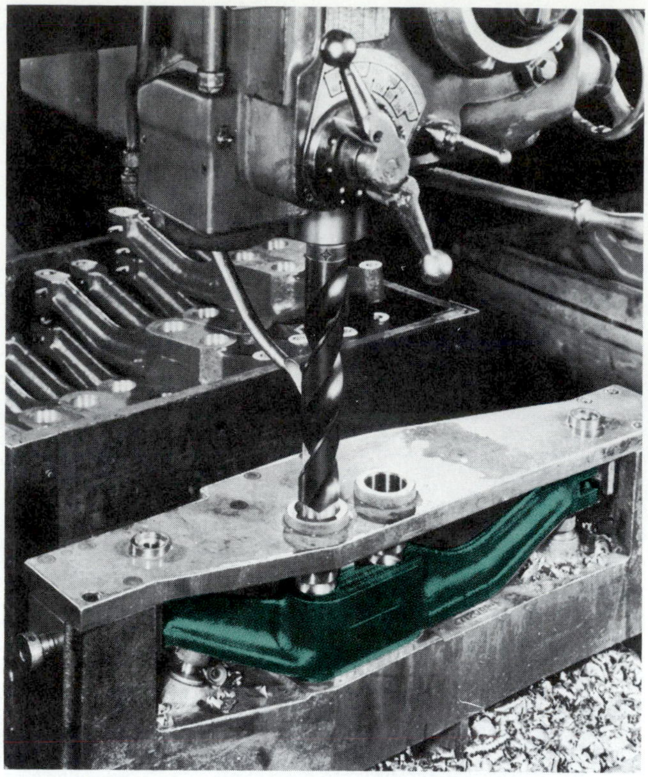

Figure 44-16 *A drill jig positions any number of identical parts so that a hole or holes may be drilled to an accurate location. (Courtesy The Cleveland Twist Drill Company.)*

REVIEW QUESTIONS

COUNTERBORING AND COUNTERSINKING

1. List the procedures for counterboring a hole.
2. Why should holes that are to be tapped be countersunk?

TAPPING

3. What is the advantage of tapping a hole by hand in a drill press?
4. Describe the procedure for tapping a hole by hand in a drill press.
5. Explain how a tapping attachment operates.

BORING

6. Explain how a drill press may be set up for boring a hole.
7. What requirements are necessary for an accurate hole with a good surface finish to be produced?

TRANSFERRING HOLE LOCATIONS

8. Name three methods of transferring the location of holes from one part to another.
9. Explain the procedure for spotting with a twist drill.
10. What are transfer punches and how are they used?
11. Describe transfer screws and explain how they are used.

SECTION 11

The Lathe

Historically, the lathe is the forerunner of all machine tools. The first application of the lathe principle was probably the potter's wheel. This machine rotated a mass of clay which enabled the clay to be formed into a cylindrical shape.

The modern lathe operates on the same basic principle. The work is held and rotated on its axis while the cutting tool is advanced along the lines of a desired cut (Fig. 1). The lathe is one of the most versatile machine tools used in industry. With suitable attachments, the lathe may be used for turning, tapering, form turning, screw cutting, facing,

STANDARD-MODERN
13 INCH SERIES 2000

Figure 1 *The main purpose of a lathe is to machine round work. (Courtesy Standard-Modern Tool Company.)*

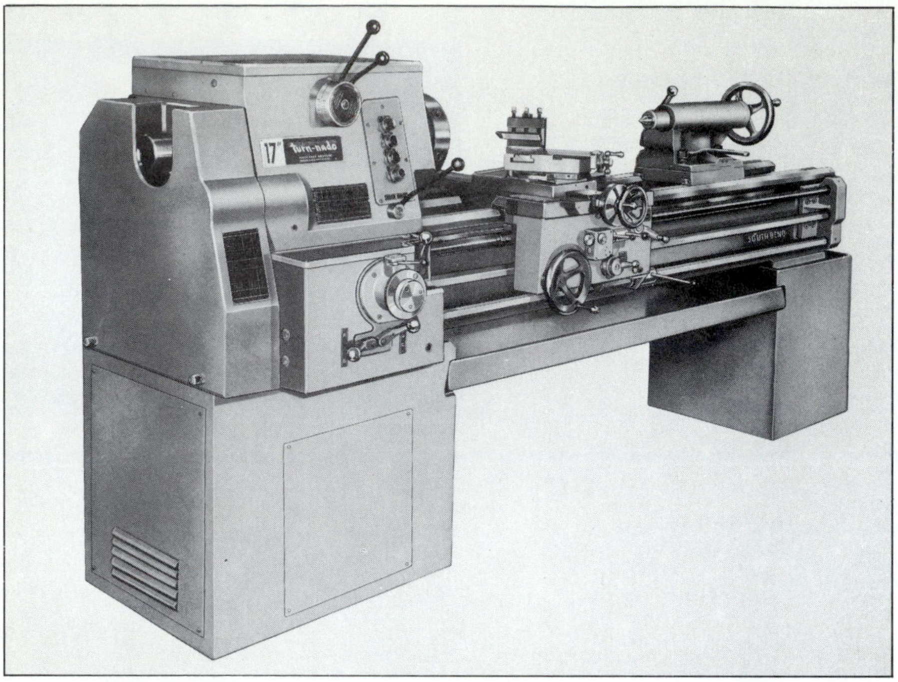
17"
SHELDON

Figure 2 *The engine lathe is the most common lathe found in a machine shop. (Courtesy South Bend Lathe Company.)*

drilling, boring, spinning, grinding, and polishing operations. Cutting operations are performed with a cutting tool fed either parallel or at right angles to the axis of the work. The cutting tool may also be fed at an angle relative to the axis of the work for machining tapers and angles.

Modern production has led to the development of many special types of lathes, such as the engine, turret, single- and multiple-spindle automatic, tracer, numerically controlled lathes, and now computer-controlled turning centers.

The *engine lathe* (Fig. 2), basically not a production lathe, is found in jobbing shops, school shops, and toolrooms.

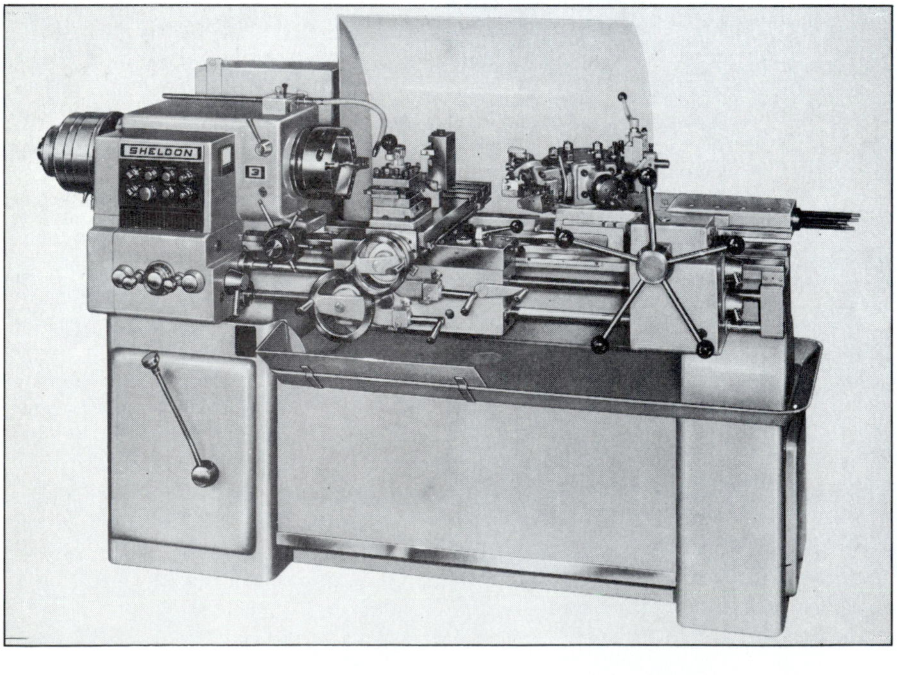
SHELDON

Figure 3 *A turret lathe is used to mass produce parts.*

When many duplicate parts are required, the *turret lathe* (Fig. 3) may be used. This lathe is equipped with a multisided toolpost called a turret to which several different cutting tools may be mounted. In use, different cutting tools are employed in a given sequence to perform a series of operations on each part. This same sequence may be repeated on many parts without having to change or reset the cutting tools.

When hundreds or thousands of identical small parts are required, they may be produced on *single-* and *multiple-spindle automatic lathes* (Fig. 4). On these machines, six or eight different operations may be performed on as many parts at the same time. Once set up, the machine will produce the parts for as long as required.

Tracer lathes (Fig. 5) are used where a few duplicate parts are required. A hydraulically operated cross-slide (and cutting tool) is controlled by a stylus bearing against a round or flat template. Tracer attachments are available for converting most engine lathes into tracer lathes.

Computerized numerically controlled lathes and *turning centers* (Fig. 6) have come into widespread use in the past several years. With these machines, the cutting-tool movements are controlled by a programmed, punched tape or computer to perform a sequence of operations automatically on the workpiece once the machine has been set up.

A very useful and popular device which may be added to an engine lathe (or any other machine) is the *digital readout system.* Engine

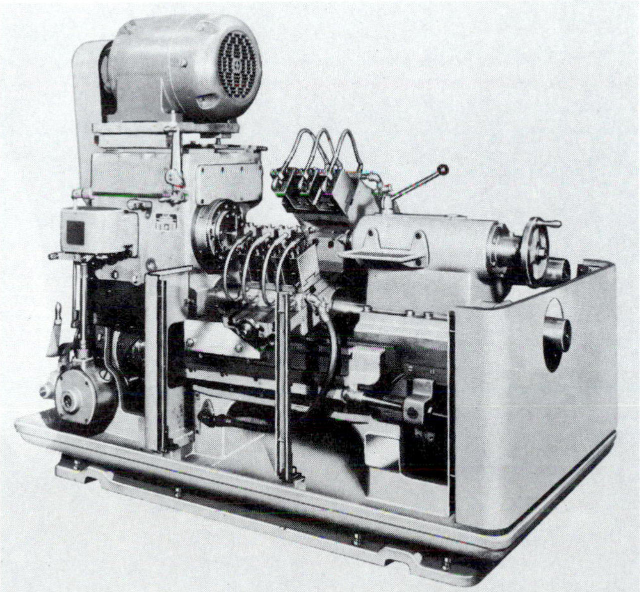

Figure 4 *A single-spindle automatic lathe.*

CINCINNATI
18

Figure 5 *A hydraulic tracer attachment mounted on a lathe. (Courtesy Cincinnati Milacron Inc.)*

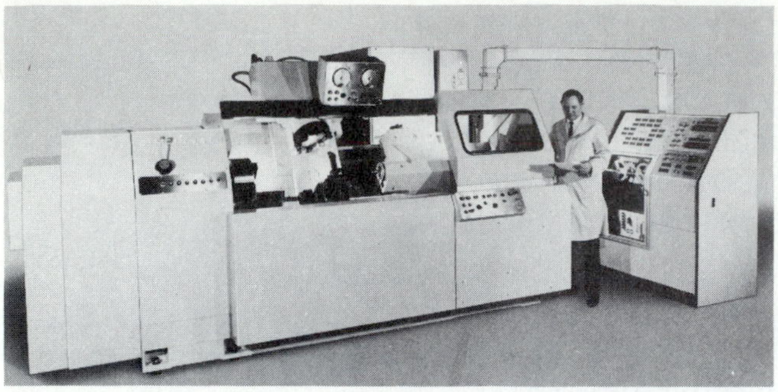

Figure 6 *A numerical control lathe. (Courtesy Cincinnati Milacron Inc.)*

+ 12.664
- 7.5870
LEBLOND

Figure 7 *A lathe equipped with a digital readout system. (Courtesy Bendix Corporation.)*

lathes equipped with these devices (Fig. 7) can produce duplicate parts to within 0.001 in. (0.02 mm) or less on the diameters and lengths of a workpiece.

Only the engine lathe, which is basic to all lathes, will be discussed in detail in the following units.

UNIT

45

Engine Lathe Parts

The majority of work turned on a lathe, other than in production shops, will be turned on an engine lathe. This lathe is an accurate and versatile machine on which many operations such as turning, tapering, form turning, threading, facing, drilling, boring, grinding, and polishing may be performed. Three common engine lathes are the tool-room, heavy-duty, and gap-bed lathes.

OBJECTIVES

After completing this unit you will be able to:

1. Identify and state the purposes of the main operative parts of the lathe
2. Set the lathe to run at any required speed
3. Set the proper feed for the cut required

LATHE SIZE AND CAPACITY

Lathe size is designated by the *largest work diameter* which can be swung over the lathe ways and generally the *maximum distance between centers* (Fig. 45-1). Some manufacturers designate the lathe size by the largest work diameter which can be swung over the ways and the overall bed length.

Lathes are manufactured in a wide range of sizes, the most common being from 9- to 30-in. swing, with a capacity of 16 in. to 12 ft between centers. A typical lathe may have a 13-in. swing, a 6-ft-long bed, and a capacity to turn work 36 in. long between centers (Fig. 45-1).

The average metric lathe used in school shops may have a 230- to 330-mm swing and have a bed length of from 500 to 3000 mm in length.

PARTS OF THE LATHE

The main parts of a lathe are the bed (and ways), headstock, quick-change gearbox, carriage, and tailstock (Fig. 45-2).

BED

The *bed* is a heavy, rugged casting made to support the working parts of the lathe. On its top section are machined ways that guide and align the major parts of the lathe.

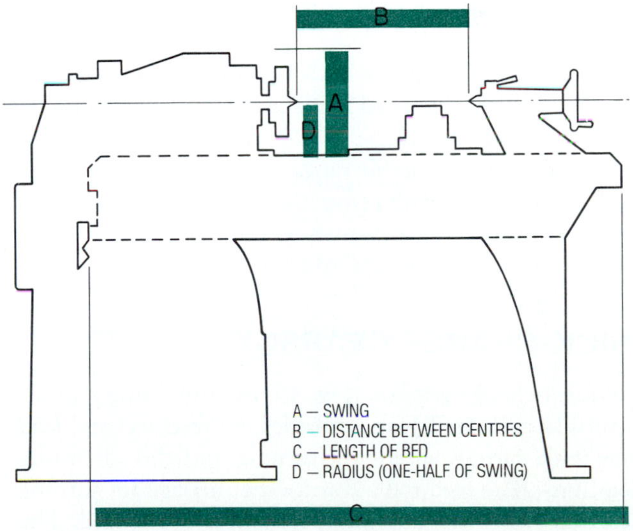
B
A
D
C
A – SWING
B – DISTANCE BETWEEN CENTRES
C – LENGTH OF BED
D – RADIUS (ONE-HALF OF SWING)

Figure 45-1 *Lathe size is indicated by the swing and the length of the bed. (Courtesy South Bend Lathe Company.)*

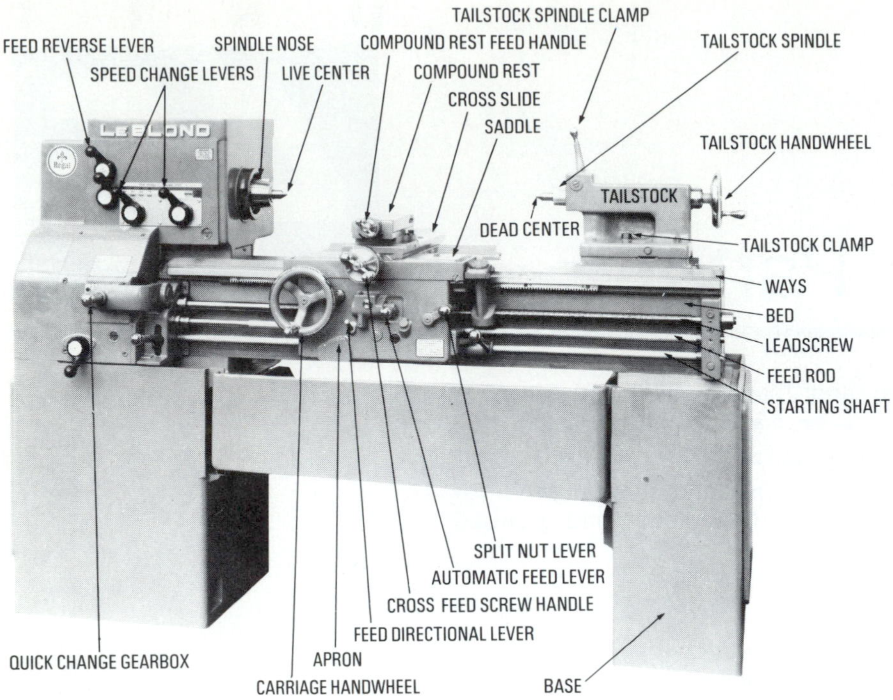
TAILSTOCK SPINDLE CLAMP
FEED REVERSE LEVER
SPINDLE NOSE
COMPOUND REST FEED HANDLE
TAILSTOCK SPINDLE
SPEED CHANGE LEVERS
LIVE CENTER
COMPOUND REST
CROSS SLIDE
SADDLE
TAILSTOCK HANDWHEEL
TAILSTOCK
DEAD CENTER
TAILSTOCK CLAMP
WAYS
BED
LEADSCREW
FEED ROD
STARTING SHAFT
SPLIT NUT LEVER
AUTOMATIC FEED LEVER
CROSS FEED SCREW HANDLE
FEED DIRECTIONAL LEVER
QUICK CHANGE GEARBOX
APRON
CARRIAGE HANDWHEEL
BASE

Figure 45-2 *The parts of an engine lathe. (Courtesy The R. K. LeBlond Machine Tool Co.)*

HEADSTOCK

The *headstock* (Fig. 45-3) is clamped on the left-hand side of the bed. The *headstock spindle,* a hollow cylindrical shaft supported by bearings, provides a drive through gears from the motor to work-holding devices. A live center, a faceplate, or a chuck can be fitted to the spindle nose to hold and drive the work. The live center has a 60° point which provides a bearing for the work to turn between centers.

Headstock spindles can be driven either by a stepped pulley and a belt, or by transmission gears in the headstock (Fig. 45-3). The lathe with a stepped pulley drive is generally called a belt-driven lathe; the gear-driven lathe is referred to as a geared-head lathe. Some light- and medium-duty lathes are equipped with a variable speed drive. On this type of drive, the speed is changed while the headstock spindle is revolving. However, when the lathe is shifted from a high- to low-speed range or vice versa, the lathe spindle must be stopped before the speed change is made. The *feed reverse lever,* mounted on the headstock, reverses the rotation of the feed rod and lead screw.

QUICK-CHANGE GEARBOX

The *quick-change gearbox* (Fig. 45-4), containing a number of different sized gears, provides the *feed rod* and *lead screw* with various speeds for turning and thread-cutting operations. The feed rod advances the carriage for turning operations when the *automatic feed lever* is engaged. The lead screw advances the carriage for thread-cutting operations when the *split-nut lever* is engaged.

CARRIAGE

The *carriage* (Fig. 45-5), consisting of three main parts, the saddle, cross-slide, and apron, is used to move the cutting tool along the lathe bed. The *saddle,* an H-shaped casting mounted on the top of the lathe ways, provides a means of mounting the cross-slide and the apron.

The *cross-slide,* mounted on top of the saddle, provides a manual or automatic cross movement for the cutting tool. The *compound rest,* fitted on top of the cross-slide, is used to support the cutting tool. It can be swiveled to any angle for taper-turning operations, and is moved manually. The cross-slide and compound rest both have graduated collars which ensure accurate cutting tool settings in thousandths

SPINDLE R.P.M.
CAUTION: STOP MOTOR WHEN SHIFTING GEARS
FEED REVERSE LEVER
50 90 167 LOW
300 540 1000 HIGH
75 135 250 LOW
450 810 1500 HIGH
SLOW RANGE
FAST RANGE

Figure 45-3 *A gear-drive headstock.*

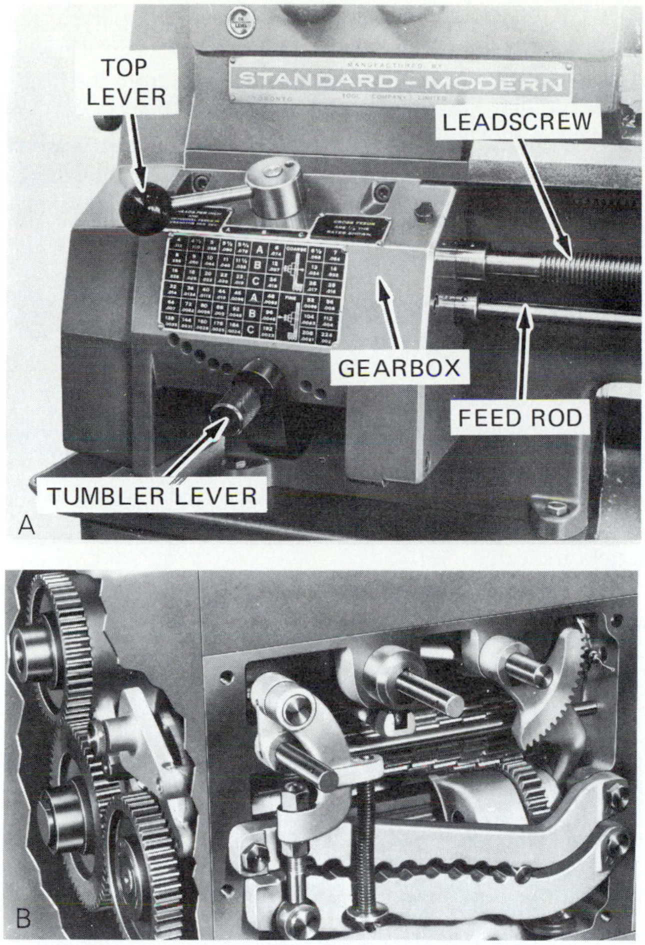
TOP
LEVER
STANDARD - MODERN
LEADSCREW
GEARBOX
FEED ROD
TUMBLER LEVER
A
B

Figure 45-4 *(A) A quick-change gearbox. (Courtesy Standard Modern Tool Company.) (B) A cutaway section of a quick-change gearbox. (Courtesy Cincinnati Lathe & Tool Company.)*

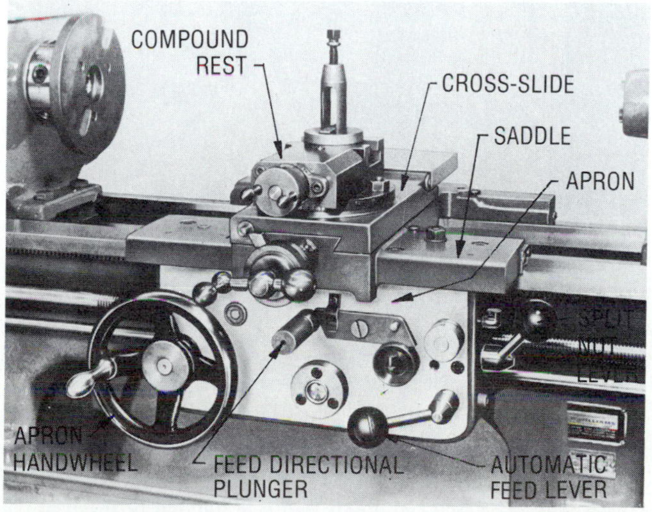
COMPOUND
REST
CROSS-SLIDE
SADDLE
APRON
APRON
HANDWHEEL
FEED DIRECTIONAL
PLUNGER
AUTOMATIC
FEED LEVER

Figure 45-5 *The main parts of an engine lathe carriage.*

of an inch or hundredths of a millimeter. The apron, fastened to the saddle, houses the gears and mechanism required to move the carriage or cross-slide automatically. A locking-off lever inside the apron prevents engaging the split-nut lever and the automatic feed lever at the same time.

The *apron handwheel* can be turned manually to move the carriage along the lathe bed. This handwheel is connected to a gear which meshes in a rack fastened to the lathe bed.

The *automatic feed lever* engages a clutch which provides automatic feed to the carriage. The *feed-change lever* can be set for longitudinal feed or for cross feed. When in the neutral position, the feed-change lever permits the split-nut lever to be engaged for thread cutting. For thread-cutting operations, the carriage is moved automatically when the split-nut lever is engaged. This causes the threads of the split-nut to engage into the threads of the revolving lead screw and move the carriage at a predetermined rate.

TAILSTOCK

The *tailstock* (Fig. 45-6), consisting of the upper and lower tailstock castings, can be adjusted for taper or parallel turning by two screws set in the base. The tailstock can be locked in any position along the bed of the lathe by the *tailstock clamp.* The *tailstock spindle* has an internal taper to receive the *dead center,* which provides support for the right-hand end of the work. Other standard tapered shank tools, such as reamers and drills, can be held in the tailstock spindle. A *spindle clamp* is used to hold the tailstock spindle in a fixed position. The *tailstock handwheel* moves the tailstock spindle in or out of the tailstock casting. It can also be used to provide a hand feed for drilling and reaming operations.

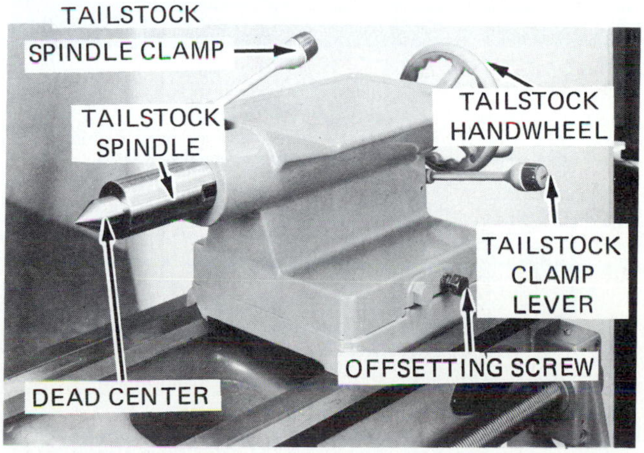
TAILSTOCK
SPINDLE CLAMP
TAILSTOCK
HANDWHEEL
TAILSTOCK
SPINDLE
TAILSTOCK
CLAMP
LEVER
OFFSETTING SCREW
DEAD CENTER

Figure 45-6 *The tailstock assembly. (Courtesy Cincinnati Milacron Inc.)*

SETTING SPEEDS ON A LATHE

Engine lathes are designed to operate at various spindle speeds for machining of different materials. These speeds are measured in revolutions per minute (r/min) and are changed by the stepped pulleys or gear levers.

On a *belt-driven lathe,* various speeds are obtained by changing the flat belt and the back gear drive.

On the *geared-head lathe* (Fig. 45-3) speeds are changed by moving the speed levers into proper positions according to the revolutions per minute chart fastened to the headstock. While shifting the lever positions, place one hand on the faceplate or chuck, and turn the lathe spindle slowly by hand. This will enable the levers to engage the gear teeth without clashing.

NOTE: Never change speeds when the lathe is running. On lathes equipped with *variable-speed drives,* the speed is changed by turning a dial or handle *while the machine is running.*

SETTING FEEDS

The feed of an engine lathe, or the distance the carriage will travel in one revolution of the spindle, depends on the speed of the feed rod or lead screw. This is controlled by the change gears in the quick-change gearbox (Fig. 45-7A). This quick-change gearbox obtains its drive from the headstock spindle through the end gear train (Fig. 45-4B). A chart mounted on the front of the quick-change gearbox indicates the various feeds and metric pitches or threads per inch which may be obtained by setting levers to the positions indicated (Fig. 45-7B).

TO SET THE FEED FOR THE APRON (CARRIAGE DRIVE)

1. Select the desired feed on the chart.
2. Move tumbler lever #4 (Fig. 45-7A) into the hole directly below the selected feed.
3. Follow the row in which the selected feed is found to the left, and set the *feed-change levers* (#1 and #2) to the letters indicated.
4. Set lever #3 to disengage the lead screw.

NOTE: Before turning on the lathe, be sure all levers are

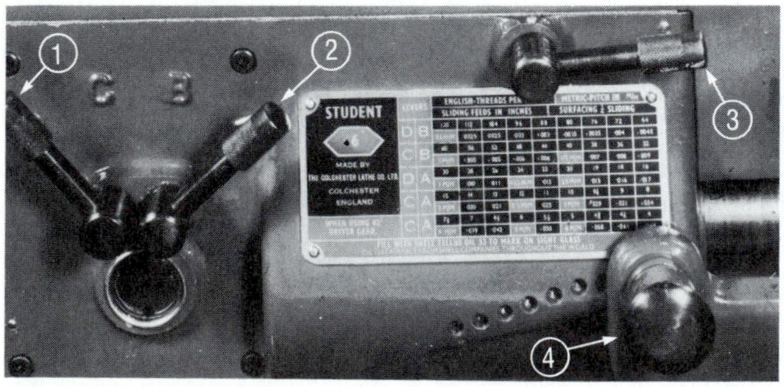
1
2
3
4

A

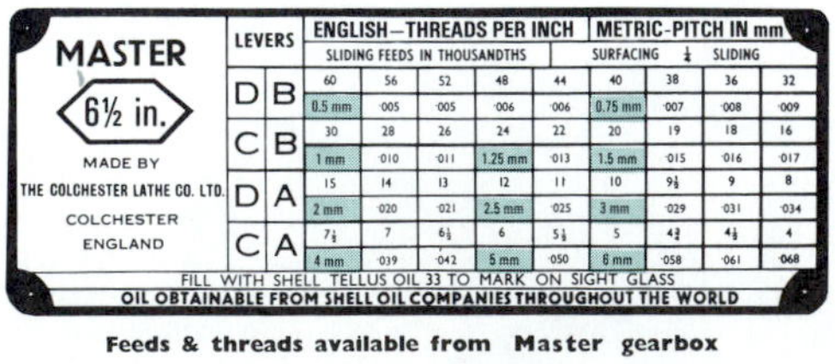

MASTER 6½ in. MADE BY THE COLCHESTER LATHE CO. LTD. COLCHESTER ENGLAND

LEVERS		ENGLISH—THREADS PER INCH					METRIC-PITCH IN mm			
		SLIDING FEEDS IN THOUSANDTHS					SURFACING ¼ SLIDING			
D	B	60	56	52	48	44	40	38	36	32
		0.5 mm	·005	·005	·006	·006	0.75 mm	·007	·008	·009
C	B	30	28	26	24	22	20	19	18	16
		1 mm	·010	·011	1.25 mm	·013	1.5 mm	·015	·016	·017
D	A	15	14	13	12	11	10	9½	9	8
		2 mm	·020	·021	2.5 mm	·025	3 mm	·029	·031	·034
C	A	7½	7	6½	6	5½	5	4¾	4½	4
		4 mm	·039	·042	5 mm	·050	6 mm	·058	·061	·068

FILL WITH SHELL TELLUS OIL 33 TO MARK ON SIGHT GLASS
OIL OBTAINABLE FROM SHELL OIL COMPANIES THROUGHOUT THE WORLD

Feeds & threads available from Master gearbox

B

Figure 45-7 *(A) A quick-change gearbox permits fast setting of feeds; (B) a quick-change gearbox chart. Note the setting for metric threads. (Courtesy Colchester Lathe Company.)*

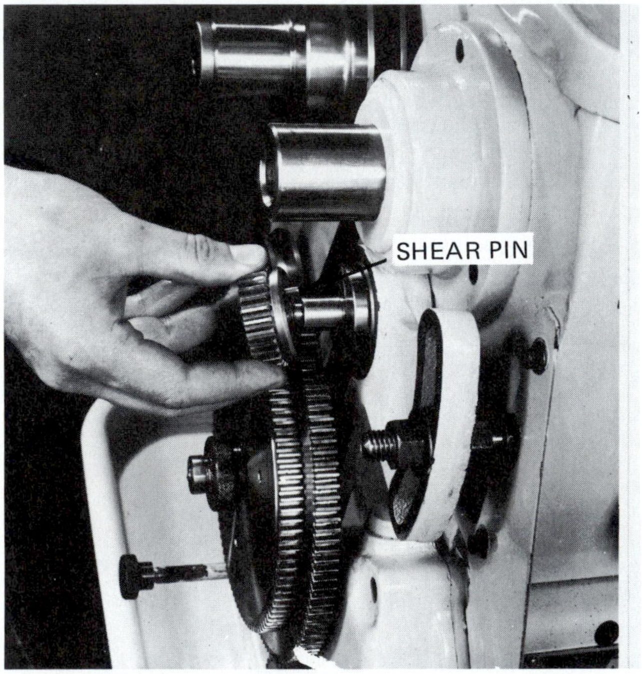
SHEAR PIN

A

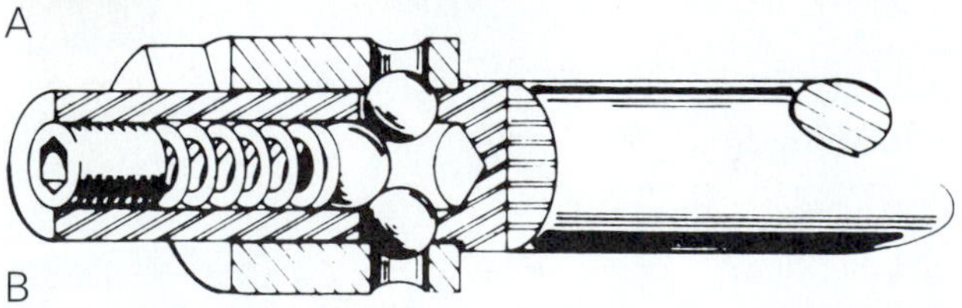

B

Figure 45-8 *(A) A shear pin in the end gear train prevents damage to the gears in case of an overload; (B) a spring-ball clutch will slip when too much strain is applied to the feed rod. (Courtesy Colchester Lathe Company.)*

fully engaged by turning the headstock spindle by hand, and see that the feed rod turns.

SHEAR PINS AND SLIP CLUTCHES

To prevent damage to the feed mechanism due to overload or sudden torque, some lathes are equipped with either *shear pins* or *slip clutches* (Fig. 45-8A, B). Shear pins, usually made of brass, may be found on the feed rod, lead screw, and the end gear train. Spring-loaded slip clutches are found only on feed rods. When the feed mechanism is overloaded, either the shear pin will break or the slip clutch will slip, causing the automatic feed to stop. This prevents damage to the gears or shafts of the feed mechanism.

REVIEW QUESTIONS

1. List the operations which can be performed on a lathe.
2. How is the size of a lathe designated?

PARTS OF THE ENGINE LATHE

3. Name the four *main* units of a lathe.
4. State the purpose of the following:
   a. Headstock spindle
   b. Lead screw and feed rod
   c. Quick-change gearbox
   d. Split-nut lever
   e. Feed-change lever
   f. Cross-slide
   g. Compound rest

SETTING SPEEDS AND FEEDS

5. List three types of lathe drives.
6. Explain how speeds on a geared-head lathe are changed.
7. List the steps to set a feed of 0.010 in. (0.25 mm).
8. What is the purpose of:
   a. A shear pin?
   b. A slip clutch?

UNIT 46

Lathe Accessories

Many lathe accessories are available to increase the versatility of the lathe and the variety of work which can be machined. Lathe accessories may be divided into two categories:

1. Work-holding, -supporting, and -driving devices
2. Cutting-tool-holding devices

Work-holding, -supporting, and -driving devices include lathe centers, chucks, faceplates, mandrels, steady and follower rests, lathe dogs, and drive plates. Cutting-tool-holding devices include various types of straight and offset toolholders, threading toolholders, boring bars, turret-type toolposts, and quick-change toolpost assemblies.

OBJECTIVES

After completing this unit you will be able to:

1. Identify and state the purpose of the common work-holding and -driving accessories
2. Identify and state the purpose of the common cutting-tool-holding accessories

WORK-HOLDING DEVICES

LATHE CENTERS

Most turning operations can be performed between centers on a lathe. Work to be turned between centers must have a center hole drilled in each end (usually 60°) to provide a bearing surface, which allows the work to turn on the centers. The centers merely support the work while the cutting operations are performed. A lathe dog, fitted into a driving plate, provides a drive for the work (Fig. 46-1).

A variety of lathe centers are used to suit various operations or workpieces. Probably the most commonly used centers in school shops were the solid 60° centers with a Morse taper shank (Fig. 46-2). These were generally made from high-speed steel or a good grade of machine steel with carbide inserts or tips. Care must be taken when using these centers to adjust and lubricate them occasionally as the work heats up and expands. If this is not done, both the center and the workpiece may be damaged. The damage to the workpiece will be the loss of concentricity, which will prevent future operations from being performed using the center holes. The lathe center must also be reground to remove the damaged section before it can be used.

The revolving *dead center* (Fig 46-3), sometimes called a live dead center, has generally replaced the solid dead center. It is commonly used to support long work held in a chuck or when work is being machined between centers. This center usually contains anti-friction bearings, which allow the center to revolve with the work. No lubrication is required between the center and the workpiece, and the center tension is not affected by workpiece expansion during the cutting action.

A *microset adjustable center* (Fig 46-4) fits into the tail-

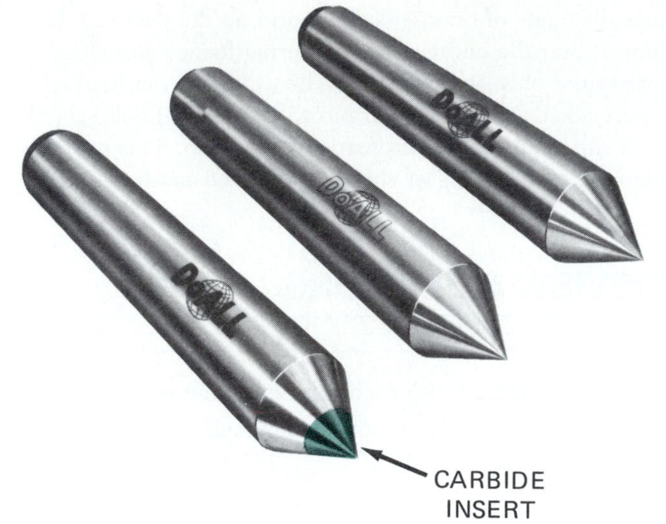

CARBIDE INSERT

Figure 46-2 *A variety of 60° lathe centers. (Courtesy DoAll Company.)*

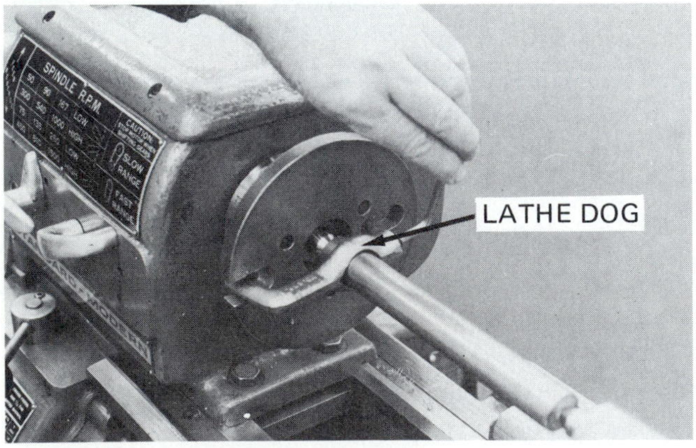

LATHE DOG

Figure 46-1 *A workpiece mounted between centers is usually driven by a lathe dog.*

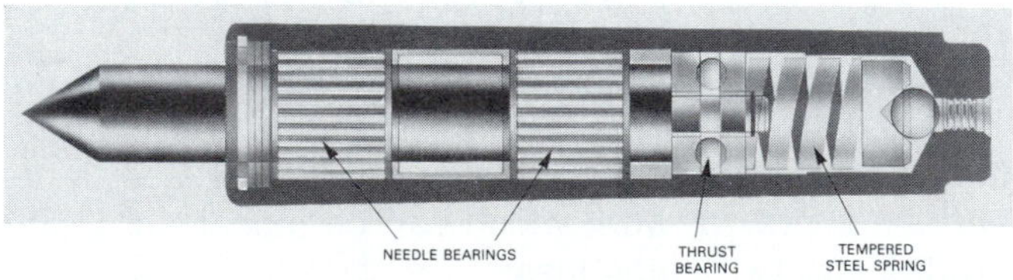

NEEDLE BEARINGS
THRUST BEARING
TEMPERED STEEL SPRING

Figure 46-3 *A revolving dead center. (Courtesy Enco Manufacturing Company.)*

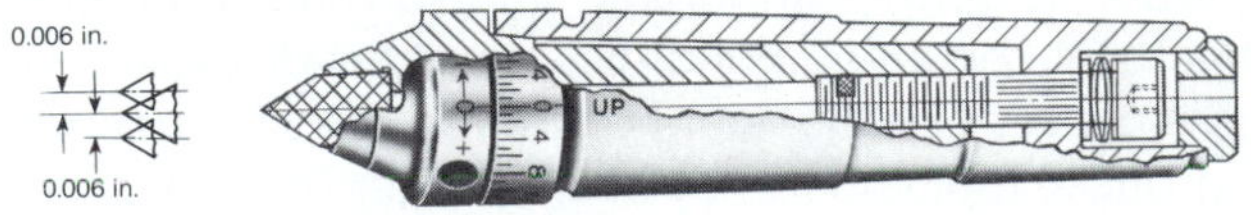

0.006 in.
0.006 in.
UP

Figure 46-4 *A microset adjustable dead center.*

stock spindle and provides a means of aligning lathe centers or producing slight tapers on work being machined between centers. An eccentric, or sometimes a dovetail slide, allows this type of center to be adjusted a limited amount to each side of center. Lathe centers are quickly and easily aligned with this type of center.

The *self-driving center* (Fig. 46-5), mounted in the headstock spindle, is used when the entire length of a workpiece is being machined in one operation and when a chuck or lathe dog could not be used to drive the work. Grooves ground around the circumference of the lathe center point provide the drive for the workpiece. The work (usually a soft material such as aluminum) is forced onto the driving center; a revolving dead center is used to support the work and hold it against the grooves of the driving center.

CHUCKS

Some workpieces, because of their size and shape, cannot be held and machined between lathe centers. Lathe chucks are used extensively for holding work for machining operations. The most commonly used lathe chucks are: the three-jaw universal, four-jaw independent, and the collet chuck.

The *three-jaw universal chuck* (Fig. 46-6) is used to hold round and hexagonal work. It grasps the work quickly and within a few thousandths of an inch or hundredths of a millimeter of accuracy, because the three jaws move simultaneously when adjusted by the chuck wrench. This simultaneous motion is caused by a scroll plate into which all three jaws fit. Three-jaw chucks are made in various sizes from 4 to 16 in. (100 to 400 mm) in diameter. They are usually provided with two sets of jaws, one for outside chucking and the other for inside chucking.

The *four-jaw independent chuck* (Fig 46-7) has four jaws, each of which can be adjusted independently by a chuck wrench. They are used to hold round, square, hexagonal, and irregularly shaped workpieces. The jaws can be reversed to hold work by the inside diameter.

Universal and independent chucks can be fitted to the three types of headstock spindles. Figure 46-8 shows a threaded spindle nose, Fig. 46-9 a tapered spindle nose, Fig. 46-10 a cam-lock spindle nose. The threaded type screws on in a clockwise direction; the tapered type is held by a lock nut that tightens on the chuck. The cam lock is held by tightening the cam locks using a T-wrench. On the taper and cam-lock types, the chuck is aligned by the taper on the spindle nose.

The *collet chuck* (Fig. 46-11) is the most accurate chuck and is used for high-precision work. Spring collets are available to hold round, square, or hexagon-shaped workpieces.

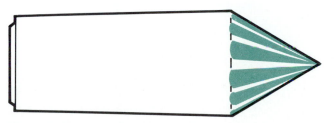

Figure 46-5 *A self-driving live center.*

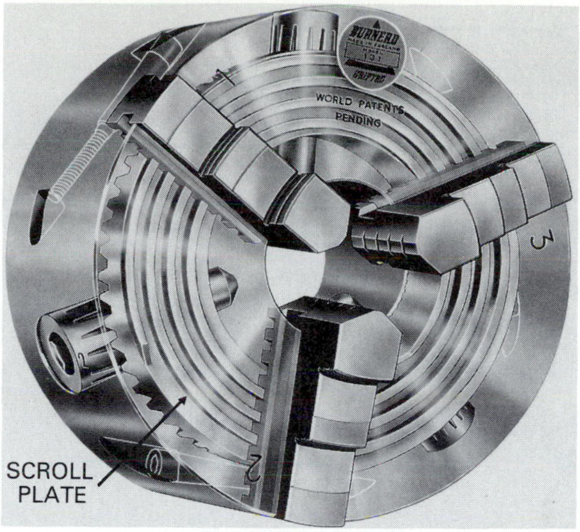
SCROLL
PLATE
WORLD PATENTS
PENDING

Figure 46-6 *A three-jaw universal geared scroll chuck.*

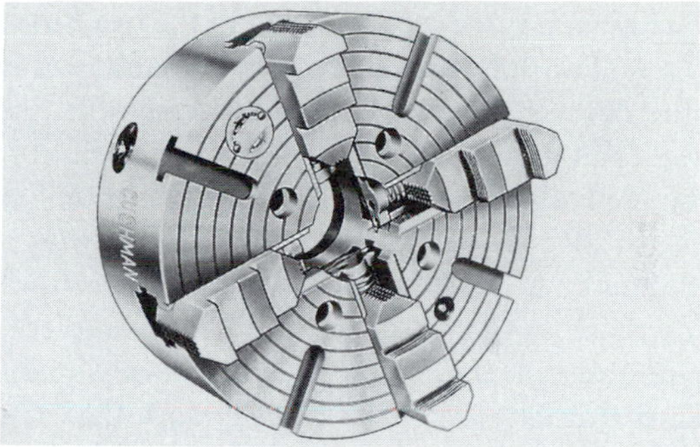

Figure 46-7 *A four-jaw independent chuck.*

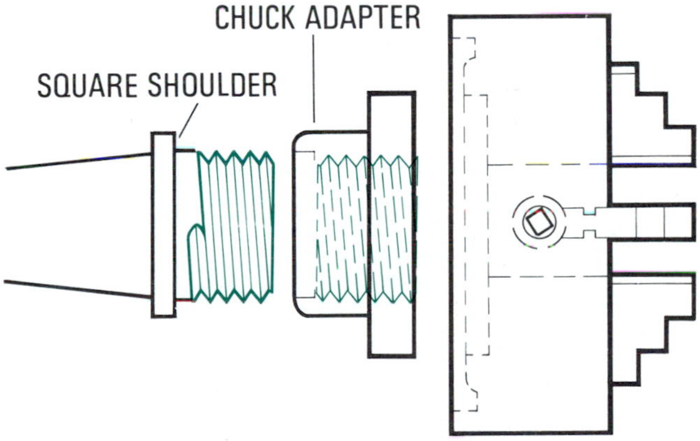
CHUCK ADAPTER
SQUARE SHOULDER

Figure 46-8 *A threaded spindle nose.*

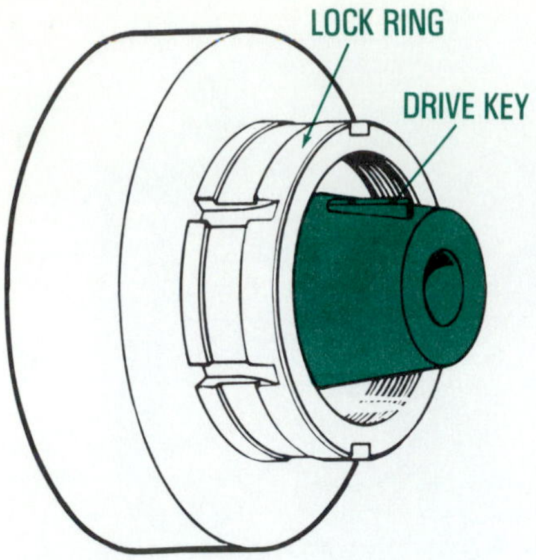
LOCK RING
DRIVE KEY

Figure 46-9 *An American standard lathe spindle nose.*

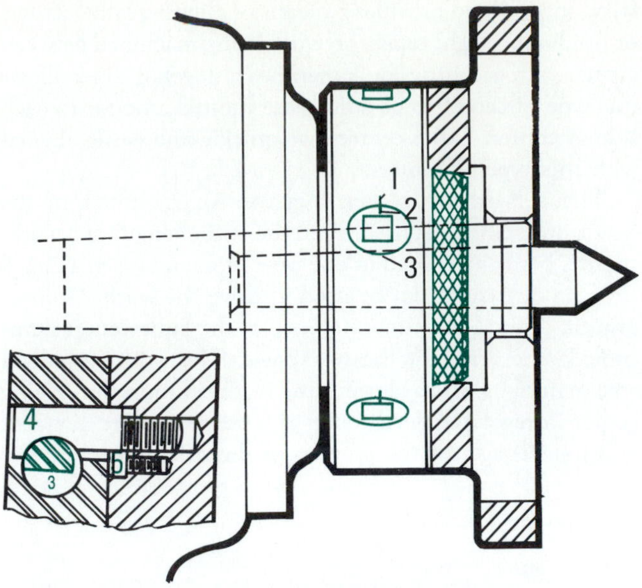
1
2
3
4
3
5

Figure 46-10 *The parts of a cam-lock spindle nose:*
1. *Registration lines on spindle nose*
2. *Registration lines on cam-lock*
3. *Cam-locks*
4. *Cam-lock mating stud on chuck or faceplate*
5. *Hollow-head retaining screw.*

Each collet has a range of only a few thousandths of an inch or hundredths of a millimeter over or under the size stamped on the collet.

A special adaptor is fitted into the taper of the headstock spindle, and a hollow draw bar having an internal thread is inserted in the opposite end of the headstock spindle. As the handwheel (and draw bar) is revolved, it draws the collet into the tapered adapter, causing the collet to tighten on the workpiece. This type of chuck is also referred to as a *spring-collet chuck.* Another form of spring-collet chuck uses a chuck wrench to tighten the collet on the workpiece. This type is mounted on the spindle nose in the same manner as standard chucks and can hold larger work than the draw-in type.

The *Jacobs collet chuck* (Fig. 46-12) has a wider range than the spring-collet chuck. Instead of a draw bar, it utilizes an impact-tightening handwheel to close the collets on the workpiece. A set of 11 Rubber-Flex collets, each having an adjustment range of almost 1/8 in. (3 mm), makes it possible to hold a wide range of work diameters. When the handwheel is turned clockwise, the rubber-flex collet is forced into a taper, causing it to tighten on the workpiece. When the handwheel is turned counterclockwise, the collet opens and releases the workpiece.

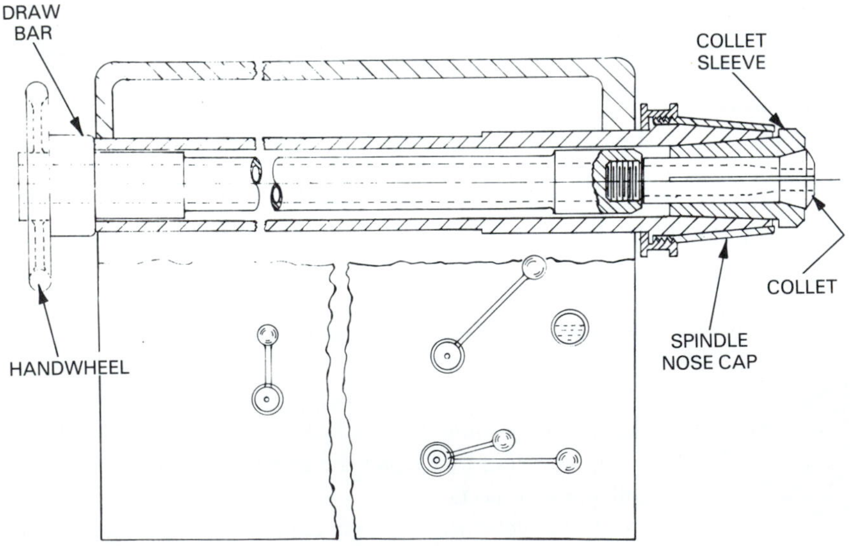
DRAW BAR
COLLET SLEEVE
COLLET
SPINDLE NOSE CAP
HANDWHEEL

Figure 46-11 *A cross-sectional view of a headstock showing the construction of a draw-in collet assembly.*

JACOBS SPINDLE NOSE LATHE CHUCK

Figure 46-12 *The Jacobs collet chuck has a wider range than other collet chucks. (Courtesy The Jacobs Manufacturing Company.)*

Magnetic chucks (Fig. 46-13) are used to hold iron or steel parts that are too thin or that may be damaged if held in a conventional chuck. These chucks are fitted to an adaptor mounted on the headstock spindle. Work is held lightly for aligning purposes by turning the chuck wrench approximately one-quarter turn. After the work has been trued, the chuck is turned to the full-on position to hold the work securely. This type of chuck is used only for light cuts and for special grinding applications.

Faceplates are used to hold work that is too large or of such a shape that it cannot be held in a chuck or between centers. Faceplates are usually equipped with several slots to permit the use of bolts to secure the work or angle plate (Fig. 46-14) so that the axis of the workpiece may be

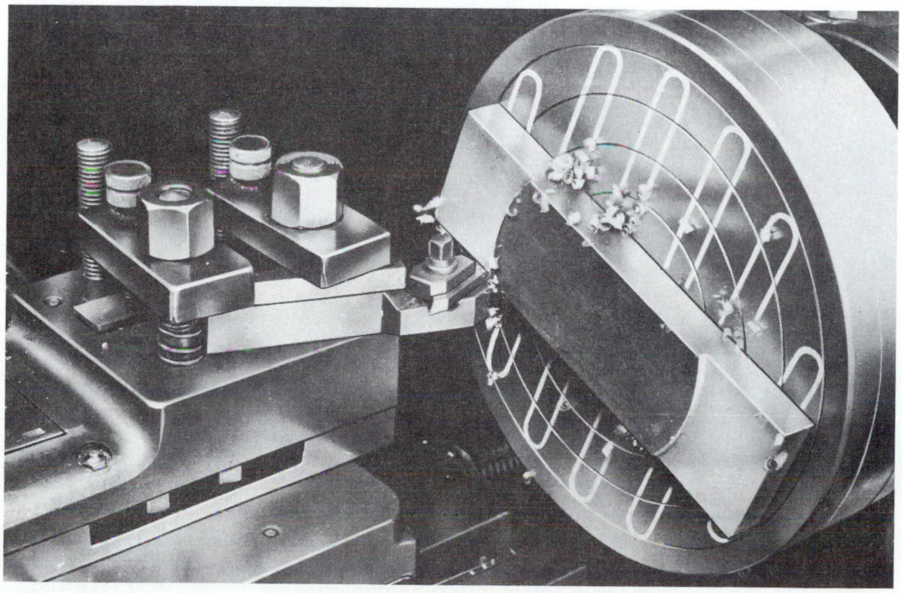

Figure 46-13 *Workpieces may be held on a magnetic chuck for turning operations.*

aligned with the lathe centers. When work is mounted off center, a counterbalance (Fig. 46-14) should be fastened to the faceplate to prevent imbalance and the resultant vibrations when the lathe is in operation.

A *steadyrest* (Fig. 46-15) is used to support long work held in a chuck or between lathe centers. It is located on, and aligned by, the ways of the lathe and may be positioned at any point along the lathe bed provided it clears the carriage travel. The three jaws, tipped with plastic, bronze, or rollers, may be adjusted to support any work diameter within the steadyrest capacity. During machining operations performed on or near the end of a workpiece, a steadyrest supports the end of work held in a chuck, when the work cannot be supported by the tailstock center. A steadyrest also supports the center of long work to prevent springing when the work is machined between centers.

A *follower rest* (Fig. 46-16), mounted on the saddle, travels with the carriage to prevent work from springing up and away from the cutting tool. The cutting tool is generally positioned just ahead of the follower rest to provide a smooth bearing surface for the two jaws of the follower rest.

A *mandrel* holds an internally machined workpiece between centers so that further machining operations are concentric with the bore. There are several types of mandrels, the most common being the *plain mandrel* (Fig. 46-17), *expanding mandrel, gang mandrel,* and *stub mandrel.*

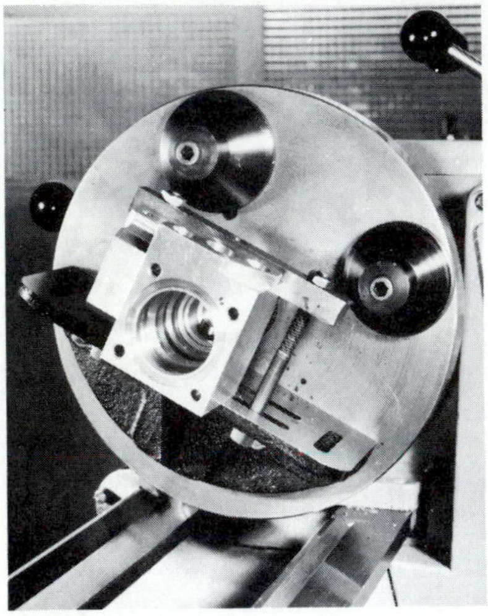

Figure 46-14 *An angle plate fastened to a faceplate is used to hold a workpiece for machining. (Courtesy Colchester Lathe & Tool Co.)*

Figure 46-15 *A steadyrest is often used to support a long or slender workpiece during machining.*

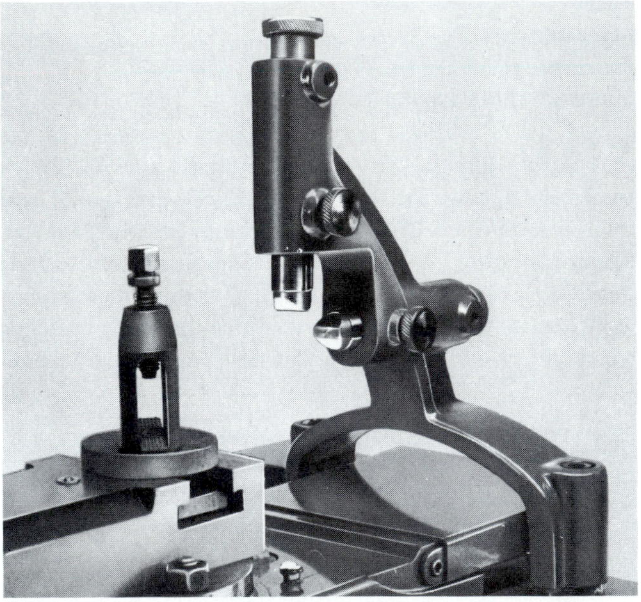

Figure 46-16 *A follower rest mounted on the saddle may be used to support a long, slender workpiece during machining.*

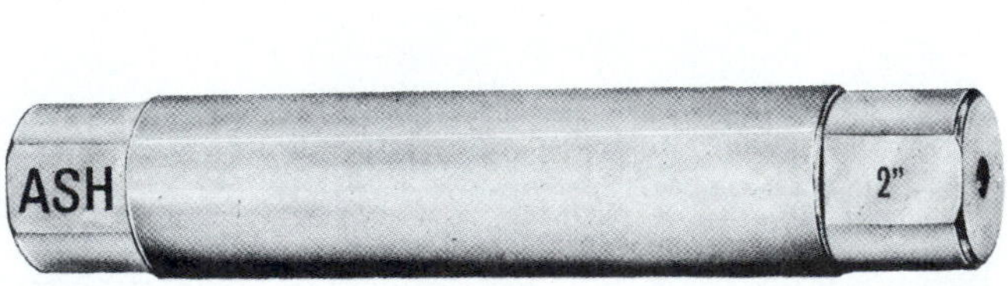
ASH
2"

Figure 46-17 *A plain mandrel. (Courtesy Ash Precision Equipment Inc.)*

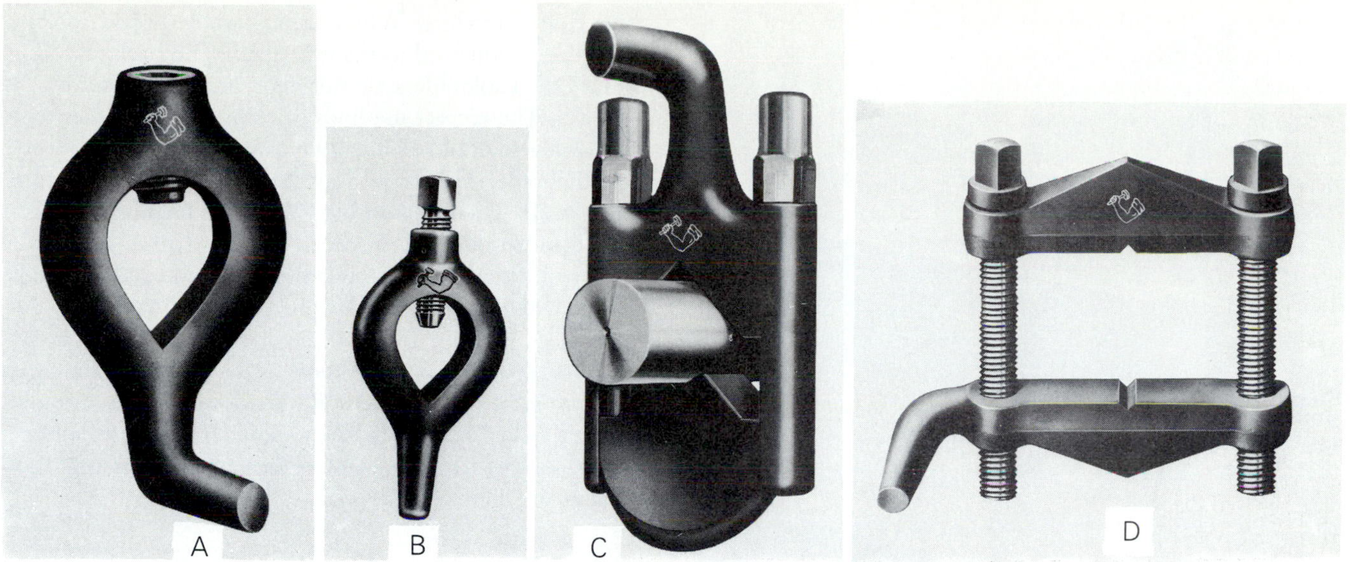
A
B
C
D

Figure 46-18 *Common types of lathe dogs: (A) standard bent-tail; (B) straight tail; (C) safety clamp; (D) clamp type. (Courtesy Armstrong Bros. Tool Co.)*

LATHE DOGS

When work is machined between centers, it is generally driven by a lathe dog. The lathe dog has an opening to receive the work and a setscrew to fasten the dog to the work. The tail of the dog fits into a slot on the driveplate and provides the drive to the workpiece. Lathe dogs are made in a variety of sizes and types to suit various workpieces.

The *standard bent-tail lathe dog* (Fig. 46-18A) is the most commonly used dog for round workpieces. These dogs are available with square-head setscrews or headless setscrews, which are safer since there is no protruding head.

The *straight-tail dog* (Fig. 46-18B) is driven by a stud in the driveplate. Since this is a more balanced type of dog than the bent-tail dog, it is used in precision turning where centrifugal force of a bent-tail dog may cause inaccuracies in the work.

The *safety clamp lathe dog* (Fig. 46-18C) may be used to hold a variety of work since it has a wide range of adjustment. It is particularly useful on finished work where the setscrew of a standard lathe dog may damage the finish.

The *clamp lathe dog* (Fig. 46-18D) has a wider range than the other types and may be used on round, square, rectangular, and odd-shaped workpieces.

CUTTING-TOOL-HOLDING DEVICES

Most toolbits used in lathe-turning operations in school shops are square and are generally held in a standard toolholder (Fig. 46-19). These toolholders are made in various

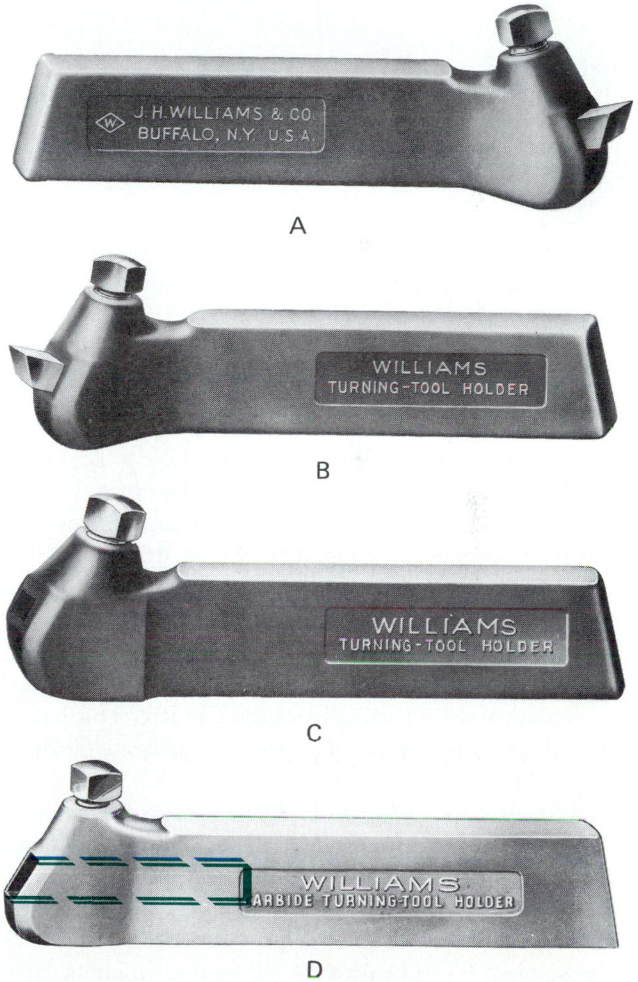
J. H. WILLIAMS & CO.
BUFFALO, N.Y. U.S.A.
A
WILLIAMS
TURNING-TOOL HOLDER
B
WILLIAMS
TURNING-TOOL HOLDER
C
WILLIAMS
ARBIDE TURNING-TOOL HOLDER
D

Figure 46-19 *Common lathe toolholders: (A) left-hand offset; (B) right-hand offset; (C) straight; (D) carbide. (Courtesy J. H. Williams and Company.)*

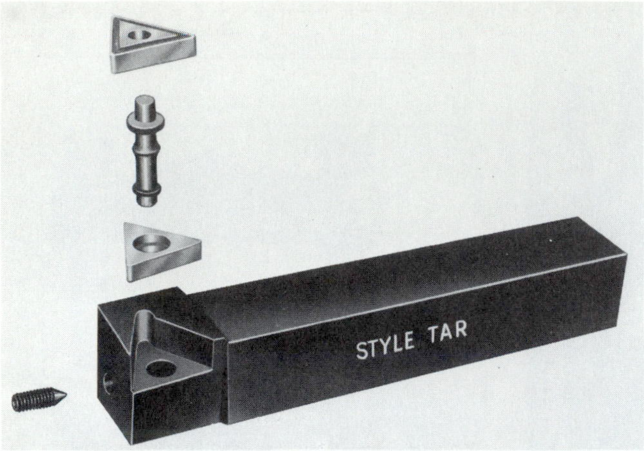
STYLE TAR

A

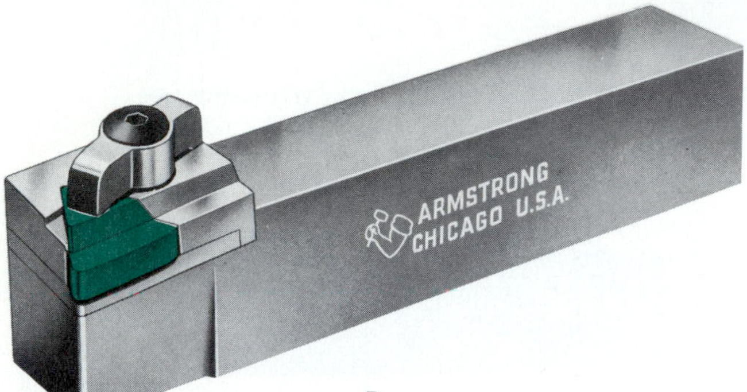
ARMSTRONG
CHICAGO U.S.A.

B

Figure 46-20 *(A) Carbide inserts are held securely in a holder by means of cam action; (B) carbide inserts are held securely by a clamp.*

styles and sizes to suit different machining operations. Toolholders for turning operations are available in three styles: left-hand offset, right-hand offset, and straight.

Each of these toolholders has a square hole to accommodate the square toolbit, which is held in place by a setscrew. The hole in the toolholder is at an angle of approximately 15 to 20° to the base of the toolholder (Fig. 46-19C). When the cutting tool is set on center, this angle provides the proper amount of back rake in relation to the workpiece.

The *left-hand offset toolholder* offset to the right (Fig. 46-19A) is designed for machining work close to the chuck or faceplate and for cutting from right to left. This type of toolholder is designated by the letter L to indicate the direction of cut.

The *right-hand offset toolholder* offset to the left (Fig. 46-19B) is designed for machining work close to the tailstock, for cutting from left to right, and for facing operations. This type of toolholder is designated by the letter R.

The *straight toolholder* (Fig. 46-19C) is a general-purpose type. It can be used for taking cuts in either direction and for general machining operations. This type of toolholder is designated by the letter S.

The *carbide toolholder* (Fig. 46-19D) has a square hole parallel to the base of the toolholder to accommodate carbide-tipped toolbits. When using carbide-tipped toolbits, hold the toolbit so that there is little or no back rake (Fig. 46-19D). Toolholders of this type are designated by the letter C. The correct methods of using this and other types of carbide toolholders are fully explained in Unit 31.

Toolholders for throwaway carbide inserts are shown in Fig. 46-20. The insert in Fig. 46-20A is held in the holder by a cam action. The insert in Fig. 46-20B is secured by means of a clamp. These types of toolholders are available for use with conventional, turret-type, and heavy-duty toolposts.

Cutting-off or *parting tools* are used when the work must be grooved or parted off. The long, thin, cutting-off blade is locked securely in the toolholder by means of either a cam lock or a locking nut. Three types of parting toolholders are shown in Fig. 46-21A–C.

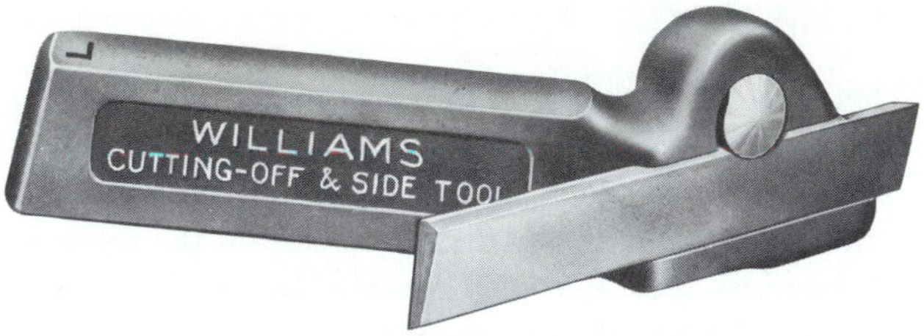
L
WILLIAMS
CUTTING-OFF & SIDE TOOL

A

S
WILLIAMS
CUTTING-OFF & SIDE TOOL

B

R
WILLIAMS
CUTTING-OFF & SIDE TOOL

C

Figure 46-21 *Types of solid cutting-off toolholders: (A) left-hand; (B) straight; (C) right-hand. (Courtesy J. H. Williams and Company.)*

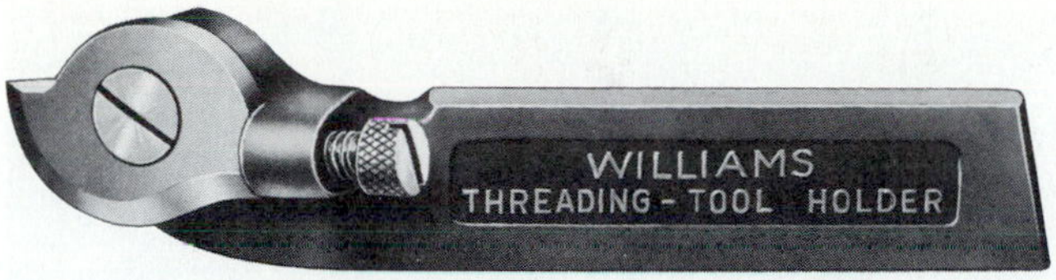
WILLIAMS
THREADING - TOOL HOLDER

Figure 46-22 *A preformed thread-cutting tool and toolholder. (Courtesy J. H. Williams and Company.)*

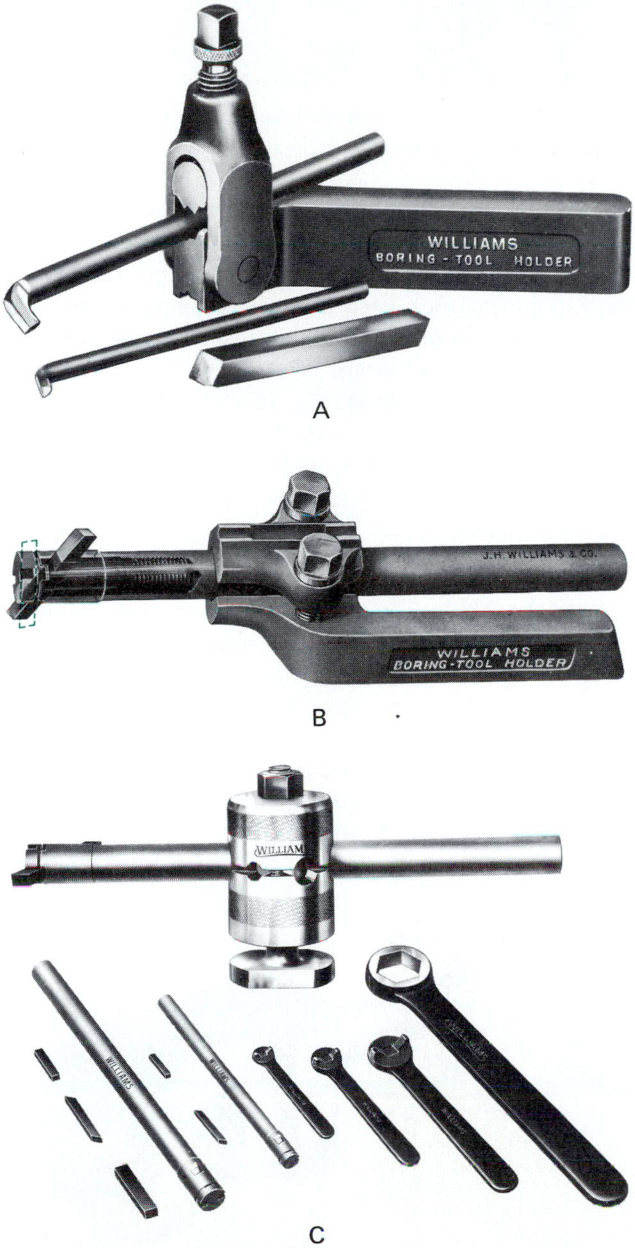
WILLIAMS
BORING - TOOL HOLDER
A
J. H. WILLIAMS & CO.
WILLIAMS
BORING-TOOL HOLDER
B
C

Figure 46-23 *(A) A light boring toolholder; (B) a medium boring toolholder; (C) a heavy-duty boring tool. (Courtesy J. H. Williams and Company.)*

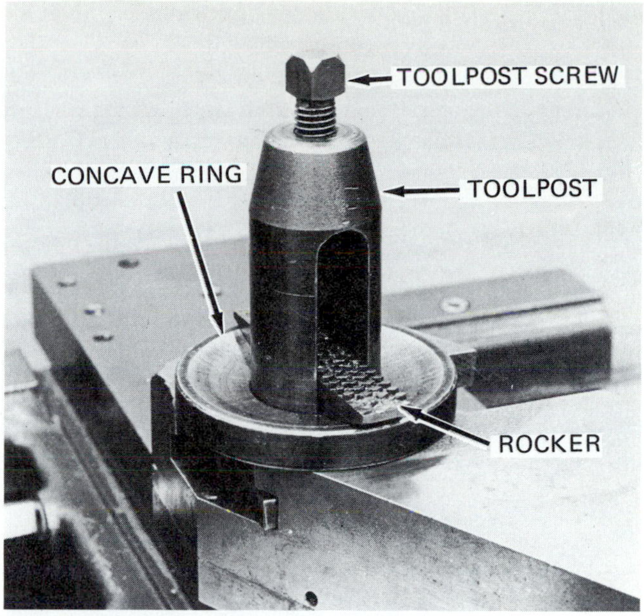
TOOLPOST SCREW
CONCAVE RING
TOOLPOST
ROCKER

Figure 46-24 *A standard toolpost.*

A *threading toolholder* (Fig. 46-22) is designed to hold a special form-relieved thread-cutting tool. Although most machinists grind their own thread-cutting tools, this tool is convenient because it has an accurately ground 60° angle. This angle is maintained throughout the life of the tool because only the top of the cutting surface is sharpened when it becomes dull.

Boring toolholders are made in several styles. A *light boring toolholder* (Fig. 46-23A) is held in a standard toolpost and is used for small holes and light cuts. The *medium boring toolholder* (Fig. 46-23B) is suitable for heavier cuts and is also held in a standard toolpost. The cutting tool may be held at 45° or 90° to the axis of the bar.

The *heavy-duty boring bar holder* (Fig. 46-23C) is mounted on the compound rest of the lathe. It has three bars of different diameters to suit the diameter of the hole being bored. With this type of boring bar, use the largest bar possible to get the maximum rigidity and to avoid chatter. The toolbit may be held at 45 or 90° to the axis.

TOOLPOSTS

The *standard* or *round toolpost* (Fig. 46-24) is generally supplied with the engine lathe. This toolpost fits into the T-slot of the compound rest. It provides a means of holding and adjusting the toolholder or cutting tool. The concave ring and the wedge or rocker provides for the adjustment of the cutting-tool height.

The *turret-type toolpost* (Fig. 46-25) provides a convenient means of holding toolbits when several operations, such as turning, threading, grooving, and parting, must be

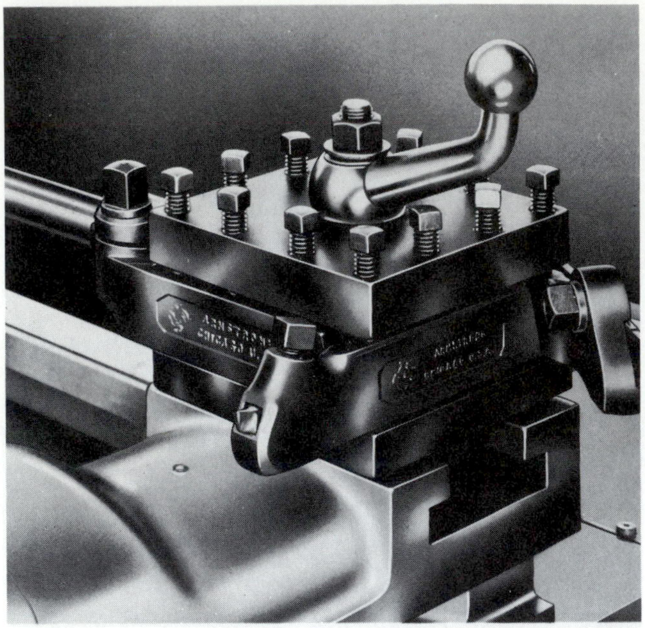

Figure 46-25 *A turret-type toolpost permits the use of up to four different cutting tools.*

performed on a workpiece. This type of toolholder is suitable for carbide-tipped tools, since no provision is made for back rake. Four different types of toolbits may be mounted in this toolholder and each may be brought into position by loosening the locking handle and rotating the turret head until the desired toolbit is in the cutting position. The handle is then turned to lock the head in place.

Quick-change toolposts and *holders* provide a quick and accurate means of setting and changing tools for different operations. The dovetailed toolholder fits onto a dovetailed toolpost, which is mounted on the compound rest (Fig. 46-26). There are several different holders to accommodate the various types of cutting tools (Fig. 46-27). The cutting tool, which is held in the holder by means of setscrews, is generally sharpened in the holder and preset (in the holder) in the toolroom. After a cutting tool becomes dull, the unit

Figure 46-26 *With a dovetailed toolpost, cutting tools can be rapidly changed and accurately set.*

can be quickly replaced with another sharpened and preset unit. This procedure ensures accuracy of size, since each preset toolbit will be in exactly the same position as the previous cutting tool. Each unit may be adjusted vertically on the toolpost by means of the knurled adjusting nut, then locked in position by means of a clamp.

The *heavy-duty toolpost* (Fig. 46-28) is also mounted in the slot of the compound rest. It is designed basically for use with carbide cutting tools and holds a single cutting tool or toolholder. This type of toolpost is used when very heavy cuts are required.

NOTE: The turret type and quick-change toolposts have replaced the standard or round toolposts in most industrial shops and in some school shops, where carbide tools are being used. These toolholders are more suitable for use with carbides. However, since many schools and jobbing shops still use high-speed steel toolbits and standard toolholders, the authors decided to continue with the use of standard toolposts since more attention must be paid to the setup using this type of toolpost.

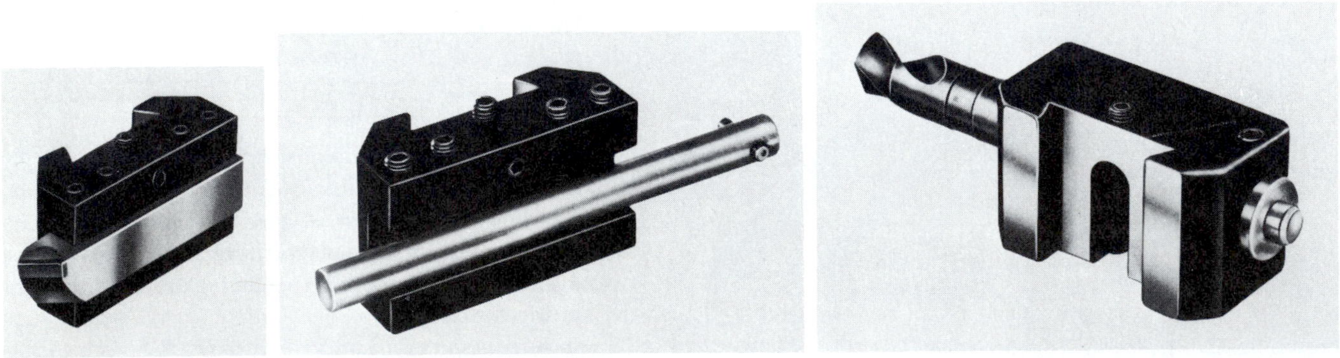

Figure 46-27 *Various types of quick-change toolholders. (Courtesy DoAll Company.)*

Figure 46-28 *A heavy-duty toolpost is used for making heavy cuts. (Courtesy Cincinnati Milacron Inc.)*

REVIEW QUESTIONS

WORK-HOLDING DEVICES

1. Name the three types of lathe centers and state the purpose of each.
2. What precautions must be taken when turning work that is supported by solid centers?
3. Describe and state the purpose of the following:
   a. Three-jaw universal chuck
   b. Four-jaw independent chuck
   c. Collet chuck
   d. Magnetic chuck
4. What advantage does a Jacobs collet chuck have over a spring-collet chuck?
5. What is the purpose of:
   a. A steadyrest? b. A follower rest?
6. Name three types of lathe dogs.
7. What is the disadvantage of the square-head setscrew in a lathe dog?
8. What is the advantage of a straight-tail dog?

CUTTING-TOOL-HOLDING DEVICES

9. Describe three types of standard toolholders and state the purpose of each.
10. How does a carbide toolholder differ from a standard toolholder?
11. State two methods by which throwaway carbide inserts are held in a toolholder.
12. What procedure should be followed when using a heavy-duty boring bar set?
13. Name and state the purpose of four types of toolposts.

UNIT 47

Cutting Speeds, Feeds, and Depth of Cut

In order to operate a lathe efficiently, the machinist must consider the importance of cutting speeds and feeds. Much time may be lost if the lathe is not set at the proper spindle speed and if the proper feed rate is not selected.

OBJECTIVES

After completing this unit you will be able to:

1. Calculate the speed at which a workpiece should revolve
2. Determine the proper feed for roughing and finishing cuts
3. Estimate the time required to machine a part in a lathe

CUTTING SPEEDS

Lathe work *cutting speed* (CS) may be defined as the rate at which a point on the work circumference travels past the cutting tool. For instance, if a metal has a CS of 90 ft/min, the spindle speed must be set so that 90 ft of the work circumference will pass the cutting tool in 1 min. Cutting speed is always expressed in feet per minute (ft/min) or in meters per minute (m/min). Do not confuse the CS of a metal with the number of turns the workpiece will make in 1 min (r/min).

Industry demands that machining operations be performed as quickly as possible; therefore correct CS must be used for the type of material being cut. If a CS is too high, the cutting tool edge breaks down rapidly, resulting in time lost to recondition the tool. With too slow a CS, time will be lost for the machining operation, resulting in low production rates. Based on research and testing by steel and cutting-tool manufacturers, the CS for high-speed steel tools listed in Table 47-1 are recommended for efficient metal removal rates. These speeds may be varied slightly to suit factors such as the condition of the machine, the type of work material, and sand or hard spots in the metal. The CS for cemented-carbide and ceramic cutting tools may be found in Units 31 and 32, Tables 31-4 and 32-3.

To calculate the lathe spindle speed in revolutions per minute (r/min), the CS of the metal and the diameter of the work must be known. The proper spindle speed can then be set by dividing the CS (in inches per minute) by the circumference of the work (in inches). The calculation for determining the spindle speed (r/min) is as follows:

INCH CALCULATIONS

$$\text{Formula:}\quad \text{r/min} = \frac{CS \times 12}{\pi D}$$

where CS = cutting speed
D = diameter of work to be turned

However, because most lathes provide only a limited number of speed settings, a simpler formula is usually used:

$$\text{r/min} = \frac{CS \times 4}{D}$$

Thus, to calculate the r/min required to rough-turn a 2-in.-diameter piece of machine steel (CS 90)

$$\text{r/min} = \frac{CS \times 4}{D} = \frac{90 \times 4}{2} = 180$$

NOTE: The recommended speeds and feeds for carbide cutting tools are found in Unit 31. The same formula for calculating spindle speed will be used for these tools.

Table 47-1 Lathe Cutting Speeds in Feet and Meters per Minute Using a High-Speed Steel Toolbit

Material	Turning and Boring				Threading	
	Rough Cut		Finish Cut			
	ft/min	m/min	ft/min	m/min	ft/min	m/min
Machine steel	90	27	100	30	35	11
Tool steel	70	21	90	27	30	9
Cast iron	60	18	80	24	25	8
Bronze	90	27	100	30	25	8
Aluminum	200	61	300	93	60	18

METRIC CALCULATIONS

$$\text{Formula:} \quad \text{r/min} = \frac{\text{CS} \times 320}{D}$$

EXAMPLE

Calculate the r/min required to turn a 45-mm-diameter piece of machine steel (CS 40 m/min).

$$\begin{aligned}\text{r/min} &= \frac{\text{CS} \times 320}{D} \\ &= \frac{40 \times 320}{45} \\ &= \frac{12{,}800}{45} \\ &= 284\end{aligned}$$

LATHE FEED

The *feed* of a lathe may be defined as the distance the cutting tool advances along the length of the work for every revolution of the spindle. For example, if the lathe is set for a 0.015-in. (0.40-mm) feed, the cutting tool will travel along the length of the work 0.015 in. (0.40 mm) for every complete turn that the work makes. The feed of an engine lathe is dependent upon the speed of the lead screw or feed rod. The speed is controlled by the change gears in the *quick-change gearbox* (Fig. 45-4A).

Whenever possible, only two cuts should be taken to bring a diameter to size: a roughing cut and a finishing cut. Since the purpose of a roughing cut is to remove excess material quickly and surface finish is not too important, a coarse feed should be used. The finishing cut is used to bring the diameter to size and produce a good surface finish, and therefore a fine feed should be used.

For general-purpose machining, a 0.010 to 0.015 in. (0.25 to 0.40 mm) feed for roughing and a 0.003 to 0.005 in. (0.07 to 0.012 mm) feed for finishing is recommended. Table 47-2 lists the recommended feeds for cutting various materials when a high-speed steel cutting tool is being used.

DEPTH OF CUT

The *depth of cut* may be defined as the depth of the chip taken by the cutting tool and is one-half the total amount removed from the workpiece in one cut. Figure 47-1 shows a 1/8-in. depth of cut being taken on a 2-in.-diameter workpiece. Note that the diameter has been reduced by 1/4 in. to 1 3/4 in. When machining a workpiece, take only one roughing and one finishing cut if possible. If much material must be removed, the roughing cut should be as deep as possible to reduce the diameter to within 0.030 to 0.040 in. (0.76 to 1.00 mm) of the size required. The depth of a rough turning cut will depend on the following factors.

- The condition of the machine
- The type and shape of the cutting tool used
- The rigidity of the workpiece, the machine, and the cutting tool
- The rate of feed

The depth of a finish turning cut will depend on the type

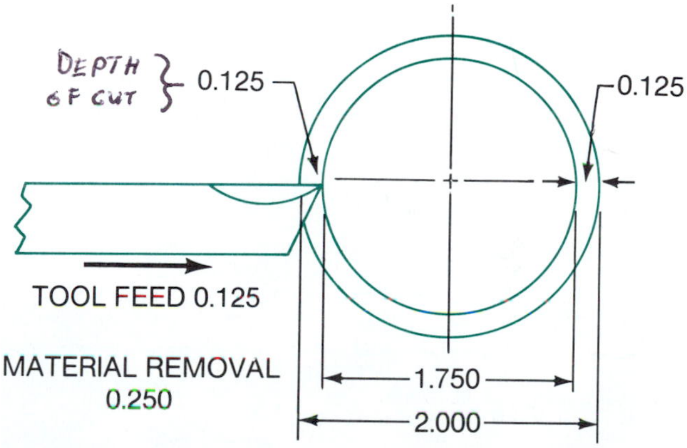

DEPTH OF CUT
0.125
0.125
TOOL FEED 0.125
MATERIAL REMOVAL
0.250
1.750
2.000

Figure 47-1 *The depth of cut on a lathe.*

Table 47-2 Feeds for Various Materials (Using a High-Speed Steel Cutting Tool)

Material	Rough Cuts in.	Rough Cuts mm	Finish Cuts in.	Finish Cuts mm
Machine steel	0.010–0.020	0.25–0.50	0.003–0.010	0.07–0.25
Tool steel	0.010–0.020	0.25–0.50	0.003–0.010	0.07–0.25
Cast iron	0.015–0.025	0.40–0.65	0.005–0.012	0.13–0.30
Bronze	0.015–0.025	0.40–0.65	0.003–0.010	0.07–0.25
Aluminum	0.015–0.030	0.40–0.75	0.005–0.010	0.13–0.25

of work and the finish required. In any case, it should not be less than 0.005 in. (0.13 mm).

GRADUATED MICROMETER COLLARS

When the diameter of a workpiece must be turned to an accurate size, *graduated micrometer collars* should be used. Graduated micrometer collars are sleeves or bushings that are mounted on the compound rest and crossfeed screws (Fig. 47-2). They assist the lathe operator to set the cutting tool accurately to remove the required amount of material from the workpiece. The collars on lathes using the inch system of measurement are usually graduated in thousandths of an inch (0.001). The micrometer collars on lathes using the metric system of measurement are usually graduated in steps of two hundredths of a millimeter (0.02 mm). The circumference of the crossfeed and compound rest screw collars on lathes using the inch system of measurement are usually divided into 100 or 125 equal divisions, each having a value of 0.001 in. Therefore, if the crossfeed screw is turned *clockwise* 10 graduations, the cutting tool will be moved 0.010 in. toward the work. Because the work in a lathe revolves, a 0.010 in. depth of cut will be taken from the entire work circumference, thereby reducing the diameter 0.020 in. (2 × 0.010 in.) (Fig. 47-3). On some lathes, however, the graduations are such that when the graduated collar is moved 10 divisions, the cutting tool will move only 0.005 in. and the diameter will be reduced by 0.010 in., or by an amount equal to the reading on the collar.

A
20
10
0
120

Figure 47-2 *Micrometer collars on the compound rest and crossfeed handles permit machining the workpiece to an accurate size. The thumbscrew, A, is used to lock the collar in place.*

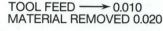

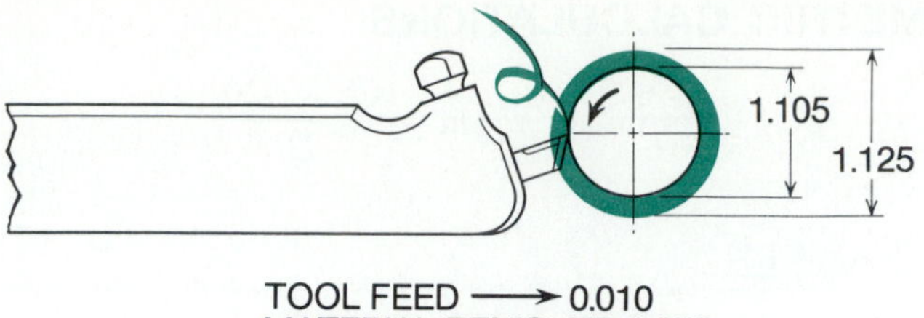
1.105
1.125
TOOL FEED → 0.010
MATERIAL REMOVED 0.020

Figure 47-3 *On machines where the workpiece revolves, the cutting tool should be set in for only half the amount to be removed from the diameter.*

NOTE: Some industrial machines may have collars with 250 or even 500 graduations, which make a much finer infeed possible.

Machine tools equipped with graduated collars generally fall into two classes:

1. *Machines in which the work revolves.* These include lathes, vertical boring mills, and cylindrical grinders.
2. *Machines in which the work does not revolve.* These include milling machines and surface grinders.

When the *circumference of work* is being cut on machines in which the work revolves, remember that since material is removed from the entire circumference, the cutting tool should be moved in only half the amount of material to be removed.

On machines in which the work does not revolve, the material removed from a workpiece is equal to the amount set on the graduated collar, because the machining is taking place only on one surface. Therefore, if a 0.010-in. depth of cut is set, 0.010 in. will be removed from the work (Fig. 47-4).

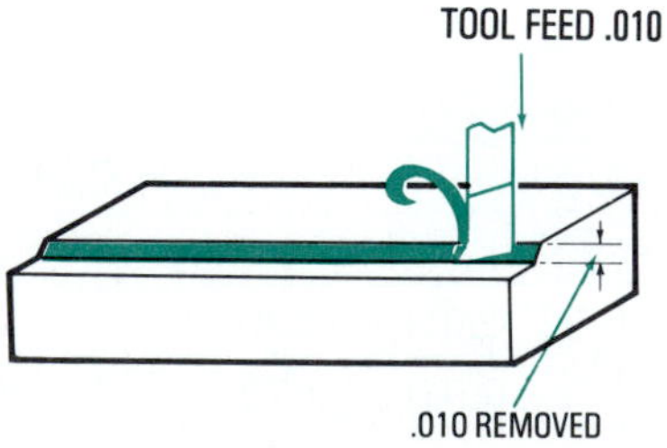
TOOL FEED .010
.010 REMOVED

Figure 47-4 *On machines where the workpiece does not revolve, the cutting tool should be set in for the amount of material to be removed.*

HINTS ON GRADUATED COLLAR USE

1. If the graduated collar has a locking screw, make sure the collar is secure before setting a depth of cut (Fig. 47-2).
2. All depths of cut must be made by feeding the cutting tool *toward the workpiece.*
3. If the graduated collar is turned past the desired setting, it must be turned backward a half-turn and then fed into the proper setting to remove the backlash (the play between the feed screw and the nut).
4. Never hold a graduated collar when setting a depth of cut. Graduated collars with friction devices can be moved easily if held when a depth of cut is being set.
5. The graduated collar on the *compound rest* can be used for accurately setting the depth of cut for the following operations.
   - *Shoulder turning*—When a series of shoulders must be spaced accurately along a piece of work, the compound rest should be set at 90° to the cross-slide. With the carriage locked in position, the graduated collar of the compound rest can be used for the spacing of shoulders to within 0.001 in. (0.02 mm on metric collars) accuracy.
   - *Facing*—The graduated collars may also be used to face a workpiece to length. When the compound rest is swung to 30°, the amount removed from the length of the work during facing will be one-half the amount of feed on the graduated collar.
   - *Machining accurate diameters*—When accurate diameters must be machined or ground in a lathe, the compound rest should be set to 84°16′ to the cross-slide. On inch-calibrated compound rests, a 0.001-in. movement would result in a 0.0001-in. infeed movement of the tool (Fig. 47-5). Similarly, on metric lathes, a 0.02-mm movement of the compound rest results in a 0.002-mm infeed movement of the cutting tool.

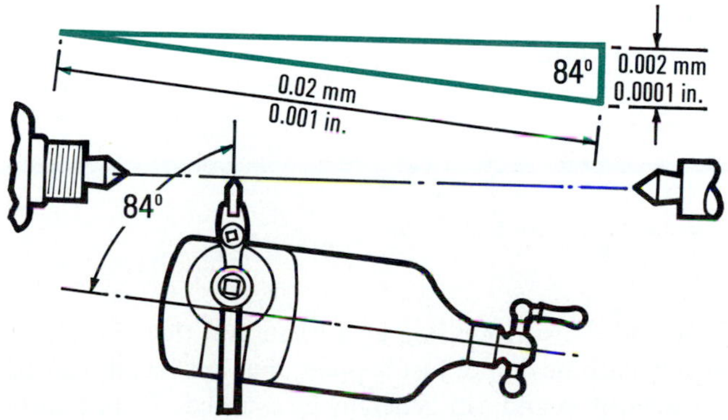

84°
0.002 mm
0.0001 in.
0.02 mm
0.001 in.
84°

Figure 47-5 *The compound rest is set at 84°16′ for making fine settings.*

CALCULATING MACHINING TIME

Any machinist should be able to estimate the time required to machine a workpiece. Factors such as spindle speed, feed, and depth of cut must be considered.

The following formula may be applied to calculate the time required to machine a workpiece.

$$\text{Time} = \frac{\text{distance}}{\text{rate}}$$

where distance = length of cut
rate = feed × r/min

EXAMPLE

Calculate the time required to machine a 2-in.-diameter machine-steel shaft 16 in. long to 1.850-in.-diameter finish size.

SOLUTION

Roughing cut—See Table 47-1:

$$\text{r/min} = \frac{CS \times 4}{D} = \frac{90 \times 4}{2} = 180$$

Roughing feed—See Table 47-2:

$$\text{Feed} = 0.020$$

$$\text{Roughing cut time} = \frac{\text{length of cut}}{\text{feed} \times \text{r/min}} = \frac{16}{0.020 \times 180} = 4.4 \text{ min}$$

Finishing cut:

$$\text{r/min} = \frac{100 \times 4}{1.850} = 216$$

$$\text{Finishing feed} = 0.003$$

$$\text{Finishing cut time} = \frac{16}{0.003 \times 216} = 24.7 \text{ min}$$

Total machine time: Roughing cut time + finishing cut time

$$= 4.4 + 24.7 = 29.1 \text{ min}$$

REVIEW QUESTIONS

CUTTING SPEEDS AND FEEDS

1. Define CS and state how it is expressed.
2. Why is proper CS important?
3. At what r/min should the lathe revolve to rough-turn a piece of cast iron 101 mm in diameter when using a high-speed steel toolbit?
4. Calculate the r/min to turn a piece of 3¾-in.-diameter machine steel using a high-speed steel toolbit.
5. Define lathe feed.

DEPTH OF CUT

6. Define depth of cut.
7. How deep should a roughing cut be?
8. What factors determine the depth of a roughing cut?
9. A 2½-in.-diameter workpiece must be machined to a 2.375-in. finished diameter. What should be the depth of:
    a. The roughing cut? b. The finishing cut?

GRADUATED MICROMETER COLLARS

10. Name and state the differences between two classes of machines that are equipped with graduated collars.
11. What precautions should be taken when setting a depth of cut?
12. What is the value of one graduation on a metric graduated micrometer collar?
13. What size will a 75-mm-diameter workpiece be after a 6.25-mm-deep cut is taken from the work?

MACHINING TIME

14. Calculate the time required to machine a 3⅛-in.-diameter piece of tool steel 14 in. long to a diameter of 3.000 in.

UNIT 48

Lathe Safety

A good worker will be aware of safety requirements in any area of a shop and will always attempt to observe safety rules. Failure to do so may result in serious injury, with the resultant loss of time and pay and in the loss of production to the company.

OBJECTIVES

After completing this unit you will be:
1. Aware of the importance of good safety standards in a shop
2. Familiar with the necessary safety precautions required to run a lathe
3. Able to point out any safety infractions by other workers

SAFETY PRECAUTIONS

The lathe, like most other machine tools, can be hazardous if not operated properly. A good lathe operator is a safe operator who realizes the importance of keeping the machine and the surrounding area clean and tidy. Accidents on any machine do not just happen; they are usually caused by carelessness and can generally be avoided. To minimize the chance of accidents when operating a lathe, the following precautions should be observed.

1. Always wear approved safety glasses. When operating a lathe, chips fly and it is important to protect your eyes.

2. Roll up your sleeves, remove your tie, and tuck in loose clothing. Short sleeves are preferable because loose clothing can get caught by revolving lathe dogs, chucks, and rotating parts of the lathe. You can be drawn into the machine and be seriously injured.

3. Never wear a ring or watch (Fig. 48-1).

   a. Rings or watches can be caught in revolving work or lathe parts and cause serious injuries.

   b. A metal object falling on the hand would bend or break a ring, causing much pain and suffering until the ring could be removed.

4. Do not operate a lathe until you fully understand its controls.

   a. Not knowing what might happen when levers or switches are turned on can be very dangerous.

   b. Be sure that you can stop the machine quickly in case something unexpected happens.

5. Never operate a machine if safety guards are removed or not closed properly.

   a. Safety guards are installed by the manufacturer to cover revolving gears, belts, or shafts.

   b. Loose clothing or a hand can be drawn into the revolving parts if guards are not replaced or closed properly.

6. Stop the lathe before you measure the work, clean, oil, or adjust the machine. Measuring revolving parts can result in broken tools or personal injury.

7. Do not use a rag to clean the work or the machine when the lathe is in operation. The rag may get caught and be drawn in along with your hand.

8. Never attempt to stop a lathe chuck or drive plate by hand. Your hand can be injured or fingers broken if they get caught in the slots and projections of the drive plate or chuck.

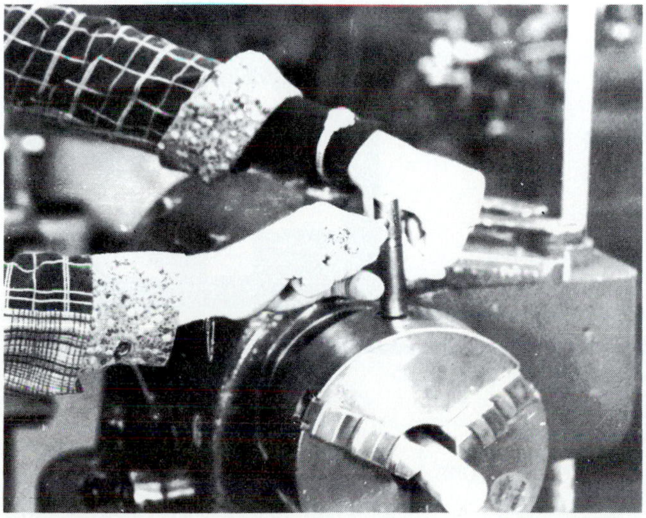

Figure 48-1 *Wearing rings and watches in a machine shop can be dangerous.*

9. Be sure that the chuck or faceplate is mounted securely before starting the lathe.

   a. If the lathe is started with a loose spindle accessory, the rotation would loosen the accessory and cause it to be thrown from the lathe.

   b. A heavy accessory, with the velocity created by the revolving spindle, can turn into a dangerous missile.

10. Always remove the chuck wrench after use (Fig. 48-2). *Never leave it in the chuck at any time!* If the lathe is started with a chuck wrench in the chuck, the following could result:

   a. The wrench could fly out and injure someone.

   b. The wrench could be jammed against the lathe bed damaging the wrench, lathe bed, chuck, and lathe spindle.

11. Move the carriage to the farthest position of the cut and revolve the lathe spindle one complete turn by hand before you start the lathe.

   a. This will ensure that all parts will clear without jamming.

   b. It will also prevent an accident and damage to the lathe.

12. Keep the floor around the machine free from grease, oil, metal cuttings, tools, and workpieces (Fig. 48-3).

Figure 48-2 *Never leave a chuck wrench in a chuck.*

Figure 48-3 *Good housekeeping can prevent tripping and slipping accidents.*

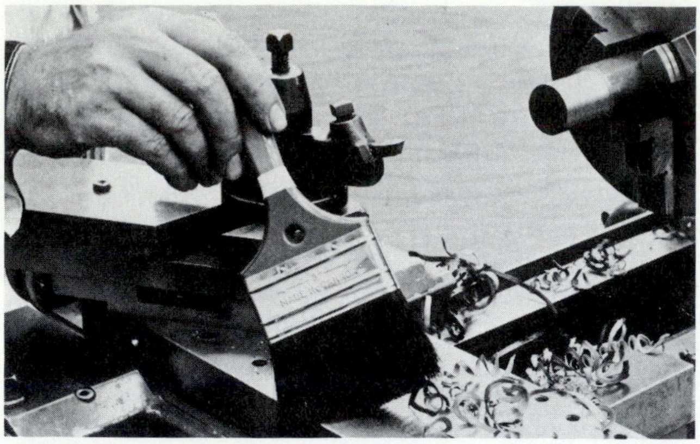

Figure 48-4 *Always remove chips with a brush or hook, never by hand.*

a. Oil and grease can cause falls, which can result in painful injuries.

b. Objects on the floor are hazards, which can cause tripping accidents.

13. Avoid horseplay at all times, especially when operating any machine tool. Horseplay can result in tripping accidents or being pushed into the revolving spindle or workpiece.

14. Always remove the chips with a brush and never with your hand or a cloth (Fig. 48-4). Steel chips are sharp and can cause cuts if handled manually or with a cloth that has chips embedded in it.

15. Whenever you are polishing, filing, cleaning, or making adjustments to the workpiece or machine, always remove the sharp toolbit from the toolholder to prevent serious cuts to your arms or hands.

REVIEW QUESTIONS

1. State three possible results of not observing the safety rules in a shop.
2. What is generally the most important cause of an accident?
3. Why are the following precautions important when operating a lathe?
   a. Wearing safety glasses
   b. Not wearing loose clothing
   c. Not wearing watches and rings
   d. Removing a chuck wrench
   e. Keeping the machine and surrounding floor area clean
   f. Stopping the lathe to measure work or clean the machine
4. Why should chips not be cleaned from a lathe with a cloth?

UNIT 49

Mounting, Removing, and Aligning Lathe Centers

Any work machined between lathe centers is generally turned for some portion of its length, then reversed, and the other end finished. It is important, when machining work between centers, that the live center runs absolutely true. If the live center does not run true, when the work is reversed to machine the opposite end, the two turned diameters will not be concentric and the part may have to be scrapped.

OBJECTIVES

After completing this unit you will be able to:
1. Properly mount and/or remove the lathe centers
2. Align the lathe centers by visual, trial-cut, and dial-indicator methods

TO MOUNT LATHE CENTERS

1. Remove any burrs from the lathe spindle, centers, or spindle sleeves (Fig. 49-1).
2. Thoroughly clean the tapers on the lathe centers and in the headstock and tailstock spindles (Fig. 49-2).

NOTE: Never attempt to clean the taper in the headstock spindle while the lathe is running.

3. Partially insert the cleaned center in the lathe spindle.
4. With a quick snap, force the center into the spindle. When mounting a tailstock center, follow the same procedure.

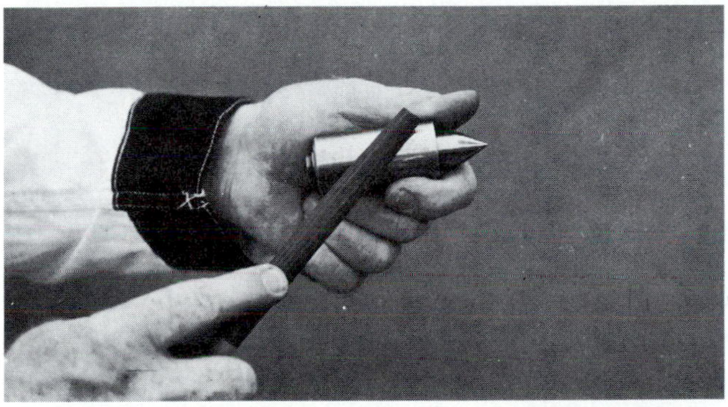

Figure 49-1 *Remove burrs from the spindle sleeve before inserting it in a lathe.*

Figure 49-2 *Clean the headstock spindle before inserting a center.*

After a center has been mounted in the headstock spindle, it should be checked for trueness. Start the lathe and observe if the center runs true. Whenever accuracy is required, check the trueness of the center with a dial indicator (Fig. 49-3). If the center is not running true and has been mounted properly, it should be ground while mounted in the lathe spindle.

TO REMOVE LATHE CENTERS

The *live center* may be removed by using a *knockout bar* that is pushed through the headstock spindle (Fig. 49-4). A

Figure 49-3 *Using a dial indicator to check the trueness of the live center.*

CLOTH
KNOCKOUT BAR

Figure 49-4 *Removing the live center with a knockout bar.*

Figure 49-5 *When removing the lathe center, hold a cloth over its point to prevent hand injury. (Courtesy Kostel Enterprises Limited.)*

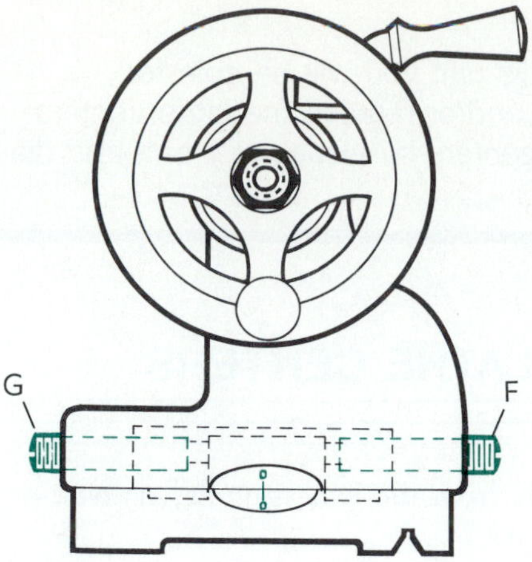

G
F

Figure 49-6 *The lines on the tailstock must be in line to produce a parallel diameter.*

slight tap is required to remove the center. When removing the live center with a knockout bar, place a cloth over the center and hold it with one hand to prevent an accident or damage to the center (Fig. 49-5).

The *dead center* can be removed by turning the tailstock handwheel to draw the spindle back into the tailstock. The end of the screw contacts the end of the dead center, forcing it out of the spindle.

ALIGNMENT OF LATHE CENTERS

To produce a parallel diameter when machining work between centers, the lathe center must be aligned; that is, the two lathe centers must be in line with each other and true with the centerline of the lathe. If the centers are not aligned, the work being machined will be tapered.

Three common methods are used to align lathe centers:

1. By aligning the centerlines on the back of the tailstock with each other (Fig. 49-6). This is only a visual check and therefore not too accurate.
2. The trial-cut method (Fig. 49-7), where a small cut is taken from each end of the work and the diameters are measured with a micrometer.
3. By using a parallel test bar and dial indicator (Fig. 49-8). This is the fastest and most accurate method of aligning lathe centers.

TO ALIGN CENTERS BY ADJUSTING THE TAILSTOCK

1. Loosen the tailstock clamp nut or lever.
2. Loosen one of the adjusting screws G or F (Fig. 49-6), depending on the direction the tailstock must be moved. Tighten the other adjusting screw until the line on the top half of the tailstock aligns exactly with the line on the bottom half.
3. Tighten the loosened adjusting screw to lock both halves of the tailstock in place.
4. Make sure that the tailstock lines are still aligned; adjust them if necessary.
5. Lock the tailstock clamp nut or lever.

TO ALIGN CENTERS BY THE TRIAL-CUT METHOD

1. Take a light cut [approximately 0.005 in. (0.12 mm)] to true the diameter, from Section A at the tailstock end for ¼ in. (6 mm) long.
2. Stop the feed and note the reading on the graduated collar of the crossfeed handle.
3. Move the cutting tool away from the work with the crossfeed handle.
4. Bring the cutting tool close to the headstock end.
5. Return the cutting tool to the same graduated collar setting as at Section A.
6. Cut a ½-in. (13-mm) length at Section B and then stop the lathe.
7. Measure both diameters with a micrometer (Fig. 49-7).
8. If both diameters are not the same size, adjust the tailstock either toward or away from the cutting tool one-half the difference of the two readings.
9. Take another light cut at A and B at the same crossfeed graduated collar setting. Measure these diameters and adjust the tailstock, if required.

TO ALIGN CENTERS USING A DIAL INDICATOR AND TEST BAR

1. Clean the lathe and work centers and mount the test bar.

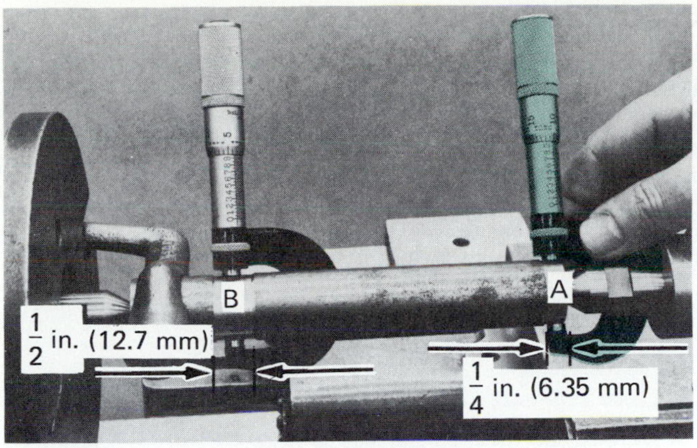
B
A
$\frac{1}{2}$ in. (12.7 mm)
$\frac{1}{4}$ in. (6.35 mm)

Figure 49-7 *A trial cut at each end of the workpiece is used to check the lathe's center alignment.*

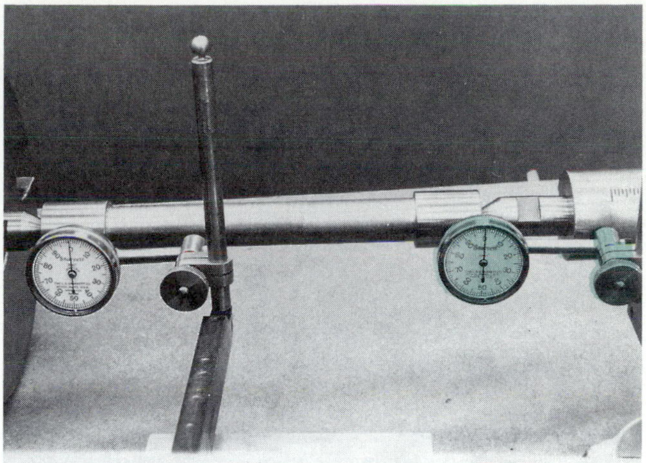

Figure 49-8 *Tailstock alignment may be accurately checked with a parallel test bar and dial indicator.*

2. Adjust the test bar snugly between centers and tighten the tailstock spindle clamp.
3. Mount a dial indicator on the toolpost or lathe carriage. Make sure that the indicator plunger is parallel to the lathe bed and that the contact point is set on center.
4. Adjust the cross-slide so that the indicator registers approximately 0.025 in. (0.65 mm) at the tailstock end.
5. Move the carriage by hand so that the indicator registers on the diameter at the headstock end (Fig. 49-8) and note the indicator reading.
6. If both indicator readings are not the same, adjust the tailstock by the adjusting screws until the indicator registers the same reading at both ends.
7. Tighten the adjusting screw which was loosened.
8. Tighten the tailstock clamp nut.
9. Adjust the tailstock spindle until the test bar is snug between the lathe centers.
10. Recheck the indicator readings at both ends and adjust the tailstock, if necessary.

REVIEW QUESTIONS

1. Briefly describe the proper procedure for mounting a lathe center.
2. How may the live center be checked for trueness after it has been mounted in the spindle?
3. What precautions should be observed when removing a live center?
4. Briefly describe how to align the lathe centers by the trial-cut method.
5. Why must the test bar be snug between centers when aligning centers with a dial indicator?

UNIT 50

Grinding Lathe Cutting Tools

Because of the number of turning operations which can be performed on a lathe, a wide variety of cutting tools is used. In order for these cutting tools to perform effectively, they must possess certain angles and clearances for the material being cut (see Table 29-2). All lathe tools cut if they have front and side clearance.The addition of side and top rake enables the chips to escape quickly from the cutting edge, making the tool cut better. All lathe cutting tools must have certain angles and clearances regardless of shape; therefore, only the grinding of a general-purpose cutting tool (Fig. 50-1) will be explained in detail.

OBJECTIVES

After completing this unit you will be able to:

1. Explain the importance of various rakes and clearance cutting tool angles
2. Grind a general-purpose cutting tool from a blank
3. Resharpen a worn cutting tool

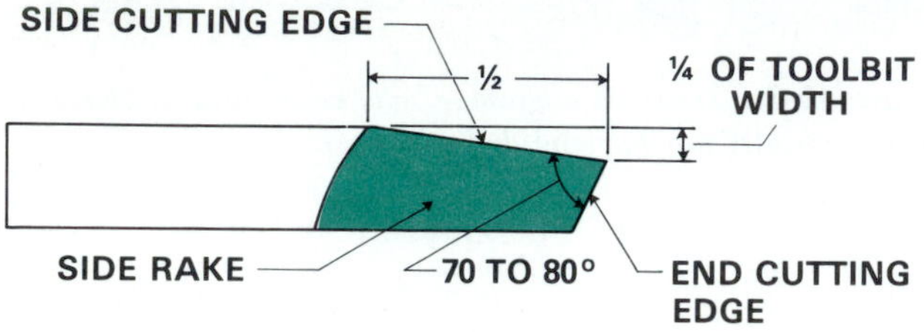
SIDE CUTTING EDGE
½
¼ OF TOOLBIT WIDTH
SIDE RAKE
70 TO 80°
END CUTTING EDGE

Figure 50-1 *The shape and dimensions of a general-purpose lathe toolbit.*

TO GRIND A GENERAL-PURPOSE TOOLBIT

1. Dress the face of the grinding wheel.
2. Grip the toolbit firmly, supporting the hands on the grinder toolrest (Fig. 50-2).
3. Hold the toolbit at the proper angle to grind the cutting edge angle. At the same time, tilt the bottom of the toolbit in toward the wheel and grind the 10° side relief or clearance angle on the cutting edge.

NOTE: The cutting edge should be approximately ½ in. (13 mm) long and should extend over about one-quarter the width of the toolbit (Fig. 50-2).

4. While grinding, move the toolbit back and forth across the face of the wheel. This accelerates grinding and prevents grooving the wheel.
5. The toolbit must be cooled frequently during the grinding operation. NEVER OVERHEAT A TOOLBIT.

NOTE: *Never* quench stellite or cemented-carbide tools and *never* grind carbides with an aluminum oxide wheel.

6. Grind the end cutting edge so that it forms an angle of a little less than 90° with the side cutting edge (Fig. 50-3). Hold the tool so that the end cutting edge angle and end relief angle of 15° are ground at the same time.
7. Using a toolbit grinding gage, check the amount of end relief when the toolbit is in the toolholder (Fig. 50-4).

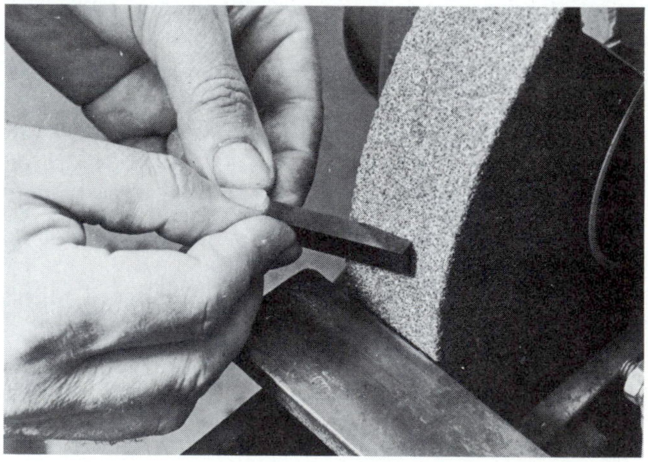

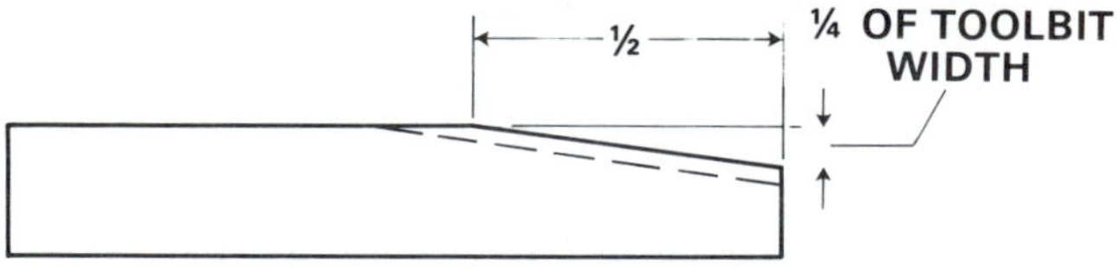
½
¼ OF TOOLBIT WIDTH

Figure 50-2 *Grinding the side cutting edge and side relief angles on a toolbit.*

8. Hold the top of the toolbit at approximately 45° to the axis of the wheel and grind the side rake to approximately 14° (Fig. 50-5).

NOTE: When grinding the side rake, *be sure that the top of the cutting edge is not ground below the top of the tool-*

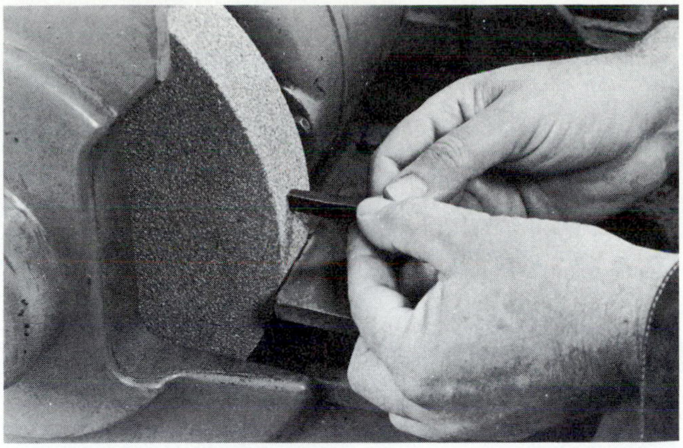

70° TO 80° POINT ANGLE

Figure 50-3 *Grinding the end relief angle on a lathe toolbit.*

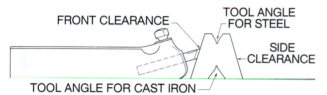

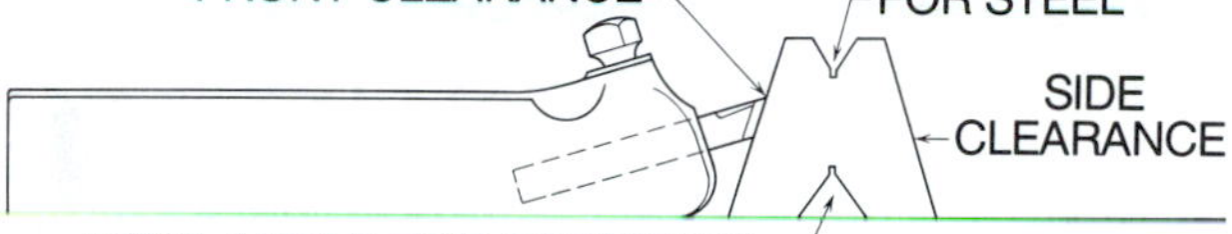
FRONT CLEARANCE
TOOL ANGLE FOR STEEL
SIDE CLEARANCE
TOOL ANGLE FOR CAST IRON

Figure 50-4 *Checking the end relief angle of a toolbit while it is in a toolholder.*

bit. If a step is ground in the top of the toolbit, a chip trap is formed, which greatly reduces the efficiency of the cutting tool.

9. Grind a slight radius on the point of the cutting tool, *being sure to maintain the same front and side clearance angle.*
10. With an oilstone, hone the cutting edge of the toolbit slightly. This will lengthen the life of the toolbit and enable it to produce a better surface finish.

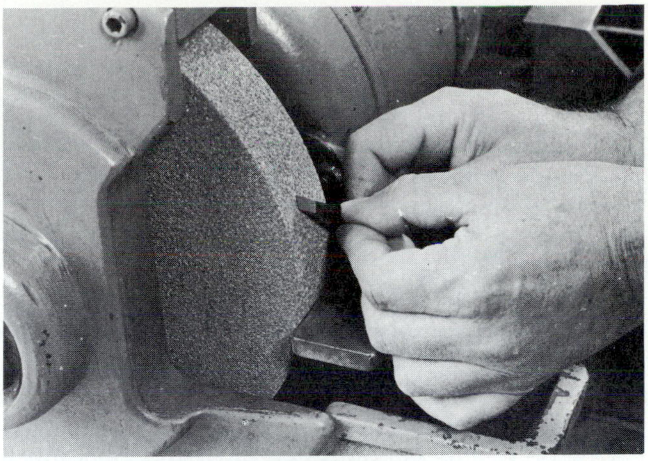

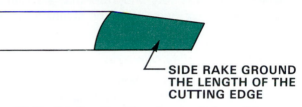

SIDE RAKE GROUND THE LENGTH OF THE CUTTING EDGE

Figure 50-5 *Grinding the side rake on a lathe toolbit. (Courtesy Kostel Enterprises Limited.)*

REVIEW QUESTIONS

1. What are two requirements that must be met to enable a lathe toolbit to cut?
2. Why should the top of the cutting edge not be ground below the top of the toolbit when the side rake is ground?
3. How should the point of a toolbit be conditioned?
4. Briefly state the procedure for grinding a general-purpose toolbit.

UNIT 51

Facing Between Centers

Work mounted between centers can be machined, removed, or set up for additional machining and still maintain the same degree of accuracy. Facing on a lathe is one of the most important machining operations in a machine shop. It is very important that the cutting tool and work be properly set up, or damage to the machine, work, and lathe centers will result.

OBJECTIVES

After completing this unit you will be able to:
1. Set up a workpiece for machining between centers
2. Set up a workpiece to face the ends
3. Face a workpiece to an exact length

TO SET UP A CUTTING TOOL FOR MACHINING

1. Move the toolpost to the *left-hand side* of the T-slot in the compound rest.
2. Mount a toolholder in the toolpost so that the setscrew in the toolholder is approximately 1 in. (25 mm) beyond the toolpost (Fig. 51-1).
3. Insert the proper cutting tool into the toolholder, having the toolbit extend ½ in. (13 mm) beyond the toolholder, but never more than twice its thickness.
4. Tighten the toolholder setscrew with only two-finger pressure on the wrench to hold the toolbit in the toolholder.

NOTE: If the toolholder setscrew is tightened too tightly, it will break the toolbit, which is very hard and brittle.

5. Set the cutting-tool point to center height. Check it against the lathe center point (Fig. 51-1).
6. Tighten the toolpost securely to prevent it from moving during a cut.

Figure 51-1 *A toolholder and toolbit being set up for a machining operation.*

TO MOUNT WORK BETWEEN CENTERS

USING A REVOLVING TAILSTOCK CENTER

1. Check the live center by holding a piece of chalk close to it while it is revolving. If the live center is not running true, the chalk will mark only the high spot.
2. If this happens, remove the live center from the headstock and clean the tapers on the center and the headstock spindle.
3. Replace the center and check for trueness.
4. Adjust the tailstock spindle until it extends about $2\frac{1}{2}$ to 3 in. (63 to 75 mm) beyond the tailstock.
5. Mount a revolving center into the tailstock spindle (Fig. 51-2).
6. Loosen the tailstock clamp nut or lever.
7. Place the lathe dog on the end of the work with the tail pointing to the left.
8. Place the end of the work with the lathe dog on the live center and slide the tailstock toward the headstock until the dead center supports the other end of the work.
9. Tighten the tailstock clamp nut or lever (Fig. 51-3).
10. Adjust the tail of the dog in the slot of the drive plate and tighten the lathe dog screw.
11. Tighten the tailstock handwheel using *thumb and finger pressure* only to snug up the work between centers.
12. Tighten the tailstock spindle clamp.
13. Move the carriage to the farthest position (left-hand end) of the cut and revolve the lathe spindle by hand to make sure that the dog does not hit the compound rest.

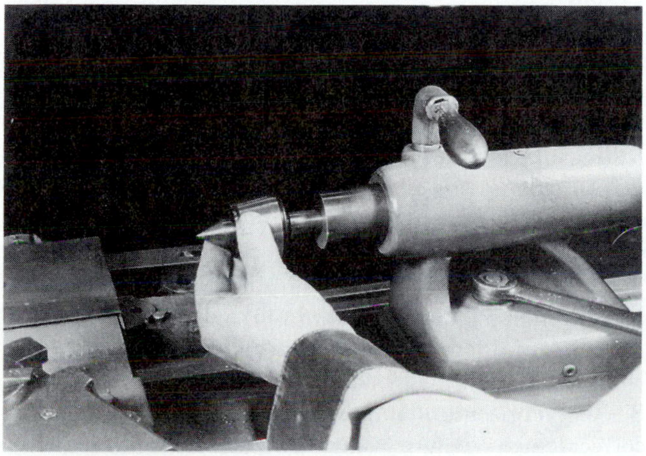

Figure 51-2 *Clean both the internal and external tapers before mounting a center in the tailstock spindle.*

Figure 51-3 *When the tailstock is in the correct position, tighten the clamp nut.*

USING A DEAD CENTER

1. Follow steps 1–4 of the revolving center method.
2. Mount the dead center in the tailstock spindle and then check the center alignment.
3. Place the dog on the end of the work and lubricate the dead center hole (Fig. 51-4).
4. Mount the work between centers and tighten the tailstock clamp nut or lever (Fig. 51-3).
5. Adjust the tail of the dog in the drive plate and tighten the set screw.
6. Turn the drive plate until the slot is in a horizontal position.
7. Hold the tail of the dog *up* in the slot and tighten the tailstock handwheel.
8. Turn the handwheel backward until the tail of the dog *just drops* and tighten the tailstock spindle clamp (Fig. 51-5).
9. Check the center tension of the work. The tail of the dog should drop of its own weight, and there should be no end play (Fig. 51-6).

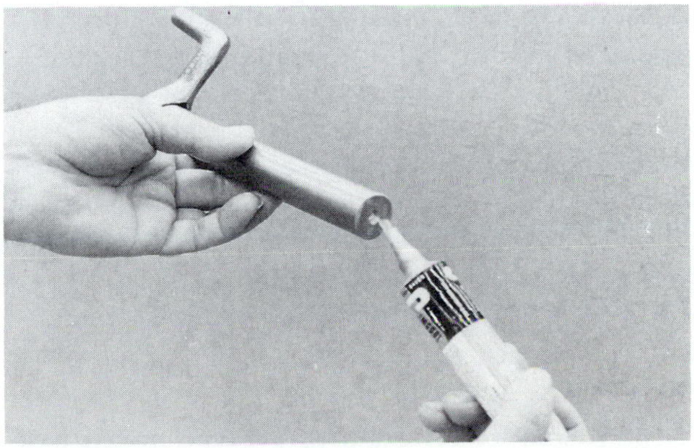

Figure 51-4 *Applying lubricant to a center hole that will be supported by a dead tailstock center.*

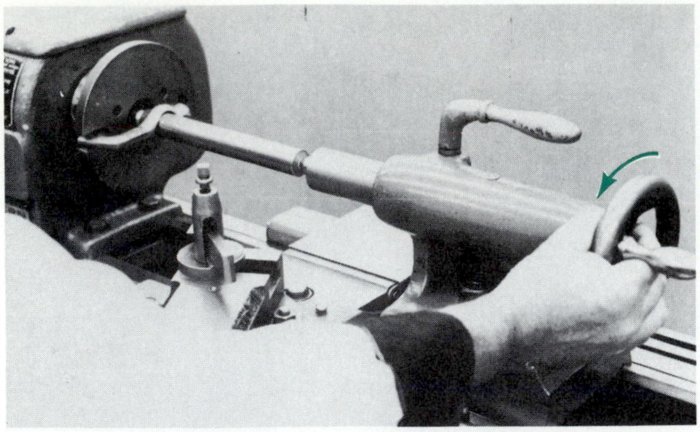

Figure 51-5 *Reverse the tailstock handwheel until the tail of the dog drops into the slot.*

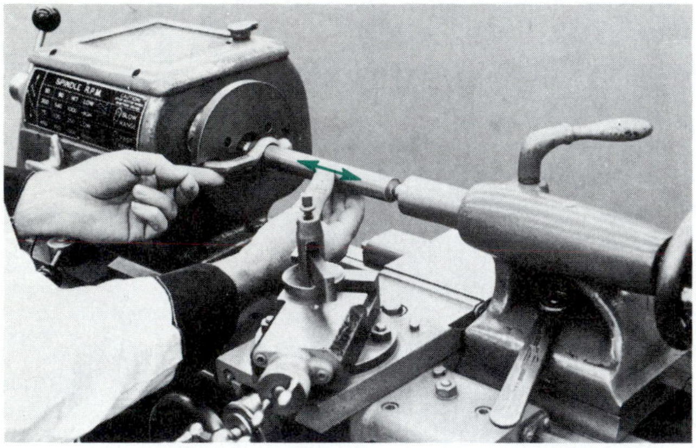

Figure 51-6 *The center tension is correct when the tail of the dog drops of its own weight and there is no end play between centers.*

FACING BETWEEN CENTERS

Workpieces to be machined are generally cut a little longer than required and then end-faced to the proper length. Facing is an operation of machining the ends of a workpiece square with its axis. To produce a flat, square surface when material is being faced between centers, the lathe centers must be in line. Work is often held in a chuck, faced to length, and center-drilled in one setup. This operation is covered in Unit 57.

The purposes of facing are:

1. To provide a true, flat surface, square with the axis of the work
2. To provide an accurate surface from which to take measurements
3. To cut the work to the required length

Figure 51-7 *Grip the toolholder short and set the toolbit to center. (Courtesy Kostel Enterprises Limited.)*

TO FACE WORK BETWEEN CENTERS

1. Move the toolpost to the *left-hand side* of the compound rest, and set the right-hand facing toolbit to the height of the lathe center point (Fig. 51-7).
2. Clean the lathe and work centers and mount the work between centers.

NOTE: Use a half-center in the tailstock if one is available (Fig. 51-8).

3. Set the facing toolbit pointing left, as shown in Fig. 51-8.

NOTE: The point of the toolbit must be closest to the work and a space must be left along the side.

4. Set the lathe to the correct speed for the diameter and type of material being cut.
5. Start the lathe and bring the toolbit as close to the center of the work as possible.
6. Move the carriage to the left, using the apron handwheel, until a small cut is started.
7. Feed the cutting tool out by turning the crossfeed handle and cut from the center outward. If the automatic crossfeed is used for feeding the cutting tool, the carriage should be locked in position (Fig. 51-9).
8. Repeat operations 5, 6, and 7 until the work is cut to the correct length. (Before facing, mark the correct length with center-punch marks and then face until the punch marks are cut in half.)

SPACE

Figure 51-8 *A half-center allows the entire surface to be faced.*

Figure 51-9 *Lock the carriage to produce a flat surface.*

NOTE: When facing, finishing cuts should begin at the center of the workpiece and feed toward the outside.

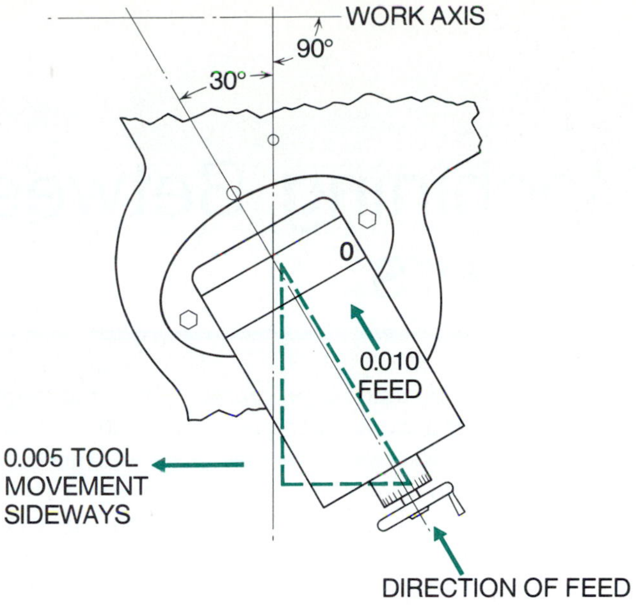
WORK AXIS
90°
30°
0
0.010 FEED
0.005 TOOL MOVEMENT SIDEWAYS
DIRECTION OF FEED

Figure 51-10 *The compound rest may be set at 30° for accurate end facing.*

FACING THE WORKPIECE TO AN ACCURATE LENGTH

When work is to be machined accurately to length, the compound rest graduated micrometer collars may be used. When facing, the compound rest may be set to 30° (to the cross-slide). The side movement of the cutting tool is always half the amount of the compound feed. For example, if the compound rest is fed in 0.010 in. (0.25 mm), the side movement of the tool, or the amount of material removed from the end of the workpiece, will be 0.005 in. (0.12 mm) (Fig. 51-10).

REVIEW QUESTIONS

MOUNTING WORK BETWEEN CENTERS

1. Explain the procedure for setting up a cutting tool for turning.
2. List the main steps for mounting work between centers.

FACING BETWEEN CENTERS

3. State three purposes of facing a workpiece.
4. Explain how a facing toolbit is set up.

UNIT

52

Machining Between Centers

In school shops or training programs, where the length of each session is fixed, much of the work machined on a lathe is mounted between centers. When both the headstock and tailstock are aligned, work can be machined, removed from the lathe at the end of the work period, and replaced for additional machining with the assurance that any machined diameter will run true (concentric) with other diameters. In training programs, it is often necessary to remove and replace work in a lathe many times before it is completed; therefore, machining between centers saves much valuable time in setting up work accurately in comparison to other work-holding methods. The most common operations performed on work mounted between centers are facing, rough and finish turning, shoulder turning, filing, and polishing.

OBJECTIVES

After completing this unit you should be able to:

1. Set up the cutting tool for turning operations
2. Turn parallel diameters to within ±0.002 in. (0.05 mm) accuracy for size
3. Produce a good surface finish by filing and polishing
4. Machine square, filleted, and beveled shoulders to within 1/64 in. (0.30 mm) accuracy

SETTING UP A CUTTING TOOL

When machining between centers, it is very important that the workpiece and cutting tool be set up properly; otherwise the workpiece could be ruined or the lathe centers damaged. Carelessness in setting up work properly could also result in work being thrown out of a lathe and causing injury to the operator.

1. Move the toolpost to the *left-hand side* of the T-slot in the compound rest.
2. Mount a toolholder in the toolpost so that the setscrew in the toolholder is approximately 1 in. (25 mm) beyond the toolpost (Fig. 52-1).

When taking heavy cuts, set the toolholder at right angles to the work (Fig. 52-2A). If the toolholder should move under pressure of the cut, the cutting tool would

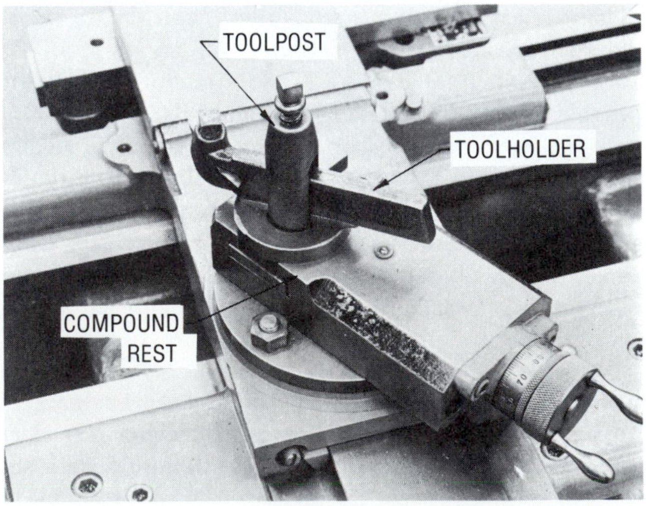
TOOLPOST
TOOLHOLDER
COMPOUND
REST
Figure 52-1 *Grip the toolholder short in the toolpost.*

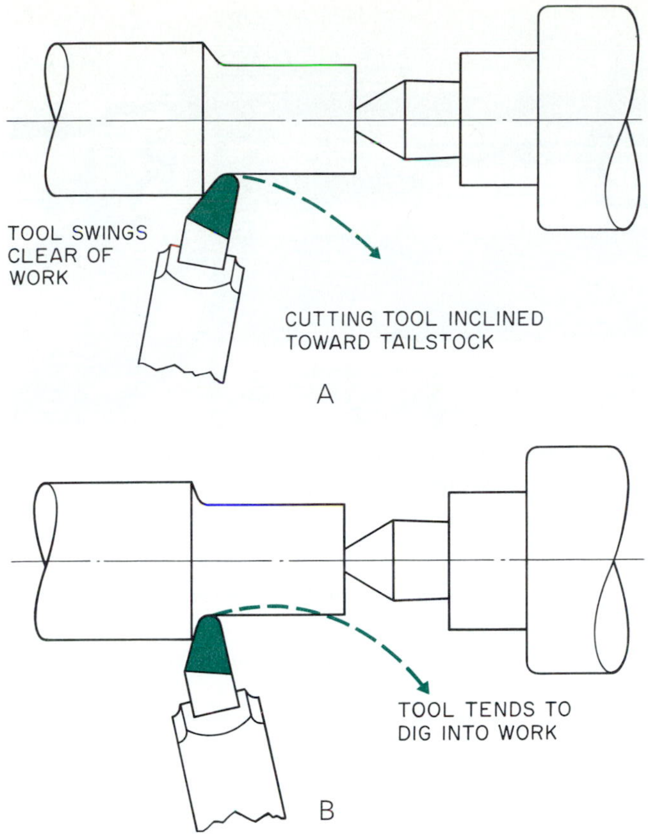
TOOL SWINGS
CLEAR OF
WORK
CUTTING TOOL INCLINED
TOWARD TAILSTOCK
A
TOOL TENDS TO
DIG INTO WORK
B

Figure 52-2 *(A) The toolholder set to prevent the toolbit from digging into the work; (B) the toolholder improperly set for making a heavy cut.*

swing away from the work and make the diameter larger.

If the toolholder were set as in Fig. 52-2B and it moved under pressure of the cut, the toolbit would swing into the work, causing the diameter to be cut undersize. A toolholder can be set as in Fig. 52-2B for light finishing cuts.

3. Insert the proper cutting tool into the toolholder, having the tool extend ½ in. (13 mm) beyond the toolholder and never more than twice its thickness.

4. Set the cutting-tool point to center height. Check it against the lathe center point (Fig. 52-3).

5. Tighten the toolpost *securely* to prevent it from moving during a cut.

MOUNTING WORK BETWEEN CENTERS

Since the procedure for mounting the work between lathe centers for machining is the same as for facing, the explanation of this operation will not be repeated. See Unit 51 for the proper procedure for mounting work between centers.

Figure 52-3 *Set the point of the toolbit even with the center.*

PARALLEL TURNING

Work is generally machined on a lathe for two reasons: to cut it to size and to produce a true diameter. Work that must be cut to size and have the same diameter along the entire length of the workpiece involves the operation of parallel turning. Many factors determine the amount of material which can be removed on a lathe at one time. Whenever possible, a diameter should be cut to size in two cuts: a roughing cut and a finishing cut (Fig. 52-4).

NOTE: To remove metal from a cylindrical piece of work and have the same diameter at each end, *the lathe centers must be in line.* (See Unit 49 for the methods of aligning centers.)

Before either the rough or finish cut is taken, the cutting tool must be set accurately for the depth of cut desired.

SETTING AN ACCURATE DEPTH OF CUT

In order to machine any diameter on a lathe accurately to a size, it is important that a trial cut be taken off the diameter

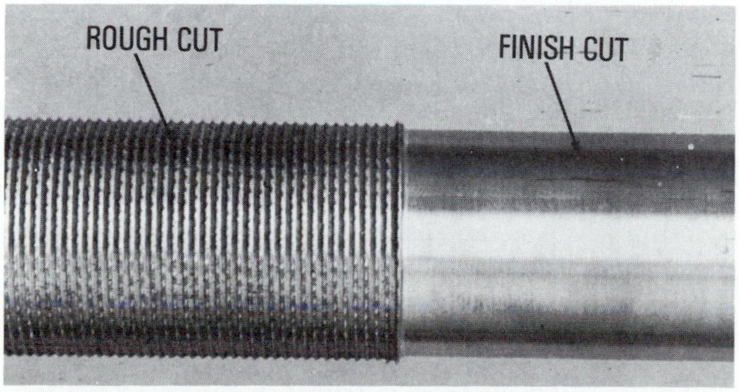
ROUGH CUT
FINISH CUT

Figure 52-4 *A rough cut and a finish cut. (Courtesy Kostel Enterprises Limited.)*

to be turned before setting *any depth of cut* on the crossfeed micrometer graduated collar. The purposes of a trial cut are to:

- Produce an accurate turned diameter, which can be measured with a micrometer
- Set the cutting-tool point to the diameter
- Set the crossfeed micrometer collar to the diameter

Only after the trial cut has been taken is it possible to set an accurate depth of cut.

To Take a Trial Cut

1. Set up the workpiece and cutting tool as for turning.
2. Set the proper speeds and feeds to suit the material being cut.
3. Start the lathe and position the toolbit over the work approximately ⅛ in. (3 mm) from the end.
4. Turn the compound rest handle *clockwise* one-quarter of a turn to remove any backlash.
5. Feed the toolbit into the work by turning the crossfeed handle clockwise until a light ring appears around the entire circumference of the work (Fig. 52-5).
6. CAUTION: *DO NOT MOVE THE CROSSFEED HANDLE SETTING.*

If the crossfeed is moved, two out of the three purposes of a trial cut are destroyed; the cutting tool and the crossfeed micrometer collar are no longer set to the diameter. Therefore the starting point of setting an accurate cut is lost.

7. Turn the carriage handwheel until the toolbit clears the end of the workpiece by about 1/16 in. (1.5 mm).
8. Turn the crossfeed handle clockwise about 0.010 in. (0.25 mm) and take a trial cut ¼ in. (6 mm) along the length of the work (Fig. 52-6).
9. Disengage the automatic feed and clear the toolbit past the end of the work with the carriage handwheel.
10. Stop the lathe.

Figure 52-5 *The first step in setting an accurate depth of cut is to machine a light ring around the circumference of the workpiece.*

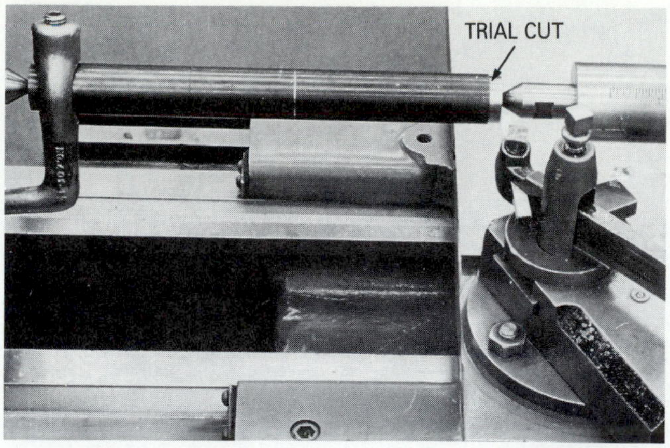
TRIAL CUT

Figure 52-6 *Make a light trial cut about ¼ in. long to clean up the diameter.*

11. Test the accuracy of the micrometer by cleaning and closing the measuring faces and then measure the trial-cut diameter (Fig. 52-7).
12. Calculate how much material must still be removed from the diameter of the work.
13. Turn the crossfeed handle clockwise one-half the amount of material to be removed (the full amount if the micrometer collar indicates material being removed from the work diameter).
14. Take another trial cut ¼ in. (6 mm) long and stop the lathe.

NOTE: If the diameter is too small, note the graduated collar setting, turn the crossfeed handle *counterclockwise* one-half turn and then clockwise to the required setting.

15. Clear the toolbit over the end of work with the carriage handwheel.
16. Measure the diameter and if necessary, readjust the crossfeed handle until the diameter is correct.
17. Machine the diameter to length.

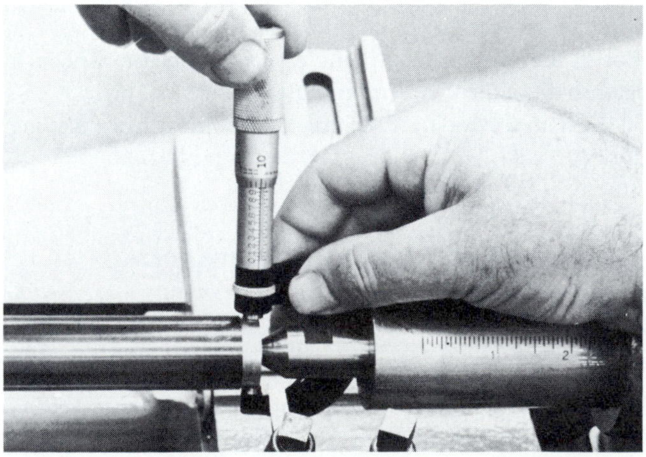

Figure 52-7 *With the cutting tool clear of the workpiece, measure the diameter.*

ROUGH TURNING

The operation of rough turning is used to remove as much metal as possible in the shortest length of time. Accuracy and surface finish are not important in this operation; therefore a 0.020- to 0.030-in. (0.50- to 0.76-mm) feed is recommended. Work is generally rough-turned to within 1/32 in. (0.80 mm) of the finished size when removing up to 1/2 in. (13 mm) from the diameter; within 1/16 in. (1.60 mm) when removing more than 1/2 in.

Proceed as follows:

1. Set the lathe to the correct speed for the type and size of material being cut (Table 47-1).
2. Adjust the quick-change gearbox for a 0.010- to 0.030-in. (0.25- to 0.76-mm) feed, depending on the depth of cut and condition of the machine.
3. Move the toolholder to the left-hand side of the compound rest and set the toolbit height to center.
4. Tighten the toolpost *securely* to prevent the toolholder from moving during the machining operation.
5. Take a light trial cut at the right-hand end of the work for 1/4-in. (6 mm) length (Fig. 52-6).
6. Measure the work and adjust the toolbit for the proper depth of cut.
7. Cut along for 1/4 in. (6 mm), stop the lathe, and check the diameter for size. The diameter should be approximately 1/32 in. (0.80 mm) over the finish size (Fig. 52-7).
8. Readjust the depth of cut, if necessary.

FINISH TURNING

Finish turning, which follows rough turning, produces a smooth surface finish and cuts the work to an accurate size. Factors such as the condition of the cutting tool, the rigidity of the machine and work, and the lathe speeds and feeds affect the type of surface finish produced.

Proceed as follows:

1. Make sure that the cutting edge of the toolbit is free from nicks, burrs, etc. It is good practice to hone the cutting edge before taking a finish cut.
2. Set the toolbit on center; check it against the lathe center point.
3. Set the lathe to the recommended speed and feed. The feed used depends upon the surface finish required.
4. Take a light trial cut 1/4 in. (6 mm) long at the right-hand end of the work to
   a. Produce a true diameter
   b. Set the cutting tool to the diameter
   c. Set the graduated collar to the diameter
5. Stop the lathe and measure the diameter.
6. Set the depth of cut for half the amount of material to be removed.
7. Cut along for 1/4 in. (6 mm), stop the lathe, and check the diameter.
8. Readjust the depth of cut, if necessary, and finish turn the diameter.

NOTE: In order to produce the truest diameter possible, finish turn work to the required size. Should it be necessary to finish a diameter by filing or polishing, *never* leave more than 0.002 to 0.003 in. (0.05 to 0.07 mm) for this operation.

FILING IN A LATHE

Work should be filed in a lathe only to remove a small amount of stock, to remove burrs, or round off sharp corners. Work should always be turned to within 0.002 to 0.003 in. (0.05 to 0.07 mm) of size, if the surface is to be filed. When larger amounts must be removed, the work should be machined, since excessive filing will produce work which is out of round and not parallel. The National Safety Council recommends filing with the left hand, so that the arms and hands can be kept clear of the revolving chuck or drive plate. Always remove the toolbit from the toolholder before filing unless the machining operation does not permit it. In this case, wind the carriage so that the toolbit is as far as possible from the area being filed.

NOTE: Before attempting to file or polish in a lathe, it is good practice to cover the lathe bed with a piece of paper to prevent filings from getting into the slides and causing excessive wear and damage to the lathe (Fig. 52-8). Cloth is

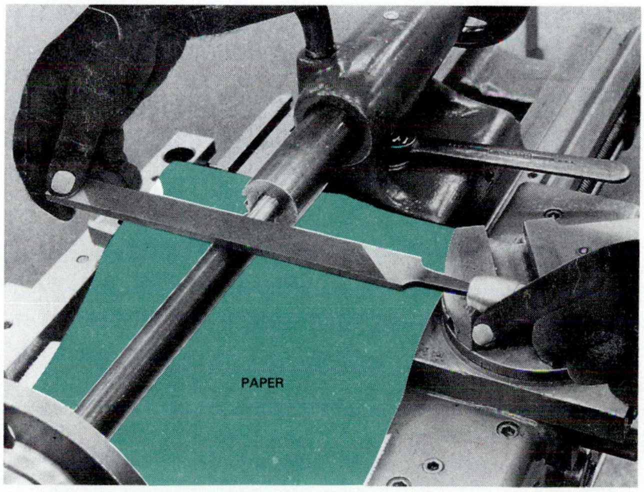
PAPER

Figure 52-8 *Always hold the file handle in your left hand to avoid injury.*

not suitable for this purpose because it tends to get caught in the revolving work or the lathe.

TO FILE IN A LATHE

1. Set the spindle speed to approximately twice that used for turning.
2. Mount the work between centers, lubricate, and carefully adjust the dead center in the workpiece. Use a revolving dead center if one is available.
3. Move the carriage as far to the right as possible and remove the toolpost.
4. Disengage the lead screw and feed rod.
5. Select a 10- or 12-in. (250 or 300 mm) *mill file* or a *long-angle lathe file.*

NOTE: Be sure that the file handle is properly secured on the tang of the file.

6. Start the lathe.
7. Grasp the file handle in the left hand and support the file point with the fingers of the right hand (Fig. 52-8).
8. Apply light pressure and push the file forward to its full length. Release the pressure on the return stroke.
9. Move the file about half the width of the file for each stroke and continue filing, using 30 to 40 strokes per minute, until the surface is finished.
10. When filing in a lathe, take the following precautions.
    a. Roll sleeves up above the elbows—short sleeves are preferable.
    b. Remove watches and rings.
    c. Never use a file without a properly fitted handle.
    d. Never apply too much pressure to the file. Excessive pressure produces out-of-roundness and causes the file teeth to clog and damage the work surface.
    e. Clean the file frequently with a file brush. Rub a little chalk into the file teeth to prevent clogging and facilitate cleaning.

POLISHING IN A LATHE

After the work surface has been filed, the finish may be improved by polishing with abrasive cloth. Proceed as follows:

1. Select the correct type and grade of abrasive cloth for the finish desired. Use a piece about 6 to 8 in. (150 to 200 mm) long and 1 in. (25 mm) wide. For ferrous metals, use aluminum oxide abrasive cloth. Silicon carbide abrasive cloth should be used for nonferrous metals.
2. Set the lathe to run at high speed.
3. Disengage the feed rod and lead screw.
4. Remove the toolpost and toolholder.
5. Lubricate and adjust the dead center. Use a revolving dead center if one is available.
6. Roll sleeves up above the elbows and tuck in any loose clothing.
7. Start the lathe.
8. Hold the abrasive cloth on the work (Fig. 52-9).
9. With the right hand, press the cloth firmly on the work while *tightly* holding the other end of the abrasive cloth with the left hand. (*Caution:* Do not let the short end of the abrasive cloth wrap around the work.)
10. Move the cloth slowly back and forth along the work.

NOTE: For normal finishes, 80- to 100-grit abrasive cloth should be used. For better finishes, use a finer grit abrasive cloth.

TURNING TO A SHOULDER

When turning more than one diameter on a piece of work, the change in diameters, or step, is known as a *shoulder.* Three common types of shoulders are illustrated in Fig. 52-10.

TO TURN A SQUARE SHOULDER

1. With the work mounted in a lathe, lay out the shoulder position from the finished end of the work. In case of filleted shoulders, allow sufficient length to permit the proper radius to be formed on the finished shoulder.
2. Place the point of the toolbit at this mark and cut a *small* groove around the circumference to mark off the length.
3. With a turning tool, rough- and finish-turn the work to within 1/16 in. (1.5 mm) of the required length (Fig. 52-11).

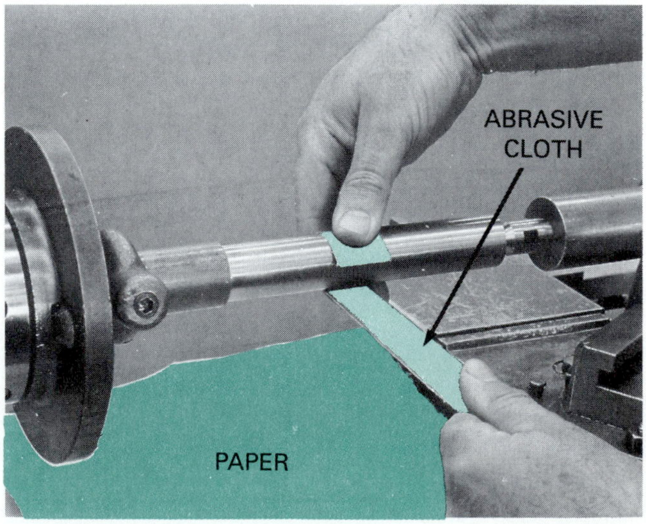
ABRASIVE
CLOTH
PAPER

Figure 52-9 *A high surface finish can be produced with abrasive cloth.*

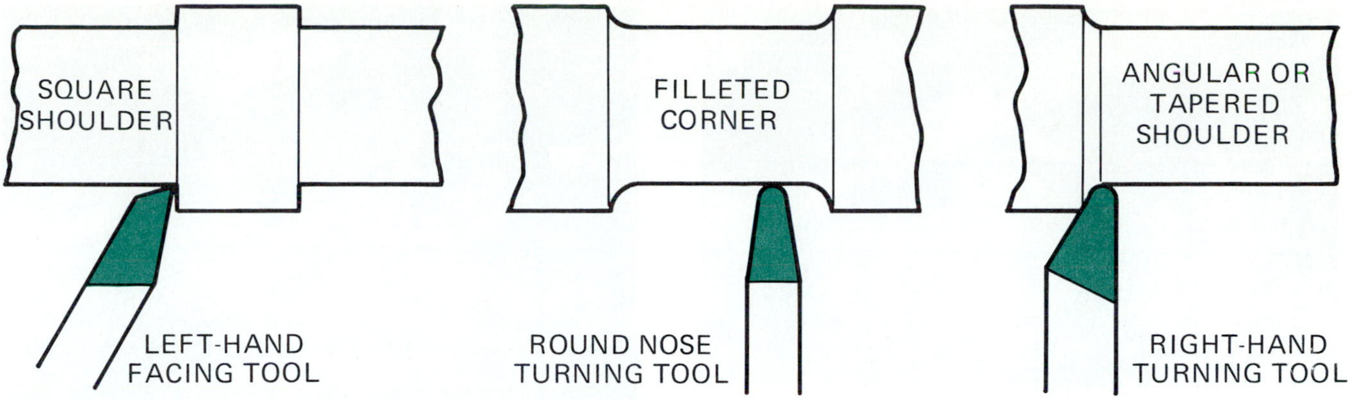
SQUARE SHOULDER
FILLETED CORNER
ANGULAR OR TAPERED SHOULDER
LEFT-HAND FACING TOOL
ROUND NOSE TURNING TOOL
RIGHT-HAND TURNING TOOL

Figure 52-10 *Types of shoulders.*

4. Set up an end-facing tool, chalk the small diameter of the work, and bring the cutting tool up until it just removes the chalk mark (Fig. 52-12).
5. Note the reading on the graduated collar of the crossfeed handle.
6. Face (square) the shoulder, cutting to the line *using hand feed.*
7. For successive cuts, return the crossfeed handle to the same graduated collar setting.

TO MACHINE A FILLETED SHOULDER

Fillets are used at a shoulder to overcome the sharpness of a corner and to strengthen the part at this point. If a filleted corner is required, a toolbit having the same radius is used to finish the shoulder. Proceed as follows:

1. Lay out the length of the shoulder with a center-punch mark or by cutting a light groove (Fig. 52-13).
2. Rough- and finish-turn the small diameter to the correct length *minus the radius to be cut.* For example, a 3-in. (75-mm) length with a 1/8-in. (3-mm) radius should be turned 2 7/8 in. (73 mm) long.
3. Mount the correct radius toolbit and set it to center.
4. Set the lathe for one-half the turning speed.
5. Coat the small diameter near the shoulder with chalk or layout dye.

Figure 52-12 *Note the reading on the graduated collar when the toolbit just touches the small diameter.*

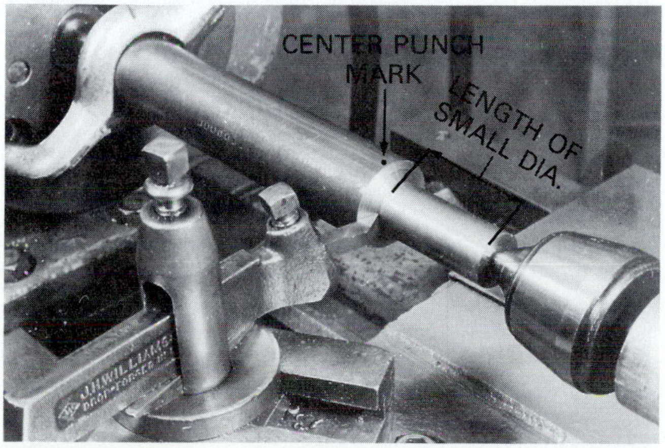
CENTER PUNCH MARK
LENGTH OF SMALL DIA.

Figure 52-11 *Turn the small diameter to within 1/16 in. of the finished length.*

CENTER PUNCH MARK

Figure 52-13 *The length of the shoulder is indicated by a center-punch mark.*

Figure 52-14 *Setting the radius toolbit to the small diameter.*

6. Start the lathe and feed the cutting tool in until it lightly marks the small diameter near the shoulder (Fig. 52-14).
7. Slowly feed the cutting tool sideways with the carriage handwheel until the shoulder is cut to the correct length.

TO MACHINE A BEVELED (ANGULAR) SHOULDER

Beveled or angular shoulders are used to eliminate sharp corners and edges, to make parts easier to handle, and to improve the appearance of the part. They are sometimes used to strengthen a part by eliminating the sharp corner of a square shoulder. Shoulders are beveled at angles ranging from 30 to 60°; however, the most common is the 45° bevel. Proceed as follows:

1. Turn the large diameter to size.
2. Lay out the position of the shoulder along the length of the workpiece.
3. Rough- and finish-turn the small diameter to size.
4. Mount a side cutting tool in the toolholder and set it to center.
5. Use a protractor and set the side cutting edge of the toolbit to the desired angle (Fig. 52-15).
6. Apply chalk or layout dye to the small diameter as close as possible to the shoulder location.
7. Set the lathe spindle to approximately one-half the turning speed.
8. Bring the point of the toolbit in until it just removes the chalk or layout dye.
9. Turn the carriage handwheel by hand to feed the cutting tool into the shoulder (Fig. 52-16).
10. Apply cutting fluid to assist the cutting action and to produce a good surface finish.
11. Machine the beveled shoulder until it is the required size.

Figure 52-15 *Using a protractor to set the toolbit side cutting edge to an angle.*

If the size of the shoulder is large and chatter occurs during cutting with the side of the toolbit, it may be necessary to cut the beveled shoulder using the compound rest (Fig. 52-17). If so, proceed as follows:

1. Set the compound rest to the desired angle.
2. Adjust the toolbit so that only the point does the cutting.
3. Machine the bevel by feeding the compound rest by hand.

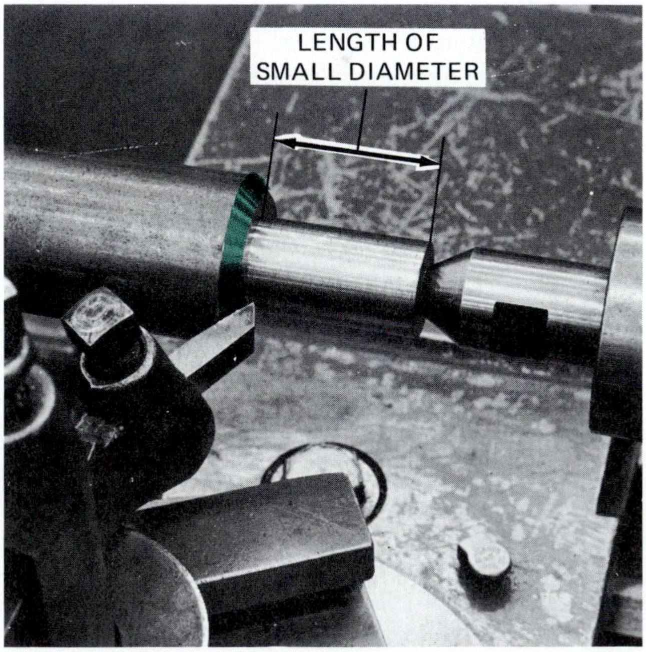
LENGTH OF
SMALL DIAMETER

Figure 52-16 *Machining a beveled shoulder with the side of a cutting tool.*

Figure 52-17 *The compound rest swiveled to cut a large beveled shoulder.*

REVIEW QUESTIONS

SETTING UP A CUTTING TOOL

1. How should the toolholder and cutting tool be set for machining between centers?
2. What might happen if the toolholder were set to the left and moved under the pressure of the cut?

PARALLEL TURNING

3. What precaution must be taken before starting a parallel turning operation?
4. Explain the procedure for setting an accurate depth of cut.
5. State the purpose of rough and finish turning.
6. How many cuts should be taken to turn a diameter to size?
7. What is the purpose of a light trial cut at the right-hand end of the work?

FILING AND POLISHING

8. How much material should be left on a diameter for filing to size?
9. How should the file be held for filing in a lathe?
10. List two of the most important things to remember about filing in a lathe.
11. List the main steps for polishing a diameter in a lathe.

SHOULDER TURNING

12. What type of cutting tool should be used to machine a square shoulder?
13. How close to finish length should a diameter which requires a filleted shoulder be cut?
14. Name two methods of cutting angular shoulders on a lathe.

UNIT 53

Knurling, Grooving, and Form Turning

Operations such as knurling, grooving, and form turning are used to alter either the shape or the finish of a round workpiece. These operations are normally performed on work mounted in a chuck; however, they can also be performed on work mounted between lathe centers if certain precautions are observed. Knurling is used to improve the surface finish on the work and provide a hand grip on the diameter. Grooving is used to provide a relief at the end of a thread or a seat for snap or O-rings. Form turning produces a concave or convex form on internal or external surfaces of a workpiece.

OBJECTIVES

After completing this unit you should be able to:

1. Set up and use knurling tools to produce diamond-shaped or straight patterns on diameters
2. Cut square, round, and V-shaped grooves on work between centers or in a chuck
3. Machine convex or concave forms on diameters freehand

KNURLING

Knurling is a process of impressing a diamond-shaped or straight-line pattern into the surface of the workpiece to improve its appearance or to provide a better gripping surface. Straight knurling is often used to increase the workpiece diameter when a press fit is required.

Diamond- and straight-pattern rolls are available in three styles: fine, medium, and coarse (Fig. 53-1).

The knurling tool (Fig. 53-2) is a toolpost-type toolholder on which a pair of hardened-steel rolls are mounted. These rolls may be obtained in diamond and straight-line patterns, and in coarse, medium, and fine pitches. Some knurling tools are made with the three various pitched rollers on one holder (Fig. 53-3).

Figure 53-1 *Fine, medium, and coarse diamond- and straight-pattern knurling rolls. (Courtesy J. H. Williams and Company.)*

TO KNURL IN A LATHE

1. Mount the work between centers and mark the required length to be knurled.

If the work is held in a chuck for knurling, the right end of the work should be supported with a revolving tailstock center.

2. Set the lathe to run at *one-quarter* the speed required for turning.
3. Set the carriage feed to *0.015 to 0.030 in.* (0.38 mm to 0.76 mm).
4. Set the center of the floating head of the knurling tool even with the dead-center point (Fig. 53-4).
5. Set the knurling tool at right angles to the workpiece and tighten it securely in this position (Fig. 53-5).
6. Start the machine and lightly touch the rolls against the work to make sure that they are tracking properly (Fig. 53-6). Adjust if necessary.
7. Move the knurling tool to the end of the work so that only half the roll face bears against the work. If the knurl

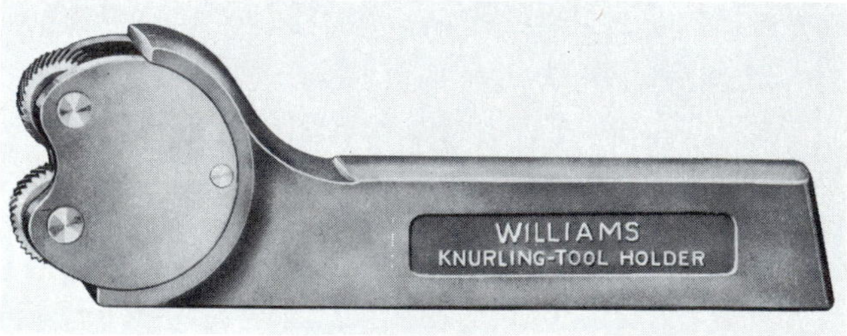
WILLIAMS
KNURLING-TOOL HOLDER

Figure 53-2 *A knurling tool with one set of rolls in a self-centering head. (Courtesy J. H. Williams and Company.)*

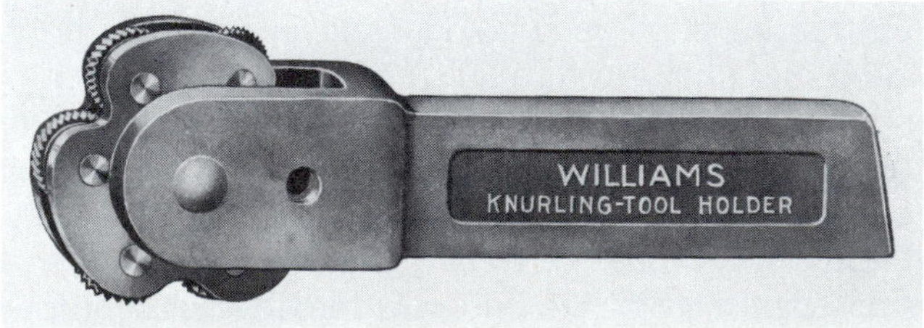
WILLIAMS
KNURLING-TOOL HOLDER

Figure 53-3 *A knurling tool with three sets of rolls in a revolving head. (Courtesy J. H. Williams and Company.)*

Figure 53-4 *Setting a knurling tool to center.*

90°

Figure 53-5 *The knurling tool set at 90° and moved near the end of the workpiece.*

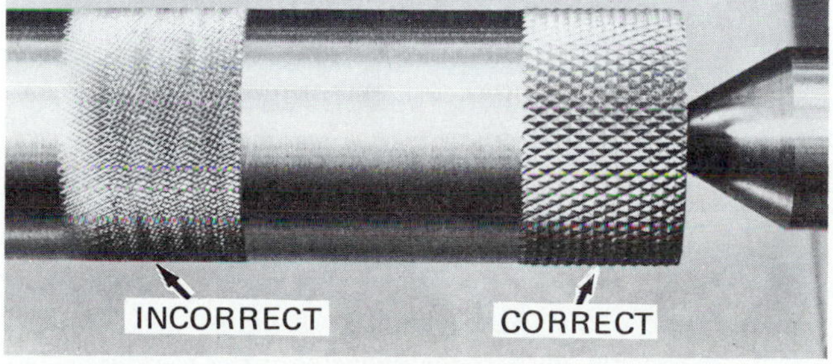
INCORRECT
CORRECT

Figure 53-6 *Correct and incorrect knurling patterns.*

does not extend to the end of the workpiece, set the knurling tool at the correct limit of the section to be knurled.

8. Force the knurling tool into the work approximately 0.025 in. (0.63 mm) and start the lathe.

OR

Start the lathe and then force the knurling tool into the work until the diamond pattern comes to a point.

9. Stop the lathe and examine the pattern. If necessary, reset the knurling tool.

   a. If the pattern is incorrect (Fig. 53-6), it is usually because the knurling tool is not set on center.

   b. If the knurling tool is on center and the pattern is not correct, it is generally due to worn knurling rolls. In this case it will be necessary to set the knurling tool off square slightly so that the corner of the knurling rolls can start the pattern.

10. Once the pattern is correct, engage the automatic carriage feed and apply cutting fluid to the knurling rolls.

11. Knurl to the proper length and depth.

NOTE: *Do not disengage the feed until the full length has been knurled; otherwise, rings will be formed on the knurled pattern* (Fig. 53-7).

12. If the knurling pattern is not to a point after the length has been knurled, reverse the lathe feed and take another pass across the work.

GROOVING

Grooving, commonly called *recessing, undercutting,* or *necking,* is often done at the end of a thread to permit full travel of the nut up to a shoulder, or at the edge of a shoulder to ensure a proper fit of mating parts. Grooves are generally square, round, or V-shaped (Fig. 53-8).

Rounded grooves are usually used where there is a strain on the part and where a square corner would lead to fracturing of the metal at this point.

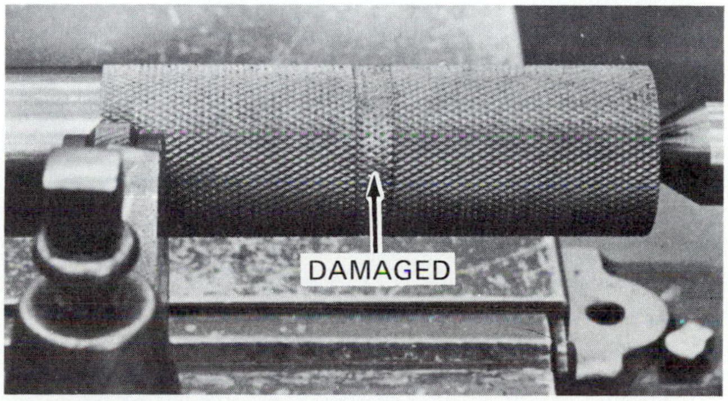
DAMAGED

Figure 53-7 *Disengaging the automatic feed will damage the knurling pattern.*

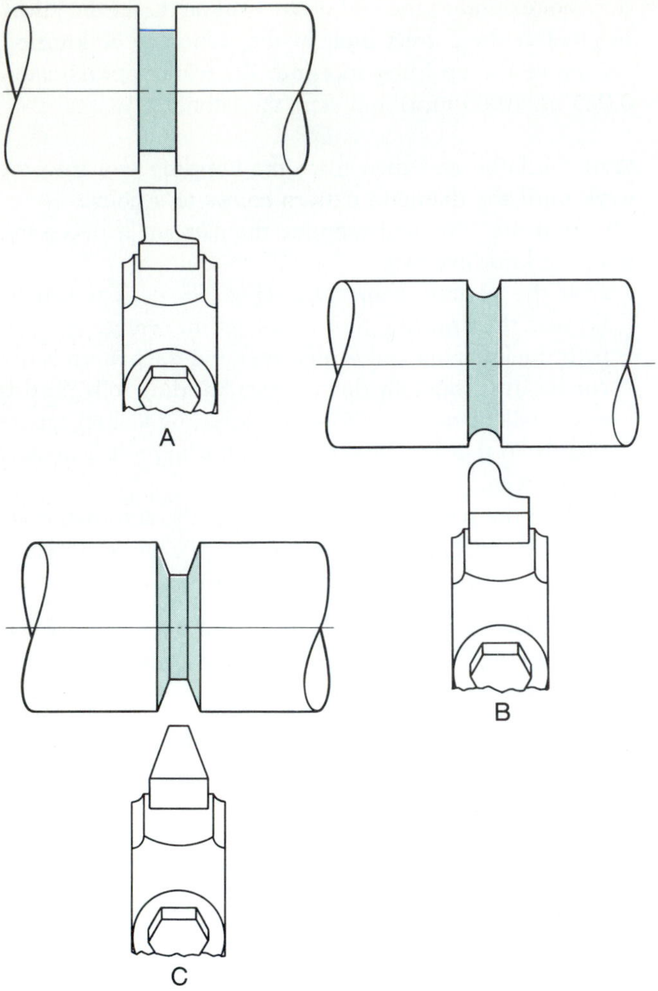
A
B
C
Figure 53-8 *Three types of common grooves.*

TO CUT A GROOVE

1. Grind a toolbit to the desired size and shape of the groove required. If a parting tool is used to cut a groove, never grind the width of the tool.
2. Lay out the location of the groove.
3. Set the lathe to half the speed for turning.
4. Mount the workpiece in the lathe.
5. Set the toolbit to center height (Fig. 53-9).
6. Locate the toolbit on the work at the position where the groove is to be cut.
7. Start the lathe and feed the cutting tool toward the work using the crossfeed handle until the toolbit lightly marks the work.
8. Hold the crossfeed handle in position and then set the graduated collar to zero (Fig. 53-10).
9. Calculate how far the crossfeed screw must be turned to cut the groove to the proper depth.
10. Feed the toolbit into the work *slowly* using the crossfeed handle.
11. Apply cutting fluid to the point of the cutting tool. To

Figure 53-9 *A grooving tool set to center.*

ensure that the cutting tool will not bind in the groove, move the carriage slightly to the left and to the right while grooving. Should chatter develop, reduce the spindle speed of the lathe.

12. Stop the lathe and check the depth of groove with outside calipers or knife-edge verniers.

NOTE: Always wear safety goggles when grooving on a lathe.

FORM TURNING ON A LATHE

It is often necessary to form irregular shapes or contours on a workpiece. Form turning may be done on a lathe by three methods:

1. Freehand
2. Form-turning tool
3. Hydraulic tracer attachment

Figure 53-10 *The graduated collar should be set to zero when the tool just touches the diameter of the workpiece. (Courtesy Kostel Enterprises Limited.)*

TURNING A FREEHAND FORM OR RADIUS

Freehand form turning probably presents the greatest problem to the beginning lathe operator. Coordination of both hands is required and practice is important in mastering this skill.

To Turn a ½-in. or 13-mm Radius on the End of a Workpiece

1. Mount the workpiece in a chuck and face the end.
2. With the work revolving, mark a line ½ in. (or 13 mm) from the end using a pencil (Fig. 53-11).
3. Mount a round-nose turning tool on center.
4. Start the lathe and adjust the toolbit in until it touches the diameter about ¼ in. (6 mm) from the end.
5. Place one hand on the crossfeed handle and the other on the carriage handwheel.
6. Turn the carriage *handwheel* (*not the handle*) to feed the toolbit slowly toward the end of the work; at the same time, turn the crossfeed handle to move the tool into the work.

NOTE: It will take practice to coordinate the movement of the carriage in relation to the crossfeed. For the first ¼ in. (6 mm) of the radius, the carriage must be moved faster than the crossfeed handle. However, for the second ¼ in. (6 mm), the crossfeed handle must be moved faster than the carriage.

7. Back the toolbit out and move the carriage to the left.
8. Take successive cuts as in step 6 until the toolbit starts to cut close to the ½-in. (13-mm) line.
9. Test the radius with a ½-in. (13-mm) gage.
10. If the radius is not correct, it may have to be recut. It is often possible to finish the cut to the required shape by filing.

NOTE: Follow the same procedure as in step 6 when cutting internal radii (Fig. 53-12). Always start at the large diameter, feeding along and in until the proper radius and diameter are obtained.

FORM TURNING TOOLS

Smaller radii and contours are conveniently formed on a workpiece by a form turning tool. The lathe toolbit is

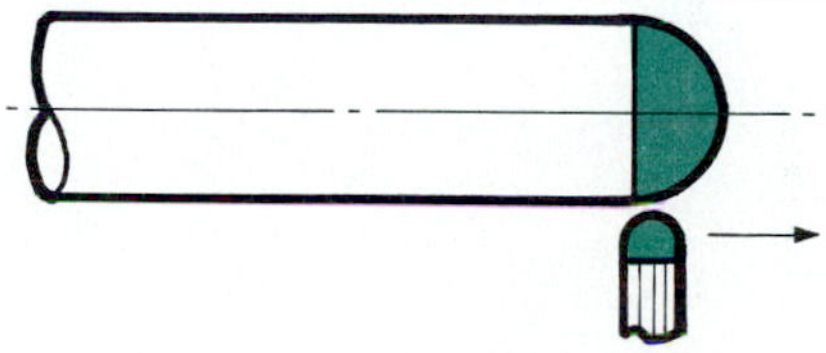

Figure 53-11 *Turning a ½-in. (13-mm) radius on the end of the workpiece.*

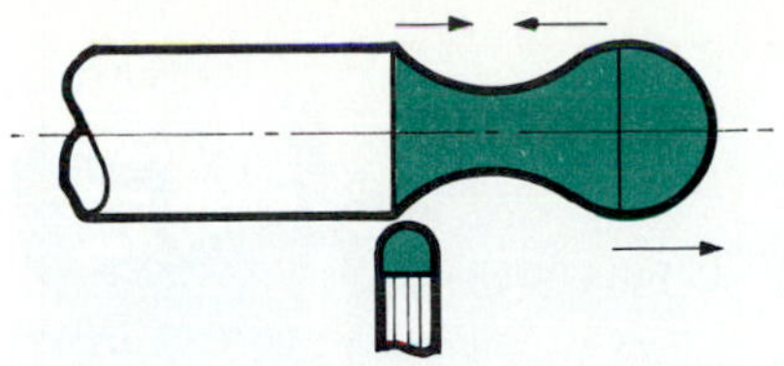

Figure 53-12 *Turning concave and convex radii on a workpiece.*

ground to the desired radius and used to form the contour on the workpiece. Toolbits may also be ground to produce a concave radius (Fig. 53-13A).

This method of forming radii and contours eliminates the need for checking with a gage or template once the toolbit is ground to the desired shape. Duplicate contours may also be formed on several workpieces when the same toolbit is used.

When producing a convex radius, it is necessary to leave a collar of the desired size on the workpiece (Fig. 53-13B).

To produce a good finish by this method, the work should be revolved slowly. The tool should be fed into the work slowly while cutting oil is applied. To eliminate chatter during the cutting operation, the cutting tool should be moved slightly back and forth (longitudinally).

FORM TURNING USING A TEMPLATE AND FOLLOWER

When only a few pieces of a special form are required, they can be accurately produced by using a template and a follower fastened to the cross-slide of the lathe. The accuracy of the template, which must be made, determines the accuracy of the form produced. Proceed as follows:

1. Make an accurate template to the form desired.
2. Mount the template on a bracket fastened to the back of the lathe (Fig. 53-14).
3. Position the template lengthwise in relation to the workpiece.
4. Mount a round-nose cutting tool in the toolpost.
5. Fasten a follower, the face of which must have the same form as the point of the cutting tool, on the cross-slide of the lathe.

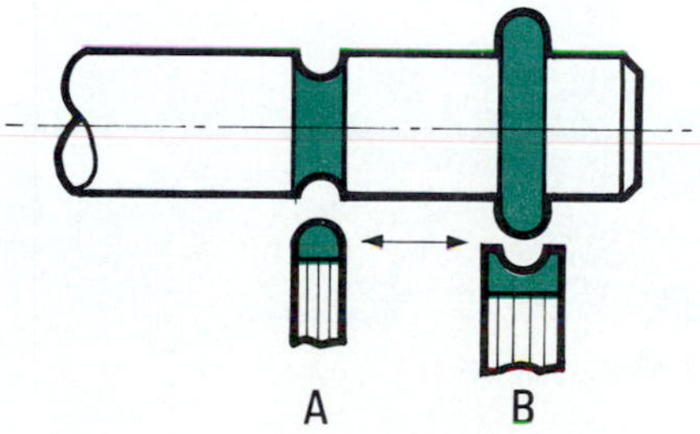
A
B

Figure 53-13 *Toolbits ground to cut (A) a concave radius; (B) a convex radius.*

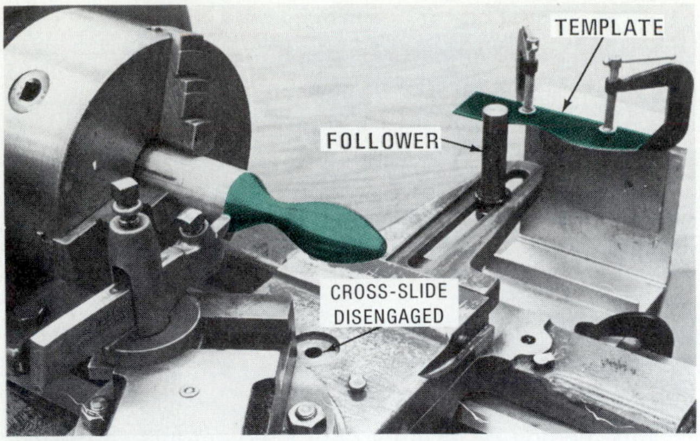
TEMPLATE
FOLLOWER
CROSS-SLIDE
DISENGAGED

Figure 53-14 *A follower on the cross-slide follows the template in order to produce a special form.*

Figure 53-15 *The cross-slide must be disconnected in order to follow the template.*

6. Rough-out the form on the workpiece freehand by keeping the follower close to the template by manually operating the carriage and cross-slide controls. While taking roughing cuts, keep the distance between the follower and the template fairly constant. For the final roughing cut, the follower should be kept within approximately $\frac{1}{32}$ in. (0.80 mm) of the template at all times.
7. Disconnect the cross-slide from the crossfeed screw (Fig. 53-15).
8. Apply light hand pressure on the cross-slide to keep the follower in contact with the template.
9. Engage the automatic carriage feed and take a finish cut from the workpiece while keeping the follower in contact with the template.

HYDRAULIC TRACER ATTACHMENT

When many duplicate parts having several radii or contours which may be difficult to produce are required, they may be easily made on a hydraulic tracer lathe or on a lathe equipped with a hydraulic tracer attachment (Fig. 53-16).

Hydraulic tracer lathes incorporate a means of moving the cross-slide by controlled oil pressure supplied by a hydraulic pump. A flat template of the desired contour of the finished piece, or a circular template identical to the finished piece, is mounted in an attachment on the lathe. Automatic control of the tool slide and duplication of the part is achieved by a stylus, which bears against the template surface. As the carriage is fed along automatically, the sty-

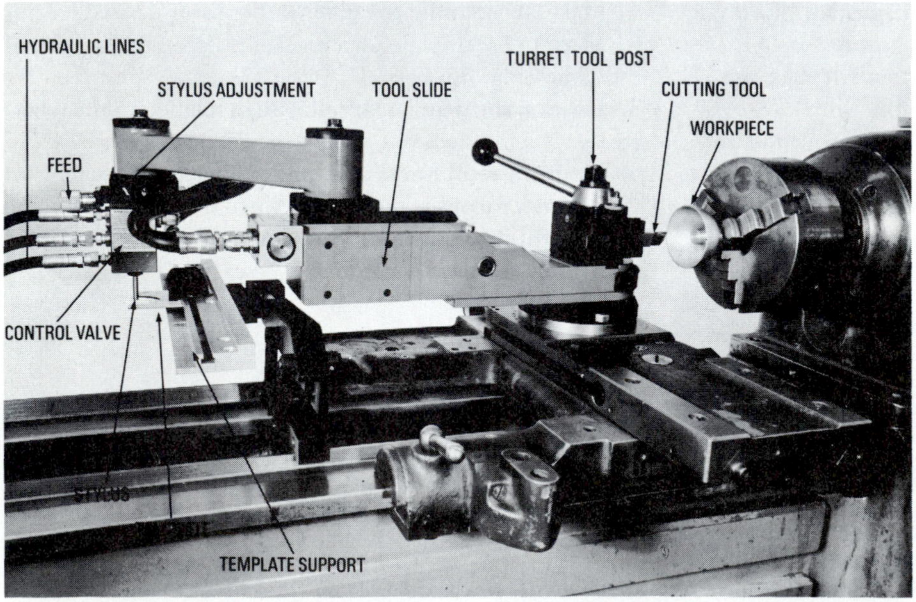
HYDRAULIC LINES
STYLUS ADJUSTMENT
TOOL SLIDE
TURRET TOOL POST
CUTTING TOOL
WORKPIECE
FEED
CONTROL VALVE
TEMPLATE SUPPORT

Figure 53-16 *A hydraulic tracer attachment being used to machine an intricate form. (Courtesy Retor Developments Ltd.)*

lus follows the contour of the template. The stylus arm actuates a control valve regulating the flow of oil into a cylinder incorporated in the tool slide base. A piston connected to the tool slide is moved in or out by the flow of oil to the cylinder. This movement causes the tool slide (and toolbit) to move in or out as the carriage moves along, duplicating the profile of a template on the workpiece.

Advantages of a Tracer Attachment

- Intricate forms, difficult to produce by other means, can be readily produced.
- Various forms, tapers, and shoulders can be produced in one cut.
- Duplicate parts can be produced rapidly and accurately.
- Accuracy and finish of the part do not depend on the skill of the operator.

Hints on the Use of a Tracer Attachment

1. The toolbit point and stylus should have the same form and radius.
2. The radius on the toolbit should be smaller than the smallest radius on the template.
3. The stylus must be set to the point on the template giving the smallest diameter of the work.
4. The centerline of the template must be parallel to the ways of the lathe.
5. The form of the template must be smooth.
6. No angle larger than 30°, or the equivalent radius, should be incorporated in the form of the template.
7. Duplicate parts produced between centers must be the same length and have the center holes drilled to the same depth.
8. Duplicate parts held in a chuck should project the same distance from the chuck jaws.
9. The included angle of the tool point should be less than the smallest angle on the template.

REVIEW QUESTIONS

KNURLING

1. Define the process of knurling.
2. Explain how to set up the knurling tool.
3. Why is it important not to disengage the feed during the knurling operation?

GROOVING

4. For what purposes are grooves used?
5. How can the depth of the cut be gaged during grooving?
6. What should be done to prevent the cutting tool from binding in a deep groove?

FORM TURNING

7. Name three methods by which form turning may be done on a lathe.
8. Briefly describe the procedure for turning a ½-in. (13-mm) radius on the end of a workpiece.
9. What is a template?
10. What types of templates may be used with a tracer lathe?
11. List three advantages of a tracer lathe or tracer attachment.
12. List six points to observe when using a tracer lathe.

UNIT 54

Tapers and Taper Turning

A *taper* may be defined as a uniform change in the diameter of a workpiece measured along its axis. Tapers in the inch system are expressed in taper per foot, taper per inch, or degrees. Metric tapers are expressed as a ratio of 1 mm per unit of length; for example, 1:20 taper would have a 1-mm change in diameter in 20 mm of length. A taper provides a rapid and accurate method of aligning machine parts and an easy method of holding tools such as twist drills, lathe centers, and reamers.

Machine tapers (those used on machines and tools) are now classified by the American Standards Association as *self-holding tapers* and *steep* or *self-releasing tapers*.

OBJECTIVES

After completing this unit you will be able to:

1. Identify and state the purpose of self-holding and self-releasing tapers
2. Cut short, steep tapers using the compound rest
3. Calculate and cut tapers on work between centers by offsetting the tailstock
4. Calculate and machine tapers with a taper attachment

SELF-HOLDING TAPERS

Self-holding tapers, when seated properly, remain in position because of the wedging action of the small taper angle. The most common forms of self-holding tapers are the Morse, the Brown and Sharpe, and the ¾-in.-per-foot machine taper. See Table 54-1.

The smaller sizes of self-holding tapered shanks are provided with a tang to help drive the cutting tool. Larger sizes employ a tang drive with the shank held in by a key or a key drive with the shank held in with a draw bolt.

STEEP TAPERS

Steep tapers (self-releasing) have a 3½ in. taper per foot (tpf). This was formerly called the standard milling machine taper. It is used mainly for alignment of milling machine arbors and accessories. A steep taper has a key drive and uses a draw-in bolt to hold it securely in the milling machine spindle.

STANDARD TAPERS

Although many of the tapers referred to in Table 54-1 are taken from the Morse and Brown and Sharpe taper series, those not listed in this table are classified as nonstandard machine tapers.

The *Morse taper,* which has approximately ⅝-in. *tpf,* is used for most drills, reamers, and lathe center shanks. Morse tapers are available in eight sizes ranging from #0 to #7.

The *Brown and Sharpe taper,* available in sizes from #4 to #12, has approximately 0.502-in. *tpf,* except #10 which has a taper of 0.516-in./ft. This self-holding taper is used on Brown and Sharpe machines and drive shanks.

The *Jarno taper,* 0.600-in. *tpf,* was used on some lathe and drill spindles in sizes from #2 to #20. The taper number indicates the large diameter in eighths of an inch and the small diameter in tenths of an inch. The taper length is indicated by the taper number divided by two.

The *standard taper pins* used for positioning and holding parts together have ¼-in. *tpf.* Standard sizes in these pins range from #6/0 to #10.

LATHE SPINDLE NOSE TAPERS

Two types of tapers are used on lathe spindle noses. The *Type D-1* has a very short tapered section (3-in. *tpf*) and is used on cam-lock spindles (Fig. 54-1). The *Type L* lathe spindle nose has a taper of 3½-in./ft and has a considerably longer taper than the Type D-1. The chuck or drive plate is held on by a threaded lock ring fitted on the spindle behind

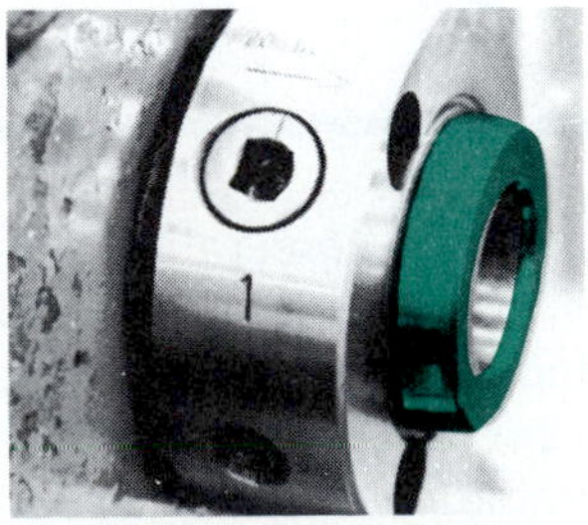

Figure 54-1 *Tapered lathe spindle nose Type D-1.*

Table 54-1 Basic Dimensions of Self-Holding Tapers

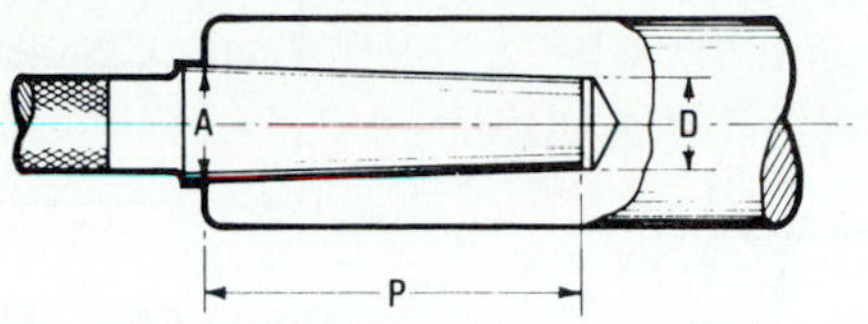
A
D
P

Number of Taper	Taper per Foot	Diameter at Gage Line (*A*)	Diameter at Small End (*D*)	Length (*P*)	Series Origin
0.239	0.502	0.2392	0.200	15/16	
0.299	0.502	0.2997	0.250	1 3/16	Brown and Sharpe taper series
0.375	0.502	0.3752	0.3125	1 1/2	
0*	0.624	0.3561	0.252	2	
1	0.5986	0.475	0.369	2 1/8	
2	0.5994	0.700	0.572	2 9/16	
3	0.6023	0.938	0.778	3 3/16	
4	0.6233	1.231	1.020	4 1/16	Morse taper series
4 1/2	0.624	1.500	1.266	4 1/2	
5	0.6315	1.748	1.475	5 3/16	
6	0.6256	2.494	2.116	7 1/4	
7	0.624	3.270	2.750	10	
200	0.750	2.000	1.703	4 3/4	
250	0.750	2.500	2.156	5 1/2	
300	0.750	3.000	2.609	6 1/4	
350	0.750	3.500	3.063	7	
400	0.750	4.000	3.516	7 3/4	
450	0.750	4.500	3.969	8 1/2	3/4-in.-*tpf* series
500	0.750	5.000	4.422	9 1/4	
600	0.750	6.000	5.328	10 3/4	
800	0.750	8.000	7.141	13 3/4	
1000	0.750	10.000	8.953	16 3/4	
1200	0.750	12.000	10.766	19 3/4	

*Taper #0 is not a part of the self-holding taper series. It has been added to complete the Morse taper series.

the taper nose. A key drive is employed in this type of taper (Fig. 54-2).

TAPER CALCULATIONS

To machine a taper, particularly by the tailstock offset method, it is often necessary to make calculations to ensure accurate results. Since tapers are often expressed in *taper*

Figure 54-2 *Tapered lathe spindle nose Type L.*

per foot, taper per inch, or *degrees,* it may be necessary to calculate any of these dimensions.

TO CALCULATE THE *tpf*

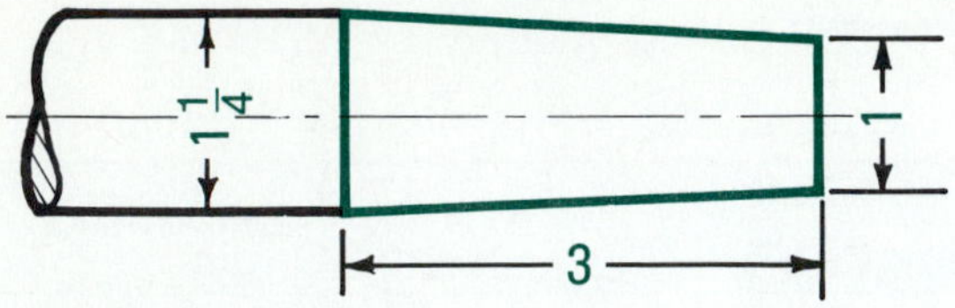

1 1/4
1
3

Figure 54-3 *The main part of an inch taper.*

To calculate the *tpf* it is necessary to know the large diameter, the small diameter, and the length of taper. The *tpf* can be calculated by applying the following formula.

$$tpf = \frac{(D - d)}{\text{length of taper}} \times 12$$

To calculate the *tpf* for the workpiece in Fig. 54-3

$$tpf = \frac{(1\tfrac{1}{4} - 1)}{3} \times 12$$
$$= \frac{1}{4} \times \frac{1}{3} \times 12$$
$$= 1$$

TO CALCULATE THE TAILSTOCK OFFSET

When calculating the tailstock offset, the *tpf* and the total length of the work must be known (Fig. 54-4).

$$\text{Tailstock offset} = \frac{tpf \times \text{length of work}}{24}$$

$$tpf = \frac{(1\tfrac{1}{8} - 1)}{3} \times 12$$
$$= \frac{1}{8} \times \frac{1}{3} \times 12$$
$$= 1/2 \text{ in.}$$

$$\text{Tailstock offset} = \frac{1/2 \times 6}{24}$$
$$= \frac{1}{2} \times \frac{1}{24} \times 6$$
$$= 1/8 \text{ in.}$$

A simplified formula can be used to calculate the tailstock offset if the taper per inch is known:

$$\text{Taper per inch} = \frac{\text{taper per foot}}{12}$$

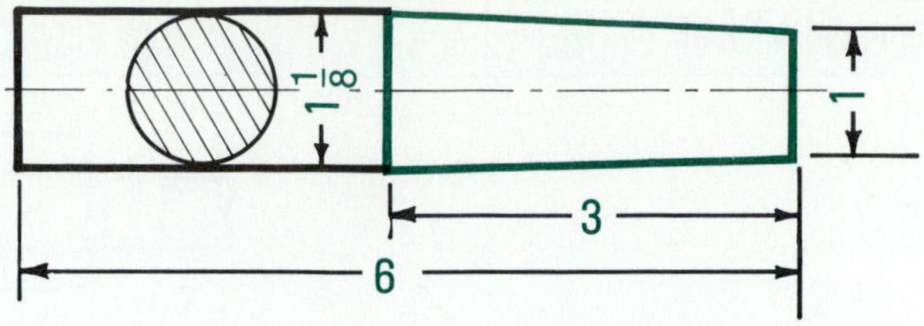

1 1/8
1
3
6

Figure 54-4 *Dimensions of a workpiece having a taper.*

$$\text{Tailstock offset} = \frac{\text{taper per inch} \times \text{OL}}{2}$$

where OL = overall length of work.

In cases where it is not necessary to find the *tpf,* the following simplified formula can be used to calculate the amount of tailstock offset.

$$\text{Tailstock offset} = \frac{\text{OL}}{\text{TL}} \times \frac{(D - d)}{2}$$

OL = overall length of work
TL = length of the tapered section
D = diameter at the large end
d = diameter at the small end

For example, to find the tailstock offset required to cut the taper for the work in Fig. 54-4

$$\text{Tailstock offset} = \frac{6}{3} \times \frac{1}{8} \times \frac{1}{2}$$
$$= 1/8 \text{ in.}$$

INCH TAPER ATTACHMENT OFFSET CALCULATIONS

Most tapers cut on a lathe with the taper attachment are expressed in *tpf.* If the *tpf* of the taper on the workpiece is not given, it may be calculated by using the following formula:

$$tpf = \frac{(D - d) \times 12}{\text{TL}}$$

EXAMPLE

Calculate the *tpf* for a taper with the following dimensions: large diameter (D), 1⅜ in.; small diameter (d), $\tfrac{15}{16}$ in.; length of tapered section (TL), 7 in.

$$tpf = \frac{(1\text{-}3/8 - 15/16) \times 12}{7}$$
$$= \frac{7/16 \times 12}{7}$$
$$= 3/4 \text{ in.}$$

METRIC TAPERS

Metric tapers are expressed as a ratio of 1 mm per unit of length. In Fig. 54-5 the work would taper 1 mm in a distance of 20 mm. This taper would then be expressed as a ratio of 1:20 and would be indicated on a drawing as "taper = 1:20."

Since the work tapers 1 mm in 20 mm of length, the diameter at a point 20 mm from the small diameter (d) will be 1 mm larger ($d + 1$).

Some common metric tapers are:

Milling machine spindle	1:3.429
Morse taper shank	1:20 (approx.)
Tapered pins and pipe threads	1:50

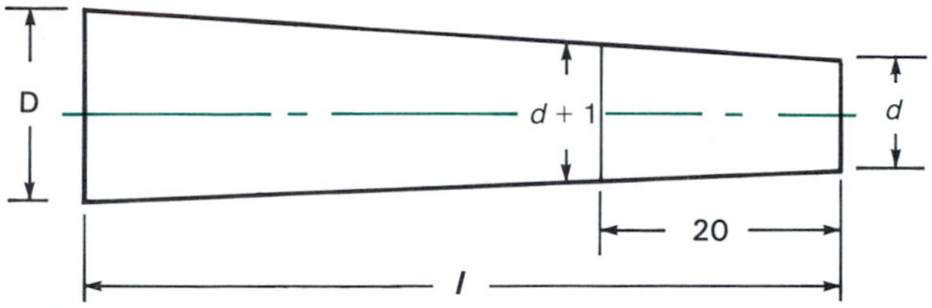
D
d + 1
d
20
l

Figure 54-5 *Characteristics of a metric taper.*

Metric Taper Calculations

If the small diameter (d), the unit length of taper (k), and the total length of taper (l) are known, the large diameter (D) may be calculated.

In Fig. 54-6, the large diameter (D) will be equal to the small diameter plus the amount of taper. The amount of taper for the unit length (k) is $(d + 1) - (d)$ or 1 mm. Therefore the amount of taper per millimeter of unit length $= 1/k$.

The *total amount of taper* is the taper per millimeter ($1/k$) multiplied by the total length of taper (l);

$$\text{Total taper} = \frac{1}{k} \times l \quad \text{or} \quad \frac{l}{k}$$

$$D = d + \text{total amount of taper}$$

$$= d + \frac{l}{k}$$

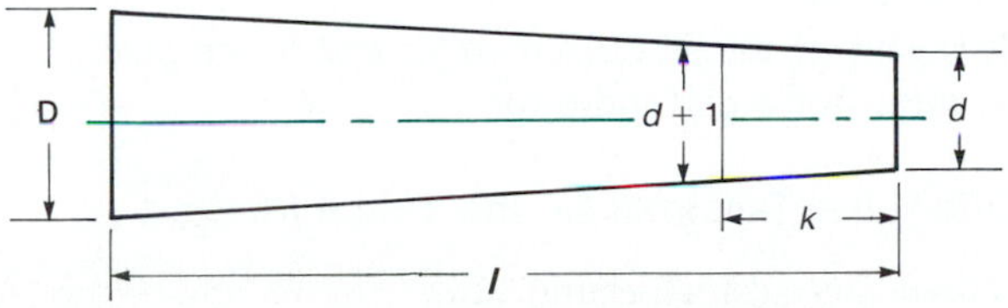
D
d + 1
d
k
l

Figure 54-6 *Dimensions of a metric taper.*

EXAMPLE

Calculate the large diameter D for a 1:30 taper having a small diameter of 10 mm and a length of 60 mm.

SOLUTION

Since the taper is 1:30, $k = 30$

$$D = d + \frac{l}{k}$$

$$= 10 + \frac{60}{30}$$

$$= 10 + 2$$

$$= 12 \text{ mm}$$

METRIC TAILSTOCK OFFSET CALCULATIONS

If the taper is to be turned by offsetting the tailstock, the amount of offset is calculated as follows. See Fig. 54-7.

$$\text{Offset} = \frac{D - d}{2 \times l} \times L$$

D = large diameter
d = small diameter
l = length of taper
L = length of work

EXAMPLE

Calculate the tailstock offset required to turn a 1:30 taper × 60 mm long on a workpiece 300 mm long. The small diameter of the tapered section is 20 mm.

SOLUTION

$$D = d + \frac{l}{k}$$

$$= 20 + \frac{60}{30}$$

$$= 20 + 2$$

$$= 22 \text{ mm}$$

$$\text{Tailstock offset} = \frac{D - d}{2 \times l} \times L$$

$$= \frac{22 - 20}{2 \times 60} \times 300$$

$$= \frac{2}{120} \times 300$$

$$= 5 \text{ mm}$$

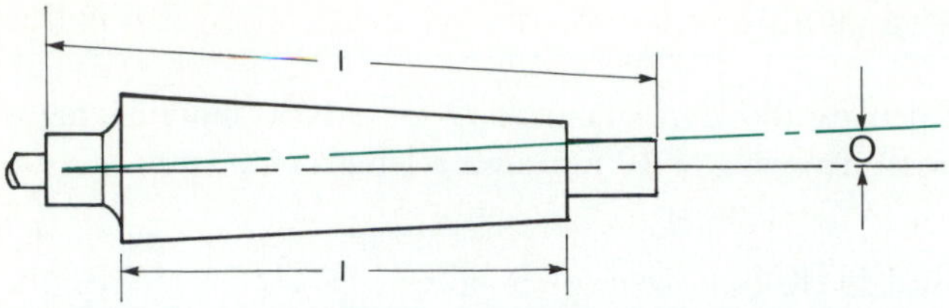
l
O
l

Figure 54-7 *Metric taper turning by the tailstock offset method.*

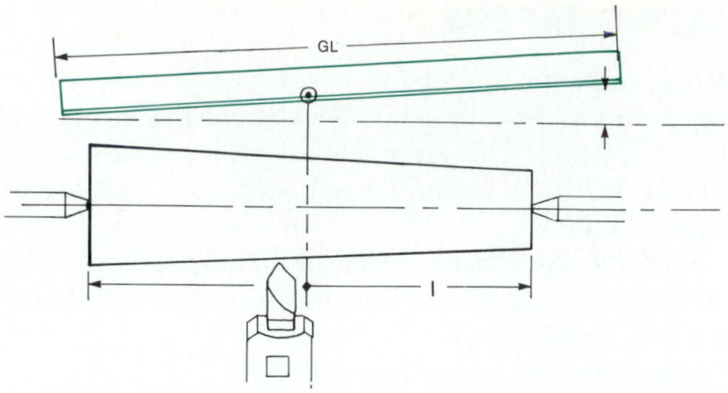
GL
l

Figure 54-8 *Metric taper turning by the taper attachment method.*

METRIC TAPER ATTACHMENT OFFSET CALCULATIONS

When the taper attachment is used to turn a taper, the amount the guide bar is set over may be determined as follows:

1. If the angle of taper is given on the drawing, set the guide bar to one-half the angle (Fig. 54-8).
2. If the angle of taper is not given on the drawing, use the following formula to find the amount of guide bar setover.

$$\text{Guide bar setover} = \frac{D - d}{2} \times \frac{GL}{l}$$

D = large diameter of taper
d = small diameter of taper
l = length of taper
GL = length of taper attachment guide bar

EXAMPLE

Calculate the amount of setover for a 500-mm-long guide bar to turn a 1:50 × 250-mm-long taper on a workpiece. The small diameter of the taper is 25 mm.

SOLUTION

$$D = d + \frac{l}{k}$$

$$= 25 + \frac{250}{50}$$

$$= 30 \text{ mm}$$

$$\text{Guide bar setover} = \frac{D - d}{2} \times \frac{GL}{l}$$

$$= \frac{30 - 25}{2} \times \frac{500}{250}$$

$$= \frac{5}{2} \times 2$$

$$= 5 \text{ mm}$$

TAPER TURNING

Taper turning on a lathe can be performed on work held between centers or in a lathe chuck. There are three methods of producing a taper:

1. By offsetting the tailstock
2. By means of a taper attachment set to the proper *tpf* or the proper taper angle of the workpiece; on inch tapers by means of the taper attachment set to the *tpf* or taper angle of the workpiece; or on metric tapers by calculating the guide bar offset
3. By adjusting the compound rest to the angle of the taper

The method used to machine any taper depends on the work length, taper length, taper angle, and number of pieces to be machined.

THE TAILSTOCK OFFSET METHOD

The tailstock offset method is generally used to cut a taper when no taper attachment is available. This involves moving the tailstock center out of line with the headstock center. However, the amount that the tailstock may be offset is limited. This method will not permit steep tapers to be turned or standard tapers to be turned on the end of a long piece of work.

Methods of Offsetting the Tailstock

The tailstock may be offset by three methods:

1. By using the graduations on the end of the tailstock (visual method)
2. By means of the graduated collar and feeler gage
3. By means of a dial indicator

To Offset the Tailstock by the Visual Method

1. Loosen the tailstock clamp nut.
2. Offset the upper part of the tailstock by loosening one

Figure 54-9 *Measuring the amount of offset with a rule.*

setscrew and tightening the other until the required amount is indicated on the graduated scale at the end of the tailstock (Fig. 54-9).

NOTE: Before machining the work make sure that both setscrews are snugged up to prevent any lateral movement of the tailstock.

To Offset the Tailstock Accurately

The tailstock may be accurately offset by using a dial indicator (Fig. 54-10).

1. Adjust the tailstock spindle to the distance it will be used in the machining setup and lock the tailstock spindle clamp.
2. Mount a dial indicator in the toolpost with the plunger in a horizontal position and on center.
3. Using the crossfeed handle, move the indicator so that it registers approximately 0.020 in. (0.50 mm) on the work, and set the indicator and crossfeed graduated collars to 0.
4. Loosen the tailstock clamp nut.
5. With the tailstock adjusting setscrews, move the tailstock until the required offset is shown on the dial indicator.
6. Tighten the tailstock setscrew that was loosened, being sure that the indicator reading does not change.
7. Tighten the tailstock clamp nut.

Figure 54-10 *Offsetting a tailstock using a dial indicator. (Courtesy Kostel Enterprises Limited.)*

The tailstock may also be offset fairly accurately by using a feeler gage between the toolpost and the tailstock spindle in conjunction with the crossfeed graduated collar (Fig. 54-11).

To Turn a Taper by the Tailstock Offset Method

1. Loosen the tailstock clamp nut.
2. Offset the tailstock the required amount.
3. Set up the cutting tool as for parallel turning.

NOTE: The cutting tool *must* be on center.

4. Starting at the small diameter, take successive cuts until the taper is 0.050- to 0.060-in. (1.27- to 1.52-mm) oversize.
5. Check the taper for accuracy using a taper ring gage, if required (see Taper Ring Gages, Unit 14).
6. Finish-turn the taper to the size and fit required.

TAPER TURNING USING THE TAPER ATTACHMENT

The use of a taper attachment for taper turning provides several advantages:

1. The lathe centers remain in alignment preventing the distortion of centers on the workpiece.
2. The setup is simple and permits changing from taper to parallel turning with no time lost to align the centers.
3. The length of the workpiece does not matter, since duplicate tapers may be turned on any length of work.

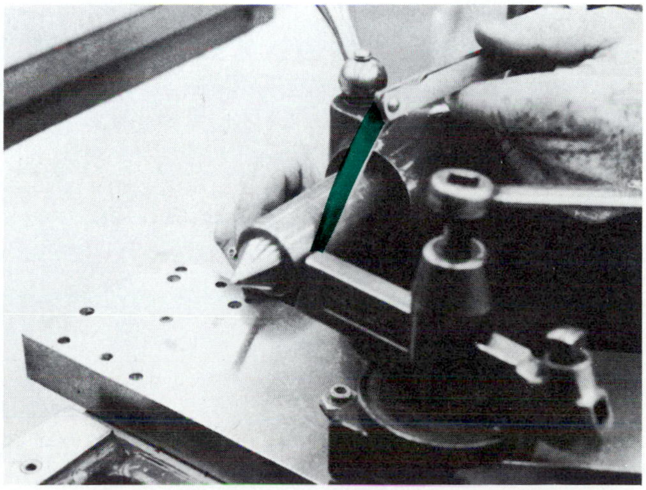

Figure 54-11 *Offsetting a tailstock using the crossfeed graduated collar and a feeler.*

4. Tapers may be produced on work held between centers, in a chuck, or in a collet.
5. Internal tapers can be produced by this method.
6. Metric taper attachments are graduated in millimeters and degrees, while inch attachments are graduated in both degrees and inches of *tpf.* This eliminates the need of lengthy calculations and set up.
7. A wider range of tapers may be produced.

There are two types of taper attachments:

1. The plain taper attachment (Fig. 54-12)
2. The telescopic taper attachment (Fig. 54-13)

When using the *plain taper attachment,* remove the binding screw which holds the cross-slide to the crossfeed screw nut. The binding screw is then used to connect the sliding block to the slide of the taper attachment. With the plain taper attachment, the depth of cut is made by using the compound rest feed handle.

When a *telescopic taper attachment* is used, the crossfeed screw is not disengaged and the depth of cut can be set by the crossfeed handle.

To Cut a Taper Using a Telescopic Taper Attachment

1. Clean and oil the guide bar B (Fig. 54-13).
2. Loosen the lock screws D^1 and D^2 and offset the end of the guide bar the required amount or, for inch attachments, set the bar to the required taper in degrees or *tpf.*
3. Tighten the lock screws.
4. Set up the cutting tool on center.
5. Set the workpiece in the lathe and mark the length of taper.
6. Tighten the connecting screw G on the sliding block E.

NOTE: If a plain taper attachment is being used, remove the binding screw in the cross-slide and use it to connect the sliding block and the connecting slide. The compound rest must also be set at right angles to the lathe bed.

7. Move the carriage until the center of the attachment is opposite the length to be tapered.
8. Lock the anchor bracket A to the lathe bed.
9. Take a cut 1/16 in. (1.5 mm) long, stop the lathe, and check the end of the taper for size.
10. Set the depth of the roughing cut to 0.050 to 0.060 in. (1.27 to 1.52 mm) oversize, and machine the taper.

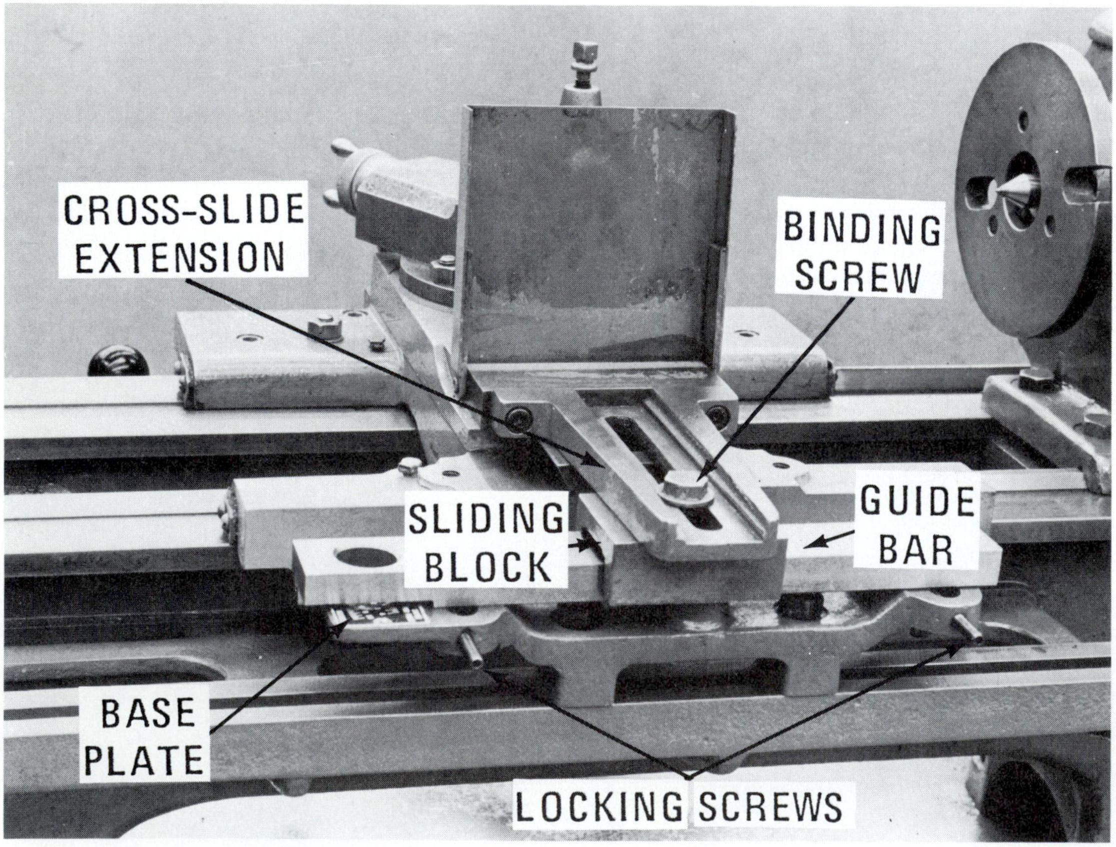
CROSS-SLIDE EXTENSION
BINDING SCREW
SLIDING BLOCK
GUIDE BAR
BASE PLATE
LOCKING SCREWS

Figure 54-12 *The parts of a plain taper attachment.*

A
D 1
E
G
B
2

Figure 54-13 *The telescopic taper attachment.*

NOTE: Start the feed about ½ in. (13 mm) before the start of the cut to remove any play in the taper attachment.

11. Readjust the taper attachment if necessary, take a light cut, and recheck the taper fit.
12. Finish-turn and fit the taper to a gage.

When standard tapers must be produced on a piece of work, a taper plug gage may be mounted between centers and the taper attachment adjusted to this angle by using a dial indicator mounted on center in the toolpost.

When an *internal taper* is cut, the same procedure is followed, except that the guide bar is set to the side of the centerline opposite to that used when turning an external taper.

When mating external and internal tapers must be cut, it is advisable first to machine the internal taper to a plug gage. The external taper is then fitted to the internal taper.

TAPER TURNING USING THE COMPOUND REST

To produce short or steep tapers stated in degrees, the compound rest method is used. The tool must be fed in by hand, using the compound rest feed handle. Proceed as follows:

1. Refer to the drawing for the amount of taper required in degrees. However, if the angle on the drawing is not given in degrees, calculate the compound rest setting as follows:

$$\text{Tan } \frac{1}{2} \text{ angle} = \frac{tpf}{24} \quad \text{or} \quad \frac{tpi}{2}$$

For example, for the workpiece illustrated in Fig. 54-3, the calculations would be:

$$tpf = \frac{1}{4} \times \frac{12}{3}$$
$$= 1 \text{ in.}$$
$$\text{Tan angle} = \frac{1}{24}$$
$$= 0.04166$$

By referring to the trigonometric tables in any handbook, it is found that one-half the angle of this taper (and the compound rest setting) is 2°23′. If a set of trigonometry tables is not available, use the simplified formula to calculate the angle of the taper and the compound rest setting. Tan angle = *tpf* × 2°23′.

2. Loosen the compound rest lock screws.

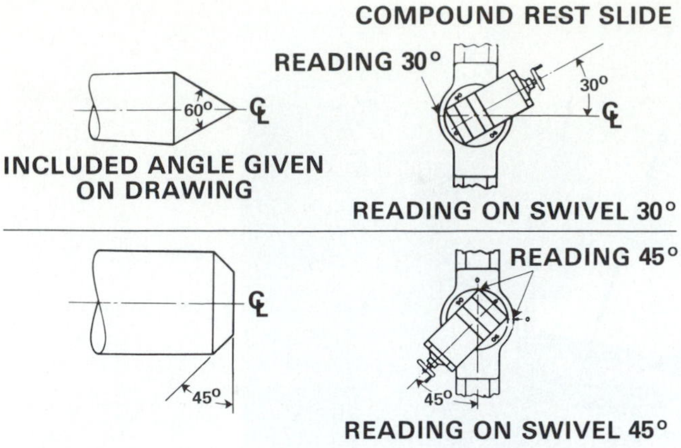
COMPOUND REST SLIDE
READING 30°
60°
℄
30°
℄
INCLUDED ANGLE GIVEN ON DRAWING
READING ON SWIVEL 30°
READING 45°
℄
45°
45°
READING ON SWIVEL 45°

Figure 54-14 *Direction to swing the compound rest for cutting various angles.*

3. Swivel the compound rest as follows:
   a. Where included angles are given on a drawing, swivel the compound rest to one-half the angle (Fig. 54-14 top).
   b. Where angles are given on one side only (Fig. 54-14 bottom), swivel the compound rest to that angle.
4. Tighten the compound rest lock screws using only two-finger pressure on the wrench to avoid stripping the lock screw threads.
5. Set the cutting tool to center with the toolholder at right angles to the taper to be cut.
6. Tighten the toolpost securely.
7. Back off the top slide of the compound rest so that there will be enough travel to machine the length of the taper.
8. Move the carriage to position the cutting tool near the start of the taper and then lock the carriage.
9. Rough turn the taper by feeding the cutting tool using the *compound rest feed handle* (Fig. 54-15).

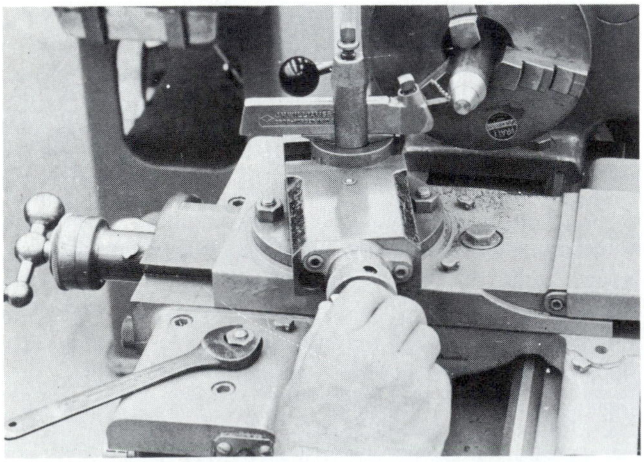

Figure 54-15 *Cutting a short taper using a compound rest.*

10. Check the taper for accuracy and readjust the compound rest setting if necessary.
11. Finish turn and check the taper for size and fit.

CHECKING A TAPER

Inch tapers can be checked by scribing two lines exactly 1 in. apart on the taper and carefully measuring the taper at these points with a micrometer (Fig. 54-16). The difference in readings will indicate the *tpi* of the workpiece. Tapers may be more accurately checked by using a sine bar (see Unit 13).

To obtain a more accurate taper, a taper ring gage is used to check external tapers. A taper plug gage is used to check internal tapers (see Unit 14).

The *taper micrometer* (Fig. 54-17) quickly and accurately measures tapers while the workpiece is still in the machine. This instrument includes an adjustable anvil and a 1 in. sine bar attached to the frame, which is adjusted by the micrometer thimble. The micrometer reading indicates the *tpi*, which can be readily converted to *tpf* or angles. The anvil can be adjusted to accommodate a wide range of work sizes.

Taper micrometers are available in various models for measuring internal tapers and dovetails (Fig. 54-18), and in bench models incorporating two indicators for quickly checking the accuracy of tapered parts.

The advantages of taper micrometers are:

- The taper accuracy can be checked while the workpiece is still in the machine.
- They provide a quick and accurate means of checking tapers.

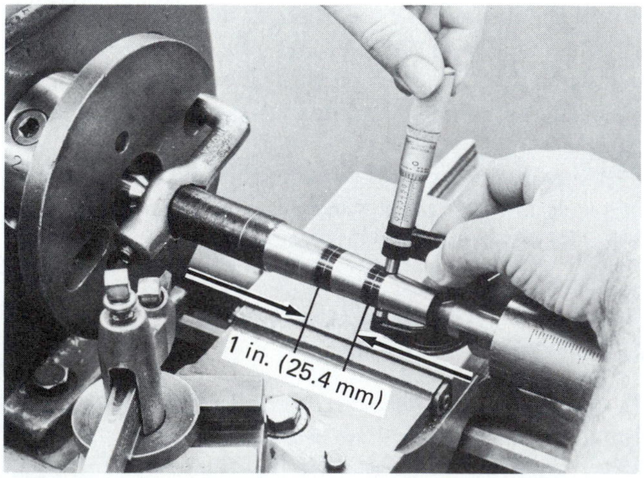
1 in. (25.4 mm)

Figure 54-16 *Checking the accuracy of a taper with a micrometer.*

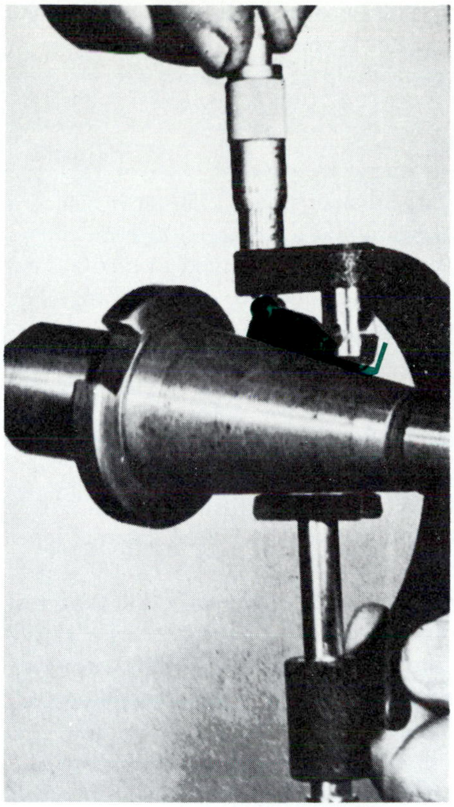

Figure 54-17 *Using a taper micrometer to check an external taper. (Courtesy Taper Micrometer Corporation.)*

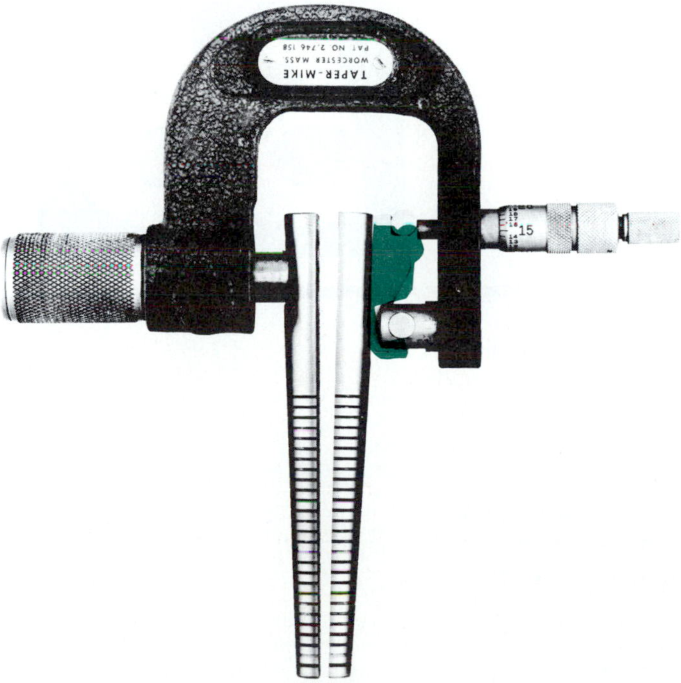

Figure 54-18 *A taper micrometer for measuring internal tapers. (Courtesy Taper Micrometer Corporation.)*

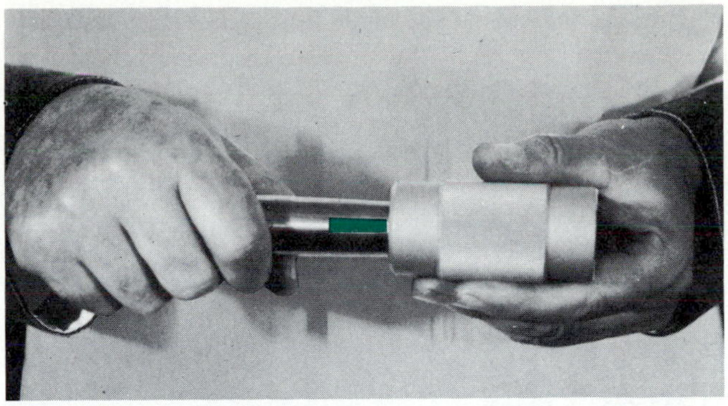

Figure 54-19 *Checking the accuracy of a taper using chalk lines.*

- They are simple to operate.
- The need for costly gaging equipment is eliminated.
- They can be used for measuring external tapers, internal tapers, and dovetails.

TO FIT AN EXTERNAL TAPER

1. Make three equally spaced lines with chalk or mechanics blue along the taper (see Unit 14, Taper Ring Gages).
2. Insert the taper into the ring gage and turn *counterclockwise* for one-half turn (Fig. 54-19).

CAUTION: Do not force the taper into the ring gage.

3. Remove the workpiece and examine the chalk marks. If the chalk has spread along the whole length of the taper, the taper is correct. If the chalk lines are rubbed from only one end, the taper setup must be adjusted.
4. Make a slight adjustment to the taper attachment and, taking trial cuts, machine the taper until the fit is correct.

TO CHECK A METRIC TAPER

1. Check the drawing for the taper required.
2. Clean the tapered section of the work and apply layout dye.
3. Lay out two lines on the taper, which are the same distance apart as the second number in the taper ratio. For example, if the taper was 1:20, the lines would be 20 mm apart.

NOTE: If the work is long enough, lay out the lines at double or triple the length of the tapered section and increase the difference in diameters by the appropriate amount. For instance, on a 1:20 taper the lines may be laid out 60 mm apart or three times the unit length of the taper. Therefore the difference in diameters would then be 3×1, or 3 mm. This will give a more accurate check of the taper.

4. Measure the diameters carefully with a metric micrometer at the two lines. The difference between these two diameters should be 1 mm for each unit of length.

5. If necessary, adjust the taper attachment setting to correct the taper.

REVIEW QUESTIONS

TAPERS

1. Define a *taper*.
2. Explain the difference between self-holding and steep tapers.
3. State the *tpf* for the following tapers:
   a. Morse
   b. Brown and Sharpe
   c. Jarno
   d. Standard taper pin
4. Describe the type D-1 and type L spindle nose and state where each is used.

TAPER CALCULATIONS

5. Calculate the *tpf* and tailstock offset for the following work:
   a. $D = 1.625$ in., $d = 1.425$ in., TL = 3 in., OL = 10 in.
   b. $D = 7/8$ in., $d = 7/16$ in., TL = 6 in., OL = 9 in.
6. Calculate the tailstock offset for the following using the simplified tailstock offset formula:
   a. $D = 3/4$ in., $d = 17/32$ in., TL = 6 in., OL = 18 in.
   b. $D = 7/8$ in., $d = 25/32$ in., TL = 3½ in., OL = 10½ in.
7. Explain what is meant by a metric taper of 1:50.
8. Calculate the large diameter of a 1:50 taper having a small diameter of 15 mm and a length of 75 mm.
9. Calculate the tailstock offset required to turn a 1:40 taper × 100 mm long on a workpiece 450 mm long. The small diameter is 25 mm.

TAPER TURNING

10. Name three methods of offsetting the tailstock for taper turning.
11. List the advantages of a taper attachment.
12. List the main steps required to cut an external taper using the taper attachment.
13. Describe a taper micrometer and state its advantages.
14. Explain in point form how to fit an external taper.
15. Calculate the amount of setover for a 480-mm-long guide bar to turn a 1:40 taper × 320 mm long on a workpiece. The small diameter of the taper is 37.5 mm.
16. At what angle should the compound rest be set to machine a workpiece with a large diameter of 1¼ in., small diameter of ¾ in., length of taper of 1 in.?

UNIT 55

Threads and Thread Cutting

Threads have been used for hundreds of years for holding parts together, making adjustments to tools and instruments, and for transmitting power and motion. A thread is basically an inclined plane or wedge that spirals around a bolt or nut. Threads have progressed throughout the ages from the early screws, which were filed by hand, to the highly accurate ball screws used on the precision machine tools of today. Although the purpose of a thread is basically the same as when the early Romans developed it, the art of producing threads has continually improved. Today threads are mass produced by taps, dies, thread rolling, thread milling, and grinding to exacting standards of accuracy and quality control. Thread cutting is a skill that every machinist should possess because it is still necessary to cut threads on an engine lathe, especially if a special size or form of thread is required.

OBJECTIVES

After completing this unit you should be able to:

1. Recognize and state the purposes of six common thread forms
2. Set up a lathe to cut inch external Unified threads
3. Set up an inch lathe to cut metric threads
4. Set up a lathe and cut internal threads
5. Set up a lathe and cut external Acme threads

THREADS

A *thread* may be defined as a helical ridge of uniform section formed on the inside or outside of a cylinder or cone. Threads are used for several purposes:

1. To fasten devices such as screws, bolts, studs, and nuts.
2. To provide accurate measurement, as in a micrometer.
3. To transmit motion. The threaded lead screw on the lathe causes the carriage to move along when threading.
4. To increase force. Heavy work can be raised with a screw jack.

THREAD TERMINOLOGY

To understand and calculate thread parts and sizes, the following definitions relating to screw threads should be known (Fig. 55-1).

- A *screw thread* is a helical ridge of uniform section formed on the inside or outside of a cylinder or cone.
- An *external thread* is cut on an external surface or cone, such as on a cap screw or a wood screw.
- An *internal thread* is produced on the inside of a cylinder or cone, such as the thread on the inside of a nut.
- The *major diameter* is the largest diameter of an external or internal thread.
- The *minor diameter* is the smallest diameter of an external or internal thread. This was formerly known as the root diameter.
- The *pitch diameter* is the diameter of an imaginary cylinder which passes through the thread at a point where the groove and thread widths are equal. The pitch diameter is equal to the major diameter minus a single depth of thread. The tolerance and allowances on threads are given at the pitch diameter line. The pitch diameter is also used to determine the outside

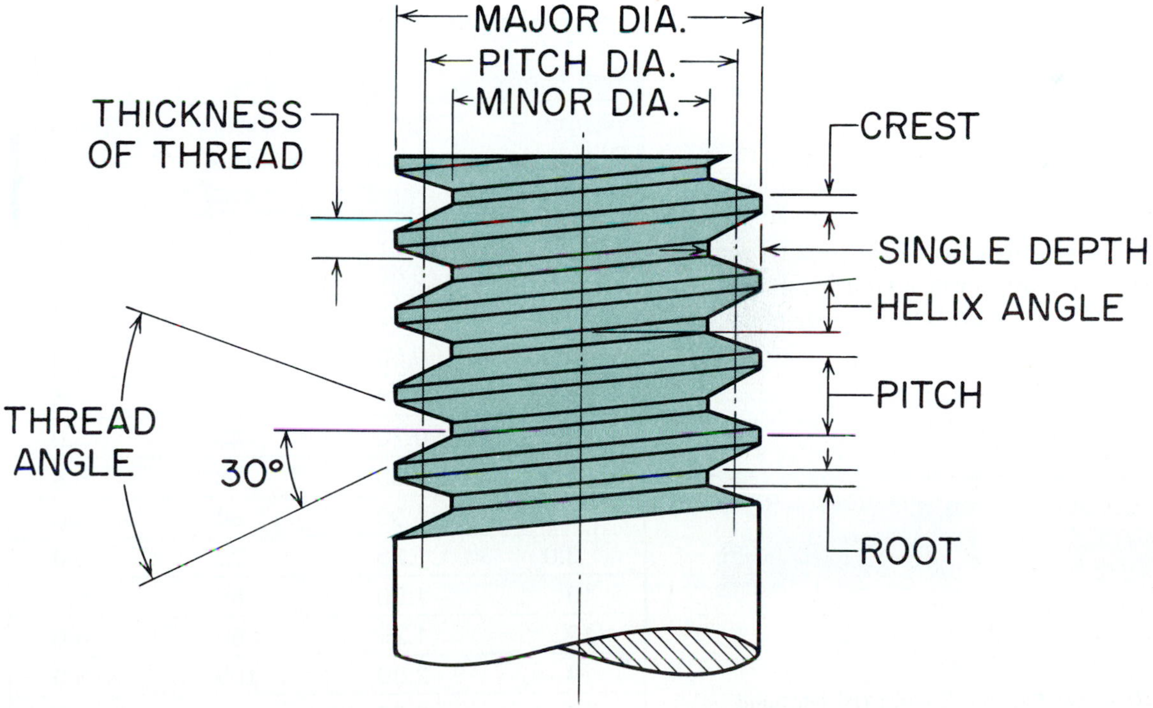

MAJOR DIA.
PITCH DIA.
MINOR DIA.
THICKNESS OF THREAD
CREST
SINGLE DEPTH
HELIX ANGLE
PITCH
THREAD ANGLE
30°
ROOT

Figure 55-1 *The parts of a screw thread.*

diameter for rolled threads. The diameter of the blank is always equal to the pitch diameter of the thread to be rolled. Thread rolling is a displacement operation and the amount of metal displaced is forced up to form the thread above the pitch line.

NOTE: Pitch diameter is not used as a basis for determining International Organization for Standardization (ISO) metric thread dimensions.

- The *number of threads per inch* is the number of crests or roots per inch of threaded section. This term does not apply to metric threads.
- The *pitch* is the distance from a point on one thread to a corresponding point on the next thread, measured parallel to the axis. Pitch is expressed in millimeters for metric threads.
- The *lead* is the distance a screw thread advances axially in one revolution. On a single start thread, the lead and the pitch are equal.
- The *root* is the bottom surface joining the sides of two adjacent threads. The root of an external thread is on its minor diameter. The root of an internal thread is on its major diameter.
- The *crest* is the top surface joining two sides of a thread. The crest of an external thread is on the major diameter; while the crest of an internal thread is on the minor diameter.
- A *flank* (side) is a thread surface which connects the crest with the root.
- The *depth of thread* is the distance between the crest and root measured perpendicular to the axis.
- The *angle of thread* is the included angle between the sides of a thread measured in an axial plane.
- The *helix angle* (lead angle) is the angle which the thread makes with a plane perpendicular to the thread axis.
- A *right-hand thread* is a helical ridge of uniform cross section onto which a nut is threaded in a clockwise direction.
  When the thread is held in a horizontal position with its axis pointing from right to left, a right-hand thread will slope *down* and to the right (Fig. 55-2A). When a right-hand thread is cut on a lathe, the toolbit advances from right to left.
- A *left-hand thread* is a helical ridge of uniform cross section onto which a nut is threaded in a counterclockwise direction.
  When the thread is held in a horizontal position with its axis pointing from right to left, the thread will slope *down* and to the left (Fig. 55-2B). When a left-hand thread is cut on a lathe the toolbit advances from left to right.

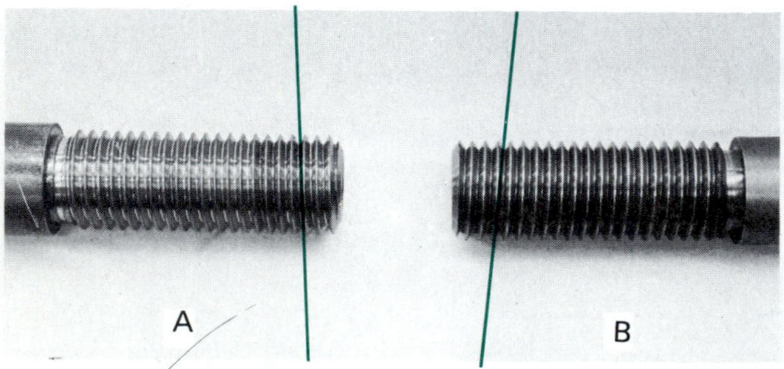

A
B

Figure 55-2 *(A) Right-hand screw threads; (B) left-hand screw threads.*

THREAD FORMS

Over the past several decades, one of the world's major industrial problems has been the lack of an international thread standard whereby the thread standard used in any country could be interchanged with that of another country. In April 1975, the ISO drew up an agreement covering a standard metric thread profile, specifying the sizes and pitches for the various threads in the new ISO Metric Thread Standard. The new series has only 25 thread sizes, which range in diameter from 1.6 to 100 mm. Countries throughout the world have been encouraged to adopt the ISO series (see Table 55-1.)

These metric threads are identified by the letter M, the nominal diameter, and the pitch. For example, a metric thread with an outside diameter of 5 mm and a pitch of 0.8 mm would be identified as follows: M 5 × 0.8.

The ISO series will not only simplify thread design but will generally produce stronger threads for a given diameter

Table 55-1 ISO Metric Pitch and Diameter Combinations

Nominal Diameter (mm)	Thread Pitch (mm)	Nominal Diameter (mm)	Thread Pitch (mm)
1.6	0.35	20	2.5
2.0	0.40	24	3.0
2.5	0.45	30	3.5
3.0	0.50	36	4.0
3.5	0.60	42	4.5
4.0	0.70	48	5.0
5.0	0.80	56	5.5
6.3	1.00	64	6.0
8.0	1.25	72	6.0
10	1.50	80	6.0
12	1.75	90	6.0
14	2.00	100	6.0
16	2.00		

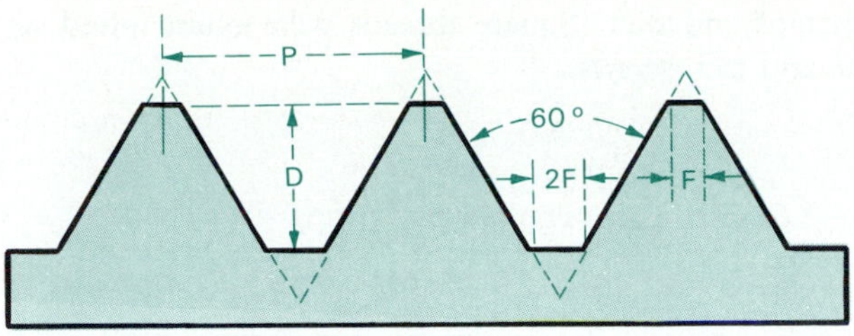

P
60°
D
2F
F

Figure 55-3 *ISO metric thread.*

and pitch and will reduce the large inventory of fasteners now required by industry.

The *ISO metric thread* (Fig. 55-3) has a 60° included angle and a crest equal to 0.125 times the pitch, similar to the National Form thread. The main difference, however, is the depth of thread (D), which is *0.6134* times the pitch. Because of these dimensions, the flat on the root of the thread (FR) is wider than the crest (FC). The root of the ISO metric thread is one-fourth of the pitch ($0.250P$).

$$D \text{ (external)} = 0.54127 \times P$$
$$\text{FC} = 0.125 \times P$$
$$\text{FR} = 0.250 \times P$$

The most commonly used thread forms in North America at the present time are:

The *American National Standard thread* (Fig. 55-4) is divided into four main series, all having the same shape and proportions: National Coarse (NC), National Fine (NF), National Special (NS), and National Pipe (NPT). This thread has a 60° angle with a root and crest truncated to one-eighth the pitch. This thread is used in fabrication, machine construction and assembly, and for components where easy assembly is desired.

The formula for calculating the depth of a 100 percent thread is $0.866/N$. However, since this thread would be very difficult to cut (especially internally) the following formula, which gives 75 percent of a thread, is generally a standard in the industry.

$$D = 0.6495 \times P \quad \text{or} \quad \frac{0.6495}{N}$$

$$F = 0.125 \times P \quad \text{or} \quad \frac{0.125}{N}$$

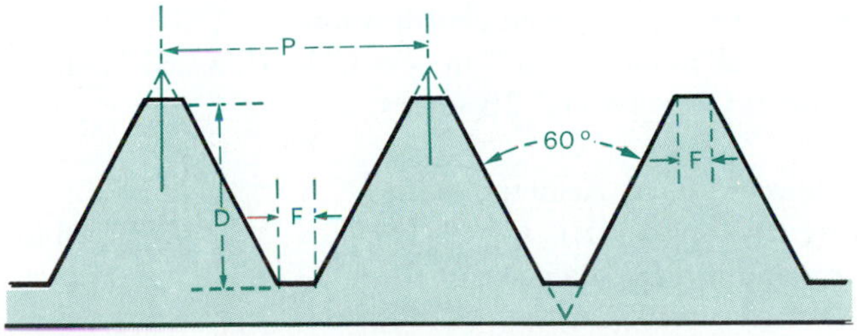

P
60°
D
F
F

Figure 55-4 *American National Standard thread.*

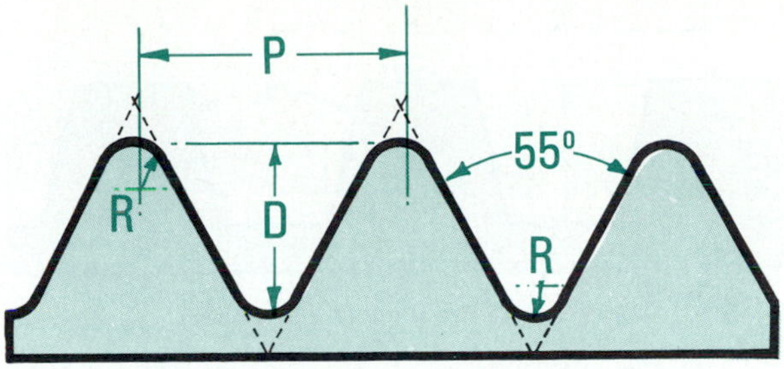

P
55°
R
D
R

Figure 55-5 *British Standard Whitworth thread.*

The *British Standard Whitworth* (BSW) *thread* (Fig. 55-5) has a 55° V-form with rounded crests and roots. This thread application is the same as for the American National form thread.

$$D = 0.6403 \times P \quad \text{or} \quad \frac{0.6403}{N}$$

$$R = 0.1373 \times P \quad \text{or} \quad \frac{0.1373}{N}$$

The *Unified thread* (Fig. 55-6) was developed by the United States, Britain, and Canada, so that equipment produced by these countries would have a standardized thread system. Until this thread was developed, many problems were created by the noninterchangeability of threaded parts being used in these countries. The Unified thread is a combination of the British Standard Whitworth and the American National form thread. This thread has a 60° angle with a rounded root, and the crest may be rounded or flat.

$$D \text{ (external thread)} = 0.6134 \times P \quad \text{or} \quad \frac{0.6134}{N}$$

$$D \text{ (internal thread)} = 0.5413 \times P \quad \text{or} \quad \frac{0.5413}{N}$$

$$F \text{ (external thread)} = 0.125 \times P \quad \text{or} \quad \frac{0.125}{N}$$

$$F \text{ (internal thread)} = 0.250 \times P \quad \text{or} \quad \frac{0.250}{N}$$

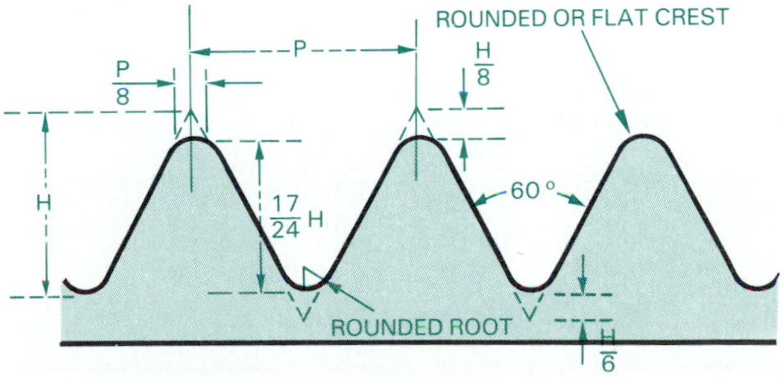

ROUNDED OR FLAT CREST
P
P/8
H/8
H
60°
17/24 H
ROUNDED ROOT
H/6

Figure 55-6 *Unified thread.*

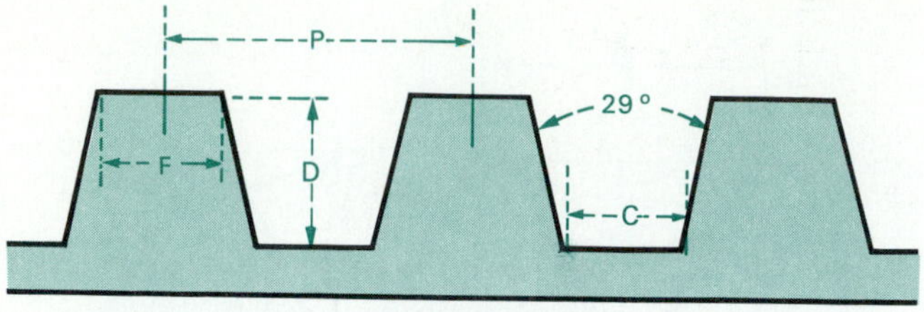
P
F
D
29°
C

Figure 55-7 *American National Acme thread.*

The *American National Acme thread* (Fig. 55-7) is replacing the square thread in many cases. It has a 29° angle and is used for feed screws, jacks, and vises.

$D =$ minimum $0.500P$
$=$ maximum $0.500P + 0.010$
$F = 0.3707P$
$C = 0.3707P - 0.0052$
(for maximum depth)

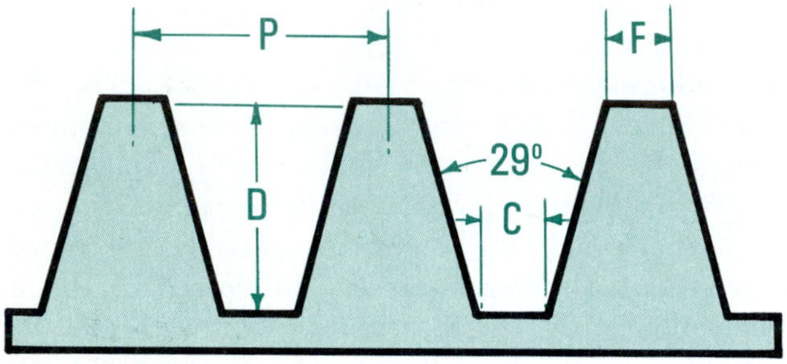
P
F
D
29°
C

Figure 55-8 *Brown and Sharpe worm thread.*

The *Brown and Sharp worm thread* (Fig. 55-8) has a 29° included angle as does the Acme thread; however, the depth is greater and the widths of the crest and root are different. This thread is used to mesh with worm gears and transmit motion between two shafts at right angles to each other but not in the same plane. The self-locking feature makes it adaptable to winches and steering mechanisms.

$D = 0.6866P$
$F = 0.335P$
$C = 0.310P$

The *square thread* (Fig. 55-9) is being replaced by the Acme thread due to the difficulty of cutting it, particularly with taps and dies. Square threads were often found on vises and jack screws.

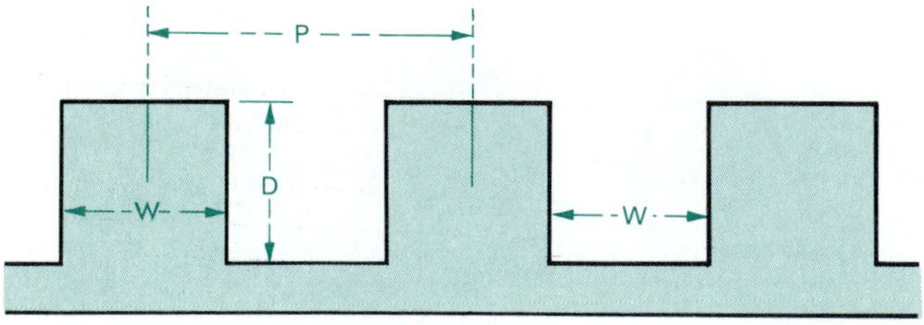
P
D
W
W

Figure 55-9 *Square thread.*

$D = 0.500P$
$F = 0.500P$
$C = 0.500P + 0.002$

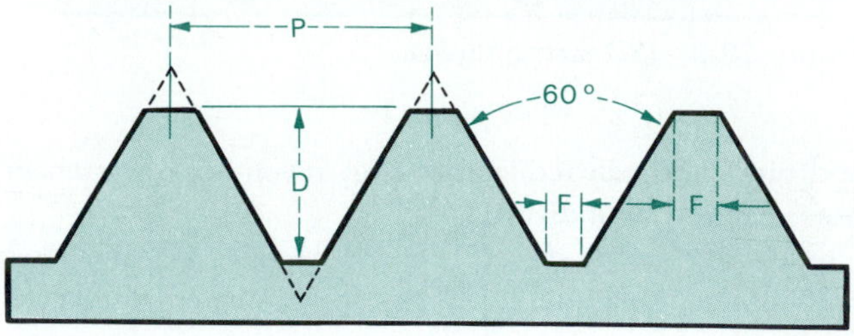
P
60°
D
F
F

Figure 55-10 *International metric thread.*

The *international metric thread* (Fig. 55-10) is a standardized thread used in Europe. This thread has a 60° included angle with a crest and root truncated to one-eighth the depth. Although this thread is used extensively throughout Europe, its use in North America has been confined mainly to spark plugs and the manufacture of instruments.

$D = 0.7035P$ (maximum)
$= 0.6855P$ (minimum)
$F = 0.125P$
$R = 0.0633P$ (maximum)
$0.054P$ (minimum)

THREAD FITS AND CLASSIFICATIONS

Certain terminology is used when referring to thread classifications and fits. To understand any thread system properly, the terminology relating to thread fits should be understood.

Fit is the relationship between two mating parts, which is determined by the amount of clearance or interference when they are assembled.

Allowance is the intentional difference in size of the mating parts or the minimum clearance between mating parts (Fig. 55-11). With threads, the allowance is the permissible difference between the largest external thread and the smallest internal thread. This produces the tightest fit acceptable for any given classification.

The allowance for a 1 in.—8 UNC (Unified National Coarse) Class 2A and 2B fit is:

Minimum pitch diameter of the internal thread (2B)	= 0.9188 in.
Maximum pitch diameter of the external thread (2A)	= 0.9168 in.
Allowance or intentional difference	= 0.002 in.

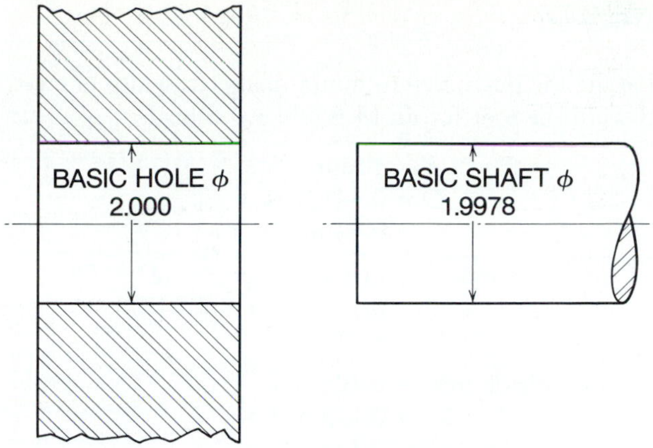
BASIC HOLE ϕ
2.000
BASIC SHAFT ϕ
1.9978

Figure 55-11 *The allowance (intentional difference) between the shaft and hole is 0.0022 in.*

Tolerance is the variation permitted in part size. The tolerance may be expressed as plus, minus, or both. The total tolerance is the sum of the plus and minus tolerances. For example, if a size is 1.000 ±0.002 (a bilateral tolerance), the total tolerance is 0.004. In the Unified and National systems, the tolerance is plus on external threads and minus on internal threads. Thus, when a thread varies from the basic or nominal size, it will ensure a freer rather than a tighter fit.

The tolerance for a 1 in.—8 UNC Class 2A thread is:

Maximum pitch diameter of the external thread (2A)	= 0.9168 in.
Minimum pitch diameter of the external thread	= 0.9100 in.
Tolerance or permitted variation	= 0.0068 in.

Limits are the maximum and minimum dimensions of a part. The limits for a 1 in.—8 UNC Class 2A thread are:

Maximum pitch diameter of the external thread (2A)	= 0.9168 in.
Minimum pitch diameter of the external thread (2A)	= 0.9100 in.

The pitch diameter of this thread must be between 0.9168 in. (upper limit) and 0.9100 in. (lower limit).

Nominal size is the designation used to identify the size of the part. For example, in the designation of 1 in.—8 UNC, the number 1 indicates a 1—in.-diameter thread.

Actual size is the measured size of a thread or part. The basic size is the size from which tolerances are set. Although the basic major diameter of a 1 in.—8 UNC Class 2A thread is 1.000 in., the actual size may vary from 0.998 to 0.983 in.

Classification of Thread Fits

With wide use of threads, it became necessary to establish certain limits and tolerances to properly identify classes of fit.

ISO Metric Tolerances and Allowances

The ISO metric screw thread tolerance system provides for allowances and tolerances defined by tolerance grades, tolerance positions, and tolerance classes.

Tolerance grades are specified numerically. For example, a medium tolerance, used on a general-purpose thread, is indicated by the number 6. Any number below 6 indicates a finer tolerance and any number above 6 indicates a greater tolerance. The tolerance for the thread at the pitch line and for the major diameter may be shown on the drawing.

Symbols are used to indicate the *allowance.* For external threads:

e indicates a large allowance
g indicates a small allowance
h indicates no allowance

For internal threads:

G indicates a small allowance
H indicates no allowance

EXAMPLE

An external metric thread may be designated as follows:

Metric	Nominal Size		Pitch		Pitch Diam. Tolerance	Outside Diam. Tolerance
↓	↓		↓		↓	↓
M	6	×	0.75	–	5*g*	6*g*

The thread fit between mating parts is indicated by the internal thread designation followed by the external thread tolerance:

$$M\ 20 \times 2 - 6H/5g\ 6g$$

Unified thread fits have been divided into three categories and the applications of each have been defined by the Screw Thread Committee. External threads are classified as 1A, 2A, and 3A, and internal threads as 1B, 2B, and 3B.

Classes 1A and 1B include those threads for work which must be readily assembled. They have the loosest fit, with no possibility of interference between the mating external and internal threads when the threads are dirty or bruised.

Classes 2A and 2B are used for most commercial fasteners. These threads provide a medium or free fit and permit power wrenching with minimum galling and seizure.

Classes 3A and 3B are used where a more accurate fit

and lead are required. No allowance is provided, and the tolerances are 75 percent of those used for 2A and 2B fits.

By classifying the tolerances of threads, the cost of threaded parts is reduced, since manufacturers may use any combination of mating threads that suit their needs. With the former system of identifying classes of tolerances (Classes 1, 2, 3, 4), it was felt, for example, that a Class 3 internal thread should be used with a Class 3 external thread.

With reference to the Unified system it should be noted that "Class" refers to tolerance or tolerance and allowance, and does not refer to fit. The fit between the mating parts is determined by the selected combination used for a specific application. For example, if a closer-than-normal fit is required, a Class 3B nut may be used on a Class 2A bolt. The basic dimensions, tolerances, and allowances for these threads may be found in any machine handbook.

THREAD CALCULATIONS

To cut a correct thread on a lathe, it is necessary first to make calculations so that the thread will have the proper dimensions. The following formulas will be helpful when calculating thread dimensions. The symbols used in these formulas are:

D = single depth of thread
P = pitch

EXAMPLE

Calculate the pitch, depth, minor diameter, and width of flat for a ¾—10 NC thread.

$$\begin{aligned} P &= \frac{1}{tpi} \\ &= \frac{1}{10} \\ &= 0.100 \text{ in.} \\ D &= 0.6495 \times P \\ &= 0.6495 \times \frac{1}{10} \\ &= 0.065 \text{ in.} \\ \text{Minor diameter} &= \text{major diameter} - (D + D) \\ &= 0.750 - (0.065 + 0.065) \\ &= 0.620 \text{ in.} \\ \text{Width of flat} &= \frac{P}{8} \\ &= \frac{1}{8} \times \frac{1}{10} \\ &= 0.0125 \text{ in.} \end{aligned}$$

EXAMPLE

What are the pitch, depth, minor diameter, width of crest, and width of root for an M 6.3 × 1.0 thread?

$$\begin{aligned} P &= 1 \text{ mm} \\ D &= 0.54127 \times 1 \\ &= 0.54 \text{ mm} \\ \text{Minor diameter} &= \text{major diameter} - (D + D) \\ &= 6.3 - (0.54 + 0.54) \\ &= 5.22 \text{ mm} \\ \text{Width of crest} &= 0.125 \times P \\ &= 0.125 \times 1 \\ &= 0.125 \text{ mm} \\ \text{Width of root} &= 0.250 \times P \\ &= 0.250 \times 1 \\ &= 0.250 \text{ mm} \end{aligned}$$

To Set the Quick-Change Gearbox for Threading

The quick-change gearbox provides a means of quickly setting the lathe for the desired pitch of the thread in number of threads per inch on inch-system lathes or millimeters on metric lathes. This unit contains a number of different sizes of gears, which vary the ratio between headstock spindle revolutions and the rate of carriage travel when thread cutting.

Proceed as follows:

1. Check the drawing for the thread pitch required.
2. From the chart on the quick-change gearbox, find the *whole number* which represents the pitch in threads per inch or in millimeters.
3. With the lathe stopped, engage the *tumbler lever* in the hole which is in line with the pitch (tpi or in millimeters) (Fig. 55-12).

SLIDING GEAR
TOP LEVER
GEAR BOX
TUMBLER LEVER
STANDARD-MODERN

Figure 55-12 *Quick-change gear mechanism is used for setting the number of threads per inch to be cut.*

4. Set the *top lever* in the proper position as indicated on the chart.
5. Engage the *sliding gear* in or out as required.

NOTE: Some lathes have two levers on the gearbox which take the place of the top lever and sliding gear, and these should be set as indicated on the chart.

6. Turn the lathe spindle by hand to ensure that the lead screw revolves.
7. Recheck the lever settings to avoid errors.

THREAD-CHASING DIAL

To cut a thread on a lathe, the lathe spindle and the lead screw must be in the same relative position for each successive cut. Most lathes have a thread-chasing dial either built into, or attached to, the carriage for this purpose. The chasing dial indicates when the split nut should be engaged with the lead screw in order for the cutting tool to follow the previously cut groove.

The thread-chasing dial is connected to a worm gear, which meshes with the threads of the lead screw (Fig. 55-13). The dial is graduated into eight divisions, four numbered and four unnumbered, and it revolves as the lead screw turns. Figure 55-14 indicates when the split-nut lever should be engaged for cutting various numbers of threads per inch (tpi).

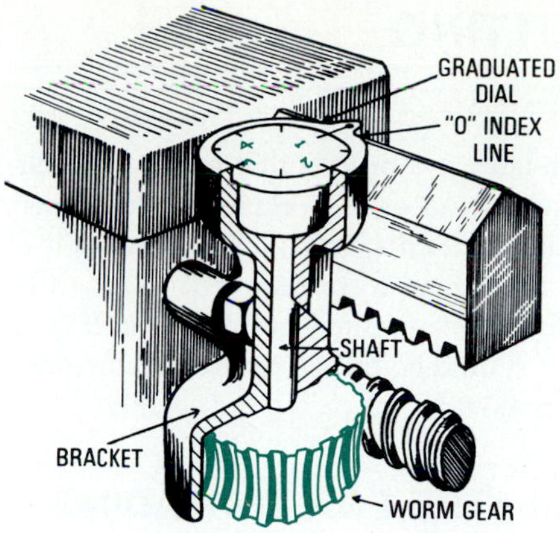

GRADUATED DIAL
"0" INDEX LINE
SHAFT
BRACKET
WORM GEAR

Figure 55-13 *Thread-chasing dial mechanism.*

- Even threads use any division.
- Odd threads must stay on either numbered or unnumbered lines; cannot use both.

THREADS PER INCH TO BE CUT	WHEN TO ENGAGE SPLIT NUT		READING ON DIAL
EVEN NUMBER OF THREADS	ENGAGE AT ANY GRADUATION ON THE DIAL	1 1 ½ 2 2 ½ 3 3 ½ 4 4 ½	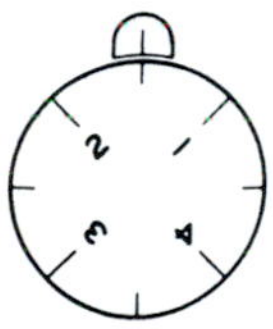
ODD NUMBER OF THREADS	ENGAGE AT ANY MAIN DIVISION	1 2 3 4	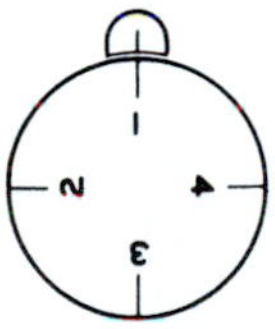
FRACTIONAL NUMBER OF THREADS	1/2 THREADS, E.G.,11 1/2 ENGAGE AT EVERY OTHER MAIN DIVISION 1 & 3, OR 2 & 4 OTHER FRACTIONAL THREADS ENGAGE AT SAME DIVISION EVERY TIME		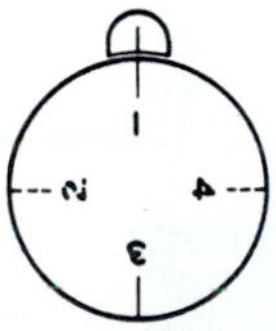
THREADS WHICH ARE A MULTIPLE OF THE NUMBER OF THREADS PER INCH IN THE LEAD SCREW	ENGAGE AT ANY TIME THAT SPLIT NUT MESHES		USE OF DIAL UNNECESSARY

Figure 55-14 *Split-nut engagement rules for thread cutting.*

THREAD CUTTING

Thread cutting on a lathe is a process that produces a helical ridge of uniform section on a workpiece. This is performed by taking successive cuts with a threading toolbit of the same shape as the thread form required. Work to be threaded may be held between centers or in a chuck. If work is held in a chuck, it should be turned to size and threaded before the work is removed.

TO SET UP A LATHE FOR THREADING (60° THREAD)

1. Set the lathe speed to about one-quarter the speed used for turning.
2. Set the quick-change gearbox for the required pitch in threads per inch or in millimeters.
3. Engage the lead screw.
4. Secure a 60° threading toolbit and check the angle using a thread center gage.
5. Set the compound rest at 29° to the right (Fig. 55-15); set it to the left for a left-hand thread.

NOTE: With the compound rest set at 29°, a slight shaving action occurs on the following edge of the thread (right side) each time that the threading tool is fed in with the compound rest handle.

6. Set the cutting tool to the height of the lathe center point.
7. Mount the work between centers. Make sure the lathe dog is tight on the work. If the work is mounted in a chuck, it must be held tightly.
8. Set the toolbit at right angles to the work, using a thread center gage (Fig. 55-16).

29°

Figure 55-15 *Compound rest set at 29° for thread cutting.*

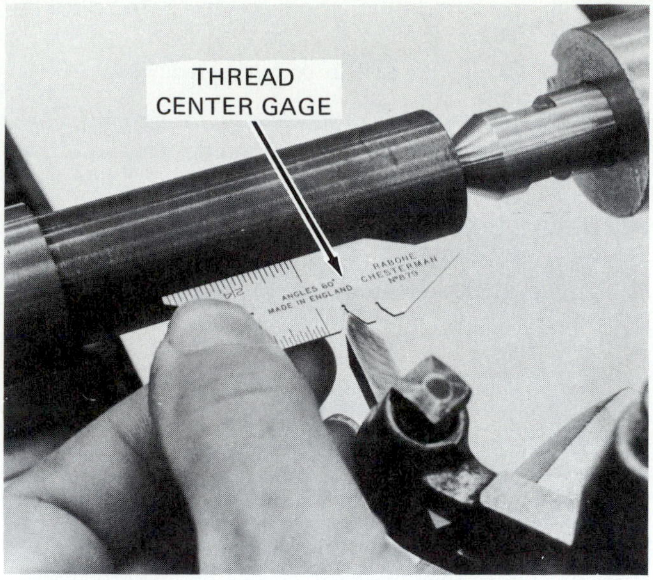
THREAD
CENTER GAGE

Figure 55-16 *Setting a threading tool square with a thread center gage.*

NOTE: Never jam a toolbit into a thread center gage. This can be avoided by aligning only the cutting edge (leading side) of the toolbit with the gage. A piece of paper on the cross-slide under the gage and toolbit makes it easier to check tool alignment.

9. Arrange the apron controls to allow the split-nut lever to be engaged.

THREAD-CUTTING OPERATION

Thread cutting is one of the more interesting operations which can be performed on a lathe. It involves manipulation of the lathe parts, coordination of the hands, and strict attention to the operation. Before proceeding to cut a thread for the first time on any lathe, take several trial passes, without cutting, in order to get the feel of the machine.

To Cut a 60° Thread

1. Check the major diameter of the work for size. It is good practice to have the diameter 0.002 in. (0.05 mm) undersize.
2. Start the lathe and chamfer the end of the workpiece with the side of the threading tool to just below the minor diameter of the thread.
3. Mark the length to be threaded by cutting a light groove at this point with the threading tool while the lathe is revolving.
4. Move the carriage until the point of the threading tool is near the right-hand end of the work.
5. Turn the *crossfeed handle* until the threading tool is

HANDLE SET
AT 3 O'CLOCK
POSITION

Figure 55-17 *The handle of the crossfeed screw set at the 3 o'clock position for thread cutting.*

close to the diameter, but stop when the handle is at the 3 o'clock position (Fig. 55-17).

6. Hold the crossfeed handle in this position and set the graduated collar to zero (0).

7. Turn the compound rest handle until the threading tool *lightly marks the work.*

8. Move the carriage to the right until the toolbit clears the end of the work.

9. Feed the compound rest *clockwise* about 0.003 in. (0.08 mm).

10. Engage the split-nut lever on the correct line of the thread-chasing dial (Fig. 55-14) and take a trial cut along the length to be threaded.

11. At the end of the cut, turn the crossfeed handle *counterclockwise* to move the toolbit away from the work and then disengage the split-nut lever.

12. Stop the lathe and check the number of *tpi* with a thread pitch gage, rule, or center gage (Fig. 55-18). If the pitch (*tpi* or millimeters) produced by the trial cut is not correct, recheck the quick-change gearbox setting.

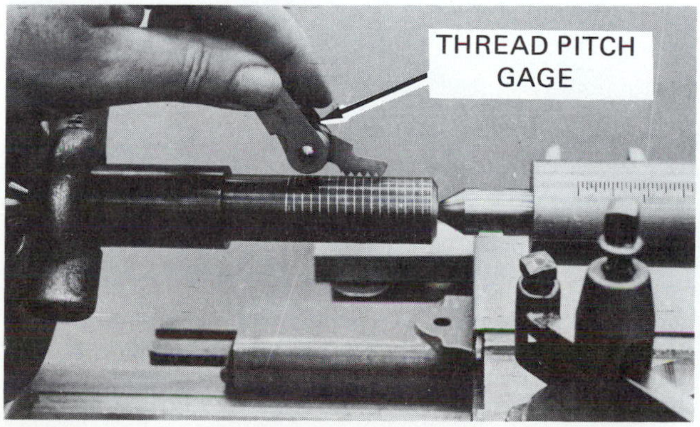
THREAD PITCH
GAGE

Figure 55-18 *Checking the number of threads per inch using a screw pitch gage.*

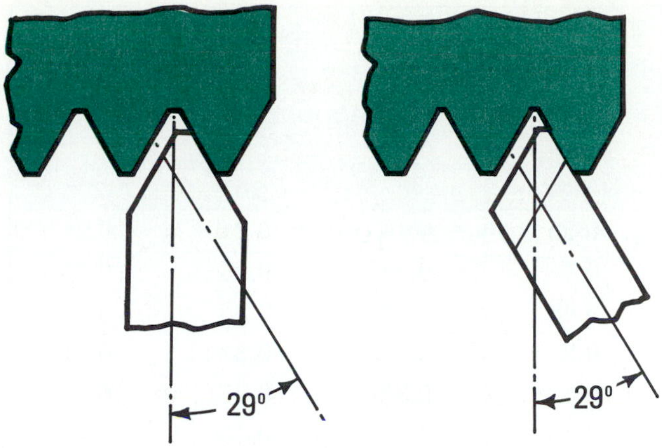
29°
29°

Figure 55-19 *When the tool is fed in at 29°, most of the cutting is done by the leading edge of the toolbit.*

13. After each cut, turn the carriage handwheel to bring the toolbit to the start of the thread and return the crossfeed handle to zero (0).

14. *Set the depth of all threading cuts with the compound rest handle* (Fig. 55-19). For National Form threads, use Table 55-2; for ISO metric threads, see Table 55-3.

15. Apply cutting fluid and take successive cuts until the top (crest) and the bottom (root) of the thread are the same width.

16. Remove the burrs from the top of the thread with a file.

Table 55-2 Depth Settings for Cutting 60° National Form Threads

	Compound Rest Setting		
TPI	0°	30°	29°
24	0.027	0.031	0.0308
20	0.0325	0.0375	0.037
18	0.036	0.0417	0.041
16	0.0405	0.0468	0.046
14	0.0465	0.0537	0.0525
13	0.050	0.0577	0.057
11	0.059	0.068	0.0674
10	0.065	0.075	0.074
9	0.072	0.083	0.082
8	0.081	0.0935	0.092
7	0.093	0.1074	0.106
6	0.108	0.1247	0.1235
4	0.1625	0.1876	0.1858

Note: When using this table for cutting National Form threads, the correct width of flat (0.125*P*) must be ground on the toolbit; otherwise, the thread will not be the correct width.

Table 55-3 Depth Setting for Cutting 60° ISO Metric Threads

Pitch (mm)	Compound Rest Setting (mm)		
	0°	30°	29°
0.35	0.19	0.21	0.21
0.4	0.21	0.25	0.24
0.45	0.24	0.28	0.27
0.5	0.27	0.31	0.31
0.6	0.32	0.37	0.37
0.7	0.37	0.43	0.43
0.8	0.43	0.50	0.49
1.0	0.54	0.62	0.62
1.25	0.67	0.78	0.77
1.5	0.81	0.93	0.93
1.75	0.94	1.09	1.08
2.0	1.08	1.25	1.24
2.5	1.35	1.56	1.55
3.0	1.62	1.87	1.85
3.5	1.89	2.19	2.16
4.0	2.16	2.50	2.47
4.5	2.44	2.81	2.78
5.0	2.71	3.13	3.09
5.5	2.98	3.44	3.40
6.0	3.25	3.75	3.71

17. Check the thread with a master nut and take further cuts, if necessary, until the nut fits the thread freely with no end play (Fig. 55-20).

There are six ways to check threads, depending on the accuracy required:

1. Master nut or screw
2. Thread micrometer
3. Three wires
4. Thread roll or snap gage

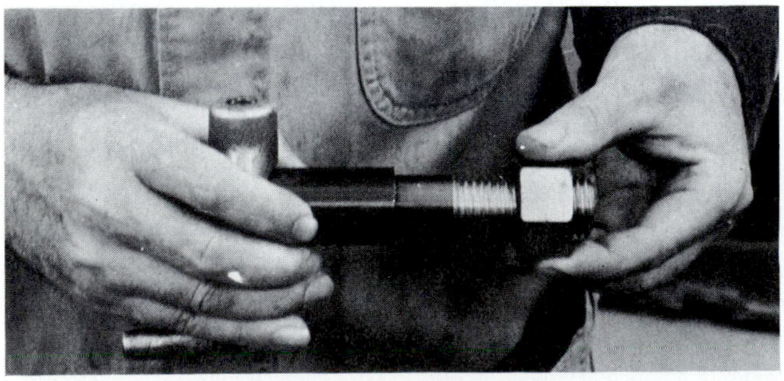

Figure 55-20 *Checking a thread with a master nut.*

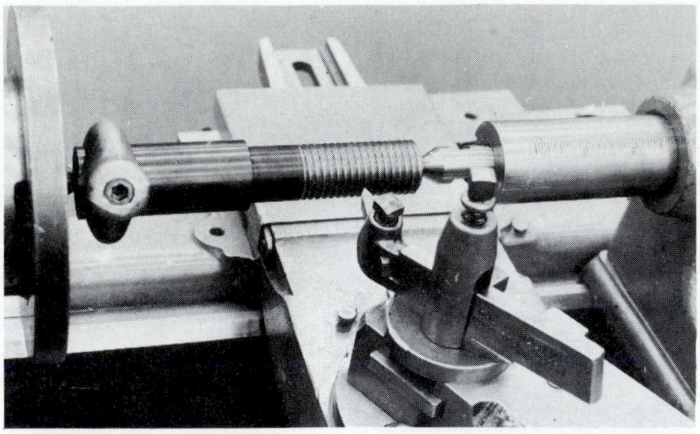

Figure 55-21 *With the split-nut lever engaged, stop the machine when the threading tool is over the thread.*

5. Thread ring or plug gage
6. Optical comparator

To Reset a Threading Tool

A threading tool must be reset whenever it is necessary to remove partly threaded work from the lathe and finish it at a later time, if the threading tool is removed for regrinding, or if the work slips under the lathe dog.

Proceed as follows:

1. Set up the lathe and work for thread cutting.
2. Start the lathe, and with the toolbit clear of the work, engage the split-nut lever on the correct line.
3. Allow the carriage to travel until the toolbit is opposite any portion of the unfinished thread (Fig. 55-21).
4. Stop the lathe, *leaving the split-nut lever engaged.*
5. Feed the toolbit into the thread groove using *only* the compound rest and crossfeed handles until the right-hand edge of the toolbit touches the rear side of the thread (Fig. 55-22).

NOTE: Do not let the cutting edge of the toolbit contact the thread at this time.

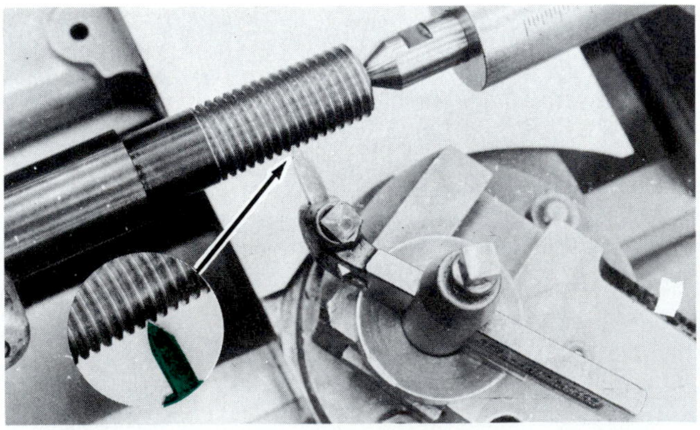

Figure 55-22 *Resetting the threading tool in a partially cut groove using only the crossfeed and compound rest handles.*

6. Set the crossfeed graduated collar to zero (0).
7. Back out the threading tool using the crossfeed handle, disengage the split-nut lever, and move the carriage until the toolbit clears the start of the thread.
8. Set the crossfeed handle back to zero (0) and take a trial cut without setting the compound rest.
9. Set the depth of cut using the compound rest handle and finish the thread to the required depth.

To Convert an Inch-Designed Lathe to Metric Threading

Metric threads may be cut on a standard quick-change gear lathe by using a pair of change gears having 50 and 127 teeth, respectively. Since the lead screw has inch dimensions and is designed to cut threads per inch, it is necessary to convert the pitch in millimeters to centimeters and then into threads per inch. To do this, it is first necessary to understand the relationship between inches and centimeters.

$$1 \text{ in.} = 2.54 \text{ cm}$$

Therefore the ratio of inches to centimeters is 1 : 2.54, or 1/2.54.

To cut a metric thread on an inch lathe, it is necessary to install certain gears in the gear train which will produce a ratio of 1/2.54. These gears are:

$$\frac{1}{2.54} \times \frac{50}{50} = \frac{50}{127} \quad \frac{\text{teeth}}{\text{teeth}}$$

In order to cut metric threads, two gears having 50 and 127 teeth must be placed in the gear train of the lathe. The 50-tooth gear is used as the spindle or drive gear and the 127-tooth gear is placed on the lead screw.

To Cut a 2-mm Metric Thread on a Standard Quick-Change Gear Lathe

1. Mount the 127-tooth gear on the lead screw.
2. Mount the 50-tooth gear on the spindle.
3. Convert the 2-mm pitch to threads per centimeter:

$$10 \text{ mm} = 1 \text{ cm}$$

$$P = \frac{10}{2} = 5 \text{ threads/cm}$$

4. Set the quick-change gearbox to 5 *tpi*. By means of the 50- and 127-tooth gears, the lathe will now cut 5 threads/cm or 2-mm pitch.
5. Set up the lathe for thread cutting. See *To Set Up a Lathe for Threading (60° Thread).*
6. Take a light trial cut. At the end of the cut, back out the cutting tool and stop the machine but *do not disengage the split nut.*
7. Reverse the spindle rotation until the cutting tool has just cleared the start of the threaded section.
8. Check the thread with a metric screw pitch gage.
9. Cut the thread to the required depth (Table 55-3).

NOTE: Never disengage the split nut until the thread has been cut to depth.

To Cut a Left-hand Thread (60°)

A left-hand thread is used to replace a right-hand thread on certain applications where the nut may loosen due to the rotation of a spindle. The procedure for cutting left-hand threads is basically the same as for right-hand threads, with a few exceptions.

1. Set the lathe speed and the quick-change gearbox for the pitch of the thread to be cut.
2. Engage the feed-direction lever so that the lead screw will revolve in the *opposite* direction to that for a right-hand thread.
3. Set the compound rest to 29° to the *LEFT* (Fig. 55-23).
4. Set up the left-hand threading tool and square it with the work.
5. Cut a groove at the left end of the section to be threaded. This gives the cutting tool a starting point.
6. Proceed to cut the thread to the same dimensions as for a right-hand thread.

Cutting a Thread on a Tapered Section

When a tapered thread, such as a pipe thread, is required on the end of a workpiece, either the taper attachment or the offset tailstock may be used for cutting the taper. The same setup is then used as for regular thread cutting. When setting up the threading tool, it is most important that it be set at 90° to the axis of the work and not square with the tapered surface (Fig. 55-24).

Figure 55-23 *Set the compound rest at 29° to the left for left-hand threads.*

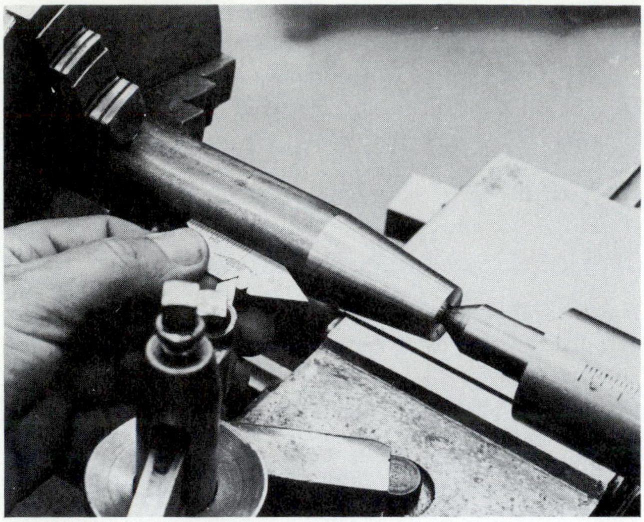

Figure 55-24 *The toolbit must be set square with the axis of the workpiece for cutting a thread on a tapered section.*

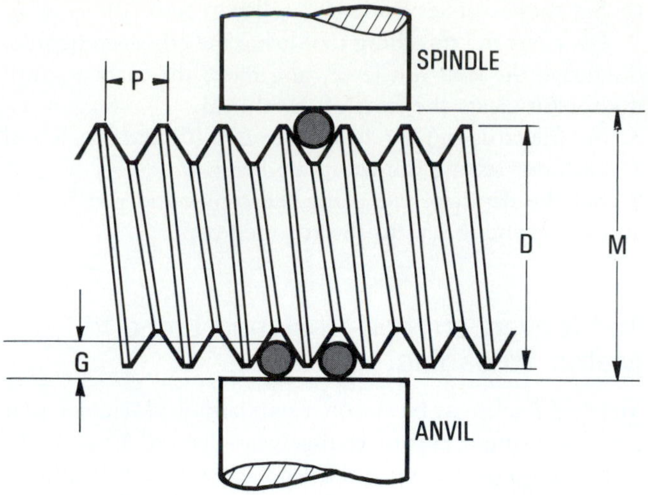
SPINDLE
P
D
M
G
ANVIL

Figure 55-25 *The three-wire method of measuring 60° threads.*

THREAD MEASUREMENT

Interchangeable manufacture demands that all parts be made to certain standards in order that, on assembly, they will fit the intended component properly. This is especially important for threaded components, and therefore the measurement and inspection of threads is important.

Threads may be measured by a variety of methods; the most common are:

1. A thread ring gage
2. A thread plug gage
3. A thread snap gage
4. A screw thread micrometer
5. A thread comparator micrometer
6. An optical comparator
7. The three-wire method

The description and use of the ring gage, plug gage, snap gage, screw thread micrometer, thread comparator micrometer, and optical comparator for checking threads are fully described in Section 5.

Three-Wire Method of Measuring Threads

The three-wire method of measuring threads is recommended by the Bureau of Standards and the National Screw Thread Commission. It is recognized as one of the best methods of checking the pitch diameter because the results are least affected by an error which may be present in the included thread angle. For threads which require an accuracy of 0.001 in. or 0.02 mm, a micrometer can be used to measure the distance over the wires. An electronic comparator should be used to measure the distance over the wires for threads requiring greater accuracy.

Three wires of equal diameter are placed in the thread; two on one side and one on the other side (Fig. 55-25). The wires used should be hardened and lapped to three times the accuracy of the thread to be inspected. A standard micrometer may then be used to measure the distance over the wires (M). Different sizes and pitches of threads require different sizes of wires. For the greatest accuracy, the *best size wire* should be used. This is one that will contact the thread at the pitch diameter (middle of the sloping sides). If the best size wire is used, the pitch diameter of the thread can be calculated by subtracting the wire constant (found in any handbook) from the measurement over the wires.

To Calculate the Measurement Over the Wires

The measurement over the wires for American National Threads (60°) can be calculated by applying the following formula.

$$M = D + 3G - \frac{1.5155}{N}$$

where M = measurement over the wires
D = major diameter of the thread
G = diameter of the wire used
N = number of *tpi*

Any of the following formulas can be used to calculate G.

$$\text{Largest size wire} = \frac{1.010}{N} \quad \text{or} \quad 1.010P$$

$$\text{Best size wire} = \frac{0.57735}{N} \quad \text{or} \quad 0.57735P$$

$$\text{Smallest size wire} = \frac{0.505}{N} \quad \text{or} \quad 0.505P$$

For the most accurate thread measurement, the best size

wire (0.57735P) should be used, since this wire will contact the thread at the pitch diameter.

EXAMPLE

Find M (measurement over the wires) for a ¾—10 NC thread.

1. Calculate G (wire size):

$$G = \frac{0.57735}{10} = 0.0577$$

2. Calculate M (measurement over the wires):

$$M = D + 3G - \frac{1.5155}{N}$$

$$= 0.750 + (3 \times 0.0577) - \frac{1.5155}{10}$$

$$= 0.750 + 0.1731 - 0.1516$$
$$= 0.9231 - 0.1516$$
$$= 0.7715$$

MULTIPLE THREADS

Multiple threads are used when it is necessary to obtain an increase in lead and a deep, coarse thread cannot be cut. Multiple threads may be double, triple, or quadruple, depending on the number of starts around the periphery of the workpiece (Fig. 55-26).

The *pitch* of a thread is always the distance from a point on one thread to the corresponding point on the next thread. The *lead* is the distance a nut advances lengthwise in one complete revolution. On a single-start thread, the pitch and lead are equal. On a double-start thread, the lead will be twice the pitch. On triple-start threads the lead will be three times the pitch.

Multiple-start threads are not as deep as single-start threads and therefore have a more pleasing appearance. For example, a double-start thread having the same lead as a single-start thread would be cut only half as deep.

Multiple threads may be cut on a lathe by

1. Using an accurately slotted drive plate or faceplate
2. Disengaging the intermediate gear of the end gear train and rotating the spindle the desired amount
3. Using the thread-chasing dial (only for double-start threads with an odd-number lead).

To Cut an 8-TPI Double Thread

1. Set up the lathe and cutting tool as for cutting a single-start thread.
2. Set the quick-change gearbox to 4 *tpi*. (The lead of this thread is ¼ in.).
3. Cut the first thread to half the depth required for 4 *tpi*.
4. Leave the crossfeed handle set to the depth of the thread and *note the reading on the compound rest graduated collar.*
5. Withdraw the threading tool from the work using the *compound rest handle.*
6. Revolve the work exactly one-half turn by either of the following methods.
   a. (1) Remove the work from the lathe with the lathe dog attached.
   (2) Replace the work in the lathe with the tail of the dog in the slot exactly opposite the one used for the first thread.

NOTE: An accurately slotted drive plate or faceplate must be used for this method of indexing. A special indexing plate may also be used for this purpose.

OR

   b. (1) Turn the lathe by hand until a tooth of the spindle gear is exactly between two teeth of the intermediate gear.
   (2) With chalk, mark both the spindle tooth and the space in the intermediate gear (Fig. 55-27).
   (3) Disengage the intermediate gear from the spindle gear.
   (4) Starting with the tooth *next* to the marked tooth on the spindle gear, count the number required for a half-revolution of the spindle. For example, if the

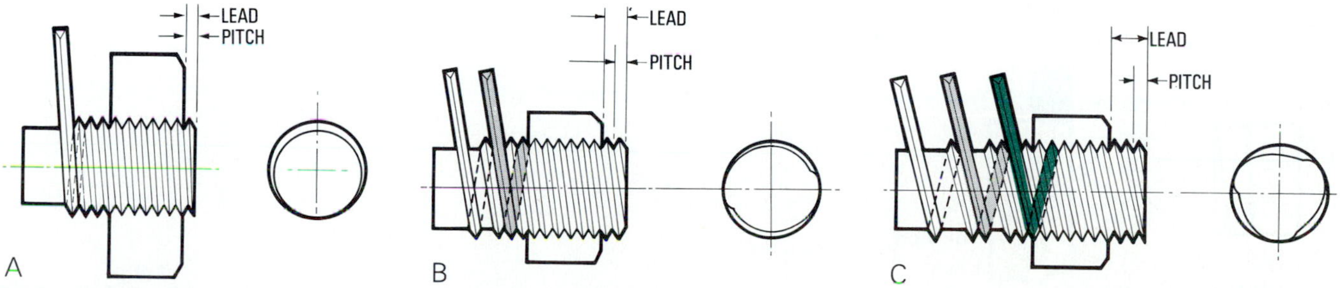

LEAD
PITCH
A
LEAD
PITCH
B
LEAD
PITCH
C

Figure 55-26 *The relationship between the pitch and the lead for a single-start and multiple-start threads: (A) single thread; (B) double thread; (C) triple thread.*

Figure 55-27 *Marking the spindle and intermediate gear before indexing the workpiece exactly one-half turn for cutting a double thread.*

spindle gear has 24 teeth, count 12 teeth and mark this one with chalk.
(5) Revolve the lathe spindle by hand one-half turn to bring the marked tooth in line with the chalk mark on the intermediate gear.
(6) Reengage the intermediate gear.

7. Reset the crossfeed handle to the same position as when cutting the first thread.
8. Cut the second thread, feeding the compound rest handle until the graduated collar is at the same setting as for the first thread.

The Thread-Chasing Dial Method of Cutting Multiple Threads

Double-start threads with an odd-numbered lead (for example, $1/5$, $1/7$, etc.) may be cut using the thread-chasing dial.

1. Take one cut on the thread by engaging the split nut at a *numbered* line on the chasing dial.
2. Without changing the depth of cut, take another cut at an *unnumbered* line on the chasing dial. The second thread will be exactly in the middle of the first thread.
3. Continue cutting the thread to depth, taking two passes (one on a numbered line, the other on an unnumbered line) for every depth-of-cut setting.

SQUARE THREADS

Square threads were often found in vise screws, jacks, and other devices where maximum power transmission was required. Because of the difficulty of cutting this thread with taps and dies, it is being replaced by Acme thread. With care, square threads can be readily cut on a lathe.

The Shape of a Square Threading Tool

The square threading tool looks like a short cutting-off tool. It differs from it in that both sides of the square threading tool must be ground at an angle to conform to the helix angle of the thread (Fig. 55-28).

The helix angle of a thread, and therefore the angle of the square threading tool, depends on two factors:

1. The helix angle changes for each *different lead* on a

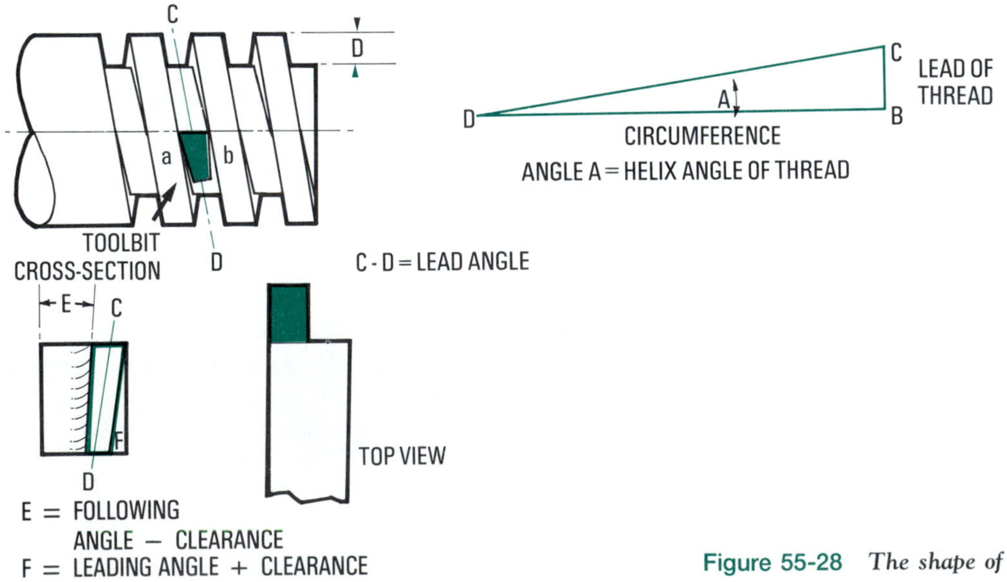

C
D
a
b
TOOLBIT
CROSS-SECTION
D
C - D = LEAD ANGLE
E
C
F
D
TOP VIEW
E = FOLLOWING
ANGLE − CLEARANCE
F = LEADING ANGLE + CLEARANCE
D
A
C
B
LEAD OF
THREAD
CIRCUMFERENCE
ANGLE A = HELIX ANGLE OF THREAD

Figure 55-28 *The shape of a square threading tool.*

given diameter. The greater the lead of the thread, the greater will be the helix angle.

2. The helix angle changes for each *different diameter* of thread for a given lead. The larger the diameter, the smaller will be the helix angle.

The helix angle of either the leading or following side of a square thread can be represented by a right-angle triangle (Fig. 55-28). The side opposite equals the *lead* of the thread, and the side adjacent equals the circumference of either the major or minor diameter of the thread. The angle between the hypotenuse and the side adjacent represents the helix angle of the thread.

To Calculate the Helix Angles of the Leading and Following Sides of a Square Thread

$$\text{Tan leading angle} = \frac{\text{lead of thread}}{\text{circumference of minor diameter}}$$

$$\text{Tan following angle} = \frac{\text{lead of thread}}{\text{circumference of major diameter}}$$

Clearance

If a square toolbit is ground to the same helix angles as the leading and following sides of the thread, it would have no clearance and the sides would rub. To prevent the tool from rubbing, it must be provided with approximately 1° clearance on each side, making it thinner at the bottom (Fig. 55-29). For the leading side of the tool, *add* 1° to the calculated helix angle. On the following side, *subtract* 1° from the calculated angle.

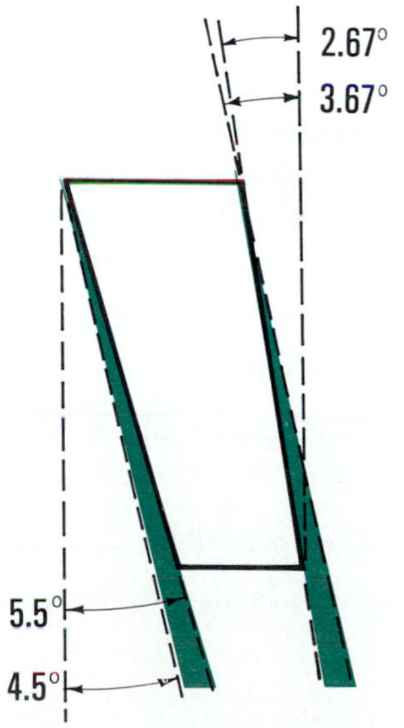

2.67°
3.67°
5.5°
4.5°

Figure 55-29 *Helix angles of the thread and clearance angles necessary for a square threading tool.*

EXAMPLE

To find the leading and following angles of a threading tool to cut a 1¼ in.—4 square thread.

SOLUTION

$$\text{Lead} = 0.250 \text{ in.}$$

$$\text{Single depth} = \frac{0.500}{4} = 0.125 \text{ in.}$$

$$\text{Double depth} = 2 \times 0.125 = 0.250 \text{ in.}$$

$$\text{Minor diameter} = 1.250 - 0.250 = 1.000 \text{ in.}$$

$$\text{Tan leading angle} = \frac{\text{lead}}{\text{minor dia. circumference}} = \frac{0.250}{1.000 \times \pi} = \frac{0.250}{3.1416} = 0.0795 \text{ in.}$$

$$\therefore \text{the angle of the thread} = 4°33'$$

$$\text{The toolbit angle} = 4°33' \text{ plus } 1° \text{ clearance} = 5°33'$$

$$\text{Tan following angle} = \frac{\text{lead}}{\text{major dia. circumference}} = \frac{0.250}{1.250\pi} = \frac{0.250}{3.927} = 0.0636 \text{ in.}$$

$$\therefore \text{the angle of the thread} = 3°38'$$

$$\text{The toolbit angle} = 3°38' \text{ minus } 1° \text{ clearance} = 2°38'$$

To Cut a Square Thread

1. Grind a threading tool to the proper leading and following angles. The width of the tool should be approximately 0.002 in. (0.05 mm) wider than the thread groove. This will allow the completed screw to fit the nut readily. Depending on the size of the thread, it may be wise to grind two tools; a roughing tool 0.015 in. (0.38 mm) undersize, and a finishing tool 0.002 in. (0.05 mm) oversize.
2. Align the lathe centers and mount the work.
3. Set the quick-change gearbox for the required number of *tpi*.
4. Set the compound rest at 30° to the right. This will

provide side movement if it becomes necessary to reset the cutting tool.

5. Set the threading tool square with the work and on center.

6. Cut the right-hand end of the work to the minor diameter for approximately 1⁄16 in. (1.58 mm) long. This will indicate when the thread is cut to the full depth.

7. If the work permits, cut a recess at the end of the thread to the minor diameter. This will provide room for the cutting tool to "run out" at the end of the thread.

8. Calculate the single depth of the thread as

$$\left(\frac{0.500}{N}\right)$$

9. Start the lathe and just touch the tool to the work diameter.

10. Set the *crossfeed graduated collar* to zero (0).

11. Set a 0.003-in. (0.08-mm) depth of cut with the *crossfeed screw* and take a trial cut.

12. Check the thread with a thread pitch gage.

13. Apply cutting fluid and cut the thread to depth, moving the *crossfeed* in from 0.002 to 0.010 in. (0.05 to 0.25 mm) for each cut. The depth of the cut will depend on the thread size and the nature of the workpiece.

NOTE: Since the thread sides are square, *all cuts* must be set using the *crossfeed screw.*

ACME THREAD

The *Acme thread* is gradually replacing the square thread because it is stronger and easier to cut with taps and dies. It is used extensively for lead screws because the 29° angle formed by its sides allows the split nut to be engaged readily during thread cutting.

The Acme thread is provided with 0.010-in. clearance for both the crest and root on all sizes of threads. The hole for an internal Acme thread is cut 0.020 in. larger than the minor diameter of the screw, and the major diameter of a tap or internal thread is 0.020 in. larger than the major diameter of the screw. This provides 0.010-in. clearance between the screw and nut on both the top and bottom.

To Cut an Acme Thread

1. Grind a toolbit to fit the end of the Acme thread gage (Fig. 55-30). Be sure to provide sufficient side clearance so that the tool will not rub while cutting the thread.

2. Grind the point of the tool flat until it fits into the slots of the gage indicating the number of threads per inch to be cut.

NOTE: If a gage is not available, the width of the toolbit point may be calculated as follows:

$$\text{Width of point} = \frac{0.3707}{N} - 0.0052 \text{ in.}$$

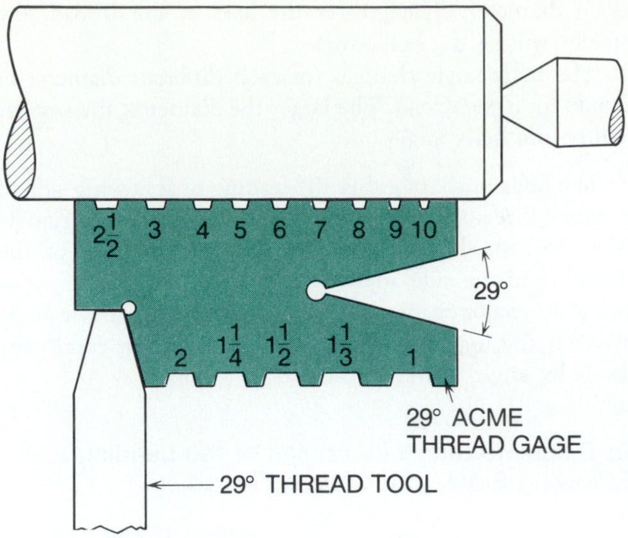
2½ 3 4 5 6 7 8 9 10
29°
2 1¼ 1½ 1⅓ 1
29° ACME
THREAD GAGE
29° THREAD TOOL
Figure 55-30 *The toolbit is squared with the workpiece by an Acme thread gage.*

3. Set the quick-change gearbox to the required number of threads per inch.

4. Set the compound rest 14½° to the right (half the included thread angle).

5. Set the Acme threading tool on center and square it with the work using the gage shown in Fig. 55-30.

6. At the right-hand end of the work, cut a section 1⁄16 in. long to the minor diameter. This will indicate when the thread is to the full depth.

7. Cut the thread to the proper depth by feeding the cutting tool, using the *compound rest.*

Measuring Acme Threads

For most purposes, the *one-wire method* of measuring Acme threads is accurate enough. A single wire or pin of the correct diameter is placed in the thread groove (Fig. 55-31) and measured with a micrometer. The thread is the correct size when the micrometer reading over the wire is the same as the major diameter of the thread and the *wire is tight in the thread.*

NOTE: It is important that the burrs be removed from the diameter before using the one-wire method.

The diameter of the wire to be used can be calculated as follows:

$$\text{Wire diameter} = 0.4872 \times \text{pitch}$$

For example, if 6 threads per inch are being cut, the wire diameter should be:

$$\text{Wire size} = 0.4872 \times \frac{1}{6}$$
$$= 0.081 \text{ in.}$$

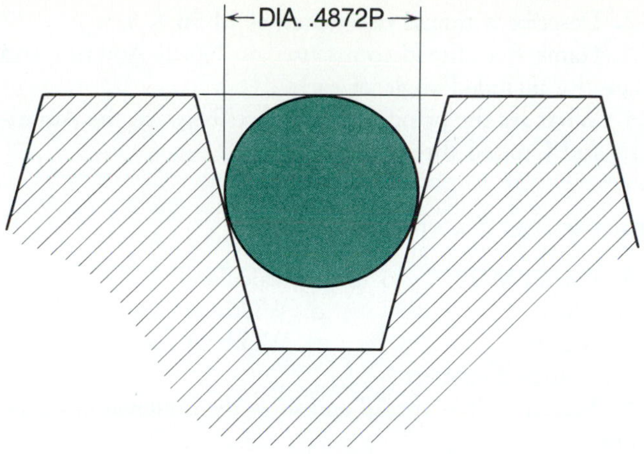
DIA. .4872P

Figure 55-31 *Using one wire to measure the accuracy of an Acme thread.*

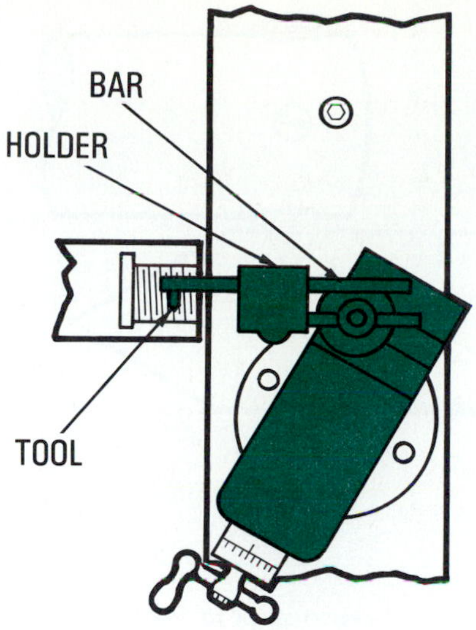
BAR
HOLDER
TOOL

INTERNAL THREADS

Most internal threads are cut with taps; however, there are times when a tap of a specific size is not available and the thread must be cut on a lathe. Internal threading, or cutting threads in a hole, is an operation performed on work held in a chuck or collet, or mounted on a faceplate. The threading tool is similar to a boring toolbit, except the shape is ground to the form of the thread to be cut.

To Cut a 1⅜ in.—6 NC Internal Thread

1. Calculate the tap drill size of the thread.

$$\text{Tap drill size} = \text{major diameter} - \frac{1}{N}$$
$$= 1.375 - \frac{1}{6}$$
$$= 1.375 - 0.166$$
$$= 1.209 \text{ in.}$$

2. Mount the work to be threaded in a chuck or collet, or on a faceplate.

3. Drill a hole approximately $\frac{1}{16}$ in. smaller than the tap drill size in the workpiece. For this thread it would be $1.209 - 0.062 = 1.147$, or a $1\frac{5}{32}$ in. hole.

4. Mount a boring tool in the lathe and bore the hole to the tap drill size (1.209 in.). The boring bar should be as large as possible and held short. The boring operation cuts the hole to size and makes it true.

5. Recess the start of the hole to the major diameter of the thread (1.375 in.) for $\frac{1}{16}$ in. length. During the thread-cutting operation, this will indicate when the thread is cut to depth.

6. If the thread does not go through the workpiece, a recess should be cut at the end of the thread to the major diameter (Fig. 55-32). This recess should be wide enough to allow the threading tool to "run out" and permit time to disengage the split-nut lever.

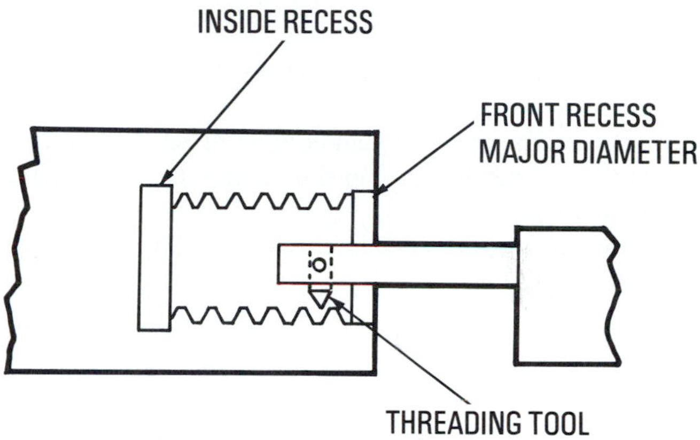
INSIDE RECESS
FRONT RECESS
MAJOR DIAMETER
THREADING TOOL

Figure 55-32 *The compound rest is set at 29° to the left for cutting right-hand internal threads.*

7. Set the compound rest at 29° to the left (Fig. 55-32); to the right for left-hand threads.

8. Mount a threading toolbit into the boring bar and set it to center.

9. Square the threading tool with a thread center gage (Fig. 55-33).

10. Place a mark on the boring bar, measuring from the threading tool, to indicate the length of hole to be threaded. This will show when the split-nut lever should be disengaged.

11. Start the lathe and turn the crossfeed handle *out* until the threading tool just scratches the internal diameter.

12. Set the crossfeed graduated collar to zero (0).

13. Set a 0.003-in. depth of cut by feeding the compound rest *out* and take a trial cut.

14. At the end of *each* cut on an internal thread, disengage the split-nut lever and feed the crossfeed handle *in* to clear the thread.

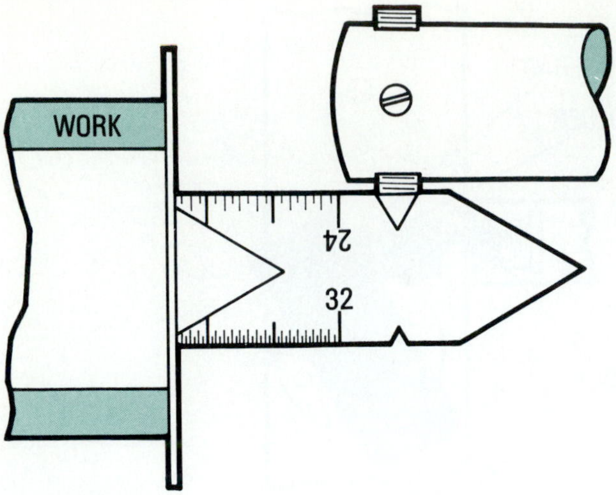
WORK
24
32

Figure 55-33 *Using a thread center gage to square the threading tool with the workpiece.*

15. Clear the threading tool from the hole and check the pitch of the thread.
16. Return the crossfeed handle back to zero (0) and set the depth of cut by turning the compound rest *out* the desired amount.
17. Cut the thread to depth; check the fit with a screw or threaded plug gage.

NOTE: The last few cuts should not be deeper than 0.001 in. each to eliminate the spring of the boring bar.

REVIEW QUESTIONS

THREADS

1. Define *thread.*
2. List four purposes of threads.

THREAD TERMINOLOGY

3. Define *pitch diameter, pitch, lead, root,* and *crest.*
4. How is the pitch designated for:
   a. UNC threads? b. Metric threads?
5. Why is the diameter of the blank for a rolled thread equal to the pitch diameter?
6. How may a right-hand thread be distinguished from a left-hand thread?

THREAD FITS AND CLASSIFICATIONS

7. Define *fit, allowance, tolerance,* and *limits.*
8. How are external UNC threads classified?
9. Name and describe three classifications of UNC fits.
10. How are thread fits designated for the ISO threads?
11. Why was the ISO metric system of threads adopted?
12. Describe a thread designated as M 56 × 5.5.
13. Name five thread forms used in North America and state the included angle of each.
14. What are the principal differences between the American and Unified threads?

THREAD CALCULATIONS

15. For an M 20 × 2.5 thread, calculate:
   a. Pitch
   b. Depth
   c. Minor diameter
   d. Width of crest
   e. Width of root
16. Sketch a UNC thread and show the dimensions of the parts.
17. How does a Brown and Sharpe worm thread differ from an Acme thread?
18. For a 1 in.—8 NC thread, calculate:
   a. Minor diameter
   b. Width of toolbit point
   c. Amount of compound rest feed
19. What is the purpose of the quick-change gearbox?
20. Describe a thread-chasing dial and state its purpose.
21. The lead screw on a lathe has 6 *tpi*; at what point or points on the thread-chasing dial may the split-nut lever be engaged to cut the following threads: 8, 9, 11½, 12?

THREAD CUTTING

22. List the main steps required to set up the lathe for cutting a 60° thread.
23. List the main steps required to cut a 60° thread.
24. It has been necessary to remove a partially threaded piece of work from the lathe and to finish it at a later time. Describe how to reset the threading tool to "pick up" the thread.

METRIC THREADS

25. a. What two change gears are required to cut a metric thread on a standard lathe?
   b. Where are these gears mounted?
26. Describe how a lathe (having a quick-change gearbox) is set up to cut a 2.5-mm thread.
27. What precaution should be taken when cutting a metric thread?

THREAD MEASUREMENT

28. Make a sketch to illustrate how a thread is measured using the three-wire method.
29. Calculate the best wire size and the measurement over the wires for the following threads:
   a. ¼ in.—20 NC
   b. ⅝ in.—11 NC
   c. 1¼ in.—7 NC

MULTIPLE THREADS

30. What is the purpose of a multiple thread?
31. If the pitch of a multiple thread is 1/8 in., what will be the lead for a double-start thread? A triple-start thread?
32. List three methods by which multiple threads may be cut.

SQUARE THREADS

33. Name two factors that affect the helix angle of a thread.
34. Calculate the leading and following angles of a square threading toolbit required to cut a 1½ in.—6 square thread.
35. List the main steps required to cut a square thread.

ACME THREADS

36. If the width of the root of an Acme thread is $0.3707P$ at minimum depth, why is the Acme threading tool ground to $0.3707P - 0.0052$?
37. Describe how an Acme thread may be measured.

INTERNAL THREADS

38. List the steps required to cut a 1¼ in.—7 UNC thread 2 in. deep in a block of steel 3 in. × 3 in. × 3 in. Show all necessary calculations.

UNIT 56

Steady Rests, Follower Rests, and Mandrels

Various accessories make it possible to machine different lengths and shapes of workpieces in a lathe. Steady and follower rests are used to support long or slender workpieces and prevent them from springing during a machining operation either between lathe centers or in a chuck. Steady rests are attached to the lathe bed and are generally set in about the middle of a workpiece. Follower rests are attached to the lathe carriage and are set up immediately to the right of the cutting tool. Since the relationship of the attachment and the cutting tool remains the same, the workpiece is supported throughout the cut.

Mandrels generally fit the hole of a thin part, such as gears, flanges, and pulleys, and they allow the outside diameter of the work to be machined between centers or in a chuck.

OBJECTIVES

After completing this unit you should be able to:

1. Set up and use a steady rest for machining a long shaft
2. Set up and use a follower rest when machining a long shaft
3. Mount and machine work on a mandrel
4. Lay out and machine an eccentric on work held between centers

STEADY REST

A *steady rest* (Fig. 56-1) is used to support long, slender work and prevent it from springing while being machined between centers. A steady rest may also be used when it is necessary to perform a machining operation on the end of a workpiece which is held in a chuck. The steady rest is fastened to the lathe bed and its three jaws are adjusted to the surface of the work to provide a supporting bearing. The jaws on a steady rest are generally made of soft material, such as fiber or brass, to prevent damaging the work surface. Other steady rests have rollers attached to the jaws to provide good support for the work.

TO SET UP A STEADY REST

1. Mount the work between centers.

   OR

   Set up and true the work in a chuck.
2. a. If the work diameter is not round, turn a true spot on the diameter (slightly wider than the steady rest jaws) at the point where the steady rest will be supporting the work. Long work in a chuck should be first supported by the tailstock center. If the diameter is rough, turn a section for the steady rest and one near the chuck to the same diameter.

   b. If it is impossible to turn a true diameter (due to the shape of the workpiece), mount and adjust a *cathead* (Fig. 56-2) on the work.
3. Move the carriage to the tailstock end of the lathe.
4. Place the steady rest on the lathe bed at the desired position. If the work diameter is turned and held in a chuck, slide the steady rest up close to the chuck.
5. Adjust the lower two jaws to the work diameter, using a paper feeler to provide clearance between the jaws and the work.
6. Slide the steady rest to the desired position and fasten it in place.

Figure 56-1 *Using a steady rest to support the end of a long workpiece held in a chuck. (Courtesy South Bend Lathe Inc.)*

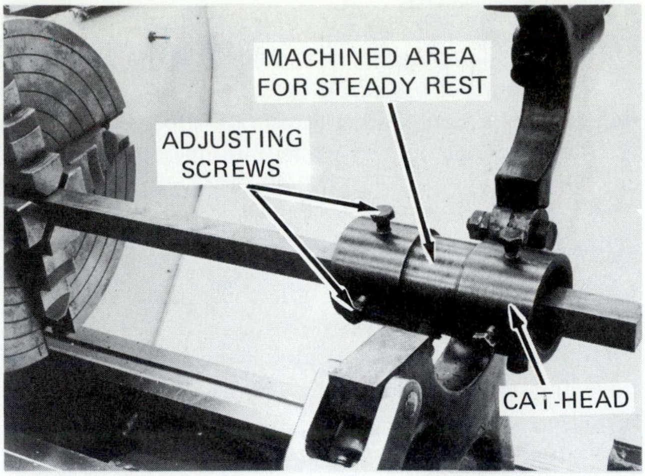
MACHINED AREA
FOR STEADY REST
ADJUSTING
SCREWS
CAT-HEAD

Figure 56-2 *The cathead can be adjusted to provide a true bearing surface for the steady rest, even if a square workpiece is being machined.*

7. Close the top section of the steady rest and adjust the top jaw, using a paper feeler.
8. Apply white or red lead to the diameter at the steady rest jaws.
9. Start the lathe and carefully adjust each jaw until it just touches the diameter.

NOTE: The white or red lead will smear just as the jaw contacts the work.

10. Tighten the lock screw on each jaw and then apply a suitable lubricant.
11. Before machining, indicate the top and front of the turned diameter at the chuck and at the steady rest to check for alignment. If the indicator reading varies, adjust the steady rest until it is correct.

TO TRUE A DAMAGED CENTER HOLE

1. Mount and true the work in a chuck and steady rest, if necessary.
2. Grind a 60° spotting tool (Fig. 56-3) and mount it on center in the tool holder.
3. Start the lathe and gradually bring the spotting tool into the damaged center hole.
4. Recut the center hole until the damaged section is removed.
5. Remove the workpiece, mount it between centers, and turn the diameter as required.

FOLLOWER REST

A *follower rest*, mounted on the saddle, moves along with the carriage to prevent work from springing up and away

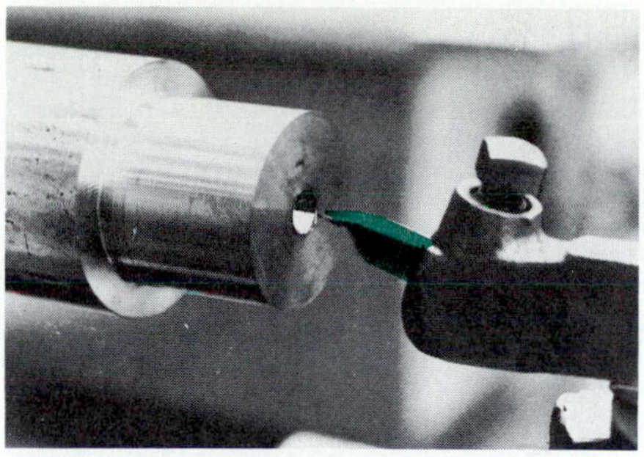

Figure 56-3 *A 60° spotting tool set up to recut a damaged center.*

from the cutting tool. The follower rest, positioned immediately behind the cutting tool, can be used to support long work for successive operations, such as thread cutting (Fig. 56-4).

TO SET UP A FOLLOWER REST

1. Mount the work between centers.
2. Fasten the follower rest to the saddle of the lathe.
3. Position the cutting tool in the toolpost so that it is just to the left of the follower rest jaws.
4. Turn the work diameter, for approximately 1½ in. (38 mm) long, to the desired size.
5. Adjust both jaws of the follower rest until they lightly contact the turned diameter.
6. Tighten the lock screw on each jaw.
7. Lubricate the work and the follower rest jaws to prevent marring the finished diameter.
8. If successive cuts are required to reduce the diameter of a workpiece, readjust the follower rest jaws as in steps 4–7.

Figure 56-4 *A follower rest used to support a long, slender workpiece between centers during thread cutting.*

MANDRELS

A *mandrel* (Fig. 56-5) is a precision tool which, when pressed into the hole of a workpiece, provides centers for a machining operation. They are especially valuable for thin work, such as flanges, pulleys, and gears, where the outside diameter must run true with the inside diameter and it would be difficult to hold the work in a chuck.

CHARACTERISTICS OF A STANDARD MANDREL

1. Mandrels are usually hardened and ground and tapered 0.006 to 0.008 in./ft (0.50 to 0.66 mm/m).
2. The nominal size is near the middle, and the small end is usually 0.001 in. (0.02 mm) under; the large end is usually 0.004 in. (0.10 mm) over the nominal size.
3. Both ends are turned smaller than the body and provided with a flat so that the lathe dog does not damage the accuracy of the mandrel.
4. The size of the mandrel is stamped on the large end.
5. The center holes, which are recessed slightly, are large enough to provide a good bearing surface and to withstand the strain caused by machining a workpiece.

TYPES OF MANDRELS

Many types of mandrels are used with various types of workpieces and machining operations. Some of the more common types of mandrels are:

- The *solid mandrel* (Fig. 56-5) is available for most of the standard hole sizes. It is a general-purpose mandrel, which may be used for a variety of workpieces.
- The *expansion mandrel* (Fig. 56-6) consists of a sleeve with four or more slots cut lengthwise fitted over a solid mandrel. A taper pin fits into the sleeve to expand it to hold work that does not have a standard size hole. Another form of expansion mandrel has a slotted bushing fitting over a tapered mandrel. Bushings of various sizes can be used with this mandrel, increasing its range.
- The *gang mandrel* (Fig. 56-7) is used to hold a number of identical parts for a machining operation. The body

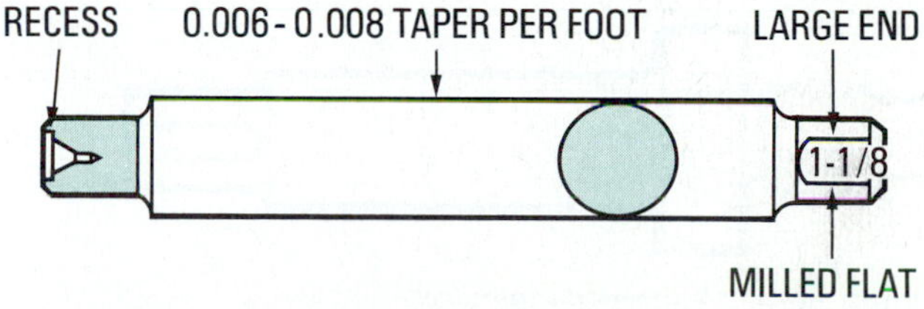
RECESS
0.006 - 0.008 TAPER PER FOOT
LARGE END
1-1/8
MILLED FLAT

Figure 56-5 *A standard solid mandrel.*

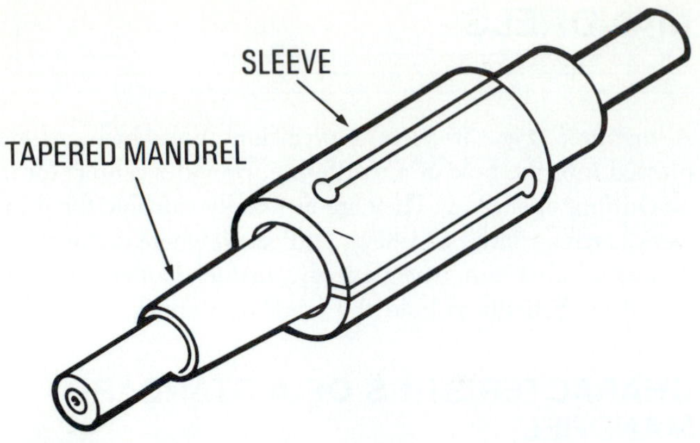
SLEEVE
TAPERED MANDREL

Figure 56-6 *An expansion mandrel.*

of the mandrel is parallel (no taper) and has a shoulder or flange on one end. The other end is threaded for a locking nut.

- The *threaded mandrel* (Fig. 56-8) is used for holding workpieces having a threaded hole. An undercut at the shoulder ensures that the workpiece will seat squarely and is not canted on the threads.
- The *taper-shank mandrel* (Fig. 56-9) may be fitted to the tapered hole in the headstock spindle. The projecting portion may be machined to any desired form to suit the workpiece. This type of mandrel is often used for small workpieces or those which have blind holes.

TO MOUNT WORK ON A PLAIN MANDREL

1. Secure a mandrel to fit the hole in the workpiece.
2. Thoroughly clean the mandrel and apply a thin film of oil on the diameter.
3. Clean and remove any burrs from the hole in the workpiece (Fig. 56-10).

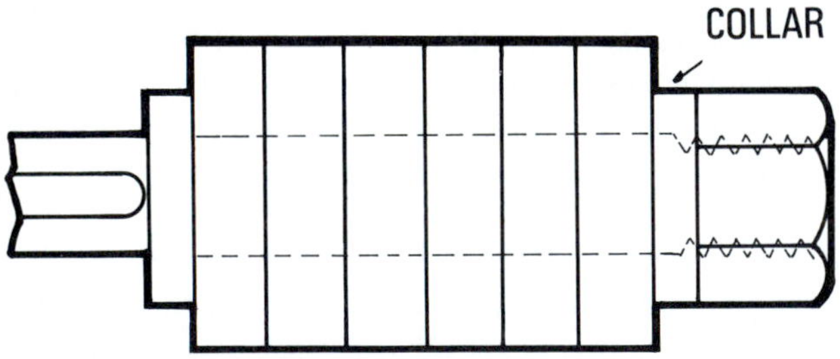
COLLAR

Figure 56-7 *A gang mandrel.*

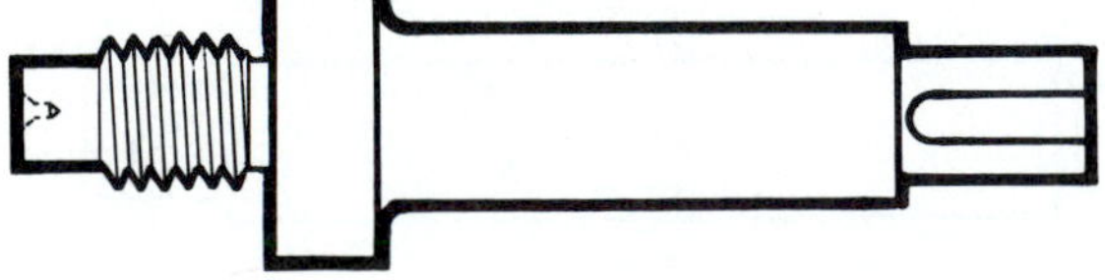

Figure 56-8 *A threaded mandrel.*

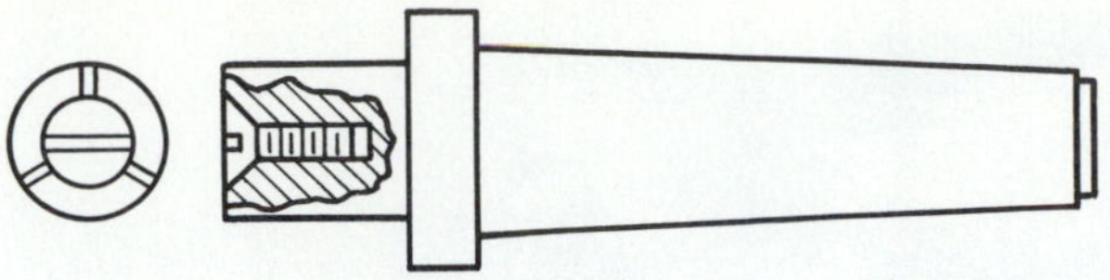

Figure 56-9 *A taper-shank mandrel.*

4. Start the small end of the mandrel into the hole (the large end has the size stamped on it) by hand.
5. Place the work on an arbor press with a machined surface down so that the hole is at right angles to the table surface (Fig. 56-11).
6. Press the mandrel firmly into the workpiece.

TO TURN WORK ON A MANDREL

Work pressed on a mandrel is held in position by friction; therefore cutting operations should be toward the large end of the mandrel. This will tend to keep the work tight on the mandrel.

Proceed as follows:

1. Fasten the lathe dog on the *large end* of the mandrel (where the size is stamped).
2. Clean the lathe and mandrel centers and then mount the work.

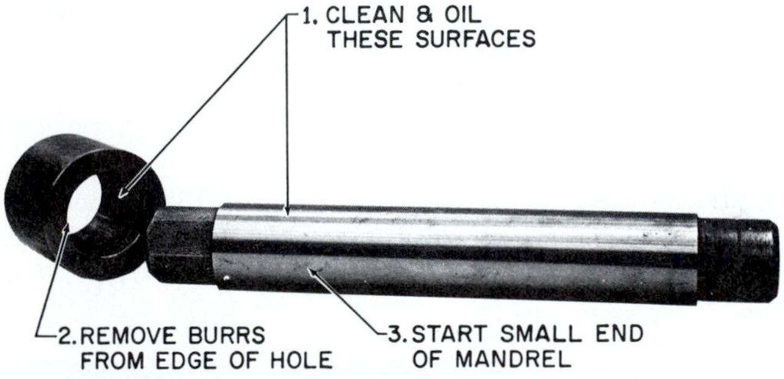
1. CLEAN & OIL THESE SURFACES
2. REMOVE BURRS FROM EDGE OF HOLE
3. START SMALL END OF MANDREL

Figure 56-10 *Preparing the mandrel and workpiece before mounting.*

Figure 56-11 *Using an arbor press to force a mandrel into a workpiece.*

3. If the entire side of the work must be faced, it is good practice to use a paper feeler between the toolbit point and mandrel for setting the toolbit. This will prevent marring or scoring the surface of the mandrel.
4. When turning the outside diameter of work, always cut toward the large end of the mandrel.
5. On large-diameter work, it is advisable to take light cuts to prevent the work from slipping on the mandrel or chattering.

ECCENTRICS

An *eccentric* (Fig. 56-12) is a shaft which may have two or more turned diameters parallel to each other but not concentric with the normal axis of the work. Eccentrics are used in locking devices, in the feed mechanism on some shapers, and in the crank shaft of automobiles, where it is necessary to *convert rotary motion into reciprocating motion,* or vice versa.

The amount of eccentricity, or *throw* of an eccentric, is the distance that a set of center holes has been offset from the normal work axis. If the center holes were offset 1/4 in. (6 mm) from the work axis, the amount of throw would be 1/4 in. (6 mm), but the total travel of the eccentric would be 1/2 in. (12 mm).

There are three types of eccentrics, which are generally cut on a lathe in the following ways.

1. When the throw enables all centers to be located on the ends of the workpiece.
2. When the throw is too small to allow all centers to be located on the workpiece at the same time.
3. When the throw is so great that all centers cannot be located on the workpiece.

TO TURN AN ECCENTRIC WITH A 0.375-in. OR 10-mm THROW

1. Place the work in a chuck and face it to length. If the center holes are to be removed later, leave the work 3/4 in. (19 mm) longer.
2. Place the work in a V-block on a surface plate and apply layout dye to both ends of the work.
3. Set a vernier height gage to the top of the work and note the vernier reading.

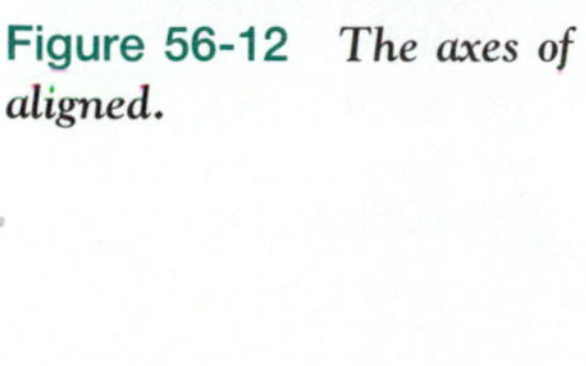

Figure 56-12 *The axes of an eccentric are parallel, not aligned.*

4. Subtract half the work diameter from the reading and set the gage to this dimension.
5. Scribe a centerline on both ends of the work.
6. Rotate the work 90° and scribe another centerline on both ends at the same height-gage setting (Fig. 56-13).
7. Lower or raise the height-gage setting 0.375 in. (10 mm) and scribe the lines for the offset centers on both ends.
8. Carefully center-punch the four scribed centers and drill the center holes in each end.
9. Mount the work in a lathe and turn the diameter with the true centers.
10. Set the work on the offset centers and turn the eccentric (center section) to the required diameter.

TO CUT AN ECCENTRIC WITH A SMALL THROW

This procedure should be followed when the centers are too close to be located on the workpiece at the same time.

1. Cut the work 3/4 in. (19 mm) longer than required.
2. Face the ends and drill one set of center holes in the lathe.
3. Mount the work between centers and turn the large diameter to size.
4. Cut off the ends to remove the center holes.
5. Lay out and drill a new set of center holes, offsetting them from center the required throw.
6. Turn the eccentric diameter to size.

TO TURN AN ECCENTRIC WITH A LARGE THROW

1. Set the work on the normal centers and turn both ends to size.
2. Secure or make a set of support blocks as shown in Fig. 56-14. The hole in the support block should fit the turned ends of the work snugly. A setscrew in each block is used to securely fasten the support blocks to the work. The number of centers required should be laid out and drilled in the support blocks (Fig. 56-14).

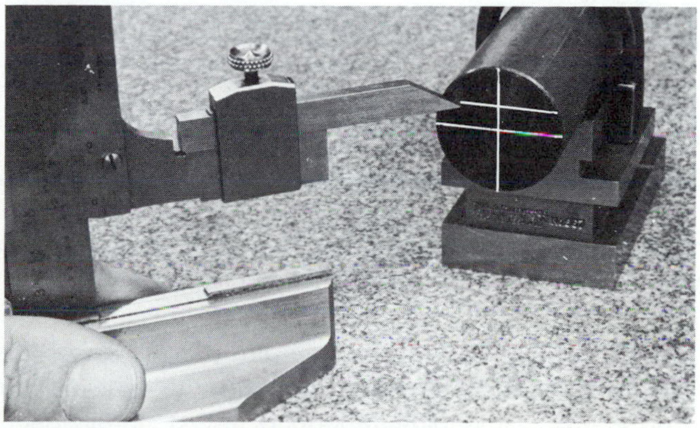

Figure 56-13 *Laying out the centers of an eccentric.*

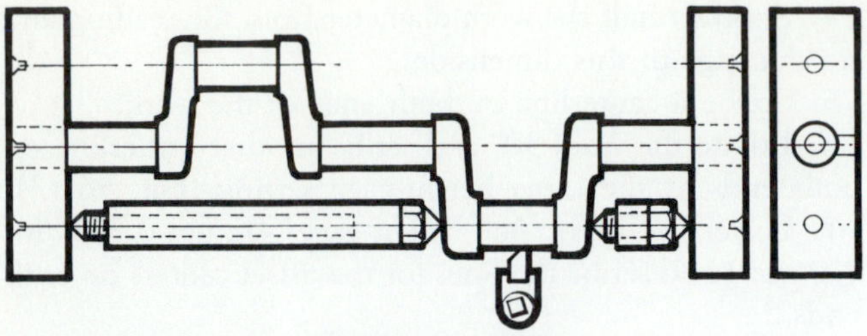

Figure 56-14 *The setup required for turning an eccentric with a large throw.*

3. Align both support blocks parallel on the work and lock them in position.
4. Counterbalance the lathe to prevent undue vibration.
5. Turn the various diameters as required.

REVIEW QUESTIONS

FOLLOWER RESTS AND STEADY RESTS

1. State the purpose of a follower rest.
2. Explain how to set up a steady rest for turning a long shaft held between centers.
3. Describe a cathead and state when it is used.
4. Explain how a damaged center hole may be trued.

MANDRELS

5. State the purpose of a mandrel.
6. Draw a 1-in. (25-mm) standard mandrel and include all specifications.
7. Name and describe four types of mandrels and state their purpose.
8. List the precautions that must be taken when a mandrel is being used (include mounting and turning).

ECCENTRICS

9. Define an *eccentric* and state its purpose.
10. Explain the difference between the throw and total travel of an eccentric.
11. A crankshaft 6 in. (150 mm) long having equal-length journals is required to produce a travel of 1½ in. (38 mm) on a piston. The journal (finished shaft) size is 1 in. (25 mm).
    a. Describe how to lay out this eccentric.
    b. What size material would be required if ⅛ in. (3 mm) is allowed for "cleaning up"? This can be calculated easily with the aid of a sketch.
12. What precautions must be taken when turning an eccentric having a large throw?

UNIT 57

Machining in a Chuck

Spindle-mounted accessories, such as chucks, centers, drive plates, and faceplates, are fitted to and driven by the headstock spindle. The most versatile and commonly used spindle accessories are the lathe chucks. The jaws of a chuck are adjustable and therefore workpieces which might be difficult to hold by another method can be held securely. Three-jaw chucks, whose jaws move simultaneously, are generally used to hold round finished diameters. Four-jaw chucks, whose jaws move independently, are generally used to hold odd-shaped pieces and when greater holding power and accuracy are required.

All the machining operations which can be performed on work between centers, such as turning, knurling, and threading, can be performed on work held in a chuck. Most chuck work is generally fairly short; however, if workpieces longer than three times their diameter must be machined in a chuck, the end of the work should be supported to prevent it from springing and being thrown out of the lathe.

OBJECTIVES

After completing this unit you should be able to:
1. Mount and machine work in a three-jaw chuck
2. Mount and machine work in a four-jaw chuck
3. Face, groove, and cut off work held in a chuck

MOUNTING AND REMOVING LATHE CHUCKS

The proper procedure for mounting and removing chucks must be carefully followed in order to not damage the lathe spindle and/or the chuck and to preserve the accuracy of the lathe. Three types of spindle noses are found on engine lathes on which lathe chucks may be mounted. They are the threaded spindle nose, the tapered spindle nose, and the cam-lock spindle nose. The procedures for mounting a chuck on each type of spindle nose follows.

TO MOUNT A CHUCK

1. Set the lathe to the slowest speed. *SHUT OFF THE ELECTRICAL SWITCH.*
2. Remove the drive plate and live center.
3. Clean all surfaces of the spindle nose and the mating parts of the chuck.

NOTE: Steel chips or dirt will destroy the accuracy of the spindle nose and the mating taper in the chuck.

4. Place a cradle block on the lathe bed in front of the spindle and place the chuck on the cradle (Fig. 57-1).
5. Slide the cradle close to the lathe spindle nose and mount the chuck.

a. *Threaded Spindle Nose Chucks*
(1) Revolve the lathe spindle *by hand* in a counterclockwise direction and bring the chuck up to the spindle. *NEVER START THE MACHINE.*
(2) If the chuck and spindle are correctly aligned, the chuck should easily thread onto the lathe spindle.
(3) When the chuck adapter plate is within 1/16 in. (1.5 mm) of the spindle shoulder, give the chuck a quick turn to seat it against the spindle shoulder.
(4) Do not jam a chuck against the shoulder too tightly, as it may damage the threads and make the chuck difficult to remove.

CHUCK CRADLE

Figure 57-1 *A properly fitted cradle block makes mounting and removal of chucks easy and safe. (Courtesy Kostel Enterprises Limited.)*

KEYWAY
KEY

Figure 57-2 *Align the keyway in the chuck with the key on the spindle.*

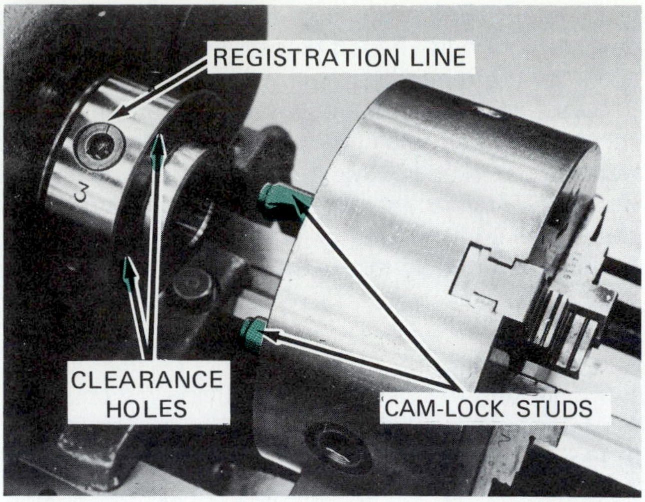
REGISTRATION LINE
CLEARANCE HOLES
CAM-LOCK STUDS

Figure 57-4 *Mounting a chuck on a cam-lock spindle nose.*

b. *Taper Spindle Nose Chucks*

(1) Revolve the lathe spindle by hand until the key on the spindle nose aligns with the keyway in the tapered hole of the chuck (Fig. 57-2).

(2) Slide the chuck onto the lathe spindle.

(3) Turn the lock ring in a counterclockwise direction until it is hand-tight.

(4) Tighten the lock ring securely with a spanner wrench by striking it sharply downward (Fig. 57-3).

c. *Cam-Lock Spindle Nose Chucks*

(1) Align the registration line of each cam lock with the registration line on the lathe spindle nose.

(2) Revolve the lathe spindle by hand until the holes in the spindle align with the cam-lock studs of the chuck (Fig. 57-4).

(3) Slide the chuck onto the spindle.

(4) Tighten each cam lock in a *clockwise* direction (Fig. 57-5).

TO REMOVE A CHUCK

The following procedures for removing a lathe chuck also apply to other spindle-mounted accessories such as drive plates and faceplates.

1. Set the lathe in the slowest speed. *STOP THE MOTOR.*
2. Place a cradle block under the chuck (Fig. 57-1).
3. Remove the chuck by the following methods.

a. *Threaded Spindle Nose Chucks*

(1) Turn the chuck until a wrench hole is in the top position.

(2) Insert the chuck wrench into the hole and pull it *sharply* toward the front of the lathe.

OR

(1) Place a block or short stick under the chuck jaw as shown in Fig. 57-6.

Figure 57-3 *Strike the spanner wrench sharply to tighten the lock ring.*

SPINDLE R.P.M.
SLOW RANGE
FAST RANGE
F

Figure 57-5 *Tighten the cam locks in a clockwise direction.*

Figure 57-6 *A hardwood block can be used to remove a chuck from a threaded spindle nose.*

(2) Revolve the lathe spindle by hand in a *clockwise* direction until the chuck is loosened on the spindle.
(3) Remove the chuck from the spindle and store it where it will not be damaged.

b. *Taper Spindle Nose Chucks*
(1) Secure the proper C-spanner wrench.
(2) Place it around the lock ring of the spindle with the handle in an upright position.
(3) Place one hand on the curve of the spanner wrench to prevent it from slipping off the lock ring (Fig. 57-7).
(4) With the palm of the other hand, *sharply* strike the handle of the wrench in a clockwise direction.
(5) Hold the chuck with one hand, and with the other hand remove the lock ring from the chuck.

NOTE: The lock ring may turn a few turns and then become tight. It may be necessary to use the spanner wrench again to loosen the taper contact between the chuck and the spindle nose.

(6) Remove the chuck from the spindle and store it with the jaws in the up position.

c. *Cam-Lock Spindle Nose Chucks*
(1) With the chuck wrench, turn each cam lock in a counterclockwise direction until its registration line coincides with the registration line on the lathe spindle nose (Fig. 57-8).
(2) Place one hand on the chuck face, and with the palm of the other hand *sharply* strike the top of the

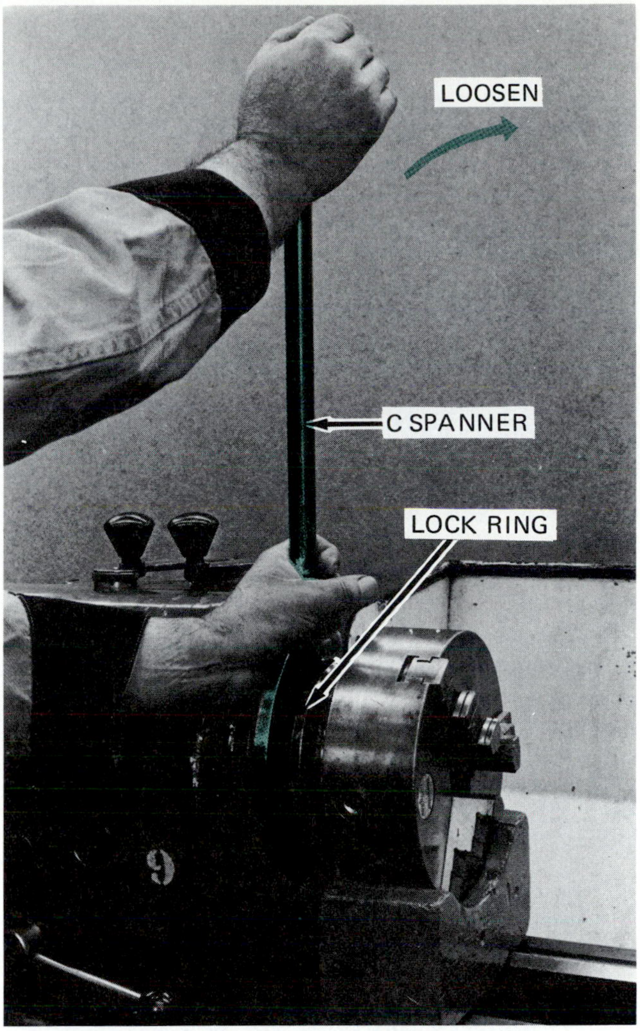
LOOSEN
C SPANNER
LOCK RING

Figure 57-7 *Using a C-spanner wrench to loosen the locking ring that holds the chuck on the taper nose spindle.*

R.P.M.
LOW
HIGH
LOW
HIGH
CAUTION
SLOW RANGE
FAST RANGE

Figure 57-8 *Loosen each cam lock until the registration lines on the lock and spindle match. (Courtesy Kostel Enterprises Limited.)*

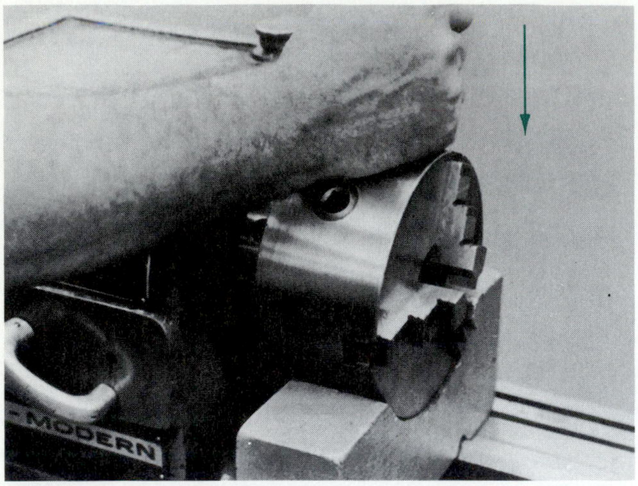
MODERN

Figure 57-9 *Strike the top of the chuck sharply with your hand to break the taper contact.*

chuck. This is necessary to break the taper contact between the chuck and the lathe spindle (Fig. 57-9).

NOTE: Sometimes it may be necessary to use a soft-faced hammer for this operation.

4. Slide the chuck clear of the spindle and place it carefully in a storage compartment.

MOUNTING WORK IN A CHUCK

In order for workpieces to be held securely for machining operations in either a three or four jaw chuck, certain procedures should be followed. Since the construction and purpose of three- and four-jaw chucks differ, the method of mounting work in each chuck also differs in certain respects.

TO MOUNT WORK IN A THREE JAW CHUCK

1. Clean the chuck jaws and the surfaces of the workpiece.
2. Use the proper size chuck wrench and open the chuck jaws slightly more than the diameter of the work.
3. Place the work in the chuck leaving no more than three times the diameter extending beyond the chuck jaws.
4. Tighten the chuck jaws with the wrench in the left hand while slowly rotating the workpiece with the right hand.
5. Tighten the chuck jaws securely by *using the chuck wrench only.*

ASSEMBLING JAWS IN A THREE-JAW UNIVERSAL CHUCK

Three-jaw chucks are supplied with two sets of chuck jaws: one set for outside gripping and one set for inside gripping (Fig. 57-10). The jaws supplied with each chuck are marked with the same serial number and should *never be used with another chuck.* When it is necessary to change the chuck jaws, the proper sequence must be followed; otherwise, the work held in the jaws will not run true.

1. Thoroughly clean the jaws and the jaw slides in the chuck.
2. Turn the chuck wrench clockwise until the start of the scroll thread is almost showing at the back edge of slide 1.
3. Insert jaw 1 (in slot 1) and press down with one hand while turning the chuck wrench clockwise with the other (Fig. 57-11).
4. After the scroll thread has engaged in the jaw, continue turning the chuck wrench clockwise until the start of the scroll is near the back edge of groove 2.
5. Insert jaw 2 and repeat steps 3 and 4 (Fig. 57-12).
6. Insert the third jaw in the same manner (Fig. 57-13).

Some chucks are equipped with one set of top jaws, which are fastened on by allen screws. In order to reverse these jaws, it is necessary to remove the screws, clean the mating parts, and replace the jaws in the reverse direction. The screws should be tightened uniformly and securely so that the jaws are not distorted.

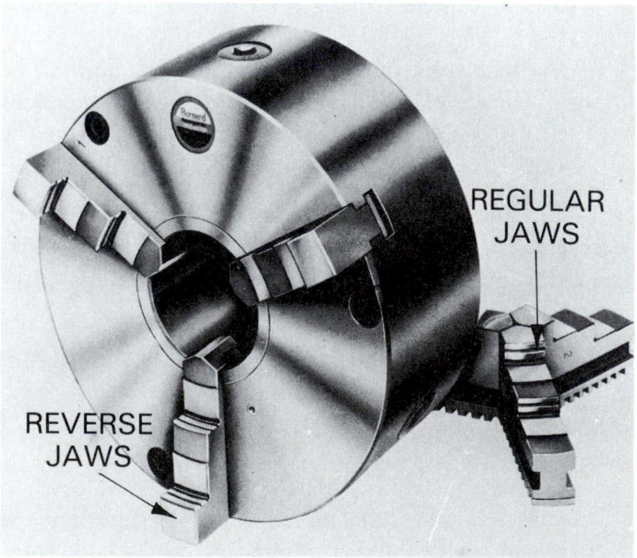
REGULAR
JAWS
REVERSE
JAWS

Figure 57-10 *A three-jaw universal chuck with reversible jaws.*

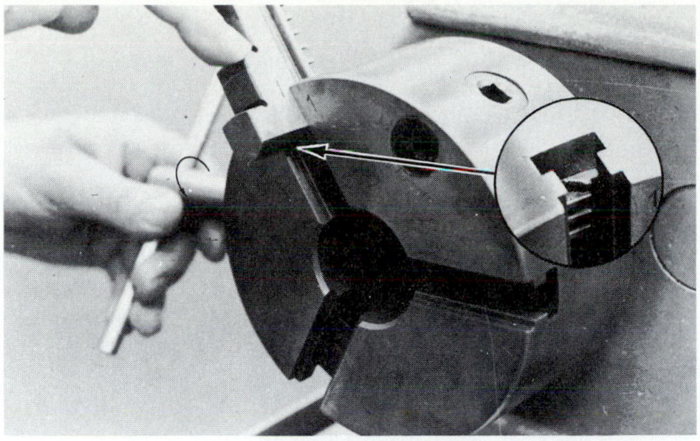

Figure 57-11 *Inserting the first jaw into the start of the scroll plate.*

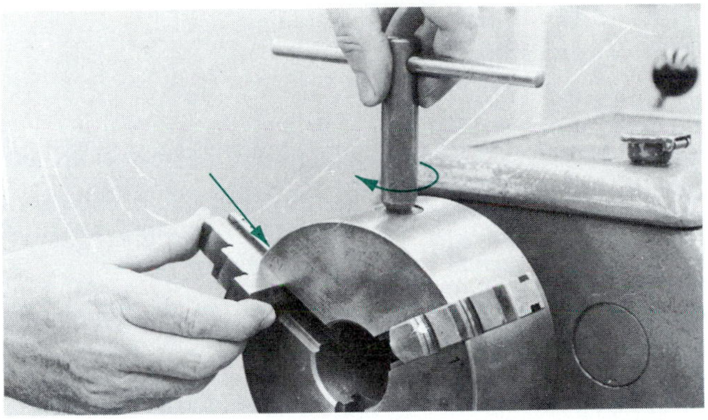

Figure 57-12 *The scroll plate turned to start the second jaw.*

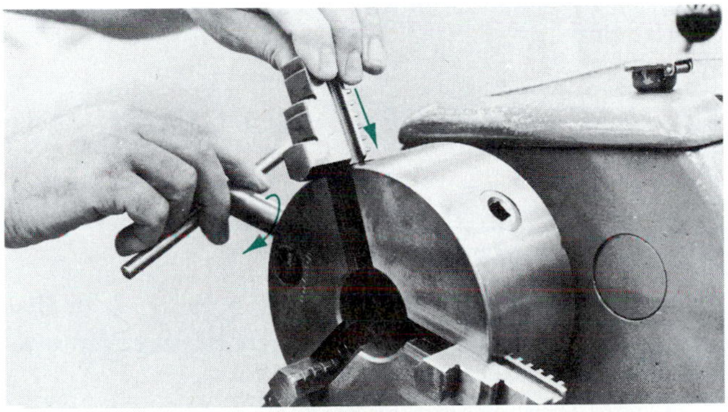

Figure 57-13 *Inserting the third jaw into a three-jaw chuck.*

TO MOUNT WORK IN A FOUR-JAW CHUCK

Work which must run absolutely true must be mounted in a four-jaw chuck because each jaw can be adjusted independently. The chuck jaws, which are reversible, can hold round, square, or irregularly shaped workpieces securely. The work can be adjusted to be either concentric or off center.

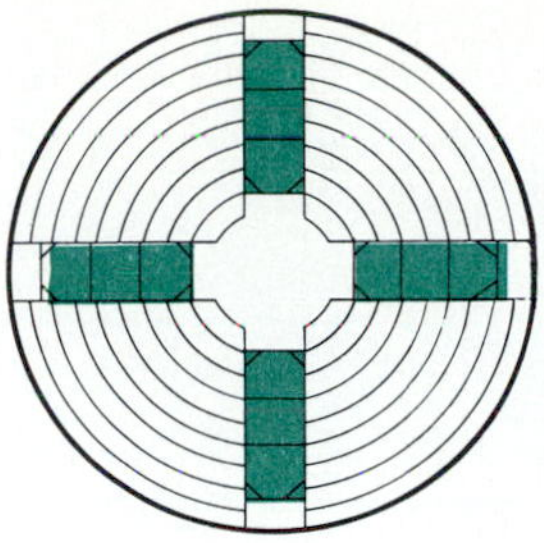

Figure 57-14 *The lines on a four-jaw chuck can be used as a guide for truing a workpiece.*

1. Measure the diameter of the work to be chucked.
2. With a chuck wrench adjust the jaws to the approximate size according to the ring marks on the face of the chuck (Fig. 57-14).
3. Set the work in the chuck and tighten the jaws snugly against the work surface.
4. True the workpiece by any of the following methods:

   a. *Chalk Method*

   (1) Start the lathe and with a piece of chalk lightly mark the high spot on the diameter (Fig. 57-15).
   (2) Stop the lathe and check the chalk mark. If it is an even, lightly marked line around the work, the work is true.
   (3) If there is only one mark, loosen the jaw *opposite* the chalk mark and tighten the jaw *next to* the chalk mark.
   (4) Continue this operation until the chalk marks

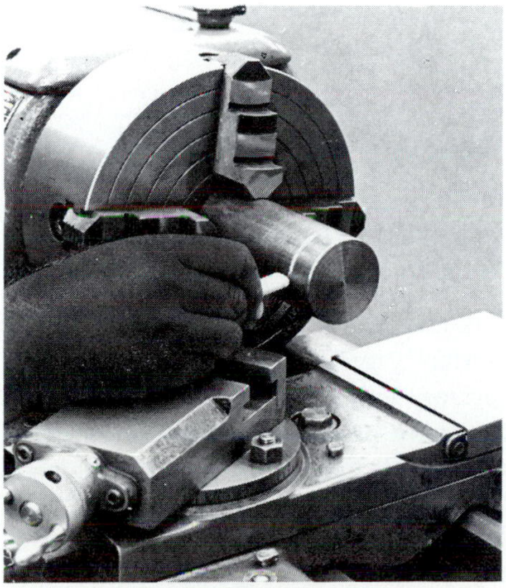

Figure 57-15 *Using chalk to indicate the high spot on a workpiece. (Courtesy Kostel Enterprises Limited.)*

Figure 57-16 *Using a surface gage to true a workpiece in a four-jaw chuck.*

Figure 57-17 *Truing a workpiece in a chuck using a dial indicator.*

lightly around the work or leaves two marks opposite each other.

b. *Surface Gage Method*

(1) Place a *surface gage* on the lathe bed and adjust the point of the scriber so that it is close to the work surface (Fig. 57-16).

(2) Revolve the lathe by hand to find the low spot on the work.

(3) Loosen the jaw *nearest* the low spot and tighten the jaw *opposite* the low spot to adjust the work closer to center.

(4) Repeat steps (2) and (3) until the work is running true.

TO TRUE WORK IN A FOUR-JAW CHUCK USING A DIAL INDICATOR

A dial indicator should be used whenever a machined diameter must be aligned to within a few thousandths of an inch or hundredths of a millimeter.

1. Mount the work and true it approximately, using either the chalk or surface gage method.
2. Mount an indicator, with a range of at least 0.100 in. (2.50 mm) in the toolpost of the lathe (Fig. 57-17).
3. Set the indicator spindle in a *horizontal position* with the contact point set to center height.
4. Bring the indicator point against the work diameter so that it registers approximately 0.020 in. (0.50 mm) and revolve the lathe *by hand.*
5. Note the highest and the lowest reading on the dial indicator.
6. Slightly loosen the chuck jaw at the lowest reading, and tighten the jaw at the high reading until the work is moved half the difference between the two indicator readings.
7. Continue to adjust *only these two opposite jaws* until the indicator registers the same at both jaws.

NOTE: Disregard the indicator readings on the work between these two jaws.

8. Adjust the other set of opposite jaws in the same manner until the indicator registers the same at any point on the work circumference.
9. Tighten all jaws evenly to secure the workpiece firmly.
10. Rotate the lathe spindle by hand and recheck the indicator reading.

TO FACE WORK HELD IN A CHUCK

The purpose of facing work in a chuck is the same as that of facing between centers: to obtain a true, flat surface and to cut the work to length.

1. True up the work in a chuck using the chalk or dial indicator method. (This is not necessary if a three-jaw universal chuck is used.)
2. Have at least an amount equal to the diameter of the work projecting from the chuck jaws.
3. Swivel the compound rest at 90° (right angles) to the cross-slide when facing a series of shoulders (Fig. 57-18).

OR

Swivel the compound rest 30° to the right if only one surface on the work must be faced (Fig. 57-19).

4. Set up the facing toolbit to the height of the dead center and pointing slightly to the left.
5. Lock the carriage in position (Fig. 57-18).
6. Set the depth of cut by using the graduated collar on the compound rest screw.

90°

Figure 57-18 *The compound rest set at 90° for facing accurate lengths.*

a. Twice the amount to be removed if the compound rest is set at 30°.

b. The same as the amount to be removed if the compound rest is set at 90° to the cross-slide.

7. Face the work to length.

TO ROUGH- AND FINISH-TURN WORK IN A CHUCK

Although much work is turned between centers in a lathe in school shops and training programs because of the ease of resetting the work accurately, most work machined in industry is held in a chuck. Since most work held in a chuck is relatively short, in many cases there is no need to support the end of the work because it is held securely enough by the lathe chuck. Workpieces that extend more than three times their diameter must be supported to prevent the work from springing during the machining operation. The most common methods of supporting the end of long workpieces is by the use of a revolving tailstock center or with a steadyrest.

The operations of rough and finish turning in a chuck are the same as turning between centers (Unit 52) and will not be detailed here. Whenever possible, a diameter should be machined to size in two cuts: one roughing and one finishing cut.

ROUGH AND FINISH TURNING

1. Mount the work securely in a chuck, with no more than three times the diameter extending beyond the chuck jaws.
2. Move the toolpost to the left of the compound rest and grip the toolholder short (Fig. 57-20).
3. Fasten a general-purpose toolbit in the toolholder and set the point to center.
4. Tighten the toolpost screw securely.
5. Set the lathe to the correct speed and feed for rough turning.
6. Take a light trial cut at the end of the work and measure the diameter (Fig. 57-21).
7. Adjust the crossfeed graduated collar to one-half the amount of metal to be removed.
8. Rough-turn the diameter to the correct length.
9. Set the lathe speed and feed for finish turning.
10. Take a light trial cut at the end of the workpiece.
11. Set the graduated collar in one-half the amount of metal to be removed.
12. Take another trial cut, measure the diameter, and if it is correct, finish-turn the diameter.

30°

Figure 57-19 *Feed the compound rest twice the amount to be faced off a surface.*

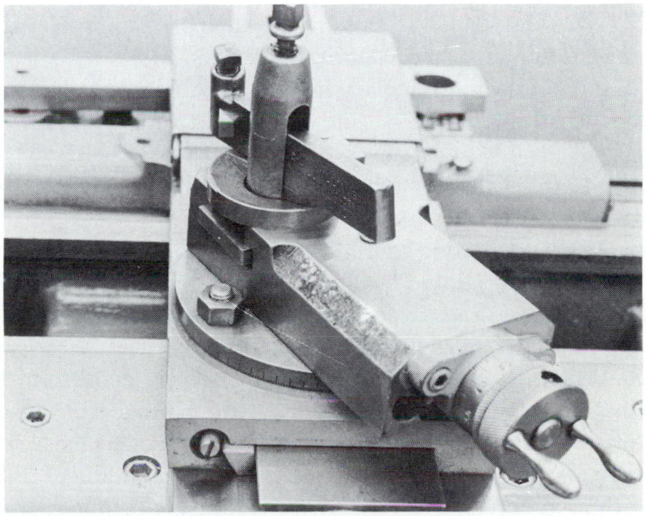

Figure 57-20 *The toolpost set on the left-hand side of the compound rest and the toolholder held short.*

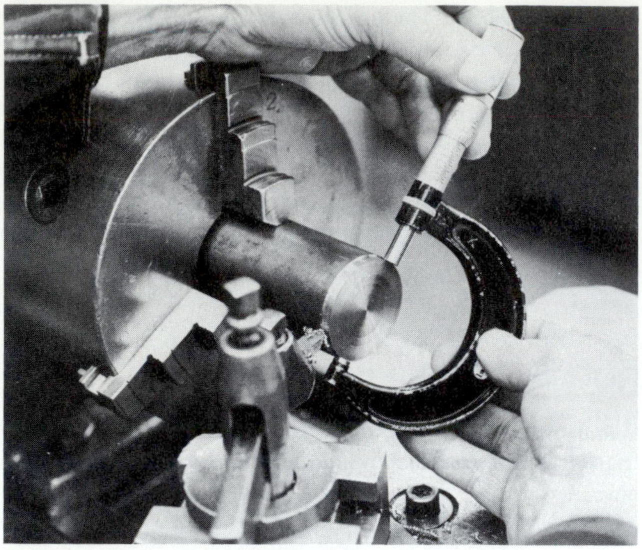

Figure 57-21 *Measuring the diameter of a trial cut.*

CUTTING OFF WORK IN A CHUCK

Cutoff tools, often called parting tools, are used for cutting off work projecting from a chuck, for grooving, and for undercutting. The inserted blade-type parting tool is the one most commonly used, and it is provided in three holders (Fig. 57-22).

PROCEDURE:

1. Mount the work in the chuck with the part to be cut off as close to the jaws as possible.
2. Mount the cutoff tool on the left-hand side of the compound rest with the cutting edge set on center (Fig. 57-23).
3. Place the holder as close to the toolpost as possible to prevent vibration and chatter.
4. Extend the cutting blade beyond the holder half the diameter of the work to be cut, plus 1/8 in. (3 mm) for clearance.
5. Set the lathe to approximately one-half the turning speed.
6. Move the cutting tool into position (Fig. 57-24).
7. Start the lathe and feed the cutoff tool into the work by hand, keeping a steady feed during the operation. Cut brass and cast iron dry, but use cutting fluid for steel.
8. When grooving or cutting off deeper than 1/4 in. (6 mm), it is good practice to move the parting tool sideways slightly. This may be accomplished by moving the carriage handwheel back and forth a few thousandths of an inch or hundredths of a millimeter during the cutting operation. This side motion cuts a little wider groove and prevents the tool from jamming.

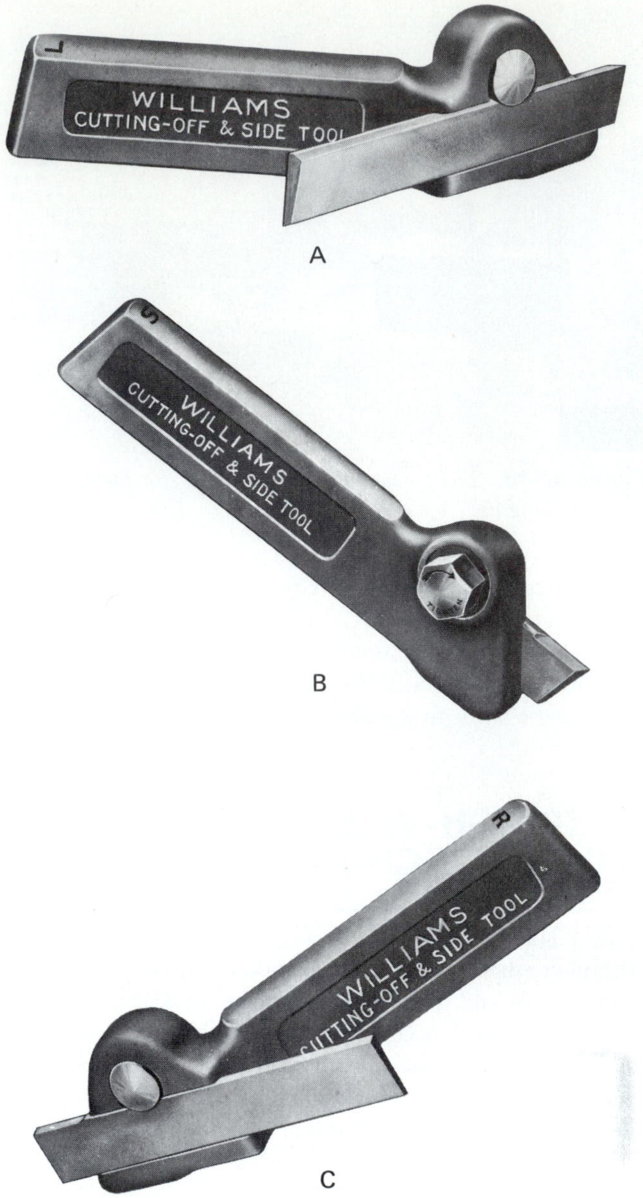
L
WILLIAMS
CUTTING-OFF & SIDE TOOL
A
S
WILLIAMS
CUTTING-OFF & SIDE TOOL
B
R
WILLIAMS
CUTTING-OFF & SIDE TOOL
C

Figure 57-22 *Inserted blade cutoff or parting tools: (A) left-hand offset; (B) straight; (C) right-hand offset.*

9. Before the cut is completed, remove the burrs from each side of the groove with a file.

NOTE: To avoid chatter, keep the tool cutting and apply cutting fluid constantly during the operation. Feed slowly when the part is almost cut off.

REVIEW QUESTIONS

MOUNTING AND REMOVING CHUCKS

1. What safety precautions should be observed before attempting to mount or remove chucks?

Figure 57-23 *The tip of the cutoff blade being set to center height.*

Figure 57-24 *The cutoff blade being moved to the correct position. (Courtesy Kostel Enterprises Limited.)*

2. Why is it necessary that the taper on the lathe spindle nose and the mating taper in the chuck be thoroughly cleaned before mounting?
3. Explain the procedure for mounting a chuck on a taper spindle nose.
4. Explain the procedure for removing a chuck from a cam-lock spindle nose.

MOUNTING WORK IN A CHUCK

5. How far may work extend beyond the chuck jaws before it must be supported?
6. Briefly list the procedure for assembling jaws in a three-jaw universal chuck.
7. For what purpose are four jaw chucks used? Explain why.
8. Briefly explain how to true a piece of work to within 0.001 in. (0.02 mm) accuracy in a four jaw chuck.

FACING IN A CHUCK

9. How should the cutting tool be set for facing?
10. How much should the compound rest graduated collar be turned in for facing:
    a. 0.050 in. with the compound rest set at 90° to the cross-slide?
    b. 0.025 in. when the compound rest is set at 30°?

ROUGH AND FINISH TURNING

11. How should the toolholder and the cutting tool be set for machining work in a chuck?
12. What is the purpose of a light trial cut at the end of a workpiece?

CUTTING OFF

13. How far should the blade extend beyond the holder for cutting off?
14. What procedure should be used for cutting grooves deeper than ¼ in. (6 mm)?

UNIT

58

Drilling, Boring, Reaming, and Tapping

Internal operations such as drilling, boring, reaming, and tapping can be performed on work being held in a chuck. Boring tools are mounted in the toolpost, while drills, reamers, and taps may be held either in a drill chuck mounted in the tailstock spindle or directly in the tailstock spindle. Since the work held in a chuck is generally machined true, these operations are usually machined concentric with the outside diameter of the workpiece.

OBJECTIVES

After completing this unit you should be able to:
1. Drill small- and large-diameter holes in a lathe
2. Ream holes to an accurate size and good surface finish
3. Bore a hole to within 0.001-in. (0.02-mm) accuracy
4. Use a tap to produce an internal thread which is concentric with the outside diameter

TO SPOT AND DRILL WORK IN A CHUCK

Spotting ensures that a drill will start in the center of the work. A spotting tool is used to make a shallow, V-shaped hole in the center of the work, which provides a guide for the drill to follow. In most cases, a hole can be spotted quickly and fairly accurately by using a center drill (Fig. 58-1). Where extreme accuracy in spotting is necessary, a spotting toolbit should be used.

PROCEDURE:

1. Mount the work true in a chuck.
2. Set the lathe to the proper speed for the type of material to be drilled.
3. Check the tailstock and make sure it is in line.
4. Spot the hole with a center drill or spotting tool.
5. Mount the twist drill in the tailstock spindle, in a drill chuck, or in a drill holder (Figs. 58-2 to 58-4).

NOTE:

a. When a tapered-shank drill is mounted directly in the tailstock spindle, a dog should be used to stop the drill from turning and scoring the tailstock spindle taper (Fig. 58-2).

b. The end of the drill may be supported with the end of a toolholder so that the drill will start on center (Fig. 58-3).

c. Tapered-shank drills are often mounted in a drill holder. The point of the drill is positioned in the hole, while the end of the holder is supported by the dead center. The handle of the holder rests on the toolholder

Figure 58-1 *Using a center drill to spot a hole.*

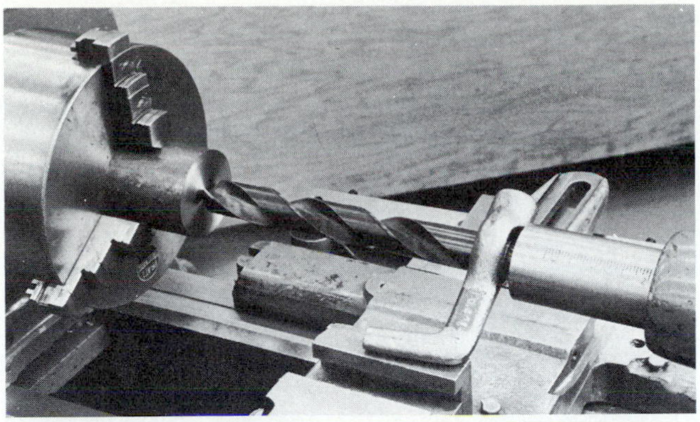

Figure 58-2 *A tapered shank-drill mounted in the spindle is prevented from turning by a lathe dog.*

Figure 58-3 *Supporting the end of a drill will prevent it from wobbling when starting a hole.*

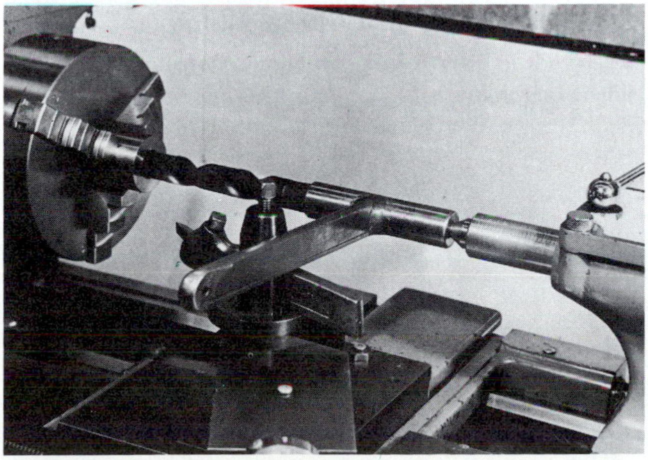

Figure 58-4 *Drilling a hole with a drill mounted in a drill holder and supported on a dead center.*

and against the toolpost to prevent the drill from turning and from pulling into the work (Fig. 58-4).

6. Start the lathe and drill to the desired depth, applying cutting fluid frequently.
7. To gage the depth of the hole, use the graduations on the tailstock spindle, or measure the depth with a steel rule (Fig. 58-5).
8. Withdraw the drill frequently to remove the chips and measure the depth of the hole.

CAUTION: Always ease the force on the feed as the drill breaks through the work.

BORING

Boring is the operation of enlarging and truing a drilled or cored hole with a single-point cutting tool. Special-diameter holes, for which no drills are available, can be produced by boring.

Holes may be drilled in a lathe; however, such holes are generally not considered accurate although the drill may have started straight. During the drilling process, the drill may become dull or hit a hard spot or blowhole in the metal, which will cause the drill to wander or run off center. If such a hole is reamed, the reamer will follow the drilled hole, and, as a result, the hole will not be straight. Therefore, if it is important that a reamed hole be straight and true, the hole should first be drilled, then bored and reamed.

TO BORE WORK IN A CHUCK

1. Mount the work in a chuck; face, spot, and drill the hole approximately 1/16 in. (1.5 mm) undersize.

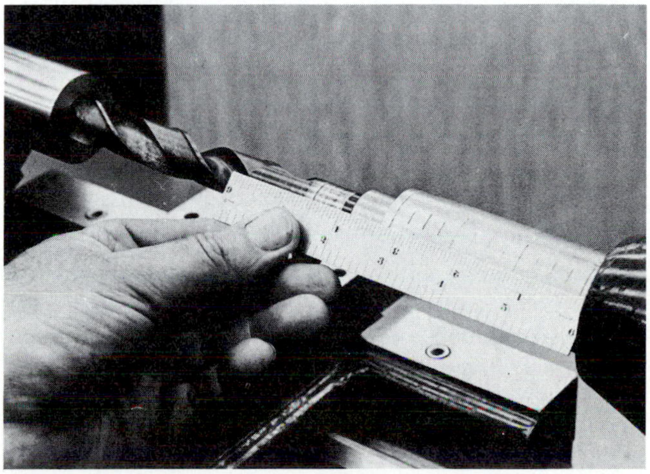

Figure 58-5 *Measuring the depth of a drilled hole using a rule.*

2. Select a boring bar as large as possible and have it extend beyond the holder only enough to clear the depth of the hole to be bored.

3. Mount the boring bar holder in the toolpost on the left-hand side of the compound rest.

4. Set the boring toolbit to center (Fig. 58-6).

5. Set the lathe to the proper speed and select a medium feed.

6. Start the lathe and bring the boring tool into contact with the inside diameter of the hole.

7. Take a light trial cut [approximately 0.005 in. (0.12 mm) or until a true diameter is produced] ¼ in. (6 mm) long at the right-hand end of the work.

8. Stop the lathe and measure the hole diameter with a telescopic gage or inside micrometer.

9. Determine the amount of material to be removed from the hole.

NOTE: Leave approximately 0.010 to 0.020 in. (0.25 to 0.50 mm) for a finish cut.

10. Set the depth of cut for half the amount of metal to be removed.

11. Start the lathe and take the roughing cut.

NOTE: If chatter or vibration occurs during machining, slow the lathe speed and increase the feed until it is eliminated.

12. Stop the lathe and bring the boring tool out of the hole without moving the crossfeed handle.

13. Set the depth of the finish cut and bore the hole to size. For a good surface finish, a fine feed is recommended.

Figure 58-6 *The boring bar should be held short and the toolbit set on center to machine inside diameters. (Courtesy R. K. LeBlond Machine Tool Company.)*

Figure 58-7 *Setup for reaming in a lathe.*

REAMING

Reaming may be performed in a lathe to quickly obtain an accurately sized hole and to produce a good surface finish. Reaming may be performed after a hole has been drilled or bored. If a true, accurate hole is required, it should be bored before the reaming operation.

TO REAM WORK IN A LATHE

1. Mount the work in a chuck; face, spot, and drill the hole to size.

For holes under ½ in. (13 mm) in diameter, drill 1/64 in. (0.40 mm) undersize; for holes over ½ in. (13 mm) in diameter drill 1/32 in. (0.80 mm) undersize. If the holes must be true, they should be bored 0.010 in. (0.25 mm) undersize.

2. Mount the reamer in a drill chuck or drill holder (Fig. 58-7).

When reaming holes ⅝ in. (16 mm) in diameter and larger, fasten a lathe dog near the reamer shank and support the tail on the compound rest to prevent the reamer from turning.

3. Set the lathe to approximately half the drilling speed.

4. Bring the reamer close to the hole and lock the tailstock in position.

5. Start the lathe, apply cutting fluid to the reamer, and slowly feed it into the drilled or bored hole with the tailstock handwheel.

6. Occasionally remove the reamer from the hole to clear chips from the flutes and apply cutting fluid.

7. Once the hole has been reamed, stop the lathe and remove the reamer from the hole.

CAUTION: *Never turn the lathe spindle or reamer backward for any reason. This will damage a reamer.*

8. Clean the reamer and store it carefully to prevent it from being nicked or damaged.

Figure 58-8 *Using a tap to cut internal threads.*

TAPPING

Tapping is one method of producing an internal thread on a lathe. The tap is aligned by placing the point of the lathe dead center in the shank end of the tap to guide it while the tap is turned by a tap wrench. A standard tap may be used for this operation; however, a gun tap is preferred because the chips are cleared ahead of the tap. When tapping a hole in a lathe, lock the spindle and turn the tap by hand (Fig. 58-8).

TO TAP A HOLE IN A LATHE

1. Mount the work in the chuck; face and center-drill.
2. Select the proper tap drill for the tap to be used.
3. Set the lathe to the proper speed.
4. Drill with the tap drill to the required depth. Use cutting fluid if required.
5. Chamfer the edge of the hole slightly larger than the tap diameter.
6. Stop the lathe and lock the spindle, or put the lathe in its lowest speed.
7. Place a taper tap in the hole and support the shank with the tailstock center.
8. With a suitable wrench, turn the tap, keeping the dead center snug into the shank of the tap by turning the tailstock handwheel.
9. Apply cutting fluid while tapping the hole (Fig. 58-8).
10. Back off the tap frequently to break the chip.
11. Remove the taper tap and finish tapping the hole with a plug or bottoming tap.

GRINDING ON A LATHE

Cylindrical and internal grinding may be done on a lathe if a proper grinding machine is not available. A toolpost grinder, mounted on a lathe, may be used for cylindrical and taper grinding as well as angular grinding of lathe centers. An internal grinding attachment for the toolpost grinder permits the grinding of straight and tapered holes. Grinding should be done on a lathe only when no other machine is available, or when the cost of performing a small grinding operation on a part would not warrant setting up a regular grinding machine. Since the work should rotate in an opposite direction to the grinding wheel, the lathe must be equipped with a reversing switch.

TO GRIND A LATHE CENTER

1. Remove the chuck or drive plate from the lathe spindle.
2. Mount the lathe center to be ground in the headstock spindle.
3. Set a slow spindle speed.
4. Swing the compound rest to 30° (Fig. 58-9) with the centerline of the lathe.
5. Protect the ways of the lathe with cloth or canvas and place a pan of water below the lathe center.
6. Mount the toolpost grinder and adjust the center of the grinding spindle to center height.
7. Mount the proper grinding wheel; true and dress.
8. Start the lathe, with the spindle revolving in reverse.
9. Start the grinder and adjust the grinding wheel until it sparks lightly against the revolving center.
10. Lock the carriage in this position.
11. Feed the grinding wheel in 0.001 in. (0.02 mm), using the crossfeed handle.
12. Move the grinder along the face of the center using the compound rest feed.
13. Check the angle of the center using a center gage, and adjust the compound rest if necessary.
14. Finish-grind the center.

Figure 58-9 *The compound rest set at 30° to grind a lathe center.*

NOTE: If a high finish is desired, polish the center with abrasive cloth at a high spindle speed.

REVIEW QUESTIONS

DRILLING

1. Why is spotting important before a hole is drilled?
2. Name three methods of holding various sizes of drills in a tailstock.
3. How can the depth of a drilled hole be gaged?

BORING

4. Define the boring process.
5. Why should a hole be bored before it is reamed?
6. How should the boring bar and cutting tool be set up for boring?
7. Briefly explain how to bore a hole to 1.750 in. (44 mm).

REAMING

8. What is the purpose of reaming in a lathe?
9. How much material should be left in a hole for reaming a:
   a. 7/16-in. hole? b. 1-in. hole?
10. Why should a reamer never be turned backward?

TAPPING

11. How is the tap started and guided so that the thread will be true to the bored hole?

SECTION

12

Milling Machines

Milling machines are machine tools used to produce one or more machined surfaces accurately on a piece of material, *the workpiece;* this is done by one or more rotary milling cutters having single or multiple cutting edges. The workpiece is held securely on the *work table* of the machine or in a holding device clamped to the table. It is then brought into contact with a revolving cutter.

The milling machine is a versatile machine tool which can handle a variety of operations normally performed by other machine tools. It is used not only for the milling of flat and irregular shaped surfaces but also for gear and thread cutting, drilling, boring, reaming, and slotting operations. Its versatility makes it one of the most important machine tools used in machine shop work.

UNIT 59

Milling Machines and Accessories

A wide variety of milling machines are required by industry to meet the job requirements of the many parts which must be machined. To make milling machines more versatile, a large variety of accessories and attachments are available so that each machine can perform more operations on each workpiece.

OBJECTIVES

After completing this unit you should be able to:
1. Recognize and explain the purposes of four milling machines
2. Know the purposes of the main operational parts of a horizontal and a vertical milling machine
3. Recognize and state the purposes of four milling machine accessories and attachments

HORIZONTAL MILLING MACHINES

In order to meet many different industrial requirements, milling machines are made in a wide variety of types and sizes. They are classified under the following headings:

1. *Manufacturing type,* in which the cutter height is controlled by vertical movement of the headstock.
2. *Special type,* designed for specific milling operations.
3. *Knee-and-column type,* in which the relationship between the cutter height and the work is controlled by vertical movement of the table.

MANUFACTURING-TYPE MILLING MACHINES

Manufacturing-type milling machines are used primarily for quantity production of identical parts. This type of machine may be either semiautomatic or fully automatic and is of simple but sturdy construction. Fixtures clamped to the table hold the workpiece for a variety of milling operations, depending upon the type of cutters or special spindle attachments used. Some of the distinctive features of manufacturing-type machines are the *automatic cycle* of cutter and work approach, the *rapid movement* during the noncutting part of the cycle, and the *automatic spindle stop.* After this machine has been set up, the operator is required only to load and unload the machine and start the automatic cycle controlled by cams and preset *trip dogs.*

Some of the more common manufacturing-type milling machines are:

- The *plain manufacturing type* (Fig. 59-1) has one horizontal spindle and one headstock. This machine is sometimes equipped with a reciprocating table cycle, which permits feeding and rapid traversing in both directions. On this type of machine, *reciprocal milling* is possible. Identical fixtures may be mounted on each end of the table. While work is being machined at one end, a new piece is being loaded into the fixture at the other end.
- The *small plain automatic knee-and-column type* (Fig. 59-2) is similar to the plain horizontal mill. It is used for milling small or medium-sized parts in different quantities. The table is operated by power and controlled automatically by trip dogs mounted on the front of the table.
- The *tracer-controlled milling machine* (Fig. 59-3) has a

CINCINNATI

Figure 59-1 *A plain manufacturing-type milling machine. (Courtesy Cincinnati Milacron Inc.)*

CINCINNATI

Figure 59-2 *A small plain automatic knee-and-column-type milling machine. (Courtesy Cincinnati Milacron Inc.)*

Figure 59-3 *A tracer-controlled milling machine. (Courtesy Cincinnati Milacron Inc.)*

hydraulic or electrical circuit designed to automatically control the relative positions of the cutter and the workpiece by a tracer stylus riding on a cam, template, or model. This machine is used for efficient, accurate reproduction of curved or irregular surfaces. If the tracer is disengaged it can be used for standard milling operations.

SPECIAL-TYPE MILLING MACHINES

Special-type milling machines are designed for individual milling operations and are used for only one particular type of job. They may be completely automatic and are used for production purposes when hundreds or thousands of similar pieces are to be machined.

MACHINING CENTERS

A special type of machine which is finding wide application in the industrial world is the horizontal numerical control machining center (Fig. 59-4). This machine is capable of handling a wide variety of work such as straight and contour milling, drilling, reaming, tapping, and boring—all in one setup. Because of the rugged construction, reliable controls, and antibacklash ball screws, they are capable of high production rates while still maintaining a high degree of accuracy. This machine and its contribution to manufacturing will be discussed in greater detail in Unit 78.

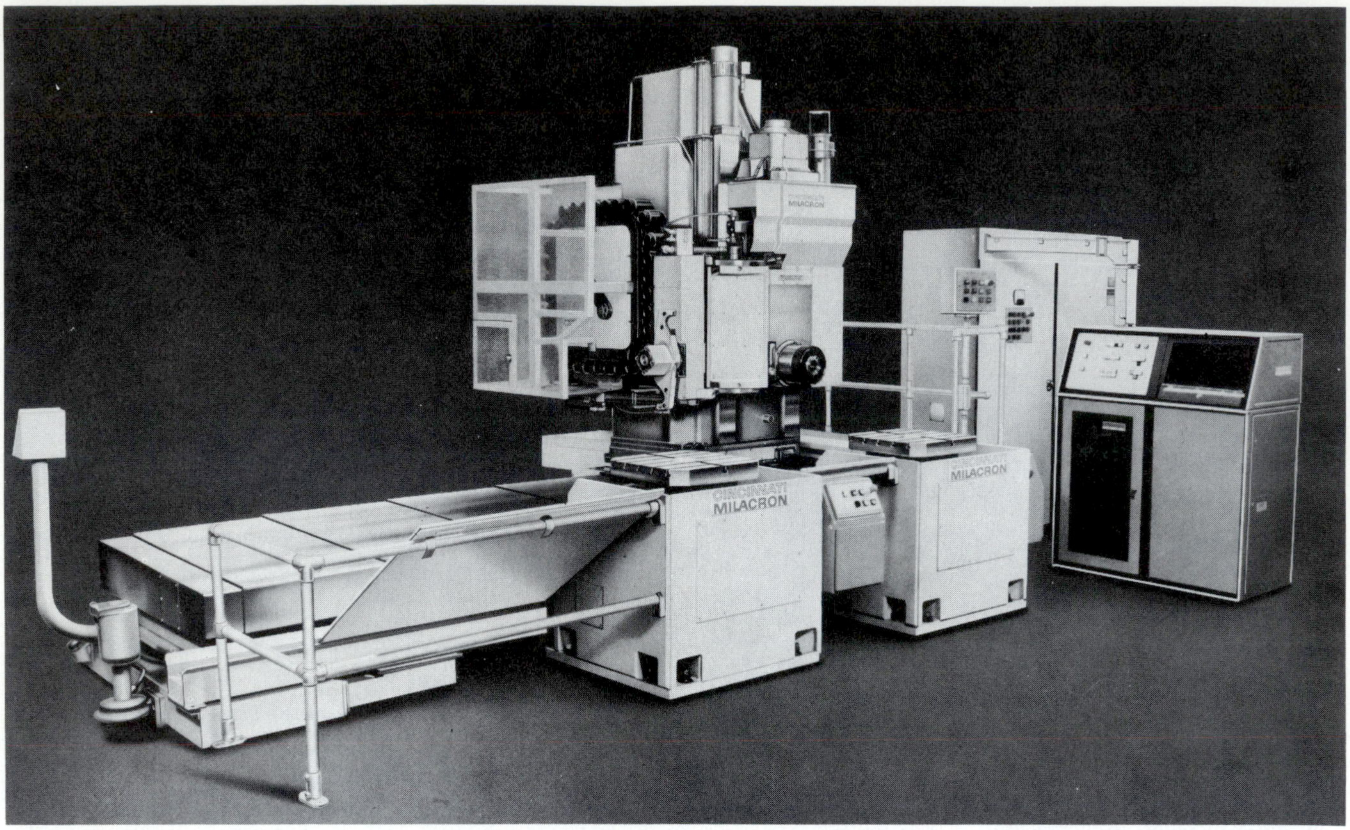

Figure 59-4 *A numerical control machining center. (Courtesy Cincinnati Milacron Inc.)*

KNEE-AND-COLUMN-TYPE MILLING MACHINES

Machines in this class fall into three categories:

1. Plain horizontal milling machines
2. Universal horizontal milling machines
3. Vertical milling machines

UNIVERSAL HORIZONTAL MILLING MACHINES

The universal horizontal milling machine is essential for advanced machine shop work. The difference between this machine and the plain horizontal mill will be dealt with in this unit.

Figure 59-5 shows the parts of a universal horizontal mill. The only difference between this mill and the plain horizontal machine is the addition of a *table swivel housing* between the table and the saddle. This housing permits the table to be swiveled 45° in either direction in a horizontal plane for such operations as the milling of helical grooves in twist drills, milling cutters, and gears.

PARTS OF THE MILLING MACHINE

- The *base* gives support and rigidity to the machine and also acts as a reservoir for the cutting fluids.
- The *column face* is a precision-machined and scraped section used to support and guide the knee when it is moved vertically.
- The *knee* is attached to the column face and may be moved vertically on the column face either manually or automatically. It houses the feed mechanism (Fig. 59-5).
- The *saddle* is fitted on top of the knee and may be moved in or out manually by means of the crossfeed handwheel or automatically by the crossfeed engaging lever (Fig. 59-6).
- The *swivel table housing,* fastened to the saddle on a universal milling machine, enables the table to be swiveled 45° to either side of the centerline.
- The *table* rests on guideways in the saddle and travels longitudinally in a horizontal plane. It supports the vise and the work (Fig. 59-6).
- The *crossfeed handwheel* is used to move the table toward or away from the column.

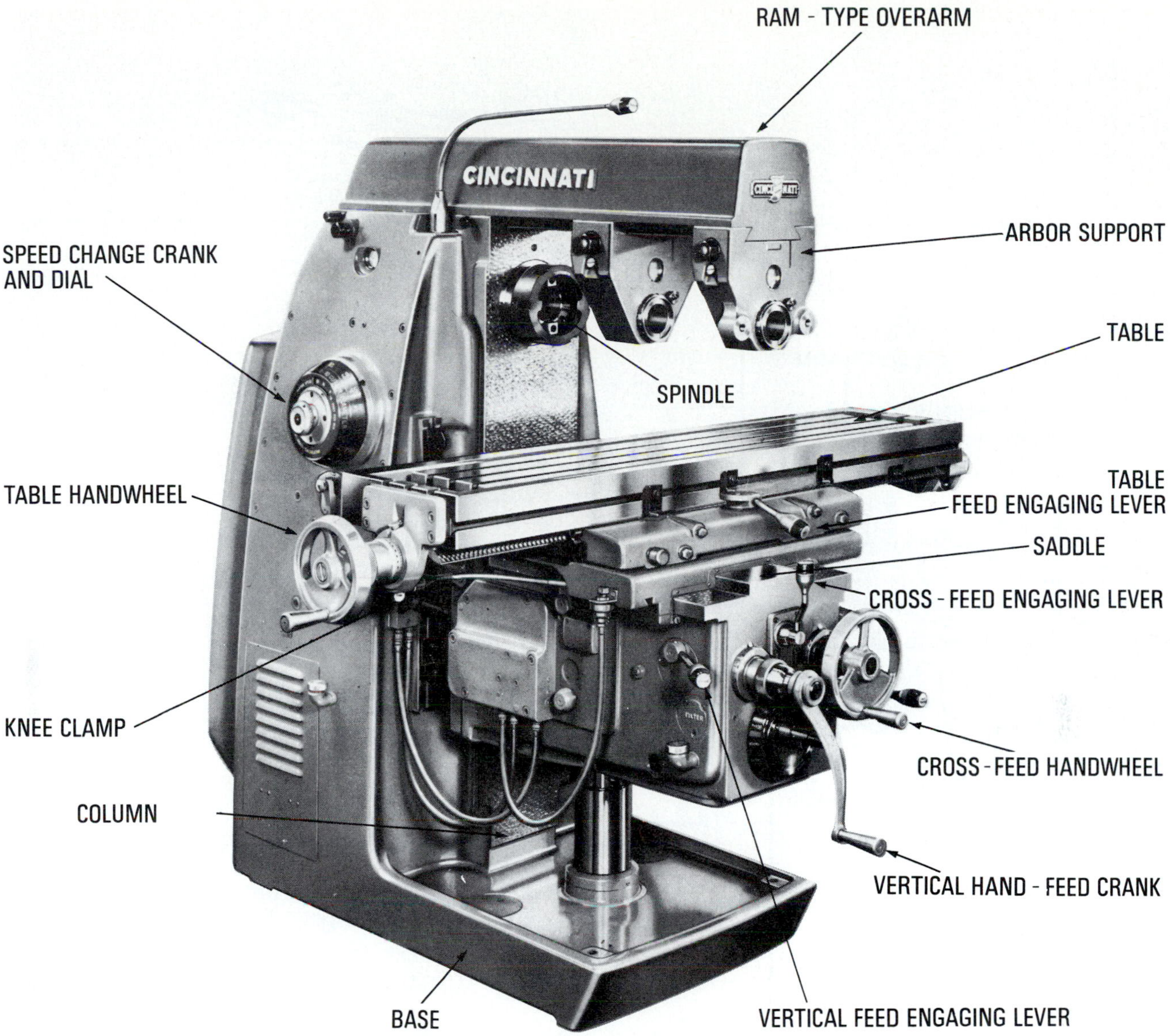
RAM - TYPE OVERARM
CINCINNATI
ARBOR SUPPORT
SPEED CHANGE CRANK AND DIAL
TABLE
SPINDLE
TABLE HANDWHEEL
TABLE FEED ENGAGING LEVER
SADDLE
CROSS - FEED ENGAGING LEVER
KNEE CLAMP
CROSS - FEED HANDWHEEL
COLUMN
VERTICAL HAND - FEED CRANK
BASE
VERTICAL FEED ENGAGING LEVER

Figure 59-5 *Universal horizontal milling machine.*

- The *table handwheel* is used to move the table horizontally back and forth in front of the column.
- The *feed dial* is used to regulate the table feeds.
- The *spindle* provides the drive for arbors, cutters, and attachments used on a milling machine.
- The *overarm* provides for correct alignment and support of the arbor and various attachments. It can be adjusted and locked in various positions, depending on the length of the arbor and the position of the cutter.
- The *arbor support* is fitted to the overarm and can be clamped at any location on the overarm. Its purpose is to align and support various arbors and attachments.
- The *elevating screw* is controlled by hand or an automatic feed. It gives an upward or downward movement to the knee and the table.
- The *spindle speed dial* is set by a crank that is turned to regulate the spindle speed. On some milling machines the spindle speed changes are made by means of two levers. When making speed changes, always check to see whether the change can be made when the machine is running or if it must be stopped.

BACKLASH ELIMINATOR

A feature on most milling machines is the addition of a *backlash eliminator*. This device, when engaged, eliminates the backlash (play) between the nut and the table lead screw, permitting the operation of *climb* (down) milling. Figure 59-7 shows a diagrammatic sketch of the Cincinnati backlash eliminator.

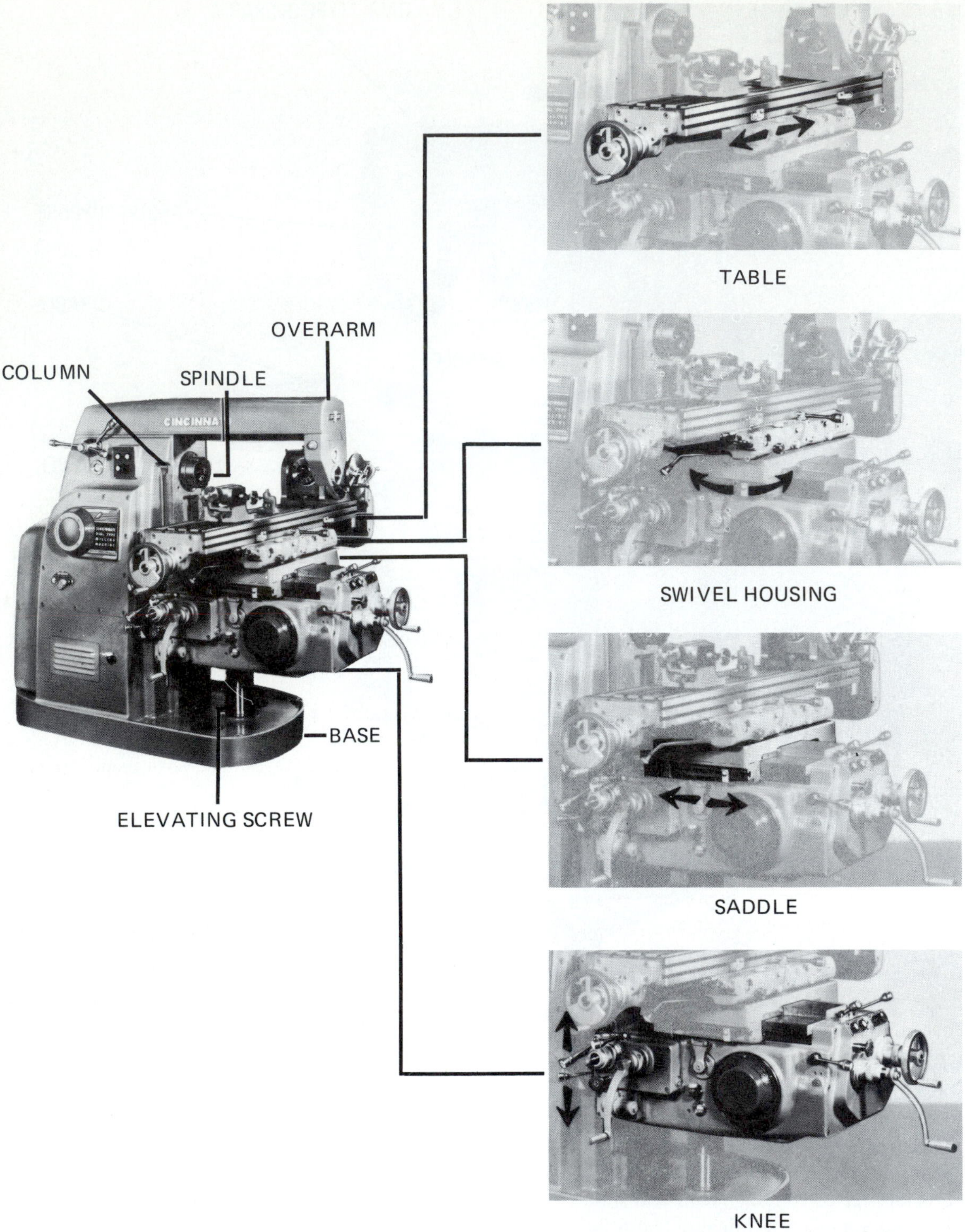
TABLE
OVERARM
COLUMN
SPINDLE
SWIVEL HOUSING
BASE
ELEVATING SCREW
SADDLE
KNEE

Figure 59-6 *The main parts of a milling machine. (Courtesy Cincinnati Milacron Inc.)*

The backlash eliminator works as follows. Two independent nuts are mounted on the lead screw. These nuts engage a common crown gear, which in turn meshes with a rack. Axial movement of the rack is controlled by the backlash eliminator engaging knob on the front of the saddle. By turning the knob in, the nuts are forced to move along the lead screw in opposite directions, removing all backlash. The nuts-gear-rack arrangement is shown in Fig. 59-7.

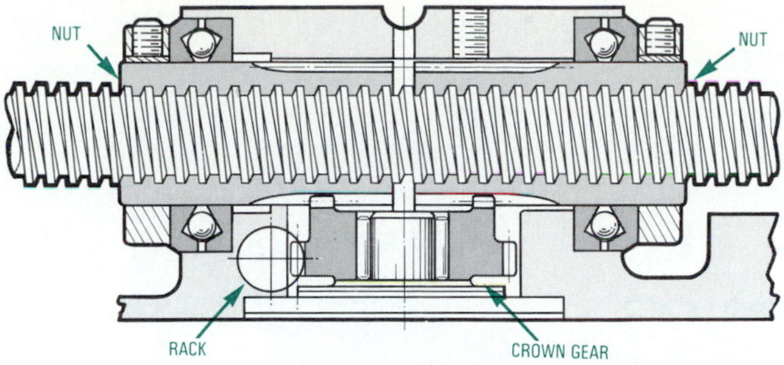
NUT
NUT
RACK
CROWN GEAR

Figure 59-7 *Cross section of a Cincinnati Milacron backlash eliminator. (Courtesy Cincinnati Milacron Inc.)*

MILLING MACHINE ACCESSORIES

A wide variety of accessories, which greatly increase its versatility and productivity, are available for the milling machine. These accessories may be classed as *fixtures* or as *attachments.*

FIXTURES

A fixture (Fig. 59-8) is a work-holding device fastened to the table of a machine or to a machine accessory, such as a rotary table. It is designed to hold workpieces that cannot be readily held in a vise or in production work when large quantities are to be machined. The fixture must be designed so that identical parts, when held in the fixture, will be positioned exactly and held securely. Fixtures may be constructed to hold one or several parts at one time and should permit the quick changing of workpieces. The work may be positioned by stops, such as pins, strips, or setscrews, and held in place by clamps, cam-lock levers, or setscrews. To produce uniform workpieces, clean the chips and cuttings from a fixture before mounting a new workpiece.

Figure 59-8 *Milling a workpiece held in a fixture. (Courtesy Cincinnati Milacron Inc.)*

ATTACHMENTS

Milling machine attachments may be divided into three classes:

1. Those designed to hold special attachments; they are attached to the spindle and column of the machine. They are the *vertical, high-speed, universal, rack milling,* and *slotting* attachments. These attachments are designed to increase the versatility of the machine.
2. *Arbors, collets,* and *adapters,* which are designed to hold standard cutters.
3. Those designed to hold the workpiece, such as a *vise, rotary table,* and *indexing* or *dividing head.*

Vertical Milling Attachment

The *vertical milling attachment* (Fig. 59-9), which may be mounted on the face of the column or the overarm, enables a plain or universal milling machine to be used as a vertical milling machine. Angular surfaces may be machined by swiveling the head, parallel to the face of the column, to any angle up to 45° on either side of the vertical position. On some models the head may be swiveled as much as 90° to either side. Vertical attachments enable the horizontal milling machine to be used for such operations as face milling, end milling, drilling, boring, and T-slot milling.

A modification of the vertical milling attachment is the *universal milling attachment,* which may be swiveled in two planes, parallel to the column and at right angles to it, for the cutting of compound angles. The vertical and universal

CINCINNATI

Figure 59-9 *A vertical milling attachment. (Courtesy Cincinnati Milacron Inc.)*

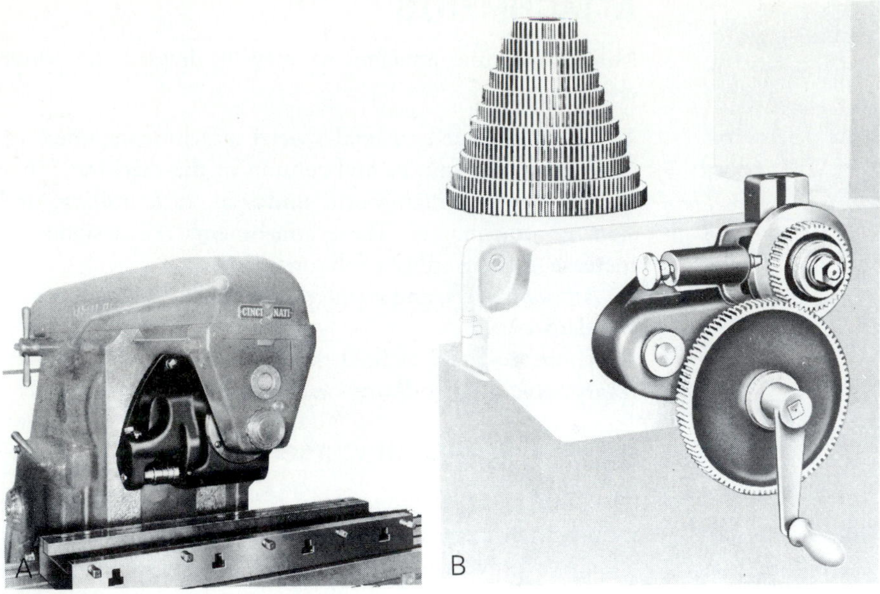
A
B

Figure 59-10 *(A) A rack milling attachment; (B) a rack indexing attachment. (Courtesy Cincinnati Milacron Inc.)*

Figure 59-11 *A slotting attachment. (Courtesy Cincinnati Milacron Inc.)*

attachments are also manufactured in *high-speed* models, which permit the efficient use of small- and medium-sized end mills and cutters for such operations as die sinking and key seating.

Rack Milling Attachment

The *rack milling attachment* (Fig. 59-10A) and the *rack indexing attachment* (Fig. 59-10B) are used to mill longer gear racks (flat gears) than could be cut with the standard horizontal milling machine. These attachments will be discussed later in the section with gear-cutting accessories.

Slotting Attachment

The *slotting attachment* (Fig. 59-11) converts the rotary motion of the spindle into reciprocating motion for cutting keyways, splines, templates, and irregularly shaped surfaces. The length of the stroke is controlled by an adjustable crank. The tool slide may be swiveled to any angle in a plane parallel to the face of the column, making the slotting attachment especially valuable in die work.

Arbors, Collets, and Adapters

Arbors, used for mounting the milling cutter, are inserted and held in the main spindle by a draw bolt or a special quick-change adapter (Fig. 59-12).

Shell-end mill arbors may fit into the main spindle or the spindle of the vertical attachment. These devices permit face milling to be done either horizontally or vertically.

Collet adapters are used for mounting drills or other tapered-shank tools in the main spindle of the machine or the vertical milling attachment.

A *quick-change adapter,* mounted in the spindle, permits such operations as drilling, boring, and milling without a change in the setup of the workpiece.

Vises

Milling machine vises are the most widely used work-holding devices for milling; they are manufactured in three styles.

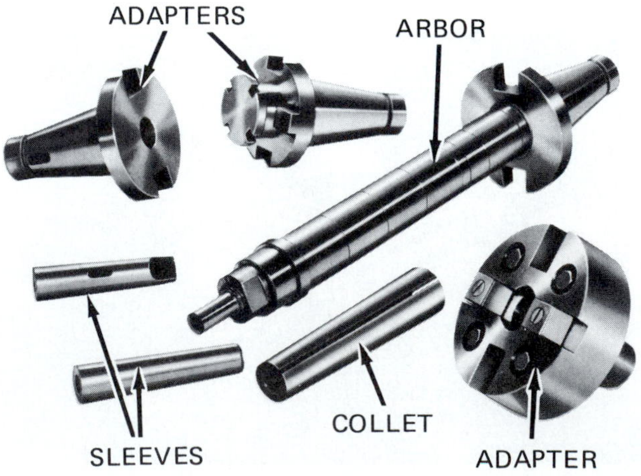
ADAPTERS
ARBOR
SLEEVES
COLLET
ADAPTER

Figure 59-12 *Arbors, collets, and adapters.*

The *plain vise* (Fig. 59-13A) may be bolted to the table so that its jaws are parallel or at right angles to the axis of the spindle. The vise is positioned quickly and accurately by keys on the bottom of the base which fit into the T-slots on the table.

The *swivel base vise* (Fig. 59-13B) is similar to the plain vise, except that it has a swivel base which enables the vise to be swiveled through 360° in a horizontal plane.

The *universal vise* (Fig. 59-13C) may be swiveled through 360° in a horizontal plane and may be tilted from 0 to 90° in a vertical plane. It is used chiefly by toolmakers, moldmakers, and diemakers, since it permits the setting of compound angles for milling.

Indexing or Dividing Head

The *indexing* or *dividing head* is a very useful accessory, permitting the cutting of bolt heads, gear teeth, ratchets, and so on. When connected to the lead screw of the milling machine, it will revolve the work as required to cut helical gears and flutes in drills, reamers, and other tools. The dividing head will be fully discussed later in the section.

REVIEW QUESTIONS

HORIZONTAL MILLING MACHINE

1. Name six operations that can be performed on a milling machine.

MANUFACTURING-TYPE MILLING MACHINE

2. List four features of a manufacturing-type milling machine.

KNEE-AND-COLUMN-MILLING MACHINE

3. What is the difference between a plain horizontal and a universal horizontal milling machine?
4. What is the purpose of the backlash eliminator?

MILLING MACHINE ACCESSORIES AND ATTACHMENTS

5. What is the purpose of a fixture?
6. List three milling machine attachments and give examples of each.

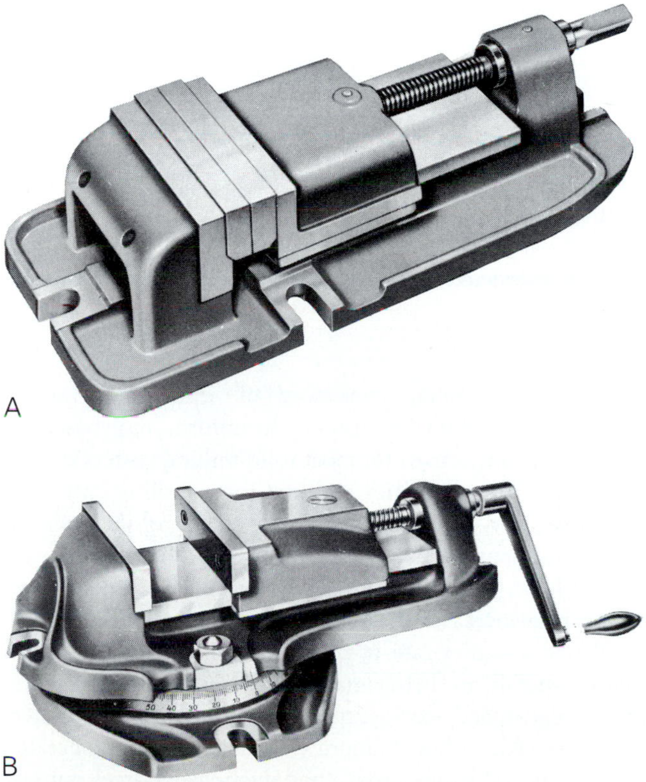
A
B

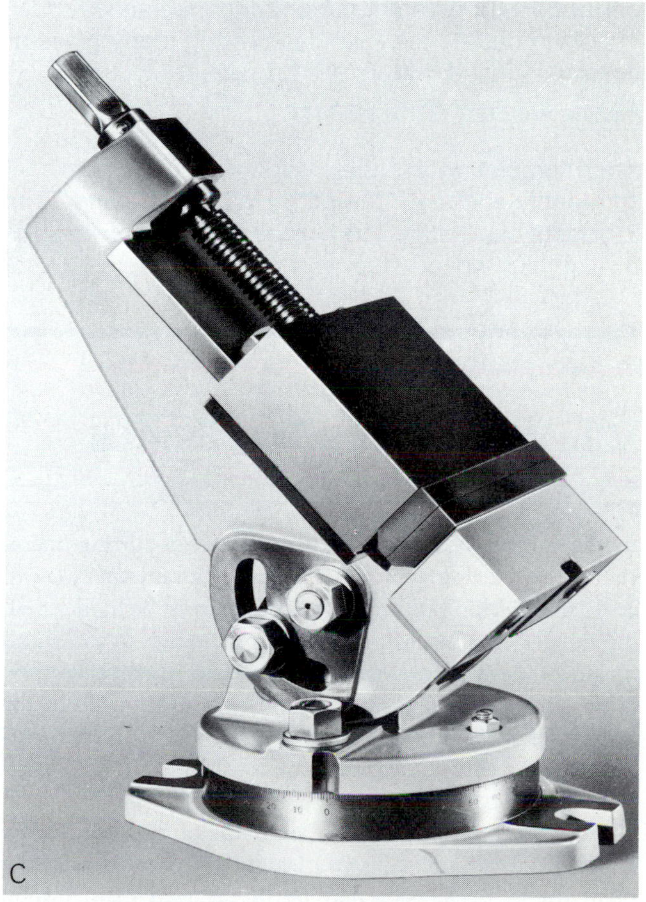
C

Figure 59-13 *(A) A plain vise (Courtesy Cincinnati Milacron Inc.); (B) a swivel base vise; (C) a universal vise.*

7. Describe the purpose of the following devices:
   a. Vertical milling attachment
   b. Rack milling attachment
   c. Slotting attachment
8. Name three methods of holding cutters on a milling machine.
9. What are the features of the:
   a. Plain vise
   b. Swivel base vise
   c. Universal vise

UNIT 60

Milling Cutters

The proper selection, use, and care of milling cutters must be practiced if the best results are to be achieved with the milling machine. The operator or apprentice must be able not only to determine the proper spindle speed for any cutter but should also constantly observe how the milling machine performs with different cutters and clearances.

Milling cutters are manufactured in many types and sizes. Only the most commonly used cutters will be discussed.

OBJECTIVES

After completing this unit you should be able to:
1. Identify and state the purposes of six standard milling cutters
2. Identify and state the purposes of four special-purpose cutters
3. Use high-speed steel and carbide cutters for proper applications

MILLING CUTTER MATERIALS

In the milling process, as with most metal-cutting operations, the cutting tool must possess certain qualities in order to function satisfactorily. Cutters must be harder than the metal being machined and strong enough to withstand pressures developed during the cutting operation. They must also be tough to resist the shock resulting from the contact of the tooth with the work. To maintain keen cutting edges, they must be able to resist the heat and abrasion of the cutting process.

Most milling cutters today are made of high-speed steel or tungsten carbide. Special-purpose cutters, made in the plant for a special job, may be made from plain carbon steel.

High-speed steel, consisting of iron with various amounts of carbon, tungsten, chromium, molybdenum, and vanadium, is used for most solid milling cutters since it possesses all the qualities required for a milling cutter. In this steel, carbon is the hardening agent, while tungsten and molybdenum enable the steel to retain its hardness up to red heat. Vanadium increases the tensile strength and chromium increases the toughness and wear resistance.

When a higher rate of production is desired and when harder metals are being machined, cemented carbides replace high-speed steel cutters. Cemented-carbide cutters (Fig. 60-1A), although more expensive, may be operated from 3 to 10 times faster than high-speed steel cutters. Cemented-carbide tips may be either brazed to a steel body or inserts may be held in place by means of locking or clamping devices (Fig. 60-1B).

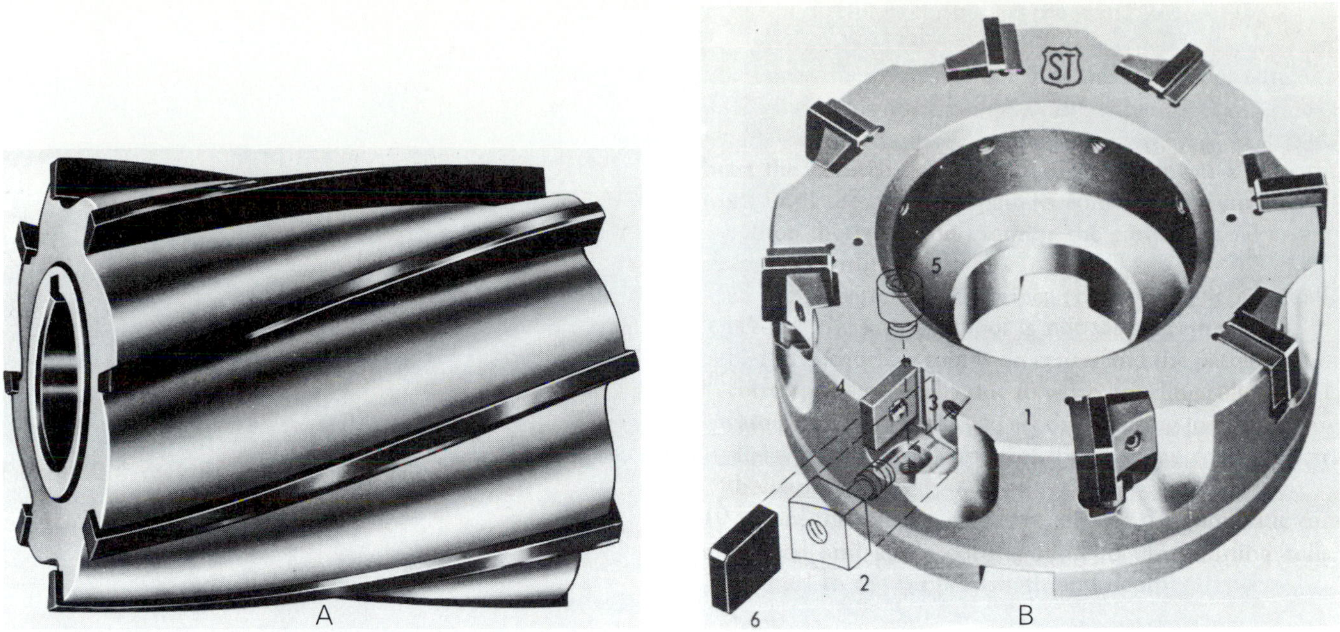
A
5
4
3
1
2
6
B

Figure 60-1 *(A) A cemented-carbide milling cutter; (B) cemented-carbide tips are held in place by a locking device.*

When cemented-carbide cutters are to be used, care must be taken to select the proper type of carbide for the job. Straight tungsten carbide is used for machining cast iron, machine steel, most nonferrous alloys, and plastics. Tantalum carbide is used to machine low- and medium-carbon steels, and tungsten–titanium carbide is used for high-carbon steels.

Although cemented carbides have many advantages as cutting tools, several disadvantages limit extensive use:

1. Cemented-carbide cutters are more costly to buy, maintain, and sharpen.
2. Efficient use of these cutters requires that machines must be rigid and have greater horsepower and speed than are required for high-speed cutters.
3. Carbide cutters are brittle, and therefore the edges break easily if misused.
4. Special grinders with silicon carbide and diamond wheels are required to sharpen carbide cutters properly.

PLAIN MILLING CUTTERS

Probably the most widely used milling cutter is the plain milling cutter, which is a cylinder made of high-speed steel with teeth cut on the periphery; it is used to produce a *flat surface.* These cutters may be of several types, as shown in Fig. 60-2.

Light-duty plain milling cutters (Fig. 60-2A), which are less than 3/4 in. (19 mm) wide, will usually have straight teeth; those over 3/4 in. (19 mm) wide have a helix angle of about 25° (Fig. 60-2B). This type of cutter is used only for light milling operations, since it has too many teeth to permit the chip clearance required for heavier cuts.

Heavy-duty plain milling cutters (Fig. 60-2C) have fewer teeth than the light-duty type, which provide for better chip clearance. The helix angle varies up to 45°. This helix angle on the teeth produces a smoother surface because of the shearing action and reduced chatter. Less power is required with this type of cutter than with straight-tooth and small-helix-angle cutters.

High-helix plain milling cutters (Fig. 60-2D) have helix angles from 45 to over 60°. They are particularly suited to the milling of wide and intermittent surfaces on contour and profile milling. Although this type of cutter is usually mounted on the milling machine arbor, it is sometimes shank-mounted with a pilot on the end and used for milling elongated slots.

STANDARD SHANK-TYPE HELICAL MILLING CUTTERS

Standard shank-type helical milling cutters (Fig. 60-3), also called *arbor-type cutters,* are used for milling forms from solid metal; for example, they are used when making yokes or forks. Shank-type milling cutters are also used for removing inner sections from solids. These cutters are inserted through a previously drilled hole and supported at

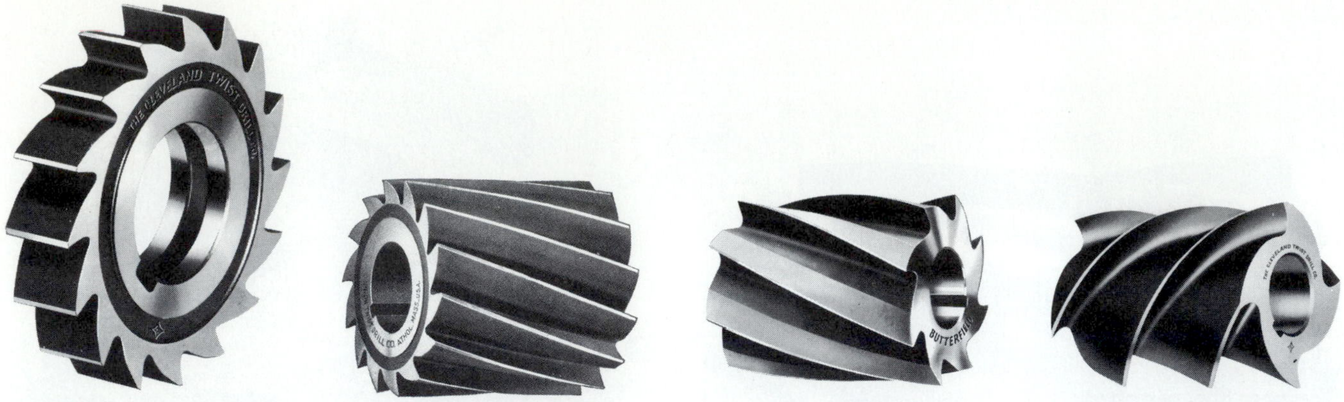
A
B
C
D

Figure 60-2 *Plain milling cutters: (A) light duty; (B) light-duty helical; (C) heavy duty; (D) high helix.*

Figure 60-3 *Standard shank-type helical milling cutter. (Courtesy The Butterfield Division, Union Twist Drill Company.)*

the outer end with type A arbor supports. Special spindle adapters are used to hold these cutters.

SIDE MILLING CUTTERS

Side milling cutters (Fig. 60-4) are comparatively narrow cylindrical milling cutters with teeth on each side as well as on the periphery. They are used for cutting slots and for face and straddle milling operations. Side milling cutters may have straight teeth (Fig. 60-4A) or staggered teeth (Fig. 60-4B). Staggered-tooth cutters have each tooth set alternately to the right and left with an alternately opposite helix angle on the periphery. These cutters have free cutting action at high speeds and feeds. They are particularly suited for milling deep, narrow slots.

Half-side milling cutters (Fig. 60-4C) are used when only one side of the cutter is required, as in end facing. These cutters are also made with interlocking faces so that two cutters may be placed side by side for slot milling. The interlocking type is more suited for slot cutting than the solid-type staggered-tooth cutter since the amount ground from the side of the cutter during regrinding may be compensated by a washer between the cutters. Half-side milling cutters have considerable rake and, therefore, are able to take heavy cuts.

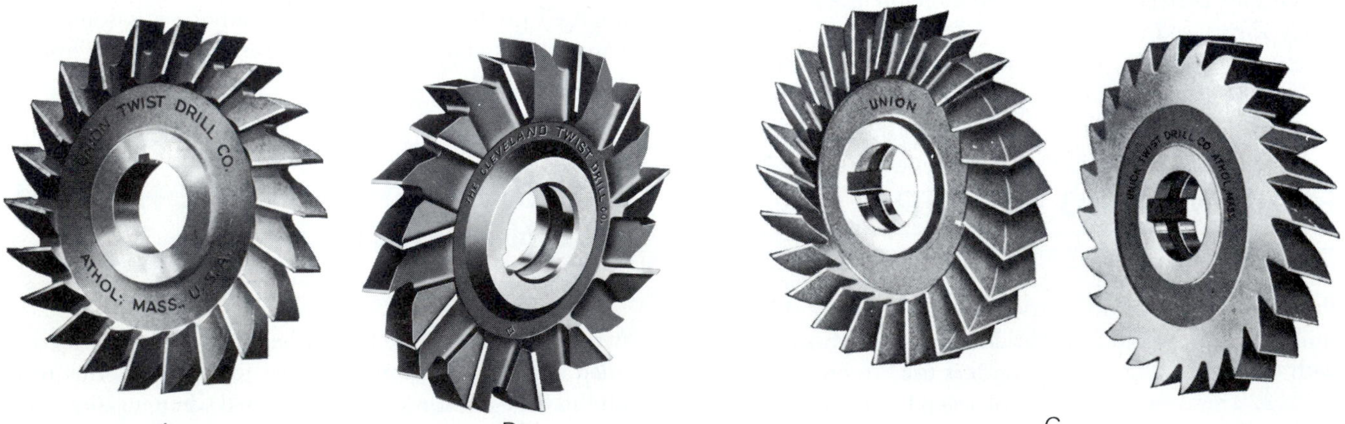
A
B
C

Figure 60-4 *Side milling cutters: (A) plain; (B) staggered tooth; (C) half side.*

FACE MILLING CUTTERS

Face milling cutters (Fig. 60-5) are generally over 6 in. (150 mm) in diameter and have *inserted teeth* held in place by a wedging device. The teeth may be of high-speed steel, cast tool steel, or they may be tipped with sintered-carbide cutting edges. The corners of this type of cutter are beveled; most of the cutting action occurs at these points and the periphery of the cutter. The face of the tooth removes a small amount of stock left by the spring of the work or cutter. To prevent chatter, only a small portion of the tooth face near the periphery is in contact with the work; the remainder is ground with a suitable clearance (8 to 10°).

This type of cutter is often used as a combination cutter, making the roughing and finishing cut in one pass. The roughing and finishing blades are mounted on the same body, with a limited number of finishing blades being set to a smaller diameter and extending slightly farther from the face than the roughing blades. The finishing blades have a slightly wider cutting face surface, which creates a better surface finish.

Face milling cutters under 6 in. (150 mm) are called *shell end mills* (Fig. 60-6). They are solid, multiple-tooth cutters with teeth on the face and the periphery. They are usually held on a stub arbor, which may be threaded or employ a key in the shank to drive the cutter. Shell end mills are more economical than large solid end mills because they are cheaper to replace when broken or worn out.

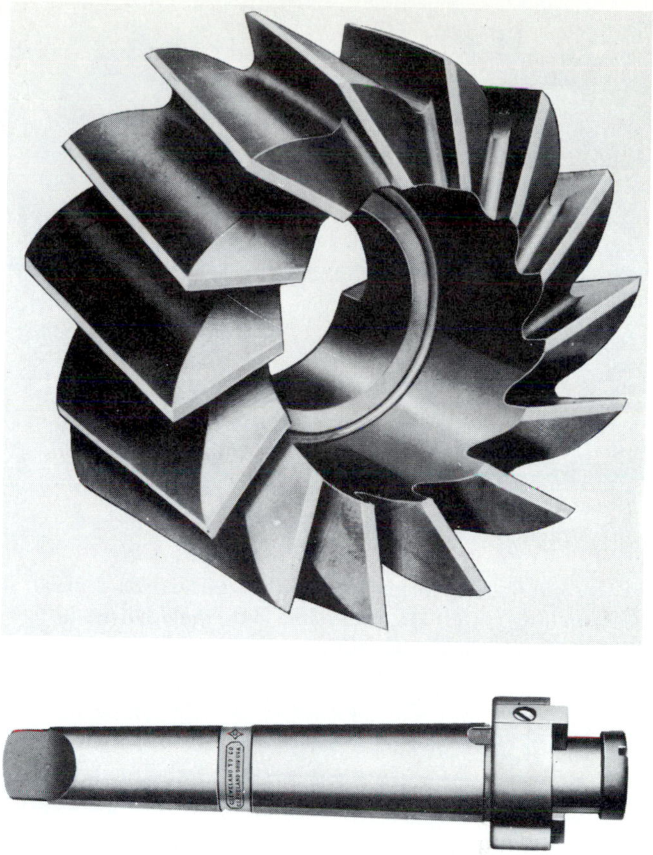

Figure 60-6 *Shell end mill and adapter.*

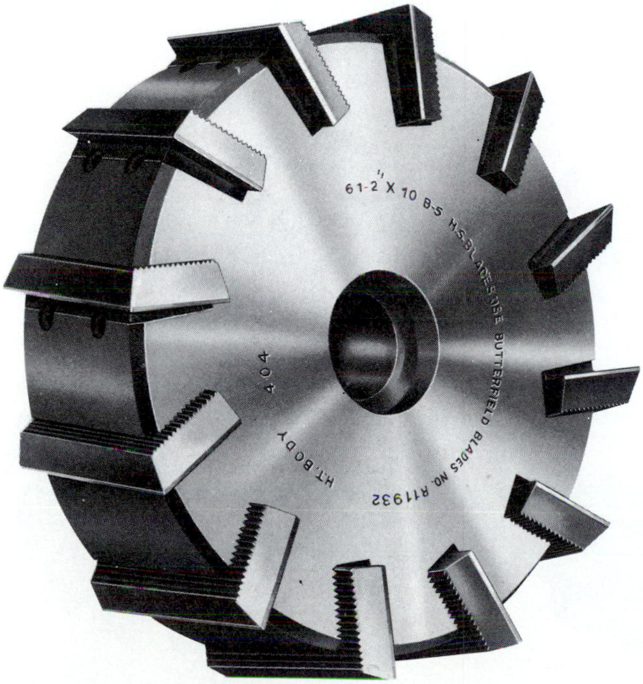
BUTTERFIELD
BLADES NO. R11932
HT. BODY
404

Figure 60-5 *A face milling cutter. (Courtesy The Butterfield Division, Union Twist Drill Company.)*

ANGULAR CUTTERS

Angular cutters have teeth that are neither parallel nor perpendicular to the cutting axis. They are used for milling angular surfaces, such as grooves, serrations, chamfers, and reamer teeth. They may be divided into two groups:

1. *Single-angle milling cutters* (Fig. 60-7A) have teeth on the angular surface and may or may not have teeth on the flat side. The included angle between the flat face and the conical face designates the cutters, such as 45 or 60° angular cutter.
2. *Double-angle milling cutters* (Fig. 60-7B) have two intersecting angular surfaces with cutting teeth on both. When these cutters have equal angles on both sides of the line at a right angle to the axis (symmetrical), they are designated by the size of the included angle. When the angles formed with this line are not the same (unsymmetrical), the cutters are designated by specifying the angle on either side of the plane or line, such as 12 to 48° double-angle milling cutter.

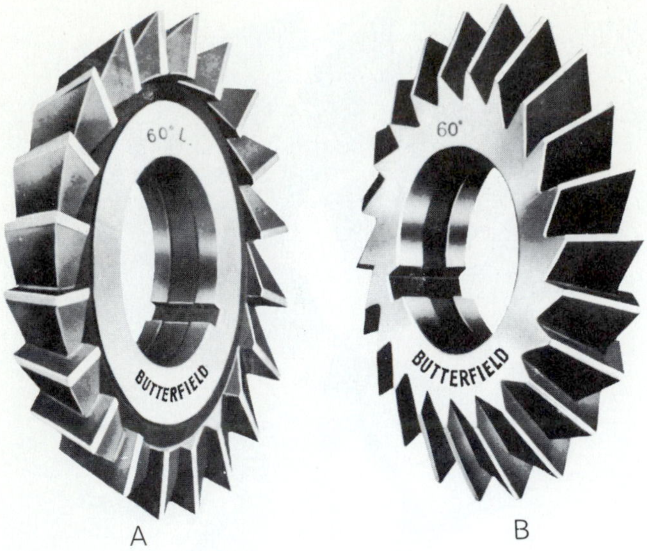
60° L
BUTTERFIELD
60°
BUTTERFIELD
A
B

Figure 60-7 *(A) Single-angle milling cutter; (B) double-angle milling cutter. (Courtesy The Butterfield Division, Union Twist Drill Company.)*

FORMED CUTTERS

Formed cutters (Fig. 60-8) incorporate the exact shape of the part to be produced, permitting exact duplication of irregularly shaped parts more economically than most other means. Formed cutters are particularly useful for the production of small parts. Each tooth of a formed cutter is identical in shape, and the clearance is machined on the full thickness of each tooth by the form or master tool in a cam-controlled relieving machine. Examples of *formed-relieved cutters* are concave, convex, and gear cutters.

Formed cutters are sharpened by grinding the tooth face. Tooth faces are radial and may have positive, zero, or negative rake, depending on the cutter application. *It is important* that the original rake on the tooth be maintained so that the profiles of the tooth and of the work are not changed. If the tooth face rake is maintained exactly, the cutter may be resharpened until the teeth are too thin for use; thus the exact, original shape of the tooth can be maintained. These cutters are sometimes produced with an angular face, which causes a shearing action and reduces chatter during the cutting process. They are, however, more difficult to sharpen.

METAL SAWS

Metal-slitting saws (Fig. 60-9) are basically thin plain milling cutters with sides relieved or "dished" to prevent rubbing or binding when in use. Slitting saws are made in widths from 1⁄32 to 3⁄16 in. (0.8 mm to 5 mm). Because of their thin cross section, they should be operated at approximately one-quarter to one-eighth of the feed per tooth used for other cutters. For nonferrous metals, their speed can be increased. Unless a special driving flange is used for slitting saws, it is *not* advisable to key the saw to the milling arbor. The arbor nut should be pulled up as tightly as possible *by hand only.* Since slitting saws are so easily broken, some operators find it desirable to "climb" or "down" mill when sawing (see Unit 63). However, to overcome the play be-

BUTTERFIELD
5/8-10° RAKE
H.S. 12
BUTTERFIELD
NO. 2-8 P. 55 TO 134 T.
UNION TWIST DRILL CO. ATHOL, MASS. U.S.A.
A
B
C

Figure 60-8 *Types of formed cutters: (A) concave; (B) convex; (C) gear tooth.*

3 X 3-32
UNION TWIST DRILL CO
ATHOL, MASS. U.S.A.

A B C

Figure 60-9 *Metal-slitting saws. (Courtesy The Butterfield Division, Union Twist Drill Company.)*

tween the lead screw and nut, the backlash eliminator should be engaged.

END MILLS

End mills have cutting teeth on the end as well as on the periphery and are fitted to the spindle by a suitable adapter. They are of two types: the *solid end mill* in which the shank and the cutter are integral, and the *shell end mill,* which, as previously stated, has a separate shank.

Solid end mills, generally smaller than shell end mills, may have either straight or helical flutes. They are manufactured with straight and tapered shanks and with two or more flutes. The two-flute type (Fig. 60-10A) has flutes which meet at the cutting end, forming two cutting lips across the bottom. These lips are of different lengths, one extending beyond the center axis of the cutter. This arrangement eliminates the center and permits the two-flute end mill to be used in a milling machine for drilling a hole to start a slot that does not extend to the edge of the metal. When a slot is being cut with a two-flute end mill, the depth of cut should not exceed one-half the diameter of the cutter. When the four-flute end mill (Fig. 60-10B) is used for slot cutting, it is usually started at the edge of the metal.

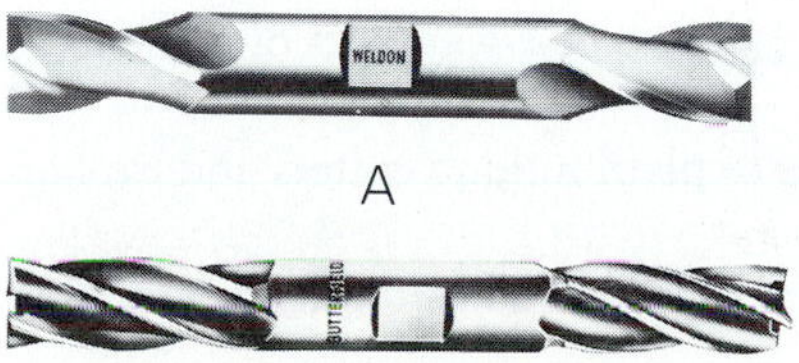
WELDON

A

B

Figure 60-10 *(A) A two-flute end mill; (B) a four-flute end mill.*

T-SLOT CUTTER

The *T-slot cutter* (Fig. 60-11A) is used to cut the wide horizontal groove at the bottom of a T-slot after the narrow vertical groove has been machined with an end mill or a side milling cutter. It consists of a small side milling cutter with teeth on both sides and an integral shank for mounting, similar to an end mill.

DOVETAIL CUTTER

The *dovetail cutter* (Fig. 60-11B) is similar to a single-angle milling cutter with an integral shank. Some dovetail cutters are manufactured with an internal thread and are mounted on a special threaded shank. They are used to form the sides of a dovetail after the tongue or the groove has been machined with another suitable cutter, usually a side milling cutter. Dovetail cutters may be obtained with 45, 50, 55 or 60° angles.

WOODRUFF KEYSEAT CUTTER

The *Woodruff keyseat cutter* (Fig. 60-11C) is similar in design to plain and side milling cutters. Smaller sizes up to approximately 2 in. (50 mm) in diameter are made with a

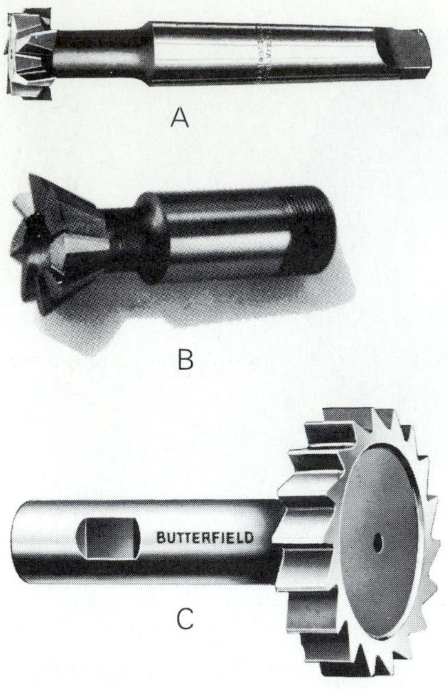
A
B
BUTTERFIELD
C

Figure 60-11 *(A) A T-slot cutter (Courtesy The Butterfield Division, Union Twist Drill Company); (B) a dovetail cutter; (C) a Woodruff keyseat cutter.*

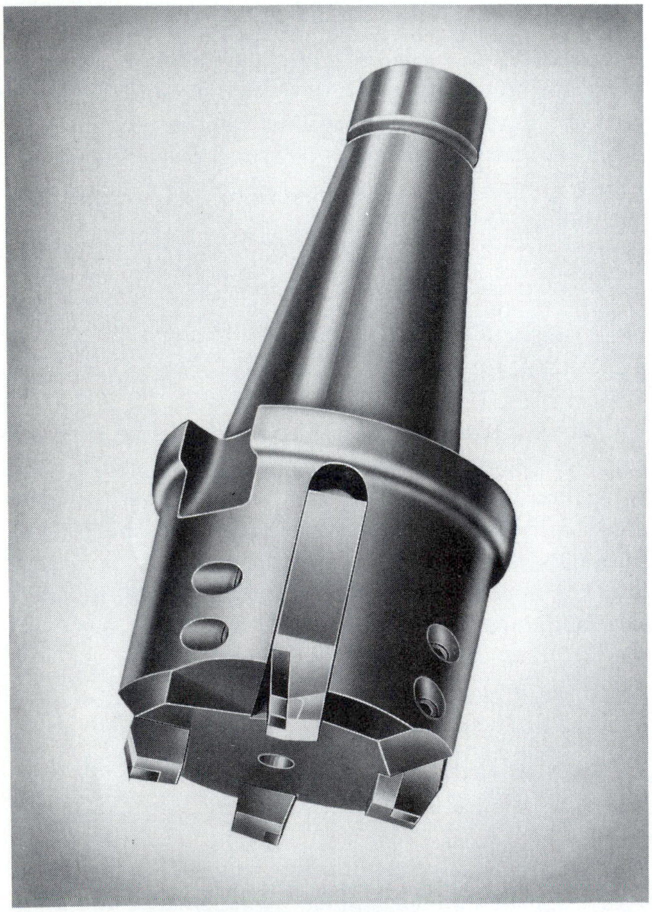

Figure 60-12 *A flycutter.*

solid shank and straight teeth; larger sizes are mounted on an arbor and have staggered teeth on both the sides and the periphery. They are used for milling semicylindrical keyseats in shafts.

Woodruff cutters are designated by a number system. The right-hand two digits give the nominal diameter in eighths of an inch, while the preceding digits give the width of the cutter in thirty-seconds of an inch. For example, a #406 cutter would have dimensions as follows:

Diameter	$06 \times \frac{1}{8} = \frac{3}{4}$ in.
Width	$4 \times \frac{1}{32} = \frac{1}{8}$ in.

FLYCUTTERS

The *flycutter* (Fig. 60-12) is a single-pointed cutting tool with the cutting end ground to the desired shape. It is mounted in a special adapter or arbor. Since all the cutting is done with one tool, a fine feed must be used. Flycutters are used in experimental work and where the high cost of a specially shaped cutter would not be warranted.

REVIEW QUESTIONS

1. Name two materials which are used to make milling cutters.
2. Name five elements that may be added to iron or steel to produce high-speed steel, and state the purpose of each additive.
3. Discuss the advantages and disadvantages of cemented-carbide tools.
4. Describe three types of plain milling cutters and explain their uses.
5. Describe three types of side milling cutters, and state where each is used.
6. List two ways in which formed cutters differ from plain milling cutters.
7. What advantage does a two-flute end mill have?

UNIT 61

Cutting Speeds, Feeds, and Depth of Cut

The most important factors affecting the efficiency of a milling operation are cutting speed, feed, and depth of cut. If the cutter is run too slowly, valuable time will be wasted, while excessive speed results in loss of time in replacing and regrinding cutters. Somewhere between these two extremes is the efficient *cutting speed* for the material being machined.

The rate at which the work is fed into the revolving cutter is important. If the work is fed too slowly, time will be wasted and cutter chatter may occur, which shortens the life of the cutter. If work is fed too fast, the cutter teeth can be broken. Much time will be wasted if several shallow cuts are taken instead of one deep or roughing cut. Therefore *speed, feed,* and *depth of cut* are three important factors in any milling operation.

OBJECTIVES

After completing this unit you will be able to:
1. Select cutting speeds and calculate the r/min for various cutters and materials
2. Select and calculate the proper feeds for various cutters and materials
3. Follow the correct procedure for taking roughing and finishing cuts

CUTTING SPEEDS

One of the most important factors affecting the efficiency of a milling operation is cutter speed. The cutting speed of a metal may be defined as *the speed, in surface feet per minute* (sf/min) *or meters per minute* (m/min) *at which the metal may be machined efficiently.* When the work is machined in a lathe, it must be turned at a specific number of revolutions per minute (r/min), *depending on its diameter,* to achieve the proper cutting speed. When work is machined in a milling machine, the *cutter* must be revolved at a specific number of r/min, *depending on its diameter,* to achieve the proper cutting speed.

Since different types of metals vary in hardness, structure, and machinability, different cutting speeds must be used for each type of metal and for various cutter materials. Several factors must be considered when the proper r/min at which to machine a metal is being determined. The most important are:

- The type of work material
- The cutter material
- The diameter of the cutter
- The surface finish required
- The depth of cut being taken
- The rigidity of the machine and work setup

The cutting speeds for the more common metals are shown in Table 61-1.

INCH CALCULATIONS

To get optimum use from a cutter, the proper speed at which the cutter should be revolved must be determined. When cutting machine steel, a high-speed steel cutter would have to achieve a surface speed of about 90 ft/min (27 m/min). Since the diameter of the cutter affects this speed, it is necessary to consider its diameter in the calculation. The following example illustrates how the formula is developed.

Table 61-1 Milling Machine Cutting Speeds

Material	High-Speed Steel Cutter		Carbide Cutter	
	ft/min	m/min	ft/min	m/min
Machine steel	70–100	21–30	150–250	45–75
Tool steel	60–70	18–20	125–200	40–60
Cast iron	50–80	15–25	125–200	40–60
Bronze	65–120	20–35	200–400	60–120
Aluminum	500–1000	150–300	1000–2000	300–600

EXAMPLE

Calculate the speed required to revolve a 3-in.-diameter high-speed steel milling cutter for cutting machine steel (90 sf/min).

SOLUTION

1. First, determine the circumference of the cutter or the distance a point on the cutter would travel in one revolution. Circumference of cutter = 3 × 3.1416.
2. To determine the proper cutter speed or *r/min,* it is necessary only to divide the cutting speed (CS) by the circumference of the cutter.

$$\text{r/min} = \frac{\text{CS (ft)}}{\text{circumference (in.)}}$$

$$= \frac{90}{3 \times 3.1416}$$

Since the numerator is in feet and the denominator in inches, the numerator must be changed to inches. Therefore,

$$\text{r/min} = \frac{12 \times \text{CS}}{3 \times 3.1416}$$

Because it is usually impossible to set a machine to the exact r/min, it is permissible to consider that 3.1416 will divide into 12 approximately 4 times. The formula now becomes

$$\text{r/min} = \frac{4 \times \text{CS}}{D}$$

Using this formula and Table 61-1, you can calculate the proper cutter or spindle speed for any material and cutter diameter.

EXAMPLE

At what speed should a 2-in.-diameter carbide cutter revolve to mill a piece of cast iron (CS 150)?

SOLUTION

$$\text{r/min} = \frac{4 \times 150}{2}$$

$$= 300$$

METRIC CALCULATIONS

The r/min at which the milling machine should be set when using metric measurements is as follows:

$$\text{r/min} = \frac{\text{CS (m)} \times 1000}{\pi \times D\ \text{(mm)}}$$

$$= \frac{\text{CS} \times 1000}{3.1416 \times D}$$

Since only a few machines are equipped with variable-speed drives, which allow them to be set to the exact calculated speed, a simplified formula can be used to calculate r/min. The π (3.1416) on the bottom line of the formula will divide into the 1000 of the top line approximately 320 times. This results in a simplified formula, which is close enough for most milling operations:

$$\text{r/min} = \frac{\text{CS (m)} \times 320}{D\ \text{(mm)}}$$

EXAMPLE

Calculate the r/min required for a 75-mm-diameter high-speed steel milling cutter when cutting machine steel (CS 30 m/min).

SOLUTION

$$\text{r/min} = \frac{30 \times 320}{75}$$

$$= \frac{9600}{75}$$

$$= 128$$

Although these formulas are helpful in calculating the cutter (spindle) speed, it should be remembered that they are approximate only and the speed may have to be altered because of the hardness of the metal and/or the machine condition. Best results may be obtained if the following rules are observed.

1. For longer cutter life, use the lower CS in the recommended range.
2. Know the hardness of the material to be machined.
3. When starting a new job, use the lower range of the CS and gradually increase to the higher range if conditions permit.
4. If a fine finish is required, reduce the feed rather than increase the cutter speed.
5. The use of coolant, properly applied, will generally produce a better finish and lengthen the life of the cutter because it absorbs heat, acts as a lubricant, and washes chips away.

MILLING FEEDS AND DEPTH OF CUT

The two other factors which affect the efficiency of a milling operation are the *milling feed,* or the rate at which the work is fed into the milling cutter, and the *depth of cut* taken on each pass.

FEED

Milling machine feed may be defined as the distance in inches (or millimeters) per minute that the work moves into the cutter. On most milling machines, the feed is regulated in inches (or millimeters) per minute and is independent of the spindle speed. This arrangement permits faster feeds for larger, slowly rotating cutters.

Feed is the rate at which the work moves into the revolving cutter, and it is measured either in inches per minute or in millimeters per minute. The *milling feed* is determined by multiplying the chip size (chip per tooth) desired, the number of teeth in the cutter, and the r/min of the cutter.

Chip or feed per tooth (CPT or FPT) is the amount of material which should be removed by each tooth of the cutter as it revolves and advances into the work. See Tables 61-2 and 61-3 for the recommended CPT for some of the more common metals.

The feed rate used on a milling machine depends on a variety of factors, such as

1. The depth and width of cut
2. The design or type of cutter
3. The sharpness of the cutter
4. The workpiece material
5. The strength and uniformity of the workpiece
6. The type of finish and accuracy required
7. The power and rigidity of the machine, the holding device, and the tooling setup.

As the work advances into the cutter, each successive tooth advances into the work an equal amount, producing chips of equal thickness. It is this thickness of the chips or the *feed per tooth,* along with the number of teeth in the cutter, which forms the basis for determining the rate of feed. The ideal rate of feed may be determined as follows.

Feed = no. of cutter teeth × feed/tooth × cutter r/min

Inch Calculations

The formula used to find the work feed in inches per minute is:

Table 61-2 Recommended Feed per Tooth (High-Speed Steel Cutters)

Material	Face Mills		Helical Mills		Slotting and Side Mills		End Mills		Form-Relieved Cutters		Circular Saws	
	in.	mm	in.	mm	in.	mm	in.	mm	in.	mm	in.	mm
Aluminum	0.022	0.55	0.018	0.45	0.013	0.33	0.011	0.28	0.007	0.18	0.005	0.13
Brass and bronze (medium)	0.014	0.35	0.011	0.28	0.008	0.20	0.007	0.18	0.004	0.10	0.003	0.08
Cast iron (medium)	0.013	0.33	0.010	0.25	0.007	0.18	0.007	0.18	0.004	0.10	0.003	0.08
Machine steel	0.012	0.30	0.010	0.25	0.007	0.18	0.006	0.15	0.004	0.10	0.003	0.08
Tool steel (medium)	0.010	0.25	0.008	0.20	0.006	0.15	0.005	0.13	0.003	0.08	0.003	0.08
Stainless steel	0.006	0.15	0.005	0.13	0.004	0.10	0.003	0.08	0.002	0.05	0.002	0.05

Table 61-3 Recommended Feed per Tooth (Cemented-Carbide-Tipped Cutters)

Material	Face Mills		Helical Mills		Slotting and Side Mills		End Mills		Form-Relieved Cutters		Circular Saws	
	in.	mm	in.	mm	in.	mm	in.	mm	in.	mm	in.	mm
Aluminum	0.020	0.50	0.016	0.40	0.012	0.30	0.010	0.25	0.006	0.15	0.005	0.13
Brass and bronze (medium)	0.012	0.30	0.010	0.25	0.007	0.18	0.006	0.15	0.004	0.10	0.003	0.08
Cast iron (medium)	0.016	0.40	0.013	0.33	0.010	0.25	0.008	0.20	0.005	0.13	0.004	0.10
Machine steel	0.016	0.40	0.013	0.33	0.009	0.23	0.008	0.20	0.005	0.13	0.004	0.10
Tool steel (medium)	0.014	0.35	0.011	0.28	0.008	0.20	0.007	0.18	0.004	0.10	0.004	0.10
Stainless steel	0.010	0.25	0.008	0.20	0.006	0.15	0.005	0.13	0.003	0.08	0.003	0.08

$$\text{Feed (in/min)} = N \times \text{CPT} \times \text{r/min}$$

N = number of teeth in the milling cutter
CPT = chip per tooth for a particular cutter and metal as given in Tables 61-2 and 61-3
r/min = revolutions per minute of the milling cutter

EXAMPLE

Find the feed in inches per minute using a 3.5 in-diameter 12-tooth helical cutter to cut machine steel (CS 80).

SOLUTION

First calculate the proper r/min for the cutter:

$$\text{r/min} = \frac{4 \times \text{CS}}{D} = \frac{4 \times 80}{3.5} = 91$$

$$\begin{aligned}\text{Feed (in./min)} &= N \times \text{CPT} \times \text{r/min} \\ &= 12 \times 0.010 \times 91 \\ &= 10.9 \quad \text{or} \quad 11\end{aligned}$$

Metric Calculations

The formula used to find work feed in millimeters per minute is the same as the formula used to find the feed in inches per minute, except that mm/min is substituted for in./min.

EXAMPLE

Calculate the feed in millimeters per minute for a 75 mm-diameter, six-tooth helical milling cutter when machining a cast iron workpiece (CS 60).

SOLUTION

First calculate the *r/min* of the cutter.

$$\text{r/min} = \frac{\text{CS} \times 320}{D} = \frac{60 \times 320}{75} = \frac{19{,}200}{75} = 256$$

$$\begin{aligned}\text{Feed (mm/min)} &= N \times \text{CPT} \times \text{r/min} \\ &= 6 \times 0.25 \times 256 \\ &= 384 \\ &= 384 \text{ mm/min}\end{aligned}$$

The calculated feeds would be possible only under ideal conditions. Under average operating conditions, the milling machine feed should be set to approximately one-third or one-half the amount calculated. The feed can then be gradually increased to the capacity of the machine and the finish desired.

Tables 61-2 and 61-3 give suggested feed per tooth for various types of milling cutters for roughing cuts under average conditions. For finishing cuts, the feed per tooth would be reduced to one-half or even one-third of the value shown.

Direction of Feed

One final consideration concerning feed is the direction in which the work is fed into the cutter. The most commonly used method of feeding is to feed the work against the rotation direction of the cutter (*conventional* or *up milling*) (Fig. 61-1). However, if the machine is equipped with a backlash eliminator, certain types of work can best be milled by *climb milling* (Fig. 61-2).

Although climb milling is not as widely used as conventional milling, it has certain advantages and disadvantages.

Advantages of Climb Milling

- It provides the best surface finish in machining most materials.

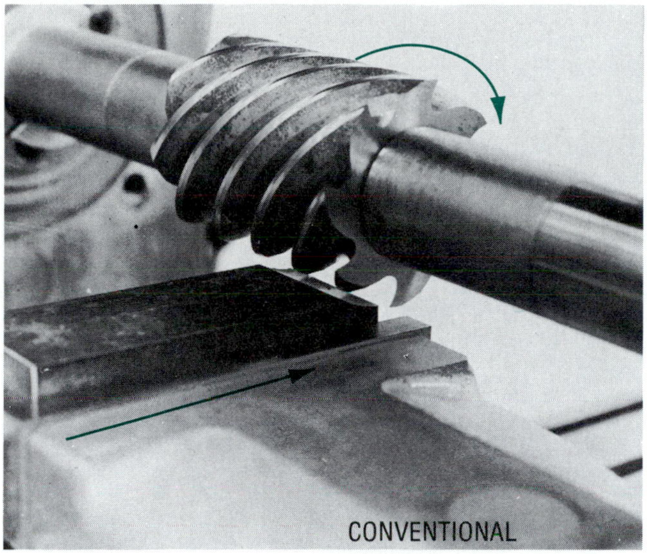
CONVENTIONAL

Figure 61-1 *Conventional or up milling.*

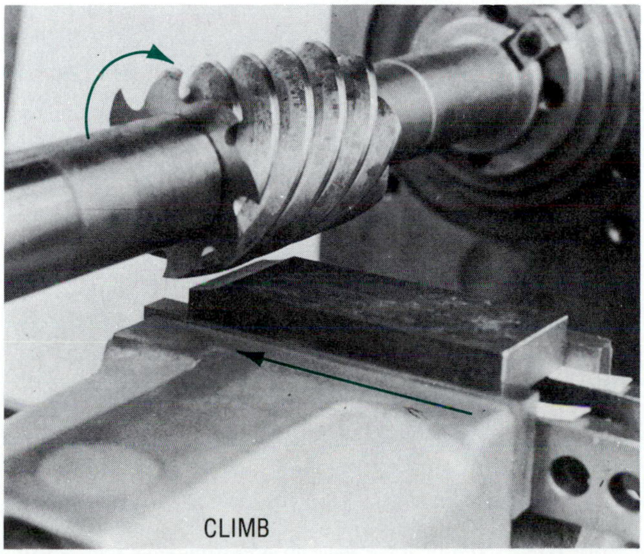
CLIMB

Figure 61-2 *Climb or down milling.*

- It is particularly suited for machining thin and hard-to-hold parts since the workpiece is forced against the table or holding device by the cutter.
- It allows work to be clamped less tightly.
- It maintains consistent parallelism and size, particularly on thin parts.
- It may be used where break-out at the edge of the workpiece could not be tolerated.
- It requires up to 20 percent less power.
- It may be used when cutting off stock or when milling deep, thin slots.

Disadvantages of Climb Milling

- *This method cannot be used unless the machine has a backlash eliminator* and the table gibs have been tightened.
- It cannot be used for machining castings or hot-rolled steel since the hard outer scale will damage the cutter.

DEPTH OF CUT

Where a smooth, accurate finish is desired, it is good milling practice to take a roughing and finishing cut. Roughing cuts should be deep, with a feed as heavy as the work and the machine will permit. Heavier cuts may be taken with helical cutters having fewer teeth, since they are stronger and have a greater chip clearance than cutters with more teeth.

Finishing cuts should be light, with a finer feed than is used for roughing cuts. The depth of the cut should be at least 1/64 in. (0.40 mm). Lighter cuts and extremely fine feeds are not advisable since the chip taken by each tooth will be thin and the cutter will often rub on the surface of the work, rather than bite into it, thus dulling the cutter. When a fine finish is required, the feed should be reduced rather than the cutter speeded up; more cutters are dulled by high speeds than by high feeds.

NOTE: To prevent damage to the finished surface (dwell marks), *never* stop the feed when the cutter is revolving over the workpiece. For the same reason, stop the cutter before returning the work to the starting position upon completion of the cut.

MILLING CUTTER FAILURE

The problem of obtaining maximum efficiency in the use of milling cutters is one which faces every user. The solution to this problem lies in constant awareness of factors that contribute to tool failure or poor performance, other than the style or type of end mill or cutter being used.

EXCESSIVE HEAT

Excessive heat (Fig. 61-3) is one of main causes of total cutting edge failure or shortened tool life. Heat, which is present in all milling operations, is caused by cutting edges rubbing on the workpiece and by chips sliding along the faces of teeth when they are separated from the part being milled. When this heat becomes excessive, the hardness of the cutting edges is affected, resulting in less resistance to wear and subsequently rapid dulling.

The heat problem is one of an ever-expanding cycle. As tool-wear lands increase, cutting temperatures rise because of increased friction. As cutting edges dull, more force is needed to part chips and the sliding pressure of chips on the faces of teeth increases, resulting in still more heat. As more heat is generated, the coolant, which is used to reduce temperature, becomes less efficient, temperatures rise higher, the hardness and strength of the cutting edges are further affected, and the cycle is repeated. Heat cannot be totally eliminated but it can be minimized by using properly designed and sharpened tools, by operating at the speeds and feeds recommended for the workpiece material, and by efficiently applying a suitable coolant.

ABRASION

Abrasion (Fig. 61-4) is a wearing-away action caused by the metallurgy of the workpiece. It dulls cutting edges and causes "wear lands" to develop around the periphery of a cutter. As dulling increases and wear lands grow wider, friction increases and greater force is required to keep the cutter cutting. The rapid rise in friction, heat, and rotational forces resulting from the abrasive nature of workpiece material can reach a point where the cutter ceases to function effectively or is totally destroyed. Since heat and abrasion are related, the suggestions for minimizing heat will also apply to abrasion. Because some materials are much more abrasive than others, it is extremely important for good tool life to follow recommendations specifically for correct CS and feeds.

SAND

Figure 61-4 *Abrasion dulls cutting edges. (Courtesy The Weldon Tool Company.)*

Figure 61-3 *Excessive heat shortens cutter life. (Courtesy The Weldon Tool Company.)*

CHIPPING OR CRUMBLING OF CUTTING EDGES

When cutting forces impose a greater load on cutting edges than their strength can withstand, small fractures occur and small areas of the cutting edges chip out (Fig. 61-5). The material that is left uncut by these chipped-out portions

Figure 61-5 *Too heavy a load on a cutting edge will cause it to chip. (Courtesy The Weldon Tool Company.)*

imposes a still greater cutting load on the following cutter tooth, aggravating the problem. This condition is progressive, and once started will lead to total cutter failure since chipped edges are dull edges that increase friction, heat, and horsepower requirements.

The major causes of chipping and fracturing of cutting edges are:

- Excessive feed per tooth (FPT)
- Poor cutter design
- Brittleness of the end mill due to improper heat treatment
- Running cutters backward
- Chattering due to a nonrigid condition in the fixture, workpiece, or machine
- Inefficient chip washout, which permits chips to be recut or compressed between workpiece surfaces and cutting edges
- Built-up edge break-away

CLOGGING

Some workpiece materials have a "gummy" composition that causes chips to be long, stringy, and compressible (Fig. 61-6). Chips from still other materials may have a tendency to cold-weld or gall to cutting edges and/or the faces of teeth. During the milling of these materials, very often the chips clog or jam into the flute area resulting in cutter breakage. This condition can be minimized by reducing the depth or width of cut, by reducing the FPT, by using tools with fewer teeth creating more chip space, and by more effectively applying a coolant under pressure that lubricates the tooth face and is directed so that it flushes out the flute area, keeping it free from chips. It might be necessary at times to use two coolant nozzles to accomplish this.

Figure 61-6 *Clogging reduces chip space and can cause a cutter to break. (Courtesy The Weldon Tool Company.)*

BUILT-UP EDGES

Built-up edges (Fig. 61-7) occur when particles of the material being machined cold-weld, gall, or otherwise adhere to the faces of teeth adjacent to the cutting edges. This process will continue until the "build-up" itself functions as the cutting edge. When this happens, more power is required and a poor surface finish is usually produced on the workpiece. Periodically, the built-up material will break away from the tooth face and go with either the chip or the workpiece. This intermittent break-away often takes with it a portion of the cutting edge, causing a complete tool breakdown with all of its attendant problems. This condition is quite prevalent on a tool such as a milling cutter because as each tooth engages the cut in an intermittent manner, the possibility of built-up material break-away and edge chipping is increased.

The built-up edge problem can often be moderated by reducing feed and/or depth of cut. However, the most effective solution to the problem is usually found in a forceful application of a good cutting fluid that gets into the area where the chip is being formed. Optimum results are gained when the coolant coats the cutting edges with a thin fluid or oxide layer that forms a cushion between the chip and the tool to prevent a built-up edge from forming.

WORK HARDENING OF THE WORKPIECE

Work hardening of the workpiece (Fig. 61-8) can cause milling cutter failure. This condition, sometimes called strain hardening, cold working, or glazing, is the result of the action of cutting edges deforming or compressing the surface of the workpiece causing a change in the work material structure that increases its hardness. Usually, this increase in hardness is evidenced by a smooth, highly glazed surface that resists the wetting action of coolants and offers

Figure 61-7 *Built-up edges result in poor cutting action. (Courtesy The Weldon Tool Company.)*

ICE WATER

Figure 61-8 *Work hardening of the workpiece can cause cutter failure. (Courtesy The Weldon Tool Company.)*

Figure 61-9 *The use of cutting fluid can reduce cratering. (Courtesy The Weldon Tool Company.)*

extreme resistance to the penetration of cutting edges. Fortunately, not all materials are subject to work hardening, but this condition can occur during machining most of the high-temperature and high-strength superalloys, all of the austenitic steels, and many of the highly alloyed carbon tool steels.

Since it is basically the rubbing action of cutting edges that causes work hardening on the surface of these materials, it is extremely important to use sharp tools operating at generous power feeds to keep rubbing contact between tool and work at a minimum. The use of proper CS is important, and climb cutting with a generous application of activated cutting oil is highly recommended. Avoid the use of dull milling cutters and light finishing cuts, and never permit a tool to dwell or rotate in contact with the workpiece without feeding.

Machines and work-holding fixtures should be massive and rigid, and tool overhang as short as possible in order to keep deflection during cut at a minimum. Should a surface that is to be milled be already glazed or workhardened from some previous cutting operation, it is very beneficial to break up the glaze by vapor honing or abrading the surface with coarse emery cloth. This will reduce the slipperiness of the surface and make it easier for tool cutting edges to bite in. Also, breaking the glaze on the surface will permit the coolant to give better wetting action, which is very helpful in prolonging tool life.

CRATERING

Cratering (Fig. 61-9) is caused by chips sliding on the tooth face adjacent to the cutting edge. This is the area of high heat and extreme abrasion due to high chip pressures. The sliding and curling of the chips erodes a narrow hollow or a groove into the tooth face. Once this cratering starts it can get progressively worse until it results in total tool failure. This problem can be minimized by efficiently applying a coolant that provides a high-pressure fluid film or a chemical oxide film on the tool to prevent metal-to-metal contact between chips and tooth face. Also beneficial are tool surface treatments properly applied that impart a high-abrasion-resisting superficial hardness to faces of the teeth.

REVIEW QUESTIONS

1. Name three factors that affect the efficiency of a milling operation.

CUTTING SPEED

2. List six factors which must be considered when selecting the proper r/min for milling.
3. At what maximum speed should a 3½ in. cemented-carbide milling cutter revolve to machine cast iron?
4. At what r/min should a 115 mm high-speed steel cutter revolve to machine a piece of tool steel?
5. What rules should be observed to obtain best results when using milling cutters?

FEED

6. Name three factors which determine milling feed.
7. Define *chip per tooth.*
8. Calculate the feed (in./min) for an eight-tooth, 3 in.-diameter cemented carbide face milling cutter to cut cast iron.
9. Calculate the feed (m/min) for a twelve-tooth 90 mm-diameter high-speed steel helical milling cutter to machine aluminum.

DEPTH OF CUT

10. What is considered to be a proper roughing cut?
11. Why is a very light cut not desirable as a finish cut?

MILLING CUTTER FAILURE

12. List seven main causes of milling cutter failure and state how each can be minimized.

UNIT 62

Milling Machine Setups

Before any operation is performed on a milling machine, it is important that the machine be properly set up. Proper setup of the machine will prolong the life of the machine and its accessories and produce accurate work. The operator should always follow safety procedures to prevent injury and to avoid damaging the machines and spoiling the workpiece.

OBJECTIVES

After completing this unit you should be able to:
1. Mount and remove a milling machine arbor
2. Mount and remove a milling cutter
3. Align the milling machine table and vise

MILLING MACHINE SAFETY

The milling machine, like any other machine, demands the total attention of the operator and a thorough understanding of the hazards associated with its operation. The following precautions should be taken during operation of a milling machine.

1. Be sure that the work and the cutter are mounted securely before taking a cut.
2. Always wear safety glasses.
3. When mounting or removing milling cutters, always hold them with a cloth to avoid being cut.
4. When setting up work, move the table as far as possible from the cutter to avoid cutting your hands.
5. Be sure that the cutter and machine parts clear the work (Fig. 62-1).
6. *Never* attempt to mount, measure, or adjust work until the cutter has *completely* stopped.
7. Keep your hands, brushes, and rags away from a revolving milling cutter *at all time.*
8. When using milling cutters, do not use an excessively heavy cut or feed. This can cause a cutter to break, and the flying pieces could cause serious injury.

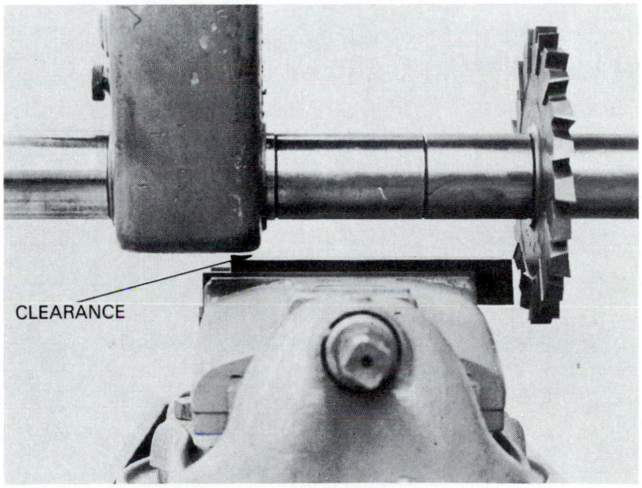
CLEARANCE
Figure 62-1 *Before making a cut be sure that the arbor and arbor support will clear the workpiece.*

9. Always use a *brush*, not a rag, to remove the cuttings after the cutter has stopped revolving.
10. Never reach over or near a revolving cutter; keep hands at least 12 in. (300 mm) from a revolving cutter.
11. Keep the floor around the machine free from chips, oil, and cutting fluid (Fig. 62-2).

MILLING MACHINE SETUPS

To prolong the life of a milling machine and its accessories and to produce accurate work, the following actions should be taken when milling machine setups are being made.

1. Prior to mounting any accessory or attachment, check to see that both the machine surface and the accessory are free from dirt and chips.
2. Do not place the tools, cutters, or parts on the milling machine table. Place them on a piece of masonite, board, or a bench kept for this purpose to prevent damaging the table or machined surfaces (Fig. 62-3).
3. When mounting cutters, be sure to use keys on all but slitting saws.

Figure 62-2 *Poor housekeeping can cause tripping and slipping accidents.*

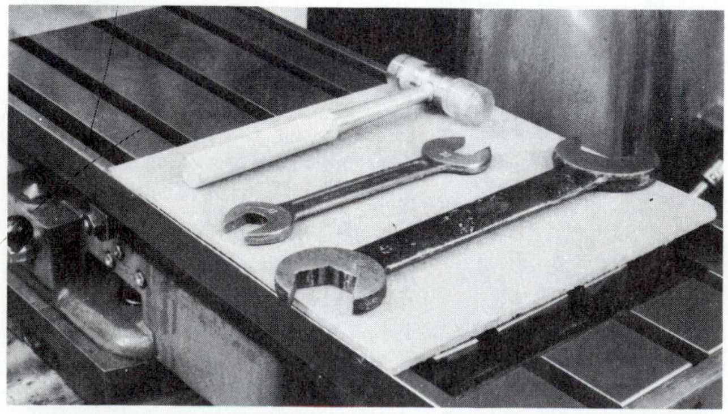

Figure 62-3 *Place tools on a piece of masonite or plywood to protect the machine table.*

4. Check that the arbor spacers and bushings are clean and free from burrs.
5. When tightening the arbor nut, take care to make it only hand tight with a wrench. *NEVER USE A HAMMER ON THE NUT OR HOLDING DEVICE ON ANY MACHINE.* The use of a hammer and wrench to tighten nuts will strip the threads and bend or damage the accessory or part.
6. When work is mounted in a vise, tighten the vise securely by hand and tap it into place with a *LEAD* OR *SOFT-FACED HAMMER.*

MILLING HINTS

The following hints should help prevent many problems that could affect the accuracy and the efficiency of a milling operation.

1. Treat milling cutters with care; they can be damaged through careless use, handling, and storage.
2. Always examine the condition of the milling cutter cutting edges before use; if necessary resharpen the cutters. *NEVER USE DULL MILLING CUTTERS.*
3. Use the correct speeds and feeds for the size of the cutter and the type of work material to be cut.

   Too high a speed quickly dulls a cutter; too low a speed reduces productivity and may break the cutter teeth.
4. For maximum cutter life, use lower cutting speeds and increase the feed rate as much as possible.
5. Mount end mills as short as possible to prevent breakage.
6. Position arbor supports as close as possible to milling cutters for maximum rigidity.
7. Climb mill wherever possible since it permits the use of higher speeds and generally improves the surface finish.

NOTE: The machine must be equipped with a backlash eliminator for climb milling.

8. Always direct the cutting force towards the solid part of the machine, vise, or fixture.
9. Make sure that the milling cutter is rotating in the proper direction for the cutting action of the teeth.
10. Use coarse tooth cutters for roughing cuts; they provide better chip clearance.
11. Stop the machine whenever there is any discoloration of the chips; it generally indicates poor cutting action. Examine the cutter.
12. Use a good supply of cutting fluid for all milling operations. This prolongs the life of the cutter and generally produces more accurate work.

MOUNTING AND REMOVING A MILLING MACHINE ARBOR

The milling arbor is used to hold the cutter during the machine operation. When mounting or removing an arbor, follow the proper procedure in order to preserve the accuracy of the machine. An improperly mounted arbor may damage the taper surfaces of the arbor or machine spindle, cause the arbor to bend, or make the cutter run out of true.

THE ARBOR ASSEMBLY

The milling cutter is driven by a key, which fits into the keyways on the arbor and cutter (Fig. 62-4). This prevents the cutter from turning on the arbor. Spacer and bearing bushings hold the cutter in position on the arbor after the nut has been tightened. The tapered end of the arbor is held securely in the machine spindle by a draw-in bar (Fig. 62-5). The outer end of the arbor assembly is supported by the bearing bushing and the arbor support.

TO MOUNT AN ARBOR

1. Clean the tapered hole in the spindle and the taper on the arbor using a clean cloth.
2. Check to make sure that there are no cuttings or burrs in the taper which would prevent the arbor from running true.
3. Check the bearing bushing and remove any burrs using a honing stone.

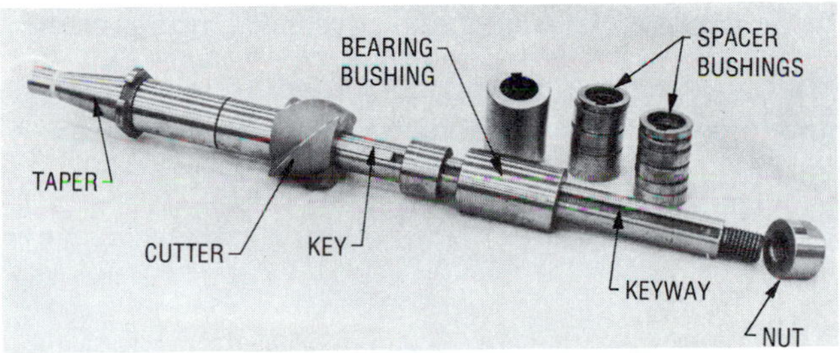
BEARING BUSHING
SPACER BUSHINGS
TAPER
CUTTER
KEY
KEYWAY
NUT

Figure 62-4 *The milling machine arbor holds and drives the milling cutter.*

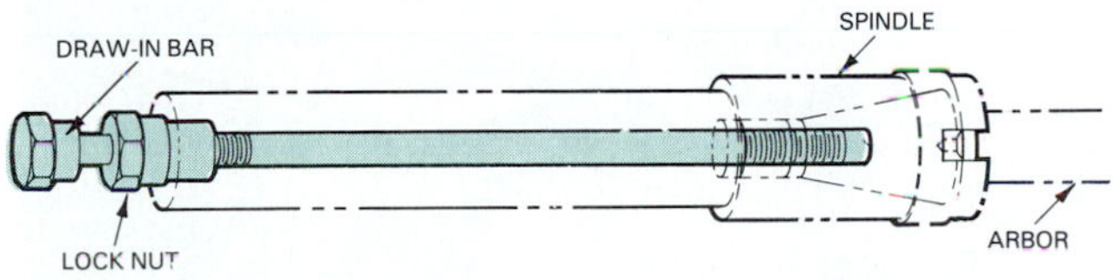
DRAW-IN BAR
SPINDLE
LOCK NUT
ARBOR

Figure 62-5 *The draw-in bar holds the arbor firmly in place in the spindle.*

Figure 62-6 *Removing a milling machine arbor.*

4. Place the tapered end of the arbor in the spindle and align the spindle driving lugs with the arbor slots.
5. Place the right hand on the draw-in bar (Fig. 62-6) and turn the thread into the arbor approximately 1 in. (25 mm).
6. Tighten the draw-in bar lock nut securely against the back of the spindle (Fig. 62-7).

TO REMOVE AN ARBOR

1. Remove the milling machine cutter.
2. Loosen the lock nut on the draw-in bar approximately two turns.
3. With a soft-faced hammer, sharply strike the end of the draw-in bar until the arbor taper is free (Fig. 62-8).
4. With one hand, hold the arbor and unscrew the draw-in bar from the arbor with the other hand (Fig. 62-6).
5. Carefully remove the arbor from the tapered spindle so as not to damage the spindle or arbor tapers.
6. Leave the draw-in bar in the spindle for further use.
7. Store the arbor in a suitable rack to prevent it from being damaged or bent (Fig. 62-9).

Figure 62-7 *Locking the arbor into the spindle with a draw-in bar.*

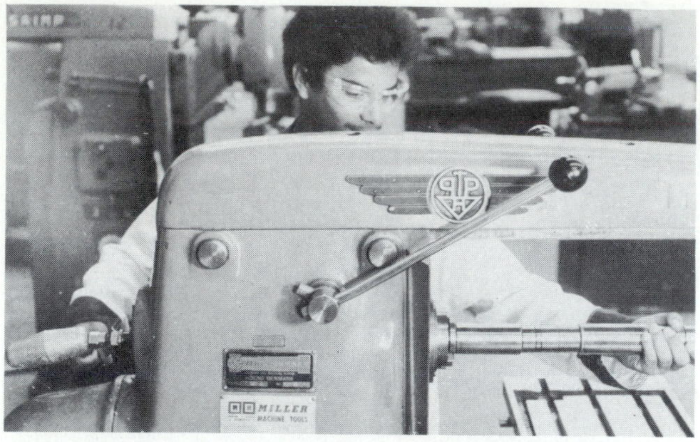

Figure 62-8 *Removing the arbor from the spindle.*

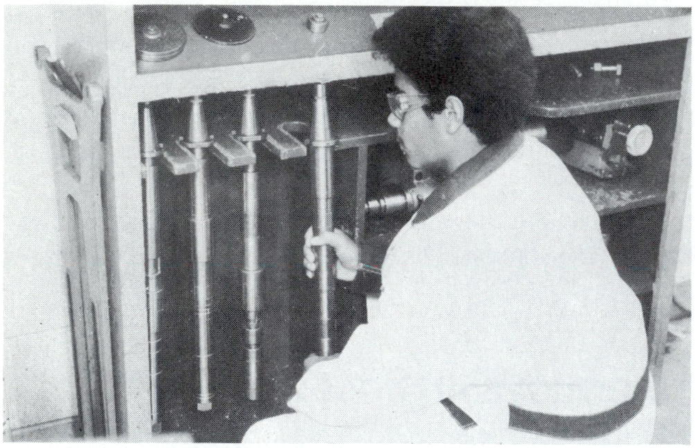

Figure 62-9 *Storing the arbor in a suitable rack.*

MOUNTING AND REMOVING MILLING CUTTERS

Milling cutters must be changed frequently to perform various operations, so it is important that certain sequences be followed in order not to damage the cutter, the machine, or the arbor.

TO MOUNT A MILLING CUTTER

1. Remove the arbor nut and collar and place them on a piece of masonite (Fig. 62-3).
2. Clean all spacing collar surfaces of cuttings and burrs.
3. Set the machine to the slowest spindle speed.
4. Check the direction of the arbor rotation by starting and stopping the machine spindle.
5. Slide the spacing collars on the arbor to the position desired for the cutter (Fig. 62-10).
6. Fit a key into the arbor keyway at the position where the cutter is to be located.
7. Hold the cutter with a cloth and mount it on the arbor,

Figure 62-10 *Use spacing collars to set the cutter in position.*

being sure that the cutter teeth point in the direction of the arbor rotation (Fig. 62-10).

8. Slide the arbor support in place and be sure that it is on a bearing bushing on the arbor.
9. Put on additional spacers leaving room for the arbor nut. *TIGHTEN THE NUT BY HAND.*
10. Lock the arbor support in position (Fig. 62-11).
11. Tighten the arbor nut firmly with a wrench, using only hand pressure.
12. Lubricate the bearing collar in the arbor support.
13. Make sure that the arbor and arbor support will clear the work (Fig. 62-1).

TO REMOVE A MILLING CUTTER

1. Be sure that the arbor support is in place and supporting the arbor on a bearing bushing before using a wrench on the arbor nut. This will prevent bending the arbor.
2. Clean all cuttings from the arbor and cutter.
3. Set the machine to the lowest spindle speed.
4. Loosen the arbor nut with a properly fitting wrench.

NOTE: Most threads on an arbor are left-hand and therefore loosen in a clockwise direction.

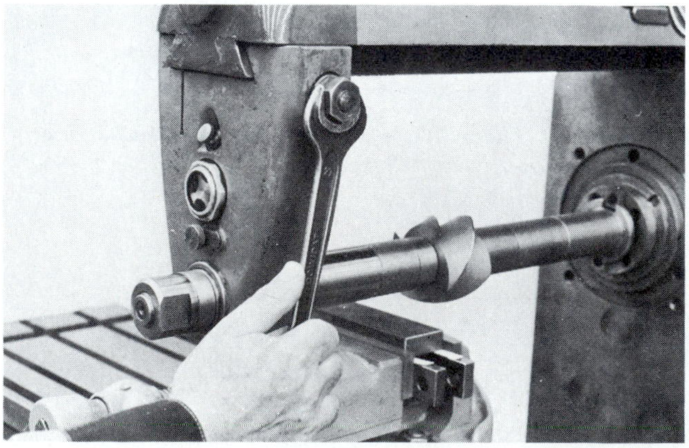

Figure 62-11 *Tightening the arbor support.*

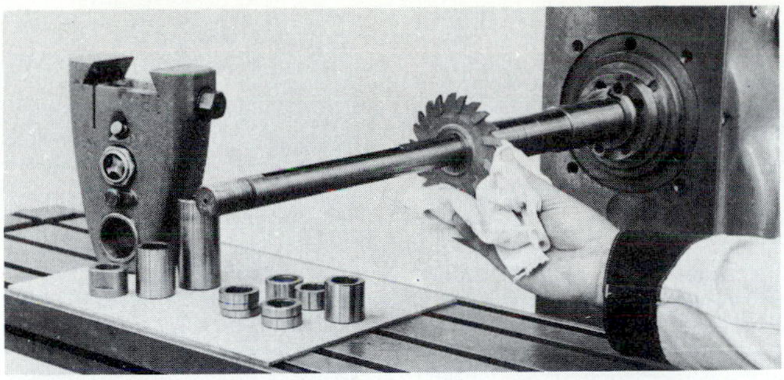

Figure 62-12 *When changing a cutter, protect the milling machine table with a piece of masonite.*

5. Loosen the arbor support and remove it from the overarm.
6. Remove the nut, spacers, and cutter. Place them on a board (Fig. 62-12) and not on the table surface.
7. Clean the spacer and nut surfaces and replace them on the arbor. *Do not use a wrench to tighten the arbor nut at this time.*
8. Store the cutter in a proper place to prevent the cutting edges from being damaged.

ALIGNING THE TABLE ON A UNIVERSAL MILLING MACHINE

If the workpiece is to be machined accurately to a layout or have cuts made square or parallel to a surface, it is always good practice to align the table of a universal milling machine prior to aligning the vise or fixture.

NOTE: If a long keyway is to be milled in a shaft, it is of utmost importance to align the table. If it is not aligned properly, the keyway will not be milled parallel to the axis of the shaft.

PROCEDURE

1. Clean the table and the face of the column thoroughly.
2. Mount a dial indicator *on the table* by means of a magnetic base or any suitable mounting device (Fig. 62-13).
3. Move the table toward the column until the dial indicator registers approximately one-quarter of a revolution.
4. Set the indicator bezel to zero (0) by turning the dial.
5. Using the table feed handwheel, move the table along the width of the column.
6. Note the reading on the dial indicator and compare it to the zero (0) reading at the other end of the column.
7. If there is any movement of the indicator hand, loosen the locking nuts on the *swivel table housing.*
8. Adjust the table for half the difference of the needle movement and lock the table housing.

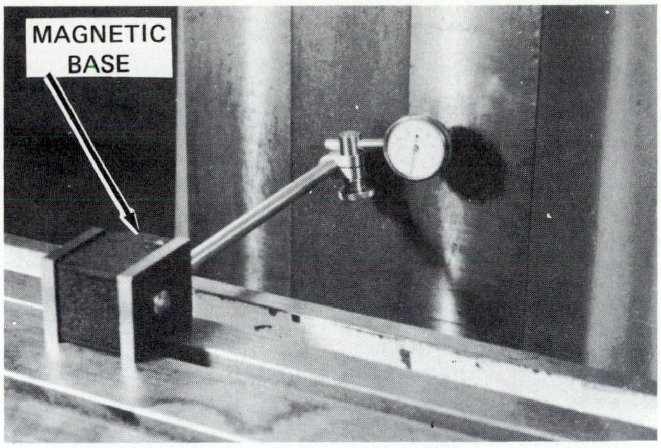
MAGNETIC
BASE

Figure 62-13 *Aligning a universal milling machine table.*

9. *Recheck* the table for alignment and adjust if necessary until there is no movement in the indicator as it is moved along the column.

NOTE: *Always indicate from the table to the face of the column, never* from the column to the table.

ALIGNING THE MILLING MACHINE VISE

Whenever accuracy is required on the workpiece, it is necessary to align the device which holds the workpiece. This may be a vise, angle plate, or a special fixture. Since most work is held in a vise, the method of aligning this accessory will be outlined.

TO ALIGN THE VISE PARALLEL TO THE TABLE TRAVEL

1. Clean the surface of the table and the base of the vise.
2. Mount and fasten the vise on the table.
3. Swivel the vise until the solid jaw is approximately parallel with the table slots.
4. Mount an indicator to the arbor or cutter (Fig. 62-14).
5. Make sure the solid jaw is clean and free from burrs.
6. Adjust the table until the indicator registers about one-quarter of a revolution against a parallel held between the jaws of the vise.
7. Set the bezel to zero (0) (Fig. 62-15A).
8. Move the table along for the length of the parallel and note the reading of the indicator (Fig. 62-15B). Compare it to the zero (0) reading at the other end of the parallel.
9. Loosen the nuts on the upper or swivel part of the vise.
10. Adjust the vise to half the difference of the indicator

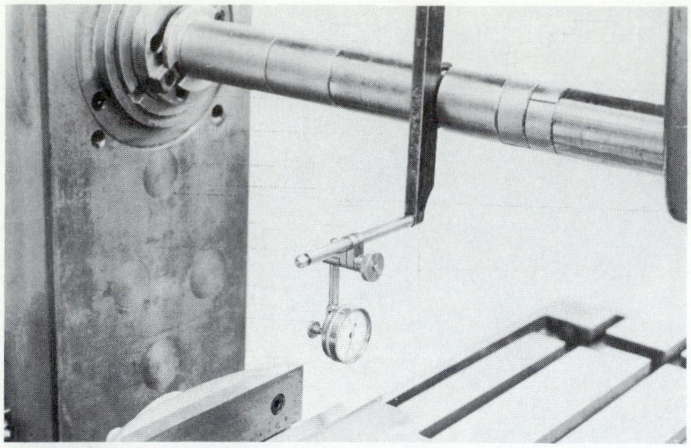

Figure 62-14 *A dial indicator mounted on the arbor for alignment purposes.*

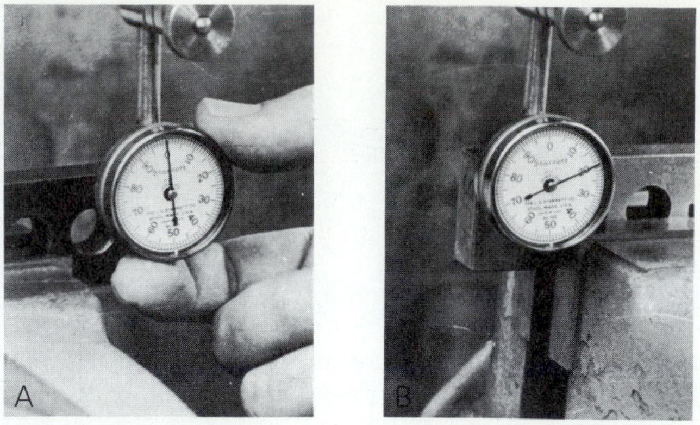
A
B

Figure 62-15 *Accurately aligning a vise parallel to table travel.*

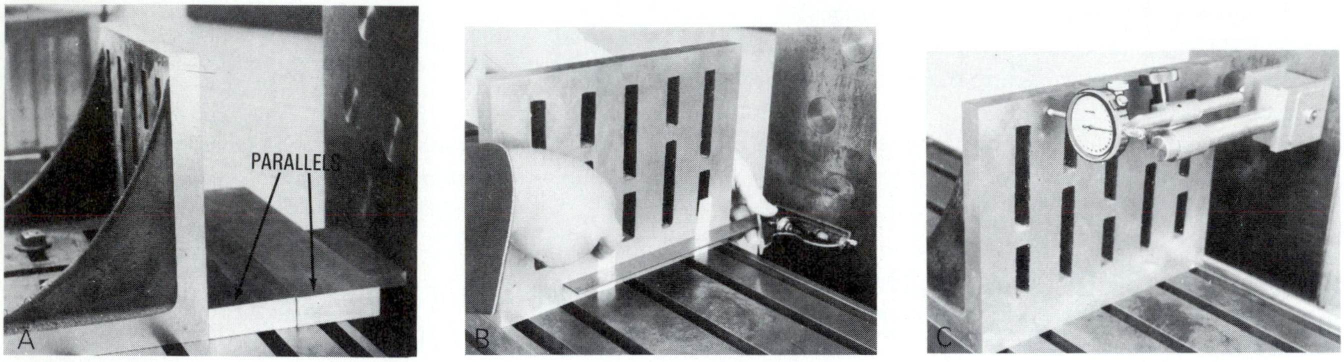
PARALLELS
A
B
C

Figure 62-16 *(A) Aligning an angle plate parallel to the column; (B) and (C) aligning an angle plate at right angles to the column.*

readings by tapping it by hand or with a *soft-faced* hammer in the appropriate direction.

NOTE: *Never tap the vise so that the parallel moves against the indicator plunger.* This will damage the indicator.

11. *Recheck* the vise for alignment and adjust if necessary until there is no movement in the indicator as it is moved along the parallel.

Methods of aligning an angle plate are shown in Fig. 62-16.

REVIEW QUESTIONS

1. Why is the proper setup of the machine important?

SAFETY

2. What can happen if an excessively heavy cut is taken or too fast a feed used?
3. How far should the hands be kept from a revolving cutter?

MOUNTING AND REMOVING ARBORS

4. What is the purpose of an arbor?
5. How is the arbor held securely in the machine spindle?
6. How is the outer end of the arbor supported?
7. How should the arbor be stored when it is not in the machine?
8. In which direction should the cutter teeth point when it is mounted on an arbor?
9. Why should the arbor support be in place before attempting to remove a cutter from the arbor?

ALIGNING THE TABLE AND VISE

10. Why is the table alignment so important to most milling operations?
11. Why must the indicator be mounted on the table and not the column or the arbor when aligning a universal table?
12. By means of suitable sketches, show how the vise may be aligned:
   a. Parallel to the line of longitudinal table travel
   b. At right angles to the milling machine column

UNIT

63

Milling Operations

The milling machine is one of the most versatile machine tools found in industry. The wide variety of operations which can be performed depends on the type of machine used, the cutter used, and the attachments and accessories available for the milling machine. The two basic types of milling are:

1. *Plain milling,* where the surface cut is parallel to the periphery of the cutter. These surfaces may either be flat or formed (Fig. 63-1).
2. *Face milling,* where the surface cut is at right angles to the axis of the cutter (Fig. 63-2).

OBJECTIVES

After completing this unit you will be able to:
1. Set a cutter to the proper depth
2. Mill a flat surface on a workpiece
3. Perform operations such as face, side, straddle, and gang milling

Figure 63-1 *Milling the surface of a workpiece held in a fixture.*

Figure 63-2 *Face milling on a manufacturing-type milling machine. (Courtesy Cincinnati Milacron Inc.)*

SETTING THE CUTTER TO THE WORK SURFACE

Before setting a depth of cut, the operator should check that the work and the cutter are properly mounted and that the cutter is revolving in the proper direction.

TO SET THE CUTTER TO THE WORK SURFACE

1. Raise the work to within ¼ in. (6 mm) of the cutter and directly under it.
2. Hold a long piece of thin paper on the surface of the work (Fig. 63-3).

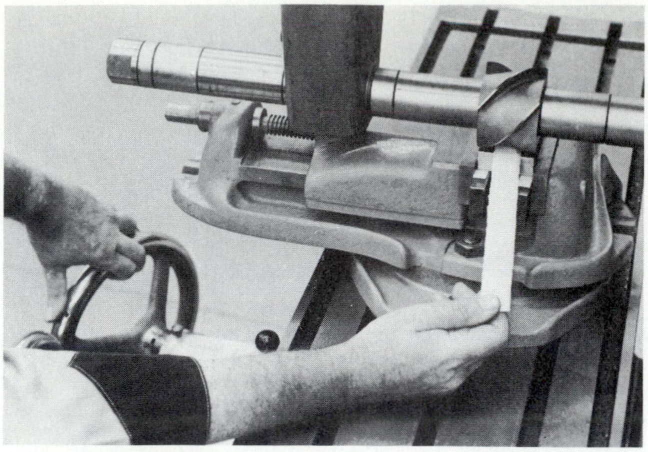

Figure 63-3 *Setting the cutter to the workpiece.*

NOTE: *Have the paper long enough to prevent the fingers from coming in contact with the revolving cutter.*

3. Start the cutter rotating.
4. With the left hand on the elevating screw handle, move the work up *slowly* until the cutter grips the paper.
5. Stop the spindle.
6. Move the machine table so that the cutter just clears the end of the workpiece (Fig. 63-4).
7. Raise the knee 0.002 in. (0.05 mm) for paper thickness.
8. Set the graduated collar on the elevating screw handle to zero (0). Do not move the elevating screw handle.
9. Move the work clear of the cutter and raise the table to the desired depth of cut.

NOTE: *If the knee is moved up beyond the desired amount, turn the handle backward one-half turn and then come up to the required line. This will take up the backlash in the thread movement.*

This method can also be used when the edge of a cutter is being set to the side of a piece of work. In this case, the paper will be placed between the side of the cutter and the side of the workpiece.

MILLING A FLAT SURFACE

One of the most common operations performed on a milling machine is that of machining a flat surface. Flat surfaces are generally machined on a workpiece with a helical milling cutter. The work may be held in a vise or clamped to a table. Proceed as follows:

1. Remove all burrs from all edges of the work with a mill file (Fig. 63-5).
2. Clean the vise and workpiece.
3. Align the vise to the column face of the milling machine, using a dial indicator.
4. Set the work in the vise, using parallels and paper feelers under each corner to make sure that the work is seated on the parallels.
5. Tighten the vise securely by hand. *DO NOT USE A HAMMER ON THE VISE WRENCH.*
6. Tap the work *lightly* at the four corners with a soft-faced hammer until the paper feelers are tight between the work and the parallels (Fig. 63-6).
7. Select a plain helical cutter wider than the work to be machined.
8. Mount the cutter on the arbor for conventional milling (Fig. 63-7).
9. Set the proper speed for the size of cutter and the type of work material.
10. Set the feed to approximately 0.003 to 0.005 (0.08 to 0.13 mm) chip per tooth (CPT).
11. Start the cutter and raise the work, using a paper feeler between the cutter and the work (Fig. 63-3).
12. Stop the spindle when the cutter just cuts the paper.
13. Raise the knee 0.002 in. (0.05 mm) for paper thickness.

Figure 63-4 *Make sure that the cutter clears the end of the workpiece before setting the depth of cut.*

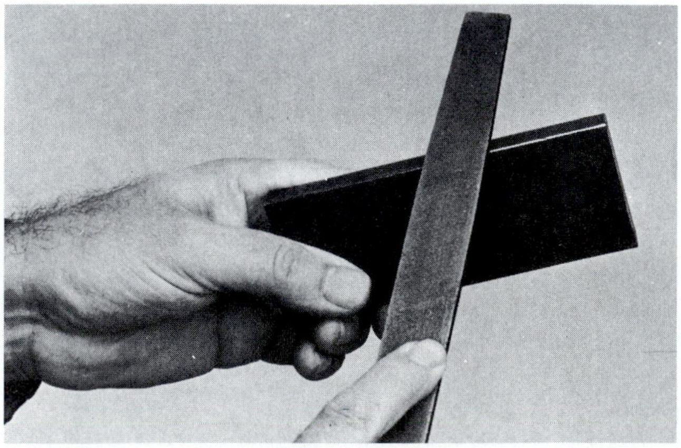

Figure 63-5 *Remove all burrs before setting a workpiece in a vise.*

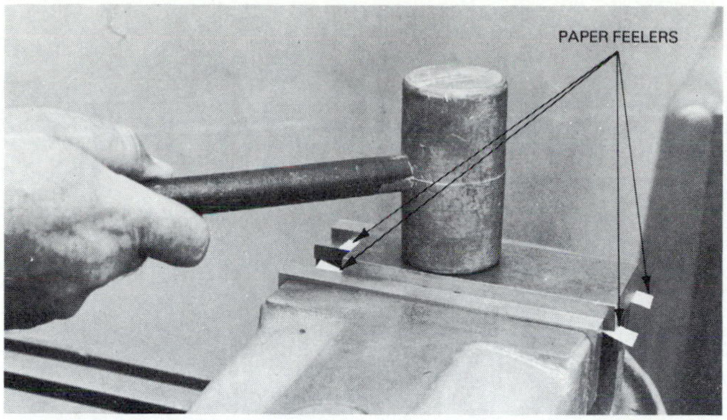
PAPER FEELERS

Figure 63-6 *Tap the workpiece down with a soft-faced hammer until all paper feelers are tight.*

Figure 63-7 *Most stock is machined by conventional milling.*

14. Set the graduated collar on the elevating screw handwheel to zero (0) (Fig. 63-8).
15. Move the work clear of the cutter and set the depth of cut using the graduated collar.
16. For roughing cuts use a depth of not less than ⅛ in. (3 mm) and 0.010 to 0.025 in. (0.25 to 0.63 mm) for finish cuts.
17. Set the table dogs for the length of cut.
18. Engage the feed and cut side 1.
19. Set up and cut the remaining sides as required.

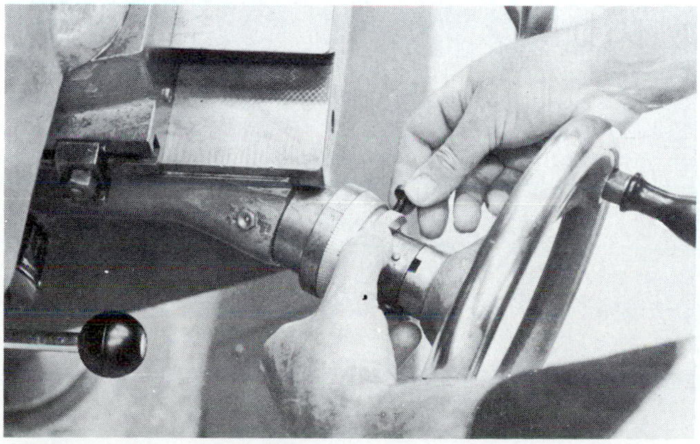

Figure 63-8 *After setting the cutter to the workpiece, set the graduated collar to zero.*

NOTE: See Unit 69 for the setup procedures if the four sides of a block must be machined.

FACE MILLING

Face milling is the process of producing a flat vertical surface at right angles to the cutter axis (Fig. 63-9). The cutters used for face milling are generally inserted-tooth cutters or shell end mills. *Face milling cutters* are made in sizes of 6 in. (150 mm) diameter and over; cutters less than 6 in. (150 mm) diameter are usually called *shell end mills.*

1. When face milling a large surface, use an inserted-tooth cutter and mount it into the spindle of the machine.
2. When milling smaller surfaces, the face milling cutter should be about 1 in. (25 mm) larger than the width of the workpiece.
3. Set the speeds and feeds for the type of cutter and the material being cut.
4. Set up the work on the milling machine, being sure that the work-holding clamps do not interfere with the cutting action.
5. Use cutting fluid if the cutter material or work will allow.
6. Take as large a roughing cut as possible to bring the surface close to within $1/32$ to $1/16$ in. (0.80 to 1.5 mm) of the finish size.
7. Set the depth of the finish cut and machine the surface to size.
8. After completing the operation, clean and store the cutter and clamping equipment in their proper places.

Figure 63-9 *Face milling produces flat, vertical surfaces.*

SIDE MILLING

Side milling is often used to machine a vertical surface on the sides or the ends of a workpiece (Fig. 63-10).

Proceed as follows:

1. Set up the work in a vise and on parallels.

NOTE: Be sure that the layout line on the surface to be cut extends about ½ in. (13 mm) beyond the edge of the vise and the parallels to prevent the cutter, vise, or parallels from being damaged.

2. Tighten the vise securely by hand. *DO NOT USE A HAMMER ON THE VISE WRENCH.*
3. Tap the work *lightly* at the four corners with a soft-faced hammer until the paper feelers are tight between the work and the parallels.
4. Mount a side milling cutter as close to the spindle bearing as possible to provide maximum rigidity when milling.
5. Set the proper speed and feed for the cutter being used.
6. Start the machine and move the table until the top corner of the work just touches the revolving cutter. Make sure that the cutter is rotating in the proper direction.
7. Set the crossfeed graduated collar to zero (0).
8. With the table handwheel, move the work clear of the cutter.
9. Set the required depth of cut with the crossfeed handle.
10. Lock the saddle to prevent movement during the cut.
11. Take the cut across the surface.

CENTERING A CUTTER TO MILL A SLOT

1. Locate the cutter as close to the center of the work as possible.
2. Using a steel square and rule (Fig. 63-11) or a gage block, adjust the work to the center by using the crossfeed screw dial.
3. Lock the saddle to prevent movement during the cut.
4. Move the work clear of the cutter and set the depth of cut.

Figure 63-10 *Side milling a vertical surface.*

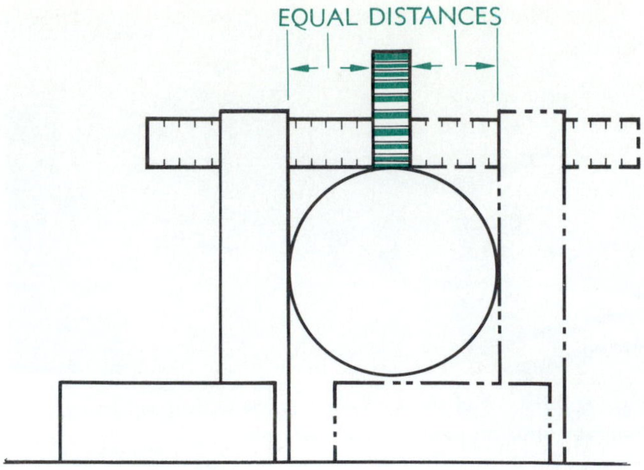
EQUAL DISTANCES

Figure 63-11 *Centering the cutter over a round shaft.*

5. Proceed to cut the slot using the same methods as for milling a flat surface.

STRADDLE MILLING

Straddle milling (Fig. 63-12) involves the use of two side milling cutters to machine the opposite sides of a workpiece parallel in one cut. The cutters are separated on the arbor by a spacer or spacers of the required length so that the distance between the inside faces of the cutters is equal to the desired size. An adjustable micrometer arbor spacer may be used to vary the distance between the two cutters and also to compensate for the wear or the regrinding of the side milling cutters. Applications of straddle milling are the milling of square and hexagonal heads on bolts.

Figure 63-12 *Straddle milling.*

The following points should be observed when straddle milling.

1. Select two sharp side milling cutters, preferably staggered tooth, of the suitable sizes.
2. Mount the cutters, with suitable arbor spacers, as close to the column as the work will permit.
3. Mount the arbor support as close to the cutters as possible to provide rigidity for the cutters and arbor.
4. Center the cutters on the workpiece in the proper location.
5. Tighten the saddle lock to prevent any movement during a cut.
6. Set the cutter to the work surface.
7. Move the table so that the cutter clears the end of the workpiece.
8. Set the depth of cut required and tighten the knee clamp.
9. Set the proper speeds and feeds for the cutter size and the type of work material; check cutter rotation.
10. Use a good supply of cutting fluid and complete the straddle milling operation in one cut.

GANG MILLING

Gang milling (Fig. 63-13) is a fast method of milling used a great deal in production work. It is performed by using two or more cutters on the arbor to produce the desired shape. The cutters may be a combination of plain and side milling cutters. If more than one helical milling cutter is used, a right- and left-hand helix cutter should be used to offset the thrust created by this type of cutter and to minimize the possibility of chatter. If several cutters are used and the slots are to be the same size, it is important that the diameter and width of each cutter be the same.

The following points should be observed to avoid problems during a gang milling operation.

1. Select cutters which are as close as possible to the same size and the same number of teeth. This will allow maximum speeds and feeds to be used.
2. Mount the correct cutters on the arbor as close to the machine column as possible.
3. Fasten the arbor support as close to the cutters as the work will permit.
4. Set the spindle speed to suit the largest-diameter cutter.
5. Be sure that the work is fastened securely and that the workholding devices will not come in contact with the cutters.
6. Use a good flow of cutting fluid to assist the cutting action and produce a good surface finish.

Figure 63-13 *Gang milling.*

SAWING AND SLITTING

Metal-slitting saws may be used for milling narrow slots and for cutting off work (Fig. 63-14). Plain slitting saws, because of their thin cross section, are rather fragile cutting tools, and if they are not used carefully they will break easily.

To get the maximum life from a slitting saw take the following precautions.

Figure 63-14 *Sawing and slitting. (Courtesy Kostel Enterprises Limited.)*

1. Never key a slitting saw on the arbor unless it is mounted in a special flanged mounting collar. If the slitting saw is keyed and jams in the work, it will rotate cutting through the key, making it difficult to remove the broken saw.

NOTE: The arbor nut must be drawn up as tightly as possible by *hand* only.

2. When selecting a slitting saw, choose one with the smallest diameter that will permit adequate clearance between the arbor collars or supports and the clamping bolts, holding device, or workpiece.
3. Mount the saw close to the column face and have the outer arbor support as close as possible to the saw.
4. Always use a sharp cutter.
5. Be sure that the table gibs are drawn up to eliminate any play between the table and the saddle.
6. Operate saws at approximately one-quarter to one-eighth the feed per tooth used for side milling cutters.
7. When sawing or slitting fairly long or deep slots, it is advisable to climb-mill to prevent the cutter from crowding sideways and breaking. The table should be fed carefully by hand and the backlash eliminator engaged.

REVIEW QUESTIONS

SETTING CUTTER TO WORK

1. Why should a long, thin piece of paper be used for setting a cutter to the work surface?
2. What should be done when the knee is moved beyond the desired amount?

MILLING A FLAT SURFACE

3. List the steps required to set work in a vise for milling.
4. Explain how the cutter should be set to the surface of the work.

FACE MILLING

5. Describe the process of face milling.
6. How large should a face milling cutter be in relation to the width of the work?

SIDE MILLING

7. For what purpose is side milling often used?
8. How should the work be set up in a vise for side milling?
9. Why should the table saddle be locked before machining?
10. Explain how a cutter can be aligned with the center of a round workpiece.

STRADDLE AND GANG MILLING

11. What is the difference between straddle and gang milling?
12. Why should the arbor support be mounted as close to the cutters as possible?

SAWING AND SLITTING

13. List five of the most important precautions for sawing or slitting.

UNIT 64

The Indexing or Dividing Head

The *indexing* or *dividing head* is one of the most important attachments for the milling machine. It is used to divide the circumference of a workpiece into equally spaced divisions when milling gears, splines, squares, and hexagons. It may also be used to rotate the workpiece at a predetermined ratio to the table feed rate to produce cams and helical grooves on gears, drills, reamers, and other parts.

OBJECTIVES

After completing this unit you will be able to:

1. Calculate and mill flats by simple and direct indexing
2. Calculate the indexing necessary with a wide-range divider
3. Calculate the indexing necessary for angular and differential indexing

INDEX HEAD PARTS

The universal dividing head set consists of the *headstock* with *index plates*, headstock *change gears* and *quadrant*, *universal chuck*, *footstock*, and the *center rest* (Fig. 64-1). A *swiveling block* mounted in the *base* enables the headstock to be tilted from 5° below the horizontal position to 10° beyond the vertical position. The side of the base and the block are graduated to indicate the angle of the setting. Mounted in the swiveling block is a *spindle*, with a *40-tooth worm wheel* attached, which meshes with a worm (Fig. 64-2). The worm, at right angles to the spindle, is connected to the *index crank*, the pin of which engages in the *index plate* (Fig. 64-3). A *direct indexing plate* is attached to the front of the spindle.

A 60° center may be inserted into the front of the spindle, and a *universal chuck* may be threaded onto the end of the spindle.

The *footstock* is used in conjunction with the headstock to support work held between centers or the end of work held in a chuck. The footstock center may be adjusted longitudinally to accommodate various lengths of work and may be raised or lowered off center. It may also be tilted out of parallel with the base when cuts are being made on tapered work.

Long, slender work held between centers is prevented from bending by the *adjustable center rest*.

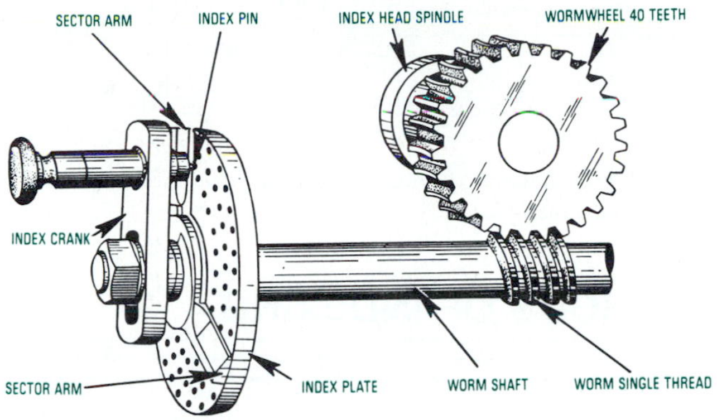
SECTOR ARM
INDEX PIN
INDEX HEAD SPINDLE
WORMWHEEL 40 TEETH
INDEX CRANK
SECTOR ARM
INDEX PLATE
WORM SHAFT
WORM SINGLE THREAD
Figure 64-2 *Section through a dividing head, showing the worm wheel and worm shaft.*

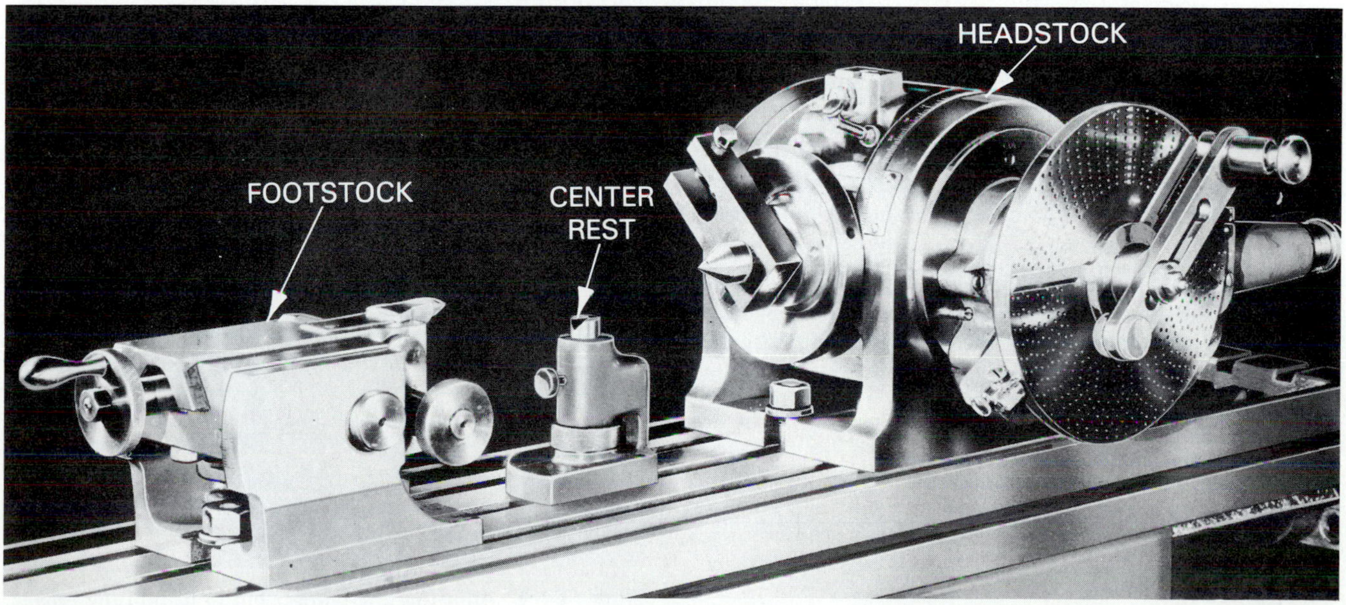
HEADSTOCK
FOOTSTOCK
CENTER REST
Figure 64-1 *A universal dividing head set.*

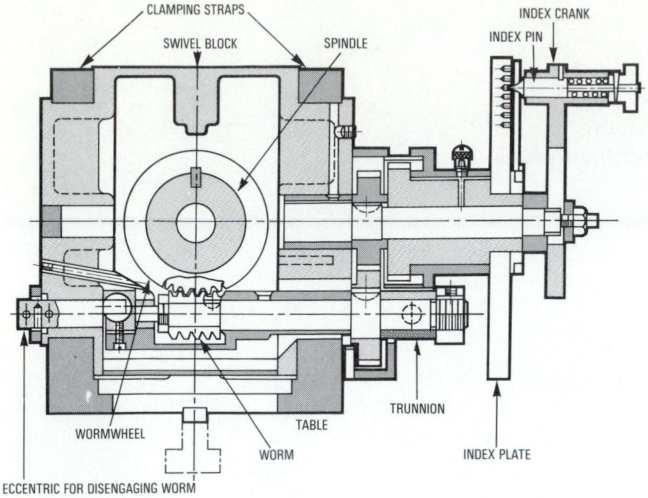
CLAMPING STRAPS
SWIVEL BLOCK
SPINDLE
INDEX CRANK
INDEX PIN
TRUNNION
TABLE
WORMWHEEL
WORM
INDEX PLATE
ECCENTRIC FOR DISENGAGING WORM

Figure 64-3 *Section through a dividing head, showing the spindle and index plate.*

METHODS OF INDEXING

The main purpose of the indexing or dividing head is to divide the workpiece circumference accurately into any number of divisions. This may be accomplished by the following indexing methods: direct, simple, angular, and differential.

DIRECT INDEXING

Direct indexing is the simplest form of indexing. It is performed by disengaging the worm shaft from the worm wheel by means of an eccentric device in the dividing head. Some direct dividing heads do not have a worm and worm wheel but rotate on bearings. The index plates contain slots, which are numbered, and a spring-loaded tongue lock is used to engage in the proper slot. Direct indexing is used for quick indexing of the workpiece when cutting flutes, hexagons, squares, and other shapes.

The work is rotated the required amount and held in place by a pin which engages into a hole or slot in the *direct indexing plate* mounted on the end of the dividing head spindle. The direct indexing plate usually contains three sets of hole circles or slots: 24, 30, and 36. The number of divisions it is possible to index is limited to numbers which are factors of either 24, 30, or 36. The common divisions that can be obtained by direct indexing are listed in Table 64-1.

EXAMPLE

What direct indexing is necessary to mill eight flutes on a reamer blank?

As the 24-hole circle is the only one divisible by 8 (the required number of divisions), it is the only circle which can be used in this case.

$$\text{Indexing} = \frac{24}{8} = 3 \text{ holes on a 24-hole circle}$$

NOTE: *NEVER* count the hole or slot in which the index pin is engaged.

Table 64-1 Direct Indexing Divisions

Plate Hole Circles														
24	2	3	4	—	6	8	—	—	12	—	—	24	—	—
30	2	3	—	5	6	—	—	10	—	15	—	—	30	—
36	2	3	4	—	6	—	9	—	12	—	18	—	—	36

To Mill a Square by Direct Indexing

1. Disengage the worm and worm shaft by turning the worm disengaging shaft lever if the dividing head is so equipped.
2. Adjust the plunger behind the index plate into the 24-hole circle or slot (Fig. 64-4).
3. Mount the workpiece in the dividing head chuck or between centers.
4. Adjust the cutter height and cut the first side.
5. Remove the plunger pin using the plunger pin lever (Fig. 64-5).
6. Turn the plate, attached to the dividing head spindle, one-half turn (12 holes or slots) and engage the plunger pin.
7. Take the second cut.

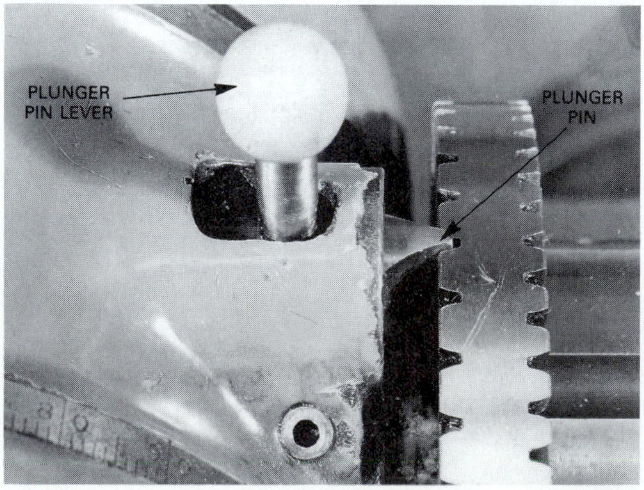
PLUNGER
PIN LEVER
PLUNGER
PIN

Figure 64-4 *Adjusting the plunger pin to fit into the proper hole circle or slot.*

Figure 64-5 *The plunger pin and the direct indexing plate are used for indexing a limited number of divisions.*

8. Measure the work across the flats and adjust the work height if required.
9. Cut the remaining sides by indexing every six holes until all surfaces are cut.

SIMPLE INDEXING

In *simple indexing,* the work is positioned by means of the crank, index plate, and sector arms. The worm attached to the crank must be engaged with the worm wheel on the dividing head spindle. Since there are 40 teeth on the worm wheel, one complete turn of the index crank will cause the spindle and the work to rotate one-fortieth of a turn. Similarly, 40 turns of the crank will revolve the spindle and work one turn. Thus there is a ratio of 40:1 between the turns of the index crank and the dividing head spindle.

To calculate the indexing or the number of turns of the crank for most divisions, it is necessary only to divide 40 by the number of divisions (N) to be cut, or

$$\text{Indexing} = \frac{40}{N}$$

EXAMPLE

The indexing required to cut eight flutes would be:

$$\frac{40}{8} = 5 \text{ full turns of the index crank}$$

If, however, it was necessary to cut seven flutes, the indexing would be

$$\frac{40}{7} = 5\tfrac{5}{7} \text{ turns}$$

Five complete turns are easily made; however, the five-sevenths of a turn involves the use of the index plate and sector arms.

INDEX PLATE AND SECTOR ARMS

The *index plate* is a circular plate provided with a series of equally spaced holes into which the index crank pin engages. The *sector arms* fit on the front of this plate and may be set to any portion of a complete turn.

To get five-sevenths of a turn, choose any hole circle (Table 64-2) which is divisible by the denominator 7, such as 21, then take five-sevenths of 21 = 15 holes on a 21-hole circle. Therefore, the indexing for seven flutes would be 40/7 = $5\tfrac{5}{7}$ turns or 5 complete turns plus 15 holes on the 21-hole circle. When extreme accuracy is required for indexing, choose the circle with the most holes.

Table 64-2 Index-Plate Hole Circles

Brown and Sharpe	
Plate 1	15-16-17-18-19-20
Plate 2	21-23-27-29-31-33
Plate 3	37-39-41-43-47-49
Cincinnati Standard Plate	
One side	24-25-28-30-34-37-38-39-41-42-43
Other Side	46-47-49-51-53-54-57-58-59-62-66

The procedure for cutting seven flutes would be as follows:

1. Mount the proper index plate on the dividing head.
2. Loosen the index crank nut and set the index pin into a hole on the 21-hole circle.
3. Tighten the index crank nut and check to see that the pin enters the hole easily.
4. Loosen the setscrew on the sector arm.
5. Place the narrow edge of the left arm against the index pin.
6. Count 15 holes on the 21-hole circle. *Do not include the hole in which the index crank pin is engaged.*
7. Move the right sector arm slightly beyond the 15th hole and tighten the sector arm setscrew.
8. Align the cutter with the workpiece.
9. Start the machine and set the cutter to the top of the work by using a paper feeler (Fig. 64-6).
10. Move the table so that the cutter clears the end of the work.
11. Tighten the friction lock on the dividing head before making each cut and loosen the lock when indexing for spaces.
12. Set the depth of cut and take the first cut.
13. After the first flute has been cut, return the table to the original starting position.
14. Withdraw the index pin and turn the crank clockwise

Figure 64-6 *Setting the cutter to the top of a workpiece using a paper feeler. (Courtesy Kostel Enterprises Limited.)*

five full turns plus the 15 holes indicated by the right sector arm. Release the index pin between the 14th and 15th holes, and gently tap it until it drops into the 15th hole.

15. Turn the sector arm *farthest from the pin* clockwise until it is against the index pin.

NOTE: It is important that the arm *farthest from the pin* be held and turned. If the arm *next* to the pin were held and turned, the spacing between both sector arms could be increased when the other arm hit the pin. This could result in an indexing error which would not be noticeable until the work was completed.

16. Lock the dividing head; then continue machining and indexing for the remaining flutes. Whenever the crank pin is moved past the required hole, *remove the backlash* between the worm and worm wheel by turning the crank *counterclockwise* approximately one-half turn and then carefully *clockwise* until the pin engages the proper hole.

ANGULAR INDEXING

When the angular distance between divisions is given instead of the number of divisions, the setup for simple indexing may be used; however, the method of calculating the indexing is changed.

One complete turn of the index crank turns the work one-fortieth of a turn, or one-fortieth of 360°, which equals 9°.

When the angular dimension is given in *degrees*, the indexing is then calculated as follows:

$$\text{Indexing in degrees} = \frac{\text{no. of degrees required}}{9}$$

EXAMPLE

Calculate the indexing for 45°.

$$\text{Indexing} = \frac{45}{9}$$

$$= 5 \text{ complete turns}$$

EXAMPLE

Calculate the indexing for 60°.

$$\text{Indexing} = \frac{60}{9}$$

$$= 6\tfrac{2}{3}$$

$$= 6 \text{ complete turns plus 12 holes on an 18-hole circle}$$

If the dimensions are given in *degrees* and *minutes*, it will be necessary to convert the degrees into minutes (number of degrees × 60′) and add this sum to the minutes required. The indexing in minutes is calculated as follows:

$$\text{Indexing in minutes} = \frac{\text{no. of minutes required}}{540}$$

EXAMPLE

Calculate the indexing necessary for 24′.

$$\text{Indexing} = \frac{24}{540}$$

$$= \frac{4}{90}$$

$$= \frac{1}{22.5}$$

The indexing for 24′ would be 1 hole on the 22.5-hole circle. As the 23-hole circle is the nearest hole circle, the indexing would be one hole on the 23-hole circle. Since in this case there is a slight error (approximately one-half minute) in indexing, it is advisable to use this method only for a few divisions if extreme accuracy is required.

EXAMPLE

Calculate the indexing for 24°30′. First convert 24°30′ into minutes:

$$\begin{aligned} (24 \times 60') &= 1440' \\ \text{Add } 30' &= \underline{\ \ 30'} \\ \text{Total} &= 1470' \end{aligned}$$

Then

$$\text{Indexing} = \frac{1470}{540}$$

$$= 2\tfrac{13}{18} \text{ turns}$$

$$= 2 \text{ complete turns plus 13 holes on an 18-hole circle}$$

When indexing for degrees and half degrees (30′), use the 18-hole circle (Brown and Sharpe).

1/2° (30′) = 1 hole on the 18-hole circle
1° = 2 holes on the 18-hole circle

When indexing for 1/3° (20′) and 2/3° (40′), the 27-hole circle should be used (Brown and Sharpe).

1/3° (20′) = 1 hole on the 27-hole circle
2/3° (40′) = 2 holes on the 27-hole circle

When indexing for minutes using a Cincinnati dividing head, note that one space on the 54-hole circle will rotate the work 10′ (1/54 × 540).

DIFFERENTIAL INDEXING

When it is impossible to calculate the required indexing by the simple indexing method, that is, when the fraction $40/N$ cannot be reduced to a factor of one of the available hole circles, it is necessary to use *differential indexing.*

With this method of indexing, the index plate must be revolved either forward or backward a part of a turn while the index crank is turned to attain the proper spacing or indexing.

In differential indexing, as in simple indexing, the index crank rotates the dividing head spindle. The spindle rotates the index plate, after the locking pin has been disengaged, by means of change gears connecting the dividing head spindle and the worm shaft (Fig. 64-7). The rotation of the plate may be either in the same direction (positive) or in the opposite direction (negative) of the index crank. This change of rotation is effected by an idler gear or gears in the gear train.

When it is necessary to calculate the indexing for a required number of divisions by the differential method, a number is chosen close to the required divisions which can be indexed by simple indexing.

To illustrate the principle of differential indexing, assume that the index crank has to be rotated one-ninth of a turn and there is only an 8-hole circle available.

If the crank is moved one-ninth of a turn, the index pin will contact the plate at a spot before the first hole on the 8-hole circle. The exact position of this spot would be the difference between one-eighth and one-ninth of a revolution of the crank. This would be

$$\frac{1}{8} - \frac{1}{9} = \frac{9-8}{72} = \frac{1}{72} \text{ of a turn } \textit{less}$$

than one-eighth of a turn, or one-seventy-second of a turn short of the first hole. Since there is no hole at this point into which the pin could engage, it is necessary to cause the plate to rotate backwards by means of change gears one-seventy-second of a turn in order that the pin will engage in a hole. At this point the index crank will be locked at exactly one-ninth of a turn.

The method of calculating the change gears (Fig. 64-7) required to rotate the plate the proper amount is as follows:

$$\text{Change gear ratio} = (A - N) \times \frac{40}{A} = \frac{\text{driver (spindle) gear}}{\text{driven (worm) gear}}$$

A = approximate number of divisions
N = required number of divisions

When the approximate number of divisions is larger than the required number, the resulting fraction is plus and the index plate must move in the same direction as the crank (clockwise). This *positive rotation* is accomplished by using an idler gear. However, if the approximate number is smaller than the required number, the resulting fraction is minus and the index plate must move in a counterclockwise direction. This *negative rotation* requires the use of two idler gears. The numerator of the fraction represents the driving (spindle) gear or gears, and the denominator represents the driven (worm) gear or gears. The gearing may be either simple or compound and the rotation is as follows:

- *Simple gearing*—One idler for a positive rotation of the index plate and two idlers for a negative rotation of the index plate
- *Compound gearing*—One idler for a negative rotation of the index plate and two idlers for a positive rotation of the index plate

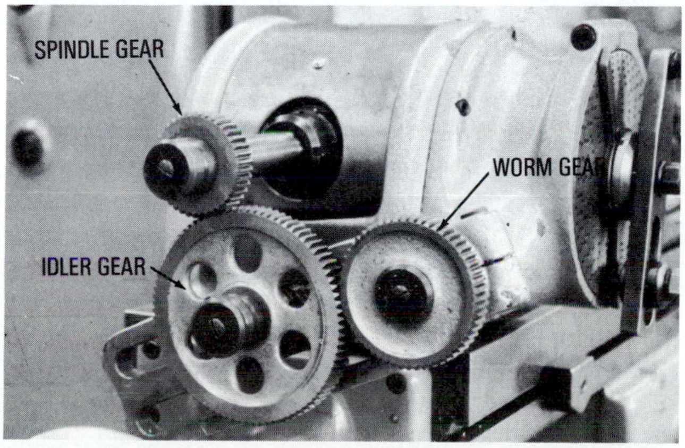
SPINDLE GEAR
WORM GEAR
IDLER GEAR
Figure 64-7 *Headstock geared for differential indexing.*

EXAMPLE

Calculate the indexing and change gears required for 57 divisions.

The change gears supplied with the dividing head are as follows: 24, 24, 28, 32, 40, 44, 48, 56, 64, 72, 86, 100.

The available index plate hole circles are as follows:

Plate 1: 15, 16, 17, 18, 19, 20
Plate 2: 21, 23, 27, 29, 31, 33
Plate 3: 37, 39, 41, 43, 47, 49

SOLUTION

1. Indexing $= \frac{40}{N} = \frac{40}{57}$

Since there is no 57-hole circle and since it is impossible to reduce this fraction to suit any hole circle, it is necessary to select an approximate number close to 57, for which simple indexing may be calculated.

2. Let the approximate number of divisions equal 56.

3. Indexing for 56 divisions $= \frac{40}{56} = \frac{5}{7}$, or 15 holes on the 21-hole circle.

4.
$$
\begin{aligned}
\text{Gear ratio} &= (A - N) \times \frac{40}{A} \\
&= (56 - 57) \times \frac{40}{56} \\
&= -1 \times \frac{40}{56} \\
&= -\frac{5}{7} \\
\text{Change gears} &= -\frac{5}{7} \times \frac{8}{8} \\
&= -\frac{40\ \text{(spindle gear)}}{56\ \text{(worm gear)}}
\end{aligned}
$$

Therefore, for indexing 57 divisions, a 40-tooth gear is mounted on the dividing head spindle, and a 56-tooth gear is mounted on the worm shaft. Since the fraction is a negative quantity and simple gearing is to be used, the index plate rotation is negative, or counterclockwise, and two idlers must be used. After the proper gears are installed, the simple indexing procedure for 56 divisions should be followed.

THE WIDE-RANGE DIVIDING HEAD

Although simple or differential indexing is satisfactory for most indexing problems, there may be certain divisions which cannot be indexed by either of these methods. Cincinnati Milacron, Inc., manufactures a *wide-range divider* that may be applied to a Cincinnati universal dividing head. With this attachment, it is possible to obtain from 2 up to 400,000 divisions.

The wide-range divider consists of a large index plate *A* (Fig. 64-8), sector arms, and crank *B* which engages in the plate *A*. This large plate contains 11-hole circles on each side. Mounted in front of the large index plate is a small index plate *C* containing a 54-hole and a 100-hole circle. The crank *D* operates through a reduction of gears having a ratio of 100:1. These gears are mounted in the housing *G*. The ratio between the worm (and the crank *B*) and the dividing head spindle is 40:1.

INDEXING FOR DIVISIONS

The ratio of the large index crank to the dividing head is 40:1, as in simple indexing. The ratio of the small index crank, which drives the large crank by planetary gearing, is 100:1. Therefore *one turn of the small crank* drives the index head spindle 1/100 of 1/40, or 1/4000 of a turn. One hole on the 100-hole circle of the small index plate *C* = 1/100 × 1/4000, or 1/400,000 of a turn. Therefore the formula for indexing *divisions* with a wide-range divider is 400,000/*N* and is applied as follows.

As the ratio of the large index crank is 40:1, any number which divides into 40 (the two numbers to the left of the short vertical line) represents full turns of the large index crank. If a 100-hole circle is used with the large crank, *one hole* on this circle will produce 1/100 or 1/40, or 1/4000 of a turn. Therefore, any numbers which divide into 4000 (the two numbers to the left of the long vertical line) are indexed on the 100-hole circle of the large plate. The numbers to the right of the long vertical line are indexed on the 100-hole circle of the small plate. Thus for the 1250 divisions, the indexing would be: 3 holes on the 100-hole circle of the large plate plus 20 holes on the 100-hole circle of the small plate.

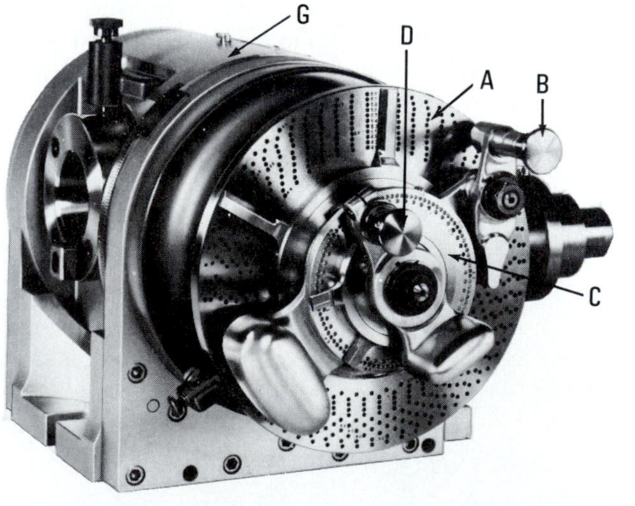
G
D
A
B
C

Figure 64-8 *A wide-range dividing head. (Courtesy Cincinnati Milacron Inc.)*

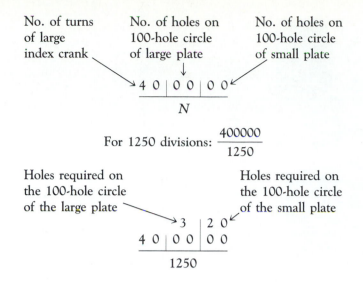

No. of turns of large index crank → 4 0 | 0 0 | 0 0 ← No. of holes on 100-hole circle of small plate

No. of holes on 100-hole circle of large plate ↓ (middle pair)

$$\frac{40\,|\,00\,|\,00}{N}$$

For 1250 divisions: $\dfrac{400000}{1250}$

Holes required on the 100-hole circle of the large plate → 3 | 2 0 ← Holes required on the 100-hole circle of the small plate

$$\frac{40\,|\,00\,|\,00}{1250} = 3\,|\,20$$

ANGULAR INDEXING WITH THE WIDE-RANGE DIVIDER

The wide-range divider is especially suited for accurate angular indexing. Indexing in degrees, minutes, and seconds is easily accomplished without the complicated calculations necessary with standard dividing heads.

For angular indexing, both the large and small index cranks are set on the 54-hole circle of each plate. Each space on the 54-hole circle of the large plate will cause the dividing head spindle to rotate 10′. Each space on the 54-hole circle of the small plate will cause the work to rotate 6″. Therefore, for indexing angles with a wide-range divider, the following formulas are used.

$$\text{Degrees} = \frac{N}{9} \text{ (indexed on the large plate)}$$

$$\text{Minutes} = \frac{N}{10} \text{ (indexed on the large plate)}$$

$$\text{Seconds} = \frac{N}{6} \text{ (indexed on the small plate)}$$

EXAMPLE

Index for an angle of 17°36′18″.

SOLUTION

1. Degrees $= \dfrac{17}{9}$

$= 1\frac{8}{9}$ turns

OR

one turn plus 48 holes on the 54-hole circle of the large index plate.

2. Minutes $= \dfrac{36}{10}$

= 3 holes on a 54-hole circle of the large index plate, leaving a remainder of 6′.

3. Convert the 6′ into seconds (6 × 60 = 360″) and add it to the 18″ still required.

4. Seconds $= \dfrac{378}{6}$

= 63 holes on a 54-hole circle of the small plate

OR

one turn and 9 holes on the 54-hole circle.

Therefore, to index for 17°36′18″ would require one turn and 51 holes (48 + 3) on the 54-hole circle of the large plate plus one turn and 9 holes on the 54-hole circle of the small plate.

LINEAR GRADUATING

The operation of producing accurate spaces on a piece of flat stock, or that of *linear graduating,* is easily accomplished on the horizontal milling machine (Fig. 64-9).

In this process the work may be clamped to the table or held in a vise, depending on the shape and size of the part. Care must be taken to align the workpiece parallel with the table travel.

To produce an *accurate* longitudinal movement of the table, the dividing head spindle is geared to the lead screw of the milling machine (Fig. 64-10).

If the dividing head spindle and the lead screw were connected with gears with equal number of teeth and the index crank turned one revolution, the spindle and lead screw on an inch milling machine would revolve one-fortieth of a revolution. This rotation of the lead screw [having 4 threads per inch (*tpi*)] would cause the table to move $\frac{1}{40} \times \frac{1}{4}$ (one turn of the lead screw) = $\frac{1}{160}$ = 0.00625 in. (0.15 mm). Thus five turns of the index crank would move the table 5 × 0.00625, or $\frac{1}{32}$ in. (0.78 mm).

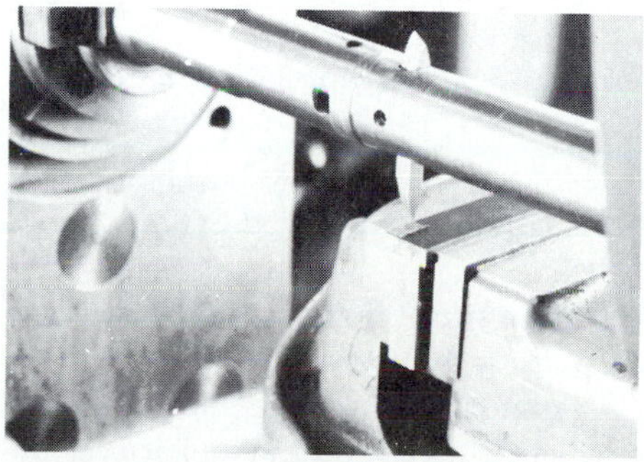

Figure 64-9 *Linear graduating.*

Figure 64-10 *Gearing required for linear graduating.*

The formula for calculating the indexing for linear graduations in thousandths of an inch is

$$\frac{N}{0.00625}$$

Very small movements of the table, such as 0.001, may be obtained by applying the formula:

$$\frac{0.001}{0.00625} = \frac{1}{6\frac{1}{4}} \text{ turns}$$

(4/25 turn), or 4 holes on the 25-hole circle.

If the lead screw of a metric milling machine had a pitch of 5 mm, one turn of the index crank would move the table one-fortieth of 5 mm, or 0.125 mm. Therefore, it would require four complete turns on the index crank to move the table 0.5 mm.

The formula for calculating the indexing for linear graduations in millimeters is

$$\frac{N}{0.125}$$

For a small movement of the table, such as 0.025 mm, apply the formula:

$$\frac{0.025}{0.125} = 1/5 \text{ turn or 5 holes on a 25-hole circle.}$$

Other suitable table movements may be obtained by using the appropriate hole circle and/or different change gears.

The point of the toolbit used for graduating is generally ground to a V-shape, although other special forms may be desired. The tool is mounted vertically in a suitable arbor which is of sufficient length to extend the toolbit over the workpiece (Fig. 64-9).

The uniformity of the length of the lines is controlled by the *accurate* movement of the crossfeed handwheel or by stops suitably mounted on the ways of the knee.

When graduating, position the starting point on the workpiece under the point of the *stationary,* vertical toolbit. The work is moved clear of the tool by the crossfeed handwheel and the proper depth is set by means of the vertical feed crank. The table is then locked in place. For a uniform width of lines to be maintained, the work must be held absolutely flat and the table height must never be adjusted.

REVIEW QUESTIONS

INDEXING OR DIVIDING HEAD

1. Name four parts of a dividing head set.
2. Name four methods of indexing that may be performed using the dividing head.
3. For what purpose is direct indexing used?

SIMPLE INDEXING

4. Explain how the ratio of 40:1 is determined on a standard dividing head.
5. Calculate the simple indexing, using a Brown and Sharpe dividing head, for the following divisions: 37, 41, 22, 34, and 120.
6. What procedure should be followed in order to set the sector arms for 12 holes on an 18-hole circle?

ANGULAR INDEXING

7. Explain the principle of angular indexing.
8. Calculate the indexing, using a Cincinnati dividing head, for the following angles: 21°, 37°, 21°30′, and 37°40′.

DIFFERENTIAL INDEXING

9. For what purpose is differential indexing used?
10. What is meant by positive rotation and negative rotation of the index plate?
11. Using a Brown and Sharpe dividing head, calculate the indexing and change gears for the following divisions: 53, 59, 101, and 175. A standard set of change gears, having the following numbers of teeth, is supplied: 24, 24, 28, 32, 40, 44, 48, 56, 64, 72, 86, 100.

WIDE-RANGE DIVIDING HEAD

12. How does the wide-range dividing head differ from a standard dividing head?
13. What two ratios are found in a wide-range dividing head?
14. Calculate the indexing for (a) 1000 and (b) 1200 divisions, using the wide-range dividing head.
15. Calculate the angular indexing for the following, using a wide-range dividing head: 20°45′, 25°15′32″.

LINEAR GRADUATING

16. How are the dividing head and the milling machine geared for linear graduating?
17. What indexing would be required to move the table 0.003 in. when using equal gearing on the dividing head and the lead screw?

UNIT 65

Gears

When it is required to transmit rotary motion from one shaft to another, several methods, such as belts, pulleys, and gears, may be used. If the shafts are parallel to each other and quite a distance apart, a flat belt and large pulleys may be used to drive the second shaft, the speed of which may be controlled by the size of the pulleys.

When the shafts are closer together, such as in the case of the sensitive drill press, a V-belt, which tends to reduce the excessive slippage of a flat belt, may be used. Here the speed of the driven shaft may be controlled by means of stepped or variable-speed pulleys. When the shafts are close together and parallel, some power may be transmitted by two rollers in contact, with one roller mounted on each shaft. Slippage is the main problem here, and the desired speed of the driven shaft could not be maintained.

The methods outlined in this unit are means by which power may be transmitted from one shaft to another, but the speed of the driven shaft may not be accurate in all cases due to slippage between the driving and driven members (belts, pulleys, or rollers). In order to eliminate slippage and produce a positive drive, gears are used.

OBJECTIVES

After completing this unit you should be able to:

1. Identify and state the purposes of six types of gears used in industry
2. Apply various formulas for calculating gear-tooth dimensions

GEARS AND GEARING

Gears are used to transmit power positively from one shaft to another by means of successively engaging teeth (in two gears). They are used in place of belt drives and other forms of friction drives when exact speed ratios and power transmission must be maintained. Gears may also be used to increase or decrease the speed of the driven shaft, thus decreasing or increasing the *torque* of the driven member.

Shafts in a gear drive or train are generally parallel. They may, however, be driven at any angle by means of suitably designed gears.

TYPES OF GEARS

Spur gears (Fig. 65-1) are generally used to transmit power between two parallel shafts. The teeth on these gears are straight and parallel to the shafts to which they are attached. When two gears of different sizes are in mesh, the larger is called the *gear* while the smaller is called the *pinion*. Spur gears are used where slow- to moderate-speed drives are required.

Internal gears (Fig. 65-2) are used where the shafts are parallel and the centers must be closer together than could be achieved with spur or helical gearing. This arrangement provides for a stronger drive since there is a greater area of contact than with the conventional gear drive. It also provides speed reductions with a minimum space requirement. Internal gears are used on heavy-duty tractors where much torque is required.

BOSTON
GEARS

Figure 65-1 *A spur gear and pinion are used for slower speeds.*

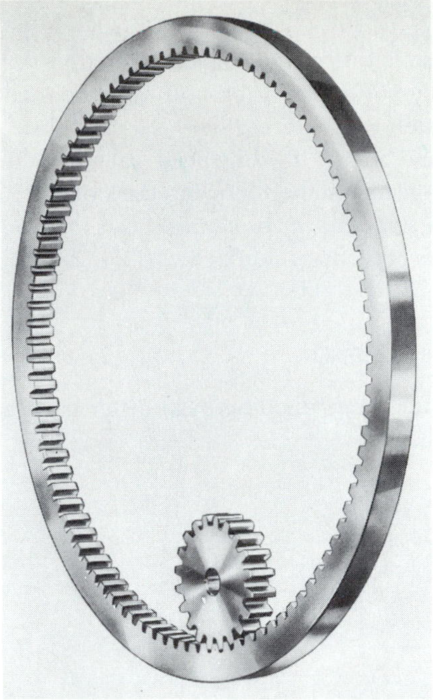

Figure 65-2 *Internal gears provide speed reductions with a minimum space requirement.*

Helical gears (Fig. 65-3) may be used to connect parallel shafts or shafts which are at an angle. Because of the progressive rather than intermittent action of the teeth, helical gears run more smoothly and quietly than spur gears. Since there is more than one tooth in engagement at any one time, helical gears are stronger than spur gears of the same size and pitch. However, special bearings (thrust bearings) are often required on shafts to overcome the end thrust produced by these gears as they turn.

On most installations where it is necessary to overcome end thrust, *herringbone gears* (Fig. 65-4) are used. This type of gear resembles two helical gears placed side by side, with one-half having a left-hand helix and the other half a right-hand helix. These gears have a smooth continuous action and eliminate the need for thrust bearings.

When two shafts are located at an angle with their axial lines intersecting at 90°, power is generally transmitted by means of *bevel gears* (Fig. 65-5A). When the shafts are at right angles and the gears are of the same size, they are called *miter gears* (Fig. 65-5B). However, it is not necessary that the shafts be only at right angles in order to transmit power. If the axes of the shafts intersect at any angle other than 90°, the gears are known as *angular bevel gears* (Fig. 65-5C). Bevel gears have straight teeth very similar to spur

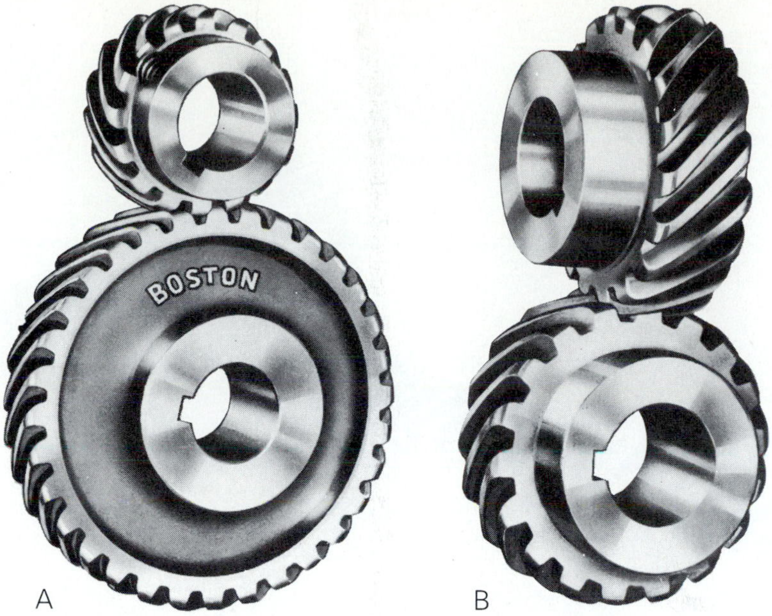
BOSTON
A
B

Figure 65-3 *Helical gears: (A) for drives that are parallel to each other; (B) for drives that are at right angles to each other.*

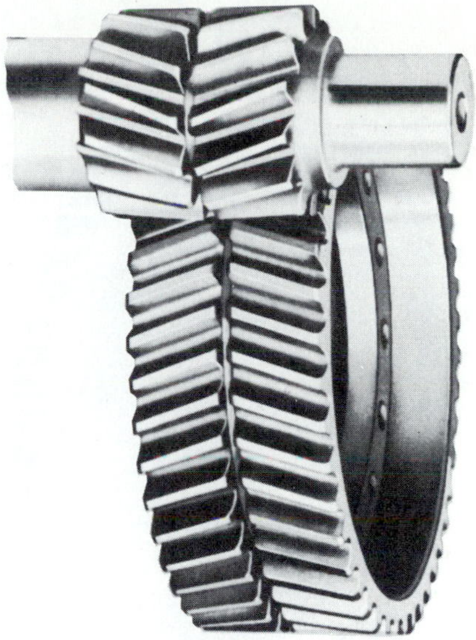

Figure 65-4 *Herringbone gears eliminate end thrust on shafts.*

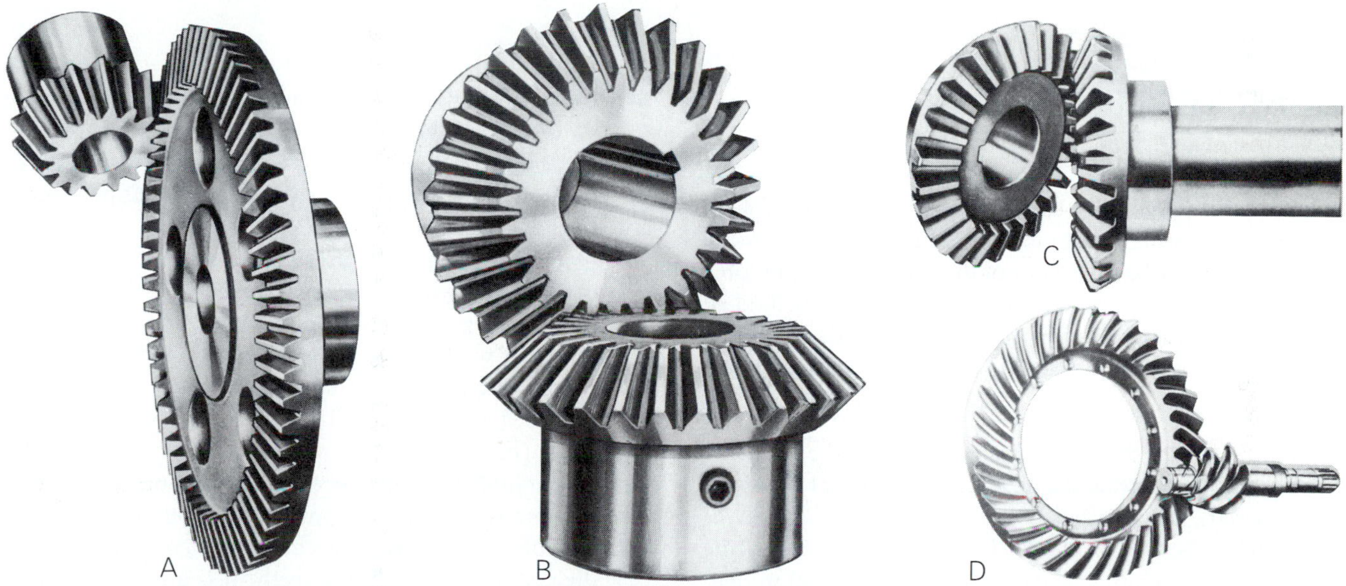
A
B
C
D

Figure 65-5 *(A) Bevel gears transmit power at 90°; (B) the driver and driven miter gears are the same size; (C) angular bevel gears are used for shafts that are not at right angles; (D) hypoid gears are used in automotive drives.*

gears. Modified bevel gears having helical teeth are known as *hypoid gears* (Fig. 65-5D). The shafts of these gears, although at right angles, are not in the same plane and, therefore, do not intersect. Hypoid gears are used in automobile drives.

When shafts are at right angles and considerable reduction in speed is required, a *worm* and *worm gear* (Fig. 65-6) may be used. The worm, which meshes with the worm gear, may be single- or multiple-start thread. A worm with a double-start thread will revolve the worm gear twice as fast as a worm with a single-start thread and the same pitch.

When it is necessary to convert rotary motion to linear motion, a *rack* and *pinion* (Fig. 65-7) may be used. The rack, which is actually a straight or flat gear, may have straight teeth to mesh with a spur gear, or angular teeth to mesh with a helical gear.

BOSTON

Figure 65-6 *A worm and worm gear is used for speed reduction.*

Figure 65-7 *A rack and pinion converts rotary motion to linear motion.*

GEAR TERMINOLOGY

A knowledge of the more common gear terms is desirable to understand gearing and to make the calculations necessary to cut a gear (Fig. 65-8). Most of these terms are applicable to either inch or metric gearing, although the method of calculating dimensions may differ. These methods, as applicable to inch and metric gear-cutting, are explained in Unit 66.

- *Addendum* is the radial distance between the pitch circle and the outside diameter or the height of the tooth above the pitch circle.
- *Center distance* is the shortest distance between the axes of two mating gears or the distance equal to one-half the sum of the pitch diameters.
- *Chordal addendum* is the radial distance measured from the top of the tooth to a point where the chordal thickness and the pitch circle intersect the edge of the tooth.
- *Chordal thickness* is the thickness of the tooth measured at the pitch circle or the length of the chord which subtends the arc of the pitch circle.
- *Circular pitch* is the distance from a point on one tooth to a corresponding point on the next tooth measured on the pitch circle.
- *Circular thickness* is the thickness of the tooth measured on the pitch circle; it is also known as the *arc thickness*.
- *Clearance* is the radial distance between the top of one tooth and the bottom of the mating tooth space.
- *Dedendum* is the radial distance from the pitch circle to the bottom of the tooth space. The dedendum is equal to the addendum plus the clearance.
- *Diametral pitch* (inch gears) is the ratio of the number of teeth for each inch of pitch diameter of the gear. For example, a gear of 10 diametral pitch and a 3-in. *pitch diameter* would have 10 × 3, or 30, teeth.
- *Involute* is the curved line produced by a point of a stretched string when it is unwrapped from a given cylinder (Fig. 65-9).
- *Linear pitch* is the distance from a point on one tooth to the corresponding point on the next tooth of a gear rack.
- *Module* (metric gears) is the pitch diameter of a gear divided by the number of teeth. It is an actual dimension, unlike diametral pitch, which is a ratio of the number of teeth to the pitch diameter.
- *Outside diameter* is the overall diameter of the gear, which is the pitch circle plus two addendums.
- *Pitch circle* is a circle which has the radius of half the pitch diameter with its center at the axis of the gear.
- *Pitch circumference* is the circumference of the pitch circle.
- *Pitch diameter* is the diameter of the pitch circle which is equal to the outside diameter minus two addendums.
- *Pressure angle* is the angle formed by a line through the point of contact of two mating teeth and tangent to the two *base circles* and a line at right angles to the center line of the gears.

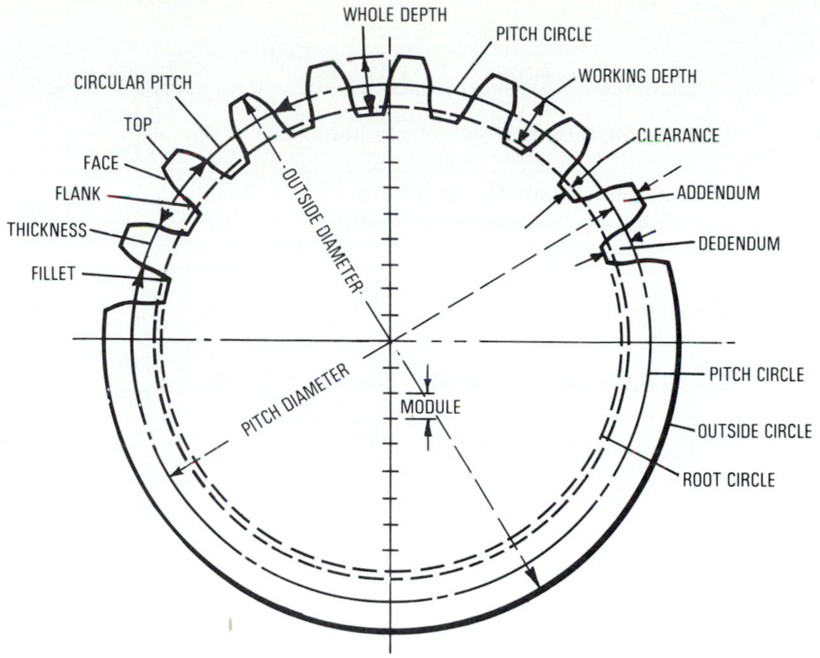
WHOLE DEPTH
PITCH CIRCLE
WORKING DEPTH
CIRCULAR PITCH
TOP
CLEARANCE
FACE
FLANK
ADDENDUM
THICKNESS
DEDENDUM
FILLET
OUTSIDE DIAMETER
PITCH DIAMETER
MODULE
PITCH CIRCLE
OUTSIDE CIRCLE
ROOT CIRCLE

Figure 65-8 *The parts of a gear.*

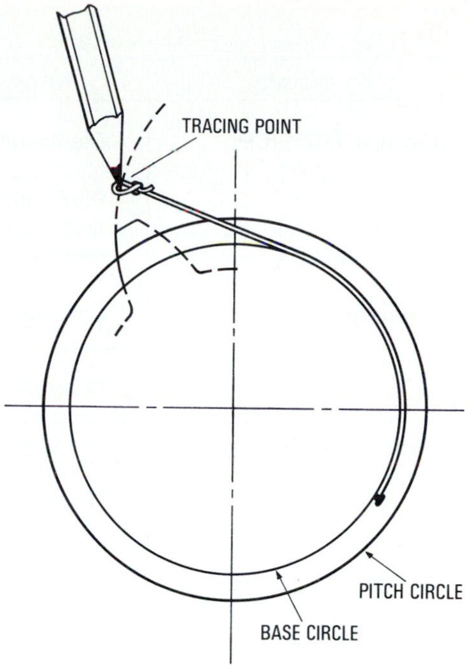
TRACING POINT
PITCH CIRCLE
BASE CIRCLE

Figure 65-9 *Generating an involute.*

- *Root circle* is the circle formed by the bottoms of the tooth spaces.
- *Root diameter* is the diameter of the root circle.
- *Tooth thickness* is the thickness of the tooth measured on the pitch circle.
- *Whole depth* is the full depth of the tooth or the distance equal to the addendum plus the dedendum.
- *Working depth* is the distance that a gear tooth extends into the tooth space of a mating gear which is equal to two addendums.

Table 65-1 Rules and Formulas for Spur Gears

To Obtain	Knowing	Rule	Formula
Addendum (*A*)	Circular pitch	Multiply the circular pitch by 0.3183	$A = CP \times 0.3183$
	Diametral pitch	Divide 1 by the diametral pitch	$A = \frac{1}{DP}$
Center distance (CD)	Circular pitch	Multiply the number of teeth in both gears by the circular pitch, and divide the product by 6.2832.	$CD = \frac{(N + n) \times CP}{6.2832}$
	Diametral pitch	Divide the total number of teeth in both gears by twice the diametral pitch.	$CD = \frac{N + n}{2 \times DP}$
Chordal (corrected) addendum (CA)	Pitch diameter addendum and number of teeth	Subtract from 1 the cosine of the result of 90° divided by the number of teeth. Multiply this result by half the pitch diameter. To this product, add the addendum.	$CA = \left[\left(1 - \cos\frac{90}{N}\right)\frac{PD}{2}\right] + A$
Chordal thickness (CT)	Pitch diameter and number of teeth	Divide 90 by the number of teeth; find the sine of this result and multiply by the pitch diameters.	$CT = \sin\frac{90}{N} \times PD$
Circular pitch (CP)	Center-to-center distance	Multiply center to center distance by 6.2832, and divide the product by the total number of teeth in both gears.	$CP = \frac{CD \times 6.2832}{N + n}$

Table 65-1 (Continued)

To Obtain	Knowing	Rule	Formula
Circular Pitch (CP)	Diametral pitch	Divide 3.1416 by the diametral pitch.	$CP = \frac{3.1416}{DP}$
	Pitch diameter and number of teeth.	Multiply pitch diameter by 3.1416 and divide by the number of teeth.	$CP + \frac{PD \times 3.1416}{N}$
Clearance (CL)	Circular pitch	Divide circular pitch by 20.	$CL = \frac{CP}{20}$
	Diametral pitch	Divide 0.157 by the diametral pitch.	$CL = \frac{0.157}{DP}$
Dedendum (*D*)	Circular pitch	Multiply the circular pitch by 0.3683.	$D = CP \times 0.3683$
	Diametral pitch	Divide 1.157 by the diametral pitch.	$D = \frac{1.157}{DP}$
Diametral pitch (DP)	Circular pitch	Divide 3.1416 by the circular pitch.	$DP = \frac{3.1416}{CP}$
	Number of teeth and outside diameter	Add 2 to the number of teeth and divide the sum by the outside diameter.	$DP = \frac{N + 2}{OD}$
	Number of teeth and pitch diameter	Divide the number of teeth by the pitch diameter.	$DP = \frac{N}{PD}$
Number of teeth (*N*)	Outside diameter and diametral pitch	Multiply the outside diameter by the diametral pitch and subtract 2.	$N = OD \times DP - 2$
	Pitch diameter and circular pitch	Multiply the pitch diameter by 3.1416 and divide by the circular pitch.	$N = \frac{PD \times 3.1416}{CP}$
	Pitch diameter and diametral pitch	Multiply the pitch diameter by the diametral pitch.	$N = PD \times DP$
Outside diameter (OD)	Number of teeth and circular pitch	Add 2 to the number of teeth and multiply the sum by the circular pitch. Divide this product by 3.1416.	$OD = \frac{(N + 2) \times CP}{3.1416}$
	Number of teeth and diametral pitch	Add 2 to the number of teeth and divide the sum by the diametral pitch.	$OD = \frac{N + 2}{DP}$
	Pitch diameter and diametral pitch	To the pitch diameter add 2 and divide by the diametral pitch.	$OD = PD + \frac{2}{DP}$
Pitch diameter (PD)	Number of teeth and circular pitch	Multiply the number of teeth by the circular pitch and divide by 3.1416.	$PD = \frac{N \times CP}{3.1416}$
	Number of teeth and diametral pitch	Divide the number of teeth by the diametral pitch.	$PD = \frac{N}{DP}$
	Outside diameter and number of teeth	Multiply the number of teeth by the outside diameter, and divide the product by the number of teeth plus 2.	$PD = \frac{N \times OD}{N + 2}$

Table 65-1 (Continued)

To Obtain	Knowing	Rule	Formula
Tooth thickness (*T*)	Circular pitch	Divide the circular pitch by 2.	$T = \frac{CP}{2}$
	Circular pitch	Multiply the circular pitch by 0.5.	$T = CP \times 0.5$
	Diametral pitch	Divide 1.5708 by the diametral pitch.	$T = \frac{1.5708}{DP}$
Whole depth (WD)	Circular pitch	Multiply the circular pitch by 0.6866.	$WD = CP \times 0.6866$
	Diametral pitch	Divide 2.157 by the diametral pitch.	$WD = \frac{2.157}{DP}$

Note: There are three commonly used gear-tooth forms which have pressure angles of 14½, 20, and 25°. The 20 and 25° tooth forms are now replacing the 14½° tooth form because the base of the tooth is wider and stronger. For the formula regarding the 20 and 25° tooth forms, see *Machinery's* or the *American Machinist's* handbooks.

REVIEW QUESTIONS

1. List six types of gears, and state where each may be used.
2. Define the following gear terms, and state the formula used to determine each. Use the formulas involving diametral pitch, *not* circular pitch, where applicable.
   a. Pitch diameter
   b. Diametral pitch
   c. Addendum
   d. Dedendum
   e. Clearance
   f. Outside diameter
   g. Number of teeth
3. Calculate the pitch diameter, outside diameter, and whole depth of tooth for the following gears.
   a. 8 DP with 36 teeth
   b. 12 DP with 81 teeth
   c. 16 DP with 100 teeth
   d. 6 DP having 23 teeth
   e. 4 DP having 54 teeth

UNIT 66

Gear Cutting

Most gears cut on a milling machine are generally used to repair or replace a gear which has been broken or lost or is no longer carried in inventory. Industry generally mass produces gears on special machines which have been designed for this purpose. The most common types of *gear-generating machines* are the *gear-shaping machines* and the *gear-hobbing machines.* It is generally more economical to buy gears from a firm which specializes in gear manufacture unless it is important that the machine be back in operation soon. At this time it is quite possible that a machinist would be called on to cut a gear in order to repair the machine quickly and get it back into production.

OBJECTIVES

After completing this unit you should be able to:

1. Select the proper cutter for any gear to be cut
2. Calculate gear tooth dimensions for inch gears
3. Calculate gear tooth dimensions for metric gears
4. Set up and cut a spur gear

INVOLUTE GEAR CUTTERS

Gear cutters are an example of formed cutters. This type of cutter is sharpened on the face and ensures exact duplication of the shape of the teeth, regardless of how far back the face of the tooth has been ground.

Gear cutters are available in many sizes, ranging from 1 to 48 diametral pitch (DP). Cutters for teeth smaller than 48 DP are available as special cutters. Comparative sizes of teeth ranging from 4 to 16 DP are shown in Fig. 66-1.

When gear teeth are cut on any gear, a cutter must be chosen to suit both the DP and the number of teeth (N). The tooth space for a small pinion cannot be of the same shape as the tooth space for a large mating gear. The teeth on smaller gears must be more "curved" to prevent binding of meshing gear teeth. Therefore, sets of gear cutters are made in a series of slightly different shapes to permit the cutting of any desired number of teeth in a gear, with the assurance that the teeth will mesh properly with those of another gear of the *same* DP.

These cutters are generally made in sets of eight and are numbered from 1 to 8 (Fig. 66-2). Notice the gradual change in shape from the #1 cutter, which has almost straight sides, to the much more curved sides of the #8 cutter. As shown in Table 66-1, the #1 cutter is used for cutting any number of teeth in a gear from 135 teeth to a rack, while the #8 cutter will cut only 12 and 13 teeth. It should be noted that in order for gears to mesh, they must be of the same DP; the cutter number permits *only* a more accurate meshing of the teeth.

Some gear cutter manufacturers have augmented the set of eight cutters with seven additional cutters in half sizes, making a total 15 cutters in the set, numbered 1, 1½, 2, 2½, etc. In the half series, a #1½ cutter would be used to cut from 80 to 134 teeth, but a 7½ cutter would cut 13 teeth only (Table 66-1).

EXAMPLE

A 10-DP gear and a pinion in mesh have 100 teeth and 24 teeth respectively. What cutters should be used to cut these gears?

CUTTER SELECTION

Since the gears are in mesh, both must be cut with a 10-DP cutter.

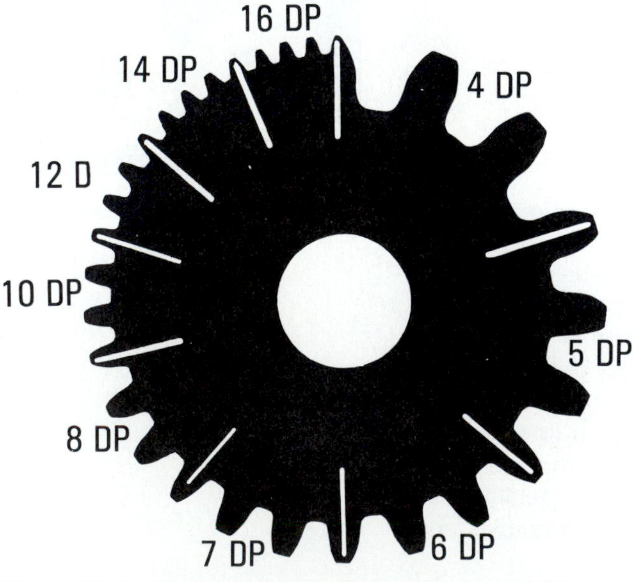
16 DP
14 DP
4 DP
12 D
10 DP
5 DP
8 DP
7 DP
6 DP

Figure 66-1 *Comparative gear tooth sizes: 4 to 16 DP.*

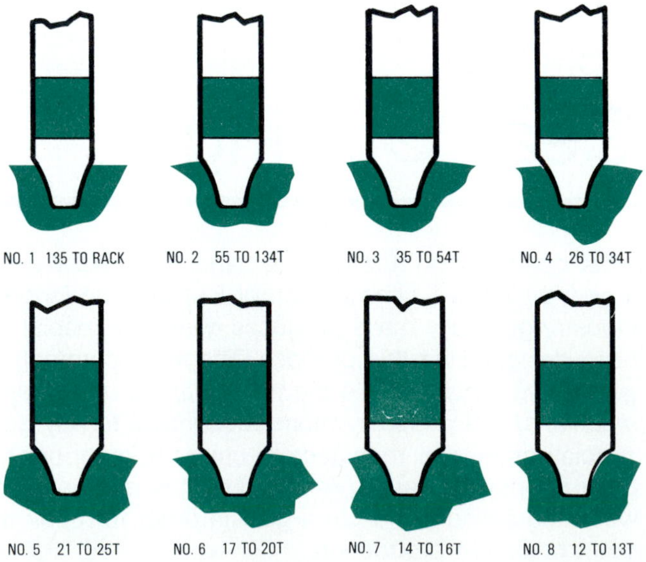
NO. 1 135 TO RACK
NO. 2 55 TO 134T
NO. 3 35 TO 54T
NO. 4 26 TO 34T
NO. 5 21 TO 25T
NO. 6 17 TO 20T
NO. 7 14 TO 16T
NO. 8 12 TO 13T

Figure 66-2 *Involute profiles for a set of gear cutters.*

Table 66-1 Involute Gear Cutters

Cutter Number	Range
1	135 teeth to a rack
1½	80 to 134 teeth
2	55 to 134 teeth
2½	42 to 54 teeth
3	35 to 54 teeth
3½	30 to 34 teeth
4	26 to 34 teeth
4½	23 to 25 teeth
5	21 to 25 teeth
5½	19 and 20 teeth
6	17 to 20 teeth
6½	15 and 16 teeth
7	14 to 16 teeth
7½	13 teeth
8	12 and 13 teeth

A #2 cutter should be used to cut the teeth on the gear, since it will cut from 55 to 134 teeth.

A #5 cutter should be used to cut the pinion, since it will cut from 21 to 25 teeth.

TO CUT A SPUR GEAR

The procedure for machining a spur gear is outlined in the following example.

EXAMPLE

A 52-tooth gear with an 8 DP is required.

PROCEDURE

1. Calculate all the necessary gear data (see Unit 65).

a. Outside diameter $= \dfrac{N+2}{DP}$

$$= \frac{54}{8}$$

$$= 6.750 \text{ in.}$$

b. Whole depth of tooth $= \dfrac{2.157}{DP}$

$$= \frac{2.157}{8}$$

$$= 0.2697 \text{ in.}$$

c. Cutter number = 3 (35 to 54 teeth)

d. Indexing (using Cincinnati standard plate)

$$= \frac{40}{N}$$

$$= \frac{40}{52}$$

$$= \frac{10}{13} \times \frac{3}{3} = 30 \text{ holes on the 39-hole circle}$$

2. Turn the gear blank to the proper outside diameter (6.750 in.)

3. Press the gear blank firmly onto the mandrel.

NOTE: If the blank was turned on a mandrel, be sure that it is tight because the heat caused by turning might have expanded the blank slightly.

4. Mount the index head and footstock, and check the alignment of the index centers (Fig. 66-3).

5. Set the dividing head so that the index pin fits into a hole on the 39-hole circle and the sector arms are set for 30 holes.

NOTE: Do not count the hole in which the pin is engaged.

6. Mount the mandrel (and workpiece), with the large end toward the indexing head, between the index centers.

NOTE:

a. The footstock center should be adjusted up tightly into the mandrel and locked in position.

b. The dog should be tightened properly on the mandrel and the tail of the dog should not bind in the slot.

c. The tail of the dog should then be locked in the driving fork of the dividing head by means of the setscrews. This will ensure that there will be no play between the dividing head and the mandrel.

d. The dog should be far enough from the gear blank to

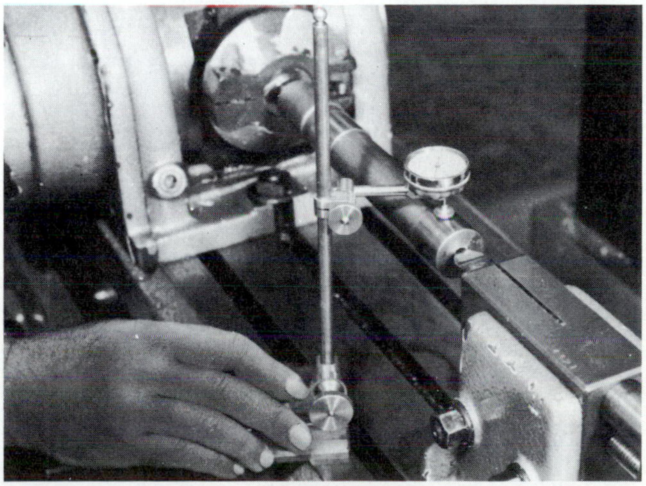

Figure 66-3 *Checking the alignment of index centers with a dial indicator.*

ensure that the cutter will not hit the dog when the gear is being cut.

7. Move the table close to the column to keep the setup as rigid as possible.

8. Mount an 8 DP#3 cutter on the milling machine arbor over the approximate center of the gear. Be sure to have the cutter rotating in the direction of the indexing head.

9. Center the gear blank with the cutter by either of the following methods:

a. Place a square against the outside diameter of the gear (Fig. 66-4). With a pair of inside calipers or a rule, check the distance between the square and the side of the cutter. Adjust the table until the distances from both sides of the gear blank to the sides of the cutter are the same.

b. A more accurate method of centralizing the cutter is to use gage blocks instead of the inside calipers or rule.

10. *LOCK THE CROSS SLIDE.*

11. Start the milling cutter and run the work under the cutter.

12. Raise the table until the cutter *just* touches the work. This can be done by using a chalk mark on the gear blank or a piece of paper between the gear blank and the cutter to indicate when the cutter is just touching the work (Fig. 66-5).

13. Set the graduated feed collar on the vertical feed to zero (0).

14. Move the work clear of the cutter by means of the longitudinal feed handle and raise the table to about two-thirds the depth of the tooth (0.180 in.); then *lock the knee clamp.*

NOTE: A special stocking cutter is sometimes used to rough out the teeth.

15. Slightly notch all gear teeth on the end of the work to check for correct indexing (Fig. 66-6).

16. Rough out the first tooth and set the automatic feed trip dog after the cutter is clear of the work.

Figure 66-4 *Centering a gear cutter and the workpiece.*

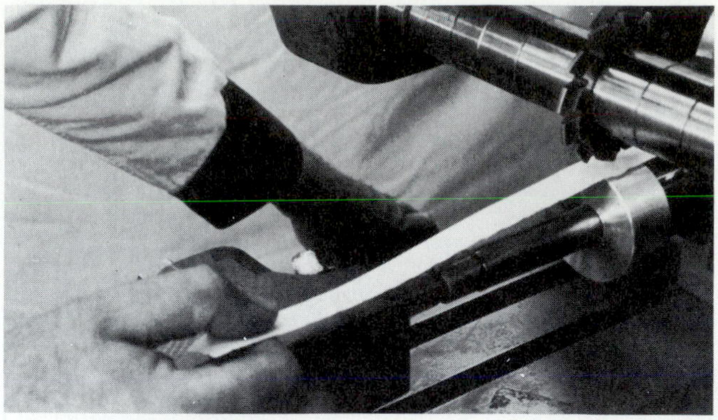

Figure 66-5 *Setting a gear cutter to the diameter of the workpiece.*

17. Return the table to the starting position.

NOTE: Clear the end of the work with a cutter.

18. Cut the remaining teeth and return the table to the starting position.

19. Loosen the knee clamp, raise the table to the full depth of 0.270 in., and *lock the knee clamp.*

NOTE: It is advisable to remove the crank from the knee-elevating shaft so that it will not be moved accidentally and change the setting.

20. Finish-cut all teeth.

NOTE: After each tooth has been cut, the cutter should be stopped before the table is returned to prevent marring the finish on the gear teeth.

METRIC GEARS AND GEAR CUTTING

Countries which have been using a metric system of measurement usually use the *module* system of gearing. The

Figure 66-6 *Notching all gear teeth eliminates errors.*

module (M) of a gear equals the pitch diameter (PD) divided by the number of teeth (N), or $M = PD/N$, whereas the DP of a gear is the ratio of N to the PD, or $DP = N/PD$. The DP of a gear is the *ratio* of the number of teeth per inch of diameter, whereas *M is an actual dimension.* Most of the terms used in DP gears remain the same for module gears; however, the method of calculating the dimensions has changed in some instances. Table 66-2 gives the necessary rules and formulas for metric spur gears.

METRIC MODULE GEAR CUTTERS

The most common metric gear cutters are available in modules ranging from 0.5 to 10 mm (Table 66-3). However, metric module gear cutters are available in sizes up to 75 mm. Any metric module size is available in a set of eight cutters, numbered from #1 to #8. The range of each cutter is the reverse of that of a DP cutter. For instance, a #1 metric module cutter will cut from 12 to 13 teeth; a #8 DP cutter will cut from 135 teeth to a rack. Table 66-3 shows the cutters available and the range of each cutter in the set.

EXAMPLE

A spur gear has PD of 60 mm and 20 teeth. Calculate:

1. Module
2. Circular pitch
3. Addendum
4. Outside diameter
5. Dedendum
6. Whole depth
7. Cutter number

SOLUTION

1. $M = \dfrac{PD}{N} = \dfrac{60}{20} = 3 \text{ mm}$

2. $CP = M \times \pi = 3 \times 3.1416 = 9.425 \text{ mm}$

3. $A = M = 3 \text{ mm}$

4. $OD = (N + 2) \times M = 22 \times 3 = 66 \text{ mm}$

5. $D = M \times 1.666 = 3 \times 1.666 = 4.998 \text{ mm}$

6. $WD = M \times 2.166 = 3 \times 2.166 = 6.498 \text{ mm}$

7. Cutter number (see Table 66-3) = 3

EXAMPLE

Two identical gears in mesh have a CD of 120 mm. Each gear has 24 teeth. Calculate:

1. Pitch diameter
2. Module
3. Outside diameter
4. Whole depth
5. Circular pitch
6. Chordal thickness

SOLUTION

1. $PD = \dfrac{2 \times CD}{2}$ (equal gears) $= \dfrac{2 \times 120}{2} = \dfrac{240}{2} = 120 \text{ mm}$

2. $M = \dfrac{PD}{N} = \dfrac{120}{24} = 5$

3. $OD = (N + 2) \times M = 26 \times 5 = 130 \text{ mm}$

4. $WD = M \times 2.166 = 5 \times 2.166 = 10.83 \text{ mm}$

5. $CP = M \times \pi = 5 \times 3.1416 = 15.708 \text{ mm}$

6. $CT = \dfrac{M \times \pi}{2} = \dfrac{5 \times 3.1416}{2} = 7.85 \text{ mm}$

Table 66-2 Rules and Formulas for Metric Module Spur Gears

To Obtain	Knowing	Rule	Formula
Addendum (*A*)	Normal module	Addendum equals module.	$A = M$
Circular pitch (CP)	Module	Multiply module by π.	$CP = M \times 3.1416$
	Pitch diameter and number of teeth	Multiply pitch diameter by π and divide by number of teeth.	$CP = \frac{PD \times 3.1416}{N}$
	Outside diameter and number of teeth	Multiply outside diameter by π and divide by number of teeth minus 2.	$CP = \frac{OD \times 3.1416}{N - 2}$
Chordal thickness (CT)	Module and outside diameter	Divide 90° by the number of teeth. Find the sine of this angle and multiply by the pitch diameter.	$CT = PD \times \sin \frac{90}{N}$
	Module	Multiply module by π and divide by 2.	$CT = \frac{M \times 3.1416}{2}$
	Circular pitch	Divide circular pitch by 2.	$CT = \frac{CP}{2}$
Clearance (CL)	Module	Multiply module by 0.166 mm.	$CL = M \times 0.166$
Dedendum (*D*)	Module	Multiply module by 1.166 mm.	$D = M \times 1.166$
Module (*M*)	Pitch diameter and number of teeth	Divide pitch diameter by the number of teeth.	$M = \frac{PD}{N}$
	Circular pitch	Divide circular pitch by π.	$M = \frac{CP}{3.1416}$
	Outside diameter and number of teeth	Divide outside diameter by number of teeth plus 2.	$M = \frac{OD}{N + 2}$
Number of teeth (*N*)	Pitch diameter and module	Divide pitch diameter by the module.	$N = \frac{PD}{M}$
	Pitch diameter and circular pitch	Multiply pitch diameter by π and divide product by circular pitch.	$N = \frac{PD \times 3.1416}{CP}$
Outside diameter (OD)	Number of teeth and module	Add 2 to the number of teeth and multiply sum of module.	$OD = (N + 2) \times M$
	Pitch diameter and module	Add 2 modules to pitch diameter.	$OD = PD + 2M$
Pitch diameter (PD)	Module and number of teeth	Multiply module by number of teeth.	$PD = M \times N$
	Outside diameter and module	Subtract 2 modules from outside diameter.	$PD = OD - 2M$
	Number of teeth and outside diameter	Multiply number of teeth by outside diameter and divide product by number of teeth plus 2.	$PD = \frac{N \times OD}{N + 2}$
Whole depth (WD)	Module	Multiply module by 2.166 mm.	$WD = M \times 2.166$
Center-to-center distance (CD)	Pitch diameters	Divide the sum of the pitch diameters by 2.	$CD = \frac{PD_1 + PD_2}{2}$

Table 66-3 Metric Module Gear Cutters

Module Size (mm)		Milling Cutter Numbers	
		Cutter No.	For Cutting
0.50	3.50		
0.75	3.75	1	12–13 teeth
1.00	4.00	2	14–16 teeth
1.25	4.50	3	17–20 teeth
1.50	5.00	4	21–25 teeth
1.75	5.50	5	26–34 teeth
2.00	6.00	6	35–54 teeth
2.25	6.50	7	55–134 teeth
2.50	7.00	8	135 teeth to rack
2.75	8.00		
3.00	9.00		
3.25	10.00		

GEAR TOOTH MEASUREMENT

To ensure that the gear teeth are of the proper dimensions, they should be measured with a gear tooth vernier caliper. The caliper should be set to the corrected addendum, a dimension which may be found in most handbooks.

Gear sizes may also be accurately checked by measuring over wires or pins of a specific diameter, which have been placed in two diametrically opposite tooth spaces of the gear (Fig. 66-7A). For gears having an odd number of teeth, the wires are placed as nearly opposite as possible (Fig. 66-7B). A measurement taken over these wires is checked against tables found in most handbooks. These tables indicate the measurement over the wires for any gear having a given number of teeth and a specific pressure angle. Since these tables are far too extensive to be printed in this book, the reader is asked to refer to any handbook for them.

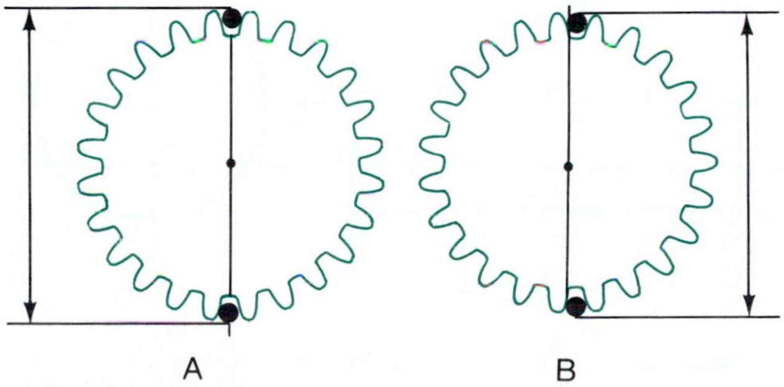
A
B

Figure 66-7 *Wires or pins are used to check gear sizes accurately.*

In order to measure inch gears accurately, the diametral pitch and number of teeth of the gear must be known. In order to measure metric gears, the module must be known. The wire or pin size to be used is determined as follows:

1. For external inch spur gears the wire or pin size is equal to 1.728 divided by the diametral pitch of the gear.
2. For internal inch spur gears, the wire size is equal to 1.44 divided by the diametral pitch of the gear.
3. Metric module gears are measured using a wire size equal to 1.728 multiplied by module of the gear. The measurement over the wires should equal the value shown in the handbook tables multiplied by module.

EXAMPLE (inch)

Determine the wire size and the measurement over the wires for a 10-DP external gear having 28 teeth and a 14.50° pressure angle.

$$\text{Wire size} = \frac{1.728}{10} = 0.1728 \text{ in.}$$

By referring to handbook tables, the size over the wires for a gear having 28 teeth and a 14.50° pressure angle should be 30.4374 in. divided by the DP. Therefore the measurement over the wires should be

$$\frac{30.4374}{10} = 3.0437 \text{ in.}$$

If the measurement is larger than this size, the PD is too large and the depth of cut will have to be increased. If it is less than the determined size, the gear is undersize. Gears having an odd number of teeth are calculated in a similar manner but using the proper tables of these gears.

REVIEW QUESTIONS

1. What cutter numbers should be used to cut the following gears?
   a. 8 DP-36 teeth
   b. 12 DP-81 teeth
   c. 16 DP-100 teeth
   d. 6 DP-23 teeth
2. Describe two methods of centering the gear blank with the cutter for machining a spur gear.
3. What precautions should be observed when mounting the gear blank between the index head and centers?
4. What is the purpose of notching before gear teeth are cut?
5. Compare the terms *module* and *pitch diameter.*
6. How does the metric module numbering system for gear cutters differ from that for diametral pitch systems?

7. For a 40-tooth spur gear, 240 mm pitch diameter, calculate:

a. Module
b. Circular pitch
c. Outside diameter
d. Addendum
e. Dedendum
f. Whole depth
g. Cutter number

GEAR TOOTH MEASUREMENT

8. State two methods of measuring gear teeth.
9. How is the wire size determined for:
   a. External gears?
   b. Internal gears?

UNIT 67

Helical Milling

The process of milling helical grooves, such as flutes in a drill, teeth in helical gears, or the worm thread on a shaft, is known as *helical milling.* It is performed on the universal milling machine by gearing the dividing head through the worm shaft to the lead screw of the milling machine.

OBJECTIVES

After completing this unit you will be able to:
1. Calculate the lead and helix angle of a helical gear
2. Set up a milling machine to machine a helix
3. Make the calculations and setup for milling a helical gear

HELICAL TERMS

The term *spiral* is often used incorrectly in place of a *helix.*

- A *helix* is a theoretical line or path generated on a *cylindrical* surface by a cutting tool which is fed lengthwise at a uniform rate, while the cylinder is also rotated at a uniform rate (Fig. 67-1). The flutes on a drill or the threads on a bolt are examples of helices.
- A *spiral* is the path generated by a point moving at a fixed rate of advance along the surface of a *rotating cone or plane* (Fig. 67-2). Threads on a wood screw and pipe threads are examples of conical spirals. Watch springs and scroll threads on a universal lathe chuck are examples of plane or flat spirals.

In order to cut either an inch or metric helix, any two of the following must be known:

1. *The lead of the helix,* which is the longitudinal distance the helix advances axially in one complete revolution of the work.
2. *The angle of the helix,* which is formed by the intersection of the helix with the axis of the workpiece.
3. *The diameter (and circumference) of the workpiece.*

In comparing two different helices, it will be noticed that the greater the angle with the centerline, the shorter will be the lead. However, if the diameter is increased but the helix

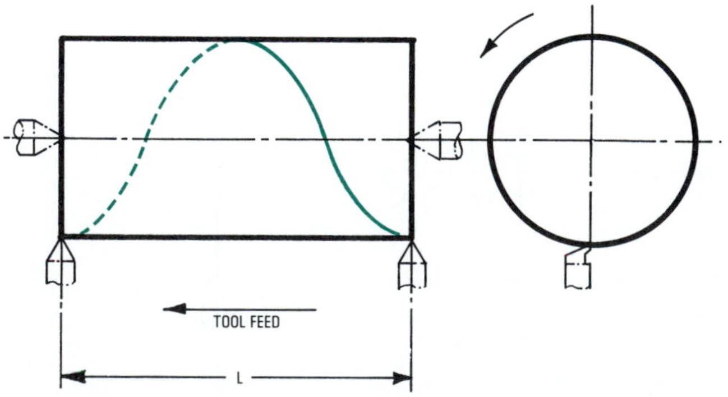
TOOL FEED
L

Figure 67-1 *A helix will be generated if the work is turned as the tool is moved along uniformly.*

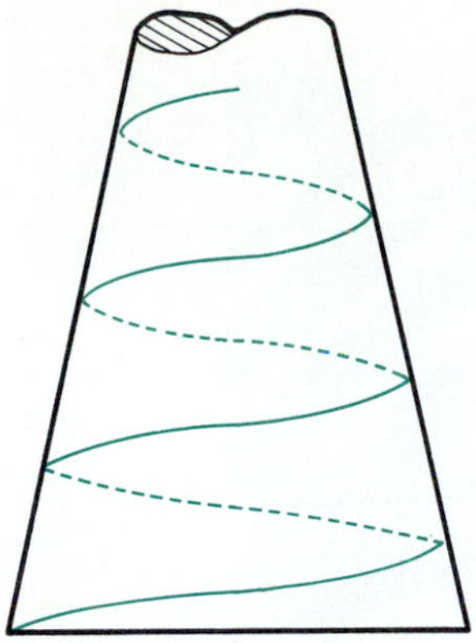

Figure 67-2 *A spiral is produced on a conical surface.*

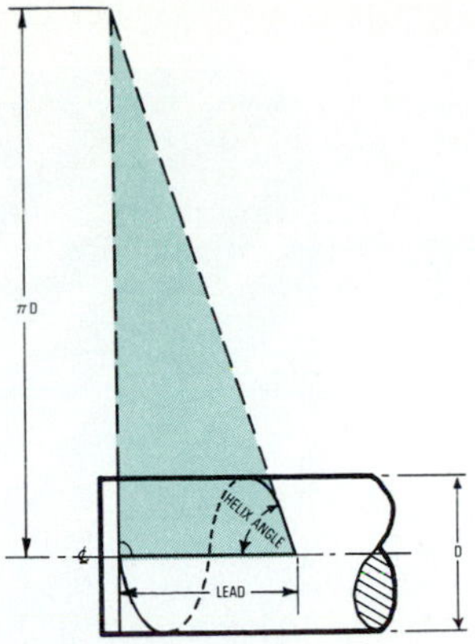
πD
HELIX ANGLE
LEAD
D

Figure 67-3 *Relationship of lead circumference to helix angle.*

angle remains the same, the greater will be the lead. Thus it is evident that the lead of a helix varies with:

1. The diameter of the work
2. The angle of the helix

The relationship between the diameter (and circumference), the helix angle, and the lead is shown in Fig. 67-3. Note that if the surface of the cylinder could be unwound to produce a flat surface, the helix would form the hypotenuse of a right-angle triangle, with the circumference forming the side opposite and the lead the side adjacent.

CUTTING A HELIX

To cut a helix on a cylinder, the following steps are necessary:

1. Swing the table in the proper direction to the angle of the helix to ensure that a groove of the same contour as the cutter is produced.
2. The work must rotate one turn while the table travels lengthwise the distance equal to the lead. This is achieved by installing the proper change gears between the worm shaft on the dividing head and on the milling machine lead screw.

DETERMINING THE HELIX ANGLE

To ensure that a groove of the same contour as the cutter is produced, the table must be swung to the angle of the helix (Fig. 67-4A). The importance of this is shown in Fig. 67-4B.

Note that when the table is not swung (Fig. 67-4B), a helix having the proper lead but an improper contour will

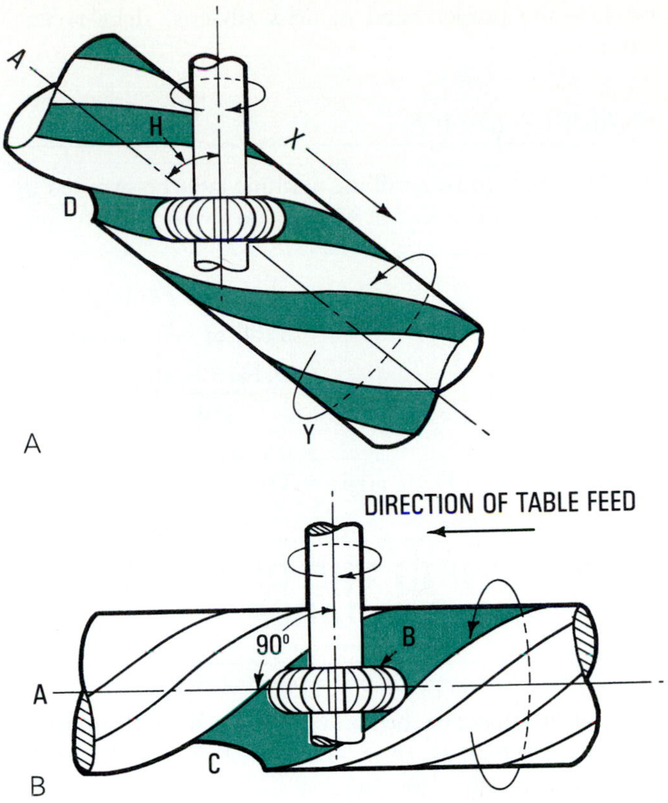
A
H
X
D
Y
A
DIRECTION OF TABLE FEED
90°
B
A
C
B

Figure 67-4 *(A) When the table is swiveled to the helix angle, the exact profile of the cutter will be generated; (B) an incorrect angle produces an incorrect profile.*

be generated. By referring to Fig. 67-3, it can easily be seen that the angle may be calculated as follows:

$$\text{Tangent of helix angle} = \frac{\text{circumference of work}}{\text{lead of helix}}$$

$$= \frac{3.1416 \times \text{diameter } (D)}{\text{lead of helix}}$$

EXAMPLE (inch)

To what angle must the milling machine table be swiveled to cut a helix having a lead of 10.882 in. on a piece of work 2 in. in diameter?

$$\text{Tangent of helix angle} = \frac{3.1416 \times D}{\text{lead of helix}}$$

$$= \frac{3.1416 \times 2}{10.882}$$

$$= \frac{6.2832}{10.882}$$

$$= 0.57739$$

$$\text{Helix angle} = 30°$$

After the helix angle has been calculated, it is necessary to determine the *direction* in which to swivel the table to produce the proper hand of helix (that is, right or left hand).

EXAMPLE (metric)

To what angle must a milling machine table be swiveled to cut a helix having a lead of 450 mm on a workpiece 40 mm in diameter?

$$\text{Tangent of helix angle} = \frac{3.1416\ D(\text{mm})}{\text{Lead of helix (mm)}}$$

$$= \frac{3.1416 \times 40}{450}$$

$$= 0.2792$$

$$\text{Helix angle} = 15°36'$$

DETERMINING THE DIRECTION TO SWING THE TABLE

In order to determine the hand of a helix, hold the cylinder on which the helix is cut in a horizontal plane with its axis running in a right–left direction.

If the helix slopes *down* and to the right, it is a right-hand helix (Fig. 67-5). A left-hand helix slopes *down* and to the left. When a *left-hand helix* is to be cut, the table of the milling machine must be swiveled in a clockwise direction (operator standing in front of the machine). A right-hand helix may be produced similarly by moving the right end of the table in toward the column or by moving it in a counterclockwise direction.

Figure 67-5 *The grooves of a right-hand helical cutter slope down to the right.*

CALCULATING THE CHANGE GEARS TO PRODUCE THE REQUIRED LEAD

To cut a helix, it is necessary to have the work move lengthwise and rotate at the same time. The amount the work (and table) travels lengthwise as the work revolves one complete revolution is the *lead*. The rotation of the work is caused by gearing the worm shaft of the dividing head to the lead screw of the machine (Fig. 67-6).

INCH CALCULATIONS

To cut a helix on an inch milling machine, it is necessary first to understand how to calculate the required change gears for any desired lead. Assume that the dividing head worm shaft is geared to the table lead screw with equal gears (for example, both having 24-tooth gears). The dividing head ratio is 40:1, and a standard milling machine lead screw has 4 threads per inch (*tpi*). The lead screw, as it revolves one turn, would revolve the dividing head spindle one-fortieth of a revolution. In order for the dividing head spindle to revolve one turn, it would be necessary for the lead screw to revolve 40 times. Thus the table would travel 40 × ¼ in., or 10 in., while the work revolves one turn. Therefore, the lead of a milling machine is said to be 10 in. when the lead screw (4 *tpi*) is connected to the dividing head (40:1 ratio) with equal gears.

In calculating the change gears required to cut any lead, the following formula may be used:

$$\frac{\text{Lead of helix}}{\text{Lead of machine (10 in.)}} = \frac{\text{product of driven gears}}{\text{product of driver gears}}$$

The ratio of gears required to produce any lead on a milling machine having a lead screw with 4 *tpi* is always equal to a

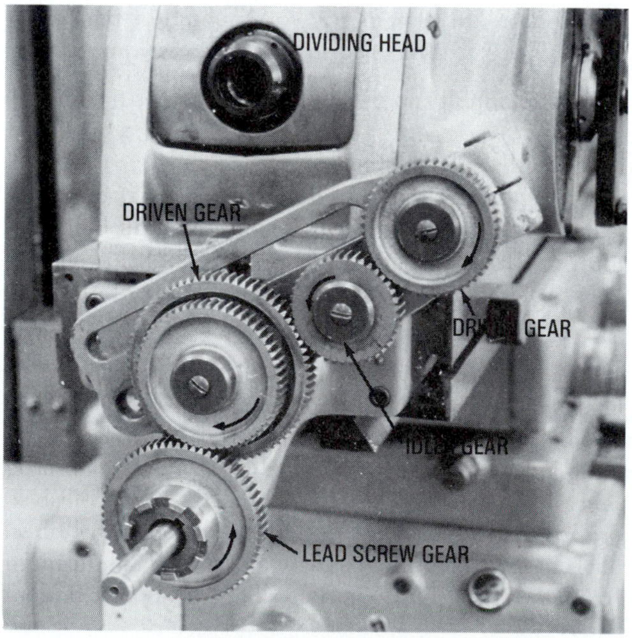
DIVIDING HEAD
DRIVEN GEAR
LEAD SCREW GEAR

Figure 67-6 *The worm shaft and the lead screw are connected for helical milling.*

fraction having the lead of the helix for the numerator and 10 for the denominator.

NOTE: The preceding formula may be inverted if preferred.

$$\frac{\text{Lead of the machine}}{\text{Lead of the helix}} = \frac{\text{product of driver gears}}{\text{product of driven gears}}$$

EXAMPLE

Calculate the change gears required to produce a helix having a lead of 25 in. on a piece of work. The available change gears have the following number of teeth: 24, 24, 28, 32, 36, 40, 44, 48, 56, 64, 72, 86, 100.

SOLUTION

$$\text{Gear ratio} = \frac{\text{lead of helix (driven gears)}}{\text{lead of machine (driver gears)}}$$

$$= \frac{25}{10}$$

Since 10- and 25-tooth gears are not supplied with standard dividing heads, it is necessary to multiply the 25:10 ratio by any number that will suit the change gears available.

$$\text{Gear ratio} = \frac{25}{10} \times \frac{4}{4}$$

$$= \frac{100 \text{ (driven gear)}}{40 \text{ (driver gear)}}$$

As both 100- and 40-tooth gears are available, simple gearing may be used.

EXAMPLE

Calculate the change gears required to produce a helix having a lead of 27 in. The available change gears are the same as those in the preceding example.

SOLUTION

$$\text{Gear ratio} = \frac{\text{lead of helix (driven gears)}}{\text{lead of machine (driver gears)}}$$

$$= \frac{27}{10}$$

Since there are no gears in the set which are multiples of both 27 and 10, it is impossible to use simple gearing. Compound gearing must therefore be used, and it becomes necessary to factor the fraction 27/10 as follows:

$$\text{Gear ratio} = \frac{27}{10}$$

$$= \frac{3}{2} \times \frac{9 \text{ (driven)}}{5 \text{ (driver)}}$$

It is now necessary to multiply both the numerator and denominator of each fraction by the same number in order to bring the ratio into the range of the gears available.

NOTE: This does not change the value of the fraction.

$$\frac{3 \times 16}{2 \times 16} = \frac{48}{32}$$

$$\frac{9 \times 8}{5 \times 8} = \frac{72}{40}$$

$$\text{The gear ratio} = \frac{48 \times 72 \text{ (driven gears)}}{32 \times 40 \text{ (driver gears)}}$$

∴ the driven gears are 48 and 72 and the driver gears are 32 and 40. The gears would be placed in the train as follows (Fig. 67-6):

Gear on worm	72	(driven)
First gear on stud	32	(driver)
Second gear on stud	48	(driven)
Gear on lead screw	40	(driver)

The preceding order is not absolutely necessary; the two driven gears may be interchanged and/or the two driver gears may be interchanged, *provided a driver gear is not interchanged with a driven gear.*

METRIC CALCULATIONS

The pitch of the lead screw on a metric milling machine is stated in millimeters. Most milling machine lead screws have a 5-mm pitch and the dividing head has a ratio of 40:1. As the lead screw revolves one turn, it would revolve the dividing head spindle one-fortieth of a revolution. In order for the dividing head spindle (and work) to revolve one full turn, the lead screw must make 40 complete revolutions. Therefore the lead of the machine would be 40 times the pitch of the lead screw.

For metric calculations, the change gears required are calculated as follows:

$$\frac{\text{Lead of helix (mm)}}{\text{Lead of machine (mm)}} = \frac{\text{product of driven gears}}{\text{product of driver gears}}$$

The normal change gears in a set are 24, 24, 28, 32, 36, 40, 44, 48, 56, 64, 72, 86, 100.

EXAMPLE

Calculate the change gears required to cut a helix having a lead of 500 mm on a workpiece using a standard set of gears. The milling machine lead screw has a pitch of 5 mm.

$$\frac{\text{Driven gears}}{\text{Driver gears}} = \frac{\text{lead of helix}}{\text{pitch of lead screw} \times 40}$$

$$= \frac{500}{5 \times 40}$$

$$= \frac{500}{200}$$

$$= \frac{5}{2} \times \frac{20}{20}$$

$$= \frac{100}{40}$$

Driven gear = 100
Driver gear = 40

DIRECTION OF SPINDLE ROTATION

Figure 67-6 illustrates the setup required to cut a left-hand helix on most machines. Note that the gear on the lead screw and the worm gear revolve in the same direction. To cut a right-hand helix, the spindle must revolve in the opposite direction, and therefore another idler must be inserted as in Fig. 67-7. The idler in this case acts neither as a driven nor a driver gear and is not considered in the calculation of the gear train. It acts merely as a means of changing the direction of rotation of the dividing head spindle. It should also be noted that the direction of spindle rotation for simple gearing will be opposite to that for compound gearing.

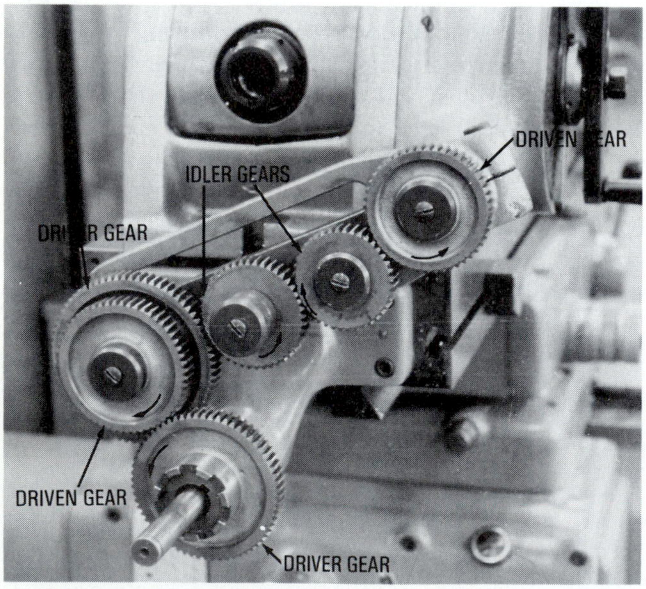
IDLER GEARS
DRIVEN GEAR
DRIVER GEAR

Figure 67-7 *A second idler reverses the direction of rotation.*

CUTTING SHORT LEAD HELICES

When it is necessary to cut leads smaller than those shown in most machinery handbooks, it is advisable to disengage the dividing head worm and wormwheel and connect the change gears directly from the table lead screw to the dividing head spindle rather than to the worm shaft. This method permits machining leads to one-fortieth of the leads shown in the handbook tables. Thus, if the machine is geared to cut a lead of 4.000 in., by connecting the worm shaft and the lead screw the same gearing would produce a lead of 1/40 × 4.000 in., or 0.100 in., when geared directly to the dividing head spindle.

EXAMPLE

A plain helical milling cutter is required to the following specifications:

Diameter: 4 in.
Number of teeth: 9
Helix: right hand
Helix angle: 25°
Rake angle: 10° positive radial rake
Angle of flute: 55°
Depth of flute: ½ in.
Length: 4 in.
Material: tool steel

PROCEDURE

1. Turn blank to sizes indicated (Fig. 67-8).
2. Apply layout die to the end of the blank and lay out as in Fig. 67-9.
3. Lay out a line on periphery to indicate direction of the *right-hand helix* (Fig. 67-10).
4. Press the cutter blank firmly on the mandrel. If a threaded mandrel is used, be sure to tighten the nut securely.
5. Mount the dividing head and footstock.
6. Calculate the indexing for nine divisions.

$$\text{Indexing} = \frac{40}{9}$$

$$= 4\tfrac{4}{9}$$

$$= 4 \text{ turns} + 8 \text{ holes on an 18-hole circle}$$

7. Set the sector arms to 8 holes on the 18-hole circle.

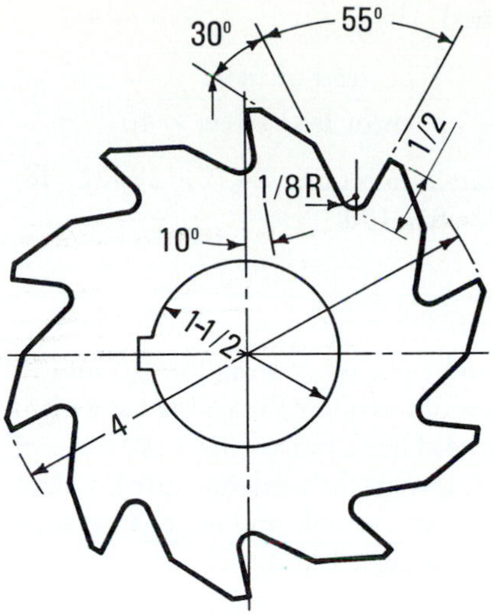
30°
55°
1/2
1/8R
10°
1-1/2
4

Figure 67-8 *Dimensions of a helical milling cutter.*

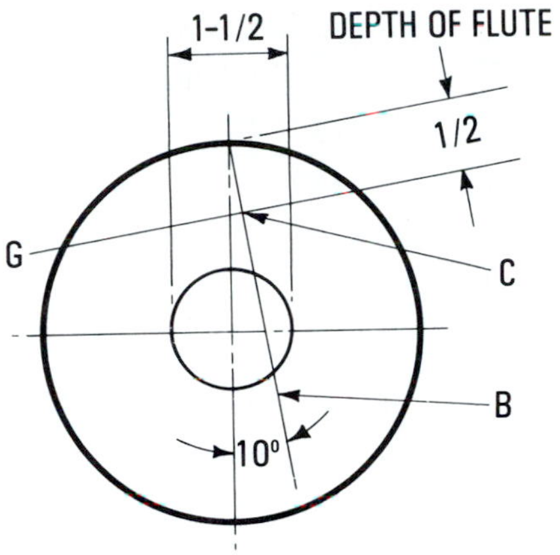
1-1/2
DEPTH OF FLUTE
1/2
G
C
B
10°

Figure 67-9 *Locating the first tooth on the cutter. (Courtesy Cincinnati Milacron Inc.)*

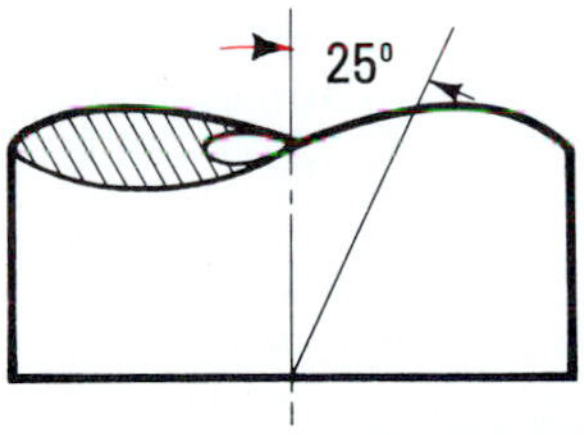
25°

Figure 67-10 *Laying out the direction of the flute. (Courtesy Cincinnati Milacron Inc.)*

NOTE: Do not count the hole in which the pin is engaged.

8. Disengage the index plate locking device.
9. Calculate the lead of the helix.

$$\text{Lead} = \frac{3.1416 \times D}{\text{tan helix angle}}$$
$$= 3.1416 \times D \text{ cot helix angle}$$
$$\left(\text{since } \frac{1}{\text{tan}} = \text{cot}\right)$$
$$= 3.1416 \times 4 \times 2.1445$$
$$= 26.949 \text{ in.}$$

10. Consult any handbook for the change gears to cut the lead closest to 26.949 in. Obviously this is 27.
11. If a handbook is not available, change gears can be calculated for the closest lead, which is 27 in.
12. Change gears required for 27-in. lead are calculated as follows:

$$\frac{\text{Required lead}}{\text{Lead of machine}} = \frac{27}{10}$$
$$= \frac{9}{5} \times \frac{3}{2}$$
$$\frac{9 \times 8}{5 \times 8} = \frac{72}{40} \qquad \frac{3 \times 16}{2 \times 16} = \frac{48}{32}$$
$$\text{Change gears} = \frac{72 \times 48 \text{ (driven gears)}}{40 \times 32 \text{ (driver gears)}}$$

13. Mount the change gears, allowing a slight clearance between mating teeth.
14. Mount the work between the centers with the large end of the mandrel against the dividing head.
15. Swivel the table 25° counterclockwise.
16. Adjust the crossfeed handwheel until the table is about 1 in. from the face of the column. This is to ensure that the table clears the column when the cutter is being machined.
17. Swing the table back to zero (0).
18. Mount a 55° double-angle cutter so that it revolves toward the dividing head, and center it approximately over the flute layout.
19. Rotate the blank until the flute layout is aligned with the cutter edge. This may be checked with a rule or straightedge (Fig. 67-11).
20. Move the blank over, using the crossfeed for the distance of M (Fig. 67-11) or until the point C (Fig. 67-9) is in line with the centerline of the cutter.

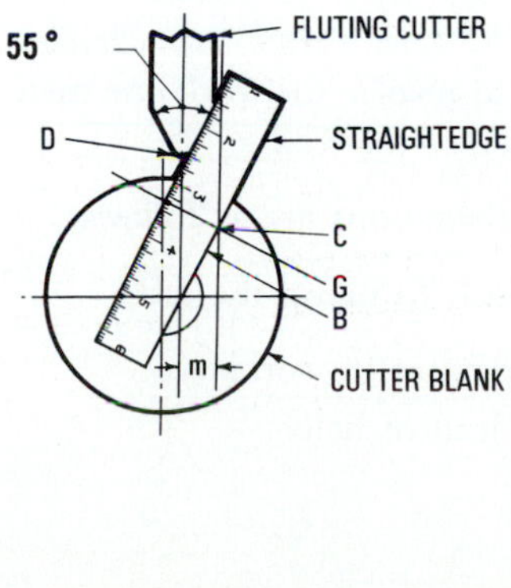
55°
FLUTING CUTTER
D
STRAIGHTEDGE
C
G
B
m
CUTTER BLANK

Figure 67-11 *Aligning the cutter blank with the cutter. (Courtesy Cincinnati Milacron Inc.)*

21. With the work clear of the cutter, set the depth to 0.500 in.
22. Rotate the table 25° counterclockwise, and lock securely (right end in toward the column).
23. Carefully cut the first tooth space, checking the accuracy of the location and the depth.
24. Index for and cut the remaining flutes.
25. Remove the fluting cutter and mount a plain helical milling cutter.
26. Rotate the work (using the index crank) until a line at 30° to the side of the flute is parallel to the table (Fig. 67-9). This may be checked by means of a surface gage. The blank, however, may be rotated by indexing an amount equal to

$$90 - \left(30 + \frac{55}{2}\right)$$
$$= 90 - 57.5$$
$$= 32.5° \; (32°30')$$

Indexing for 32°30′:

$$\begin{aligned} 32° \times 60' &= 1920' \\ 30' &= \underline{\;\;30'} \\ 32°30' &= 1950' \\ &= \frac{1950}{540} \\ &= \frac{330}{540} \\ &= 3\tfrac{11}{18} \\ &= 3 \text{ turns} + 11 \text{ holes on the 18-hole circle} \end{aligned}$$

27. Adjust the workpiece under the cutter.
28. With the cutter rotating, raise the table until the width of the land on the workpiece is about 1/32 in. wide.
29. Cut the secondary clearance (30° angle) on all teeth of the workpiece.

HELICAL MILLING CALCULATIONS (METRIC)

For any helical milling calculation is it necessary to determine:

1. The angle at which the table must be swiveled to produce the proper helix angle
2. The change gears required to revolve the work one turn as the work travels the distance of the lead

For metric helices, these calculations are as follows:

1. Helix angle or angle to which to swivel the table

$$\text{Tangent } \angle = \frac{\text{circumference of workpiece}}{\text{lead of helix}}$$

2. Change gears required

$$\frac{\text{Driven gears}}{\text{Driver gears}} = \frac{\text{lead of helix}}{\text{pitch of lead screw} \times 40}$$

The normal change gears in a set are: 24, 24, 28, 32, 36, 40, 44, 48, 56, 64, 72, 86, 100.

EXAMPLE

A lead of 480 mm is to be cut on a workpiece 40 mm in diameter. The lead screw of the milling machine has a pitch of 5 mm. The dividing head has a ratio of 40:1 (40 turns of the crank are required to revolve the work one turn). Calculate the angle at which to set the table and the change gears necessary to produce the required lead.

SOLUTION

1. *Helix angle*:

$$\begin{aligned} \text{Tangent helix angle} &= \frac{\text{circumference of work}}{\text{lead of helix}} \\ &= \frac{3.1416 \times 40}{480} \\ &= 0.26180 \\ \text{Helix angle} &= 14°40' \end{aligned}$$

2. *Change gears*:

$$\begin{aligned} \frac{\text{Driven gears}}{\text{Driver gears}} &= \frac{\text{lead of helix}}{\text{pitch of lead screw} \times 40} \\ &= \frac{480}{5 \times 40} \\ &= \frac{480}{200} \\ &= \frac{12}{5}\,\frac{(6 \times 2)}{(5 \times 1)} \end{aligned}$$

$$\frac{6}{5} \times \frac{8}{8} = \frac{48}{40} \qquad \frac{2}{1} \times \frac{28}{28} = \frac{56}{28}$$

$$\text{Gears} = \frac{48}{40} \times \frac{56}{28}$$

$$\text{Driven gears} = 48 \text{ and } 56$$
$$\text{Driver gears} = 40 \text{ and } 28$$

HELICAL GEARING

Helical gearing requires a thorough knowledge of all the aspects of spur gearing, as well as the formulas and procedures used in helical milling. In helical gear cutting two new terms are encountered: *normal circular pitch* (NCP) and *normal diametrical pitch* (NDP).

The NCP is the distance from a point on one tooth to a

corresponding point on the next tooth, measured on the pitch circle at right angles to the face of the tooth. The *circular pitch* (CP) is measured on the face of the gear in a plane at right angles to the axis of the gear. Figure 67-12 illustrates the relationship between the CP and the NCP. Note that a right-angle triangle is formed, with a line representing the CP as the hypotenuse and a line representing the NCP as the side opposite. By simple geometry, it may be proved that the angle a is equal to the angle a_1, which is the angle of the helix. Therefore, in the triangle illustrated in Fig. 67-12,

$$\text{Angle } a = \frac{\text{NCP}}{\text{CP}}$$

Therefore, the relationship between NCP and CP is exactly proportional to the length of these lines. The number of teeth in a helical and a spur gear of the same size and pitch will also be proportional to the length of these lines.

It is obvious that the CP will increase as the helix angle increases; therefore, the greater the helix angle, the fewer will be the teeth in a helical gear, compared to a spur gear of the same pitch diameter and diametral pitch (DP).

Since most gearing is calculated on the DP system, it will be necessary to convert the CP into DP terms. Referring to the spur gearing formulas,

$$\text{DP} = \frac{3.1416}{\text{CP}} \quad \text{or} \quad \text{CP} = \frac{3.1416}{\text{DP}}$$

Since normal diametral pitch (NDP) bears the same relation to NCP as DP does to CP, the following formulas will apply:

$$\text{NDP} = \frac{3.1416}{\text{NCP}}$$

$$\text{NCP} = \frac{3.1416}{\text{NDP}}$$

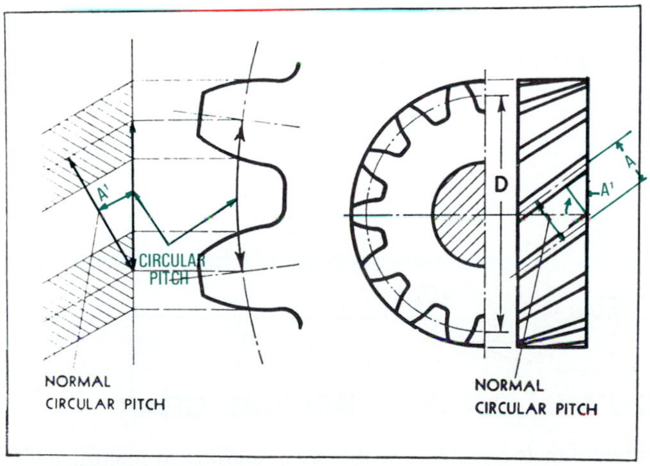

CIRCULAR PITCH
A'
D
A'
A
NORMAL CIRCULAR PITCH
NORMAL CIRCULAR PITCH

Figure 67-12 *Relationships between circular pitch and normal circular pitch.*

Referring to Fig. 67-12, note that

$$\text{Cosine helix angle} = \frac{\text{NCP}}{\text{CP}}$$

This may be converted to DP as follows:

$$\begin{aligned}\text{Cosine helix angle} &= \frac{\text{NCP}}{\text{CP}}\\ &= \frac{(3.1416/\text{NDP})}{(3.1416/\text{DP})}\\ &= \frac{3.1416}{\text{NDP}} \times \frac{\text{DP}}{3.1416}\\ &= \frac{\text{DP}}{\text{NDP}}\end{aligned}$$

Other formulas may be derived as follows:

$$\text{NDP} = \frac{\text{DP}}{\cos \text{ helix angle}}$$

$$\text{DP} = \text{NDP} \times \cos \text{ helix angle}$$

CUTTER SELECTION

The type of cutter used for cutting helical gears is the same as that used for spur gears. However, the DP of a spur gear and the cutter now becomes the NDP of the helical gear. The thickness of the cutter at the pitch line for cutting helical gears should be equal to one-half the NCP.

When a spur gear is being cut, the number of the cutter used is dependent on the number of teeth being cut (Table 66-1). However, when a helical gear is being cut, this table does not apply, since the shape of the tooth would be changed because of the helix angle of the gear teeth. To determine the proper cutter for machining a helical gear, it is necessary to consider the number of teeth being cut and the helix angle of the tooth. The cutter number may be determined by dividing the number of teeth (N) by the cube of the cosine of the angle. Thus,

$$\text{Cutter number} = \frac{N}{(\cos a)^3}$$

EXAMPLE

Determine the cutter number required to cut 38 teeth on a helical gear having an angle of 45°.

$$\begin{aligned}\text{Number of teeth for which to select the cutter} &= \frac{N}{(\cos a)^3}\\ &= \frac{38}{(0.7071)^3}\\ &= \frac{38}{0.3534}\\ &= 108\end{aligned}$$

Therefore, a #2 cutter would be used since it cuts from 55 to 134 teeth.

HELICAL GEAR CALCULATIONS AND FORMULAS

Many of the calculations and formulas that apply to spur gears also apply to helical gears. Others may be calculated from Table 67-1 as required. It is well to remember the following when making helical gears:

1. The DP of a spur gear is the NDP of a helical gear.
2. The NDP of the helical gear is the DP of the cutter
3. The cutter number must be calculated using the formula:

$$\text{Number of teeth for which to select a cutter} = \frac{N}{(\cos a)^3}$$

The alignment chart (Table 67-2) will be helpful when it is necessary to determine cutter numbers for helical gears. Using a straightedge, align the actual number of teeth in the left-hand column with the helix angle in the right-hand column. Read the cutter number from the center column where the straightedge crosses the line.

Metric Helical Gear Calculations

Most metric module helical gear calculations are similar to those for DP helical gears and module spur gears. There are, however, certain changes which must be understood before any calculations are attempted. Refer to Fig. 67-12 to more clearly understand the following terms.

The *circular pitch* (CP) is the distance from one tooth to a corresponding point on the next tooth, measured on the pitch circle. The *real (pitch) module* (*M*) is calculated on the pitch circle.

The *normal circular pitch* (NCP) is the distance from one tooth to a corresponding point on the next tooth, measured at right angles to the tooth face. The *normal (pitch) module* (nM) is calculated on this distance.

For all other calculations, see Unit 66 on metric spur gears and helical milling. Table 67-3 covers the rules and formulas for calculating metric module helical gears.

EXAMPLE

A helical metric 5 (normal) module gear having 38 teeth must be cut. The pitch diameter (PD) of the gear is 200 mm. Calculate:

1. Real module (*M*)
2. Normal module (nM)
3. Circular pitch (CP)
4. Normal circular pitch (NCP)
5. Addendum (*A*)
6. Outside diameter (OD)
7. Depth of tooth (WD)
8. Helix angle (∠)
9. Lead (*L*)
10. Indexing

SOLUTION

1. $M = \frac{200}{38}$
   $= 5.26$ mm
2. nM = 5
3. $\text{CP} = M \times 3.1416$
   $= 5.26 \times 3.1416$
   $= 16.52$ mm
4. $\text{NCP} = \text{nM} \times 3.1416$
   $= 5 \times 3.1416$
   $= 15.70$ mm
5. $A = \text{nM}$
   $= 5$ mm
6. $\text{OD} = \text{PD} + 2\text{nM}$
   $= 200 + 10$
   $= 210$ mm
7. $\text{WD} = \text{nM} \times 2.166$
   $= 10.83$ mm
8. $\text{Cos} \angle = \frac{\text{nM}}{M}$
   $= \frac{5}{5.2632}$
   $= 0.950$
   $= 18°11'$
9. $L = \text{PD} \times 3.1416 \times \cot \angle$
   $= 200 \times 3.1416 \times 3.0445$
   $= 1913$ mm
10. $\text{Indexing} = \frac{40}{N}$
    $= \frac{40}{38}$
    $= 1\frac{2}{38}$
    $= 1\frac{1}{19}$
    = 1 turn of crank + 1 hole in the 19-hole circle

TO MILL AN INCH HELICAL GEAR

The following problem will serve as a guide in making helical gear calculations. If the steps are followed in sequence, little difficulty should be encountered.

Table 67-1 Rules for Calculating Inch Helical Gears

No. of Rule	To Find Rule	Formula
Addendum (*A*)	Divide by the normal diametral pitch	$A = \frac{1}{NDP}$
Cutter number (CN)	Divide the number of teeth by the cube of the cosine of the helix angle.	$CN = \frac{N}{(\cos \angle)^3}$
Helix angle of teeth (∠)	Divide the normal circular pitch by the circular pitch. The quotient equals the cosine of the helix angle.	$\cos \angle = \frac{NCP}{CP}$
	Divide the diametral pitch by the normal diametral pitch. The quotient equals the cosine of the helix angle.	$\cos \angle = \frac{DP}{NDP}$
	Divide the number of teeth by the product of the normal diametral pitch and the pitch diameter. The quotient equals the cosine of the helix angle.	$\cos \angle = \frac{N}{NDP \times PD}$
Lead of the helix (∠)	Divide the product of the number of teeth and the circular pitch by the tangent of the helix angle.	$L = \frac{N \times CP}{\tan \angle}$
	Multiply the pitch diameter by 3.1416 times the cotangent of the helix angle.	$L = 3.1416 \times PD \cot \angle$
Normal diameter pitch (NDP)	Divide the number of teeth by the cosine of the helix angle, add 2, and divide the sum by the outside diameter.	$NDP = \left(\frac{N}{\cos \angle} + 2\right) \div OD$
Normal circular pitch (NCP)	Multiply the circular pitch by the cosine of the helix angle.	$NCP = CP \cos \angle$
Outside diameter of gear blank (OD)	Add 0.6366 times the normal circular pitch to the pitch diameter.	$OD = 0.6366 \times NCP + PD$
Pitch diameter	Multiply the product of the circular pitch and the number of teeth by 0.3183 and divide the product by the cosine of the helix angle.	$PD = \frac{0.3183 \times CP \times N}{\cos \angle}$
	Divide the number of teeth by the product of the normal diameter pitch and the cosine of the helix angle.	$PD = \frac{N}{NDP \cos \angle}$
Thickness of tooth at pitch line (*T*)	Divide 1.571 by the normal diametral pitch.	$T = \frac{1.571}{NDP}$
Whole depth of tooth (WD)	Divide 2.157 by the normal diametral pitch.	$WD = \frac{2.157}{NDP}$

PROBLEM

It is required to cut two mating helical gears having a 3:2 ratio, 12-DP pitch, and 3.000-in. center-to-center distance.

Calculations

Pitch Diameters

The sum of the PDs is 3.000 in. × 2 = 6.000 in. Since the total of the ratios is 3 + 2 = 5, the large gear will occupy three-fifths of the total distance of the pitch diameters (6.000 in.), while the pinion will be two-fifths of the total distance. That is,

$$\begin{aligned} \text{PD of gear} &= 3/5 \times 6.0 = 3.600 \\ \text{PD of pinion} &= 2/5 \times 6.0 = \underline{2.400} \\ & \qquad\qquad\quad 6.000 \text{ in. (proof)} \end{aligned}$$

Table 67-2 Helical GearCutter (Inch) Nomograph

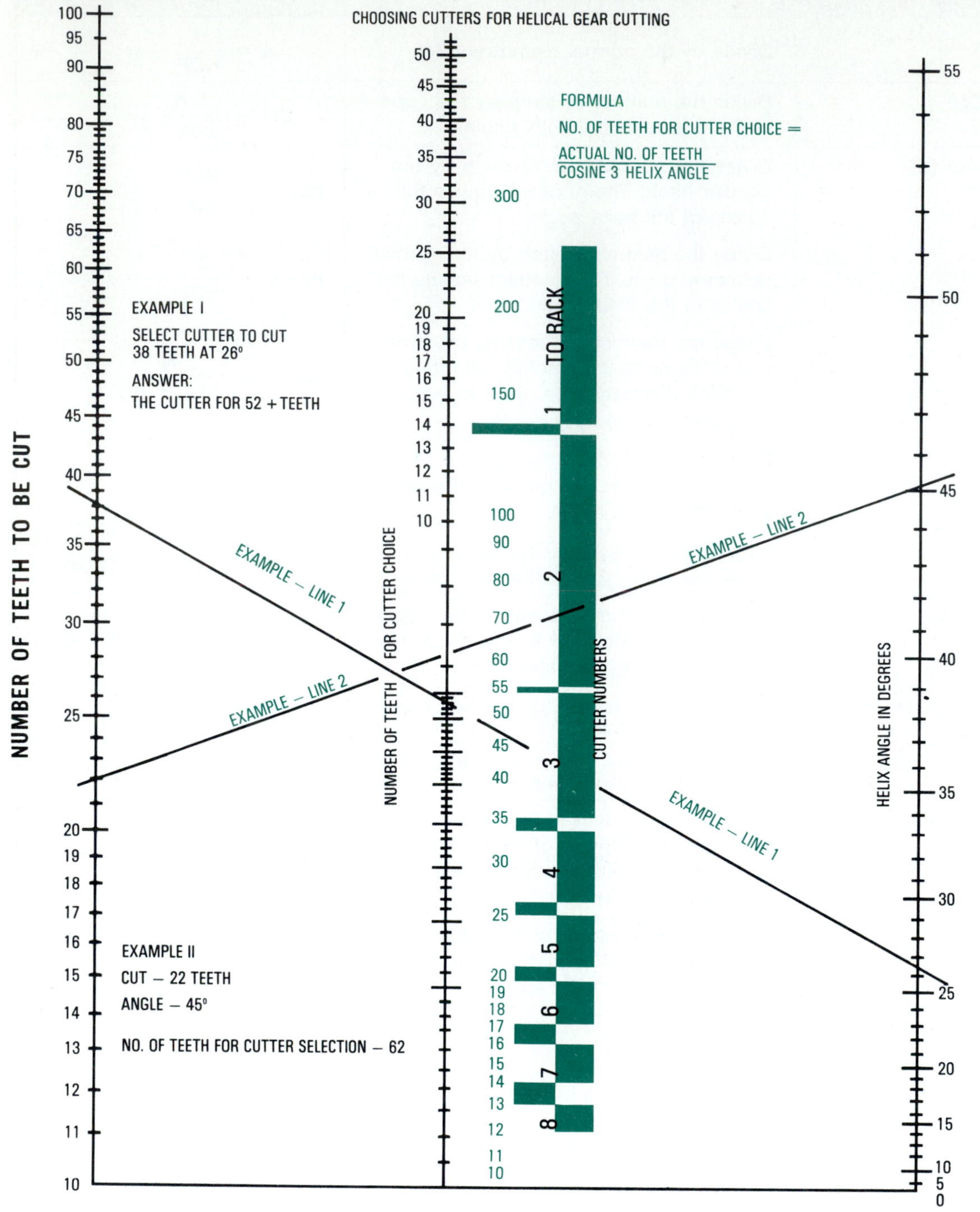
CHOOSING CUTTERS FOR HELICAL GEAR CUTTING
FORMULA
NO. OF TEETH FOR CUTTER CHOICE =
ACTUAL NO. OF TEETH
COSINE 3 HELIX ANGLE
NUMBER OF TEETH TO BE CUT
NUMBER OF TEETH FOR CUTTER CHOICE
CUTTER NUMBERS
HELIX ANGLE IN DEGREES
TO RACK
EXAMPLE I
SELECT CUTTER TO CUT
38 TEETH AT 26°
ANSWER:
THE CUTTER FOR 52 + TEETH
EXAMPLE II
CUT – 22 TEETH
ANGLE – 45°
NO. OF TEETH FOR CUTTER SELECTION – 62
EXAMPLE – LINE 1
EXAMPLE – LINE 2

Number of Teeth

$$N = DP \times PD$$
$$N \text{ in gear} = 3.6 \times 12 = 43.2$$
$$N \text{ in pinion} = 2.4 \times 12 = 28.8$$

Since it is impossible to cut 43.2 or 28.8 teeth, it is necessary to change the number of teeth in each gear to two whole numbers which have a 3:2 ratio. The teeth of the gear are cut at an angle, and there will be fewer teeth in a helical gear than in a corresponding spur gear. It is therefore necessary to select two whole numbers which are *less* than 43.2 and 28.8 and which have a ratio of 3:2.

The largest numbers under 43.2 and 28.8 having this ratio are 42 and 28. Therefore:

$$N \text{ in gear} = 42$$
$$N \text{ in pinion} = 28$$

Table 67-3 Rules and Formulas for Module Helical Gears

To Obtain	Knowing	Rule	Formula
Addendum (*A*)	Normal module	Addendum equals normal module.	$A = \text{nM}$
Dedendum (*D*)	Normal module	Multiply normal module by 1.166.	$D = \text{nM} \times 1.166$
Circular pitch (CP)	Module	Multiply the module by π.	$\text{CP} = M \times 3.1416$
	Normal circular pitch and helix angle	Divide the normal circular pitch by the cosine of the helix angle.	$\text{CP} = \dfrac{\text{NCP}}{\cos \angle}$
	Normal module and helix angle	Multiply the normal module by π and divide the product by the cosine of the helix angle.	$\text{CP} = \dfrac{\text{nM} \times 3.1416}{\cos \angle}$
Normal circular pitch (NCP)	Normal module	Multiply the normal module by π.	$\text{NCP} = \text{nM} \times 3.1416$
	Circular pitch and helix angle	Multiply the circular pitch by the cosine of the helix angle.	$\text{NCP} = \text{CP} \times \cos \angle$
	Pitch diameter, helix angle, and number of teeth	Multiply the pitch circumference by the cosine of the helix angle and divide the product by the number of teeth.	$\text{NCP} = \dfrac{\text{PD} \times 3.1416 \times \cos \angle}{N}$
Helix angle of teeth ($\angle$)	Normal module and module	Normal module divided by the module equals the cosine of the helix angle.	$\cos \angle = \dfrac{\text{nM}}{M}$
	Number of teeth, normal module, and pitch diameter	The number of teeth multiplied by the normal module and divided by the pitch diameter equals the cosine of the helix angle.	$\cos \angle = \dfrac{N \times \text{nM}}{\text{PD}}$
Lead of helix (*L*)	Pitch diameter and helix angle	Multiply the circumference of the pitch circle by the cotangent of the helix angle.	$L = \text{PD} \times 3.1416 \times \cot \angle$
	Number of teeth, module, and helix angle	Multiply the number of teeth by the module times 3.1416. Multiply the product by the cotangent of the helix angle.	$L = N \times M \times 3.1416 \times \cot \angle$
Normal module (nM)	Module and helix angle	Multiply the module by the cosine of the helix angle.	$\text{nM} = M \times \cos \angle$
	Normal circular pitch	Divide the normal circular pitch by π.	$\text{nM} = \dfrac{\text{NCP}}{3.1416}$
	Circular pitch and helix angle	Multiply the circular pitch by the cosine of the helix angle and divide the product by π.	$\text{M} = \dfrac{\text{CP} \times \cos \angle}{3.1416}$
Module (*M*)	Normal module and helix angle	Divide normal module by the cosine of the helix angle.	$M = \dfrac{\text{nM}}{\cos \angle}$
	Circular pitch	Divide the circular pitch by π	$M = \dfrac{\text{CP}}{3.1416}$
	Normal circular pitch and helix angle	Divide the normal circular pitch by π times the cosine of the helix angle.	$M = \dfrac{\text{NCP}}{3.1416 \times \cos \angle}$

Table 67-3 (Continued)

To Obtain	Knowing	Rule	Formula
Outside diameter (OD)	Pitch diameter and normal module	Add 2 normal modules to the pitch diameter.	$OD = PD + 2nM$
	Normal module, number of teeth, and helix angle	Divide the number of teeth by the cosine of the helix angle and add 2 to the quotient. Multiply this number by the normal module.	$OD = nM\left(\frac{N}{\cos \angle} + 2\right)$
Pitch diameter (PD)	Normal module and number of teeth	Multiply the number of teeth by the normal module and divide the product by the cosine of the helix angle.	$PD = \frac{N \times nM}{\cos \angle}$
Number of teeth (N)	Pitch diameter and circular pitch	Multiply the pitch diameter by π and divide the product by the circular pitch.	$N = \frac{PD \times 3.1416}{CP}$
	Pitch diameter and module	Divide the pitch diameter by the module.	$N = \frac{PD}{M}$
	Pitch diameter, normal module, and helix angle	Multiply the pitch diameter by the cosine of the helix angle and divide the product by the normal module.	$N = \frac{PD \times \cos \angle}{nM}$
Whole depth of tooth (WD)	Normal module	Multiply the normal module by 2.166.	$WD = nM \times 2.166$

Diametral Pitch

$$DP = \frac{N}{PD}$$

$$\text{DP of gear} = \frac{42}{3.6}$$

$$= 11.666 \text{ teeth}$$

OR

$$\text{DP of pinion} = \frac{28}{2.4}$$

$$= 11.666 \text{ teeth (proof)}$$

It has now been established that the DP, or the number of teeth per inch of PD of each gear, is 11.666. However, a 12-DP cutter, which will cut 12 teeth per inch of PD, must be used. This will necessitate swinging the table to the proper helix angle to cut the gears.

Helix Angle

$$\text{Cosine helix angle} = \frac{DP}{NDP}$$

$$= \frac{11.666}{12}$$

$$= 0.97216$$

$$\therefore \text{helix angle} = 13°33'$$

Outside Diameter

$$OD = PD + \frac{2}{DP}$$

$$\text{OD of gear} = 3.600 + \frac{2}{12}$$

$$= 3.766 \text{ in.}$$

$$\text{OD of pinion} = 2.400 + \frac{2}{12}$$

$$= 2.566 \text{ in.}$$

Leads

$$\text{Lead} = PD \times 3.1416 \cot \text{helix angle}$$

$$\text{Lead of gear} = 3.6 \times 3.1416 \cot 13°33'$$
$$= 3.6 \times 3.1416 \times 4.1493$$
$$= 46.926 \text{ in.}$$

$$\text{Lead of pinion} = 2.4 \times 3.1416 \times 4.1493$$
$$= 31.285 \text{ in.}$$

Change Gears

The change gears may be determined from the table supplied with the dividing head or most machinery handbooks. To determine the change gears for the gear having a lead of

46.926 in., check a handbook for the closest lead, which is 46.880 in. The gears for this lead are as follows:

$$\frac{\text{Driven gears}}{\text{Driver gears}} = \frac{100 \times 72}{32 \times 48}$$

The error in this lead is 46.926 − 46.880 = 0.046 in. in approximately 47 in. Change gears for the pinion having a lead of 31.285 in.; the closest lead is 31.270 in.

$$\frac{\text{Driven gears}}{\text{Driver gears}} = \frac{86 \times 64}{40 \times 44}$$

The error in this lead is 31.285 − 31.270 = 0.015 in. in approximately 31 in.

NOTE: If a chart or handbook is not available, the change gears may be calculated by *continued fractions*, which are dealt with later in this unit where the examples for this pinion are shown.

Cutter Number

$$\begin{aligned}\text{Number of teeth in gear for which to select a cutter} &= \frac{\text{number of teeth in gear}}{(\cos \text{helix angle})^3} \\ &= \frac{42}{(0.97216)^3} \\ &= 45.71 \text{ teeth}\end{aligned}$$

A #3 cutter would be used since it will cut from 35 to 54 teeth.

$$\begin{aligned}\text{Number of teeth in pinion for which to select a cutter} &= \frac{28}{(0.97216)^3} \\ &= 30.47 \text{ teeth}\end{aligned}$$

A #4 cutter would be selected since it will cut from 26 to 34 teeth.

The alignment chart (Table 67-2) may be used to determine the proper cutter number. This will eliminate the need to calculate the required number of the cutter.

CUTTING THE GEARS

The setup for cutting the gears is exactly the same as that for helical milling. Having made all the necessary calculations for the gear and the pinion, list all the pertinent information, including the depth of tooth and indexing for both gears, on a separate sheet of paper.

PROCEDURE

1. Mount the dividing head at the end of the milling machine table.
2. Mount the work between centers.
3. Set the indexing crank and the sector arms.
4. Disengage the index plate locking device.
5. Mount the change gears as required.
6. Swing the table to the helix angle and in the proper direction.

NOTE: When cutting mating helical gears for parallel shafts, it is always necessary to cut one gear with a left-hand

Helical Gear Data		
	Gear	Pinion
DP	12	12
PD	3.600 in.	2.400 in.
OD	3.766 in.	2.566 in.
N	42	28
Cutter number	3	4
Indexing (Cincinnati plate)	20 holes on 21-hole circle	1 turn 9 holes on 21-hole circle
Depth of tooth	0.1797 in.	0.1797 in.
Helix angle	13°33′ (right-hand)	13°33′ (left-hand)
Direction to swing table	counterclockwise	clockwise
Lead	46.926 in.	31.285 in.
Change gears—Worm gear	100	86
First stud gear	32	40
Second stud gear	72	64
Lead screw gear	48	44

helix and the other with a right-hand helix so that the gears will mesh.

7. Move the table, using the crossfeed handle, to within 1 in. (25 mm) of the column to ensure clearance between the table and column.
8. Swing the table back to zero (0).
9. Mount the cutter on the arbor over the approximate center of the work.
10. Center the work with the cutter.
11. Raise the table until the cutter just touches the work.
12. Set the vertical feed screw collar to zero (0).
13. With the work clear of the cutter, raise the table to the proper depth of the tooth.
14. Swing the table in the proper direction.

NOTE: When the table is swung in the other direction to cut the opposite helix, it will be necessary to add an idler to the gear train in order to reverse the direction of rotation of the work.

15. Nick each tooth to ensure the proper indexing.
16. Cut all teeth.

NOTE: Lower the table one full turn of the handwheel before returning it to the starting position so that the cutter will not damage the finish on the teeth.

CONTINUED FRACTIONS

Most handbooks contain a table of inch change gears for milling various inch leads. If there are no tables available or if the lead is not within the range shown in the tables, the change gears may be calculated by means of *continued fractions* or *successive quotients.*

To calculate the change gears by this method, it is necessary to form a fraction using the required lead and the lead of the machine.

EXAMPLE

Calculate the gears required to produce a lead of 5.189 in. on a helical gear. The lead of the milling machine is 10 in. The available change gears are as follows: 24, 24, 28, 32, 36, 40, 44, 48, 56, 64, 72, 86, 100.

SOLUTION

1. Gear ratio $= \dfrac{\text{required lead}}{\text{lead of machine}}$

$$= \frac{\text{product driven}}{\text{product drivers}}$$

$$= \frac{5.189}{10}$$

$$= \frac{5189}{10000}$$

2. Since this fraction cannot be factored, the required gears may be obtained by dividing the smaller number into the larger number to obtain the successive quotients. Continue to divide the remainder into the previous divisor until the remainder is zero (0) (Fig. 67-13).
3. Arrange the successive quotients on a chart (Fig. 67-14). In the left-hand columns, place the figures 1 and 0 as in Fig. 67-15. The numbers are arranged in this way only when the gear ratio is a proper fraction, that is, 5189/10 000, or when the lead of the helix is less than the lead of the machine.

If the ratio is an improper fraction, such as that produced by the lead required for the larger helical gear, which has a lead greater than the lead of the machine (as in the foregoing problem, 15 566/10 000), the numbers must be arranged as in Fig. 67-16.

4. After the numbers have been arranged on the chart, the convergents are found by multiplying each quotient (A) by the number in the box (B), the upper row of convergents, which is located below and one square to the left of the quotient. To this product add the number in the next box to the left (C) (Fig. 67-17). The result is then placed in the convergent box below the quotient used (A). For example.

$$A \times B + C \text{ (in the upper section)} = 1 \times 0 + 1 = 1$$

5. Move to the next quotient in the upper row and follow the same procedure.

NOTE: Always multiply the number in the upper row by the number in the box below and to the left. To this product, add the value of the next box (second row) to the left. For example,

$$12 \times 1 + 1 = 13$$

which would be placed below the 12 (Fig. 67-18). Continue this process until the original number (5189) is arrived at (Fig. 67-19).

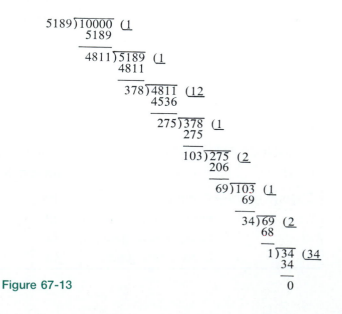

```
5189)10000 (1
     5189
     ----
     4811)5189 (1
          4811
          ----
           378)4811 (12
               4536
               ----
                275)378 (1
                    275
                    ---
                    103)275 (2
                        206
                        ---
                         69)103 (1
                             69
                             --
                             34)69 (2
                                68
                                --
                                 1)34 (34
                                   34
                                   --
                                    0
```

Figure 67-13

		1	1	12	1	2	1	2	34	QUOTIENTS
1	0									
0	1									
		1	1	12	1	2	1	2	34	QUOTIENTS

Figure 67-14

1	0	
0	1	

Figure 67-15

0	1
1	0

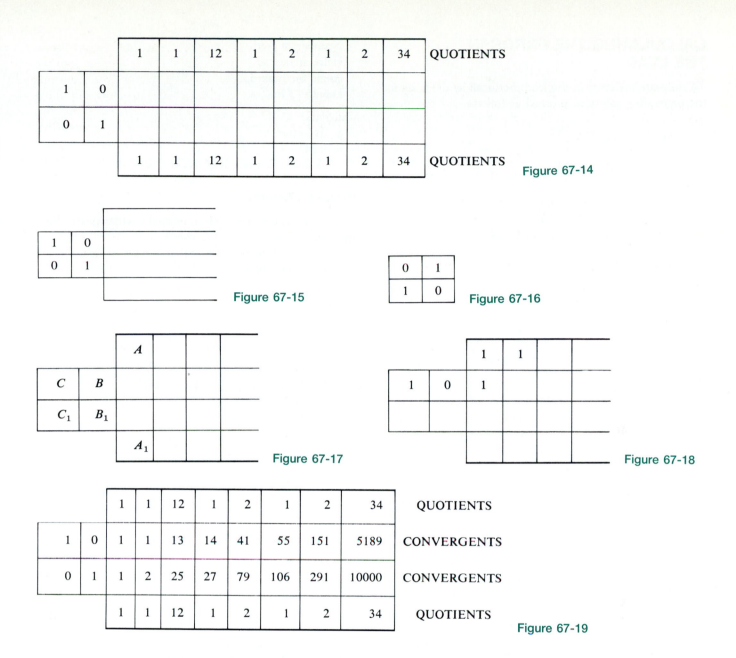

Figure 67-16

		A			
C	*B*				
C_1	B_1				
		A_1			

Figure 67-17

		1	1		
1	0	1			

Figure 67-18

		1	1	12	1	2	1	2	34	QUOTIENTS
1	0	1	1	13	14	41	55	151	5189	CONVERGENTS
0	1	1	2	25	27	79	106	291	10000	CONVERGENTS
		1	1	12	1	2	1	2	34	QUOTIENTS

Figure 67-19

6. Follow the same procedure for the lower set of convergents, which should end with the original number (10 000). The calculation is proved correct when the original number appears in the runs of convergents. Examination of Fig. 67-19 reveals several sets of fractions in the convergent squares. These are really ratios which become more accurate as they move to the right until the original numbers finally appear.

7. Starting at the right, examine each ratio until one is found that will factor into a set of numbers which can be fitted into the range of change gears available. In this case (Fig. 67-19), the ratio or fraction is 14/27.

8. Factor this fraction as follows:

$$\frac{14}{27} = \frac{7 \times 2}{9 \times 3}$$

9. These fractions may then be multiplied by a fraction having the value of 1 to suit the change gears available.

$$\frac{7}{9} \times \frac{8}{8} = \frac{56}{72}$$

$$\frac{2}{3} \times \frac{16}{16} = \frac{32}{48}$$

Therefore, the gears used to cut a lead reasonably close to 5.189 in. are:

$$\frac{56 \times 32}{72 \times 48} = \frac{\text{driven gears}}{\text{driver gears}}$$

It should be noted that the farther to the right that the factorable fraction lies, the more accurate will be the lead produced by the gears.

CALCULATING THE ERROR IN THE LEAD

To calculate the error in the lead produced by the gears in the preceeding solution, proceed as follows.

$$\begin{aligned}\text{Lead produced} &= \frac{56}{72} \times \frac{32}{48}\\ &= \frac{14}{27} \times 10 \text{ (lead of machine)}\\ &= 5.1895 \text{ in.}\\ \text{Desired lead} &= 5.189\\ \text{Error} &= 5.189 - 5.185\\ &= 0.004 \text{ in.}\end{aligned}$$

REVIEW QUESTIONS

HELICAL MILLING

1. Define:

a. Helix
b. Spiral
c. Lead
d. Angle of helix

2. Make a sketch to illustrate the relationship between the lead, circumference, and helix angles.

3. List two factors which affect the lead of a helix.

4. To what angle must the table be swung to cut the following helices?

a. Lead 10.290 in., diameter of workpiece 3¼ in.
b. Lead 12.000 in., diameter of workpiece 2¾ in.
c. Lead = 600 mm, work diameter = 100 mm
d. Lead = 232 mm, work diameter = 25 mm

5. How may a right- and left-hand helix be recognized, and in which direction should the table be swiveled for each?

6. a. Calculate the change gears to cut the following leads:

(1) 6.000 in.
(2) 7.500 in.
(3) 9.600 in.

b. The lead screw of a milling machine has a pitch of 5 mm. The available change gears are 24, 24, 28, 32, 36, 40, 44, 48, 56, 64, 72, 86, 100. Calculate the change gears required for leads of 800 mm and 560 mm.

7. It is required to make a helical milling cutter having the following specifications.

Diameter: 3.475 in.
Helix: left-hand
Rake angle: 5° positive
Depth of flute: 0.5 in.
Material: tool steel
Number of teeth: 7
Helix angle: 20°
Angle of flute: 55°
Length: 3 in.

Calculate:

a. indexing
b. lead
c. change gears required to cut this lead

HELICAL GEARING

8. Compare circular pitch to normal circular pitch. Illustrate by means of a suitable sketch.

9. a. The following information applies to helical gears having parallel shafts.

	Pitch	Ratio	Center Distance
(1)	3	3:1	10.5 in.
(2)	10	4:1	5.5 in.
(3)	6	3:1	7.5 in.
(4)	8	12:5	6.0625 in.

For the above gears and pinions calculate:

(1) Pitch diameter
(2) Number of teeth
(3) Diametral pitch
(4) Helix angle
(5) Outside diameter
(6) Lead
(7) Change gears
(8) Cutter number to be used

NOTE: A standard set of change gears is supplied.

b. A 34-tooth helical metric 6 module gear with a 600 mm lead is to be cut. The pitch diameter of the gear is 180 mm and the pitch of the lead screw is 5 mm. Calculate:

(1) Module
(2) Circular pitch
(3) Normal circular pitch
(4) Addendum
(5) Outside diameter
(6) Depth of tooth
(7) Helix angle
(8) Indexing

CONTINUED FRACTIONS

10. Using continued fractions, calculate the change gears required to produce the following leads, when a standard set of change gears is supplied.

a. 20.570 in.
b. 31.360 in.
c. 9.778 in.
d. 6.667 in.

UNIT 68

Cam, Rack, Worm, and Clutch Milling

The variety of attachments available make the milling machine a very versatile machine tool. Besides the standard operations generally performed on a milling machine, with the proper setups and attachments, it is possible to cut cams, racks, worms, and clutches. Although a machinist may not be called on frequently to perform them, it is wise to be familiar with these operations so that you will be able to cut the form required.

OBJECTIVES

After completing this unit you will be able to:
1. Calculate and cut a uniform-motion cam
2. Set up the machine and cut a rack
3. Understand how a worm is cut
4. Set up the machine and cut a clutch

CAMS AND CAM MILLING

A *cam* is a device generally applied to a machine to change rotary motion into straight-line or reciprocating motion and to transmit this motion to other parts of the machine through a follower. The cam shaft on an automobile engine incorporates several cams, which control the opening and closing of the intake and exhaust valves. Many machine operations, especially on automatic machines, are controlled by cams, which transmit the desired motion to the cutting tool through a follower and some type of push rod.

Cams are also used to transform linear motion into a reciprocating motion of the follower. Cams of this type are called *plate,* or *bar,* cams or are often referred to as *templates.* Templates are often used on tracer-type milling machines and lathes where parts must be produced to the profile of the template.

Cams may also be used as locking devices. Extensive applications are found in jig and fixture design and in quick-locking clamps.

CAMS USED TO IMPART MOTION

Cams used to impart motion are generally found on machines and may be of two types: the *positive* and the *nonpositive.*

Positive-type cams, such as the cylindrical and grooved plate (Fig. 68-1), control the follower at all times. That is, the follower remains engaged in the groove on the face or the periphery of the cam and uses no other means to maintain this engagement.

Example of the *nonpositive-type cams* are the plate, toe and wiper, and crown (Fig. 68-2). In the nonpositive types, the cam pushes the follower in a given direction and depends on some external force, such as gravity or springs, to keep the follower bearing against the cam surface.

Followers may be of several types:

- The *roller type* (Fig. 68-3A) has the least frictional drag and requires little or no lubrication.
- The *tapered roller type* (Fig. 68-3B) is used with grooved plate or cylindrical cams.
- The *flat,* or *plunger, type* (Fig. 68-3C) is used to transmit large forces and requires lubrication.
- The *knife-edge,* or *pointed, type* (Fig. 68-3D) is used on more intricate cams as it permits sharp contours to be followed more readily than with a roller cam.

CAM MOTIONS

There are three standard types of motions imparted by cams to followers and machine parts:

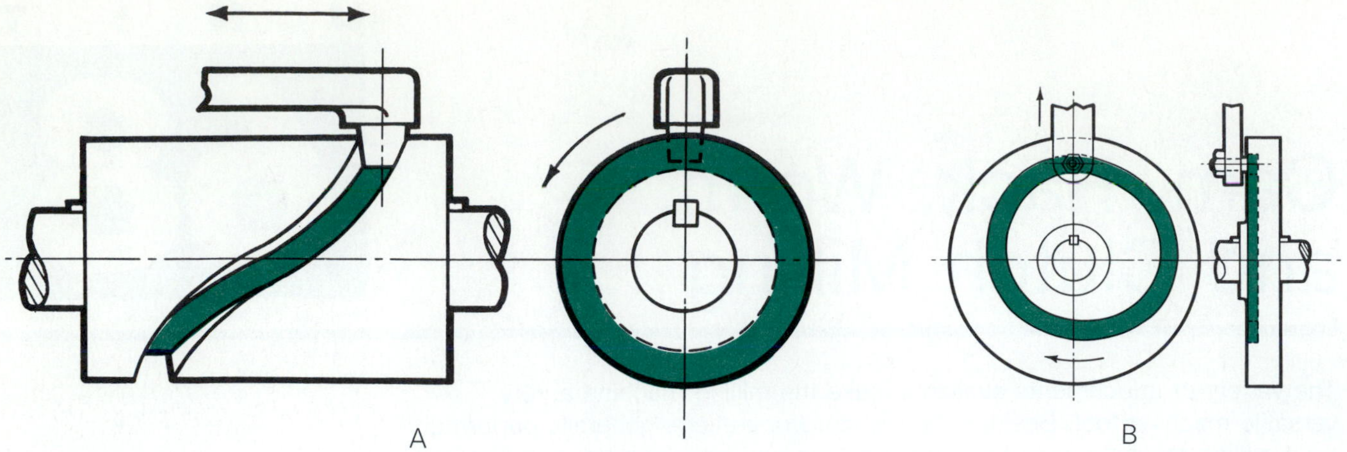
A
B

Figure 68-1 *(A) Cylindrical or drum cam with a tapered roller follower; (B) grooved-plate cam with a roller follower.*

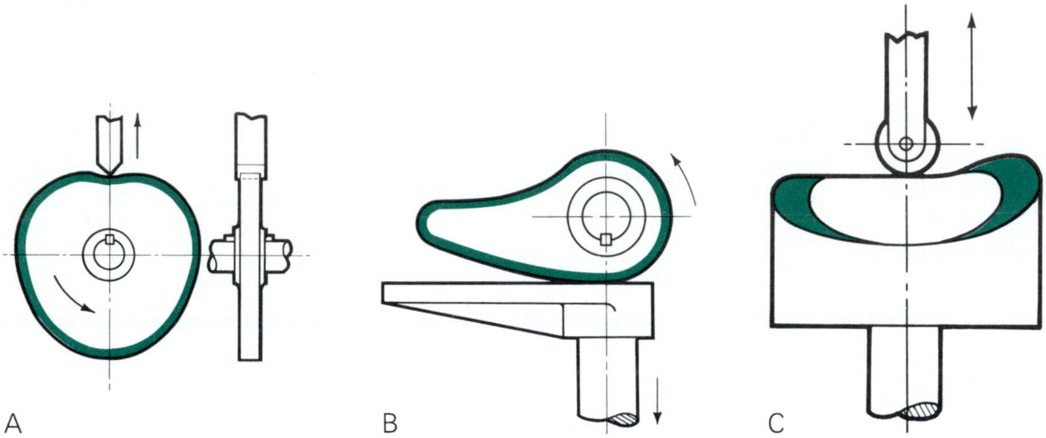
A
B
C

Figure 68-2 *(A) Plate cam with a knife-edge follower; (B) toe and wiper cam; (C) crown cam.*

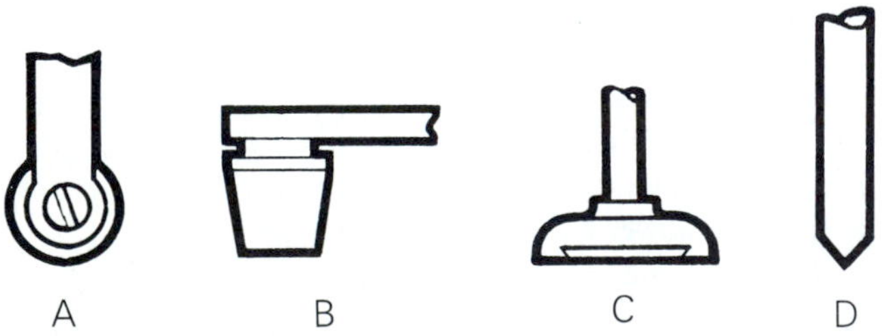
A
B
C
D

Figure 68-3 *Types of cam followers: (A) roller; (B) tapered roller; (C) flat; (D) knife edge or pointed.*

1. Uniform motion
2. Harmonic motion
3. Uniformly accelerated and decelerated motion

The *uniform-motion cam* moves the follower at the same rate from the beginning to the end of the stroke. Since the movement goes from zero to full speed instantaneously and ends in the same abrupt way, there is a distinct shock at the beginning and the end of the stroke. Machines using this type of cam must be rigid and sturdy enough to withstand this constant shock.

The *harmonic-motion cam* provides a smooth start and stop to the cycle. It is used when uniformity of motion is not essential and where high speeds are required.

The *uniformly accelerated and decelerated cam* moves the follower slowly at first, then accelerates or decelerates at a uniform rate. It then gradually decreases in speed, permitting the follower to come to a slow stop before reversal takes place. This type is considered the smoothest of the three motions and is used on high-speed machines.

RADIAL CAM TERMS

A *lobe* is a projecting part of the cam which imparts a reciprocal motion to the follower. Cams may have one or several lobes, depending on the application to the machine (Figs. 68-4 and 68-5).

The *rise* is the distance one lobe will raise or lower the follower as the cam revolves.

The *lead* is the total travel which would be imparted to the follower in one revolution of a uniform-rise cam, having only one lobe in 360°. In Fig. 68-5, the lead for a double-lobe cam is twice the lead of a single-lobe cam having the same rise. It is the lead of the cam and not the rise that controls the gear selection in cam milling.

Figure 68-4 *Single-lobe uniform-rise cam.*

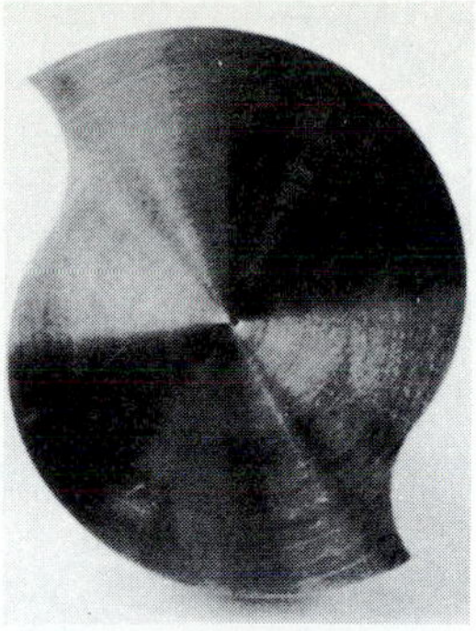

Figure 68-5 *Double-lobe uniform-rise cam.*

Uniform rise is the rise generated on a cam which moves inward at an even rate around the cam, assuming the shape of an Archimedes spiral. This is caused by uniform feed and rotation of the work when a cam is being machined.

CAM MILLING

In the majority of plate cams which do not have a uniform rise, the cam must be laid out and machined by incremental cuts. By this method, the blank is rotated through an angular increment and the cut is taken to the layout line or a predetermined point. This process is repeated until the outline of the cam is produced as closely as possible. The ridges left between each successive cut are then removed by filing and polishing (Fig. 68-6).

Uniform-rise cams may be produced in the milling machine with a vertical head, by the combined *uniform rotation* of the cam blank, held in the spindle of a dividing head, and the *uniform feed* of the table.

When a cam is machined by this method, the work and the vertical head are usually swung at an angle so that the axis of the work and the axis of the mill are parallel (Fig. 68-7).

If the work and the vertical milling attachment are maintained in a vertical position, only a cam having the same lead for which the machine is geared can be cut. When the work and the vertical milling attachment are inclined, any desired lead may be produced, providing that the desired lead is *less* than the lead for which the machine is geared. In other words, the required lead to be cut on the cam must always be less than the forward feed of the table during one revolution of the work.

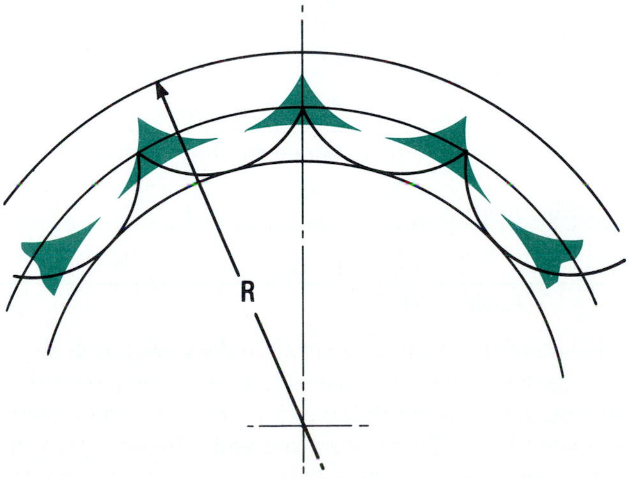

R

Figure 68-6 *Ridges left by incremental cuts.*

CINCINNATI

Figure 68-7 *Machine set up for cam milling using a short-lead milling attachment.*

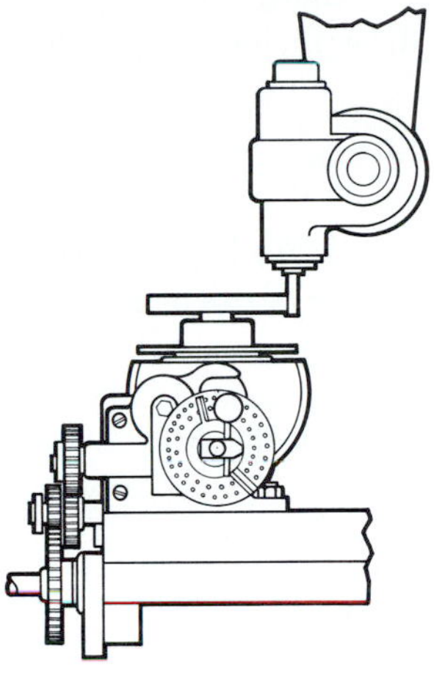

Figure 68-8 *Vertical head and workpiece set at 90°.*

The principle involved in swinging the head is as follows: If a milling machine is set up to cut a cam and has equal gears on the dividing head and the lead screw, with the work and the vertical attachment in a vertical position (Fig. 68-8), the table would advance 10 in. while the work revolved one turn. An Archimedes spiral would be generated, and the cam would have a lead of 10 in., which is also a rise of 10 in. in 360°.

Rotate both the work and cutter so that they are parallel to the table and reading zero (Fig. 68-9). If a cut is taken, the work will move along the length of the end mill (which in this case would have to be 10 in.), and a circle having no lead would be generated.

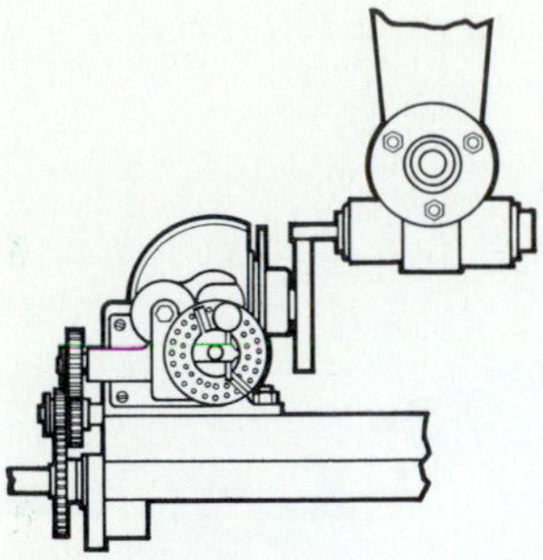

Figure 68-9 *Vertical attachment and workpiece at zero.*

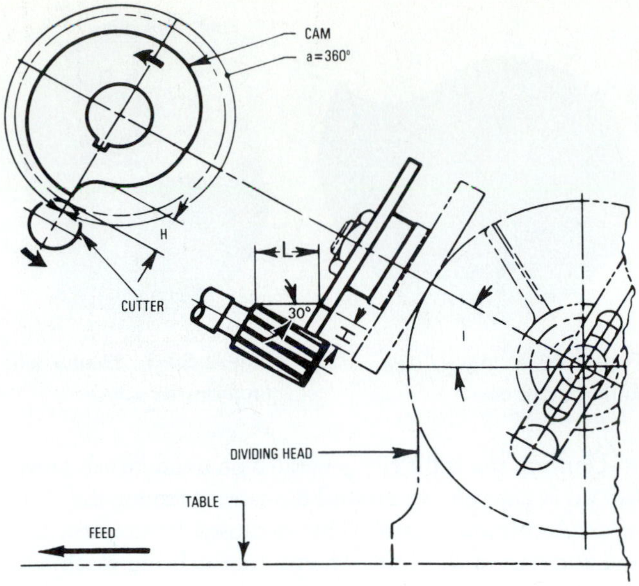
CAM
a = 360°
H
CUTTER
L
30°
H
I
DIVIDING HEAD
TABLE
FEED

Figure 68-10 *Relation between cam lead and table lead. (Courtesy Cincinnati Milacron Inc.)*

From these two examples, it can be seen that it is possible to mill any lead between zero (0) and that for which the machine is geared, if the cutter and the work are inclined to any given angle between 0 and 90°. If this method were not used, it would be necessary to have a different change gear combination for every lead to be cut. This would be impossible because of the large number of change gears required and the time required to change the gears for each different lead.

Calculations Required

From the drawing determine the *lead* of the lobe or lobes of the cam; that is, determine the amount of rise of each lobe if it were continued for the full circumference of the cam.

If the space occupied by the lobe is indicated in degrees on the drawing, the lead would be calculated as follows:

$$\text{Lead} = \frac{\text{rise of lobe in inches} \times 360}{\text{number of degrees of circumference occupied by the lobe}}$$

However, if the circumference is divided into 100 equal parts, it will be necessary to calculate the lead as follows:

$$\text{Lead} = \frac{\text{rise of lobe in inches} \times 100}{\text{percent of circumference occupied by the lobe}}$$

EXAMPLE

A uniform-rise cam having a rise of 0.375 in. in 360° is to be cut. Calculate the required lead, the inclination of the work, and the vertical head.

SOLUTION

$$\text{Lead of cam} = 0.375 \text{ in.}$$

The machine and the dividing head should then be geared to 0.375 in. This is impossible, since the shortest lead which can be cut with regular change gears on a milling machine is generally 0.670 in.

NOTE: Any handbook will contain these milling tables.

To cut a lead of 0.375 in., it will be necessary to gear the machine to something more than 0.375 in.; in this case it will be 0.670 in. By consulting a handbook, it will be noted that the gears required for a 0.670-in. lead are 24, 86, 24, 100. It will also be necessary to swing the work and the vertical head to a definite angle. Figure 68-10 illustrates how this is calculated.

In the diagram,

L = lead to which the machine is geared
H = lead of the cam
i = angle of inclination of the dividing head spindle in degrees

$$\sin i = \frac{H}{L}$$

Therefore,

$$\text{Sine angle of inclination} = \frac{\text{lead of cam}}{\text{lead to which machine is geared}}$$

$$\text{Sine angle} = \frac{0.375}{0.670}$$

$$= 0.55970$$

$$\text{Angle} = 34°2'$$

Figure 68-10 illustrates that when the work travels along the distance L (0.670), it will rotate one turn and reach a point on the cutter which is 0.375 in. higher than the starting point. This will produce a rise and a lead of 0.375 in. on the cam.

EXAMPLE

It is required to cut a uniform-rise cam having three lobes, each lobe occupying 120° and each having a rise of 0.150 in.

PROCEDURE

1. Lead of cam $= \dfrac{0.150 \times 360}{120}$

$= 0.450$ in.

2. The smallest lead over 0.450 in. to which a machine can be geared is 0.670 in. (handbook).
3. Change gears required to produce a lead of 0.670 in.

$$= \frac{24}{86} \times \frac{24 \text{ (driven)}}{100 \text{ (drivers)}}$$

4. Mount the dividing head, and connect the worm shaft and the lead screw with the above gears. *Disengage the index plate locking device.*
5. Scribe three marks 120° apart on the periphery of the cam.
6. Mount the work in the dividing head chuck.
7. Mount an end mill of sufficient length in the vertical attachment.
8. Calculate the offset of the work and the vertical head:

$$\text{Sine of angle} = \frac{0.450}{0.670}$$
$$= 0.67164$$
$$\therefore \text{angle} = 42°12'$$

9. Swivel the dividing head to 42°12′.
10. Swivel the vertical milling attachment to 90° − 42°12′ = 47°48′.
11. Centralize the work and the cutter.
12. Rotate work with the index crank until one scribed mark on the cam blank is exactly at the bottom dead center.
13. Adjust the table until the cutter is touching the lower side of the work and the center of the cutter is in line with the mark.

NOTE: When the cutter is below the work, the setup will be more rigid and there will be no chance of chips obscuring any layout lines.

14. Set the vertical feed collar to zero (0).
15. Start the machine.
16. Using the index head crank, rotate the work one-third turn or until the second scribed line is in line with the forward edge of the cutter.

NOTE: If the machine is geared to a lead over 2½ in., the automatic feed may be used. If it is geared to a lead of less than 2½ in., a short lead attachment should be used or the table should be fed along manually by means of the index head crank.

17. Lower the table slightly and disengage the gear train, or disengage the dividing head worm.
18. Return the table to the starting position.
19. Rotate the work until the next line on the circumference is in line with the center of the cutter.
20. Reengage the gear train or the dividing head worm.
21. Cut the second lobe.
22. Repeat steps 17, 18, 19, and 20, and cut the third lobe.

NOTE: When calculating the offset as in step 8, the lead of the machine is often set to the whole number nearest to *twice* the cam lead. When the lead of the machine is exactly twice the lead of the cam, the offset will always equal 30° because the lead of the machine, which is the hypotenuse (Fig. 68-10), is twice the lead of the cam. Thus the sine of the angle of inclination is equal to 1/2 or 0.500, which then makes the angle equal to 30°.

RACK MILLING

A *rack,* in conjunction with a gear (pinion), is used to convert rotary motion into longitudinal motion. Racks are found on lathes, drill presses, and many other machines in a shop. A rack may be considered as a spur gear which has been straightened out so that the teeth are all in one plane. The circumference of the pitch circle of this gear would now become a straight line, which would just touch the pitch circle of a gear meshing with the rack. Thus the pitch line of a rack is the distance of one addendum (1/DP) below the top of the tooth.

The pitch of a rack is measured in linear (circular) pitch, which is obtained by dividing 3.1416 by the diametral pitch; that is:

$$\text{Rack pitch} = \frac{3.1416}{\text{DP}}$$

The method used to cut a rack will depend generally on the length of the rack. If the rack is reasonably short [10 in. (250 mm) or less], it may be held in the milling machine vise in a position parallel to the cutter arbor. On short racks, the teeth may be cut by accurately moving the cross-slide of the machine an amount equal to the circular pitch of the gear and then moving the table longitudinally to cut each tooth. If the rack is longer than the cross travel of the milling machine table, it must be held longitudinally on the table and is generally held in a special fixture.

The milling cutter is held in a *rack milling attachment* (Fig. 68-11). When cutting a straight tooth, the cutter is held at 90° to the position used when cutting a spur gear.

It is possible to mount slotting or narrow side milling cutters on the rack milling attachment for milling opera-

Figure 68-11 *Cutting the teeth on a helical rack using the rack milling and indexing attachments. (Courtesy Cincinnati Milacron Inc.)*

tions that can be handled more easily by using the machine crossfeed.

RACK INDEXING ATTACHMENT

When cutting a rack using the rack milling attachment, the table is often moved (indexed) for each tooth by means of the rack indexing attachment (Fig. 68-11). This consists of an indexing plate with two diametrically opposed notches and a locking pin. Two change gears selected from a set of 14 are mounted as shown in Fig. 68-11. Different combinations of change gears permit the machine table to be moved accurately in increments, corresponding to the linear (circular) pitch of the rack, by making either a half-turn or one complete turn of the plate. For indexing requiring one complete turn only, provision is made to close off one of the slots, thus preventing any error in indexing.

This attachment permits the indexing of all diametral pitches from 4 to 32, as well as all circular pitches from $\frac{1}{8}$ in. to $\frac{3}{4}$ in., varying by sixteenths. The following table movements can also be produced: $\frac{1}{7}$, $\frac{1}{6}$, $\frac{1}{5}$, $\frac{2}{7}$, $\frac{1}{3}$, and $\frac{2}{5}$ in.

WORMS AND WORM GEARS

Worms and *worm gears* are used when a great ratio reduction is required between the driving and driven shafts. A worm is a cylinder on which is cut a single- or multiple-start Acme-type thread. The angle of this thread ranges from a 14.5 to 30° pressure angle. As the lead angle of the worm increases, the greater the pressure angle should be on the side of the thread. The teeth on a worm gear are machined on a peripheral groove, which has a radius equal to half the root diameter of the worm. The drive ratio between a worm and worm gear assembly depends on the number of teeth in the worm gear and whether the worm has a single- or double-start thread. Thus if a worm gear had 50 teeth, the ratio would be 50:1, providing the worm has a single-start thread. If it had a double-start thread, the ratio would be 50:2, or 25:1.

TO MILL A WORM

Worms are often cut on a milling machine with a rack milling attachment and a thread milling cutter (Fig. 68-12). The setup of the cutter is similar to that for rack milling. The work is held between index centers and is rotated by suitable gears between the worm shaft and the lead screw of the milling machine. This is similar to the setup for helical milling. A short lead attachment is generally used when a worm is being milled because the thread usually has a short lead. If a short lead attachment is not available, the work may be rotated and the table moved lengthwise by means of the index crank on the dividing head.

PROCEDURE

1. Calculate all dimensions of the thread, that is, lead, pitch, depth, and angle of thread.

NOTE: The angle of the thread is calculated using the pitch diameter.

2. Mount the worm blank between the dividing head centers located at the end of the milling machine table.
3. Determine the proper gears for the lead, and mount them so that they connect the worm shaft and the lead screw.

Figure 68-12 *Milling machine set up for machining a worm. (Courtesy Cincinnati Milacron Inc.)*

4. Disengage the index plate locking device.
5. Mount the proper thread milling cutter on the rack milling attachment.
6. Swing the rack milling attachment to the required helix angle of the worm thread and in the proper direction for the lead of the worm.
7. Center the work under the cutter.
8. Raise the work up to the cutter.
9. Move the work clear of the cutter, and raise the table to the required depth of thread.
10. Cut the thread using the automatic feed or by turning the index crank handle to feed the table manually.

CLUTCHES

Positive-drive clutches are used extensively to drive or disconnect gears and shafts in machine gearboxes. The headstocks on most lathes use clutches, machined on the hubs of gears, to engage or disengage gears to provide different spindle speeds. The positive drive on this type of clutch is produced by means of interlocking teeth or projections on the driving and driven parts and does not rely on friction drive as in the case of friction-type clutches.

Three forms of positive-drive clutches are shown in Fig. 68-13. The *straight tooth clutch* (Fig. 68-13A) permits rotation in either direction. This type is more difficult to engage since the mating teeth and grooves must be in perfect alignment before engagement is possible.

The *inclined-tooth clutch* (Fig 68-13B) provides an easier means of engaging or disengaging the driving and driven members because of an 8 or 9° angle machined on the faces of the teeth. Since this type of clutch tends to disengage more rapidly, it must be provided with a positive means of locking it in engagement. Clutches of this type permit the shafts to run in either direction without backlash.

The *sawtooth clutch* (Fig. 68-13C) permits drive in only one direction but is more easily engaged than the other two types of clutches. The angle of the teeth in this type is generally 60°.

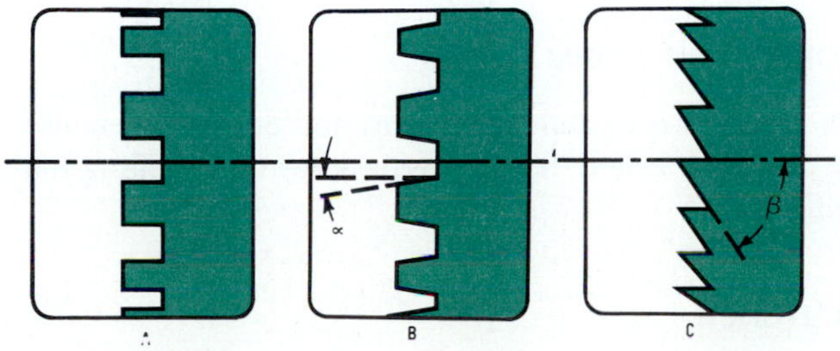

A
B
C

Figure 68-13 *Types of clutch teeth: (A) straight tooth; (B) inclined tooth; (C) saw tooth.*

TO MACHINE A STRAIGHT-TOOTH CLUTCH HAVING THREE TEETH

NOTE: This method applies to all clutches having an odd number of teeth.

1. Mount the dividing head on the milling machine table.
2. Mount a three-jaw chuck in the dividing head.
3. Mount the workpiece in the chuck.

NOTE: The spindle and the chuck of the dividing head may be positioned either horizontally or vertically. For this operation, it is assumed to be in a vertical position.

4. Set the sector arms to the proper indexing, that is, 40/3 = 13⅓ turns or 13 turns and 13 holes on a 39-hole circle.
5. Mount a side milling cutter on the milling machine arbor. The cutter should be no wider than the narrowest or innermost part of the groove.
6. Set the proper spindle speed and table feed.
7. Start the cutter and adjust the work until the edge nearest the front of the machine *just* touches the inner side of the cutter (Fig. 68-14). Set the crossfeed graduated feed collar to zero (0).
8. Move the table longitudinally until the work is clear of the cutter.
9. Move the table laterally half the diameter of the work plus about 0.001 in. (0.02 mm) for clearance. Lock the saddle in this position.
10. Set the depth of cut, and lock the knee clamps.
11. Take a cut across the full width of the workpiece as in Fig. 68-15.
12. Return the table to starting position.
13. Index for the next tooth and take the second cut as shown in Fig. 68-16.
14. Return the table to starting position.
15. Index for the next tooth.
16. Take the third cut as shown in Fig. 68-17.

Figure 68-14 *Adjusting a workpiece to the cutter.*

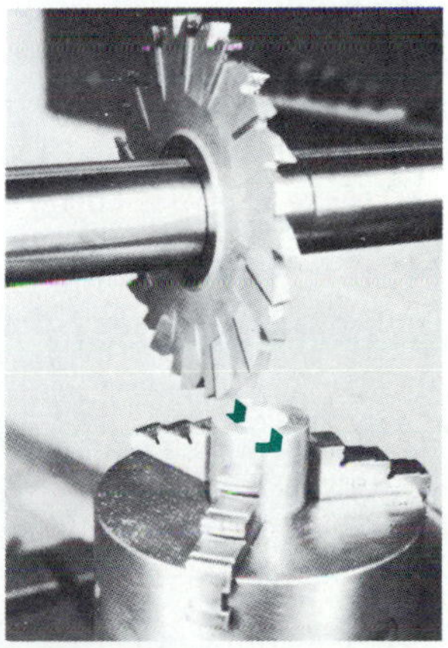

Figure 68-15 *Cutting the first tooth.*

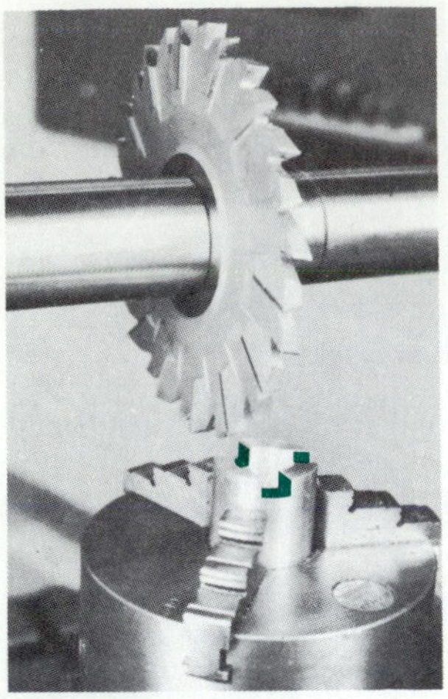

Figure 68-16 *Cutting the second tooth.*

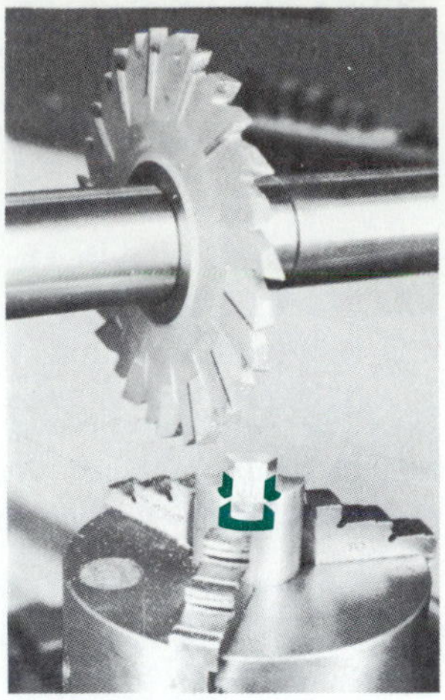

Figure 68-17 *Cutting the third tooth.*

TO MACHINE A STRAIGHT-TOOTH CLUTCH HAVING FOUR TEETH

When machining a straight- or inclined-tooth clutch having an even number of teeth, it is necessary to machine one side of each tooth first and then machine the second side of each. This obviously requires more time; therefore, clutches should be designed with an odd number of teeth to reduce machining time and the chance of error.

PROCEDURE

1. Mount the work as in the previous example and set the proper indexing.
2. Start the cutter and adjust the workpiece until the edge nearest the front of the machine just touches the inner edge of the cutter (Fig. 68-14).
3. With the work clear of the cutter, move the saddle over half the diameter of the work plus the thickness of the cutter minus 0.001 in. (0.02 mm) for clearance.
4. Adjust the work until the cutter is over the center hole of the clutch. It is advisable to cut from the center to the outside of the clutch in order to minimize the vibration.
5. Set the depth and lock the knee clamp.
6. Take the first cut.

NOTE: When cutting an even number of clutch teeth, it is absolutely necessary that the cut be made through one wall only.

7. Index for and cut the remaining teeth on the one side. (Indexing = 10 turns for each tooth.)
8. Revolve the work one-eighth of a turn (five turns of the index crank).
9. Touch the opposite side of the work to the other side of the cutter, that is, the outside of the cutter to the inner edge of the work.
10. Repeat operations 3, 4, 5, 6, and 7 until all the teeth have been cut. If any pie-shaped pieces of metal remain in the tooth spaces, they must be removed with an additional cut through the center of the space.

REVIEW QUESTIONS

CAMS AND CAM MILLING

1. Define a *cam*.
2. Name four types of cams.
3. Name four types of followers and state how each is used.
4. List three cam motions, and describe the type of motion which is imparted to the follower in one revolution of the cam.
5. In a single-lobe cam and a double-lobe cam, what is the relationship of the lead to the rise?
6. Calculate:
   a. Lead of cam
   b. The change gears required to produce this lead
   c. The angle at which to swing the dividing head
   d. The angle at which to swing the vertical head for each of the following examples of uniform rise:
      (1) a single-lobe cam having a rise of 0.125 in. in 360°
      (2) a two-lobe cam, each lobe having a rise of 0.187 in. in 180°
      (3) a three-lobe cam, each lobe having a rise of 0.200 in. in 120°

RACK MILLING

7. Define *rack* and state its purpose.
8. A 10-DP gear is in mesh with a rack. The gear has 42 teeth. The rack is 1 in. thick from the top of the tooth to the bottom of the rack. Calculate the distance from the center of the gear to the bottom of the rack.
9. Calculate the linear pitch of a 5-pitch rack, an 8-pitch rack, and a 14-pitch rack.

WORMS AND WORM GEARS

10. Define *worm* and *worm gear,* and describe their use.
11. Describe briefly how a worm is cut on a milling machine.

CLUTCHES

12. List three types of clutches, and state the application of each.

13. Why are clutches with an odd number of teeth preferred to those with an even number of teeth?

14. After the outside surface of a 3-in.-diameter (or 76-mm) clutch blank has been touched up to a ½ in.-wide (or 13-mm) side cutter, how far must the table be moved over when cutting:

a. A clutch having five teeth?

b. A clutch having six teeth?

The Vertical Milling Machine—Construction and Operation

UNIT 69

Much of the work on a milling machine is best done with a vertical milling attachment. The time involved in setting up a vertical attachment prevents the milling machine from being used for other milling operations at the same time; therefore, the vertical milling machine has become popular in industry. This machine offers versatility not found in any other machine. Some of the operations which can conveniently be carried out on these machines are face milling, end milling, keyway cutting, dovetail cutting, T-slot and circular slot cutting, gear cutting, drilling, boring, and jig boring. Because of the machine construction (vertical spindle), many of the facing operations can be done with a fly cutter, which reduces the cost of cutters considerably. Also, since most cutters are much smaller than for the horizontal mill, the cost of the cutters for the same job is usually much less for a vertical milling machine.

OBJECTIVES

After completing this unit you will be able to:

1. Name and state the purposes of the main operative parts
2. Align the vertical head and the vise to within an accuracy of ±0.001 in. (0.02 mm)
3. Mill flat and angular surfaces
4. Mill a block square and parallel
5. Mill keyways in a shaft

TYPES OF VERTICAL MILLING MACHINES

The *standard vertical milling machine* (Fig. 69-1) has all the construction features of a plain horizontal milling machine, except that the cutter spindle is mounted in a vertical position. The spindle head on most vertical milling machines may be swiveled, which readily permits the machining of angular surfaces. The spindles on most vertical milling machines have a short travel, which facilitates step milling and the drilling and boring of holes.

The cutters used are of the end mill or shell end mill type. This type of machine is particularly suited to the use of the rotary table, permitting the machining of circular grooves and the positioning of holes which have been laid out with angular measurements.

The *ram-type vertical milling machine* (Fig. 69-2) is

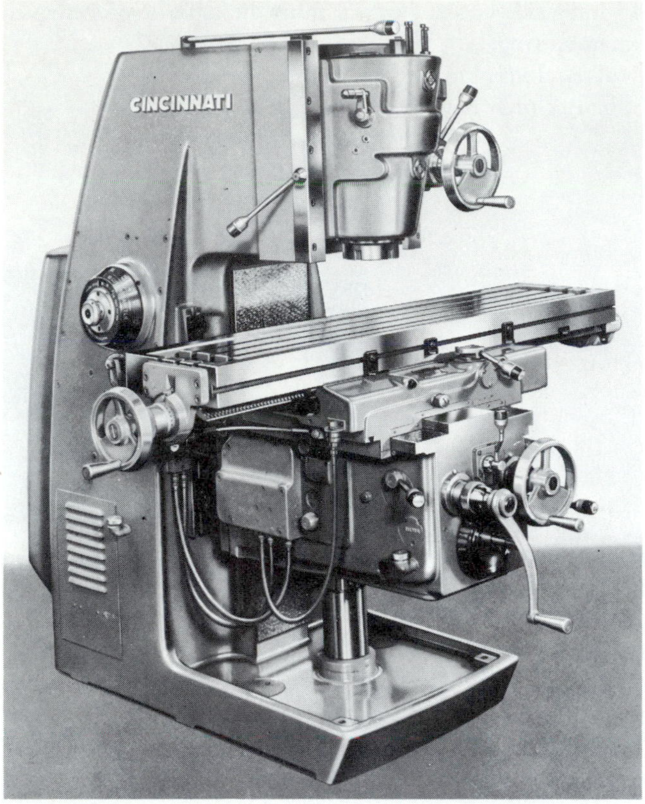
CINCINNATI

Figure 69-1 *A standard vertical milling machine.*

lighter than the standard vertical machine. Because of its simplicity and the ease of setup, it has become increasingly popular. Machines of this type are generally used for lighter types of milling machine work.

PARTS OF THE RAM-TYPE VERTICAL MILL

The *base* is made of ribbed cast iron. It may contain a coolant reservoir.

The *column* is often cast integrally with the base. The machined face of the column provides the ways for vertical movement of the knee. The upper part of the column is machined to receive a *turret* on which the overarm is mounted.

The *overarm* is round, or of the ram type, as illustrated in Fig. 69-2. It may be adjusted toward or away from the column to increase the capacity of the machine.

The *head* is attached to the end of the ram. Provision is made to swivel the head in one plane. On universal-type machines, the head may be swiveled in two planes. Mounted on top of the head is the *motor* which provides drive to the *spindle,* usually through V-belts. Spindle speed changes are effected by means of a variable-speed pulley and crank or by belt changes and a reduction gear. The spindle may be fed by means of a hand lever, a handwheel, or by automatic power feed. Most machines are equipped with a micrometer quill stop for precision drilling and boring to depth.

The *knee* moves up and down on the face of the column and supports the saddle and the table. The knee in this type of machine does *not* contain the gears for the automatic feed as in the horizontal milling machine and the standard vertical milling machine. The automatic feed on most of these types of machines is not a standard feature and is usually added as an accessory. It is an external device and controls only the longitudinal feeds of the table. Most cutting on the vertical milling machine is done by end mill cutters; therefore, it is not necessary to swing the table even when cutting a helix. As a result, vertical milling machines are equipped with plain tables only.

ALIGNING THE VERTICAL HEAD

Proper alignment of the head is of the utmost importance when machining holes or when face milling. If the head is not at an angle of 90° to the table, the holes will not be square with the work surface when the cutting tool is fed by hand or automatically. When face milling, the machined surface will be stepped if the head is not square with the table. Although all heads are graduated in degrees and some have vernier devices used for setting the head, it is wise to check the spindle alignment as follows:

PROCEDURE

1. Mount a dial indicator on a suitable rod, bent at 90° and held in the spindle (Fig. 69-3).
2. Position the indicator over the front of the table.

NOTE: If a flat ground cylindrical plate is available, place it on the clean table and position the indicator near the outer edge.

3. Carefully lower the spindle until the indicator button touches the table and the dial indicator registers about one-quarter revolution, then set the bezel to zero (0) (Fig. 69-4). Lock the spindle in place.
4. Carefully rotate the vertical mill spindle 180° by hand until the button bears on the opposite side of the table (Fig. 69-5). Compare the readings.
5. If there is any discrepancy in the readings, loosen the locking nuts on the swivel mounting and adjust the head until the indicator registers one-half the difference between the two readings. Tighten the locking nuts.
6. Recheck the accuracy of the head and adjust if necessary.
7. Rotate the vertical mill spindle 90°, and set the dial indicator as in step 3.

HEAD
QUILL FINE FEED HANDWHEEL
QUILL FEED HAND LEVER
RAM
SPINDLE
TABLE
COLUMN
TABLE TRAVERSE HANDWHEEL
CROSSFEED HANDWHEEL
VERTICAL TRAVERSE LEVER
KNEE
BASE

Figure 69-2 *A ram-type vertical milling machine.*

8. Rotate the machine spindle 180°, and check the reading at the other end of the table.
9. If the two readings do not coincide, repeat step 5 until the readings are the same.
10. Tighten the locking nuts on the swivel mount.
11. Recheck the readings and adjust if necessary.

NOTE: When readings are taken, it is important that the indicator button does not catch in the T-slots of the table. To prevent this, it is advisable to work from the high reading first and then rotate to the low reading. It should be apparent that the longer the rod used, the more accurate the setting will be.

Figure 69-3 *An indicator assembly fastened to a rod in a chuck.*

Figure 69-4 *Setting the indicator to 0 on the righthand side of the table.*

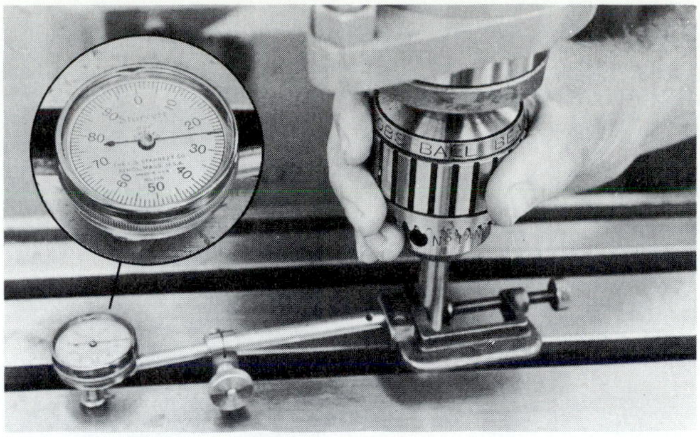

Figure 69-5 *The spindle rotated 180°.*

ALIGNING THE VISE

When the vise is aligned on a vertical milling machine, the dial indicator may be attached to the quill or the head by any convenient means, such as clamps (Fig. 69-6) or magnetic base. The same method of alignment should be followed as was outlined in unit 62 for aligning the vise on a horizontal milling machine.

TO MACHINE A FLAT SURFACE

1. Clean the vise and mount the work securely in the vise, on parallels if necessary (Fig. 69-7).
2. Check that the vertical head is square with the table.
3. If possible, select a cutter which will just overlap the edges of the work. This will then require only one cut to be taken to machine the surface. If the surface to be machined is fairly narrow, an end mill slightly larger in diameter than the width of the work should be used. If the surface is large and requires several passes, a shell end mill or a suitable fly cutter should be used.

NOTE: It is not advisable to use a facing cutter which is too wide since the head may be thrown out of alignment if the cutter should jam.

4. Set the proper spindle speed for the size and type of cutter and the material being machined; check cutter rotation.
5. Tighten the quill clamps.
6. Start the machine, and adjust the table until the end of the work is under the edge of the cutter.
7. Raise the table until the work surface just touches the cutter. Move the work clear of the cutter.
8. Raise the table about 1/32 in. (0.80 mm) and take a trial cut for approximately 1/4 in. (6 mm).
9. Move the work clear of the cutter, *stop the cutter,* and measure the work (Fig. 69-8).
10. Raise the table the desired amount, and lock the knee clamp.
11. Mill the surface to size using the automatic feed if the machine is so equipped.

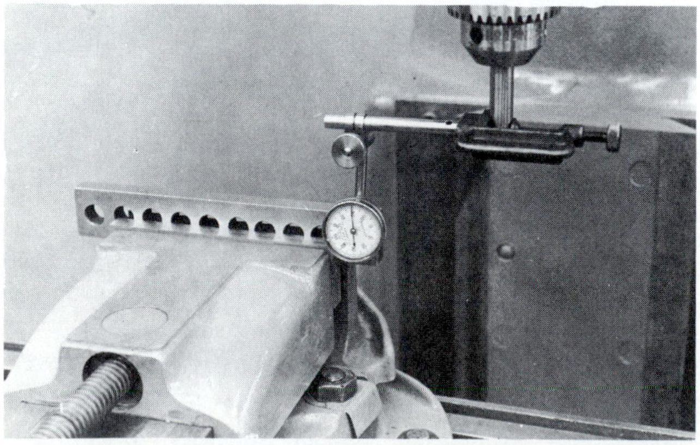

Figure 69-6 *Indicator assembly for checking vise alignment.*

Figure 69-7 *Tap the workpiece down with a soft-faced hammer until all feelers are tight.*

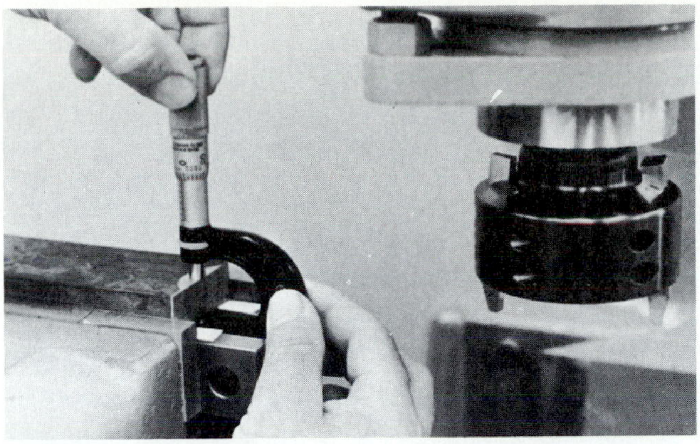

Figure 69-8 *Measuring a workpiece after a light trial cut.*

MACHINING A BLOCK SQUARE AND PARALLEL

In order to mill the four sides of a piece of work so that its sides are square and parallel, it is important that each side be machined in a definite order. It is very important that dirt and burrs be removed from the work, vise, and parallels, since they can cause inaccurate work.

MACHINING SIDE 1

1. Clean the vise thoroughly and remove all burrs from the workpiece.
2. Set the work on parallels in the center of the vise with the largest surface (side 1) facing up (Fig. 69-9).
3. Place short paper feelers under each corner between the parallels and the work.
4. Tighten the vise securely.
5. With a soft-faced hammer, tap the workpiece down until all paper feelers are tight.
6. Mount a flycutter in the milling machine spindle.
7. Set the machine for the proper speed for the size of the cutter and the material to be machined.
8. Start the machine and raise the table until the cutter just touches near the right-hand end of side 1.
9. Move work clear of the cutter.
10. Raise the table about 0.030 in. (0.76 mm) and machine side 1 using a steady feed rate.
11. Take the work out of the vise and remove all burrs from the edges with a file.

MACHINING SIDE 2

12. Clean the vise, work, and parallels thoroughly.
13. Place the work on parallels, if necessary, with side 1 against the solid jaw and side 2 up (Fig 69-10).
14. Place short paper feelers under each corner between the parallels and the work.
15. Place a round bar between side 4 and the movable jaw.

NOTE: The round bar must be in the *center* of the amount of work being held inside the vise jaws.

16. Tighten the vise securely and tap the work down until the paper feelers are tight.
17. Follow steps 8 to 11 and machine side 2.

MACHINING SIDE 3

18. Clean the vise, work, and parallels thoroughly.
19. Place side 1 against the solid vise jaw with side 2 resting on parallels if necessary (Fig. 69-11).
20. Push the parallel to the left so that the right edge of the work extends about 1/4 in. (6 mm) beyond the parallel.
21. Place short paper feelers under each end or corner between the parallels and the work.
22. Place a round bar between side 4 and the movable jaw being sure that the round bar is in the center of the amount of work held *inside the vise.*

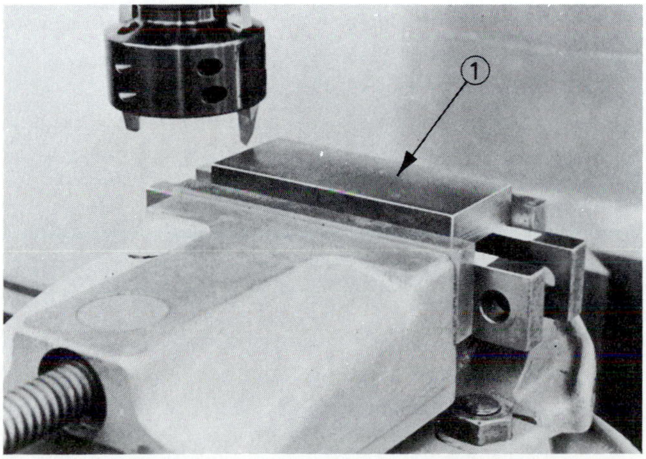
1

Figure 69-9 *Side 1 or the surface with the largest area should be facing up.*

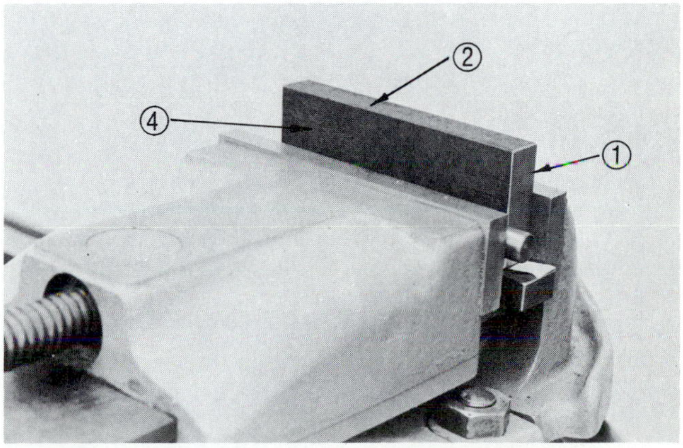
2
4
1

Figure 69-10 *The finished side (side 1) should be placed against the cleaned solid jaw for machining side 2.*

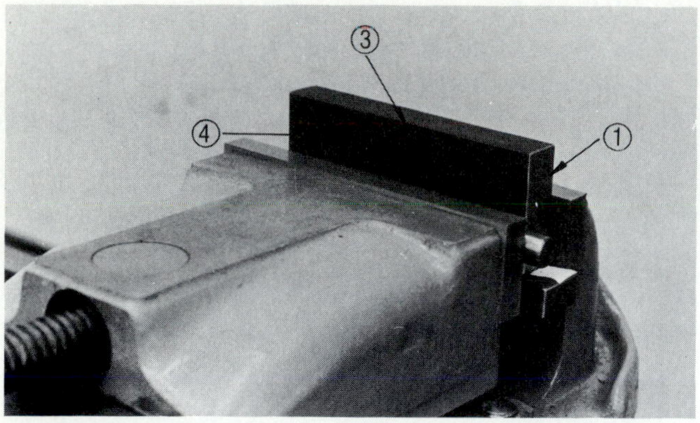
3
4
1

Figure 69-11 *Place side 1 against the solid jaw and side 2 down for machining side 3.*

23. Tigten the vise securely and tap the work down until the paper feelers are tight.
24. Start the machine and raise the table until the cutter just touches near the right-hand end of side 3.
25. Move the work clear of the cutter and raise the table about 0.010 in. (0.25 mm).
26. Take a trial cut about ¼ in. (6 mm) long, stop the machine, and measure the width of the work (Fig. 69-12).
27. Raise the table the required amount and machine side 3 to correct width.
28. Remove the work and file off all burrs.

MACHINING SIDE 4

29. Clean the vise, work, and parallels thoroughly.
30. Place side 1 down on the parallels with side 4 up and tighten the vise securely (Fig. 69-13).

NOTE: With three finished surfaces, the round bar is not required when machining side 4.

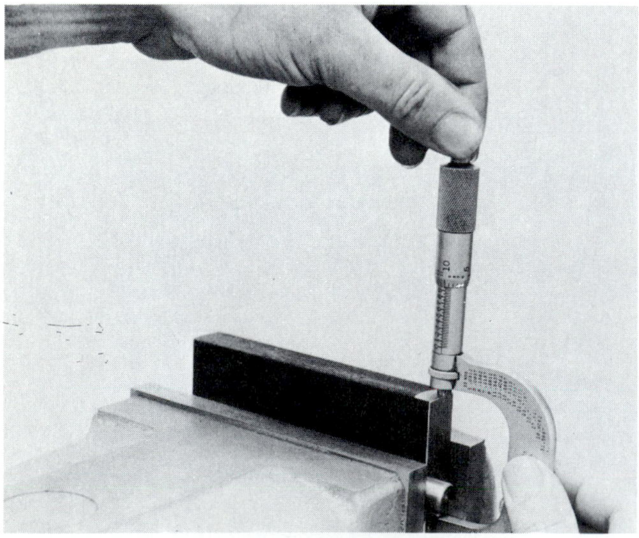

Figure 69-12 *Measuring the trial cut on side 3.*

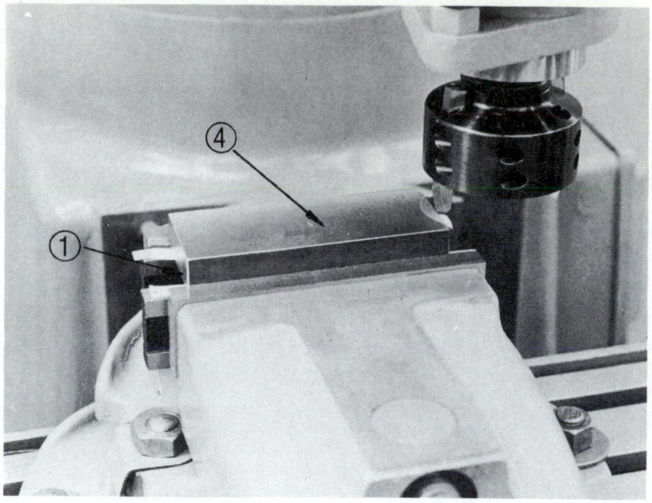
4
1

Figure 69-13 *The workpiece is set up properly for machining side 4. Note the paper feelers between the parallels and the workpiece.*

31. Place short paper feelers under each corner between the parallels and work (Fig. 69-13).
32. Tighten the vise securely.
33. Tap the work down until the paper feelers are tight.
34. Follow steps 24 to 27 and machine side 4 to the correct thickness.

MACHINING THE ENDS SQUARE

Two common methods are used to square the ends of workpieces in a vertical mill. Short pieces are generally held vertically in the vise and are machined with an end mill or flycutter (Fig. 69-14). Long pieces are generally held flat in

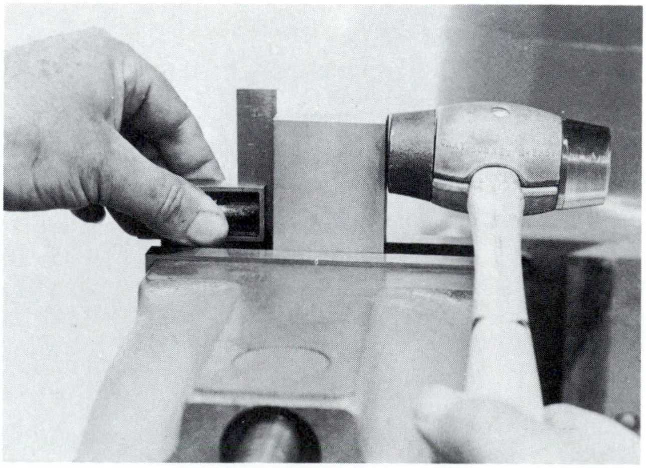

Figure 69-14 *Setting the workpiece edge square for machining its ends.*

Figure 69-15 *One end machined square with the side of the workpiece.*

the vise with one end extending past the end of the vise. The end surface is cut square with the body of an end mill.

PROCEDURE FOR SHORT WORK

1. Set the work in the center of the vise with one of the ends up and tighten the vise lightly.
2. Hold a square down firmly on top of the vise jaws and bring the blade into light contact with the side of the work.
3. Tap the work until its edge is aligned with the blade of the square (Fig. 69-14).
4. Tighten the vise securely and recheck the squareness of the side.
5. Take about 0.030 in.-deep (0.76 mm) cut and machine the end square (Fig. 69-15).
6. Remove the burrs from the end of the machined surface.
7. Clean the vise and set the machined end on paper feelers in the bottom of the vise.
8. Tigten the vise securely and tap the work down until the paper feelers are tight.
9. Take a trial cut from the end until the surface cleans up.
10. Measure the length of the workpiece with a depth micrometer (Fig. 69-16).
11. Raise the table the required amount and machine the work to length.

TO MACHINE AN ANGULAR SURFACE

1. Lay out the angular surface.
2. Clean the vise.
3. Align the vise with the direction of the feed. *This is of the utmost importance.*
4. Mount the work on parallels in the vise.

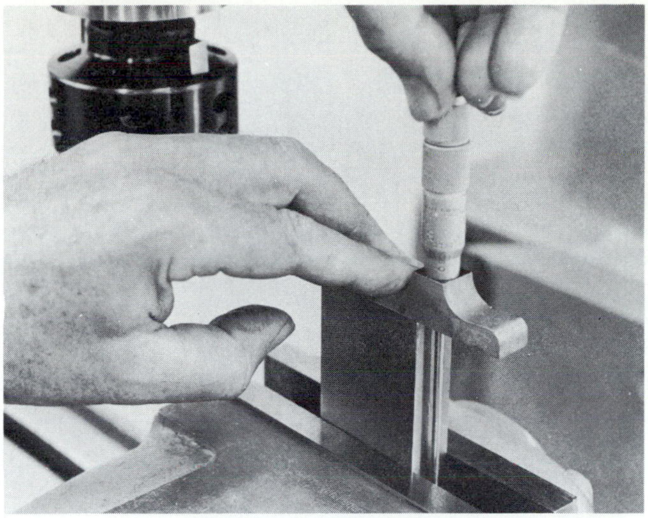

Figure 69-16 *Measuring the length of the workpiece with a depth micrometer.*

5. Swivel the vertical head to the required angle (Fig. 69-17).
6. Tighten the quill clamp.
7. Start the machine and raise the table until the cutter touches the work. Carefully raise the table until the cut is of the desired depth.

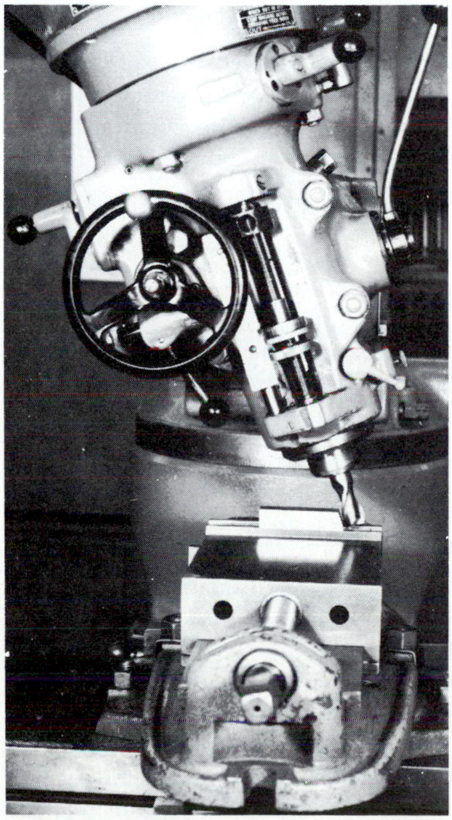

Figure 69-17 *Head swiveled to machine an angle.*

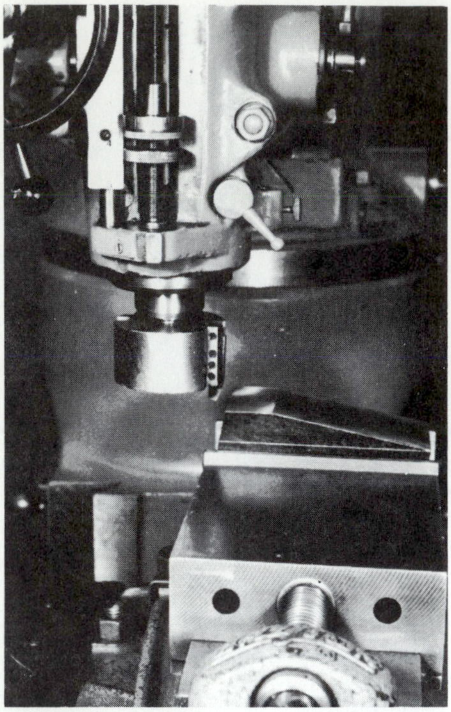

Figure 69-18 *Machining an angle by adjusting the workpiece.*

8. Take a trial cut for about ½ in. (13 mm).
9. Check the angle with a protractor.
10. If the angle is correct, continue the cut.

NOTE: It is always advisable to feed the work into the rotation of the cutter, rather than with the rotation of the cutter, which may draw the work into the cutter and cause damage to the work, the cutter, or both.

11. Machine to the required depth, taking several cuts if necessary.

Alternate Method

Angles may sometimes be milled by leaving the head in a vertical position and setting the work on an angle in the vise (Fig. 69-18). This will depend on the shape and size of the workpiece.

PROCEDURE

1. Check that the vertical head is square with the table.
2. Clean the vise.
3. Lock the quill clamp.
4. Set the workpiece in the vise with the layout line parallel to the top of the vise jaws and about ¼ in. (6 mm) above them.
5. Adjust the work under the cutter so that the cut will start at the narrow side of the taper and progress into the thicker metal.
6. Take successive cuts of about 0.125 to 0.150 in. (3 to 4 mm), or until the cut is about 1/32 in. (0.80 mm) above the layout line.
7. Check to see that the cut and the layout line are parallel.
8. Raise the table until the cutter just touches the layout line.
9. Clamp the knee at this setting.
10. Take the finishing cut.

CUTTING SLOTS AND KEYSEATS

Slots and keyseats with one or two blind ends may be cut in shafts more easily on a vertical milling machine, using a two- or three-fluted end mill, than with a horizontal mill and a side facing cutter.

PROCEDURE

1. Lay out the position of the keyseat on the shaft, and scribe reference lines on the end of the shaft (Fig. 69-19).
2. Secure the workpiece in a vise on a parallel. If the shaft is long, it may be clamped directly to the table by placing it in one of the table slots or in V-blocks.
3. Using the layout lines on the end of the shaft, set up the shaft so that the keyseat layout is in the proper position on top of the shaft.
4. Mount a two- or three-fluted end mill of a diameter equal to the width of the keyseat, in the milling machine spindle.

NOTE: If the keyseat has two blind ends, a two- or three-lip end mill must be used, since they may be used as a drill to start the slot. If the slot is at one end of the shaft (one blind end only), a four-fluted end mill may be used, although a two- or three-lip end mill will give better chip clearance.

5. Center the workpiece by carefully touching the cutter up to one side of the shaft (Fig. 69-20). This may also be

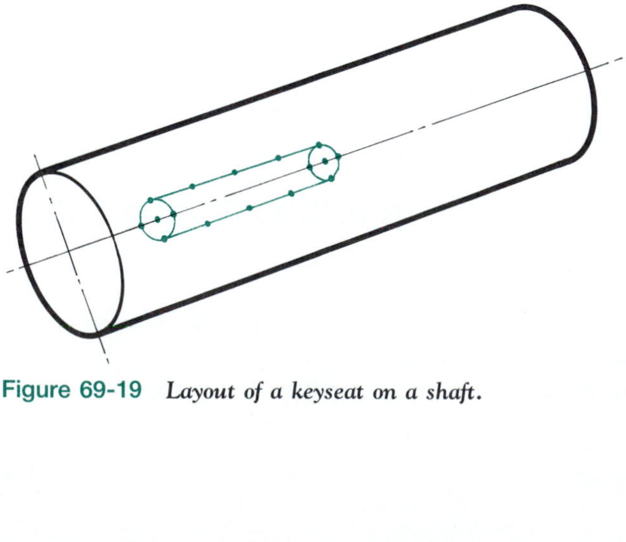

Figure 69-19 *Layout of a keyseat on a shaft.*

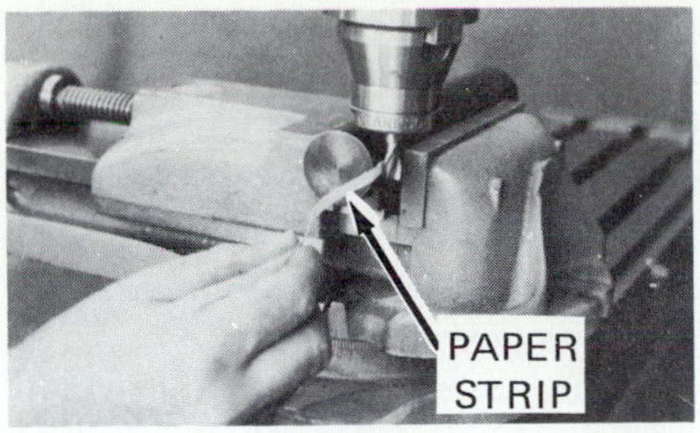
PAPER
STRIP

Figure 69-20 *Setting a cutter to the side of the workpiece.*

done by placing a piece of thin paper between the shaft and the cutter.

NOTE: Paper may be made to adhere to shafts or work surfaces by wetting it with coolant or oil before applying it to the surface. This eliminates the necessity of holding paper between the cutter and the work, thus making it a safer operation.

6. Lower the table until the cutter clears the workpiece.
7. Move the table over an amount equal to half the diameter of the shaft plus half the diameter of the cutter plus the thickness of the paper (Fig. 69-21). For example, if a 1/4 in. (6-mm) slot is required in a 2 in. (50-mm) shaft and the thickness of the paper used is 0.002 in. (0.05 mm), the table would be moved over 1.000 + 0.125 + 0.002 = 1.127 in. (25 + 3 + 0.05 = 28.05 mm).
8. If the keyseat being cut has two blind ends, adjust the work until the end of the keyseat is aligned with the edge of the cutter.
9. Feed the cutter down (or the table up) until the cutter *just* cuts to its full diameter. If the keyseat has one blind end only, the work is adjusted so that this cut is taken at the end of the work. The work would now be moved clear of the cutter.
10. Adjust the depth of cut to one-half the thickness of the key, and machine the keyseat to the proper length (Fig. 69-22).

Figure 69-21 *Cutter centered on the layout.*

Figure 69-22 *Keyseat milled to depth and length.*

MEASURING THE DEPTH OF KEYSEATS

If the keyseat is at the end of a shaft, the proper depth of the keyseat is checked by measuring diametrically from the bottom of the keyseat to the opposite side of the shaft. This distance may be calculated as follows (Fig. 69-23):

$$M = D - \left(\frac{W}{2} + h\right)$$

$$h = 0.5(D - \sqrt{D^2 - W^2})$$

M = the distance from the bottom of the keyseat to the opposite side of the shaft
D = diameter of the shaft
W = width of the keyseat
h = height of the segment above the width of the keyseat

EXAMPLE

Calculate the measurement M if the shaft is 50 mm in diameter and a 6-mm key is to be used.

$$\begin{aligned} h &= 0.5(D - \sqrt{D^2 - W^2}) \\ &= 0.5(50 - \sqrt{2500 - 36}) \\ &= 0.5(50 - \sqrt{2464}) \\ &= 0.5(50 - 49.64) \\ &= 0.5(0.36) \\ &= 0.18 \text{ mm} \end{aligned}$$

$$\begin{aligned} M &= D - \left(\frac{W}{2} + h\right) \\ &= 50 - \left(\frac{6}{2} + 0.18\right) \\ &= 50 - (3 + 0.18) \\ &= 50 - 3.18 \\ &= 46.82 \text{ mm} \end{aligned}$$

WOODRUFF KEYS

Woodruff keys are used when keying shafts and mating parts (Fig. 69-24A). Woodruff keyseats can be cut more

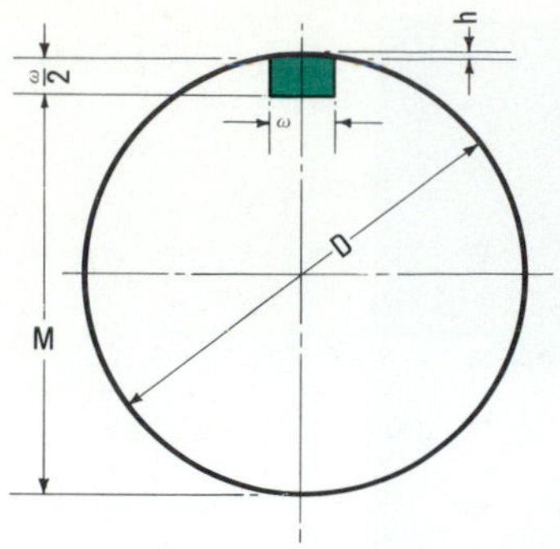
h
D
M

Figure 69-23 *Keyseat calculations.*

quickly than square keyseats, and the key should not require any fitting after the keyseat has been cut. These keys are semicircular in shape and can be purchased in standard sizes, which are designated by E numbers. They can be conveniently made from round bar stock of the required diameter.

Woodruff keyseat cutters (Fig. 69-24B) have shank diameters of $\frac{1}{2}$ in. for cutters up to $1\frac{1}{2}$ in. in diameter. The shank is undercut adjacent to the cutter to permit the cutter to go into the proper depth. The sides of the cutter are slightly tapered toward the center to permit clearance during cutting. Cutters over 2 in. in diameter are mounted on an arbor.

The size of the cutter is stamped on the shank. The last two digits on the number indicate the nominal diameter in eighths of an inch. The digit or digits preceding the last two numbers indicate the nominal width of the cutter in thirty-seconds of an inch. Thus a cutter marked 608 would be $8 \times \frac{1}{8}$, or 1 in. in diameter, and $6 \times \frac{1}{32}$, or $\frac{3}{16}$ in. wide. The key would be a semicircular cross section to fit the groove exactly.

To Cut a Woodruff Keyseat

1. Align the spindle of the vertical milling machine 90° to the table.
2. Lay out the position of the keyseat.
3. Set the shaft in the vise of the milling machine or on V-blocks. Be sure that the shaft is level (parallel to the table).
4. Mount the cutter of the proper size in the spindle.
5. Start the cutter, and touch the bottom of the cutter to the top of the workpiece. Set the vertical graduated feed collar to zero (0) and check cutter rotation.
6. Move the work clear of the cutter. Raise the table half the diameter of the work plus half the thickness of the cutter. Lock the knee at this setting.
7. Position the center of the slot with the center of the cutter. Lock the table in this position.
8. Touch the revolving cutter to the work. Use a strip of paper between the cutter and the work if desired. Set the crossfeed screw collar to zero (0).
9. Cut keyseat to the proper depth.

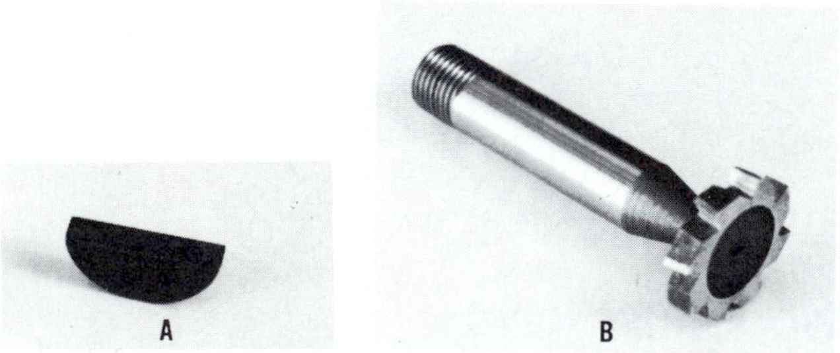
A
B

Figure 69-24 *Woodruff key and keyseat cutters.*

NOTE: Keyseat proportions may be found in any handbook.

REVIEW QUESTIONS

1. Why has the vertical machine been so readily accepted by industry?
2. Name six operations which can be performed in a vertical mill.
3. Describe:
   a. A standard vertical mill
   b. A ram-type vertical mill
4. State the purpose of the following parts:
   a. Column
   b. Overarm
   c. Head
   d. Knee

ALIGNING THE VERTICAL HEAD AND VISE

5. Why is it necessary to align the vertical head square with the table?
6. Describe briefly how the vertical head may be aligned with the table surface.
7. Why is it important that the vise be aligned with the table travel?
8. Describe one method of aligning the vise parallel to the table travel.

MACHINING A FLAT SURFACE

9. What precautions should be observed before machining a flat surface?
10. What type of cutter should be used to machine a large surface?
11. How can the edge of a flat piece be machined square with a flat finished surface?

MACHINING A BLOCK SQUARE AND PARALLEL

12. How should the work be set in a vise to machine side 1?

13. How should the work be set in a vise to machine side 2?

14. Draw a sketch of the work setup required for machining side 3 and side 4.

MACHINING ANGULAR SURFACES

15. Describe briefly two methods of machining angular surfaces.

CUTTING SLOTS AND KEYWAYS

16. What type of cutter should be used to cut slots or keyways having one or two blind ends?

17. How can round work be held for machining slots and keyways?

18. Explain one method of aligning the end mill with the center of a shaft.

19. How may the depth of a keyway be measured?

20. Calculate the measurement for a ½-in.-wide keyseat from the bottom of a 2-in.-diameter shaft to the bottom of the keyseat (½ × ½ key).

21. Describe a Woodruff key and state its purpose.

22. Explain the marking 810 found on a Woodruff key cutter.

23. Explain how to center a Woodruff key cutter with a shaft.

UNIT 70

Special Milling Operations

The versatility of the vertical milling machine is further increased by the use of specially shaped cutters and machine accessories. Operations such as milling of T-slots and dovetails are possible because special milling cutters are manufactured for each purpose. The rotary table accessory permits the milling of radii and circular slots. A boring head mounted in the machine spindle and measuring rods on the table allow the vertical milling machine to be used for the accurate jig boring of holes.

OBJECTIVES

After completing this unit you will be able to:
1. Set up and use the rotary table to mill a circular slot
2. Set up and mill internal and external dovetails
3. Jig bore holes on a vertical mill

THE ROTARY TABLE

The *rotary table* or the *circular milling attachment* (Fig. 70-1) can be used on plain universal vertical milling machines and slotters. It may provide rotary motion to the workpiece in addition to the longitudinal and vertical motion provided by the machine. With this attachment, it is possible to cut radii, circular grooves, and circular sections not possible by other means. The drilling and boring of holes which have been designated by angular measurements, as well as other indexing operations, are easily accomplished with this accessory. This attachment is also suitable for use with the slotting attachment on a milling machine.

Rotary tables may be of two types: those having hand feed and those having power feed. The construction of these is basically the same, the only exception being the automatic feed mechanism.

Figure 70-1 *A hand feed rotary table.*

ROTARY TABLE CONSTRUCTION

The rotary table unit has a *base,* which is bolted to the milling machine table. Fitting into the base is the *rotary table,* on the bottom of which is mounted a *worm gear* (Fig. 70-1). A *worm shaft* mounted in the base meshes with and drives the worm gear. The worm shaft may be quickly disengaged when rapid rotation of the table is required, as when setting work concentric with the table. A *handwheel* is mounted on the outer end of the worm shaft. The bottom edge of the table is graduated in half degrees. On most rotary table units, there is a *vernier scale* on the handwheel collar, which permits setting to within two minutes of a degree. The table has T-slots cut into the top surface to permit the clamping of work.

A hole in the center of the table accommodates test plugs to permit easy centering of the table with the machine spindle. Work may be centered with the table by means of test plugs or arbors.

Some rotary tables have an *indexing attachment* instead of a handwheel (Fig. 70-2). This attachment is often supplied as an accessory to the standard rotary table. It not only serves the same purpose as a handwheel, but also permits the indexing of work with dividing head accuracy. The worm and wormwheel ratio of rotary tables is not necessarily 40:1 as on most dividing heads. The ratio may be 72:1, 80:1, 90:1, 120:1, or any other ratio. Larger ratios are usually found on larger tables. The method of calculating the indexing is the same as for the dividing head, except that the number of teeth in the wormwheel is used rather than 40 as in the dividing head calculations.

EXAMPLE

Calculate the indexing for five equally spaced holes on a circular plate using a rotary table with an 80:1 ratio.

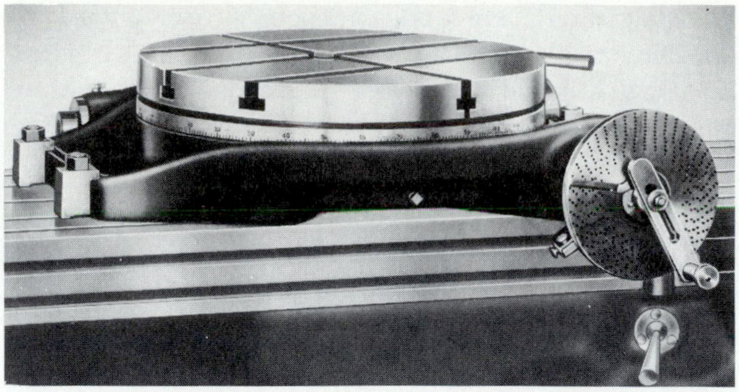

Figure 70-2 *A rotary table with an indexing attachment. (Courtesy Cincinnati Milacron Inc.)*

$$\text{Indexing} = \frac{\text{number of teeth in wormwheel}}{\text{number of divisions required}}$$

$$= \frac{80}{5}$$

$$= 16 \text{ turns}$$

Accurately spaced holes on circles and segments (such as clutch teeth and teeth in gears too large to be held between index centers) are some of the applications of this type of rotary table. It may be supplied with a power feed mechanism.

The third type of rotary table has provision for power rotation of the table (Fig. 70-3). Here the worm shaft is connected to the milling machine lead screw drive gear by a special shaft and an end gear train. The rate of rotation is controlled by the feed mechanism of the milling machine. When operated by power, the rotation of the table may be controlled by the operator using a feed lever attached to the unit or by trip dogs located on the periphery of the table. The operation of the table may be controlled by the handwheel or the automatic feed lever as required.

Figure 70-3 *A rotary table with power feed. (Courtesy Cincinnati Milacron Inc.)*

This type of attachment is particularly suited for production work and continuous milling operations when large numbers of small identical parts are required. In this operation, the parts are mounted in suitable fixtures on the table, and the rotary feed moves the parts under the cutter. After the piece has passed beneath the cutter, the finished piece is removed and replaced with an unfinished workpiece.

To Center the Rotary Table with the Vertical Mill Spindle

1. Square the vertical head with the machine table.
2. Mount the rotary table on the milling machine.
3. Place a test plug in the center hole of the rotary table.
4. Mount an indicator with a grasshopper leg in the machine spindle.
5. With the indicator just clearing the top of the test plug, rotate the machine spindle by hand and approximately align the plug with the spindle.
6. Bring the indicator into contact with the diameter of the plug, and rotate the spindle by hand.
7. Adjust the machine table by the longitudinal and crossfeed handles until the dial indicator registers no movement.
8. Lock the machine table and saddle, and recheck the alignment.
9. Readjust if necessary.

To Center a Workpiece with the Rotary Table

Often it is necessary to perform a rotary table operation on several identical workpieces, each having a machined hole in the center. To align each workpiece quickly, a special plug can be made to fit the center hole of the workpiece and the hole in the rotary table. Once the machine spindle has been aligned with the rotary table, each succeeding piece can be aligned quickly and accurately by placing it over the plug.

If there are only a few pieces, which would not justify the manufacture of a special plug, or if the workpiece does not have a hole through its center, the following method can be used to center the workpiece on the rotary table.

1. Align the rotary table with the vertical head spindle.
2. Lightly clamp the workpiece to the rotary table in the approximate center.

NOTE: *Do not* move the crossfeed or longitudinal feed handles.

3. Disengage the rotary table worm mechanism.
4. Mount an indicator in the machine spindle or on the milling machine table, depending upon the workpiece.
5. Bring the indicator into contact with the surface to be indicated, and revolve the rotary table by hand.
6. With a soft metal bar, tap the work (away from the indicator movement) until no movement is registered on the indicator in a complete revolution of the rotary table.
7. Clamp the workpiece tightly, and recheck the accuracy of the setup.

NOTE: If a center-punch mark must be aligned, a wiggler instead of an indicator is mounted in the milling machine spindle and the punch mark is positioned under the point of the wiggler.

RADIUS MILLING

When it is required to mill the ends on a workpiece to a certain radius or to machine circular slots having a definite radius, a certain sequence should be followed. Figure 70-4 illustrates a typical setup.

PROCEDURE

1. Align the vertical milling machine spindle at 90° to the table.
2. Mount a circular milling attachment (rotary table) on the milling machine table.
3. Center the rotary table with the machine spindle, using a test plug in the table and a dial indicator in the spindle.

Figure 70-4 *Milling a circular slot in a workpiece.*

4. Set the longitudinal feed dials and the crossfeed dial to zero (0).
5. Mount the work on the rotary table, aligning the center of the radial cuts with the center of the table. A special arbor may be used for this. Another method is to align the center of the radial cut with a wiggler mounted in the machine spindle.
6. Move either the crossfeed or the longitudinal feed (whichever is more convenient) an amount equal to the radius required.
7. *Lock both the table and the saddle,* and remove the handles if convenient.
8. Mount the proper end mill.
9. Rotate the work, using the rotary table feed handwheel, to the starting point of the cut.
10. Set the depth of cut and machine the slot to the size indicated on the drawing, using hand or power feed.

MILLING A T-SLOT

T-slots are machined in the tops of machine tables and accessories to receive bolts for clamping workpieces. They are machined in two operations.

PROCEDURE

1. Consult a handbook for the T-slot dimensions.
2. Lay out the position of the T-slot.
3. Square the vertical milling machine spindle with the machine table.
4. Mount the work on the milling machine. If the work is to be held in a vise, the vise jaw must be aligned with the table travel. If the work is clamped to the table, the position of the slot must be aligned with the table travel.
5. Mount an end mill having a diameter slightly larger than the diameter of the bolt body. The size of the end mill to be used is shown in the T-slot tables.
6. Machine the center slot to the proper depth of the T-slot, using the end mill.
7. Remove the end mill, and mount the proper T-slot cutter.
8. Set the T-slot cutter depth to the bottom of the slot.
9. Machine the lower part of the slot.

MILLING DOVETAILS

Dovetails are used to permit reciprocating motion between two elements of a machine. They are composed of an external or an internal part and are adjusted by means of a gib. Dovetails may be machined on a vertical milling machine or on a horizontal mill equipped with a vertical milling attachment. A dovetail cutter is a special single-angle end mill type of cutter ground to the angle of the dovetail required.

Milling an Internal Dovetail

1. Refer to a handbook for the method of measuring a dovetail.
2. Check the measurements of the workpiece in which the dovetail is to be cut. Remove all burrs.
3. Lay out the position of the dovetail.
4. Mount the workpiece in a vise and clamp it on a rotary table, or if the work is long, bolt it directly to the machine table.
5. Indicate the side of the workpiece or the slot layout to see that it is parallel to the line of table travel.
6. Mount an end mill of a diameter narrower than the center section of the dovetail (Fig. 70-5A).
7. Start the end mill, and touch up to the side of the work, after checking cutter rotation.
8. Set the crossfeed dial to zero (0).
9. Move the work over until the end mill is in the center of the dovetail. In this case, it will be the distance from the side of the workpiece to the center of the dovetail plus half the diameter of the cutter (plus the thickness of paper, if used).
10. Lock the saddle in this position, and set the crossfeed dial to zero (0).
11. Touch the edge of the cutter to the top of the work.
12. Move the work clear of the cutter, and set the depth of cut. Lock the knee in this position. The depth of this slot should be 0.030 to 0.050 in. (0.76 to 1.27 mm) deeper than the bottom of the dovetail to prevent drag and to provide clearance for dirt and chips between the mating dovetail parts.
13. Mill the channel to the width of the cutter (Fig. 70-5A).
14. Move the work over an amount equal to half the difference between the machined slot size and the size of the dovetail at the top. *Check for backlash.*
15. Take this finish cut along the one side of the work.
16. Check the width of the slot.
17. Move the work over to the finished width of the top of the dovetail. *Check for backlash.*
18. Cut the second side and check the width of the slot.
19. Return the saddle to zero (0).
20. Mount a dovetail cutter.
21. Set the depth for a roughing cut. This should be about 0.005 to 0.010 in. (0.12 to 0.25 mm) less than the finish depth.
22. Calculate the width of the dovetail at the bottom.
23. Move the work over 0.010 in. (0.25 mm) less than the finished size of this side. This will leave enough for the finish cut. Note the readings on the crossfeed dial.
24. Rough out the angle on the first side.
25. Move the work over to the other side the same amount from the centerline, and rough-cut the other side.
26. Set the cutter to the proper depth.
27. Machine the bottom surface (both sides) of the dovetail to the finished depth.

A
B

Figure 70-5 *Milling a dovetail: (A) center section roughed out; (B) both sides of the dovetail machined.*

28. Using two rods, measure the dovetail for size.
29. Move the table over half the difference between the rough dovetail and the finished size. *Check for backlash.*
30. Take the finish cut on one side.
31. Move the table over the required amount, and finish the other side (Fig. 70-5B).
32. Check the finished size of the dovetail.

NOTE: If the work is mounted centrally on a rotary table, it is possible to rotate the work a half-turn (180°) after step 18 and take the same cuts on each side of the block for each successive step.

To Mill an External Dovetail

1. Center the cutter over the dovetail position.
2. Remove as much material as possible from each side of the external dovetail; that is, cut it to the largest size of the dovetail. In this operation it will be necessary to note the readings from the centerline and remove the backlash for each side.
3. Mount a dovetail cutter, and center it with the workpiece.
4. Move the work over one-half the width of the dovetail plus half the diameter of the cutter. Allow 0.010 in. (0.25 mm) for a finish cut.
5. Take this roughing cut.
6. Move the work over an equal amount to the other side of the centerline, taking up the backlash.
7. Rough-cut the second side.
8. Adjust the work over, and take the finish cut on the one side.
9. Measure the dovetail using two rods.
10. Adjust for the finish cut on the second side, and take this cut.
11. Measure the width of the finished dovetail.

NOTE: If the work is mounted centrally on a rotary table, the work may be rotated a half-turn and the same cuts taken on each side.

JIG BORING ON A VERTICAL MILLING MACHINE

If a jig borer is not available, the vertical milling machine may be used for jig boring purposes. When the vertical milling machine is used for accurate hole location, the same setup, locating, and machining methods used in a jig borer apply. The coordinate system of hole location is used in each case. Since this topic is covered extensively in Unit 72, the reader is advised to refer to this unit before jig boring on a vertical milling machine.

Since the vertical milling machine does not have the same lead screw accuracy as a jig borer, it must have some external measuring system to ensure the accuracy of the table setting. Measuring rods and dial indicators, a vernier scale, or optical measuring devices, such as digital readout boxes, may be used for this purpose.

Inch precision end-measuring rods (Fig. 70-6) are generally supplied in sets of 11 rods, including two micrometer heads capable of measuring from 4 to 5 in. to an accuracy of 0.0001. These micrometer heads (Fig. 70-6B) are usually furnished with a red identifying ring on one and a black

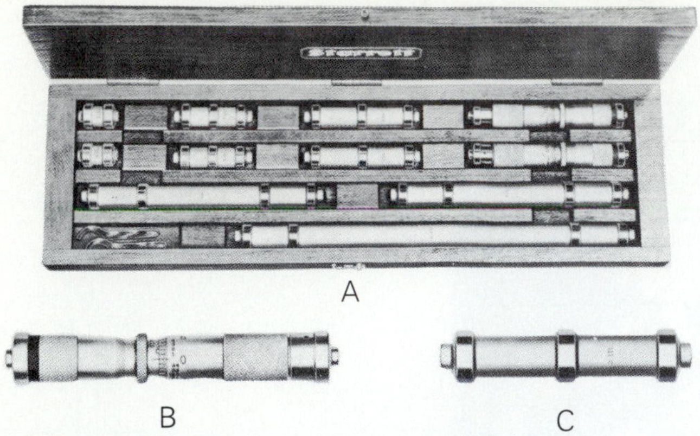
A
B
C

Figure 70-6 *Precision end-measuring rods: (A) a set of measuring rods; (B) a micrometer head; (C) a 3-in. measuring rod. (Courtesy The L. S. Starrett Company.)*

identifying ring on the other. One is used for longitudinal settings and the other for transverse settings. The solid rods are made of hardened and ground tool steel and have several concentric collars (Fig. 70-6C). The nine rods in a standard set include two each in lengths of 1, 2, and 6 in. and one 12 in. rod. Other rods are available in lengths of 4, 5, 7, 8, 10, and 15 in. The rod ends (and the micrometer head ends) are hardened, ground, and precision lapped, providing extreme accuracy.

The rods are held in V-shaped troughs, one mounted on the milling machine table and the other on the saddle (Fig. 70-7). A stop rod is mounted at one end of the trough, while at the other end a long-range dial indicator graduated in 0.0001 is mounted.

A *metric precision end-measuring rod* set consists of two micrometer heads capable of measuring to an accuracy of 0.002 mm and a number of solid measuring rods.

A standard measuring rod set consists of two rods in the following lengths: 25, 50, 75, 150, and 300 mm. Other rods are available in lengths of 100, 125, 175, 200, 250, and 375 mm.

The table is positioned accurately from the *X* and *Y* coordinates by adding (or subtracting) specified lengths or buildups of various rod combinations and micrometer settings.

TO POSITION THE MILLING MACHINE TABLE USING MEASURING RODS

1. Set the spindle to the edge of the work. See Unit 72 for methods of locating an edge.
2. Clean the trough and ends of the stop rods and indicator rods.
3. Check the indicator for free operation.
4. Place the required number of rods, including the micrometer head, in the trough to take up the space between the stop rod and the indicator rod.
5. Adjust the micrometer until it has extended enough to cause the indicator needle to move a half-turn.

Figure 70-7 *Using measuring rods to position the table accurately. (Courtesy The L. S. Starrett Company.)*

6. Lock the table.
7. Set the indicator bezel to zero (0).
8. Increase the rod and micrometer buildup by the length of the measurement between the side of the workpiece and the hole location.
9. Move the table along more than this required distance.
10. Insert the rods and the micrometer head.
11. Move the table back until the needle moves the half-turn and registers zero (0).
12. Lock the table.
13. Recheck the setting and adjust if necessary.

DIGITAL READOUT BOXES

Electronically controlled measuring devices, suitably mounted on the table and the saddle, indicate the table travel to an accuracy of 0.0005 or 0.0001 in. (0.012 or 0.002 mm). A digital readout box (Fig. 70-8) resembles the odometer on an automobile dashboard and indicates the distance traveled through a series of numbers visible through a glass front on the box. This arrangement permits quick and accurate setting of the machine table in two directions: horizontal and transverse, or on the *X* and *Y* axes.

VERNIER SCALE

The milling machine may be equipped with scales on the table and saddle with pointers suitably mounted on the machine. The scales are generally graduated in increments of tenths of an inch or 2.0 mm. A vernier arrangement is mounted adjacent to the feed screw collars, which permits the reading in 0.0001 in. or 0.002 mm.

Figure 70-8 *A digital readout box. (Courtesy The Bendix Corporation.)*

NOTE: Settings on this machine must be made in one direction only in order to remove the backlash.

Vertical milling machines may be used to position holes without the use of any of the aforementioned equipment; however, this method is not too accurate.

The table is positioned by means of the graduated feed collars. This method may be used if the accuracy of the location must not be less than 0.002 in. or 0.05 mm. Again, in order to eliminate backlash, all settings must be made in one direction only.

VERTICAL MILLING MACHINE ATTACHMENTS

The versatility of the vertical milling machine may be further increased by the use of certain attachments.

The *rack milling attachment* permits the machining of racks and broaches on the vertical mill. The spindle of this attachment is fitted into the spindle of the machine while the housing is clamped to the machine quill. It operates on the same principle as a rack milling attachment on the horizontal milling machine.

The *slotting attachment* (Fig. 70-9) is generally fitted to the back end of the overarm, which may be rotated 180° to permit the use of this device. The slotting attachment operates independently of the machine drive on most vertical milling machines. It transmits a reciprocating motion to a single-pointed tool by means of a motor-driven eccentric. This attachment may be used for cutting keyways and slotting out small blanking dies.

DIE SINKING

One important application of the vertical milling machine is that of *die sinking.* Dies used in drop forging and die casting have impressions or cavities cut in them by means of die sinking. The operation of machining a die cavity on a vertical mill is generally done by hand control of the machine, using various end-mill-type cutters. This is usually followed by considerable filing, scraping, and polishing to produce the highly finished, properly curved and contoured surfaces required on the die.

Figure 70-9 *Cutting an internal keyway with a slotting attachment. (Courtesy Cincinnati Milacron Inc.)*

Complicated shapes and patterns on molds and dies are made more conveniently and accurately on a vertical milling machine equipped with tracer control. In addition to the regular cutting head, the machine is equipped with a tracer head. Electrical discharge machining is finding ever-increasing use in the manufacture of molds and dies. Intricate shapes and forms, which would be difficult or impossible to produce on conventional machines, can be quickly and easily reproduced in even supertough, space-age metals.

On a tracer-controlled machine (Fig. 70-10) the form of a master or pattern is transferred to the workpiece by means of a hydraulic tracer unit actuated by a stylus or tracer finger. This contacts the master and moves the cutter up or down with the vertical or side travel of the stylus.

Since both the work and the master are fastened to the table, they both travel at the same rate. As the pattern moves under the stylus, the workpiece assumes the identical form of the master. Some machines may be equipped with ratio arms or devices, and the master can be made considerably larger than the finished workpiece. If a master is made 10 times larger, the arms on the machine are set at a 10:1 ratio and then the part is cut. An error of 0.010 in. (0.25 mm) on the master will only result in a 0.001 in. (0.025 mm) error on the workpiece.

Only a light pressure on the stylus, in contact with the master or template, is necessary to deflect the tracer arm

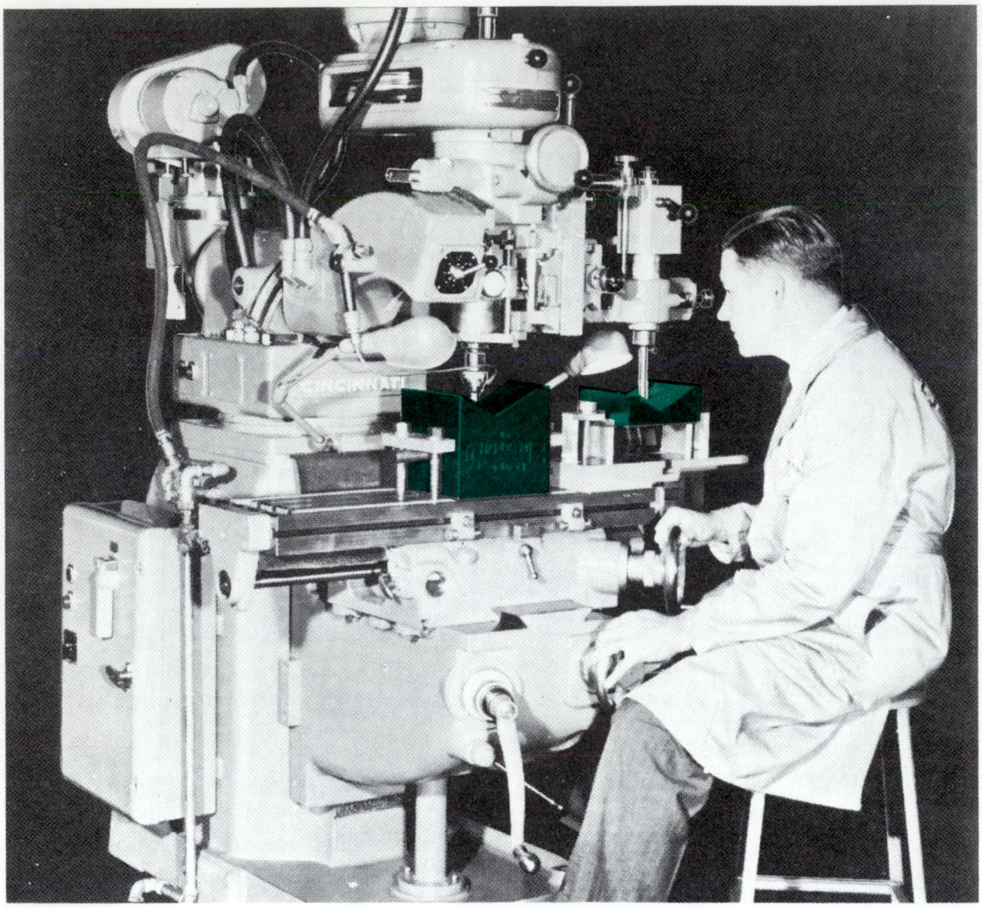

Figure 70-10 *A tracer-controlled vertical milling machine. (Courtesy Cincinnati Milacron Inc.)*

and actuate the control valve, which controls the movement of the cutting head.

REVIEW QUESTIONS

ROTARY TABLE

1. For what purpose may a rotary table be used?
2. Describe briefly the construction of a rotary table.
3. What is the purpose of the hole in the center of a rotary table?
4. What common ratios are found on rotary tables?
5. Describe briefly how a rotary table may be centered with the vertical mill spindle.
6. Explain how a number of identical parts, having a hole in the center, can be quickly aligned on a rotary table.

RADIUS AND T-SLOT MILLING

7. Explain how a large radius may be cut using a rotary table.
8. What purpose do T-slots serve?
9. List the two operations necessary in order to cut a T-slot.

MILLING DOVETAILS

10. What is the purpose of a dovetail?
11. What is the procedure for machining the center section of an internal dovetail?
12. Explain how the first angular side of a dovetail is cut.
13. How can an internal dovetail be measured accurately for size?

JIG BORING ON A VERTICAL MILL

14. Name and describe three types of measuring systems used on vertical mills for jig boring.
15. Name three methods of locating an edge.
16. How may a table be moved exactly 2.6836 in. (68.163 mm) from one location to another by using measuring rods?

VERTICAL MILLING MACHINE ATTACHMENTS

17. What are the purposes of the rack milling and slotting attachments?
18. How can a vertical mill be used for various die-sinking operations:
    a. manually?   b. automatically?
19. Explain the purpose of ratio arms or devices in die sinking.

SECTION

13

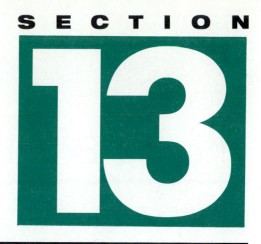

The Jig Borer and Jig Grinder

The jig borer was developed primarily to overcome the toolmaker's perpetual problem of accurately locating and producing holes. The jig borer is especially useful in the manufacture of jigs and fixtures when there must be an accurate dimensional relationship between the locators aligning the workpiece and the bushing holes which are used to provide accurate hole locations. It is an invaluable machine tool in the manufacture of simple, compound, progressive, and lamination dies, which require great accuracy between a variety of locating parts. Holes in the punch pad and die plate, pilot holes, and bushing holes which align the stripper plate are obvious "naturals" for a jig borer. Such holes can be produced quickly and accurately. By interchanging the tools in the spindle of the jig borer, operations such as drilling, boring, reaming, and counterboring can be readily performed.

The jig grinder was developed because of the need for accurately located holes in work material which had to be hardened. Much work produced in toolrooms must be hardened so that the part does not wear too quickly and will provide years of service. During the hardening process, material has a tendency to warp and distort; therefore, the holes must be jig-ground to reposition the holes accurately.

UNIT

71

The Jig Borer

A jig borer (Fig. 71-1), although similar to a vertical milling machine, is much more accurate and built closer to the floor so that the worker can operate it while seated. The precision-ground lead screws controlling the table movements are capable of infinitely fine incremental divisions and permit simultaneous measurement and positioning to within an accuracy of 0.0001 in. (0.002 mm) over the table length. This machine must be not only rugged for the heavy cuts necessary for roughing purposes but also sensitive for the more accurate finishing cuts.

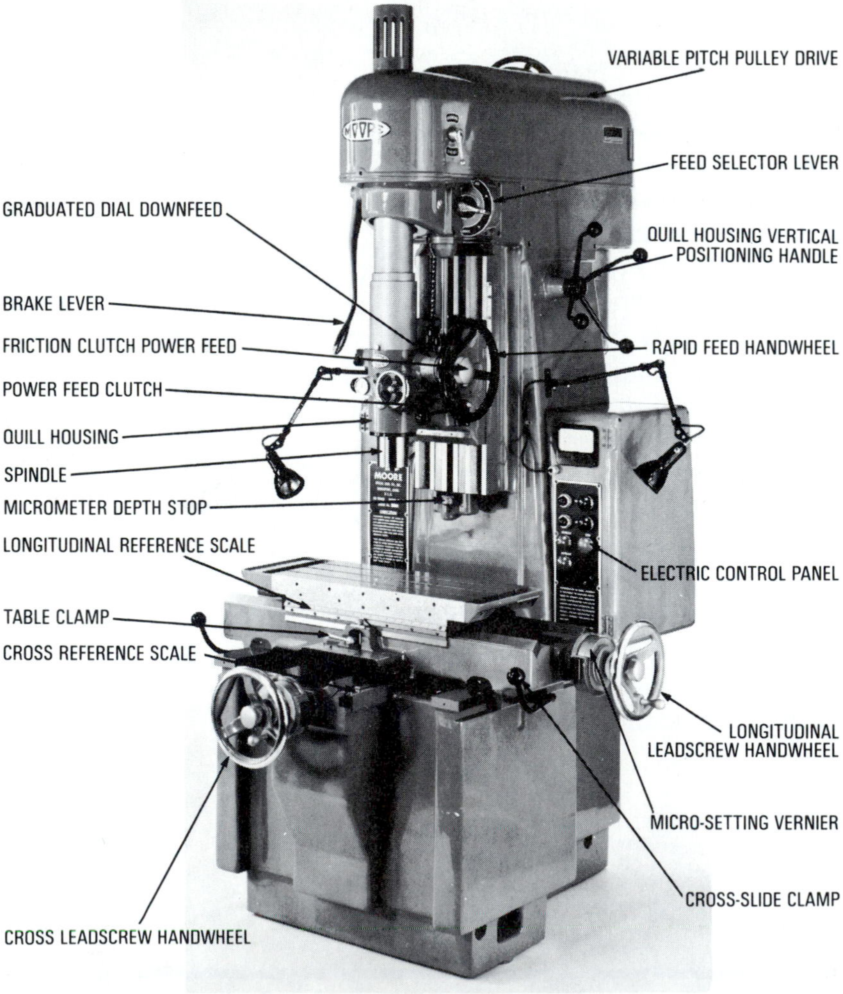
VARIABLE PITCH PULLEY DRIVE
FEED SELECTOR LEVER
GRADUATED DIAL DOWNFEED
QUILL HOUSING VERTICAL POSITIONING HANDLE
BRAKE LEVER
FRICTION CLUTCH POWER FEED
RAPID FEED HANDWHEEL
POWER FEED CLUTCH
QUILL HOUSING
SPINDLE
MICROMETER DEPTH STOP
LONGITUDINAL REFERENCE SCALE
ELECTRIC CONTROL PANEL
TABLE CLAMP
CROSS REFERENCE SCALE
LONGITUDINAL LEADSCREW HANDWHEEL
MICRO-SETTING VERNIER
CROSS-SLIDE CLAMP
CROSS LEADSCREW HANDWHEEL

Figure 71-1 *A #3 Moore precision jig borer. (Courtesy Moore Special Tool Company Inc.)*

OBJECTIVES

After completing this unit you will be able to:
1. Identify and state the purposes of the main operative parts of a jig borer
2. Use various accessories and work-holding devices for setting up and boring holes

JIG BORER PARTS

- The *variable-pitch pulley drive* is operated by pushing a button on the *electric control panel* to provide the spindle with a variable-speed range from 60 to 2250 revolutions per minute (r/min).
- The *quill housing* can be raised or lowered to accommodate various sizes of work if first the *quill housing clamp* is loosened and then the *quill housing vertical positioning handle* is turned.
- The *brake lever* is manually operated to stop the rotation of the spindle. It is especially useful while various tools are being removed or replaced in the spindle.
- The *rapid feed handwheel* allows the spindle to be raised or lowered rapidly by hand.
- The *friction clutch* may be used to engage or disengage the handfeed of the quill.
- The *graduated downfeed dial,* by means of a vernier, reads the distance of vertical spindle travel in hundredths of a millimeter or in thousandths of an inch for inch-designed dials.
- The *adjustable stop for hole depths* can be adjusted to allow the spindle to move to a predetermined depth for drilling or boring a hole.
- The *spindle* revolves inside the quill and supplies drive for the cutting tools. An internal taper in the spindle allows a variety of tools to be rapidly and accurately interchanged.
- The *reference scales* (longitudinal and crossfeed) serve as reference points in moving the table into position. They determine the position of the starting or reference point of the job.
- The *graduated dials,* with microsetting verniers on the *longitudinal* and *crossfeed screw handwheels,* allow the table to be positioned quickly and accurately to within 0.002 mm or to within 0.0001 in. for inch-graduated dials.

TO INSERT SHANKS IN THE SPINDLE

Toolroom work, with a variety of hole sizes, requires frequent changing of tools (Fig. 71-2). The tapered hole in the spindle allows this to be done quickly and accurately if certain precautions are taken.

1. The taper shank on the tool being inserted and the hole in the spindle must be *perfectly clean;* otherwise the tapers will be damaged and inaccuracy will occur.
2. Protect the taper shanks from finger perspiration, especially if the shank will be in the machine for some time. This may cause both the shank and the spindle to rust.
3. Avoid inserting shanks too tightly, especially when the spindle is warmer than the inserted shank. When the shank warms up, it expands and may jam in the spindle.
4. When removing or replacing tools in the spindle, apply the brake firmly with the left hand. The wrench should be held carefully to prevent it from slipping out of the hand and damaging the machine table.

ACCESSORIES AND SMALL TOOLS

A wide variety of accessories enable a jig borer to meet three basic requirements: *accuracy, versatility,* and *productivity.* Only accessories concerned with drilling, boring, and reaming will be dealt with in this unit.

DRILLING

Key-type and keyless chucks are used to hold the smaller-sized straight-shank spotting tools, drills, and reamers. Special collets are used to hold the larger straight-shank spotting tools, drills, and reamers. A setscrew in the collet is tightened against a flat on the tool shank, providing a positive grip, which eliminates twisting and scuffing of the tool shank.

BORING

Single-point boring, the most accurate method of attaining locational accuracy in jig boring, makes it necessary to have a wide variety of boring tools. The most commonly used boring tools are a solid boring bar, a swivel block boring chuck, an offset boring chuck, and a DeVlieg microbore boring bar.

The *solid boring bar* (Fig. 71-3) is fitted with an adjusting screw which, when adjusted, advances the toolbit over a

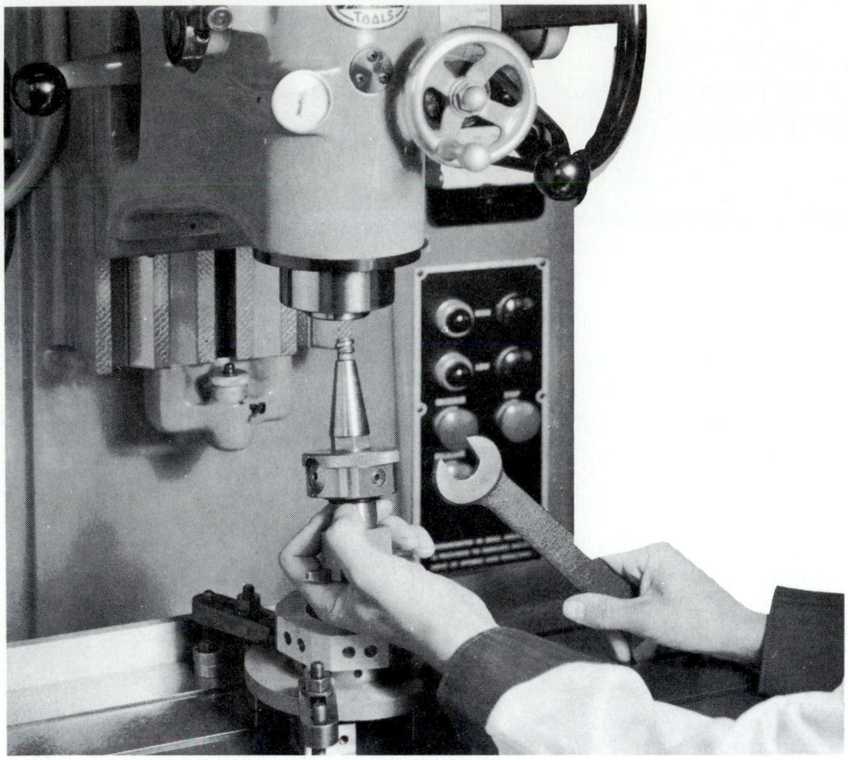

Figure 71-2 *Replacing a tool in the jig borer spindle. (Courtesy Moore Special Tool Company Inc.)*

relatively short range. Solid boring bars are rigid, making them especially useful in boring deep holes; a number of solid boring bars may be left set at a specific size for repetitive boring.

The *swivel block boring chuck* (Fig. 71-4) provides a greater range of adjustment than do other types in proportion to its diameter and better visibility to the operator while boring. One disadvantage of this type of chuck is that since the tool swings in an arc, the graduations for adjusting the tool travel vary depending upon the length of cutting tool used.

The *offset boring chuck* (Fig. 71-5) is a versatile tool which permits the cutting tool to be moved outward at 90° to the spindle axis of the machine. This allows the use of a wide variety of cutting tools without altering the value of the adjusting graduations. This chuck makes it possible to perform operations such as boring, counterboring, facing, undercutting, and machining outside diameters.

The *DeVlieg microbore boring bar* (Fig. 71-6) is equipped with a micrometer vernier adjustment, which permits the cutting tool to be accurately adjusted to within a tenth of a thousandth.

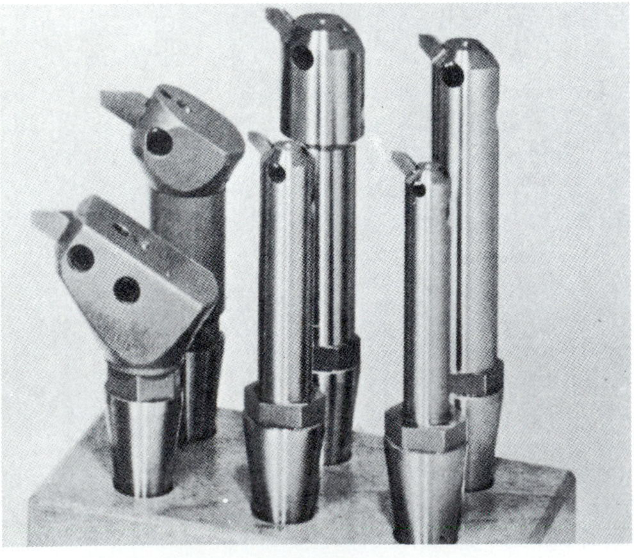

Figure 71-3 *A set of solid boring tools.*

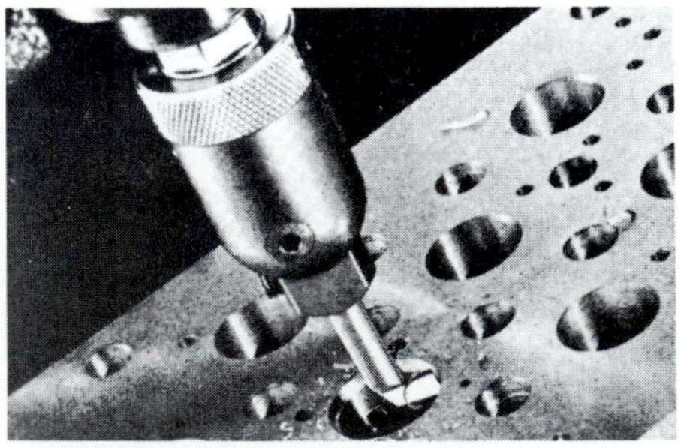

Figure 71-4 *A swivel block boring chuck. (Courtesy Moore Special Tool Company Inc.)*

Figure 71-5 *A rugged, versatile boring chuck designed for making heavy cuts.*

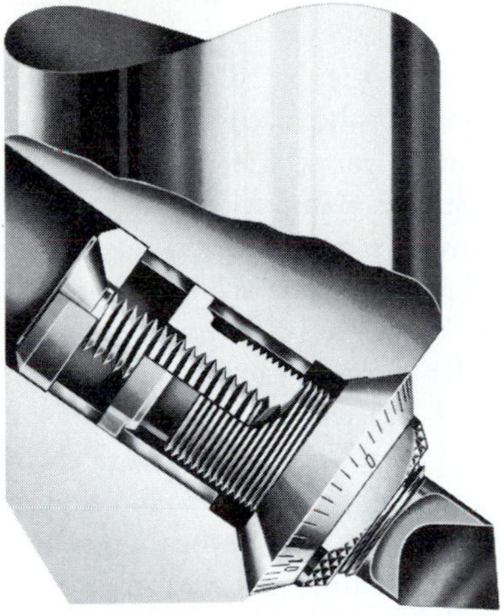

Figure 71-6 *A DeVleig microbore boring bar. (Courtesy DeVleig Machine Company.)*

Single-Point Boring Tools

Since single-point boring is the most accurate method of generating accurate hole location, a wide variety of cutting tools is available for this operation. This toolbits shown in Fig. 71-7 are generally used for small holes; however, with the necessary chuck attachments, larger holes may be bored. These toolbits are available in high-speed steel and also with brazed, cemented-carbide tips.

Collets and Chucks

An assortment of collets and chucks (Fig. 71-8) is available for a jig borer spindle to hold straight-shank spotting tools, drills, and precision end-cutting reamers.

REAMERS

Two types of reamers, the rose, or fluted, and the precision end cutting, are used in jig boring for bringing a hole to size

Figure 71-7 *Single-point boring tools. (Courtesy Moore Special Tool Company Inc.)*

quickly. The *rose,* or *fluted reamer,* is used after a hole has been bored and provides an accurate method of sizing a hole. If handled carefully and used to remove only about 0.001 to 0.003 in. (0.02 to 0.07 mm), these reamers maintain fairly accurate hole sizes.

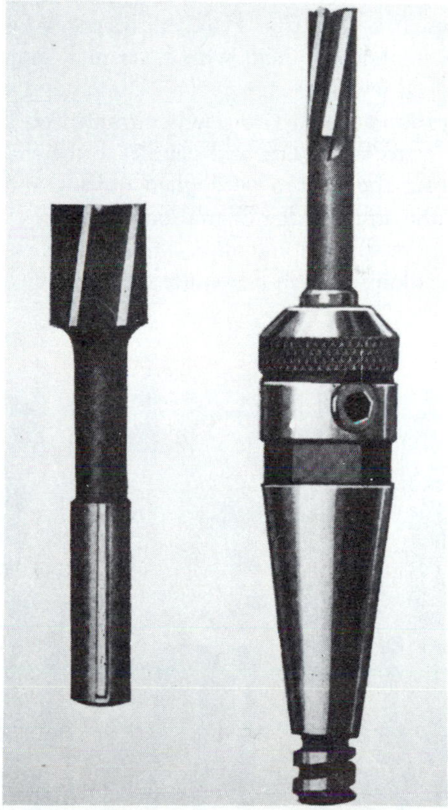

Figure 71-8 *A hardened and ground collet holding a precision end-cutting reamer. (Courtesy Moore Special Tool Company Inc.)*

Precision end-cutting reamers, with short, sturdy shanks (Fig. 71-8), provide the fastest method of locating and sizing holes to within an accuracy of ±0.0005 in. (0.01 mm). The end-cutting reamer, held rigidly and running true with the spindle, acts like a boring tool and reamer, locating and sizing the hole at the same time. End-cutting reamers eliminate the use of boring tools if the accuracy of the hole diameter and location does not require closer tolerance than ±0.0005 in. (0.01 mm).

Figure 71-10 *A precision vise. (Courtesy Moore Special Tool Company Inc.)*

WORK-HOLDING DEVICES

A wide variety of work-holding devices is necessary to fasten work for jig boring. Parallel setup blocks, matched parallel setup angle irons, matched box parallels, and extension parallels help to set up and align a variety of work shapes. To prevent machining into the table surface, most work held on a jig borer is mounted on parallels or another suitable device. *Bolts* and *strap clamps* (Fig. 71-9) provide an efficient and convenient method of holding most types of work. The heel rests for these clamps are made of brass to avoid marring the table and will build up to any height from approximately ⅜ to 9 in. (9.5 to 225 mm).

The *precision vise* (Fig. 71-10) is a valuable accessory for holding work too small to be held with bolts and strap clamps. It has stepped jaws, which serve as parallels and a V-slot for holding round work. The vise is mounted on a base plate, which is ground square and parallel to the stationary jaw, allowing the vise to be aligned quickly and accurately against the straightedge of the machine.

V-blocks are used to support and align cylindrical workpieces. If the work is long enough to require the use of two V-blocks (Fig. 71-11), it is important that a matched set be used to ensure parallelism with the table surface. The work must also be aligned parallel with the table travel. Indicate along the diameter of the cylinder and tap the V-blocks to correct any error.

The *microsine plate* (Fig. 71-12), based on the sine bar principle, is used to hold work for the machining of angular holes. It can be set to any angle from 0 to 90° by the use of the proper gage block buildup. A clamp rod is used to prevent the microsine plate from moving during machining operations. The surface of the sine plate is large enough to allow the rotary table (Fig. 71-12) to be mounted for work requiring compound angular setups and spacings.

The *rotary table* (Fig. 71-12) can be mounted on the machine table and then used for spacing holes accurately in a circle. The table is accurately graduated around its circumference in half-degree divisions and, by means of a vernier on the handwheel dial, settings accurate to within ±12″ or less can be made. When holes are required at 90° to the axis of the work, the *Moore Precision Rotary Table,* constructed so that it may be mounted in a vertical position on the table of the jig borer, may be used. A lapped bushing

Figure 71-9 *Bolts, strap clamps, and heel rests. (Courtesy Moore Special Tool Company Inc.)*

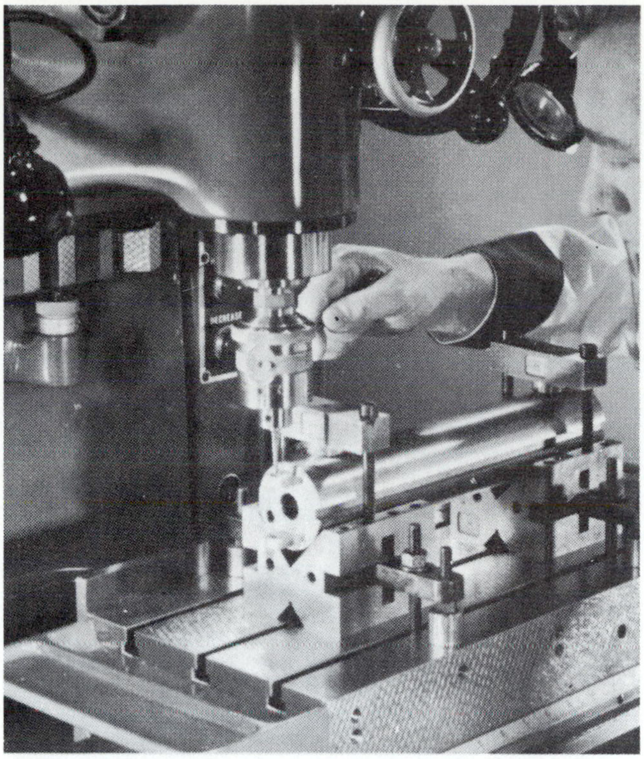

Figure 71-11 *Cylindrical work set up in V-blocks. (Courtesy Moore Special Tool Company Inc.)*

in the center of the table may be used for aligning workpieces centrally. On some rotary tables, the handwheel can be removed and replaced with an index plate attachment. This is especially valuable where a large number of holes or graduations are required, since it eliminates errors resulting from calculation of the angles.

REVIEW QUESTIONS

JIG BORER

1. For what purpose were jig borers developed?
2. For what type of work are they especially valuable?
3. Name the operations which can be performed on a jig borer.
4. Explain the difference between a jig borer and a vertical milling machine.

JIG BORER PARTS

5. State the purpose of each of the following: quill housing, vertical positioning handle, brake lever, rapid feed handwheel, adjustable stop for hole depths, and spindle.
6. What purpose do reference scales serve in jig boring?
7. Explain why longitudinal and crossfeed settings may be set quickly and accurately to 0.0001 in. (0.002 mm).
8. List three precautions that should be observed while inserting shanks in the machine spindle.

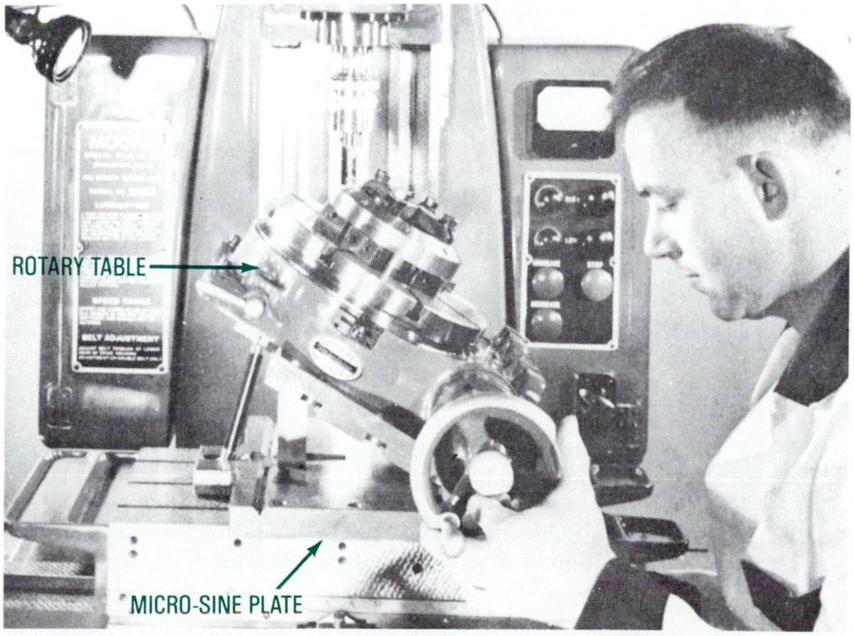
ROTARY TABLE
MICRO-SINE PLATE

Figure 71-12 *Combining the microsine plate and the rotary table for compound angle setups and indexing spacing. (Courtesy Moore Special Tool Company Inc.)*

ACCESSORIES AND SMALL TOOLS

9. Name four common boring tools and explain the advantages of each.
10. Name two types of reamers used in jig boring and explain the advantages of each.

WORK-HOLDING DEVICES

11. Explain why a precision vise is a very valuable jig borer accessory.
12. How should a long, cylindrical piece of work be set up and aligned?
13. State the purpose of:
   a. A microsine plate   b. A rotary table
14. Explain how the versatility of the rotary table may be increased.

UNIT 72

Jig-Boring Holes

Since the beginning of the machine age, the problem of locating holes accurately has plagued the toolmaker. Before the development of accurate locating and measuring machines, the toolmaker was faced with a tedious and costly, though reasonably accurate, method of locating holes. The *jig borer,* first developed in 1917, now provides the toolmaker with a means of quickly and accurately locating holes to within an accuracy of 0.000090 in. over 18 in. of length (0.002 mm over 450 mm of length). It is used to finish-bore holes in material left soft or to rough-bore holes in work which will later be hardened and jig-ground.

OBJECTIVES

After completing this unit you will be able to:
1. Set up and align workpieces for jig boring
2. Calculate hole locations using the coordinate system
3. Jig-bore and inspect holes

TO SET UP WORK

Several methods of setting up work are used in jig boring. The most common are:

1. Setting the work parallel to the machine table and aligning one edge of the workpiece with the table travel
2. Mounting the work on a sine plate for angular machining
3. Mounting the work on a rotary table for the angular spacing of holes in a circle.

The most common type of workpiece is flat and rectangular and is one of the easiest to set up. The basic requirement for this type of work is that *two edges be accurately ground at right angles to each other.* This greatly assists in aligning the workpiece and also provides an accurate surface for aligning the center of the spindle with an edge.

The work is usually placed on parallels to avoid cutting

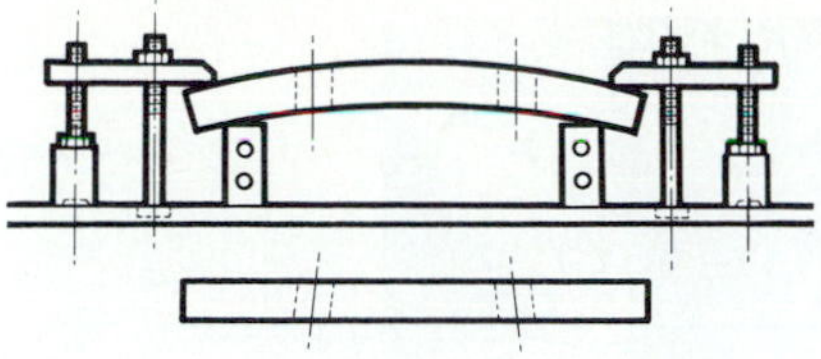

Figure 72-1 *The spring caused by improper clamping will result in holes machined out of square. (Courtesy Moore Special Tool Company Inc.)*

the table and is fastened by means of bolts and strap clamps. It is important that the supporting parallels be directly under the clamps to prevent the work from being sprung or bowed, as in Fig. 72-1.

The edge of the workpiece may be aligned with the table travel by setting one of the two ground edges against the straightedge of the machine. Should it be desirable to have the work closer to the center of the table, parallel setup blocks or gage blocks may be used between the straightedge and the work (Fig. 72-2). To ensure correct alignment, it is good practice to check the work with a dial indicator.

Figure 72-3 *Aligning the edge of the work parallel to table travel with an indicator. (Courtesy Moore Special Tool Company Inc.)*

TO ALIGN THE WORK EDGE WITH AN INDICATOR

On work that does not allow the use of the straightedge, alignment may be accomplished by holding an indicator in the machine spindle (Fig. 72-3).

PROCEDURE

1. Clamp the work lightly to the table.
2. Hold the indicator shank in a collet or drill chuck in the machine spindle.

Figure 72-2 *Setup blocks spacing the workpiece from the straightedge, parallel to table travel. (Courtesy Moore Special Tool Company Inc.)*

3. Bring the indicator against the ground edge of the work and make it register approximately 0.020 in. (0.50 mm).
4. Turn the machine spindle slightly in each direction and stop when the indicator registers the highest reading. This establishes a 90° relationship between the indicator point and the work edge.
5. Set the indicator dial to zero (0).
6. Turn the jig borer table handwheel to move the workpiece edge past the indicator point. In this way, the indicator will show how much error there is in the work alignment.
7. Then gently tap the work until the indicator shows no movement while the work is moved past the indicator.

NOTE: The work should *always be tapped away from the indicator,* not toward it, to prevent damage to the indicator.

8. When no indicator movement is shown, tighten the clamping nuts and *recheck the alignment* with the indicator.

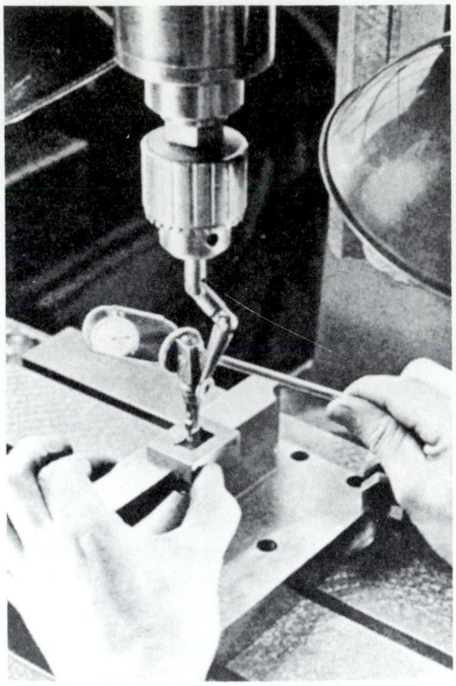

Figure 72-4 *Picking up the edge with an edgefinder and indicator. The mirror assists reading of the indicator when it is turned away from the operator. (Courtesy Moore Special Tool Company Inc.)*

METHODS OF LOCATING AN EDGE

Once the work has been set parallel with the table travel, it is necessary to align the center of the jig borer spindle to some reference or starting point. The reference points vary greatly and may include a hole, pin, boss, scribed line, slot, or contour or an edge. The two most commonly used reference points are a hole or an edge.

BACKLASH

When a reference point is being picked up, table dials set, or the table positioned, *the movement must always be made in the direction of the arrows on the dials* (Fig. 72-9) to eliminate errors as the result of *backlash.* If the crossfeed or longitudinal dial is turned past the required setting, it is necessary to back the dial away from the setting approximately one-quarter turn to eliminate backlash, and then reset it by turning it in the proper direction. When it is necessary to move the table backward (in the opposite direction of the dial arrows), turn the dial past the setting required by at least one-quarter turn, and then make the final setting in the direction of the arrows.

PICKING UP AN EDGE WITH AN EDGEFINDER

The *edgefinder* (Fig. 72-4) is a valuable accessory for picking up an edge. It is constructed so that the surface of the edgefinder which is held against the edge of the work is exactly in the center of the slot used for indicating purposes.

PROCEDURE

1. Hold the edgefinder firmly against the work. On some workpieces, it may be possible to hold the edgefinder to the work with the aid of a clamp.
2. Adjust the indicator holder so that the indicator point just touches one edge of the slot.
3. Bring the indicator into contact with one side of the slot, and have it register approximately 0.010 to 0.020 in. (0.25 to 0.50 mm).
4. Make sure that the indicator is at right angles to the edge by *slightly turning* the jig borer spindle backward and forward. Stop when the indicator registers its highest reading and *set the indicator dial to zero (0).*
5. Turn the machine spindle one-half turn, stopping when the indicator registers its highest reading on the other edge (Fig. 72-4).
6. Turn the table handwheel in the direction of the arrow one-half the difference between the two indicator readings.
7. Repeat steps 4, 5, and 6 until both indicator readings are exactly the same.
8. When both indicator readings are identical, the center of the machine spindle is located on the edge of the work.

Figure 72-5 *A spring-loaded edgefinder. (Courtesy Brown & Sharpe Mfg. Co.)*

Work may be quickly located to within 0.0005 in. (0.01 mm) by means of a spring-loaded edgefinder (Fig. 72-5) mounted in a drill chuck or collet. This device, which has a movable head the same diameter as the shank, moves approximately 1/32 in. (0.80 mm) off center when the edgefinder is rotated. As the edgefinder is brought against the edge of the workpiece, the eccentricity of the head decreases. When the head of the edgefinder runs concentric with the body, the edge of the workpiece will be half the diameter of the head from the centerline of the spindle.

TO PICK UP AN EDGE WITHOUT AN EDGEFINDER

If an edgefinder is not available, it is still possible to pick up an edge fairly accurately by using the method illustrated in Fig. 72-6.

PROCEDURE

1. Mount the indicator into the machine spindle.
2. Adjust the indicator holder so that the indicator point is as close as possible to the center of the machine spindle.
3. Bring the indicator into contact with the edge of the work, and have it register approximately 0.010 to 0.020 in. (0.25 to 0.50 mm).
4. Make sure the indicator is at right angles to the edge by *slightly turning* the jig borer spindle backward and forward. Stop when the indicator registers its lowest reading and *set the indicator dial to zero (0).*
5. Raise the machine spindle to clear the top of the work and rotate the spindle 180° (one-half turn).
6. Place a gage block against the work edge (Fig. 72-6), and turn the machine spindle slightly backward and forward. Stop when the indicator registers its lowest reading.
7. Turn the table handwheel in the direction of the arrow one-half the difference between the two indicator readings.
8. Repeat steps 4, 5, 6, and 7 until both indicator readings are exactly the same.

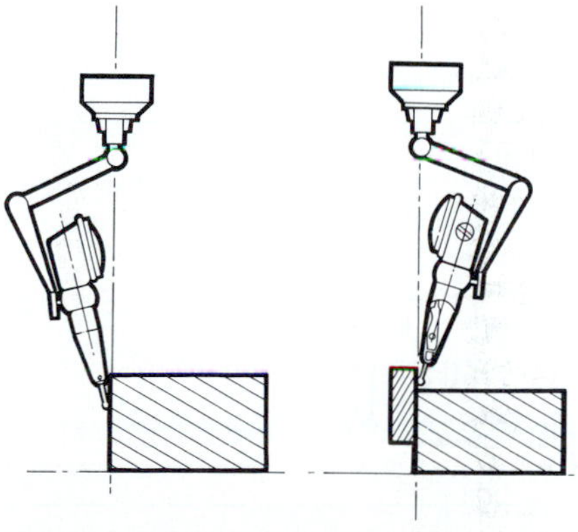

Figure 72-6 *Picking up an edge without an edgefinder. The indicator is set against the workpiece, raised, rotated 180°, and set against a gage block held against the edge. (Courtesy Moore Special Tool Company Inc.)*

TO LOCATE WITH A LINEFINDER

A method which is not very accurate but is sometimes used to pick up an edge or a scribed line, is to locate it with a *linefinder* or *wiggler.*

PROCEDURE

1. Mount the wiggler (Fig. 72-7) into a drill chuck held in the machine spindle.
2. Start the machine and make the wiggler run true by bringing a finger into contact with the wobbling point.
3. Move the table so that the point of the wiggler is as close as possible to the reference line or edge.
4. Use a magnifying glass to make the final settings as accurate as possible.

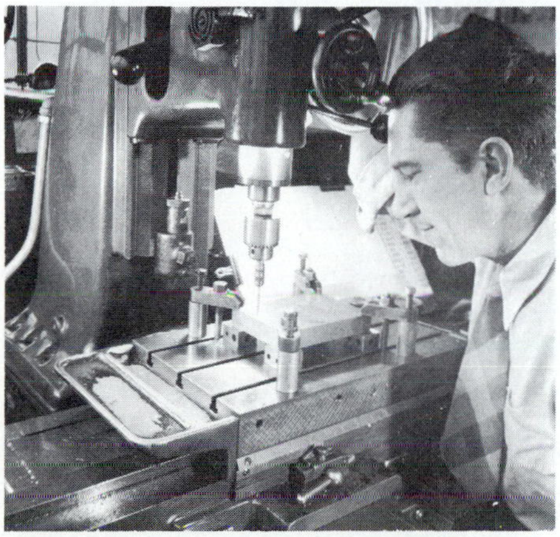

Figure 72-7 *Picking up an edge with a linefinder. (Courtesy Moore Special Tool Company Inc.)*

TO LOCATE WITH A MICROSCOPE

There are times when the reference point on the work does not suit the pickup methods of an indicator or linefinder. When small or partial holes, irregular contours, slots, and punch marks are used as reference points, a *locating microscope* (Fig. 72-8) is used. This microscope has a 40× magnification, great enough to permit 0.0001 in. (0.002 mm) to be seen. The reticle reference on the microscope consists of a number of concentric circles and two pairs of crossed centerlines for picking up a wide variety of reference points.

THE COORDINATE LOCATING SYSTEM

The *coordinate locating system* is the most efficient method of establishing hole locations for jig-boring operations. This system eliminates the tedious and often inaccurate methods of hole location, such as buttoning and layout. It also permits the best sequence of operations, such as spotting, drilling, and rough boring all holes on a workpiece before the finish-boring operation. The coordinate locating system provides the best conditions for maintaining precision hole location in jig boring.

The two general types of coordinates used are:

1. *Rectangular coordinates,* for dimensions given in straight lines
2. *Polar coordinates,* used with rotary table for holes on circles and showing angles and distances from a zero line or center

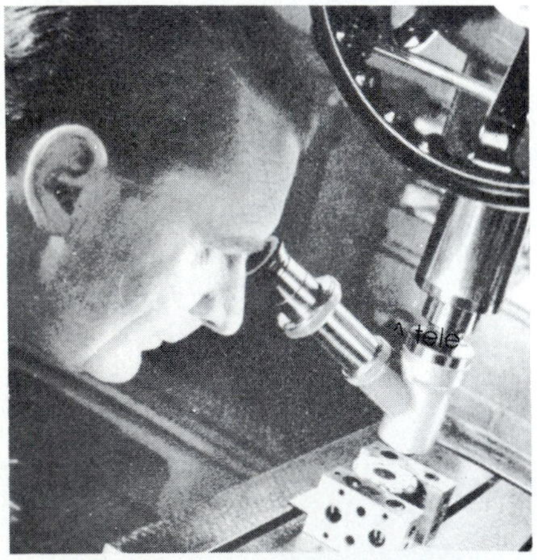

Figure 72-8 *The Moore locating microscope permits optical pickups where conventional means are impractical. (Courtesy Moore Special Tool Company Inc.)*

RECTANGULAR COORDINATES

The rectangular coordinate system consists of establishing the relationship of the work to be jig-bored to a pair of crossed ordinates, or *zero lines.* On the Moore jig borer, the zero lines are at the upper left-hand corner of the table. These crossed ordinates, sometimes referred to as the *X* and *Y* axes, represent zero (0) readings on both the crossfeed and longitudinal reference scales of the jig borer. Figure 72-9 shows the relationship of the reference scales and the lead screw dials with the coordinate dimensions. (Most jig borers on the market are capable of providing readouts in both inches and millimeters.)

After the work is fastened to the table and the spindle aligned with *both edges* of the work, the reference scales (longitudinal and crossfeed) should be set to the nearest inch line. For example, if the crossfeed reference scale is set at 4.000 in. and the longitudinal reference scale is set at 6.000 in., the figures 4.000 in. and 6.000 in. are immediately written on the corresponding edge of the drawing in the correct relationship to the table movement.

Figure 72-10 shows a conventionally dimensioned drawing on which some of the dimensions are from an edge, and others are from hole to hole. Figure 72-11 shows the same drawing with rectangular coordinate dimensions. The 4.000 in. on the upper right-hand corner and the 6.000 in. on the upper left-hand corner correspond to the settings on the crossfeed and longitudinal reference scales. All dimensions are added to these two points and are labeled on the diagram correspondingly (Fig. 72-11). For example, the

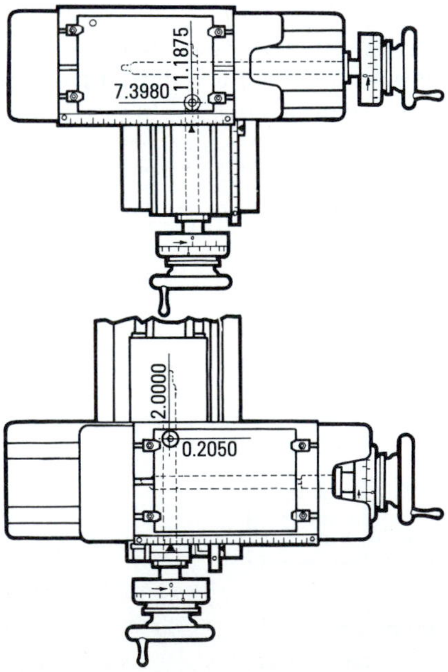

11.1875
7.3980
2.0000
0.2050

Figure 72-9 *The relationship of reference scales and dials to coordinate dimensions. (Courtesy Moore Special Tool Company Inc.)*

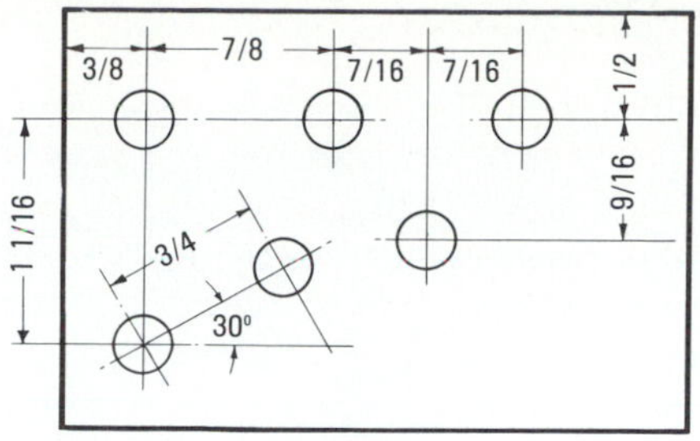

3/8
7/8
7/16
7/16
1/2
9/16
1 1/16
3/4
30°

Figure 72-10 *A conventionally dimensioned drawing. (Courtesy Moore Special Tool Company Inc.)*

three holes on top are all 0.500 in. from the edge which now becomes 4.500 in. from the reference point on the scale. The distance between these three and the hole immediately below is 0.5625 in., which now becomes 5.0625 in. (4.000 + 0.500 + 0.5625) from the reference point.

POLAR COORDINATES

The polar coordinate system is the conventional dimensioning system used to indicate on a circle the location of holes which are all the same distance from a common center. Such dimensions may be given either in the form of angles between holes, or by the number of equally spaced holes required on the circle. To convert the number of equally spaced holes into an angular value between each hole, divide this number into 360°. Great care must be taken in these calculations to avoid error, especially if the number is not evenly divisible into 360°.

Once polar coordinates are calculated, they lend themselves directly to use with the rotary table. Here also, if many holes, the number of which does not divide evenly into 360°, are required, great care must be taken in the indexing of the rotary table since a small error between each hole can result in a sizable cumulative error. The example in Fig. 72-12 shows the problems which can be encountered in calculating the angular spacing when the number is not evenly divisible into 360°.

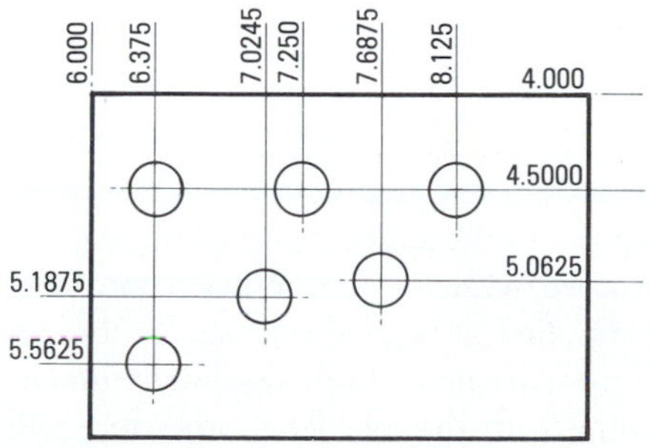

6.000
6.375
7.0245
7.250
7.6875
8.125
4.000
4.5000
5.0625
5.1875
5.5625

Figure 72-11 *The same drawing with rectangular coordinates. (Courtesy Moore Special Tool Company Inc.)*

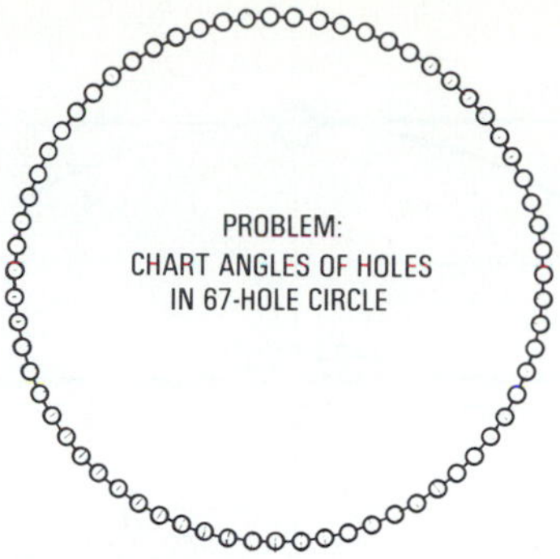

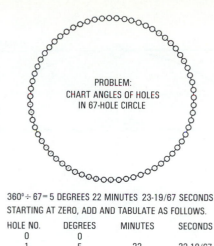

PROBLEM:
CHART ANGLES OF HOLES
IN 67-HOLE CIRCLE

360° ÷ 67 = 5 DEGREES 22 MINUTES 23-19/67 SECONDS
STARTING AT ZERO, ADD AND TABULATE AS FOLLOWS.

HOLE NO.	DEGREES	MINUTES	SECONDS
0	0		
1	5	22	23-19/67
ADD	5	22	23-19/67
2	10	44	46-38/67
ADD	5	22	23-19/67
3	16	7	9-57/67
ADD	5	22	23-19/67
4	21	29	33-9/67
ADD	5	22	23-19/67
5	26	51	56-28/67
ADD	5	22	23-19/67
6	32	14	19-47/67
	etc.	etc.	etc.

Figure 72-12 *Steps required to calculate angles between holes on a circle. To prevent cumulative error, each small fraction of an angle must be added every time. (Courtesy W. J. and J. D. Woodworth.)*

The Woodworth tables (see Appendix) were developed by W. J. Woodworth and J. D. Woodworth to eliminate the problems and inaccuracies encountered in the use of polar coordinates. The tables establish rectangular coordinates for each hole from an upper and left-hand tangent line, enabling holes to be accurately located without the use of a rotary table. A complete set of coordinate tables for up to 100 holes may be found in the book *Holes, Contours and Surfaces*, published by the Moore Special Tool Company, Bridgeport, Connecticut.

To Calculate Rectangular Coordinates Using Woodworth Tables

1. Determine the coordinates of line *A* (left tangent) and line *B* (upper tangent). These two coordinates would be taken from the reference scales on the jig borer table.
2. Multiply the *A* factor by the diameter of the circle for the location of each hole from the left tangent line (Fig. 72-13).
3. Add the result of the calculation to the coordinate of the *A* line.
4. Multiply the *B* factor by the diameter of the circle for

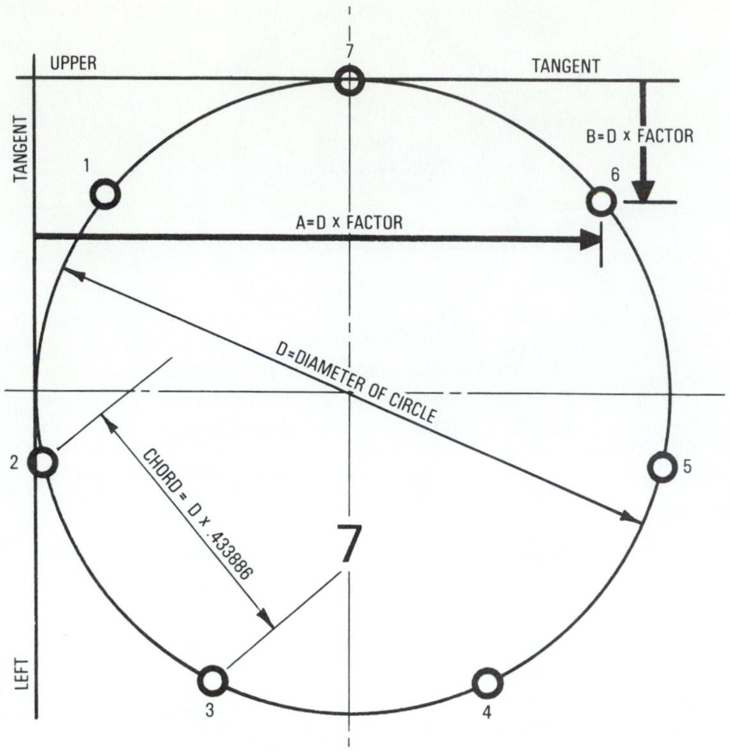
UPPER
TANGENT
TANGENT
7
B=D x FACTOR
1
6
A=D x FACTOR
D=DIAMETER OF CIRCLE
2
5
CHORD = D x .433886
7
LEFT
3
4

→	Factor for A	Factor for B		↓	Angle of Hole Degree	Minute	Second
1	.109084	1	.188255	1	51	25	42-6/7
2	.012536	2	.611261	2	102	51	25-5/7
3	.283058	3	.950484	3	154	17	8-4/7
4	.716942	4	.950484	4	205	42	51-3/7
5	.987464	5	.611261	5	257	8	34-2/7
6	.890916	6	.188255	6	308	34	17-1/7
7	.500000	7	.000000	7	360	0	0

Figure 72-13 *Rectangular coordinate factors and angles for seven evenly spaced holes. (Courtesy W. J. and J. D. Woodworth.)*

the location of each hole from the upper tangent line (Fig. 72-13).

5. Add the result of the calculation to the coordinate of the *B* line.

PREFIGURING COORDINATES

In order to save operator time and secure the greatest amount of production from a jig borer, many firms are supplying drawings on which dimensions are already given in coordinate readings. In this way, after the job is clamped to the table, the operator is required only to set the reference scales to correspond to the starting point indicated on the drawing and proceed with the job. When coordinate dimensions are not supplied, it is good practice for the operator to convert all measurements into coordinate dimensions before proceeding with the job. This procedure will not only save time, but also prevent many errors occurring as a result of alternation between jig-boring operations and locational calculations.

MAKING SETTINGS

Some jig borer manufacturers use measuring rods and indicators while others, such as the Moore Special Tool Company, use accurate lead screws with vernier readings on the dial to move the table to the required location. Regardless of the system used, it is important to remember to make *all locational settings* by turning the table handwheels *in one direction only* to avoid errors resulting from backlash. On Moore jig borers, this direction is indicated by an arrow on the crossfeed and longitudinal dials. Whenever it is necessary to move the table in the opposite direction, the table should be moved past the setting by approximately one-quarter turn and then brought to the desired setting by the handwheel being turned in the direction of the arrows. This procedure will eliminate error resulting from backlash.

JIG-BORING PROCEDURE

Because of the wide variety of work encountered in jig boring, no standard procedure would always apply. However, the following sequence should be followed whenever possible.

1. Set up and carefully align the work parallel to the table travel.
2. Align the center of the machine spindle with the reference point on the work. Set the reference scales of the machine to the nearest major measurement line.
3. Calculate the coordinate location of all holes and mark them on the drawing of the workpiece.
4. Spot the location of all holes lightly with a center drill or spotting tool.
5. Respot all the holes to a depth which will provide a good guide for future drilling operations. This operation is a means of rechecking the initial spotting operation to ensure that errors have not been made in reading the scale or dials.
6. Drill holes over ½ in. (13 mm) in diameter to within 1/32 in. (0.80 mm) of finish size. For holes less than ½ in. (13 mm) in diameter, drill to within 1/64 in. (0.40 mm) of finish size.

NOTE: When drilling large holes, it is advisable to enlarge the hole in ¼-in. (6-mm) steps to avoid undue drilling pressure, which may move the work location.

7. Recheck the work alignment, as well as the alignment of the spindle with the reference points, to make sure that it has not shifted during the roughing operation. If any error is evident, it will be necessary to repeat steps 1 and 2 before proceeding.
8. Rough-bore all holes to within 0.003 to 0.005 in. (or 0.07 to 0.12 mm) of the size required. Work requiring precise accuracy should be allowed to return to *room temperature* before the finish-boring operation.
9. Finish-bore all holes to the required size.
10. Mount an indicator in the machine spindle and inspect the accuracy of the jig boring before removing the work from the machine.

NOTE: If the accuracy of the work is not required to closer tolerances than ±0.0005 in. (0.01 mm), a precision end-cutting reamer may be used to finish the hole, a procedure which eliminates steps 8 and 9.

MEASUREMENT AND INSPECTION OF HOLES

MEASURING HOLE SIZE

A wide variety of instruments is available for measuring hole sizes to various accuracies.

Telescopic and small hole gages are used in the same manner as inside calipers. However, because they are more rigid, greater measuring accuracy is possible. Figure 72-14 shows a telescopic gage being used to measure the hole diameter.

A *plug gage* (Fig. 72-15) is more accurate than either calipers or telescopic gages for checking a hole diameter. It has a serious limitation in that it is effective only in checking a hole whose size is exactly that of the gage. Since a plug gage does not relate hole size in measurement units, it is impossible to determine how much material must still be removed from the hole.

Flat leaf taper gages (Fig. 72-16) provide a rapid means of checking hole sizes, especially during the rough-boring operation. The most common set consists of 36 leaf taper gages, allowing holes from 0.095 to 1.005 in. (2.4 to

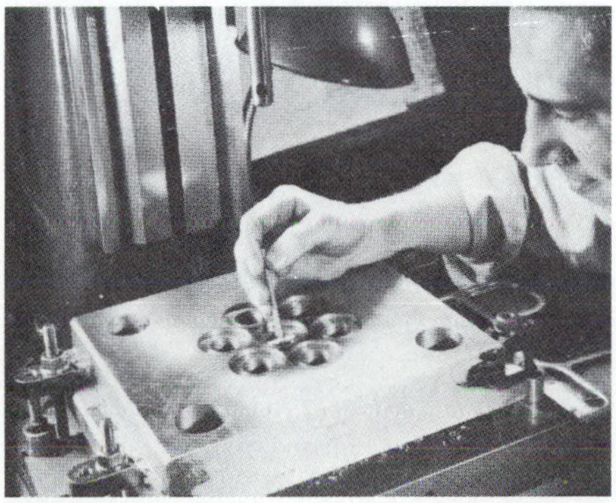

Figure 72-14 *A telescopic gage used to measure hole size. (Courtesy Moore Special Tool Company Inc.)*

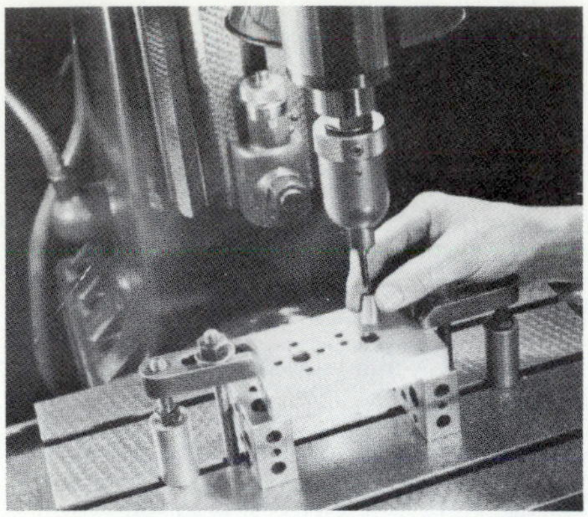

Figure 72-15 *A plug gage is an accurate means of checking final hole size. (Courtesy Moore Special Tool Company Inc.)*

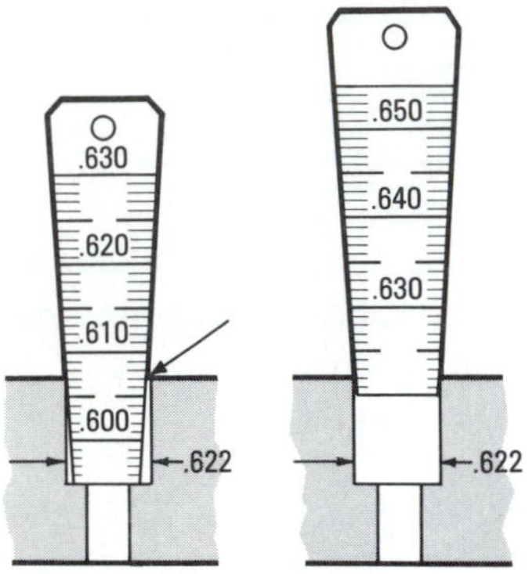

.630
.620
.610
.600
.622
.650
.640
.630
.622

Figure 72-16 *Hole size is read directly from the leaf taper gage at the point of contact with the edges of the hole. (Courtesy Moore Special Tool Company Inc.)*

25 mm) to be measured. Each leaf taper gage is 1½ in. (38 mm) long, marked off in 0.001 in. (0.025 mm) graduations, and has a range of 0.030 in. (0.76 mm). Because these gages are tapered, they cannot determine whether a hole is bell-mouthed or tapered and cannot measure to within a ten-thousandth.

Vernier calipers and various types of *inside micrometers* are instruments capable of measuring hole sizes accurately to within thousandths of an inch or hundredths of a millimeter. They are direct-reading instruments and, therefore, do not require their settings to be transferred to another instrument.

The *internal indicator gage* (or *dial bore gage*) (Fig. 72-17) is an extremely accurate instrument capable of measuring hole sizes to 0.0001 in. (0.002 mm). This gage is set for a particular size against either a standard ring gage or a micrometer, and then the indicator dial is revolved until the needle is on the zero line. The gage is then inserted into the hole and the hole size is compared to the indicator setting. Errors, such as taper, bell mouth, or out-of-roundness, can be easily checked with this type of instrument.

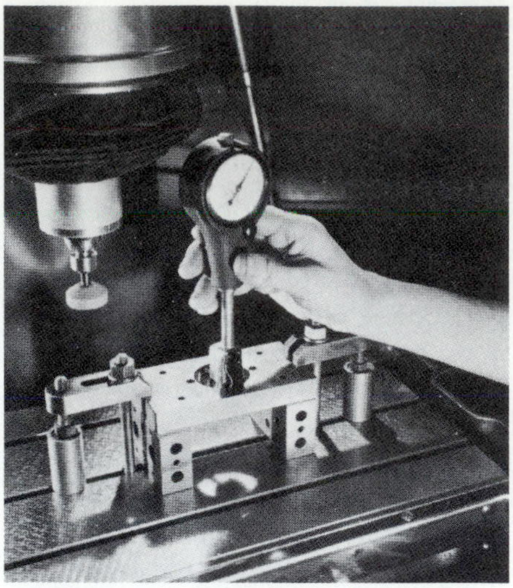

Figure 72-17 *The internal indicator gage provides extremely accurate measurements. (Courtesy Moore Special Tool Company Inc.)*

INSPECTING HOLE LOCATION

After the work has been jig-bored, it is important to verify the accuracy of the hole locations. Many methods of inspecting hole locations may be employed; the method chosen depends upon the accuracy required.

It is good practice to use the *jig borer* to inspect the accuracy of hole locations before removing the work from the machine. This is one of the most accurate and convenient methods of inspection since the work is already set up in the machine. An *indicator* (Fig. 72-18) is mounted in the machine spindle and the original reference point is then picked up. From this reference point, the table is moved to the various hole locations, which are checked for accuracy by having the indicator entered into the hole and the machine spindle revolved. The accuracy of the hole location is shown by the amount the indicator needle varies. A locating microscope, mounted in the jig borer spindle, is especially valuable in inspecting locations given from reference points, such as small or partial holes, contours, or slots.

The advantages of using the jig borer for inspecting hole locations are:

1. The work is already set up, saving time and eliminating errors which could occur in setting up work again.
2. The same coordinate dimensions which were used for jig boring are used for inspection.

Figure 72-18 *An indicator mounted in a spindle for inspection purposes. (Courtesy Moore Special Tool Company Inc.)*

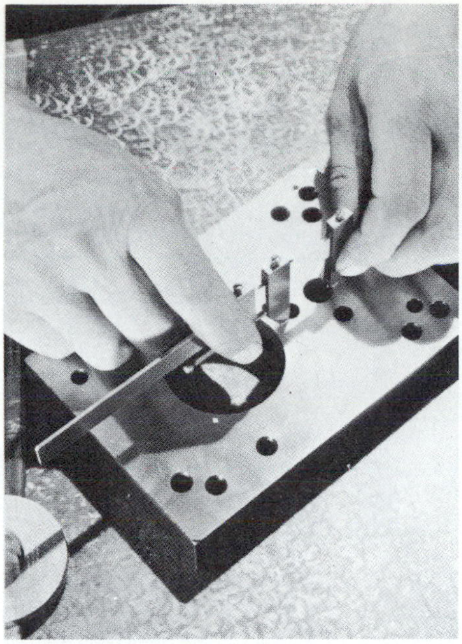

Figure 72-19 *Measuring the distance between two holes with a vernier caliper. (Courtesy Moore Special Tool Company Inc.)*

3. The machine's measuring system is just as accurate as most measuring standards.
4. The indicator mounted in the machine spindle may be used to check the location, out-of-roundness, bell mouth, or taper of a hole.
5. Work which was bored to polar coordinates can be inspected with the use of rectangular coordinates; the accuracy of rotary table calculations and settings will also be checked.

Vernier calipers (Fig. 72-19) provide a rapid, but not very accurate method of measuring the distance between two holes. The accuracy in using vernier calipers is subject to:

1. Improper tension on the instrument
2. Mistakes in reading the vernier caliper
3. Errors in angular alignment, which are not easily noticed

An *outside micrometer* may be used to measure the distance between tightly fitting plugs which have been inserted into the holes. The accuracy of this method is subject to:

1. Holes being out of square with each other
2. Looseness between the plug and the hole
3. Burrs or dirt in the hole or on the plug

A fairly accurate method of inspecting hole location is by means of *gage blocks and a dial indicator* (Fig. 72-20). The work is clamped to an angle plate with the finished edge resting on the surface plate, and the distance from the work edge to the hole surface or inserted plug is calculated. A gage block buildup for the calculated dimension is set up and the dial indicator set to the blocks. The indicator is then passed over the surface of the plug or into the hole, comparing this location with the gage block buildup.

The *measuring machine,* specially developed for inspecting hole locations to less than 0.0001 in. (0.002 mm), is

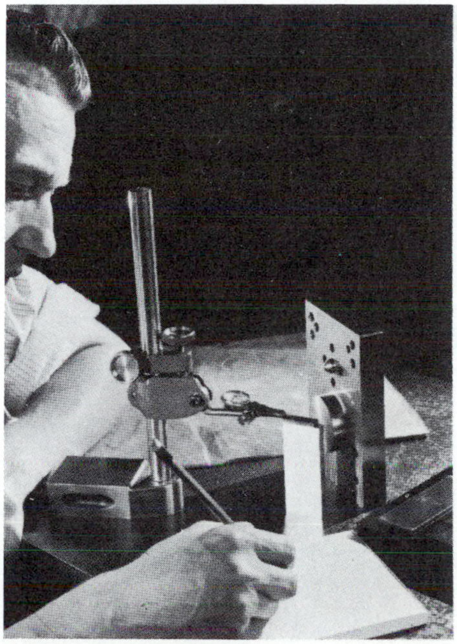

Figure 72-20 *Measuring the distance from the edge of the workpiece to a plug fitted in a hole. (Courtesy Moore Special Tool Company Inc.)*

Figure 72-21 *The Moore measuring machine. (Courtesy Moore Special Tool Company Inc.)*

the most accurate instrument used for inspecting purposes. The Moore measuring machine (Fig. 72-21) incorporates the same basic principles used in jig borers for accurately establishing hole locations.

REVIEW QUESTIONS

SETTING UP WORK

1. Name three methods of setting up work on a jig borer.
2. What requirements are necessary for rectangular work before it is set up in a machine? Explain why they are necessary.
3. What precautions should be taken in clamping?
4. Briefly describe two methods of aligning the edge of a workpiece parallel to the table travel.
5. Why should the work always be tapped away from an indicator, not toward it?

METHODS OF LOCATING AN EDGE

6. What must be done to eliminate errors resulting from backlash?
7. If the dial is turned past the required setting, how is backlash eliminated?
8. Name four methods used to pick up an edge or reference point.
9. Describe a locating microscope and state the purpose for which it is used.

THE COORDINATE LOCATING SYSTEM

10. Define and state the purposes of the two types of coordinates.
11. Explain the principle of the coordinate locating system and how it is used in jig boring.
12. What purpose do the reference scales serve in the coordinate locating system?

MAKING SETTINGS

13. Why should all locational settings be made in the same direction?
14. What must be done when it is necessary to move the table in the direction opposite to that used for locational settings?

JIG-BORING PROCEDURE

15. Why is it advisable to calculate all the coordinate dimensions before proceeding with jig boring?
16. What is the purpose of respotting all holes after the initial spotting operation?
17. What precaution should be observed after the drilling operation to ensure accurate hole locations?
18. When is it advisable to use precision end-cutting reamers?

MEASUREMENT AND INSPECTION OF HOLES

19. List four instruments used to measure hole size.
20. Describe a precision instrument which can accurately measure hole size.
21. State the advantages of using the internal indicator gage for checking holes.
22. List the advantages of using the jig borer for inspection purposes.

UNIT 73

The Jig Grinder

The need for accurate hole locations in hardened material led to the development of the jig grinder in 1940. While it was originally developed to position and grind accurately straight or tapered holes, many other uses have been found for the jig grinder over the years. The most important of these has been the grinding of contour forms, which may include a combination of radii, tangents, angles, and flats (Fig. 73-1).

The advantages of jig grinding are:

1. Holes distorted during the hardening process can be accurately brought to correct size and position.
2. Holes and contours requiring taper or draft may be ground. Mating parts, such as punches and dies, can be finished to size, eliminating the tedious job of hand fitting.
3. Because more accurate fits and better surface finishes are possible, the service life of the part is greatly prolonged.
4. Many parts requiring contours can be made in a solid form, rather than in sections, as was formerly necessary.

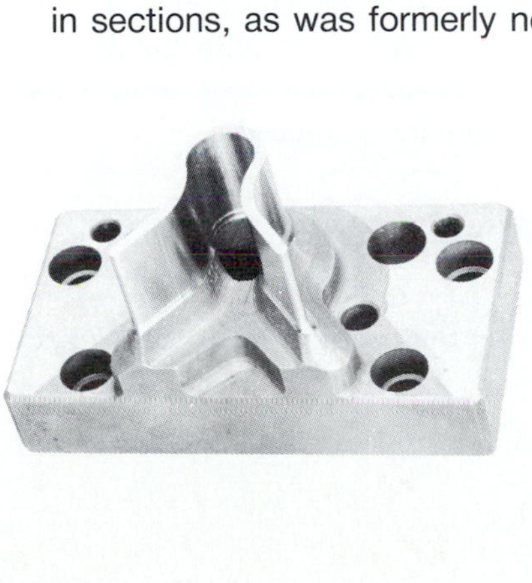

Figure 73-1 *A flanged punch represents an ideal example of jig grinding. (Courtesy Moore Special Tool Company Inc.)*

OBJECTIVES

After completing this unit you will be able to:

1. Select the wheels and grinding methods required for jig-grinding holes
2. Set up work and jig-grind a straight hole to within a tolerance of ±0.0002 in. (0.005 mm)

JIG GRINDER PARTS

The *jig grinder* (Fig. 73-2) is similar to a jig borer, both having precision-ground lead screws capable of positioning the table within 0.0001 in. (0.002 mm) accuracy over its entire length. Both are vertical spindle machines and are based on the same basic cutting principle encountered in single-point boring. The main difference between these two machines is in the spindles.

The jig grinder is equipped with a high-speed pneumatic turbine grinding spindle for holding and driving the grinding wheel. The spindle construction permits outfeed grinding (Fig. 73-3), and also the grinding of tapered holes (Fig. 73-4).

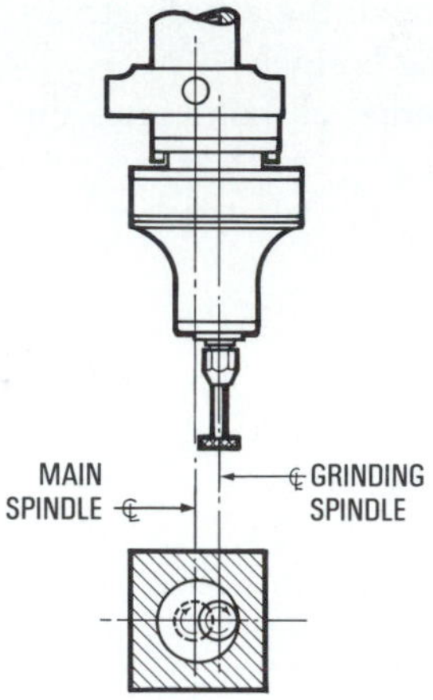
MAIN SPINDLE ℄
℄ GRINDING SPINDLE

Figure 73-3 *The grinding spindle may be offset from the main spindle. Lower view shows the planetary path of rotation.*

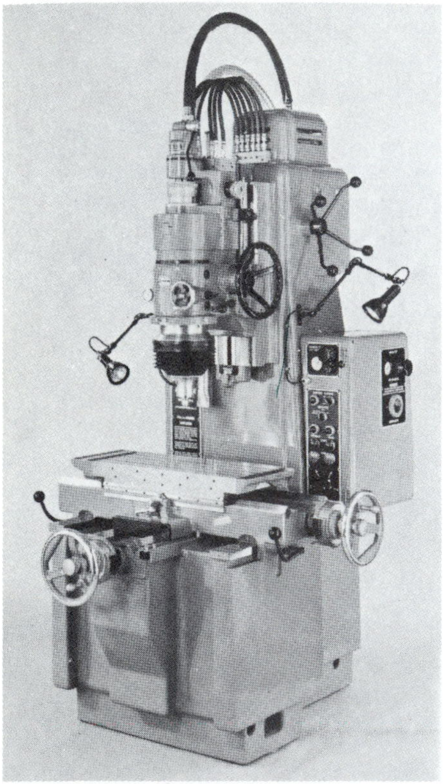

Figure 73-2 A #3 Moore jig grinder. *(Courtesy Moore Special Tool Company Inc.)*

GRINDING HEAD OUTFEED

A horizontal dovetail slide connects the grinding head to the main spindle of the jig grinder. The grinding head may be offset from the center of the main spindle to grind various size holes. The amount of eccentricity (offset) of the grinding head can be *accurately* controlled by the internally threaded *outfeed dial,* which is mounted on the nonrotating yoke at the top of the jig-grinding spindle (Fig. 73-5). The dial is graduated in steps of 0.0001 in. (0.002 mm) permitting accurate control of the hole size during grinding.

Coarse adjustment of the grinding wheel position is attained by a fine pitch-adjusting screw within the dovetail slide (Fig. 73-5). This coarse-adjusting screw is accessible only when the machine spindle is stopped.

DEPTH-MEASURING DEVICES

The Moore jig grinder has three distinct features for controlling and measuring the depth of holes (Fig. 73-6A and B):

1. The *adjustable positive stop* is on the left-hand end of

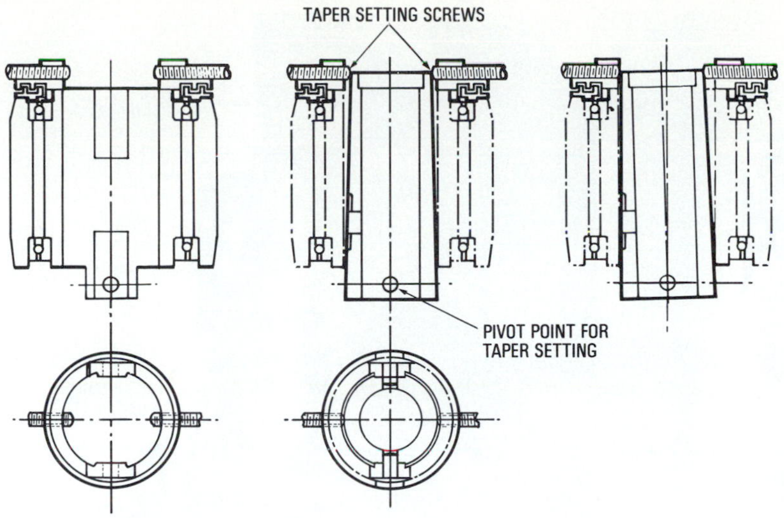
TAPER SETTING SCREWS
PIVOT POINT FOR
TAPER SETTING

Figure 73-4 *Main spindle assembly.*

the pinion shaft. Microadjustments can be made by a limiting screw.

2. The *graduated dial* on the downfeed handwheel indicates the travel of the quill. It can be set to zero (0) at any position and reads the travel depth in steps of 0.001 in. (0.02 mm).

3. The *micrometer stop* (Fig. 73-6), fastened to the column of the grinder, controls hole depth.

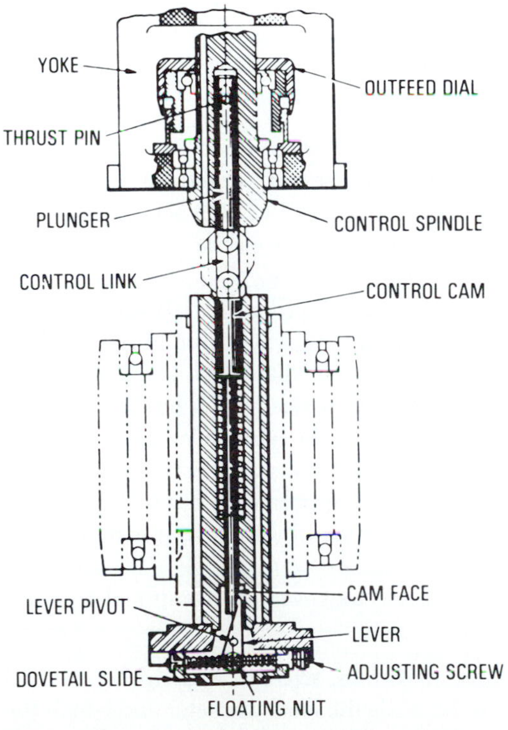
YOKE
OUTFEED DIAL
THRUST PIN
PLUNGER
CONTROL SPINDLE
CONTROL LINK
CONTROL CAM
CAM FACE
LEVER PIVOT
LEVER
DOVETAIL SLIDE
ADJUSTING SCREW
FLOATING NUT

Figure 73-5 *Assembly for controlling size by outfeed. Dial setting outfeed is graduated in tenths. (Courtesy Moore Special Tool Company Inc.)*

DIAMOND DRESSING ARM

Jig grinders must rapidly dress grinding wheels without disturbing the setup and location of the grinding spindle. The diamond dressing arm (Fig. 73-7) may be quickly swung into the approximate grinding wheel location and then locked into position. The final approach to the grinding wheel is done by a fine-adjusting knurled screw, which advances the diamond through the dressing arm.

GRINDING METHODS

The removal of material from a hole with a conventional grinding wheel is carried out by two methods: outfeed and plunge grinding. Each method has its advantages, and at times both can be used effectively to grind the same hole. Small holes, less than ¼ in. (6 mm) in diameter, can be effectively ground by using diamond-charged mandrels. Holes larger in diameter than the normal machine range can be ground effectively if an extension plate (Fig. 73-8) is used between the grinding spindle and the main spindle. With the use of an extension plate, holes up to 9 in. (225 mm) in diameter may be ground.

OUTFEED GRINDING

Outfeed grinding is similar to internal grinding where the wheel is fed radially into the work with passes as fine as 0.0001 in. (0.002 mm) at a time. The cutting action takes place with the periphery of the grinding wheel. Outfeed grinding is generally used to remove small amounts of stock when high finish and accurate hole size are required.

DOWNFEED GRADUAL DIAL AND VERNIER
OUTFEED GRADUATED DIAL
ADJUSTABLE POSITIVE STOP
MICROMETER STOP

Figure 73-6 *Depth-measuring devices. (Courtesy Moore Special Tool Company Inc.)*

PLUNGE GRINDING

Plunge grinding with a grinding wheel can be compared to the cutting action of a boring tool. The grinding wheel is fed radially to the desired diameter and then into the work (Fig. 73-9). Cutting is done with the bottom corner of the wheel only. It is a rapid method of removing excess stock, and if the wheel is properly dressed, it produces satisfactory finishes for some jobs. The sharp cutting action which results from the small contact area of the wheel keeps the work cooler than outfeed grinding.

Figure 73-7 *Dressing a wheel with a diamond dressing arm.*

Diamond-charged mandrels (Fig. 73-10) are used instead of conventional grinding wheels for grinding holes of less than 1/4 in. (6 mm) in diameter. These mandrels should be made of cold-rolled steel which has been turned to the correct size and shape in relation to the hole to be ground. The grinding end of the mandrel is placed in diamond dust and is tapped sharply with a small hardened hammer to embed the diamond dust in the mandrel surface.

The advantages of this tool over a conventional grinding wheel are:

1. Mandrels have maximum strength and rigidity.
2. Mandrels can be made the ideal diameter and length for each hole.
3. The velocity required for efficient grinding is approximately one-quarter of that for a wheel.
4. The cost per hole is less due to the greater efficiency.

Figure 73-8 *Grinding a large hole using an extension plate. (Courtesy Moore Special Tool Company Inc.)*

Figure 73-9 *Hole size may be increased by 1/16 in. (1.59 mm) in one cut by plunge grinding.*

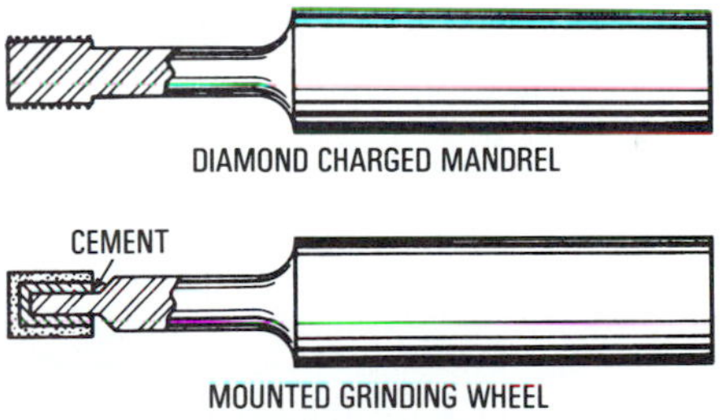
DIAMOND CHARGED MANDREL
CEMENT
MOUNTED GRINDING WHEEL

Figure 73-10 *The strength and rigidity of a diamond-charged mandrel exceed those of a mounted grinding wheel. (Courtesy Moore Special Tool Company Inc.)*

GRINDING WHEELS

SELECTION

Selection of the proper wheel is necessary for satisfactory grinding performance. Since many specific factors influence the selection of the grinding wheel to be used, a few general principles will be helpful.

1. The shank or mandrel of mounted wheels should be as short as possible to assure rigidity.
2. Wherever possible, the grinding wheel diameter should be approximately three-quarters of the diameter of the hole to be ground.
3. Widely spaced abrasive grains in the bond increase the penetrating power of the wheel.
4. When soft, low-tensile-strength materials are being ground, a hard abrasive grain with a fairly strong or hard bond should be used.
5. For grinding high-alloy hardened steels, a hard abrasive grain in a soft or weak bond is recommended.

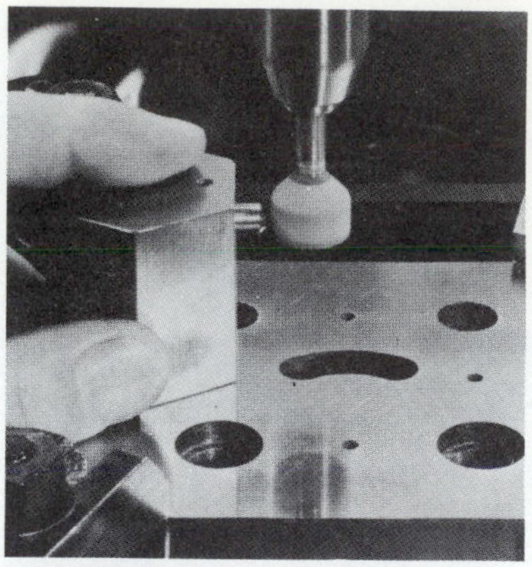

Figure 73-11 *The hand-held diamond dresser is convenient and effective.*

WHEEL SPEED

The majority of grinding wheels are operated most efficiently at about 6000 surface feet per minute (sf/min) (1828 m/min). Diamond-charged mandrels used for small hole grinding should be operated at approximately 1500 sf/min (457 m/min). The spindle speed can be varied for the different types and diameters of wheels used; three grinding heads are available for the Moore jig grinder. With these heads, a range from 12,000 to 60,000 revolutions per minute (r/min) is possible. The speed of each head may be varied within its range by adjustment of the pressure regulator, which controls its air supply.

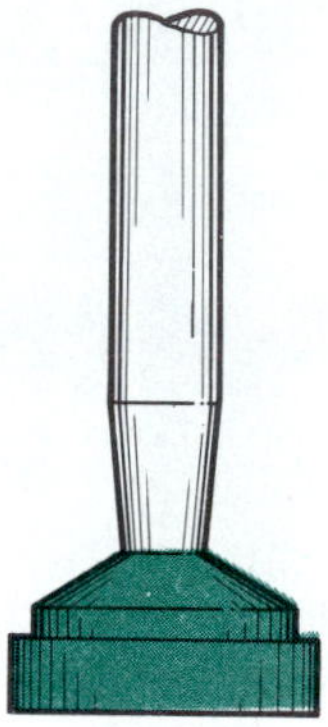

Figure 73-12 *The width of the wheel face is reduced to avoid excessive side pressure during grinding. (Courtesy Moore Special Tool Company Inc.)*

WHEEL DRESSING

For a grinding wheel to perform efficiently, it must be dressed or trued properly. An improperly dressed wheel will tend to produce the following conditions.

1. Poor surface finish on the hole
2. Surface burns
3. Holes which are out-of-round
4. Taper or bell-mouth holes
5. Locational error

Care in dressing a grinding wheel can prevent many of these undesirable conditions from developing. To develop the best cutting characteristics of the wheel, use the following recommended techniques for dressing a grinding wheel on a jig grinder.

1. While the wheel is running at a reduced rate, dress the top and bottom face with an abrasive stick held in the hand.
2. Dress the diameter of the wheel with a *sharp* diamond (Fig. 73-11).
3. Repeat steps 1 and 2 with the wheel at the proper operating speed.
4. Relieve the upper portion of the diameter (Fig. 73-12) so that approximately ¼ in. (6 mm) of the cutting face remains.
5. The bottom face of the wheel should be concaved slightly with an abrasive stick for grinding to a shoulder to the bottom of a hole.

In *outfeed grinding,* only the diameter of the wheel should be dressed when required. When *plunge grinding,* dress the bottom face of the wheel with an abrasive stick.

GRINDING ALLOWANCES

Many factors determine the amount of material which should be left in a hole for the jig grinding operation. Some of the more common factors are:

1. Type of surface finish in the bored hole
2. Size of the hole
3. Material of the workpiece
4. Distortions which occur during the hardening process

It is difficult to set specific rules on the amount of material which should be left for grinding, because of the many variable factors involved. However, general rules which would apply in most cases are:

1. Holes of up to ½ in. (13 mm) in diameter should be

0.005 to 0.008 in. (0.12 to 0.20 mm) undersize for the grinding operation.
2. Holes of over ½ in. (13 mm) in diameter should be 0.010 to 0.015 in. (0.25 to 0.40 mm) undersize for the grinding operation.

SETTING UP WORK

When setting up work for jig grinding, take care to avoid distortion of the workpiece or machine table due to clamping pressures. Keep the following points in mind while mounting the work.

1. When bolts or strap clamps are used, keep the bolts as close as possible to the work.
2. Strap clamps should be placed exactly over the parallels supporting the work. Distortion of the work can occur if a strap clamp is tightened over a part of the work not supported by parallels.
3. Bolts should *not* be tightened any more than is required to hold the workpiece. There is less pressure exerted during jig grinding than during jig boring.
4. Do not clamp work too tightly in the precision vise clamp since this may spring the stationary jaw, dislocating the aligned edge of the work.
5. Set up work on parallels high enough to allow the bottom of the hole being ground to be measured.

TO LOCATE THE WORKPIECE

The same basic methods as in jig boring are used to locate accurately a workpiece on the jig grinder; the straightedge, edge-finder, and indicator are used, for example. Distortion of the workpiece during the heat-treating process may necessitate "juggling" during the setting-up process to ensure that all holes will "clean up." The workpiece may be set up parallel to the table travel by three methods:

1. Indicate an edge of the workpiece.
2. Set the work against the table straightedge; then check the alignment with an indicator.
3. On a heat-treated piece, indicate two or more holes and set up the work to suit the average location of a group of holes.

GRINDING SEQUENCE

When a series of holes in a workpiece must be accurately related to each other, consideration must be given to the sequence of grinding operations. The following sequence is suggested when a number of different holes are required, for example, straight, tapered, or blind or holes with shoulders.

1. Rough-grind all holes first. When a high degree of accuracy is required, allow the work to cool to room temperature before proceeding to finish-grind.
2. Finish-grind all holes which can be ground with the same grinding head. This avoids continually changing grinding heads.
3. Holes whose relationship to others is most important should be ground in one continuous period of time.
4. Grind holes with shoulders or steps only once to avoid having to make accurate depth settings twice.

TO GRIND A TAPERED HOLE

The grinder spindle can be set for grinding tapers in either direction by loosening one adjusting screw and tightening the other. For most taper hole grinding, it is accurate enough to set the grinder spindle to the degrees indicated on the taper-setting plate (Fig. 73-13).

When an *extremely accurate* angular setting is required, the following steps are suggested.

1. Convert the angle into thousandths taper per inch (or into hundredths of a millimeter per 25 mm of length) by mathematical calculations.
2. Mount an indicator in the machine spindle.
3. Set an angle plate or master square on the machine table.
4. Move the indicator through 1 in. (or 25 mm) of vertical movement as read on the downfeed dial.
5. Set the adjusting screws until the desired taper is attained.

NOTE: If the adjusting screws are too tight, they will bind the vertical movement of the spindle; if too loose, the machine may grind out of round.

By reversing this procedure, one can accurately reset the machine spindle for straight-hole grinding.

Most tapered holes also have a straight section, and the taper is ground to a certain distance from the top. Sometimes it is difficult to see exactly where the taper begins because of the high finish in the hole and the gradual advance of most tapers. Two methods are generally used to show where the tapered section begins:

1. Apply layout dye to the top portion of the hole with a pipe cleaner. The dye will be removed from the taper portion during the grinding operation, allowing the length of the straight hole to be measured.
2. On holes too small or difficult to see and measure, it is recommended that the taper be ground first to dimension *X* in Fig. 73-14. This involves the use of the formula in Fig. 73-14 in order to calculate what size the hole would be at

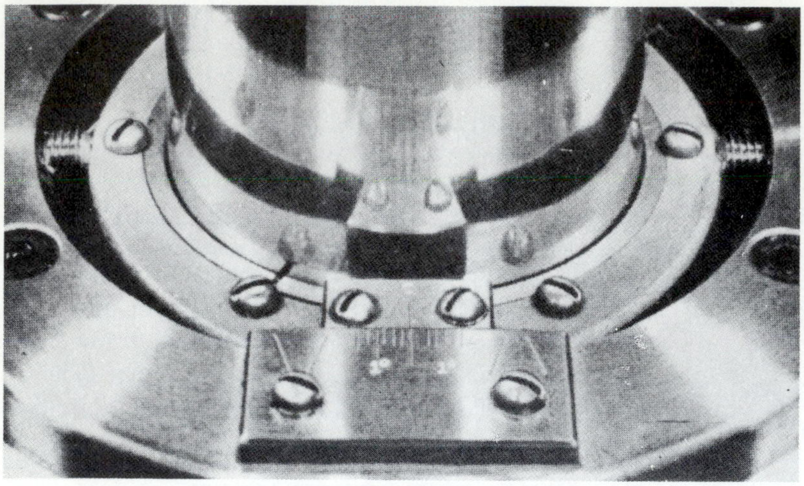

Figure 73-13 *The taper angle is indicated on the taper-setting plate. (Courtesy Moore Special Tool Company Inc.)*

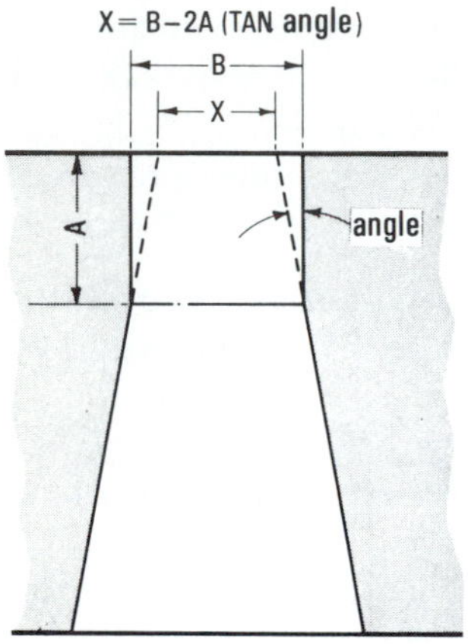
X= B−2A (TAN angle)
B
X
A
angle

Figure 73-14 *Formula for calculating X at the top of a tapered hole to produce the desired length of section A. (Courtesy Moore Special Tool Company Inc.)*

dimension *X*. Once the tapered hole has been correctly ground to size, the straight hole is ground to the proper diameter. This will automatically produce the proper length of the straight hole *A*.

TO GRIND SHOULDERED HOLES

Many times it is necessary to grind not only the diameter of a hole, but also the bottom of blind or shouldered holes. Grinding shouldered holes presents a few problems not encountered in straight or taper grinding, and the following suggestions are offered.

1. Select the proper size grinding wheel for the hole size (Fig. 73-15).
2. Make the bottom of the wheel slightly concave with an abrasive stick.
3. Set the depth stop so that the wheel just touches the bottom or shoulder of the hole.
4. Rough-grind the sides and shoulder of the hole at the same time. This eliminates leaving a slight step near the bottom of the hole.
5. Dress the wheel and proceed to finish-grind the hole.

Jig-Grinding Hints

1. Calculate all coordinate hole locations first.
2. Clamp work just enough to hold it in place. Clamping too tightly may cause distortion.

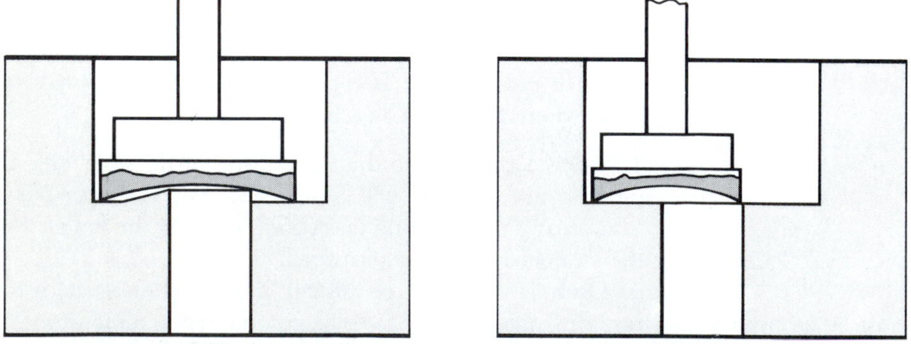

Figure 73-15 Left: *Grinding wheel is too large and cannot grind a flat surface.* Right: *Wheel is small enough to clear the opposite side of the hole.*

3. Select a grinding wheel which is three-quarters the diameter of the hole to be ground.
4. A wheel with widely spaced grains should be selected for rough grinding.
5. Relieve the wheel diameters so that only ¼ in. (6 mm) of the cutting face remains.
6. Never use a glazed wheel for grinding.
7. Rough-grind all holes by plunge grinding.
8. Allow the work to cool before finish grinding.
9. Finish-grind holes with a freshly dressed wheel by outfeed grinding.

REVIEW QUESTIONS

1. Why was the jig grinder developed?
2. State the advantages of jig grinding.

JIG GRINDING PARTS

3. How is the jig grinder spindle constructed to allow for the grinding of taper holes?
4. Explain how the grinding wheel may be positioned to the hole diameter.
5. What three methods may be used for controlling and measuring the depth of a hole?
6. Name two methods of dressing a wheel on a jig grinder.

GRINDING METHODS

7. Compare outfeed and plunge grinding.
8. How may large holes be ground on a jig grinder?
9. State the advantages of diamond-charged mandrels over grinding wheels for small hole grinding.

GRINDING WHEELS

10. List four general principles which should be followed in selecting a grinding wheel.
11. At what speed should the following be operated:
   a. Grinding wheels?
   b. Diamond-charged mandrels?
12. What undesirable conditions will an improperly dressed wheel cause?
13. Explain the procedure for dressing a grinding wheel.

GRINDING ALLOWANCES

14. Name the factors which determine the amount of material that should be left in a hole for jig grinding.
15. State the grinding allowance which would apply in most cases for holes:
   a. Under ½ in. (13 mm)
   b. Over ½ in. (13 mm)

SETTING UP WORK

16. Explain how bolts and clamps should be placed when work is being set up.
17. Why is it important that bolts not be tightened too tightly?
18. Name three methods which are used to set up work parallel to the table travel.

GRINDING SEQUENCE

19. List the sequence suggested when grinding a variety of holes.
20. Explain how the grinding head may be set to an accurate angle.
21. Calculate the *X* dimension for a 3⁄16 in.-diameter hole with a 0.200 in. straight section. The included angle of the tapered hole is 2° (see Fig. 75-14).
22. Explain the procedure for grinding shouldered holes.

SECTION

14

Computer-Age Machining

No single invention since the industrial revolution has made such an impact on society as the computer. Today computers can guide and direct spaceships to the moon and outer space and bring them back safely to earth. They route long-distance telephone calls, schedule and control train and plane operations, determine weather forecasts, and produce an instantaneous report of your bank balance. In chain stores, the cash register (connected with a central computer) totals bills, posts sales, and updates the inventory with every entry. These are but a few of the applications of the computer in our society.

During the past two to three decades, basic computers were applied to machine tools to program and control the machine operations. These devices have been steadily improved until they are now highly sophisticated units capable of totally controlling the programming, maintenance, trouble shooting, and operation of a single machine, a group of machines, or soon even a complete manufacturing plant.

UNIT 74

The Computer

Ever since primitive people became aware of the concept of quantity, people have used some device to count and perform calculations. Primitive people used their fingers, toes, and stones to count (Fig. 74-1). In about 4000 B.C., the abacus, really the first computer, was developed in the Orient. It uses the principle of moving beads on several wires to make calculations (Fig. 74-2). The abacus is very accurate when properly used. It may still be found in some of the older and smaller Oriental businesses today.

OBJECTIVES

After completing this unit you will be able to:

1. Describe generally the development of computers over the ages
2. Explain briefly the effect of computers on everyday life

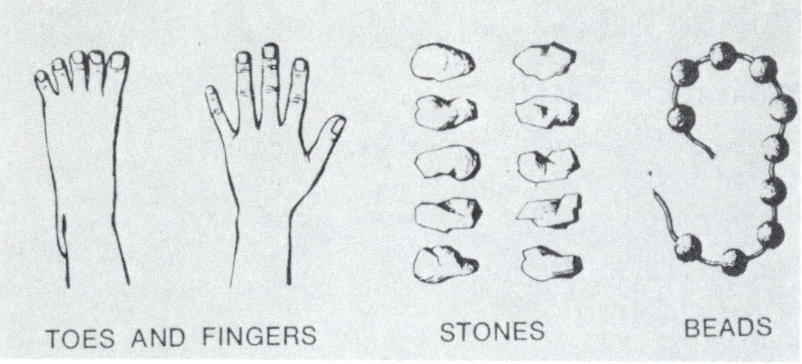
TOES AND FINGERS
STONES
BEADS

Figure 74-1 *Primitive means of counting.*

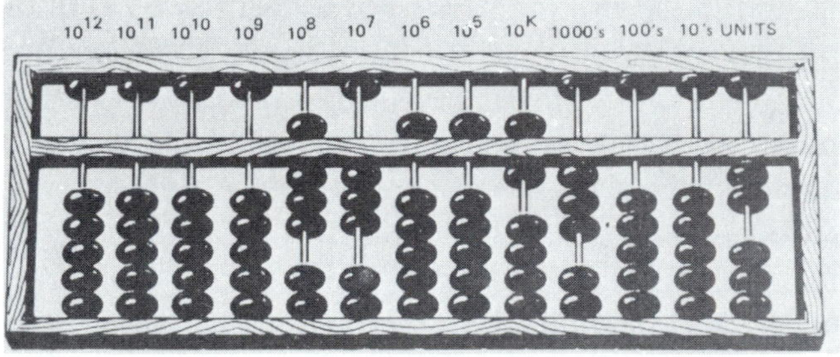
10^12 10^11 10^10 10^9 10^8 10^7 10^6 10^5 10^K 1000's 100's 10's UNITS

Figure 74-2 *The abacus was the first real computer.*

HISTORY OF THE COMPUTER

In 1642, the first mechanical calculator was constructed by a Frenchman named Blaise Pascal. It consisted of eight wheels or dials each with the numbers 0 to 9 and each wheel representing units, tens, hundreds, thousands, etc. It could, however, only add or subtract. Multiplication or division was done by repeated additions or subtractions.

In 1671, a German mathematician added the capability of multiplication and division. However, this advanced machine could only do arithmetical problems.

Charles Babbage, a 19th-century English mathematician, produced a machine called the difference engine that could rapidly and accurately calculate long lists of various functions, including logarithms.

In 1804, a French mechanician, J. M. Jacquard, introduced a punch card system to direct the operations of a weaving loom. In the United States, Herman Hollerith introduced the use of punched cards to record personal information, such as age, sex, race, and marital status, for the 1890 United States census (Fig. 74-3). The information was encoded on cards and read and tabulated by electric sensors. This use of punched cards led to the development of the early office machines for the tabulation of data.

In the 1930s, a German named Konrad Zuse built a

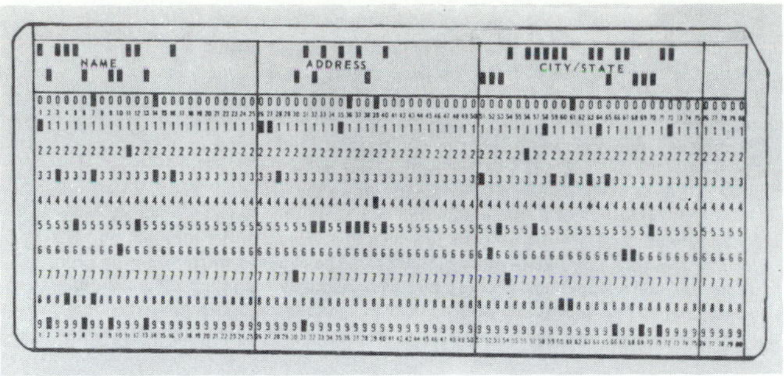
NAME
ADDRESS
CITY/STATE

Figure 74-3 *Punched cards were our first method of data processing.*

simple computer, which among other things, was used to calculate wing designs for the German aircraft industry.

A mathematician named George Stibitz produced a similar device in 1939 for the Bell Telephone Laboratories in the United States. This machine was capable of doing calculations over telephone wires; thus was born the first remote data processing machine.

During World War II, the British built a computer called the Colossus I, which helped break the German military codes.

The earliest digital computers used electro-mechanical on–off switches or relays. The first large computer, the Mark I, assembled at Harvard University by IBM, could multiply two 23-digit numbers in about 5 seconds—a very slow feat compared to today's machines.

In 1946, the world's first electronic digital computer, the ENIAC (electronic *n*umerical *i*ntegrating *a*utomatic *c*omputer) was produced. It contained more than 19 000 vacuum tubes, weighed almost 30 tons, and occupied more than 15 000 ft^2 of floor space. It was a much faster computer—able to add two numbers in $1/5000$ of a second. A machine of this size had many operational problems, particularly with tubes burning out and with circuit wiring.

In 1947, the first transistor was produced by the Bell Laboratories. These were used as switches to control the flow of electrons. They were much smaller than vacuum tubes, had fewer failures, gave off less heat, and were much cheaper to make. Computers were then assembled using transistors but still the problem of extensive hand wiring existed. This problem led to the development of the printed circuit.

In the late 1950s, Kilby of Texas Instruments and Noyce of Fairchild discovered that any number of transistors along with the connections between them could be etched on a small piece of silicon (about $1/4 \times 1/4 \times 1/32$ in. thick). These chips, called *integrated circuits* (ICs), contained entire sections of the computer, such as a logic circuit or a memory register. These chips have been further improved until today thousands of transistors and circuits are crammed into this tiny silicon chip (Fig. 74-4). The only problem with this advanced chip was that the circuits were

Figure 74-4 *Thousands of bits of information can be stored on a tiny silicon chip. (Courtesy Rockwell International.)*

rigidly fixed and the chips could only do the duties for which they were designed.

In 1971, the Intel Corporation produced the microprocessor—a chip which contained the entire *central processing unit* (CPU) for a single computer. This single chip could be programmed to do any number of tasks from steering a spacecraft to operating a watch or controlling the new personal computers.

THE ROLE OF THE COMPUTER

Although today's computers seem to amaze us (the older generation particularly), they have become part of everyday life. They will become an even greater influence in the years to come.

We are amazed that some computers today can perform 1 million calculations per second because of the thousands of transistors and circuits jammed into the tiny chips (ICs). Computer scientists can foresee the day when 1 billion transistors or electronic switches (with the necessary connections) will be crowded into a single chip. A single chip will have a memory large enough to store the text of 200 long novels. Advances of this type will decrease the size of computers considerably.

Some computer scientists feel that by 1990 the prototype of a thinking computer incorporating *artificial intelligence* (AI) will be introduced. They feel that the commercial product will follow about 5 years later. This machine will be able to recognize natural speech and written langauge and will be able to translate and type documents automatically. Once a verbal command is given, the computer will act, unless it does not understand the command. At this point the machine will ask questions until it is able to form its own judgments and act. It will also learn by recalling and studying its errors.

Computers today are being used in larger medical centers to catalog all known diseases with their symptoms and known cures. This knowledge is more than any doctor could remember. Doctors now are able to patch their own computer into the central computer and get an immediate and accurate diagnosis of the patient's problem, thus saving many hours and even days of awaiting the results of routine tests.

Because of the computer, children will learn more at a younger age and, as may be expected, future generations will have a much broader and deeper knowledge than those of past generations. It was said that in the past we doubled our knowledge every 25 years. Now with the computer, the amount of knowledge is said to double every 3 years. With this greater knowledge, the human race will explore and develop new sciences and areas that are unknown to us today (much the same as the computer has affected the older people of this era).

In other areas, the computer has been and will continue to be used to forecast weather; guide and direct planes, spaceships, missiles, and military artillery; and to monitor industrial environments.

In everyday life, everyone is and will continue to be affected by the computer. Department store computers list and total your purchases, at the same time keeping the inventory up to date and advising the company of people's buying habits. Thus the computer permits the company to buy more wisely. Credit bureau computers know how much every adult owes, to whom, and how the debt is being repaid. School computers record students' courses, grades, and other information. Hospital and medical records are kept on anyone who has been admitted to a hospital.

Police agencies have access to a national computer which can produce the police records of any known offender. The census bureau and tax department of any country have information on all its citizens on computer.

On the office front, computers have relieved the accountants of the drudgery of repetitious jobs such as payroll processing. In the future it is expected that many office workers will work at home on a company computer. This will eliminate the necessity of traveling long distances to work and the need for baby-sitting services required by so many young working families today.

In air defense control systems, the position and course of all planes from the network of radar stations are fed into the computer along with the speed and direction of each. The information is stored and the future positions of the planes are calculated.

In the manufacturing industry the computer has contributed to the efficient manufacture of all goods. It appears that the impact of the computer will be even greater in the years to come. Computers will continue to improve productivity through *computer-aided design* (CAD) by which the design of a product can be researched, fully developed, and tested before production begins (Fig. 74-5). *Computer-assisted manufacturing* (CAM) results in less scrap and more reliability through the computer control of the machining sequence and the cutting speeds and feeds.

Figure 74-5 *CAD systems are invaluable to engineers who research and design products.*

Robots, which are computer-controlled, are being used by industry to an increasing extent. Robots can be programmed to paint cars, weld, feed forges, load and unload machinery, assemble electric motors, and perform dangerous and boring tasks formerly done by humans.

REVIEW QUESTIONS

1. Name three methods of counting used by primitive people.
2. What was the first computer ever developed?
3. For what purpose were the first punched cards used in the United States?
4. How many transistors and circuits can be found on a silicon chip?
5. How are computers used in medical centers?
6. How are computers affecting the manufacturing industry?

UNIT

75

Numerical Control

Numerical control (NC) may be defined as a method of accurately controlling the operation of a machine tool by a series of coded instructions, consisting of numbers, letters of the alphabet, and symbols that the *machine control unit* (MCU) can understand. These instructions are converted into electrical pulses of current, which the machine motors and controls follow to carry out machining operations on a workpiece. The numbers, alphabet, and symbols are coded instructions that refer to specific distances, positions, functions, or motions, which the machine tool can understand as it machines the workpiece.

A basic form of numerical control has existed since the first print or sketch of a part was dimensioned by a drafter. The numbers on a print convey the information to the machine operator who proceeds to transform these numbers manually into movements of the machine tool. A numerical control machine is supplied with detailed information regarding the part by means of punched tape (Fig. 75-1). The machine decodes this punched information and electronic devices activate the various motors on the machine tool, causing them to follow specific instructions. The measuring and recording devices incorporated into numerical control machine tools ensure that the part being manufactured will be accurate. Numerical control machines minimize the possibility of human error, which existed before their development.

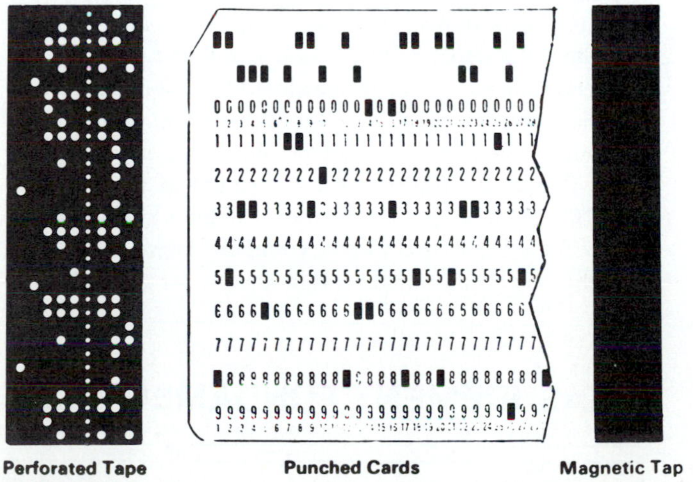
Perforated Tape
Punched Cards
Magnetic Tap

Figure 75-1 *Types of numerical control input program media.*

OBJECTIVES

After completing this unit you should be able to:
1. Identify the types of systems and controls used in numerical control
2. List the steps involved in producing a part by numerical control
3. Discuss the advantages and disadvantages of numerical control

NUMERICAL CONTROL THEORY

Numerical control is really an efficient method of reading prints and conveying this information to the motors which control the speeds, feeds, and various motions of the machine tool. The designer's information is punched on a tape, which is then put into the machine tool reader. Here holes are scanned by beams of light or small mechanical wires or fingers (Fig. 75-2). The beams of light are connected to an electric circuit. Each time a hole in the tape appears beneath the beam of light, a specific circuit is activated, sending signals to start or stop motors and control various functions of the machine tool.

TYPES OF NUMERICAL CONTROL SYSTEMS

Open loop and *closed loop* are the two main types of control systems used for numerical control machine tools.

OPEN LOOP SYSTEM

In the *open loop system* (Fig. 75-3), the tape is fed into a *tape reader,* which decodes the information on the tape and stores it until the machine is ready to use it and then converts it into electric pulses or signals. These signals are sent to the *machine control unit,* which energizes the *servo control units.* The servo control units direct the *servomotors* to perform certain functions according to the information supplied by the tape. The amount each servomotor will move the leadscrew of the machine depends on the number of electric pulses it receives from the servo control unit.

This type of system is fairly simple; however, since there is no means of checking to determine whether the servomotor has performed its function correctly, it is not generally used where an accuracy greater than 0.001 in. (0.02 mm) is required. The open loop system may be compared with a gun crew that has made all the calculations necessary to hit a distant target but does not have an observer to confirm the accuracy of the shot.

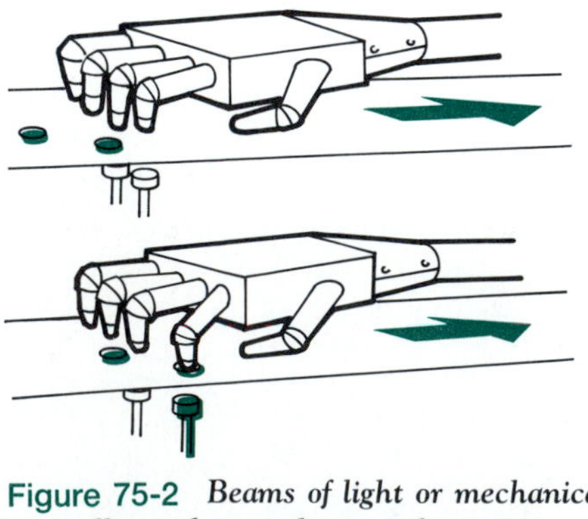

Figure 75-2 *Beams of light or mechanical fingers are generally used to read tape information.*

CLOSED LOOP SYSTEM

The *closed loop system* (Fig. 75-4) can be compared with the same gun crew that now has an observer to confirm the accuracy of the shot. The observer relays the information regarding the accuracy of the shot to the gun crew, which then makes the necessary adjustments to hit the target.

The closed loop system is similar to the open loop system, with the exception that a *feedback unit* (Fig. 75-4) is introduced in the electric circuit. This feedback unit, generally called a *transducer,* compares the amount the machine table has been moved by the servomotor with the signal sent by the control unit. The control unit instructs the servomotor to make whatever adjustments are necessary until both the signal from the control unit and the one from the servounit are equal. In the closed loop system, 10 000 electric pulses are required to move the machine slide 1 in. (25 mm). Therefore, in this type of system, one pulse will cause 0.0001 in. (0.002 mm) movement of the machine slide. Closed loop numerical control systems are very accurate because the accuracy of the command signal is recorded and there is an automatic compensation for error.

MEASUREMENT FUNDAMENTALS

Numerical control data processing (numbers, alphabet, and symbols) is done in a computer or machine control unit

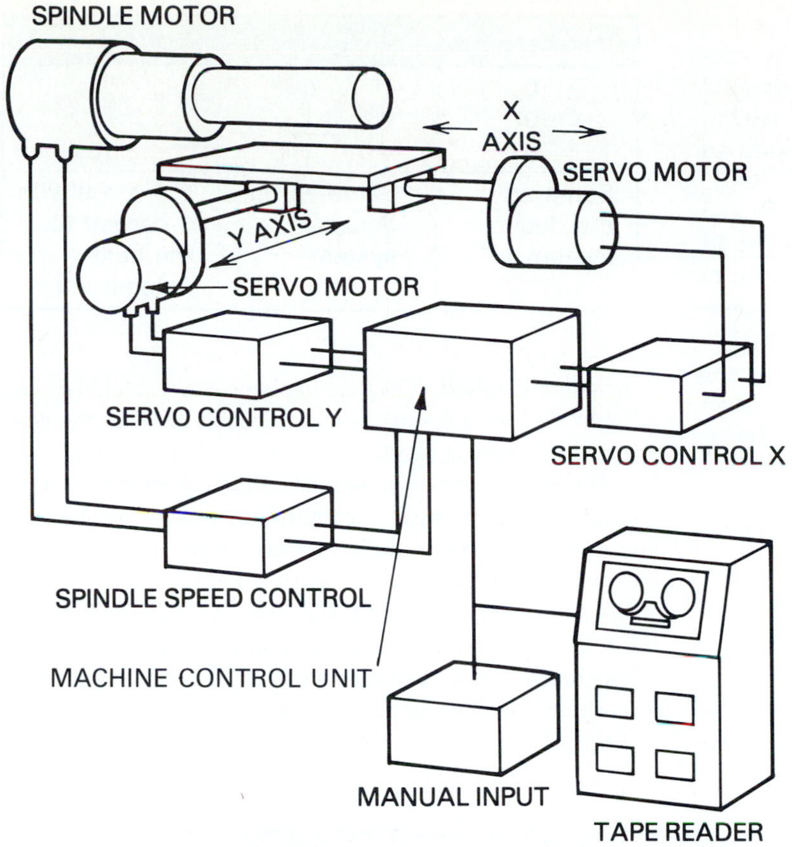
SPINDLE MOTOR
X
AXIS
SERVO MOTOR
Y AXIS
SERVO MOTOR
SERVO CONTROL Y
SERVO CONTROL X
SPINDLE SPEED CONTROL
MACHINE CONTROL UNIT
MANUAL INPUT
TAPE READER

Figure 75-3 *An open loop numerical control system.*

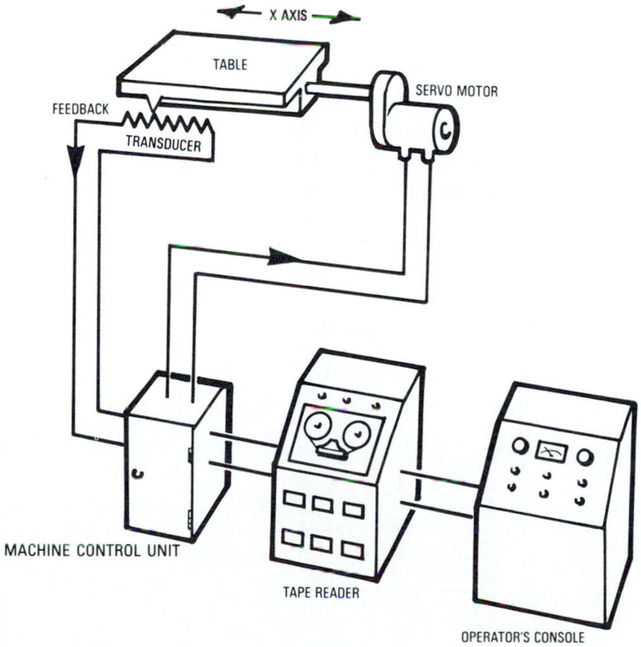
X AXIS
TABLE
SERVO MOTOR
FEEDBACK
TRANSDUCER
MACHINE CONTROL UNIT
TAPE READER
OPERATOR'S CONSOLE

Figure 75-4 *A closed loop numerical control system showing the X axis only.*

(MCU) by adding, subtracting, multiplying, dividing, and comparing. The computer can be programmed to recognize *A* before *B,* or any other facts required. Subtracting is done by adding negative values, multiplying is done through a series of additions, and dividing is done by a series of subtractions. The computer is capable of handling numbers very quickly and the addition of two simple numbers may take only a billionth of a second (a nano second).

BINARY NUMBERS

Primitive people always used their ten fingers and ten toes to count numbers. This method evolved into our present decimal or Arabic system from which *base ten* or *the power of ten* was derived. However, computers and machine control units use the *binary system* to recognize numerical values. A knowledge of the binary system is not essential for the machine operator, since both the computer and the MCU can recognize the standard decimal (Arabic) system and convert it into binary data. To provide an understanding of the binary system, let's compare it with the decimal system:

- In the *decimal system* the value of each digit depends on where it is placed in relation to the other digits in a number. The number one (1) by itself is worth 1; however, if it is placed to the left of one zero (0) or two zeros (00) it is worth 10 or 100, respectively. In order to add or subtract numbers, they must first be arranged in their proper place columns. In the decimal system, each position to the left of a decimal point means an increase in the power of 10.

- The *binary system* uses only two digits—zero (0) and one (1)—and is based on the power of two (2) instead of ten (10), as in the decimal system. Each position to the left means an increase in the power of 2. For example, $2^1 = 2$, $2^2 = 4\ (2 \times 2)$, $2^3 = 8\ (2 \times 2 \times 2)$, $2^4 = 16\ (2 \times 2 \times 2 \times 2)$ and $2^5 = 32$. Therefore, any numerical value can be made by using only the two digits 1 and 0. Since there are only two digits, the binary system is often called the *on* or *off* system. For example, 1 = *ON* and 0 = *OFF.* On numerical punched tape, a hole represents a 1 and no hole represents a 0.

Table 75-1 illustrates the difference between the decimal and binary systems. Examples are shown to clarify both systems.

The rules of addition in the binary system are somewhat different from those of the decimal system because only two digits are involved. They are explained in Table 75-2. See Table 75-3 for a comparison of addition between the decimal and binary systems.

Binary numbers are essential to the computer's and the machine control unit's capabilities to process information at a very high speed. Binary notations are used in electric circuits because they are stable in either of two conditions: *ON* OR *OFF, POSITIVE* OR *NEGATIVE, CHARGED* OR *DISCHARGED,* etc. Since a punched tape either has a hole or no hole at a specific position, the tape reader on a machine tool decodes this information and converts it into electrical pulses.

Table 75-2 Binary Addition Rules

0 +0 = 0	0 +1 = 1	1 +1 = 10
Same as the decimal system	Same as the decimal system	1 + 1 = 0 with 1 carried to the next column left

Table 75-1 Comparison of the Decimal and Binary Systems

	Decimal System (Power of 10)		Binary System (Power of 2)			
	Place	Place	Place	Place	Place	Place
	10	1	8	4	2	1
Digits		0				0
		1				1
		2			1	0
		3			1	1
		4		1	0	0
		5		1	0	1
		6		1	1	0
		7		1	1	1
		8	1	0	0	0
		9	1	0	0	1
	1	0	1	0	1	0

To convert the number 55 from decimal to binary:

1. Subtract the largest possible power of two from the number.
2. Mark a one in each column used.
3. Keep subtracting the largest possible power of two from the remainder.
4. Mark a zero in each column not used.
5. Keep subtracting (power of two) until the remainder is zero.

Example:

Decimal

	10	1
	5	5
−	3	2
	2	3
−	1	6
		7
−		4
		3
−		2
		1
−		1
		0

Binary

32	16	8	4	2	1
1	1	0	1	1	1

Table 75-3 A Comparison of Addition in the Decimal and Binary Systems

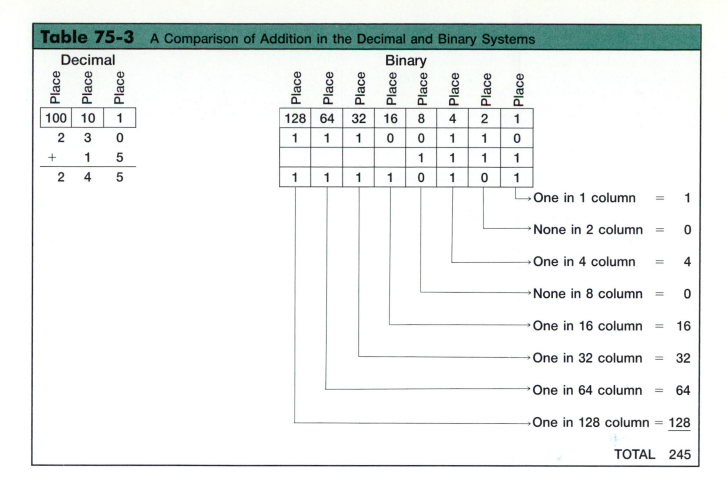

Decimal

	100 Place	10 Place	1 Place
	2	3	0
+		1	5
	2	4	5

Binary

128 Place	64 Place	32 Place	16 Place	8 Place	4 Place	2 Place	1 Place
1	1	1	0	0	1	1	0
				1	1	1	1
1	1	1	1	0	1	0	1

One in 1 column	=	1
None in 2 column	=	0
One in 4 column	=	4
None in 8 column	=	0
One in 16 column	=	16
One in 32 column	=	32
One in 64 column	=	64
One in 128 column	=	128
TOTAL		245

Computers and machine control units accept information in the decimal system commonly used in industry and convert this information into binary form. The programmer does not have to make any conversions from one system to the other.

CARTESIAN COORDINATE SYSTEM

Almost everything that can be produced on a conventional machine tool can be produced on a numerical control machine tool with its many advantages. During production of a part, machine tool movements involve two basic classifications of positioning: point-to-point (straight-line movements) and continuous-path (contouring movements).

The *Cartesian,* or *rectangular coordinate, system* was discovered by the French mathematician and philosopher René Descartes. With this system, any specific point can be described in mathematical terms from any other point along three perpendicular axes. This concept fits machine tools perfectly, since their construction is generally based on three axes of motion (*X, Y, Z*) plus an axis of rotation (Fig. 75-5). Using a plain vertical milling machine as an example, the *X*-axis is the horizontal movement (right or left) of the table, the *Y*-axis is the table cross-movement (toward or away from the column), and the *Z*-axis is the vertical movement of the knee or the spindle. Numerical control systems rely heavily on the use of rectangular coordinates because the programmer can locate every point on a job precisely.

For locating points on a workpiece, two straight inter-

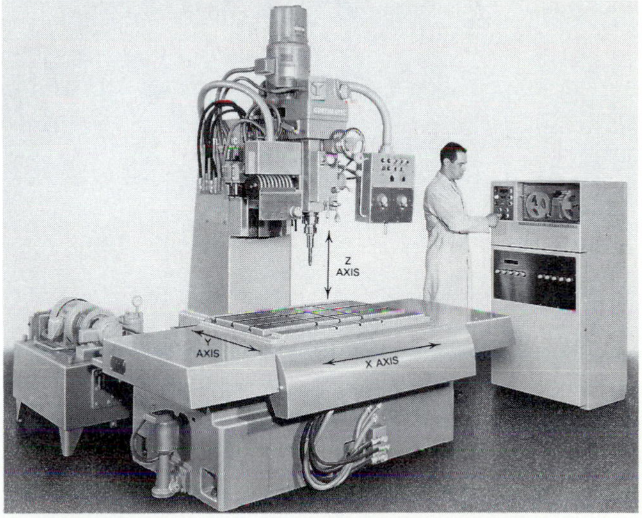

Z AXIS
Y AXIS
X AXIS

Figure 75-5 *The X, Y, and Z axes of a NC drilling machine give three-dimensional coordinates with the Z axis in a vertical relationship to the X and Y axes.*

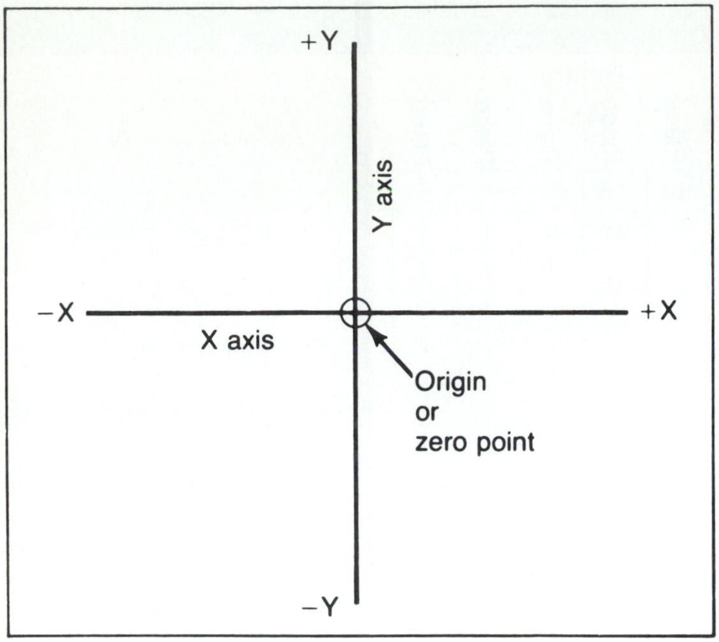

+Y
Y axis
−X
+X
X axis
Origin
or
zero point
−Y

Figure 75-6 *Intersecting lines form right angles and establish the zero (0) point.*

secting lines, one vertical and one horizontal are used. These lines must be at right angles to each other and the point where they cross is called the *origin* or *zero point* (Fig. 75-6).

The three dimensional coordinate planes are shown in Fig. 75-7. The *X*- and *Y*-planes (axes) are horizontal and represent horizontal machine table motions. The *Z*-plane (axis) represents vertical tool motion. The plus (+) and minus (−) signs indicate the direction from the zero point (origin) along the axis of movement. The four quadrants formed when the *XY*-axes cross are numbered in a counterclockwise direction (Fig. 75-8). All positions located in quadrant 1 would be positive *X* (+*X*) and positive *Y*(+*Y*). In the second quadrant, all positions would be negative *X*(−*X*) and positive *Y*(+*Y*). In the third quadrant, all locations would be negative *X* (−*X*) and negative *Y*(−Y). In

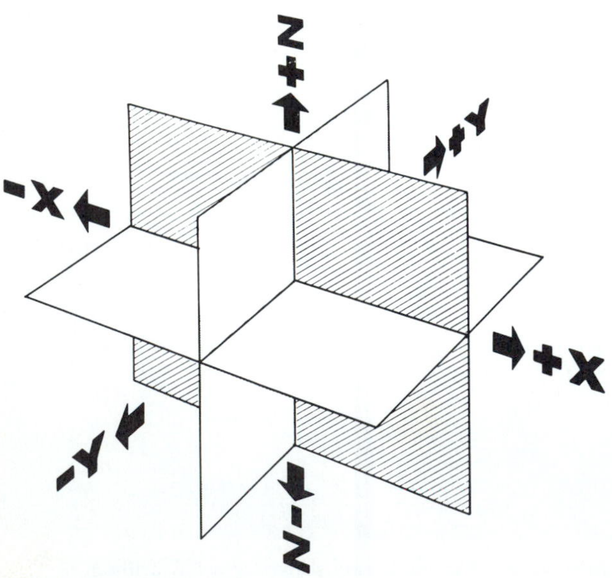

+Z
+Y
-X
+X
-Y
-Z

Figure 75-7 *The three-dimensional coordinate planes (axes) used in numerical control work.*

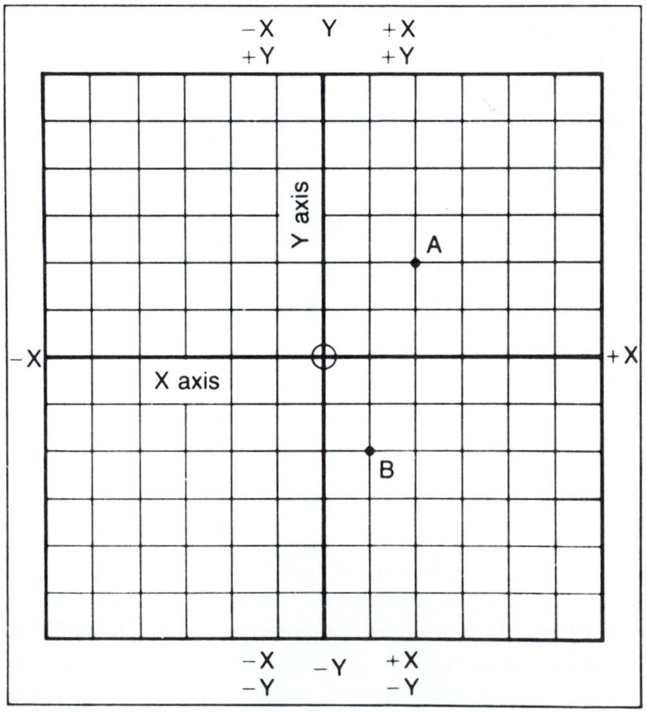

−X
+Y
Y
+X
+Y
Y axis
A
−X
+X
X axis
B
−X
−Y
−Y
+X
−Y

Figure 75-8 *The quadrants formed when the* X *and* Y *axes are used to accurately locate points from the* XY *zero (origin point).*

the fourth quadrant, all locations would be positive X ($+X$) and negative Y($-Y$).

In Fig. 75-8, point A would be 2 units to the right of the Y-axis and 2 units above the X-axis. Assume that each unit equals 1 in.; the location of point A would be (X: +2.000, Y: +2.000). For point B, the location would be (X: +1.000, Y: −2.000). In numerical control programming it is not necessary to indicate plus (+) values since these are automatically assumed; however, the minus (−) values must be indicated. For example, the locations of both A and B would be indicated as follows:

A	*X*2.000	*Y*2.000
B	*X*1.000	*Y*−2.000

TYPES OF CONTROL

Numerical control programming falls into two distinct categories: point-to-point control and continuous-path, or contouring, control (Fig. 75-9). At one time the difference between the two categories was very distinct, but now most control units are able to handle both point-to-point and continuous path machining. A knowledge of both programming methods is necessary for an understanding of the applications that each has in numerical control.

POINT-TO-POINT POSITIONING

Point-to-point positioning is used when it is necessary to accurately locate the spindle, or the workpiece mounted on the machine table, at one or more specific locations to perform operations such as drilling, reaming, boring, tapping, or punching (Fig. 75-10). It is the process of positioning from one coordinate (X–Y) position or location to another, performing the machining operation, and continuing this pattern until all the operations have been completed at all programmed locations.

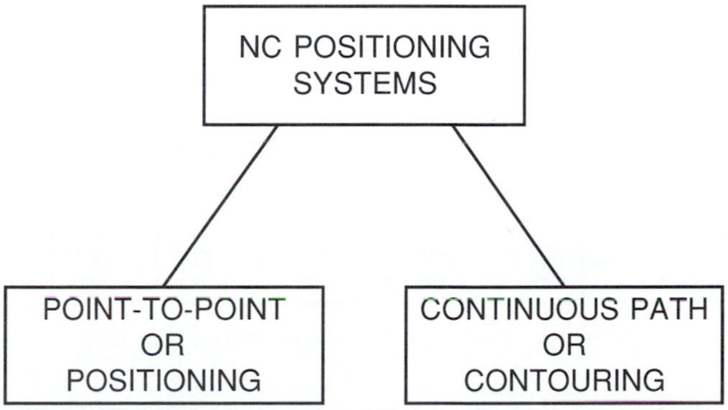
NC POSITIONING SYSTEMS
POINT-TO-POINT OR POSITIONING
CONTINUOUS PATH OR CONTOURING

Figure 75-9 *The two types of numerical control positioning systems.*

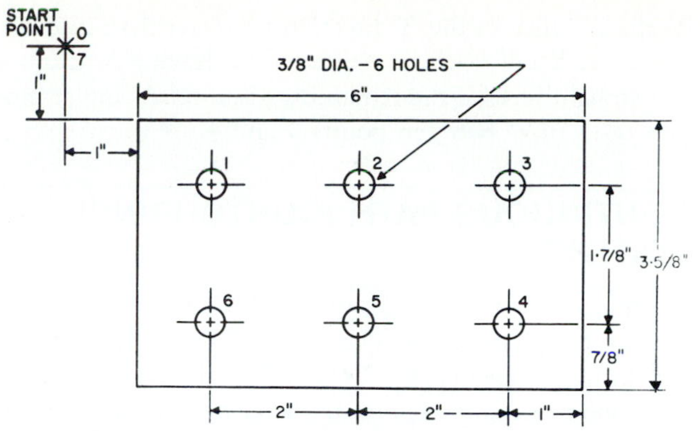
START POINT
0
7
1"
3/8" DIA. – 6 HOLES
6"
1"
1
2
3
4
5
6
1-7/8"
3-5/8"
7/8"
2"
2"
1"

Figure 75-10 *Point-to-point positioning is used when drilling a series of holes in various locations.*

Drilling machines or point-to-point machines are ideally suited for positioning the machine tool (in this case a drill) to an exact location or point, performing the machining operation (drilling a hole), and then moving to the next location where another hole would be drilled. By simply identifying each point or hole location in the program, this operation can be repeated as many times as required.

Point-to-point machining moves from one point to another as fast as possible (rapid traverse) while the cutting tool is above the work surface. Rapid traverse is used to quickly position the cutting tool or workpiece between each location point before a cutting action is started. The rate of rapid traverse is usually between 150 and 400 in./min (38 to 100 m/min). Both X- and Y-axes move simultaneously and at the same rate during rapid traverse. This results in a movement along a 45° angle line until one axis is reached; then there is a straight-line movement to the other axis.

In Fig. 75-11, point 1 to point 2 is a straight line, and the machine moves only on the X-axis, but points 2 and 3 require that motion on both the X- and Y-axes take place.

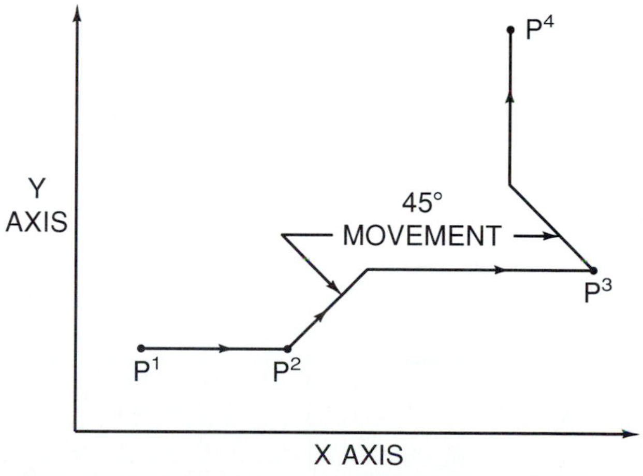
Y AXIS
X AXIS
45° MOVEMENT
P1
P2
P3
P4

Figure 75-11 *In point-to-point positioning, the machine table rapids simultaneously on the XY axis until one axis location is reached.*

As the distance in the X direction is greater than in the Y direction, Y will reach its position first, leaving X to travel in a straight line for the remaining distance. A similar motion takes place between points 3 and 4.

CONTINUOUS-PATH (CONTOURING) CONTROL

Continuous-path or *contouring* machining involves work such as that produced on a lathe or milling machine, where the cutting tool is in contact with the workpiece as it travels from one programmed point to the next. Continuous-path positioning (contouring) is the ability to control motion along two or more machine axes simultaneously to keep a constant cutter–workpiece relationship. The programmed information on the NC tape must accurately position the cutting tool from one point to the next and follow a predefined, accurate path at a programmed feed rate in order to produce the form or contour required (Fig. 75-12A and B).

The method by which contouring machine tools move from one programmed point to the next is called *interpolation.* This ability to merge individual axis points into a predefined tool path is built into most of today's machine control units (MCU). There are five methods of interpolation: linear, circular, helical, parabolic, and cubic. All contouring controls provide linear interpolation, and most controls are capable of both linear and circular interpolation. Helical, parabolic, and cubic interpolation are used by industries that manufacture parts having complex shapes, such as aerospace parts and dies for car bodies.

PROGRAMMING SYSTEMS

Two types of programming modes are used for numerical control: the incremental system and the absolute system. Both systems have applications in numerical control programming, and neither system is the best one to use all the time. Most controls on machine tools built today are capable of handling either incremental or absolute programming.

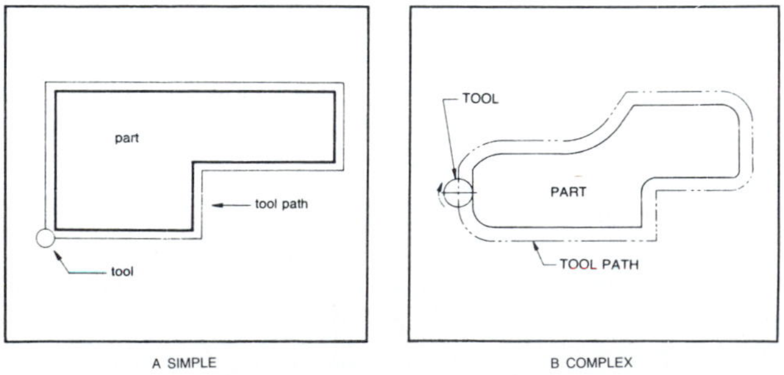
part
tool path
tool
TOOL
PART
TOOL PATH
A SIMPLE
B COMPLEX

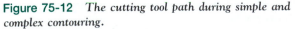

Figure 75-12 *The cutting tool path during simple and complex contouring.*

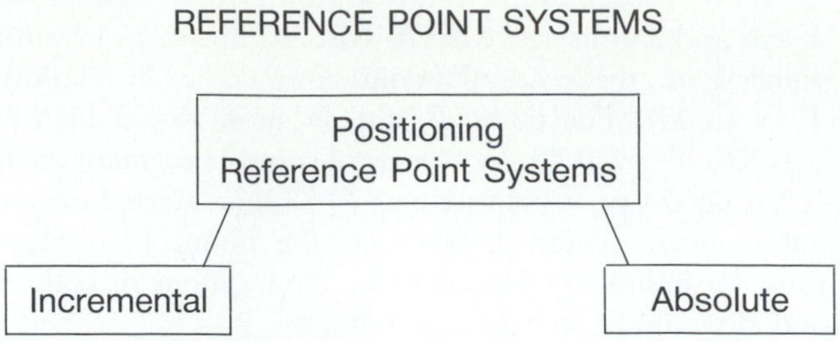
REFERENCE POINT SYSTEMS
Positioning
Reference Point Systems
Incremental
Absolute

INCREMENTAL SYSTEM

In the *incremental system,* dimensions or positions are given from previously known points. An example of incremental instructions would be a person who delivers newspapers to certain houses on a street. He or she could be given instructions to deliver a newspaper to the first house on a street which is 60 feet from the corner (Fig. 75-13). The second house that should receive a paper is 120 feet from the first house; the third house is 60 feet from the second, and so on. All distances are expressed in terms of the previous known point. Incremental dimensioning on a job print is shown in Fig. 75-14. As you will note, the dimensions for each hole are given from the previous hole. One disadvantage of incremental positioning or program-

PAPER

Figure 75-13 *The incremental system being used to locate the houses that require newspapers.*

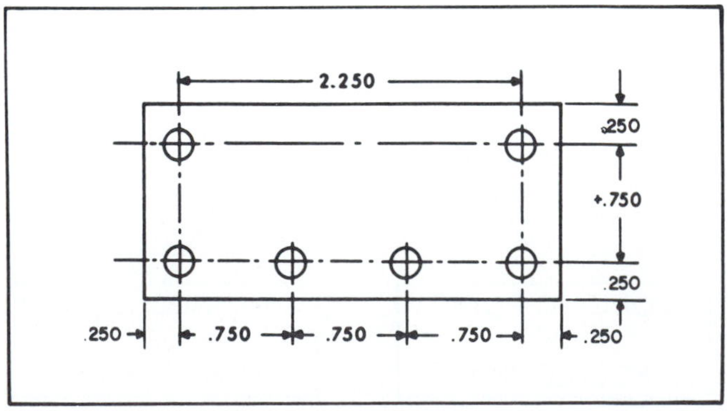
2.250
.250
.750
.250
.250
.750
.750
.750
.250

Figure 75-14 *A workpiece dimensioned in the incremental system mode.*

ming is that if an error has been made in any location, this error will be carried over to all the locations after that point.

Incremental program locations are always given as the distance and direction from the immediately preceding point (Fig. 75-14). Command codes which tell the machine to move the table, spindle, and knee will be explained using a vertical milling machine as an example:

- A *plus X* (+*X*) command causes the machine table to move to the right of the last point.
- A *minus X* (−*X*) command causes the machine table to move to the left of the last point.
- A *plus Y* (+*Y*) command causes the machine table to move toward the column.
- A *minus Y* (−*Y*) command causes the machine table to move away from the column.
- A *plus Z* (+*Z*) command causes the cutting tool or spindle to move up or away from the workpiece.
- A *minus Z* (−*Z*) command causes the cutting tool to move down or into the workpiece.

In incremental programming the G91 command indicates to the computer and MCU that programming is to be in the incremental mode.

ABSOLUTE SYSTEM

In the *absolute system,* all dimensions or positions are given from a zero or reference point. For instance, the person delivering newspapers could be given instructions to use the street corner as a zero or reference point. Newspapers then would be delivered to the house 60 feet from the corner (Fig. 75-15), the second house 180 feet from the corner, the third house 240 feet from the corner, and so on. As you will note, all distances have been given from the corner, which is the zero or reference point. In Fig. 75-16, the same workpiece is used as in Fig. 75-14, except that all dimensions are given from the zero or reference point. Therefore, in the absolute system of dimensioning or programming, an error in any dimension is still an error, but it is not carried over to any other location.

Absolute program locations are always given from a single fixed zero or origin point (Fig. 75-16). The zero or origin point may be a position on the machine table, such as the corner of the work table, or at any specific point on the workpiece. Since each point or location on the workpiece is given as a certain distance from the zero or reference point:

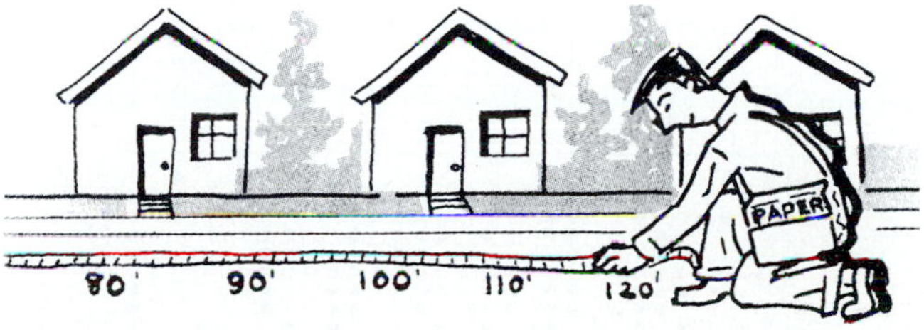

80'
90'
100'
110'
120'
PAPERS

Figure 75-15 *The absolute system locates all houses requiring a newspaper as the distance each is from the street corner.*

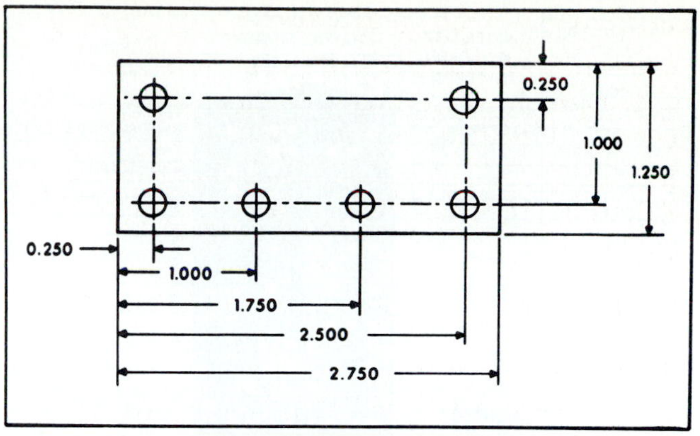

0.250
1.000
1.250
0.250
1.000
1.750
2.500
2.750

Figure 75-16 *A workpiece dimensioned in the absolute system mode. NOTE: All dimensions are given from a known point of reference.*

- A *plus X* (+*X*) commands the machine table to move to the right of the zero or origin point.
- A *minus X* (−*X*) commands the machine table to move to the left of the zero or origin point.
- A *plus Y* (+*Y*) commands the machine table to move toward the column from the zero or reference point.
- A *minus Y* (−*Y*) commands the machine table to move away from the column from the zero or reference point.

In absolute programming the G90 command indicates to the computer and MCU that the programming is to be in the absolute mode.

INPUT MEDIA

With the development of numerical control, a variety of input media has been used to convey information from the drawing to the machine. The most common types of input media used are magnetic tape, punched cards, punched tape, and magnetic disks. Most machines now use a standard 1-in.-wide punched tape, which may be made of paper, Mylar, or foil.

TAPE CODING

The Electronics Industries Association adopted the binary-coded decimal system for standardizing information punched on tapes used for numerical control. The standard coding system used for thc 1 in. (25 mm) wide, 8-channel tape is illustrated in Fig. 75-17.

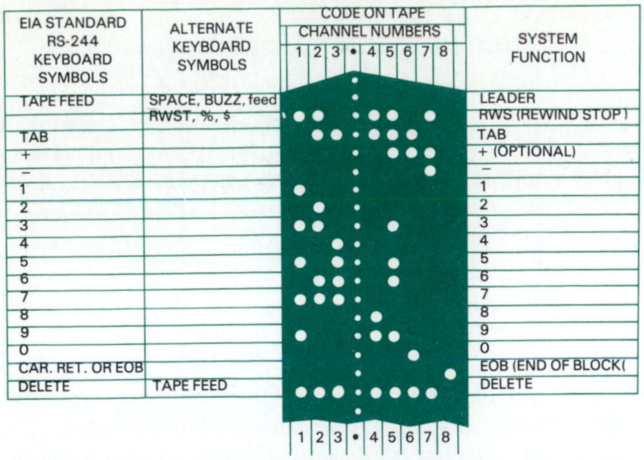
EIA STANDARD RS-244 KEYBOARD SYMBOLS
ALTERNATE KEYBOARD SYMBOLS
CODE ON TAPE
CHANNEL NUMBERS
1 2 3 • 4 5 6 7 8
SYSTEM FUNCTION
TAPE FEED
SPACE, BUZZ, feed
LEADER
RWST, %, $
RWS (REWIND STOP)
TAB
TAB
+
+ (OPTIONAL)
−
−
1
1
2
2
3
3
4
4
5
5
6
6
7
7
8
8
9
9
0
0
CAR. RET. OR EOB
EOB (END OF BLOCK(
DELETE
TAPE FEED
DELETE
1 2 3 • 4 5 6 7 8

Figure 75-17 *The EIA (Electronics Industries Association) standard coding for a 1-inch wide eight-channel tape.*

The numbers 1 to 8 on the top of the tape represent the number of each channel and have no relationship to the holes punched in the tape.

1. The *sprocket holes* between channels 3 and 4 are used to drive the tape through the machine. The sprocket holes are set off center so that it is impossible to put the reverse side of the tape into the machine.
2. *Channels 1, 2, 3, and 4* are used for *numerical data,* such as dimensions, speeds, and feeds. These channels have numerical values of 1, 2, 4, and 8, respectively.
3. *Channel 5,* marked CH, is called the *parity check.* An odd number of holes must appear in each row; otherwise, the tape reader will stop the machine. If an even number of holes must be punched in one row to get the correct information on the tape, an additional hole must be punched in channel 5. The odd parity check tests for any errors or mechanical failures in tape preparation.
4. *Channel 6* always represents a zero (0).
5. *Channel 7,* marked X, is used to select a letter to identify various machine operations and is not used for dimensions. Each letter (a to z) in the left-hand, vertical column represents a certain machine function or operation, such as drilling, boring, or reaming. When channels 6 and 7 are punched together, in conjunction with channels 1, 2, 3, and 4, operations "a to i" will be selected. When only channel 7 is punched, in conjunction with channels 1, 2, 3, and 4, operations "j to r" will be selected. If only channel 6 is punched, in conjunction with channels 1, 2, 3, and 4, operations "s to z" will be selected.
6. *Channel 8,* marked EL (or EOB), represents the *end of a line* or end of a *block* of information. It must always be at the beginning of a tape and at the end of each block of information.

The TAB code, punched in channels 2, 3, 4, 5, and 6, is used to separate each operation or dimension. The remaining codes in Fig. 75-17, beginning with a period (.) and ending with lowercase letters, are all acceptable codes that are used in numerical control tapes.

EXAMPLES OF PUNCHED INFORMATION

1. Assume that the numerical value of 1 must be recorded. A hole must be punched in channel 1 because it has a numerical value of 1.
2. Assume that the numerical value of 5 must be recorded. A hole must be punched in channels 3 and 1 because channel 3 has a numerical value of 4 and channel 1 has a numerical value of 1. These two totaled equal a numerical value of 5.

NOTE: Since only two holes are punched across this row, an extra hole must be punched in the parity check channel (5) because the tape reader will not recognize an even number of holes (Fig. 75-17). If a hole is not punched in the parity check channel, the tape reader will stop the machine.

3. Operation "g" would be recorded by punching holes in channels 1, 2, 3, 6, and 7.

TAPE FORMAT

Each block of information must contain five complete words or pieces of information. If five complete words are not included in each block, the tape reader will not recognize the information on the tape, and therefore will not activate the control unit. Figure 75-18 shows a punched tape containing one complete block of information consisting of five complete words. From left to right, the information punched on the tape is as follows:

1. A hole is punched in channel 8 (EOB), which represents the end or the beginning of a line or block of information.
2. The *first word* of the block represents the number of the operation or sequence number. H001 represents the first operation on the tape. The *letter H* is recorded by punching holes in channels 4, 6, and 7 (Fig. 75-18). The *two zeros (0)* are recorded by punching channel 6 in two successive

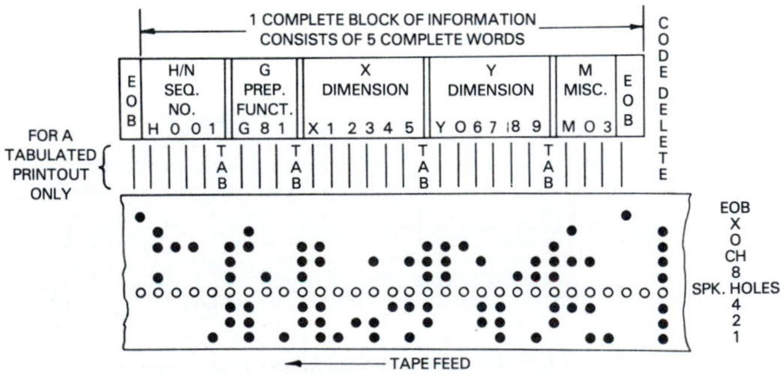
1 COMPLETE BLOCK OF INFORMATION
CONSISTS OF 5 COMPLETE WORDS
CODE DELETE
EOB
H/N SEQ. NO.
G PREP. FUNCT.
X DIMENSION
Y DIMENSION
M MISC.
EOB
H 0 0 1
G 8 1
X 1 2 3 4 5
Y 0 6 7 8 9
M 0 3
FOR A TABULATED PRINTOUT ONLY
TAB
TAB
TAB
TAB
EOB
X
0
CH
8
SPK. HOLES
4
2
1
TAPE FEED

Figure 75-18 *A complete block of information punched on a tape.*

rows. The *1* is recorded by punching channel 1, which has a numerical value of 1.

3. The *TAB* code is used to separate each word into a block of information.

4. The *second word* in the block represents a preparatory function. G81 is a drill cycle on Cincinnati Milacron Numerical Control systems. The *letter G* is recorded by punching channels 1, 2, 3, 6, and 7. The *8* is recorded by punching channel 4, which has a numerical value of 8. The *1* is recorded by punching channel 1, which has a numerical value of 1.

5. The *third word* represents the distance the table slide must move from the *X*-axis. The information contained in the third word of Fig. 75-18 is X12345. As standard tapes on closed loop systems program all dimensions to ten-thousandths of an inch, the machine will move in a positive direction 1.2345 along the *X*-axis.

6. The *fourth word* represents the distance the table slide must move from the *Y*-axis. In Fig. 75-18, Y06789 represents a positive table movement of 0.6789 along the *Y*-axis.

7. The *fifth word* represents a miscellaneous machining function. M03 would start the spindle revolving clockwise.

PROGRAM, TAPE, AND MACHINE PREPARATION

Although numerical control systems differ greatly in detail and complexity according to the manufacturer, all have basically the same elements. Regardless of the type of input media used, all operate a machine tool the same way. A complete sequence of operations beginning with the programmer transferring the information from a drawing to a program sheet until the finished part is removed from the machine is illustrated in Fig. 75-19.

1. A programmer reviews the part drawing, determines the sequence of operations required, and records this information on a program sheet (Fig. 75-19A).
2. A typist transfers the information contained on the program sheet to the tape using a special tape-punching machine (Fig. 75-19B). To ensure that the typist has not made an error in punching the information on the tape, a second typist may also type the same program. These two tapes are then compared, and if they are identical it is assumed that the program is correct.
3. The punched tape and a copy of the program sheet are then handed to the machine operator. After positioning the part to be machined on the machine table according to the instructions contained on the program sheet, the operator threads the tape into the tape reader (Figs. 75-19C and 75-20).
4. The tape reader, which automatically advances the tape, is then started.
5. As each block of information is decoded by the tape reader, it sends the necessary information to the *machine control unit* (MCU).
6. After the end of each machining operation (Fig. 75-19D), the *feedback switch* informs the MCU that the previous operation has been completed.
7. The MCU instructs the *command memory unit* to transfer the tape instructions for the next operation to the *indexer.*
8. The indexer starts the servomotors, which in turn move the machine table slides the required amount; and the next machining operation is performed, and so on, until all machining operations have been completed (Fig. 75-19E).

NUMERICAL CONTROL PERFORMANCE

Great advances in numerical control have been made since it was first introduced in the mid 1950s as a means of guiding machine tools through various motions automatically, without human assistance. The early machines were capable only of point-to-point positioning (straight-line motions), were very costly, and required highly skilled technicians and mathematicians to produce the tape programs.

Great advances in numerical control resulted from new technology in the electronics industry. The development of transistors, solid-state circuitry, integrated circuits, and the computer chip have made it possible to program machine tools to perform tasks undreamed of as little as a decade ago. Not only have the machine tools and controls been dramatically improved, but the cost has continually dropped. Numerical control machines are now within the financial reach of small manufacturing shops and educational institutions. Their wide acceptance throughout the world is based on their accuracy, reliability, repeatability, and productivity (Fig. 75-21).

PRODUCTIVITY

The goal of industry has always been to produce better products at competitive or lower prices in order to gain a larger share of the market. Soaring production in other countries has increased the competition for world and domestic markets. To meet this competition, manufacturers must continue to cut manufacturing costs and build better quality products. Meeting this competition means that there must be greater output per worker, greater output per machine, and greater output for each dollar of capital investment. These factors alone justify using numerical control—and also for automating our factories. It provides us with the opportunity to produce goods of better quality, faster, and at lower cost.

Let's compare the costs of producing a typical part man-

A

D

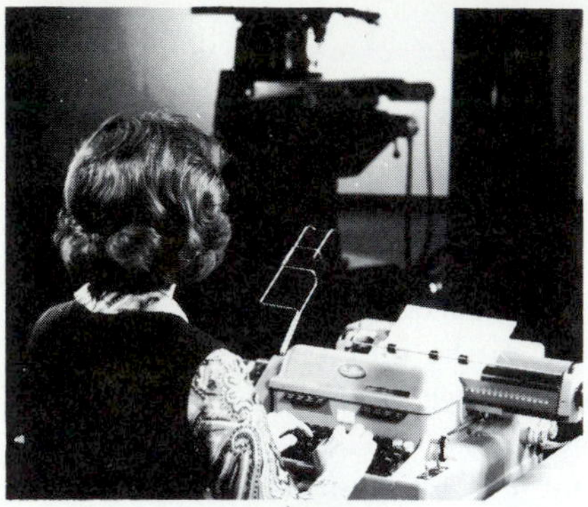

B

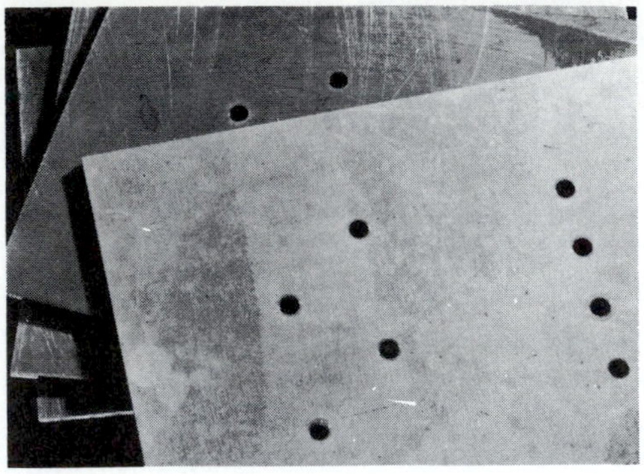

E

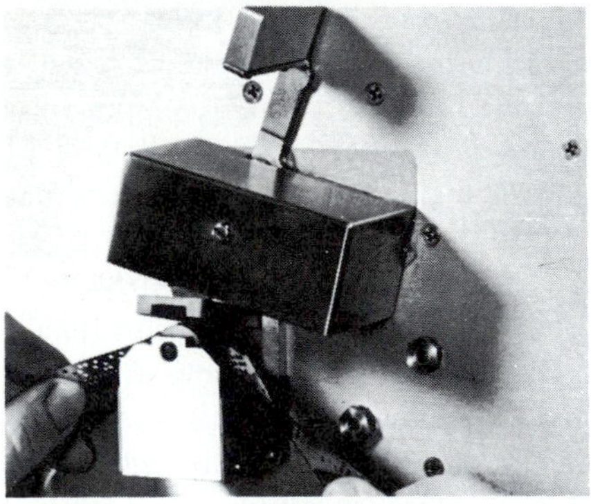

C

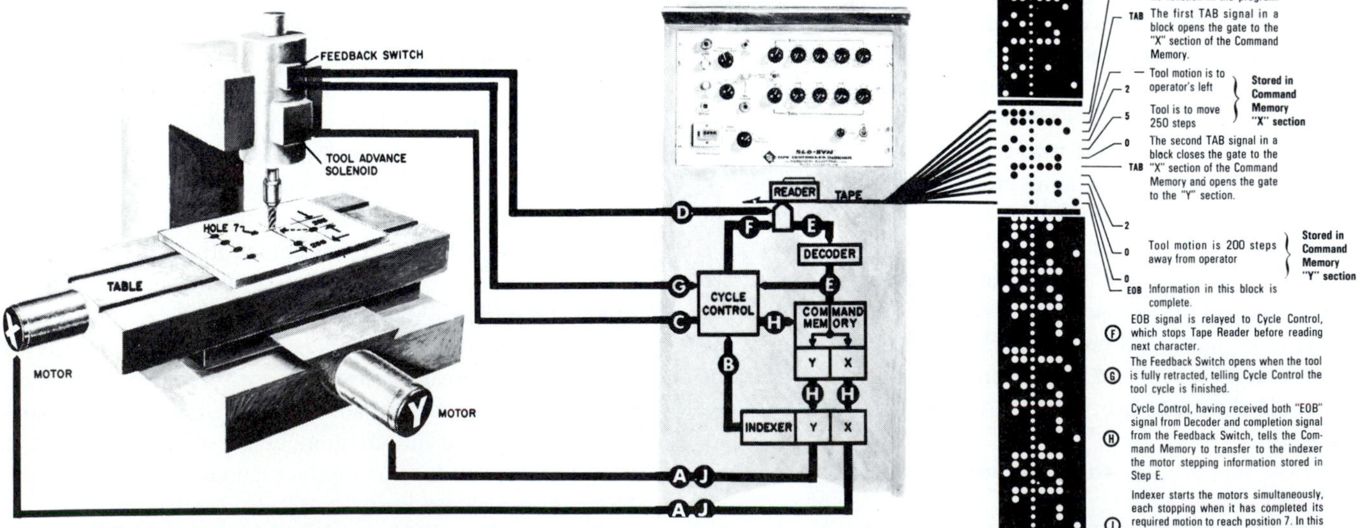

FEEDBACK SWITCH
TOOL ADVANCE SOLENOID
HOLE 7
TABLE
MOTOR
MOTOR
Y
READER
TAPE
DECODER
CYCLE CONTROL
COMMAND MEMORY
Y
X
INDEXER
Y
X
The example below illustrates how one cycle, or block, of taped information is processed and acted upon by a tape controlled indexer.
Ⓐ Assume the Table is in motion, moving to position 6 as directed by the indexer.
Ⓑ When the Indexer completes both the X and Y counts, it returns a signal to the Cycle Control.
Ⓒ The Cycle Control then energizes the Tool Advance Solenoid.
Ⓓ When the tool advances, the Feedback Switch closes, telling the Tape Reader to read the next block of information.
Ⓔ Tape information goes to the Decoder. In this case:
7 The sequence number is for reference only and performs no function in the program.
TAB The first TAB signal in a block opens the gate to the "X" section of the Command Memory.
— Tool motion is to operator's left
2
5 Tool is to move 250 steps
Stored in Command Memory "X" section
0 The second TAB signal in a block closes the gate to the
TAB "X" section of the Command Memory and opens the gate to the "Y" section.
2
0 Tool motion is 200 steps away from operator
0
Stored in Command Memory "Y" section
EOB Information in this block is complete.
Ⓕ EOB signal is relayed to Cycle Control, which stops Tape Reader before reading next character.
Ⓖ The Feedback Switch opens when the tool is fully retracted, telling Cycle Control the tool cycle is finished.
Ⓗ Cycle Control, having received both "EOB" signal from Decoder and completion signal from the Feedback Switch, tells the Command Memory to transfer to the indexer the motor stepping information stored in Step E.
Ⓙ Indexer starts the motors simultaneously, each stopping when it has completed its required motion to reach position 7. In this case, the "Y" motor will stop first, having moved 200 steps. The "X" motor will continue until it has moved 250 steps.
Ⓚ The cycle is complete, return to Step B.

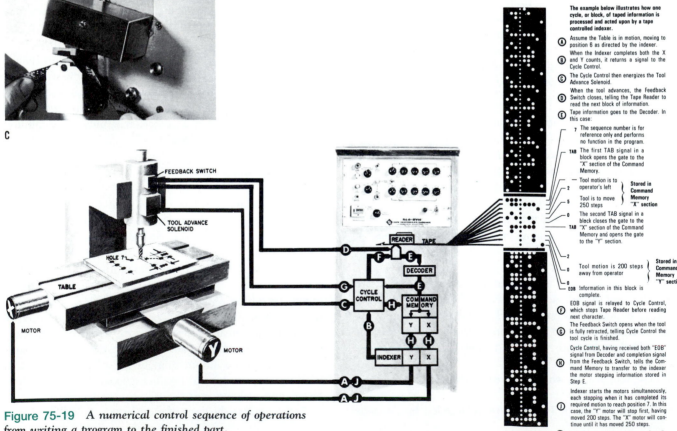

Figure 75-19 *A numerical control sequence of operations from writing a program to the finished part.*

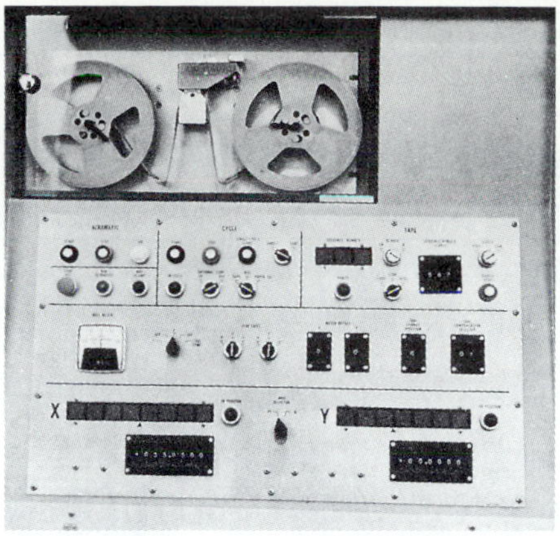
X
Y

Figure 75-20 *A Cincinnati Milacron Acromatic tape control reader.*

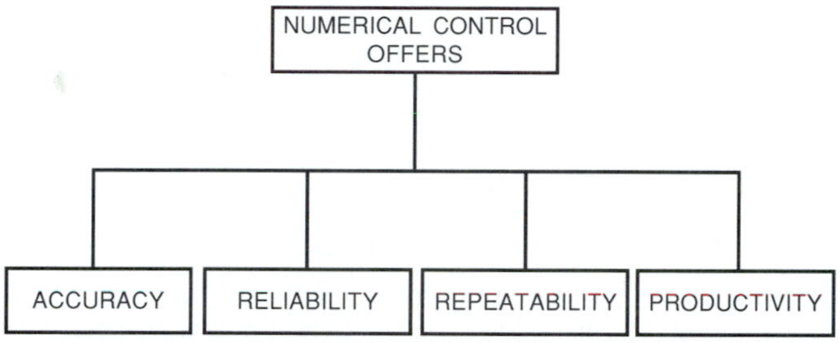
NUMERICAL CONTROL OFFERS
ACCURACY
RELIABILITY
REPEATABILITY
PRODUCTIVITY

Figure 75-21 *Numerical control has been widely accepted because of the many advantages it has to offer industry.*

ually and by numerical control. The part shown in Fig. 75-22 requires twenty-five holes of four different sizes, involving the operations of spotting, drilling, countersinking, counterboring, reaming, and drilling. Eleven different cutting tools are required for these operations and the locational tolerance for the holes is ±0.001 in. (0.025 mm).

Let's assume that a firm received an order for 700 of

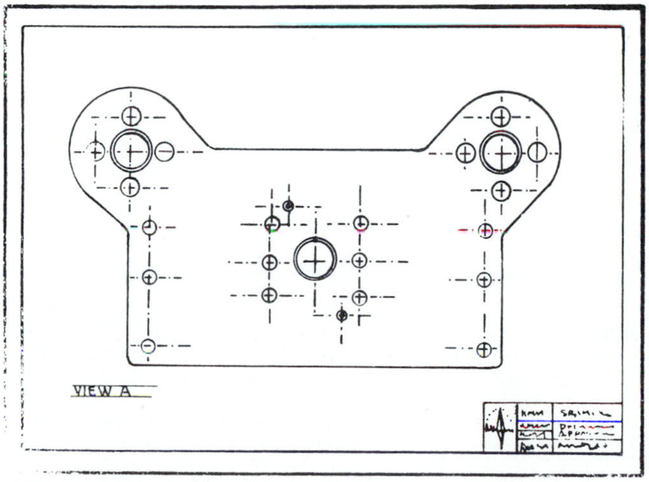
VIEW A

Figure 75-22 *A typical part which must be produced for a customer.*

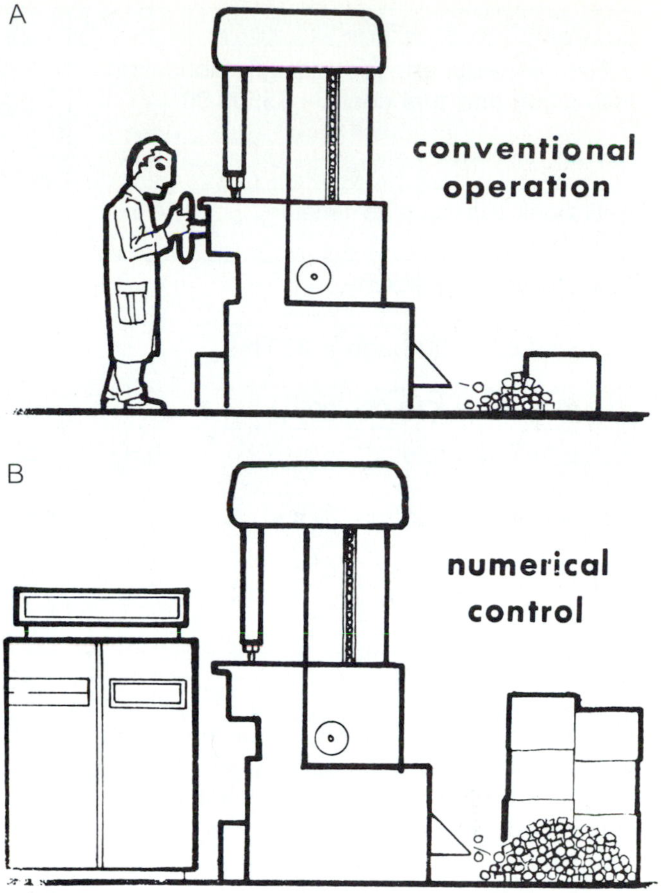
A
conventional operation
B
numerical control

Figure 75-23 *(A) Producing the part by conventional machining methods. (B) Producing the part by numerical control results in increased production.*

these parts, which had to be supplied to the customer in lots of 35. Therefore, because the delivery date is uncertain, it is quite likely that the firm may produce these as required and would need 20 lots or manufacturing runs. For a comparison of this production see Fig. 75-23A and 75-23B.

The savings of $14,947 (page 536) is fairly consistent and reflects increased productivity along with savings in production costs. Obviously, making a general rule regarding the savings that numerical control offers is difficult; some jobs will show greater savings while others will show less.

ADVANTAGES

Numerical control has been applied to a variety of machines and has gained wide acceptance by industry. Machine tools such as lathes, turret drills, and milling and boring machines are some of the more common types of equipment which employ numerical control. Figure 75-24 lists some of the advantages of numerical control equipment.

1. The machine has greater flexibility, since one machine can act as a drill, mill, and turret lathe.

Manually	Numerical Control	Savings
2 Fixtures required. Tool design, fabrication, inspection and trial costs = $5500.00	1 simple fixture $125.00 Tape preparation 2 hrs. @ $40.00/hr. $80.00 = $205.00	$5295.00
4–6 week tooling lead time	1 day for alignment, fixture, and tape preparation	
Floor-to-floor time/part 30 minutes × 700 parts = 350 hrs. @ $40.00/hr. = $14,000.00	Floor-to-floor time/part 10 minutes × 700 parts = 117 hrs. @ $40.00/hr. = $4680.00	$9320.00
Setup time/job lots 30 minutes × 20 lots = 10 hrs. @ $40.00 = $400.00	Setup time/job lots 5 minutes × 20 lots = 1.7 hrs. @ $40.00 = $68.00	$332.00 Total $14,947.00

2. Once the program has been set up, there is usually a 20 to 30 percent increase in productivity.

3. The reliability of the system eliminates the human error associated with manual operation, thereby reducing scrap loss.

4. Special jigs and fixtures usually required for positioning are eliminated, since the machine can locate positions quickly and accurately.

5. The time required for setting up and locating the workpiece is reduced.

6. Complex operations can be performed with ease.

7. Single parts or production runs can be made with minimum effort and cost.

8. The program can be quickly changed by inserting a new tape in the machine.

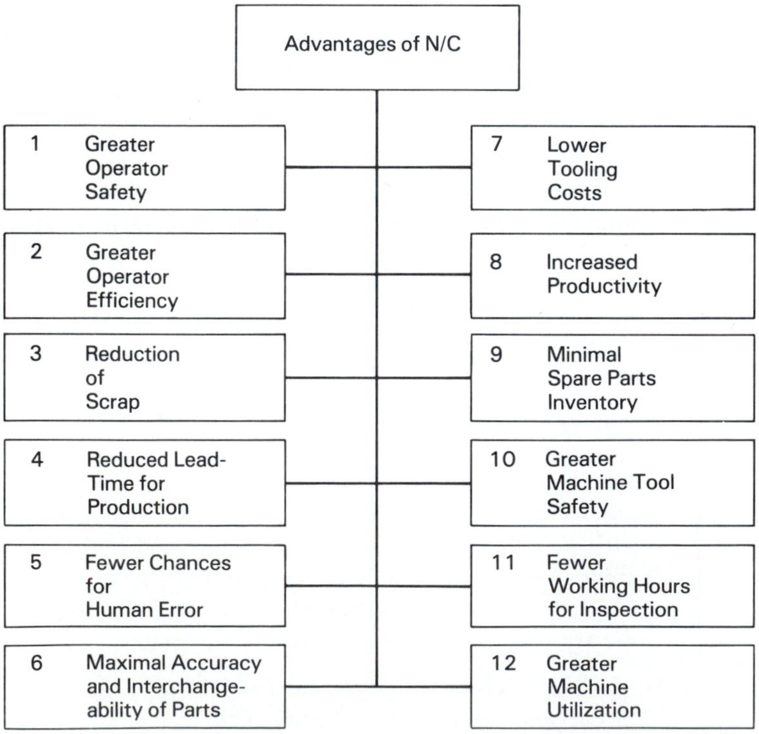
Advantages of N/C
1 Greater Operator Safety
2 Greater Operator Efficiency
3 Reduction of Scrap
4 Reduced Lead-Time for Production
5 Fewer Chances for Human Error
6 Maximal Accuracy and Interchange-ability of Parts
7 Lower Tooling Costs
8 Increased Productivity
9 Minimal Spare Parts Inventory
10 Greater Machine Tool Safety
11 Fewer Working Hours for Inspection
12 Greater Machine Utilization

Figure 75-24 *Advantages of numerical control.*

9. Inspection costs are reduced because of the reliability of the system.
10. Once the program and tooling have been tested, the equipment does not require a highly skilled operator.

DISADVANTAGES

1. The initial cost of numerical control machines is higher than that of conventional machines.
2. Personnel trained in electronics are required to service the equipment.
3. Additional floor space is required for this equipment.
4. Personnel must be trained in the programming and operation of this equipment.

RECENT ADVANCES IN NUMERICAL CONTROL

Although numerical control has provided industry with a method of increasing productivity while maintaining a high degree of accuracy, many changes have been made in machining procedures, resulting in lower manufacturing costs.

With advancing technology, more efficient use of NC has been developed. The continued developments in microelectronics and computer technology have made this possible by shrinking the size of machine control hardware while providing greater capacity and capabilities. Most machine tools being made today are controlled by computers or other computerlike devices, such as programmable controllers.

Since NC was first introduced, computers have always been on the sidelines helping to create parts programs. The direct control of machine tools by computers was just a matter of time. *Computer numerical control* bypasses the intermediate step of putting program data for any parts onto punched cards, punched tape, or magnetic disks.

TYPES OF COMPUTER CONTROL

There are two different types of computer control for NC machines: *computer numerical control* (*CNC*) and *direct numerical control* (*DNC*).

COMPUTER NUMERICAL CONTROL

There are four basic elements in a *computer numerical control* system:

1. A *general purpose computer,* which gathers and stores the required signals.
2. A *control unit,* which establishes contact and directs the flow of information between the computer and the machine control unit.
3. The *machine logic,* which receives information and passes it on to the machine control unit.
4. The *machine control unit,* which contains the servo units, speed and feed controls, and machine operations such as spindle and table movements and toolchanger.

Most NC systems being built are CNC systems. The housings for these systems have tended to look much like those of the older NC systems, and the controls operate and are used similarly. The control unit is mounted on the machine or stands alongside it (Fig. 75-25).

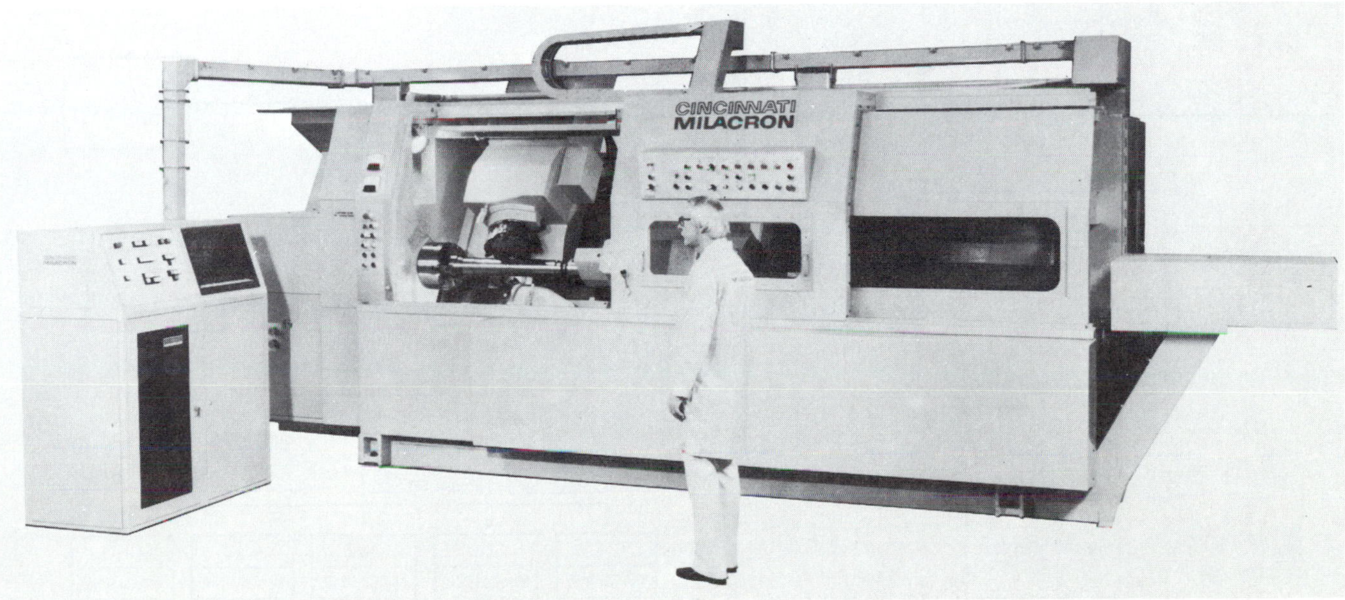
CINCINNATI
MILACRON

Figure 75-25 *A computer controlled turning center. (Courtesy Cincinnati Milacron Inc.)*

The CNC system (Fig. 75-26), built around a powerful minicomputer, contains much more memory capacity and has many more features to assist in programming, program editing at the machine, setup of the machine, operation of it, and maintenance. Many of these features are sets of machine and control instructions stored in memory, which can be called into use by the part program or by the machine user.

Most CNC systems still have tape readers, with the part programs still being prepared in an office off-line unit and delivered to the machine in the form of punched tape. In some cases, however, the tape is read once, with the part program being stored in memory for repetitive machining.

In the more recent CNC machines, minicomputers and then microcomputers have been incorporated into the controls. This permits the machine operator to manually input the data required to produce the part. These data are stored in the computer memory for the production of further parts.

The main advantage of this type of system is its ability to operate in a "live" or "conversational" mode, with direct communication between the machine and the computer. This capability enables a programmer to make program changes at the machine or even develop a program on the spot. His or her input to the computer is transformed almost instantly into machine motions. Therefore, the results of changes to the program can be observed immediately and revisions made if necessary. This new concept of machine control allows programs to be tried out, corrected, and revised in a fraction of the time that was required with tape systems.

Advantages of CNC

- It is more flexible in that modifications can be made to the program rather than making a completely new tape as with older conventional (hand-wired) controls.
- It can diagnose programs and before the part is produced detect machine and control malfunctions as or before they occur.
- It can be integrated with DNC systems in highly sophisticated manufacturing systems.
- It increases productivity because of the ease of programming.
- It makes corrections on the first part possible, which cuts costs on the entire batch.
- It is practical, and even profitable, to produce short-run lots.

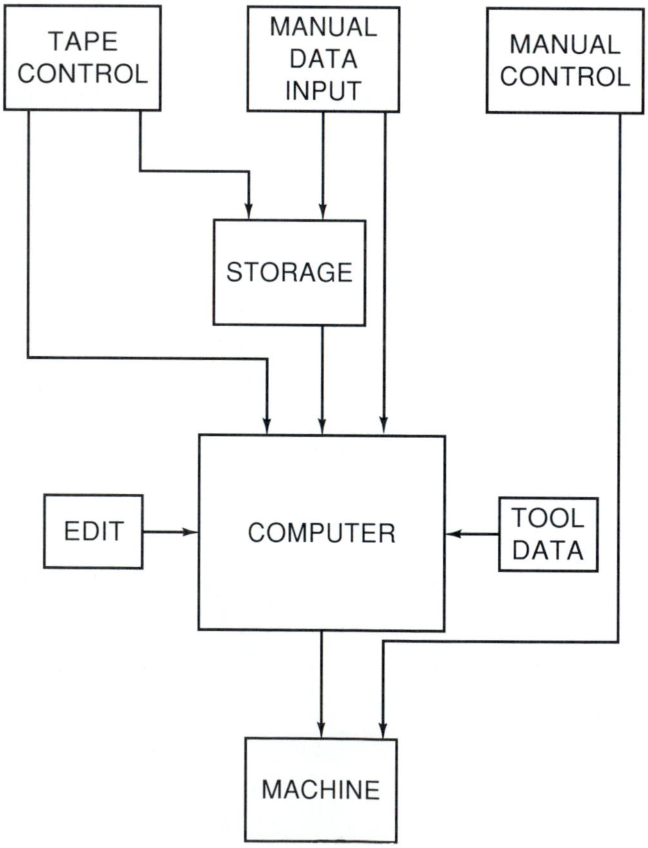
TAPE CONTROL
MANUAL DATA INPUT
MANUAL CONTROL
STORAGE
EDIT
COMPUTER
TOOL DATA
MACHINE

Figure 75-26 *A CNC system of machine control.*

DIRECT NUMERICAL CONTROL

In this type of system (Fig. 75-27), a number of CNC-equipped machines are controlled remotely from a mainframe computer, which handles the scheduling of work and downloads into the machine's memory a complete program

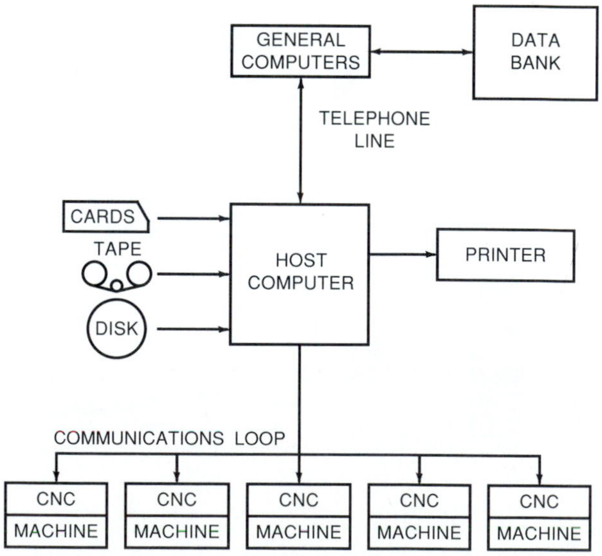
GENERAL COMPUTERS
DATA BANK
TELEPHONE LINE
CARDS
TAPE
DISK
HOST COMPUTER
PRINTER
COMMUNICATIONS LOOP
CNC MACHINE
CNC MACHINE
CNC MACHINE
CNC MACHINE
CNC MACHINE

Figure 75-27 *A typical DNC system.*

when a new part is to be machined. It bypasses the control's tape reader. If automatic work handling is provided at the machines, the computer can control automatic cycling of each machine. In this case, a machine operator is needed only for initial setup and troubleshooting.

Since most CNC machines now come with a minicomputer or microcomputer, it is possible for each machine to be operated individually by CNC, should the mainframe computer break down.

Advantages of DNC

- The often troublesome tape reader on the machine is bypassed.
- A single computer can control many machine tools at the same time.
- Considerable time is saved in eliminating program errors or revising the program. The programmer can make revisions or corrections on a typewriter-style keyboard right at the machine tool.
- Programming is faster, simpler, and more flexible.
- The computer can record any desired production, machining, or time data.
- The key control unit can be kept in a processing room away from adverse shop conditions.
- When three or more machines are DNC controlled, the initial cost is lower than for conventional NC.
- Operating costs are lower than with NC.

REVIEW QUESTIONS

1. Define *numerical control.*
2. Briefly describe an open loop system.
3. What is the basic difference between an open loop system and a closed loop system?
4. What is the accuracy of a machine fitted with:
   a. An open loop system
   b. A closed loop system
5. Name two types of controls used to position and control the cutting tool.
6. Briefly explain the principle of the binary system.
7. How can any point on a job be located using the Cartesian coordinate system?
8. Define *origin* or *zero point.*
9. Label a neat diagram of the four quadrants showing the positive (+) and negative (−) locations.
10. Explain the difference between point-to-point and continuous-path systems.
11. Define and give an example of absolute and incremental programming.
12. What feature on a standard tape prevents it from being mounted incorrectly?
13. What is *parity check* and what is its purpose?
14. Explain the purpose of TAB and EL code.
15. List briefly the steps required to produce a part by numerical control.
16. Why have there been such great advances in numerical control in a very short period of time?
17. List three factors which are justification for installing a numerical control system.
18. List six important advantages of numerical control.

CNC AND DNC

19. List four advantages of CNC.
20. State the difference between the CNC and DNC systems.

UNIT 76

Computer-Aided Design

The advent of the computer proved a boon to the design engineer in that it simplified the long and tedious calculations often required for a project. As time went on, it was felt that the computer could be further extended in the design area, and in the 1960s a new system called *computer-aided design* (CAD) was introduced. Computer-aided design allows the designer or engineer to produce finished engineering drawings from simple pencil sketches or from models, and to modify the drawings if they appear not to be functional. From three-drawing orthographic views, it is possible to transform drawings into a three-dimensional view and with appropriate computer software to show the anticipated performance of the part. This has proved extremely valuable to designers, engineers, and drafters. This new system has streamlined the design and drafting processes and its use is increasing by an estimated 40 percent per year.

OBJECTIVES

After completing this unit you will be able to:
1. Explain the principle and purpose of CAD
2. Name and state the purposes of the main parts of a CAD system
3. State six advantages of CAD

CAD COMPONENTS

Computer-aided design is a television-like system that produces a picture from electronic signals relayed from the computer. Most systems in larger plants use two levels of computers. Smaller desk-top computers, each of which has its own television-like *cathode ray tube* (CRT), are connected to the large mainframe, or host, computer. By adding to this system a keyboard, a light pen, or an electronic tablet and a plotter, the operator can direct the computer to produce any drawing or view required (Fig. 76-1).

The desk-top computer is used by the operator to insert all the required data for the job. Because the memory bank of the desk-top will be much smaller than that of the host computer, only the commonly used data required by the operator will be stored in its memory bank. Additional and more complex information, as well as final part information, will be stored in the memory of the larger host computer. The desk-top computer may call on the host computer to perform tasks beyond the smaller computer's capability. The operator can at any time, through the desk-top computer, recall information stored in the host computer.

DESIGNING THE PART

The operator may start off with a pencil sketch and, using the light pen or an electronic tablet, can produce a properly

Figure 76-1 *A total computer-aided design (CAD) system. (Courtesy Bausch & Lomb.)*

scaled drawing of the part on the screen and also record it in the computer's memory. The computer will calculate the coordinates of the ends of the line, the points of intersection, as well as the required arcs, circles, radii, etc., producing a view of the evolving part on the CRT as each feature is added. Basically the operator provides the input and the computer does the calculating.

If any design changes are necessary at this time, the designer makes the changes on the screen with the light pen, and these changes are automatically made in the computer's memory. In most systems, CRT screens can display multiple orthographic views of a design (front, top, and side) in combination with a three-dimensional isometric view. These views are frequently displayed simultaneously on a split screen, and any design changes made on one view are always automatically added to the other views. The designer is able to create and change parts and lines on the CRT with a light pen, an electro-mechanical cursor, or an electronic tablet (Fig. 76-2). It should also be noted that the operator can rotate any drawing to any position desired so that he or she can better study it or compare it to another component.

Most systems have a typewriter-like keyboard for entering the text and commands into the computer. Some systems also provide a list or *menu* of standard commands containing drawing functions for specifying the size and location of lines, arcs, texts, cross-hatching, standard drafting symbols, and other items (Fig. 76-3). These menus are designed to increase the speed at which drawing data can be entered. Any desired function contained in the menu is entered with the push of a single button. Menus may also be displayed on the screen or on a data tablet and items may be selected from these with the electronic or light pen.

Figure 76-2 *An electronic tablet being used to make design changes in a part illustrated on the CRT. (Courtesy Bausch & Lomb.)*

Figure 76-3 *Standard drawing functions are contained in the menu and can be added to a drawing at the push of a button. (Courtesy Bausch & Lomb.)*

DESIGN DATA AND TESTING

The information stored in the computer's memory provides a database from which the computer can extract and process data for future needs. For instance, it, or other computers linked with it, can calculate the volume, mass, center of gravity, and other features of the part to be produced. It can also perform stress analysis, provide English-to-metric conversion of the figures, generate a bill of material, and even produce a numerical control (NC) tape and the instructions for machining the part.

After the design has been completed, an engineer (or designer) can ask the computer questions regarding the anticipated performance of the part. For instance, the engineer who designs a tower which must support a weight of 220 000 lb (100 000 kg, or 100 tonnes) can ascertain from the computer if the structure will support this weight. The computer will tell if it will support this weight or if it will break and at what point. If the information displayed shows that the structure will not support this weight, the designer merely makes the required modifications on the screen (using the light pen) and then asks the same question of the computer. This process is repeated until an affirmative answer is received from the computer.

Once the details of a drawing have been entered into the system, the operator, engineer, or designer can make changes easily and quickly to any area of the drawing without having to redraw the original. With this equipment, the accuracy of the drawing is greatly improved, which permits a quick check of critical components for clearance and tolerance while the "product" is still in the drawing stage.

TESTING THE FUNCTION OF COMPONENTS

The operator can also assemble parts electronically by laying the drawing for one component over the other and zoom in on the critical areas to check visually for clearance, interference, etc. By using CAD, engineers are able to make performance analyses on the part. Here they can check the part for stress and strain and determine if it will stand up in use. At this point it is possible to alter the structure and strengthen any weak areas. For example, the part may be thickened at one point or a larger radius added where two surfaces join. Any changes made will also be recorded in the computer's memory. At this time it is also possible to estimate the cost of the finished part, run a weight analysis, and even prepare an NC tape for the machining process—all while the part is still in the drawing stage.

After the operator has carefully examined the drawing on the display screen and is satisfied that the detail is accurate, the data may then be directed to the *plotter* to produce the finished drawing. The plotter will hold the drafting paper and produce permanent drawings in ink or ball-point pen. Some systems offer a photoplotter, which projects a highly accurate light beam onto photosensitive paper to produce a drawing with sharp, high-contrast lines.

CAD ADVANTAGES

Let's look at another example of the value of CAD. Imagine that an aircraft design engineer is studying the image of an aircraft fuselage on the CRT. By pushing a button he or she could display the landing gear. As the engineer studies the components, some interference may be noted between the extended landing gear and the fuselage. The engineer will then press a key on the keyboard and with the light pen alter the shape of the fuselage, permitting the landing gear to operate freely. By the proper use of CAD equipment, it is possible to correct an error in less than a minute that would have taken months to resolve by the standard procedures formerly used, such as making a prototype or model of the plane.

Following is a list of CAD advantages.

- Greater productivity of drafting personnel
- Less drawing production time
- Better drawing change procedure
- Greater drawing and design accuracy
- Greater detail in layouts
- Superior drawing appearance
- Greater parts standardization
- Better factory assembly procedures
- Less scrap

REVIEW QUESTIONS

1. What is CAD?
2. Name the main components of a CAD system.
3. Briefly explain how CAD can be used to design a part.
4. Define *menu* and state its purpose.
5. What design and testing data can be obtained from a CAD system?
6. Why is CAD so important to designing a new product?

UNIT 77

Chucking and Turning Centers

Extensive studies during the mid-1960s showed that about 40 percent of all metal cutting operations were performed on lathes. Up until that time, most work was done on engine or turret lathes, which were not very efficient by present-day standards. Intensive research led to the development of numerically controlled turning centers and chucking lathes, which could produce work of almost any configuration automatically and much more efficiently (Fig. 77-1A). In recent years these have been updated to more powerful computer-controlled units capable of greater precision and higher production rates than their predecessors.

OBJECTIVES

After completing this unit you will be able to:

1. State the purposes and functions of chucking and turning centers

Figure 77-1 *(A) Turning and chucking centers are versatile and very productive. (Courtesy Cincinnati Milacron Inc.)*

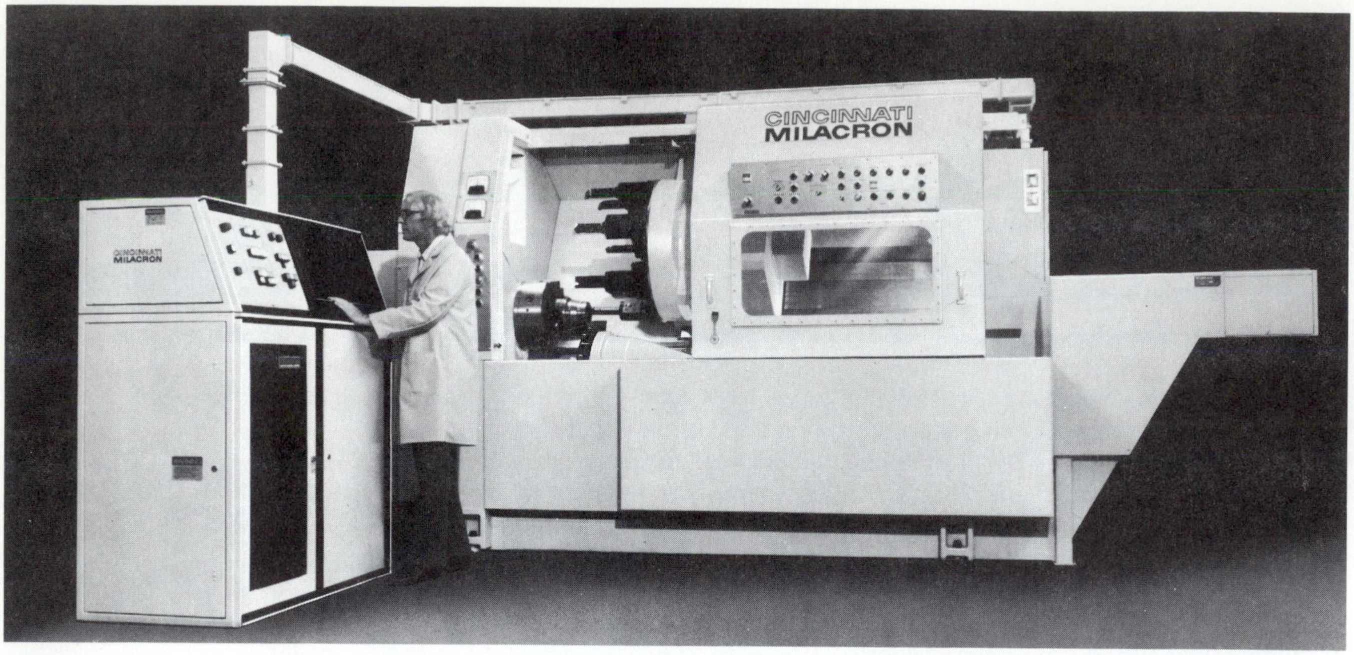
CINCINNATI
MILACRON

Figure 77-1 *(B) A chucking center is designed to increase productivity. (Courtesy Cincinnati Milacron Inc.)*

2. Identify the applications of computer numerical control (CNC) for turning centers
3. Name the machining operations which may be performed simultaneously

CNC CHUCKING CENTERS

Chucking centers (Fig. 77-1B), designed to machine most work held in a chuck, are available in a variety of chuck sizes from 8 to 36 in. in diameter. Because there are many similar types of machines made by different manufacturers and all perform similar functions, only one chucking lathe, a four-axis model, will be described.

This chucking center has two turrets, operating independently on separate slides, to machine the workpiece simultaneously. While the seven-tool upper turret is machining the inside diameter (ID), the lower turret, which also mounts seven different cutting tools, may be machining the outside diameter (OD) (Fig. 77-2). However, if the workpiece requires mainly internal operations, both turrets can work on the inside of the workpiece simultaneously (Fig. 77-3). This type of operation is suitable for large-diameter parts which require boring, chamfering, threading, internal radii, or retaining grooves.

For parts with mostly OD operations, the upper turret can be equipped with turning tools so that both turrets can

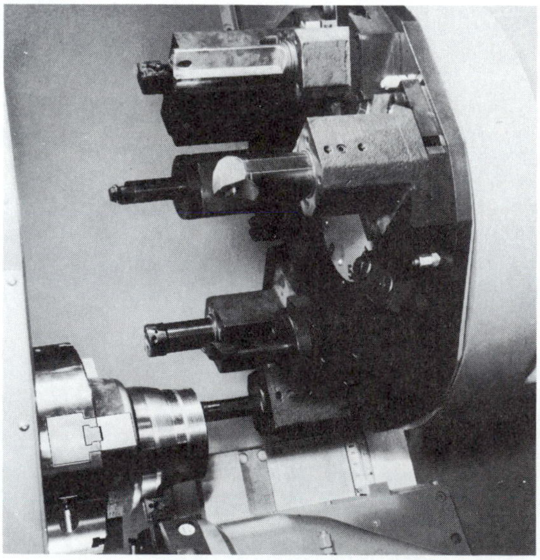

Figure 77-2 *Turning and boring operations may be carried on simultaneously. (Courtesy Cincinnati Milacron Inc.)*

Figure 77-3 *Both turrets being used for internal operations. (Courtesy Cincinnati Milacron Inc.)*

Figure 77-5 *A long shaft being supported by a center mounted in the upper turret. (Courtesy Cincinnati Milacron Inc.)*

machine the OD. Figure 77-4 shows a diameter and a chamfer being machined at the same time,

When longer parts must be machined, the right-hand end of the shaft may be supported with a center mounted in the upper turret, while the lower turret performs the external machining operations (Fig. 77-5). Other operations which may be performed simultaneously are turning and facing (Fig. 77-6) and internal threading (Fig. 77-7).

Other chucking centers are chiefly two-axis models. They may have a single disklike turret on which both ID and OD tools are mounted, or they may have two turrets (usually on the same slide). In the latter case, one turret is normally designed for OD tools and the other for ID tools. Whatever the arrangement, the two-axis control will drive only one turret at a time.

CONSTRUCTION

Because of the high spindle speeds [up to 2240 revolutions per minute (r/min)] and the high horsepower requirements, machines of this type must be rugged to absorb the high cutting forces. The bed and machine frame is a heavy one-piece casting. To permit easy chip removal and easy loading and unloading of the workpieces, the bed is slanted 30° from the vertical plane.

Figure 77-4 *Both turrets being used for external turning. (Courtesy Cincinnati Milacron Inc.)*

Figure 77-6 *Simultaneous turning and facing operations using both turrets. (Courtesy Cincinnati Milacron Inc.)*

Figure 77-7 *Cutting internal and external threads simultaneously. (Courtesy Cincinnati Milacron Inc.)*

Figure 77-8 *A gage for presetting tools for turning centers. (Courtesy Cincinnati Milacron Inc.)*

Direct-current servo drives provide accurate positioning and travel of the cutting tools. The servo motors position the turret cross-slides by means of high-precision preloaded ball screws, which ensures slide positioning repeatability to ±0.0002 in. These turret slides operate at speeds from 0 to 200 in./min (0 to 5080 mm/min) and 400 in./min (10 160 mm/min) on rapid traverse.

TOOLING

Both the upper turret and the lower turret can accommodate seven different types of tools. The toolholders for machining ODs are located in the lower turret and are qualified (preset) and only require changing the carbide insert when the tool must be replaced. Tools for machining the ID are mounted in a dovetailed block and preset off the machine by means of a tool-setting gage (Fig. 77-8). The dovetailed block is then mounted in the upper turret, ensuring proper tool point positioning every time.

COMPUTER NUMERICAL CONTROL

The control console shown in Fig. 77-9 is the "brains" of the chucking center. It features a minicomputer, a cathode ray tube (CRT) (video display), a tape reader, part program storage, part program edit, and maintenance diagnostic display.

The minicomputer performs control logic, mechanism control and input–output control. The CRT provides a visual display of slide positions, the spindle operating condition, the sequence numbers, preparatory functions, system fault conditions (diagnostics), operator instructions, and keyboard data.

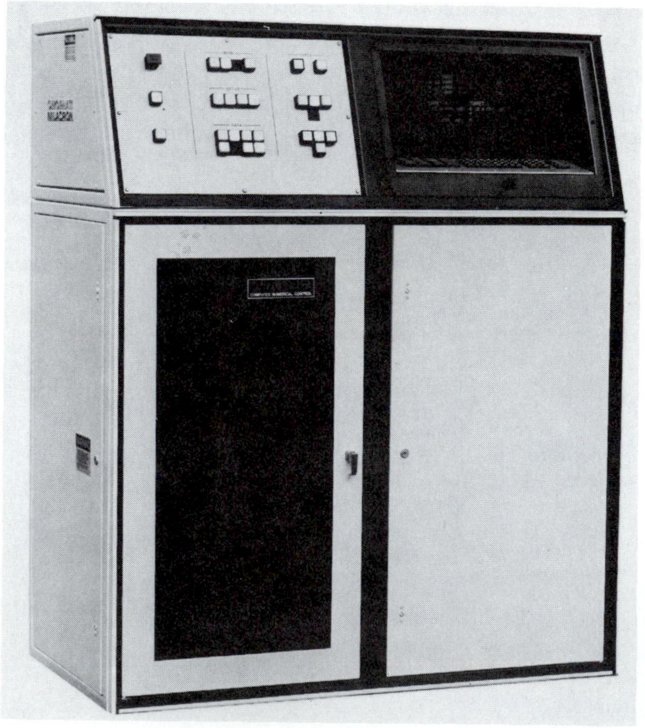

Figure 77-9 *A chucking center control panel contains the computer, video display, and machine control functions. (Courtesy Cincinnati Milacron Inc.)*

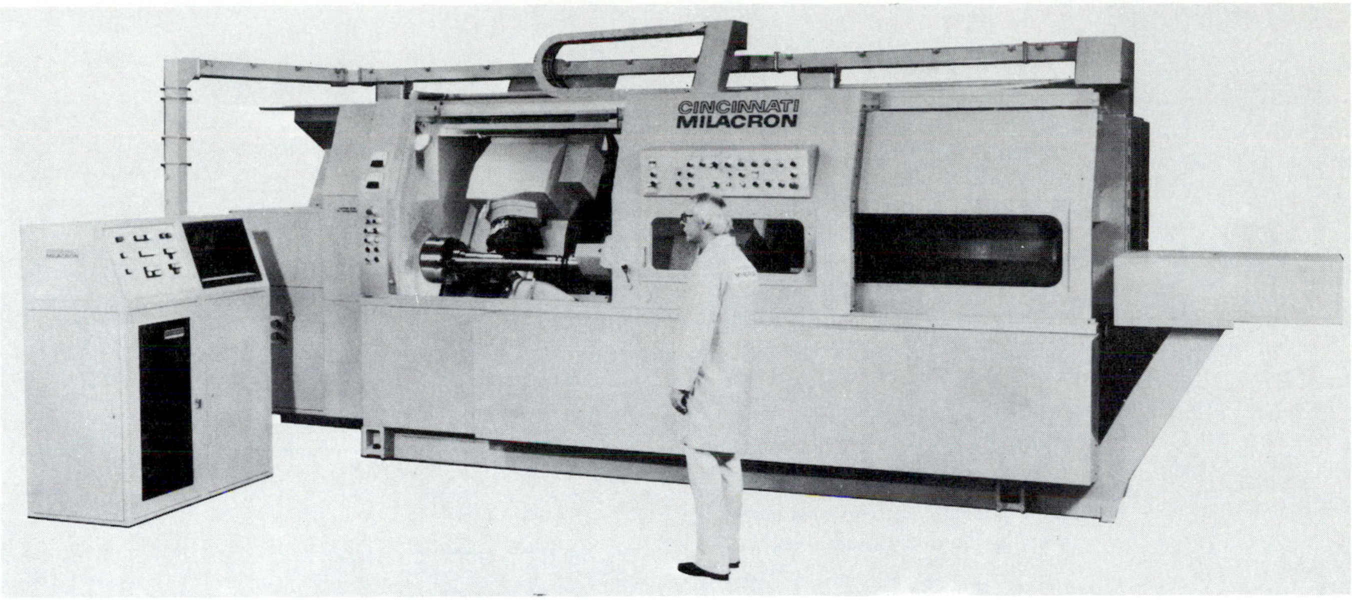
CINCINNATI
MILACRON

Figure 77-10 A CNC *turning center is designed for maximum efficiency on shaft-type workpieces. (Courtesy Cincinnati Milacron Inc.)*

The keyboard input provides the means of communicating with the CNC system. It is used to enter setup and tooling data on new parts programs or for correcting tooling data. The keyboard is also used to run diagnostic checks on the systems with the results (regarding oil pressure, spindle operating conditions, etc.) being displayed on the CRT.

Part program storage can store the data from up to 300 ft (91.4 m) of part program tape. A part program edit feature allows the program to be modified at time of use. Other features include a constant surface speed monitor which checks the part diameter and controls the spindle speed accordingly.

CNC TURNING CENTERS

Computer numerical control turning centers (Fig. 77-10), while not unlike chucking centers, are designed mainly for machining shaft-type workpieces, which are supported by a chuck and a heavy-duty tailstock center.

On four-axis machines two opposed turrets, each capable of holding seven different tools, are mounted on separate cross-slides, one above and one below the centerline of the work. Because the turrets balance the cutting forces applied to the work, extremely heavy cuts can be taken on a workpiece when it is supported by the tailstock. The dual turrets also lend themselves to other operations such as:

- Roughing and finishing cuts in one pass
- Machining different diameters on a shaft simultaneously (Fig. 77-11)
- Finish turning and threading simultaneously
- Cutting two different sections of a shaft at the same time

Because the lower turret is designed to accept ID tools, parts may be gripped in the chuck and be machined inside and outside at the same time (Fig. 77-12).

Figure 77-11 *Both turrets being used to machine a shaft having different diameters. (Courtesy Cincinnati Milacron Inc.)*

Figure 77-12 *The inside and outside diameters of a part being machined at the same time. (Courtesy Cincinnati Milacron Inc.)*

When the turning center is equipped with a steady rest, operations such as facing and threading may be performed on the end of a shaft.

A bar-feeding mechanism (Fig. 77-13), supplied as an option, permits the machining of shafts and parts from bar stock, providing it is smaller than the diameter of the spindle through-hole.

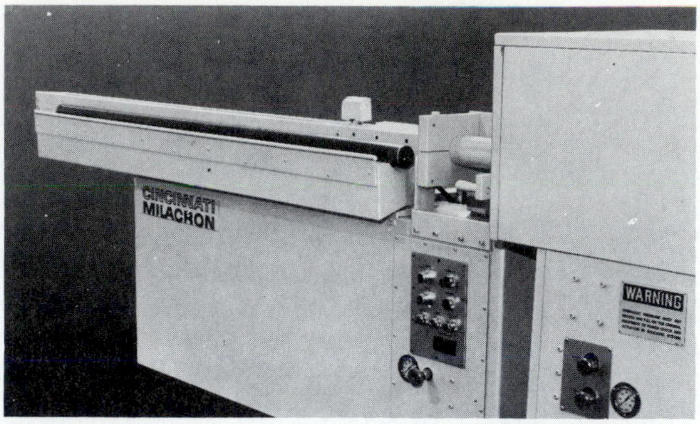
MILACRON
WARNING

Figure 77-13 *A bar-feeding attachment allows parts to be manufactured from bar stock. (Courtesy Cincinnati Milacron Inc.)*

REVIEW QUESTIONS

1. Why were chucking and turning centers developed?
2. Describe the construction of the two turrets of the chucking lathe.
3. What is the purpose of the slanted bed on a chucking or turning center?
4. State the purpose of the following CNC parts:
   a. Minicomputer b. CRT c. Keyboard
5. What is the main purpose of a turning center?
6. Why can extremely heavy cuts be taken on a turning center?

UNIT 78

Machining Centers

In the 1960s industrial surveys showed that smaller machine components requiring several operations took a long time to complete. The reason was that the part had to be sent to several machines before it was finished and often had to wait a week or more at each machine before being processed. In some cases, parts requiring many different operations spent as many as 20 weeks in the shop before completion. Another startling fact was that during all the time in the shop, the part was only in a machine for 5 percent of the time, and while it was in the machine only 30 percent of the time was spent in machining. Therefore, a part was machined for only 30 percent of 5 percent or 1.5 percent of its time in the shop. In contrast, larger parts like large machine castings took much less time because all the machining was usually done on one machine.

Also in the 1960s, there was much "operator intervention" during the machining process. The operator had to watch the performance of the cutting tool and change spindle speeds and feeds to suit the job and the machine. The operator was frequently changing cutters as they became dull and always had to set depths of roughing and finishing cuts. All of these problems were recognized by machine tool manufacturers, and in the late 1960s and early 1970s they began designing machines which would perform several operations and probably do about 90 percent of the machining on one machine. One of the results of this research was the *machining center* and the later more elaborate version called the *processing center.* The machines can perform the operations of drilling, milling, boring, tapping, and precision profiling very efficiently.

OBJECTIVES

After completing this unit you will be able to:

1. Describe the development of the machining center
2. Identify the types and construction of machining centers
3. Explain the operation of the machining center and the processing center

CONSTRUCTION AND OPERATION

There are two main types of machining centers: the *horizontal* and the *vertical spindle* types. Only the horizontal spindle type will be discussed in this unit. There are two main types of horizontal spindle machining centers: the traveling-column type and the fixed-column type.

The *traveling-column type* (Fig. 78-1) is equipped with one or usually two tables on which the work is mounted. With this type the column (and the cutter) move toward

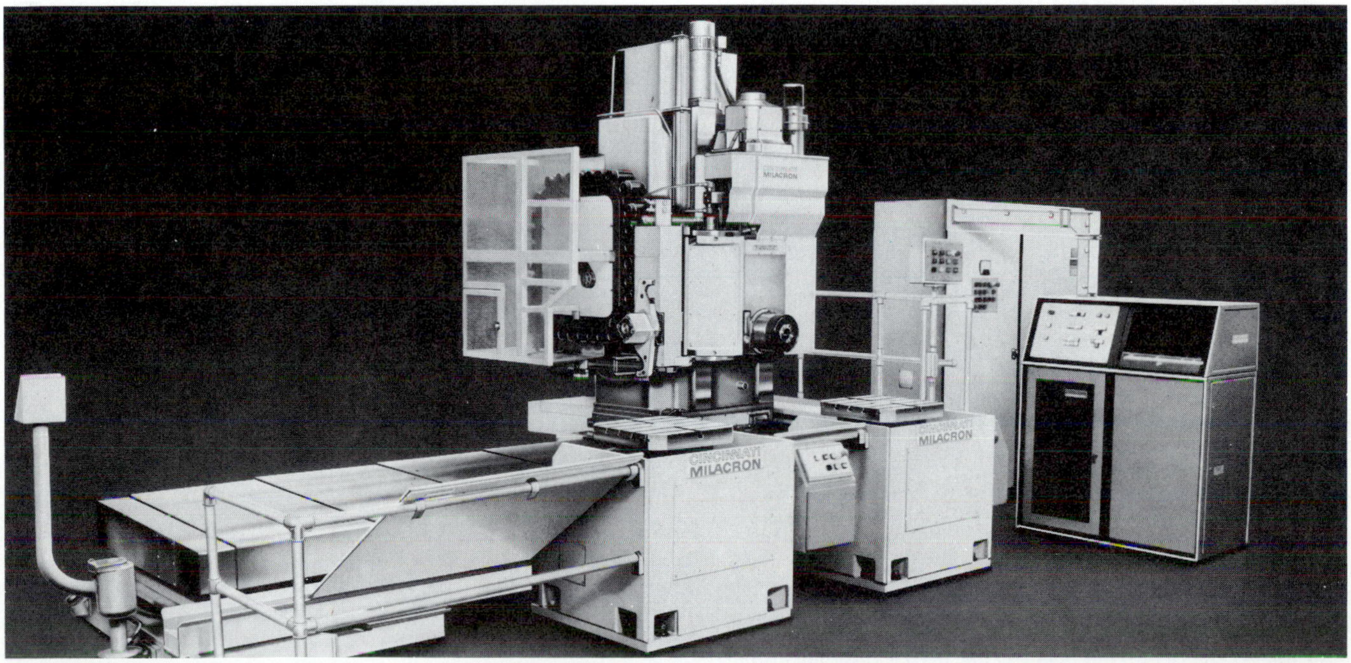
MILACRON
CINCINNATI
MILACRON
MILACRON

Figure 78-1 *A traveling-column-type machining center. (Courtesy Cincinnati Milacron Inc.)*

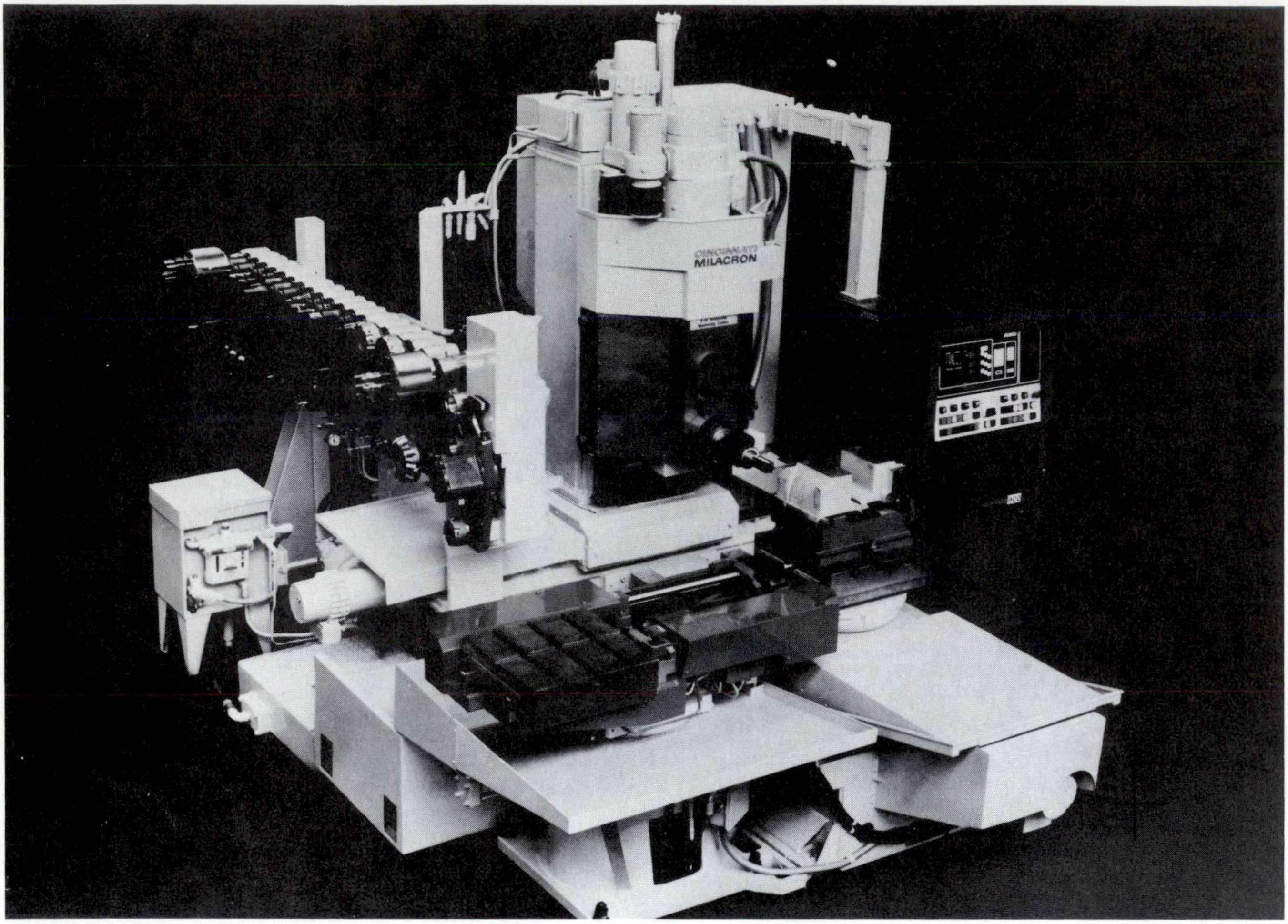
CINCINNATI
MILACRON

Figure 78-2 *A fixed-column-type machining center. (Courtesy Cincinnati Milacron Inc.)*

the work, and while it is machining the work on one table, the operator is changing the workpiece on the other table.

The *fixed-column type* (Fig. 78-2) is equipped with a pallet shuttle. The pallet is a removable table on which the workpiece is mounted. With this type, after one workpiece has been machined (Fig. 78-3), the pallet (and workpiece) is forced off the receiver onto the shuttle. The shuttle is then rotated, bringing a new pallet into position for the

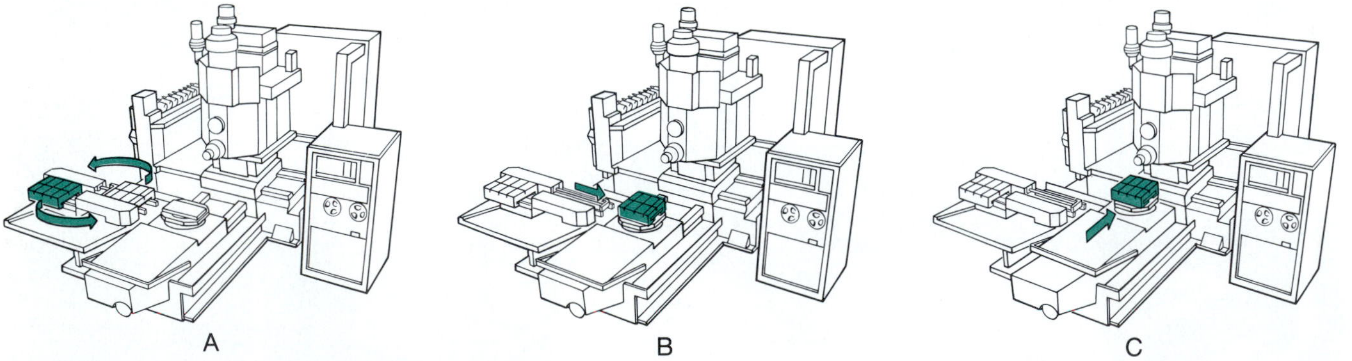
A
B
C

Figure 78-3 *(A) A new pallet ready for loading while the other pallet is ready for unloading; (B) pallet being shuttled onto the receiver; (C) the pallet being moved into machining position. (Courtesy Cincinnati Milacron Inc.)*

shuttle and the finished work pallet into position for unloading.

In Fig. 78-3B the new pallet is shuttled and clamped onto the receiver, while in Fig. 78-3C, the pallet and workpiece are positioned into the programmed machining position. The cycle for changing and shuttling the pallets off and onto the receiver takes only 20 s.

TOOL CHANGER

All machining centers are equipped for automatic, numerically controlled tool changing, which is much faster and more reliable than manual tool changing. The tools (up to 90, depending on the type of machine) are held in a storage chain (matrix) and each is identified by either the tool number or the storage pocket number. This information is stored in the computer's memory. While one operation is being performed on the workpiece, the tool required for the next operation is moved to the pick-up position (Fig. 78-4A), where the tool-change arm removes and holds it.

Immediately on completion of the machining cycle, the tool-change unit swings 90° to the tool-change position (Fig. 78-4B). The tool-change arm then rotates 90° and removes the cutting tool from the spindle. It then rotates 90° and places the new cutting tool in the spindle after which it returns the old cutting tool to its position in the tool carrier (Fig. 78-4C). This whole operation is completed in about 11 s.

COMPUTER

The computer numerical control (CNC) system is built around a minicomputer which is either mounted on the machine or in a separate unit beside the machine. Most computer controls have similar features, such as a CRT or display unit and an alphanumeric keyboard for passing additional information to the computer. There will also be storage for 15 to 20 programs for machining different parts as well as program editing capabilities. Most computer controls now are capable of adjusting automatic backlash and lead screw compensation, as well as indicating any malfunction in the machine or any errors in the program's functions. This diagnostic display informs the operator of any problems in the machine or control and, by referring to a maintenance manual, the user normally can correct the problem. On some controls, diagnostic units can be tied in by long-distance telephone to a mainframe computer in the machine or control manufacturer's plant. There the problem is assessed immediately and instructions given by phone to the user.

In unmanned machining centers, the computer controls the total program. It programs the parts onto the machine and up to the spindle. It selects the proper program for the part and then carries out the machining program and controls the loading and unloading often performed by robots.

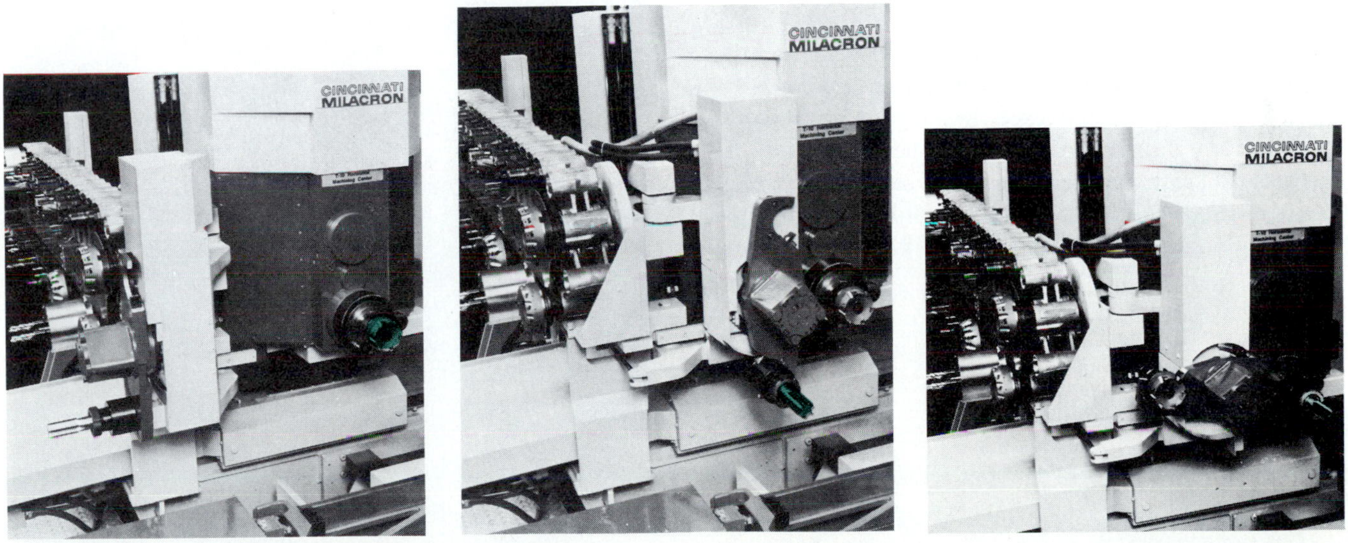
CINCINNATI MILACRON
CINCINNATI MILACRON
CINCINNATI MILACRON

A B C

Figure 78-4 *(A) The cutting tool required being removed from the storage chain; (B) the tool-change unit swings 90° to change tools in the spindle; (C) the old cutting tool has been removed and replaced with the tool required for the next operation. (Courtesy Cincinnati Milacron Inc.)*

MACHINING CENTER ACCESSORIES

TORQUE CONTROL MACHINING

A feature that is fast becoming popular is that of *torque control machining,* the torque being calculated from measurements at the spindle drive motor (Fig. 78-5). This device will increase productivity by preventing or sensing damage to the cutting tool. The torque is measured when the machine is turning but not cutting, and this value is stored in the computer's memory.

As the machining operation begins, the stored value is subtracted from the torque reading at the motor. This will give the net cutting torque, which is compared to the programmed torque or limits stored in the computer (or on NC tape). If the net cutting torque exceeds the programmed torque limits, the computer will act by reducing the feed rate, turning on the coolant, or even stopping the cycle. The feed rate will be lowered whenever the horsepower required exceeds the rated motor capacity or the programmed code value.

The system display of three yellow lights advises the operator of the operational conditions in the machine at the time. A left-hand yellow light indicates that the torque control unit is in operation. The middle yellow light indicates that the horsepower limits are being exceeded. The right-hand light comes on when the feed rate drops below 60 percent of the programmed rate. The meter (Fig. 78-5) indicates the cutting torque (or operational feed rates) as a percent of the programmed feed rate.

As the tool gets dull, the torque will increase and the machine will back off on the feed and ascertain the problem. It could be that there is excessive material on the workpiece or a very dull or broken tool. If the tool is dull, the machine will finish the operation, and a new backup tool of the same size will be selected from the storage chain when that operation is to be performed again. If the torque is too great, the machine will stop the operation on the workpiece and program the next piece into position for machining.

100
1000
10000
50
25
100
0
FT.-LB.
% FEEDRATE

Figure 78-5 *Torque control raises or lowers the feed rate depending on the depth of cut at any time during the cutting cycle. (Courtesy Cincinnati Milacron Inc.)*

PRECISION SURFACE-SENSING PROBE

The *precision probe* (Fig. 78-6) is an extremely sensitive stylus used for setting up and checking workpieces on the machining center. The probe is stored in the tool storage matrix (Fig. 78-7) and is loaded in the spindle, when required, to determine if the work is mounted and positioned properly. If there is not enough (or too much) material on the workpiece for proper machining, or if the workpiece is not positioned properly on the pallet, the probe will reject the part and the computer will call up the next workpiece.

The probe may also be used to verify that the proper fixture has been selected, or for checking the amount of core shift in castings before machining. It can also be used to calibrate the accuracy of the machine and to inspect the accuracy of the finished workpiece. One company had previously taken 25 h to inspect a complex part using standard inspection equipment at the bench. The probe performed the same inspection while the part was still mounted on the machine in 12 min. This type of precision surface-sensing

PROBE

Figure 78-6 *A precision sensing probe is used to quickly check the position of the workpiece. (Courtesy Cincinnati Milacron Inc.)*

PROBE

Figure 78-7 *A probe may be loaded into the automatic tool changer for use whenever required. (Courtesy Cincinnati Milacron Inc.)*

probe may also be used for setting up and checking parts which are to be machined in a turning center.

THE PROCESSING CENTER

The Cincinnati Milacron T-10 machining center may be equipped with an eight-pallet automatic work changer (AWC) that includes a manually indexible load–unload work station to permit loading and unloading while the machine is running. With this addition, eight alike or different parts may be loaded on the pallets and presented to the cutting tool(s) for the machining operations required (Fig. 78-8). The operations for each part are programmed into the computer, and if the workpiece has been set up properly and accepted by the probe, the computer will select the proper program and the machine will perform the necessary operations on it. This setup permits the machining of eight different parts automatically.

ADVANTAGES OF MACHINING CENTERS

1. Increased productivity.
2. This type of machine can machine to closer tolerances. This means:

- Fewer finishing operations required, such as grinding or scraping
- Less chance of error because fewer machines are being used
- Less inspection required
- Less paperwork needed
- Lower rejection or scrap rate

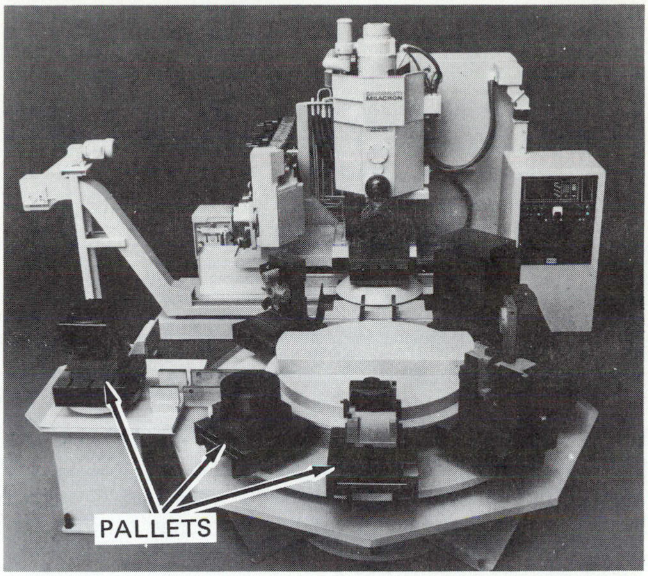
PALLETS

Figure 78-8 *The processing center is a machine designed for the factory of the future. (Courtesy Cincinnati Milacron Inc.)*

3. Reduction in inventory because there is less floor time between operations.

REVIEW QUESTIONS

1. What factors led to the development of the machining center?

CONSTRUCTION AND OPERATION

2. Name two types of horizontal spindle machining centers and briefly state the principle of each.
3. Describe how the required tool is mounted in the spindle.
4. What is the purpose of:
   a. A cathode ray tube? b. A diagnostic display?

MACHINING CENTER ACCESSORIES

5. Briefly describe the operation of the torque-control device.
6. List three factors that may cause the precision surface-sensing probe to reject a workpiece.
7. State four other uses of the probe.

UNIT

79

Robotics

Over the past two decades, industry has realized that in order to be competitive in world markets, it had to increase productivity and reduce manufacturing costs. Since the source of skilled workers was dwindling and it was difficult to find people who would perform tasks which were considered to be monotonous, physically difficult, or environmentally unpleasant, industry has found it necessary to automate many manufacturing processes. The development of the computer has made it possible for industry to produce reliable machine tools and robots, which are making manufacturing processes more productive and reliable, thereby improving their user's competitive position in the world market.

Over the past 15 years, industry has slowly introduced robots to various aspects of manufacturing. With the need to be competitive in world markets, new control systems, more dependable and lower-priced computers, and more and better robots are being developed in the industrial world. Today's industrial robots are tough, tireless, very accurate, and are helping to make human drudgery a thing of the past in hundreds of factories throughout the world, and at the same time improving productivity and reducing manufacturing costs. The word *robot* is derived from a Czechoslovakian word *robota* meaning "work." The word *robot* has been used for many years in science fiction to mean a "mechanical being." Today's industrial robots still have a long way to go before they are capable of performing like the science-fiction robot, but they are capable of performing tasks which are monotonous, physically difficult, or present health or safety hazards to human workers.

OBJECTIVES

After completing this unit you will be able to:

1. State the purpose of the industrial robot in manufacturing
2. Describe the construction and operation of an industrial robot
3. Explain the applications and benefits of robots in manufacturing

THE INDUSTRIAL ROBOT

The *industrial robot* of today (Fig. 79-1) is basically a single-arm device that can manipulate parts or tools through a sequence of operations or motions as programmed by the computer. These operations or sequences may or may not necessarily be repetitive, because the robot, through the computer, has the ability to make logical decisions. The robot can be applied to many different operations in industry and is capable of being taken from one operation and easily "taught" to do another. This ability is based on the concept of *flexible automation,* where one machine is capable of economically performing many different operations with a minimum amount of special engineering or debugging. This differs from the "hard," or ded-

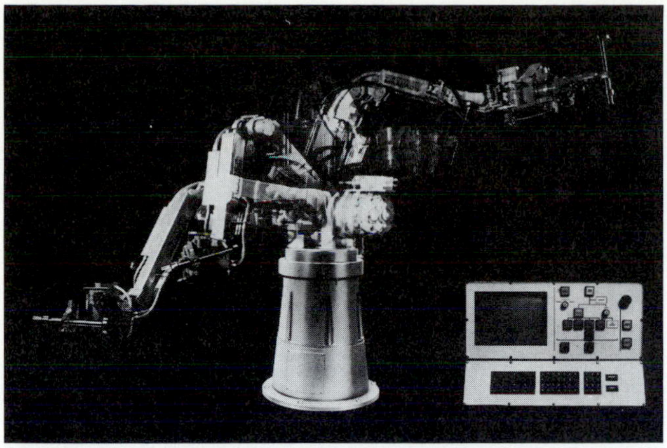

Figure 79-1 *Industrial robots perform back-breaking, boring, or repetitive jobs. (Courtesy Cincinnati Milacron Inc.)*

icated, system, which is designed only to perform one operation or task.

INDUSTRIAL ROBOT REQUIREMENTS

In order for an industrial robot to be most effective, it must meet three criteria: be flexible for many applications, be reliable, and be easy to teach.

APPLICATION FLEXIBILITY

The ideal industrial robot should be capable of performing many different operations. There are two general categories of applications where robots are being used extensively: handling and processor applications.

In *handling applications,* robots must be equipped with some type of device, such as a pair of gripping fingers or vacuum cups, in order to pick up or move the material. They can be used to load or unload machine tools, forging presses, injection-molding machines, and even inspecting or gaging systems. These applications also include the handling of materials (moving them from one station to another), retrieving parts from storage areas or conveyor systems, packaging, and palletizing (placing parts on a pallet for transporting to other stations).

In *processor applications,* robots carry out operations such as spot welding, seam welding, spray painting, metallizing, cleaning, or any other operation where the robot can operate a tool to carry out a manufacturing process.

RELIABILITY

The industrial robot must be very reliable since it will be working in conjunction with some very sensitive machinery, which would be left idle when a robot breaks down. The robot should not only be reliable but also have the ability to diagnose a problem quickly when it occurs and, if possible, take corrective action.

EASE OF TEACHING

An industrial robot is taught by example by a human operator who must lead it through the sequence of steps or operations needed for each specific application. This programmed teaching routine is necessary to teach the robot to make logical decisions based on information generated by two-way communication with its working environment.

Robots which are easy to teach do not require that the worker know how to program them or how their control system operates. For simplicity, the human worker needs to know only two things: what the robot has to do to perform an operation, and what the worker has to do to *teach the robot.*

INDUSTRIAL ROBOT DESIGN

A robot system is a means of moving a pair of grippers or some other form of end effector and directing it through a certain sequence of motions. At certain points in the sequence, the system directs some function to occur, such as "close the grippers" or "start the machine tool cycle."

The Tomorrow Tool (T3), manufactured by Cincinnati Milacron Inc., is a hydraulically powered, servo-operated, and computer-controlled industrial robot system which has been designed for maximum flexibility. The arm has six rotary axes of movement: arm sweep, shoulder swivel, and elbow extension, along with pitch, yaw, and roll of the wrist (Fig. 79-2). The elbow-axis is driven by a piston–cylinder arrangement; all other axes are powered by direct-coupled

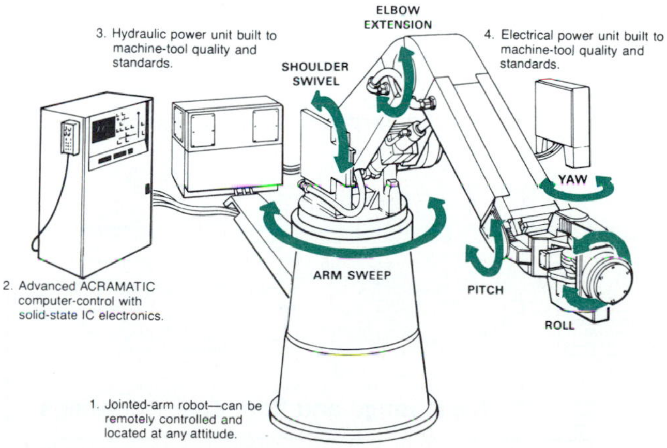
ELBOW EXTENSION
3. Hydraulic power unit built to machine-tool quality and standards.
4. Electrical power unit built to machine-tool quality and standards.
SHOULDER SWIVEL
YAW
ARM SWEEP
PITCH
ROLL
2. Advanced ACRAMATIC computer-control with solid-state IC electronics.
1. Jointed-arm robot—can be remotely controlled and located at any attitude.

Figure 79-2 *The robot's arm has six rotary axes of movement. (Courtesy Cincinnati Milacron Inc.)*

rotary actuators. This arm provides a mechanical system which has no gears or linear slides and a minimum number of rotating parts. This produces a very reliable system, which creates very few maintenance problems.

Because of the construction design, the arm can reach over 150 in. (3800 mm) vertically upward and extend out horizontally almost 100 in. (2500 mm) (Fig. 79-3). It can also operate through 240° of arm sweep. Standard arms are capable of lifting up to a 100 lb (45-kg) load located at 10 in. (250 mm) from the roll-axis plate, and heavy-duty arms can lift up to 225 lb (100 kg). The arm can repeat for position anywhere in its range within an accuracy of ±0.050 in. (1.27 mm). However, with the addition of a special system option, the repeatability can be improved to ±0.025 in. (0.64 mm).

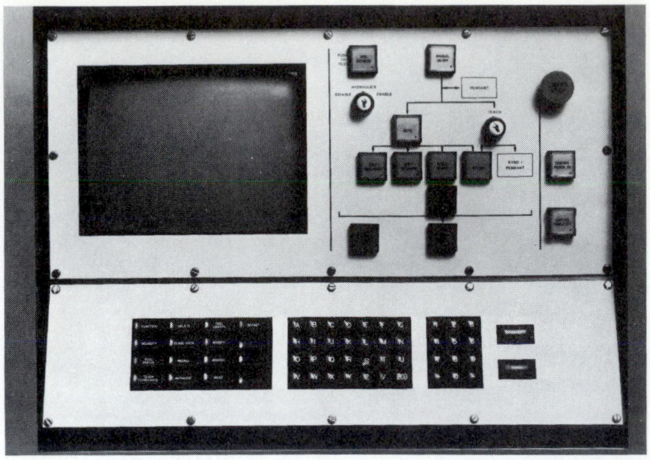

Figure 79-4 *The robot control unit consists of a computer, video display, and control panel. (Courtesy Cincinnati Milacron Inc.)*

THE CONTROL SYSTEM

The purpose of an industrial robot's system is to direct the robot through its movements and sequences of operations and to provide a means for the operator to teach the robot the motions and sequences required to perform various operations.

The control unit used for robots is based on a minicomputer because it is very flexible and allows the operator to teach the robot. The robot's control unit (Fig. 79-4) consists of a minicomputer, a CRT and keyboard, a control panel, axis servo-control electronics, servo test panel, a teach pendant, and terminals for communication with other equipment (Fig. 79-5).

In order to operate effectively, the robot's control system should have the following characteristics:

1. When a robot is operating, it should move from position to position in a straight line with all six axes moving proportionally so that they all start and finish at the same time.
2. The robot should be able to communicate between itself and the control unit in order to start and complete cycles, make logical decisions on its next move, and verify that an operation has been completed.
3. The robot should be able to diagnose problems as they occur in order to reduce downtime and minimize loss of production.
4. During the teaching mode, the human teacher should be able to guide the tool center point (TCP) along a rectilinear (*XYZ*) coordinate system or even a cylindrical system (Fig. 79-6).
5. The human teacher should be able to edit a previously taught program by easily adding or deleting points if necessary. Therefore, channels of communication must exist between the teacher and the control.

MODES OF OPERATION FOR ROBOTS

The T3 Cincinnati Milacron robot has three modes of operation: manual, teach, and automatic. The mode desired is selected on the control console.

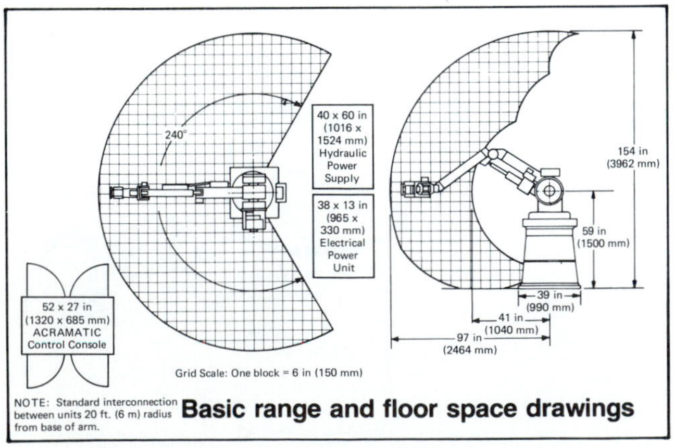

240°
40 x 60 in (1016 x 1524 mm) Hydraulic Power Supply
38 x 13 in (965 x 330 mm) Electrical Power Unit
52 x 27 in (1320 x 685 mm) ACRAMATIC Control Console
154 in (3962 mm)
59 in (1500 mm)
39 in (990 mm)
41 in (1040 mm)
97 in (2464 mm)
Grid Scale: One block = 6 in (150 mm)
NOTE: Standard interconnection between units 20 ft. (6 m) radius from base of arm.
Basic range and floor space drawings

Figure 79-3 *The robot arm is designed to operate over a wide range. (Courtesy Cincinnati Milacron Inc.)*

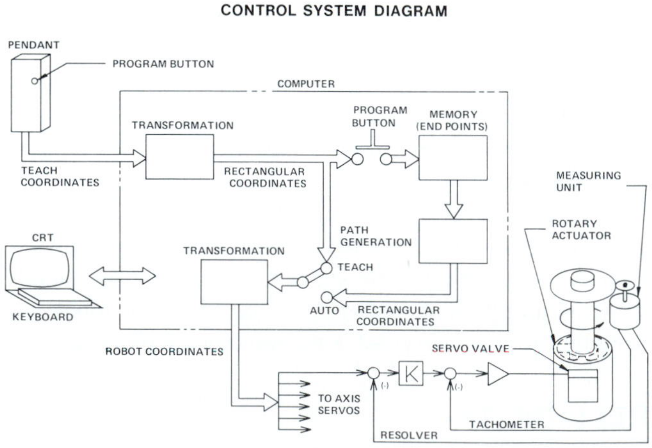

CONTROL SYSTEM DIAGRAM
PENDANT
PROGRAM BUTTON
COMPUTER
TRANSFORMATION
PROGRAM BUTTON
MEMORY (END POINTS)
TEACH COORDINATES
RECTANGULAR COORDINATES
MEASURING UNIT
ROTARY ACTUATOR
CRT
PATH GENERATION
TRANSFORMATION
TEACH
KEYBOARD
AUTO
RECTANGULAR COORDINATES
ROBOT COORDINATES
SERVO VALVE
TO AXIS SERVOS
TACHOMETER
RESOLVER

Figure 79-5 *Role of the computer in the control system. (Courtesy Cincinnati Milacron Inc.)*

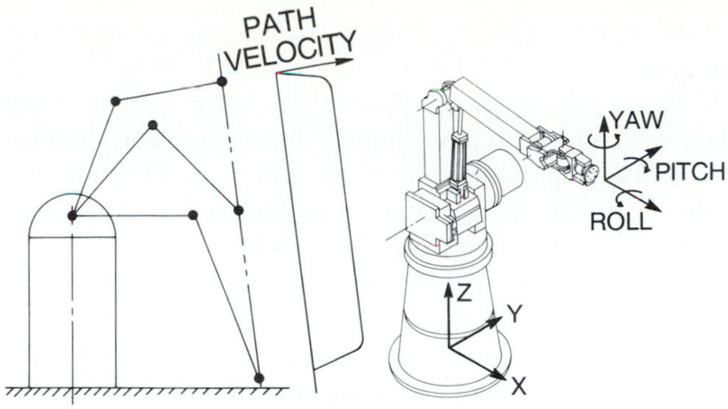
PATH
VELOCITY
YAW
PITCH
ROLL
Z
Y
X

Figure 79-6 *The key operating concepts of a robot: (A) controlled path—straight line; (B) rectangular and angular coordinates. (Courtesy Cincinnati Milacron Inc.)*

1. In the *manual mode,* the human worker can move each axis of the arm independently of all the others with the use of the teach pendant (Fig. 79-7). The purpose of the manual mode is to allow the operator to align the arm to its "home" position. Home is a unique location where the six coordinates of the robot's arm are known to the control system. Before the human operator can program the robot through the teach mode or enter the automatic mode, the arm must be aligned to home. A special display on the control console and arrows on the robot arm can be used for this alignment.

2. In the *teach mode,* the operator can teach the robot arm to perform a certain task or operation. The axes of the robot arm can be commanded to move in a coordinated fashion.

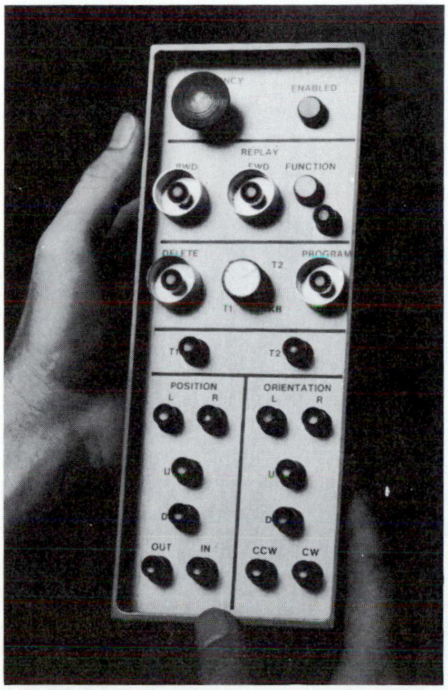
ENABLED
REPLAY
FUNCTION
POSITION
ORIENTATION
OUT
IN
CCW
CW

Figure 79-7 *The teach pendant allows the robot to be taught easily. (Courtesy Cincinnati Milacron Inc.)*

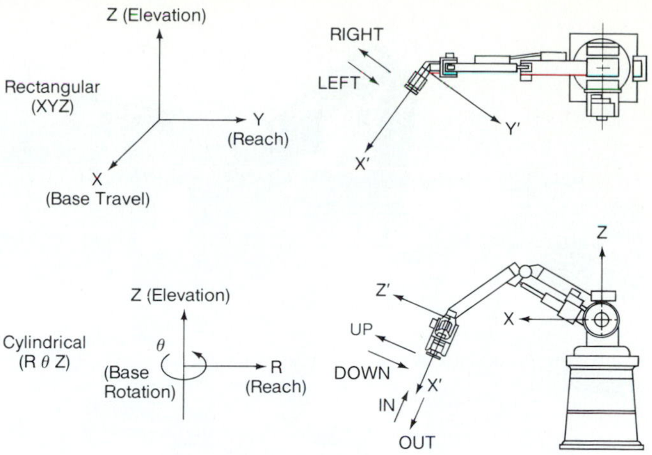
Z (Elevation)
Rectangular
(XYZ)
Y
(Reach)
X
(Base Travel)
RIGHT
LEFT
Y'
X'
Z (Elevation)
Cylindrical
(R θ Z)
θ
(Base
Rotation)
R
(Reach)
Z
Z'
X
UP
DOWN
IN
X'
OUT

Figure 79-8 *The teach coordinate systems. (Courtesy Cincinnati Milacron Inc.)*

The three coordinate system options available are cylindrical, rectangular, and hand (Fig. 79-8), and the way that coordination occurs is a function of the teach coordinate system chosen.

Movement within a given coordinate system is controlled by the positional pushbuttons on the pendant. For the rectangular system, two positional buttons will cause the tool center point (TCP) to move left or right; two other buttons cause the TCP to move up and down, and two more buttons cause the TCP to move in and out. The pushbuttons on the teach pendant make it very simple for the human operator to "teach" the robot to perform specific tasks using any coordinate system.

3. In the *automatic mode,* the robot performs its taught cycle continuously until either the "end of cycle stop" or "interrupt" buttons on the computer console are depressed. The end of cycle stop button makes the arm stop the next time it reaches the last point in the cycle. The interrupt button makes the robot arm stop immediately. An error code appears on the control console display whenever something goes wrong with the system or the human worker issues an unacceptable command.

INDUSTRIAL APPLICATIONS FOR ROBOTS

As mentioned previously, the industrial robot is finding many applications where tasks are monotonous, physically difficult, or environmentally unpleasant for the human worker. Because of the flexibility of the robot arm and the ease of teaching it to do new tasks, the robot is continually finding new industrial applications. The most common robot applications are:

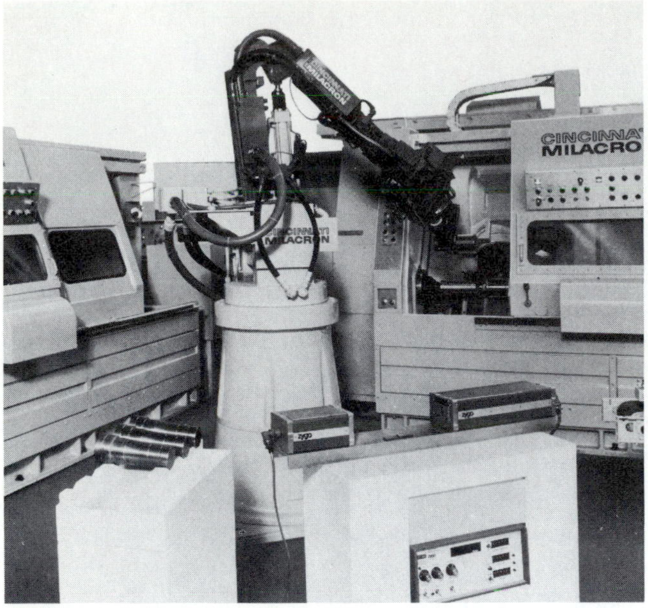

Figure 79-9 *A robot being used to load and unload machine tools. (Courtesy Cincinnati Milacron Inc.)*

1. *Loading and unloading machine tools.* Robots are finding wide use in loading and unloading machine tools, forging presses, and injection-molding machines (Fig. 79-9). It is not uncommon to find one robot loading and unloading two or more machines at the same time.
2. *Welding.* Computer-controlled industrial robots can spot-weld car bodies as they are moving on a conveyor. Unique tracking capabilities enable the robot to spot the welds without stopping the line. Seam welding is another application where robots are finding wide use (Fig. 79-10). The Cincinnati Milacron Milatrac seam-tracking device monitors the weld path "through the arc" and adjusts the robot's position to compensate for variations in the start point or seam path to produce consistently accurate weld weaves.
3. *Moving heavy parts.* Robots are being used to move heavy parts from an input station, through various machining and gaging stations, to an exit conveyor. They can also bring materials from storage areas, package and palletize, inspect, sort, and assemble (Fig. 79-11).
4. *Spray painting.* The painting of automobile bodies is an example of a job which would be environmentally unpleasant for a human worker. The robot can be "taught" (programmed) to follow the contour of an automobile body or part and apply a consistently even coat of paint to the surface.
5. *Assembly.* Assembly jobs such as stringing a wire harness around a series of pegs is an example of a task which is very monotonous for the human worker. The robot arm is fitted with a spool of wire and a feeding mechanism, and then taught to proceed around pegs and through closely spaced slots to produce a harness.
6. *Machining operations.* Some repetitive machining operations are ideally suited for the computer-controlled industrial robot. Figure 79-12 illustrates a robot being used to drill airplane parts. As another example, the robot puts vent holes in casting molds.

BENEFITS OF COMPUTER-CONTROLLED INDUSTRIAL ROBOTS

The computer-controlled industrial robot has many advantages and features which make it a very important industrial manufacturing tool.

Figure 79-10 *Robots are used extensively for welding operations. (Courtesy Cincinnati Milacron Inc.)*

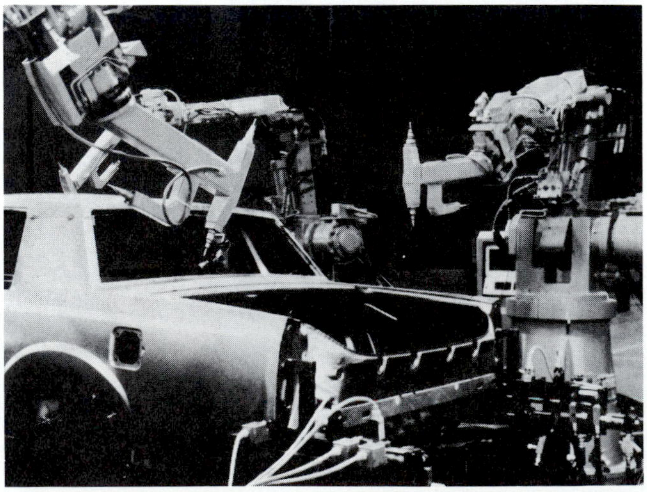

Figure 79-11 *Car bodies are checked for assembly accuracy by robots. (Courtesy Cincinnati Milacron Inc.)*

Figure 79-12 *A robot performing a drilling operation on airplane parts.*

1. The *controlled path algorithm,* made possible by the computer, controls the position, orientation, velocity, and acceleration of the path of the TCP at all times.
2. The robot can easily be taught because the robot can be moved in different coordinate systems with the teach pendant. The console display provides communication between the teacher and the control.
3. The computer can store system software and taught data in its memory. This allows easy editing of the taught program at any time, especially when there are changes in the application.
4. The memory of the computer can be easily expanded to allow for the inclusion of more taught points or extra functions.
5. The robot can be interfaced with other equipment and can receive and interpret signals from various sensor devices. It can also provide valuable data to a higher level computer on manufacturing conditions, quantities, and problems.
6. The system can easily be modified to suit changes in production. The robot computer can be linked to a supervisory computer, which can direct the robot to convert to other operations as required.
7. The computer is able to monitor the robot and the equipment it services, and diagnostic messages appear on the console screen to indicate various error conditions. If necessary, the computer may direct the robot to stop operations.

ROBOT SAFETY

Robots, like other pieces of machinery, must be treated with respect in order to prevent accidents. Although robots are taking over hazardous jobs, it is still good practice to take various safety precautions.

1. The robot working area should be enclosed by some form of barrier to prevent people from entering while the robot is working. Entrance gates, if any, should have controls which automatically stop the robot if anyone enters.
2. Emergency stop buttons, which stop all robot action and cut off the power, should be readily accessible outside the working range of the robot.
3. Robots should be programmed, serviced, and operated only by trained workers who fully understand and can anticipate the robot's action.
4. Care should be exercised during the programming cycle if it is necessary to have people within the robot's working area.
5. The work performed by a robot should not present a health or safety hazard to human workers in nearby areas.
6. Hydraulic and electrical cables should be located so that they cannot be damaged by the operation of the robot or any related equipment.

The six-axis computer-controlled industrial robot has found a variety of uses in industry. As more sophisticated robots are developed, it is quite possible that the robots of the future will play an ever-increasing role in manufacturing. Under development are auditory systems, vision systems, and sensors which will enable a robot to see, hear, and feel.

REVIEW QUESTIONS

1. Why were robots developed?
2. Define an *industrial robot.*
3. Name three criteria of an industrial robot.
4. Name the six rotary axes of movement of a robot arm.
5. List the components in a robot control unit.
6. What three modes can be used to operate robots?
7. List six applications of industrial robots.
8. Name three important safety precautions which should be observed with robots.

UNIT

80

Manufacturing Systems

A manufacturing system may be defined as a collection of two or more manufacturing units, interconnected with material-handling machinery, under the supervision of one or more dedicated, supervisory (executive) computers.

There are two types of manufacturing systems: the random (or flexible) system and the dedicated system.

The *random system* consists of standard numerical control (NC) machine tools linked together with appropriate handling devices and total system control. This system is designed for low-volume, random-order machining of parts of any size or shape. This type of system is quite flexible and will permit changes in production of the parts simply by reprogramming.

The *dedicated system* consists of specialized machines and/or some standard NC machines linked together with less-flexible handling devices and complete system control which may be a computer or a programmable controller. Because the specialized machines are capable of performing only one certain operation and cannot be changed and/or because the handling devices require operations to be done in a fixed sequence, the machine (and the system) is said to be dedicated. Since some of the machines are specialized and do not provide for frequent tool changes, and because the handling system lacks adaptability, the dedicated system requires medium-volume production runs.

OBJECTIVES

After completing this unit you will be able to:

1. State the purpose and function of a flexible manufacturing system
2. State the purpose and function of a dedicated manufacturing system

FLEXIBLE MANUFACTURING SYSTEMS

The fully flexible manufacturing system is in effect a *random system* because it consists of standard NC machine modules linked together with a flexible parts-handling device and total-system computer control.

The flexible manufacturing system shown in Fig. 80-1 consists of two turning centers served by a centrally located robot and a suitable conveyor. Two or more similar NC machines that are grouped to handle workpieces of similar size and shape and are controlled by a central computer are usually called a *cell.* When a robot is used, its computer serves as a system supervisor and is linked to each of the machine computer control units.

OPERATION

The rough part is removed from the conveyor by the programmed robot and mounted in the lathe. Proper positioning is assured by the *probe,* which checks surfaces and will later measure outside and inside diameters in process and

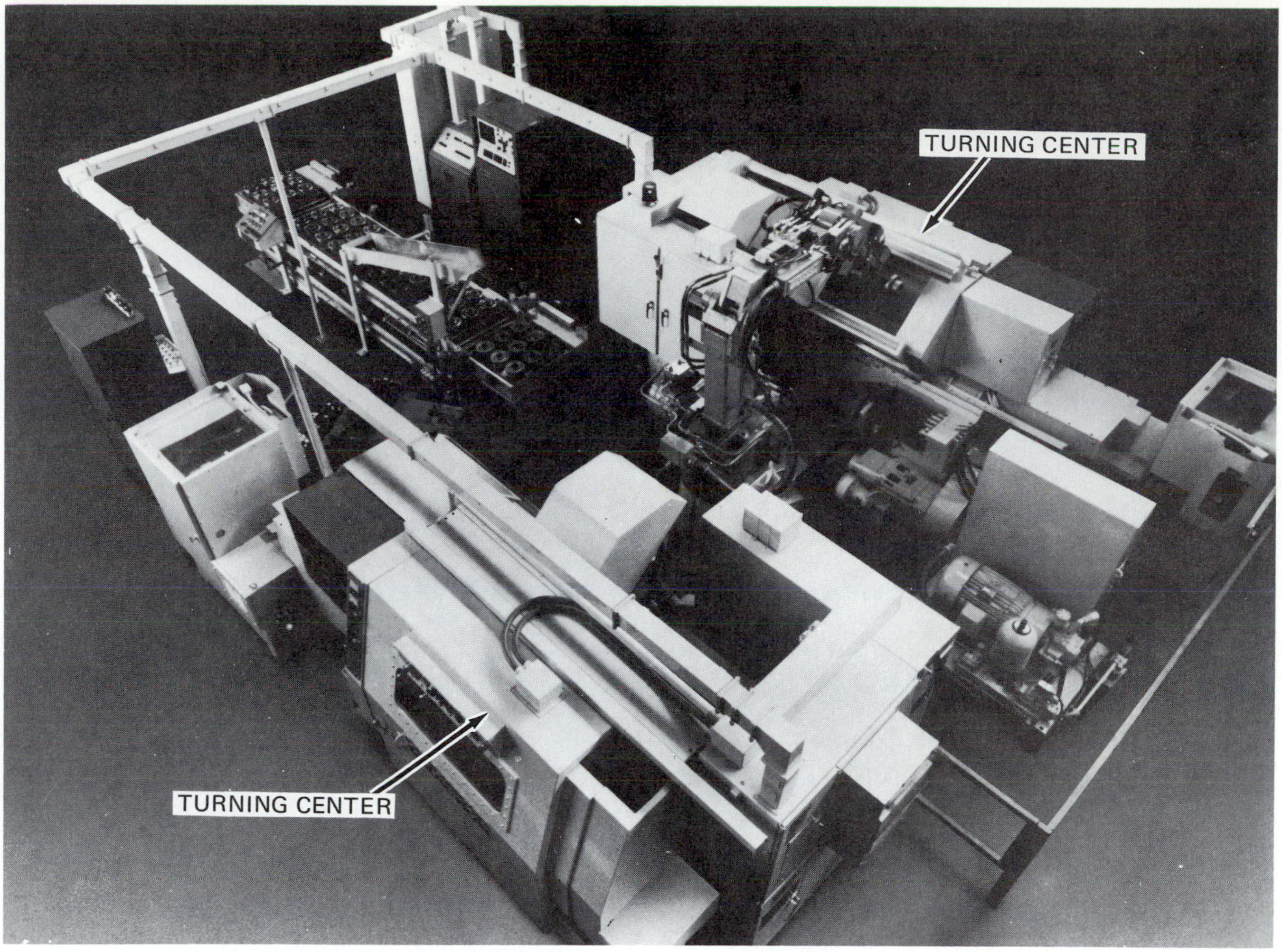
TURNING CENTER
TURNING CENTER

Figure 80-1 *A flexible manufacturing system with two turning centers, a robot, and a conveyor system. (Courtesy Cincinnati Milacron Inc.)*

instruct the computer to make appropriate tool adjustments.

As the machining process continues, the cutting tools will become dull. A *tool cycle time monitor* incorporated in the computer keeps track of tool usage and will signal for a backup tool when the expected life of the tool expires. After the particular machining cycle has been completed, the dull cutting tool is automatically replaced with a sharp one from the tool chain matrix. The old tool is automatically replaced in the tool magazine rack.

To ensure continuous machining with the fewest possible problems, a *machining monitor* constantly measures the torque at the machine spindle. This feature will detect dull or broken tools and other malfunctions in the lathe. Should a problem occur, the monitor will retract the cutting tool and shut off the machine.

After the part has been finished, it will be removed from the lathe by the robot and placed in the storage facility or on an exit conveyor. As an intermediate step, after the part has been machined it may be presented by the robot to be checked by a *post-process gaging control.* Here the final part size is checked by laser for accuracy of size. Should any out-of-tolerance dimensions be detected, the gaging control will send feedback signals through the computer to the machine which automatically adjusts the tool to the correct position.

Flexible machining systems are not limited to turning centers. Figure 80-2 shows a flexible manufacturing system for automating the production of rectangular or box-type parts. Here two traveling-column machining centers automatically machine the parts, which are delivered to the stations by computer-controlled carts. Replacement tooling is also delivered to the machines when required by these carts. Flexible manufacturing systems are usually used on the second and third shifts where it is difficult to obtain skilled help.

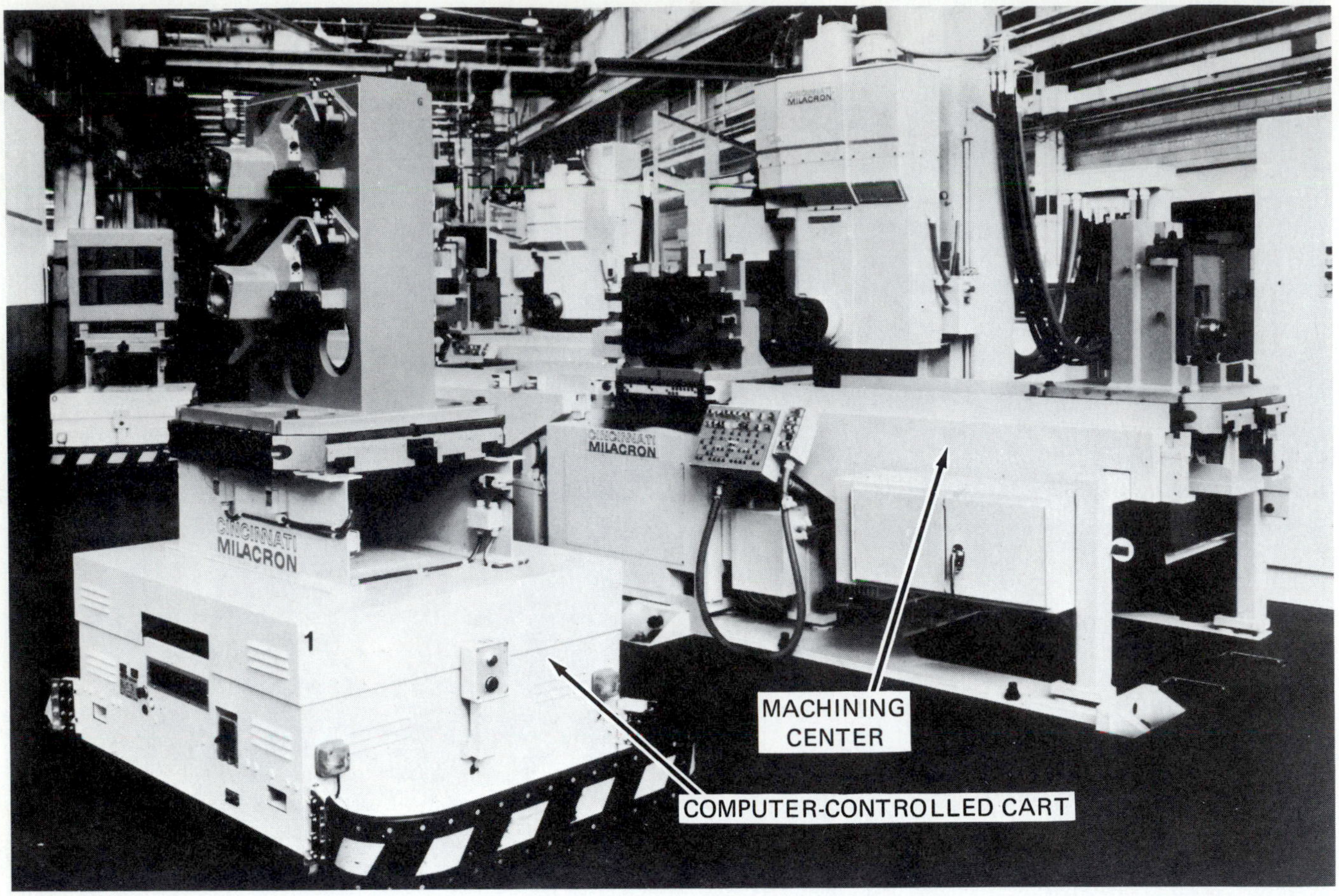
CINCINNATI
MILACRON
1
CINCINNATI
MILACRON
MACHINING
CENTER
COMPUTER-CONTROLLED CART

Figure 80-2 *A flexible manufacturing system automated for the production of boxlike parts. (Courtesy Cincinnati Milacron Inc.)*

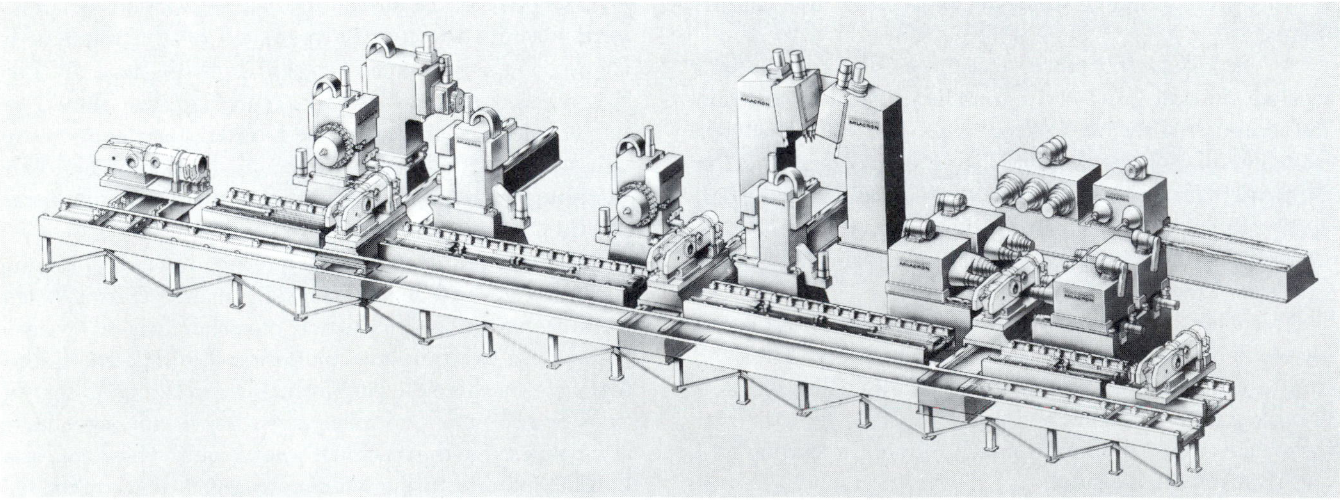

Figure 80-3 *Dedicated manufacturing systems perform only specialized operations. (Courtesy Cincinnati Milacron Inc.)*

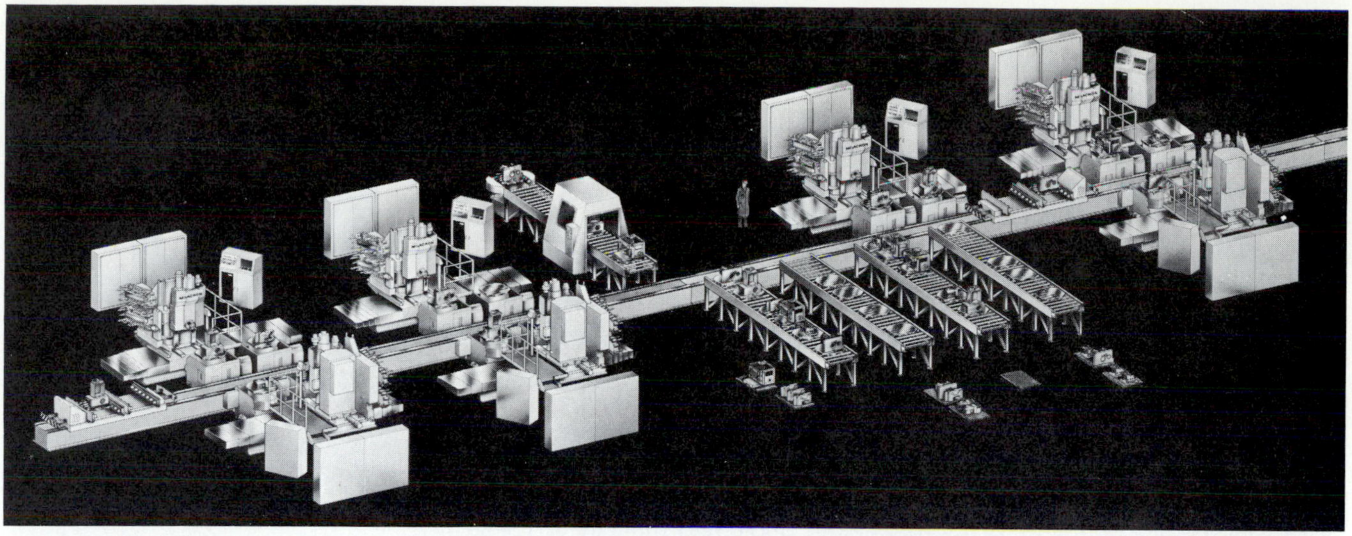

Figure 80-4 *A dedicated manufacturing system that combines flexible and standard machines. (Courtesy Cincinnati Milacron Inc.)*

DEDICATED MANUFACTURING SYSTEMS

A dedicated system often includes machines which are designed to perform only specialized operations on a part and additional standard NC machine tools all linked together with appropriate parts-handling equipment. The total system is controlled by a computer. Dedicated systems are designed for "medium-volume" production of a part or a restricted family of parts (restricted by shape and size).

The system illustrated in Fig. 80-3 is composed of three machines (stations), which are capable of performing only specialized operations (dedicated). The workpieces, each weighing 2000 lb, are mounted on pallets and are moved automatically from one station to the next for the various operations required.

Another type of dedicated system is shown in Fig. 80-4. Here standard machining centers are combined with special (dedicated) machines for washing and inspecting the parts. The workpieces are shuttled from station to station on pallets. This system is fully automated and produces and inspects several similar and dissimilar parts. As a part is moved into position for machining, the probe and the computer will select the proper machining program for the part. The washing and gaging stations in this system are dedicated.

Machines which are grouped to make up dedicated systems will have features and controls similar to those outlined for flexible systems. However, the work moves from station to station in a fixed sequence. Pallets tend to be less universal than handling devices in flexible or random-order systems, and system software is less sophisticated.

REVIEW QUESTIONS

1. Define a *manufacturing system.*
2. What is the difference between a random and a dedicated system?

FLEXIBLE MANUFACTURING SYSTEMS

3. Describe the composition of a flexible manufacturing system.
4. Explain the function of the following components:
   a. Robot
   b. Probe
   c. Tool cycle monitor
   d. Machining monitor

DEDICATED MANUFACTURING SYSTEMS

5. How does the dedicated system differ from the flexible manufacturing system?
6. Explain what occurs as a part is moved into position for machining.

UNIT 81

Factories of the Future

One of the large problems in manufacturing is that of "batch manufacturing," whereby small quantities of hundreds of different parts are produced. At least three-quarters of all industrial production falls into this category. Large quantities of one item can be produced most efficiently on a transfer (production) line, or by a series of dedicated machines. These processes, however, are not suitable to the manufacture of small numbers of one part (batch). This problem, along with the small percentage of time that machines are being used (often for only one shift), and the lack of skilled personnel for the second and third shifts, has led manufacturers to look to more efficient production methods.

OBJECTIVES

After completing this unit you will:

1. Have an overview of the factory of tomorrow
2. Understand the impact of the computer on our society in general and on manufacturing in particular

VISION OF THE FUTURE

Industry has taken steps in the past few years to overcome production problems. Computers, the key to the new technology, have been applied to the automation of equipment to increase machine use, reduce the complexity of batch production, and improve productivity.

In order to achieve optimum efficiency in manufacturing, machine tools should be run 8760 h per year. Planners are now looking for ways to achieve this goal and some of the means to implement this follow. The factory of the future will be operated at full production with very few people present to produce a great variety of different parts in small batches (Fig. 81-1). Many objects now made of metal will probably be made of synthetic materials and by different processes.

Machines will be grouped together in small clusters or *cells* with *automatic handling* of all materials in and between different cells. Machines will be *computer-controlled,* which will greatly improve the accuracy and consistency of production. The computer's memory permits the repetition of a series of complicated operations with the push of a button. These programs can be stored in the computer for future use. *Automatic tool changing* (Fig. 81-2) will be common on all machines. With these improvements, it will be possible to remove the operator and employ the person as a monitor or supervisor to control all the production in the cell. The operator will be required to step into the cell only in case of a breakdown. With the addition of sensing devices, which will warn of impending machine failures, the cell will be able to perform batch operations continuously without human intervention.

In order to satisfy production needs, several cells will be set up, each connected with the others. These in turn will be connected with areas of the plant such as material storage, transport, and data communications with the executive and mainline or central computers. These several cells will be controlled by one person in the executive computer room.

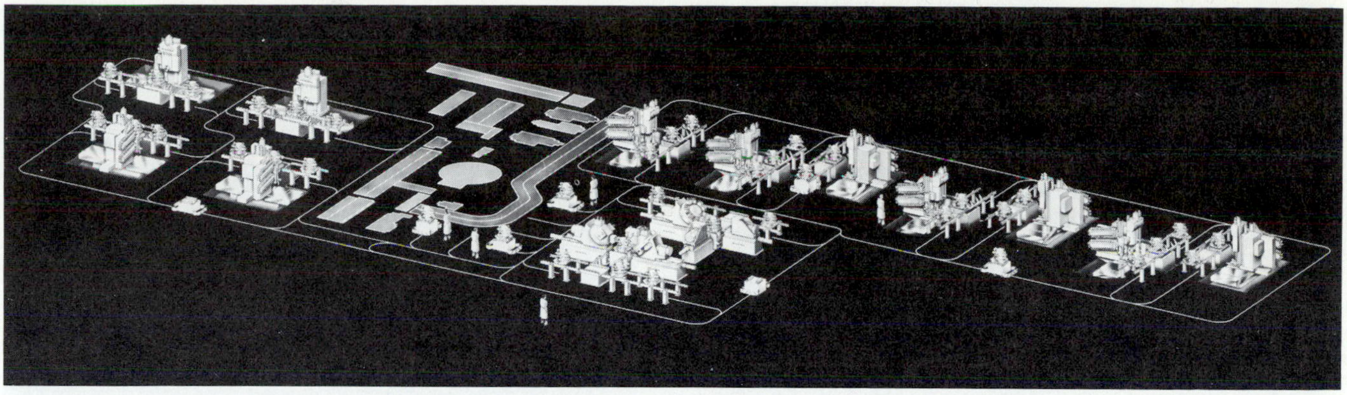

Figure 81-1 *A designer's view of the factory of the future. (Courtesy Cincinnati Milacron Inc.)*

COMPLETE MANUFACTURING SYSTEMS

In order to make the system complete, all functions in the manufacturing process, that is, *design, planning, inventory control, scheduling,* and *shop floor control,* must be able to communicate with each other automatically. In this way they will all share the common data base and all other current information. Each of the cell systems can take any independent action required, based on this shared information. The mainline or central computer reports to the staff and calls attention to problems which may require attention.

An overview of the factory of the future is shown in Fig. 81-3. There are three general phases of manufacturing: design, planning, and production.

DESIGN

The *design process* will be improved by computer-aided design (CAD), whose main purposes are drafting and design. The design engineer may make or modify design drawings by using the CRT and the light pen (Fig. 81-4). Color graphics will be added to make three-dimensional shapes easier to comprehend and to reduce the chance of design error. Analysis pertaining to the function of the final product will be carried out by CAD.

PLANNING

All operations relating to production planning will be computerized. *Capacity Resource Planning* (CRP) will determine equipment and labor requirements. By the use of animation, parts can be moved from one cell to another (on the CRT) to ascertain any problems and determine machine utilization.

Figure 81-2 *Automatic tool changing will play an increasingly important role in manufacturing systems. (Courtesy Cincinnati Milacron Inc.)*

DESIGN

CAD

PLANNING FOR MANUFACTURING

CAPACITY PLANNING | MRP | CAPP | PART PROGRAMMING | TOOL DESIGN

MANUFACTURING

SHOP FLOOR CONTROL | PROCESS AUTOMATION

Figure 81-3 *Top-down diagram of the computer-integrated factory. (Courtesy Cincinnati Milacron Inc.)*

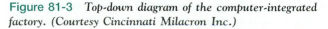

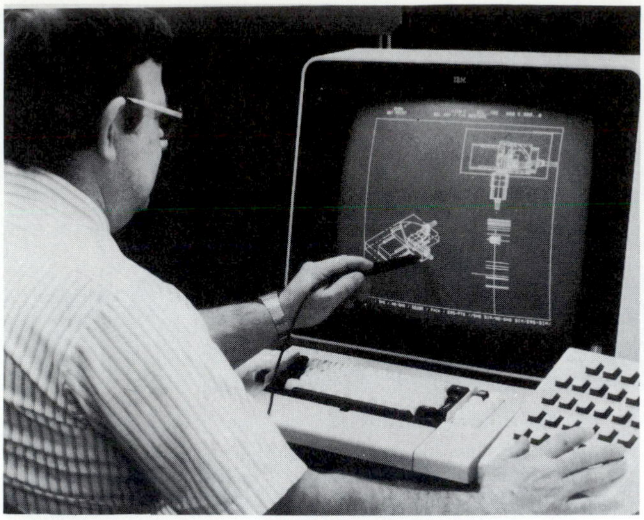

Figure 81-4 *A light pen being used to change the design of a part in a CAD system. (Courtesy Cincinnati Milacron Inc.)*

PROBE

Figure 81-5 *In the future the probe will be widely used to locate, realign, and inspect workpieces. (Courtesy Cincinnati Milacron Inc.)*

PRODUCTION

In order to further increase productivity, *group technology* (GT), whereby parts are classified and coded by similarities in design and processing, will be employed. Computer-aided process planning will be employed. Here, with input from CAD and from the database along with the use of GT, the computer will be able to determine the sequence of operations, as well as time standards and cost estimates.

Computer-aided parts programming will take input from other sources and produce instructions for individual machine programming. These programs will be stored in general memory and be available to shop floor computers whenever required. During the final planning process, cutting-tool movements can be simulated by solid modeling on the computer. The movements are recorded in the computer which calculates the numerical description of the path.

TOOL DESIGN

Special tooling and mold design will be improved by the use of CAD, which can compare the proposed design against the design of the part to see if it is functional. On the shop floor, all operations, such as machine and tool scheduling, management, and monitoring, will be totally computer controlled.

Sensing devices linked to the computer are essential for machine operation without operators. The *probe* (Fig. 81-5) will be widely used to locate, realign, and inspect work. Advanced sensors, which will give each tool the sense of touch for positioning work, will no doubt be developed in the near future. Damage to the machine and tools will be controlled by means of *torque-control* devices mounted at the motor drive. These will warn the computer of excessive cutting torque and tool wear. Should other machine problems develop, a *diagnostic system* in the machine computer will warn the executive and central computers of these and they will take appropriate action.

Another area controlled by the computer will be *material-handling* methods. Computer-controlled *programmable robots* (Fig. 81-6) will play a large part in the factory of the future. Robots will be used to perform dangerous and tedious tasks now performed by people. They may also be used for loading and unloading machines and performing other tasks that fixed automation cannot do.

Material-handling devices, such as wire-guided carts (Fig. 81-7), carousels, and conveyors, will be computerized and integrated into the whole system to supply materials

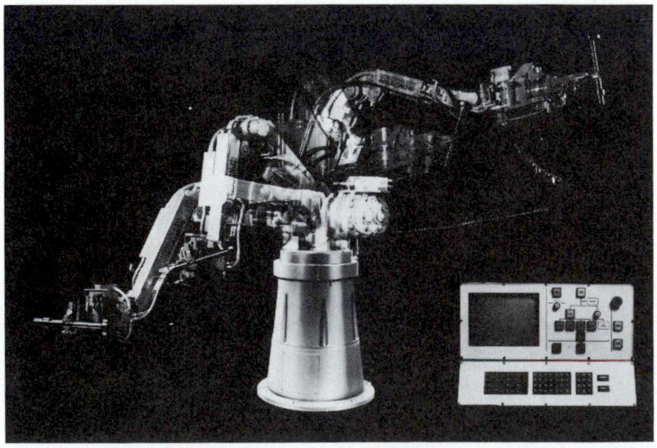

Figure 81-6 *Ever-increasing uses of robots will be found in the factory of the future. (Courtesy Cincinnati Milacron Inc.)*

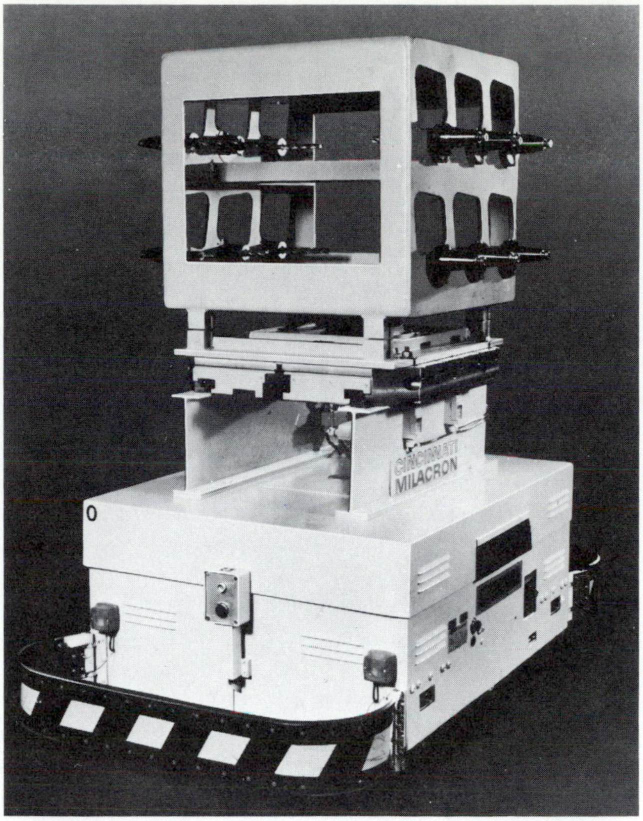
MILACRON

Figure 81-7 *Computer-controlled, wire-guided carts will transport tools and materials in the factory of the future. (Courtesy Cincinnati Milacron Inc.)*

and tools to the machines and remove the finished products.

Inspection and quality control will also be computerized. Constant feedback to the rest of the system will be maintained and any trend which may be developing in part sizes and other characteristics will be detected early and appropriate computer adjustments will be made to the machine which produced the part.

CONCLUSION

Each of these elements of factory automation is in use in some factories today, but not on a total basis in any one factory. When total integration is obtained, factories will be able to operate 8760 h per year virtually unattended.

REVIEW QUESTIONS

1. Why is the computer so important to the factory of the future?
2. How will the factory of the future be operated?
3. Define *cell*.
4. What functions must be able to communicate with each other so that a manufacturing system will be complete?
5. State the purpose of:
   a. Capacity resource planning
   b. Computer-aided parts programming
   c. The probe
   d. The diagnostic system
   e. Wire-guided carts

SECTION 15

Grinding

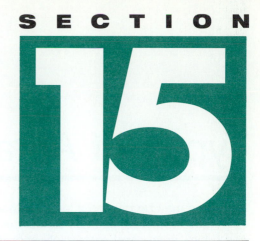

Early humans used abrasive stones to sharpen tools and produce a smooth surface. Over the years, the manufacture of abrasives and grinding tools has developed gradually to produce stronger and better abrasives. The use of abrasives in modern industry has contributed much to our mass production methods. Because of abrasives and modern grinding machines, it has been possible to produce parts and products to the close tolerances and high surface finish required by industry and the machine tool trade.

In order to function properly, an abrasive must have certain characteristics:

1. It must be harder than the material being ground.
2. It must be strong enough to withstand grinding pressures.
3. It must be heat-resistant so that it does not become dull at grinding temperatures.
4. It must be friable (capable of fracturing) so that when the cutting edges become dull, they will break off and present new sharp surfaces to the material being ground.

UNIT

82

Types of Abrasives

Abrasives may be divided into two classes: natural and artificial.

Natural abrasives, such as sandstone, garnet, flint, emery, quartz, and corundum, were used extensively prior to the early part of the 20th century. However, they have been almost totally replaced by manufactured abrasives with their inherent advantages. One of the best natural abrasives is diamond but, because of the high cost of industrial diamonds (bort), its use in the past was limited mainly to grinding cemented carbides and glass and sawing concrete, marble, limestone, and granite. However, due to the introduction of synthetic or manufactured diamonds, industrial natural diamonds will become cheaper in cost and will be used on many more grinding applications.

Manufactured abrasives are used extensively because their grain size, shape, and purity can be closely controlled. This uniformity of grain size and shape, which ensures that each grain does its share of work, is not possible with natural abrasives. There are several types of manufactured abrasives: aluminum oxide, silicon, carbide, boron carbide, cubic boron nitride, and manufactured diamond.

OBJECTIVES

After completing this unit you will be able to:

1. Describe the manufacture of aluminum oxide and silicon carbide abrasives
2. Select the proper grinding wheel for each type of work material
3. Discuss the applications of grinding wheels and abrasive products

ALUMINUM OXIDE

Aluminum oxide is probably the most important abrasive, since about 75 percent of the grinding wheels manufactured are made of this material. It is generally used for high-tensile-strength materials, including all ferrous metals except cast iron.

Aluminum oxide is manufactured with various degrees of purity for different applications, the hardness and brittleness increases as the purity increases. Regular aluminum oxide (Al_2O_3) is about 94.5 percent pure and is a tough abrasive capable of withstanding abuse. It is a grayish color and is used for grinding strong, tough materials, such as steel, malleable and wrought iron, and tough bronzes.

Aluminum oxide of about 97.5 percent purity is more brittle and not as tough as the regular aluminum oxide. This gray abrasive is used in the manufacture of grinding wheels for centerless, cylindrical, and internal grinding of steel and cast iron.

The purest form of aluminum oxide for grinding wheels is a white material which produces a sharp cutting edge when fractured. It is used for grinding the hardest steels and stellite and for die and gage grinding.

MANUFACTURE OF ALUMINUM OXIDE

Bauxite ore, from which aluminum oxide is made, is mined by the open-pit method in Arkansas and in Guyana, Suriname, and French Guiana (South America). The bauxite ore is usually calcined (reduced to powder form) in a large furnace, where most of the water is removed. The calcined bauxite is then loaded into a cylindrical, unlined steel shell about 5 ft (1.5 m) in diameter by 5 ft (1.5 m) deep (Fig. 82-1). This furnace is open at the top and lined with carbon bricks at the bottom. Two or three electrodes of carbon or graphite project into the open top of the furnace. During the operation, the outside of the furnace is cooled by a circumferential spray of water. The furnace is half filled with a mixture of bauxite, coke screenings, and iron borings in the proper proportions. The coke is used to reduce the impurities in the ore to their metals, which combine with the iron and sink to the bottom of the furnace. The electrodes are then lowered onto the top of the charge, a starting batch of coke screenings is placed between the electrodes, and the current is applied. The coke is rapidly heated to incandescence and the fusion of the bauxite starts. After a small amount of bauxite is fused it becomes the conductor and carries the current. After the fusion of the bauxite has begun, more bauxite and coke screenings are added, and the process continues until the shell is full. During the entire operation, the height of the electrodes is automatically adjusted to maintain a constant rate of power input. When the furnace is full, the power is shut off, and the furnace is allowed to cool for about 12 h. The fused ingot is removed from the shell and allowed to cool for about a week. The ingot is broken up and fed into crushers; the material then is washed, screened, and graded to size.

Figure 82-1 *Aluminum oxide is produced in an arc-type furnace. (Courtesy The Carborundum Company.)*

SILICON CARBIDE

Silicon carbide is suited for grinding materials which have a low tensile strength (aluminum, brass, and bronze) and high density, such as cemented carbides, stone, and ceramics. It is also used for cast iron and most nonferrous and nonmetal materials. It is harder and tougher than aluminum oxide. Silicon carbide may vary in color from green to black. Green silicon carbide is used mainly for grinding cemented carbides and other hard materials. Black silicon carbide is used for grinding cast iron and soft nonferrous metals, such as aluminum, brass, and copper. It is also suitable for grinding ceramics.

MANUFACTURE OF SILICON CARBIDE

A mixture of silica sand and high-purity coke is heated in an electric resistance furnace (Fig. 82-2). Sawdust is added to produce porosity in the finished product and to permit the escape of the large volume of gas formed during the operation. Sodium chloride (salt) is added to assist in removing some of the impurities.

The furnace is a brick-lined rectangle about 50 ft (15 m) long, 8 ft (2.50 m) wide, and 8 ft (2.50 m) high. It is open at the top. One or more electrodes protrude from each end of the furnace. The mixture of sand, coke, and sawdust is loaded into the furnace to the height of the electrodes. A granular core of coke is placed around the electrodes for the full length of the furnace. The core is covered with more sand, coke, and sawdust mixture and heaped to the top of

Figure 82-2 *Silicon carbide is produced in a resistance-type furnace. (Courtesy The Carborundum Company.)*

the furnace. The current is then applied to the furnace and the voltage closely regulated to maintain the desired rate of power input. The time required for this operation is about 36 h.

After the furnace has cooled for about 12 h, the brick sidewalls are removed and the unfused mixture falls to the floor; the silicon carbide ingot can then cool more rapidly. After cooling for several days, the ingot is broken up and the silicon carbide removed. Care must be taken to remove the pure silicon carbide since the outer layer has not been properly fused and is not usable. The inner core surrounding the electrodes is also unusable since it is only graphitized coke.

The resultant silicon carbide is then crushed, treated with acid and alkalis to remove any remaining impurities, screened, and graded to size.

ZIRCONIA-ALUMINUM OXIDE

One of the more recent additions to the abrasive family is *zirconia-aluminum oxide.* This material, containing about 40 percent zirconia, is made by fusing zirconium oxide and aluminum oxide at extremely high temperatures [3450°F (1900°C)]. This is the first *alloy abrasive* ever produced. It is used for heavy-duty rough and finish grinding in steel mills, snagging in foundries, and in metal fabrication plants for the rapid rough and finish grinding of welds. The performance of zirconia-aluminum abrasive is superior to that of standard aluminum oxide for rough grinding and snagging operations because the actions of the two grain types are quite different.

During any grinding process, much frictional heat is generated, which breaks down the organic bond (rubber, shellac, or phenalic) on the wheel or the coated disk and lets the wheel "self-dress." Standard aluminum oxide abrasive wheels will, at first, penetrate the surface of the workpiece and remove the metal effectively. Soon the abrasive grains will become dull, the wheel will lose its penetrating power and ride on the surface of the work. Frictional heat builds up, which softens the organic bond, and the entire grain is expelled after about only 25 to 30 percent of its life. With the newer zirconia-alumina abrasive, the grain action is quite different. The cutting action starts in the same manner, but when the abrasive grain starts to dull, microfracturing takes place just below the abrasive grain surface. As the small particles break off, sharp new cutting points are generated while the grain remains secured in the bond. As the small broken-off particles of abrasive are ejected, they take with them much of the heat generated by the grinding action. Due to the cooler cutting action of this abrasive, sharp new cutting edges are produced many times before the grain is finally expelled from the bond and only after about 75 to 80 percent of the abrasive grain has been used. Wheels and disks made with zirconia-alumina base will last from two to five times longer than standard aluminum oxide wheels, depending on the application.

Zirconia-alumina offers several advantages over standard abrasives for heavy-duty grinding:

- Higher grain strength
- Higher impact strength
- Longer grain life
- Maintains its shape and cutting ability under high pressure and temperature
- Higher production per wheel or disk
- Less operator time spent changing wheels or disks

BORON CARBIDE

Another of the new abrasives is *boron carbide.* It is harder than silicon carbide and, next to diamond, it is the hardest material manufactured. Boron carbide is not suitable for use in grinding wheels and is used only as a loose abrasive and a relatively cheap substitute for diamond dust. Because of its extreme hardness, it is used in the manufacture of precision gages and sand blast nozzles. Crush dressing rolls made of boron carbide have proved superior to tungsten carbide rolls for the dressing of grinding wheels on multiform grinders. Boron carbide is also widely accepted as an abrasive used in ultrasonic machining applications.

MANUFACTURE OF BORON CARBIDE

Boron carbide is produced by dehydrated boric acid being mixed with high-quality coke. The mixture is heated in a horizontal steel cylinder, which is completely enclosed except for a hole in each end to accommodate a graphite electrode and vent holes to prevent the escape of the gases formed. The outside of the furnace is sprayed with water to prevent the shell from melting. During the heating process, air must be excluded from the furnace. This is done by dampening the mixture with kerosene, which will volatize and expel the air from the furnace as it is heated. A high current at low voltage is applied for about 24 h, after which the furnace is cooled. The resulting product, boron carbide, is a hard, black, lustrous material.

CUBIC BORON NITRIDE

One of the more recent developments in the abrasive field has been the introduction of cubic boron nitride. This synthetic abrasive has hardness properties between silicon carbide and diamond. The crystal known as Borazon CBN (cubic boron nitride) was developed by the General Electric

Company in 1969. This material is capable of grinding high-speed steel with ease and accuracy and is superior to diamond in many applications.

Cubic boron nitride is about twice as hard as aluminum oxide and is capable of withstanding grinding temperatures of up to 2500°F (1371°C) before breaking down. CBN is cool-cutting and chemically resistant to all inorganic salts and organic compounds. Because of the extreme hardness of this material, grinding wheels made of CBN are capable of maintaining very close tolerances. These wheels require very little dressing and are capable of removing a constant amount of material across the face of a large work surface without having to compensate for wheel wear. Because of the cool-cutting action of CBN wheels, there is little or no surface damage to the work surface.

Manufacture

Cubic boron nitride is synthesized in crystal form from hexagonal boron nitride ("white graphite") with the aid of a catalyst, heat, and pressure (Fig. 82-3A). The combination of extreme heat [2725°F (1496°C)] and tremendous pressure (47 540 000 psi) on the hexagonal boron nitride and the catalyst produces a strong, hard, blocky, crystalline structure with sharp corners known as CBN (Fig. 82-3B).

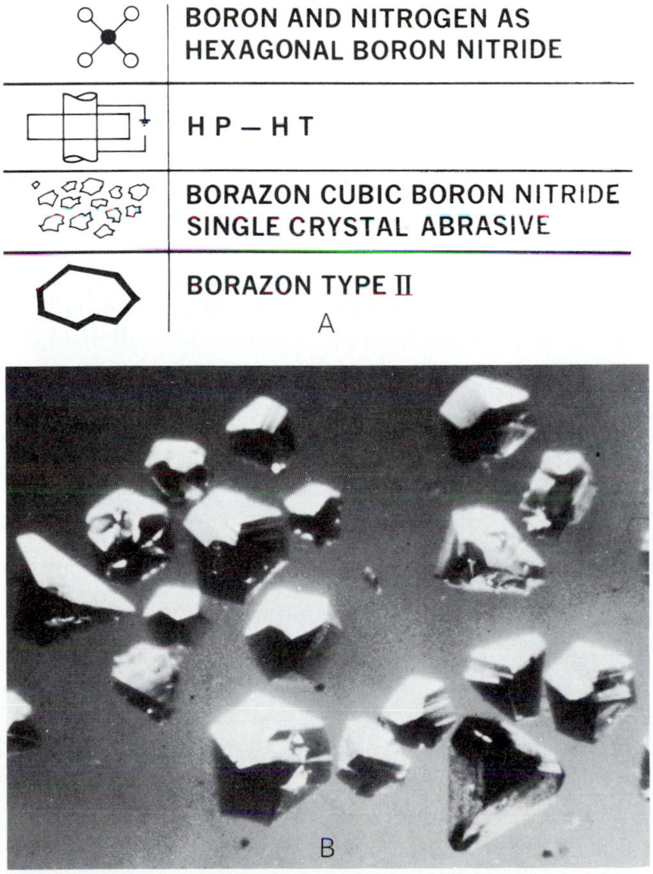
BORON AND NITROGEN AS HEXAGONAL BORON NITRIDE
H P — H T
BORAZON CUBIC BORON NITRIDE SINGLE CRYSTAL ABRASIVE
BORAZON TYPE II
A
B

Figure 82-3 *(A) Making Borazon (cubic boron nitride); (B) crystals of cubic boron nitride. (Courtesy General Electric Company.)*

There are two types of CBN:

- *Borazon CBN* is an uncoated abrasive which can be used on plated mandrels and in metal-bonded grinding wheels. This type of wheel is used for general-purpose grinding and for internal grinding on hardened steel.
- *Borazon Type II CBN* is nickel-plated grains of cubic boron nitride used in resin bonds for general-purpose dry and wet grinding of hardened steel. Uses of these wheels range from resurfacing blanking dies to the sharpening of high-speed steel end mills.

MANUFACTURED DIAMONDS

Diamond, the hardest substance known, was primarily used in machine shop work for truing and dressing grinding wheels. Because of the high cost of natural diamonds, industry began to look for cheaper, more reliable sources. In 1954, the General Electric Company, after four years of research, produced Man-Made® diamonds in their laboratory. In 1957, the General Electric Company, after more researching and testing, began the commercial production of these diamonds.

Many forms of carbon were used in experiments to manufacture diamonds. After much experimentation with various materials, the first success came when carbon and iron sulfide in a granite tube closed with tantalum disks were subjected to a pressure of 66 536 750 psi and temperatures between 2550 and 4260°F (1400 and 2350°C). Various diamond configurations are produced by using other metal catalysts such as chromium, manganese, tantalum, cobalt, nickel, or platinum in place of iron. The temperatures used must be high enough to melt the metal saturated with carbon and start the diamond growth.

DIAMOND TYPES

Because the temperature, pressure, and catalyst-solvent can be varied, it is possible to produce diamonds of various sizes, shapes, and crystal structure best suited to a particular need.

Type RVG Diamond

The RVG type of manufactured diamond is an elongated, friable crystal with rough edges (Fig. 82-4A). The letters RVG indicate that this type may be used with a resinoid or vitrified bond and is used for grinding ultrahard materials such as tungsten carbide, silicon carbide, and space-age alloys. The RVG diamond may be used for wet and dry grinding.

Type MBG-II Diamond

A tough and blocky-shaped crystal, the MBG-II type is not as friable as the RVG type and is used in metal-bonded grinding (MBG) wheels (Fig. 82-4B). It is used for grinding cemented carbides, sapphires, and ceramics as well as in electrolytic grinding.

Type MBS Diamond

The MBS type is a blocky, extremely tough crystal with a smooth, regular surface, which is not very friable (Fig. 82-4C). It is used in metal-bonded saws (MBS) to cut concrete, marble, tile, granite, stone, and masonry materials.

Diamonds may be coated with nickel or copper to provide a better holding surface in the bond and to prolong the life of the wheel.

CERAMIC ALUMINUM OXIDE

In 1988, a new abrasive product, ceramic aluminum oxide, known as SG abrasive, was introduced by the Norton Company. This new material greatly outperforms conventional aluminum oxide wheels in the grinding of tough alloys and other hard ferrous and nonferrous metals.

Aluminum oxide grains are made from fused aluminum oxide, which is then crushed to the desired particle size. This produces a grain having very few crystal particles. With this type of grain as much as one-fifth of the grinding surface of the grain can be lost when a crystal particle breaks away after it becomes dull.

On the other hand, Norton SG abrasive is made by a nonfused process. Thousands of submicron-sized particles are sintered to provide a single abrasive grain of more uniform shape and size, having vastly more cutting edges which remain sharp as they fracture. This self-sharpening feature combats friction heat caused by the dulling of the standard aluminum oxide grains.

Norton SG abrasive is harder than aluminum oxide and zirconia alumina but not as hard or as long lasting as cubic boron nitride or other superabrasive products, which often require special machines for their use. Wheels made of SG abrasive are well suited to CNC grinding because of their cool-grinding feature and their resistance to loading and wear. This combination results in fewer wheel changes, less wheel dressing, higher productivity, and therefore lower labor costs.

When selecting the proper SG wheel for the job, remember that a higher surface finish is created by a dull grain. Since SG grains remain sharp far longer, a finer grit size than that of the aluminum oxide wheels should be selected.

Advantages of SG Abrasives Over Conventional Abrasives

- They outlast conventional wheels from five to ten times.
- Metal removal rates are doubled.
- Heat damage to the surface of very thin workpieces is reduced.
- Grinding cycle time is reduced.
- Dressing time is reduced as much as 80 percent.

ABRASIVE PRODUCTS

After the abrasive has been produced, it is formed into products such as grinding wheels, coated abrasives, polishing and lapping powders, and abrasive sticks, all of which are used extensively in machine shops.

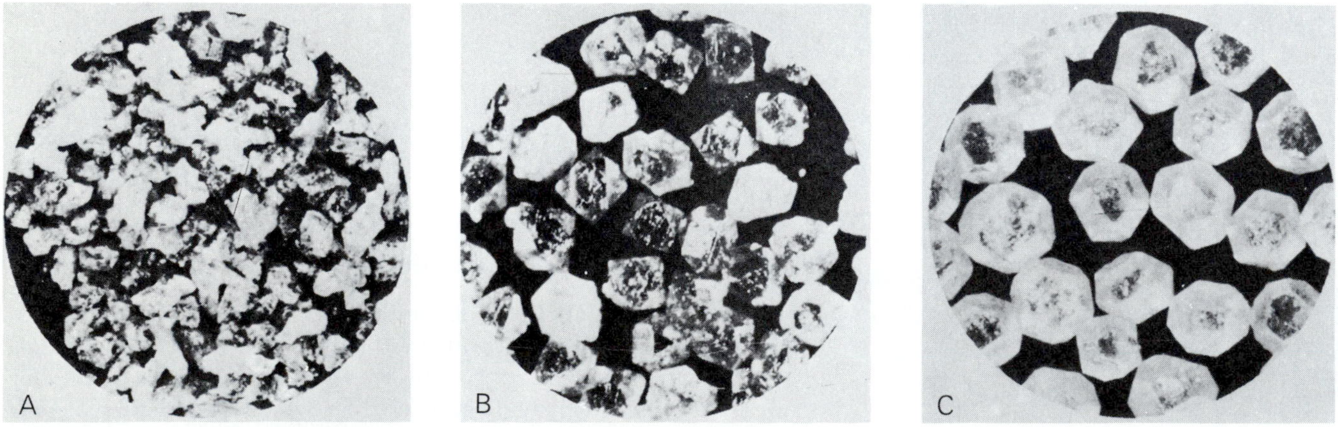
A
B
C

Figure 82-4 *(A) Type RVG diamond is used to grind ultrahard materials; (B) type MBG-II is a tough diamond crystal used in metal-bonded grinding wheels; (C) type MBS is a very tough, large crystal used in metal-bonded saws.*

GRINDING WHEELS

Grinding wheels, the most important products made from abrasives, are composed of abrasive material held together with a suitable bond. The basic functions of grinding wheels in a machine shop are:

1. Generation of cylindrical, flat, and curved surfaces
2. Removal of stock
3. Production of highly finished surfaces
4. Cutting-off operations
5. Production of sharp edges and points

For grinding wheels to function properly, they must be hard and tough, and the wheel surface must be capable of gradually breaking down to expose new sharp cutting edges to the material being ground.

The material components of a grinding wheel are the *abrasive grain* and the *bond;* however, there are other physical characteristics, such as *grade* and *structure,* that must be considered in grinding wheel manufacture and selection.

ABRASIVE GRAIN

The abrasive used in most grinding wheels is either *aluminum oxide* or *silicon carbide.* The function of the abrasive is to remove material from the surface of the work being ground. Each abrasive grain on the working surface of a grinding wheel acts as a separate cutting tool and removes a small metal chip as it passes over the surface of the work. As the grain becomes dull, it fractures and presents a new, sharp cutting edge to the material. The fracturing action reduces the heat of friction which would be caused if the grain became dull, producing a relatively cool cutting action. As a result of hundreds of thousands of individual grains all working on the surface of a grinding wheel, a smooth surface can be produced on the workpiece.

One important factor to consider in grinding wheel manufacture and selection is the *grain size.* After the abrasive ingot, or pig, is removed from the electric furnace, it is crushed, and the abrasive grains are cleaned and sized by passing them through screens, which contain a certain number of meshes, or openings, per inch. A #8 grain size would pass through a screen having eight meshes per linear inch and would be approximately 1/8 in. across. The sizing of the abrasive grain is rather an important operation since undersize grains in a wheel will fail to do their share of the work, while oversize grains will scratch the surface of the work.

Commercial grain sizes are classified as follows:

Very Coarse	Coarse	Medium
6	14	30
8	16	36
10	20	46
12	24	54
		60

Fine	Very Fine	Flour Size
70	150	280
80	180	320
90	220	400
100	240	500
120		600

Relative grain sizes are shown in Fig. 82-5. The factors affecting the selection of grains sizes are:

1. *The type of finish desired.* Coarse grains are best suited for rapid removal of metal. Fine grains are used for producing smooth and accurate finishes.
2. *The type of material being ground.* Generally coarse grains are used on soft material, while fine grains are used for hard materials.
3. *The amount of material to be removed.* Where a large amount of material is to be removed and surface finish is not important, a coarse-grain wheel should be used. For finish grinding, a fine-grain wheel is recommended.
4. *The area of contact between the wheel and the workpiece.* If the area of contact is wide, a coarse-grain wheel is generally used. Fine-grain wheels are used when the area of contact between the wheel and the work is small.

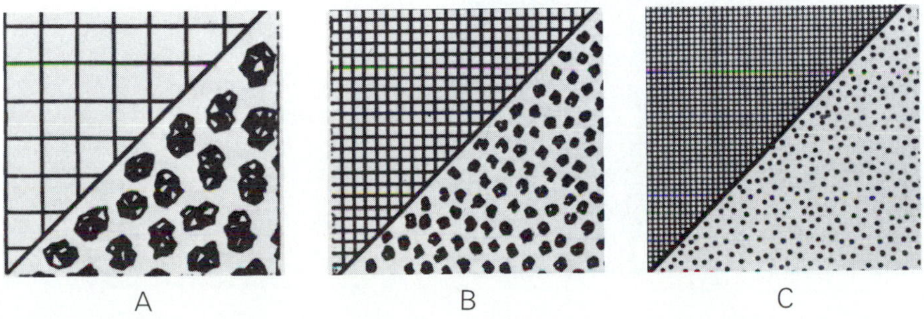
A
B
C

Figure 82-5 *Relative abrasive grain sizes: (A) 8-grain; (B) 24-grain; (C) 60-grain.*

BOND TYPES

The function of the bond is to hold the abrasive grains together in the form of a wheel. There are six common bond types used in grinding wheel manufacture: vitrified, resinoid, rubber, shellac, silicate, and metal.

Vitrified Bond

Vitrified bond is used on most grinding wheels. It is made of clay or feldspar, which fuses at a high temperature and when cooled forms a glassy bond around each grain. Vitrified bonds are strong but break down readily on the wheel surface to expose new grains during the grinding operation. This bond is particularly suited to wheels used for the rapid removal of metal. Vitrified wheels are not affected by water, oil, or acid and may be used in all types of grinding operations. Vitrified wheels should be operated between 6300 and 6500 sf/min (1920 and 1980 m/min).

Resinoid Bond

Synthetic resins are used as bonding agents in resinoid wheels. The majority of resinoid wheels generally operate at 9500 sf/min (2895 m/min); however, the modern trend is toward greater power and faster speeds for faster stock removal. Special resinoid wheels are manufactured to operate at speeds of 12 500 to 22 500 sf/min (3810 to 6858 m/min) for certain applications. These wheels are cool-cutting and remove stock rapidly. They are used for cutting-off operations, snagging, and rough grinding, as well as for roll grinding.

Rubber Bond

Rubber-bonded wheels produce high finishes such as those required on ball bearing races. Because of the strength and flexibility of this type of wheel, it is used for thin cutoff wheels. Rubber-bonded wheels are also used as regulating wheels on centerless grinders.

Shellac Bond

Shellac-bonded wheels are used for producing high finishes on parts such as cutlery, cam shafts, and paper-mill rolls. They are not suitable for rough or heavy grinding.

Silicate Bond

Silicate-bonded wheels are not used to any extent in industry. Silicate bond is used principally for large wheels and for small wheels where it is necessary to keep heat generation to a minimum. The bond (silicate of soda) releases the abrasive grains more rapidly than does the vitrified bond.

Metal Bond

Metal bonds (generally nonferrous) are used on diamond wheels and for electrolytic grinding operations where the current must pass through the wheel.

GRADE

The grade of a grinding wheel may be defined as the degree of strength with which the bond holds the abrasive particles in the bond setting. If the bond posts are very strong (Fig. 82-6C), that is, if they retain the abrasive grains in the wheel during the grinding operation, the wheel is said to be of a hard grade. If the grains are released rapidly during the grinding operation, the wheel is classified as a soft grade (Fig. 82-6A).

The selection of the proper grade of wheel is important. Wheels which are too hard do not release the grains readily; consequently, the grains become dull and do not cut effectively. This is known as glazing. Wheels which are too soft release the grain too quickly, and the wheel will wear rapidly.

One of the most difficult features to determine in selecting a grinding wheel for each job is the grade. Trial and error are generally used to decide what grade works best. The characteristics of wheels that are too hard, or too soft, are listed in Table 82-1 as a guide to help select the wheel grade.

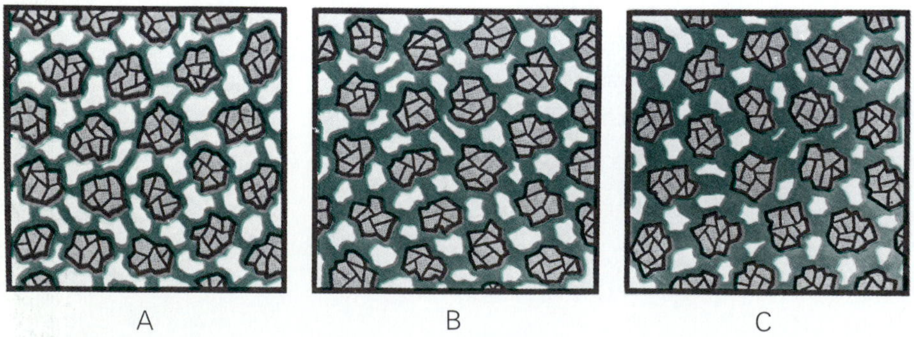
A
B
C

Figure 82-6 *Grinding wheel grades: (A) weak bond posts; (B) medium bond posts; (C) strong bond posts.*

Table 82-1 Wheel Grade Faults

Hard-Wheel Characteristics	Soft-Wheel Characteristics
Glaze: Grain wears flat and cannot be discharged from bond post.	*Breaks down too fast:* Wheel is too soft for the grinding operation.
Loading: Material being ground deposits on the wheel face.	*Surface finish gets worse:* Grain breaks from the wheel too rapidly—before the surface can get smooth.
Burn: Flat abrasive grains rub the work, causing burn marks which could result in grinding cracks.	*Cuts freely:* The wheel does not glaze and, therefore, the cutting action is good.
Squeal: Hard wheels emit a high-pitched sound called squealing.	*Sparks out quickly:* The free-cutting wheel sparks out because of the low cutting pressure and low machine distortion.
Doesn't cut freely: Cutting action is slow, wheel-to-work pressures are high, and wheel will not spark out.	*Chatter:* The chatter pattern is usually wide and surface finish is poor.
Inaccurate work: High cutting pressures and heat distort the workpiece.	*Sizing difficult:* Because of the rapid grain loss, maintaining workpiece size is difficult.
Chatter: Finely spaced chatter marks are often produced in the work.	*Scratches, "fish tails":* Grains breaking out from the wheel roll between the wheel and work, causing surface scratches.

It is well to remember that all abrasive grains are hard and that the hardness of a wheel refers to grade (the strength of the bond) and not to the hardness of the grain. Wheel grade symbols are indicated alphabetically ranging from A (softest) to Z (hardest). The grade selected for a particular job depends on the following factors.

1. *Hardness of the material.* A hard wheel is generally used on soft material and soft grades on hard materials.
2. *Area of contact.* Soft wheels are used where the area of contact between the wheel and the workpiece is large. Small areas of contact require harder wheels.
3. *Condition of the machine.* If the machine is rigid, a softer grade of wheel is recommended. Light-duty machines or machines with loose spindle bearings require harder wheels.
4. *The speed of the grinding wheel and the workpiece.* The higher the wheel speed in relation to the workpiece, the softer the wheel should be. Wheels which revolve slowly wear faster; therefore, a harder wheel should be used at slow speeds.
5. *Rate of feed.* Higher rates of feed require the use of harder wheels since the pressure on the grinding wheel is greater than the slower feeds.
6. *Operator characteristics.* An operator who removes the material quickly requires a harder wheel than one who removes the material more slowly. This is particularly evident in offhand grinding and where piece-work programs are involved.

STRUCTURE

The structure of a grinding wheel is the space relationship of the grain and bonding material to the voids that separate them. In brief, it is the *density* of the wheel.

If the spacing of the grains is close, the structure is dense (Fig. 82-7A). If the spacing of the grains is relatively wide, the structure is open (Fig. 82-7C).

Selection of the wheel structure depends on the type of work required. Wheels with open structures (Fig. 82-8) provide greater chip clearance than those with dense structures and remove material faster than dense wheels.

The structure of grinding wheels is indicated by numbers ranging from 1 (dense) to 15 (open). Selection of the proper wheel structure is affected by the following factors.

1. *The type of material being ground.* Soft materials will require greater chip clearance; therefore, an open wheel should be used.
2. *Area of contact.* The greater the area of contact, the more open should be the structure to provide better chip clearance.
3. *Finish required.* Dense wheels will give a better, more accurate finish.

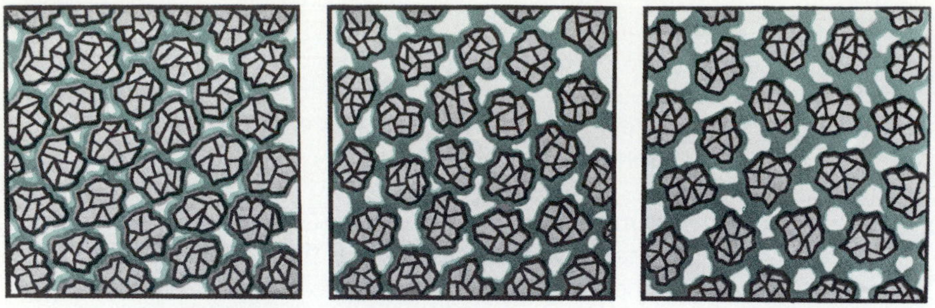
a
b
c

Figure 82-7 *Grinding wheel structure: (A) dense; (B) medium; (C) open.*

4. *Method of cooling.* Open structure wheels provide a better supply of coolant for machines using "through the wheel" coolant systems.

In summary, Table 82-2 will serve as a guide to the factors that must be considered in selecting a grinding wheel.

GRINDING WHEEL MANUFACTURE

Most grinding wheels used for machine shop operations are manufactured with vitrified bonds; therefore, the manufacture of only this type of wheel will be discussed. The main operations in the manufacture of vitrified grinding wheels are as follows.

Mixing

The correct proportions of abrasive grain and bond are carefully weighed and thoroughly mixed in a rotary power mixing machine (Fig. 82-9). A certain percentage of water is added to moisten the mix.

Molding

The proper amount of this mixture is placed in a steel mold of the desired wheel shape and compressed in a hydraulic press (Fig. 82-10) to form a wheel slightly larger than the finished size. The amount of pressure used varies with the size of the wheel and the structure required.

Table 82-2 Factors to Be Considered in the Selection of a Grinding Wheel

Grinding Factors	Wheel Considerations				
	Abrasive Type	Grain Size	Bond	Grade	Structure
Material to be ground (high or low tensile strength; hard or soft)	X	X		X	X
Type of operation (cylindrical, centerless, surface, cutoff, snagging, etc.)			X		
Machine characteristics (rugged or light; loose or tight bearings)				X	
Wheel speed (slow or fast)			X	X	
Rate of feed (slow or rapid)				X	
Area of contact (large or small)		X		X	X
Operator characteristics				X	
Amount of stock to be removed (light cut or heavy cut)			X	X	X
Finish required		X	X		X
Use of coolant (wet or dry grinding)			X	X	X

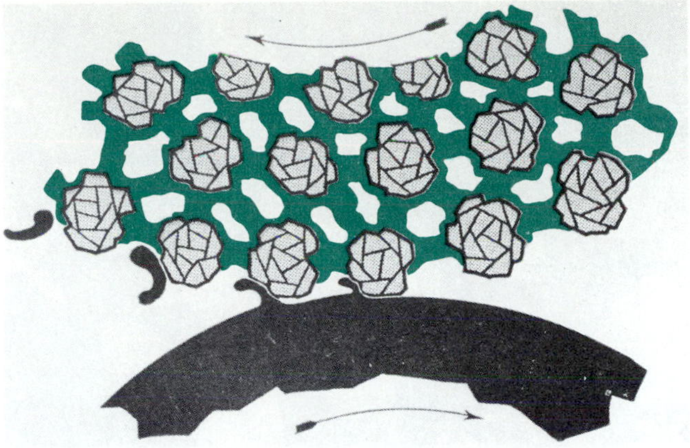

Figure 82-8 *An open-structure wheel provides greater chip clearance. (Courtesy The Carborundum Company.)*

Figure 82-9 *Mixing abrasive grain and bond. (Courtesy The Carborundum Company.)*

Shaving

Although the majority of wheels are molded to shape and size, some machines require special wheel shapes and recesses. These are shaped or shaved to size in the green, or unburned, state on a shaving machine, which resembles a potter's wheel.

Firing (Burning)

The "green" wheels are carefully stacked on cars and are moved slowly through a long kiln 250 to 300 ft (76 to 90 m) long. The temperature of the kiln is held at approximately 2300°F (1260°C). This operation, which takes about five days, causes the bond to melt and form a glassy case around each grain; the product is a hard wheel.

Figure 82-10 *Molding grinding wheels using a hydraulic press. (Courtesy Cincinnati Milacron Inc.)*

Truing

The cured wheels are mounted in a special lathe and turned to the required size and shape by hardened-steel conical cutters, diamond tools, or special grinding wheels.

Bushing

The arbor hole in a grinding wheel is fitted with a lead or plastic-type bushing to fit a specific spindle size. The edges of the bushing are then trimmed to the thickness of the wheel.

Balancing

To remove vibration which may occur while a wheel is revolving, each wheel is balanced. Generally, small, shallow holes are drilled in the "light" side of the wheel and filled with lead to assure proper balance.

Speed Testing

Wheels are rotated in a special, heavy, enclosed case and revolved at speeds at least 50 percent above normal operating speeds. This ensures that the wheel will not break under normal operating speeds and conditions.

STANDARD GRINDING WHEEL SHAPES

Nine standard grinding wheel shapes have been established by the United States Department of Commerce, the Grinding Wheel Manufacturers, and the Grinding Machine Manufacturers. Dimensional sizes for each of the shapes have also been standardized. Each of these nine shapes is identified by a number as shown in Table 82-3.

MOUNTED GRINDING WHEELS

Mounted grinding wheels (Fig. 82-11) are driven by a steel shank mounted in the wheel. They are produced in a variety of shapes for use with jig grinders, internal grinders, portable grinders, toolpost grinders, and flexible shafts. They are manufactured in both aluminum oxide and silicon carbide types.

GRINDING WHEEL MARKINGS

The standard marking system chart (Fig. 82-12) is used by the manufacturers to identify grinding wheels. This information is found on the blotter of all small- and medium-sized grinding wheels. It is stenciled on the side of larger wheels.

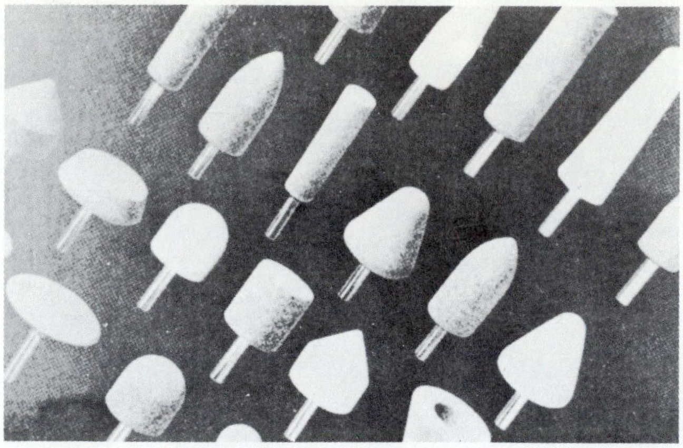

Figure 82-11 *A variety of mounted grinding wheels. (Courtesy The Carborundum Company.)*

Table 82-3 Common Grinding Wheel Shapes and Applications

Shape	Name	Applications
	Straight (Type 1)	Cylindrical, centerless, internal, cutter, surface, and offhand grinding operations.
	Cylinder (Type 2)	Surface grinding on horizontal and vertical spindle grinders.
	Tapered (both sides) (Type 4)	Snagging operations. The tapered sides lessen the chance of the wheel breaking.
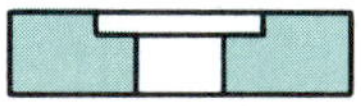	Recessed (one side) (Type 5)	Cylindrical, centerless, internal, and surface grinders. The recess provides clearance for the mounting flange.
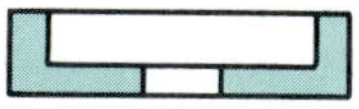	Straight cup (Type 6)	Cutter and tool grinder and surface grinding on vertical and horizontal spindle machines.
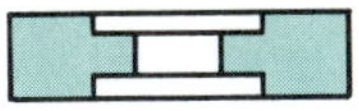	Recessed (both sides) (Type 7)	Cylindrical, centerless, and surface grinders. The recesses provide clearance for mounting flanges.
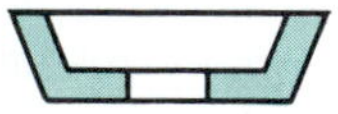	Flaring cup (Type 11)	Cutter and tool grinder. Used mainly for sharpening milling cutters and reamers.
	Dish (Type 12)	Cutter and tool grinder. Its thin edge permits it to be used in narrow slots.
	Saucer (Type 13)	Saw gumming, gashing, milling cutter teeth.

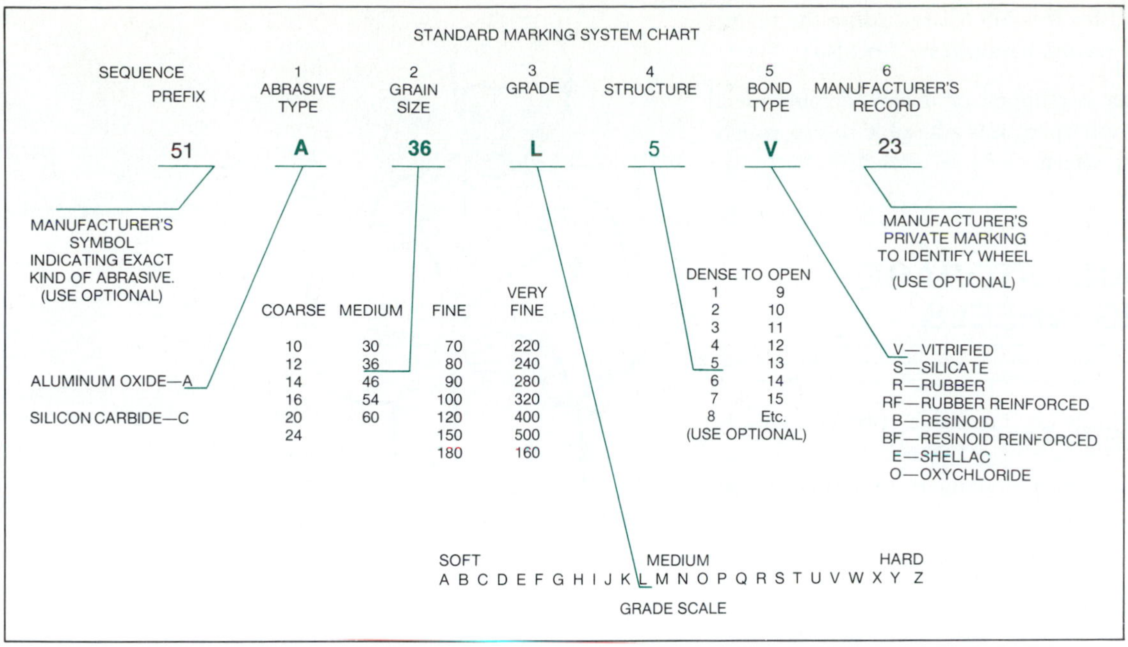
STANDARD MARKING SYSTEM CHART
SEQUENCE
PREFIX
1
ABRASIVE
TYPE
2
GRAIN
SIZE
3
GRADE
4
STRUCTURE
5
BOND
TYPE
6
MANUFACTURER'S
RECORD
51
A
36
L
5
V
23
MANUFACTURER'S SYMBOL INDICATING EXACT KIND OF ABRASIVE. (USE OPTIONAL)
ALUMINUM OXIDE—A
SILICON CARBIDE—C
COARSE
MEDIUM
FINE
VERY FINE
10 12 14 16 20 24
30 36 46 54 60
70 80 90 100 120 150 180
220 240 280 320 400 500 160
DENSE TO OPEN
1 2 3 4 5 6 7 8
9 10 11 12 13 14 15 Etc.
(USE OPTIONAL)
MANUFACTURER'S PRIVATE MARKING TO IDENTIFY WHEEL
(USE OPTIONAL)
V—VITRIFIED
S—SILICATE
R—RUBBER
RF—RUBBER REINFORCED
B—RESINOID
BF—RESINOID REINFORCED
E—SHELLAC
O—OXYCHLORIDE
SOFT
MEDIUM
HARD
A B C D E F G H I J K L M N O P Q R S T U V W X Y Z
GRADE SCALE

Figure 82-12 *Straight grinding wheel marking system chart. (Courtesy Grinding Wheel Institute.)*

The six positions shown in the standard sequence are followed by all manufacturers of grinding wheels. The prefix shown is a manufacturer's symbol and is not always used by all grinding wheel producers.

NOTE: This marking system is used only for aluminum oxide and silicon carbide wheels; it is not used for diamond wheels.

SELECTING A GRINDING WHEEL FOR A SPECIFIC JOB

From the foregoing information, the machinist should be able to select the proper wheel for the job required.

EXAMPLE

It is required to rough surface grind a piece of SAE 1045 steel using a straight wheel. Coolant is to be used.

Type of abrasive Because steel is to be ground, *aluminum oxide* should be used.

Size of grain Since the surface is not precision-finished, a medium grain can be used—about *46 grit.*

Grade A *medium-grade* wheel which will break down reasonably well should be selected. Use grade *J.*

Structure Since this steel is of medium hardness, the wheel should be of medium density—about *7.*

Bond type Since the operation is standard surface grinding and since coolant is to be used, a *vitrified* bond should be selected.

After the various factors have been considered, an A46-J7-V grinding wheel should be selected to rough-grind SAE 1045 steel.

NOTE: These specifications do not include the manufacturer's prefix or the manufacturer's records.

EXAMPLE

It is required to finish-grind a high-speed steel milling cutter on the cutter and tool grinder.

Type of abrasive Since the cutter is steel, an *aluminum oxide* wheel should be used.

Size of grain Since the milling cutter must have a smooth finish, a medium to fine grain should be used. About a *60 grit* is recommended for this type of operation.

Grade It is important that a cool-cutting wheel be used to prevent burning the cutting edge of the cutter. A wheel which breaks downs reasonably well will permit cool grinding. Use a medium-soft grade such as *J.*

Structure In order to produce a smooth cut, a medium-dense wheel should be used. For this application, use a *#6.*

Bond type Because most cutter and tool grinders are designed for standard speeds, a *vitrified* bond should be used. When the speed is excessive for the wheel size, a *resinoid* bond should be used.

The wheel selected for this job (disregarding the manufacturer's prefix and records) should be A60-J6-V.

NOTE: If the cutter is chipped or if considerable metal must be removed to resharpen, it is advisable to first rough-grind using a 46-grit wheel.

HANDLING AND STORAGE OF GRINDING WHEELS

Since all grinding wheels are breakable, proper handling and storage is important. Damaged wheels can be dangerous if used; therefore, the following rules should be observed.

1. Do not drop or bump grinding wheels.
2. Always store wheels properly on shelves or in bins provided. Flat and tapered wheels can be stored on edge, while large cup wheels and cylindrical wheels should be stored on the flat sides with a suitable layer of packing between each wheel.
3. Thin organic (resinoid and rubber) bonded wheels should be laid flat on a horizontal surface, away from excessive heat, to prevent warping.
4. Small cup and small internal grinding wheels may be stored separately in boxes, bins, or drawers.

INSPECTION OF WHEELS

After wheels have been received, they should be inspected to see that they have not been damaged in transit. For further assurance that wheels have not been damaged, they should be suspended and tapped lightly with a screwdriver handle for small wheels (Fig. 82-13) or with a wooden mallet for larger wheels. If vitrified or silicate wheels are sound, they give a clear, metallic ring. Organic-bonded wheels give a duller ring, and cracked wheels do not produce a ring. Wheels must be dry and free from sawdust before testing; otherwise the sound will be deadened.

Figure 82-13 *Testing a grinding wheel for cracks.*

DIAMOND WHEELS

Diamond wheels are used for grinding cemented carbides and hard vitreous materials, such as glass and ceramics. Diamond wheels are manufactured in a variety of shapes, such as straight, cup, dish, and thin cutoff wheels.

Wheels of ½ in. (13 mm) diameter or less have diamond particles throughout the wheel. Wheels larger than ½ in. (13 mm) are made with a diamond surface on the grinding face only. The diamonds for this purpose are made in grain sizes ranging from 100 to 400. The proportions of the diamond and bond mixture vary with the application. This diamond concentration is identified by the letters A, B, or C. The C concentration contains four times the number of diamonds in an A concentration. This mixture is coated on the grinding face of the wheel in thicknesses ranging from 1/32 to ¼ in. (0.80 to 6 mm).

BONDS

There are three types of bonds available for diamond wheels: resinoid, metal, and vitrified.

Resinoid-bonded wheels give a maximum cutting rate and require very little dressing. These wheels remain sharp for a long time and are well suited to grinding carbides.

A recent development in the manufacture of resinoid-bonded diamond wheels has been the coating of the diamond particles with nickel plating by means of an electroplating process. This process is carried out before the diamonds are mixed with the resin. It reduces the tendency of the diamonds to chip and results in cooler-grinding, longer-lasting wheels.

Metal bonds, generally nonferrous, are particularly suited to offhand grinding and cutting-off operations. This type of wheel holds its form extremely well and does not wear on radius work or on small areas of contact.

Vitrified-bonded wheels remove stock rapidly but require frequent cleaning with a boron carbide abrasive stick to prevent the wheel from loading. These wheels are particularly suited for offhand and surface grinding of cemented carbides.

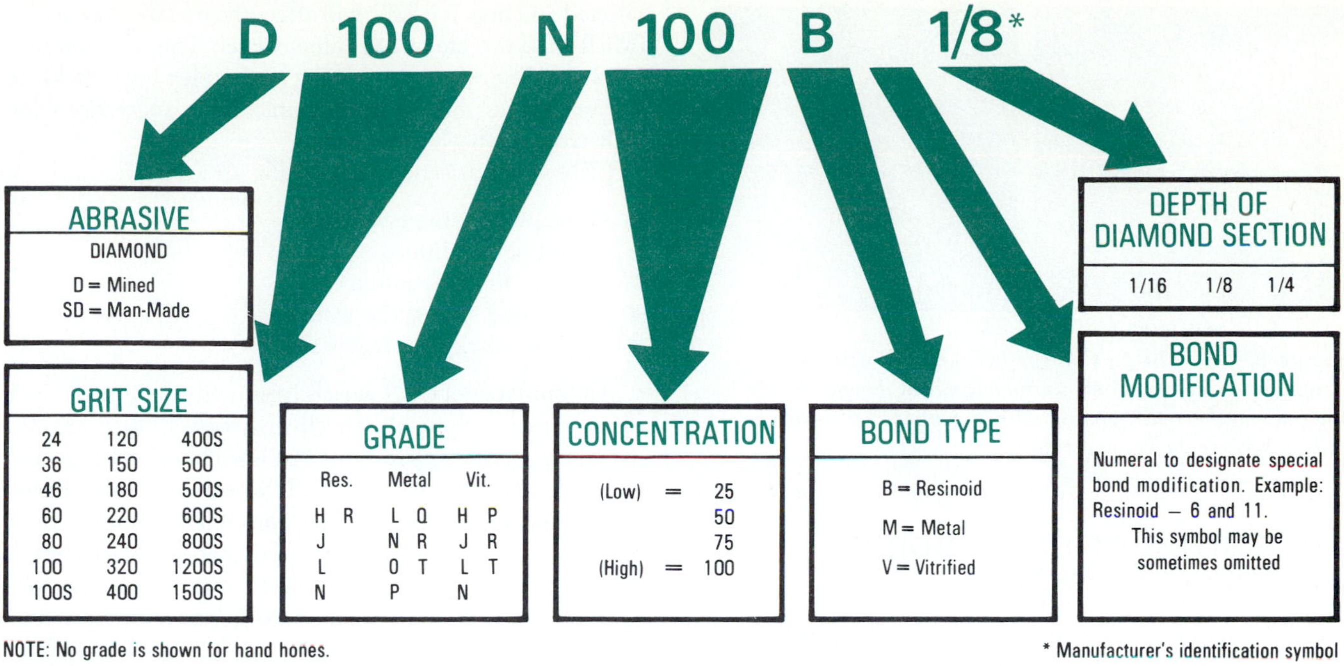
D 100 N 100 B 1/8*
ABRASIVE
DIAMOND
D = Mined
SD = Man-Made
GRIT SIZE
24 120 400S
36 150 500
46 180 500S
60 220 600S
80 240 800S
100 320 1200S
100S 400 1500S
GRADE
Res. Metal Vit.
H R L Q H P
J N R J R
L O T L T
N P N
CONCENTRATION
(Low) = 25
50
75
(High) = 100
BOND TYPE
B = Resinoid
M = Metal
V = Vitrified
DEPTH OF DIAMOND SECTION
1/16 1/8 1/4
BOND MODIFICATION
Numeral to designate special bond modification. Example: Resinoid – 6 and 11.
This symbol may be sometimes omitted
NOTE: No grade is shown for hand hones.
* Manufacturer's identification symbol

Figure 82-14 *Diamond grinding wheel marking system chart. (Courtesy The Norton Company of Canada Ltd.)*

DIAMOND WHEEL IDENTIFICATION

The method used to identify diamond wheels differs from that used for other grinding wheels (Fig. 82-14).

CUBIC BORON NITRIDE WHEELS

Cubic boron nitride (CBN) grinding wheels are now recognized as superior cutting tools for grinding difficult-to-machine metals. From their initial use in toolrooms and cutter grinding applications CBN wheels are really making their presence felt in production grinding operations where the alternative had been to use less costly conventional abrasives, which wear out at much faster rates. The applications for CBN wheels range from the elementary regrinding of high-speed steel cutting tools to the ultra-high-speed grinding of hardened steel components in the automotive industries.

CBN grinding wheels have more than twice the hardness of conventional abrasives for grinding difficult-to-grind ferrous metals (Fig. 82-15). Hardness in an abrasive is meaningless if the abrasive is too brittle to withstand the machining pressures and the heat of production grinding. The CBN abrasive crystal has the toughness to match its hardness so that its cutting edges stay sharp longer with much slower wear rates than those of conventional abrasives.

On difficult-to-grind materials, conventional grinding wheels dull quickly and, as a result, generate high frictional heat. As the abrasive grains dull, the material removal rate falls and it is difficult to maintain part accuracy and geometry. The CBN wheel's prolonged cutting capacity and high thermal conductivity help prevent uncontrolled heat buildup and therefore reduce the chances of wheel glazing and workpiece metallurgical damage (Fig. 82-16). The CBN abrasive is also thermally and chemically stable at temperatures above 1832°F (1000°C), that is, well above the temperatures generally reached in grinding. This means reduced grinding wheel wear, with easier production of precision workpiece geometry and accuracy.

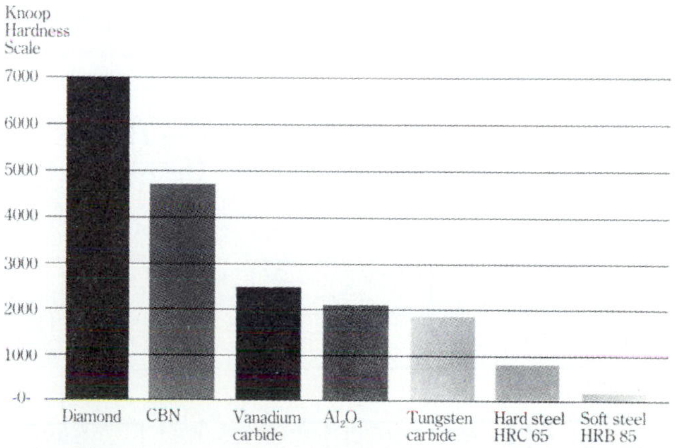
Knoop Hardness Scale
7000
6000
5000
4000
3000
2000
1000
-0-
Diamond
CBN
Vanadium carbide
Al_2O_3
Tungsten carbide
Hard steel HRC 65
Soft steel HRB 85

Figure 82-15 *Hardness of various metals and abrasives. (Courtesy The Norton Company.)*

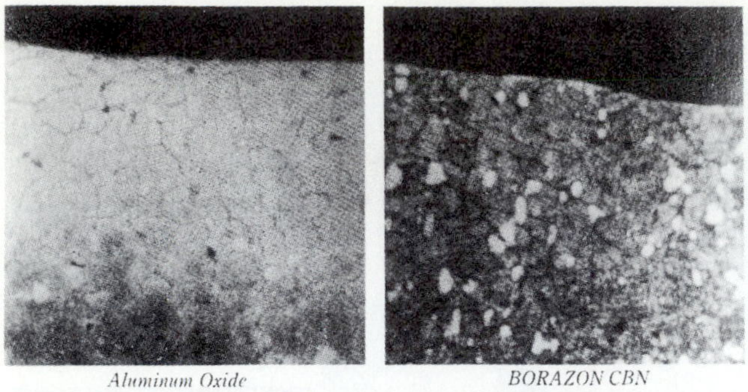
Aluminum Oxide
BORAZON CBN

Figure 82-16 *The part on the left, ground with an aluminum oxide wheel, shows metallurgical damage, while the part on the right, ground with a cubic boron nitride (CBN) wheel, shows no damage.*

PROPERTIES OF CUBIC BORON NITRIDE (CBN) WHEELS

Cubic boron nitride is a man-made material, which is not found in nature. CBN grinding wheels contain the four main properties necessary to grind extremely hard or abrasive materials at high metal-removal rates: hardness, abrasion resistance, compressive strength, and thermal conductivity (Fig. 82-17).

1. *Hardness*—Cubic boron nitride, next to diamond in hardness, is about twice as hard as silicon carbide and aluminum oxide.
2. *Abrasion resistance*—CBN wheels maintain their sharpness much longer, thereby increasing productivity while at the same time producing parts which are dimensionally accurate.
3. *Compressive strength*—The high compressive strength of CBN crystals gives them excellent qualities to withstand the forces created during high metal-removal rates.
4. *Thermal conductivity*—CBN wheels have excellent thermal conductivity, which allows greater heat dissipation (transfer), especially when hard, abrasive, tough materials are being ground at high material-removal rates.

WHEEL SELECTION

Any successful grinding operation depends to a large extent on choosing the right wheel for the job. The type of wheel selected and how it is used will affect the metal-removal rate (MRR) and the life of a grinding wheel. The selection of a CBN grinding wheel can be a very complex task, and it is always wise to follow the manufacturer's suggestions for each type of wheel.

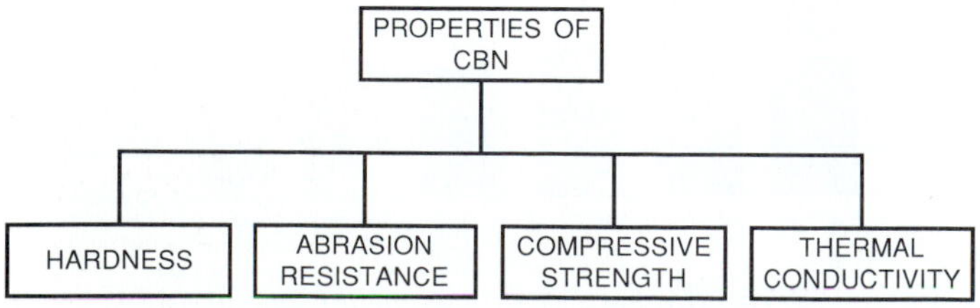
PROPERTIES OF CBN
HARDNESS
ABRASION RESISTANCE
COMPRESSIVE STRENGTH
THERMAL CONDUCTIVITY

Figure 82-17 *The properties of CBN grinding wheels.*

CBN wheel selection is generally affected by:

- Type of grinding operation
- Grinding conditions
- Surface finish requirements
- Shape and size of the workpiece
- Type of workpiece material.

All four types of CBN wheels (resin, vitrified, metal, and electroplated) are highly effective; however, they are designed for specific applications and must be selected accordingly. There is no one type of CBN wheel that is suitable for all grinding operations and, therefore, for the best grinding performance, the characteristics of the wheel abrasive must be matched to the requirements of the specific grinding job. Abrasive characteristics such as concentration, size, and toughness must be considered in the selection of a wheel because they affect metal-removal rates, wheel life, and surface finish of any grinding operation. However, the wheel manufacturer understands the type of CBN abrasive available and will provide the most suitable type for the job.

Wheel Selection Guidelines

Cubic boron nitride (CBN) grinding wheels are available in a complete range of shapes and sizes, including wheels with straight or formed faces, ring wheels, disk wheels, flaring cup wheels, mounted wheels, mandrels, and hones. Individually engineered wheels are also available to suit specific superabrasive machining systems or specific grinding operations. As with all wheels which use high-cost abrasives, the CBN wheel is constructed with a precision preformed core with the abrasive portion on the grinding face of the wheel. The abrasive portion is usually between 1⁄16 to 1⁄4 in. (1.5 to 6 mm) in depth, and this information is shown on the identification label of the wheel. There generally is a CBN wheel readily available to suit any grinding operation.

The wheel manufacturer will provide the right CBN abrasive for the bond system selected. The manner in which the wheel is actually used, however, will determine whether ideal wear conditions are achieved. Always be sure that the actual operating conditions are within the range of the wheel's capabilities. In order to successfully use CBN grinding wheels, follow these general guidelines:

1. *Select the bond.* Refer to Table 82-4 for a good first choice in selecting bonds for all major applications.
2. *Specify normal wheel diameters and widths.* When an aluminum oxide wheel is being replaced with a CBN wheel, the CBN wheel should be the same diameter and width as the wheel it replaces.
3. *Choose the largest abrasive mesh size that produces the*

Table 82-4 Grinding Performance and Specifications of CBN Bond Systems

Item	Wheel Bond Type			
	Resin	Metal	Vitreous	Electroplated
Rim depth	0.079 in. (2 mm) or more	0.079 in. (2 mm) or more	0.079 in. (2 mm) or more	One abrasive layer
Wheel life	Limited	Long	Long	Limited
Grinding ratio	Medium	High	Medium	N/A
Cutting action	Good to excellent	Good	Excellent	Good to excellent
Material-removal rate	High	Limited	Medium to high	High
Form holding	Good	Excellent	Good	Good
Precautions for use	None	Rigid machines recommended	Least resistance to damage	None
Grinding	Wet or dry	Wet	Wet	Wet preferred; can be used dry

desired finish. For any given set of grinding operations, a wheel containing coarse (large mesh size) CBN abrasive will have a longer life than a wheel containing fine (small mesh size) abrasive. However, the wheel with fine abrasive will produce better surface finishes at lower downfeed rates.
4. *Choose wheels with the optimum abrasive concentration.* Although low concentration wheels will generally do the job, they may not always be the most cost effective. Always select the highest concentration which the grinding machine has the power to drive effectively.

COATED ABRASIVES

Coated abrasives (Fig. 82-18) consist of a flexible backing (cloth or paper) to which abrasive grains have been bonded. Garnet, flint, and emery (natural abrasives) are being replaced by aluminum oxide and silicon carbide in the manufacture of coated abrasives. This is due to the greater toughness and the more uniform grain size and shape of manufactured abrasives.

Coated abrasives serve two purposes in the machine shop: metal grinding and polishing. Metal grinding may be done on a belt or disk grinder and up until the last few years was a rapid, nonprecision method of removing metal. Coarse-grit coated abrasives are used for rapid removal of metal, whereas fine grits are used for polishing.

Emery, a natural abrasive which is black in appearance, is used to manufacture coated abrasives, such as emery cloth and emery paper. Since the grains are not as sharp as artificial abrasives, emery is generally used for polishing metal by hand.

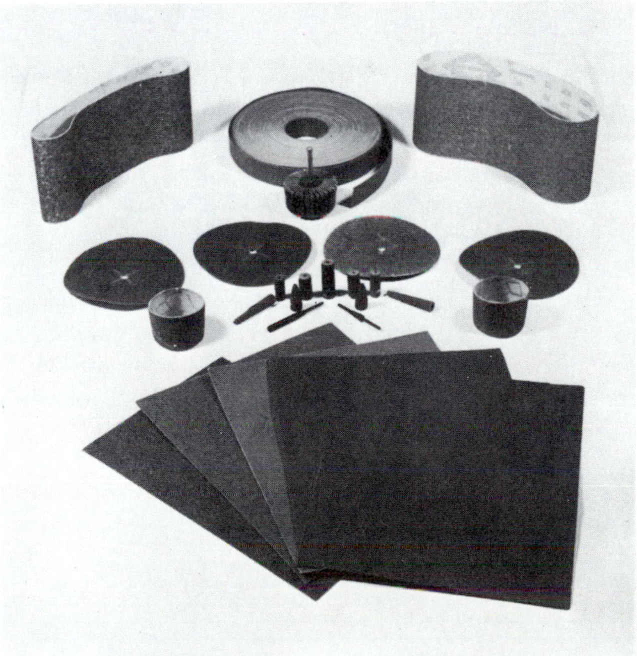

Figure 82-18 *A variety of coated abrasives. (Courtesy The Norton Company.)*

SELECTION OF COATED ABRASIVES

Aluminum Oxide

Aluminum oxide, gray in appearance, is used for high-tensile-strength materials, such as steels, alloy steels, high-carbon steels, and tough bronzes. Aluminum oxide is characterized by the long life of its cutting edges.

For *hand operations,* 60 to 80 grit is used for fast cutting (roughing), while 120 to 180 grit is recommended for finishing operations.

For *machine operations,* such as on belt and disk grinders, 36 to 60 grit is used for roughing, while 80 to 120 grit is recommended for finishing operations.

Silicon Carbide

Silicon carbide, bluish-black in appearance, is used for low-tensile-strength materials, such as cast iron, aluminum, brass, copper, glass, and plastics. The selection of grit size for hand and machine operations is the same as for aluminum oxide–coated abrasives.

COATED ABRASIVE MACHINING

Over the past few years, coated abrasive machining has become widely used in industry. With improved abrasives and bonding materials, better grain structure, more uniform belt splicing, and new polyester belt backing, abrasive belt machining is being used to a much greater extent. Some types of work which had formerly been performed by milling, turning, and abrasive wheel grinding in machine shops, steel mills, steel fabrication plants, and foundries, are now being done more efficiently by belt or disk grinding operations. Coated abrasive operations are now capable of grinding to less than 0.001 in. (0.03 mm) tolerance and with a surface finish of 10 to 20 μin. (0.3 to 0.5 μm).

As a result of greatly improved coated abrasive products, heavier and better machines requiring much higher horsepower are being produced in this field. The basic machines have become more automated and now include automatic loaders, feeders, and unloaders. Belt machining has been applied very successfully to the centerless grinding concept where it is used extensively for bar and tube grinding in steel and metalworking plants. In many lumber mills, belt abrasive machining is now used to finish the lumber to size, an operation that previously was done by planing.

REVIEW QUESTIONS

1. What characteristics must an abrasive have in order for it to function properly?

TYPES OF ABRASIVES

2. Name four natural abrasives.
3. What is "bort" and for what purpose is it used?
4. Name three manufactured abrasives and state why they are used extensively today.
5. For what purposes are aluminum oxide and silicon carbide abrasives used?
6. Describe the manufacture of aluminum oxide.
7. Describe the manufacture of silicon carbide.
8. List three uses of zirconia-alumina abrasive.
9. Describe the cutting action of zirconia-alumina abrasive.
10. List four advantages of zirconia-alumina abrasive wheels.
11. How does the manufacture of boron carbide differ from that of the other manufactured abrasives?
12. List the uses for boron carbide.
13. List five advantages of cubic boron nitride grinding wheels.
14. Name two types of Borazon® (cubic boron nitride) and state where each is used.
15. List three types of manufactured diamonds and state where each is used.
16. a. Name two materials used to coat diamonds.
    b. What is the purpose of coating diamonds?

GRINDING WHEELS

17. What are the basic functions of a grinding wheel?
18. Describe the function of each abrasive grain.
19. How is grain size determined and why is it important?
20. What factors affect the selection of the proper grain size?
21. What is the function of a bond and how does it affect the grade of a grinding wheel?
22. Name six types of bonds and state the purpose of each type in the manufacture of grinding wheels.
23. Why is the selection of the grade of wheel important to the grinding operation?
24. What factors should be considered when the grade of wheel is being selected for a particular job?
25. Define the *structure* of a grinding wheel and state how this is indicated.
26. What factors affect the selection of the proper wheel structure?

GRINDING WHEEL MANUFACTURE

27. Describe briefly the manufacture of a vitrified grinding wheel.

STANDARD GRINDING WHEEL SHAPES

28. Describe the following grinding wheels and state their purposes: Types 1, 5, 6, 11.
29. For what purpose are mounted grinding wheels used?

GRINDING WHEEL MARKINGS

30. Explain the meaning of the following grinding wheel markings: A80-G8-S, C-60-L4-V.
31. What wheel should be selected for grinding:
  a. Machine steel?
  b. Cemented carbide?
  c. Cast iron?

INSPECTION OF WHEELS

32. Explain why it is important to inspect a grinding wheel before it is mounted on a grinder.
33. How may grinding wheels be inspected?

DIAMOND WHEELS

34. For what purpose are diamond wheels used?
35. Explain *diamond concentration* in a grinding wheel.
36. Name three types of bonds used in diamond wheels and state the purpose of each.
37. Define the following diamond wheel markings: D 120–N 100–B 1/8.

COATED ABRASIVES

38. Name the three common abrasives used in the manufacture of coated abrasives and state their purposes.
39. What grit size is recommended for:
  a. Hand operations?
  b. Machine operations?

UNIT 83

Surface Grinders and Accessories

Grinding is an important part of the machine tool trade. Improved grinding machine construction has permitted the production of parts to extremely fine tolerances with improved surface finishes and accuracy. Because of

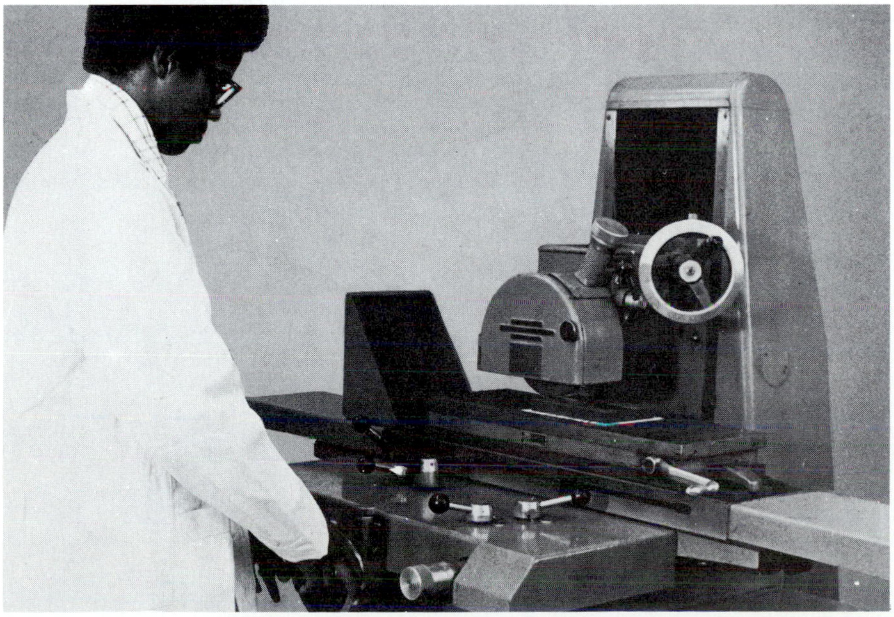

the dimensional accuracy obtained by grinding, interchangeable manufacture has become commonplace in most industries.

Grinding has also, in many cases, eliminated the need for conventional machining. With the development of new abrasives and better machines, the rough part is often finished in one grinding operation, thus eliminating the need for other machining processes. The role of grinding machines has changed over the years; initially they were used on hardened work and for truing hardened parts which had been distorted by heat treating. Today, grinding is applied extensively to the production of unhardened parts where high accuracy and surface finish are required. In many cases, modern grinding machines permit the manufacture of intricate parts faster and more accurately than do other machining methods.

OBJECTIVES

After completing this unit you should be able to:

1. Name four methods of surface grinding and state the advantage of each
2. True and dress a grinding wheel
3. Select the proper grinding wheel to be used for each type of work material

THE GRINDING PROCESS

In the grinding process the workpiece is brought into contact with a revolving grinding wheel. Each small abrasive grain on the periphery of the wheel acts as an individual cutting tool and removes a chip of metal (Fig. 83-1). As the abrasive grains become dull, the pressure and heat created between the wheel and the workpiece cause the dull face to break away, leaving new sharp cutting edges.

Regardless of the grinding method used, whether it be cylindrical, centerless, or surface grinding, the grinding process is the same, and certain general rules will apply in all cases:

1. Use a silicon carbide wheel for low-tensile-strength material and an aluminum oxide wheel for high-tensile-strength materials.
2. Use a hard wheel on soft materials and a soft wheel on hard materials.
3. If the wheel is too hard, increase the speed of the work or decrease the speed of the wheel to make it act as a softer wheel.
4. If the wheel appears too soft or wears rapidly, decrease the speed of the work or increase the speed of the wheel, but not above its recommended speed.
5. A glazed wheel will affect the finish, accuracy, and metal-removal rate. The main causes of wheel glazing are:
   a. The wheel speed is too fast.
   b. The work speed is too slow.
   c. The wheel is too hard.
   d. The grain is too small.
   e. The structure is too dense, which causes the wheel to load.
6. If a wheel wears too quickly, the cause may be any of the following.
   a. The wheel is too soft.
   b. The wheel speed is too slow.
   c. The work speed is too fast.
   d. The feed rate is too great.
   e. The face of the wheel is too narrow.
   f. The surface of the work is interrupted by holes or grooves.

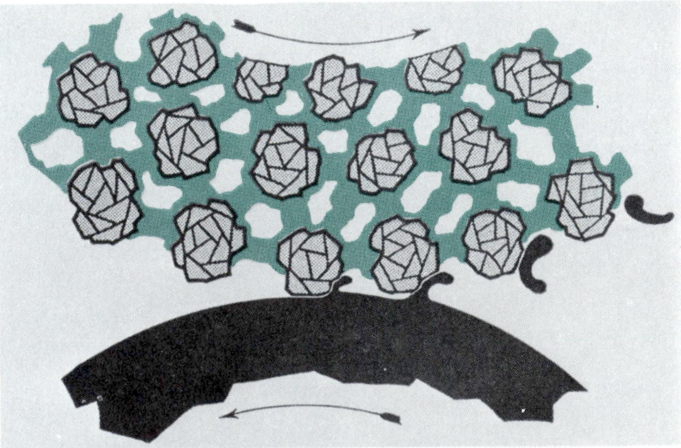

Figure 83-1 *Cutting action of abrasive grains. (Courtesy The Carborundum Company.)*

SURFACE GRINDING

Surface grinding is a technical term referring to the production of flat, contoured, and irregular surfaces on a piece of work which is passed against a revolving grinding wheel.

TYPES OF SURFACE GRINDERS

There are four distinct types of surface-grinding machines (Fig. 83-2), all of which provide a means of holding the metal and bringing it into contact with the grinding wheel.

The horizontal spindle grinder with a reciprocating table (Fig. 83-2A) is probably the most common type of surface grinder used in the toolroom. The work is reciprocated (moved back and forth) under the grinding wheel, which is fed down to provide the desired depth of cut. Feed is obtained by a transverse movement of the table at the end of each stroke.

The horizontal spindle grinder with a rotary table (Fig. 83-2B) is often found in toolrooms for the grinding of flat circular parts. The surface pattern it produces makes it particularly suitable for grinding parts which must rotate in contact with each other. The work is held on the magnetic chuck of a rotating table and passed under a grinding wheel. Feed is obtained by the transverse movement of the wheelhead. This type of machine permits faster grinding of circular parts since the wheel is always in contact with the workpiece.

The vertical spindle grinder with a rotary table (Fig. 83-2C) produces a finished surface by grinding with the face of the wheel rather than the periphery, as in horizontal spindle machines. The surface pattern appears as a series of intersecting arcs. Vertical spindle grinders have a higher metal removal rate than the horizontal-type spindle machines. It is probably the most efficient and accurate form of grinder for the production of flat surfaces.

The vertical spindle grinder with a reciprocating table (Fig. 83-2D) grinds on the face of the wheel while the work is moved back and forth under the wheel. Because of its vertical spindle and greater area of contact between the wheel and the work, this machine is capable of heavy cuts. Material up to ½ in. (13 mm) thick may be removed in one pass on larger machines of this type. Provision is made on most of these grinders to tilt the wheelhead a few degrees from the vertical. This permits greater pressure where the rim of the wheel contacts the workpiece and results in faster metal removal. When the wheelhead is vertical and grinding is done on the face of the wheel, the surface pattern

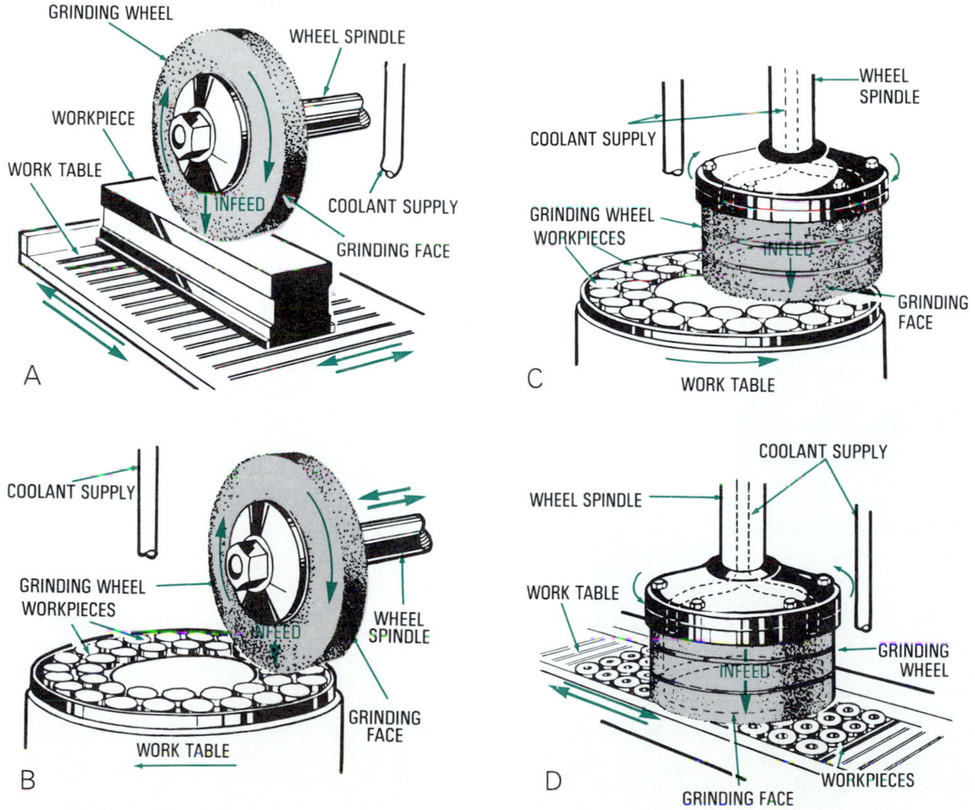

GRINDING WHEEL
WHEEL SPINDLE
WORKPIECE
WORK TABLE
INFEED
COOLANT SUPPLY
GRINDING FACE
A
COOLANT SUPPLY
WHEEL
SPINDLE
GRINDING WHEEL
WORKPIECES
INFEED
GRINDING
FACE
WORK TABLE
C
COOLANT SUPPLY
GRINDING WHEEL
WORKPIECES
INFEED
WHEEL
SPINDLE
GRINDING
FACE
WORK TABLE
B
COOLANT SUPPLY
WHEEL SPINDLE
WORK TABLE
INFEED
GRINDING
WHEEL
WORKPIECES
GRINDING FACE
D

Figure 83-2 *(A) Horizontal spindle grinder with a reciprocating table; (B) horizontal spindle grinder with a rotary table; (C) vertical spindle grinder with a rotary table; (D) vertical spindle grinder with a reciprocating table. (Courtesy The Carborundum Company.)*

produced is a series of uniform intersecting arcs. If the wheelhead is tilted, it produces a semicircular pattern.

The Horizontal Spindle Reciprocating Table Surface Grinder

The horizontal spindle reciprocating table surface grinder is the most commonly used and will be discussed in detail. Machines of this type may be either hand or hydraulically operated (Fig. 83-3).

Parts of a Hydraulic Surface Grinder

The *base* is generally of heavy cast-iron construction. It usually contains the hydraulic reservoir and pump used to operate the table and power feeds. The top of the base has accurately machined ways to receive the saddle.

The *saddle* may be moved in or out across the ways, manually or by automatic feed.

The *table* is mounted on the top of the saddle. The ways for the table are at right angles to those on the base. Thus the table reciprocates across the upper ways on the saddle, while the saddle (and table) moves in or out on the ways of the base.

The *column,* mounted on the back of the frame, contains the ways for the *spindle housing* and *wheelhead.* The wheel feed handwheel provides a means of moving the wheelhead vertically to set the depth of cut.

The reciprocating action of the table may be controlled manually by the *table traverse handwheel* or by the *hydraulic control valve lever.*

The direction of the table is reversed when one of the *stop dogs* mounted on the side of the table strikes the *table traverse reverse lever.*

The table may be fed toward or away from the column manually by means of the *crossfeed handwheel* or automatically by the *power crossfeed control.* This operation moves the work laterally under the wheel.

GRINDING WHEEL CARE

In order to ensure the best results in any surface-grinding operation, proper care of the grinding wheel must be taken.

1. When not in use, all grinding wheels should be properly stored.
2. Wheels should be tested for cracks prior to use.
3. Select the proper type of wheel for the job.
4. Grinding wheels should be properly mounted and operated at the recommended speed.

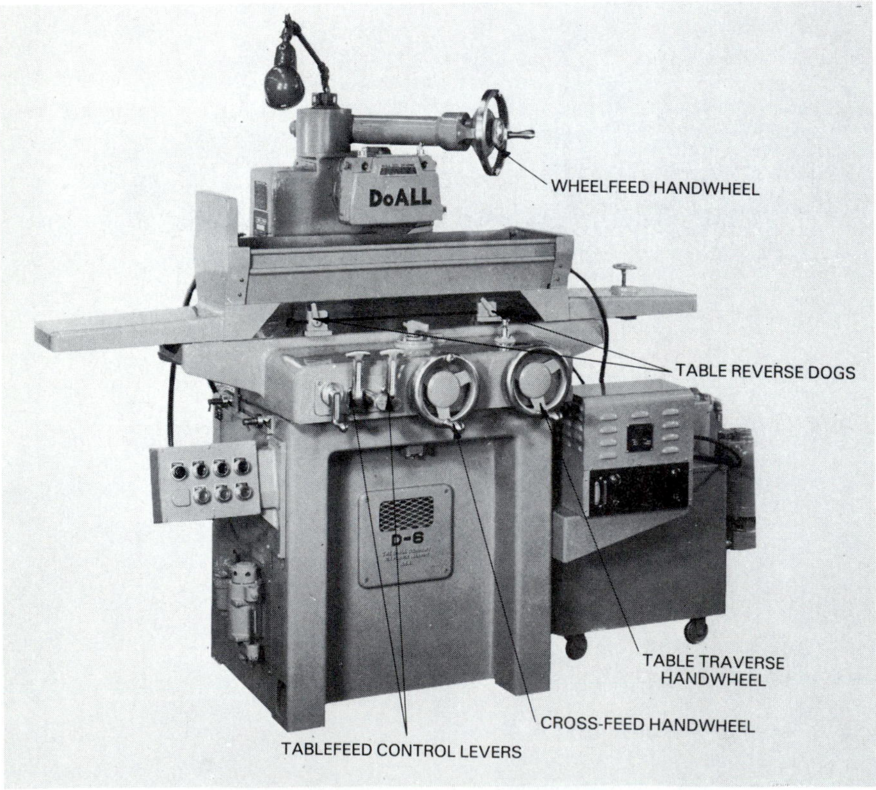
DoALL
WHEELFEED HANDWHEEL
TABLE REVERSE DOGS
D-6
TABLE TRAVERSE HANDWHEEL
CROSS-FEED HANDWHEEL
TABLEFEED CONTROL LEVERS

Figure 83-3 *A hydraulic surface grinder. (Courtesy The DoAll Co.)*

MOUNTING A GRINDING WHEEL

After the correct wheel has been selected for the job, proper mounting of the grinding wheel ensures the best grinding performance.

PROCEDURE

1. Test the wheel to see that it is not cracked by ring testing with the handle of a screwdriver or hammer.
2. Clean the grinding wheel adapter.
3. Mount the adapter through the wheel and tighten the threaded flange (Fig. 83-4).
   a. Be sure that the blotter is on each side of the wheel prior to mounting. A perforated blotter should be used for through-the-wheel coolant. A rubber washer is sometimes used in place of the blotter on some grinders.
   b. The wheel should be a good fit on the adapter or spindle. If it is too tight or too loose, the wheel should not be mounted.
   c. To comply with the Wheel Manufacturer's Safety Code, the diameter of the flanges should not be less than one-third the diameter of the wheel.
4. Tighten the wheel adapter flanges only enough to hold wheel firmly. If it is tightened too much, it may damage the flanges or break the wheel.

BALANCING A GRINDING WHEEL

Proper balance of a mounted grinding wheel is very important since improper balance will greatly affect the surface finish and accuracy of the work. Excessive imbalance creates vibration which will damage the spindle bearings.

There are two methods of balancing a wheel:

1. *Static balancing.* On some grinders, the wheel is balanced off the machine with the use of a balancing stand and arbor. Counterweights in the wheel flange must be correctly positioned in order to balance the grinding wheel.

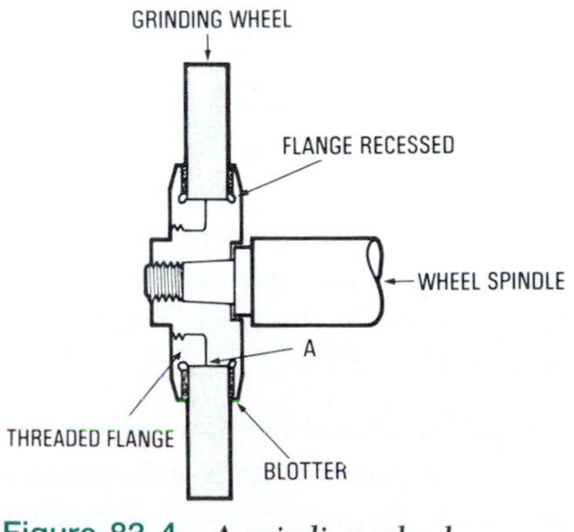
GRINDING WHEEL
FLANGE RECESSED
WHEEL SPINDLE
A
THREADED FLANGE
BLOTTER

Figure 83-4 *A grinding wheel properly mounted on the grinder spindle.*

2. *Dynamic balancing.* Most new grinding machines are equipped with ball-bearing balancing devices, which automatically balance a wheel in a matter of seconds while it is revolving on the grinder.

After the wheel has been mounted on the adapter it should be balanced, if provision is made in the adapter for balancing.

TO BALANCE A GRINDING WHEEL

1. Mount the wheel and adapter on the surface grinder and true the wheel with a diamond dresser.
2. Remove the wheel assembly and mount a special tapered balancing arbor in the hole of the adapter.
3. Place the wheel and arbor on a balancing stand (Fig. 83-5) which has been leveled.
4. Allow the wheel to rotate until it stops. This will indicate that the heavy side is at the bottom. Mark this point with chalk.
5. Rotate the wheel and stop it at three positions, one-quarter, one-half, and three-quarters of a turn, to check the balance. If the wheel moves from any of these positions, it is not balanced.
6. Loosen the setscrews in the wheel counterbalances, in the grooved recess of the flange, and move the counterbalances opposite the chalk mark (Fig. 83-6).
7. Check the wheel in the four positions mentioned in steps 4 and 5.

DoALL

Figure 83-5 *A grinding wheel balancing stand. (Courtesy The DoAll Co.)*

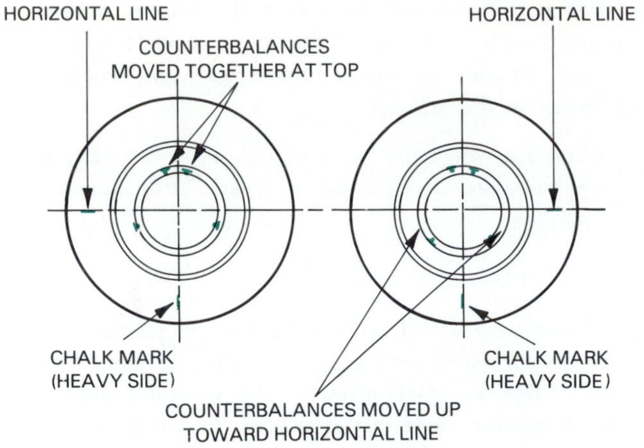
HORIZONTAL LINE
HORIZONTAL LINE
COUNTERBALANCES
MOVED TOGETHER AT TOP
CHALK MARK
(HEAVY SIDE)
CHALK MARK
(HEAVY SIDE)
COUNTERBALANCES MOVED UP
TOWARD HORIZONTAL LINE

Figure 83-6 *Adjusting counterbalances to balance a grinding wheel.*

8. Move the counterbalances around the groove an equal amount on each side of the centerline and check for balance again.
9. Continue to move the balances away from the heavy side until the wheel remains stationary at any position.
10. Tighten the counterbalances in place.

TRUING AND DRESSING A GRINDING WHEEL

After mounting a grinding wheel, it is necessary to true the wheel to ensure that it will be concentric with the spindle.

Truing is the process of making a grinding wheel round and concentric with its spindle axis and to produce the required form or shape on the wheel (Fig. 83-7). This procedure involves the grinding or wearing away of a portion of the abrasive section of a grinding wheel in order to produce the desired form or shape.

Dressing a wheel is the operation of removing the dull grains and metal particles. This operation exposes sharp cutting edges of the abrasive grains to make the wheel cut better (Fig. 83-7). A dull, glazed, or loaded wheel should be dressed for the following reasons.

1. To reduce the heat generated between the work surfaces and the grinding wheel

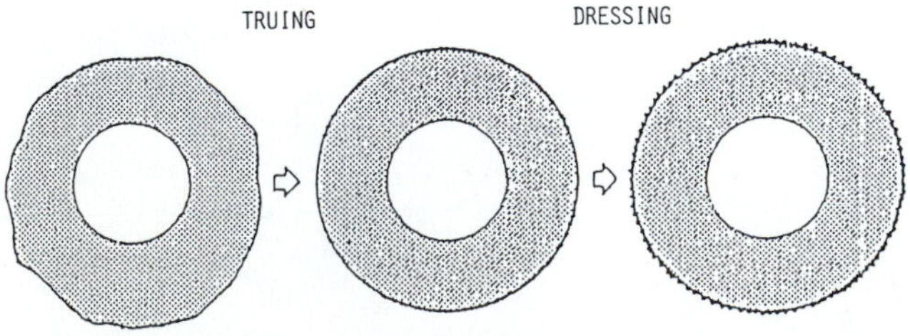
TRUING
DRESSING

Figure 83-7 *Truing makes the wheel round and true with its axis; dressing sharpens the wheel.*

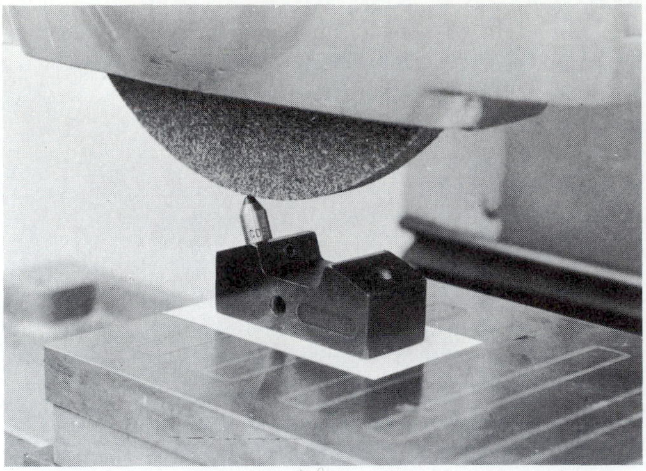

Figure 83-8 *A diamond dresser used to dress a grinding wheel. (Courtesy Kostel Enterprises Limited.)*

2. To reduce the strain on the grinding wheel and the machine
3. To improve the surface finish and accuracy of the work
4. To increase the rate of metal removal

An industrial diamond, mounted in a suitable holder on the magnetic chuck, is generally used to true and dress a grinding wheel (Fig. 83-8).

TO TRUE AND DRESS A GRINDING WHEEL

1. Check the diamond for wear and, if necessary, turn it in the holder to expose a sharp cutting edge to the wheel.

NOTE: Most diamonds are mounted in the holder at an angle of 5 to 15° from the vertical to prevent the possibility of chatter and of the diamond digging into the wheel. It also permits the diamond to wear on an angle so that a sharp cutting edge may be obtained by merely turning the diamond in the holder.

2. Clean the magnetic chuck thoroughly with a cloth and wipe it with the palm of the hand to remove all grit and dirt.
3. Place a piece of paper, slightly larger than the base of the diamond holder, on the left-hand end of the magnetic chuck. This prevents scratching of the chuck when the diamond holder is being removed.
4. Place the diamond holder on the paper, covering as many magnetic inserts as possible, and energize the chuck. The diamond should be pointing in the same direction as the grinding wheel rotation (Fig. 83-9).

NOTE: The diamond dresser should be mounted on the left-hand end of the chuck to prevent flying grit, created during the dressing operation, from pitting and damaging the surface of the magnetic chuck.

5. Raise the wheel above the height of the diamond.

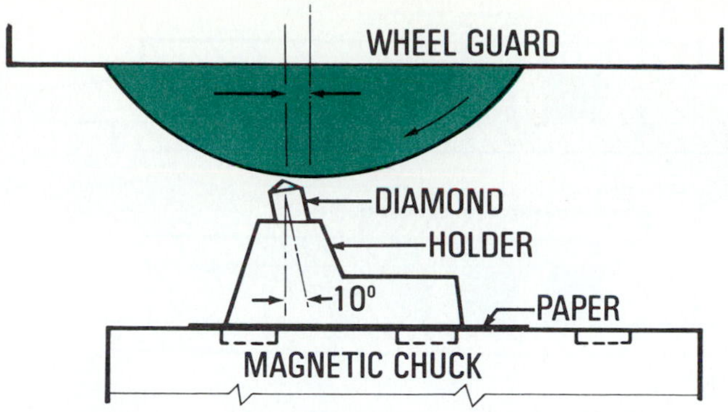
WHEEL GUARD
DIAMOND
HOLDER
10°
PAPER
MAGNETIC CHUCK

Figure 83-9 *Proper positioning of the diamond dresser in relation to the grinding wheel.*

6. Move the table longitudinally so that the diamond is offset approximately ¼ in. (6 mm) to the left of the centerline of the wheel.
7. Adjust the table laterally so that the diamond is positioned under the high point on the face of the wheel (Fig. 83-10). This is important since grinding wheels will wear more quickly on the edges of the wheel, leaving the center of the face higher than the edges.
8. Start the wheel revolving and carefully lower the wheel until the high point touches the diamond.
9. Move the table laterally, using the crossfeed handwheel to feed the diamond across the face of the wheel.
10. Lower the grinding wheel about 0.001 to 0.002 in. (0.02 to 0.05 mm) per pass, and rough-dress the face of the wheel until it is flat and has been dressed all around the circumference.
11. Lower the wheel 0.0005 in. (0.01 mm) and take several passes across the face of the wheel. The rate of crossfeed will vary with the structure of the wheel. A rule of thumb is to use a fast crossfeed with coarse wheels and a slow, but regular, crossfeed for fine, closely spaced grains.

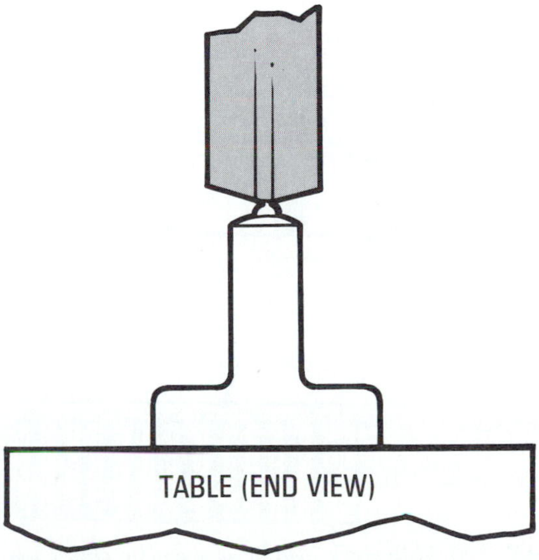
TABLE (END VIEW)

Figure 83-10 *Positioning the diamond under the high point of the wheel.*

The following additional points may be helpful when truing or dressing a grinding wheel.

1. To minimize wear on the diamond, rough-dress the grinding wheel with an abrasive stick.
2. If coolant is to be used during the grinding operation, it is advisable to use coolant when dressing the wheel. This will protect the diamond and the wheel from excessive heat.
3. A loaded wheel is indicated by a discoloration on the periphery or grinding wheel face. When dressing the wheel, take off sufficient material to remove completely any discoloration on the wheel face.
4. If rapid removal of metal is more important than the surface finish, do not finish-dress the wheel. After the wheel has been rough-dressed, some operators will take a final pass of 0.001 to 0.002 in. (0.02 to 0.05 mm) at a high rate of feed. The rough surface produced by this operation will remove the metal more rapidly than a finish-dressed wheel.

WORK-HOLDING DEVICES

THE MAGNETIC CHUCK

In some surface-grinding operations the work may be held in a vise, on V-blocks, or bolted directly to the table. However, most of the ferrous work ground on a surface grinder is held on a *magnetic chuck,* which is clamped to the table of the grinder.

Magnetic chucks may be of two types: the electromagnetic chuck and the permanent magnetic chuck.

The *electromagnetic chuck* uses electromagnets to provide the holding power. It has the following advantages.

- The holding power of the chuck may be varied to suit the area of contact and the thickness of the work.
- A special switch neutralizes the residual magnetism in the chuck, permitting the work to be removed easily.

The *permanent magnetic chuck* provides a convenient means of holding most workpieces to be ground. The holding power is provided by means of permanent magnets. The principle of operation is the same for both electromagnetic and permanent magnetic chucks.

PERMANENT MAGNETIC CHUCK CONSTRUCTION (FIG. 83-11)

The *base plate* provides a base for the chuck and a means of clamping it to the table of the grinder.

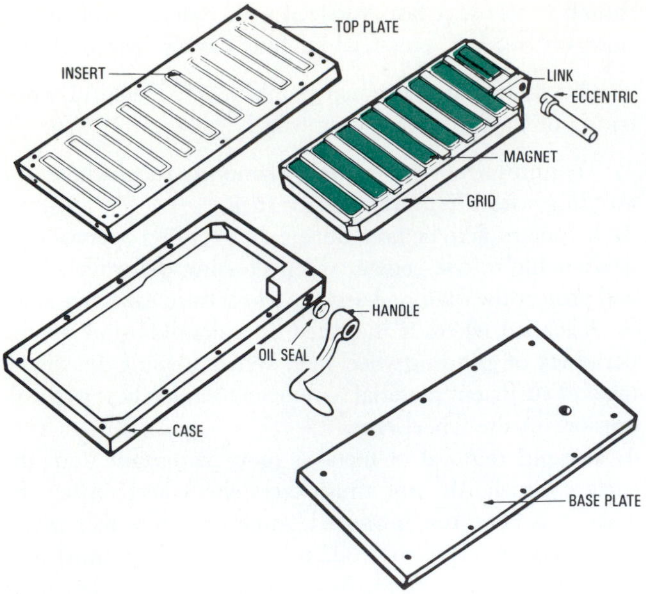
TOP PLATE
INSERT
LINK
ECCENTRIC
MAGNET
GRID
HANDLE
OIL SEAL
CASE
BASE PLATE

Figure 83-11 *Construction of a permanent magnetic chuck. (Courtesy James Neill & Co. Ltd.)*

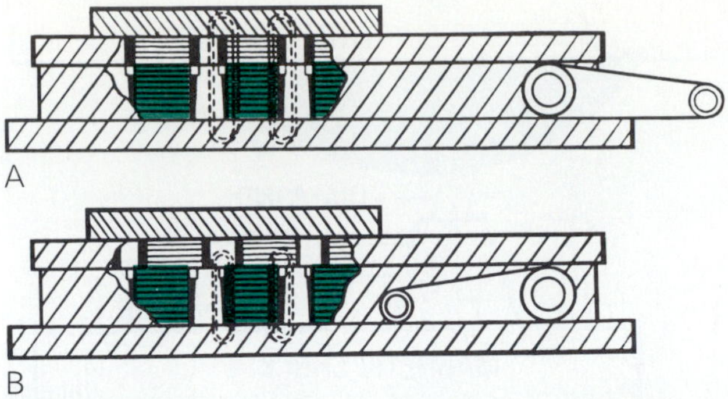
A
B

Figure 83-12 *(A) Magnetic chuck in the on position; (B) magnetic chuck in the off position. (Courtesy James Neill & Co. Ltd.)*

The *grid* or *magnetic pack* houses the *magnets* and the *grid conductor bars*. It is moved longitudinally by a handle when the chuck is placed in the *ON* or *OFF* position.

The *case* houses the grid assembly and permits longitudinal movement of the grid. It also provides an oil reservoir for the lubrication of the moving parts.

The *top plate* contains *inserts* or *pole pieces,* which are separated magnetically from the surrounding plate by means of white metal. This separation provides the poles necessary to conduct the magnetic lines of flux.

When the work is placed on the face of the chuck (top plate) and the handle moved to the *on* position (Fig. 83-12A), the grid conductor bars and the inserts in the top plate are in line. This permits the magnetic flux to pass through the work, holding it onto the top plate.

When the handle is rotated 180° to the *off* position (Fig. 83-12B), the grid assembly moves the grid conductor bars and the inserts out of line. In this position, the magnetic lines of flux enter the top plate and inserts but not the work.

MAGNETIC CHUCK ACCESSORIES

Often it is not possible to hold all work on the chuck. The size, shape, and type of work will dictate how the work should be held for surface grinding. The holding power of a magnetic chuck is dependent on the size of the workpiece, the area of contact, and the thickness of the workpiece. A highly finished piece will be held better than a poorly machined workpiece.

Very thin workpieces will not be held too securely on the face of a magnetic chuck because there are too few magnetic lines of force entering the workpiece (Fig. 83-13A).

An *adapter plate* (Fig. 83-13B) is used to securely hold thin work [less than 1/4 in. (6 mm)]. The plate's alternate layers of steel and brass convert the wider pole spacing of the chuck to finer spacing with more but *weaker* flux paths. This method is particularly suited to small, thin pieces and reduces the possibility of distortion when thin work is being ground.

Magnetic chuck blocks (Fig. 83-14) provide a means of extending the flux paths to hold workpieces that cannot be held securely on the chuck face. V-blocks may be used to hold round or square stock for *light* grinding.

NOTE: Set the chuck blocks so that the maximum number of magnetic loops pass through the workpiece. They

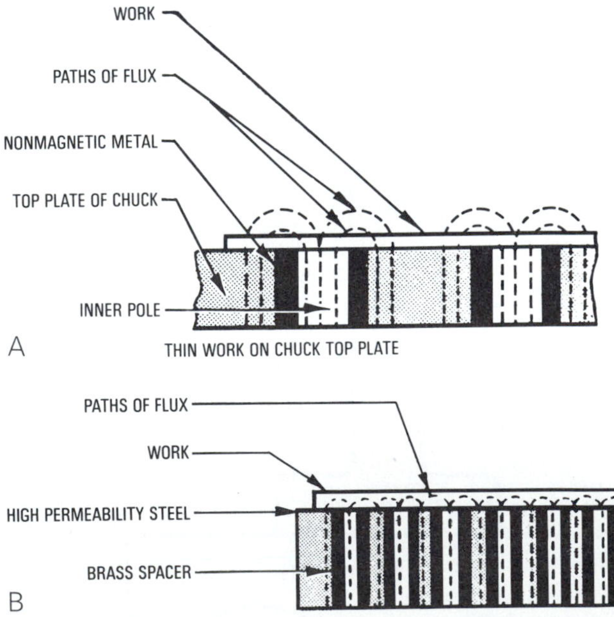
WORK
PATHS OF FLUX
NONMAGNETIC METAL
TOP PLATE OF CHUCK
INNER POLE
A
THIN WORK ON CHUCK TOP PLATE
PATHS OF FLUX
WORK
HIGH PERMEABILITY STEEL
BRASS SPACER
B

Figure 83-13 *(A) Thin work held on a magnetic chuck top plate attracts fewer magnetic lines of force; (B) thin work held on an adapter plate. (Courtesy James Neill & Co. Ltd.)*

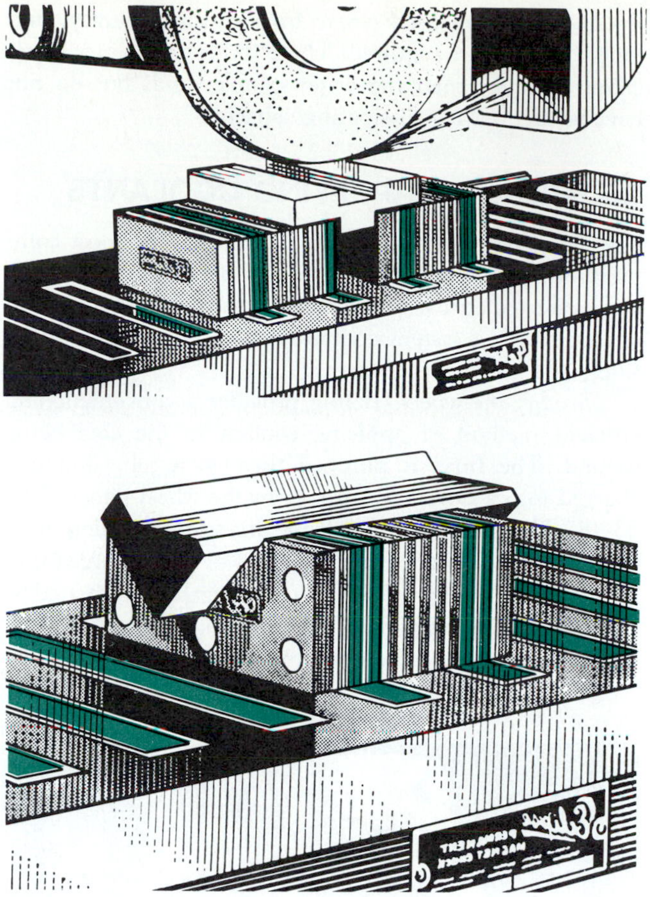

Figure 83-14 *Applications of magnetic chuck blocks. (Courtesy James Neill & Co. Ltd.)*

must also be placed so that the laminations are in line with the inner poles or inserts.

If the following points are observed, magnetic chuck blocks will last longer.

1. Clean thoroughly before and after use.
2. Store in a covered wooden box.
3. Check frequently for accuracy and burrs.
4. If regrinding is necessary to restore accuracy, take light cuts with a dressed wheel. Use coolant when grinding to prevent the magnetic chuck blocks from heating (even slightly).

Sine chuck. When it is required to grind an angle on a workpiece, the work may be set up with a sine bar and clamped to an angle plate. Often a *sine chuck* (Fig. 83-15), which is a form of a magnetic sine plate, is used to hold the work. The buildup for the angles is the same as for the sine bar. Compound sine chucks, which have two plates hinged at right angles to each other, are available for grinding compound angles.

Magna-vise clamps (Fig. 83-16) may be used when the workpiece does not have a large bearing area on the chuck, or when the work is nonmagnetic. These magnetically actuated clamps consist of comblike bars attached to a solid bar by a piece of spring steel. When work is held with these clamps, the solid bar of one clamp is placed against the backing plate of the magnetic chuck. The work is placed on the chuck surface between the toothed edges of two clamps as shown. The toothed edges on the bars in contact with the work should be above the magnetic chuck face. When the chuck is energized, the jaws of the clamps are brought down toward the face of the chuck, locking the work in place.

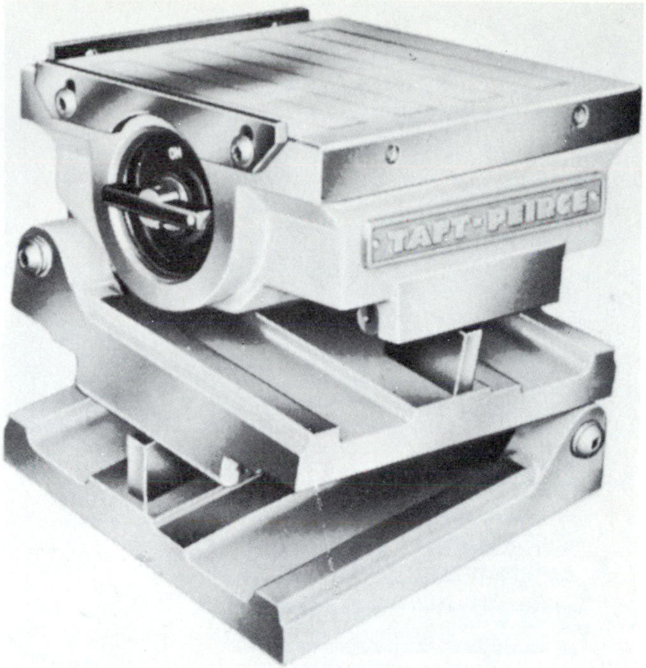
TAFT-PEIRCE

Figure 83-15 *A compound sine chuck. (Courtesy The Taft-Pierce Manufacturing Company.)*

Double-faced tape is often used for holding thin, nonmagnetic pieces on the chuck for grinding. The tape, having two adhesive sides, is placed between the chuck and the work, causing the work to be held securely enough for light grinding.

Special fixtures are often used to hold nonmagnetic materials and odd-shaped workpieces, particularly when a large number of workpieces must be ground.

GRINDING FLUIDS

Although work is ground dry in many cases, most machines have provision for applying grinding fluids or coolants. Grinding fluids serve four purposes.

1. *Reduction of grinding heat,* which affects work accuracy, surface finish, and wheel wear.
2. *Lubrication* of the surface between the workpiece and the grinding wheel, which results in a better surface finish.

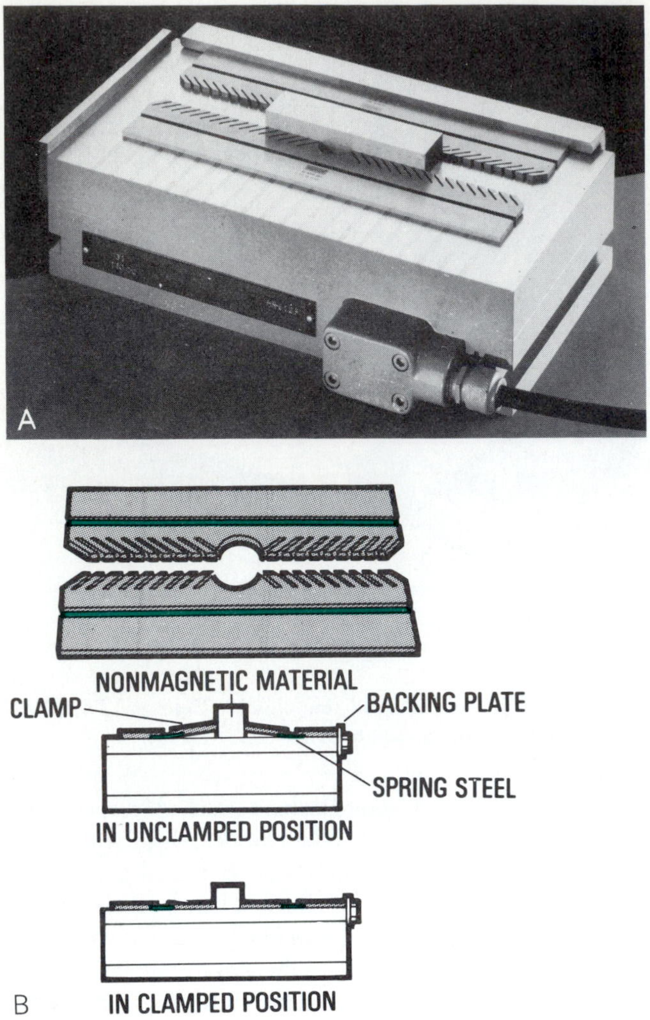
A
NONMAGNETIC MATERIAL
CLAMP
BACKING PLATE
SPRING STEEL
IN UNCLAMPED POSITION
B
IN CLAMPED POSITION

Figure 83-16 *(A) Nonmagnetic workpiece being held for grinding with magna-vise clamps. (Courtesy Magna Lock Corporation); (B) magna-vise clamps used to hold workpiece on a magnetic chuck. (Courtesy Brown & Sharpe Mfg. Co.)*

3. *Removal of swarf* (small metal chips and abrasive grains) from the cutting area.
4. *Control of grinding dust,* which may present a health hazard.

TYPES OF GRINDING FLUIDS

1. *Soluble oil and water* when mixed form a milky solution which provides excellent cooling, lubricating, and rust-resistant qualities. This solution is generally applied by flooding the surface of the work.
2. *Soluble chemical grinding fluids and water* when mixed form a grinding fluid, which may be used with flood cooling or "through-the-wheel" cooling systems. The chemical grinding fluid contains rust inhibitors and bactericides to minimize odors and skin irritation.
3. *Straight oil grinding fluids,* generally applied by the flood system, are used where high finish, accuracy, and long wheel life are required. These fluids have better lubricating qualities than the water-soluble fluids but do not have as high a heat-dissipating capacity.

METHODS OF APPLYING COOLANTS

The *flood system* (Fig. 83-17) is probably the most common form of coolant application. Bt this method, the coolant is directed onto the workpiece by a nozzle and is recirculated through a system containing a reservoir, a pump, a filter, and a control valve.

Through-the-wheel cooling provides a convenient and efficient method of applying coolant to the area being ground. The fluid is pumped through a tube and discharged into a dovetailed groove in the wheel flange (Fig. 83-18). Holes through the flange and corresponding holes in the rubber washer permit the fluid to be discharged into the porous grinding wheel. The centrifugal force, created by the high-speed rotation of the wheel, forces the fluid through the wheel onto the area of contact between the

Figure 83-17 *Many grinding operations use the flood system to keep the workpiece cool. (Courtesy DoAll Company.)*

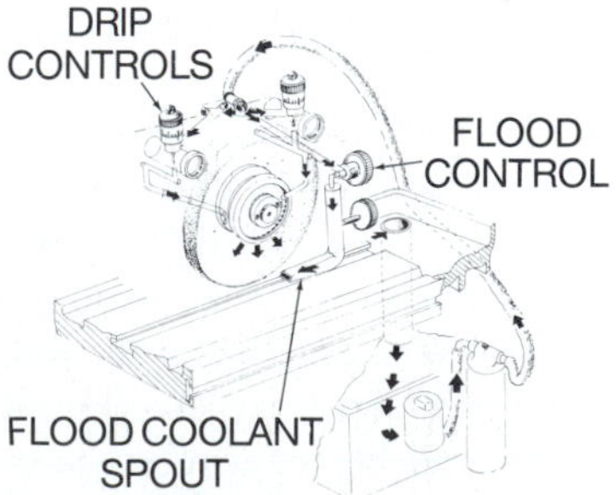
DRIP CONTROLS
FLOOD CONTROL
FLOOD COOLANT SPOUT

Figure 83-18 *Through-the-wheel system applies coolant to the point of grinding contact. (Courtesy DoAll Company.)*

wheel and the work. Some machines have a coolant reservoir above the grinding wheel guard which feeds the coolant into the wheel flange groove by gravity.

The *mist cooling system,* which supplies coolant in the form of a mist, uses the atomizer principle. Air passes through a line containing a T-connection, which leads to the coolant reservoir. The velocity of the air as it passes through the T-connection draws a small amount of coolant from the reservoir and discharges it through a small nozzle in the form of vapor. The nozzle is directed to the point of contact between the work and the wheel. The air and the vapor, as it evaporates, cause the cooling action. The force of the air also blows away the swarf.

Figure 83-19 *Poor surface finish may be caused by a clogged wheel or by using the wrong grinding wheel.*

SURFACE FINISH

The finish produced by a surface grinder is important, and the factors affecting it should be considered. Some parts that are ground do not require a fine surface finish, and time should not be spent producing fine finishes if they are not required.

The following factors affect the surface finish.

- **Material being ground.** Soft material, such as brass and aluminum, will not permit as high a finish as harder ferrous materials. A much finer finish can be produced on hardened-steel workpieces than can be produced on soft steel or cast iron.
- **Amount of material being removed.** If a large amount of material is to be removed, a coarse-grit, open-structure wheel should be used. This will not produce as fine a finish as a fine-grit, dense wheel.
- **Grinding wheel selection.** A wheel containing abrasive grains which are friable (fracture easily) will produce a better finish than a wheel made up of tough grains. A fine-grit, dense-structure wheel produces a smoother surface than a coarse-grit, open wheel. A grinding wheel which is too soft releases the abrasive grains too easily, causing them to roll between the wheel and the work, creating deep scratches in the work (Fig. 83-19).
- **Grinding wheel dressing.** An improperly dressed wheel will leave a pattern of scratches on the work. Care should be taken when finish dressing the wheel to move the diamond slowly across the wheelface. Always dress the wheel sufficiently to expose new abrasive grains and ensure that all glazing or foreign particles have been removed from the periphery of the wheel. New grinding wheels *which have not been properly balanced and trued* will produce a chatter pattern on the surface of the work.
- **Condition of the machine.** A light machine or one with loose spindle bearings will not produce the accuracy and fine surface finish possible in a rigid machine with properly adjusted spindle bearings. Also, to ensure optimum accuracy and surface finish, the machine should be kept clean.
- **Feed.** Coarse feeds tend to produce a rough finish. If "feed lines" persist when a fine feed is used, the wheel edges should be rounded slightly with an abrasive stick.

REVIEW QUESTIONS

GRINDING MACHINES AND PROCESS

1. Discuss how grinding has contributed to interchangeable manufacture.
2. How has the role of grinding changed over the years?
3. Outline the action that takes place during a grinding operation.
4. List five important rules which apply to any grinding operation.

SURFACE GRINDING

5. Define surface grinding.
6. Name four different types of surface grinders and briefly outline the principle of each.

PARTS OF THE HYDRAULIC SURFACE GRINDER

7. Name five *main* parts of the hydraulic surface grinder.
8. Name and state the purposes of five controls found on the hydraulic surface grinder.

GRINDING WHEEL CARE

9. List four points to be observed in grinding wheel care.

TO MOUNT A GRINDING WHEEL

10. List the steps required to mount a grinding wheel.
11. Why is a blotter on each side of the wheel necessary when a grinding wheel is being mounted?

TO BALANCE A GRINDING WHEEL

12. Why is the proper balance of a grinding wheel essential?
13. Describe briefly the procedure for balancing a grinding wheel.

TO TRUE AND DRESS A GRINDING WHEEL

14. Define truing and dressing.
15. Why are most diamond dressers mounted at an angle of 10 to 15° to the base?
16. Explain how a grinding wheel should be finish-dressed for:
    a. rough grinding b. finish grinding

WORK-HOLDING DEVICES

17. List the advantages and disadvantages of an electromagnetic chuck.
18. Describe the construction and operation of a permanent magnetic chuck.

MAGNETIC CHUCK ACCESSORIES

19. How are thin workpieces held for surface grinding? Why is this necessary?
20. What precautions must be observed when using magnetic chuck blocks?
21. Describe and state the purpose of magna-vise clamps.

GRINDING FLUIDS

22. State four purposes of grinding fluids.
23. Name and describe three methods of applying coolant.

SURFACE FINISH

24. List any five factors that affect the surface finish on the part being ground. How do these factors affect the surface finish?

UNIT 84

Surface-Grinding Operations

The surface grinder is used primarily for grinding flat surfaces on hardened or unhardened workpieces. Since the workpiece can be held by various methods and the wheel face can be shaped by dressing, it is possible to perform such operations as form, angular, and vertical grinding. Surface grinding brings the work to close tolerances and produces a high surface finish.

Good surface-grinding results depend on several factors, such as the proper mounting of the work, and the proper wheel selection for the job. A knowledge of the controls and grinding safety practices are of the utmost importance before anyone attempts to use a surface grinder.

OBJECTIVES

After completing this unit you should be able to:

1. Set up various workpieces for grinding
2. Observe the safety rules to operate the grinder
3. Grind flat vertical and angular surfaces

MOUNTING THE WORKPIECE FOR GRINDING

The size, shape, and type of work will determine the method by which the work should be held for surface grinding.

FLAT WORK OR PLATES

1. Remove all burrs from the surface of the work.
2. Clean the chuck surface with a clean cloth and then wipe the palm of the hand over the surface to remove dirt.
3. Place a piece of paper slightly larger than the workpiece in the center of the magnetic chuck face.
4. Place the work on top of the paper, and be sure to straddle as many magnetic inserts as possible.
5. If the workpiece is warped and rocks on the chuck face, shim the work where necessary to prevent rocking. This will avoid distortion when the work is removed from the magnetic chuck.
6. Turn the handle to the *ON* position.
7. Check the work to see that it is held securely by trying to remove the workpiece.

THIN WORKPIECES

Thin workpieces tend to warp because of the heat created during the grinding operation. To minimize the amount of heat generated, it is advisable to mount the workpiece at an angle of approximately 15 to 30° from the side of the chuck (Fig. 84-1). This reduces the length of time the wheel is in contact with the work, which in turn reduces the amount of heat generated per pass. If an adapter is available, it should be used, and the work mounted at an angle.

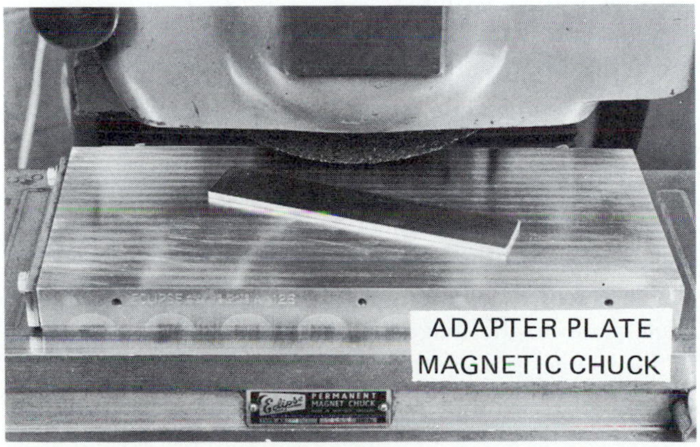
ADAPTER PLATE
MAGNETIC CHUCK

Figure 84-1 *A small, thin workpiece should be set at an angle on the adapter plate to minimize warpage caused by grinding heat.*

SHORT WORKPIECES

Work which does not straddle three magnetic poles will generally not be held firmly enough for grinding. It is advisable to straddle as many poles as possible and to set parallels or steel pieces around the work to prevent it from moving during the grinding operation (Fig. 84-2). The parallels or steel pieces should be slightly thinner than the workpiece to provide maximum support.

GRINDER SAFETY

When operating any type of grinder, it is important that certain basic, time-tested safety precautions be observed. Generally, the safest grinding practice is also the most efficient.

1. Before mounting a grinding wheel, ring test the wheel to check for defects.
2. Be sure that the grinding wheel is properly mounted on the spindle.
3. See that the wheel guard covers at least one-half the wheel.
4. Make sure that the magnetic chuck has been turned on by trying to remove the work.
5. See that the grinding wheel clears the work before starting a grinder.
6. Be sure that the grinder is operating at the correct speed for the wheel being used.
7. When starting a grinder, always stand to one side of the wheel and make sure no one is in line with the grinding wheel in case it breaks on startup.

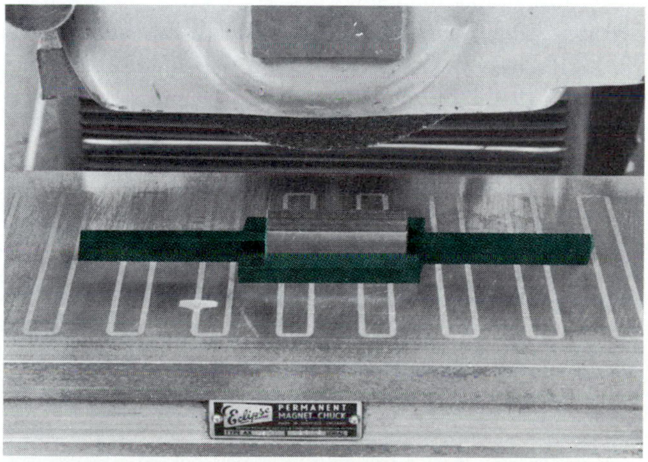

Figure 84-2 *Steel blocks or parallels are placed around a short workpiece to prevent it from moving during grinding. (Courtesy Kostel Enterprises Limited.)*

8. Never attempt to clean the magnetic chuck or mount and remove work until the wheel has stopped completely.
9. *Always* wear safety glasses when grinding.

GRINDING OPERATIONS

The most common operation performed on a surface grinder is the grinding of flat (horizontal) surfaces. Regardless of the type of grinding operation, it is important that the correct wheel be mounted and the work held securely.

TO GRIND A FLAT (HORIZONTAL) SURFACE

1. Remove all burrs and dirt from the workpiece and the face of the magnetic chuck.
2. Mount the work on the chuck, placing a piece of paper between the chuck and the workpiece.

NOTE: Paper is used so that the work may be easily lifted off the magnetic chuck rather than sliding it. Sliding traps the abrasive grains between the work and the chuck face, which will scratch and damage the chuck.

3. Check to see that the work is held firmly.
4. Set the table reverse dogs so that the center of the grinding wheel clears each end of the work by approximately 1 in. (25 mm).
5. Set the crossfeed for the type of grinding operation—roughing cuts, 0.030 to 0.050 in. (0.76 to 1.27 mm); finishing cuts, 0.005 to 0.020 in. (0.12 to 0.50 mm).
6. Bring the work under the grinding wheel by hand, *having about $\frac{1}{8}$ in. (3 mm) of the wheel edge over the work* (Fig. 84-3).
7. Start the grinder and lower the wheelhead until the wheel just sparks the work.
8. The wheel may have been set on a low spot of the work. It is good practice, therefore, to always raise the wheel about 0.005 in. (0.12 mm).
9. Start the table traveling automatically and feed the entire width of the work under the wheel to check for high spots.
10. Lower the wheel for every cut until the surface is completed—roughing cuts 0.001 to 0.003 in. (0.02 to 0.07 mm), finishing cuts 0.0005 to 0.001 in. (0.01 to 0.02 mm).

NOTE: If the surface finish of the work is not satisfactory, refer to Table 84-1 for the possible cause and suggested remedies.

11. Release the magnet and remove the workpiece, by raising one edge, to break the magnetic attraction. This will prevent scratching the chuck surface.

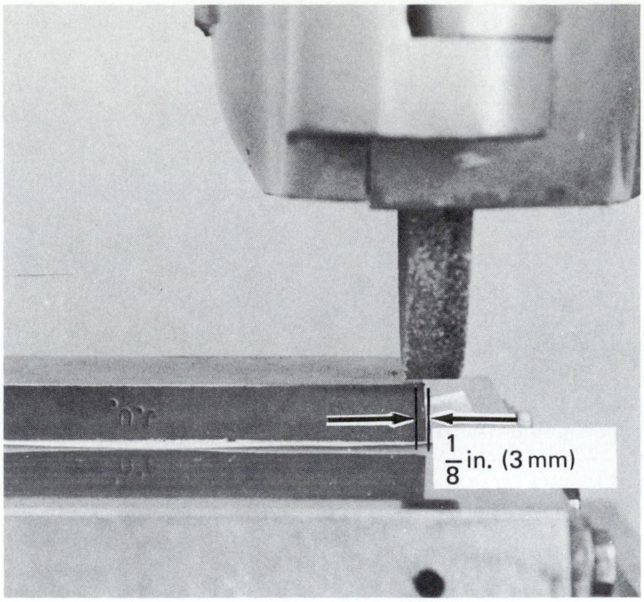

$\frac{1}{8}$ in. (3 mm)

Figure 84-3 *The edge of the wheel should overlap the workpiece by about 1/8 in. (3 mm). (Courtesy Kostel Enterprises Limited.)*

NOTE: Cutting fluid should be used whenever possible to aid the grinding action and keep the work cool.

TO GRIND THE EDGES OF A WORKPIECE

Much work which is machined on a surface grinder must have the edges ground square and parallel so that these edges may be used for further layout or machining operations.

Work which is to be ground all over should be machined to about 0.010 in. (0.25 mm) over the finished size for each surface. The large, flat surfaces are usually ground first, which then permits them to be used as reference surfaces for further setups.

When the four edges of a workpiece must be ground, clamp the work to an angle plate so that two adjacent sides may be ground square without moving the workpiece.

Setting Up the Workpiece

1. Clean and remove all burrs from the workpiece, the angle plate, and the magnetic chuck.
2. Place a piece of paper which is slightly larger than the angle plate on the magnetic chuck.
3. Place one end of the angle plate on the paper (Fig. 84-4).
4. Place a flat-ground surface of the workpiece against the angle plate so that the top and one edge of the workpiece project about 1/2 in. (13 mm) beyond the edges of the angle plate (Fig. 84-4).

NOTE: Be sure that the one edge of the work does not project beyond the base of the angle plate. If the work is smaller than the angle plate, a suitable parallel must be used to bring the top surface beyond the end of the angle plate.

5. Hold the work firmly against the angle plate and turn on the magnetic chuck.
6. Clamp the work to the angle plate and set the clamps so that they will not interfere with the grinding operation.

NOTE: Place a piece of soft metal between the clamp and the work to prevent marring the finished surface.

7. Turn off the magnetic chuck and carefully place the base of the angle plate on the magnetic chuck (Fig. 84-5).
8. Carefully fasten two more clamps on the end of the workpiece to hold the work securely.

Table 84-1 Surface-Grinding Problems, Causes, and Remedies

Grinding Problem	Cause	Possible Remedy
Burning or discoloration	Wheel is too hard.	Use a softer, free-cutting wheel. Decrease wheel speed. Increase work speed. Coarse-dress the wheel. Take lighter cuts and dress the wheel frequently. Use coolant directed at point of contact between the wheel and the work.
Burnished work surface (work is highly polished in irregular patches)	Wheel is glazed.	Dress the wheel. Use a coarser-grit wheel. Use a softer wheel. Use a more open-structure wheel.
Chatter or wavy pattern	Wheel is out of balance. Wheel is out of round. Spindle bearing is too loose. Wheel is too hard. Glazing of wheel.	Rebalance. True and dress. Adjust or replace bearings. Use a softer wheel, coarser grit, or more open structure. Increase table speed. Redress the wheel.
Scratches on the work surface	Grinding wheel is too soft. (Abrasive grains break off too readily and catch between the wheel and work surface.) Wheel is too coarse. Loose particles of swarf fall onto the work from the wheel guard. Dirty coolant carries dirt particles onto the work surface. Feed lines.	Use a harder wheel. Use a finer-grit wheel. Clean the grinding wheel guard when changing a wheel. Clean the coolant tank and replace the coolant. Slightly round the edges of the wheel.

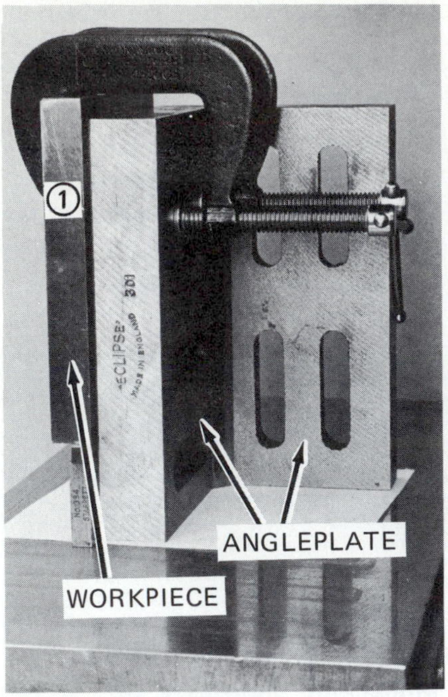
1
ANGLEPLATE
WORKPIECE

Figure 84-4 *The workpiece may be clamped to an angle plate for grinding the edges square. (Courtesy Kostel Enterprises Limited.)*

Grinding the Edges of a Workpiece Square and Parallel

After the work has been properly set up on the magnetic chuck, the following procedure should be used for grinding the four edges of the workpiece.

1. Raise the wheelhead so that it is about ½ in. (13 mm) above the top of the work.

1

Figure 84-5 *A workpiece set up for grinding the first edge. (Courtesy Kostel Enterprises Limited.)*

2. Set the table reverse dogs so that each end of the work clears the grinding wheel by about 1 in. (25 mm).
3. With the work under the center of the wheel, turn the crossfeed handle until *about $\frac{1}{8}$ in. (3 mm) of the wheel edge overlaps the edge of the work* (Fig. 84-3).
4. Start the grinding wheel and lower the wheelhead until the wheel just sparks the work.
5. Move the work clear of the wheel with the crossfeed handle.
6. Raise the wheel about 0.005 to 0.010 in. (0.12 to 0.25 mm) in case the wheel had been set to a low spot on the work.
7. Check for high spots by feeding the table by hand so that the entire length and width of the work passes under the wheel. Raise the wheel if necessary.
8. Engage the table reverse lever and grind the surface until all marks are removed. The depth of cut should be 0.001 to 0.003 in. (0.02 to 0.07 mm) for roughing cuts and 0.0005 to 0.001 in. (0.01 to 0.02 mm) for finishing cuts.
9. Stop the machine and remove the clamps from the right-hand end of the work.
10. Turn off the magnetic chuck and remove the angle plate and workpiece as one unit. Be careful not to jar the work setup.
11. Clean the chuck and the angle plate.
12. Place the angle plate (with the attached workpiece) on its end with the surface to be ground at the top (Fig. 84-6).

2
1

Figure 84-6 *Angle plate and work set for grinding the second edge of the workpiece at 90° to the first edge. (Courtesy Kostel Enterprises Limited.)*

13. Fasten two clamps to the right-hand side of the workpiece and the angle plate.
14. Remove the original clamps from the top of the setup.
15. Repeat steps 1 to 8 and grind the second edge.
16. Remove the assembly from the chuck and remove the workpiece from the angle plate.

Grinding the Third and Fourth Edges

When two adjacent sides have been ground, they are then used as reference surfaces to grind the other two sides square and parallel.

1. Clean the workpiece, the angle plate, and the magnetic chuck thoroughly and remove any burrs.
2. Place a clean piece of paper on the magnetic chuck
3. Place a ground edge of the workpiece on the paper.
   a. If the workpiece is at least 1 in. (25 mm) thick and long enough to span three magnetic poles on the chuck, and no more than 2 in. (50 mm) high, no angle plate is required (Fig. 84-7).
   b. If the work is less than 1 in. (25 mm) thick and does not span three magnetic poles, it should be fastened to an angle plate (Fig. 84-8).
      (1) Place a ground edge on the paper and place an angle plate no higher than the workpiece against the workpiece.

NOTE: A suitable parallel may be required to raise the edge of the work above the edge of the angle plate.

      (2) Turn on the chuck and carefully clamp the work to the angle plate.
4. Grind the third edge to the required size.
5. Repeat operations 1 to 3 and grind the fourth edge.

Figure 84-7 *A workpiece having a sufficient bearing surface may be set on the chuck for finishing the remaining edges. (Courtesy Kostel Enterprises Limited.)*

Figure 84-8 *An angle plate may be required to finish the third and fourth edge. (Courtesy Kostel Enterprises Limited.)*

TO GRIND A VERTICAL SURFACE

Although most grinding performed on a surface grinder is the grinding of flat horizontal surfaces, it is often necessary to grind a vertical surface (Fig. 84-9). Extreme care must be taken in the setup of the workpiece when grinding the vertical surface. Before grinding a vertical surface, it is necessary

PARALLELS

Figure 84-9 *The setup for vertical grinding.*

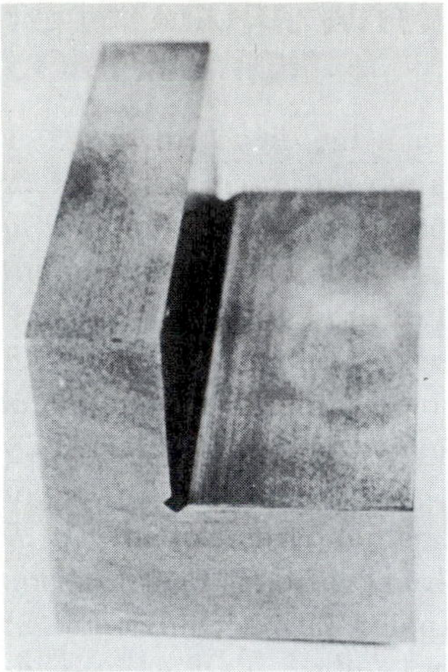

Figure 84-10 *Corner of the workpiece is relieved to provide clearance and prevent the corner of the wheel from breaking down. (Courtesy Kostel Enterprises Limited.)*

to relieve the corner of the work (Fig. 84-10) to ensure clearance for the edge of the wheel.

PROCEDURE

1. Mount the proper grinding wheel; true, dress, and balance as required.
2. Dress the side of the wheel to give it a slight clearance (Fig. 84-11).
3. Clean the surface of the magnetic chuck and mount the work.
4. With an indicator, align the edge of the work parallel to the table travel.

OR

Place the work against the stop bar of the magnetic chuck which has been aligned. If the work cannot be set against the stop bar, parallels may be used to position the work on the magnetic chuck (Fig. 84-9).

5. Turn on the magnetic chuck and test to see that the work is held securely.
6. Set the reversing dogs, allowing sufficient table travel to permit clearance for the wheel at each end of the stroke.
7. Bring the side of the wheel close to the vertical surface to be ground.
8. Lower the wheel to within 0.002 to 0.005 in. (0.05 to 0.12 mm) of the flat or horizontal surface which has been finish ground.
9. Start the table traveling *slowly* and feed the wheel across until it just sparks the vertical surface.
10. Rough-grind the vertical surface to within 0.002 in. (0.05 mm) of size by feeding the table in approximately 0.001 in. (0.02 mm) maximum per pass.
11. Redress the side of the wheel if necessary.
12. Finish-grind the vertical surface by feeding the table approximately 0.0005 in. (0.01 mm) maximum per pass.

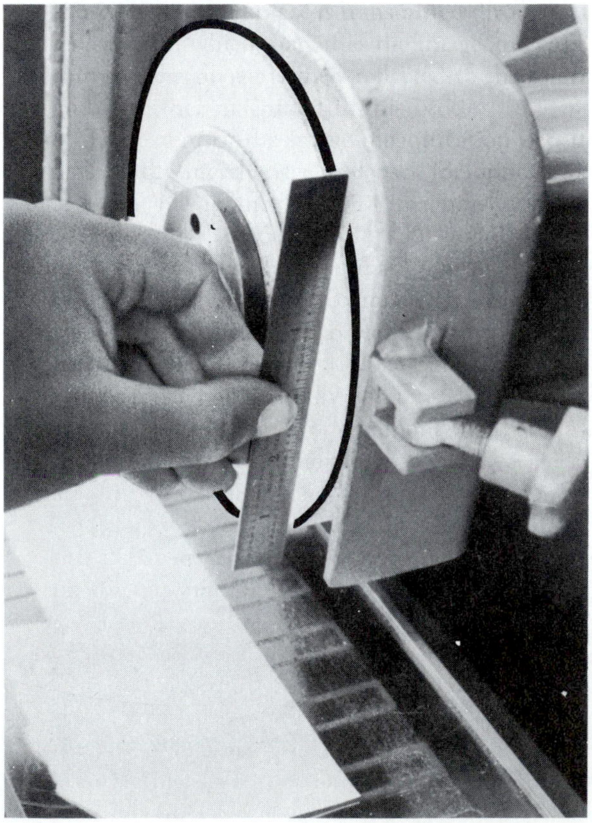

Figure 84-11 Left: *dressing the side of a wheel for grinding a vertical surface.* Right: *the side of the wheel should be slightly concave for grinding a vertical surface.*

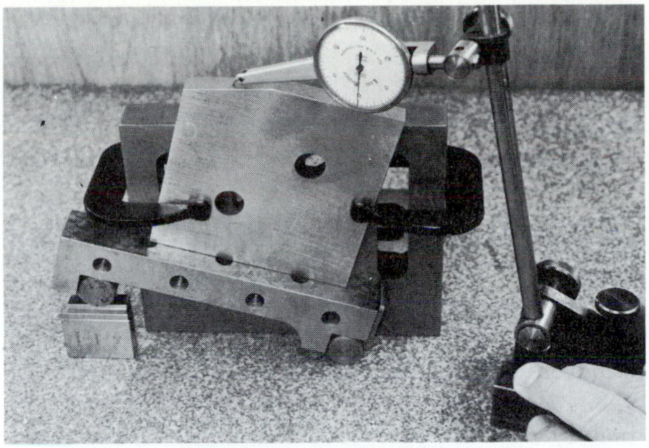

Figure 84-12 *The workpiece may be set to a sine bar for grinding an accurate angle.*

TO GRIND AN ANGULAR SURFACE

When it is necessary to grind an angular surface, the work may be held at an angle by a sine bar and angle plate (Fig. 84-12), a sine chuck (Fig. 84-13), or an adjustable angle vise (Fig. 84-14). When work is held by any of these methods, it is ground with a flat-dressed wheel.

Angular surfaces may also be ground by holding the work flat and dressing the grinding wheel to the required angle with a sine dresser (Fig. 84-15).

When a sine dresser is not available, a parallel set to the desired angle by means of a sine bar may be clamped to an angle plate. This setup is then placed on a magnetic chuck beneath the grinding wheel (Fig. 84-16).

DoALL
Magna-Lock

Figure 84-13 *Grinding an accurate angle using a sine chuck. (Courtesy DoAll Company.)*

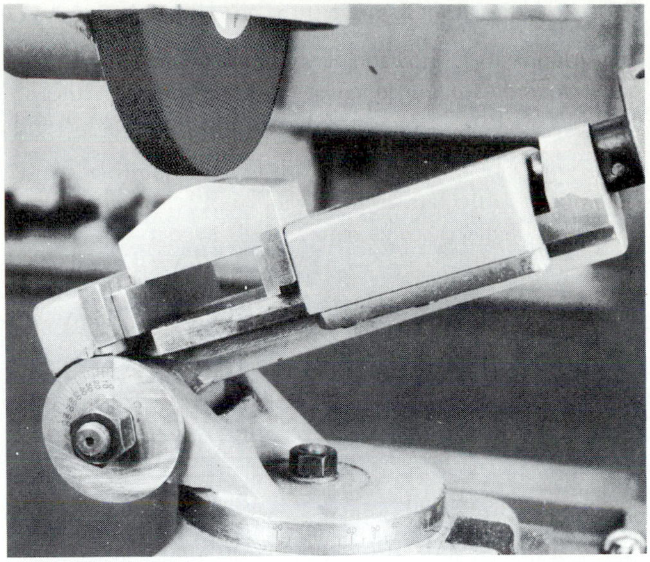

Figure 84-14 *Workpiece held for grinding in an adjustable angle vise.*

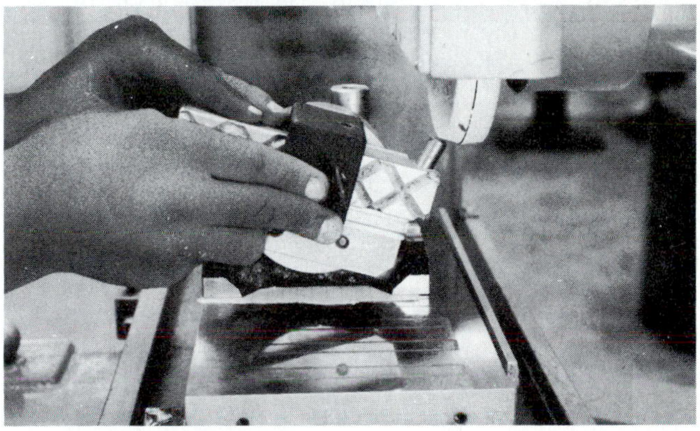

Figure 84-15 *Dressing a wheel to an angle using a sine dresser.*

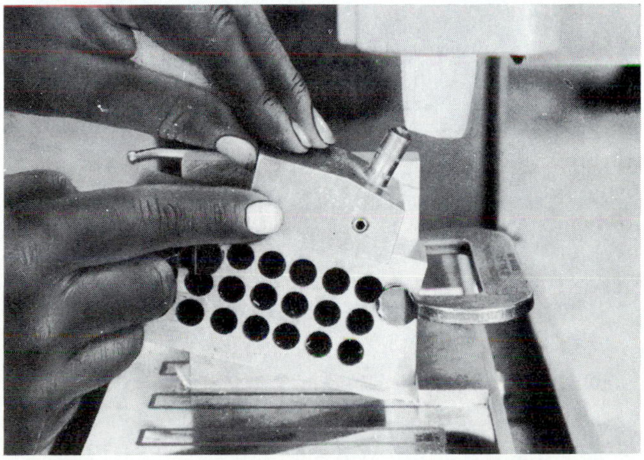

Figure 84-16 *Dressing a wheel using a parallel set to an angle.*

FORM GRINDING

Form grinding refers to the production of curved and angular surfaces produced by means of a specially dressed wheel. The reverse form or contour required on the workpiece is dressed on the grinding wheel (Fig. 84-17). Contours and radii may be produced on the grinding wheel by means of a radius wheel dresser (Fig. 84-18A).

TO DRESS A CONVEX RADIUS ON A GRINDING WHEEL

1. Mount the radius dresser (Fig. 84-18A) squarely on a clean magnetic chuck.
2. Set both stops on the radius dresser so that it can only be rotated one-quarter of a turn. The two stops should be 90° apart.
3. Fasten the diamond height-setting bar in the radius dresser. The bottom surface of the height-setting bar is the center of the radius dresser.
4. Place a gage block buildup using wear blocks on each side, equal to the radius required on the grinding wheel, between the height-setting bar and the diamond point.
5. Raise the diamond until it just touches the gage blocks (Fig. 84-18B) and then lock it in this position.

NOTE: When dressing a *concave radius,* the diamond point must be set *above* the center of the radius dresser a distance equal to the radius desired.

6. Move the table longitudinally until the diamond is under the center of the grinding wheel (Fig. 84-18A).
7. Lock the table to prevent longitudinal movement.
8. Rotate the arm of the radius dresser one-quarter of a turn, so that the diamond is in a horizontal position.
9. Start the machine spindle and, using the crossfeed handle, bring the diamond in until it just touches the side of the grinding wheel.
10. Lock the table cross-slide in this position.

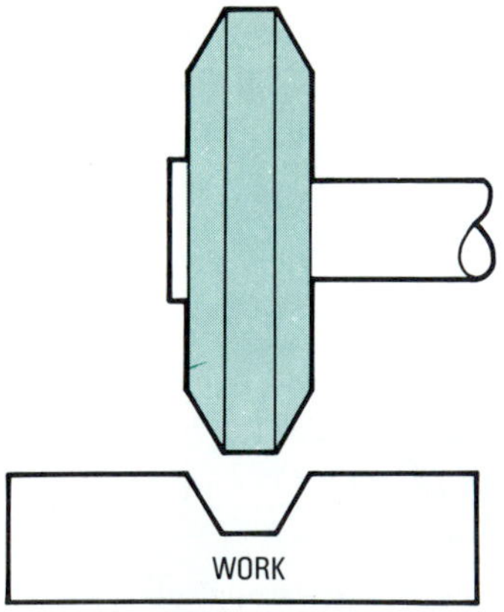
WORK

Figure 84-17 *For form grinding, the wheel is dressed to the reverse profile of the form required.*

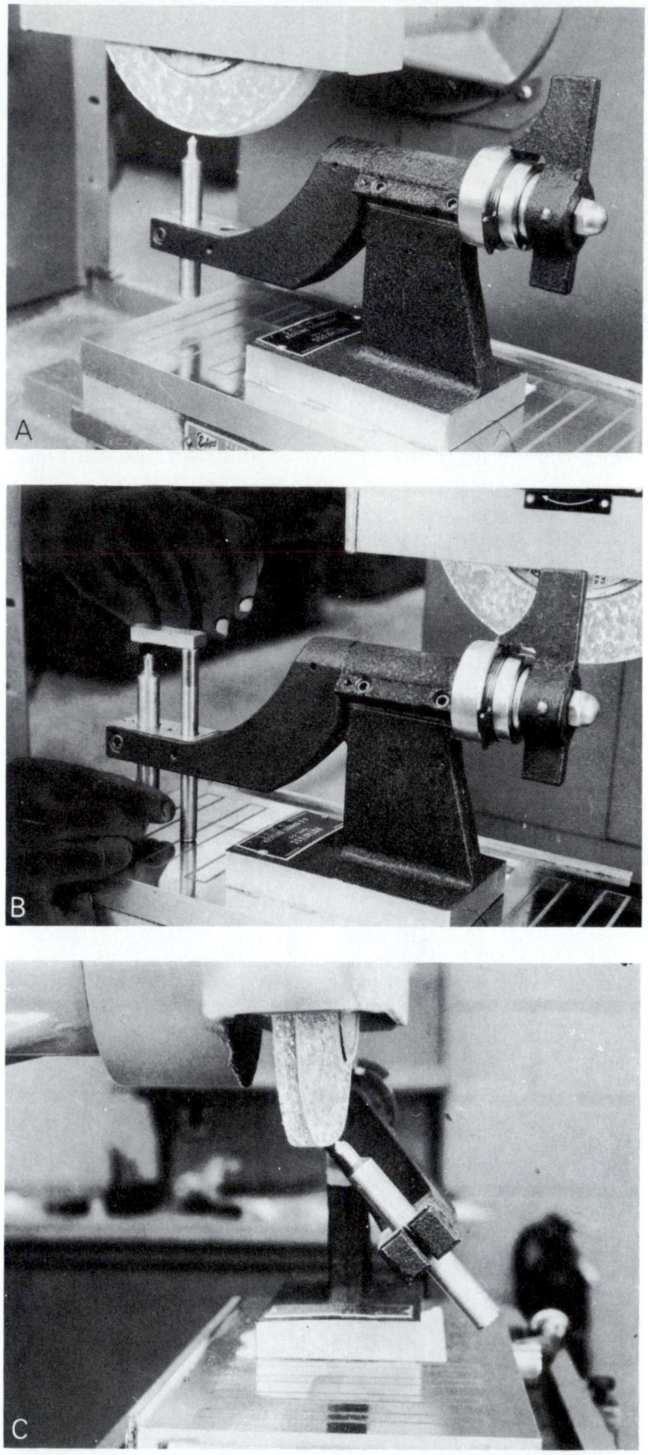
A
B
C

Figure 84-18 *(A) A radius wheel dresser mounted on a magnetic chuck; (B) gage blocks being used to set a diamond to the correct height to dress the radius on a grinding wheel; (C) a convex radius being dressed on a grinding wheel.*

11. Stop the grinder and raise the wheel until it clears the diamond.
12. Start the grinder, and while slowly rotating the diamond back and forth through the 90° arc, lower the wheel until it just touches the diamond.
13. Feed the wheel down approximately 0.002 to 0.003 in. (0.05 to 0.07 mm) for every rotation of the dresser.
14. Continue to dress the radius until the periphery of the wheel just touches the diamond when it is in a vertical position. This indicates that the radius is completely formed (Fig. 84-18C).
15. Stop the grinder, raise the wheel, and remove the radius dresser.

SPECIAL FORMS

When complex profiles must be ground on long production runs, the wheel may often be crush-formed. A tool-steel or carbide roll, having the desired form or contour of the finished workpiece, is forced into the slowly revolving grinding wheel [200 to 300 sf/min (60 to 90 m/min)] (Fig. 84-19). The grinding wheel assumes the reverse form of the crushing roll. The wheel is then used to grind the form or contour on the workpiece (Fig. 84-20). Grinding wheels may be crush-formed to tolerances as close as ±0.002 in. (0.05 mm), and to radii as small as 0.005 in. (0.12 mm), depending on the grit size and structure of the wheel. As the wheel is used, it will gradually wear out of tolerance and the form must be redressed, using the crush roll. When the crush roll wears, as a result of many redressings, it must be reground to the original tolerance.

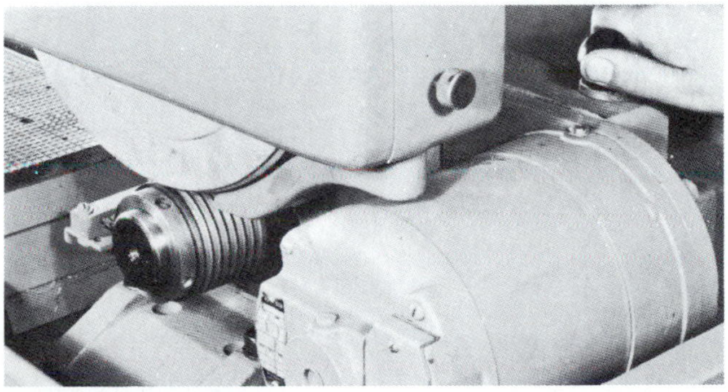

Figure 84-19 *Crush-dressing a wheel to grind serrations, with a roll mounted on the machine.*

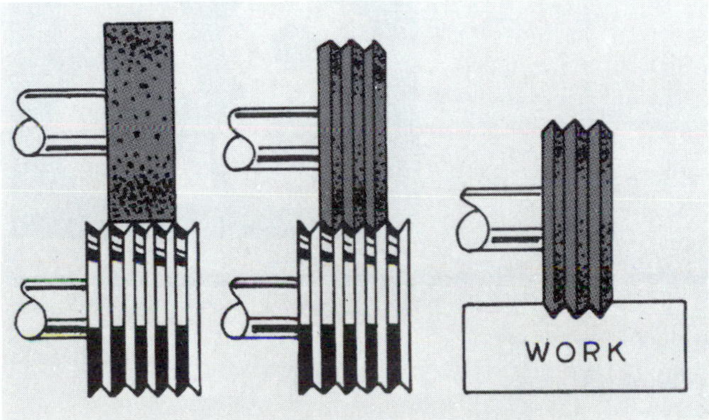
WORK

Figure 84-20 *Principle of crush-form grinding.*

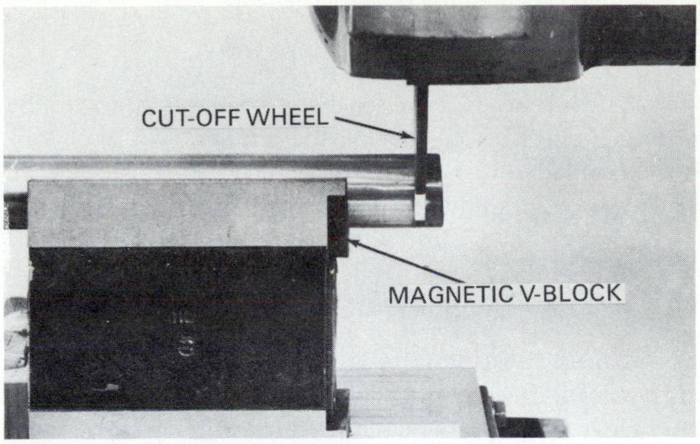
CUT-OFF WHEEL
MAGNETIC V-BLOCK

Figure 84-21 *Cutting off a workpiece held in a magnetic V-block on a surface grinder. (Courtesy Kostel Enterprises Limited.)*

Some surface grinders are not designed for crush-form dressing of the wheel. It is not advisable to perform crush-form dressing on any machine equipped with a ball bearing spindle, since the bearings are subjected to a considerable load in crush dressing and may be damaged. Machines equipped with roller bearings have proved quite satisfactory for crush-dressing operations.

CUTTING-OFF OPERATIONS

The surface grinder may be used for cutting off hardened materials by using thin cutoff wheels. The work may be clamped in a fixture or vise and positioned below the wheel (Fig. 84-21). For thin, short pieces, the wheelhead may be fed straight down to cut off the work. If longer pieces are to be cut, the work is properly mounted and the table is reciprocated as in normal grinding, while the wheel is fed down. Diamond wheels may also be mounted for cutting-off operations on carbides.

REVIEW QUESTIONS

MOUNTING THE WORKPIECE FOR GRINDING

1. Describe briefly the procedure for mounting the following for grinding:
   a. Flat work
   b. Short work

GRINDER SAFETY

2. How should a wheel be checked for defects?
3. Why is it necessary to see that no one is in line with the grinding wheel before starting the grinder?
4. List five grinder safety rules which you consider to be most important and explain the reason for the selection of each.

TO GRIND A FLAT SURFACE

5. Explain how the grinding wheel should be set to the surface of the work.
6. How should work be removed from a magnetic chuck?

TO GRIND THE EDGES OF A WORKPIECE

7. Why are the edges of workpieces often ground square and parallel?
8. How much allowance should be left on a surface for grinding?
9. What precautions should be observed before mounting any workpiece or accessory on the magnetic chuck?
10. What precautions should be observed when mounting a workpiece on an angle plate for grinding two adjacent sides?
11. After the first side of a workpiece has been ground, how is the work set to grind the second side square to the first?
12. What precautions should be observed when setting the grinding wheel to the work surface?

TO GRIND A VERTICAL SURFACE

13. Outline briefly the procedure for grinding a vertical surface.
14. When grinding a vertical surface, why is it necessary first to relieve the corner between the two adjacent surfaces?
15. How is a grinding wheel dressed when grinding a vertical surface? Why is this necessary?

TO GRIND AN ANGULAR SURFACE

16. By what methods may work be held for grinding an angle when a flat dressed wheel is being used?
17. Name two methods of dressing a wheel to an angle.

FORM GRINDING

18. Name two methods of dressing contours and radii on grinding wheels.
19. Describe the principle of crush-form dressing.
20. On what types of surface grinders should crush-form dressing be performed?

SURFACE GRINDING PROBLEMS, CAUSES, AND REMEDIES

21. List three causes and three remedies for each of the following grinding problems.
    a. Chatter or wavy pattern
    b. Scratches on the work surface

UNIT 85

Cylindrical Grinders

When the diameter of a workpiece must be ground accurately to size and to a high surface finish, this may be done on a *cylindrical* grinder. There are two types of machines suitable for cylindrical grinding—the *center type* and the *centerless type*—each with its special applications.

OBJECTIVES

After completing this unit you should be able to:

1. Set up and grind work on a cylindrical grinder
2. Internal-grind on a universal cylindrical grinder
3. Identify and state the principles of three methods of centerless grinding

CENTER-TYPE CYLINDRICAL GRINDERS

Work which is to be finished on a cylindrical grinder is generally held between centers but may also be held in a chuck. There are two types of cylindrical grinders (center type), the *plain* and the *universal.* The plain-type grinder is generally a manufacturing-type of machine. The universal cylindrical grinder (Fig. 85-1) is more versatile, since both the wheelhead and headstock may be swiveled.

PARTS OF THE UNIVERSAL CYLINDRICAL GRINDER

The *base* is of heavy cast-iron construction to provide rigidity. The top of the base is machined to form the *ways* for the table.

The *wheelhead* is mounted on a cross-slide at the back of the machine. The ways on which it is mounted are at right angles to the table ways, permitting the wheelhead to be fed toward the table and the work, either automatically or by hand. On universal machines, the wheelhead may be

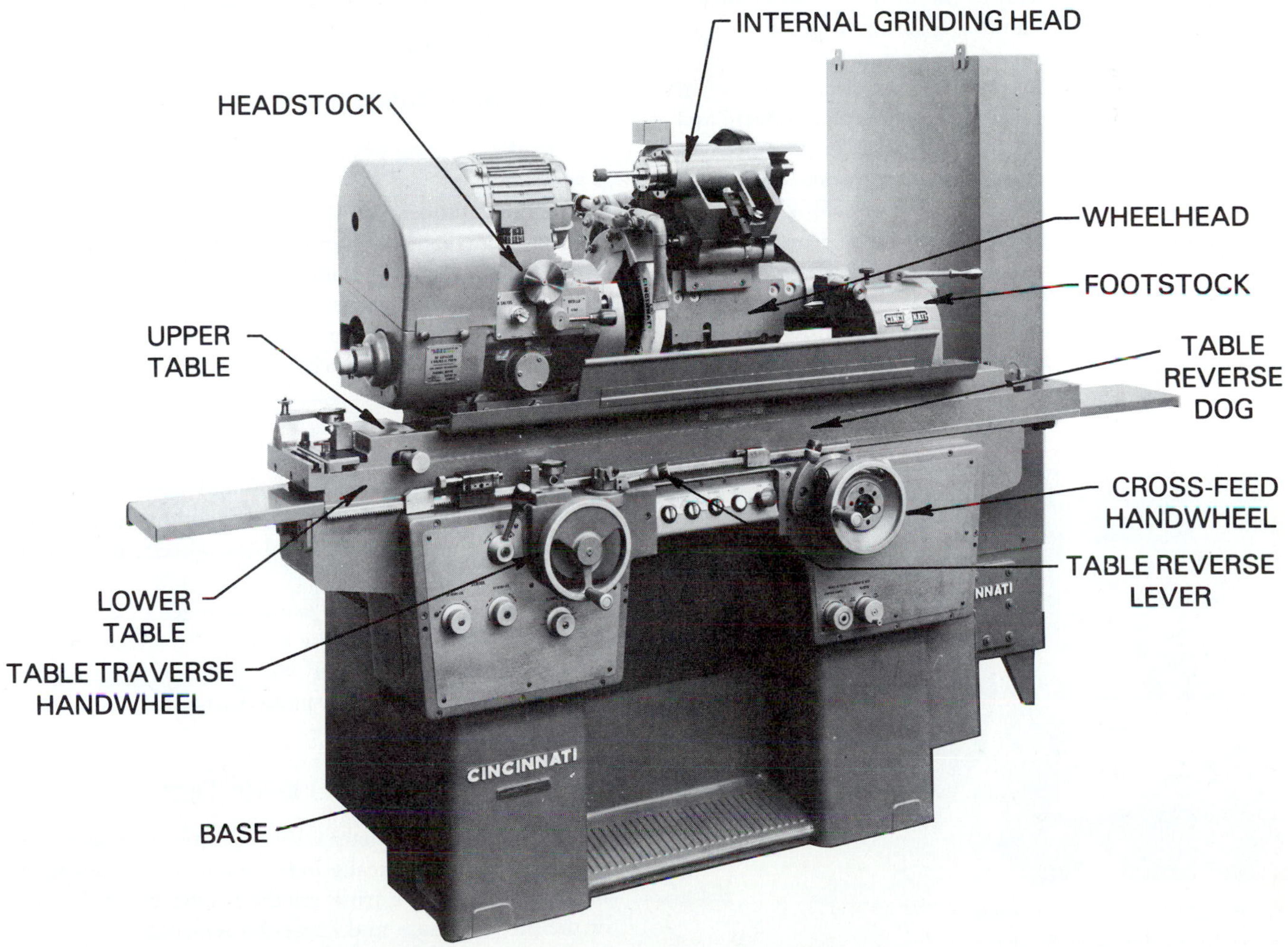

INTERNAL GRINDING HEAD
HEADSTOCK
WHEELHEAD
FOOTSTOCK
UPPER TABLE
TABLE REVERSE DOG
CROSS-FEED HANDWHEEL
TABLE REVERSE LEVER
LOWER TABLE
TABLE TRAVERSE HANDWHEEL
BASE
CINCINNATI

Figure 85-1 *The parts of a universal cylindrical grinder. (Courtesy Cincinnati Milacron Inc.)*

swiveled to permit the grinding of steep tapers by plunge grinding.

The *table,* mounted on the ways, is driven back and forth by hydraulic or mechanical means. The reversal of the table is controlled by *trip dogs.* The table is composed of the *lower table,* which rests on the ways, and the *upper table,* which may be swiveled for grinding tapers and alignment purposes. The *headstock* and *footstock,* used to support work held between centers, are mounted on the table.

The *headstock* unit is mounted on the left end of the table and contains a motor for rotating the work. A dead center is mounted in the headstock spindle. When work is mounted between centers, it is rotated on *two dead centers* by means of a dog and a drive plate, which is attached to, and revolves with, the headstock spindle. The purpose of a dead center in the headstock is to overcome any spindle inaccuracies (looseness, burrs, etc.) which may be transferred to the workpiece. Grinding work on two dead centers results in truer diameters, which are concentric with the centerline of the work. Work may also be held in a chuck which is mounted on the spindle nose of the headstock.

The *footstock* supports the right end of the work and is adjustable along the length of the table. The dead center, on which the work is mounted, is spring-loaded to provide the proper center tension on the workpiece.

The *backrest* or *steadyrest* provides support for long, slender work and prevents it from springing. Outward and downward movement of the workpiece is prevented by means of adjustable supports in front of and below the workpiece. It may be positioned anywhere along the length of the table.

The *center rest,* which resembles a lathe steadyrest, may be mounted at any point on the table. It is used to support the right end of the work when external grinding is confined to the end of the workpiece. Long workpieces on which internal grinding is to be performed are also supported on the end by the center rest.

An *internal grinding attachment* may be mounted on the wheelhead on most machines for internal grinding. It is usually driven by a separate motor.

A *diamond wheel dresser* may be clamped to the table to dress the grinding wheel as required. On some types of grinders, diamond dressers may be permanently mounted on the footstock.

The *coolant system* is built into all cylindrical grinders to provide dust control, temperature control, and a better surface finish on the workpiece.

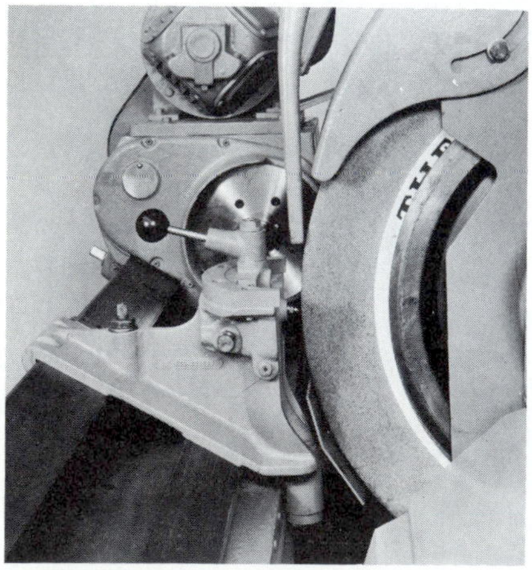

Figure 85-2 *Dressing a grinding wheel on a cylindrical grinder. (Courtesy Cincinnati Milacron Inc.)*

MACHINE PREPARATION FOR GRINDING

MOUNTING THE WHEEL

All precautions, such as wheel balancing and mounting procedures, used on surface grinders should be observed for cylindrical grinding.

To True and Dress the Wheel

1. Start the grinding wheel to permit the spindle bearings to warm up.
2. Mount the proper diamond in the holder and clamp it to the table. The diamond should be mounted at an angle of 10 to 15° to the wheelface and should be held on or slightly below the centerline of the wheel (Fig. 85-2).
3. Adjust the wheel until the diamond almost touches the high point of the wheel, which is usually in the center of the wheelface.
4. Turn on the coolant if this is to be used for the grinding operation.
5. Feed the wheel into the diamond about 0.001 in. (0.02 mm) per pass, and move it back and forth across the wheelface at a medium rate until the wheelface has been completely dressed.
6. Finish-dress the wheel by using an infeed of 0.0005 in. (0.01 mm) and a slow traverse feed. When a very fine finish is required, the diamond should be traversed slowly across the wheelface for a few passes without any additional infeed.

To Parallel-Grind an Outside Diameter

Grinding a parallel diameter is the most common operation performed on a cylindrical grinder. If any undesirable work characteristics occur during grinding, refer to Table 85-1 for the possible cause and suggested remedies.

1. Lubricate the machine as required.

2. Start the grinding wheel to warm up the spindle bearings. This will ensure the utmost accuracy when grinding.

3. True and dress the grinding wheel if required and then shut off the spindle.

4. Clean the machine centers and the center holes of the work. If the grinder centers are damaged, they must be reground. On hardened-steel workpieces, the center holes should be honed or lapped to ensure the utmost accuracy.

5. Align the headstock and footstock centers with a test bar and indicator.

6. Lubricate the center holes with a suitable lubricant.

7. Set the headstock and footstock for the proper length of work so that the center of the work will be over the center of the table.

8. Mount the work between centers with the dog mounted loosely on the left end of the work.

9. Tighten the dog on the end of the work and engage the drive plate pin in the fork of the dog.

10. Adjust the table dogs so that the wheel will overrun each end of the work by about one-third of the width of the wheelface. If grinding must be done up to a shoulder, the table traverse must be carefully set so that it reverses just before the wheel touches the shoulder.

11. Set the grinder to the proper speed for the wheel being used. Some machines are provided with a means of increasing the wheel speed as the wheel becomes smaller. If this is not done, the wheel will act softer and wear quickly.

12. Set the work speed for the diameter and type of material being ground. Proper work speed is very important. A slow speed causes heating and distortion of the work. High work speeds will cause the wheel to act softer and break down quickly.

13. Set the headstock spindle to rotate the work in an opposite direction to that of the grinding wheel. When grinding, the sparks should be directed down toward the table.

14. If the machine is so equipped, set the automatic infeed for each table reversal. Also set the dwell or "tarry" time, which permits the wheel to clear itself at each end of the stroke.

15. Select and set the desired table traverse or speed. This should be such that the table will move one-half to two-thirds of the wheel width per revolution of the work. Finish grinding is done at a slower rate of table traverse.

16. Start the table and move the wheel up to the workpiece until it just sparks.

17. Engage the wheel feed clutch lever and grind until the work is cleaned up.

18. Check the work for taper and adjust if necessary.

19. Determine the amount of material to be removed and set the feed index for this amount. The infeed of the wheel will stop automatically when the work is at the proper diameter.

20. Stop the machine with the wheel clear of the work and measure the size of the workpiece. If necessary, make a correction on the index setting and grind the piece to size.

TO GRIND A TAPERED WORKPIECE

Tapered work, held between centers, is ground in the same manner as parallel work, except that the table is swiveled to half the included angle of the taper. The taper should be checked for accuracy after the workpiece is cleaned up and the table adjusted if necessary.

Short, steep tapers may be ground on work held in a chuck or between centers by swiveling the wheelhead to the desired angle and plunge-grinding the tapered suface with the face of the wheel.

Table 85-1 Cylindrical Grinding Faults, Causes, and Remedies

Fault	Cause	Possible Remedy
"Barber Pole" finish	Work loose on centers.	Adjust center tension.
	Centers are a poor fit in the spindle.	Use properly fitting centers.
Burnt work, cracked surfaces	Grinding wheel not trued.	True and dress the wheel.
	Grinding wheel too hard.	Use a softer wheel.
	Grinding wheel structure too dense.	Use a more open wheel.
	Incorrect bond.	Consult manufacturer's handbook.
	Grinding wheel too fast.	Adjust the speed.
	Work revolving too slowly.	Increase the speed.
	Too heavy a cut.	Reduce the depth of cut.
	Insufficient coolant.	Increase coolant supply.
	Wrong type of coolant.	Try other type.
	Dull diamond dresser.	Turn diamond in holder.

Table 85-1 Cylindrical Grinding Faults, Causes, and Remedies *(Continued)*

Fault	Cause	Possible Remedy
Chatter marks	Too heavy a cut.	Try a lighter cut.
	Grinding wheel too hard.	Use a softer wheel.
		Increase work speed.
		Reduce wheel speed.
	Work too slender.	Use a steadyrest.
	Vibrations in machine.	Locate the vibrations and correct.
	External vibrations transferred to machine.	Isolate the machine to prevent vibrations.
Diamond truing lines	Fast dressing feed.	Slow dressing feed.
	Diamond too sharp.	Use a cluster-type nib.
Feed lines	Wrong wheel structure.	Change to suit.
	Improper traverse speed when finish grinding.	Change work speed and wheel speed.
	Coolant not directed properly.	Adjust nozzle.
	Improperly adjusted steadyrest.	Check and adjust.
Intermittent cutting action	Work too tight on centers.	Adjust center tension.
	Work too hot.	Use coolant.
	Work out of balance.	Counterbalance as required.
Out-of-round work	Work center holes damaged or dirty.	Hone and lap centers.
	Work loose on centers.	Adjust the center tension.
	Loose machine centers.	Clean and reset.
	Machine centers worn.	Regrind centers.
	Loose gibs on the table.	Adjust the gibs.
Rough finish	Wheel too coarse.	Use a finer wheel.
	Wheel too hard.	Use a softer wheel.
	Diamond too sharp.	Use a cluster-type nib.
	Wheel rough-dressed.	Finish-dress wheel.
	Table traverse too fast.	Slow to suitable feed.
	Work speed too fast.	Slow to suitable speed.
	Work springing.	Use a steadyrest.
Wavy marks (long)	Wheel out of balance.	Redress and balance wheel.
	Coolant has been directed against a stationary wheel.	Always shut off coolant well before stopping the wheel.
Wavy marks (short)	Vibrations caused by unmatched belts.	Replace belts with a matched set.
		Check motor and pulleys for balance.

PLUNGE GRINDING

When a short tapered or parallel surface is to be ground on a workpiece, it may be plunge-ground by feeding the wheel into the revolving work, with the table remaining stationary. The grinding wheel may be fed in automatically to the setting on the feed index. It then dwells for a suitable time to permit "spark out" and retracts automatically. In the case of tapered work, the wheelhead must be swung to half the included angle. The length of the surface to be ground must be no longer than the width of the grinding wheel face.

INTERNAL GRINDERS

Internal grinding may be defined as the accurate finishing of holes in a workpiece by a grinding wheel. Although internal grinders were originally designed for hardened workpieces, they are now used extensively for finishing holes to size and accuracy in soft material.

Production internal grinding is done on internal grinders designed exclusively for this type of work. The wheel is fed into the work automatically until the hole reaches the required diameter. When the hole is finished to size, the wheel is withdrawn from the hole and automatically dressed before the next hole is ground.

Internal grinding may also be performed on the universal cylindrical grinder, the cutter and tool grinder, and the lathe. Since these machines are not designed primarily for internal grinding, they are not as efficient as the standard internal grinder. For most internal grinding operations, the work is rotated in a chuck mounted on the workhead spindle. Work may also be mounted on a faceplate, collet chucks, or special fixtures. When work is too large to be rotated, internal diameters may be ground by using a planetary grinder. In this operation, the grinding wheel is guided in a circular motion about the axis of the hole and is fed out to the required diameter. The work or the grinding head is fed parallel to the wheel spindle to provide a smooth, uniform surface.

INTERNAL GRINDING ON A UNIVERSAL CYLINDRICAL GRINDER

Although the universal cylindrical grinder is not designed primarily as an internal grinder, it is used extensively in toolrooms for this purpose. On most universal cylindrical grinders, the internal grinding attachment is mounted on the wheelhead column and is easily swung into place when required. One advantage of this machine is that the outside and inside diameters of a workpiece may often be finished in one setup. Although the grade of the wheel used for internal grinding will depend on the type of work and the rigidity of the machine, the wheels used for internal grinding are generally softer than those for external grinding, for the following reasons.

1. There is a larger area of contact between the wheel and the workpiece during the internal grinding operation.
2. A soft wheel requires less pressure to cut than a hard wheel; thus the spindle pressure and spring is reduced.

To Grind a Parallel Internal Diameter on a Universal Cylindrical Grinder

If any problems occur during internal grinding, refer to Table 85-2 for the possible cause and suggested remedies.

1. Mount the workpiece in a universal chuck or a collet chuck or on a faceplate. Care must be taken not to distort thin workpieces.
2. Swing the internal grinding attachment into place and mount the proper spindle in the quill. For maximum rigidity, the spindle should be as large as possible, with the shortest overhang.
3. Mount the proper grinding wheel, as large as possible, for the job.
4. Adjust the spindle height until its center is in line with the center axis of the hole in the workpiece.
5. True and dress the grinding wheel.
6. Set the wheelspeed to 5000 to 6500 sf/min (1520 to 1980 m/min).
7. Set the workspeed to 150 to 200 sf/min (45 to 60 m/min).
8. Adjust the table dogs so that *only* one-third of the wheel width overlaps the ends of the work at each end of the stroke. On blind holes, the dog should be set to reverse the table just as the wheel clears the undercut at the bottom of the hole.

NOTE: In order to prevent bell-mouthing, the wheel must *never* overlap the end of the work by more than one-half the width of the wheel.

9. Start the work and the grinding wheel.
10. Touch the grinding wheel to the diameter of the hole.
11. Turn on the coolant.
12. Grind until the hole just cleans up, feeding the wheel in no more than 0.002 in. (or 0.05 mm) per table reversal.
13. Check the hole size and set the automatic feed (if the machine is so equipped) to disengage when the work is roughed to within 0.001 in. (0.02 mm) of size.
14. Reset the automatic infeed to 0.0002 in. (0.005 mm) per table reversal.
15. Finish-grind the work and let the wheel spark out.
16. Move the table longitudinally and withdraw the wheel and spindle from the workpiece.
17. Check the hole diameter and finish grind if necessary.

Table 85-2 Internal Grinding Problems, Causes, and Remedies

Problem	Cause	Remedy
Bell-mouthed hole	Stroke is too long and wheel overlaps hole too much.	Reduce overlap of wheel at each end of hole.
	Centerlines of workpiece and wheel spindle are at different heights.	Align the centers before setting up work.
Burning or discoloration of work	Wheel too hard.	Use a softer wheel.
		Increase work speed.
		Decrease diameter of wheel.
		Use a narrower wheel.
		Coarse dress the wheel.
	Insufficient coolant.	Increase coolant supply and direct it at the point of grinding contact.
Chatter marks	Worn spindle bearings.	Adjust if possible or replace.
	Belt slipping.	Adjust tension.
	Defective belts.	Replace the complete set of belts.
	Wheel out of balance.	Balance the wheel.
	Wheel not true.	True and dress.
	Wheel too hard.	Use a softer wheel.
	Incorrect work speed.	Adjust.
Feed lines or spirals	Improper dressing.	Dress the wheel carefully, using a sharp diamond.
	Wheel too hard.	Use a softer wheel.
	Edges of wheel too sharp.	Round edges slightly with an abrasive stick.
	Feed too coarse.	Reduce feed on final passes.
	Wheelhead is tipped or swung.	Align wheelhead and spindle.
Out-of-round hole	Work is distorted during mounting in the chuck or holding device.	Use extreme care when mounting work.
	Work overheated during rough grinding.	Reduce depth of cut and feed. If work is mounted on a faceplate, loosen each clamp slightly and retighten evenly.
Scratches on ground surface	Wheel too soft and abrasive grains are caught between work surface and wheel.	Use a harder wheel.
	Improperly dressed wheel.	Carefully dress the wheel.
	Dirty coolant deposits particles between the wheel and the work.	Clean coolant tank and replace coolant.
	Wheel too coarse.	Use a finer wheel.
Tapered hole	Workhead set at a slight angle.	Align workhead.
	Wheel too soft to hold size.	Use a harder wheel.
	Feed too fast.	Reduce feed.

Table 85-2 Internal Grinding Problems, Causes, and Remedies *(Continued)*		
Problem	**Cause**	**Remedy**
Wheel glazing	Wheel too hard.	Use a softer wheel. Increase work speed.
	Wheel too dense.	Use a more open wheel.
	Improper dressing.	Use a sharp diamond and rough-dress.
Wheel loading	Wheel too hard.	Use a softer wheel. Increase work speed. Increase traverse feed.
	Wheel too fine.	Use a coarser grain.
	Dirty coolant.	Clean coolant system and replace coolant.
	Truing diamond is dull.	Use a sharp diamond and coarse-dress the wheel.

TO GRIND A TAPERED HOLE

The same procedures and precautions should be followed for grinding tapered holes as for grinding parallel holes. However, the workhead must be set to one-half the included angle of the taper. It is *very* important that the centerlines of the grinding wheel and the hole be set at the same height in order to produce the correct taper.

CENTERLESS GRINDERS

The production of cylindrical, tapered, and multidiameter workpieces may be achieved on a centerless grinder (Fig. 85-3). As the name suggests, the work is not supported on centers but rather by a work rest blade, a regulating wheel, and a grinding wheel (Fig. 85-4).

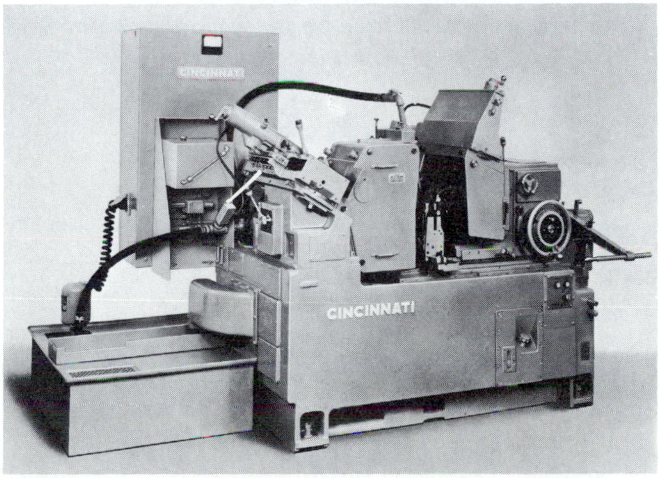
CINCINNATI
CINCINNATI

Figure 85-3 *A centerless grinder. (Courtesy Cincinnati Milacron Inc.)*

On a centerless grinder, the work is supported on the work rest blade, which is equipped with suitable guides for the type of workpiece. The rotation of the grinding wheel forces the workpiece onto the work rest blade and against the regulating wheel, while the regulating wheel controls the speed of the work and the longitudinal feed movement. To provide longitudinal feed to the work, the regulating wheel is set at a slight angle. The rate of feed may be varied by changing the angle and the speed of the regulating wheel. The regulating and grinding wheels rotate in the same direction, and the center heights of these wheels are fixed. Because the centers are fixed, the diameter of the workpiece is controlled by the distance between the wheels and the height of the work rest blade.

The higher the workpiece is placed above the centerlines of the wheels, the faster it will be ground cylindrical. However, there is a limit to the height at which it may be placed, since the work will eventually be lifted periodically from the work rest blade. There is one exception to placing the work above center: when removing slight bends in long, small-diameter work; in this case the center of the piece is placed

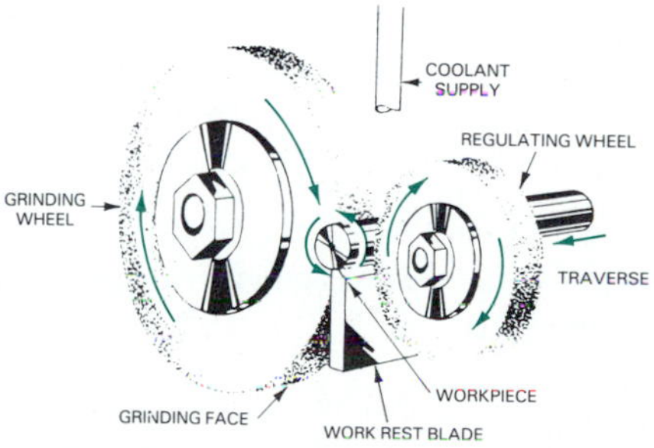
COOLANT SUPPLY
REGULATING WHEEL
GRINDING WHEEL
TRAVERSE
WORKPIECE
GRINDING FACE
WORK REST BLADE

Figure 85-4 *Principle of the centerless grinder.*

below the centerline of the wheels and the rate of traverse is high. This operation eliminates whipping and chattering that might result from bent work and is used primarily for straightening the workpiece. After the work has been straightened, it is ground in the normal manner above centers.

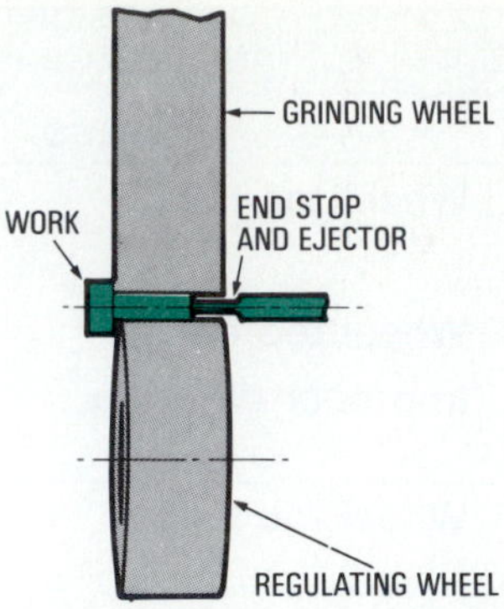

GRINDING WHEEL
WORK
END STOP
AND EJECTOR
REGULATING WHEEL

Figure 85-6 *Principle of infeed centerless grinding. (Courtesy Cincinnati Milacron Inc.)*

METHODS OF CENTERLESS GRINDING

There are three methods of centerless grinding: thru-feed, infeed, and endfeed.

THRU-FEED CENTERLESS GRINDING

Thru-feed centerless grinding (Fig. 85-5) consists of feeding the work between the grinding and regulating wheels. The cylindrical surface is ground as the work is fed by the regulating wheel past the grinding wheel. The speed at which the work is fed across the grinding wheel is controlled by the speed and angle of the regulating wheel.

INFEED CENTERLESS GRINDING

Infeed centerless grinding, a form of plunge grinding, is used when the work being ground has a shoulder or head (Fig. 85-6). Several diameters of a workpiece may be finished simultaneously by infeed grinding. Tapered, spherical, and other irregular profiles are ground efficiently by this method.

With infeed grinding, the work rest blade and the regulating wheel are clamped in a fixed relation to each other. The work is placed on the rest, against the regulating wheel, and is fed into the grinding wheel by moving the infeed lever through a 90° arc. When the lever is at the full end of the travel, the predetermined size has been reached and the part is the desired size. When the lever is reversed, the regulating wheel and work rest move back and the part is ejected either manually or automatically.

If the part to be ground is longer than the wheels, one end is supported on the work rest and the other on rollers mounted on the machine.

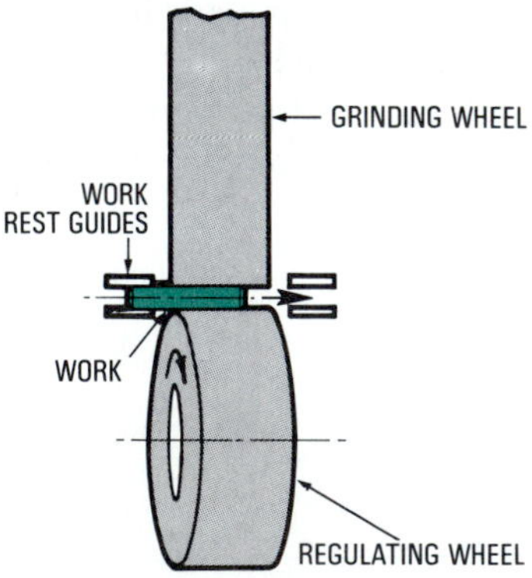

GRINDING WHEEL
WORK
REST GUIDES
WORK
REGULATING WHEEL

Figure 85-5 *Principle of thru-feed centerless grinding. (Courtesy Cincinnati Milacron Inc.)*

ENDFEED CENTERLESS GRINDING

The *endfeed method* (Fig. 85-7) is used mainly for grinding tapered work. The grinding wheel, the regulating wheel, and the work rest all remain in a fixed position. The work is then fed in from the front, manually or mechanically, up to a fixed stop. When the machine is prepared for endfeed grinding, the grinding wheel and the regulating wheel are often dressed to the required taper. In some cases where only a few parts are required, only the regulating wheel may be dressed.

ADVANTAGES OF CENTERLESS GRINDING

- There is no limit to the length of work being ground.
- There is no axial thrust on the workpiece, permitting

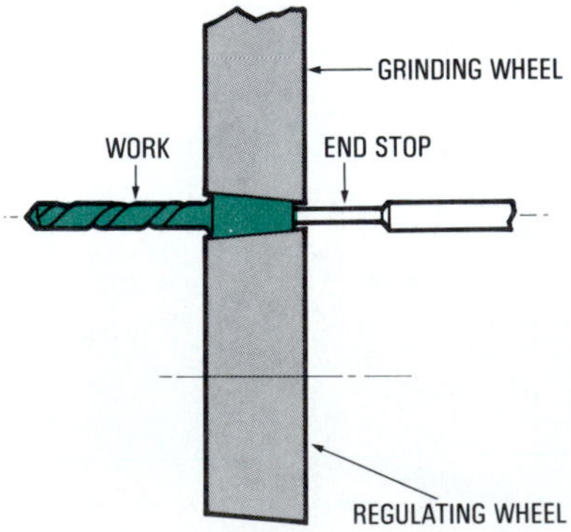

GRINDING WHEEL
WORK
END STOP
REGULATING WHEEL

Figure 85-7 *Principle of endfeed centerless grinding. (Courtesy Cincinnati Milacron Inc.)*

the grinding of long workpieces, which would be distorted by other methods.

- For truing purposes, less stock is required on the workpiece than if the work is held between centers. This is due to the fact that the work "floats" in the centerless grinder. Work held between centers may run eccentrically and require more stock for truing up.
- Because there is less stock to be removed, there is less wheel wear and less grinding time is required.

REVIEW QUESTIONS

CYLINDRICAL GRINDERS

1. Name two types of cylindrical grinders.
2. List the main parts of a cylindrical grinder and state the purpose of each.
3. What precautions should be observed when truing and dressing the grinding wheel of a cylindrical grinder?

TO PARALLEL-GRIND THE OUTSIDE DIAMETER

4. List six precautions which must be taken to ensure the utmost accuracy during parallel grinding the outside diameter of a workpiece.

TO GRIND A TAPERED WORKPIECE

5. List three ways in which tapers may be ground on a universal cylindrical grinder.

CYLINDRICAL GRINDING FAULTS, CAUSES, AND REMEDIES

6. List three causes and three remedies for the following cylindrical grinding faults.

   a. Burnt work
   b. Chatter marks
   c. Feed lines
   d. Work out-of-round
   e. Rough finish
   f. Wavy marks

INTERNAL GRINDING

7. Describe internal grinding and name four machines on which it may be performed.

INTERNAL GRINDING ON A UNIVERSAL CYLINDRICAL GRINDER

8. Why are grinding wheels used for internal grinding generally softer than those used for external grinding?
9. What precautions must be taken when setting up and grinding a parallel hole in a workpiece?

TO GRIND A TAPERED HOLE

10. At what height should the grinding wheel be set for grinding a taper? Why is this necessary?

INTERNAL GRINDING PROBLEMS, CAUSES, AND REMEDIES

11. List four important problems encountered in internal grinding. State at least two causes and two remedies for each of the problems listed.

CENTERLESS GRINDING

12. Describe the principle of centerless grinding. Illustrate by means of a suitable sketch.
13. How is the work held during centerless grinding to make it cylindrical as quickly as possible?

METHODS OF CENTERLESS GRINDING

14. Name three methods of centerless grinding and illustrate each of these methods by a suitably labeled sketch.
15. List four advantages of centerless grinding.

UNIT 86

Universal Cutter and Tool Grinder

The universal cutter and tool grinder is designed primarily for the grinding of cutting tools such as milling cutters, reamers, and taps. Its universal feature and various attachments permit a variety of other grinding operations to be performed. Other operations which may be performed are internal, cylindrical, taper, and surface grinding, single-point tool grinding, and cutting-off operations. Most of the latter operations require additional attachments or accessories.

OBJECTIVES

After completing this unit you should be able to:

1. Identify and state the purposes of the main parts of a cutter and tool grinder
2. Grind clearance angles on helical and staggered tooth cutters
3. Grind a form-relieved cutter
4. Set up the grinder for cylindrical and internal grinding

PARTS OF THE UNIVERSAL CUTTER AND TOOL GRINDER (FIG. 86-1)

The *base* is of heavy, cast-iron, boxlike construction that provides rigidity. The top of the base is machined to provide the *ways* (which are generally hardened) for the saddle.

The *wheelhead* is mounted on a column at the back of the base. It may be raised or lowered by the wheelhead handwheels located on either side of the base. The wheelhead may be swiveled through 360°. The wheelhead spindle is mounted in antifriction bearings and is tapered and threaded at both ends to receive grinding wheel collets. The spindle speed may be varied by stepped pulleys to suit the size of the wheel being used.

The *saddle* is mounted on the ways of the base and is moved in and out by the crossfeed handwheels located at the front and back of the machine. The upper part of the saddle has machined and hardened ways at right angles to the ways on top of the base.

The *table* is composed of two units, the *upper* and *lower* table. The lower table, mounted on the upper ways of the saddle, rests and moves on antifriction bearings. The upper table is fastened to the lower table and may be swiveled for grinding tapers. The table unit (upper and lower tables) may be moved longitudinally by three *table traverse knobs,* one located at the front of the machine and two at the back. The table may also be traversed slowly by means of the *slow table traverse crank.* The table may be locked in place laterally and longitudinally with locking screws.

Stop dogs, mounted in a T-slot on the front of the table, control the length of the table traverse. Each dog has a positive stop pin on one side and a spring-loaded plunger on the other. They are reversible to provide for a positive or a cushioned stop for the table, as desired.

ACCESSORIES AND ATTACHMENTS

The *right-* and *left-hand tailstocks* are mounted in the T-slot of the upper table and support the work for certain

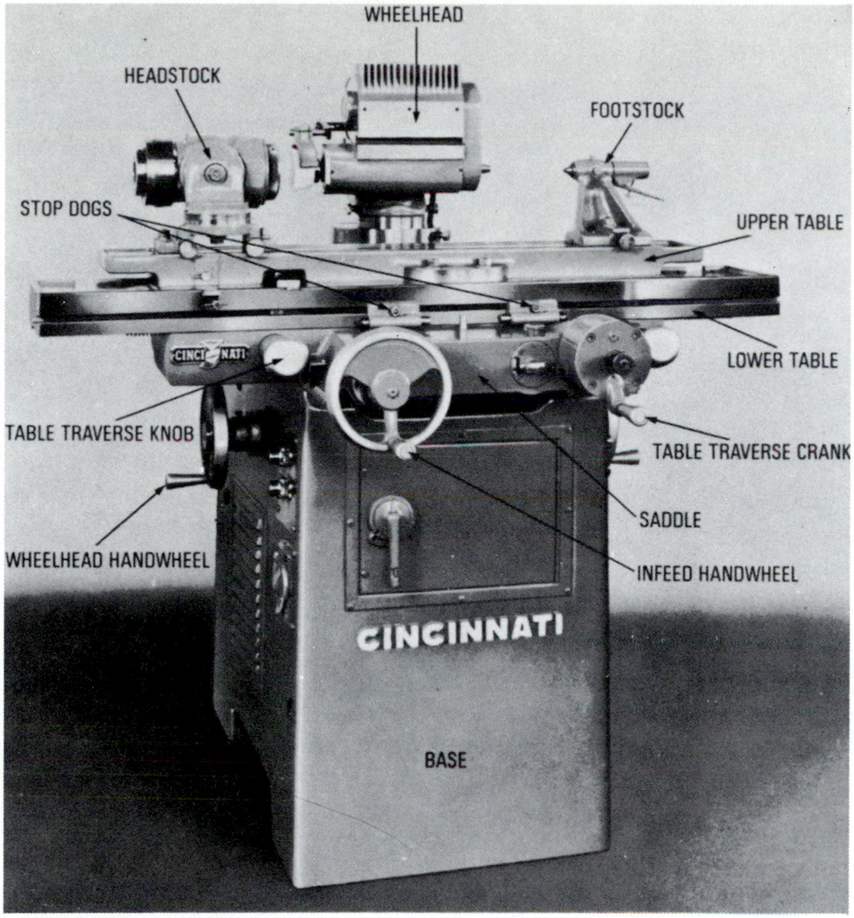
WHEELHEAD
HEADSTOCK
FOOTSTOCK
STOP DOGS
UPPER TABLE
LOWER TABLE
TABLE TRAVERSE KNOB
TABLE TRAVERSE CRANK
SADDLE
WHEELHEAD HANDWHEEL
INFEED HANDWHEEL
CINCINNATI
BASE

Figure 86-1 *A universal cutter and tool grinder. (Courtesy Cincinnati Milacron Inc.)*

grinding operations. They may be placed at any point along the table.

The *universal workhead* (Fig. 86-1) or headstock is mounted on the left side of the table and used for supporting end mills and face mills for grinding. It may also be equipped with a pulley and motor (motorized headstock; see Fig. 86-19) and used for cylindrical grinding. A chuck may be mounted in the workhead to hold work for internal and cylindrical grinding, as well as cutting-off operations (see Fig. 86-20).

The *centering gage* (see Fig. 86-13) is used to align quickly the tailstock center with the center of the wheelhead spindle. It is also used to align the cutter tooth on center in some grinding setups.

The *adjustable tooth rest* supports the cutter tooth and may be fastened to the wheelhead or table, depending on the type of cutter being ground. Another form of tooth rest is the *universal micrometer flicker type,* which has a micrometer adjustment for small vertical movements of the tooth rest.

Plain tooth rest blades (Fig. 86-2A) are used for grinding straight-tooth milling cutters.

Rounded tooth rest blades (Fig. 86-2B) are used for sharpening shell end mills, small end mills, taps, and reamers.

Offset tooth rest blades (Fig. 88-2C) are a universal type suitable for most applications, such as coarse-pitch helical milling cutters and large face mills with inserted blades.

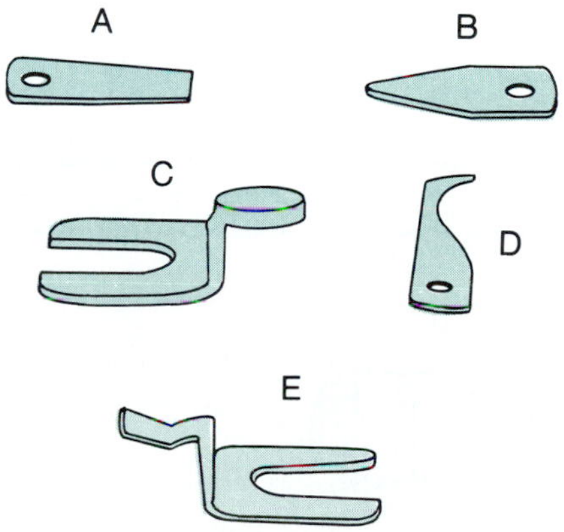
A
B
C
D
E

Figure 86-2 *Various shapes of tooth rest blades.*

Hook or *L-shaped tooth rest blades* (Fig. 86-2D) are used for sharpening slitting saws, straight-tooth plain milling cutters with closely spaced teeth, and end mills.

Inverted V-tooth rest blades (Fig. 86-2E) are used for grinding the periphery of staggered-tooth cutters.

Cutter grinding mandrels and arbors (Fig. 86-3A, B). It is most important when grinding a milling cutter to hold it in the same manner as it is held for milling. For example, shell end mills should be sharpened on the same arbor as that used for milling.

Plain milling and *side facing cutters,* which are held on the standard milling machine arbor, should be held on a grinding mandrel (Fig. 86-3A) or a cutter grinding arbor (Fig. 86-3B).

A *grinding mandrel* rather than a lathe mandrel should be used to hold the cutter. This is necessary since a lathe mandrel will hold the cutter only at one end. The straight length of a grinding mandrel is a sliding fit into the cutter, and the slightly tapered end will hold the cutter securely for grinding.

Where considerable cutter grinding is done, a cutter grinder arbor will be useful.

MILLING CUTTER NOMENCLATURE

To grind cutters correctly, the cutter parts and their function should be understood. Milling cutter parts are shown in Fig. 86-4. A brief description of the various parts follows.

- *Primary clearance* is the clearance ground on the land adjacent to the tooth face. It is the angle formed between the slope of the land and a line tangent to the periphery. Primary clearance prevents the land behind the cutting edge from rubbing on the work. The amount of primary clearance on a cutter will vary with the type of material being cut.
- *Secondary clearance* is ground behind the primary clearance and gives additional clearance to the cutter behind the tooth face. When grinding the clearance on a milling cutter, always grind the primary clearance first. The secondary clearance is then used to control the width of the land.
- The *cutting edge* is formed by the intersection of the face of the tooth with the land. This angle formed by the face of the tooth and the primary clearance is called the *angle of keenness.*

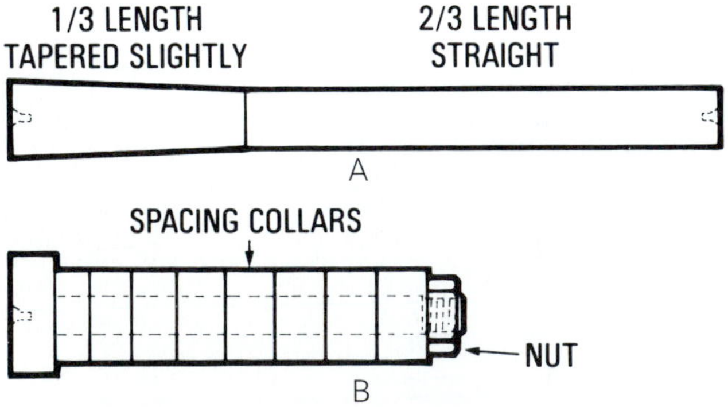

1/3 LENGTH
TAPERED SLIGHTLY
2/3 LENGTH
STRAIGHT
A
SPACING COLLARS
NUT
B

Figure 86-3 *(A) A cutter grinding mandrel; (B) a cutter grinding arbor.*

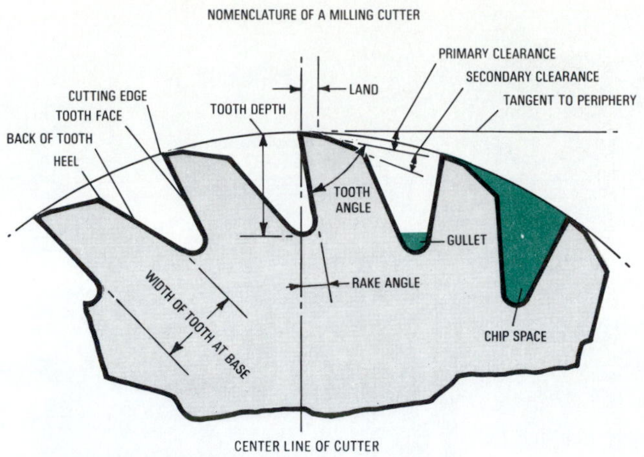

NOMENCLATURE OF A MILLING CUTTER
PRIMARY CLEARANCE
SECONDARY CLEARANCE
TANGENT TO PERIPHERY
LAND
CUTTING EDGE
TOOTH FACE
TOOTH DEPTH
BACK OF TOOTH
HEEL
TOOTH ANGLE
GULLET
RAKE ANGLE
WIDTH OF TOOTH AT BASE
CHIP SPACE
CENTER LINE OF CUTTER

Figure 86-4 *Milling cutter nomenclature.*

On side milling cutters, the cutting edges may be on one or both sides as well as on the periphery. When the teeth are straight, the cutting edge engages along the full width of the tooth at the same moment. This creates a gradual buildup of pressure as the tooth cuts into the work and a sudden release of this pressure as the tooth breaks through, causing a vibration or chatter. This type of cutter produces a poor finish and does not retain its sharp cutting edge as long as a helical cutter does.

When the teeth are helical, the length of the cutting edge contacting the work varies with the helix angle. The number of teeth in contact with the work will vary with the size of the surface being machined, the number of teeth in the cutter, the cutter diameter, and the helix angle. Helical cutters produce a shearing action on the material being cut, which reduces vibration and chatter.

The *helix angle,* sometimes called the *shear angle,* is the angle formed by the angle of the teeth and the centerline of the cutter. It may be measured with a protractor or by bluing the edge of the cutter teeth and rolling the cutter against a straightedge over a sheet of paper (Fig. 86-5). The marks left by the teeth can easily be measured in relation to the axis of the cutter to determine the helix angle.

The *land* is the narrow surface behind the cutting edge on the primary clearance produced when the secondary

Figure 86-5 *One method of measuring the helix angle of a milling cutter. (Courtesy Kostel Enterprises Limited.)*

clearance is ground on the cutter. The width of the land varies from about 1/64 in. (0.40 mm) on small cutters to about 1/16 in. (1.5 mm) on large cutters. On face mills, the land is more correctly called the *face edge.*

The *tooth angle* is the included angle between the face of the tooth and the land caused by grinding the primary clearance. This angle should be as large as possible to provide maximum strength at the cutting edge and better dissipation of heat generated during the cutting process.

The *tooth face* is the surface on which the metal being cut forms a chip. This face may be flat, as in straight-tooth plain milling cutters and inserted face-tooth mills, or curved, as in helical milling cutters.

CUTTER CLEARANCE ANGLES

To perform efficiently, a milling cutter must be ground to the correct clearance angle. The proper clearance angle on a milling cutter may be determined only by the "cut-and-try" method. The clearance angle will be influenced by such factors as finish, the number of pieces per sharpening, the type of material, and the condition of the machine. Excessive primary clearance produces chatter, causing the cutter to dull quickly.

A general rule followed by a large machine tool manufacturer for grinding cutter clearance angles on high-speed steel cutters is 5° primary clearance plus an additional 5° for the secondary clearance. Thus the primary clearance is 5° and the secondary clearance is 10° for cutting machine steel. Carbide cutters used on machine steel are ground to 4° primary clearance plus an additional 4°, or a total of 8° for the secondary clearance.

Table 86-1 provides a rule of thumb for grinding cutter clearance angles on high-speed steel milling cutters. Table 86-2 gives the angles for carbide cutters. It should be remembered that this is only a guide. If the cutter does not perform satisfactorily with these angles, adjustments will have to be made to suit the job.

Table 86-1 Clearance Angles for High-Speed Steel Cutters

Material to Be Machined	Primary Clearance Angle	Secondary Clearance Angle
High-carbon and alloy steels	3–5°	6–10°
Machine steel	3–5°	6–10°
Cast iron	4–7°	7–12°
Medium and hard bronze	4–7°	7–12°
Brass and soft bronze	10–12°	13–17°
Aluminum, magnesium, and plastics	10–12°	13–17°

Table 86-2 Primary Clearance Angles for Cemented-Carbide Cutters

Type of Cutter	Periphery			Chamfer			Face		
	Steel	Cast Iron	Aluminum	Steel	Cast Iron	Aluminum	Steel	Cast Iron	Aluminum
Face or side	4–5°	7°	10°	4–5°	7°	10°	3–4°	5°	10°
Slotting	5–6°	7°	10°	5–6°	7°	10°	3°	5°	10°
Sawing	5–6°	7°	10°	5–6°	7°	10°	3°	5°	10°

METHODS OF GRINDING CLEARANCE ON CUTTERS

Clearance may be ground on cutters by clearance, hollow, and circle grinding. The type of cutter being ground will determine the method used.

CLEARANCE GRINDING

Clearance grinding (Fig. 86-6) produces a flat surface on the land. A 4 in. (100 mm) flared-cup wheel is used for this method and is offset slightly to permit long cutters to clear the opposite side of the wheel. When clearance grinding, the tooth rest may be set between the center and the top of the wheel, but never below center. The higher the tooth rest is placed, the less will be the clearance between the cutter and the opposite edge of the wheel. When clearance grinding, the tooth rest may be attached to the table or the wheelhead, depending on the type of cutter being ground. For straight-tooth cutters, it may be mounted on the table, but for helical teeth it must be mounted on the wheelhead.

Figure 86-6 *Setup for clearance grinding using a flaring-cup wheel.*

2730

Figure 86-7 *The periphery of the wheel is used for hollow grinding. (Courtesy Cincinnati Milacron Inc.)*

HOLLOW GRINDING

The land produced by hollow grinding (Fig. 86-7) is concave. A 6 in. diameter (150 mm) dish wheel or a 6-in.-diameter cutoff wheel is desirable. The cutoff wheel generally produces a better finish and breaks down more slowly because of the resinoid bond. Since all grinding wheels break down in use, it is better to grind diagonally opposite teeth in rotation and to take light cuts. In hollow grinding, the wheel and cutter centers must be aligned. Then, clearance is obtained by raising or lowering the wheel, depending on the method used to set up the cutter.

CIRCLE GRINDING

Circle grinding (Fig. 86-8) provides only a minute amount of clearance and is used mainly for reamers. The reamer is mounted between centers and is rotated *backward* so that the heel of the tooth contacts the wheel first. As the tooth

Figure 86-8 *The workpiece is revolved against the grinding wheel for circle grinding.*

rotates against the grinding wheel, the pressure of the wheel causes the cutter to spring back slightly as each cut progresses. Thus a very small amount of clearance is produced between the cutting edge and the heel of the tooth. The grinding wheel should be set on center for circle grinding. Secondary clearance must be obtained by clearance or hollow grinding. Circle grinding is also used to obtain concentricity of milling cutters prior to clearance or hollow grinding.

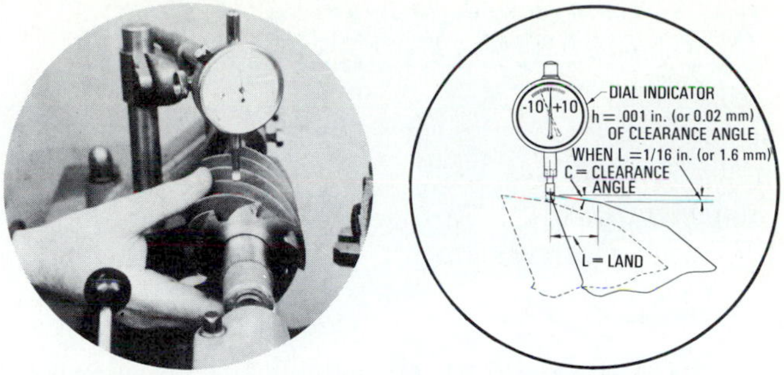
-10
+10
DIAL INDICATOR
h = .001 in. (or 0.02 mm)
OF CLEARANCE ANGLE
WHEN L = 1/16 in. (or 1.6 mm)
C = CLEARANCE
ANGLE
L = LAND

Figure 86-9 *Measuring the clearance angle with a dial indicator.*

METHODS OF CHECKING CUTTER CLEARANCE ANGLES

There are four methods of determining tooth clearance on a milling cutter:

1. Dial indicator
2. Brown and Sharpe cutter clearance gage
3. Starrett cutter clearance gage

TO CHECK CUTTER CLEARANCE WITH A DIAL INDICATOR

When a dial indicator is being used, clearance is determined by the movement of the indicator needle from the front to the back of the cutter land (Fig. 86-9). The basic rule used to determine the clearance by this method is as follows.

For a land of 1/16 in. (or 1.5 mm) width, 1° of clearance is equivalent to 0.001 in. (or 0.02 mm) on the dial indicator. Thus 4° of clearance on a 1/16-in. (or 1.5 mm) land would register 0.004 in. (or 0.10 mm) on the dial indicator. The cutter diameter does not affect the measurement.

TO CHECK CUTTER CLEARANCE WITH A BROWN AND SHARPE CLEARANCE GAGE

When the Brown and Sharpe clearance gage is being used (Fig. 86-10), the inside surfaces of the hardened arms (which are at 90°) are placed on top of two teeth of the cutter. The cutter is revolved sufficiently to bring the face of the tooth into contact with the angle ground on the end of the hardened center blade. The clearance angle of the tooth should correspond with the angle marked on the end of the blade. Two gage blades are furnished with each gage and are stamped at each end with the diameters of the cutters for which they are intended. This cutter clearance gage measures all cutters from ½ to 8 in. (13 to 200 mm) in diameter, except those with less than eight teeth.

CHECKING THE CUTTER CLEARANCE WITH A STARRETT CUTTER CLEARANCE GAGE

The Starrett gage (Fig. 86-11) may be used to check the clearance on all types of inch cutters from 2 to 30 in. (50 to 750 mm) in diameter, and on small cutters and end mills from ½ to 2 in. (13 to 15 mm), providing the teeth are evenly spaced. This gage consists of a frame graduated from 0 to 30°, a fixed foot, and a beam. An adjustable foot slides along the beam extension. A blade, which may be adjusted angularly and vertically, is used to check the angle of the land on a tooth. When in use, the feet are positioned on two alternate teeth of the cutter with the gage at right angles to the tooth face. The adjustable blade is then lowered onto the top of the middle tooth and adjusted until the angle corresponds to the angle of the land being checked. The land angle is indicated on the protractor on the top of the frame.

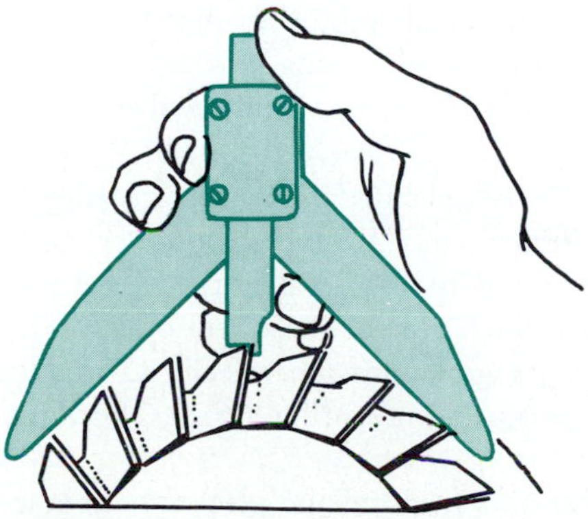

Figure 86-10 *Checking the clearance angle with a Brown and Sharpe cutter clearance gage.*

Figure 86-11 *Checking the clearance angle with a Starrett cutter clearance gage.*

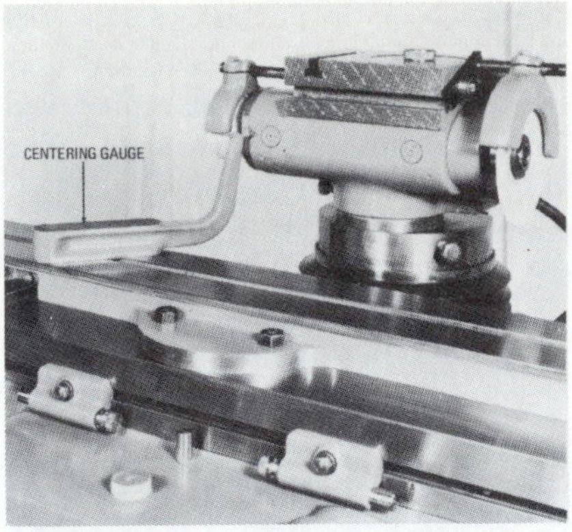
CENTERING GAUGE

Figure 86-12 *Setting the wheelhead spindle to center height with a centering gage. (Courtesy Cincinnati Milacron Inc.)*

CUTTER GRINDING OPERATIONS AND SETUPS

It is most important that milling cutters be ground properly and to the correct clearance angles. Otherwise, the cutter will not cut efficiently and its life will be shortened considerably.

TO GRIND A PLAIN HELICAL MILLING CUTTER

Primary Clearance

1. Mount a parallel-ground test bar between the tailstock centers and check the alignment with an indicator. This will ensure that the table travel is parallel to the grinding edge of the wheel.
2. Remove the test bar.
3. Mount a 4-in. (100-mm) flaring-cup wheel (A 60-L 5-V BE) on the grinding head spindle so that the wheel rotates in a counterclockwise direction.
4. Adjust the machine so that the wheel revolves at the proper speed.
5. True the face of the wheel and dress the cutting edge so that it is no more than 1/16 in. (1.5 mm) wide.
6. Swivel the wheelhead to 89° so that the wheel will touch the cutter on the left side of the wheel only.
7. Using a centering gage, adjust the wheelhead spindle to the height of the tailstock centers (Fig. 86-12). Lock the wheelhead spindle.
8. Mount the cutter on a mandrel and place it temporarily between the footstock centers on the machine table.
9. Set up the tooth rest, on which an offset tooth rest blade (Fig. 86-2C) has been mounted, on the wheelhead housing. Adjust the top of the tooth rest to approximately center height.
10. Move the table until the cutter is near the tooth rest.
11. Adjust the tooth rest between two teeth at the approximate helix angle of the cutter teeth.
12. Chalk or blue the top of the tooth rest blade.
13. Move the cutter over the tooth rest blade and rotate it until the tooth rests on top of the blade.
14. While holding the tooth face against the rest, traverse the table back and forth to mark the point where the tooth bears on the tooth rest blade.
15. Remove the mandrel and cutter from between the centers.
16. Using the centering gage, adjust the tooth rest so that the *center* of the marked bearing point on the tooth rest is at center height and also in the center of the grinding surface of the wheel (Fig. 86-13). The tooth rest blade must be as close to the wheel as possible without touching it.

NOTE: At this point, the center of the grinding spindle, the footstock centers, and the bearing point on the tooth rest blade are in line.

17. Place a dog on the end of the grinding mandrel and mount the work between the tailstock centers.

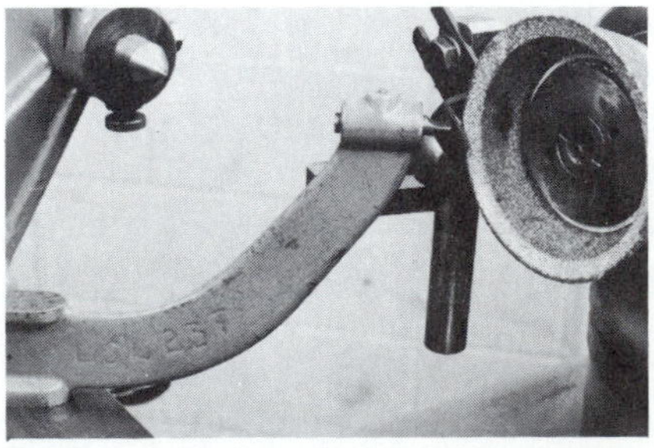

Figure 86-13 *Centering the bearing point on the tooth rest.*

18. Adjust a cutter tooth onto the top of the tooth rest blade.

19. Set the cutter clearance setting dial (Fig. 86-14) to zero (0) and lock. Adjust the dog into the pin of the cutter clearance gage.

20. Set the wheelhead graduated collar to zero (0).

21. Loosen the wheelhead lock and the cutter clearance setting dial lock.

22. Holding the cutter tooth on the tooth rest blade, carefully *lower* the wheelhead until the required clearance is shown on the cutter clearance dial. See Table 86-3 for the wheelhead adjustment for the required angle. When using a flaring-cup wheel, the distance to lower the wheelhead may also be calculated by either of the following methods:

a. Distance = 0.0087 × clearance angle × cutter dia.

b. Distance = sine of clearance angle × cutter dia. ÷ 2

Table 86-3 Vertical Wheelhead Adjustment for Cutter Clearance Angles

Cutter Diameter (in.)	Clearance Angle and Distance							
	4°		5°		6°		7°	
	in.	mm	in.	mm	in.	mm	in.	mm
½	0.017	0.45	0.022	0.55	0.026	0.65	0.031	0.80
¾	0.026	0.65	0.033	0.85	0.040	1.00	0.046	1.15
1	0.035	0.90	0.044	1.10	0.053	1.35	0.061	1.55
1¼	0.044	1.10	0.055	1.40	0.066	1.65	0.077	1.95
1½	0.053	1.35	0.066	1.65	0.079	2.00	0.092	2.35
1¾	0.061	1.55	0.076	1.95	0.092	2.35	0.108	2.75
2	0.070	1.75	0.087	2.20	0.105	2.65	0.123	3.10
2½	0.087	2.20	0.109	2.75	0.131	3.30	0.153	3.90
2¾	0.097	2.45	0.120	3.05	0.144	3.65	0.168	4.25
3	0.105	2.65	0.131	3.30	0.158	4.00	0.184	4.65
3½	0.122	3.10	0.153	3.90	0.184	4.65	0.215	5.45
4	0.140	3.55	0.195	4.95	0.210	5.35	0.245	6.20
4½	0.157	4.00	0.197	5.00	0.237	6.00	0.276	7.00
5	0.175	4.45	0.219	5.55	0.263	6.70	0.307	7.80
5½	0.192	4.85	0.241	6.10	0.289	7.35	0.338	8.60
6	0.210	5.35	0.262	6.65	0.315	8.00	0.368	9.35
6½	0.226	5.75	0.283	7.20	0.339	8.60	0.396	10.05
7	0.244	6.20	0.305	7.45	0.365	9.25	0.426	10.80
7½	0.261	6.60	0.326	8.30	0.392	9.95	0.457	11.60
8	0.278	7.05	0.348	8.85	0.418	10.60	0.487	12.35
8½	0.296	7.50	0.370	9.40	0.444	11.25	0.519	13.20
9	0.313	7.95	0.392	9.95	0.470	11.95	0.548	13.90
9½	0.331	8.40	0.413	10.50	0.496	12.60	0.579	14.70
10	0.348	8.85	0.435	11.05	0.522	13.25	0.609	15.45
11	0.383	9.70	0.479	12.15	0.574	14.55	0.670	17.00
12	0.418	10.60	0.522	13.25	0.626	15.90	0.731	18.55
13	0.452	11.50	0.566	14.35	0.679	17.25	0.792	20.10
14	0.487	12.35	0.609	15.45	0.731	18.55	0.853	21.65
15	0.522	13.25	0.653	16.60	0.783	19.90	0.914	23.20
16	0.557	14.15	0.696	17.65	0.835	21.20	0.974	24.75

Note: Inch and metric wheelhead adjustments are approximate equivalents.

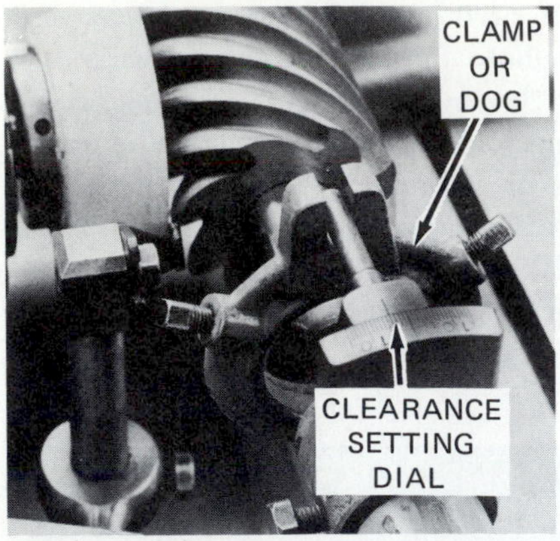
CLAMP
OR
DOG
CLEARANCE
SETTING
DIAL

Figure 86-14 *The cutter clearance dial indicates the amount of clearance being ground on the cutter. (Courtesy Kostel Enterprises Limited.)*

If the cutter is being hollow ground, the distance to lower the wheelhead is

$$\text{Distance} = 0.0087 \times \text{clearance angle} \times \text{wheel dia.}$$

23. Remove the dog from the end of the mandrel and unlock the table.
24. Adjust the table stops so that the wheel clears the cutter sufficiently at each end to permit indexing for the next tooth.
25. Start the grinding wheel.
26. Carefully feed the cutter in until it just touches the wheel.
27. Standing at the rear of the machine, turn the table traverse knob with the left hand. At the same time, with the right hand, hold the arbor firmly enough to keep the cutter tooth on the tooth rest (Fig. 86-6).
28. Grind one tooth for the full length and return to the starting position, being careful *at all times* to keep the tooth tight against the tooth rest.
29. Traverse the table until the cutter is clear of the tooth rest and rotate the cutter until the diagonally opposite tooth comes in line with the tooth rest blade.
30. Grind this tooth without changing the infeed setting.
31. Check for taper by measuring both ends of the cutter with a micrometer.
32. Remove any taper, if necessary, by loosening the holding nuts on the upper table and adjusting the table.
33. Grind the remaining teeth.
34. Finish-grind all teeth by using a 0.0005-in. (0.01-mm) depth of cut.
35. If the land is over 1/16 in. (1.5 mm) for larger cutters grind the secondary clearance.

To Grind the Secondary Clearance of a Plain Helical Milling Cutter

1. Reset the dog on the mandrel as in step 17 for grinding the primary clearance.
2. Loosen the clearance dial setscrew.
3. Hold the cutter tooth against the tooth rest and lower the wheelhead until the required secondary clearance is shown on the clearance setting dial.
4. Lock the dial, remove the dog, and proceed to grind the secondary clearance in the same manner as for primary clearance.
5. Grind the secondary clearance until the land is the required width.

TO GRIND A STAGGERED TOOTH CUTTER

To grind the primary clearance on the periphery of a staggered-tooth cutter (Fig. 86-15) proceed as follows:

1. Carry out steps 1 to 7 as for grinding primary clearance on a plain helical milling center.
2. Mount a staggered-tooth cutter tooth rest blade (Fig. 86-2E) in the holder and mount the unit on the wheelhead.
3. Place the high point of the inverted V *exactly* in the center of the width of the grinding wheel cutting face and at center height.
4. Place the centering gage on the table and adjust the wheelhead height until the highest point of the tooth rest blade is at center height.
5. Mount the cutter between centers with the dog loosely

TOOTH REST
CLEARANCE
SETTING
DIAL

Figure 86-15 *Setup for clearance grinding a staggered-tooth cutter. (Courtesy Kostel Enterprises Limited.)*

on the mandrel and adjust the table until one cutter tooth rests on the blade. Lock the table in position.

6. Set the cutter clearance dial (Fig. 86-14) to zero (0) and tighten the dog on the mandrel.

7. Loosen the cutter clearance dial lock and the wheelhead lock.

8. Lightly holding the cutter tooth onto the tooth rest blade, lower the wheelhead until the required clearance shows on the clearance setting dial.

9. Remove the clearance setting dog and unlock the table.

10. Set the stop dogs so that the wheel clears both sides of the cutter enough to allow indexing for the next tooth.

11. Start the grinding wheel.

12. Adjust the saddle until the cutter just touches the grinding wheel.

13. Grind one tooth and move the cutter clear of the tooth rest.

14. Rotate the next tooth, which is offset in the opposite direction, onto the tooth rest and grind it on the return stroke.

15. After grinding two teeth, check them with a dial indicator to see whether they are the same height. If not, adjust the blade slightly toward the high side and grind the next two teeth. Repeat the process until the teeth are within 0.0003 in. (0.007 mm).

FLICKER-TYPE
TOOTH REST

Figure 86-16 *A flicker-type tooth rest is used for grinding the clearance on the sides of the teeth. (Courtesy Kostel Enterprises Limited.)*

Secondary Clearance

Because it is necessary to provide adequate chip clearance when milling deep slots, a secondary clearance of 20 to 25° on staggered-tooth cutters is recommended. It is also suggested that enough secondary clearance be ground to reduce the width of the land to approximately 1/32 in. (0.80 mm). This will permit regrinding of the primary clearance at least once without the need for grinding the secondary clearance.

To Grind the Secondary Clearance on a Staggered-Tooth Cutter

1. Remove the tooth rest from the wheelhead and mount it on the table between the tailstocks. A universal micrometer flicker-type tooth rest and a straight blade (Fig. 86-16) should be used to permit the cutter to be rotated.

2. Place the centering gage on the table and bring the *center* of one tooth to center height. Mark this tooth with layout dye or chalk.

3. Locate the dog on the clearance setting dial pin and tighten it on the mandrel.

4. Rotate the cutter to the desired amount of clearance using the clearance setting dial.

5. Adjust the tooth rest under, or on the side of, the marked tooth.

6. Swivel the table sufficiently to the right or the left (depending on the helix angle of the tooth being ground) to grind a straight land.

7. Grind the secondary clearance on this tooth until the land is 1/32 in. (0.80 mm) wide.

8. Grind all remaining teeth having the same slope or helix.

9. Swivel the table in the opposite direction and follow steps 6, 7, and 8 to set up and grind the remaining teeth.

Side Clearance

The side of the teeth of any milling cutter should not be ground unless absolutely necessary, since this reduces the width of the cutter. If the teeth must be ground, proceed as follows:

1. Mount the cutter on a stub arbor in the workhead (Fig. 86-16).

2. Mount a flaring-cup wheel.

3. Tilt the workhead to the desired primary clearance angle. This is generally 2 to 4°. The secondary clearance is about 12°.

4. Place the centering gage on the wheelhead and adjust one tooth of the cutter until it is on center and level. Clamp the workhead spindle.

5. Mount the tooth rest on the workhead using a flicker-type rest and a plain blade.

6. Raise or lower the wheelhead so that the grinding wheel contacts only the tooth resting on the blade.

7. Grind the primary clearance on all teeth.

8. Tilt the workhead to the required angle for the secondary clearance and grind all teeth.

TO GRIND A FORM-RELIEVED CUTTER

Unlike other types of milling cutters, form-relieved cutters are ground on the face of the teeth rather than on the periphery; otherwise, the form of the cutter will be changed when it is sharpened.

When grinding formed cutters for the first time, grind the backs of the teeth before grinding the cutting face, to ensure that all teeth are the same thickness. This is necessary because the locating pawl on the grinding fixture bears against *the back of the tooth* when the cutter is being ground.

PROCEDURE

1. Swing the wheelhead so that the spindle is at 90° to the table travel.
2. Mount a dish wheel and the proper wheel guard.
3. Mount a gear cutter sharpening attachment on the table to the left of the grinding wheel (Fig. 86-17).
4. Place the gear cutter on the stud of the attachment so that the back of each tooth may be ground.

NOTE: This operation is necessary only when the cutter is being sharpened for the first time.

5. Place the centering gage on the wheelhead and adjust the wheelhead until the center of the tooth face is on center.
6. Move the table in until the back edge of a tooth is near the grinding wheel. At the same time, rotate the cutter until the back of the tooth is parallel with the face of the wheel (Fig. 86-18).
7. Engage the edge of the pawl on the face of the tooth and clamp the pawl in place (Fig. 86-18).
8. Grind the back of this tooth.
9. Move the table to the left so that the cutter is clear of the grinding wheel.
10. Index the cutter so that the pawl will bear against the next tooth. Hold the tooth face against the pawl when grinding.
11. Grind the backs of all teeth.
12. Reverse the cutter on the stud and adjust the pawl against the back of the tooth, after the face of the tooth has been brought to bear against the centering gage fastened to the attachment. Swing the centering gage out of the way.
13. Adjust the saddle to bring the face of one tooth in line with the grinding wheel. Thereafter, adjust the saddle only to compensate for wheel wear.
14. Loosen one setscrew and tighten the other to rotate the cutter against the grinding wheel.
15. Grind one tooth, traverse the table, and index for the next tooth.
16. Grind all tooth faces.

GEAR CUTTER
SHARPENING ATTACHMENT

Figure 86-17 *Grinder setup to sharpen a form-relieved cutter.*

MANDREL

Figure 86-18 *The back of the tooth is set parallel to the wheel and held in place by the pawl. (Courtesy Cincinnati Milacron Inc.)*

CYLINDRICAL GRINDING

With the aid of a motorized workhead, the cutter and tool grinder may be used for cylindrical and plunge grinding. Work may be ground between centers or held in a chuck, depending on the type of work.

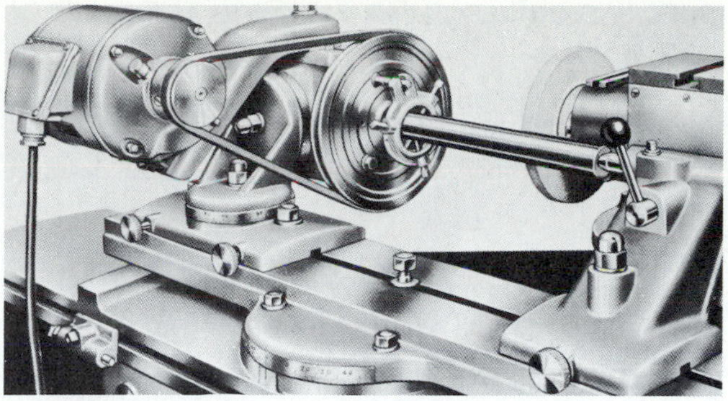

Figure 86-19 *Setup for cylindrical grinding on a cutter and tool grinder.*

To Grind Work Parallel Between Centers

1. Mount the motorized workhead on the left end of the table (Fig. 86-19).
2. Examine the centers of the machine and the work to see that they are in good condition.
3. Using the centering gage on the wheelhead, adjust the wheelhead to the tailstock center height.
4. Mount a 6 in. (150 mm) straight grinding wheel on the wheelhead spindle so that the wheel rotates downward at the front of the wheel. This will deflect the sparks downward.
5. Mount a parallel, hardened and ground test bar between centers.
6. Using a dial indicator, align the centers for height and then align the side of the bar parallel with the table travel. Remove the bar and indicator.
7. Mount the work between centers.
8. Set the stop dogs so that the wheel overlaps the work by one-third the width of the wheelface at each end.
9. Start the grinding wheel and workhead. The workpiece should revolve in the opposite direction to that of the grinding wheel.
10. Bring the revolving work up until it just touches the grinding wheel.
11. Traverse the table slowly and clean up the workpiece. The traverse speed should be such that the work travels approximately one-quarter the width of the wheel for each revolution of the work.
12. Measure each end of the workpiece for size and taper. If a taper exists, adjust as required.
13. After the work is parallel, set the crossfeed graduated collar to zero (0).
14. Feed the work into the grinding wheel approximately 0.001 in. (0.02 mm) per pass until the work is within 0.001 in. (0.02 mm) of finished size. Use 0.0002-in. (0.005-mm) cuts for finishing.
15. Feed in the work until the graduated collar indicates it is the proper size.

NOTE: Since work expands during grinding, it should never be measured for accurate size when warm.

16. Traverse the table several times to permit the wheel to spark out.

The same procedure is followed for taper grinding, except that the table must be swung to half the angle of the taper. After the work is cleaned up, the taper should be carefully checked for size and accuracy, and adjustments made as required. When taper grinding, it is most important that the center height of the wheel and the workpiece be in line.

INTERNAL GRINDING

Light internal grinding may be performed on the tool and cutter grinder by mounting the internal-grinding attachment on the wheelhead. The workpiece is held in a chuck mounted on the motorized workhead (Fig. 86-20).

1. Mount a test bar in the workhead spindle and align it both vertically and horizontally. When grinding a tapered hole, the workhead spindle must be aligned vertically and then swung to half the included angle of the taper.
2. Mount the internal-grinding attachment on the workhead.
3. Center the grinding wheel spindle using the centering gage.
4. Mount the proper grinding wheel on the spindle.
5. Mount a chuck on the motorized workhead.
6. Mount the work in the chuck. Care must be taken not to distort the workpiece by gripping it too tightly.
7. Set the rotation of the workhead in the opposite direction to that of the grinding spindle.
8. Start the grinding wheel and the workpiece.
9. Carefully bring the wheel into the hole of the workpiece.

Figure 86-20 *Setup for internal grinding on a cutter and tool grinder. (Courtesy Cincinnati Milacron Inc.)*

10. Set the table travel so that only one-third of the wheel overlaps the hole at each end.
11. Clean up the inside of the hole and check for size, parallelism, and bell-mouthing. Correct as required.
12. Set the crossfeed graduated collar to zero (0) and determine the amount of material to be removed.
13. Feed the grinding wheel in about 0.0005 in. (0.01 mm) per pass.
14. When work is close to the finished size, let the wheel spark out to improve the finish and remove the spring from the spindle.
15. Finish-grind the hole to size.

REVIEW QUESTIONS

PARTS OF THE UNIVERSAL CUTTER AND TOOL GRINDER

1. Name four main parts of the universal cutter and tool grinder.
2. How many controls are there for each of the following parts of the universal cutter and tool grinder?
   a. Wheelhead b. Saddle c. Table
3. Name five accessories used with the universal cutter and tool grinder and state the purpose of each.
4. Name five types of tooth rest blades and indicate the purpose for which each is used.

CUTTER GRINDING MANDRELS AND ARBORS

5. Make a suitable sketch of a cutter grinding mandrel and a cutter grinding arbor.
6. How does a cutter grinding mandrel differ from a lathe mandrel?

MILLING CUTTER NOMENCLATURE

7. Make a suitable sketch of at least two teeth on a milling cutter and indicate the following parts.
   a. Primary clearance angle
   b. Secondary clearance angle
   c. Cutting angle
   d. Land
   e. Tooth angle
   f. Tooth face
8. Why are cutters with widely spaced teeth more efficient than those with closely spaced teeth?
9. What factors influence the cutter clearance angle?

METHODS OF GRINDING CLEARANCE ON MILLING CUTTERS

10. Name and describe briefly three methods of grinding clearance on cutting tools.

CHECKING CUTTER CLEARANCE WITH A DIAL INDICATOR

11. Describe the principle by which cutter clearance is checked using a dial indicator.
12. Name two other methods of checking cutter clearance angles.

CUTTER GRINDING OPERATIONS AND SETUPS

13. Describe how the tooth edge of a milling cutter is placed on center.
14. Describe how the correct cutter clearance is set on the machine using the clearance setting dial.
15. Explain two methods of determining how far to lower the wheelhead, for the proper cutter clearance angle.
16. How far would the wheelhead be lowered to grind the proper clearance on the following cutters?
   a. 5° clearance angle on a 3-in.-diameter cutter using a 6-in. straight wheel.
   b. 5° clearance angle on a 75-mm-diameter cutter using a flaring-cup wheel
17. List four precautions to be taken for grinding a helical milling cutter.
18. Where is the tooth rest mounted when grinding cutters having:
   a. Helical teeth? b. Straight teeth?

TO GRIND A STAGGERED-TOOTH CUTTER

19. How does the grinding of a staggered-tooth milling cutter differ from the grinding of a plain helical milling cutter?

TO GRIND A FORM-RELIEVED CUTTER

20. On what surface are form-relieved cutters ground when they are being sharpened? Explain why.
21. Why is it advisable to grind the backs of the teeth on a new gear cutter?
22. What type of grinding wheel is used for sharpening gear cutters?

CYLINDRICAL GRINDING

23. List the main steps required to *set up* the cutter and tool grinder for cylindrical grinding.
24. How should the stop dogs be set when cylindrical grinding?
25. How fast should the table be traversed when cylindrical grinding on a cutter and tool grinder?
26. What precautions must be taken when setting up the machine for taper grinding?

INTERNAL GRINDING

27. What precautions should be taken when setting up the work for internal grinding?
28. List briefly the steps required to grind the inside diameter of a bushing using a cutter and tool grinder.

SECTION

16

Metallurgy

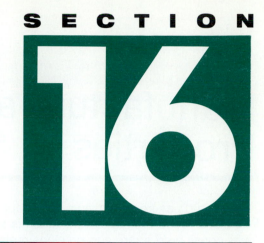

Understanding the properties and heat treatment of metals has become increasingly important to machinists during the past two decades. Study of metal properties and development of new alloys have facilitated reduction in mass and increase in strength of machines, automobiles, aircraft, and many present-day commodities.

The most commonly used metals today are ferrous metals, or those which contain iron. The composition and properties of ferrous materials may be changed by the addition of various alloying elements during manufacture, to impart the desired qualities to the material. Cast iron, machine steel, carbon steel, alloy steel, and high-speed steel are all ferrous metals, each having different properties.

UNIT

87

Manufacture and Properties of Steel

Metals such as iron, aluminum, and copper are among the most common elements found in nature. Iron, which is found in most parts of the world, was considered a rare and precious metal in ancient times. Throughout the ages, iron was transformed into steel, and today it is one of our more versatile metals. Almost every product made today contains some steel or was manufactured by tools made of steel. Steel can be made hard enough to cut glass, pliable as the steel in a paper clip, flexible as steel in springs, or strong enough to withstand enormous stress. Metals may be shaped in several ways: casting, forging, rolling and bending, drawing and forming, cutting, and joining. To better understand this versatile metal, a knowledge of its properties and manufacture is desirable.

OBJECTIVES

After completing this unit you should be able to:
1. Identify six properties of metals
2. Explain the processes by which iron and steel are made
3. Describe the effect of alloying elements on steel

PHYSICAL PROPERTIES OF METALS

Physical metallurgy is the science concerned with the physical and mechanical properties of metals. The properties of metals and alloys are affected by three variables:

1. *Chemical properties*—those that a metal attains through the addition of various chemical elements
2. *Physical properties*—those that are not affected by outside forces, such as color, density, conductivity, or melting temperature
3. *Mechanical properties*—those that are affected by outside forces such as rolling, forming, drawing, bending, welding, and machining.

To better understand the use of the various metals, one should become familiar with the following terms:

- *Brittleness* (Fig. 87-1A) is that property of a metal which permits no permanent distortion before breaking. Cast iron is a brittle metal; it will break rather than bend under shock or impact.
- *Ductility* (Fig. 87-1B) is the ability of the metal to be permanently deformed without breaking. Metals such as copper and machine steel, which may be drawn into wire, are ductile materials.
- *Elasticity* (Fig. 87-1C) is the ability of a metal to return to its original shape after any force acting upon it has been removed. Properly heat-treated springs are good examples of elastic materials.
- *Hardness* (Fig. 87-1D) may be defined as the resistance to forceable penetration or plastic deformation.
- *Malleability* (Fig. 87-1E) is that property of a metal which permits it to be hammered or rolled into other sizes and shapes.
- *Tensile strength* (Fig. 87-1F) is the maximum amount of pull that a material will withstand before breaking. It is expressed as the number of pounds per square inch (on inch testers) or in kilograms per square centimeter (on metric testers) of pull required to break a bar hav-

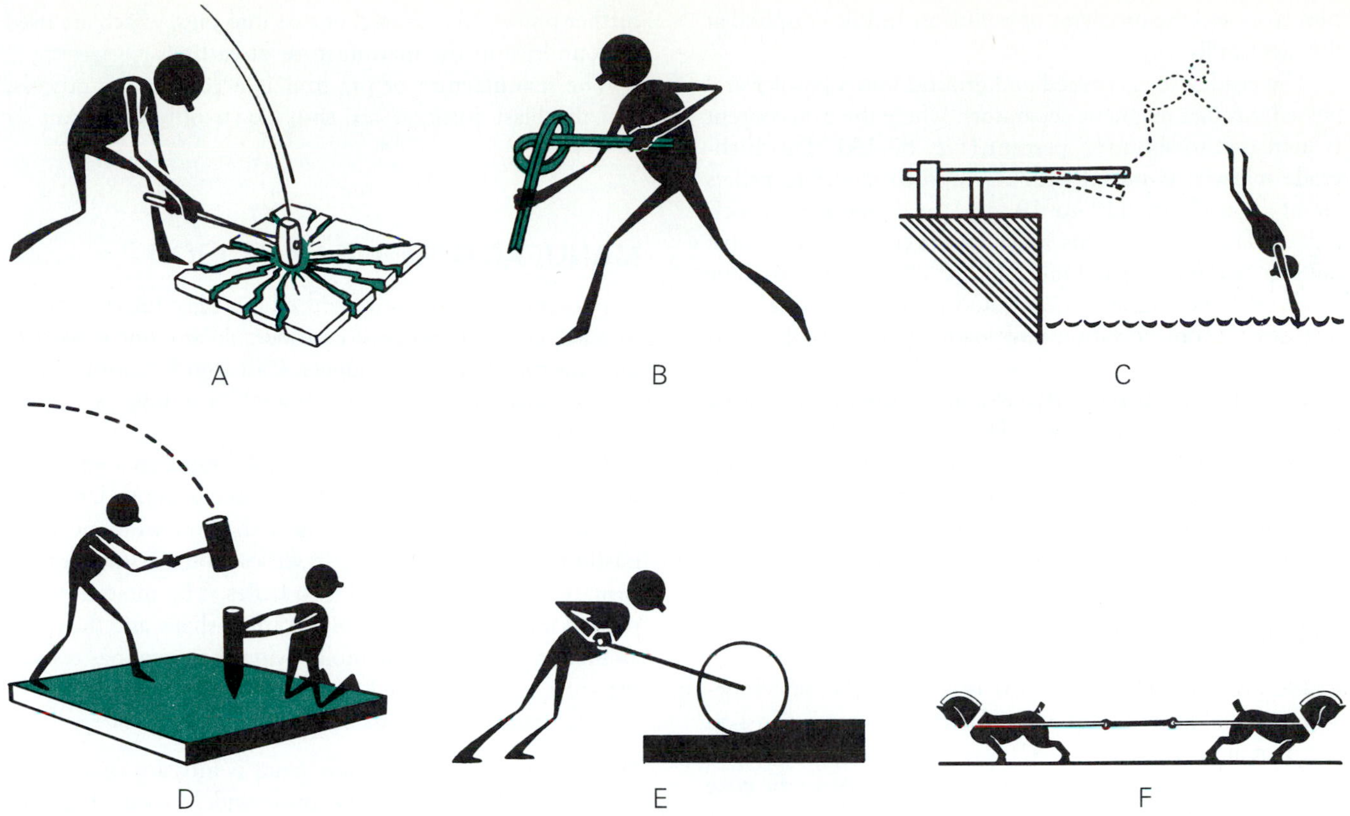
A
B
C
D
E
F

Figure 87-1 *(A) Brittle metals will not bend but break easily; (B) ductile metals are easily deformed; (C) elastic metals return to their original shape after the load is removed; (D) hard metals resist penetration; (E) malleable metals may be easily formed or shaped; (F) tensile strength is the amount that a metal will resist a direct pull. (Courtesy Linde Division, Union Carbide Corporation.)*

ing a one-square-inch or one-square-centimeter cross section.

- *Toughness* is the property of a metal to withstand shock or impact. Toughness is the opposite condition to brittleness.

MANUFACTURE OF FERROUS METALS

PIG IRON

Production of pig iron in the blast furnace (see Fig. 87-3) is the first step in the manufacture of cast iron or steel.

Raw Materials

Iron ore is the chief raw material used to make iron and steel. The main sources of iron ore in North America are at Steep Rock, the Ungava district near the Quebec–Labrador border, and the Mesabi range, situated at the western end of Lake Superior. The most important iron ores are:

- *Hematite,* which contains about 70 percent iron and varies in color from black to brick red.
- *Limonite,* a brownish ore similar to hematite, contains water. When the water has been removed by roasting, the ore resembles hematite.
- *Magnetite,* a rich, black ore, contains a higher percentage of iron than any other ore but is not found in large quantities.
- *Taconite,* a low-grade ore containing about 25 to 30 percent iron, must be specially treated before it is suitable for reduction into iron.

Pelletizing Process

Low-grade iron ores are uneconomical to use in the blast furnace and, as a result, go through a pelletizing process, where most of the rock is removed and the ore is brought to a higher iron concentration. Some steelmaking firms are now pelletizing most ores at the mine to reduce transportation costs and the problems of pollution and slag disposal at the steel mills.

The crude ore is crushed and ground into a powder and passed through magnetic separators, where the iron content is increased to about 65 percent (Fig. 87-2A). This high-grade material is mixed with clay and formed into pellets about ½ to ¾ in. (13 to 19 mm) in diameter in a pelletizer. The pellets at this stage are covered with coal dust and sintered (baked) at 2354°F (1290°C) (Fig. 87-2B). The resultant hard, highly concentrated pellets will remain intact during transportation and loading into the blast furnace.

Coal, after being converted to coke, is used to supply the heat to reduce the iron ore. The burning coke produces carbon monoxide, which removes the oxygen from the iron ore and reduces it to a spongy mass of iron.

Limestone is used as a flux in the production of pig iron to remove the impurities from the iron ore.

MANUFACTURE OF PIG IRON

In the blast furnace (Fig. 87-3) iron ore, coke, and limestone are fed into the top of the furnace by means of a skip car. Hot air at 1000°F (537°C) is fed into the bottom of the furnace through the bustle pipe and tuyeres. After the coke is ignited, the hot air makes it burn vigorously. Carbon monoxide, produced by the burning coke, combines with the oxygen in the iron ore, reducing it to a spongy mass of iron. The iron gradually seeps down through the charge and collects in the bottom of the furnace. During this process, the decomposed limestone acts as a flux and unites with the impurities (silica and sulfur) in the iron ore to form a slag, which also seeps to the bottom of the furnace. Since the slag is lighter, it floats on the top of the molten iron. Every 6 h, the furnace is tapped. The slag is drawn off first and then the molten iron is poured into ladles. The iron may be further processed into steel or cast into *pigs,* which are used by foundries in the manufacture of castings.

The manufacture of pig iron is a continuous process, and the blast furnaces are shut down only for repair or rebricking.

MANUFACTURE OF CAST IRON

Most of the pig iron manufactured in a blast furnace is used to make steel. However, a considerable amount is used to manufacture cast-iron products. Cast iron is manufactured in a cupola furnace, which resembles a huge stovepipe (Fig. 87-4).

Layers of coke, solid pig iron, and scrap iron are fed into the top of the furnace. After the furnace is charged, the fuel is ignited and air is forced in near the bottom to aid combustion. As the iron melts, it settles to the bottom of the furnace, where it is tapped into ladles. The molten iron is poured into sand molds of the required shape and the metal assumes the shape of the mold. After the metal has cooled, the castings are removed from the molds.

The principal types of cast-iron castings are:

- *Gray-iron castings,* made from a mixture of pig iron and steel scrap, are the most widely used. They are made into a wide variety of products, including bathtubs, sinks, and parts for automobiles, locomotives, and machinery.
- *Chilled-iron castings* are made by pouring molten metal into metal molds so that the surface cools rapidly. The surface of such castings becomes very hard, and the castings are used for crusher rolls or other products requiring a hard, wear-resistant surface.
- *Alloyed castings* contain certain amounts of alloys such as chromium, molybdenum, and nickel. Castings of

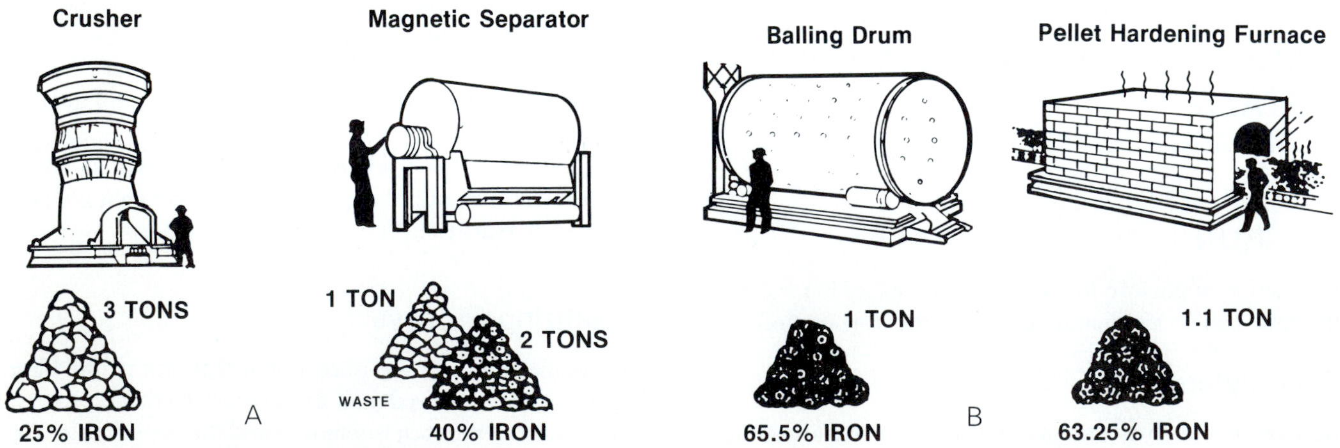
Crusher
Magnetic Separator
Balling Drum
Pellet Hardening Furnace
3 TONS
1 TON
2 TONS
WASTE
1 TON
1.1 TON
25% IRON
A
40% IRON
65.5% IRON
B
63.25% IRON

Figure 87-2 *(A) The first step in pelletizing separates the iron ore from the rock; (B) iron ore pellets are hardened in the furnace.*

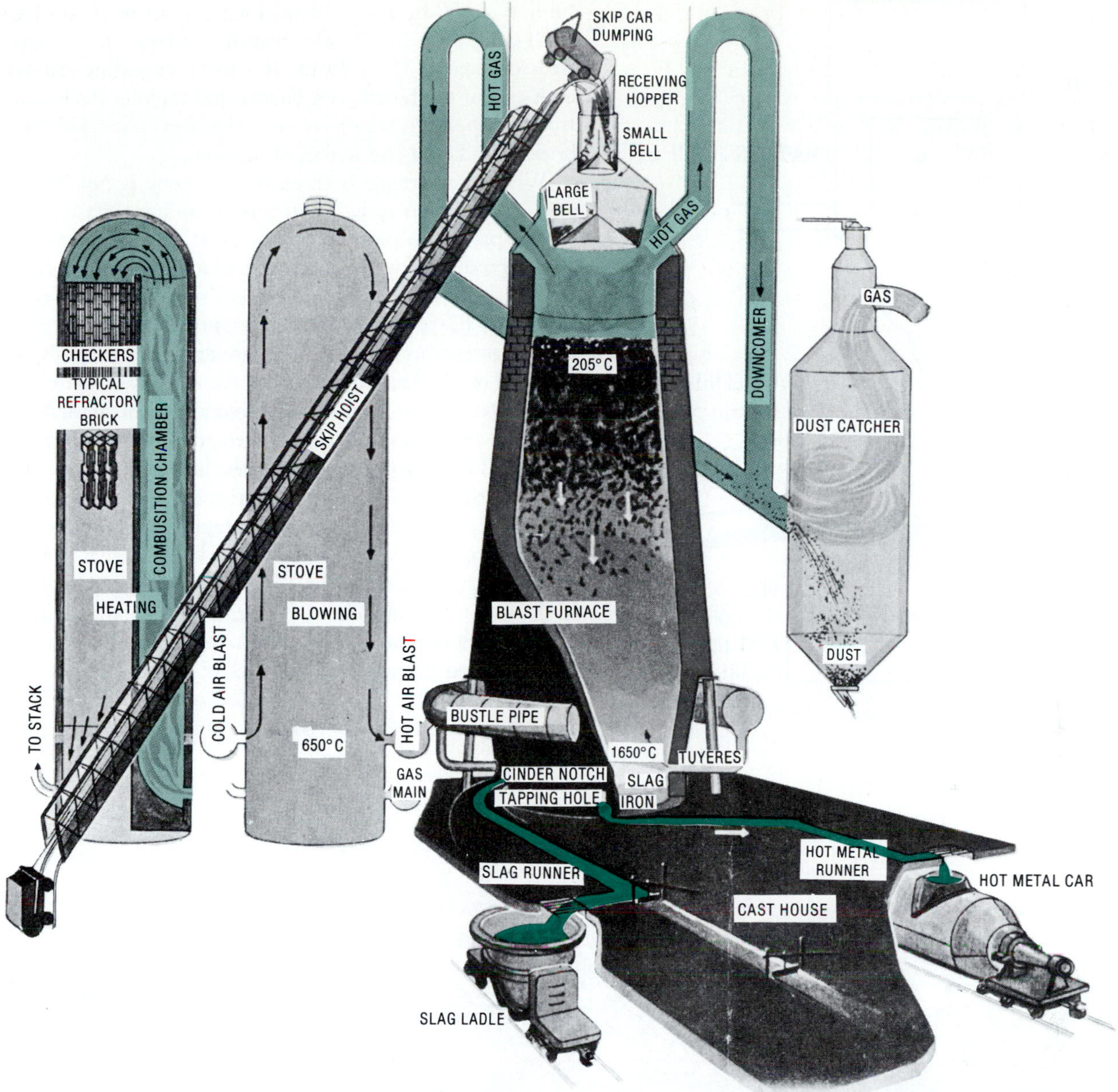
SKIP CAR
DUMPING
HOT GAS
RECEIVING
HOPPER
SMALL
BELL
LARGE
BELL
HOT GAS
GAS
DOWNCOMER
205° C
CHECKERS
TYPICAL
REFRACTORY
BRICK
SKIP HOIST
COMBUSTION CHAMBER
DUST CATCHER
STOVE
STOVE
HEATING
BLOWING
BLAST FURNACE
COLD AIR BLAST
HOT AIR BLAST
DUST
TO STACK
BUSTLE PIPE
650° C
1650° C
TUYERES
GAS
MAIN
CINDER NOTCH
SLAG
TAPPING HOLE
IRON
HOT METAL
RUNNER
SLAG RUNNER
HOT METAL CAR
CAST HOUSE
SLAG LADLE

Figure 87-3 *A blast furnace produces pig iron. (Courtesy American Iron and Steel Institute.)*

this type are used extensively by the automobile industry.

- *Malleable castings* are made from a special grade of pig iron and foundry scrap. After these castings have solidified, they are annealed in special furnaces. This makes the iron malleable and resistant to shock.

MANUFACTURE OF STEEL

Before molten pig iron from the blast furnace can be converted into steel, most of the impurities must be burned out. This may be done in one of four types of furnaces: the open-hearth furnace, the Bessemer converter, the basic oxygen process furnace, and the electric furnace.

Open-Hearth Process

For several decades, the *open-hearth furnace* was the "work horse" of the steel industry. At one time, almost 90 percent of the steel produced in North America was made in the open-hearth furnace. Although some steel companies may still use the modified open-hearth furnace, this process is gradually giving way to more efficient methods of producing steel. The furnace is a large, rectangular, brick struc-

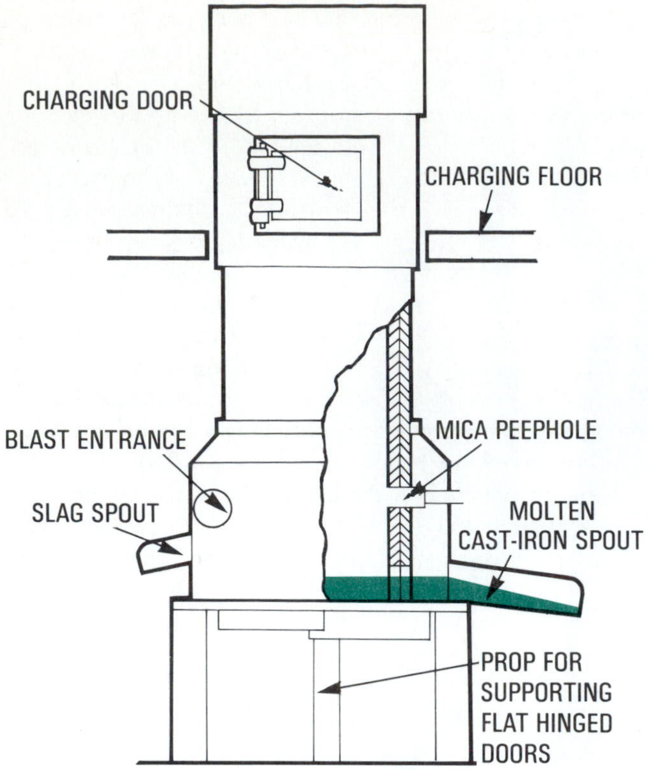
CHARGING DOOR
CHARGING FLOOR
BLAST ENTRANCE
MICA PEEPHOLE
SLAG SPOUT
MOLTEN
CAST-IRON SPOUT
PROP FOR
SUPPORTING
FLAT HINGED
DOORS

Figure 87-4 *A cupola furnace is used to make cast iron.*

ture, 40 to 50 ft (12 to 15 m) long and 15 to 17 ft (4 to 5 m) wide (Fig. 87-5). The *hearth* is a large dish-shaped structure about 25 to 35 in. (63 to 88 cm) deep. Above each end of the hearth is a *burner* and an inlet for hot air from the *checkers,* which are large chambers filled with firebrick placed in a checkerboard fashion.

When the furnace is charged, limestone is put in and steel scrap is then added. After the scrap has been melted, molten pig iron is poured into the hearth. Large tongues of flame, produced by the union of fuel and the hot gas from the checkers, sweep over the molten metal from the right side (Fig. 87-5) and burn the impurities out of it. The burned gases from this operation are drawn off through the checkers on the left side and heat these checkers. At intervals of about 10 to 15 min, the direction of the air and the flame are reversed and pass from left to right over the molten metal. During this operation, the limestone unites with the impurities and rises to the top as slag.

After 7 to 12 h, the furnace is tapped and the molten steel is run into a ladle, which is just large enough to hold the batch (heat) of steel; the slag overflows through a spout in the ladle. Alloying materials, such as silicon, manganese, or other elements, are then added to the molten steel to give it the desired properties. The steel is poured, or *teemed,* into ingot molds and after solidification the ingots are removed from the molds and transferred to soaking pits.

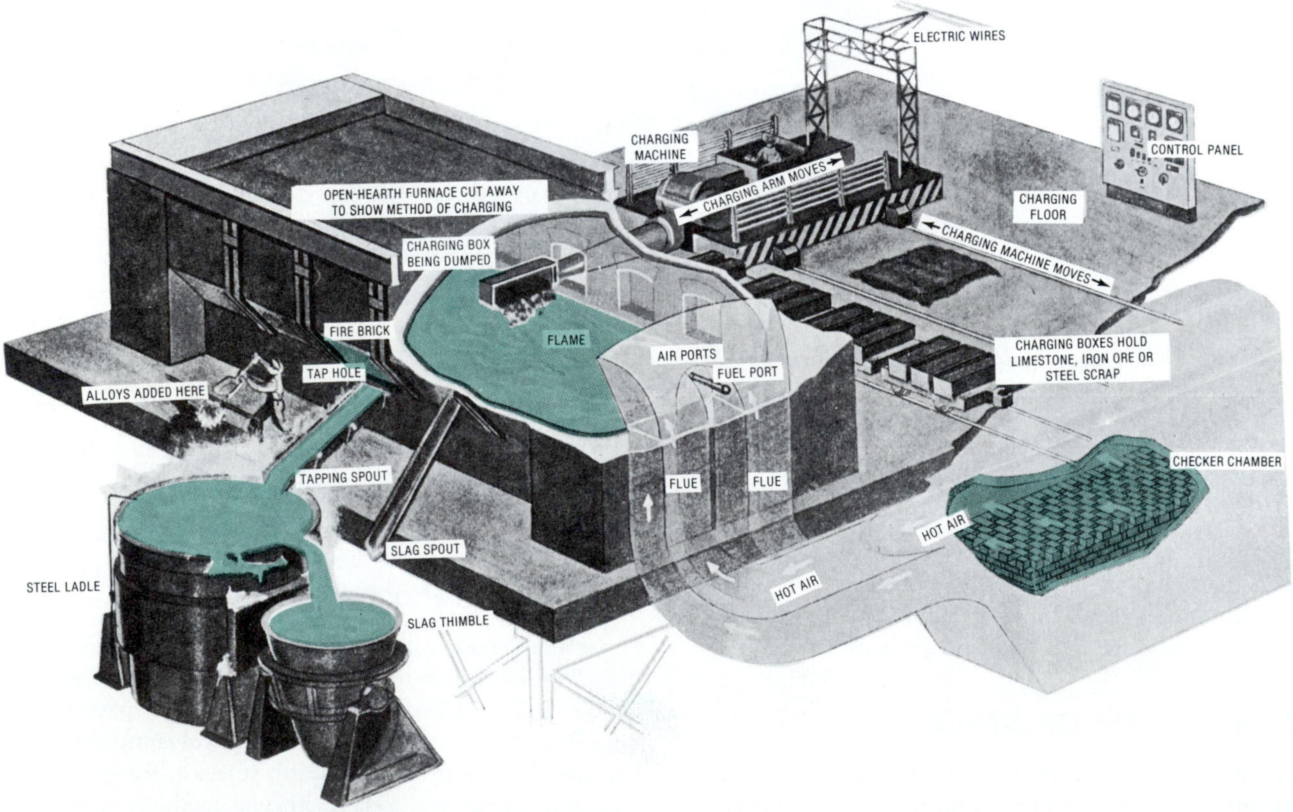
ELECTRIC WIRES
CHARGING MACHINE
CONTROL PANEL
OPEN-HEARTH FURNACE CUT AWAY TO SHOW METHOD OF CHARGING
CHARGING ARM MOVES
CHARGING FLOOR
CHARGING BOX BEING DUMPED
CHARGING MACHINE MOVES
FIRE BRICK
FLAME
AIR PORTS
FUEL PORT
CHARGING BOXES HOLD LIMESTONE, IRON ORE OR STEEL SCRAP
ALLOYS ADDED HERE
TAP HOLE
CHECKER CHAMBER
TAPPING SPOUT
FLUE
FLUE
HOT AIR
SLAG SPOUT
HOT AIR
STEEL LADLE
SLAG THIMBLE

Figure 87-5 *A schematic cross section of an open-hearth furnace. (Courtesy American Iron and Steel Institute.)*

There they are brought to a uniform temperature prior to rolling into the desired shapes and sizes. Open-hearth furnaces have a capacity of up to 550 tons (498 t) of steel.

Since the introduction of the basic oxygen furnace, steelmakers have found that the addition of oxygen to any steelmaking process will increase production considerably. Consequently, many open-hearth furnaces have been updated with the addition of an oxygen lance (Fig. 87-5), which directs almost pure oxygen at a high velocity onto the top of the molten steel in the hearth. The oxygen burns the impurities out of the steel much quicker than was possible by the straight open-hearth process. As a result of this modification, the production of these furnaces has been greatly increased.

Bessemer Converter

At one time the Bessemer converter was used to produce almost all the steel made. Now, however, the open-hearth furnace introduced in 1908 produces most of the steel, and the Bessemer converter makes less than 3 percent of it.

Basic Oxygen Process

One of the recent developments in steel-making is the basic oxygen process. The basic oxygen furnace (Fig. 87-6) resembles a Bessemer converter but does not have the air chamber and tuyeres at the bottom to admit air through the charge. Instead of air being forced through the molten metal, as in the Bessemer process, a high-pressure stream of pure oxygen is directed onto the top of the molten metal.

The furnace is tilted and is first charged with scrap metal (about 30 percent of the total charge). Molten pig iron is poured into the furnace, after which the required fluxes are added. An oxygen lance with a water-cooled hood is then lowered into the furnace until the tip is within 60 to 100 in. (152 to 254 cm) above the surface of the molten metal, depending on the blowing qualities of the iron and the type of scrap used. The oxygen is turned on and flows at the rate of 5000 to 6000 ft^3/min (141 to 169 m^3/min) at a pressure of 140 to 160 psi (965 to 1103 kPa).

The introduction of the oxygen causes the temperature of the molten steel (batch) to rise, at which time lime and fluospar may be added to help carry off the impurities in the form of slag. A recent development in this process is the *Lance Bubbling Equilibrium* (LBE). With this process, inert gases such as argon and nitrogen are introduced by lances through the bottom of the furnace. These gases bubble up through the molten metal, increasing the contact between the metal and the slag, and speed mixing. This results in an increase in yield and a reduction in alloying

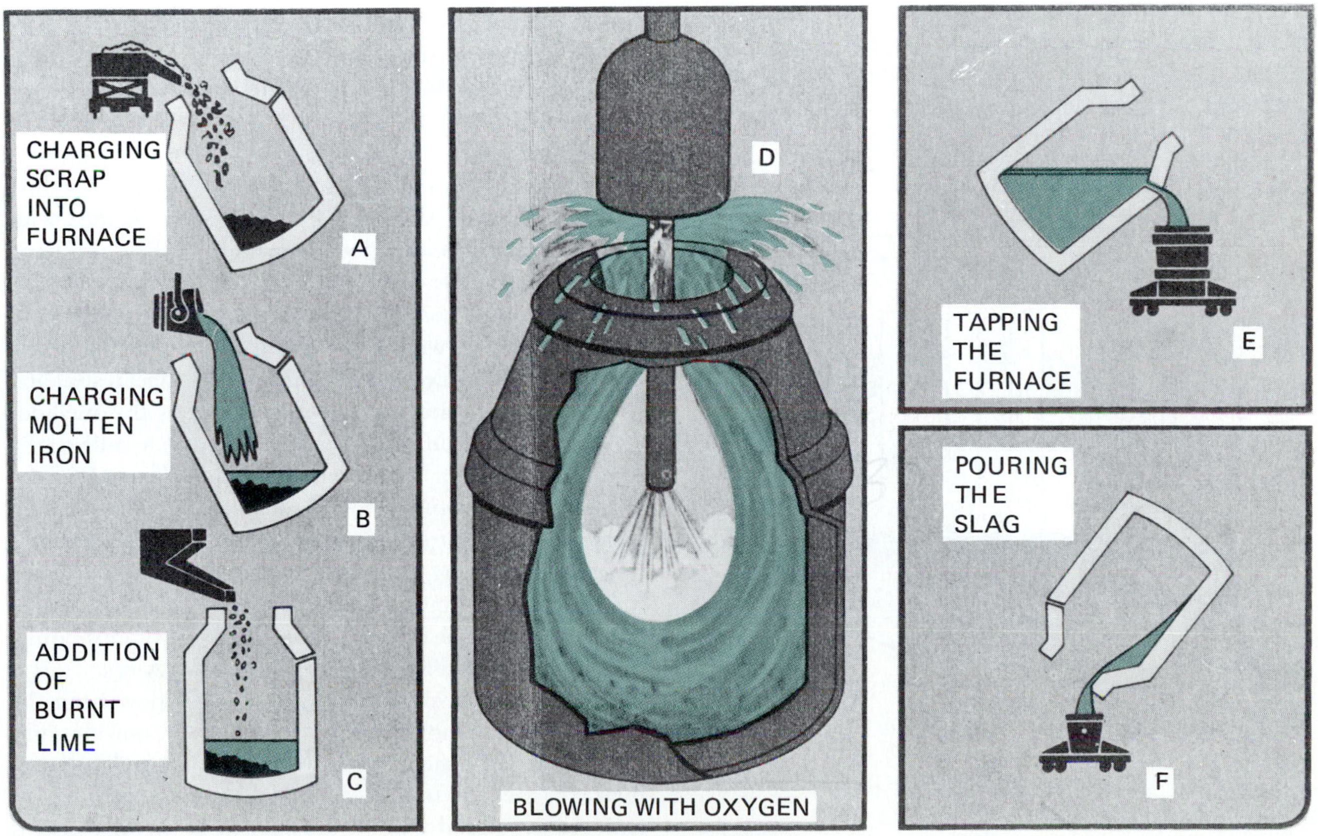
CHARGING SCRAP INTO FURNACE
A
CHARGING MOLTEN IRON
B
ADDITION OF BURNT LIME
C
D
BLOWING WITH OXYGEN
TAPPING THE FURNACE
E
POURING THE SLAG
F

Figure 87-6 *The basic oxygen process. (Courtesy Inland Steel Corporation.)*

elements, such as aluminum or silicon, which are used to reduce the oxygen and carbon content in the steel.

The force of the oxygen starts a high-temperature churning action and burns out the impurities. After all the impurities have been burned out, there will be a noticeable drop in the flame and a definite change in sound. The oxygen is then shut off and the lance removed.

The furnace is now tilted (Fig. 87-6) and the molten steel flows into a ladle or taken directly to the strand casting machine. The required alloys are added, after which the molten metal is teemed into ingots or formed into slabs. The refining process takes only about 50 min and about 300 tons (272 t) of steel can be made per hour.

Electric Furnace

The electric furnace (Fig. 87-7) is used primarily to make fine alloy and tool steels. The heat, the amount of oxygen, and atmospheric conditions can be regulated at will in the electric furnace; this furnace is therefore used to make steels that cannot be readily produced in any other way.

Carefully selected steel scrap, containing smaller amounts of the alloying elements than are required in the finished steel, is loaded into the furnace. The three carbon electrodes are lowered until an arc jumps from them to the scrap. The heat generated by the electric arcs gradually melts all the steel scrap. Alloying materials such as chromium, nickel, and tungsten, are then added to make the type of alloy steel required. Depending on the size of the furnace, it takes from 4 to 12 h to make a heat of steel. When the metal is ready to be tapped, the furnace is tilted forward and the steel flows into a large ladle. From the ladle, the steel is teemed into ingots.

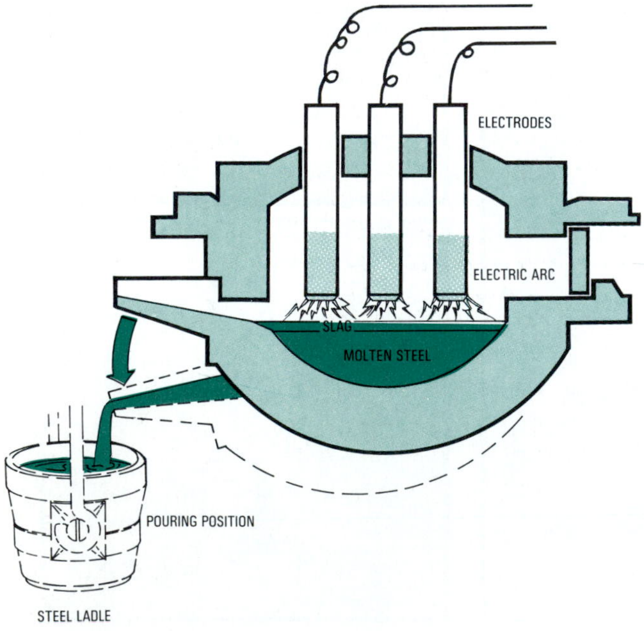
ELECTRODES
ELECTRIC ARC
SLAG
MOLTEN STEEL
POURING POSITION
STEEL LADLE

Figure 87-7 *Diagram of the electric furnace.*

Steel Processing

After the steel has been properly refined in any of the furnaces, it is tapped into ladles where alloying elements and deoxidizers may be added. The molten steel may then be teemed into *ingots* weighing as much as 20 tons (18 t), or it may be formed directly into slabs by the *continuous-casting process* (Fig. 87-8).

The steel is teemed into ingot molds and allowed to solidify. The ingot molds are then removed or *stripped* and the hot ingots are placed in soaking pits at 2200°F (1204°C) for up to 1.5 h to make them a uniform temperature throughout. The ingots are then sent to the rolling mills where they are rolled and reduced in cross section to form blooms, billets, or slabs.

Blooms are generally rectangular or square and are larger than 36 in^3 in cross section. They are used to manufacture structural steel and rails.

Billets may be rectangular or square, but are less than 36 in^2 in cross-sectional area. They are used to manufacture steel rods, bars, and pipes.

Slabs are usually thinner and wider than billets. They are used to manufacture plate, sheet, and strip steel.

Strand or Continuous Casting

Strand or continuous casting (Fig. 87-8) is the most modern and efficient method of converting molten steel into semifinished shapes such as slabs, blooms, and billets. This process eliminates ingot teeming, stripping, soaking, and rolling. It also produces higher quality steel, reduces energy consumption, and has increased overall productivity by over 13 percent.

Continuous casting is fast becoming the major method of producing steel slabs and billets. This, combined with the much faster production of molten steel by the basic oxygen process, has greatly improved the efficiency of steel production. In 1982, 31 percent of the steel produced in the United States was by the strand or continuous casting method. In the same year, 53 percent of European steel and 85 percent of Japanese steel was produced by this method.

Molten steel from the furnace is taken in a ladle to the top of the strand or continuous caster and poured into the tundish. The *tundish* provides an even pool of molten metal to be fed into the casting machine. It also acts as a reservoir permitting the empty ladle to be removed and the full ladle to be poured without interrupting the flow of molten metal to the caster. The steel is stirred continuously by a nitrogen lance or by electromagnetic devices.

The molten steel drops in a controlled flow from the tundish into the mold section. Cooling water quickly chills the outside of the metal to form a solid skin, which becomes thicker as the steel strand descends through the cooling system. As the strand reaches the bottom of the machine, it becomes solid throughout. The solidified steel is moved in a

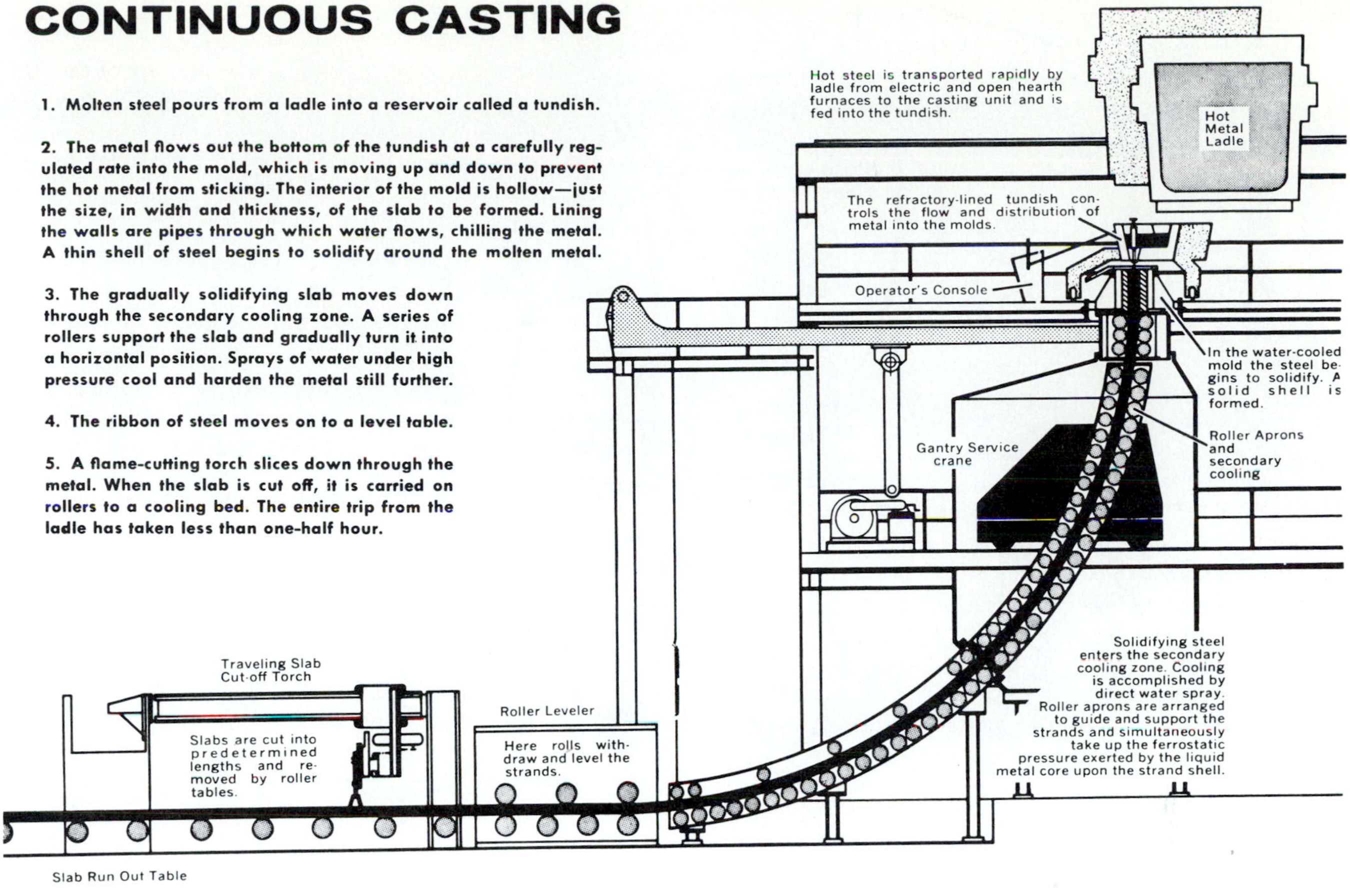
CONTINUOUS CASTING
1. Molten steel pours from a ladle into a reservoir called a tundish.
2. The metal flows out the bottom of the tundish at a carefully regulated rate into the mold, which is moving up and down to prevent the hot metal from sticking. The interior of the mold is hollow—just the size, in width and thickness, of the slab to be formed. Lining the walls are pipes through which water flows, chilling the metal. A thin shell of steel begins to solidify around the molten metal.
3. The gradually solidifying slab moves down through the secondary cooling zone. A series of rollers support the slab and gradually turn it into a horizontal position. Sprays of water under high pressure cool and harden the metal still further.
4. The ribbon of steel moves on to a level table.
5. A flame-cutting torch slices down through the metal. When the slab is cut off, it is carried on rollers to a cooling bed. The entire trip from the ladle has taken less than one-half hour.
Hot steel is transported rapidly by ladle from electric and open hearth furnaces to the casting unit and is fed into the tundish.
Hot Metal Ladle
The refractory-lined tundish controls the flow and distribution of metal into the molds.
Operator's Console
In the water-cooled mold the steel begins to solidify. A solid shell is formed.
Roller Aprons and secondary cooling
Gantry Service crane
Solidifying steel enters the secondary cooling zone. Cooling is accomplished by direct water spray. Roller aprons are arranged to guide and support the strands and simultaneously take up the ferrostatic pressure exerted by the liquid metal core upon the strand shell.
Traveling Slab Cut-off Torch
Slabs are cut into predetermined lengths and removed by roller tables.
Roller Leveler
Here rolls withdraw and level the strands.
Slab Run Out Table

Figure 87-8 *The continuous-casting process for making steel blooms or slabs.*

gentle curve by bending rolls until it reaches a horizontal position. The strand is then cut by a traveling cutting torch into required lengths. In some strand casting machines, the solidified steel is cut when it is in the vertical position. The slab or billet then topples to the horizontal position and is taken away.

Continuous casters are capable of producing strands at up to 15 ft per min (4 m per min). This process is designed for high tonnage operations or small batches, as required. When small batches are required, the molten steel is generally produced by methods other than the basic oxygen or open-hearth furnaces.

After the metal has been made into blooms or slabs by either of the aforementioned processes, it is further rolled into billets and then, while still hot, into the desired shape, such as round, flat, square, hexagonal, etc. (Fig. 87-9). These rolled products are known as *hot-rolled steel* and are easily identified by the bluish-black scale on the outside.

Hot-rolled steel may be further processed into *cold-rolled* or *cold-drawn steel* by removing the scale in an acid bath and passing the metal through rolls or dies of the desired shape and size.

One of the newest processes in the production of sheet steel is the *Continuous Annealing Line* (CAL), which produces high-strength and ultra-high-strength cold-rolled steel without increased weight, mainly for the automotive industry. This process is capable of altering the steel strength without changing its chemistry. The sheet steel produced by this method is virtually free from defects such as edge damage or sticker breaks. Continuous annealing reduces the time from the 5 to 6 days required for batch annealing to a mere 10 min. It also permits superior steel to be made with lesser amounts of costly alloying elements. Advanced computer controls permit the heating–cooling sequence to "lock in" the grain structure at the desired strength level. It can produce sheet steel with a yield strength of up to 220 000 psi.

Vacuum Processing of Molten Steel

Steel used in space and nuclear projects is often processed and solidified in a vacuum to remove oxygen, nitrogen, and hydrogen, and thus produce a high-quality steel.

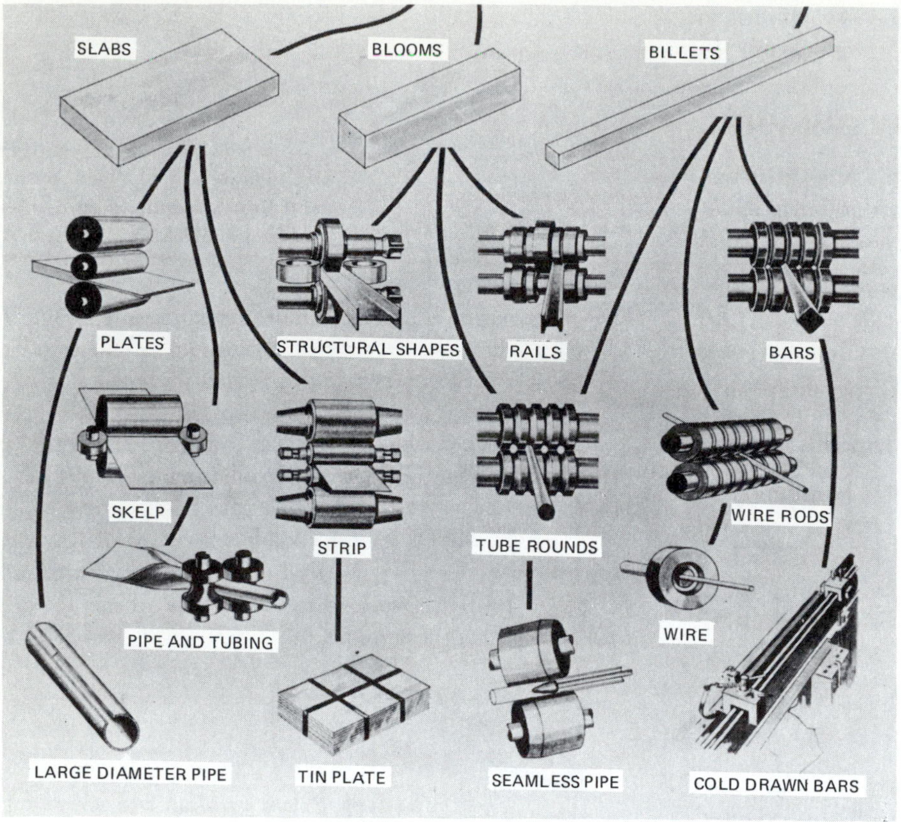
SLABS
BLOOMS
BILLETS
PLATES
STRUCTURAL SHAPES
RAILS
BARS
SKELP
STRIP
TUBE ROUNDS
WIRE RODS
PIPE AND TUBING
WIRE
LARGE DIAMETER PIPE
TIN PLATE
SEAMLESS PIPE
COLD DRAWN BARS

Figure 87-9 *Various steel shapes produced by rolling.*

CHEMICAL COMPOSITION OF STEEL

Although iron and carbon are the main elements in steel, certain other elements may be present in varying quantities. Some are present because they are difficult to remove, and others are added to impart certain qualities to the steel. The elements found in plain carbon steel are carbon, manganese, phosphorus, silicon, and sulfur.

Carbon is the element which has the greatest influence on the property of the steel, since it is the hardening agent. The hardness, hardenability, tensile strength, and wear resistance will be increased as the percentage of carbon is increased up to about 0.83 percent. After this point has been reached, additional carbon does not noticeably affect the hardness of the steel but increases wear resistance and hardenability.

Manganese, when added in small quantities (0.30 to 0.60 percent) during the manufacture of steel, acts as a deoxidizer or purifier. Manganese helps to remove the oxygen which, if it remained, would make the steel weak and brittle. Manganese also combines with sulfur, which in most cases is considered an undesirable element in steel. The addition of manganese increases the strength, toughness, hardenability, and shock resistance of steel. It will also slightly lower the critical temperature and increase ductility.

When manganese is added in quantities above 0.60 percent, it is considered an alloying element and will impart certain properties to the steel. When 1.5 to 2 percent manganese is added to high-carbon steel it will produce deep-hardening, nondeforming steel, which must be quenched in oil. Hard, wear-resistant steels, suitable for use in power shovel scoops, rock crushers, and grinding mills, are produced when up to 15 percent manganese is added to high-carbon steel.

Phosphorus is generally considered an undesirable element in carbon steel when present in amounts over 0.6 percent, since it will cause the steel to fail under vibration or shock. This condition is termed *cold-shortness.* Small amounts of phosphorus (about 0.3 percent) tend to eliminate blow holes and decrease shrinkage in the steel. Phosphorus and sulfur may be added to low-carbon steel (machine steel) to improve the machinability.

Silicon, present in most steels in amounts from 0.10 to 0.30 percent, acts as a deoxidizer and makes steel sound when it is cast or hot-worked. Silicon, when added in larger amounts (0.60 to 2 percent) is considered an alloying element. It is never used alone or simply with carbon; some deep-hardening element, such as manganese, molybdenum, or chromium, is usually added with silicon. When added as an alloying element, silicon increases the tensile strength, toughness, and hardness penetration of steel.

Sulfur, generally considered an impurity in steel, causes the steel to crack during working (rolling or forging) at high temperatures. This condition is known as *hot-shortness.* Sulfur may be added purposely to low-carbon steel in quantities ranging from 0.07 to 0.30 percent, to increase its machinability. Sulfurized, free-cutting steel is known as *screw stock* and is used in automatic screw machines.

CLASSIFICATION OF STEEL

Steel may be classified into two groups: carbon steels and alloy steels.

PLAIN CARBON STEELS

Plain carbon steels may be classified as those which contain only carbon and no other major alloying element. They are divided into three categories: low-carbon steel, medium-carbon steel, and high-carbon steel.

Low-carbon steel contains from 0.02 to 0.30 percent carbon by mass. Because of the low carbon content, this type of steel cannot be hardened, but can be case-hardened. *Machine steel* and *cold-rolled steel,* which contain from 0.08 to 0.30 percent carbon, are the most common low-carbon steels. These steels are commonly used in machine shops for the manufacture of parts that do not have to be hardened. Such items as bolts, nuts, washers, sheet steel, and shafts are made of low-carbon steel.

Medium-carbon steel contains from 0.30 to 0.60 percent carbon and is used where greater tensile strength is required. Because of the higher carbon content, this steel can be hardened, which makes it ideal for steel forgings. Tools such as wrenches, hammers, and screwdrivers are drop-forged from medium-carbon steel and later heat treated.

High-carbon steel, also known as *tool steel,* contains over 0.60 percent carbon and may range as high as 1.7 percent. This type of steel is used for cutting tools, punches, taps, dies, drills, and reamers. It is available in hot-rolled stock or in finish-ground flat stock and drill rod.

ALLOY STEELS

Often certain steels are needed which have special characteristics that a plain carbon steel would not possess. It is then necessary to choose an *alloy steel.*

Alloy steel may be defined as steel containing other elements, in addition to carbon, which produce the desired qualities in the steel. The addition of alloying elements may impart one or more of the following properties to the steel.

1. Increase in tensile strength
2. Increase in hardness
3. Increase in toughness
4. Alteration of the critical temperature of the steel
5. Increase in wear abrasion
6. Red hardness
7. Corrosion resistance

High-Strength, Low-Alloy Steels

A recent development in the steelmaking industry is that of the high-strength, low-alloy (HSLA) steels. These steels, containing a maximum carbon content of 0.28 percent and small amounts of vanadium, columbium, copper, and other alloying elements, offer many advantages over the regular low-carbon construction steels. Some of these advantages are:

- Higher strength than medium-carbon steels.
- Less expensive than other alloy steels.
- Strength properties are "built into" the steel, and no further heat treating is needed.
- Bars of smaller cross sections can do the work of larger regular-carbon steel bars.
- Higher hardness, toughness (impact strength), and fatigue failure limits than carbon steel bars.
- May be used unpainted because it develops a protective oxide coating on exposure to the atmosphere.

Effects of the Alloying Elements

Elements such as chromium, cobalt, manganese, molybdenum, nickel, phosphorus, silicon, sulphur, tungsten, and vanadium may be added to steel to give it a wide range of desired properties. The properties imparted to the steel by these elements are given in Table 87-1.

REVIEW QUESTIONS

1. What effect has the study of metals had on modern living?

PHYSICAL PROPERTIES OF METALS

2. Compare hardness, brittleness, and toughness.

MANUFACTURING OF PIG IRON

3. Briefly describe the manufacture of pig iron.

Table 87-1 The Effect of Alloying Elements on Steel

Effect	Carbon	Chromium	Cobalt	Lead	Manganese	Molybdenum	Nickel	Phosphorus	Silicon	Sulfur	Tungsten	Vanadium
Increases tensile strength	X	X			X	X	X					
Increases hardness	X	X										
Increases wear resistance	X	X			X		X				X	
Increases hardenability	X	X			X	X	X					X
Increases ductility					X							
Increases elastic limit		X				X						
Increases rust resistance		X					X					
Increases abrasion resistance		X			X							
Increases toughness		X				X	X					X
Increases shock resistance		X					X					X
Increases fatigue resistance												X
Decreases ductility	X	X										
Decreases toughness			X									
Raises critical temperature		X	X								X	
Lowers critical temperature					X		X					
Causes hot-shortness										X		
Causes cold-shortness								X				
Imparts red-hardness			X			X					X	
Imparts fine grain structure					X							X
Reduces deformation					X		X					
Acts as deoxidizer					X				X			
Acts as desulphurizer					X							
Imparts oil-hardening properties		X			X	X	X					
Imparts air-hardening properties					X	X						
Eliminates blow holes								X				
Creates soundness in casting									X			
Facilitates rolling and forging					X				X			
Improves machinability				X						X		

MANUFACTURING OF CAST IRON

4. Explain how cast iron is manufactured.
5. Name four types of cast iron and state one purpose for each.

THE OPEN-HEARTH PROCESS

6. What is the purpose of the checkers in an open-hearth furnace?
7. How are the impurities removed from the molten metal in an open-hearth furnace?

8. What modification has been made to most open-hearth furnaces and what has been its result?

THE BESSEMER PROCESS AND THE BASIC OXYGEN PROCESS

9. What are the chief differences between the Bessemer and the basic oxygen processes?
10. Briefly describe the basic oxygen process.

THE CHEMICAL COMPOSITION OF STEEL

11. What effect does carbon in excess of 0.83 percent have on steel?
12. What is the purpose of adding small quantities of manganese to steel?
13. What effect will the addition of larger quantities of manganese (1.5 to 2.0 percent) have on steel?
14. Why is phosphorus considered an undesirable element in steel?
15. How will steel be affected by the addition of silicon:
    a. In small amounts? b. In large amounts?
16. Why is sulfur considered an undesirable element in steel?
17. Why are sulfur and phosphorus sometimes added to steel?

CLASSIFICATION OF STEEL

18. State the carbon content of and two uses for:
    a. Low-carbon steel
    b. High-carbon steel
19. List six properties that alloying elements may impart to steel.

UNIT 88

Heat Treatment of Steel

In order for a steel component to function properly, it is often necessary to heat treat it. Heat treating is the process of heating and cooling a metal in its solid state in order to obtain the desired changes in its physical properties. One of the most important mechanical properties of steel is its ability to be hardened to resist wear and abrasion or be softened to improve ductility and machinability. Steel may also be heat treated to remove internal stresses, reduce grain size, or increase its toughness. During manufacture, certain elements are added to steel to produce special results when the metal is properly heat treated.

OBJECTIVES

After completing this unit you should be able to:
1. Select the proper grade of tool steel for a workpiece
2. Harden and temper a carbon-steel workpiece
3. Case-harden a piece of machine steel

HEAT-TREATING EQUIPMENT

The heat treating of metal is carried out in specially controlled furnaces, which may employ gas, oil, or electricity to provide heat. These furnaces must also be equipped with certain safety devices, as well as control and indicating devices to maintain the temperature required for the job. All furnace installations should be equipped with a fume hood and exhaust fan to take away any fumes resulting from the heat-treating operation, or in the case of a gas installation, to exhaust the gas fumes.

In most gas installations, the exhaust fan, when running, will actuate the air switch in the exhaust duct. The air switch in turn operates a solenoid valve, which permits the main gas valve to be opened. Should the exhaust fan fail for any reason, the air switch will also fail and the main gas supply will close down.

The furnace temperature is controlled by a thermocouple and an indicating pyrometer (Fig. 88-1). After the furnace has been started, the desired temperature is set on the indicating pyrometer. This pyrometer is connected on one side to a thermocouple, and on the other side to a solenoid valve, which controls the flow of gas to the furnace.

The thermocouple is made up of two dissimilar metal wires twisted together and welded at the end. The thermocouple is generally mounted in the back of the furnace in a refectory tube to prevent damage and oxidation to the thermocouple wires.

As the temperature in the furnace rises, the thermocouple becomes hot, and due to the dissimilarity of the wires, a small electrical current is produced. This current is conducted to the pyrometer on the wall and causes the pyrometer needle to indicate the temperature of the furnace. When the temperature in the furnace reaches the amount set on the pyrometer, a solenoid valve connected to the gas supply is actuated and the flow of gas to the furnace is restricted. When the furnace temperature drops below the temperature indicated on the pyrometer, the solenoid valve opens, permitting a full flow of gas.

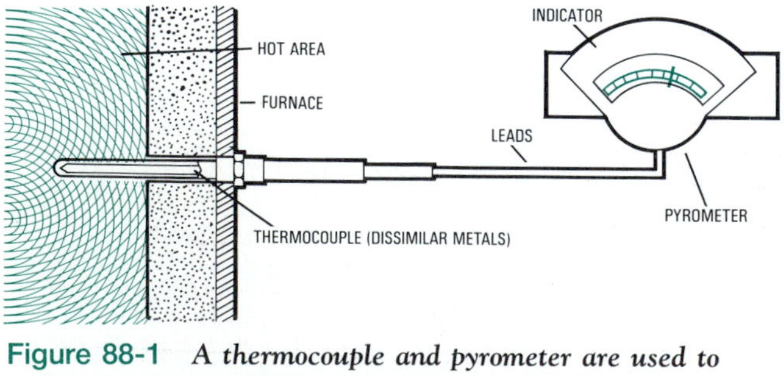
INDICATOR
HOT AREA
FURNACE
LEADS
PYROMETER
THERMOCOUPLE (DISSIMILAR METALS)

Figure 88-1 *A thermocouple and pyrometer are used to indicate and control the temperature of a heat-treating furnace.*

TYPES OF FURNACES

For most heat-treating operations, it is advisable to have a *low-temperature furnace* capable of temperatures up to 1300°F (704°C), a *high-temperature furnace* capable of temperatures up to 2500°F (1371°C), and a *pot-type furnace* (Fig. 88-2). The pot-type furnace may be used for hardening and tempering by immersing the part to be heat treated in the molten heat-treating medium, which may be salt or lead. Cyanide-type mixtures are used for case-hardening operations. One advantage of this type of furnace is that the parts, when being heated, do not come into contact with the air. This eliminates the possibility of oxidation or scaling. The temperature of the pot furnace is controlled by the same method as the high-temperature and low-temperature furnaces.

HEAT-TREATMENT TERMS

Plain carbon steel is composed of alternate layers of iron and iron carbide. In this unhardened state, it is known as pearlite (Fig. 88-3A). Before getting involved in the theory of heat treating, it is well to study and understand a few terms connected with this topic.

- *Heat treatment*—the heating and subsequent cooling of metals to produce the desired mechanical properties.
- *Decalescence point*—that temperature at which carbon steel, when being heated, transforms from pearlite to austenite; this is generally at about 1330°F (721°C) for 0.83 percent carbon steel.
- *Recalescence point*—that temperature at which carbon

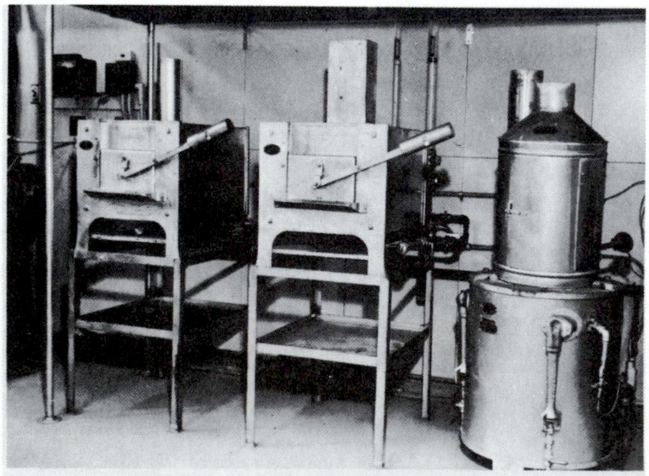

A Low temperature B High temperature C Pot type

Figure 88-2 *Types of heat-treating furnaces. (Courtesy Charles A. Hones, Inc.)*

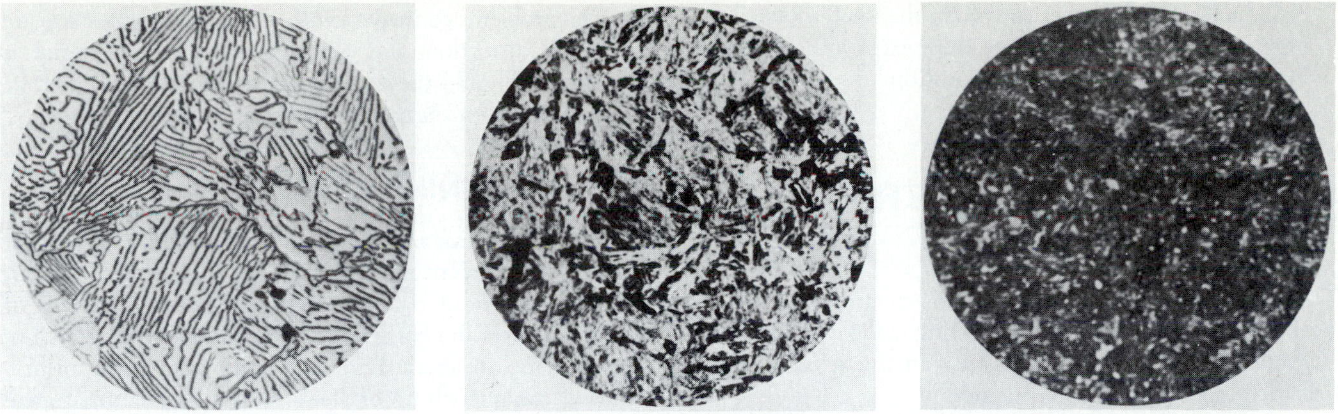

A B C

Figure 88-3 *(A) Pearlite is usually the condition of carbon steel before heat treating; (B) martensite is the condition of the hardened carbon steel; (C) martensite structure of steel can be altered by tempering.*

steel, when being slowly cooled, transforms from austenite to pearlite.

- *Lower critical temperature point*—the lowest temperature at which steel may be quenched in order to harden it. This temperature coincides with the decalescence point.
- *Upper critical temperature point*—the highest temperature at which steel may be quenched in order to attain maximum hardness and the finest grain structure.
- *Critical range*—the temperature range bounded by the upper and the lower critical temperatures.
- *Hardening*—the heating of steel above its lower critical temperature and quenching in the proper medium (water, oil, or air) to produce martensite (Fig. 88-3B).
- *Tempering* (drawing)—reheating hardened steel to a desired temperature below its lower critical temperature, followed by any desired rate of cooling. Tempering removes the brittleness and toughens the steel. Steel in this condition is called tempered martensite (Fig. 88-3C).
- *Annealing* (full)—heating metal to just above its upper critical point for the required period of time, followed by slow cooling in the furnace, lime, or sand. Annealing will soften the metal, relieve the internal stresses and strains, and improve its machinability.
- *Process annealing*—heating the steel to just below the lower critical temperature, followed by any suitable cooling method. This process is often used on metals which have been work-hardened. Process annealing will soften it sufficiently for further cold working.
- *Normalizing*—heating the steel to just above its upper critical temperature and cooling it in still air. Normalizing is done to improve the grain structure and remove the stresses and strains. In general, it brings the metal back to its normal state.
- *Spheroidizing*—the heating of steel to just below the lower critical temperature for a prolonged period of time followed by cooling in still air. This process produces a grain structure with globular-shaped particles (spheroids) of cementite rather than the normal needlelike structure, which improves the machinability of the metal.
- *Alpha iron*—the state in which iron exists below the lower critical temperature. In this state, the atoms form a body-centered cube.
- *Gamma iron*—the state in which iron exists in the critical range. In this state the molecules form face-centered cubes. Gamma iron is nonmagnetic.
- *Pearlite*—a laminated structure of ferrite [iron, and cementite (iron carbide)]; usually the condition of steel before heat treatment (Fig. 88-3A).
- *Cementite*—a carbide of iron (Fe_3C), which is the hardener in steel.
- *Austenite*—a solid solution of carbon in iron, which exists between the lower and upper critical temperatures.
- *Martensite*—the structure of fully hardened steel obtained when austenite is quenched. Martensite is characterized by its needlelike pattern (Fig. 88-3B).
- *Tempered martensite*—the structure obtained after martensite has been tempered (Fig. 88-3C). Tempered martensite was formerly known as *troosite* and *sorbite*.
- *Eutectoid steel*—steel containing just enough carbon to dissolve completely in the iron when the steel is heated to its critical range. Eutectoid steel contains from 0.80 to 0.85 percent carbon. This may be likened to a saturated solution of salt in water.
- *Hypereutectoid steel*—steel containing more carbon than will completely dissolve in the iron when the steel is heated to the critical range. This is similar to a supersaturated solution.
- *Hypoeutectoid steel*—steel containing less carbon than

can be dissolved by the iron when the steel is heated to the critical range. Here there is an excess of iron. This is similar to an unsaturated solution.

SELECTION OF TOOL STEEL

The proper selection and proper heat treatment of a tool steel are both essential if the part being made is to perform efficiently. Problems that may arise in the selection and heat treatment of tool steel include:

1. It may not be tough or strong enough for the job.
2. It may not offer sufficient abrasion resistance.
3. It may not have sufficient hardening penetration.
4. It may warp during heat treatment.

Because of these problems, the steel producers have been forced to manufacture many types of alloy steels to cover the range of most jobs.

To select the correct steel for a job, consult the handbook provided by the steel company for the specifications and application of the various steels it produces. This handbook will also outline the proper heat treatment of the particular material chosen. These instructions should be carefully followed.

Tool steels are generally classified as water-hardening, oil-hardening, air-hardening, or high-speed steels. They are usually identified by each manufacturer by a trade name such as Alpha 8, Keewatin, Nutherm, or Nipigon, which are some of the trade names of the Atlas Steel Company's products.

WATER-HARDENING TOOL STEEL

Water-hardening tool steels generally contain from 0.50 to 1.3 percent carbon, along with small amounts (about 0.20 percent) of silicon and manganese. The addition of silicon facilitates the forging and rolling of the material, while manganese helps to make the steel more sound when it is first cast into the ingot. Further addition of silicon (above 0.20 percent) will reduce the grain size and increase the toughness of water-hardening steel.

Most water-hardening steels achieve the maximum hardness for a depth of about ⅛ in. (3 mm); the inner core remains softer but still tough. Chromium or molybdenum is sometimes added to increase the hardenability (hardness penetration), toughness, and wear resistance of water-hardening steels. Water-hardening steels are heated to around 1450 to 1500°F (787 to 815°C) during the hardening process. These steels are used where a dense, fine-grained outer casing with a tough inner core is required. Typical applications are drills, taps, reamers, punches, jig bushings, and dowel pins.

The problems connected with water-hardening steels are those of distortion and cracking when the material is quenched. Should these problems occur, it would then be wise to select an oil-hardening steel.

OIL-HARDENING STEELS

A typical *oil-hardening steel* contains about 0.90 percent carbon, 1.6 percent manganese, and 0.25 percent silicon. The addition of manganese in quantities of 1.5 percent or more increases the hardenability (hardness penetration) of the steel up to about 1 in. (25 mm) from each surface. During the quenching of steels with a higher manganese content, the hardening is so rapid that a less severe quenching medium (oil) must be used. The use of oil as a quenching medium retards the cooling rate and reduces the stresses and strains in the steel, which cause warping and cracking. Chromium and nickel, in varying quantities, may be added to oil-hardening steel to increase its hardness and wear resistance. Higher hardening temperatures, from 1500 to 1550°F (815 to 843°C), are required for these latter alloy steels.

Often, due to the intricate shape of the part, it may not be possible to eliminate warping or cracking during the quench, and it will be necessary to select air-hardening steel for the particular part. Typical applications of oil-hardening steels are blanking, forming, and punching dies, precision tools, broaches, and gages.

AIR-HARDENING STEELS

Due to the slower cooling rate of *air-hardening steels,* the stresses and strains which cause cracking and distortion are kept to a minimum. Air-hardening steels are also used on parts having large cross sections, where full hardness throughout could not be obtained by using water- or oil-hardening steels.

A typical air-hardening steel will contain about 1.00 percent carbon, 0.70 percent manganese, 0.20 percent silicon, 5.00 percent chromium, 1.00 percent molybdenum, and 0.20 percent vanadium. Air-hardening steels require higher hardening temperatures—from 1600 to 1775°F (871 to 968°C).

Typical applications of this steel are large blanking, forming, trimming, and coining dies, rolls, long punches, precision tools, and gages.

HIGH-SPEED STEELS

High-speed steels are used in the manufacture of cutting tools such as drills, reamers, taps, milling cutters, and lathe cutting tools. The analysis of a typical high-speed steel could be as follows: 0.72 percent carbon, 0.25 percent manganese, 0.20 percent silicon, 4 percent chromium, 18 percent tungsten, and 1 percent vanadium. Tools made of high-speed steel will retain their hardness and cutting edges even when operating at red heat.

During heat treatment, high-speed steels must be preheated slowly to 1500 to 1600°F (815 to 871°C) in a neutral atmosphere and then transferred to another furnace and quickly brought up to 2300 to 2400°F (1260 to 1315°C). They are generally quenched in oil, but small, intricate sections may be air cooled.

CLASSIFICATION OF STEEL

In order to ensure that the composition of various types of steel remains constant and that a certain type of steel will meet the required specifications, the Society of Automotive Engineers (SAE) and the American Iron and Steel Institute (AISI) have devised similar methods of identifying different types of steel, and both are widely used.

THE SAE–AISI CLASSIFICATION SYSTEMS

The systems designed by the SAE and the AISI are similar in most respects. They both use a series of four or five numbers to designate the type of steel.

The first digit in these series indicates the predominant alloying element. The last two digits (or sometimes three in certain corrosion- or heat-resisting alloys) indicate the average carbon content in points (hundredths of 1 percent, or 0.01%) (Fig. 88-4).

The main difference in the two systems is that the AISI system indicates the steel-making process used by the following prefixes:

A—basic open-hearth alloy steel
B—acid-Bessemer carbon steel
C—basic open-hearth carbon steel
D—acid-open-hearth carbon steel
E—electric furnace steel

In the classification charts, the various types of steels are indicated by the first number in the series as follows:

1. Carbon
2. Nickel
3. Nickel–chrome
4. Molybdenum
5. Chromium
6. Chromium–vanadium
8. Triple alloy
9. Manganese–silicon

Table 88-1 indicates the SAE classification of the various steels and alloys. The number 7 does not appear on the chart. It formerly represented tungsten steel, which is no longer listed in this chart since it is now considered a special steel.

EXAMPLES OF STEEL IDENTIFICATION

Determine the types of steel indicated by the following numbers: 1015, A2340, 4170.

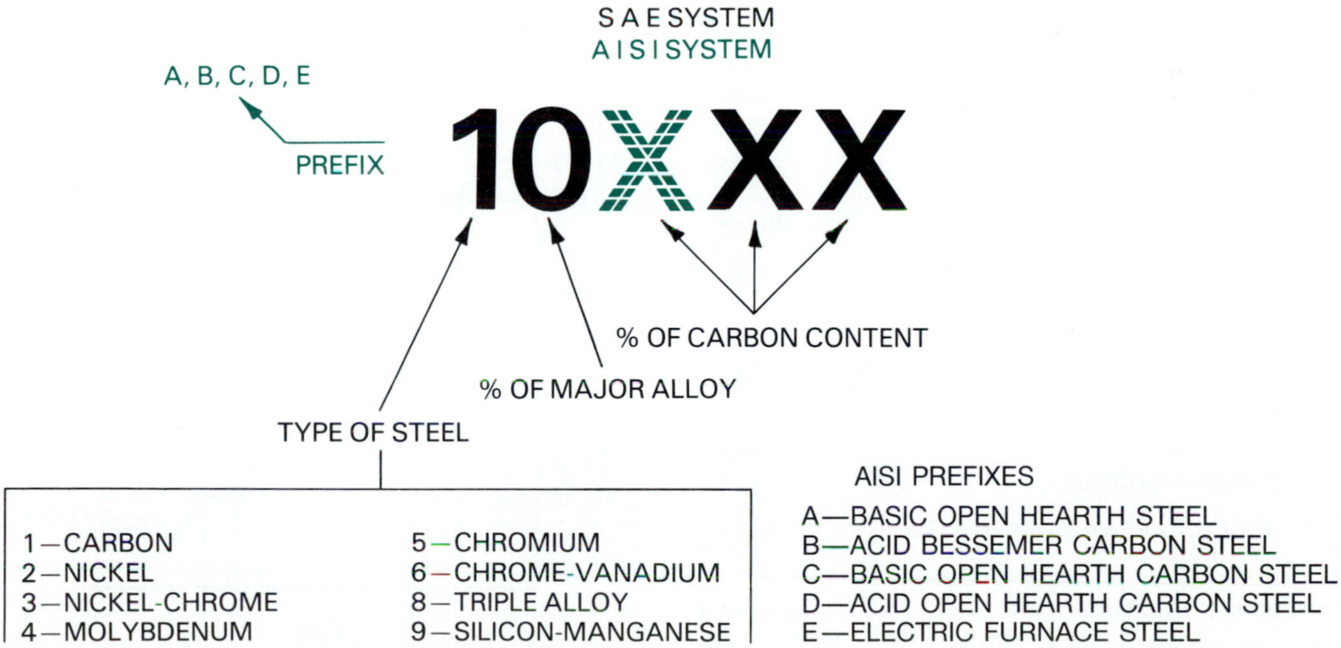
S A E SYSTEM
A I S I SYSTEM
A, B, C, D, E
PREFIX
10XXX
% OF CARBON CONTENT
% OF MAJOR ALLOY
TYPE OF STEEL
1—CARBON
2—NICKEL
3—NICKEL-CHROME
4—MOLYBDENUM
5—CHROMIUM
6—CHROME-VANADIUM
8—TRIPLE ALLOY
9—SILICON-MANGANESE
AISI PREFIXES
A—BASIC OPEN HEARTH STEEL
B—ACID BESSEMER CARBON STEEL
C—BASIC OPEN HEARTH CARBON STEEL
D—ACID OPEN HEARTH CARBON STEEL
E—ELECTRIC FURNACE STEEL

Figure 88-4 *The SAE and AISI classification systems.*

1015—1 indicates plain carbon steel.
—0 indicates there are no major alloying elements.
—15 indicates there is between 0.01 and 0.20 percent carbon content.

NOTE: This steel would naturally contain small quantities of manganese, phosphorus, and sulfur.

A2360—A indicates an alloy steel made by the basic open-hearth process.
—23 indicates the steel contains 3.5 percent nickel (see Table 88-1).
—60 indicates 0.60 percent carbon content.

4170—41 indicates a chromium–molybdenum steel.
—70 indicates 0.70 percent carbon content.

Table 88-1 SAE Classification of Steels

CARBON STEELS	1xxx
Plain carbon	10xx
Free-cutting (resulfurized screw stock)	11xx
Free-cutting manganese	X13xx
HIGH-MANGANESE	T13XX
NICKEL STEELS	2xxx
0.50% nickel	20xx
1.50% nickel	21xx
3.50% nickel	23xx
5.00% nickel	25xx
NICKEL–CHROMIUM STEELS	3xxx
1.25% nickel, 0.60% chromium	31xx
1.75% nickel, 1.00% chromium	32xx
3.50% nickel, 1.50% chromium	33xx
3.00% nickel, 0.80% chromium	34xx
Corrosion- and heat-resisting steels	30xxx
MOLYBDENUM STEELS	4xxx
Chromium–molybdenum	41xx
Chromium–nickel–molybdenum	43xx
Nickel–molybdenum	46xx and 48xx
CHROMIUM STEELS	5xxx
Low-chromium	51xx
Medium-chromium	52xxx
CHROMIUM–VANADIUM STEELS	6xxx
TRIPLE-ALLOY STEELS (nickel, chromium, molybdenum)	8xxx
MANGANESE–SILICON STEELS	9xxx

HEAT TREATMENT OF CARBON STEEL

The proper performance of a steel part depends not only on the correct selection of steel but also upon the correct heat-treating procedure and an understanding of the theory behind it. When steel is heated from room temperature to the upper critical temperature and then quenched, several changes take place in the steel. These may be more easily understood if the changes which take place in water, from the frozen state until it is transformed into steam, are considered.

By referring to Fig. 88-5 it is noted that water exists as a solid at or below 32°F (0°C). If the ice is heated, the temperature will remain 32°F (0°C) until the ice completely melts. If the water is heated further, it will turn into steam at 212°F (100°C). Again the water remains at this temperature for a short time before turning into steam. It should also be noticed that if the process were reversed and the steam cooled, it would form water at 212°F (100°C) and ice at 32°F (0°C). The points where water transforms to another state are known as the *critical points* of water.

Steel, like water, has critical points which, when determined, will lead to successful heat treatment of the metal.

TO DETERMINE THE CRITICAL POINTS OF 0.83 PERCENT CARBON STEEL

A simple experiment may be performed to illustrate the critical points and the changes that take place in a piece of carbon steel when heated and slowly cooled.

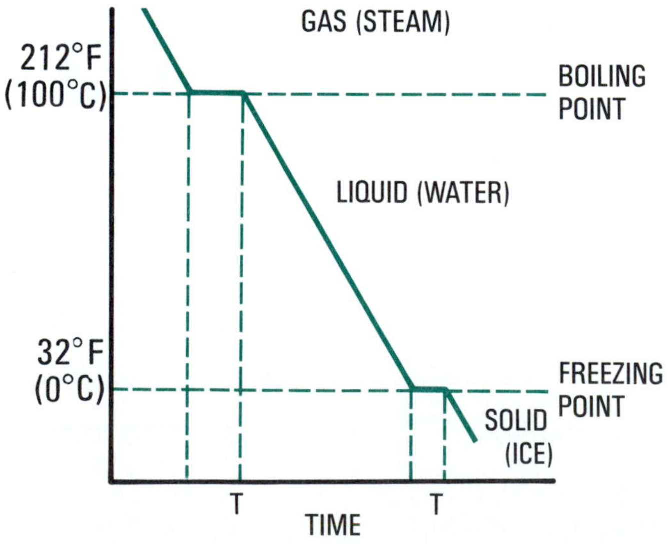

GAS (STEAM)
212°F
(100°C)
BOILING POINT
LIQUID (WATER)
32°F
(0°C)
FREEZING POINT
SOLID (ICE)
T
T
TIME

Figure 88-5 *The critical points of water.*

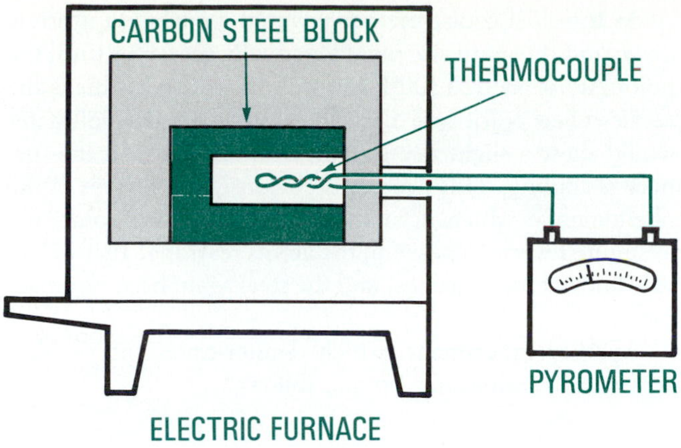
CARBON STEEL BLOCK
THERMOCOUPLE
PYROMETER
ELECTRIC FURNACE

Figure 88-6 *Setup to determine the critical points of steel.*

PROCEDURE

1. Select a piece of 0.83 percent (eutectoid) carbon steel about 1½ in. × 1½ in. × 2 in. (38 mm × 38 mm × 50 mm) long and drill a small hole in one end for most of the length.
2. Insert a thermocouple in the hole and seal the end of the hole with fireclay.
3. Place the block in a furnace and run the thermocouple wire to a voltmeter (Fig. 88-6).
4. Light the furnace and set the temperature for about 1425°F (773°C) on the pyrometer.
5. Plot the readings of the voltmeter needle at regular time intervals (Fig. 88-7).

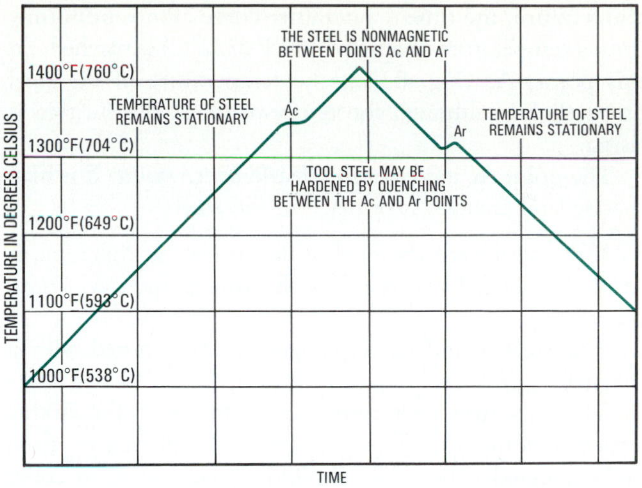
THE STEEL IS NONMAGNETIC
BETWEEN POINTS Ac AND Ar
1400°F(760°C)
TEMPERATURE OF STEEL
REMAINS STATIONARY
Ac
TEMPERATURE OF STEEL
REMAINS STATIONARY
Ar
1300°F(704°C)
TOOL STEEL MAY BE
HARDENED BY QUENCHING
BETWEEN THE Ac AND Ar POINTS
1200°F(649°C)
1100°F(593°C)
1000°F(538°C)
TEMPERATURE IN DEGREES CELSIUS
TIME

Figure 88-7 *A graph illustrating the critical points of steel.*

6. When the furnace reaches 1425°F (773°C), shut it down and let it cool.
7. Continue to plot the readings until the temperature in the furnace drops to approximately 1000°F (537°C).

OBSERVATIONS AND CONCLUSIONS

Steel at room temperature consists of laminated layers of ferrite (iron) and cementite (iron carbide). This structure is called pearlite (Fig. 88-8). As the steel is heated from room

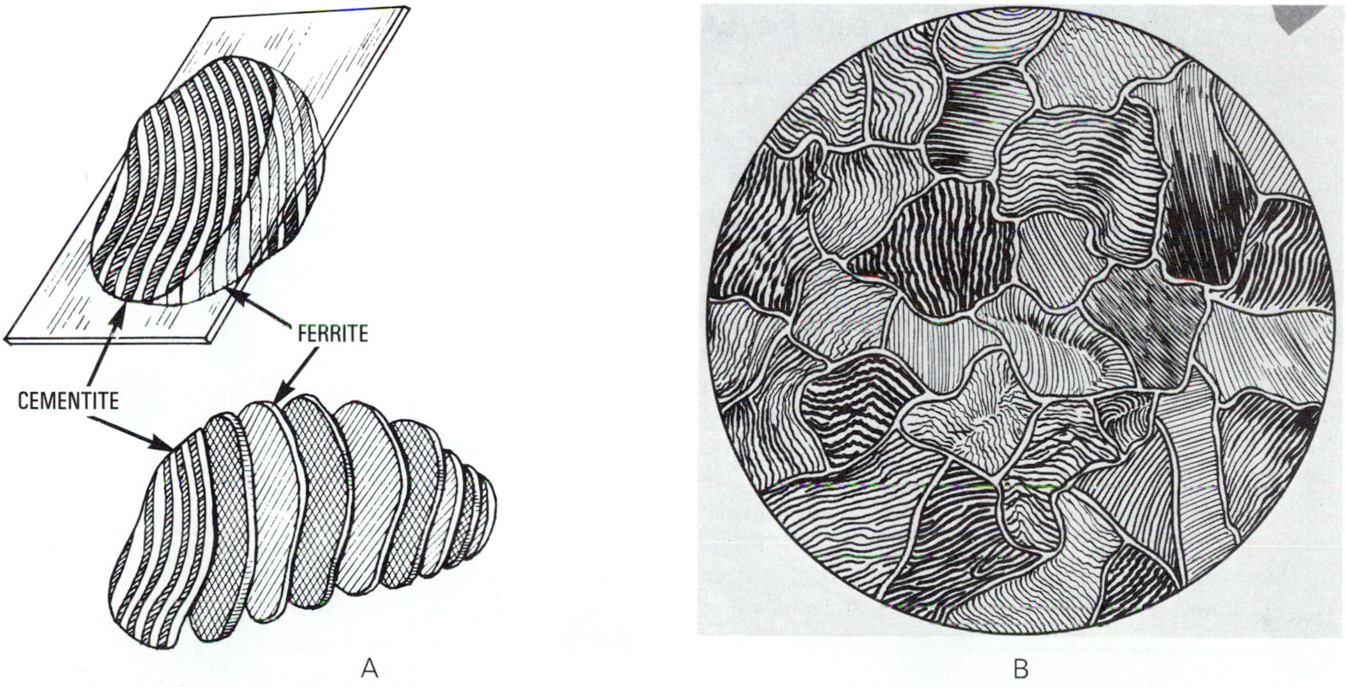
FERRITE
CEMENTITE
A
B

Figure 88-8 *(A) A pearlite grain is composed of alternate layers of iron (ferrite) and iron carbide (cementite); (B) a photomicrograph of high-carbon steel, showing the pearlite grains surrounded by iron carbide (white lines). (Courtesy Linde Division, Union Carbide Corporation.)*

temperature, the time/temperature curve climbs uniformly until a temperature of about 1333°F (722°C) is reached. At this point, Ac (Fig. 88-7), the temperature of the steel drops slightly although the temperature of the furnace is rising.

The point Ac indicates the *decalescence point.* It is here that several changes take place in the steel:

1. If the steel were observed in the furnace at this time, it would be noticed that the dark shadows in the steel disappear.
2. The steel would be nonmagnetic when tested with a magnet.
3. These changes were caused by a change in the atomic structure of the steel. The atoms rearrange themselves from body-centered cubes (Fig. 88-9A) to face-centered cubes (Fig. 88-9B). When the atoms are rearranged, the energy (heat) required for this change is drawn from the metal; thus a slight drop in the temperature of the workpiece is recorded at the decalescence point. The layers of iron carbide completely dissolve in the iron to form a solid solution known as *austenite.* Thus the decalescence point marked the transformation point from *pearlite* to *austenite,* or from body-centered cubes to face-centered cubes.
4. It is at this point that the steel, if quenched in water, would also show the first signs of hardening.
5. If the steel could be examined under a microscope, it would be noticed that the grain structure starts to get smaller. As the curve progresses upward past Ac, the grain size would become progressively smaller until the upper critical temperature [1425°F (773°C)] is reached.

As the steel cools, the curve would continue uniformly down and the grain size would gradually get larger until the point Ar, at about 1300°F (704°C), is reached. This is the *recalescence point* and here the needle on the voltmeter would show a slight rise in temperature, although the furnace is cooling. This process is obviously the reverse of the phenomenon which occurs at the decalescence point; the austenite reverts to pearlite, the atoms rearrange themselves into body-centered cubes, and the steel again becomes magnetic.

Another experiment, which demonstrates the decalescence and recalescence points, follows.

Decalescence Point

1. Place a magnet on a firebrick.
2. Select a ½ to ⅝ in. (13 to 15 mm) round piece of 0.90 to 1.00 carbon steel and place it on the magnet (Fig. 88-10).
3. Place a can of cold water under the magnet ends.
4. Heat the piece held to the magnet using a small flame.

NOTE: Do not allow the flame to come into contact with the magnet.

5. When the temperature reaches its critical point, the steel will drop into the water and become hardened.

Recalescence Point

1. Remove the can of water from under the magnet.

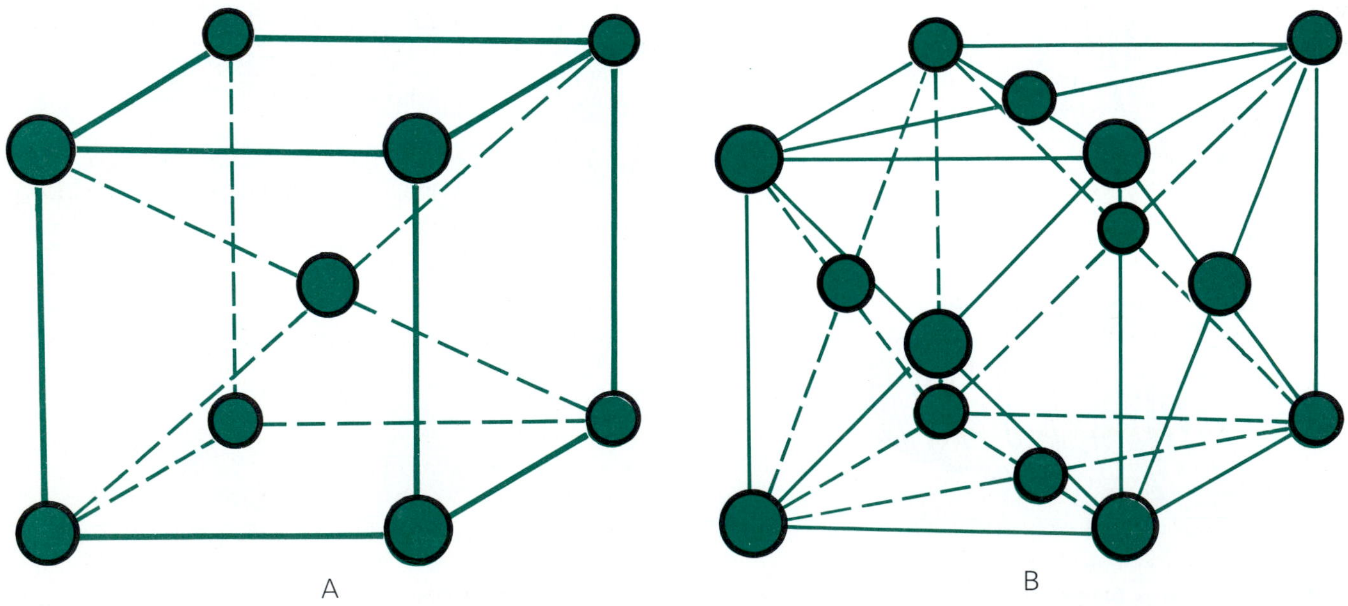

A
B

Figure 88-9 *(A) Arrangement of the atoms in a body-centered cube; (B) arrangement of the atoms in a face-centered cube.*

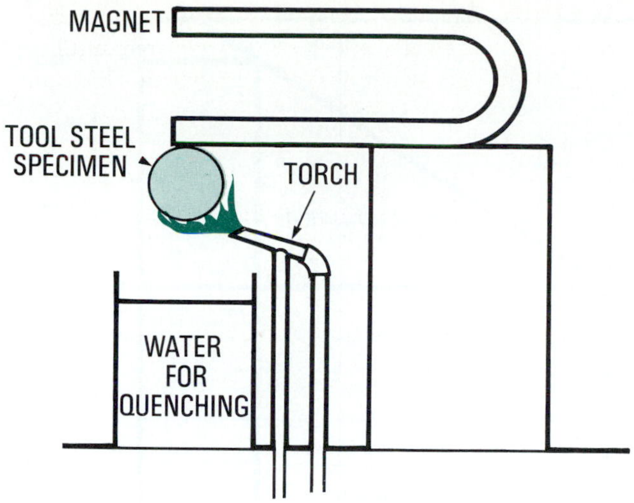
MAGNET
TOOL STEEL SPECIMEN
TORCH
WATER FOR QUENCHING

Figure 88-10 *Decalescence-point experiment.*

2. Place a flat plate under the work held on the magnet.
3. Heat the steel until it drops from the magnet onto the plate.
4. When the steel cools, it will become attracted by the magnet.

Summary

When the steel loses its magnetic value (decalescence point), it drops into the water and the change in the steel is trapped or stopped. The steel then hardens because it does not have time to revert to another state.

Conversely, when the steel is not quenched but is allowed to cool gradually from the decalescence point, it regains its magnetic value when it has cooled slightly (recalescence point). The steel does not change from magnetic to nonmagnetic; it merely acquires temporary characteristics of being attracted or *not* attracted to the magnet.

HARDENING OF 0.83 PERCENT CARBON STEEL

Once the critical temperature of a steel has been determined, the proper quenching temperature, which is about 50°F (27°C) over the upper critical temperature, can be determined. Not all steels have the same critical temperature; Fig. 88-11 indicates that the critical temperature of a steel drops as the carbon content increases up to 0.83 percent, after which it does not change. As a result, steels containing over 0.83 percent carbon (hypereutectoid) only need be heated to just above the lower critical temperature Ac_1 (Fig. 88-11) to obtain maximum hardness. This makes it possible to use a lower hardening temperature for hypereutectoid steels, thus decreasing the possibility of warping. The increase in the carbon content beyond 0.83 percent will not increase the hardness of the steel; however, it does increase the wear resistance of the steel considerably.

In order for the steel to be hardened properly, it must be heated uniformly to about 50°F (27°C) above the upper critical temperature and held at this temperature long enough to allow sufficient carbon to dissolve and form a solid solution, which permits maximum hardness. At this point the steel will have the smallest grain size and, when quenched, will produce the maximum hardness.

The critical temperature of a steel is also affected by alloying elements such as manganese, nickel, chromium, cobalt, and tungsten.

QUENCHING

When steel has been properly heated throughout, it is quenched in brine, water, or oil (depending on the type of steel), to cool it rapidly. During this operation, the austenite is transferred into *martensite,* a hard brittle metal. Because the steel is cooled rapidly, the austenite is prevented from passing through the recalescence point (Ar), as in the case of slow cooling, and the small grain size of the austenite is retained in the martensite (Fig. 88-12).

The rate of cooling affects the hardness of steel. If a water-hardening steel is quenched in oil, it cools more slowly and does not attain the maximum hardness. On the other hand, if an oil-hardening steel is cooled too quickly by quenching it in water, it may crack. Cracking may also occur when the quenching medium is too cold.

The method of quenching greatly affects the stresses and strains set up in the metal, which cause warping and cracking. For this reason, long, flat pieces should be held vertically above the quenching medium and plunged straight into the liquid. After immersion, the part should be moved around in a figure 8 motion. This keeps the liquid at a uniform temperature and prevents air pockets forming on the steel, which would affect the uniformity of hardness.

METCALF'S EXPERIMENT

This simple experiment demonstrates the effect that various degrees of heat have on the grain structure, hardness, and strength of tool steel.

1. Select a piece of SAE 1090 (tool steel) about ½ in. (13 mm) in diameter and about 4 in. (100 mm) long.
2. With a sharp, pointed tool, cut shallow grooves approximately ½ in. (13 mm) apart.
3. Number each section (Fig. 88-13).
4. Heat the bar with an oxyacetylene torch, bringing section 1 to a white heat.
5. Keep section 1 at white heat, and heat sections 4 and 5 to a cherry red. *Do not* apply heat to sections 6 to 8.
6. Quench in cold water or brine.
7. Test each section with the edge of a file for hardness.
8. Break off the sections and examine the grain structure under a microscope.

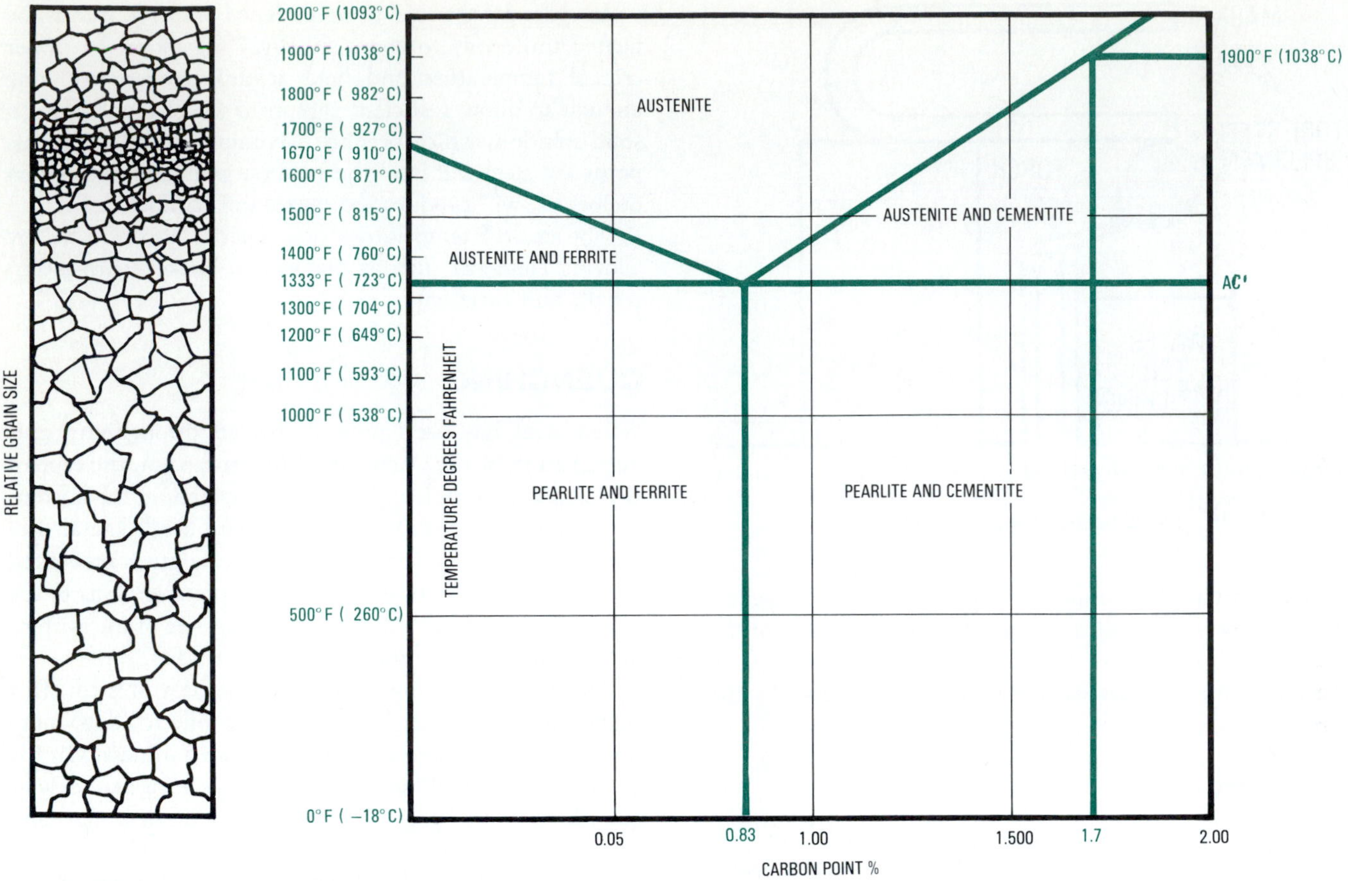
2000° F (1093° C)
1900° F (1038° C)
1800° F (982° C)
1700° F (927° C)
1670° F (910° C)
1600° F (871° C)
1500° F (815° C)
1400° F (760° C)
1333° F (723° C)
1300° F (704° C)
1200° F (649° C)
1100° F (593° C)
1000° F (538° C)
500° F (260° C)
0° F (−18° C)
RELATIVE GRAIN SIZE
TEMPERATURE DEGREES FAHRENHEIT
AUSTENITE
AUSTENITE AND CEMENTITE
AUSTENITE AND FERRITE
PEARLITE AND FERRITE
PEARLITE AND CEMENTITE
1900° F (1038° C)
AC'
0.05
0.83
1.00
1.500
1.7
2.00
CARBON POINT %

Figure 88-11 *Iron carbide diagram illustrating the relationship between the carbon content of steel and critical temperatures.*

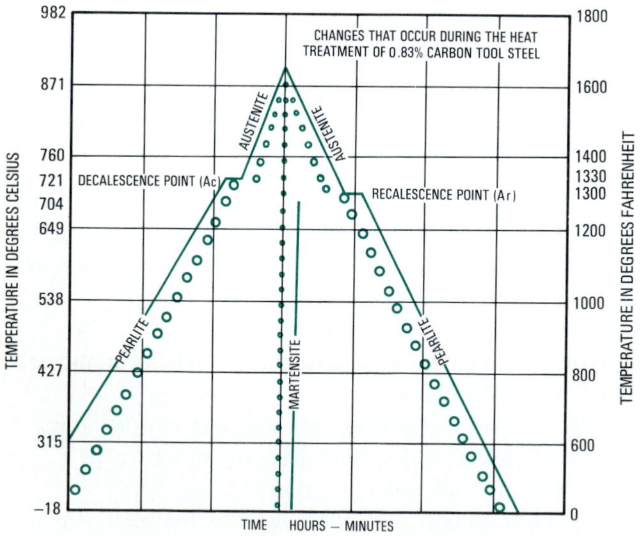
CHANGES THAT OCCUR DURING THE HEAT TREATMENT OF 0.83% CARBON TOOL STEEL
982
871
760
721
704
649
538
427
315
−18
1800
1600
1400
1330
1300
1200
1000
800
600
0
TEMPERATURE IN DEGREES CELSIUS
TEMPERATURE IN DEGREES FAHRENHEIT
DECALESCENCE POINT (Ac)
RECALESCENCE POINT (Ar)
AUSTENITE
AUSTENITE
PEARLITE
PEARLITE
MARTENSITE
TIME HOURS – MINUTES

Figure 88-12 *Carbon steel, when quenched at the upper critical temperature, retains the smallest grain size.*

Results

Sections 1 and 2 have been overheated. They break easily and the grain structure is very coarse (Fig. 88-14A).

Section 3 requires more force to break and the grain structure is somewhat finer. Sections 4 and 5 have the greater strength and resistance to shock. These sections have the finest grain structure (Fig. 88-14B).

Sections 6 to 8, where the metal was underheated, require the greatest force to break, and bending occurs. Note that the grain structure becomes coarser toward section 8 (Fig. 88-14C). This section is the original structure of unheated steel (pearlite).

TEMPERING

Tempering is the process of heating a hardened carbon or alloy steel below its lower critical temperature and cooling it

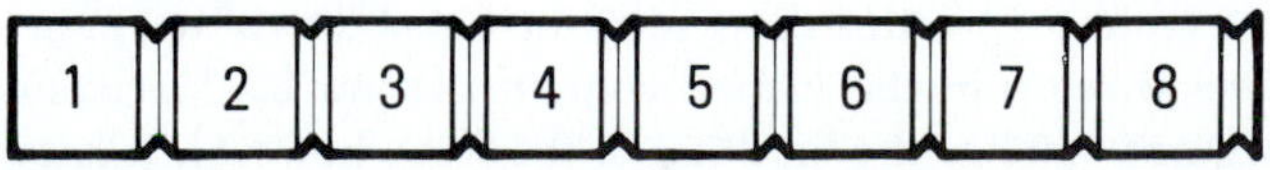
1
2
3
4
5
6
7
8

Figure 88-13 *A test piece for Metcalf's experiment.*

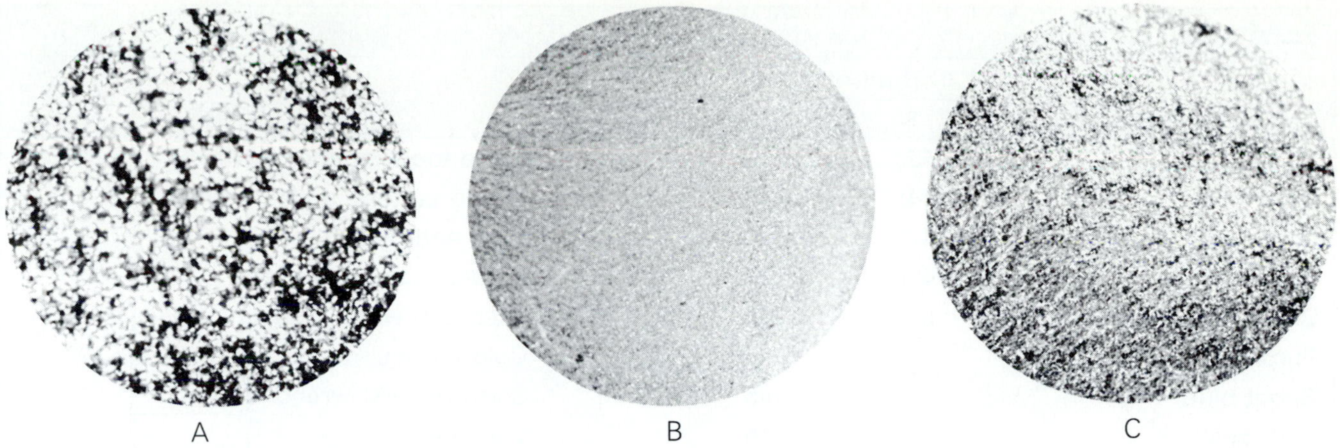
A
B
C

Figure 88-14 *(A) Toolsteel overheated; (B) toolsteel heated to the proper temperature; (C) toolsteel underheated. (Courtesy Kostel Enterprises Limited.)*

by quenching in a liquid or in air. This operation removes many of the stresses and strains set up when the metal was hardened. Tempering imparts toughness to the metal but decreases the hardness and tensile strength. The tempering process modifies the structure of the martensite, changing it to *tempered martensite,* which is somewhat softer and tougher than martensite.

The tempering and drawing temperature is not the same for each type of steel and is affected by several factors:

1. The toughness required for the part
2. The hardness required for the part
3. The carbon content of the steel
4. The alloying elements present in the steel

The hardness obtained after tempering depends on the temperature used and the length of time the workpiece is held at this temperature. Generally, hardness decreases and toughness increases as the temperature is increased (Fig. 88-15).

As the length of the tempering time is increased for a specific part, the hardness of the metal decreases. On the other hand, if the tempering time is too short, the stresses and strains set up by hardening are not totally removed and the metal will be brittle. The cross-sectional size of the workpiece affects its tempering time. The tempering time and temperatures for various steels are always supplied in the steel manufacturer's handbook; these recommended times and temperatures should be followed to obtain the best results.

Tempering Colors

When a piece of steel is heated from room temperature to a red heat, it passes through several color changes, caused by the oxidation of the metal. These color changes indicate the approximate temperature of the metal and are often used as a guide when tempering (Table 88-2).

ANNEALING

Annealing is a heat-treating operation used to soften metal and to improve its machinability. Annealing also relieves the internal stresses and strains caused by previous operations, such as forging or rolling.

PROCEDURE

1. Set the pyrometer approximately 30°F (16°C) above the upper critical temperature and start the furnace (Fig. 88-16).

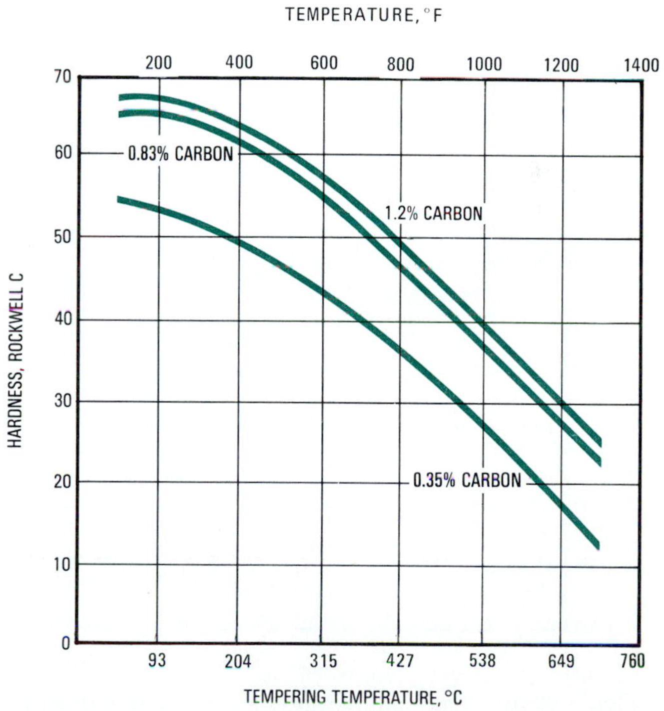
TEMPERATURE, °F
200
400
600
800
1000
1200
1400
70
60
50
40
30
20
10
0
0.83% CARBON
1.2% CARBON
0.35% CARBON
HARDNESS, ROCKWELL C
93
204
315
427
538
649
760
TEMPERING TEMPERATURE, °C

Figure 88-15 *As tempering heat increases, hardness decreases.*

Table 88-2 Tempering Colors and Approximate Temperatures for Carbon Steel

Color	Temperature °F	Temperature °C	Use
Pale yellow	430	220	Lathe tools, shaper tools
Light straw	445	230	Milling cutters, drills, reamers
Dark straw	475	245	Taps and dies
Brown	490	255	Scissors, shear blades
Brownish-purple	510	265	Axes and wood chisels
Purple	525	275	Cold chisels, center punches
Bright blue	565	295	Screwdrivers, wrenches
Dark blue	600	315	Woodsaws

2. Place the part in the furnace. After the required temperature has been reached, allow it to soak for 1 h per inch (25 mm) of workpiece thickness.
3. Shut off the furnace and allow the part to cool slowly in the furnace; or remove the part from the furnace and pack it immediately in lime or ashes and leave it covered for several hours, depending on the size, until it is cool.

NORMALIZING

Normalizing is performed on metal to remove internal stresses and strains and to improve its machinability.

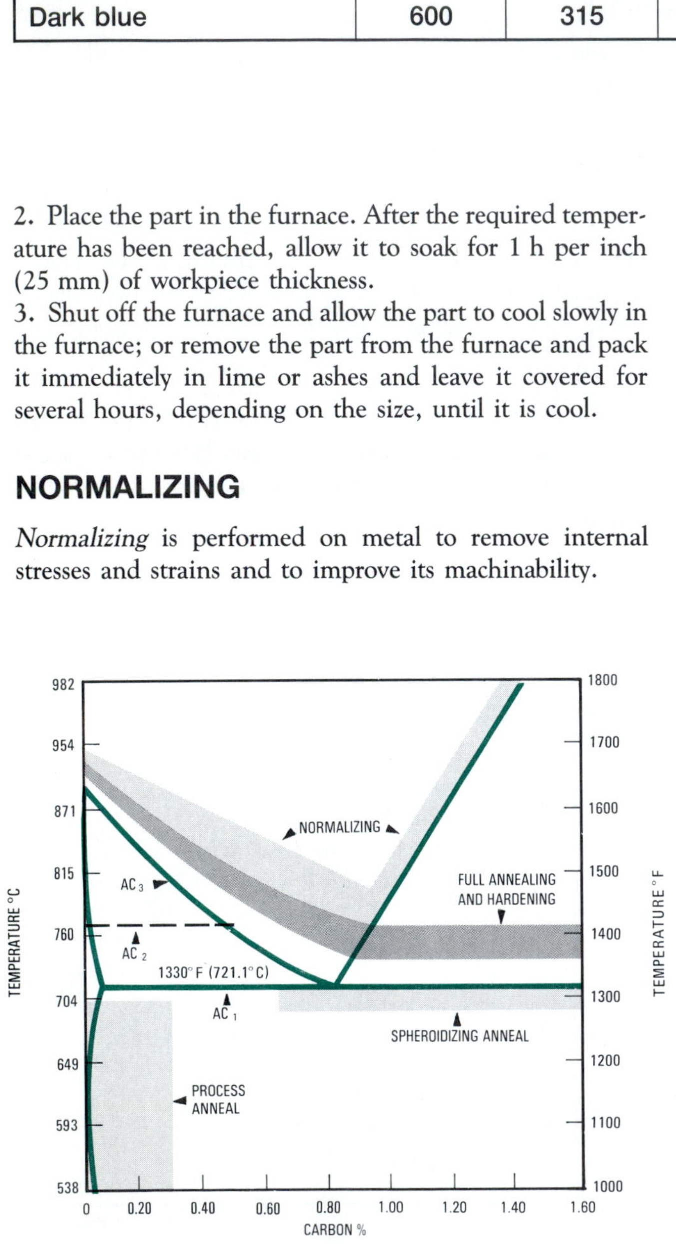

982
954
871
815
760
704
649
593
538
1800
1700
1600
1500
1400
1300
1200
1100
1000
TEMPERATURE °C
TEMPERATURE °F
NORMALIZING
FULL ANNEALING AND HARDENING
AC 3
AC 2
AC 1
1330° F (721.1° C)
SPHEROIDIZING ANNEAL
PROCESS ANNEAL
0
0.20
0.40
0.60
0.80
1.00
1.20
1.40
1.60
CARBON %

Figure 88-16 *Temperature ranges for various heat-treating operations.*

PROCEDURE

1. Set the pyrometer approximately 30°F (16°C) above the upper critical temperature of the metal and start the furnace (Fig. 88-16).
2. Place the part in the furnace. After the required temperature has been reached, allow the workpiece to soak for 1 h per inch (25 mm) of thickness.
3. Remove the part from the furnace and allow it to cool slowly in still air. Thin workpieces may cool too rapidly and may harden if normalized in air. It may be necessary to pack them in lime to retard the cooling rate.

SPHEROIDIZING

Spheroidizing is a process of heating metal for an extended period to just below the lower critical temperature. This process produces a special kind of grain structure whereby the cementite particles become spherical in shape. Spheroidizing is generally done on high-carbon steel to improve the machinability.

PROCEDURE

1. Set the pyrometer approximately 30°F (16°C) below the lower critical temperature of the metal and start the furnace (Fig. 88-16).
2. Place the part in the furnace and allow it to soak for several hours at this temperature.
3. Shut down the furnace and let the part cool slowly to about 1000°F (537°C).
4. Remove the part from the furnace and cool it in still air.

CASE-HARDENING METHODS

When hardened parts are required, they may be made from carbon steel and heat treated to the specifications. Often these parts can be made more cheaply from machine steel and then case-hardened. This process produces a hard outer case with a soft inner core, which is often preferable to through-hard parts made from carbon steel. Case hardening may be performed by several methods, such as carburizing, carbonitriding, and nitriding processes.

CARBURIZING

Carburizing is a process whereby low-carbon steel, when heated with some carbonaceous material, absorbs carbon into its outer surface. The depth of penetration depends on the time, the temperature, and the carburizing material used. Carburizing may be performed by three methods: pack carburizing, liquid carburizing, and gas carburizing.

Pack carburizing is generally used when a hardness depth of penetration of 0.060 in. (1.5 mm) or more is required. The parts to be carburized are packed with a carbonaceous material such as activated charcoal in a sealed steel box.

PROCEDURE

1. Place a 1- to 1½ in. (25 to 38 mm) layer of carbonaceous material in the bottom of a steel box which will fit into the furnace.
2. Place the parts to be carburized in the box, leaving about 1½ in. (38 mm) between parts.
3. Pack the carburizer around the parts and cover the parts with about 1½ in. (38 mm) of material.
4. Tap the sides of the box to settle the material and to pack it around the workpieces. This will exclude most of the air.
5. Place a metal cover over the box and seal around the joint with fireclay.
6. Place the box in the furnace and bring the temperature up to about 1700°F (926°C).
7. Leave the box in the furnace long enough to give the required penetration. The rate of penetration is generally about 0.007 to 0.008 in./h (0.17 to 0.20 mm/h), however this decreases as the depth of penetration increases. The proper time (and temperature) for any depth of penetration is usually given in the literature supplied by the manufacturer of the carburizing material (Fig. 88-17).
8. Shut down the furnace and leave the box in the furnace until it cools. This may take 12 to 16 h.
9. Remove the box from the furnace and take out the parts and clean them.

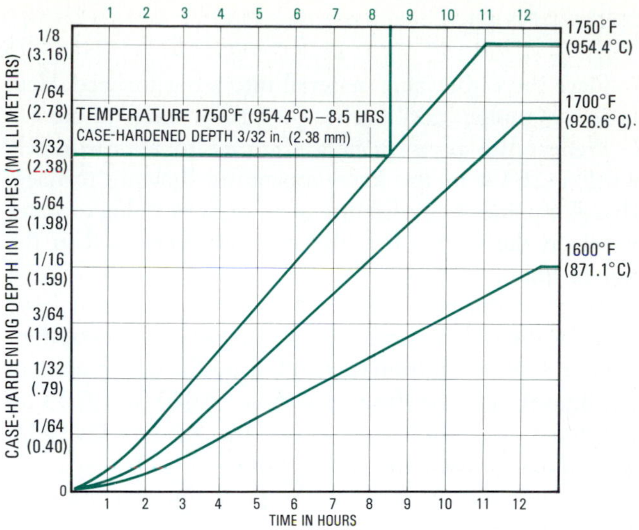
CASE-HARDENING DEPTH IN INCHES (MILLIMETERS)
1/8 (3.16)
7/64 (2.78)
3/32 (2.38)
5/64 (1.98)
1/16 (1.59)
3/64 (1.19)
1/32 (.79)
1/64 (0.40)
0
1 2 3 4 5 6 7 8 9 10 11 12
TIME IN HOURS
TEMPERATURE 1750°F (954.4°C)–8.5 HRS
CASE-HARDENED DEPTH 3/32 in. (2.38 mm)
1750°F (954.4°C)
1700°F (926.6°C)
1600°F (871.1°C)

Figure 88-17 *Relationship between temperature, time, and depth of case.*

NOTE: The surfaces of the workpieces have now been transformed into a thin layer of carbon steel, which is soft due to the slow cooling of the parts in the carburizing material.

10. Heat the parts to the proper critical temperature in a furnace and quench in oil or water, depending on the shape of the parts and the hardness required. The parts are now surrounded with a hard layer of carbon steel and have a soft inner core (Fig. 88-18).

Liquid carburizing is generally used to produce a thin layer of carbon steel on the outside of low-carbon steel parts. The parts are not usually machined after liquid carburization.

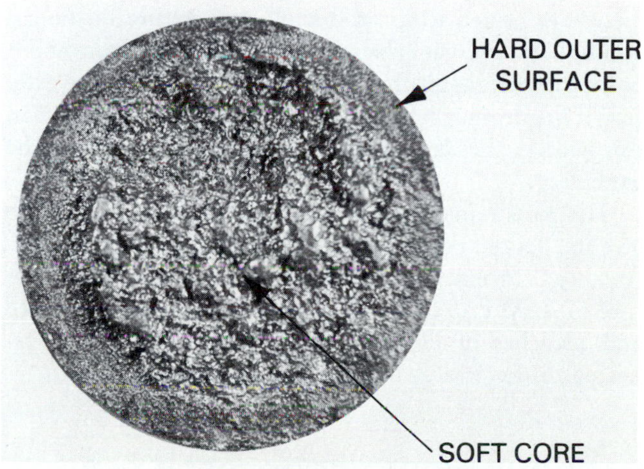
HARD OUTER SURFACE
SOFT CORE

Figure 88-18 *A broken cross section of a piece of case-hardened machine steel, showing the hardened outer case.*

PROCEDURE

1. Place the carburizing material into a pot furnace. Heat until it is molten and reaches the proper temperature.
2. Preheat the parts to be carburized to approximately 800°F (426°C) in the low-temperature drawing furnace. This eliminates the possibility of an explosion due to water or oil on the work when the parts are immersed in the molten liquid.
3. Suspend the parts in the liquid carburizer and leave them for the time required to give the desired penetration. The depth of penetration (depending on the temperature of the liquid) may be from 0.015 to 0.020 in. (0.38 to 0.50 mm) for the first hour and about 0.010 in. (0.25 mm) for each succeeding hour.
4. Use dry tongs to remove the parts; quench the parts immediately in water.

CAUTION: Since most liquid carburizers contain cyanide, extreme care must be taken when these materials are being used.

- Avoid letting *any moisture* come into contact with the liquid carburizer. Such contact will cause an explosion.
- Heat the jaws of the tongs before using, to remove any moisture or oil.
- Avoid inhaling the fumes; they are toxic.
- Wear protective clothing (gloves, face and arm shields) when removing and quenching parts.

Gas carburizing, like pack carburizing, is used on parts where over 0.060 in. (1.5 mm) depth of case-hardening is required and where it is necessary to grind the parts after carburizing. This method is generally done by specialized heat-treating firms, since it requires special types of furnaces.

The parts are placed in a sealed drum into which natural gas or propane is introduced and circulated. The workpieces are heated to a carburizing temperature in the gas atmosphere. The gas exhausts at one end of the drum and is burned to prevent air from entering the chamber. The exterior of the drum is heated by a source such as gas or oil. In this process the carbon from the gas is absorbed by the workpiece.

The parts remain in the drum for the time required to give the desired penetration. Depths of 0.020 to 0.030 in. (0.50 to 0.76 mm) are obtained in about 4 h at a temperature of 1700°F (926°C). The parts may then be removed and quenched or allowed to cool, after which they are reheated to the critical temperature and quenched.

CARBONITRIDING PROCESSES

Carbonitriding is a process whereby both carbon and nitrogen are absorbed by the surface of a steel workpiece when it is heated to the critical temperature to produce a hard, shallow outer case. Carbonitriding may be done by liquid or gas methods.

Cyaniding (liquid carbonitriding) is a process that uses a salt bath composed of cyanide-carbonate-chloride salts with varying amounts of cyanide, depending on the application. Liquid cyaniding is generally carried out in a pot-type furnace, and since cyanide fumes are poisonous, extreme care must be taken when using this method.

The parts are suspended in a liquid cyanide bath, which must be at a temperature above the lower critical point of the steel being used. The depth of penetration is about 0.005 to 0.010 in. (0.12 to 0.25 mm) in 1 h at 1550°F (843°C). A depth of about 0.015 in. (0.38 mm) may be obtained in 2 h at the same temperature. The parts may then be quenched in water or oil, depending on the steel being used. After the parts have been hardened, they should be thoroughly washed to remove all traces of the cyanide salt.

Carbonitriding (gas cyaniding) is carried out in a special furnace similar to the gas carburizing furnace. The workpieces are put into the inner drum of the furnace. A mixture of ammonia and a carburizing gas is introduced into and circulated through this chamber, which is heated externally to a temperature of 1350 to 1700°F (732 to 926°C). During this process, the workpiece absorbs carbon from the carburizing gas and nitrogen from the ammonia.

The parts are removed from the furnace and quenched in oil, which gives the parts maximum hardness and minimum distortion. The depth of case-hardening produced by this method is relatively shallow. Depths of about 0.030 in. (0.76 mm) are obtained in 4 to 5 h at a temperature of 1700°F (926°C).

NITRIDING PROCESSES

Nitriding is used on certain alloy steels to provide maximum hardness. Most carbon alloy steels can be hardened only to about 62 Rockwell C by conventional means, whereas readings of 70 Rockwell C may be obtained on certain vanadium and chromium alloy steels using a nitriding process. Nitriding may be done in a protected atmosphere furnace or in a salt bath.

In *gas nitriding* the parts to be nitrided are placed in an airtight drum, which is heated externally to a temperature of 900 to 1150°F (482 to 621°C). Ammonia gas is circulated through the chamber. The ammonia, at this temperature, decomposes into nitrogen and hydrogen. The nitrogen penetrates the outer surface of the workpiece and combines with the alloying elements to form hard nitrides. Gas nitriding is a slow process requiring approximately 48 h to obtain a case-hardened depth of 0.020 in. (0.50 mm). Because of the low operating temperatures used in this process, and since no quenching of the part is required, there is little or no distortion. This method of increasing hardness is used on parts that have been hardened and ground. No further finishing is required on such parts.

Salt bath nitriding is carried out in a salt bath containing nitriding salts. The hardened part is suspended in the molten nitriding salt, which is held at a temperature from 900 to 1100°F (482 to 593°C) depending on the application. Parts such as high-speed taps, drills, and reamers are nitrided to increase surface hardness, which improves durability.

SURFACE HARDENING OF MEDIUM-CARBON STEELS

When selected areas of a part are to be surface-hardened, to increase wear resistance and retain a soft inner core, the part must have a medium- or high-carbon content. It may be surface hardened by flame or induction hardening, depending on the size of the part and its application. In both processes, the steel must contain carbon, since no external carbon is added as in other case-hardening methods.

INDUCTION HARDENING

In *induction hardening* the part is surrounded by a coil through which a high-frequency electrical current is passed. The current heats the surface of the steel to above the critical temperature in a few seconds. An automatic spray of water, oil, or compressed air is used to quench and harden the part, which is held in the same position as for heating. Since only the surface of the metal is heated, the hardness is localized at the surface. The depth of hardness is governed by the current frequency and heating-cycle duration.

The current frequencies vary from 1 kHz to 2 MHz. Higher frequencies produce shallow hardening depths. Lower frequencies produce hardening depths up to ¼ in. (6 mm).

Induction hardening may be used for the selective hardening of gear teeth, splines, crank shafts, camshafts, and connecting rods.

FLAME HARDENING

Flame hardening is used extensively to harden ways on lathes and other machine tools, as well as gear teeth, splines, crank shafts, etc.

The surface of the metal is heated very rapidly to above the critical temperature and is hardened quickly by a quenching spray. Large surfaces, such as lathe ways, are heated by a special oxyacetylene torch which is moved automatically along the surface, followed by a quenching spray. Smaller parts are placed under the flame, and spray-quenched automatically.

Flame-hardened parts should be tempered immediately to remove strains created by the hardening process. Large surfaces are tempered by a special low-tempering torch, which follows the quenching nozzle as it moves along the work. The depth of flame hardening varies from 1/16 to ¼ in. (1.5 to 6 mm), depending on the speed at which the surface is brought up to the critical temperature.

REVIEW QUESTIONS

HEAT-TREATING EQUIPMENT

1. Describe a thermocouple and explain how it functions.
2. Sketch a furnace installation showing the furnace, thermocouple, and pyrometer.
3. What is the purpose of:
   a. The pyrometer?
   b. The solenoid valve?
   c. The air switch?

HEAT-TREATMENT TERMS

4. What is the difference between the decalescence point and the recalescence point of a piece of steel?
5. At what point in relation to the upper and lower critical temperatures are the following heat-treating operations performed?
   a. Hardening
   b. Tempering
   c. Annealing
   d. Normalizing
   e. Spheroidizing
6. What is the difference between hypereutectoid and hypoeutectoid steel?

WATER-HARDENING TOOL STEEL

7. State two problems which often occur with water-hardening steel.

OIL-HARDENING TOOL STEEL

8. Why is oil used to advantage as a quenching medium?

AIR-HARDENING STEELS

9. What are the advantages of air-hardening steels?
10. What elements will give red-hardness quality to a high-speed steel toolbit?
11. Explain the difference in the hardening procedures between plain carbon steel and high-speed steel.

THE SAE–AISI CLASSIFICATION SYSTEMS

12. How does the AISI system differ from the SAE system of identifying steel?
13. What is the composition of the following steels?
a. 2340 b. 1020 c. E4340

HEAT TREATMENT OF CARBON STEEL

14. Explain how the critical points of water are determined.
15. List five changes that take place in steel when it is heated to (and above) the decalescence point from room temperature.
16. Describe the changes that take place in a piece of steel when it is cooled to the recalescence point from the upper critical temperature.

HARDENING OF 0.83 PERCENT CARBON STEEL

17. What is the advantage of using hypereutectoid steel?
18. To what temperature should steel be heated prior to quenching, in order to produce the best results?

QUENCHING

19. Name two quenching media and state the purpose of each.
20. Describe the proper method of quenching long, thin workpieces.

TEMPERING

21. What is the purpose of tempering a piece of steel?
22. What factors will affect the temperature at which a piece of steel is tempered?
23. What will happen to the steel if the tempering time is:
a. Too long? b. Too short?

ANNEALING, NORMALIZING, AND SPHEROIDIZING

24. What is the difference between annealing, normalizing, and spheroidizing?

CARBURIZING

25. Describe briefly how pack carburizing is performed.
26. What precautions should be taken when the liquid carburizing process is being used?
27. Describe the gas carburizing process.

CARBONITRIDING PROCESSES

28. Explain the difference between the carburizing and the carbonitriding processes.

SURFACE HARDENING OF CARBON STEELS

29. What type of steel must be used for the flame-hardening and induction-hardening processes?

INDUCTION HARDENING

30. Describe briefly the induction-hardening principle.

FLAME HARDENING

31. Why is it desirable to temper parts immediately after flame hardening?
32. How may this be done on large surfaces?

UNIT 89

Testing of Metals and Nonferrous Metals

After heat treating certain tests may be performed to determine the properties of the metal. These tests fall into two categories:

1. *Nondestructive testing,* whereby the test may be performed without damaging the sample
2. *Destructive testing,* whereby a sample of the material is broken to determine the qualities of the metal

Nonferrous metals find a variety of uses in the machine trade. Since nonferrous metals contain little or no iron, they are used as bearings to prevent two like metal parts from being in contact with each other, where rust and corrosion is a factor, and where the weight of the product is important.

OBJECTIVES

After completing this unit you should be able to:
1. Explain three methods of hardness testing
2. Perform a Rockwell C hardness test on a workpiece
3. Perform tensile strength and impact tests on a workpiece
4. Describe several nonferrous metals used in industry

HARDNESS TESTING

Hardness testing is the most common form of nondestructive testing and is used to determine the hardness of a metal. Hardness in steel may be defined as its capacity to resist wear and deformation. The term *hardness* applied to metal is relative and indicates some of its properties. For example, if a piece of steel is hardened, tensile strength increases, but ductility is reduced. If the hardness of a metal is known, the properties and performance of the metal can be predicted accurately.

Two types of testing machines are used to measure the hardness of a metal:

1. Those which measure the depth of penetration made by a penetrator under a known load. Rockwell, Brinell, and Vickers hardness testers are examples of this type.
2. Those which measure the height of rebound of a small mass dropped from a known height. The scleroscope is based on this principle.

ROCKWELL HARDNESS TESTER

The Rockwell hardness tester (Fig. 89-1) indicates the hardness value by the depth that a penetrator advances into the metal under a known pressure. A 120° conical diamond penetrator (brale) is used for testing hard materials. A 1/16- or 1/8 in. diameter (1.5 or 3-mm) ball is used as a penetrator for soft materials (Fig. 89-2)

Rockwell hardness is designated by various letters and numbers. The scales are indicated by the letters A, B, C, and D. The C scale, which is the outside scale on the dial, is used in conjunction with the 120° diamond penetrator and a 330-lb (150-kg) major load for testing hardened metals. The B scale, or the red inner scale, is read when the 1/16-in. (1.5-mm) ball penetrator is used along with the 220-lb (100-kg) load for testing soft metals. The other letters, A

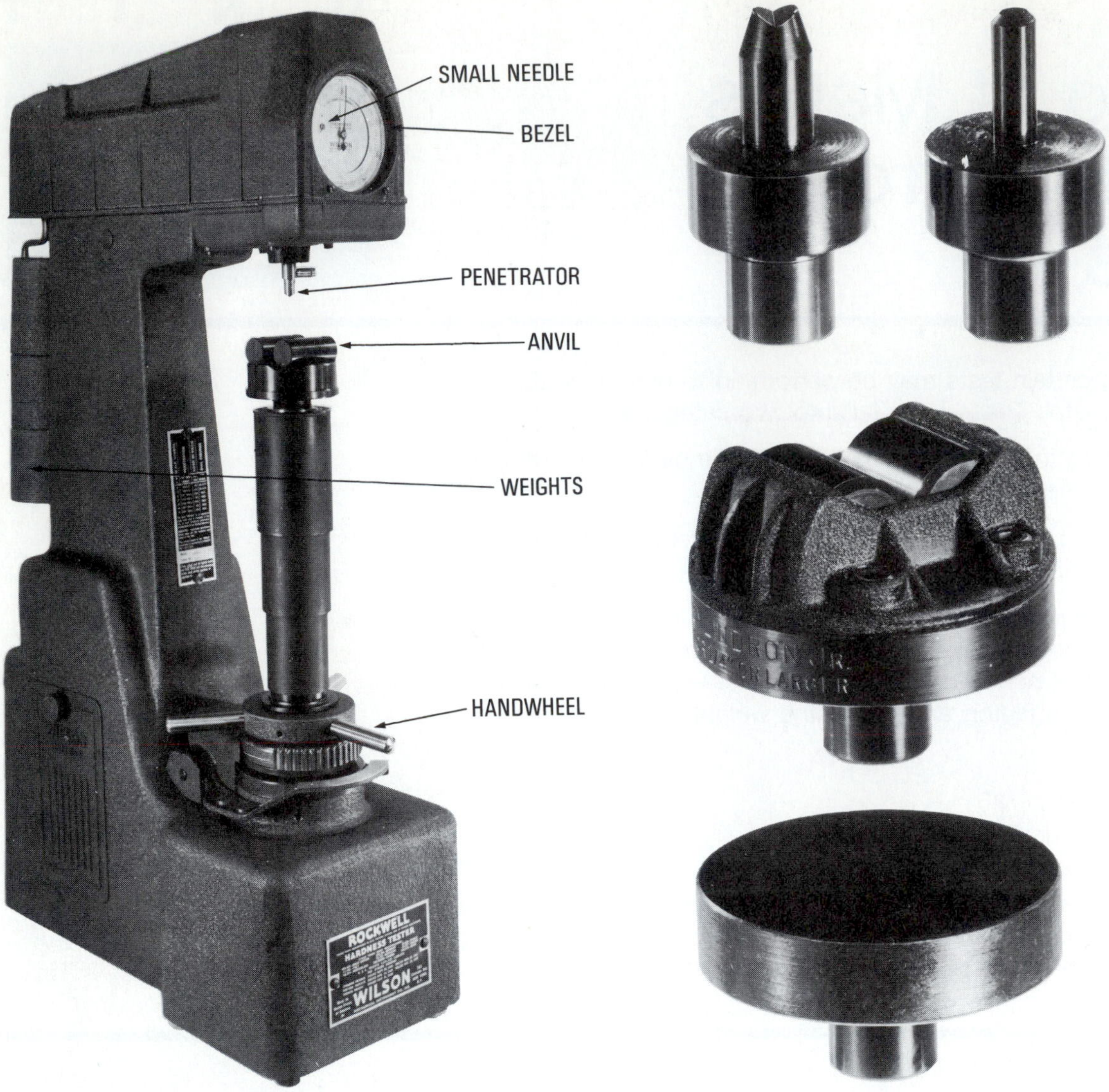
SMALL NEEDLE
BEZEL
PENETRATOR
ANVIL
WEIGHTS
HANDWHEEL

Figure 89-1 *Rockwell hardness tester, showing various anvils. (Courtesy Wilson Instrument Division, American Chain and Cable Company.)*

and D, are special scales and are not used as often as the B and C scales. Rockwell superficial hardness scales are used for testing the hardness of thin materials and case-hardened parts.

To Perform a Rockwell C Hardness Test

Although the various Rockwell-type testers may differ slightly in construction, they all operate on the same principle.

PROCEDURE

1. Select the proper penetrator for the material to be tested. Use a 120° diamond for hardened materials. Use a ¹⁄₁₆-in. (1.5-mm) ball for soft steel, cast iron, and nonferrous metals.
2. Mount the proper anvil for the shape of the part being tested.
3. Remove the scale or oxidation from the surface on which the test is to be made. Usually an area of about ½ in. (13 mm) in diameter is sufficient.
4. Place the workpiece on the anvil and apply the minor load (10 kg) by turning the handwheel until the small needle is in line with the red dot on the dial.
5. Adjust the bezel (outer dial) to zero (0).
6. Apply the major load (150 kg).
7. After the large hand stops, remove the major load.
8. When the hand ceases to move backward, note the hardness reading on the C scale (black). This reading indicates the difference in penetration of the brale between the minor and major loads (Fig. 89-3) and indicates the Rockwell C (Rc) hardness of the material.
9. Release the minor load and remove the specimen.

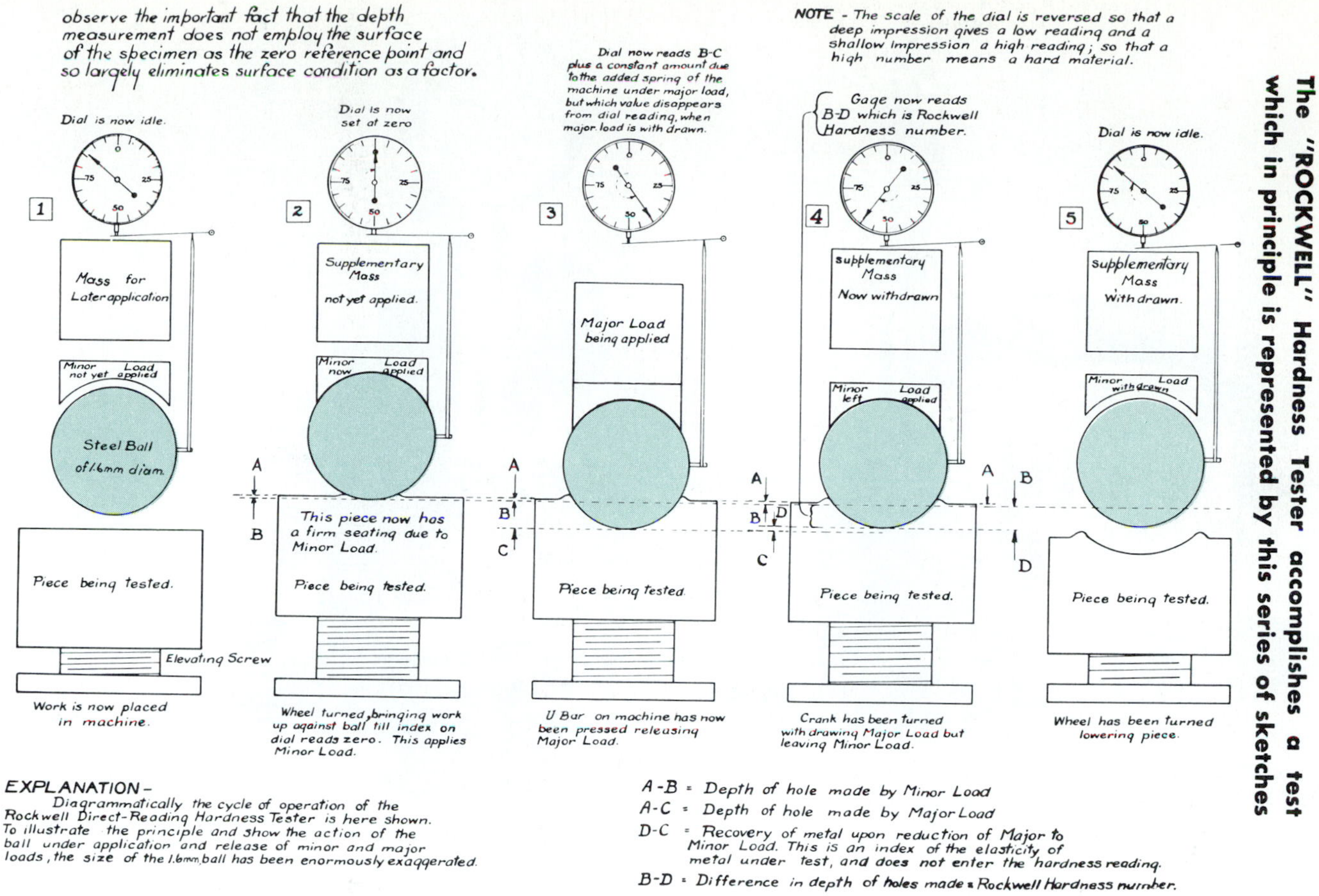

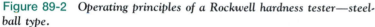

The "ROCKWELL" Hardness Tester accomplishes a test which in principle is represented by this series of sketches
observe the important fact that the depth measurement does not employ the surface of the specimen as the zero reference point and so largely eliminates surface condition as a factor.
NOTE - The scale of the dial is reversed so that a deep impression gives a low reading and a shallow impression a high reading; so that a high number means a hard material.
Dial is now idle.
Dial is now set at zero
Dial now reads B-C plus a constant amount due to the added spring of the machine under major load, but which value disappears from dial reading, when major load is withdrawn.
Gage now reads B-D which is Rockwell Hardness number.
Dial is now idle.
1
2
3
4
5
Mass for Later application
Supplementary Mass not yet applied.
Major Load being applied
Supplementary Mass Now withdrawn
Supplementary Mass Withdrawn.
Minor Load not yet applied
Minor Load now applied
Minor Load left applied
Minor Load withdrawn
Steel Ball of 1/16 mm diam.
This piece now has a firm seating due to Minor Load.
Piece being tested.
Elevating Screw
Work is now placed in machine.
Wheel turned, bringing work up against ball till index on dial reads zero. This applies Minor Load.
U Bar on machine has now been pressed releasing Major Load.
Crank has been turned withdrawing Major Load but leaving Minor Load.
Wheel has been turned lowering piece.
EXPLANATION -
Diagrammatically the cycle of operation of the Rockwell Direct-Reading Hardness Tester is here shown. To illustrate the principle and show the action of the ball under application and release of minor and major loads, the size of the 1/16mm ball has been enormously exaggerated.
A-B = Depth of hole made by Minor Load
A-C = Depth of hole made by Major Load
D-C = Recovery of metal upon reduction of Major to Minor Load. This is an index of the elasticity of metal under test, and does not enter the hardness reading.
B-D = Difference in depth of holes made = Rockwell Hardness number.

Figure 89-2 *Operating principles of a Rockwell hardness tester—steel-ball type.*

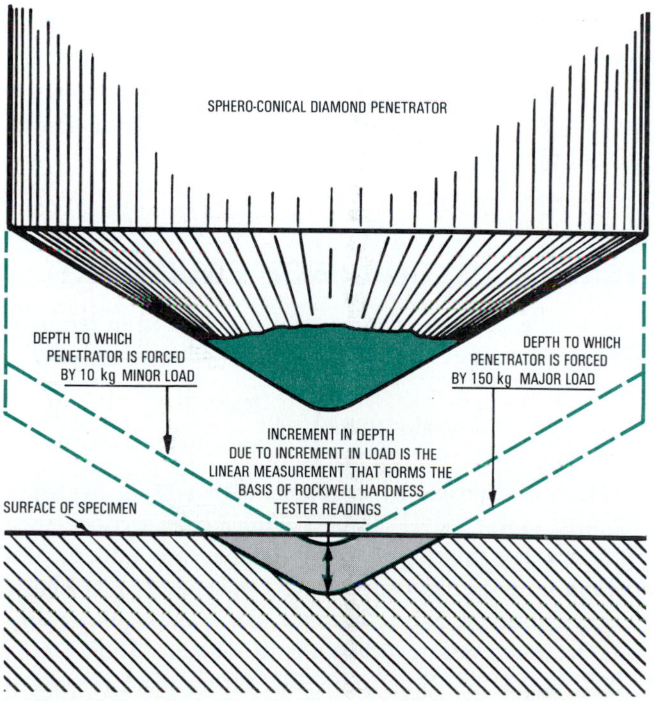

SPHERO-CONICAL DIAMOND PENETRATOR
DEPTH TO WHICH PENETRATOR IS FORCED BY 10 kg MINOR LOAD
DEPTH TO WHICH PENETRATOR IS FORCED BY 150 kg MAJOR LOAD
INCREMENT IN DEPTH DUE TO INCREMENT IN LOAD IS THE LINEAR MEASUREMENT THAT FORMS THE BASIS OF ROCKWELL HARDNESS TESTER READINGS
SURFACE OF SPECIMEN

Figure 89-3 *Operating principle of a Rockwell hardness tester—diamond-cone type.*

NOTE: For accurate results, two or three readings should be taken and averaged. If either the brale or the anvil has been removed and replaced, a "dummy run" should be made to properly seat these parts before performing a test. A test piece of known hardness should be tested occasionally to check the accuracy of the instrument.

BRINELL HARDNESS TESTER

The Brinell hardness tester is operated by pressing a 10-mm hardened steel ball under a load of 3000 kg into the surface of the specimen and measuring the diameter of the impression with a microscope. When the diameter of the impression and the applied load is known, the Brinell hardness number (BHN) can be obtained from Brinell hardness tables. The Brinell hardness value is determined by dividing the load in kilograms which was applied to the penetrator by the area of the impression (in square millimeters).

A standard load of 500 kg is used for testing nonferrous metals. The impression made in softer metals is larger and the Brinell hardness number is lower.

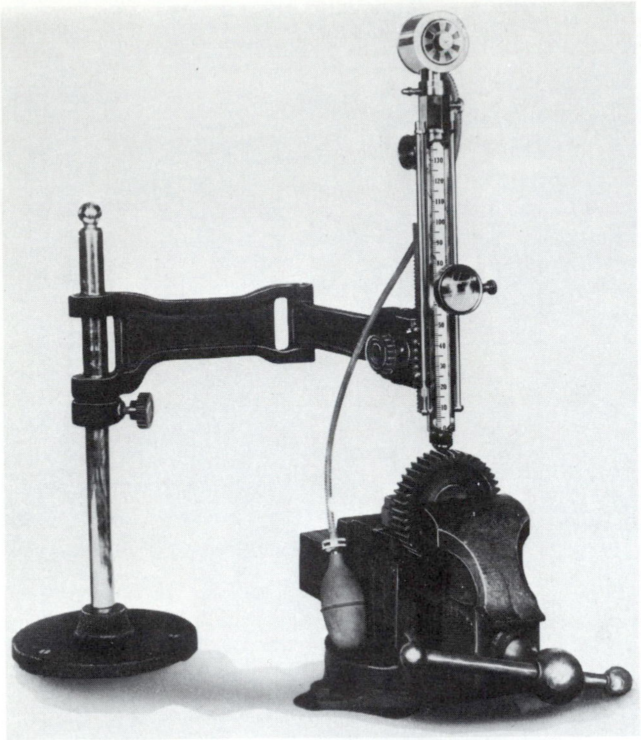

Figure 89-4 *A vertical scale scleroscope. (Courtesy The Shore Instrument and Manufacturing Company.)*

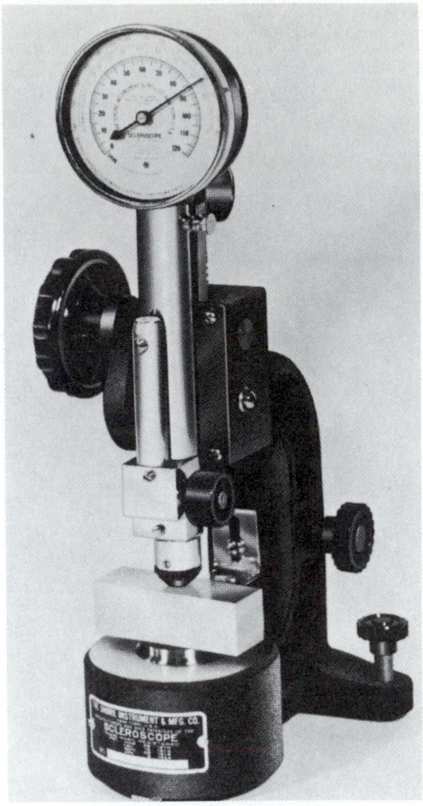

Figure 89-5 *A dial recording scleroscope.*

SCLEROSCOPE HARDNESS TESTER

The scleroscope hardness tester is operated on the principle that a small diamond-tipped hammer, when dropped from a fixed height, will rebound higher from a hard surface than from a softer one. The height of the rebound is converted to a hardness reading.

Scleroscopes are available in several models. On some models, the hardness reading is taken directly from the vertical barrel or tube (Fig. 89-4). On others, the hardness numbers are marked on the dial (Fig. 89-5). These models may also show the corresponding Brinell and Rockwell numbers. See Table 14 in the appendix for hardness conversion chart.

DESTRUCTIVE TESTING

There is a close relationship between the various properties of a metal; for example, the tensile strength of a metal increases as the hardness increases, and the ductility decreases as the hardness increases. Thus the tensile strength of a metal may be determined with reasonable accuracy if the hardness and the composition of the metal are known. A more accurate method of determining the tensile strength of a material is that of tensile testing. The tensile strength, or the maximum amount of pull (force) that a material can withstand before breaking, is determined on a tensile testing machine (Fig. 89-6).

A sample of the metal is pulled or stretched in a machine until the metal fractures. This test indicates not only the tensile strength but also the elastic limit, the yield point, the percentage of area reduction, and the percentage of elongation of the material.

TENSILE TESTING (INCH EQUIPMENT)

The tensile strength of a material is expressed in terms of pounds per square inch and is calculated as follows:

$$\text{Tensile strength} = \frac{\text{load, lb}}{\text{area, in}^2}$$

Machines capable of extremely high tension loads are required to test a steel sample having a cross-sectional area of 1 in^2. For this reason, most samples are reduced to a definite cross-sectional area which can be used conveniently in calculating the tensile strength. For example, most samples are machined to 0.505 in. in diameter (Fig. 89-7), which is 0.2 in^2. Smaller tensile testing machines use a sample that has been machined to 0.251 in. in diameter (Fig. 89-8), which is an area of $\frac{1}{20}$ in^2.

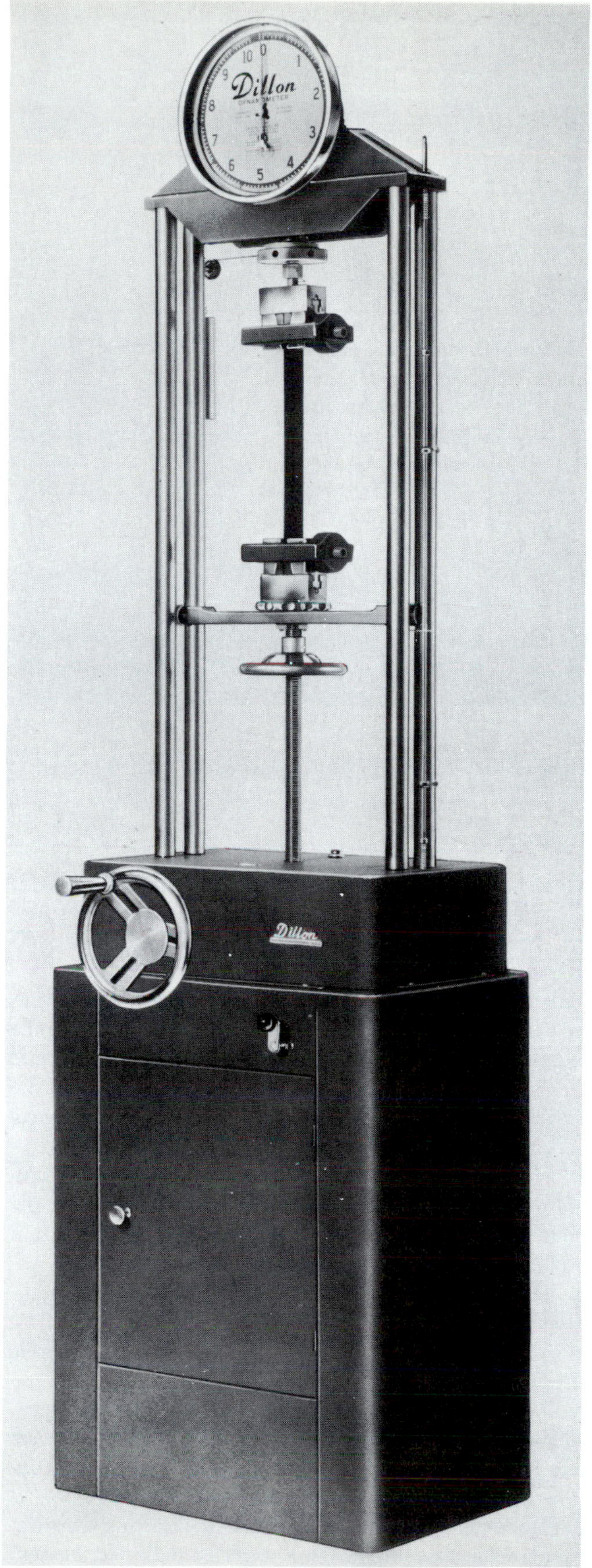
Dillon
DYNAMOMETER
Dillon

Figure 89-6 *A tensile testing machine.*

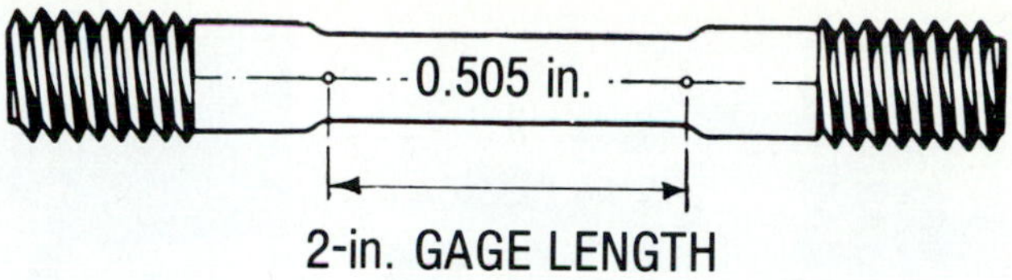
0.505 in.
2-in. GAGE LENGTH

Figure 89-7 *A test sample 0.505 in. in diameter.*

For a load of 10 000 lb using 0.505-in.-diameter sample (0.2 in^2), the tensile strength is:

$$\text{Tensile strength} = \frac{10\,000}{0.2} = 50\,000 \text{ psi}$$

For a load of 3000 lb using a 0.251-in.-diameter sample 1/20 in^2), the tensile strength is:

$$\text{Tensile strength} = \frac{3000}{1/20} = 60\,000 \text{ psi}$$

TO DETERMINE THE TENSILE STRENGTH OF STEEL

1. Turn a sample of the steel to be tested to the dimensions shown in Fig. 89-8, and place on it two center-punch marks exactly 2 in. apart.
2. Mount the specimen in the machine (Fig. 89-9A), and make sure that the jaws grip the sample properly by jogging the motor switch until the black needle just starts to move. If necessary, remove any tension by reversing the motor.
3. Turn the red hand back until it bears against the black hand on the dial.
4. Set a pair of dividers to the center-punch marks on the sample (2 in.).
5. Start the machine and apply the load to the specimen.
6. Observe and record the readings at which there are any changes in the uniform movement of the needle.

NOTE: At this time it is possible to determine the elastic limit of the metal, or the maximum stress which can be developed without causing permanent deformation. This is done by checking the distance between the two center-punch marks with the preset dividers. Increased loads must be applied to the specimen and removed several times. After each load has been removed, check that the distance

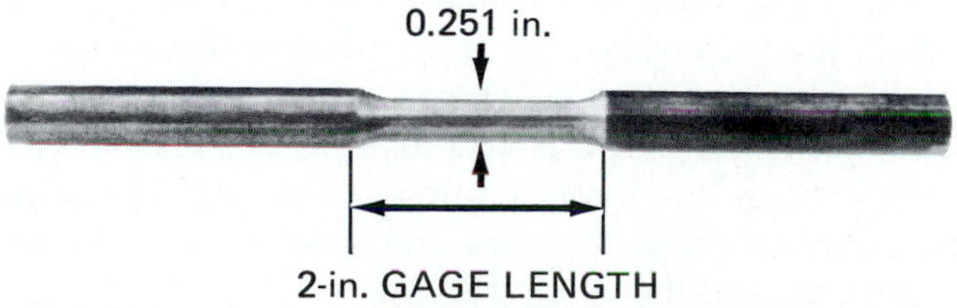
0.251 in.
2-in. GAGE LENGTH

Figure 89-8 *A test sample 0.251 in. in diameter.*

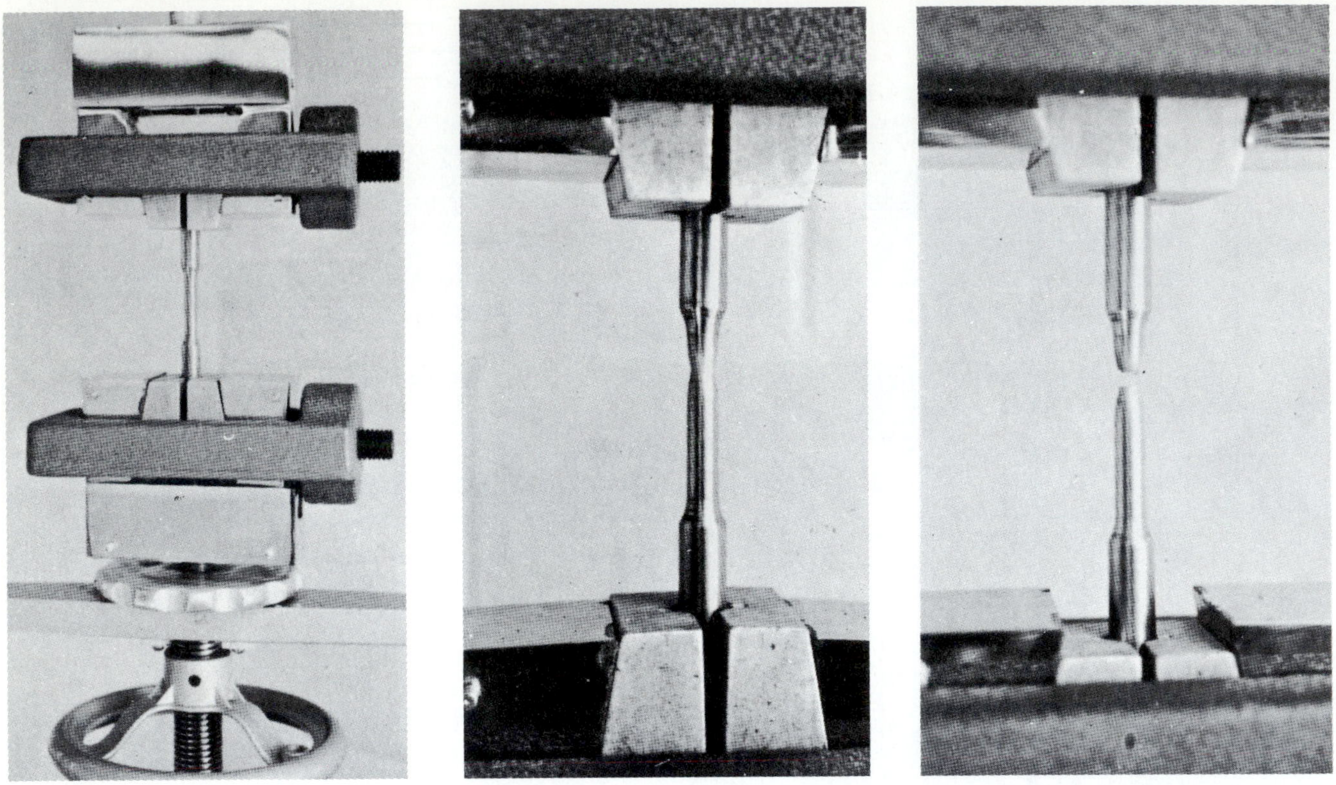

A B C

Figure 89-9 *(A) A 0.251-in.-diameter specimen mounted and ready for testing; (B) specimen "necking down"; (C) fractured specimen.*

between the center-punch marks remains at 2 in. When this distance increases even slightly, the elastic limit of the metal has been reached. An *extensometer* may also be used to indicate the elastic limit.

7. Continue to exert the pull on the sample until it "necks down" (Fig. 89-9B) and finally breaks (Fig. 89-9C). From this procedure, several properties of the metal may be determined (Fig. 89-10).
8. Remove the sample pieces, place the broken ends together, and clamp them in this position.
9. Measure the distance between the center-punch marks to determine the amount of elongation.
10. Measure the diameter of the specimen at the break to determine the reduction in diameter.

Observations (Refer to Fig. 89-10)

1. The needle continued to move uniformly until about 3600 lb showed on the scale, after which it slowed down slightly. The point at which it began to slow down indicated the *proportional limit.* It is here that the metal reached its *elastic limit* and no longer returned to its original size or shape. At this point, the stress and strain were no longer proportional and the curve changed.

The *yield point* was reached just beyond the proportional limit and here the metal started to stretch or yield; the strain increased without a corresponding stress increase.
2. The needle continued to move slowly up to about 6300 lb and then it remained stationary.
3. After a short time, the metal began to show a reduction in diameter (necking down), at which time the needle began to show a rapid drop. The highest travel of the needle indicated the *ultimate strength* or the *tensile strength* of the metal. This was the maximum pull to which the metal may be subjected before breaking.
4. The needle continued to move backward (leaving the red hand stationary) and suddenly the metal broke at the point where it had necked down. The point at which it broke is known as the *breaking stress.*
5. The position of the red hand was at 6300 lb. This indicated the load required to break a cross-sectional area of $\frac{1}{20}$ in^2. The ultimate strength of the metal or the tensile strength was $6300 \times 20 = 126\,000$ lb.
6. When the pieces were placed back together, clamped, and measured, the distance between the center-punch marks was about 2.185 in. This was an elongation of about 9 percent.
7. When the diameter of the metal at the fracture was measured, it was found to be 0.170 in. This indicated a reduction in area of 0.080 in., or about 32 percent.

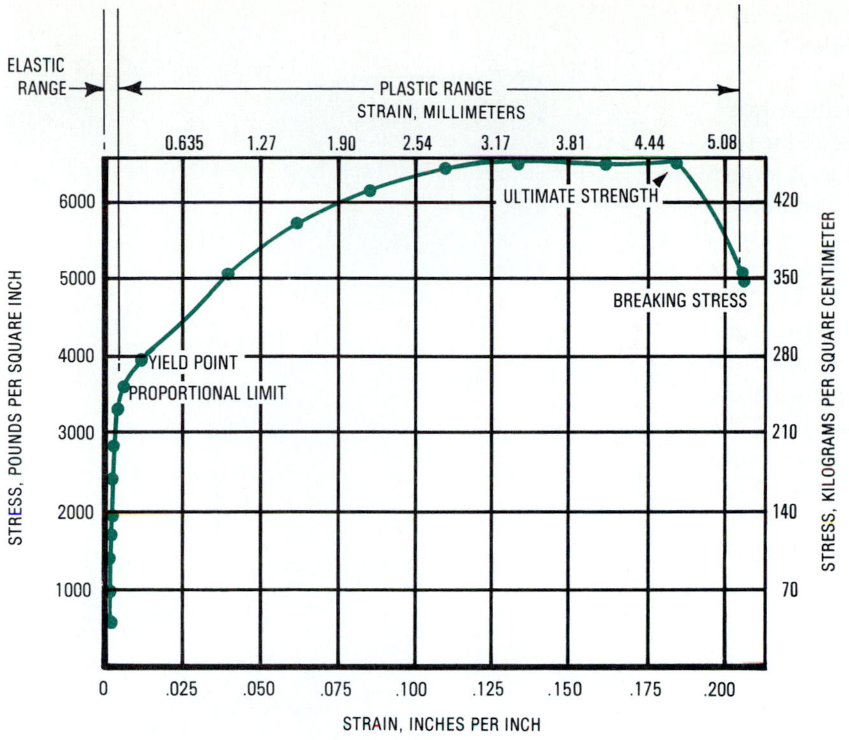
ELASTIC RANGE
PLASTIC RANGE
STRAIN, MILLIMETERS
0.635 1.27 1.90 2.54 3.17 3.81 4.44 5.08
ULTIMATE STRENGTH
BREAKING STRESS
YIELD POINT
PROPORTIONAL LIMIT
6000 5000 4000 3000 2000 1000
420 350 280 210 140 70
STRESS, POUNDS PER SQUARE INCH
STRESS, KILOGRAMS PER SQUARE CENTIMETER
0 .025 .050 .075 .100 .125 .150 .175 .200
STRAIN, INCHES PER INCH

Figure 89-10 *Stress–strain graph.*

TENSILE TESTING (METRIC EQUIPMENT)

Metric tensile testers are graduated in kilograms per square centimeter and the cross-sectional area of the sample is calculated in square centimeters or square millimeters. Metric extensometers are graduated in millimeters.

The calculations involved in determining the tensile strength, percentage elongation, and area of reduction are the same as for inch calculations.

Although not the accepted SI unit of pressure, most metric tensile testing machines available and in use at the time of publication were graduated in kilograms per square centimeter (kg/cm^2). For conversion to pascals, use the formula 1 kg/cm^2 = 980.6 Pa.

EXAMPLE

A sample piece is 1.3 cm in diameter. The ultimate pull exerted on it during a tensile test was 4650 kg. What was the tensile strength of this metal?

$$\text{Tensile strength} = \frac{\text{load, kg}}{\text{area, cm}^2} = \frac{4650}{1.327} = 3504\ \text{kg/cm}^2$$

PERCENTAGE OF ELONGATION

If the punch marks were 50 mm apart at the start of the pull and 54 mm after the sample was broken, what was the percentage elongation?

$$\%\ \text{of elongation} = \frac{\text{amount of elongation}}{\text{original length}} \times 100 = \frac{4}{50} \times 100 = 8\%$$

PERCENTAGE REDUCTION OF AREA

The original diameter was 1.3 cm and after breaking, the diameter was 0.95 cm. What was the percentage of reduction of area?

$$\text{Amount of reduction} = 1.3 - 0.95 = 0.35\ \text{cm} = \frac{0.35}{1.3} \times 100 = 27\%$$

IMPACT TESTS

The toughness of metal, or its ability to withstand a sudden shock or impact may be measured by the *Charpy impact test* or the *Izod test.* In both tests a 10 mm-square specimen is used; it may be notched or grooved, depending on the test used.

Both tests use a swinging pendulum of a fixed mass, which is raised to a standard height, depending on the type of specimen. The pendulum is released, and as it swings through an arc, it strikes the specimen in the pendulum's path.

In the Charpy test (Fig. 89-11), the specimen is mounted in a fixture and supported at both ends. The V or notch is placed on the side opposite to the direction of the pendulum's swing. When the pendulum is released, the knife edge strikes the sample in the center, reducing the travel of the pendulum. The difference in height of the pendulum at the beginning and end of the stroke is shown on the gage and this indicates the amount of energy used to fracture the specimen.

The Izod test (Fig. 89-12) is similar in principle to the Charpy test. One end of the work is gripped in a clamp with the notched side toward the direction of the pendulum's swing. The amount of energy required to break the specimen is recorded on the scale (Fig. 89-13).

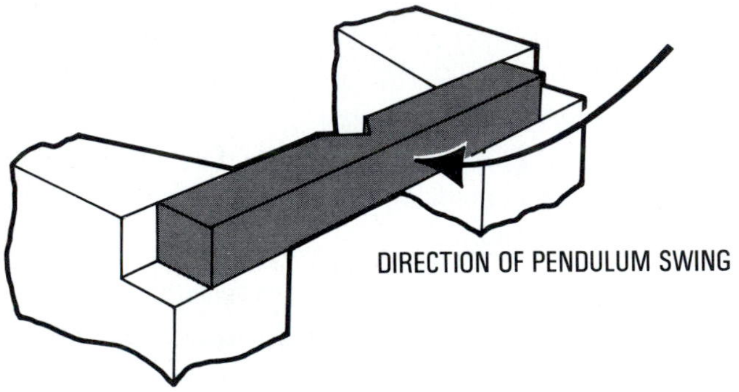
DIRECTION OF PENDULUM SWING

Figure 89-11 *Principle of the Charpy impact test.*

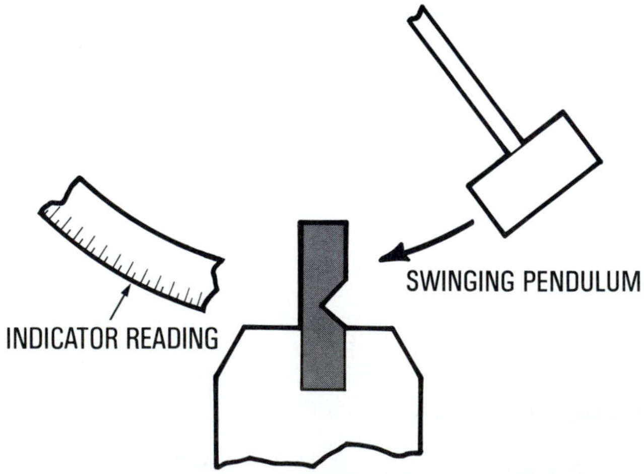
SWINGING PENDULUM
INDICATOR READING

Figure 89-12 *Principle of the Izod impact test.*

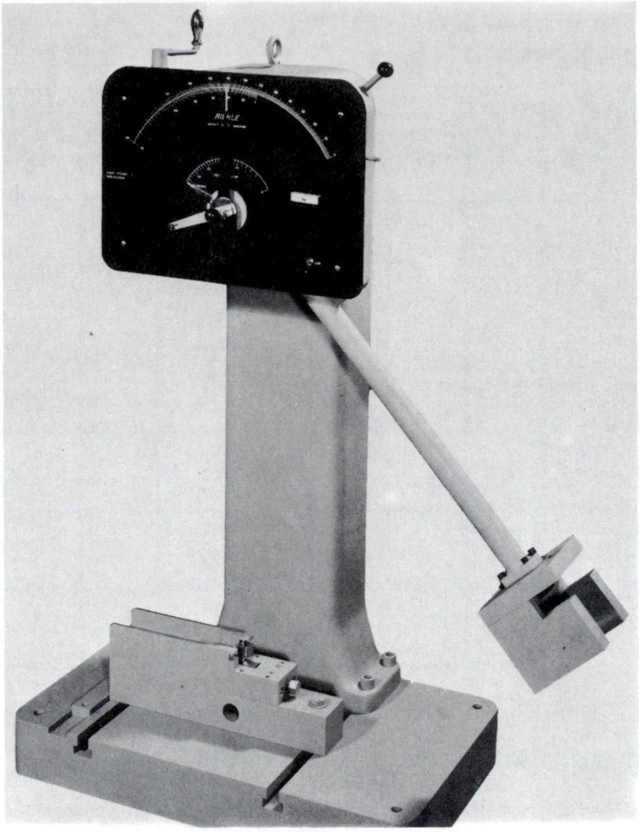

Figure 89-13 *Impact testing machine. (Courtesy Ametek Testing Equipment.)*

NONFERROUS METALS AND ALLOYS

Nonferrous metals, as the name implies, contain little or no iron and are generally nonmagnetic. Since all pure nonferrous metals do not offer the qualities required for industrial applications, they are often combined to produce alloys having the desired qualities for a particular job. The most widely used nonferrous metals for industrial use are aluminum, copper, lead, magnesium, nickel, tin, and zinc.

ALUMINUM

Aluminum is a light, soft, white metal produced from bauxite ore. It is resistant to atmospheric corrosion and is a good conductor of electricity and heat. It is malleable and ductile and can easily be machined, forged, rolled, and extruded. It has a low melting point of 1220°F (660°C) and can be cast easily. It is used extensively in transportation vehicles of all types, the construction industry, transmission lines, cooking utensils, and hardware.

Aluminum is not generally used in a pure state, since it

is too soft and weak. It is alloyed with other metals to form strong alloys that are used extensively in industry.

Aluminum-Base Alloys

Duralumin, an alloy of 95 percent aluminum and 4 percent copper, 0.05 percent manganese, and 0.05 percent magnesium, is widely used in the aircraft and transportation industries. This is a naturally aging alloy; that is, it hardens as it ages. Due to this peculiarity, duralumin must be kept at a subzero temperature until ready for use. When it is brought to room temperature, the hardening process begins.

Other alloys contain varying amounts of copper, magnesium, and manganese.

Aluminum–manganese alloys have good formability, good resistance to corrosion, and good weldability. These alloys are used for utensils, gasoline and oil tanks, pressure vessels, and piping.

Aluminum–silicon alloys are easily forged and cast. They are used for forged automotive pistons, intricate castings, and marine fittings.

Aluminum–magnesium alloys have good corrosion resistance and moderate strength. They are used for architectural extrusions and automotive gas and oil lines.

Aluminum–silicon–magnesium alloys have excellent corrosion resistance, are heat treatable, and are easily worked and cast. They are used for small boats, furniture, bridge railings, and architectural applications.

Aluminum–zinc alloys contain zinc, magnesium, and copper, with smaller amounts of manganese and chromium. They have high tensile strength, good corrosion resistance, and may be heat treated. They are used for aircraft structural parts when great strength is required.

COPPER

Copper is a heavy, soft, reddish-colored metal refined from copper ore (copper sulfide). It has high electrical and thermal conductivity, good corrosion resistance, and strength and is easily welded, brazed, or soldered. It is very ductile and is easily drawn into wire and tubing. Since copper work-hardens readily, it must be heated at about 1200°F (648°C) and quenched in water to anneal.

Because of its softness, copper does not machine well. The long chips produced in drilling and tapping tend to clog the flutes of the cutting tool and must be cleared frequently. Sawing and milling operations require cutters with good chip clearance. Coolant should be used to minimize heat and aid the cutting action.

Copper-Base Alloys

Brass, an alloy of copper and zinc, has good corrosion resistance and is easily formed, machined, and cast. There are several forms of brass. *Alpha* brasses containing up to 36 percent zinc are suitable for cold working. *Alpha + beta* brasses containing 54 to 62 percent copper are used in hot working of this alloy. Small amounts of tin or antimony are added to alpha brasses to minimize the pitting effect of salt water on this alloy. Brass alloys are used for water and gasoline line fittings, tubing, tanks, radiator cores, and rivets.

Bronze originally referred to an alloy of copper and tin, has now been extended to include all copper-base alloys, except copper–zinc alloys, which contain up to 12 percent of the principal alloying element.

Phospor–bronze contains 90 percent copper, 10 percent tin, and a very small amount of phosphorus, which acts as a hardener. This metal has high strength, toughness, and corrosion resistance and is used for lock washers, cotter pins, springs, and clutch discs.

Silicon–bronze (a copper–silicon alloy) contains less than 5 percent silicon and is the strongest of the work-hardenable copper alloys. It has the mechanical properties of machine steel and the corrosion resistance of copper. It is used for tanks, pressure vessels, and hydraulic pressure lines.

Aluminum–bronze (a copper–aluminum alloy) contains between 4 and 11 percent aluminum. Other elements, such as iron, nickel, manganese, and silicon, are added to aluminum bronzes. Iron (up to 5 percent) increases the strength and refines the grain. Nickel, when added (up to 5 percent), has effects similar to those of iron. Silicon (up to 2 percent) improves machinability. Manganese promotes soundness in casting.

Aluminum–bronzes have good corrosion resistance and strength and are used for condenser tubes, pressure vessels, nuts, and bolts.

Beryllium–bronze (copper and beryllium), containing up to about 2 percent beryllium, is easily formed in the annealed condition. It has high tensile strength and fatigue strength in the hardened condition. Beryllium–bronze is used for surgical instruments, bolts, nuts, and screws.

LEAD

Lead is a soft, heavy metal that has a bright, silvery color when freshly cut but turns gray quickly when exposed to air. It has a low melting point, low strength, low electrical conductivity, and high corrosion resistance. It is used extensively in the chemical and plumbing industries. Lead is also added to bronzes, brasses, and machine steel to improve their machinability.

LEAD ALLOYS

Antimony and *tin* are the most common alloying elements of lead. Antimony, when added to lead (up to 14 percent), increases its strength and hardness. This alloy is used for battery plates and cable-sheathing.

The most common lead–tin alloy is solder, which may be composed of 40 percent tin and 60 percent lead, or 50 percent of each. Antimony is sometimes added as a hardener.

Lead–tin–antimony alloys are used as type metals in the printing industry.

MAGNESIUM

Magnesium is a lightweight element which, when alloyed, produces a light, strong metal used extensively in the aircraft and missile industries. Magnesium plates are used to prevent corrosion by salt water in underwater fittings on ship hulls. Magnesium rods, when inserted in galvanized domestic water tanks, will prolong the life of the tank. Other uses for this metal are in photographic flash bulbs and thermite bombs.

NICKEL

Nickel, a whitish metal, is noted for its resistance to corrosion and oxidation. It is used extensively for electroplating, but its most important application is in the manufacture of stainless and alloy steels.

Nickel Alloys

Nickel–chromium–iron-base alloys (containing about 60 percent nickel, 16 percent chromium, and 24 percent iron) are widely used for electric heating elements in toasters, percolators, and water heaters.

Monel metal, containing about 60 percent nickel, 38 percent copper, and small amounts of manganese or aluminum, is a tough, ductile metal with good machining qualities. It is corrosion-resistant, nonmagnetic, and is used in valve seats, chemical marine pumps, and in nonmagnetic aircraft parts.

Hasteloy, containing about 87 percent nickel, 10 percent silicon, and 3 percent copper, is widely used in the chemical industry because of its noncorrosive qualities.

Inconel, a strong, tough alloy containing about 76 percent nickel, 16 percent chromium, and 8 percent iron is used in food-processing equipment, milk pasteurizers, exhaust manifolds for aircraft, and in heat-treating furnaces and equipment.

TIN

Pure *tin* has a silvery-white appearance, good corrosion resistance, and melts at about 450°F (232°C). It is used as a coating on other metals, such as iron, to form tin plate.

Tin Alloys

As already mentioned, tin when alloyed with lead forms solder.

Babbitt is an alloy containing tin, lead, and copper.

Pewter, another tin-base alloy, is composed of 92 percent tin and 8 percent antimony and copper.

ZINC

Zinc is a coarse, crystalline, brittle metal used mainly for die-casting alloys and as a coating for sheet steel, chain, wire, screws, and piping. Zinc alloys, containing approximately 90 to 95 percent zinc, 5 percent aluminum, and small amounts of copper or magnesium, are widely used in the die-casting field to produce automotive parts, building hardware, padlocks, and toys.

BEARING METALS

Bearing metals may be divided into two groups: leaded bronzes and babbitt.

Leaded Bronzes

The composition of bronze bearings varies according to their use. Bearings used to support heavy loads contain about 80 percent copper, 10 percent tin, and 10 percent lead. For lighter loads and faster speeds, the lead content is increased. A typical bearing of this type might contain 70 percent copper, 5 percent tin, and 25 percent lead.

Babbitt

Babbitt bearing materials may be divided into two groups: lead base and tin base.

Lead-base babbitt may contain 75 percent lead, 10 percent tin, and 15 percent antimony, depending on the application. A small amount of arsenic is often added to permit the bearing to carry heavier loads. Applications of lead-base babbitt bearings are in automotive connecting rods, main and crankshaft bearings, and diesel engine bearings.

Tin-base babbitt may contain up to 90 percent tin with copper and antimony added, to 65 percent tin, 15 percent antimony, 2 percent copper, and 18 percent lead. Since tin has become less plentiful, tin-base babbitts are used in high-grade bearing applications, such as steam turbines.

REVIEW QUESTIONS

TESTING OF METALS

1. Explain the difference between nondestructive and destructive testing.
2. What information can be determined by hardness testing?

ROCKWELL HARDNESS TESTER

3. Name two scales which are found on a Rockwell-type hardness tester and state the penetrator used in each case.
4. Briefly describe how a Rockwell C hardness test is performed.

BRINELL HARDNESS TESTER

5. Compare the principles of the Brinell hardness tester and the Rockwell hardness tester.

SCLEROSCOPE

6. Describe the principle of the scleroscope.

DESTRUCTIVE TESTING

7. What effect does hardening have on the tensile strength and ductility of a metal?
8. Explain the principle of tensile testing.
9. What other properties may be determined by a tensile test?
10. How is the tensile strength of a metal calculated?
11. Explain the procedure for performing a tensile test.
12. How may the elastic limit of a metal be determined?
13. Define: *proportional limit, yield point, ultimate strength,* and *breaking stress.*
14. Describe the principle of impact testing.

NONFERROUS METALS AND ALLOYS

15. Define *nonferrous metal.*
16. Name four nonferrous metals commonly used as base metals in alloys.
17. Name four aluminum-base alloys and state one application of each.
18. Name two types of brasses and state the composition of each.
19. What is the difference between brass and bronze alloys?
20. Why is lead added to steel?
21. Where are magnesium alloys used extensively?
22. Name three nickel-base alloys and state two applications of each.
23. Name two tin-base alloys.
24. For what purpose are zinc alloys used extensively?
25. What are the three metals used for bearing alloys?
26. What is the basic difference between bronze and babbitt?
27. Explain the use of the following:
    a. Lead-base babbitt
    b. Tin-base babbitt

SECTION 17

Hydraulics

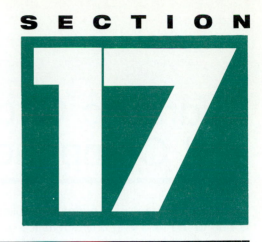

A hydraulic system is a method of transmitting force or motion by applying pressure on a confined liquid. It can be used to substitute liquid linkage between moving parts for mechanical linkages, such as shafts, gears, and connecting rods. The science of hydraulics dates back several thousand years to when dams, sluice gates, and waterwheels were first used to control water flow for domestic use and irrigation purposes.

Hydraulic power is used in nearly every phase of industry. Machine tools, cars, airplanes, jacks, missiles, and boats are only a few of the areas where hydraulics is applied. Hydraulic systems, because of their extreme versatility, are adaptable to almost every type of machine tool to provide feeds and drives.

No other medium combines the same degree of accuracy, positiveness, and flexibility of control with the ability to transmit a maximum of motion in a minimum of bulk and mass. Oil is the fluid most frequently used in hydraulic systems because it has all these characteristics. Oil also lubricates and gives anticorrosion properties to the mechanical parts of the system.

UNIT 90

Hydraulic Circuits and Components

The wide acceptance of hydraulic applications in the machine tool industry is due to the following characteristics of fluids.

1. A fluid is one of the most versatile means of transmitting power and modifying motion.
2. A fluid is infinitely flexible, yet as unyielding as steel, due to its incompressibility.
3. A fluid adjusts readily to changes in diameter and shape.
4. A fluid can be divided to perform work in different areas simultaneously. The hydraulic brake system in an automobile is an example of this.

OBJECTIVES

After completing this unit you should be able to:
1. State Pascal's law and the hydraulic principle
2. Identify and state the purposes of the main parts in a hydraulic system
3. Explain the application of constant- and variable-volume systems as applied to machines

THE HYDRAULIC PRINCIPLE

The extensive use of hydraulics is due to the fact that liquids possess two properties which make them especially useful in power transmission. Liquids cannot be compressed and they have the ability to multiply force. In the 17th century, the French scientist Pascal studied the physical properties of liquids and formulated the basic law of hydraulics. Pascal's law states that: *pressure at any one point in a static liquid is the same in every direction, and pressure exerted on an enclosed liquid is transmitted undiminished in every direction, and acts with equal force on equal areas* (Fig. 90-1).

The two types of hydraulic systems are the *hydrostatic* and the *hydrodynamic*. A basic comparison of these two principles is shown in Fig. 90-2.

HYDROSTATIC PRINCIPLE

Most industrial systems work on the hydrostatic principle, which can be easily understood by referring to Fig. 90-1. Suppose a piston with a cross-sectional area of 1 square inch (in^2) is tightly fitted in a liquid-filled cylinder. If a 1-lb force is applied to the piston, a pressure of 1 psi is created and this pressure is transmitted to all points in the cylinder. If there were no variations of pressure in the cylinder due to gravity, frictional losses, etc., a gage installed anywhere in the cylinder would record a pressure of 1 psi (Fig. 90-1).

The hydrostatic principle is often used to increase force or pressure. This can best be understood by connecting the piston–cylinder combination in Fig. 90-1 to a piston–

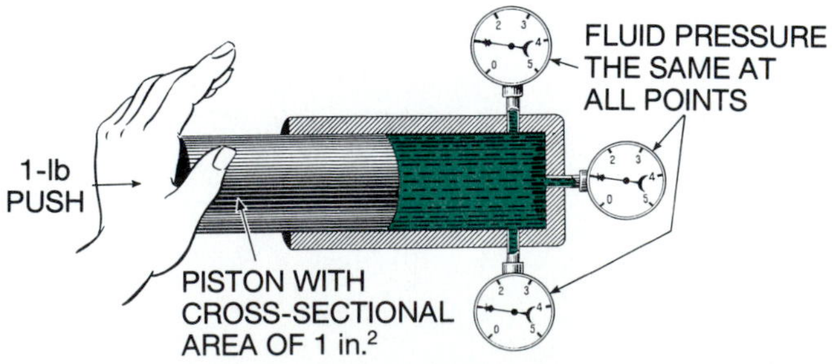
FLUID PRESSURE
THE SAME AT
ALL POINTS
1-lb
PUSH
PISTON WITH
CROSS-SECTIONAL
AREA OF 1 in.²
Figure 90-1 *An illustration of Pascal's law.*

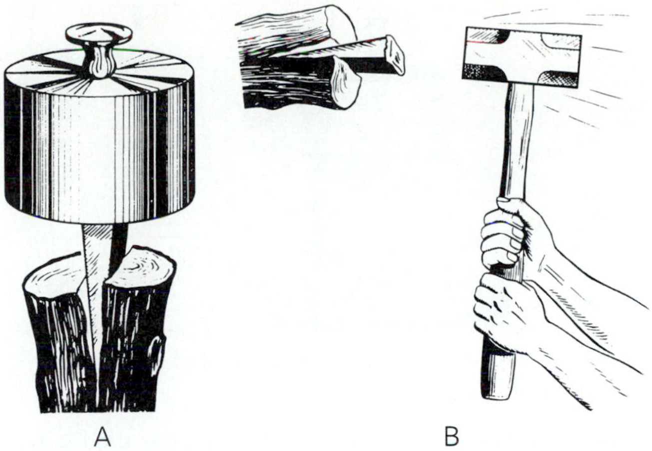
A
B

Figure 90-2 *(A) The static principle: A heavy mass applies constant pressure to split a log. (B) The dynamic principle: The force of a sledge hammer is used on the wedge to split the log.*

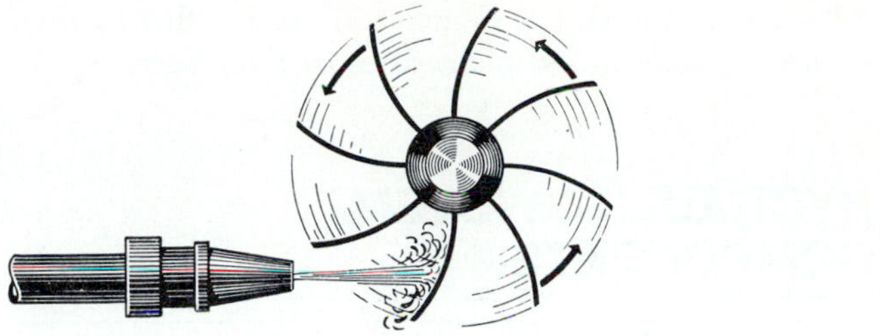

Figure 90-4 *Hydrodynamic force is used to turn a paddle wheel.*

cylinder combination which has a cross-sectional area of 10 in^2 (Fig. 90-3). By applying a 1-lb force on the small piston, this 1-lb force is transmitted to the 10 in^2 piston area, creating a total force of 10 lb (Fig. 90-3). This increase in force is gained at the expense of the distance moved. For example, it would be necessary to move the small piston 10 in. in order to move the large piston 1 in. By applying pressure on the large piston, it is possible to reduce the total force and increase the distance the small piston moves.

Nearly all industrial hydraulic systems based on the hydrostatic principle have pumps which apply constant pressure to the fluid. By using various devices to control the pressure, rate, and direction of flow, extremely flexible and versatile machines can be designed.

HYDRODYNAMIC PRINCIPLE

Dynamic force uses *kinetic energy* stored in a moving body to perform work. A log can be split *dynamically* by hitting the wedge with a sledge hammer (Fig. 90-2B). The amount of work performed depends on the speed and mass of the sledge hammer. Increasing either, or both, increases the amount of force exerted.

Hydraulic fluid can be used dynamically by directing a stream of fluid against a paddle wheel (Fig. 90-4). Through impact, much of the kinetic energy of the moving fluid is conveyed to the paddle wheel. Hydraulic couplings and torque converters work on the hydrodynamic principle.

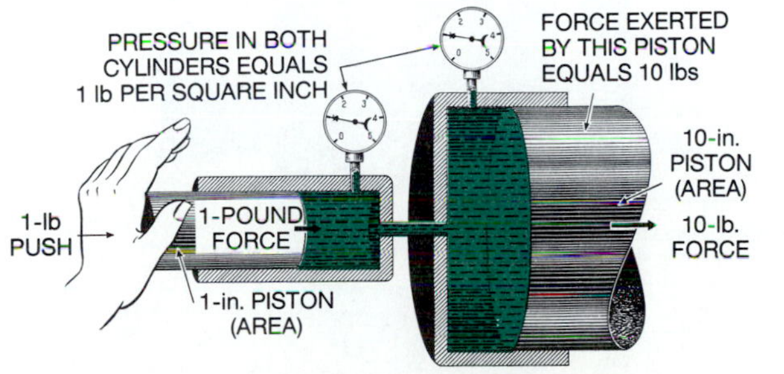
PRESSURE IN BOTH
CYLINDERS EQUALS
1 lb PER SQUARE INCH
FORCE EXERTED
BY THIS PISTON
EQUALS 10 lbs
10-in.
PISTON
(AREA)
10-lb.
FORCE
1-lb
PUSH
1-POUND
FORCE
1-in. PISTON
(AREA)

Figure 90-3 *The hydrostatic principle is used to increase force or pressure.*

FUNDAMENTAL HYDRAULIC CIRCUIT

All hydraulic systems are basically simple and have six essential elements (Fig. 90-5):

1. A reservoir for the oil supply
2. A pump to move the oil under pressure
3. One or more control valves to regulate the flow
4. A piston and cylinder (hydraulic motor) to convert hydraulic power into mechanical power
5. Pipe or tubing to connect the various parts
6. Hydraulic fluid, which is the "life blood" of the system

There is no such thing as an all-hydraulic system. At certain points in any hydraulic circuit, mechanical devices must be used to control the fluid. A hydraulic system is a

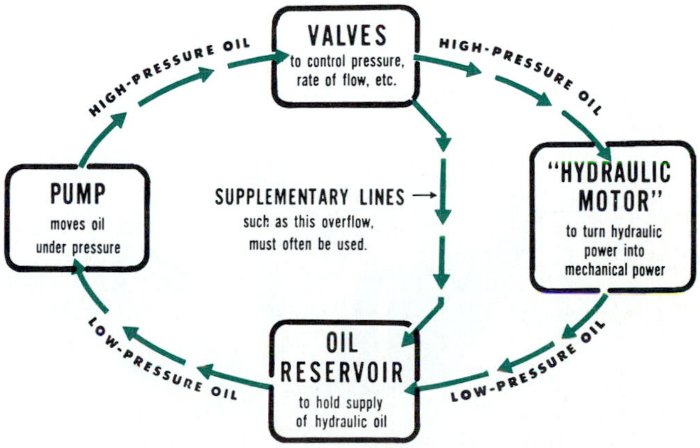
VALVES
to control pressure,
rate of flow, etc.
HIGH-PRESSURE OIL
HIGH-PRESSURE OIL
PUMP
moves oil
under pressure
SUPPLEMENTARY LINES →
such as this overflow,
must often be used.
"HYDRAULIC
MOTOR"
to turn hydraulic
power into
mechanical power
LOW-PRESSURE OIL
LOW-PRESSURE OIL
OIL
RESERVOIR
to hold supply
of hydraulic oil

Figure 90-5 *A fundamental hydraulic circuit.*

collection of mechanical elements joined together by *fluid connecting rods* and in many cases by *fluid levers.*

HYDRAULIC SYSTEM COMPONENTS

Certain basic parts must be used in every hydraulic system; other parts are sometimes used to refine the system or to cope with special conditions. Each basic part will be discussed in detail, to show how a hydraulic system operates.

OIL RESERVOIR

Every hydraulic system requires a reservoir for storing the oil used in the system. A separate tank (Fig. 90-6) is generally used for large hydraulic systems. On smaller machines, the oil reservoir may be incorporated in the base of the machine. The built-in reservoirs are generally compact but are hard to get at and difficult to clean.

PUMPS

The pump is an important part of a hydraulic system; its purpose is to transmit force by causing oil to flow. A simple example of a hydraulic pump may be found in the hydraulic brake system of a car. Depressing the brake pedal causes the piston in the master brake cylinder to move toward the discharge end, causing the fluid under pressure to flow to the four wheel cylinders. The force applied at one end is transmitted from one point (the brake pedal) to four points (the wheel cylinders). As soon as the brake shoes contact the brake drums, oil stops flowing. The primary purpose of the master cylinder, which is a very simple pump, is to transmit force hydraulically. In more complex hydraulic systems, the pump must provide a steady flow of liquid under pressure. Therefore the pump must provide a means of transmitting both force and motion.

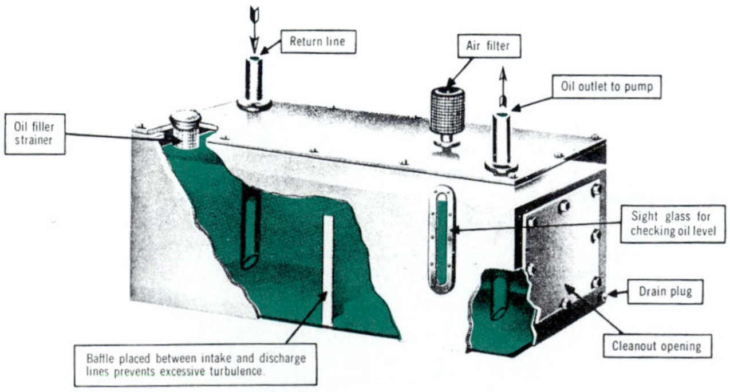
Return line
Air filter
Oil outlet to pump
Oil filler strainer
Sight glass for checking oil level
Drain plug
Cleanout opening
Baffle placed between intake and discharge lines prevents excessive turbulence.

Figure 90-6 *The structure of a hydraulic oil reservoir.*

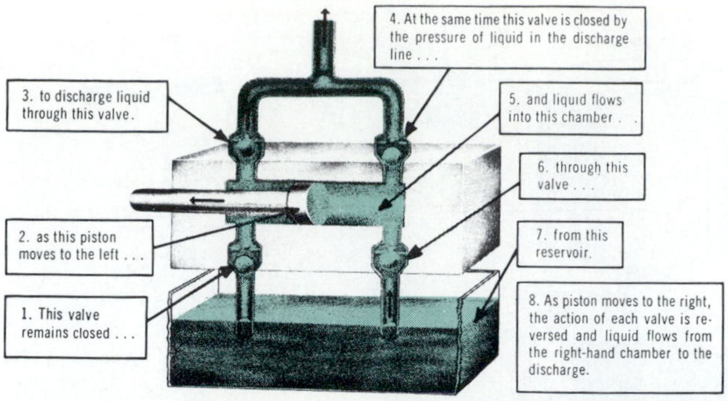
4. At the same time this valve is closed by the pressure of liquid in the discharge line . . .
3. to discharge liquid through this valve.
5. and liquid flows into this chamber . .
6. through this valve . . .
2. as this piston moves to the left . . .
7. from this reservoir.
1. This valve remains closed . . .
8. As piston moves to the right, the action of each valve is reversed and liquid flows from the right-hand chamber to the discharge.

Figure 90-7 *Principle of a reciprocating hydraulic pump.*

Some of the more common types of pumps used in hydraulic systems will be described.

Reciprocating pumps (Fig. 90-7) have limited use because the oil pulsates too much to provide smooth operation. However, by using several multipiston types, a fairly steady flow can be provided. Reciprocating pumps are quite rugged and work well under adverse conditions such as low temperatures.

Gear pumps (Fig. 90-8) are simple in design. Because of their almost nonpulsating flow, they are well suited to applications requiring smooth operation. Simple gear pumps have spur gears; however, for quieter and smoother operation, some types contain helical or herringbone gears.

Internal gear pumps (Fig. 90-9) are basically the same as gear pumps except that they have an internal gear. They are generally not as efficient as gear pumps and lose their efficiency rapidly as soon as mating parts become worn.

Screw pumps (Fig. 90-10) are usually used to transfer oil from one point to another. They are seldom used to provide hydraulic power. Screw pumps are simple, rugged, and capable of operating at high speeds.

Vane pumps (Fig. 90-11) are used extensively to provide hydraulic power because they produce a reasonably steady

2. which is picked up and carried in these spaces to outlet side of pump . . .
3. where mating teeth displace oil and force it through outlet.
1. Parting gears create slight suction here that helps to draw in oil . . .

Figure 90-8 *Principle of a gear pump.*

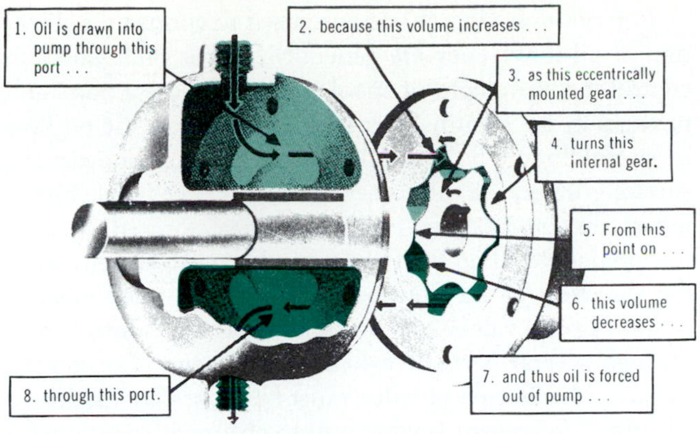

1. Oil is drawn into pump through this port . . .
2. because this volume increases . . .
3. as this eccentrically mounted gear . . .
4. turns this internal gear.
5. From this point on . . .
6. this volume decreases . . .
7. and thus oil is forced out of pump . . .
8. through this port.

Figure 90-9 *Principle of an internal pump.*

flow. Wear of parts does not affect the efficiency of these pumps greatly, since the vanes can always maintain close contact with the ring in which they rotate.

Radial-piston pumps (Fig. 90-12) are compact, rugged, and used in applications requiring a flexible source of hydraulic power. Through the use of ring-shifting mechanisms, the output of radial-piston pumps can be varied between zero and full volume.

Centrifugal pumps (Fig. 90-13) provide a nonpulsating flow and are used in moving large quantities of oil. This pump cannot be overloaded because the oil in the pump begins to rotate with the rotor as soon as the output pressure becomes too great. The output pressure is difficult to control; therefore centrifugal pumps are not generally used to provide hydraulic power.

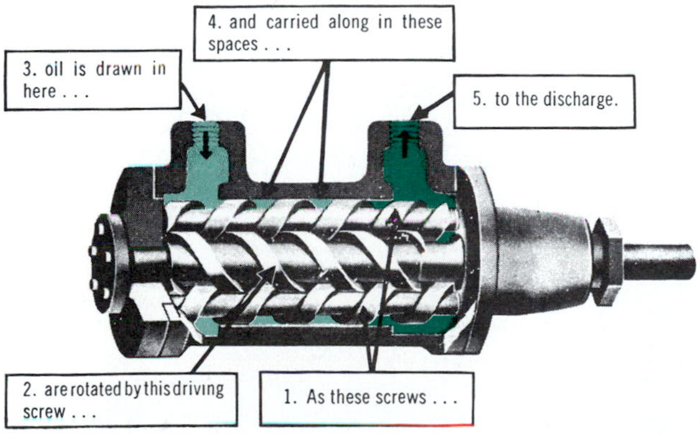

4. and carried along in these spaces . . .
3. oil is drawn in here . . .
5. to the discharge.
2. are rotated by this driving screw . . .
1. As these screws . . .

Figure 90-10 *Principle of a screw pump.*

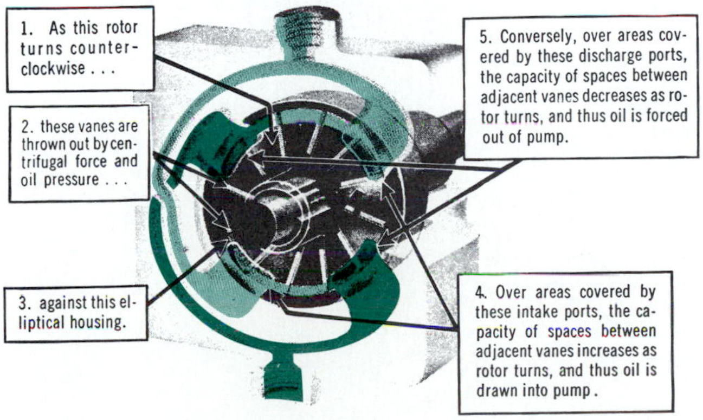

1. As this rotor turns counter-clockwise . . .
2. these vanes are thrown out by centrifugal force and oil pressure . . .
3. against this elliptical housing.
4. Over areas covered by these intake ports, the capacity of spaces between adjacent vanes increases as rotor turns, and thus oil is drawn into pump.
5. Conversely, over areas covered by these discharge ports, the capacity of spaces between adjacent vanes decreases as rotor turns, and thus oil is forced out of pump.

Figure 90-11 *Principle of a vane pump.*

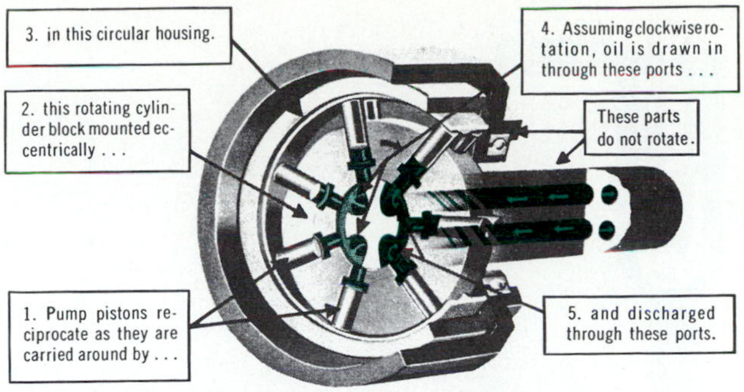

3. in this circular housing.
4. Assuming clockwise rotation, oil is drawn in through these ports . . .
2. this rotating cylinder block mounted eccentrically . . .
These parts do not rotate.
1. Pump pistons reciprocate as they are carried around by . . .
5. and discharged through these ports.

Figure 90-12 *Principle of a radial-piston pump.*

VALVES

Once the pump begins to move the oil under pressure, valves are usually required *to control pressure* and *to direct and control oil flow.* Valves were the only method of controlling flow and pressure until recently. Pumps that have the means of varying both flow and pressure have been successfully developed during the past few years and are proving popular. However, valves are still the most important devices for providing flexibility in a hydraulic system, and a few of the more common types will be discussed.

Ball valves (Fig. 90-14) are used in small reciprocating pumps and in small oil lines to permit oil flow in one direction only. Ball valves are usually spring-loaded and will open when the pressure in the system becomes greater than pressure exerted by the spring. Spring-loaded ball valves are often used as pressure-relief valves. They are placed in a line connecting the hydraulic pump with the oil reservoir to prevent the buildup of excessive pressure in the system.

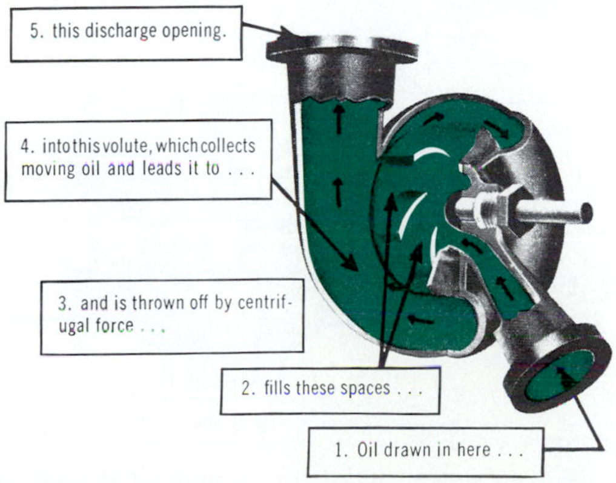

5. this discharge opening.
4. into this volute, which collects moving oil and leads it to . . .
3. and is thrown off by centrifugal force . . .
2. fills these spaces . . .
1. Oil drawn in here . . .

Figure 90-13 *Principle of a centrifugal pump.*

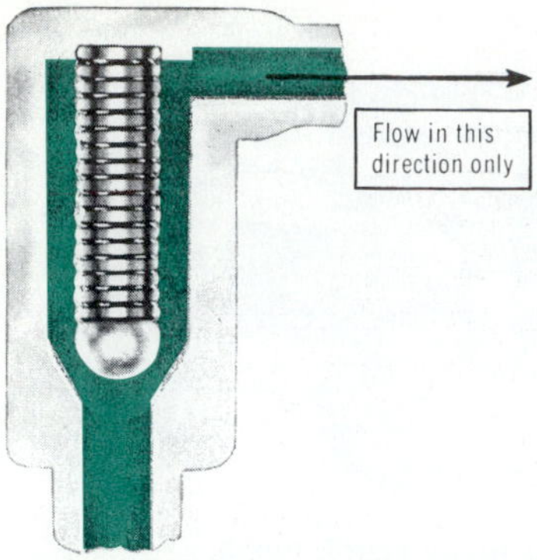
Flow in this
direction only
Figure 90-14 *A ball valve permits flow in one direction only.*

Rotary valves (Fig. 90-15) are used to control the direction of oil flow. They are generally used as pilot valves to control the movement of spool valves. The extra ports and passages in the rotating part of the valve allow the oil flow in several lines to be controlled by one valve. In a hydraulic surface grinder, the reciprocating and crossfeed table movements are controlled by one rotary valve.

Spool valves (Fig. 90-16) are used extensively to control the direction of oil flow because of their quick positive action. Spool valves are versatile and can control flow through several parts of a hydraulic system. The mating surfaces of this type of valve must be accurately machined and fitted to prevent leakage and to ensure efficient operation. Oil grooves are often machined around the pistons to

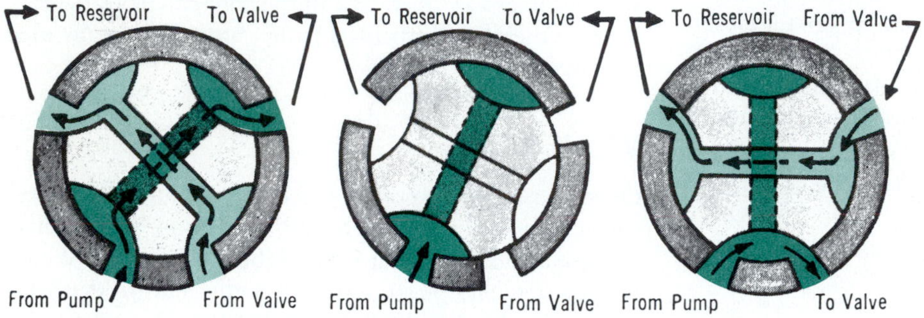
To Reservoir
To Valve
From Pump
From Valve
To Reservoir
To Valve
From Pump
From Valve
To Reservoir
From Valve
From Pump
To Valve
Figure 90-15 *A rotary valve may control the flow of oil to the spool valve or to the cylinder.*

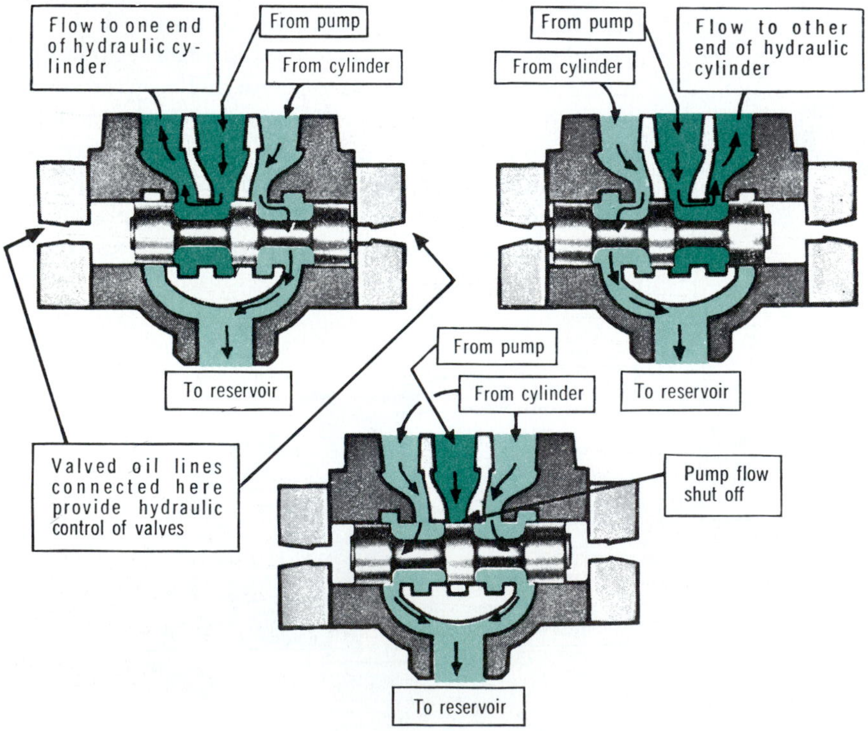
Flow to one end of hydraulic cylinder
From pump
From cylinder
To reservoir
From pump
From cylinder
Flow to other end of hydraulic cylinder
To reservoir
From pump
From cylinder
Valved oil lines connected here provide hydraulic control of valves
Pump flow shut off
To reservoir
Figure 90-16 *A spool valve controls the flow of oil to and from the cylinder.*

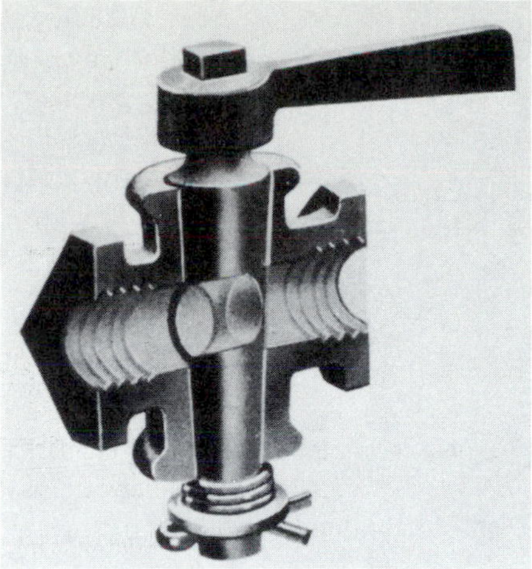

Figure 90-17 *A cock valve.*

keep the spool centered and lubricated. Spool valves can be operated mechanically, electrically, or hydraulically.

Cock valves (Fig. 90-17) are simple valves used to bleed cylinders, eliminate air pockets, and control the operation of pressure gages. They are generally built in small sizes and are used in systems using moderate pressure.

Needle valves (Fig. 90-18) provide a means of regulating oil flow precisely. They are operated manually and can be used in any position from fully opened to fully closed. Their design prevents an abrupt change in oil flow; they are therefore often used in lines connecting sensitive devices which could be damaged by a sudden surge of pressure.

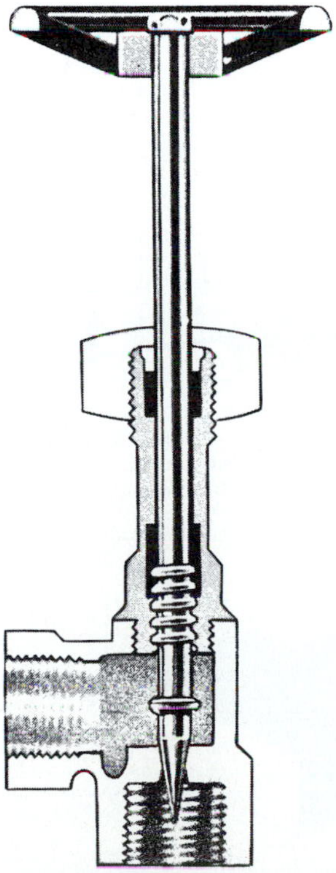

Figure 90-18 *A needle valve.*

HYDRAULIC MOTOR

A hydraulic motor, generally a simple piston-and-cylinder combination, converts the oil flow into mechanical motion. Hydraulic motors alone can produce only two types of motion: straight-line (piston and cylinder type) and rotary (torque converter). When more complex motions are required, mechanical elements must be used in conjunction with the hydraulic motor.

The *piston motor* (Fig. 90-19) is the most commonly used hydraulic motor. Its design is simple and it can be made to withstand any pressure. When oil flows into the cylinder from the right side, the piston is moved to the left a certain distance in a given time. When the oil flow is reversed, the piston moves to the right at a slow rate because, with no piston rod to take up some of the cylinder volume, more oil must flow into the cylinder to move the piston the same distance. This type of piston motor, commonly called a *differential piston motor* (Fig. 90-19), is often used in machines which require a powerful working stroke and a fast, but less powerful, return stroke.

THE HYDRAULIC SYSTEM

A hydraulic system is produced when the various components such as reservoir, pump, valve, and motor are combined and joined together with the necessary pipe, tubing, and fittings. The hydraulic system illustrated in Fig. 90-20 is quite simple; however, when it becomes necessary to vary speeds, motions, etc., in a fixed or variable sequence, the complexity of the system increases. The most complex system still contains the basic components found in a simple system. Two types of hydraulic systems, the constant volume and the variable volume, are generally used.

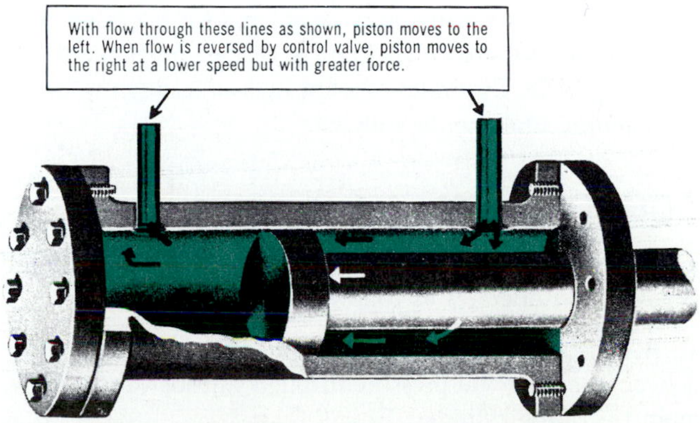
With flow through these lines as shown, piston moves to the left. When flow is reversed by control valve, piston moves to the right at a lower speed but with greater force.

Figure 90-19 *A simple hydraulic piston motor.*

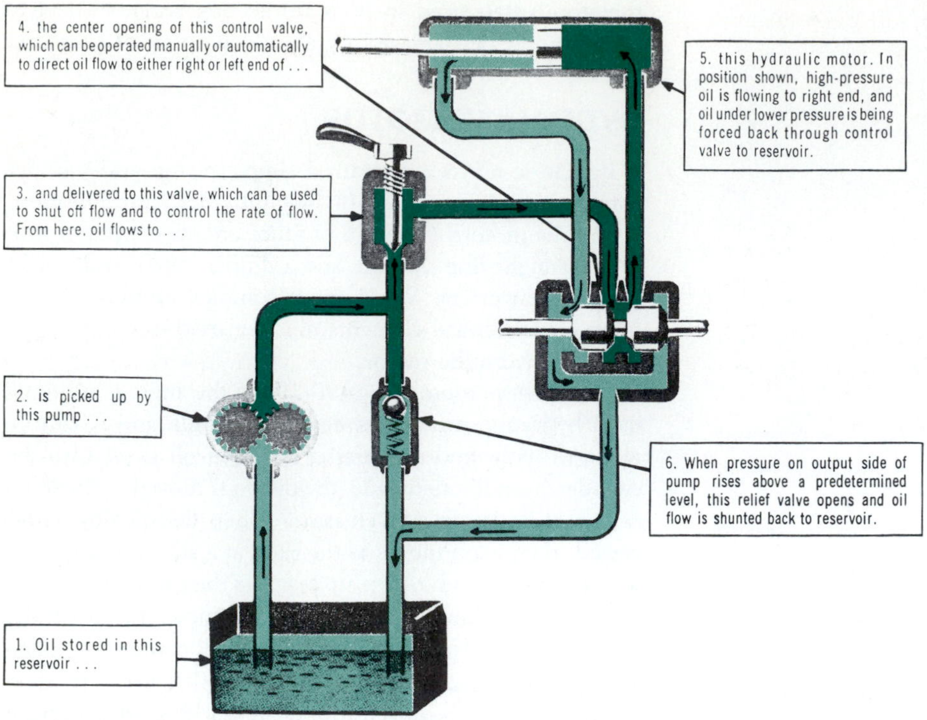
4. the center opening of this control valve, which can be operated manually or automatically to direct oil flow to either right or left end of . . .
5. this hydraulic motor. In position shown, high-pressure oil is flowing to right end, and oil under lower pressure is being forced back through control valve to reservoir.
3. and delivered to this valve, which can be used to shut off flow and to control the rate of flow. From here, oil flows to . . .
2. is picked up by this pump . . .
6. When pressure on output side of pump rises above a predetermined level, this relief valve opens and oil flow is shunted back to reservoir.
1. Oil stored in this reservoir . . .

Figure 90-20 *A simple hydraulic system.*

Constant-volume systems contain pumps which have a fixed output. The pump operates whenever the machine is in operation. If the full flow of oil delivered by the pump is not required in the system, it is bypassed through valves back to the oil reservoir.

Variable-volume systems contain pumps whose output can be varied to suit different requirements. Combinations of mechanical, electrical, and pneumatic controls may be used to regulate the flow of oil. Variable-volume systems generally do not require certain bypass and flow-control valves, which are needed in a constant-volume system. Variable-volume systems are widely used when the output power and speed vary over a wide range.

Many machine tools use both the constant-volume and the variable-volume systems. Each system may be used individually to operate different parts of a machine, or they may be combined to alternately control certain portions of the operating cycle. Examples of a few hydraulic systems found on machine tools are as follows:

CONSTANT-VOLUME SYSTEM

The constant-volume hydraulic system of a surface grinder (Fig. 90-21) operates as follows:

1. A constant-volume or a constant-discharge gear pump delivers moderate oil pressure through various valves to the main operating cylinder.
2. The oil acts on the piston to reciprocate the work table.
3. Excess oil pressure is released through a spring-loaded relief valve and returns to the reservoir.
4. At each reversal of the work table, cams actuate the reversing valve, directing the oil through a wheel-infeed cylin-

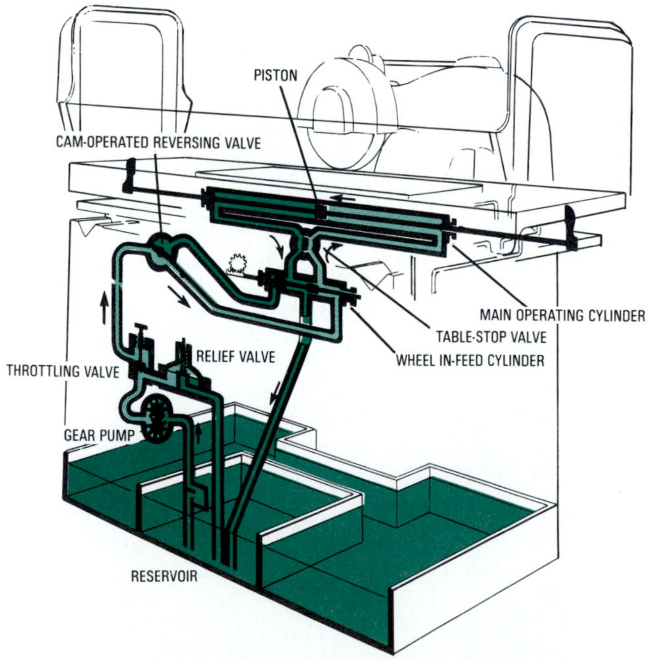
PISTON
CAM-OPERATED REVERSING VALVE
MAIN OPERATING CYLINDER
TABLE-STOP VALVE
WHEEL IN-FEED CYLINDER
RELIEF VALVE
THROTTLING VALVE
GEAR PUMP
RESERVOIR

Figure 90-21 *The constant-volume hydraulic system of a surface grinder.*

der, then alternately to opposite ends of the main operating cylinder.
5. The wheel-infeed cylinder also actuates the ratchet mechanism of the wheel feed.
6. The rate of table travel is controlled by adjusting the throttling valve.

VARIABLE-VOLUME SYSTEM

The milling machine requires a slow feed during the cutting stroke, followed by a rapid return stroke which requires very little oil pressure. The lock-feed hydraulic system of a milling machine using three pumps is illustrated in Fig. 90-22. A brief explanation of the operation of this system follows.
1. The simple gear pump draws oil from the reservoir and supplies it at low pressure to the booster pump.
2. Oil from the booster pump, now under high pressure, travels through a reversing valve to one end of the working cylinder.
3. As the piston moves forward, the machine table, attached to the piston, also moves at a slow, steady rate.
4. A variable-discharge metering pump, connected to both ends of the cylinder, draws a specific amount of oil from the cylinder *ahead* of the motion of the piston.
5. The metering pump holds back and regulates the advance of the piston, which determines the rate of table travel.
6. When the reversing valve is actuated, the discharge and suction ports of the metering pump are interconnected.

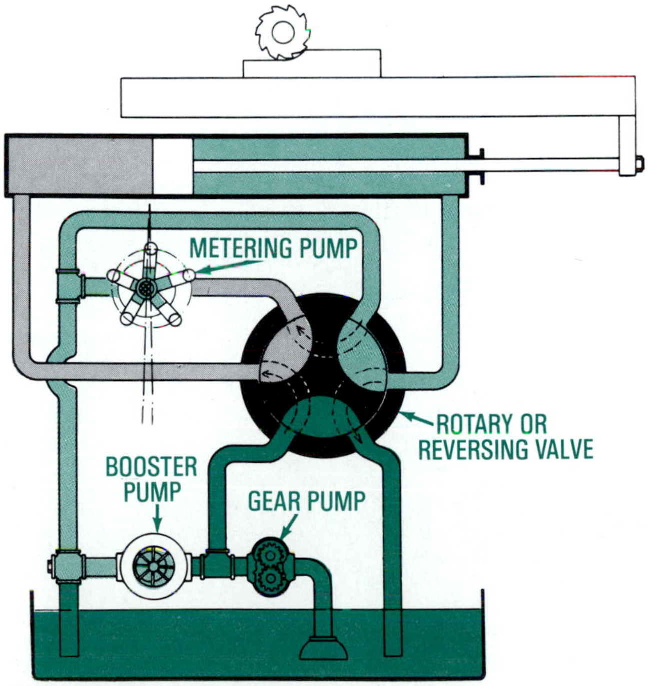
METERING PUMP
ROTARY OR
REVERSING VALVE
BOOSTER
PUMP
GEAR PUMP

Figure 90-22 *The variable-volume system of a manufacturing-type milling machine.*

7. The booster pump then discharges oil into the reservoir and the gear pump supplies low-pressure oil to the cylinder for the rapid return stroke.

HYDRAULIC COUPLINGS AND TORQUE CONVERTERS

In the past few years, great strides have been made in perfecting two *hydrodynamic* devices: the hydraulic coupling and the hydraulic torque converter. Hydraulic couplings and hydraulic torque converters are not new but have only recently come into widespread use because of refinements in both.

A *hydraulic coupling,* sometimes called a *fluid clutch* (Fig. 90-23), is used to join two mechanisms hydraulically. An engine or motor at the power-input end may be joined hydraulically to another mechanism at the output end, without the use of mechanical connections. A hydraulic coupling transfers the torque (twist) delivered to it very smoothly, and at the same time protects the mechanisms against damage from vibration or shock loads.

A *hydraulic torque converter,* sometimes called a *hydraulic gearbox* (Fig. 90-24), is both a clutch and transmission source. It provides a means of coupling two mechanical elements and also of varying the ratio of the input torque to the output torque. Hydraulic torque converters provide a smooth flow of power, along with high torque at low speed, low torque at high speed, or any combination of the two within the capacity of the equipment.

HYDRAULIC OIL REQUIREMENTS

A hydraulic system contains an assembly of valves, cams, operating cylinders, etc., which all have close fitting tolerances. To obtain maximum efficiency from the system and protect the intricate moving parts, it is important that the proper oil be selected for use in hydraulic systems. A hydraulic oil must perform two important functions:

1. It must transmit power efficiently.
2. It must lubricate adequately.

For a hydraulic oil to perform these two important functions satisfactorily, it must have the following characteristics.

1. *Proper viscosity,* to ensure a ready flow of oil at all times and to minimize leakage.
2. *Film strength and lubricity,* to provide adequate lubrication between closely fitted sliding parts.
3. *Resistance to oxidation,* to prevent intricate moving

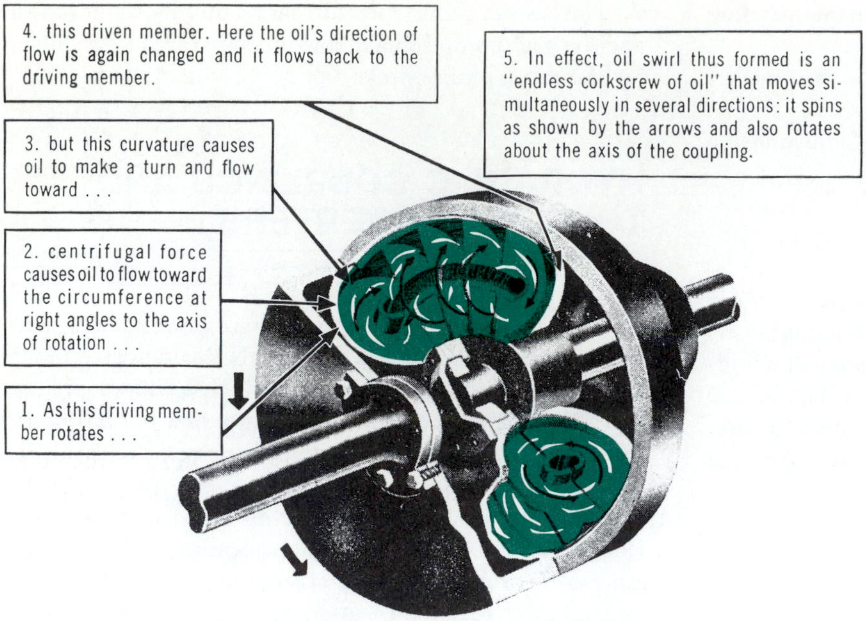

4. this driven member. Here the oil's direction of flow is again changed and it flows back to the driving member.
5. In effect, oil swirl thus formed is an "endless corkscrew of oil" that moves simultaneously in several directions: it spins as shown by the arrows and also rotates about the axis of the coupling.
3. but this curvature causes oil to make a turn and flow toward . . .
2. centrifugal force causes oil to flow toward the circumference at right angles to the axis of rotation . . .
1. As this driving member rotates . . .

Figure 90-23 *Principle of a hydraulic coupling.*

parts from damage due to oxidation of the oil (sludge formation).

4. *Resistance to emulsification,* to separate quickly from any water which may enter the system.

5. *Resistance to corrosion and rusting,* to prevent damage to the closely fitting parts.

6. *Resistance to foaming,* to free itself readily of air drawn into the hydraulic system.

VISCOSITY

Viscosity is one of the most important qualities that a hydraulic oil must possess. Viscosity should be thought of as a measure of the resistance of oil to flow. Low-viscosity oils flow freely; high-viscosity oils flow sluggishly. The viscosity of oil changes with the temperature; low temperatures cause the oil to flow sluggishly, while heat causes the oil to run freely.

The viscosity of oil has a direct bearing on the efficient transmission of power and the adequate lubrication of the parts in a hydraulic system. If hydraulic oil does not have the right viscosity, it cannot perform satisfactorily. Viscosity of an oil affects the operation of a hydraulic system in the following ways.

1. If an oil is too light (low viscosity), it usually results in:
   a. Excessive leakage
   b. Lower efficiency at the pump
   c. Increased wear on parts
   d. Loss of oil pressure
   e. Lack of positive hydraulic control
   f. Lower overall efficiency

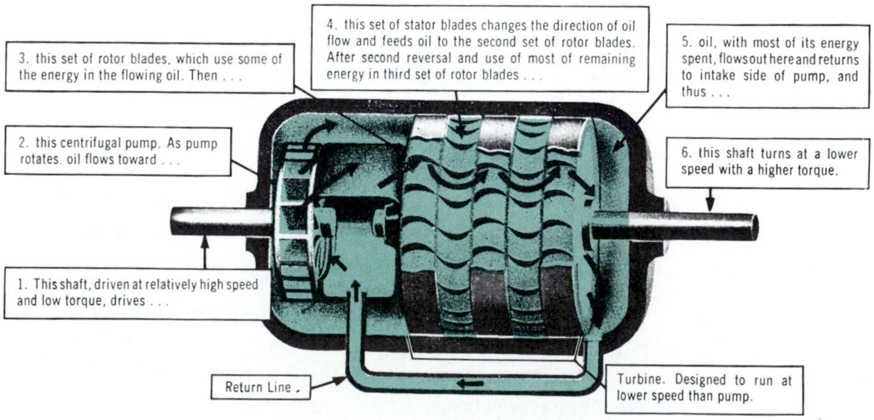

3. this set of rotor blades, which use some of the energy in the flowing oil. Then . . .
4. this set of stator blades changes the direction of oil flow and feeds oil to the second set of rotor blades. After second reversal and use of most of remaining energy in third set of rotor blades . . .
5. oil, with most of its energy spent, flows out here and returns to intake side of pump, and thus . . .
2. this centrifugal pump. As pump rotates, oil flows toward . . .
6. this shaft turns at a lower speed with a higher torque.
1. This shaft, driven at relatively high speed and low torque, drives . . .
Return Line.
Turbine. Designed to run at lower speed than pump.

Figure 90-24 *Principle of a hydraulic torque converter.*

2. If an oil is too heavy (high viscosity), it usually results in:

   a. Increased pressure drop
   b. Higher oil temperatures
   c. Sluggish operation
   d. Lower mechanical efficiency
   e. Higher power consumption

FILM STRENGTH AND LUBRICITY

When parts of the hydraulic pump, cylinder, and control valves become worn, loss of pressure, leakage, and less accurate control may result. An important quality of a hydraulic oil should be its ability to minimize wear under the *thin-film* or *boundary lubrication* conditions that exist between the closely fitted sliding parts. A suitable oil maintains strong films that resist being squeezed out or pushed aside.

Both light (low-viscosity) and heavy (high-viscosity) oils have antiwear or film-strength qualities. High-viscosity oils have these qualities to a greater extent than low-viscosity oils. Heavy oils, because of their greater resistance to leakage, tend to resist displacement from lubricated surfaces better than light oils. Therefore, an oil with as high a viscosity as possible, which fulfills all other requirements, should be used to minimize wear.

RESISTANCE TO OXIDATION

Hydraulic oil comes into contact with warm air in the reservoir. Because of this contact, the oil tends to oxidize, that is, combine chemically with the oxygen of the air. The tendency to oxidize is greatly increased at high operating temperatures and pressures, and by excessive agitation or splashing of oil.

A slight oxidation is not harmful, but if an oil has poor resistance to this chemical change, oxidation may become too great. If this occurs, substantial amounts of both soluble and insoluble oxidation products are created and the oil gradually increases in viscosity.

Insoluble oxidation products may be deposited in the form of gum and sludge in the oil passages, pump parts, and valve parts, where it will restrict the flow of oil. This may slow down the movement of pump vanes and spool- and piston-type valves and cause seated valves to leak.

RESISTANCE TO EMULSIFICATION

Regardless of the precautions taken, some water gets into hydraulic systems through leaks and the condensation of atmospheric moisture. Most of the condensation occurs above the oil in the reservoir as the machine cools during idle periods. Any appreciable amount of water in a hydraulic system will promote rusting and decrease the lubricating value of the oil. This will eventually cause leakage, erratic pump action, and a breakdown in the system.

Where the percentage of water in a system is high, it is very important to use a hydraulic oil that has the ability to separate quickly and completely from the water. Oils that have exceptionally high oxidation resistance retain good water-separating abilities for a long time.

RESISTANCE TO CORROSION AND RUSTING

Water and oxygen can cause rusting of ferrous metal parts in a hydraulic system. Air is always present, oxygen is available, and some water is usually present; these conditions promote rusting. The danger of rusting is greatest when the machine is idle and surfaces usually covered with oil are left unprotected.

Rusting causes surface deterioration of metal parts, and rust particles may eventually be carried into the hydraulic system. This contributes to the formation of sludge-like deposits which can cause serious abrasions and interfere with the operation of cylinders, valves, and pumps.

To guard against rusting, it is important that a hydraulic system be kept as free from water as possible. High-quality hydraulic fluids contain a rust inhibitor, which forms a film that resists displacement by water, thereby protecting the surfaces from contact with the water.

RESISTANCE TO FOAMING

Oil in a hydraulic system usually dissolves a certain amount of air, which causes a foaming (bubbling) action. Air can enter a system by oil falling from a considerable height into the reservoir, through leaks, or by suction of the pump. A hydraulic system can absorb the amount of air bubbles normally present; however, if an excessive amount of air is present and the bubbles persist at operating temperatures, pumping will be noisy and irregular, and the action of motors and controls will be erratic.

Foaming can be controlled by taking the following precautions.

1. Use of a premium-quality hydraulic oil which has good resistance to foaming.
2. The oil returning from the system should not fall from a great height into the reservoir.
3. Vent valves should be installed at high points in the hydraulic system.
4. The reservoir should be properly designed to allow the air to free itself readily at the oil surface.

ADVANTAGES OF HYDRAULIC SYSTEMS

The development of suitable hydraulic oils and the precision tolerances maintained by manufacturers have led to the wide use of hydraulics in machine tool manufacture. The reason for this widespread use is that hydraulic power transmission offers many advantages, including the following.

- A complex combination of cams, gears, levers, etc., can often be replaced by a simpler combination of pumps, valves, lines, and hydraulic motors.
- Speeds and feeds can be varied infinitely without stopping the machine.
- Smooth, steady cutting action, with a minimum amount of chatter or vibration, is easily obtained.
- Even, positive motion is provided at all loads.
- The rate of oil pressure or flow can easily be controlled by simple valves.
- Relief valves prevent overloads which could otherwise cause extensive damage.
- Power costs and friction losses can be held to a minimum.
- All moving parts in the system are constantly lubricated by the hydraulic oil.
- The operation of a hydraulic system is comparatively quiet.
- Rapid reversals at the end of each reciprocating stroke, such as on a surface grinder, are cushioned (smooth and shockless).
- Component parts can be located close to the moving members because the power is transmitted through pipes which can run in any direction.
- Intricate operations can be made almost completely automatic.

REVIEW QUESTIONS

1. Define *hydraulics.*
2. List four reasons why hydraulic applications have found wide acceptance in the machine tool industry.

THE HYDRAULIC PRINCIPLE

3. Name the two important properties that liquids possess.
4. State and illustrate Pascal's law.
5. Explain the hydrostatic principle and state the purpose for which it is used.
6. Compare the hydrostatic and the hydrodynamic principles. Illustrations should be used to supplement the answer.
7. What hydraulic components operate on the hydrodynamic principle?

THE FUNDAMENTAL HYDRAULIC CIRCUIT

8. Name the six essential elements of a hydraulic system.

HYDRAULIC SYSTEM COMPONENTS

9. Give an example of a simple hydraulic pump and explain how it operates.
10. What purpose does a pump serve in a hydraulic system?
11. Describe briefly the operation of the following.
    a. Gear pumps
    b. Vane pumps
    c. Radial-piston pumps
12. What purpose do valves serve in a hydraulic system?
13. For what purpose are the following used?
    a. Ball valves
    b. Rotary valves
    c. Spool valves
    d. Needle valves
14. Explain how a simple hydraulic motor operates.

THE HYDRAULIC SYSTEM

15. Compare a constant-volume and a variable-volume hydraulic system.
16. Explain briefly the operation of the constant-volume hydraulic system of a surface grinder.
17. Explain briefly the operation of the variable-volume hydraulic system of a milling machine.

HYDRAULIC COUPLINGS AND TORQUE CONVERTERS

18. State the purpose of a hydraulic coupling and briefly explain its operation.
19. State the purpose of a hydraulic torque converter and explain how it operates.

HYDRAULIC OIL REQUIREMENTS

20. What two important functions must a hydraulic oil perform?
21. List six characteristics that a hydraulic oil should possess in order to perform satisfactorily.
22. Discuss the importance of viscosity.
23. What would happen if either too light or too heavy an oil were used in a hydraulic system?
24. What effect does oxidation have on a hydraulic system?
25. Explain how water may enter a hydraulic system.
26. What effect does water have on a hydraulic system?
27. Discuss the effects of rusting and explain how it can be controlled.
28. What effect does foaming have on a hydraulic system?
29. How can foaming be minimized?

ADVANTAGES OF HYDRAULIC SYSTEMS

30. List six important advantages of hydraulic power transmission.

SECTION

18

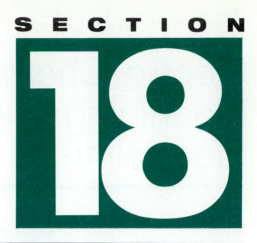

Special Processes

With the introduction of unique mechanisms and exotic materials in recent years, it has been necessary to develop new methods of efficiently machining metals. Parts made out of cemented carbide or difficult-to-machine metals were previously shaped by the costly process of diamond wheel grinding. *Electro-chemical machining, electrical discharge machining,* and *electrolytic grinding* are three methods which have been developed recently. All remove metal by some form of electrical discharge.

Production of other hard-to-machine parts has been made possible by high-energy forming using either the explosive or electromagnetic process. Another widely used nonmachining process is that of powder metallurgy, whereby parts are made from compressed, heat-treated powder metals. Powder-metallurgy parts have compared mechanically with machined parts and are usually cheaper to produce.

In recent years, the laser has proved invaluable for many scientific, industrial, and medical applications. It is widely used to cut holes in any material as well as for special welding applications not possible by other methods. Extremely fine measurements as well as accurate machine part alignment is possible with the laser beam.

Electro-Chemical Machining and Electrolytic Grinding

UNIT 91

Electro-chemical machining, more commonly called ECM, differs from conventional metal-cutting techniques in that electrical and chemical energy are used as the cutting tools. This process machines metal easily, regardless of the work hardness, and is characterized by its "chipless" operation. A nonrotating tool the shape of the cavity required is the *cutting tool;* therefore, square or difficult-to-machine shapes can easily be cut in a workpiece. The wear on the cutting tool is hardly noticeable since the tool *never* touches the work. Electro-chemical machining is particularly suitable for producing round through holes, square through holes, round or square blind holes, simple cavities which have straight, parallel sides, and planing operations. Electro-chemical machining is especially valuable when metals that exceed a hardness of 42 Rockwell C (400 BHN) are machined. Sharp corners, flat bottom sections, or true radii are difficult to maintain because of the slight overcut which occurs during this process. A significant advantage of ECM is that the surfaces and edges of workpieces are not deformed and are left burr-free.

Although *electrolytic grinding* is based on the principle of ECM, its application is different in that a grinding wheel is used instead of an electrode.

OBJECTIVES

After completing this unit you should be able to:

1. Describe the principle and purposes of ECM and electrolytic grinding
2. Identify the various components and their functions in each of these systems

ELECTRO-CHEMICAL MACHINING

THE PROCESS

For years, metal in a solution has been transferred from one metal to another by means of electro-plating baths. Since *electro-chemical machining* (ECM) evolved from this process, it may be wise to examine the electroplating principle (Fig. 91-1):

1. Two bars of unlike metal are immersed in an electrolyte solution.

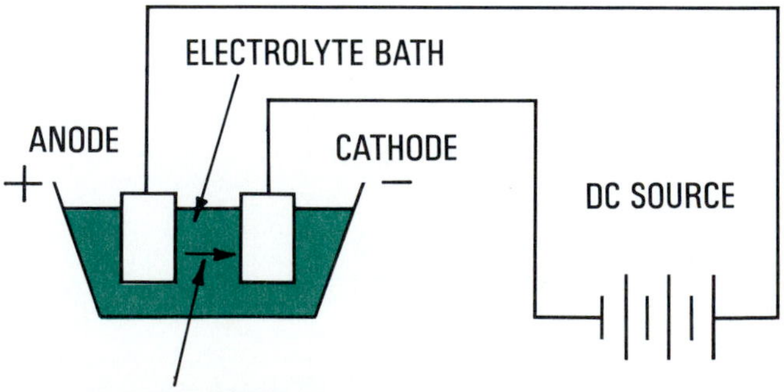
ELECTROLYTE BATH
ANODE
+
CATHODE
−
DC SOURCE
PLATING CURRENT

Figure 91-1 *Principle of the electroplating process.*

2. One bar is fastened to a negative lead on a battery while the other is fastened to the positive lead.
3. When the circuit is closed, direct current passes through the electrolyte between the two bars of metal.
4. Chemical reaction transfers metal from one bar to the other.

Electro-chemical machining differs from the plating process in that an electro-chemical reaction *dissolves metal from a workpiece* into an electrolyte solution. A direct current is passed through an electrolyte solution between the electrode tool (the shape of the cavity desired), which is negative, and the workpiece, which is positive. This causes metal to be removed ahead of the electrode tool as the tool is fed toward the work. Chemical reaction caused by the direct current in the electrolyte dissolves the metal from the workpiece (Fig. 91-2).

The *electrode* for ECM is not simple bar of metal, but a precision insulated tool which has been made to a specific shape and exact size. The electrode (tool) and workpiece, although located within 0.002 to 0.003 in. (0.05 to 0.07 mm) of each other, never contact each other. The *electrolyte solution* is a controlled, swiftly flowing stream which carries the current. The *direct current* used may at times be as high as 10 000 A/in^2 (1550 A/cm^2) of work material. All these factors affect the operation of the electro-chemical machining process and will be discussed in greater detail.

THE ELECTROLYTE

The electrolyte is a solution of water to which salt, mineral acid, caustic-potash, or caustic soda has been added to increase the electrical conductivity. A weak supply of the electrolyte solution will result in two disadvantages:

- Metal removal rates will be low.
- Excessive heat will destroy the effectiveness of the solution.

Electrical energy which starts a chemical reaction in the electrolyte solution results in the formation of gas between the tool and workpiece and dissolves metal from the workpiece. The gas escapes into the atmosphere while the dissolved metal is carried away in the solution. As there is some resistance to current flow and because chemical reactions occur, heat is generated in the machining area. The electrolyte is introduced into the machining area in great quantities to dissipate the heat and wash away the dissolved metal. Filters placed in an ECM system remove the dissolved metal from the electrolyte and ensure a fresh flow of solution to the machining area.

The electrolyte is introduced to the machining area *through* the electrode tool; therefore, the amount of flow is affected by the electrode length, diameter, and shape. As much flow as possible is desirable, and some applications have been known to use as much as 200 gal/min (757 l/min) at pressures up to 300 psi (2068 kPa).

The Electrode

The *electrode tool,* which is always the *negative terminal* of the electrical circuit, is an insulated tool made to the size and shape of the cavity desired. The electrolyte solution is fed to the machining area by a hole through the center of the electrode. Because it is necessary that the electrolyte flow completely around the electrode and allowance be made for the overcut which occurs during the ECM process, the tool is made about 0.005 in (0.12 mm) per side smaller than the hole it produces. The periphery of the electrode is insulated (Fig. 91-3) to prevent the sides from cutting as the tool extends deeper into the hole.

One of the chief purposes of the electrode tool is to impart its shape to the workpiece. For example, a square electrode will produce a square hole, and a round electrode will produce a round hole. The shape of the cavity which can be produced through ECM is limited only by the shape which can be cut on the electrode. The material used to make the electrode tool should possess the following characteristics.

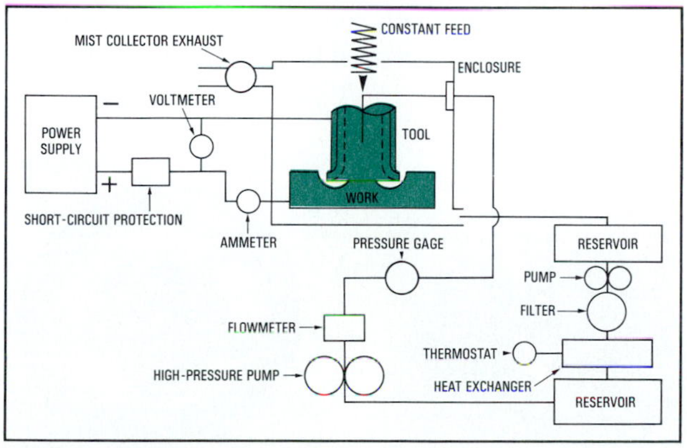

MIST COLLECTOR EXHAUST
CONSTANT FEED
ENCLOSURE
VOLTMETER
POWER SUPPLY
TOOL
WORK
SHORT-CIRCUIT PROTECTION
AMMETER
PRESSURE GAGE
RESERVOIR
PUMP
FILTER
FLOWMETER
THERMOSTAT
HIGH-PRESSURE PUMP
HEAT EXCHANGER
RESERVOIR

Figure 91-2 *Schematic diagram of a typical electro-chemical machining system.*

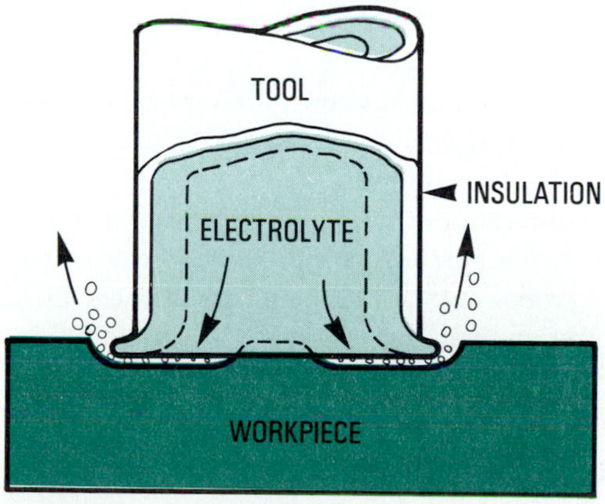

TOOL
INSULATION
ELECTROLYTE
WORKPIECE

Figure 91-3 *Electrolyte flowing through the electrode tool.*

1. It must be machinable.
2. It must be rigid.
3. It must be a good conductor of electricity.
4. It must be able to resist corrosion.

Copper, brass, and stainless steel have generally been found to be good electrode material for ECM.

Copper, an excellent conductor of electricity, is not recommended for manufacturing electrodes for thin-walled sections or deep holes. Because of its softness and tendency to bend, it is difficult to machine in long or thin sections.

Brass, while not as good a conductor of electricity as copper, can be used for electrode material with good results. The greater strength of brass, its ease in machining, and its lower cost make it an ideal material for most electrodes.

Stainless steel is used when large volume flow and high pressures of electrolyte are necessary. It is stronger than the other two materials, but its high initial cost and difficulties encountered in machining limit its use as an electrode material.

Metal Removal

In the electro-chemical process, the distance between the electrode and the work (the machining gap) (Fig. 91-3) is important. In order to encourage efficient electrical transmission, the tool and the work must be as close as possible to each other yet never come into contact. For most conditions, this gap will vary from 0.001 to 0.003 in (0.02 to 0.07 mm). Because of the high current levels used, serious damage to both the electrode tool and the work will occur if there is any physical contact between them.

The rate of metal removal is directly proportional to the current passing between the tool and the workpiece. Current densities ranging from a maximum of 1000 A/in^2 (6452 A/cm^2) to a maximum of 10 000 A/in^2 (64 520 A/cm^2) have been used in ECM applications. The use of high current will result in a high rate of metal removal, while low current will result in a low rate of metal removal.

The amount of overcut (the difference between the tool size and the hole produced) depends on cutting conditions and may vary from 0.008 to 0.012 in. (0.20 to 0.30 mm). Once the amount of overcut is known for a given tool, hole sizes will be repeated to within at least 0.0015 to 0.005 in. (0.03 to 0.01 mm) of roundness.

The rate of penetration varies with the type of operation being performed, the type of work material, the cross section of the electrode, and the current density used. ECM rates of penetration may vary from 0.250 to 0.430 in. (6.35 to 10.92 mm).

ADVANTAGES OF ECM

Electro-chemical machining is one of the metal-cutting processes which has contributed to the machining of space-age metals. Some of its characteristics and advantages are:

- Metal of any hardness can be machined.
- It competes with drilling and some through-hole milling operations, especially if the work exceeds 42 Rockwell C hardness.
- No heat is created during the machining process; therefore, there is no work distortion.
- It machines metal without tool rotation.
- Tool wear is insignificant, as the tool never touches the work.
- Because the tool never touches the work, thin and fragile sections can be machined without distortion.
- The workpiece is left burr-free.
- Intricate forms, difficult to machine by other processes, can be produced easily (Fig. 91-4).
- It is suitable for production-type work where multiple holes or cavities may be machined at the same time.
- Surface finishes of 25 μin. (0.63 μm) or less may be obtained.

ELECTROLYTIC GRINDING

Electrolytic grinding has been a boon to the machining of thin, fragile, metal products and tough, difficult-to-machine space-age alloys. In electrolytic grinding, the metal is removed from the work surface by a combination of electro-chemical action and the action of a metal-bonded abrasive grinding wheel (Fig. 91-5). Approximately 90 percent of the metal removed from the work surface is the result of this electro-chemical deplating action, while the remaining 10 percent is "wiped away" by the grinding wheel. The electrolytic grinding process is very similar to the process used for ECM.

THE PROCESS

The metal-bonded grinding wheel and the electrically conductive workpiece are both connected to a *direct current* power supply and are separated by the protruding abrasive

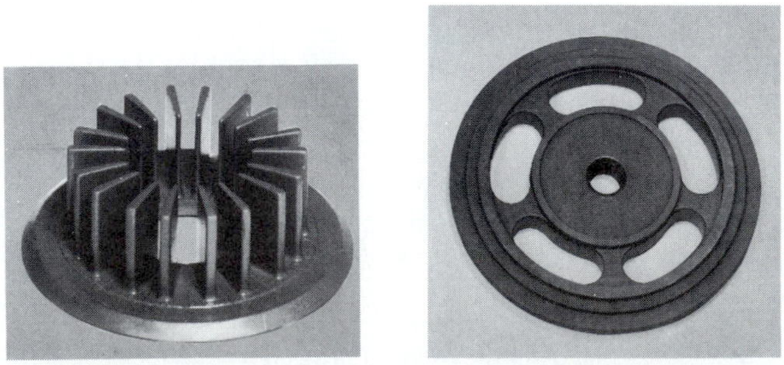

Figure 91-4 *Electro-chemical machining applications.*

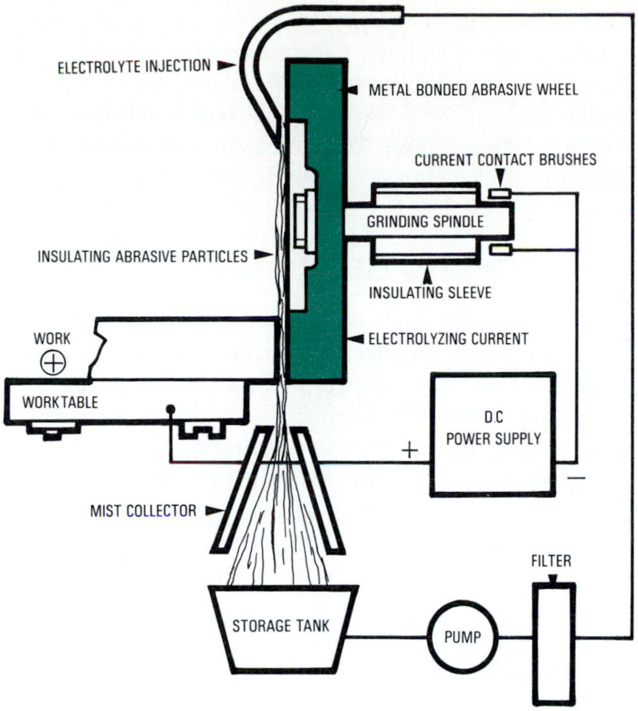

ELECTROLYTE INJECTION
METAL BONDED ABRASIVE WHEEL
CURRENT CONTACT BRUSHES
GRINDING SPINDLE
INSULATING ABRASIVE PARTICLES
INSULATING SLEEVE
WORK
ELECTROLYZING CURRENT
WORKTABLE
DC
POWER SUPPLY
+
−
MIST COLLECTOR
FILTER
STORAGE TANK
PUMP

Figure 91-5 *Electrolytic grinding principle. (Courtesy Cincinnati Milacron Inc.)*

particles of the wheel. An electrolytic grinding solution (electrolyte) is injected into the gap between the wheel and the work, completing the electrical circuit and producing the necessary deplating action which decomposes the work material. This decomposed material is removed by the action of the revolving grinding wheel and is washed away in the solution. In electrolytic grinding, the wheel never comes into actual contact with the work material.

THE GRINDING WHEEL

Metal-bonded, electrically conductive, abrasive grinding wheels are used for the electrolytic grinding process. Brass-, bronze-, and copper-bonded wheels which can be dressed to various forms are common in electrolytic grinding. The grinding wheel is the cathode (−) of the electrical circuit. The wheel spindle is connected to a *direct current* power supply through a set of contact brushes and is isolated from the rest of the machine by an insulating sleeve (Fig. 91-5). Metal-bonded diamond wheels are recommended for the grinding of tungsten carbides. Metal-bonded aluminum oxide wheels are used for grinding all other electrically conductive materials.

The wheel abrasive performs an important function in the electrolytic grinding process. The abrasive particles, protruding about 0.0005 to 0.001 in. (0.01 to 0.02 mm) beyond the metal-bonded wheel, act as nonconductive spacers, maintaining the necessary gap between the wheel and the workpiece. They also provide thousands of little pockets filled with electrolyte solution which completes the electrical circuit. The accuracy of the working gap (the distance between the metal bond of the wheel and the workpiece) is determined by the amount these abrasive particles protrude from the wheel (cathode).

TRUING

It is important that the grinding wheel run true within 0.0005 in. (0.01 mm) to maintain grinding accuracy and also to provide maximum passage for the electrolyte solution to achieve maximum stock removal. Metal-bonded aluminum oxide wheels can be dressed or trued with a single-point diamond or a commercial brake-truing attachment. Diamond wheels should be trued by the use of a dial indicator. After each dressing or truing operation, the electrolytic machining process is reversed by switching the electrical leads on the power supply, causing a deplating action of the metal wheel bond. This removes a slight amount of the metal wheel bond and causes the abrasive particles to protrude.

THE WORKPIECE

The work, which must be electrically conductive, is the anode (+) of the circuit. It is electrically connected to the *direct current* power supply through the table of the machine. Any material, even tungsten carbide and the difficult-to-machine space-age alloys, can be readily ground by this process, providing that the material is an electrical conductor.

THE CURRENT

A direct current of relatively low voltage (approximately 4 to 16 V) and high amperage (300 to 1000 A or higher) is used in the electrolytic grinding process. The amount of current flow depends on the size of the area on which the cutting action is occurring. A rule of thumb for stock removal rates is 0.010 in^3/min (0.16 cm^3/min) for every 100 A of electrolytic grinding current. Each work material has a saturation point which limits the current it can accept. Machine tool manufacturers of electrolytic grinders generally provide charts with their machines showing the stock-removal rates for various materials.

THE ELECTROLYTE

The electrolyte, generally a saline solution, serves two important functions:

1. It acts as a current conductor between the work and the wheel.
2. It combines chemically with the decomposed work material.

As the current flows from the workpiece (positive pole) through the electrolytic solution to the grinding wheel (negative pole), pockets of the solution act as electro-chemical cells decomposing the work surface. The current combines with the electrolyte to form a soft oxide film on the work surface, which will not allow the current to flow. The wiping action of the revolving grinding wheel removes these oxides, allowing the current to flow again. The rate of this decomposition is in direct proportion to the amount of current flowing from the work to the wheel.

As the solution becomes contaminated with metal particles, it is flushed away into a storage tank, then passed through filters to remove these particles. A freshly filtered solution is necessary for efficient electrolytic grinding. An uneven supply of the electrolyte will result in excessive wheel wear, while a weak supply retards the metal removal rates.

The machine surfaces coming into contact with the electrolyte must be chrome-plated to overcome the corrosive action of the electrolytic solution.

SURFACE FINISH

The surface finish obtained by electrolytic grinding ranges from 8 to 20 μin. (0.2 to 0.5 μm) when steels and various alloys are ground. As a rule, the higher the alloy, the better is the surface finish obtainable. When surface or traverse grinding tungsten carbide, a surface finish of 10 to 12 μin. (0.25 to 0.3 μm) may be expected. A surface finish of 8 to 10 μin. (0.2 to 0.25 μm) is generally obtained from plunge grinding carbides. These surface finishes can be obtained at maximum stock-removal rates with no finish cut required.

ELECTROLYTIC GRINDING METHODS

Grinding methods, such as cylindrical, form, plunge, surface, and traverse, all lend themselves to electrolytic grinding (Fig. 91-6). With all types of grinding methods, one important fact must be kept in mind: that is the area of work–wheel contact along the cutting path of the wheel should never exceed ¾ in. (19 mm) under normal conditions. If it is necessary to exceed this ¾-in. (19-mm) dimension, a much slower feed rate must be used or the electrolyte must be supplied to the grinding area by auxiliary methods.

Cylindrical Grinding

All the advantages of electrolytic grinding apply to cylindrical grinding, except high stock-removal rates. The reason for the lower stock-removal rate is that the area of contact between the work and wheel is small, allowing little current

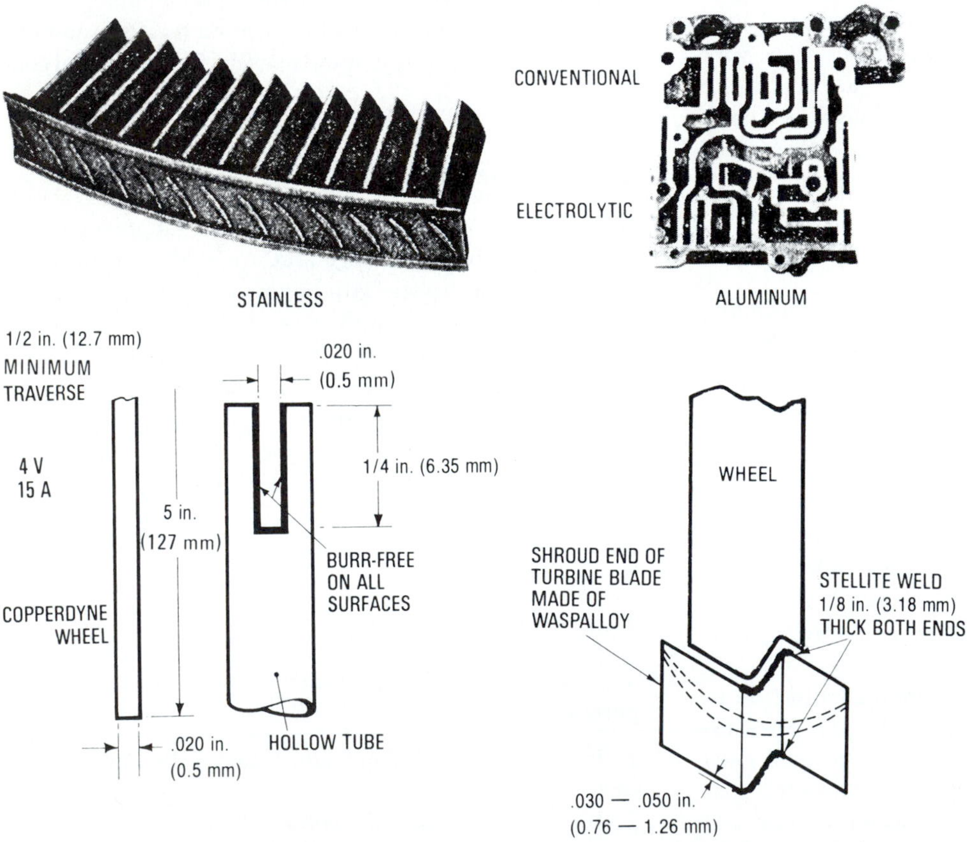

CONVENTIONAL
ELECTROLYTIC
STAINLESS
ALUMINUM
1/2 in. (12.7 mm)
MINIMUM
TRAVERSE
4 V
15 A
5 in.
(127 mm)
COPPERDYNE
WHEEL
.020 in.
(0.5 mm)
.020 in.
(0.5 mm)
1/4 in. (6.35 mm)
BURR-FREE
ON ALL
SURFACES
HOLLOW TUBE
WHEEL
SHROUD END OF
TURBINE BLADE
MADE OF
WASPALLOY
STELLITE WELD
1/8 in. (3.18 mm)
THICK BOTH ENDS
.030 — .050 in.
(0.76 — 1.26 mm)

Figure 91-6 *Types of electrolytic grinding. (Courtesy Avco Bay State Abrasive Company.)*

to flow. It is recommended that plunge grinding be used to rough out work to within a few thousandths of size and then that work be finished by traverse grinding.

Form Grinding

Metal-bonded aluminum oxide wheels are usually employed for form-grinding operations. They are easily formed to the desired shape by a conventional diamond truing attachment. Formed diamond wheels for grinding carbides are expensive to change from form to form and difficult to true. Therefore, metal-bonded abrasive wheels are generally used for the production grinding of carbides.

Plunge Grinding

When an area is plunge-ground, there is no traverse motion of the wheel, and the wheel is fed straight into the surface. Either the face or the side of the wheel may be used for plunge grinding. Single-point tools, face-milling cutters, straight side-milling cutters, and any other single-plane surfaces within the range of the wheel sizes may be effectively ground by this method. Fine manual feeds should be used for grinding individual pieces; automatic feeds are recommended for production grinding.

Surface Grinding

Since the area of contact in surface grinding varies with the depth of cut for each wheel diameter, the full depth of cut per pass is recommended for maximum results. Always use the largest wheel possible, depending on the capacity of the machine. The rate of traverse is not dependent on the width of the wheel as long as there is sufficient current available. Whenever possible, the use of automatic power feeds is advisable, since too slow a feed rate results in excessive overcut and too fast a feed results in excessive wheel wear. At no time should the work and wheel contact exceed ¾ in. (19 mm). Machine tool manufacturers generally supply feed rate charts with their machines. The charts show grinding current, wheel size, depth of cut, and area of contact.

Traverse Grinding

Most cutter-sharpening operations are performed by traverse grinding with a cup- or flare-type wheel. Helical side milling cutters and others that are impractical to plunge-grind are sharpened by this method.

Edge grinding with cup wheels is impractical as the work–wheel contact is small and allows little current to flow. For efficiency in traverse grinding, the wheel is swiveled slightly, or an angle of 1 to 2° is dressed on the wheel surface to increase the area of work–wheel contact. Traverse feed rates depend on the swivel angle, width of the wheel face, and the amount of current available.

ADVANTAGES OF ELECTROLYTIC GRINDING

Electrolytic grinding offers many advantages over conventional machining methods for metal removal.

- It saves wheel costs, especially in metal-bonded diamond wheels, since only 10 percent of the metal is removed by abrasive action.
- There is a high ratio of stock removal in relation to wheel wear.
- Higher production rates are possible because of the longer runs between wheel truing.
- No heat is generated during the grinding operation; therefore, there is no burning or heat distortion of the work.
- The process is burr-free, thereby eliminating deburring operations.
- Thin, fragile workpieces can be ground without distortion since the wheel never touches the work.
- Tungsten carbide and supertough alloys can be ground quickly and with ease.
- Exotic materials, such as zirconium and beryllium, can be cut regardless of their hardness, fragility, or thermal sensitivity.
- Dissimilar metals can be ground, providing they are electrically conductive.
- Cutters may be reground in one pass, eliminating the necessity of finishing cuts.
- No stresses are created in the work material.
- No work hardening occurs during this process.

DISADVANTAGES AND LIMITATIONS

While there are many advantages of electrolytic grinding, there are also some disadvantages or limitations.

- Only electrically conductive work can be ground.
- Grinding wheels, especially the metal-bonded diamond wheels, are more expensive than regular wheels.
- Inside corners cannot be ground sharper than a 0.010 to 0.015 in. (0.25 to 0.40 mm) radius because of the overcut occurring during the electrochemical action.
- Accuracy is possible to within 0.0005 in. (0.01 mm) only.
- The electrolytic solution is corrosive; therefore, machine parts coming into contact with the electrolyte must be chrome-plated.
- The wheel contact should not exceed ¾ in. (19 mm).

CONCLUSION

At present, this method of electrical machining simplifies production methods and provides savings in both machining time and grinding wheel costs. As new types of metal-bonded wheels, power supplies, and electrolytes are developed, this grinding process should find many more applications.

REVIEW QUESTIONS

ELECTRO-CHEMICAL MACHINING

1. How does ECM differ from conventional machining techniques?
2. Describe briefly the process of ECM.
3. Name four suitable electrolytes.
4. Explain fully the purpose of the electrolyte solution.
5. Explain what part the electrolyte plays in the ECM process.
6. What are the characteristics of a good electrode tool?
7. Name three electrode materials suitable for ECM and state the advantage of each.
8. Define:
   a. Machining gap
   b. Overcut
   c. Rate of penetration
9. State seven main advantages of ECM.

ELECTROLYTIC GRINDING

10. Describe briefly the electrolytic grinding process.
11. Describe the type and function of the electrolytic grinding wheel.
12. What purpose does the wheel abrasive serve in electrolytic grinding?
13. Discuss the importance and operation of truing a grinding wheel.
14. What function does the workpiece serve in the electrolytic grinding circuit?
15. What type of current is used for this process?
16. Name two functions of the electrolyte solution.
17. Explain what occurs to the electrolyte when the current flows.
18. What surface finishes can be obtained by electrolytic grinding?
19. Discuss plunge and surface grinding.
20. Name eight advantages of the electrolytic grinding process.
21. List four disadvantages of electrolytic grinding.

UNIT 92

Electrical Discharge Machining

Electrical discharge machining, commonly known as EDM, is a process that is used to remove metal through the action of an electrical discharge of short duration and high current density between the tool and the workpiece (Fig. 92-1A). This principle of removing metal by an electric spark has been known for quite some time. In 1889, Paschen explained the phenomenon and devised a formula which would predict its arcing ability in various materials. The EDM process can be compared to a miniature version of a lightning bolt striking a surface, creating a localized intense heat, and melting away the work surface.

OBJECTIVES

After completing this unit you should be able to:

1. Explain the principle of electrical discharge machining
2. Name and state the purposes of the main components of an electrical discharge machine
3. State the advantages and applications of the electrical discharge process

USE OF EDM

Electrical discharge machining has proved especially valuable in the machining of supertough, electrically conductive materials, such as the new space-age alloys. These metals would have been difficult to machine by conventional methods. EDM has made it relatively simple to machine intricate shapes that would be impossible to produce with conventional cutting tools. This machining process is continually finding further applications in the metal-cutting industry. It is being used extensively in the plastics industry to produce cavities of almost any shape in the steel molds.

PRINCIPLE OF EDM

Electrical discharge machining is a controlled metal-removal technique whereby an electric spark is used to cut (erode) the workpiece, which takes a shape opposite to that of the cutting tool or electrode (Fig. 92-1B). The *cutting tool (electrode)* is made from electrically conductive material, usually carbon. The electrode, made to the shape of the cavity required, and the workpiece are both submerged in a *dielectric fluid,* which is generally a light lubricating oil. This dielectric fluid should be a nonconductor (or poor conductor) of electricity. A *servo mechanism* maintains a gap of about 0.0005 to 0.001 in. (0.01 to 0.02 mm) between the electrode and the work, preventing them from coming into contact with each other. A *direct current* of low voltage and high amperage is delivered to the electrode at the rate of approximately 20 000 hertz (HZ). These electrical energy impulses vaporize the oil at this point. This permits the spark to jump the gap between the electrode and the workpiece through the dielectric fluid (Fig. 92-2). Intense heat is created in the localized area of the spark impact; the metal melts and a small particle of molten metal is expelled from the surface of the workpiece. The dielectric fluid, which is constantly being circulated, carries away the eroded particles of metal during the off-cycle of

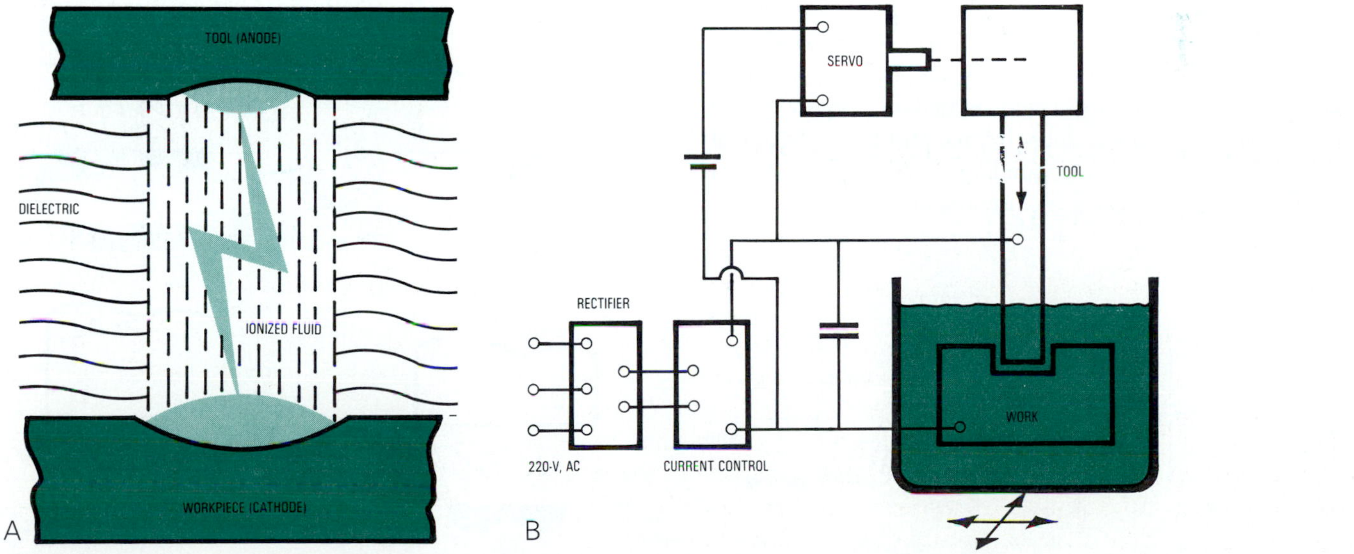
TOOL (ANODE)
DIELECTRIC
IONIZED FLUID
WORKPIECE (CATHODE)
A
SERVO
TOOL
RECTIFIER
220-V, AC
CURRENT CONTROL
WORK
B

Figure 92-1 *(A) A controlled spark removes metal during electrical discharge machining (EDM); (B) basic elements of an electrical discharge system.*

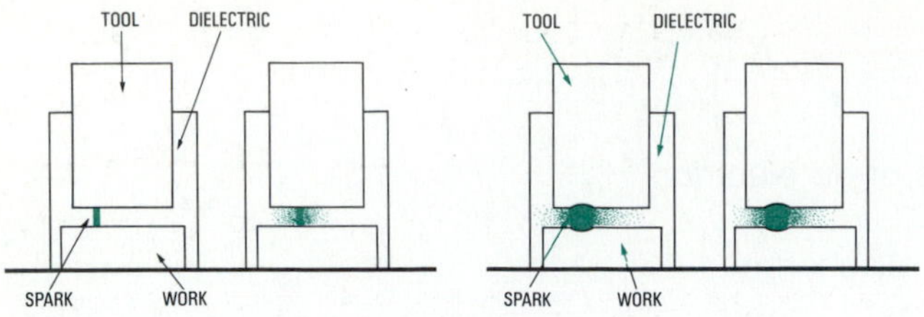
TOOL
DIELECTRIC
SPARK
WORK
TOOL
DIELECTRIC
SPARK
WORK

Figure 92-2 *Stages of single spark.*

the pulse and also assists in dissipating the heat caused by the spark. This process continues at the rate of over 20 000 cycles per second.

TYPES OF EDM CIRCUITS

Several types of electrical discharge power supply have been used for EDM. Although there are many differences between them, each type is used for the same basic purpose, that is, the precise, economical removal of metal by electrical spark erosion.

The two most common types of power supplies are the:

1. Resistance–capacitance power supply
2. Pulse-type power supply

Resistance–capacitance power supply, also known as the *relaxation-type power supply,* was widely used on the first EDM machines. It is still the power supply used on many of the machines of foreign manufacture.

As illustrated in Fig. 92-3A, the capacitor is charged through a resistance from a direct current voltage source that is generally fixed. As soon as the voltage across the capacitor reaches the breakdown value of the dielectric fluid in the gap, a spark occurs. A relatively high voltage (125 V), high capacitance of over 100 μF (microfarads) for roughing cuts, low spark frequency, and high amperage are characteristics of the resistance–capacitance power supply.

In resistance–capacitance circuits, an increase in metal-removal rates depends more on larger amperage and capacitance than on increasing the number of discharges per second. The combination of low frequency, high voltage, high capacitance, and high amperage results in:

1. A rather coarse surface finish
2. Large overcut around the electrode (tool)
3. Larger metal particles being removed and more space being required to flush out particles

The advantages of the resistance–capacitance power supply are:

- The circuit is simple and reliable.
- It works well at low amperages, especially with milliampere currents required for holes under 0.005 in. (0.12 mm) in diameter.

Pulse-type power supply is used almost exclusively by American manufacturers. It is similar to the resistance–capacitance type; however, vacuum tubes or solid-state devices are used to achieve an extremely fast pulsing switch effect (Fig. 92-3B). The pulse width and intervals may also be accurately controlled by switching devices. The switching is extremely fast and the discharges per second are 10 or more times greater than with the resistance–capacitance power supply at low frequencies. The results of more discharges per second are illustrated in Fig. 92-4. With more discharges per second, and using the same current (10 A), it is clear that smaller craters are created, producing a finer surface finish while still maintaining the same metal-removal rate.

Pulse-type power supply circuits are usually operated on low voltages (70 to 80 V), high frequency (sparks at the rate

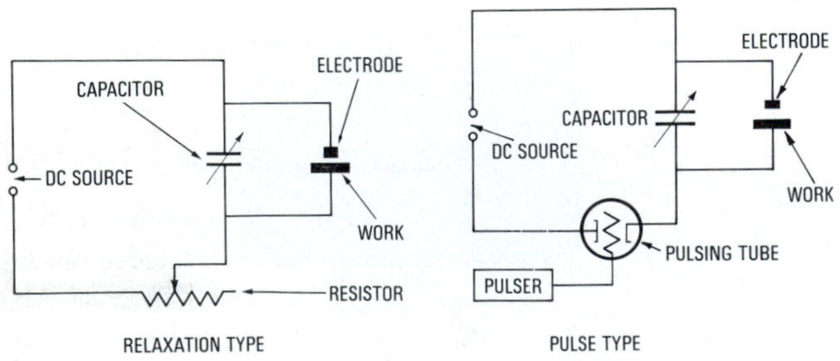
CAPACITOR
ELECTRODE
DC SOURCE
WORK
RESISTOR
RELAXATION TYPE
ELECTRODE
CAPACITOR
DC SOURCE
WORK
PULSING TUBE
PULSER
PULSE TYPE

Figure 92-3 *Electrical discharge system power supply circuits.*

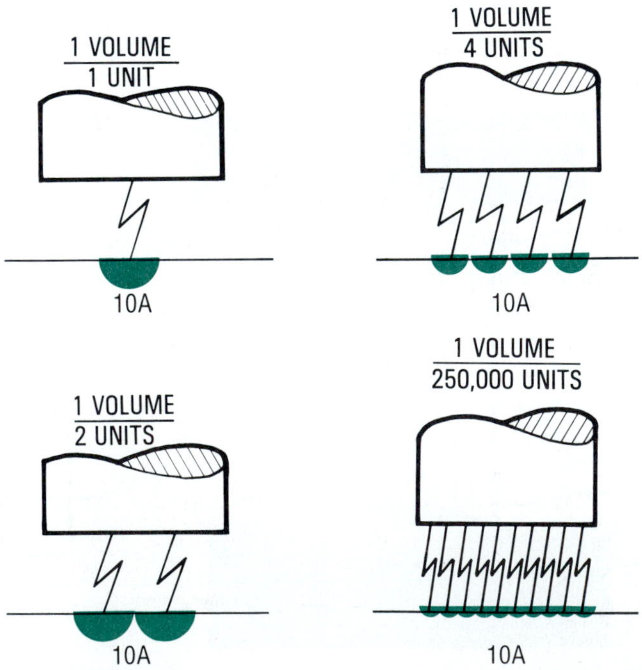
1 VOLUME / 1 UNIT
10A
1 VOLUME / 4 UNITS
10A
1 VOLUME / 2 UNITS
10A
1 VOLUME / 250,000 UNITS
10A

Figure 92-4 *Effects on cratering and surface finish using various frequencies of spark discharge.*

of 260 000 Hz), low capacitance (50 μF or less), and low-energy spark levels.

The main advantages of the pulse-type circuit are:

- It is extremely versatile and can be accurately controlled for roughing and finishing cuts.
- Better surface finish is produced as less metal is removed per spark since there are many sparks per unit of time.
- There is less overcut around the electrode (tool).

THE ELECTRODE

The *electrode* in EDM is formed to the shape of the cavity desired. As in conventional machining, some materials have better cutting and wearing qualities than others. Electrode materials must, therefore, have the following characteristics.

1. Be good conductors of electricity and heat
2. Be easily machined to shape at a reasonable cost
3. Produce efficient metal removal from the workpiece
4. Resist deformation during the erosion process
5. Exhibit low electrode (tool) wear rates

Much experimentation has been carried out to find a good, economical material for the manufacture of EDM electrodes. The more common electrode materials are graphite, copper, copper graphite, copper tungsten, brass, and steel. None of these electrode materials has general-purpose application; each machining operation dictates the selection of the electrode material. *Yellow brass* has been used primarily as electrode material for pulse-type circuits because of its good machinability, electrical conductivity, and relatively low cost. *Copper* produces better results in the resistance–capacitance circuits where higher voltages are employed.

High-density and high-purity carbon, or *graphite,* is a relatively new electrode material, which is gaining wide acceptance. It is commercially available in various shapes and sizes, is relatively inexpensive, can be machined easily, and makes an excellent electrode. Its tool wear rate is much less and its high metal-removal rate is almost double that of any other electrode material. Development of superior graphite electrodes has increased the use of graphite electrodes because of their low cost and ease of fabrication.

THE EDM PROCESS

The use of EDM is increasing as more and more applications are found for this process. As new, important technological advances in equipment and application techniques become available, more industries are adopting the process. It is necessary, therefore, to discuss the various aspects of EDM in more detail.

THE SERVO MECHANISM

It is important that there be no physical contact between the electrode (tool) and the workpiece; otherwise, arcing will occur, causing damage to both the electrode and the workpiece. Electrical discharge machines are equipped with a servo control mechanism that automatically maintains a constant gap of approximately 0.0005 to 0.001 in. (0.01 to 0.02 mm.) between the electrode and the workpiece. The mechanism also advances the tool into the workpiece as the operation progresses and senses and corrects any shorted condition by rapidly retracting and returning the tool. Precise control of the gap is essential to a successful machining operation. If the gap is too large, ionization of the dielectric fluid does not occur and machining cannot take place. If the gap is too small, the tool and workpiece may weld together.

Precise gap control is accomplished by a circuit in the power supply, which compares the average gap voltage to a preselected reference voltage. The difference between the two voltages is the input signal, which tells the servo mechanism how far and how fast to feed the tool and when to retract it from the workpiece.

When chips in the spark gap reduce the voltage below a critical level, the servo mechanism causes the tool to withdraw until the chips are flushed out by the dielectric fluid. The servo system should not be too sensitive to "short-lived" voltages caused by chips being flushed out; otherwise, the tool would be constantly retracting, thereby seriously affecting machining rates.

Servo feed control mechanisms can be used to control the vertical movement of the electrode (tool) for sinking cavities. It can also be applied to the table of the machine for work requiring horizontal movement of the electrode (tool).

CUTTING CURRENT (AMPERAGE)

The EDM power supply provides the direct current electrical energy for the electrical discharges which occur between the tool and the workpiece. As the pulse-type power supply is the more commonly used in North America, only the characteristics of this type will be listed.

Characteristics of Pulse-Type Circuits

1. Low voltages (normally about 70 V, which drops to about 20 V after the spark is initiated)
2. Low capacitance (about 50 μF or less)

3. High frequencies (usually 20 000 to 30 000 Hz but may be as high as 260 000 Hz)
4. Low-energy spark levels

THE DISCHARGE PROCESS

Upon application of sufficient electrical energy between the electrode (cathode) and the workpiece (anode), the dielectric fluid changes into a gas, allowing a heavy discharge of current to flow through the ionized path and strike the workpiece. The energy of this discharge vaporizes and decomposes the dielectric fluid surrounding the column of electric conduction. As the conduction continues, the diameter of the discharge column expands and the current increases. The heat between the electrode and the work surface causes a small pool of molten metal to be formed on the work surface. When the current is stopped, usually only for microseconds, the molten metal particles solidify and are washed away by the dielectric fluid.

These electrical discharges occur at the rate of 20 000 to 30 000 Hz between the electrode and the workpiece. Each discharge removes a minute amount of metal. Since the voltage during discharge is constant, the amount of metal removed will be proportional to the amount of charge between the electrode and the work. For fast metal removal, high amounts of current should be delivered as quickly as possible to melt the maximum amount of metal. This, however, produces large craters in the workpiece, resulting in rough surface finish. To obtain smaller craters and therefore finer surface finishes, smaller charges of energy can be used. This results in slower stock removal rates. If the current is maintained but the frequency (number of hertz) is increased, this also results in smaller craters and better surface finish. The surface finish is proportional to the number of electrical discharges (cycles) per second (Fig. 92-4).

THE DIELECTRIC FLUID

The dielectric fluid used in the electrical discharge process serves several main functions:

1. Serves as an insulator between the tool and the workpiece until the required voltage is reached
2. Vaporizes (ionizes) to initiate the spark between the electrode and the workpiece
3. Confines the spark path to a narrow channel
4. Flushes away the metal particles to prevent shorting
5. Acts as a coolant for both the electrode and the workpiece

Types of Dielectrics

Many types of fluids have been used as dielectrics, resulting in various rates of metal removal. They must be able to ionize (vaporize) and deionize rapidly and have a low viscosity that will allow them to be pumped through the narrow machining gap. The most commonly used EDM fluids, proven to be satisfactory dielectrics, have been various petroleum products, such as light lubricating oils, transformer oils, silicon-base oils, and kerosene. These all perform reasonably well, especially with graphite electrodes, and are reasonable in cost. In certain cases, dielectrics such as carbon tetrachloride and certain compressed gases have been used.

The selection of the dielectric is important to the EDM process since it affects the metal-removal rate and electrode wear. Industry is continually searching for new and better dielectric fluids for this process. A fluid consisting of triethylene glycol, water, and monoethyl ether of ethylene glycol has been used in research with superior results, especially with metallic electrodes. It is quite likely that many new dielectric fluids will be developed to improve the EDM process.

Methods of Circulating Dielectrics

The dielectric fluid must be circulated under constant pressure if it is to flush away efficiently the metal particles and assist in the machining process. The pressure used generally begins with 5 psi (34 kPa) and is increased until optimum cutting is attained. Too much dielectric fluid will remove the chips before they can assist in the cutting action and thereby cause slower machining rates. Too little pressure will not remove the chips quickly enough and thereby cause short circuits.

Four methods are generally used to circulate the dielectric fluid. All must use fine filters in the system to remove the metal particles so that they are not recirculated.

Down through the Electrode (Fig. 92-5A)

A hole (or holes) is drilled through the electrode, and the dielectric fluid is forced through the electrode and between it and the workpiece. This rapidly flushes away the metal particles from the machining area. On cavities, a small standing slug or core remains and must be ground away after the machining operation has been completed.

Up through the Workpiece (Fig. 92-5B)

Another common method is to cause the fluid to be circulated up through the workpiece. This type of flushing is limited to through-hole cutting applications and to cavities having holes for core or ejector pins.

Vacuum Flow (Fig. 92-5C)

A negative pressure (vacuum) is created in the gap, which causes the dielectric to flow through the normal 0.001 in. (0.02 mm) clearance between the electrode and the workpiece. The flow can be either up through a hole in the electrode or down through a hole in the workpiece. Vacuum flow has several advantages over other methods: It

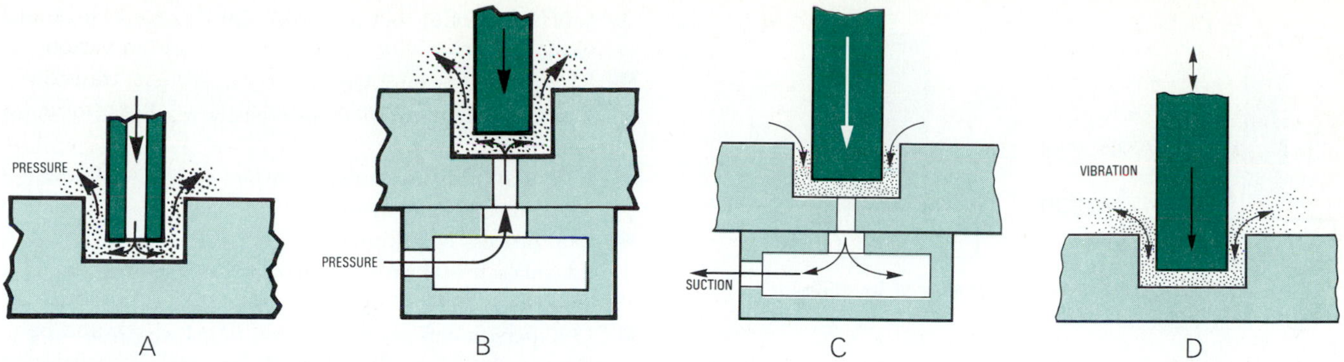
PRESSURE
A
PRESSURE
B
SUCTION
C
VIBRATION
D

Figure 92-5 *(A) Circulating dielectric fluid down through the electrode; (B) circulating dielectric fluid up through the workpiece; (C) circulating the dielectric fluid down through the workpiece by suction; (D) circulating the dielectric fluid by vibration.*

improves machining efficiency, reduces smoke and fumes, and helps to reduce or eliminate taper in the workpiece.

Vibration (Fig. 92-5D)

A pumping and sucking action is used to cause the dielectric to disperse the chips from the spark gap. The vibration method is especially valuable for very small holes, deep holes, or blind cavities where it would be impractical to use other methods.

METAL-REMOVAL RATES

Metal-removal rates for EDM are somewhat slower than with conventional machining methods. The rate of metal removal is dependent on the following factors.

1. Amount of current in each discharge
2. Frequency of the discharge
3. Electrode material
4. Workpiece material
5. Dielectric flushing conditions

The normal metal-removal rate is approximately 1 in^3 (16 cm^3) of work material per hour for every 20 A of machining current. However, metal removal rates of up to 15 in^3/h (245 cm^3/h) are possible for roughing cuts with special power supplies.

ELECTRODE (TOOL) WEAR

During the discharge process, the electrode (tool) as well as the workpiece is subject to wear or erosion. As a result, it is difficult to hold close tolerances as the tool gradually loses its shape during the machining operation. At times it is necessary to use as many as five electrodes to produce a cavity of the required shape and tolerance. For through-hole operations, stepped electrodes are used to produce roughing and finishing cuts in one pass.

Fortunately, the rate at which the tool wears is considerably less than that of the workpiece. An *average wear ratio* of the workpiece to the electrode is 3:1 for metallic tools, such as copper, brass, zinc alloys, etc. With graphite electrodes, this wear ratio can be greatly improved to 10:1.

Much research and development remains to be done to reduce the wear ratio of the electrode. *Reverse-polarity machining,* a relatively new development, promises to be a major breakthrough in reducing electrode wear. With this method, molten metal from the workpiece is deposited on a graphite electrode about as fast as the electrode is worn away. Thus minute electrode wear is continually being replaced by a deposit of the work material. Reverse-polarity machining operates best on low spark-discharge frequencies and high amperage. It improves the metal-removal rates and greatly reduces electrode wear.

OVERCUT

Overcut is the amount the cavity in the workpiece is cut larger than the size of the electrode used in the machining process. The distance between the surface of the work and the surface of the electrode (overcut) is equal to the length of the sparks discharged, which is constant over all areas of the electrode.

The amount of overcut in EDM ranges from 0.0002 to 0.007 in. (0.005 to 0.1 mm) and is dependent on the amount of gap voltage. As illustrated in Fig. 92-6, the overcut distance increases with the increased gap voltage. Therefore, the amount of overcut can be controlled fairly accurately. The amount of overcut is generally varied to suit the metal-removal rate and the surface finish required, which in turn determines the size of the chip removal.

Most manufacturers of EDM machines provide overcut charts to show the amount of clearance produced with various currents. The charts make it possible to determine accurately the electrode size required to machine an opening to within 0.0001 in. (0.002 mm).

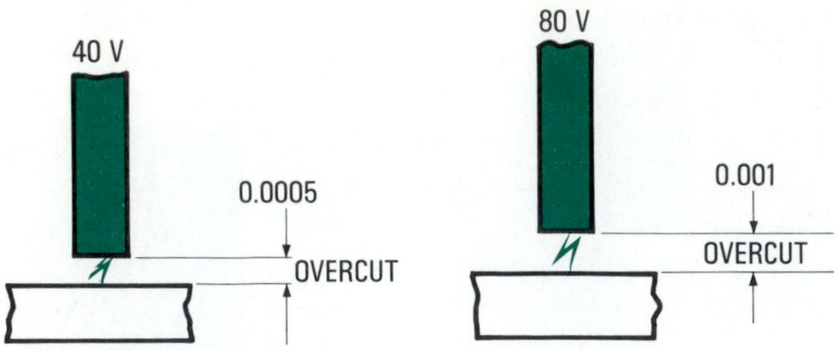
40 V
0.0005
OVERCUT
80 V
0.001
OVERCUT

Figure 92-6 *Example of overcut produced by different voltages in EDM.*

The size of the chips removed is an important factor in setting the amount of overcut because:

1. Chips in the space between the electrode and the work serve as conductors for the electrical discharges.
2. Large chips produced with higher amperages require a larger gap to enable them to be flushed out effectively.

Therefore, overcut depends upon the gap voltage and the chip size, which vary with the amperage used.

SURFACE FINISH

In the last few years, major advances have been made with regard to the surface finishes that can be produced. With low metal-removal rates, surface finishes of 2 to 4 μin. (0.05 to 0.10 μm) are possible. With high metal-removal rates [as much as 15 in^3/h (245 cm^3/h)] finishes of 1000 μin. (24 μm) are produced.

The type of finish required determines the number of amperes which can be used and the capacitance, frequency, and voltage setting. For fast metal removal (roughing cuts), high amperage, low frequency, high capacitance, and minimum gap voltage are required. For slow metal removal (finish cut) and good surface finish, low amperage, high frequency, low capacitance, and the highest gap voltage are required.

ADVANTAGES OF EDM

Electrical discharge machining has many advantages over conventional machining processes, including the following.

- Any material that is electrically conductive can be cut, regardless of its hardness. EDM is especially valuable for cemented carbides and the new supertough space-age alloys that are extremely difficult to cut by conventional means.
- Work can be machined in a hardened state, thereby overcoming the deformation caused by the hardening process.
- Broken taps or drills can readily be removed from workpieces.
- It does not create stresses in the work material since the tool (electrode) never comes into contact with the work.
- The process is burr-free.
- Thin, fragile sections can be easily machined without deforming.
- Secondary finishing operations are generally eliminated for many types of work.
- The process is automatic in that the servo mechanism advances the electrode into the work as the metal is removed.
- One person can operate several EDM machines at one time.
- Intricate shapes, impossible to produce by conventional means, are cut out of a solid with relative ease (Fig. 92-7A and B).
- Better dies and molds can be produced at lower cost.
- A die punch can be used as the electrode to reproduce its shape in the matching die plate, complete with the necessary clearance.

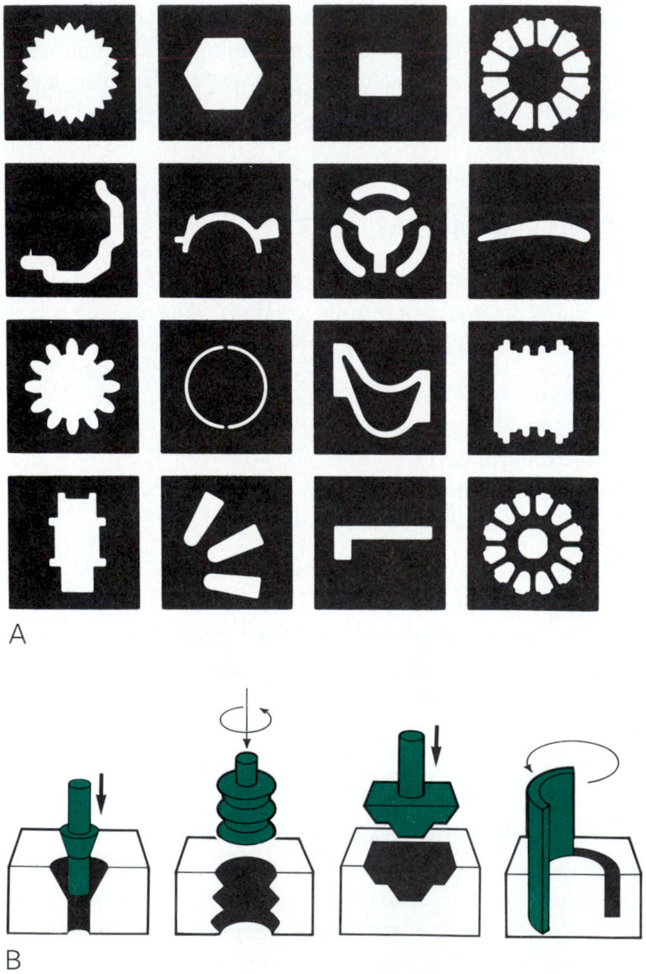
A
B

Figure 92-7 *(A) Examples of work produced by EDM; (B) tool movement required to produce cavities of various shapes.*

LIMITATIONS OF EDM

Electrical discharge machining has found many applications in the machine tool trade. However, it does have some limitations:

- Metal-removal rates are low.
- The material to be machined must be electrically conductive.
- Cavities produced are slightly tapered, but can be controlled for most applications to as little as 0.0001 in. (0.002 mm) in every ¼ in. (6 mm).
- Rapid electrode wear can become costly in some types of EDM equipment.
- Electrodes smaller than 0.003 in. (0.07 mm) in diameter are impractical.
- The work surface is damaged to a depth of 0.0002 in. (0.005 mm) but it is easily removed.
- A slight case hardening occurs. This, however, may be classed as an advantage in some instances.

THE WIRE-CUT EDM MACHINE

Electrical discharge machining has advanced quickly with the addition of computer numerical control (CNC). Today EDM is used for a wide variety of precision metalworking applications which would have been almost impossible just a few years ago. Cutting tolerances, cutting speeds, and surface finish quality have been greatly improved.

Another application of electrical discharge machining is the wire-cut EDM machine (Fig. 92-8). Unlike other EDM applications which use an electrode in the shape and size of the cavity or hole required, this machine generally uses a thin brass or copper wire as the electrode, making it possible to cut most shapes and contours from flat plate materials.

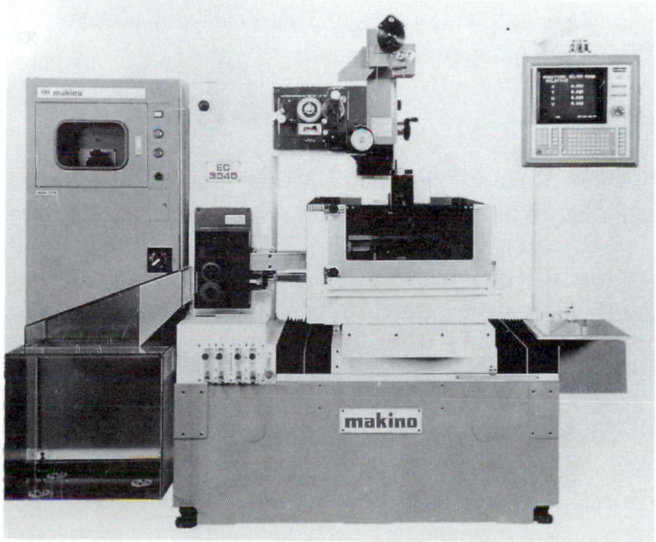
makino
EC
3040

Figure 92-8 *The wire-cut electrical discharge machine is used for machining complex forms. (Courtesy LeBlond-Makino Machine Tool Co.)*

Wire-cut EDM can do things older technologies cannot do as well, as quickly, as inexpensively, and as accurately. Most parts can now be programmed and produced as a solid, rather than in sections and then assembled as a unit as was necessary previously. Wire-cut EDM is capable of producing complex shapes such as tapers, involutes, parabolas, ellipses, etc. (Fig. 92-9). This process is now commonly used to machine tungsten carbide, difficult-to-machine material, polycrystalline diamond, polycrystalline cubic boron nitride, and pure molybdenum.

THE PROCESS

The wire-cut EDM machine uses CNC to move the workpiece along the *X*- and *Y*-axes (backward, forward, and sideways) in a horizontal plane toward a vertically moving wire electrode (Fig. 92-10). The wire electrode does not contact the workpiece but operates in a stream of dielectric fluid (usually deionized water), which is directed to the spark area between the work and the electrode. When the machine is in operation, the dielectric fluid in the spark area breaks down, forming a gas that permits the spark to jump between the workpiece and the electrode. The eroded material caused by the spark is then washed away by the dielectric fluid.

Figure 92-9 *Tapers, involutes, parabolas, elipses, and many other shapes can be cut easily on wire-cut electrical discharge machines. (Courtesy LeBlond-Makino Machine Tool Co.)*

Figure 92-10 *The workpiece is moved along the X- and Y-axes by numerical control to cut the desired shape. (Courtesy LeBlond-Makino Machine Tool Co.)*

Movement of the wire is controlled continuously in minimum increments of 0.00001 in. (0.0002 mm) in two- to five-axis positions. Along with the *XY*-axis, the head can be tilted up to 30° (*UV*-axis) for cutting tapered sections and can be raised or lowered (*Z*-axis) to suit various workpiece thicknesses. Any contour, within the size of the machine, can be cut very accurately. During the cutting action, an arc gap of 0.001 to 0.002 in. (0.02 to 0.05 mm) is maintained between the workpiece and the wire electrode.

OPERATING SYSTEMS

The four main operating systems, or components, of the wire-cut electrical discharge machines are the servo mechanism, the dielectric fluid, the electrode, and the machine control unit.

The Servo Mechanism

The EDM servo mechanism controls the cutting current levels, the feed rate of the drive motors, and the traveling speed of the wire. The servo mechanism automatically maintains a constant gap of approximately 0.001 to 0.002 in. (0.02 to 0.05 mm) between the wire and the workpiece. It is important that there be no physical contact between the wire (electrode) and the workpiece; otherwise, arcing will occur, which could damage the workpiece and break the wire. The servo mechanism also advances the workpiece into the wire as the operation progresses, senses the work–wire spacing, and slows or speeds up the drive motors, as required, to maintain the proper arc gap. Precise control of the gap is essential to a successful machining operation. If the gap is too large, the dielectric fluid between the wire (electrode) and the workpiece will not break down into a gas, a spark cannot be conducted between the wire and the workpiece, and therefore machining cannot take place. If the gap is too small, the wire will touch the workpiece causing the wire to short and break.

The Dielectric Fluid

One of the most important factors in a successful EDM operation is the removal of the particles (chips) from the working gap. Flushing these particles out of the gap with the dielectric fluid will produce good cutting conditions, while poor flushing will cause erratic cutting and poor machining conditions.

The dielectric fluid in the wire-cut EDM process is usually deionized water. This is tap water that is circulated through an ion-exchange resin. The deionized water makes a good insulator, while untreated water is a conductor and is not suitable for the electrical discharge machining process. The amount of deionization of the water determines its resistance. For most operations, the lower the resistance, the faster will be the cutting speed. However, the resistance of the dielectric fluid should be much higher when carbides and high-density graphites are being cut.

The dielectric fluid used in the wire-cut EDM process serves several functions:

1. It helps to initiate the spark between the wire (electrode) and the workpiece.
2. It serves as an insulator between the wire and the workpiece.
3. It flushes away the particles of disintegrated wire and workpiece to prevent shorting.
4. It acts as a coolant for both the wire and the workpiece.

The dielectric fluid must be circulated under constant pressure if it is to flush away the particles and assist in the machining process.

The Electrode

The electrode in wire-cut EDM may be a spool of brass, copper, tungsten, molybdenum, or zinc wire ranging from 0.002 to 0.012 in. (0.05 to 0.30 mm) in diameter and from 2 to 100 lb (0.90 to 45.36 kg) in weight. The electrode continuously travels from a supply spool to a takeup spool, so that new wire is always in the spark area. With this type of electrode, the wear on the wire does not affect the accuracy of the cut because new wire is being constantly fed past the workpiece at rates from a fraction of an inch to several inches per minute. Both the electrode wear and the material-removal rate from the workpiece depend on factors such as the material's electrical and thermal conductivity, its melting point, and the duration and intensity of the electrical pulses. As in conventional machining, some materials have better cutting and wearing qualities than others;

therefore, electrode materials must have the following characteristics:

1. Be a good conductor of electricity.
2. Have a high melting point.
3. Have a high tensile strength.
4. Have good thermal conductivity.
5. Produce efficient metal removal from the workpiece.

The Machine Control Unit

The control unit for the wire-cut EDM can be separated into three individual operator panels:

- The control panel for setting cutting conditions (servo mechanism)
- The control panel for machine setup and the data required to produce the part (numerical control)
- The control panel for manual data input (MDI) and a cathode ray tube display (Fig. 92-11)

Although some of the newer wire-cut electrical discharge machines eliminate some of these controls and incorporate them as part of the machine's automatic cutting cycle, a knowledge of what is being controlled during the cutting cycle should give the operator a better overall understanding of the wire-cut machine and cutting process.

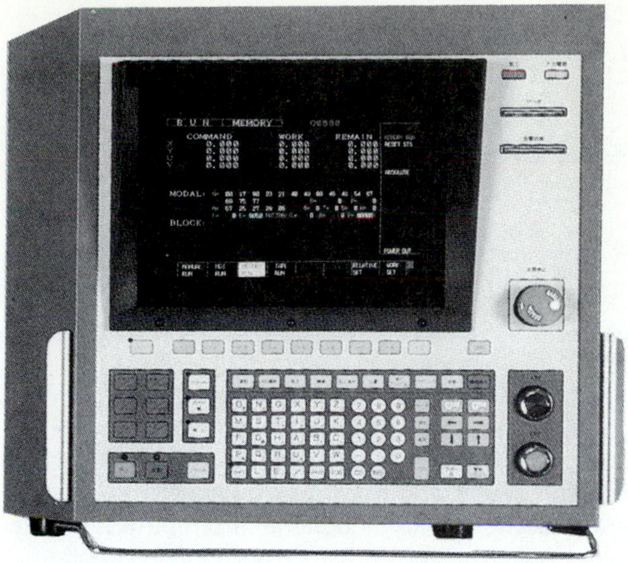

Figure 92-11 *The control panel is used to set the cutting conditions of the wire-cut electrical discharge machine. (Courtesy LeBlond-Makino Machine Tool Co.)*

CONCLUSION

Since EDM was first put to practical use, its applications have been expanding by virtue of its ability to perform economically jobs which are extremely difficult or impossible to perform by conventional machining methods. Research and development are continuing to improve tool wear ratios, to increase metal-removal rates without the surface finish suffering, and to improve power supplies and machine tool components. Wider recognition of this process and its possibilities will lead to increasing the importance of EDM in the future.

REVIEW QUESTIONS

1. Describe briefly the principle of EDM.
2. Explain the operation and advantages of the pulse-type power supply.
3. List the characteristics of a good electrode material.
4. Explain why graphite is gaining wide acceptance as an electrode material.
5. What is the purpose of the servo mechanism and how does it operate?
6. Briefly explain what occurs during the discharge process.
7. What purpose does a dielectric serve?
8. Discuss the four methods of circulating dielectric fluids and state the advantages of each method.
9. What factors affect the metal-removal rate?
10. Explain the principle of reverse-polarity machining.
11. Define *overcut* and explain how it can be controlled.
12. What surface finishes are possible by EDM and how are they achieved?
13. List six main advantages of the EDM process.
14. List four limitations of EDM.

UNIT

93

Forming Processes

Forming large metal parts has become important as a result of the missile and aerospace program. Small parts may be easily formed to shape in conventional presses; however, these presses do not have the capacity or the force necessary to form materials with higher elastic strength or large parts made of common material. Formed parts as large as 10 ft (3.04 m) in diameter were required by the aerospace program. To produce them in conventional presses would have been economically impractical due to the cost and size of the press required.

High-energy-rate (explosive) forming has received considerable attention in recent years because of its ability to apply great amounts of pressure [as much as 100 000 psi (689 MPa) or higher] by controlled explosions to form metal to shape. In this type of metal forming, only the cavity part of the die is required, and the force created by the controlled explosion replaces the die punch. Although there are many variations of explosive forming, only the chemical explosive, electric spark discharge, and electromagnetic methods will be briefly discussed to give the reader an insight into the principles involved.

One of the most widely used forming processes is that of powder metallurgy. *Powder metallurgy* is the process of producing metal parts by:

1. Blending powdered metals and alloys
2. Compressing the powders in a die which is the shape of the part to be produced
3. Subjecting the shaped form to elevated temperatures (sintering) which causes the metal particles to "weld" together and form a solid part

The process of fabricating useful objects by forming powders is not new. The ancient Egyptians practiced a form of powder metallurgy, and since that time experimentation has continued to perfect the process. Since World War I, research and development of this process have led to considerable progress. Today powder metallurgy is used to produce cams, gears, self-oiling bearings, light filaments, levers, cemented-carbide tools, automotive filters, etc., which were previously produced by conventional machining methods (Fig. 93-A).

OBJECTIVES

After you have completed this unit you should be able to:

1. Describe the processes of explosive and electromagnetic forming
2. Explain how the powder metallurgy process can be used to produce parts without machining

BRICATORS

EXPLOSIVE FORMING

Chemical explosives have proved to be a compact and inexpensive source of energy. Only a small amount of explosive is required to create a large force to form metal. For example, 2 oz (57 g) of explosives in a 5 ft diameter (1.5 m) water tank can produce the same force as a hydraulic press rated at 1000 tons (907 t). If 5 lb (2268 g) of explosives were used, the resulting force would far exceed the capacity of any existing press.

THE PROCESS

1. The workpiece (blank) to be formed is clamped to the top of the die cavity by a clamping ring (Fig. 93-1).

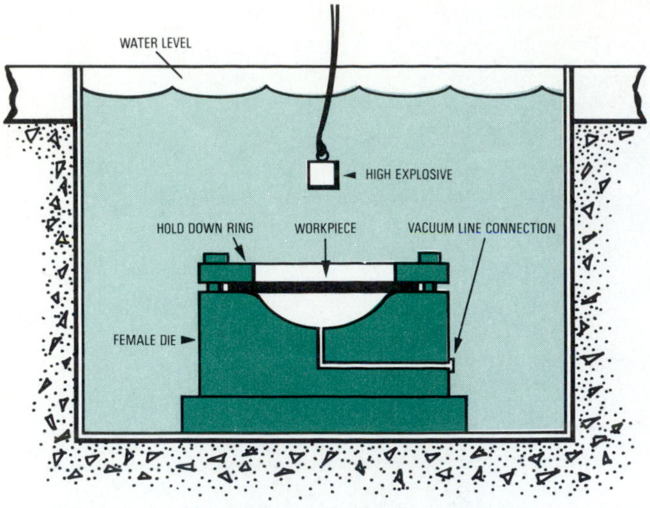
WATER LEVEL
HIGH EXPLOSIVE
HOLD DOWN RING
WORKPIECE
VACUUM LINE CONNECTION
FEMALE DIE

Figure 93-1 *Explosive forming setup.*

2. The explosive is mounted over the blank at a predetermined distance away from its surface.
3. The entire assembly is lowered into a water tank.
4. A vacuum is created in the die (the space between the workpiece and the die cavity) (Fig. 93-1).
5. The explosive charge is set off.
6. The shock wave which is created causes the metal to form to the shape of the die cavity.

The detonation of the explosive under water creates a shock wave that strikes the workpiece, forcing the metal into contact with the die. The pressures exerted on the work are high; however, the intensity and duration of the pressure can be controlled so that only slightly more energy than required is exerted. If too much pressure is exerted, it may cause the workpiece to rupture.

ADVANTAGES

Many advantages are offered by the use of explosives for forming large metal sections.

- Since only one-half the die is required, the cost of die manufacture is reduced.
- The cost of the equipment required is relatively low.
- Parts difficult or impossible to form by mechanical means can be formed.
- A better surface finish on the work is created than is possible by conventional die sets.
- Better forming accuracy is possible because there is little or no springback in the workpiece.
- Annealing operations required for deep forming by conventional means are eliminated.

LIMITATIONS AND DISADVANTAGES

Certain limitations or disadvantages must be considered when using explosives to form metals:

- Employees must be trained in the safe use of explosives.
- Insurance rates are generally higher because of the use of explosives.
- The forming must be done in a remote area, which increases transportation and handling costs.
- Elevated temperatures, which could assist in forming the workpiece, cannot be used because of the cooling effects of water.

CONCLUSION

The use of explosives has provided industry with a relatively inexpensive method of forming large metal parts. This process should be considered for parts difficult or impossible to form by conventional means or if costs are lower.

ELECTRIC SPARK DISCHARGE FORMING

Shock waves, comparable to those produced by explosives, created by an underwater electrical discharge of high voltage have been used successfully for metal forming. The amount of electrical energy required for forming workpieces depends on the following factors.

1. Diameter and depth of the die cavity
2. Distance from the spark to the surface of the water
3. Width of the spark gap
4. Thickness of the workpiece

Electric spark discharge (electrospark) forming cannot and was not intended to compete with conventional presses in the forming of simple parts. It is generally used for forming parts which would be difficult or impossible to manufacture by conventional methods. Operations such as forming, blanking, piercing, embossing, flaring, trimming, and forming tubular parts either by expansion or contraction are possible with electrospark forming.

THE PROCESS

1. The piece to be formed is fastened securely to the top of the die by means of a clamp ring (Fig. 93-2).
2. The die is then submerged in water.
3. A vacuum is created in the die cavity.
4. The electrodes are positioned at a predetermined distance above the workpiece (Fig. 93-2).

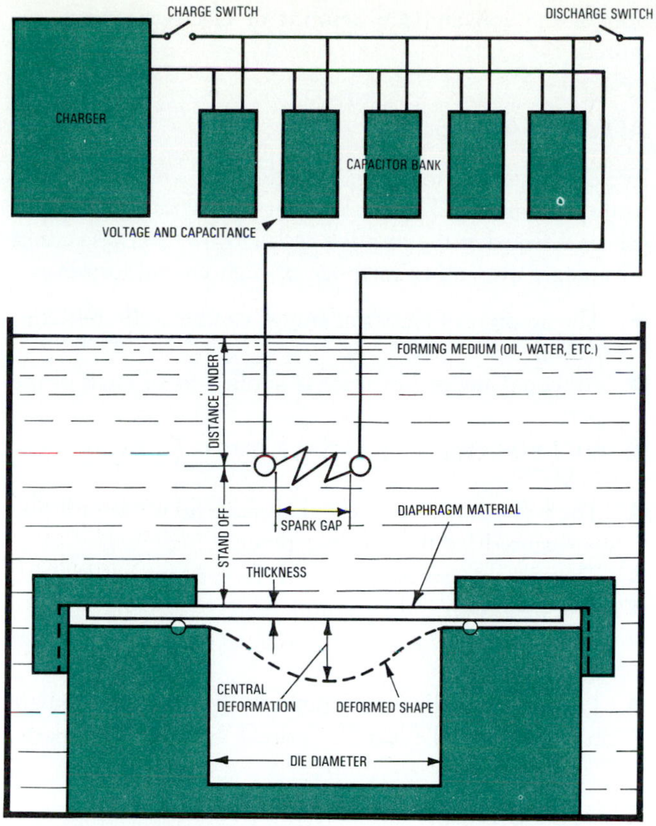
CHARGE SWITCH
DISCHARGE SWITCH
CHARGER
CAPACITOR BANK
VOLTAGE AND CAPACITANCE
FORMING MEDIUM (OIL, WATER, ETC.)
DISTANCE UNDER
STAND OFF
SPARK GAP
DIAPHRAGM MATERIAL
THICKNESS
CENTRAL
DEFORMATION
DEFORMED SHAPE
DIE DIAMETER

Figure 93-2 *Electrospark forming principle.*

5. The stored energy from the high-voltage capacitor bank is released between the submerged electrodes.
6. The spark discharge between the electrodes creates a shock wave, causing the workpiece to form to the shape of the die cavity.

The release of high energy between the submerged electrodes heats and vaporizes a thin channel of water, rapidly expanding it and creating a shock wave. When this shock wave reaches the workpiece, there is an imbalance of forces on the workpiece caused by low pressure in the die cavity as a result of the vacuum and high pressure of the shock wave. Because of this imbalance of forces, the work is forced in the direction of the low pressure and assumes the shape of the die cavity.

Often the path of the spark is controlled by connecting the space between the two electrodes with a thin wire. During discharge, the spark is confined to a specific path; however, the wire is vaporized and must be replaced for each part formed.

CONCLUSION

Most advantages and disadvantages listed for explosive forming apply also to electrospark forming. The differences between the two processes are:

1. The cost of the equipment required to initiate the discharge is considerably higher than with explosive forming.
2. Electrospark forming is a much safer process and does not have to be performed in a remote area.
3. Higher production rates are possible than with explosive forming.
4. The path of the discharge can be more accurately controlled in electrospark forming.
5. The amount of discharge is limited to the capacity of the electrical power bank.

ELECTROMAGNETIC FORMING

Electromagnetic forming is one of the newest methods of high-energy metal forming. The magnetic field of force, created by passing a high current through a coil around either the outside or inside of a workpiece, is used to form parts to the desired shape. The best results with this method of high-energy metal forming have been obtained when forming parts made of copper and aluminum, which are good electrical conductors. Electromagnetic forming is used primarily for swaging or expanding tubular shapes. It may also be used for embossing, punching, forming, and shrinking operations (Fig. 93-3).

THE PROCESS

In electromagnetic forming, the electrical energy from the capacitor bank (Fig. 93-4) is passed through a coil rather than between two electrodes as is the case in electrospark forming. A large magnetic field builds up around the coil, inducing a voltage in the workpiece. The resultant high current builds up its own magnetic field. These two magnetic fields of force are opposite in direction and repel each other, causing the deformation of the workpiece. If the coil is placed on the inside of a tubular workpiece, the magnetic force will cause the workpiece to bulge and assume the shape of the die cavity. Workpieces may be shrunk onto formed mandrels by placing the coil around the outside of

Figure 93-3 *Examples of electromagnetic forming.*

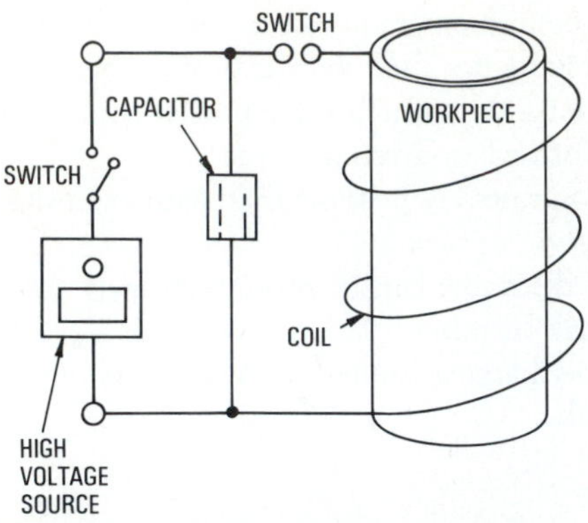
SWITCH
CAPACITOR
WORKPIECE
SWITCH
COIL
HIGH VOLTAGE SOURCE
Figure 93-4 *Electromagnetic forming principle.*

the piece to be formed (Fig. 93-4). Field-shaping coils may be used to concentrate the magnetic field of force when irregular shapes must be formed.

Experiments have been conducted using variations of electrical energy. The best results for electromagnetic forming have been obtained with a capacitor bank of low voltage and high capacitance. The amount of work deformation increases in relation to the amount of stored electrical energy released. Deformation of the workpiece decreases when the distance between the coil and the work is increased. Parts made of poor or nonconductive materials can be formed by placing a conductive foil around or inside the parts to be formed. The foil acts as would a good electrically conductive material and causes the workpiece to assume the shape of the die cavity or formed mandrel.

ELECTROMAGNETIC FORMING FACTORS

Although electromagnetic forming is a relatively new method of metal forming, it has proved invaluable for many operations which were difficult or impossible to perform by other methods. The following factors must be carefully considered since they affect the efficiency of the electromagnetic forming process.

1. The amount of electrical energy employed must be sufficient to form the part completely.
2. The coil should be designed so that it is stronger than the part to be formed; otherwise, the electromagnetic force may deform the coil rather than the workpiece.
3. The size of the wire and the number of turns in the coil are very important because they determine the strength of the electromagnetic field which can be created. The coil must be placed at a specific distance from the part to be formed.
4. The electrical conductivity of the work material is an important factor in this process.
5. The thickness of the work material determines the location of the coil and the amount of electrical energy required.

ADVANTAGES

In electromagnetic forming, the size of the work that can be formed is controlled by the amount of electrical power that can be directed to the forming coil. However, it has certain advantages over other methods of high-energy forming:

- The amount of electrical energy can be accurately controlled.
- An equal amount of force is applied to all areas of the part.
- No forces are set up unless a part is in the magnetic field.
- The work may be preheated because no water or liquid is required for the forming process.
- There are no moving parts in the forming equipment.
- The operation can be automated.
- Forming can be performed in a vacuum or in an inert atmosphere.
- It provides a low-cost method of assembling where bands of metal must be formed around other parts.

CONCLUSION

Electromagnetic forming, although a relatively new process, has been used for a great variety of industrial applications ranging from the aerospace program to the assembly of electronic components. As electromagnetic pulse equipment of greater capacities becomes available, this method of metal forming should assume more importance in manufacturing processes.

POWDER METALLURGY

Powder metallurgy is a radical departure from the conventional machining process (Fig. 93-5). With conventional machining methods, a piece of steel or bar stock larger than the finished part is selected and the required form is machined. The material cut away in the machining process is in the form of steel chips, which are considered *scrap loss* (Fig. 93-6). In powder metallurgy, the correct types of powders are blended together and formed into the required shape in a die. Since machining is not required and only the required amount of powder is used for each part, scrap loss is minute.

The unique features of the powder metallurgy process are:

1. Handling of molten metals is not involved.
2. Machining or finishing operations are rarely required.

Figure 93-5 *Examples of parts produced by powder metallurgy. (Courtesy The Powder Metallurgy Parts Manufacturing Association.)*

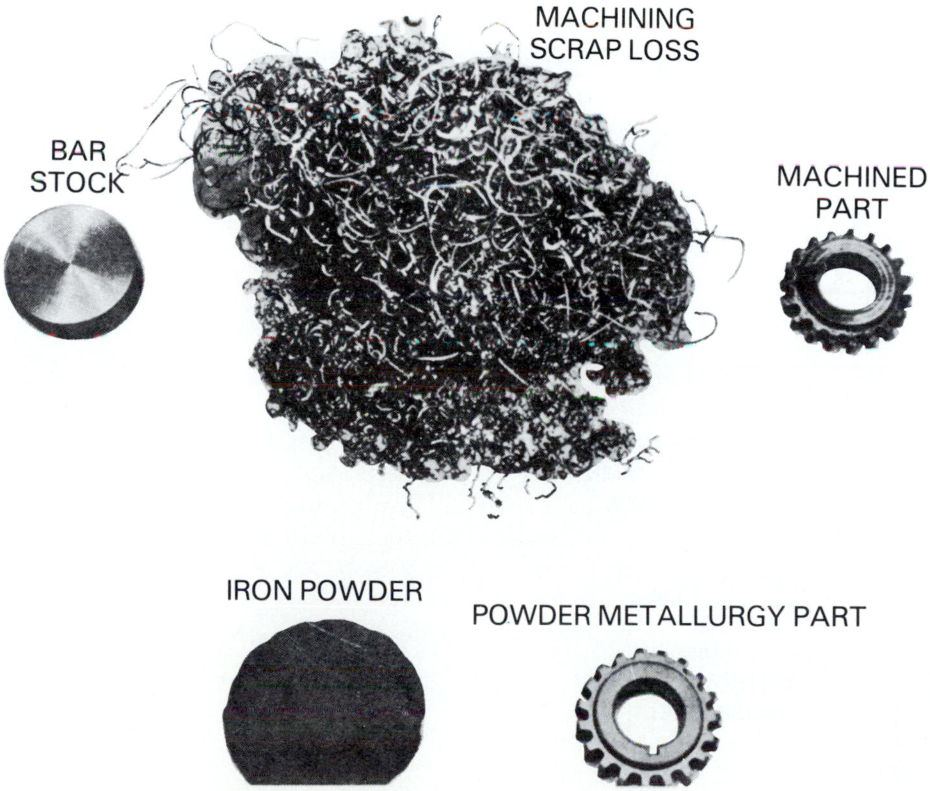

MACHINING
SCRAP LOSS
BAR
STOCK
MACHINED
PART
IRON POWDER
POWDER METALLURGY PART

Figure 93-6 *Comparison of conventional machining and powder metallurgy processes.*

3. It permits rapid mass production of steel and other high-melting metal shapes.
4. Beryllium, molybdenum, and tungsten parts can be produced easily, which would be impractical or uneconomical by other methods.
5. It permits combinations of metal and nonmetals as well as alloys which are not possible by any other method.
6. It permits accurate control over the density or the porosity of the finished part.

THE POWDER METALLURGY PROCESS

A part made by the powder metallurgy process begins with metal powders and goes through four main stages before it becomes a finished product (Fig. 93-7). These include securing raw powders, blending, compacting, and sintering.

Raw Powders

Almost any type of metal can be produced in powder form; however, only a few have the desired characteristics and properties necessary for economical production. Iron- and copper-base powders are the two main types which lend themselves well to the powder metallurgy process. Aluminum, nickel, silver, and tungsten powders are not widely used; however, they have some important applications.

Some of the more common methods of manufacturing powders are:

1. *Atomization* or metal spraying is a means of producing powders from low-melting temperature metals, such as aluminum, lead, tin, and zinc. This process produces powder particles irregular in shape.
2. *Electrolytic deposition* is the common method used to produce copper, iron, silver, and tantalum powders.

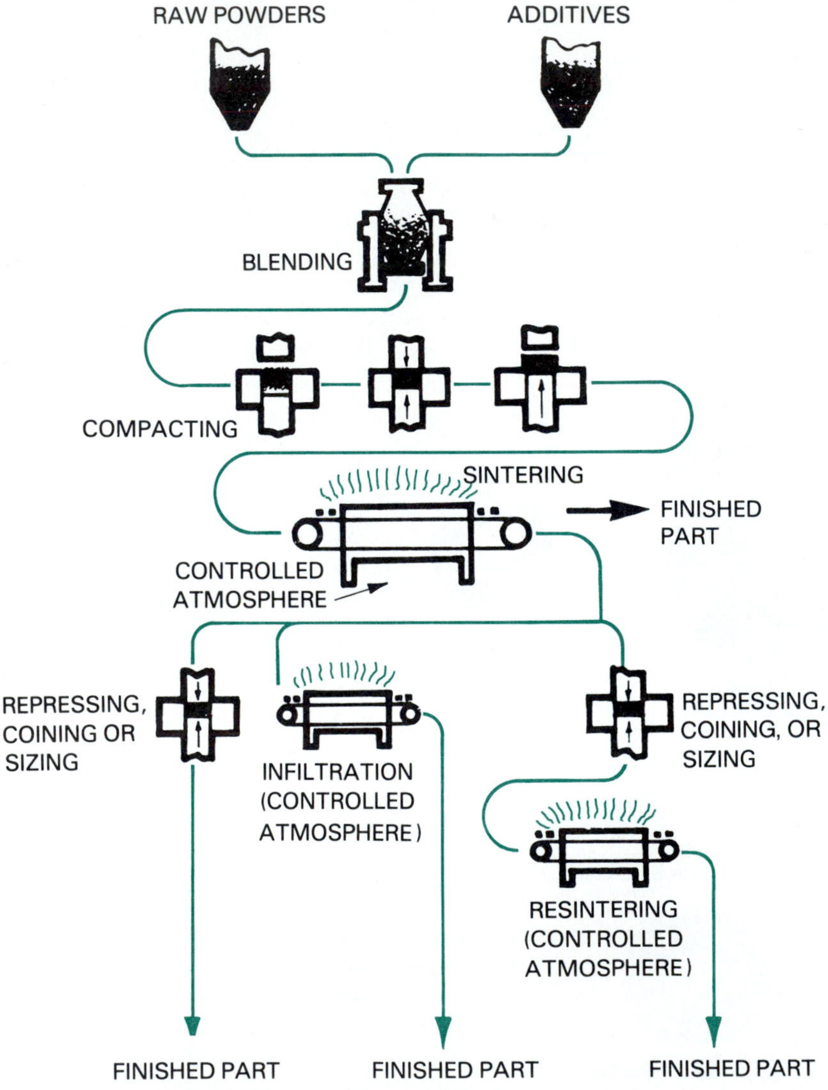
RAW POWDERS
ADDITIVES
BLENDING
COMPACTING
SINTERING
FINISHED PART
CONTROLLED ATMOSPHERE
REPRESSING, COINING OR SIZING
INFILTRATION (CONTROLLED ATMOSPHERE)
REPRESSING, COINING, OR SIZING
RESINTERING (CONTROLLED ATMOSPHERE)
FINISHED PART
FINISHED PART
FINISHED PART

Figure 93-7 *The powder metallurgy process.*

3. *Granulation* is used to convert a few metals into powder. The metal is stirred rapidly while it is being cooled. This process depends on the formation of oxides on the metal particles during stirring.
4. *Machining* produces coarse powders and is primarily used for producing magnesium powders.
5. *Milling* involves using various types of crushers, rollers, or presses to break the metal into powders.
6. *Reduction* is used to transform metal oxides to powder form by contact with gas at temperatures below their melting point. Cobalt, iron, molybdenum, nickel, and tungsten powders are produced by the reduction process.
7. *Shotting* is the operation of passing molten metal through an orifice or sieve and having the particles drop into water. Most metals can be powdered by this method; however, the particles are generally large.

Blending

The powder for a specific product must be carefully selected to ensure economical production and so that the finished part will have the required characteristics. The following information regarding the powder should be considered during its selection.

1. Shape of the powder particle
2. Particle size
3. Flowability (its ability to flow readily)
4. Compressibility (its ability to maintain a form)
5. Sintering ability (its ability to fuse or bond)

Once the correct powders have been selected, the mass of each is carefully determined in the proportion required for the finished component, and a die lubricant, such as powdered graphite, zinc stearate, or stearic acid, is added. The purpose of this lubricant is to assist the flow of powder in the die, prevent scoring of the die walls, and permit easy ejection of the compressed part. This composition is then carefully mixed or blended (Fig. 93-8) to ensure homogeneous grain distribution in the finished product.

Compacting (Briquetting)

The powder blend is then fed into a precision die which is the shape and size of the finished product. The die generally consists of a die shell, an upper punch, and a lower punch (Fig. 93-8). These dies are usually mounted on hydraulic or mechanical presses where pressures from about 3000 psi (20 MPa) to as much as 200 000 psi (1379 MPa) are used to compress the powder. Soft powder particles can be pressed or keyed together readily and therefore do not require as high a pressure as the harder particles. The density and hardness of the finished product increases with the amount of pressure used to compact or briquette the powder. However, for every case there is an optimum pressure above which very little improvement in qualities and properties can be obtained.

Sintering

After compacting, the "green part" must be heated sufficiently to effect permanent cohesion of the metal particles into a solid. This operation of heating is known as *sintering*. The parts are passed through controlled protective-atmosphere furnaces, which are maintained at a temperature approximately one-third below the melting point of the principal powder. The carefully controlled atmosphere and temperature during the sintering operation permits particle bonding and recrystallization across the particle interfaces. Most sintering is done in a hydrogen atmosphere, producing parts with no scaling or discoloration.

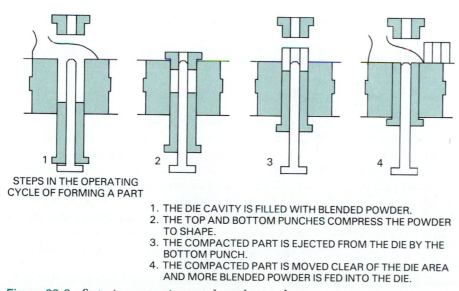

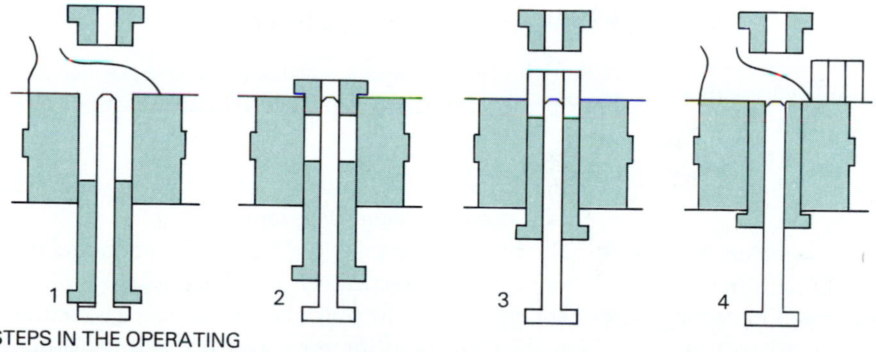
1
2
3
4
STEPS IN THE OPERATING
CYCLE OF FORMING A PART
1. THE DIE CAVITY IS FILLED WITH BLENDED POWDER.
2. THE TOP AND BOTTOM PUNCHES COMPRESS THE POWDER TO SHAPE.
3. THE COMPACTED PART IS EJECTED FROM THE DIE BY THE BOTTOM PUNCH.
4. THE COMPACTED PART IS MOVED CLEAR OF THE DIE AREA AND MORE BLENDED POWDER IS FED INTO THE DIE.

Figure 93-8 *Steps in compacting metal powders to shape.*

The temperature within the furnaces varies from 1600 to 2700°F (871 to 1482°C), depending on the type of metal powder being sintered. The time required to sinter a part varies, depending on its shape and size; normally, however, it is from 15 to 45 min.

The purpose of the sintering operation is to bond the powder particles together to form a strong, homogeneous part having the desired physical characteristics.

FINISHING OPERATIONS

After sintering, most parts are ready for service. However, some parts requiring very close tolerances or other qualities may require some additional operations, such as sizing and repressing, impregnation, infiltration, plating, heat treating, and machining.

Sizing and Repressing

Parts requiring close tolerances or increased density must be *sized* or *repressed*. This involves putting the part into a die, which is similar to the one used for compacting, and repressing it. This operation improves the surface finish and dimensional accuracy, increases the density, and gives added strength to the part.

Impregnation

The process of filling the pores of a sintered part with a lubricant or nonmetallic material is called *impregnation*. Oilite bearings are a good example of a part impregnated with oil to overcome the necessity for constant lubrication and maintenance. Parts may be impregnated by a vacuum process or by soaking the parts in oil for several hours.

Infiltration

Infiltration is the process of filling the pores of a part with a metal or alloy having a lower melting point than the sintered piece. The purpose of this operation is to increase the density, strength, hardness, impact resistance, and ductility of the manufactured part.

Slugs of the material to be infiltrated are placed on the compacted parts and passed through the sintering furnace. Because of its lower melting point, the infiltrated material melts and penetrates the pores of the compacted part by capillary action.

Plating, Heat Treating, Machining

Depending on the type of material, its application, and requirements, some powder metallurgy parts may be plated, heat treated, machined, brazed, or welded.

ADVANTAGES OF POWDER METALLURGY

The use of powder metallurgy is increasing rapidly, with many parts being made better and more economically than with previous manufacturing methods. Some of the advantages of the powder metallurgy process are:

- Close dimensional tolerances of ±0.001 in. (0.02 mm) and smooth finishes can be obtained without costly secondary operations.
- Complex or unusually shaped parts, which would be impractical to obtain by any other method, can be produced (Fig. 93-9).
- It is capable of producing porous bearings and cemented-carbide tools.
- The pores of a part can be infiltrated with other metals.
- Surfaces of great wear resistance can be produced.
- The porosity of the product can be accurately controlled.
- Products made of extremely pure metals can be produced.
- There is little waste in this process.
- The operation can be automated and employ unskilled labor, keeping costs low.
- The physical properties of the product can be minutely controlled.
- Duplicate parts are accurate and unvarying.
- Powders made from difficult-to-machine alloys can be formed into parts which would be difficult to produce by machining processes.

DISADVANTAGES OF POWDER METALLURGY

Although there are many advantages of the powder metallurgy process, there are also certain disadvantages or limitations:

- Metal powders are quite expensive and must be carefully stored to avoid deterioration.
- The cost of the presses and furnaces is high, and therefore it is not practical for short-run jobs.
- The size of the part which can be produced is controlled by the capacity of the press available and the compression ratio of the powders.
- Parts requiring sharp corners or abrupt changes in thickness are very difficult to produce.
- Parts requiring internal threads, grooves, or undercuts are impossible to produce.
- Some of the low-melting-temperature powders may cause difficulties during the sintering operation.

Figure 93-9 *Examples of products available through powder metallurgy. (Courtesy The Powder Metallurgy Parts Manufacturing Association.*

Powder Metallurgy Part Design Rules

1. The shape of the part must permit it to be ejected from the die. The following cannot be molded and must be machined later: undercuts, grooves, internal threads, diamond knurls, holes at right angles, and reverse tapers.
2. Avoid part designs that would require powder to flow into thin walls, narrow splines, or sharp corners. Parts with these characteristics can be produced only with extreme difficulty and should be avoided whenever possible.
3. The die should be designed to provide the maximum strength to all the components.
4. The length of the part should not be longer than 2½ times the diameter.
5. The part should be designed with as few steps or diameters as possible.

REVIEW QUESTIONS

HIGH-ENERGY FORMING

1. Explain why high-energy metal forming has become increasingly important in recent years.
2. Name three methods of explosive forming.
3. Explain briefly what occurs when a chemical explosive is detonated under water.
4. Name four advantages of explosive forming.
5. What are the disadvantages or limitations of using explosives for forming metals?

ELECTRIC SPARK DISCHARGE FORMING

6. How does electric spark discharge forming differ from explosive forming?
7. Describe briefly the process of electrospark forming.
8. Explain what occurs when high energy is released between two submerged electrodes.
9. Why is it necessary to create a vacuum in the die cavity for explosive and electrospark forming?

ELECTROMAGNETIC FORMING

10. On what principle does electromagnetic forming operate.
11. List four important factors which should be considered for electromagnetic forming.
12. Name five advantages of electromagnetic forming.

13. List the four steps in the powder metallurgy process.
14. Name six products which are produced by powder metallurgy.
15. Compare conventional machining methods with powder metallurgy, and list four unique features of the process.
16. Name and briefly describe four methods of producing metal powders.
17. Describe the process of blending and state its importance to the finished product.
18. Describe the compacting operation with regard to the type of die used, pressures required, and the qualities obtained.
19. Fully describe and state the purposes of sintering.
20. Define and state the purpose of:
   a. Sizing and repressing
   b. Impregnation
   c. Infiltration
21. List six important advantages of powder metallurgy.
22. State four disadvantages or limitations of powder metallurgy.

UNIT 94

The Laser

Since the first operational laser was built in 1960, laser technology has advanced at a rapid rate and is one of the most versatile developments at our disposal. The word laser is an abbreviation for *light amplification by stimulated emission of radiation.* Lasers can be used to cut holes in any material, including diamond, or to perform microscopic surgery on the eye and many other parts of the human body. Lasers may also be used to guide missiles and satellites, to trigger thermonuclear fusion, and to measure distances from milicrons (millionth of a meter) to hundreds of thousands of miles with extreme accuracy. For instance, scientists using lasers can determine the distance from the earth to the moon at any given time, to within an accuracy of about 1 ft.

The laser beam is a very narrow, intense beam of monocolored light which can be controlled over a wide range of temperatures, ranging from one that would feel slightly warm to one which is several times hotter than the surface of the sun at the point of focus. Some lasers can instantaneously produce temperatures of 75 000°F (41 650°C).

OBJECTIVES

After completing this unit you should be able to:

1. Identify and briefly describe the operation of four types of lasers
2. State the applications and advantages of industrial lasers
3. Describe the purpose and operation of the interferometer

TYPES OF LASERS

There are several types of lasers and each has a particular use, depending on the job required. Most lasers may be classified as solid, gas, liquid, and semiconductor. Because solid, gas, and liquid lasers are based on the same principle, only these will be discussed. The main parts of a laser (Fig. 94-1) are basically the same for all types of lasers. These include a *power supply,* a *lasing medium* (suitably contained), and a pair of *precisely aligned mirrors.* One of the mirrors has a totally reflective surface, while the other is only partially reflective (about 96 percent).

SOLID LASERS

The power source charge of *solid lasers* fires the large flash tube in a manner similar to a photographic electronic flash. The flash tube emits a burst of light sending or "pumping" energy into the lasing medium, which could be a ruby rod. This excites the chromium atoms in the ruby rod to a high energy level. As these atoms fall back to their initial state, they emit heat and bundles of light energy called *photons.* The photons produced will strike other atoms causing other photons of identical wavelengths to be produced (Fig. 94-2). These photons are reflected back and forth (oscillate) along the rod by the mirrored, square surface on each end of the rod. Consequently, the number of photons increase and produce more energy until some of them pass through the partially reflective mirror producing a laser beam, which then passes through a lens onto the point of focus on the work (Fig. 94-3).

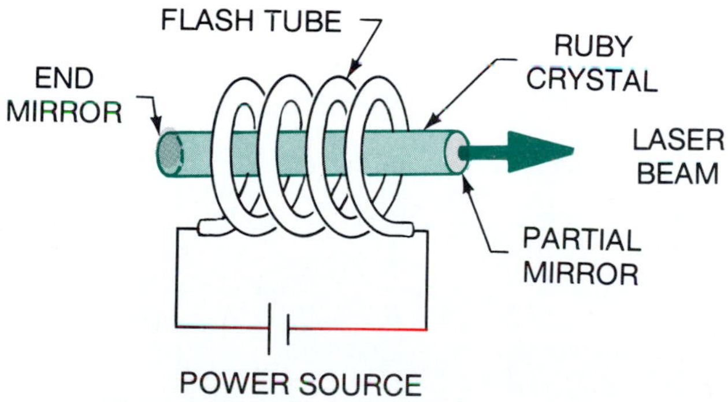
FLASH TUBE
END MIRROR
RUBY CRYSTAL
LASER BEAM
PARTIAL MIRROR
POWER SOURCE

Figure 94-1 *The solid ruby laser is used in surgery, drilling, measurement, and spot welding.*

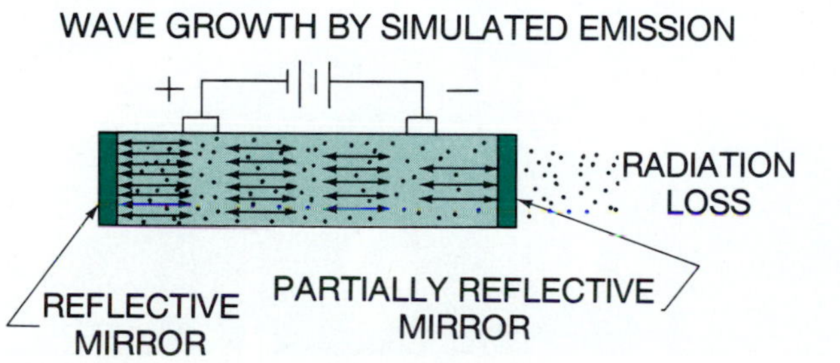
WAVE GROWTH BY SIMULATED EMISSION
+
−
RADIATION LOSS
REFLECTIVE MIRROR
PARTIALLY REFLECTIVE MIRROR

Figure 94-2 *Light waves traveling along the axis of the rod grow by stimulated emission of photons.*

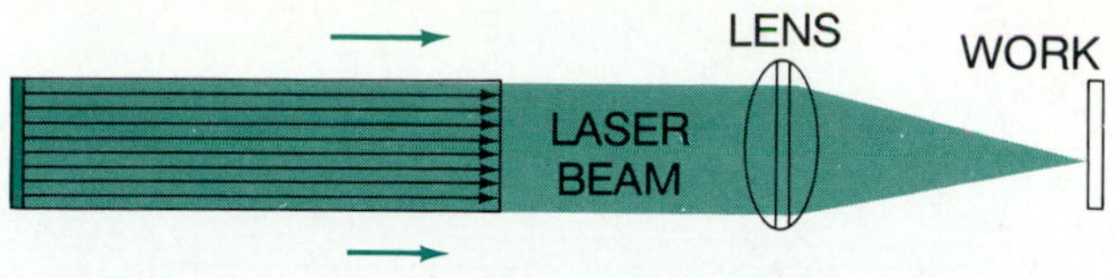
LENS
WORK
LASER BEAM

Figure 94-3 *Parallel light waves pass through the partially reflective mirror and the lens.*

Solid lasers have proved very useful in the fields of surgery, atomic fusion, drilling diamond dies, measurement, and spot welding.

GAS LASERS

Probably the most common *gas laser* is the *helium–neon* (HeNe) *laser.* The gas, mixture of 10 parts helium to 1 part of neon, is contained in a sealed glass unit with angled ends to reduce reflection loss. As the power-pulsed circuit closes, an electron discharge takes place within the tube, exciting the helium atoms to higher energy levels. These atoms collide with the neon atoms bringing them up to the same level. When the neon atoms drop back to a lower level, they emit heat and red photons of laser light. The action continues to multiply and the atoms oscillate back and forth in the tube until some of the light escapes through the partial mirror in the form of a red laser beam (Fig. 94-4). HeNe lasers are small, relatively inexpensive, and relatively safe. Consequently they are used in laboratories and schools for laser experiments.

Another form of gas laser is the *carbon dioxide* (CO_2) *laser* which uses CO_2 in the sealed tube. This device is capable of much higher output and is more efficient than the HeNe laser. If the CO_2 gas is pumped through the laser cavity or tube and cooled in a heat exchanger, it can operate continuously at very high power. Large units of this type are

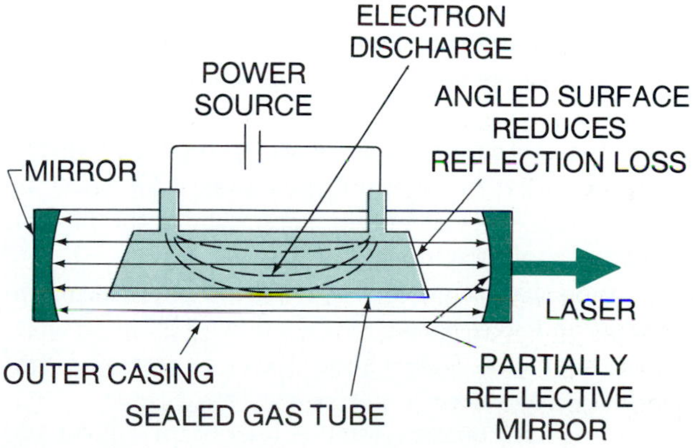
ELECTRON DISCHARGE
POWER SOURCE
ANGLED SURFACE REDUCES REFLECTION LOSS
MIRROR
LASER
OUTER CASING
SEALED GAS TUBE
PARTIALLY REFLECTIVE MIRROR

Figure 94-4 *A sealed-tube gas laser.*

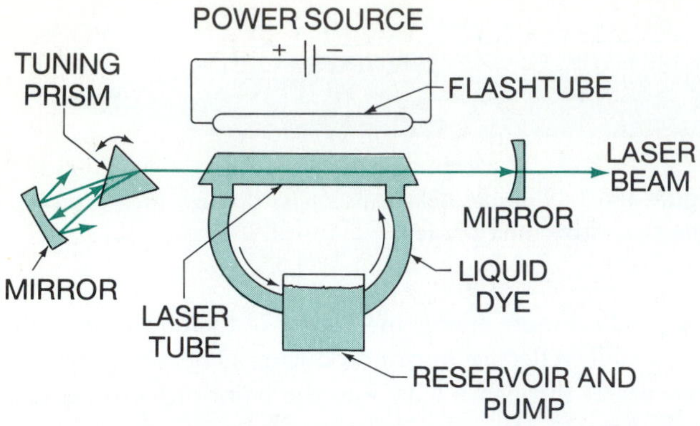
POWER SOURCE
TUNING PRISM
FLASHTUBE
LASER BEAM
MIRROR
LIQUID DYE
MIRROR
LASER TUBE
RESERVOIR AND PUMP

Figure 94-5 *Liquid lasers are finding wide use in chemical laboratories.*

capable of vaporizing *any* material. CO_2 lasers are used for cutting organic materials such as rubber and leather, thus eliminating the need for punch presses. They are also used for hole production, perforating, welding, and oxygen-assisted metal cutting.

LIQUID LASERS

Liquid lasers employ an organic dye in a solvent as the lasing medium (Fig. 94-5). The liquid is circulated and the flash tube pumps the dye molecules to high energy levels and photons are produced. The process continues as the adjustable external mirrors create a feedback and the atoms oscillate in the tube until some of the light passes through the partial mirror as a laser beam. Liquid lasers are particularly suited to chemical analysis because of the tuning prism which can be rotated and which in turn produces a choice of several different colors and corresponding wavelengths.

INDUSTRIAL APPLICATIONS OF LASERS

Lasers may be used for:

1. Cutting holes in any known substance. Ruby lasers are used to produce precise holes in diamonds for wire drawing dies.
2. Producing holes in paper, plastic, rubber, etc. where a poor finish is left by conventional means. CO_2 lasers are used to punch holes in plastic pipes, baby bottle nipples, and aerosol can spray tips.
3. Cutting shallow grooves and engraving measuring instruments and steel parts.
4. Cutting metal. A laser beam leaves no tears or jagged edges like a saw does.
5. Cutting hard quartz crystal. A laser beam will cut 100 times faster than a diamond wheel.
6. Removing very narrow sections of unwanted metal and printed circuits.
7. Welding which would be difficult by any other means, such as welding the gears to the synchronizing mechanism in an automobile transmission.
8. Machining microscopic parts.
9. Heat treating, such as hardening the surfaces of gears or cylinders. The intense heat required for this is generated almost instantaneously. The beam may be aimed to harden specific areas and no further polishing is needed.
10. Measuring with a very high degree of accuracy: within 0.000006 in. per foot of length (Fig. 94-6). (See Unit 17 for the principle of the laser interferometer.)

ADVANTAGES OF LASERS

- Capable of producing a very narrow heat-affected zone, which is suitable for welding next to a glass–metal seal.
- Can be used in welding delicate electronic components that could not withstand resistance welding.
- Can be used to weld in a vacuum.
- Capable of producing holes in any known material.
- Capable of producing extremely small and accurate holes.
- There is no need to clamp a workpiece in position since there is no torque involved, as in conventional drilling.
- Ideal for tack-welding operations.

CINCINNATI
MILACRON
WARNING

Figure 94-6 *A laser being used to check the alignment of machine parts. (Courtesy Cincinnati Milacron Inc.)*

- Very useful in welding parts, as no other material or equipment is needed.
- Can weld dissimilar metals easily.
- Particularly suited for drilling paper, plastics, rubber, etc., where a poor finish is left by conventional means.
- Can be used for localized heat treating on finished parts.
- Useful for cutting shallow grooves and engraving letters and numbers in steel and measuring instruments.

REVIEW QUESTIONS

TYPES OF LASERS

1. Name four types of lasers.
2. Name the main parts of any laser.
3. Describe the operation of the solid laser.
4. Where are HeNe gas lasers used? Why?
5. State four uses for a CO_2 laser.
6. For what purpose are liquid lasers generally used?

INDUSTRIAL APPLICATIONS

7. State three major industrial laser applications.

ADVANTAGES OF LASERS

8. List six of the most important advantages of lasers.

Appendix of Tables

TABLE 1: DECIMAL INCH, FRACTIONAL INCH, AND MILLIMETER EQUIVALENTS

Decimal inch	Fractional inch	Millimeter	Decimal inch	Fractional inch	Millimeter
.015625	1/64	0.397	.515625	33/64	13.097
.03125	1/32	0.794	.53125	17/32	13.494
.046875	3/64	1.191	.546875	35/64	13.891
.0625	1/16	1.588	.5625	9/16	14.288
.078125	5/64	1.984	.578125	37/64	14.684
.09375	3/32	2.381	.59375	19/32	15.081
.109375	7/64	2.778	.609375	39/64	15.478
.125	1/8	3.175	.625	5/8	15.875
.140625	9/64	3.572	.640625	41/64	16.272
.15625	5/32	3.969	.65625	21/32	16.669
.171875	11/64	4.366	.671875	43/64	17.066
.1875	3/16	4.762	.6875	11/16	17.462
.203125	13/64	5.159	.703125	45/64	17.859
.21875	7/32	5.556	.71875	23/32	18.256
.234375	15/64	5.953	.734375	47/64	18.653
.25	1/4	6.350	.75	3/4	19.05
.265625	17/64	6.747	.765625	49/64	19.447
.28125	9/32	7.144	.78125	25/32	19.844
.296875	19/64	7.541	.796875	51/64	20.241
.3125	5/16	7.938	.8125	13/16	20.638
.328125	21/64	8.334	.828125	53/64	21.034
.34375	11/32	8.731	.84375	27/32	21.431
.359375	23/64	9.128	.859375	55/64	21.828
.375	3/8	9.525	.875	7/8	22.225
.390625	25/64	9.922	.890625	57/64	22.622
.40625	13/32	10.319	.90625	29/32	23.019
.421875	27/64	10.716	.921875	59/64	23.416
.4375	7/16	11.112	.9375	15/16	23.812
.453125	29/64	11.509	.953125	61/64	24.209
.46875	15/32	11.906	.96875	31/32	24.606
.484375	31/64	12.303	.984375	63/64	25.003
.5	1/2	12.700	1.	1	25.400

TABLE 2:

CONVERSION OF INCHES TO MILLIMETERS						CONVERSION OF MILLIMETERS TO INCHES					
Inches	Millimeters	Inches	Millimeters	Inches	Millimeters	Millimeters	Inches	Millimeters	Inches	Millimeters	Inches
.001	0.025	.290	7.37	.660	16.76	0.01	.0004	0.35	.0138	0.68	.0268
.002	0.051	.300	7.62	.670	17.02	0.02	.0008	0.36	.0142	0.69	.0272
.003	0.076	.310	7.87	.680	17.27	0.03	.0012	0.37	.0146	0.70	.0276
.004	0.102	.320	8.13	.690	17.53	0.04	.0016	0.38	.0150	0.71	.0280
.005	0.127	.330	8.38	.700	17.78	0.05	.0020	0.39	.0154	0.72	.0283
.006	0.152	.340	8.64	.710	18.03	0.06	.0024	0.40	.0157	0.73	.0287
.007	0.178	.350	8.89	.720	18.29	0.07	.0028	0.41	.0161	0.74	.0291
.008	0.203	.360	9.14	.730	18.54	0.08	.0031	0.42	.0165	0.75	.0295
.009	0.229	.370	9.40	.740	18.80	0.09	.0035	0.43	.0169	0.76	.0299
.010	0.254	.380	9.65	.750	19.05	0.10	.0039	0.44	.0173	0.77	.0303
.020	0.508	.390	9.91	.760	19.30	0.11	.0043	0.45	.0177	0.78	.0307
.030	0.762	.400	10.16	.770	19.56	0.12	.0047	0.46	.0181	0.79	.0311
.040	1.016	.410	10.41	.780	19.81	0.13	.0051	0.47	.0185	0.80	.0315
.050	1.270	.420	10.67	.790	20.07	0.14	.0055	0.48	.0189	0.81	.0319
.060	1.524	.430	10.92	.800	20.32	0.15	.0059	0.49	.0193	0.82	.0323
.070	1.778	.440	11.18	.810	20.57	0.16	.0063	0.50	.0197	0.83	.0327
.080	2.032	.450	11.43	.820	20.83	0.17	.0067	0.51	.0201	0.84	.0331
.090	2.286	.460	11.68	.830	21.08	0.18	.0071	0.52	.0205	0.85	.0335
.100	2.540	.470	11.94	.840	21.34	0.19	.0075	0.53	.0209	0.86	.0339
.110	2.794	.480	12.19	.850	21.59	0.20	.0079	0.54	.0213	0.87	.0343
.120	3.048	.490	12.45	.860	21.84	0.21	.0083	0.55	.0217	0.88	.0346
.130	3.302	.500	12.70	.870	22.10	0.22	.0087	0.56	.0220	0.89	.0350
.140	3.56	.510	12.95	.880	22.35	0.23	.0091	0.57	.0224	0.90	.0354
.150	3.81	.520	13.21	.890	22.61	0.24	.0094	0.58	.0228	0.91	.0358
.160	4.06	.530	13.46	.900	22.86	0.25	.0098	0.59	.0232	0.92	.0362
.170	4.32	.540	13.72	.910	23.11	0.26	.0102	0.60	.0236	0.93	.0366
.180	4.57	.550	13.97	.920	23.37	0.27	.0106	0.61	.0240	0.94	.0370
.190	4.83	.560	14.22	.930	23.62	0.28	.0110	0.62	.0244	0.95	.0374
.200	5.08	.570	14.48	.940	23.88	0.29	.0114	0.63	.0248	0.96	.0378
.210	5.33	.580	14.73	.950	24.13	0.30	.0118	0.64	.0252	0.97	.0382
.220	5.59	.590	14.99	.960	24.38	0.31	.0122	0.65	.0256	0.98	.0386
.230	5.84	.600	15.24	.970	24.64	0.32	.0126	0.66	.0260	0.99	.0390
.240	6.10	.610	15.49	.980	24.89	0.33	.0130	0.67	.0264	1.00	.0394
.250	6.35	.620	15.75	.990	25.15	0.34	.0134	......	......	......	......
.260	6.60	.630	16.00	1.000	25.40						
.270	6.86	.640	16.26	......	......						
.280	7.11	.650	16.51	......	......						

Courtesy Automatic Electric Company

TABLE 3: LETTER DRILL SIZES

Letter	mm	in.	Letter	mm	in.
A	5.9	.234	N	7.7	.302
B	6.0	.238	O	8.0	.316
C	6.1	.242	P	8.2	.323
D	6.2	.246	Q	8.4	.332
E	6.4	.250	R	8.6	.339
F	6.5	.257	S	8.8	.348
G	6.6	.261	T	9.1	.358
H	6.7	.266	U	9.3	.368
I	6.9	.272	V	9.5	.377
J	7.0	.277	W	9.8	.386
K	7.1	.281	X	10.1	.397
L	7.4	.290	Y	10.3	.404
M	7.5	.295	Z	10.5	.413

TABLE 4: DRILL GAUGE SIZES

No.	mm	inch	No.	mm	inch	No.	mm	inch
1	5.80	.2280	34	2.80	.1110	66	0.84	.0330
2	5.60	.2210	35	2.80	.1100	67	0.81	.0320
3	5.40	.2130	36	2.70	.1065	68	0.79	.0310
4	5.30	.2090	37	2.65	.1040	69	0.74	.0292
5	5.20	.2055	38	2.60	.1015	70	0.71	.0280
6	5.20	.2040	39	2.55	.0995	71	0.66	.0260
7	5.10	.2010	40	2.50	.0980	72	0.64	.0250
8	5.10	.1990	41	2.45	.0960	83	0.61	.0240
9	5.00	.1960	42	2.40	.0935	74	0.57	.0225
10	4.90	.1935	43	2.25	.0890	75	0.53	.0210
11	4.90	.1910	44	2.20	.0860	76	0.51	.0200
12	4.80	.1890	45	2.10	.0820	77	0.46	.0180
13	4.70	.1850	46	2.05	.0810	78	0.41	.0160
14	4.60	.1820	47	2.00	.0785	79	0.37	.0145
15	4.60	.1800	48	1.95	.0760	80	0.34	.0135
16	4.50	.1770	49	1.85	.0730	81	0.33	.0130
17	4.40	.1730	50	1.80	.0700	82	0.32	.0125
18	4.30	.1695	51	1.70	.0670	83	0.31	.0120
19	4.20	.1660	52	1.60	.0635	84	0.29	.0115
20	4.10	.1610	53	1.50	.0595	85	0.28	.0110
21	4.00	.1590	54	1.40	.0550	86	0.27	.0105
22	4.00	.1570	55	1.30	.0520	87	0.25	.0100
23	3.90	.1540	56	1.20	.0465	88	0.24	.0095
24	3.90	.1520	57	1.10	.0430	89	0.23	.0091
25	3.80	.1495	58	1.05	.0420	90	0.22	.0087
26	3.70	.1470	59	1.05	.0410	91	0.21	.0083
27	3.70	.1440	60	1.00	.0400	92	0.20	.0079
28	3.60	.1405	61	0.99	.0390	93	0.19	.0075
29	3.50	.1360	62	0.97	.0380	94	0.18	.0071
30	3.30	.1285	63	0.94	.0370	95	0.17	.0067
31	3.00	.1200	64	0.92	.0360	96	0.16	.0063
32	2.95	.1160	65	0.89	.0350	97	0.15	.0059
33	2.85	.1130						

TABLE 5: TAP DRILL SIZES

Nominal Diameter mm	Thread Pitch mm	Tap Drill Size mm	Nominal Diameter mm	Thread Pitch mm	Tap Drill Size mm
1.60	0.35	1.20	20.00	2.50	17.50
2.00	0.40	1.60	24.00	3.00	21.00
2.50	0.45	2.05	30.00	3.50	26.50
3.00	0.50	2.50	36.00	4.00	32.00
3.50	0.60	2.90	42.00	4.50	37.50
4.00	0.70	3.30	48.00	5.00	43.00
5.00	0.80	4.20	56.00	5.50	50.50
6.30	1.00	5.30	64.00	6.00	58.00
8.00	1.25	6.80	72.00	6.00	66.00
10.00	1.50	8.50	80.00	6.00	74.00
12.00	1.75	10.20	90.00	6.00	84.00
14.00	2.00	12.00	100.00	6.00	94.00
16.00	2.00	14.00			

TABLE 6: ISO METRIC PITCH & DIAMETER COMBINATIONS

Nominal Dia. (mm)	Thread Pitch (mm)	Nominal Dia. (mm)	Thread Pitch (mm)
1.6	0.35	20	2.5
2	0.40	24	3.0
2.5	0.45	30	3.5
3	0.50	36	4.0
3.5	0.60	42	4.5
4	0.70	48	5.0
5	0.80	56	5.5
6.3	1.00	64	6.0
8	1.25	72	6.0
10	1.50	80	6.0
12	1.75	90	6.0
14	2.00	100	6.0
16	2.00		

TABLE 7: TAP DRILL SIZES AMERICAN NATIONAL FORM THREAD

NC National Coarse			NF National Fine		
Tap Size	Threads per inch	Tap Drill Size	Tap Size	Threads per inch	Tap Drill Size
# 5	40	#38	# 5	44	#37
# 6	32	#36	# 6	40	#33
# 8	32	#29	# 8	36	#29
#10	24	#25	#10	32	#21
#12	24	#16	#12	28	#14
1/4	20	# 7	1/4	28	# 3
5/16	18	F	5/16	24	I
3/8	16	5/16	3/8	24	Q
7/16	14	U	7/16	20	25/64
1/2	13	27/64	1/2	20	29/64
9/16	12	31/64	9/16	18	33/64
5/8	11	17/32	5/8	18	37/64
3/4	10	21/32	3/4	16	11/16
7/8	9	49/64	7/8	14	13/16
1	8	7/8	1	14	15/16
1-1/8	7	63/64	1-1/8	12	1-3/64
1-1/4	7	1-7/64	1-1/4	12	1-11/64
1-3/8	6	1-7/32	1-3/8	12	1-19/64
1-1/2	6	1-11/32	1-1/2	12	1-27/64
1-3/4	5	1-9/16			
2	4-1/2	1-25/32			
NPT NATIONAL PIPE THREAD					
1/8	27	11/32	1	11-1/2	1-5/32
1/4	18	7/16	1-1/4	11-1/2	1-1/2
3/8	18	19/32	1-1/2	11-1/2	1-23/32
1/2	14	23/32	2	11-1/2	2-3/16
3/4	14	15/16	2-1/2	8	2-5/8

The major diameter of an NC or NF number size tap or screw = (N × .013) + .060
EXAMPLE: The major diameter of a #5 tap equals
(5 × .013) + .060 = .125 diameter

TABLE 8:

THREE WIRE THREAD MEASUREMENT
(60° Metric Thread)

M = PD + C PD = M − C

M = Measurement over wires
PD = Pitch diameter
C = Constant

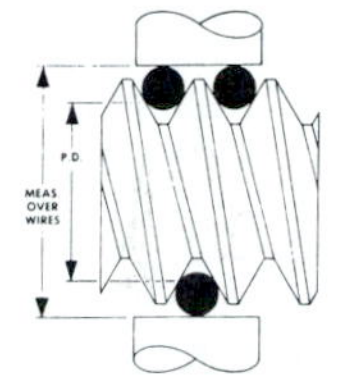

Pitch		Best Wire Size		Constant	
mm	Inches	mm	Inches	mm	Inches
0.2	.00787	0.1155	.00455	0.1732	.00682
0.225	.00886	0.1299	.00511	0.1949	.00767
0.25	.00934	0.1443	.00568	0.2165	.00852
0.3	.01181	0.1732	.00682	0.2598	.01023
0.35	.01378	0.2021	.00796	0.3031	.01193
0.4	.01575	0.2309	.00909	0.3464	.01364
0.45	.01772	0.2598	.01023	0.3897	.01534
0.5	.01969	0.2887	.01137	0.4330	.01705
0.6	.02362	0.3464	.01364	0.5196	.02046
0.7	.02756	0.4041	.01591	0.6062	.02387
0.75	.02953	0.4330	.01705	0.6495	.02557
0.8	.03150	0.4619	.01818	0.6928	.02728
0.9	.03543	0.5196	.02046	0.7794	.03069
1.0	.03937	0.5774	.02273	0.8660	.03410
1.25	.04921	0.7217	.02841	1.0825	.04262
1.5	.05906	0.8660	.03410	1.2990	.05114
1.75	.06890	1.0104	.03978	1.5155	.05967
2.0	.07874	1.1547	.04546	1.7321	.06819
2.5	.09843	1.4434	.05683	2.1651	.08524
3.0	.11811	1.7321	.06819	2.5981	.10229
3.5	.13780	2.0207	.07956	3.0311	.11933
4.0	.15748	2.3094	.09092	3.4641	.13638
4.5	.17717	2.5981	.10229	3.8971	.15343
5.0	.19685	2.8868	.11365	4.3301	.17048
5.5	.21654	3.1754	.12502	4.7631	.18753
6.0	.23622	3.4641	.13638	5.1962	.20457
7.0	.27559	4.0415	.15911	6.0622	.23867
8.0	.31496	4.6188	.18184	6.9282	.27276
9.0	.35433	5.1962	.20457	7.7942	.30686
10.0	.39370	5.7735	.22730	8.6603	.34095

TABLE 9: MORSE TAPERS

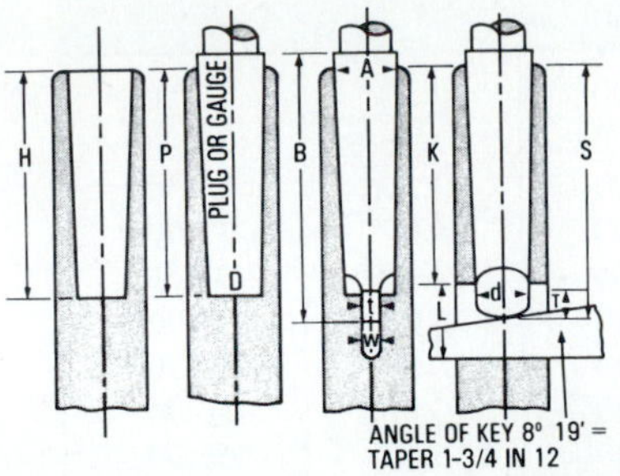
H
P
PLUG OR GAUGE
D
B
A
K
S
t
W
L
d
T
ANGLE OF KEY 8° 19' =
TAPER 1-3/4 IN 12

Number of taper	Diameter of plug at small end	Diameter at end of socket	Whole length of shank	Shank depth	Depth of hole	Standard plug depth	Thickness of tongue	Length of tongue	Diameter of tongue	Width of keyway	Length of keyway	End of socket to keyway	Taper per foot
	D	A	B	S	H	P	t	T	d	w	L	K	
0	.252	.356	2-11/32	2-7/32	2-1/32	2	5/32	1/4	.235	.160	9/16	1-15/16	.6246
1	.369	.475	2-9/16	2-7/16	2-3/16	2-1/8	13/64	3/8	.343	.213	3/4	2-1/16	.5986
2	.572	.700	3-1/8	2-15/16	2-5/8	2-9/16	1/4	7/16	17/32	.260	7/8	2-1/2	.5994
3	.778	.938	3-7/8	3-11/16	3-1/4	3-3/16	5/16	9/16	23/32	.322	1-3/16	3-1/16	.6023
4	1.020	1.231	4-7/8	4-5/8	4-1/8	4-1/16	15/32	5/8	31/32	.478	1-1/4	3-7/8	.6232
5	1.475	1.748	6-1/8	5-7/8	5-1/4	5-3/16	5/8	3/4	1-13/32	.635	1-1/2	4-15/16	.6315
6	2.116	2.494	8-9/16	8-1/4	7-3/8	7-1/4	3/4	1-1/8	2	.760	1-3/4	7	.6256
7	2.750	3.270	11-1/4	11-5/8	10-1/8	10	1-1/8	1-3/8	2-5/8	1.135	2-5/8	9-1/2	.6240

Note: All measurements are in inches

TABLE 10: STANDARD MILLING MACHINE TAPER

Milling Machine Spindles

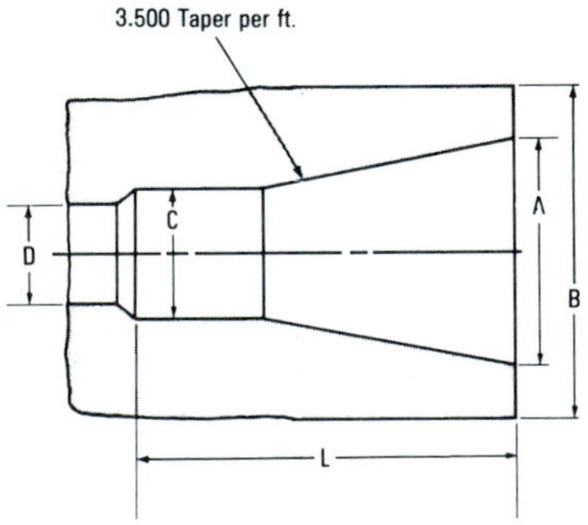
3.500 Taper per ft.
C
D
A
B
L

Milling Machine Arbors

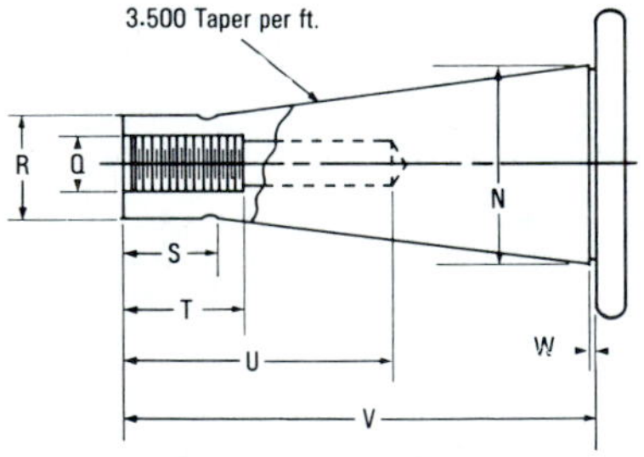
3.500 Taper per ft.
R
Q
N
S
T
U
V
W

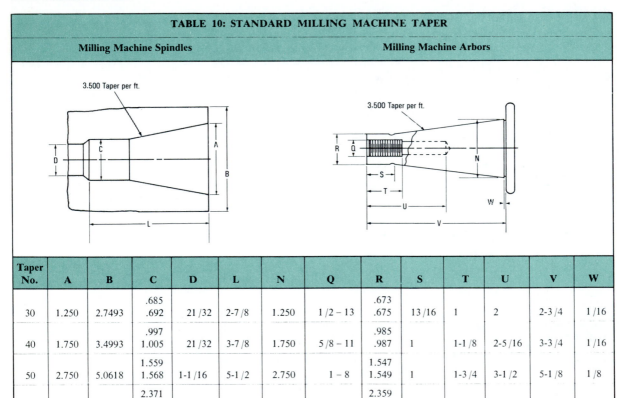

Taper No.	A	B	C	D	L	N	Q	R	S	T	U	V	W
30	1.250	2.7493	.685 .692	21/32	2-7/8	1.250	1/2 – 13	.673 .675	13/16	1	2	2-3/4	1/16
40	1.750	3.4993	.997 1.005	21/32	3-7/8	1.750	5/8 – 11	.985 .987	1	1-1/8	2-5/16	3-3/4	1/16
50	2.750	5.0618	1.559 1.568	1-1/16	5-1/2	2.750	1 – 8	1.547 1.549	1	1-3/4	3-1/2	5-1/8	1/8
60	4.250	8.718	2.371 2.381	1-3/8	8-5/8	4.250	1-1/4 – 7	2.359 2.361	1-3/4	2-1/4	4-1/4	8-5/16	1/8

Note: All measurements are in inches

TABLE 11: TAPERS AND ANGLES

Taper per Foot	Included Angle		With Center Line		Taper per Inch	Taper per Inch from Center Line
	Degree	Minute	Degree	Minute		
1/8	0	36	0	18	.010416	.005208
3/16	0	54	0	27	.015625	.007812
1/4	1	12	0	36	.020833	.010416
5/16	1	30	0	45	.026042	.013021
3/8	1	47	0	53	.031250	.015625
7/16	2	05	1	02	.036458	.018229
1/2	2	23	1	11	.041667	.020833
9/16	2	42	1	21	.046875	.023438
5/8	3	00	1	30	.052084	.026042
11/16	3	18	1	39	.057292	.028646
3/4	3	35	1	48	.062500	.031250
13/16	3	52	1	56	.067708	.033854
7/8	4	12	2	06	.072917	.036458
15/16	4	28	2	14	.078125	.039063
1	4	45	2	23	.083330	.041667
1-1/4	5	58	2	59	.104166	.052083
1-1/2	7	08	3	34	.125000	.062500
1-3/4	8	20	4	10	.145833	.072917
2	9	32	4	46	.166666	.083333
2-1/2	11	54	5	57	.208333	.104166
3	14	16	7	08	.250000	.125000
3-1/2	16	36	8	18	.291666	.145833
4	18	56	9	28	.333333	.166666
4-1/2	21	14	10	37	.375000	.187500
5	23	32	11	46	.416666	.208333
6	28	04	14	02	.500000	.250000

Courtesy Morse Twist Drill & Machine Co.

TABLE 12: ALLOWANCES FOR FITS

RUNNING FITS

Shaft Diameter	For Shafts with Speeds Under 600 r/min Ordinary Working Conditions	For Shafts with Speeds Over 600 r/min. Heavy Pressure – Severe Working Conditions
Up to 1/2	–.0005 to –.001	–.0005 to –.001
1/2 to 1	–.00075 to –.0015	–.001 to –.002
1 to 2	–.0015 to –.0025	–.002 to –.003
2 to 3-1/2	–.002 to –.003	–.003 to –.004
3-1/2 to 6	–.0025 to –.004	–.004 to –.005

SLIDING FITS

Shaft Diameter	For Shafts with Gears, Clutches or Similar Parts which Must be Free to Slide
Up to 1/2	–.0005 to –.001
1/2 to 1	–.00075 to –.0015
1 to 2	–.0015 to –.0025
2 to 3-1/2	–.002 to –.003
3-1/2 to 6	–.0025 to –.004

PUSH FITS

Shaft Diameter	For Light Service where Part is Keyed to Shaft and Clamped Endwise – No Fitting	With Play Eliminated – Part Should Assemble Readily – Some Fitting and Selecting May be Required
Up to 1/2	Standard to –.00025	Standard to +.00025
1/2 to 3-1/2	Standard to –.0005	Standard to +.0005
3-1/2 to 6	Standard to –.00075	Standard to +.00075

DRIVING FITS

Shaft Diameter	For Permanent Assembly of Parts so Located that Driving Cannot be Done Readily	For Permanent Assembly and Severe Duty and Where There is Ample Room for Driving
Up to 1/2	Standard to +.00025	+.0005 to +.001
1/2 to 1	+.00025 to +.0005	+.0005 to +.001
1 to 2	+.0005 to +.00075	+.0005 to +.001
2 to 3-1/2	+.0005 to +.001	+.00075 to +.00125
3-1/2 to 6	+.0005 to +.001	+.001 to +.0015

FORCED FITS

Shaft Diameter	For Permanent Assembly and Very Severe Service – Hydraulic Press Used for Larger Parts
Up to 1/2	+.00075 to +.001
1/2 to 1	+.001 to +.002
1 to 2	+.002 to +.003
2 to 3-1/2	+.003 to +.004
3-1/2 to 6	+.004 to +.005

Note: All measurements are in inches

TABLE 13: RULES FOR FINDING DIMENSIONS OF CIRCLES, SQUARES, ETC.

D is diameter of stock necessary to turn shape desired.
E is distance "across flats," or diameter of inscribed circle.
C is depth of cut into stock turned to correct diameter.

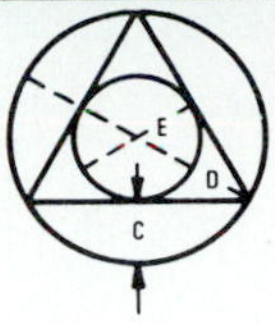

E
D
C

Triangle

E = side × .57735
D = side × 1.1547 = 2E
Side = D × .866
C = E × .5 = D × .25

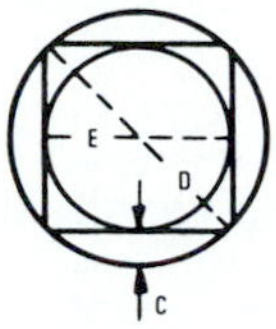

E
D
C

Square

E = side = D × .7071
D = side × 1.4142 = diagonal
Side = D × .7071
C = D × .14645

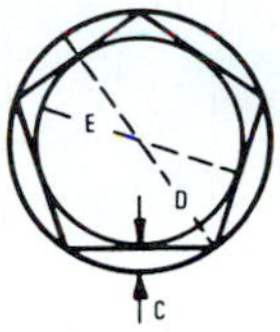

E
D
C

Pentagon

E = side × 1.3764 = D × .809
D = side × 1.7013 = E × 1.2361
Side = D × .5878
C = D × .0955

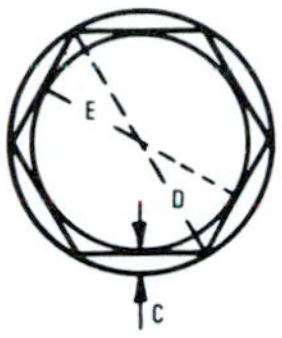

E
D
C

Hexagon

E = side × 1.7321 = D × .866
D = side × 2 = E × 1.1547
Side = D × .5
C = D × .067

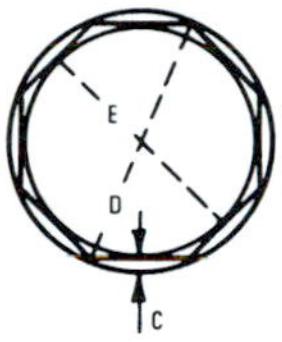

E
D
C

Octagon

E = side × 2.4142 = D × .9239
D = side × 2.6131 = E × 1.0824
Side = D × .3827
C = D × .038

Courtesy Morse Twist Drill & Machine Co.

TABLE 14: HARDNESS CONVERSION CHART

10 mm Ball 3000 kg	120° Cone 150 kg	1/16 in. Ball 100 kg	Model C	Mpa	10 mm Ball 3000 kg	120° Cone 150 kg	1/16 in. Ball 100 kg	Model C	Mpa
Brinell	Rockwell C	Rockwell B	Shore Scleroscope	Tensile Strength	Brinell	Rockwell C	Rockwell B	Shore Scleroscope	Tensile Strength
800	72	......	100	......	276	30	105	42	938
780	71	......	99	......	269	29	104	41	910
760	70	......	98	......	261	28	103	40	889
745	68	......	97	2530	258	27	102	39	876
725	67	......	96	2460	255	26	102	39	862
712	66	......	95	2413	249	25	101	38	848
682	65	......	93	2324	245	24	100	37	820
668	64	......	91	2248	240	23	99	36	807
652	63	......	89	2193	237	23	99	35	793
626	62	......	87	2110	229	22	98	34	779
614	61	......	85	2062	224	21	97	33	758
601	60	......	83	2013	217	20	96	33	738
590	59	......	81	2000	211	19	95	32	717
576	57	......	79	1937	206	18	94	32	703
552	56	......	76	1862	203	17	94	31	689
545	55	......	75	1848	200	16	93	31	676
529	54	......	74	1786	196	15	92	30	662
514	53	120	72	1751	191	14	92	30	648
502	52	119	70	1703	187	13	91	29	634
495	51	119	69	1682	185	12	91	29	627
477	49	118	67	1606	183	11	90	28	621
461	48	117	66	1565	180	10	89	28	614
451	47	117	65	1538	175	9	88	27	593
444	46	116	64	1510	170	7	87	27	579
427	45	115	62	1441	167	6	87	27	565
415	44	115	60	1407	165	5	86	26	558
401	43	114	58	1351	163	4	85	26	552
388	42	114	57	1317	160	3	84	25	538
375	41	113	55	1269	156	2	83	25	524
370	40	112	54	1255	154	1	82	25	517
362	39	111	53	1234	152		82	24	510
351	38	111	51	1193	150		81	24	510
346	37	110	50	1172	147	......	80	24	496
341	37	110	49	1158	145	......	79	23	490
331	36	109	47	1124	143	......	79	23	483
323	35	109	46	1089	141	......	78	23	476
311	34	108	46	1055	140	......	77	22	476
301	33	107	45	1020	135	......	75	22	462
293	32	106	44	993	130	......	72	22	448
285	31	105	43	965	......	......	......	......	......

TABLE 15: SOLUTIONS FOR RIGHT-ANGLED TRIANGLES

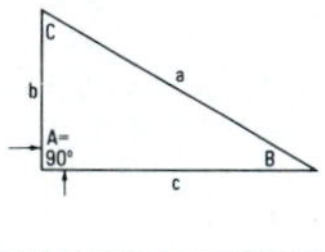	Sine ∠ = $\frac{\text{Side opposite}}{\text{Hypotenuse}}$	Cosecant ∠ = $\frac{\text{Hypotenuse}}{\text{Side opposite}}$
	Cosine ∠ = $\frac{\text{Side adjacent}}{\text{Hypotenuse}}$	Secant ∠ = $\frac{\text{Hypotenuse}}{\text{Side adjacent}}$
	Tangent ∠ = $\frac{\text{Side opposite}}{\text{Side adjacent}}$	Cotangent ∠ = $\frac{\text{Side adjacent}}{\text{Side opposite}}$

Knowing		Formulas to find	
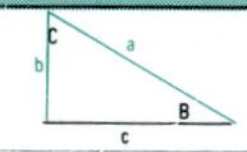	Sides a & b	$c = \sqrt{a^2 - b^2}$	$\sin B = \frac{b}{a}$
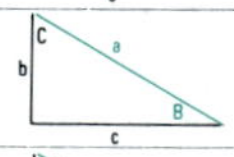	Side a & angle B	$b = a \times \sin B$	$c = a \times \cos B$
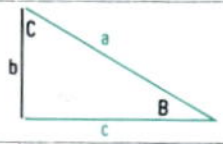	Sides a & c	$b = \sqrt{a^2 - c^2}$	$\sin C = \frac{c}{a}$
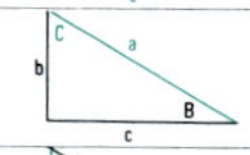	Side a & angle C	$b = a \times \cos C$	$c = a \times \sin C$
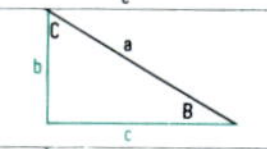	Sides b & c	$a = \sqrt{b^2 + c^2}$	$\tan B = \frac{b}{c}$
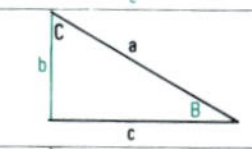	Side b & angle B	$a = \frac{b}{\sin B}$	$c = b \times \cot B$
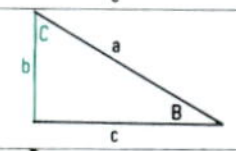	Side b & angle C	$a = \frac{b}{\cos C}$	$c = b \times \tan C$
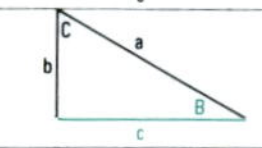	Side c & angle B	$a = \frac{c}{\cos B}$	$b = c \times \tan B$
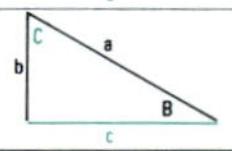	Side c & angle C	$a = \frac{c}{\sin C}$	$b = c \times \cot C$

TABLE 16: SINE BAR CONSTANTS (5 in. BAR)
(Multiply Constants by Two for a 10 in. Sine Bar)

Min.	0°	1°	2°	3°	4°	5°	6°	7°	8°	9°	10°	11°	12°	13°	14°	15°	16°	17°	18°	19°	Min.
0	.00000	.08725	.17450	.26170	.34880	.43580	.52265	.60935	.69585	.78215	.86825	.95405	1.0395	1.1247	1.2096	1.2941	1.3782	1.4618	1.5451	1.6278	0
2	.00290	.09015	.17740	.26460	.35170	.43870	.52555	.61225	.69875	.78505	.87110	.95690	.0424	.1276	.2124	.2969	.3810	.4646	.5478	.6306	2
4	.00580	.09310	.18030	.26750	.35460	.44155	.52845	.61510	.70165	.78790	.87395	.95975	.0452	.1304	.2152	.2997	.3838	.4674	.5506	.6333	4
6	.00875	.09600	.18320	.27040	.35750	.44445	.53130	.61800	.70450	.79080	.87685	.96260	.0481	.1332	.2181	.3025	.3865	.4702	.5534	.6361	6
8	.01165	.09890	.18615	.27330	.36040	.44735	.53420	.62090	.70740	.79365	.87970	.96545	.0509	.1361	.2209	.3053	.3893	.4730	.5561	.6388	8
10	.01455	.10180	.18905	.27620	.36330	.45025	.53710	.62380	.71025	.79655	.88255	.96830	1.0538	1.1389	1.2237	1.3081	1.3921	1.4757	1.5589	1.6416	10
12	.01745	.10470	.19195	.27910	.36620	.45315	.54000	.62665	.71315	.79940	.88540	.97115	.0566	.1417	.2265	.3109	.3949	.4785	.5616	.6443	12
14	.02035	.10760	.19485	.28200	.36910	.45605	.54290	.62955	.71600	.80230	.88830	.97405	.0594	.1446	.2293	.31[illegible]	.3977	.4813	.5644	.6471	14
16	.02325	.11055	.19775	.28490	.37200	.45895	.54580	.63245	.71890	.80515	.89115	.97690	.0623	.1474	.2322	.3165	.4005	.4841	.5672	.6498	16
18	.02620	.11345	.20065	.28780	.37490	.46185	.54865	.63530	.72180	.80800	.89400	.97975	.0651	.1502	.2350	.3193	.4033	.4868	.5699	.6525	18
20	.02910	.11635	.20355	.29070	.37780	.46475	.55155	.63820	.72465	.81090	.89685	.98260	1.0680	1.1531	1.2378	1.3221	1.4061	1.4896	1.5727	1.6553	20
22	.03200	.11925	.20645	.29365	.38070	.46765	.55445	.64110	.72755	.81375	.89975	.98545	.0708	.1559	.2406	.3250	.4089	.4924	.5755	.6580	22
24	.03490	.12215	.20940	.29655	.38360	.47055	.55735	.64400	.73040	.81665	.90260	.98830	.0737	.1587	.2434	.3278	.4117	.4952	.5782	.6608	24
26	.03780	.12505	.21230	.29945	.38650	.47345	.56025	.64685	.73330	.81950	.90545	.99115	.0765	.1615	.2462	.3306	.4145	.4980	.5810	.6635	26
28	.04070	.12800	.21520	.30235	.38940	.47635	.56315	.64975	.73615	.82235	.90830	.99400	.0793	.1644	.2491	.3334	.4173	.5007	.5837	.6663	28
30	.04365	.13090	.21810	.30525	.39230	.47925	.56600	.65265	.73905	.82525	.91120	.99685	1.0822	1.1672	1.2519	1.3362	1.4201	1.5035	1.5865	1.6690	30
32	.04655	.13380	.22100	.30815	.39520	.48210	.56890	.65550	.74190	.82810	.91405	.99970	.0850	.1700	.2547	.3390	.4228	.5063	.5893	.6718	32
34	.04945	.13670	.22390	.31105	.39810	.48500	.57180	.65840	.74480	.83100	.91690	1.0016	.0879	.1729	.2575	.3418	.4256	.5091	.5920	.6745	34
36	.05235	.13960	.22680	.31395	.40100	.48790	.57470	.66130	.74770	.83385	.91975	.0054	.0907	.1757	.2603	.3446	.4284	.5118	.5948	.6772	36
38	.05525	.14250	.22970	.31685	.40390	.49080	.57760	.66415	.75055	.83670	.92260	.0082	.0935	.1785	.2631	.3474	.4312	.5146	.5975	.6800	38
40	.05820	.14540	.23265	.31975	.40680	.49370	.58045	.66705	.75345	.83960	.92545	1.0110	1.0964	1.1813	1.2660	1.3502	1.4340	1.5174	1.6003	1.6827	40
42	.06110	.14835	.23555	.32265	.40970	.49660	.58335	.66995	.75630	.84245	.92835	.0139	.0992	.1842	.2688	.3530	.4368	.5201	.6030	.6855	42
44	.06400	.15125	.23845	.32555	.41260	.49950	.58625	.67280	.75920	.84530	.93120	.0168	.1020	.1870	.2716	.3558	.4396	.5229	.6058	.6882	44
46	.06690	.15415	.24135	.32845	.41550	.50240	.58915	.67570	.76205	.84820	.93405	.0196	.1049	.1898	.2744	.3586	.4423	.5257	.6085	.6909	46
48	.06980	.15705	.24425	.33135	.41840	.50530	.59200	.67860	.76495	.85105	.93690	.0225	.1077	.1926	.2772	.3614	.4451	.5285	.6113	.6937	48
50	.07270	.15995	.24715	.33425	.42130	.50820	.59490	.68145	.76780	.85390	.93975	1.0253	1.1106	1.1955	1.2800	1.3642	1.4479	1.5312	1.6141	1.6964	50
52	.07565	.16285	.25005	.33715	.42420	.51105	.59780	.68435	.77070	.85680	.94260	.0281	.1134	.1983	.2828	.3670	.4507	.5340	.6168	.6991	52
54	.07855	.16580	.25295	.34010	.42710	.51395	.60070	.68720	.77355	.85965	.94550	.0310	.1162	.2011	.2856	.3698	.4535	.5368	.6196	.7019	54
56	.08145	.16870	.25585	.34300	.43000	.51685	.60355	.69010	.77645	.86250	.94835	.0338	.1191	.2039	.2884	.3726	.4563	.5395	.6223	.7046	56
58	.08435	.17160	.25875	.34590	.43290	.51975	.60645	.69300	.77930	.86540	.95120	.0367	.1219	.2068	.2913	.3754	.4591	.5423	.6251	.7073	58
60	.08725	.17450	.26170	.34880	.43580	.52265	.60935	.69585	.78215	.86825	.95405	1.0395	1.1247	1.2096	1.2941	1.3782	1.4618	1.5451	1.6278	1.7101	60

Courtesy Brown & Sharpe Mfg. Co.

Table 16 continued

TABLE 16: SINE BAR CONSTANTS (5 in. BAR)
(Multiply Constants by Two for a 10 in. Sine Bar)

Min.	20°	21°	22°	23°	24°	25°	26°	27°	28°	29°	30°	31°	32°	33°	34°	35°	36°	37°	38°	39°	Min.
0	1.7101	1.7918	1.8730	1.9536	2.0337	2.1131	2.1918	2.2699	2.3473	2.4240	2.5000	2.5752	2.6496	2.7232	2.7959	2.8679	2.9389	3.0091	3.0783	3.1466	0
2	.7128	.7945	.8757	.9563	.0363	.1157	.1944	.2725	.3499	.4266	.5025	.5777	.6520	.7256	.7984	.8702	.9413	.0114	.0806	.1488	2
4	.7155	.7972	.8784	.9590	.0390	.1183	.1971	.2751	.3525	.4291	.5050	.5802	.6545	.7280	.8008	.8726	.9436	.0137	.0829	.1511	4
6	.7183	.8000	.8811	.9617	.0416	.1210	.1997	.2777	.3550	.4317	.5075	.5826	.6570	.7305	.8032	.8750	.9460	.0160	.0852	.1534	6
8	.7210	.8027	.8838	.9643	.0443	.1236	.2023	.2803	.3576	.4342	.5100	.5851	.6594	.7329	.8056	.8774	.9483	.0183	.0874	.1556	8
10	1.7237	1.8054	1.8865	1.9670	2.0469	2.1262	2.2049	2.2829	2.3602	2.4367	2.5126	2.5876	2.6619	2.7354	2.8080	2.8798	2.9507	3.0207	3.0897	3.1579	10
12	.7265	.8081	.8892	.9697	.0496	.1289	.2075	.2855	.3627	.4393	.5151	.5901	.6644	.7378	.8104	.8821	.9530	.0230	.0920	.1601	12
14	.7292	.8108	.8919	.9724	.0522	.1315	.2101	.2881	.3653	.4418	.5176	.5926	.6668	.7402	.8128	.8845	.9554	.0253	.0943	.1624	14
16	.7319	.8135	.8946	.9750	.0549	.1341	.2127	.2906	.3679	.4444	.5201	.5951	.6693	.7427	.8152	.8869	.9577	.0276	.0966	.1646	16
18	.7347	.8162	.8973	.9777	.0575	.1368	.2153	.2932	.3704	.4469	.5226	.5976	.6717	.7451	.8176	.8893	.9600	.0299	.0989	.1669	18
20	1.7374	1.8189	1.8999	1.9804	2.0602	2.1394	2.2179	2.2958	2.3730	2.4494	2.5251	2.6001	2.6742	2.7475	2.8200	2.8916	2.9624	3.0322	3.1012	3.1691	20
22	.7401	.8217	.9026	.9830	.0628	.1420	.2205	.2984	.3755	.4520	.5276	.6025	.6767	.7499	.8224	.8940	.9647	.0345	.1034	.1714	22
24	.7428	.8244	.9053	.9857	.0655	.1447	.2232	.3010	.3781	.4545	.5301	.6050	.6791	.7524	.8248	.8964	.9671	.0369	.1057	.1736	24
26	.7456	.8271	.9080	.9884	.0681	.1473	.2258	.3036	.3807	.4570	.5327	.6075	.6816	.7548	.8272	.8988	.9694	.0392	.1080	.1759	26
28	.7483	.8298	.9107	.9911	.0708	.1499	.2284	.3061	.3832	.4596	.5352	.6100	.6840	.7572	.8296	.9011	.9718	.0415	.1103	.1781	28
30	1.7510	1.8325	1.9134	1.9937	2.0734	2.1525	2.2310	2.3087	2.3858	2.4621	2.5377	2.6125	2.6865	2.7597	2.8320	2.9035	2.9741	3.0438	3.1125	3.1804	30
32	.7537	.8352	.9161	.9964	.0761	.1552	.2336	.3113	.3883	.4646	.5402	.6149	.6889	.7621	.8344	.9059	.9764	.0461	.1148	.1826	32
34	.7565	.8379	.9188	.9991	.0787	.1578	.2362	.3139	.3909	.4672	.5427	.6174	.6914	.7645	.8368	.9082	.9788	.0484	.1171	.1849	34
36	.7592	.8406	.9215	2.0017	.0814	.1604	.2388	.3165	.3934	.4697	.5452	.6199	.6938	.7669	.8392	.9106	.9811	.0507	.1194	.1871	36
38	.7619	.8433	.9241	.0044	.0840	.1630	.2414	.3190	.3960	.4722	.5477	.6224	.6963	.7694	.8416	.9130	.9834	.0530	.1216	.1893	38
40	1.7646	1.8460	1.9268	2.0070	2.0867	2.1656	2.2440	2.3216	2.3985	2.4747	2.5502	2.6249	2.6987	2.7718	2.8440	2.9153	2.9858	3.0553	3.1239	3.1916	40
42	.7673	.8487	.9295	.0097	.0893	.1683	.2466	.3242	.4011	.4773	.5527	.6273	.7012	.7742	.8464	.9177	.9881	.0576	.1262	.1938	42
44	.7701	.8514	.9322	.0124	.0920	.1709	.2492	.3268	.4036	.4798	.5552	.6298	.7036	.7766	.8488	.9200	.9904	.0599	.1285	.1961	44
46	.7728	.8541	.9349	.0150	.0946	.1735	.2518	.3293	.4062	.4823	.5577	.6323	.7061	.7790	.8512	.9224	.9928	.0622	.1307	.1983	46
48	.7755	.8568	.9376	.0177	.0972	.1761	.2544	.3319	.4087	.4848	.5602	.6348	.7085	.7815	.8535	.9248	.9951	.0645	.1330	.2005	48
50	1.7782	1.8595	1.9402	2.0204	2.0999	2.1787	2.2570	2.3345	2.4113	2.4874	2.5627	2.6372	2.7110	2.7839	2.8559	2.9271	2.9974	3.0668	3.1353	3.2028	50
52	.7809	.8622	.9429	.0230	.1025	.1814	.2596	.3371	.4138	.4899	.5652	.6397	.7134	.7863	.8583	.9295	.9997	.0691	.1375	.2050	52
54	.7837	.8649	.9456	.0257	.1052	.1840	.2621	.3396	.4164	.4924	.5677	.6422	.7158	.7887	.8607	.9318	3.0021	.0714	.1398	.2072	54
56	.7864	.8676	.9483	.0283	.1078	.1866	.2647	.3422	.4189	.4949	.5702	.6446	.7183	.7911	.8631	.9342	.0044	.0737	.1421	.2095	56
58	.7891	.8703	.9510	.0310	.1104	.1892	.2673	.3448	.4215	.4975	.5727	.6471	.7207	.7935	.8655	.9365	.0067	.0760	.1443	.2117	58
60	1.7918	1.8730	1.9536	2.0337	2.1131	2.1918	2.2699	2.3473	2.4240	2.5000	2.5752	2.6496	2.7232	2.7959	2.8679	2.9389	3.0091	3.0783	3.1466	3.2139	60

Courtesy Brown & Sharpe Mfg. Co.

Table 16 continued

TABLE 16: SINE BAR CONSTANTS (5 in. BAR)
(Multiply Constants by Two for a 10 in. Sine Bar)

Min.	40°	41°	42°	43°	44°	45°	46°	47°	48°	49°	50°	51°	52°	53°	54°	55°	56°	57°	58°	59°	Min.
0	3.2139	3.2803	3.3456	3.4100	3.4733	3.5355	3.5967	3.6567	3.7157	3.7735	3.8302	3.8857	3.9400	3.9932	4.0451	4.0957	4.1452	4.1933	4.2402	4.2858	0
2	.2161	.2825	.3478	.4121	.4754	.5376	.5987	.6587	.7176	.7754	.8321	.8875	.9418	.9949	.0468	.0974	.1468	.1949	.2418	.2873	2
4	.2184	.2847	.3499	.4142	.4774	.5396	.6007	.6607	.7196	.7773	.8339	.8894	.9436	.9967	.0485	.0991	.1484	.1965	.2433	.2888	4
6	.2206	.2869	.3521	.4163	.4795	.5417	.6027	.6627	.7215	.7792	.8358	.8912	.9454	.9984	.0502	.1007	.1500	.1981	.2448	.2903	6
8	.2228	.2890	.3543	.4185	.4816	.5437	.6047	.6647	.7235	.7811	.8377	.8930	.9472	4.0001	.0519	.1024	.1517	.1997	.2464	.2918	8
10	3.2250	3.2912	3.3564	3.4206	3.4837	3.5458	3.6068	3.6666	3.7254	3.7830	3.8395	3.8948	3.9490	4.0019	4.0536	4.1041	4.1533	4.2012	4.2479	4.2933	10
12	.2273	.2934	.3586	.4227	.4858	.5478	.6088	.6686	.7274	.7850	.8414	.8967	.9508	.0036	.0553	.1057	.1549	.2028	.2494	.2948	12
14	.2295	.2956	.3607	.4248	.4879	.5499	.6108	.6706	.7293	.7869	.8433	.8985	.9525	.0054	.0570	.1074	.1565	.2044	.2510	.2963	14
16	.2317	.2978	.3629	.4269	.4900	.5519	.6128	.6726	.7312	.7887	.8451	.9003	.9543	.0071	.0587	.1090	.1581	.2060	.2525	.2978	16
18	.2339	.3000	.3650	.4291	.4921	.5540	.6148	.6745	.7332	.7906	.8470	.9021	.9561	.0089	.0604	.1107	.1597	.2075	.2540	.2992	18
20	3.2361	3.3022	3.3672	3.4312	3.4941	3.5560	3.6168	3.6765	3.7351	3.7925	3.8488	3.9039	3.9579	4.0106	4.0621	4.1124	4.1614	4.2091	4.2556	4.3007	20
22	.2384	.3044	.3693	.4333	.4962	.5581	.6188	.6785	.7370	.7944	.8507	.9058	.9596	.0123	.0638	.1140	.1630	.2107	.2571	.3022	22
24	.2406	.3065	.3715	.4354	.4983	.5601	.6208	.6805	.7390	.7963	.8525	.9076	.9614	.0141	.0655	.1157	.1646	.2122	.2586	.3037	24
26	.2428	.3087	.3736	.4375	.5004	.5621	.6228	.6824	.7409	.7982	.8544	.9094	.9632	.0158	.0672	.1173	.1662	.2138	.2601	.3052	26
28	.2450	.3109	.3758	.4396	.5024	.5642	.6248	.6844	.7428	.8001	.8562	.9112	.9650	.0175	.0689	.1190	.1678	.2154	.2617	.3066	28
30	3.2472	3.3131	3.3779	3.4417	3.5045	3.5662	3.6268	3.6864	3.7448	3.8020	3.8581	3.9130	3.9667	4.0193	4.0706	4.1206	4.1694	4.2169	4.2632	4.3081	30
32	.2494	.3153	.3801	.4439	.5066	.5683	.6288	.6883	.7467	.8039	.8599	.9148	.9685	.0210	.0722	.1223	.1710	.2185	.2647	.3096	32
34	.2516	.3174	.3822	.4460	.5087	.5703	.6308	.6903	.7486	.8058	.8618	.9166	.9703	.0227	.0739	.1239	.1726	.2201	.2662	.3111	34
36	.2538	.3196	.3844	.4481	.5107	.5723	.6328	.6923	.7505	.8077	.8636	.9184	.9720	.0244	.0756	.1255	.1742	.2216	.2677	.3125	36
38	.2561	.3218	.3865	.4502	.5128	.5744	.6348	.6942	.7525	.8096	.8655	.9202	.9738	.0262	.0773	.1272	.1758	.2232	.2692	.3140	38
40	3.2583	3.3240	3.3886	3.4523	3.5149	3.5764	3.6368	3.6962	3.7544	3.8114	3.8673	3.9221	3.9756	4.0279	4.0790	4.1288	4.1774	4.2247	4.2708	4.3155	40
42	.2605	.3261	.3908	.4544	.5169	.5784	.6388	.6981	.7563	.8133	.8692	.9239	.9773	.0296	.0807	.1305	.1790	.2263	.2723	.3170	42
44	.2627	.3283	.3929	.4565	.5190	.5805	.6408	.7001	.7582	.8152	.8710	.9257	.9791	.0313	.0823	.1321	.1806	.2278	.2738	.3184	44
46	.2649	.3305	.3950	.4586	.5211	.5825	.6428	.7020	.7601	.8171	.8729	.9275	.9809	.0331	.0840	.1337	.1822	.2294	.2753	.3199	46
48	.2671	.3326	.3972	.4607	.5231	.5845	.6448	.7040	.7620	.8190	.8747	.9293	.9826	.0348	.0857	.1354	.1838	.2309	.2768	.3213	48
50	3.2693	3.3348	3.3993	3.4628	3.5252	3.5866	3.6468	3.7060	3.7640	3.8208	3.8765	3.9311	3.9844	4.0365	4.0874	4.1370	4.1854	4.2325	4.2783	4.3228	50
52	.2715	.3370	.4014	.4649	.5273	.5886	.6488	.7079	.7659	.8227	.8784	.9329	.9861	.0382	.0891	.1386	.1870	.2340	.2798	.3243	52
54	.2737	.3391	.4036	.4670	.5293	.5906	.6508	.7099	.7678	.8246	.8802	.9347	.9879	.0399	.0907	.1403	.1886	.2356	.2813	.3257	54
56	.2759	.3413	.4057	.4691	.5314	.5926	.6528	.7118	.7697	.8265	.8820	.9364	.9896	.0416	.0924	.1419	.1902	.2371	.2828	.3272	56
58	.2781	.3435	.4078	.4712	.5335	.5947	.6548	.7138	.7716	.8283	.8839	.9382	.9914	.0433	.0941	.1435	.1917	.2387	.2843	.3286	58
60	3.2803	3.3456	3.4100	3.4733	3.5355	3.5967	3.6567	3.7157	3.7735	3.8302	3.8857	3.9400	3.9932	4.0451	4.0957	4.1452	4.1933	4.2402	4.2858	4.3301	60

Courtesy Brown & Sharpe Mfg. Co.

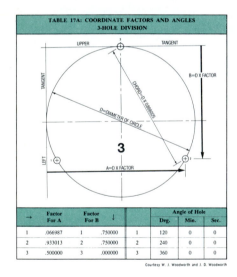

TABLE 17A: COORDINATE FACTORS AND ANGLES 3-HOLE DIVISION

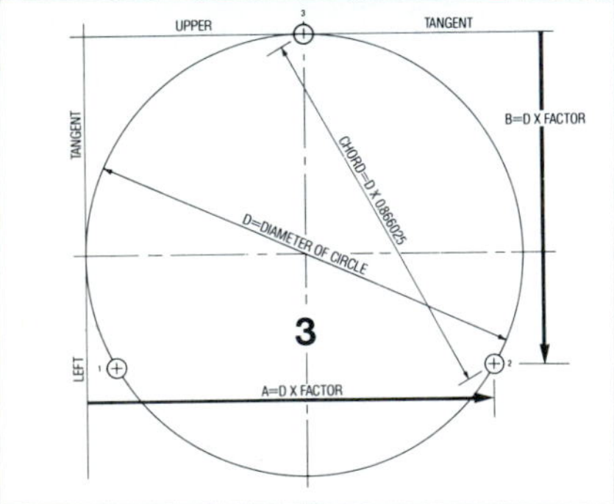

UPPER
TANGENT
TANGENT
B=D X FACTOR
CHORD=D X 0.866025
D=DIAMETER OF CIRCLE
3
LEFT
A=D X FACTOR

→	Factor For A		Factor For B ↓		Angle of Hole Deg.	Min.	Sec.
1	.066987	1	.750000	1	120	0	0
2	.933013	2	.750000	2	240	0	0
3	.500000	3	.000000	3	360	0	0

Courtesy W. J. Woodworth and J. D. Woodworth

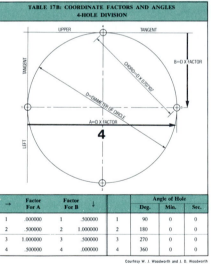

TABLE 17B: COORDINATE FACTORS AND ANGLES 4-HOLE DIVISION

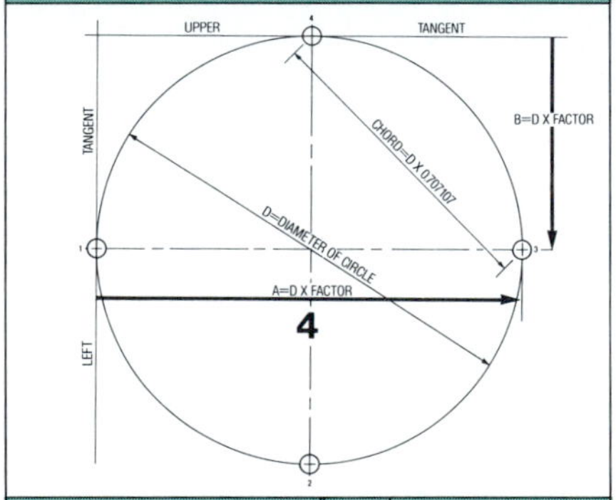

UPPER
TANGENT
TANGENT
B=D X FACTOR
CHORD=D X 0.707107
D=DIAMETER OF CIRCLE
A=D X FACTOR
4
LEFT

→	Factor For A		Factor For B ↓		Angle of Hole Deg.	Min.	Sec.
1	.000000	1	.500000	1	90	0	0
2	.500000	2	1.000000	2	180	0	0
3	1.000000	3	.500000	3	270	0	0
4	.500000	4	.000000	4	360	0	0

Courtesy W. J. Woodworth and J. D. Woodworth

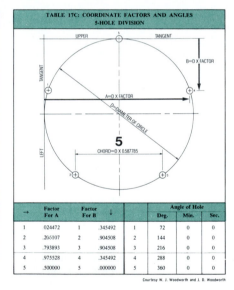

TABLE 17C: COORDINATE FACTORS AND ANGLES 5-HOLE DIVISION

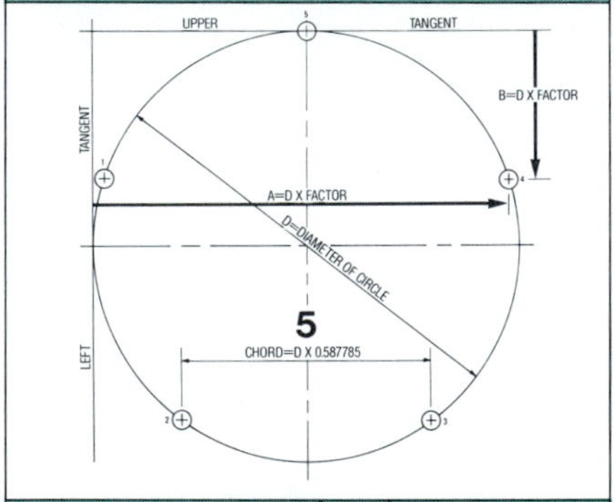

UPPER
TANGENT
TANGENT
B=D X FACTOR
A=D X FACTOR
D=DIAMETER OF CIRCLE
5
CHORD=D X 0.587785
LEFT

→	Factor For A		Factor For B ↓		Angle of Hole Deg.	Min.	Sec.
1	.024472	1	.345492	1	72	0	0
2	.206107	2	.904508	2	144	0	0
3	.793893	3	.904508	3	216	0	0
4	.975528	4	.345492	4	288	0	0
5	.500000	5	.000000	5	360	0	0

Courtesy W. J. Woodworth and J. D. Woodworth

TABLE 17D: COORDINATE FACTORS AND ANGLES 6-HOLE DIVISION

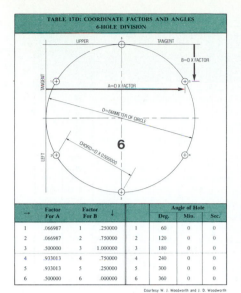

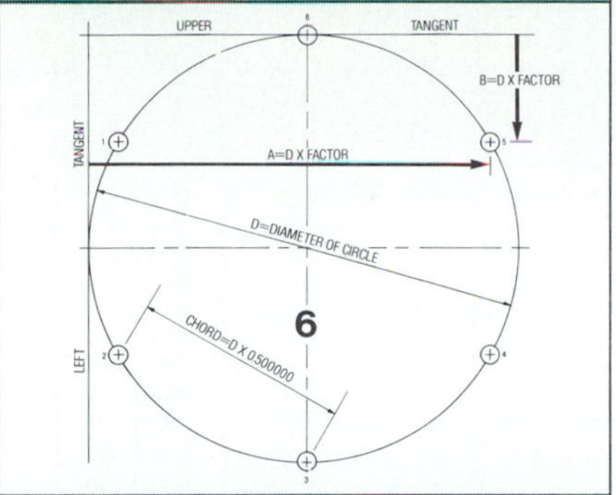
UPPER
TANGENT
B=D X FACTOR
TANGENT
A=D X FACTOR
D=DIAMETER OF CIRCLE
6
CHORD=D X 0.500000
LEFT

→	Factor For A		Factor For B ↓		Angle of Hole Deg.	Min.	Sec.
1	.066987	1	.250000	1	60	0	0
2	.066987	2	.750000	2	120	0	0
3	.500000	3	1.000000	3	180	0	0
4	.933013	4	.750000	4	240	0	0
5	.933013	5	.250000	5	300	0	0
6	.500000	6	.000000	6	360	0	0

Courtesy W. J. Woodworth and J. D. Woodworth

TABLE 17E: COORDINATE FACTORS AND ANGLES 7-HOLE DIVISION

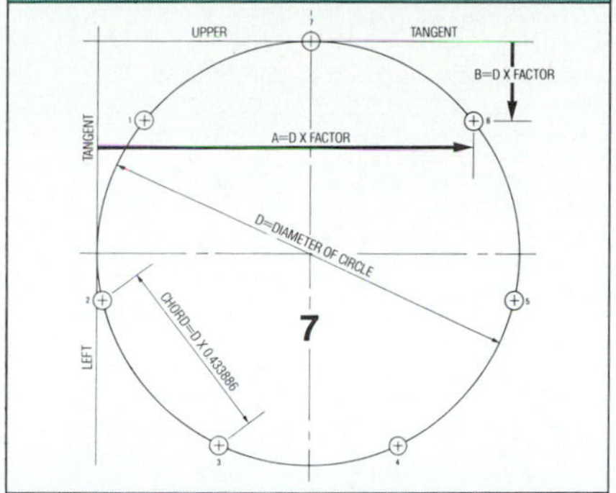
UPPER
TANGENT
B=D X FACTOR
TANGENT
A=D X FACTOR
D=DIAMETER OF CIRCLE
7
CHORD=D X 0.433886
LEFT

→	Factor For A		Factor For B ↓		Angle of Hole Deg.	Min.	Sec.
1	.109084	1	.188255	1	51	25	42-6/7
2	.012536	2	.611261	2	102	51	23-5/7
3	.283058	3	.950484	3	154	17	8-4/7
4	.716942	4	.950484	4	205	42	51-3/7
5	.987464	5	.611261	5	257	8	34-2/7
6	.890916	6	.188255	6	308	34	17-1/7
7	.500000	7	.000000	7	360	0	0

Courtesy W. J. Woodworth and J. D. Woodworth

TABLE 17F: COORDINATE FACTORS AND ANGLES 8-HOLE DIVISION

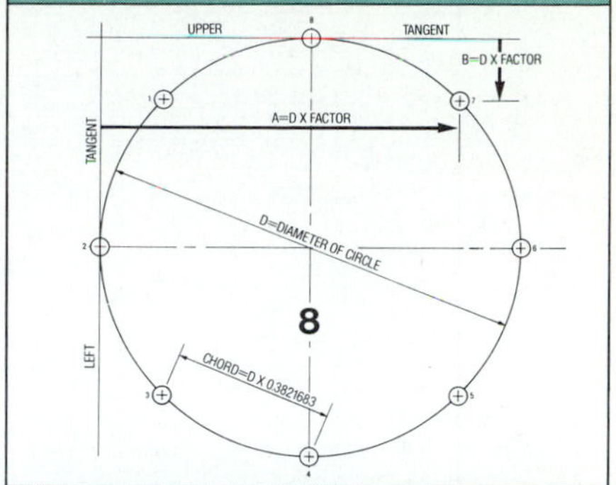
UPPER
TANGENT
B=D X FACTOR
TANGENT
A=D X FACTOR
D=DIAMETER OF CIRCLE
8
CHORD=D X 0.3821683
LEFT

→	Factor For A		Factor For B ↓		Angle of Hole Deg.	Min.	Sec.
1	.146447	1	.146447	1	45	0	0
2	.000000	2	.500000	2	90	0	0
3	.146447	3	.853553	3	135	0	0
4	.500000	4	1.000000	4	180	0	0
5	.853553	5	.853553	5	225	0	0
6	1.000000	6	.500000	6	270	0	0
7	.853553	7	.146447	7	315	0	0
8	.500000	8	.000000	8	360	0	0

Courtesy W. J. Woodworth and J. D. Woodworth

TABLE 17G: COORDINATE FACTORS AND ANGLES 9-HOLE DIVISION

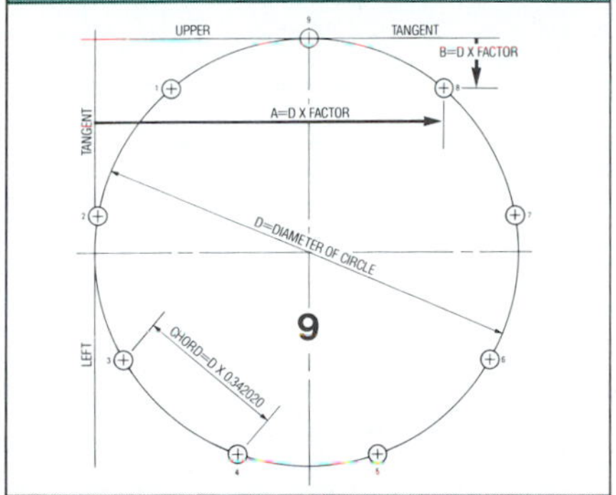
UPPER
TANGENT
B=D X FACTOR
TANGENT
A=D X FACTOR
D=DIAMETER OF CIRCLE
9
CHORD=D X 0.342020
LEFT

→	Factor For A		Factor For B ↓		Angle of Hole Deg.	Min.	Sec.
1	.178606	1	.116978	1	40	0	0
2	.007596	2	.413176	2	80	0	0
3	.066987	3	.750000	3	120	0	0
4	.328990	4	.969846	4	160	0	0
5	.671010	5	.969846	5	200	0	0
6	.933013	6	.750000	6	240	0	0
7	.992404	7	.413176	7	280	0	0
8	.821394	8	.116978	8	320	0	0
9	.500000	9	.000000	9	360	0	0

Courtesy W. J. Woodworth and J. D. Woodworth

TABLE 17H: COORDINATE FACTORS AND ANGLES 10-HOLE DIVISION

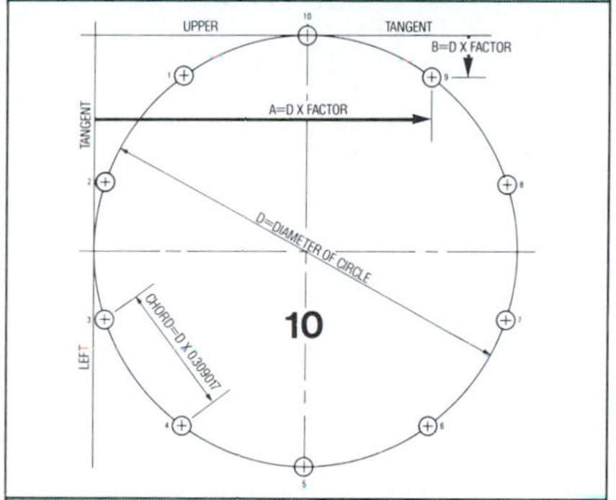
UPPER
TANGENT
B=D X FACTOR
TANGENT
A=D X FACTOR
D=DIAMETER OF CIRCLE
10
CHORD=D X 0.309017
LEFT

→	Factor For A		Factor For B ↓		Angle of Hole Deg.	Min.	Sec.
1	.206107	1	.095492	1	36	0	0
2	.024472	2	.345492	2	72	0	0
3	.024472	3	.654508	3	108	0	0
4	.206107	4	.904508	4	144	0	0
5	.500000	5	1.000000	5	180	0	0
6	.793893	6	.904508	6	216	0	0
7	.975528	7	.654508	7	252	0	0
8	.975528	8	.345492	8	288	0	0
9	.793893	9	.095492	9	324	0	0
10	.500000	10	.000000	10	360	0	0

Courtesy W. J. Woodworth and J. D. Woodworth

TABLE 17I: COORDINATE FACTORS AND ANGLES 11-HOLE DIVISION

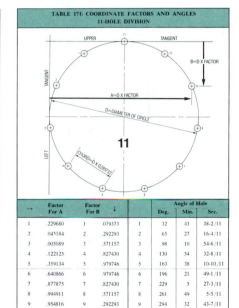

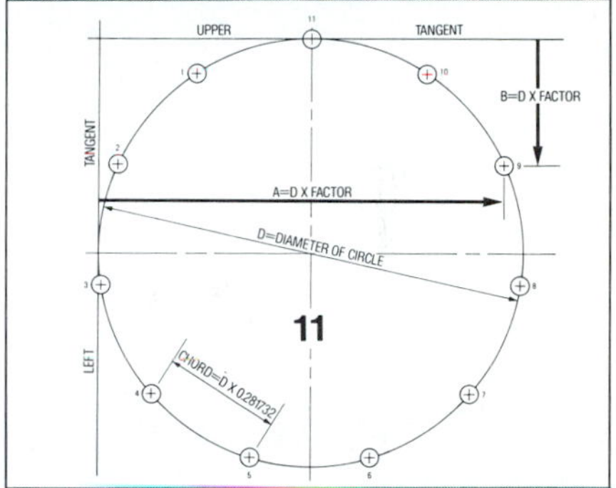
UPPER
TANGENT
B=D X FACTOR
TANGENT
A=D X FACTOR
D=DIAMETER OF CIRCLE
11
CHORD=D X 0.281732
LEFT

→	Factor For A		Factor For B ↓		Angle of Hole Deg.	Min.	Sec.
1	.229680	1	.079373	1	32	43	38-2/11
2	045184	2	.292293	2	65	27	16-4/11
3	.005089	3	.571157	3	98	10	54-6/11
4	.122125	4	.827430	4	130	54	32-8/11
5	.359134	5	.979746	5	163	38	10-10/11
6	.640866	6	.979746	6	196	21	49-1/11
7	.877875	7	.827430	7	229	5	27-3/11
8	.994911	8	.571157	8	261	49	5-5/11
9	.954816	9	.292293	9	294	32	43-7/11
10	.770320	10	.079373	10	327	16	21-9/11
11	.500000	11	.000000	11	360	0	0

Courtesy W. J. Woodworth and J. D. Woodworth

NATURAL TRIGONOMETRIC FUNCTIONS

	0°						1°						2°						3°						
′	sin	cos	tan	cot	sec	cosec	sin	cos	tan	cot	sec	cosec	sin	cos	tan	cot	sec	cosec	sin	cos	tan	cot	sec	cosec	′
0	.00000	1.0000	.00000	Infinite	1.0000	Infinite	.01745	.99985	.01745	57.290	1.0001	57.299	.03490	.99939	.03492	28.636	1.0006	28.654	.05234	.99863	.05241	19.081	1.0014	19.107	60
1	.00029	.0000	.00029	3437.7	.0000	3437.7	.01774	.99984	.01775	56.350	.0001	56.359	.03519	.99938	.03521	28.399	.0006	28.417	.05263	.99861	.05270	18.975	.0014	19.002	59
2	.00058	.0000	.00058	1718.9	.0000	1718.9	.01803	.99984	.01804	55.441	.0001	55.450	.03548	.99937	.03550	28.166	.0006	28.184	.05292	.99860	.05299	18.871	.0014	18.897	58
3	.00087	.0000	.00087	1145.9	.0000	1145.9	.01832	.99983	.01833	54.561	.0002	54.570	.03577	.99936	.03579	27.937	.0006	27.955	.05321	.99858	.05328	18.768	.0014	18.794	57
4	.00116	.0000	.00116	859.44	.0000	859.44	.01861	.99983	.01862	53.708	.0002	53.718	.03606	.99935	.03608	27.712	.0006	27.730	.05350	.99857	.05357	18.665	.0014	18.692	56
5	.00145	1.0000	.00145	687.55	.0000	687.55	.01891	.99982	.01891	52.882	.0002	52.891	.03635	.99934	.03638	27.490	.0007	27.508	.05379	.99855	.05387	18.564	.0014	18.591	55
6	.00174	.0000	.00174	572.96	.0000	572.96	.01920	.99981	.01920	52.081	.0002	52.090	.03664	.99933	.03667	27.271	.0007	27.290	.05408	.99854	.05416	18.464	.0015	18.491	54
7	.00204	.0000	.00204	491.11	.0000	491.11	.01949	.99981	.01949	51.303	.0002	51.313	.03693	.99932	.03696	27.056	.0007	27.075	.05437	.99852	.05445	18.365	.0015	18.393	53
8	.00233	.0000	.00233	429.72	.0000	429.72	.01978	.99980	.01978	50.548	.0002	50.558	.03722	.99931	.03725	26.845	.0007	26.864	.05466	.99850	.05474	18.268	.0015	18.295	52
9	.00262	.0000	.00262	381.97	.0000	381.97	.02007	.99980	.02007	49.816	.0002	49.826	.03751	.99930	.03754	26.637	.0007	26.655	.05495	.99849	.05503	18.171	.0015	18.198	51
10	.00291	0.99999	.00291	343.77	.0000	343.77	.02036	.99979	.02036	49.104	.0002	49.114	.03781	.99928	.03783	26.432	.0007	26.450	.05524	.99847	.05532	18.075	.0015	18.103	50
11	.00320	.99999	.00320	312.52	.0000	312.52	.02065	.99979	.02066	48.412	.0002	48.422	.03810	.99927	.03812	26.230	.0007	26.249	.05553	.99846	.05562	17.980	.0015	18.008	49
12	.00349	.99999	.00349	286.48	.0000	286.48	.02094	.99978	.02095	47.739	.0002	47.750	.03839	.99926	.03842	26.031	.0007	26.050	.05582	.99844	.05591	17.886	.0016	17.914	48
13	.00378	.99999	.00378	264.44	.0000	264.44	.02123	.99977	.02124	47.085	.0002	47.096	.03868	.99925	.03871	25.835	.0007	25.854	.05611	.99842	.05620	17.793	.0016	17.821	47
14	.00407	.99999	.00407	245.55	.0000	245.55	.02152	.99977	.02153	46.449	.0002	46.460	.03897	.99924	.03900	25.642	.0008	25.661	.05640	.99841	.05649	17.701	.0016	17.730	46
15	.00436	.99999	.00436	229.18	.0000	229.18	.02181	.99976	.02182	45.829	.0002	45.840	.03926	.99923	.03929	25.452	.0008	25.471	.05669	.99839	.05678	17.610	.0016	17.639	45
16	.00465	.99999	.00465	214.86	.0000	214.86	.02210	.99975	.02211	45.226	.0002	45.237	.03955	.99922	.03958	25.264	.0008	25.284	.05698	.99837	.05707	17.520	.0016	17.549	44
17	.00494	.99999	.00494	202.22	.0000	202.22	.02240	.99975	.02240	44.638	.0002	44.650	.03984	.99921	.03987	25.080	.0008	25.100	.05727	.99836	.05737	17.431	.0016	17.460	43
18	.00524	.99999	.00524	190.98	.0000	190.99	.02269	.99974	.02269	44.066	.0002	44.077	.04013	.99919	.04016	24.898	.0008	24.918	.05756	.99834	.05766	17.343	.0017	17.372	42
19	.00553	.99998	.00553	180.93	.0000	180.93	.02298	.99974	.02298	43.508	.0003	43.520	.04042	.99918	.04045	24.718	.0008	24.739	.05785	.99832	.05795	17.256	.0017	17.285	41
20	.00582	.99998	.00582	171.88	.0000	171.89	.02326	.99973	.02327	42.964	.0003	42.976	.04071	.99917	.04075	24.542	.0008	24.562	.05814	.99831	.05824	17.169	.0017	17.198	40
21	.00611	.99998	.00611	163.70	.0000	163.70	.02356	.99972	.02357	42.433	.0003	42.445	.04100	.99916	.04104	24.367	.0008	24.388	.05843	.99829	.05853	17.084	.0017	17.113	39
22	.00640	.99998	.00640	156.26	.0000	156.26	.02385	.99971	.02386	41.916	.0003	41.928	.04129	.99915	.04133	24.196	.0008	24.216	.05872	.99827	.05883	16.999	.0017	17.028	38
23	.00669	.99998	.00669	149.46	.0000	149.47	.02414	.99971	.02415	41.410	.0003	41.423	.04158	.99913	.04162	24.026	.0009	24.047	.05902	.99826	.05912	16.915	.0017	16.944	37
24	.00698	.99997	.00698	143.24	.0000	143.24	.02443	.99970	.02444	40.917	.0003	40.930	.04187	.99912	.04191	23.859	.0009	23.880	.05931	.99824	.05941	16.832	.0018	16.861	36
25	.00727	.99997	.00727	137.51	.0000	137.51	.02472	.99969	.02473	40.436	.0003	40.448	.04217	.99911	.04220	23.694	.0009	23.716	.05960	.99822	.05970	16.750	.0018	16.779	35
26	.00756	.99997	.00756	132.22	.0000	132.22	.02501	.99969	.02502	39.965	.0003	39.978	.04246	.99910	.04249	23.532	.0009	23.553	.05989	.99820	.05999	16.668	.0018	16.698	34
27	.00785	.99997	.00785	127.32	.0000	127.32	.02530	.99968	.02531	39.506	.0003	39.518	.04275	.99908	.04279	23.372	.0009	23.393	.06018	.99819	.06029	16.587	.0018	16.617	33
28	.00814	.99997	.00814	122.77	.0000	122.78	.02559	.99967	.02560	39.057	.0003	39.069	.04304	.99907	.04308	23.214	.0009	23.235	.06047	.99817	.06058	16.507	.0018	16.538	32
29	.00843	.99996	.00844	118.54	.0000	118.54	.02589	.99966	.02589	38.618	.0003	38.631	.04333	.99906	.04337	23.058	.0009	23.079	.06076	.99815	.06087	16.428	.0018	16.459	31
30	.00873	.99996	.00873	114.59	.0000	114.59	.02618	.99966	.02618	38.188	.0003	38.201	.04362	.99905	.04366	22.904	.0009	22.925	.06105	.99813	.06116	16.350	.0019	16.380	30
31	.00902	.99996	.00902	110.89	.0000	110.90	.02647	.99965	.02648	37.769	.0003	37.782	.04391	.99903	.04395	22.752	.0010	22.774	.06134	.99812	.06145	16.272	.0019	16.303	29
32	.00931	.99996	.00931	107.43	.0000	107.43	.02676	.99964	.02677	37.358	.0003	37.371	.04420	.99902	.04424	22.602	.0010	22.624	.06163	.99810	.06175	16.195	.0019	16.226	28
33	.00960	.99995	.00960	104.17	.0000	104.17	.02705	.99963	.02706	36.956	.0004	36.969	.04449	.99901	.04453	22.454	.0010	22.476	.06192	.99808	.06204	16.119	.0019	16.150	27
34	.00989	.99995	.00989	101.11	.0000	101.11	.02734	.99963	.02735	36.563	.0004	36.576	.04478	.99900	.04483	22.308	.0010	22.330	.06221	.99806	.06233	16.043	.0019	16.075	26
35	.01018	.99995	.01018	98.218	.0000	98.223	.02763	.99962	.02764	36.177	.0004	36.191	.04507	.99898	.04512	22.164	.0010	22.186	.06250	.99804	.06262	15.969	.0019	16.000	25
36	.01047	.99994	.01047	95.489	.0000	95.495	.02792	.99961	.02793	35.800	.0004	35.814	.04536	.99897	.04541	22.022	.0010	22.044	.06279	.99803	.06291	15.894	.0020	15.926	24
37	.01076	.99994	.01076	92.908	.0000	92.914	.02821	.99960	.02822	35.431	.0004	35.445	.04565	.99896	.04570	21.881	.0010	21.904	.06308	.99801	.06321	15.821	.0020	15.853	23
38	.01105	.99994	.01105	90.463	.0001	90.469	.02850	.99959	.02851	35.069	.0004	35.084	.04594	.99894	.04599	21.742	.0010	21.765	.06337	.99799	.06350	15.748	.0020	15.780	22
39	.01134	.99993	.01134	88.143	.0001	88.149	.02879	.99958	.02880	34.715	.0004	34.729	.04623	.99893	.04628	21.606	.0011	21.629	.06366	.99797	.06379	15.676	.0020	15.708	21
40	.01163	.99993	.01164	85.940	.0001	85.946	.02908	.99958	.02910	34.368	.0004	34.382	.04652	.99892	.04657	21.470	.0011	21.494	.06395	.99795	.06408	15.605	.0020	15.637	20
41	.01193	.99993	.01193	83.843	.0001	83.849	.02937	.99957	.02939	34.027	.0004	34.042	.04681	.99890	.04687	21.337	.0011	21.360	.06424	.99793	.06437	15.534	.0021	15.566	19
42	.01222	.99992	.01222	81.847	.0001	81.853	.02967	.99956	.02968	33.693	.0004	33.708	.04711	.99889	.04716	21.205	.0011	21.228	.06453	.99791	.06467	15.464	.0021	15.496	18
43	.01251	.99992	.01251	79.943	.0001	79.950	.02996	.99955	.02997	33.366	.0004	33.381	.04740	.99888	.04745	21.075	.0011	21.098	.06482	.99790	.06496	15.394	.0021	15.427	17
44	.01280	.99992	.01280	78.126	.0001	78.133	.03025	.99954	.03026	33.045	.0004	33.060	.04769	.99886	.94774	20.946	.0011	20.970	.06511	.99788	.06525	15.325	.0021	15.358	16
45	.01309	.99991	.01309	76.390	.0001	76.396	.03054	.99953	.03055	32.730	.0005	32.745	.04798	.99885	.04803	20.819	.0011	20.843	.06540	.99786	.06554	15.257	.0021	15.290	15
46	.01338	.99991	.01338	74.729	.0001	74.736	.03083	.99952	.03084	32.421	.0005	32.437	.04827	.99883	.04832	20.693	.0012	20.717	.06569	.99784	.06583	15.189	.0022	15.222	14
47	.01367	.99991	.01367	73.139	.0001	73.146	.03112	.99951	.03113	32.118	.0005	32.134	.04856	.99882	.04862	20.569	.0012	20.593	.06598	.99782	.06613	15.122	.0022	15.155	13
48	.01396	.99990	.01396	71.615	.0001	71.622	.03141	.99951	.03143	31.820	.0005	31.836	.04885	.99881	.04891	20.446	.0012	20.471	.06627	.99780	.06642	15.056	.0022	15.089	12
49	.01425	.99990	.01425	70.153	.0001	70.160	.03170	.99950	.03172	31.528	.0005	31.544	.04914	.99879	.04920	20.325	.0012	20.350	.06656	.99778	.06671	14.990	.0022	15.023	11
50	.01454	.99989	.01454	68.750	.0001	68.757	.03199	.99949	.03201	31.241	.0005	31.257	.04943	.99878	.04949	20.205	.0012	20.230	.06685	.99776	.06700	14.924	.0022	14.958	10
51	.01483	.99989	.01484	67.402	.0001	67.409	.03228	.99948	.03230	30.960	.0005	30.976	.04972	.99876	.04978	20.087	.0012	20.112	.06714	.99774	.06730	14.860	.0023	14.893	9
52	.01512	.99988	.01513	66.105	.0001	66.113	.03257	.99947	.03259	30.683	.0005	30.699	.05001	.99875	.05007	19.970	.0012	19.995	.06743	.99772	.06759	14.795	.0023	14.829	8
53	.01542	.99988	.01542	64.858	.0001	64.866	.03286	.99946	.03288	30.411	.0005	30.428	.05030	.99873	.05037	19.854	.0013	19.880	.06772	.99770	.06788	14.732	.0023	14.765	7
54	.01571	.99988	.01571	63.657	.0001	63.664	.03315	.99945	.03317	30.145	.0005	30.161	.05059	.99872	.05066	19.740	.0013	19.766	.06801	.99768	.06817	14.668	.0023	14.702	6
55	.01600	.99987	.01600	62.499	.0001	62.507	.03344	.99944	.03346	29.882	.0005	29.899	.05088	.99870	.05095	19.627	.0013	19.653	.06830	.99766	.06846	14.606	.0023	14.640	5
56	.01629	.99987	.01629	61.383	.0001	61.391	.03374	.99943	.03375	29.624	.0006	29.641	.05117	.99869	.05124	19.515	.0013	19.541	.06859	.99764	.06876	14.544	.0024	14.578	4
57	.01658	.99987	.01658	60.306	.0001	60.314	.03403	.99942	.03405	29.371	.0006	29.388	.05146	.99867	.05153	19.405	.0013	19.431	.06888	.99762	.06905	14.482	.0024	14.517	3
58	.01687	.99986	.01687	59.266	.0001	59.274	.03432	.99941	.03434	29.122	.0006	29.139	.05175	.99866	.05182	19.296	.0013	19.322	.06918	.99760	.06934	14.421	.0024	14.456	2
59	.01716	.99985	.01716	58.261	.0001	58.270	.03461	.99940	.03463	28.877	.0006	28.894	.05204	.99864	.05212	19.188	.0013	19.214	.06947	.99758	.06963	14.361	.0024	14.395	1
60	.01745	.99985	.01745	57.290	.0001	57.299	.03490	.99939	.03492	28.636	.0006	28.654	.05234	.99863	.05241	19.081	.0014	19.107	.06976	.99756	06993	14.301	.0024	14.335	0
′	cos	sin	cot	tan	cosec	sec	cos	sin	cot	tan	cosec	sec	cos	sin	cot	tan	cosec	sec	cos	sin	cot	tan	cosec	sec	′
	89°						88°						87°						86°						

Courtesy Bethlehem Steel Corporation

NATURAL TRIGONOMETRIC FUNCTIONS

′	4° sin	4° cos	4° tan	4° cot	4° sec	4° cosec	5° sin	5° cos	5° tan	5° cot	5° sec	5° cosec
0	.06976	.99756	.06993	14.301	1.0024	14.335	.08715	.99619	.08749	11.430	1.0038	11.474
1	.07005	.99754	.07022	14.241	.0025	14.276	.08744	.99617	.08778	11.392	.0038	11.436
2	.07034	.99752	.07051	14.182	.0025	14.217	.08773	.99614	.08807	11.354	.0039	11.398
3	.07063	.99750	.07080	14.123	.0025	14.159	.08802	.99612	.08837	11.316	.0039	11.360
4	.07092	.99748	.07110	14.065	.0025	14.101	.08831	.99609	.08866	11.279	.0039	11.323
5	.07121	.99746	.07139	14.008	.0025	14.043	.08860	.99607	.08895	11.242	.0039	11.286
6	.07150	.99744	.07168	13.951	.0026	13.986	.08889	.99604	.08925	11.205	.0040	11.249
7	.07179	.99742	.07197	13.894	.0026	13.930	.08918	.99601	.08954	11.168	.0040	11.213
8	.07208	.99740	.07226	13.838	.0026	13.874	.08947	.99599	.08983	11.132	.0040	11.176
9	.07237	.99738	.07256	13.782	.0026	13.818	.08976	.99596	.09013	11.095	.0040	11.140
10	.07266	.99736	.07285	13.727	.0026	13.763	.09005	.99594	.09042	11.059	.0041	11.104
11	.07295	.99733	.07314	13.672	.0027	13.708	.09034	.99591	.09071	11.024	.0041	11.069
12	.07324	.99731	.07343	13.617	.0027	13.654	.09063	.99588	.09101	10.988	.0041	11.033
13	.07353	.99729	.07373	13.563	.0027	13.600	.09092	.99586	.09130	10.953	.0041	10.998
14	.07382	.99727	.07402	13.510	.0027	13.547	.09121	.99583	.09159	10.918	.0042	10.963
15	.07411	.99725	.07431	13.457	.0027	13.494	.09150	.99580	.09189	10.883	.0042	10.929
16	.07440	.99723	.07460	13.404	.0028	13.441	.09179	.99578	.09218	10.848	.0042	10.894
17	.07469	.99721	.07490	13.351	.0028	13.389	.09208	.99575	.09247	10.814	.0043	10.860
18	.07498	.99718	.07519	13.299	.0028	13.337	.09237	.99572	.09277	10.780	.0043	10.826
19	.07527	.99716	.07548	13.248	.0028	13.286	.09266	.99570	.09306	10.746	.0043	10.792
20	.07556	.99714	.07577	13.917	.0029	13.235	.09295	.99567	.09335	10.712	.0043	10.758
21	.07585	.99712	.07607	13.146	.0029	13.184	.09324	.99564	.09365	10.678	.0044	10.725
22	.07614	.99710	.07636	13.096	.0029	13.134	.09353	.99562	.09394	10.645	.0044	10.692
23	.07643	.99707	.07665	13.046	.0029	13.084	.09382	.99559	.09423	10.612	.0044	10.659
24	.07672	.99705	.07694	12.996	.0029	13.034	.09411	.99556	.09453	10.579	.0044	10.626
25	.07701	.99703	.07724	12.947	.0030	12.985	.09440	.99553	.09482	10.546	.0045	10.593
26	.07730	.99701	.07753	12.898	.0030	12.937	.09469	.99551	.09511	10.514	.0045	10.561
27	.07759	.99698	.07782	12.849	.0030	12.888	.09498	.99548	.09541	10.481	.0045	10.529
28	.07788	.99696	.07812	12.801	.0030	12.840	.09527	.99545	.09570	10.449	.0046	10.497
29	.07817	.99694	.07841	12.754	.0031	12.793	.09556	.99542	.09599	10.417	.0046	10.465
30	.07846	.99692	.07870	12.706	.0031	12.745	.09584	.99540	.09629	10.385	.0046	10.433
31	.07875	.99689	.07899	12.659	.0031	12.698	.09613	.99537	.09658	10.354	.0046	10.402
32	.07904	.99687	.07929	12.612	.0031	12.652	.09642	.99534	.09688	10.322	.0047	10.371
33	.07933	.99685	.07958	12.566	.0032	12.606	.09671	.99531	.09717	10.291	.0047	10.340
34	.07962	.99682	.07987	12.520	.0032	12.560	.09700	.99528	.09746	10.260	.0047	10.309
35	.07991	.99680	.08016	12.474	.0032	12.514	.09729	.99525	.09776	10.229	.0048	10.278
36	.08020	.99678	.08046	12.429	.0032	12.469	.09758	.99523	.09805	10.199	.0048	10.248
37	.08049	.99675	.08075	12.384	.0032	12.424	.09787	.99520	.09834	10.168	.0048	10.217
38	.08078	.99673	.08104	12.339	.0033	12.379	.09816	.99517	.09864	10.138	.0048	10.187
39	.08107	.99671	.08134	12.295	.0033	12.335	.09845	.99514	.09893	10.108	.0049	10.157
40	.08136	.99668	.08163	12.250	.0033	12.291	.09874	.99511	.09922	10.078	.0049	10.127
41	.08165	.99666	.08192	12.207	.0033	12.248	.09903	.99508	.09952	10.048	.0049	10.098
42	.08194	.99664	.08221	12.163	.0034	12.204	.09932	.99505	.09981	10.019	.0050	10.068
43	.08223	.99661	.08251	12.120	.0034	12.161	.09961	.99503	.10011	9.9893	.0050	10.039
44	.08252	.99659	.08280	12.077	.0034	12.118	.09990	.99500	.10040	9.9601	.0050	10.010
45	.08281	.99656	.08309	12.035	.0034	12.076	.10019	.99497	.10069	9.9310	.0050	9.9812
46	.08310	.99654	.08339	11.992	.0035	12.034	.10048	.99494	.10099	9.9021	.0051	9.9525
47	.08339	.99652	.08368	11.950	.0035	11.992	.10077	.99491	.10128	9.8734	.0051	9.9239
48	.08368	.99649	.08397	11.909	.0035	11.950	.10106	.99488	.10158	9.8448	.0051	9.8955
49	.08397	.99647	.08426	11.867	.0035	11.909	.10134	.99485	.10187	9.8164	.0052	9.8672
50	.08426	.99644	.08456	11.826	.0036	11.868	.10163	.99482	.10216	9.7882	.0052	9.8391
51	.08455	.99642	.08485	11.785	.0036	11.828	.10192	.99479	.10246	9.7601	.0052	9.8112
52	.08484	.99639	.08514	11.745	.0036	11.787	.10221	.99476	.10275	9.7322	.0053	9.7834
53	.08513	.99637	.08544	11.704	.0036	11.747	.10250	.99473	.10305	9.7044	.0053	9.7558
54	.08542	.99634	.08573	11.664	.0037	11.707	.10279	.99470	.10334	9.6768	.0053	9.7283
55	.08571	.99632	.08602	11.625	.0037	11.668	.10308	.99467	.10363	9.6493	.0053	9.7010
56	.08600	.99629	.08632	11.585	.0037	11.628	.10337	.99464	.10393	9.6220	.0054	9.6739
57	.08629	.99627	.08661	11.546	.0037	11.589	.10366	.99461	.10422	9.5949	.0054	9.6469
58	.08658	.99624	.08690	11.507	.0038	11.550	.10395	.99458	.10452	9.5679	.0054	9.6200
59	.08687	.99622	.08719	11.468	.0038	11.512	.10424	99455	.10481	9.5411	.0055	9.5933
60	.08715	.99619	.08749	11.430	.0038	11.474	.10453	99452	.10510	9.5144	.0055	9.5668
′	85° cos	85° sin	85° cot	85° tan	85° cosec	85° sec	84° cos	84° sin	84° cot	84° tan	84° cosec	84° sec

′	6° sin	6° cos	6° tan	6° cot	6° sec	6° cosec	7° sin	7° cos	7° tan	7° cot	7° sec	7° cosec	′
0	.10453	.99452	.10510	9.5144	1.0055	9.5668	.12187	.99255	.12278	8.1443	1.0075	8.2055	60
1	.10482	.99449	.10540	.4878	.0055	.5404	.12216	.99251	.12308	.1248	.0075	.1861	59
2	.10511	.99446	.10569	.4614	.0056	.5141	.12245	.99247	.12337	.1053	.0076	.1668	58
3	.10540	.99443	.10599	.4351	.0056	.4880	.12273	.99244	.12367	.0860	.0076	.1476	57
4	.10568	.99440	.10628	.4090	.0056	.4620	.12302	.99240	.12396	.0667	.0076	.1285	56
5	.10597	.99437	.10657	9.3831	.0057	9.4362	.12331	.99237	.12426	8.0476	.0077	8.1094	55
6	.10626	.99434	.10687	.3572	.0057	.4105	.12360	.99233	.12456	.0285	.0077	.0905	54
7	.10655	.99431	.10716	.3315	.0057	.3850	.12389	.99229	.12485	.0095	.0078	.0717	53
8	.10684	.99428	.10746	.3060	.0057	.3596	.12418	.99226	.12515	7.9906	.0078	.0529	52
9	.10713	.99424	.10775	.2806	.0058	.3343	.12447	.99222	.12544	.9717	.0078	.0342	51
10	.10742	.99421	.10805	9.2553	.0058	9.3092	.12476	.99219	.12574	7.9530	.0079	8.0156	50
11	.10771	.99418	.10834	.2302	.0058	.2842	.12504	.99215	.12603	.9344	.0079	7.9971	49
12	.10800	.99415	.10863	.2051	.0059	.2593	.12533	.99211	.12633	.9158	.0079	.9787	48
13	.10829	.99412	.10893	.1803	.0059	.2346	.12562	.99208	.12662	.8973	.0080	.9604	47
14	.10858	.99409	.10922	.1555	.0059	.2100	.12591	.99204	.12692	.8789	.0080	.9421	46
15	.10887	.99406	.10952	9.1309	.0060	9.1855	.12620	.99200	.12722	7.8606	.0080	7.9240	45
16	.10916	.99402	.10981	.1064	.0060	.1612	.12649	.99197	.12751	.8424	.0081	.9059	44
17	.10944	.99399	.11011	.0821	.0060	.1370	.12678	.99193	.12781	.8243	.0081	.8879	43
18	.10973	.99396	.11040	.0579	.0061	.1129	.12706	.99189	.12810	.8062	.0082	.8700	42
19	.11002	.99393	.11069	.0338	.0061	.0890	.12735	.99186	.12840	.7882	.0082	.8522	41
20	.11031	.99390	.11099	9.0098	.0061	9.0651	.12764	.99182	.12869	7.7703	.0082	7.8344	40
21	.11060	.99386	.11128	8.9860	.0062	.0414	.12793	.99178	.12899	.7525	.0083	.8168	39
22	.11089	.99383	.11158	.9623	.0062	.0179	.12822	.99174	.12928	.7348	.0083	.7992	38
23	.11118	.99380	.11187	.9387	.0062	8.9944	.12851	.99171	.12958	.7171	.0084	.7817	37
24	.11147	.99377	.11217	.9152	.0063	.9711	.12879	.99167	.12988	.6996	.0084	.7642	36
25	.11176	.99373	.11246	8.8918	.0063	8.9479	.12908	.99163	.13017	7.6821	.0084	7.7469	35
26	.11205	.99370	.11276	.8686	.0063	.9248	.12937	.99160	.13047	.6646	.0085	.7296	34
27	.11234	.99367	.11305	.8455	.0064	.9018	.12966	.99156	.13076	.6473	.0085	.7124	33
28	.11262	.99364	.11335	.8225	.0064	.8790	.12995	.99152	.13106	.6300	.0085	.6953	32
29	.11291	.99360	.11364	.7996	.0064	.8563	.13024	.99148	.13136	.6129	.0086	.6783	31
30	.11320	.99357	.11393	8.7769	.0065	8.8337	.13053	.99144	.13165	7.5957	.0086	7.6613	30
31	.11349	.99354	.11423	.7542	.0065	.8112	.13081	.99141	.13195	.5787	.0087	.6444	29
32	.11378	.99350	.11452	.7317	.0065	.7888	.13110	.99137	.13224	.5617	.0087	.6276	28
33	.11407	.99347	.11482	.7093	.0066	.7665	.13139	.99133	.13254	.5449	.0087	.6108	27
34	.11436	.99344	.11511	.6870	.0066	.7444	.13168	.99129	.13284	.5280	.0088	.5942	26
35	.11465	.99341	.11541	8.6648	.0066	8.7223	.13197	.99125	.13313	7.5113	.0088	7.5776	25
36	.11494	.99337	.11570	.6427	.0067	.7004	.13226	.99121	.13343	.4946	.0089	.5611	24
37	.11523	.99334	.11600	.6208	.0067	.6786	.13254	.99118	.13372	.4780	.0089	.5446	23
38	.11551	.99330	.11629	.5989	.0067	.6569	.13283	.99114	.13402	.4615	.0089	.5282	22
39	.11580	.99327	.11659	.5772	.0068	.6353	.13312	.99110	.13432	.4451	.0090	.5119	21
40	.11609	.99324	.11688	8.5555	.0068	8.6138	.13341	.99106	.13461	7.4287	.0090	7.4957	20
41	.11638	.99320	.11718	.5340	.0068	.5924	.13370	.99102	.13491	.4124	.0090	.4795	19
42	.11667	.99317	.11747	.5126	.0069	.5711	.13399	.99098	.13520	.3961	.0091	.4634	18
43	.11696	.99314	.11777	.4913	.0069	.5499	.13427	.99094	.13550	.3800	.0091	.4474	17
44	.11725	.99310	.11806	.4701	.0069	.5289	.13456	.99090	.13580	.3639	.0092	.4315	16
45	.11754	.99307	.11836	8.4489	.0070	8.5079	.13485	.99086	.13609	7.3479	.0092	7.4156	15
46	.11783	.99303	.11865	.4279	.0070	.4871	.13514	.99083	.13639	.3319	.0092	.3998	14
47	.11811	.99300	.11895	.4070	.0070	.4663	.13543	.99079	.13669	.3160	.0093	.3840	13
48	.11840	.99296	.11924	.3862	.0071	.4457	.13571	.99075	.13698	.3002	.0093	.3683	12
49	.11869	.99293	.11954	.3655	.0071	.4251	.13600	.99071	.13728	.2844	.0094	.3527	11
50	.11898	.99290	.11983	8.3449	.0071	8.4046	.13629	.99067	.13757	7.2687	.0094	7.3372	10
51	.11927	.99286	.12013	.3244	.0072	.3843	.13658	.99063	.13787	.2531	.0094	.3217	9
52	.11956	.99283	.12042	.3040	.0072	.3640	.13687	.99059	.13817	.2375	.0095	.3063	8
53	.11985	.99279	.12072	.2837	.0073	.3439	.13716	.99055	.13846	.2220	.0095	.2909	7
54	.12014	.99276	.12101	.2635	.0073	.3238	.13744	.99051	.13876	.2066	.0096	.2757	6
55	.12042	.99272	.12131	8.2434	.0073	8.3039	.13773	.99047	.13906	7.1912	.0096	7.2604	5
56	.12071	.99269	.12160	.2234	.0074	.2840	.13802	.99043	.13935	.1759	.0097	.2453	4
57	.12100	.99265	.12190	.2035	.0074	.2642	.13831	.99039	.13965	.1607	.0097	.2302	3
58	.12129	.99262	.12219	.1837	.0074	.2446	.13860	.99035	.13995	.1455	.0097	.2152	2
59	.12158	.99258	.12249	.1640	.0075	.2250	.13888	.99031	.14024	.1304	.0098	.2002	1
60	.12187	.99255	.12278	8.1443	.0075	8.2055	.13917	.99027	.14054	7.1154	.0098	7.1853	0
′	83° cos	83° sin	83° cot	83° tan	83° cosec	83° sec	82° cos	82° sin	82° cot	82° tan	82° cosec	82° sec	′

NATURAL TRIGONOMETRIC FUNCTIONS

′	8°						9°						10°						11°						′
′	sin	cos	tan	cot	sec	cosec	sin	cos	tan	cot	sec	cosec	sin	cos	tan	cot	sec	cosec	sin	cos	tan	cot	sec	cosec	′
0	.13917	.99027	.14054	7.1154	1.0098	7.1853	.15643	.98769	.15838	6.3137	1.0215	6.3924	.17365	.98481	.17633	5.6713	1.0154	5.7588	.19081	.98163	.19438	5.1445	1.0187	5.2408	60
1	.13946	.99023	.14084	.1004	.0099	.1704	.15672	.98764	.15868	.3019	.0125	.3807	.17393	.98476	.17663	.6616	.0155	.7493	.19109	.98157	.19468	.1366	.0188	.2330	59
2	.13975	.99019	.14113	.0854	.0099	.1557	.15701	.98760	.15898	.2901	.0125	.3690	.17422	.98471	.17693	.6520	.0155	.7398	.19138	.98152	.19498	.1286	.0188	.2252	58
3	.14004	.99015	.14143	.0706	.0099	.1409	.15730	.98755	.15928	.2783	.0126	.3574	.17451	.98465	.17723	.6425	.0156	.7304	.19166	.98146	.19529	.1207	.0189	.2174	57
4	.14032	.99010	.14173	.0558	.0100	.1263	.15758	.98750	.15958	.2665	.0126	.3458	.17479	.98460	.17753	.6329	.0156	.7210	.19195	.98140	.19559	.1128	.0189	.2097	56
5	.14061	.99006	.14202	7.0410	.0100	7.1117	.15787	.98746	.15987	6.2548	.0127	6.3343	.17508	.98455	.17783	5.6234	.0157	5.7117	.19224	.98135	.19589	5.1049	.0190	5.2019	55
6	.14090	.99002	.14232	.0264	.0101	.0972	.15816	.98741	.16017	.2432	.0217	.3228	.17537	.98450	.17813	.6140	.0157	.7023	.19252	.98129	.19619	.0970	.0191	.1942	54
7	.14119	.98998	.14262	.0117	.0101	.0827	.15844	.98737	.16047	.2316	.0128	.3113	.17565	.98445	.17843	.6045	.0158	.6930	.19281	.98124	.19649	.0892	.0191	.1865	53
8	.14148	.98994	.14291	6.9972	.0102	.0683	.15873	.98732	.16077	.2200	.0128	.2999	.17594	.98440	.17873	.5951	.0158	.6838	.19309	.98118	.19680	.0814	.0192	.1788	52
9	.14176	.98990	.14321	.9827	.0102	.0539	.15902	.98727	.16107	.2085	.0129	.2885	.17622	.98435	.17903	.5857	.0159	.6745	.19338	.98112	.19710	.0736	.0192	.1712	51
10	.14205	.98986	.14351	6.9682	.0102	7.0396	.15931	.98723	.16137	6.1970	.0129	6.2772	.17651	.98430	.17933	5.5764	.0159	5.6653	.19366	.98107	.19740	5.0658	.0193	5.1636	50
11	.14234	.98982	.14380	.9538	.0103	.0254	.15959	.98718	.16167	.1856	.0130	.2659	.17680	.98425	.17963	.5670	.0160	.6561	.19395	.98101	.19770	.0581	.0193	.1560	49
12	.14263	.98978	.14410	.9395	.0103	.0112	.15988	.98714	.16196	.1742	.0130	.2546	.17708	.98419	.17993	.5578	.0160	.6470	.19423	.98095	.19800	.0504	.0194	.1484	48
13	.14292	.98973	.14440	.9252	.0104	6.9971	.16017	.98709	.16226	.1628	.0131	.2434	.17737	.98414	.18023	.5485	.0161	.6379	.19452	.98090	.19831	.0427	.0195	.1409	47
14	.14320	.98969	.14470	.9110	.0104	.9830	.16045	.98704	.16256	.1515	.0131	.2322	.17766	.98409	.18053	.5393	.0162	.6288	.19480	.98084	.19861	.0350	.0195	.1333	46
15	.14349	.98965	.14499	6.8969	.0104	6.9690	.16074	.98700	.16286	6.1402	.0132	6.2211	.17794	.98404	.18083	5.5301	.0162	5.6197	.19509	.98078	.19891	5.0273	.0196	5.1258	45
16	.14378	.98961	.14529	.8828	.0105	.9550	.16103	.98695	.16316	.1290	.0132	.2100	.17823	.98399	.18113	.5209	.0163	.6107	.19537	.98073	.19921	.0197	.0196	.1183	44
17	.14407	.98957	.14559	.8687	.0105	.9411	.16132	.98690	.16346	.1178	.0133	.1990	.17852	.98394	.18143	.5117	.0163	.6017	.19566	.98067	.19952	.0121	.0197	.1109	43
18	.14436	.98952	.14588	.8547	.0106	.9273	.16160	.98685	.16376	.1066	.0133	.1880	.17880	.98388	.18173	.5026	.0164	.5928	.19595	.98061	.19982	.0045	.0198	.1034	42
19	.14464	.98948	.14618	.8408	.0106	.9135	.16189	.98681	.16405	.0955	.0134	.1770	.17909	.98383	.18203	.4936	.0164	.5838	.19623	.98056	.20012	4.9969	.0198	.0960	41
20	.14493	.98944	.14648	6.8269	.0107	6.8998	.16218	.98676	.16435	6.0844	.0134	6.1661	.17937	.98378	.18233	5.4845	.0165	5.5749	.19652	.98050	.20042	4.9894	.0199	5.0886	40
21	.14522	.98940	.14677	.8131	.0107	.8861	.16246	.98671	.16465	.0734	.0135	.1552	.17966	.98373	.18263	.4755	.0165	.5660	.19680	.98044	.20073	.9819	.0199	.0812	39
22	.14551	.98936	.14707	.7993	.0107	.8725	.16275	.98667	.16495	.0624	.0135	.1443	.17995	.98368	.18293	.4665	.0166	.5572	.19709	.98039	.20103	.9744	.0200	.0739	38
23	.14579	.98931	.14737	.7856	.0108	.8589	.16304	.98662	.16525	.0514	.0136	.1335	.18023	.98362	.18323	.4575	.0166	.5484	.19737	.98033	.20133	.9669	.0201	.0666	37
24	.14608	.98927	.14767	.7720	.0108	.8454	.16333	.98657	.16555	.0405	.0136	.1227	.18052	.98357	.18353	.4486	.0167	.5396	.19766	.98027	.20163	.9594	.0201	.0593	36
25	.14637	.98923	.14796	6.7584	.0109	6.8320	.16361	.98652	.16585	6.0296	.0136	6.1120	.18080	.98352	.18383	5.4396	.0167	5.5308	.19794	.98021	.20194	4.9520	.0202	5.0520	35
26	.14666	.98919	.14826	.7448	.0109	.8185	.16390	.98648	.16615	.0188	.0137	.1013	.18109	.98347	.18413	.4308	.0168	.5221	.19823	.98016	.20224	.9446	.0202	.0447	34
27	.14695	.98914	.14856	.7313	.0110	.8052	.16419	.98643	.16644	.0080	.0137	.0906	.18138	.98341	.18444	.4219	.0169	.5134	.19851	.98010	.20254	.9372	.0203	.0375	33
28	.14723	.98910	.14886	.7179	.0110	.7919	.16447	.98638	.16674	5.9972	.0138	.0800	.18166	.98336	.18474	.4131	.0169	.5047	.19880	.98004	.20285	.9298	.0204	.0302	32
29	.14752	.98906	.14915	.7045	.0111	.7787	.16476	.98633	.16704	.9865	.0138	.0694	.18195	.98331	.18504	.4043	.0170	.4960	.19908	.97998	.20315	.9225	.0204	.0230	31
30	.14781	.98901	.14945	6.6911	.0111	6.7655	.16505	.98628	.16734	5.9758	.0139	6.0588	.18223	.98325	.18534	5.3955	.0170	5.4874	.19937	.97992	.20345	4.9151	.0205	5.0158	30
31	.14810	.98897	.14975	.6779	.0111	.7523	.16533	.98624	.16764	.9651	.9139	.0483	.18252	.98320	.18564	.3868	.0171	.4788	.19965	.97987	.20375	.9078	.0205	.0087	29
32	.14838	.98893	.15004	.6646	.0112	.7392	.16562	.98619	.16794	.9545	.9140	.0379	.18281	.98315	.18594	.3780	.0171	.4702	.19994	.97981	.20406	.9006	.0206	.0015	28
33	.14867	.98889	.15034	.6514	.0112	.7262	.16591	.98614	.16824	.9439	.0140	.0274	.18309	.98309	.18624	.3694	.0172	.4617	.20022	.97975	.20436	.8933	.0207	4.9944	27
34	.14896	.98884	.15064	.6383	.0113	.7132	.16619	.98609	.16854	.9333	.0141	.0170	.18338	.98304	.18654	.3607	.0172	.4532	.20051	.97969	.20466	.8860	.0207	.9873	26
35	.14925	.98880	.15094	6.6252	.0113	6.7003	.16648	.98604	.16884	5.9228	.0141	6.0066	.18366	.98299	.18684	5.3521	.0173	5.4447	.20079	.97963	.20497	4.8788	0208	4.9802	25
36	.14953	.98876	.15123	.6122	.0114	.6874	.16677	.98600	.16914	.9123	.0142	5.9963	.18395	.98293	.18714	.3434	.0174	.4362	.20108	.97957	.20527	.8716	.0208	.9732	24
37	.14982	.98871	.15153	.5992	.0114	.6745	.16705	.98595	.16944	.9019	.0142	.9860	.18424	.98288	.18745	.3349	.0174	.4278	.20136	.97952	.20557	.8644	.0209	.9661	23
38	.15011	.98867	.15183	.5863	.0115	.6617	.16734	.98590	.16973	.8915	.0143	.9758	.18452	.98283	.18775	.3263	.0175	.4194	.20165	.97946	.20588	.8573	.0210	.9591	22
39	.15050	.98862	.15213	.5734	.0115	.6490	.16763	.98585	.17003	.8811	.0143	.9655	.18481	.98277	.18805	.3178	.0175	.4110	.20193	.97940	.20618	.8501	.0210	.9521	21
40	.15068	.98858	.15243	6.5605	.0115	6.6363	.16791	.98580	.17033	5.8708	.0144	5.9554	.18509	.98272	.18835	5.3093	.0176	5.4026	.20222	.97934	.20648	4.8430	.0211	4.9452	20
41	.15097	.98854	.15272	.5478	.0116	.6237	.16820	.98575	.17063	.8605	.0144	.9452	.18538	.98267	.18865	.3008	.0176	.3943	.20250	.97928	.20679	.8359	.0211	.9382	19
42	.15126	.98849	.15302	.5350	.0116	.6111	.16849	.98570	.17093	.8502	.0145	.9351	.18567	.98261	.18895	.2923	.0177	.3860	.20279	.97922	.20709	.8288	.0212	.9313	18
43	.15155	.98845	.15332	.5223	.0117	.5985	.16878	.98565	.17123	.8400	.0145	.9250	.18595	.98256	.18925	.2839	.0177	.3777	.20307	.97916	.20739	.8217	.0213	.9243	17
44	.15183	.98840	.15362	.5097	.0117	.5860	.16906	.98560	.17153	.8298	.0146	.9150	.18624	.98250	.18955	.2755	.0178	.3695	.20336	.97910	.20770	.8147	.0213	.9175	16
45	.15212	.98836	.15391	6.4971	.0118	6.5736	.16935	.98556	.17183	5.8196	.0146	5.9049	.18652	.98245	.18985	5.2671	.0179	5.3612	.20364	.97904	.20800	4.8077	.0214	4.9106	15
46	.15241	.98832	.15421	.4845	.0118	.5612	.16964	.98551	.17213	.8095	.0147	.8950	.18681	.98240	.19016	.2588	.0179	.3530	.20393	.97899	.20830	.8007	.0215	.9037	14
47	.15270	.98827	.15451	.4720	.0119	.5488	.16992	.98546	.17243	.7994	.0147	.8850	.18709	.98234	.19046	.2505	.0180	.3449	.20421	.97893	.20861	.7937	.0215	.8969	13
48	.15298	.98823	.15481	.4596	.0119	.5365	.17021	.98541	.17273	.7894	.0148	.8751	.18738	.98229	.19076	.2422	.0180	.3367	.20450	.97887	.20891	.7867	.0216	.8901	12
49	.15328	.98818	.15511	.4472	.0119	.5243	.17050	.98536	.17303	.7794	.0148	.8652	.18767	.98223	.19106	.2339	.0181	.3286	.20478	.97881	.20921	.7798	.0216	.8833	11
50	.15356	.98814	.15540	6.4348	.0120	6.5121	.17078	.98531	.17333	5.7694	.0149	5.8554	.18795	.98218	.19136	5.2257	.0181	5.3205	.20506	.97875	.20952	4.7728	.0217	4.8765	10
51	.15385	.98809	.15570	.4225	.0120	.4999	.17107	.98526	.17363	.7594	.0150	.8456	.18824	.98212	.19166	.2174	.0182	.3124	.20535	.97869	.20982	.7659	.0218	.8967	9
52	.15413	.98805	.15600	.4103	.0121	.4878	.17136	.98521	.17393	.7495	.0150	.8358	.18852	.98207	.19197	.2092	.0182	.3044	.20563	.97863	.21012	.7591	.0218	.8630	8
53	.15442	.98800	.15630	.3980	.0121	.4757	.17164	.98516	.17423	.7396	.0151	.8261	.18881	.98201	.19227	.2011	.0183	.2963	.20592	.97857	.21043	.7522	.0219	.8563	7
54	.15471	.98796	.15659	.3859	.0122	.4637	.17193	.98511	.17453	.7297	.0151	.8163	.18909	.98196	.19257	.1929	.0184	.2883	.20620	.97851	.21073	.7453	.0220	.8496	6
55	.15500	.98791	.15689	6.3737	.0122	6.4517	.17221	.98506	.17483	5.7199	.0152	5.8067	.18938	.98190	.19287	5.1848	.0184	5.2803	.20649	.97845	.21104	4.7385	.0220	4.8429	5
56	.15528	.98787	.15719	.3616	.0123	.4398	.17250	.98501	.17513	.7101	.0152	.7970	.18967	.98185	.19317	.1767	.0185	.2724	.20677	.97839	.21134	.7317	.0221	.8362	4
57	.15557	.98782	.15749	.3496	.0123	.4279	.17279	.98496	.17543	.7004	.0153	.7874	.18995	.98179	.19347	.1686	.0185	.2645	.20706	.97833	.21164	.7249	.0221	.8296	3
58	.15586	.98778	.15779	.3376	.0124	.4160	.17307	.98491	.17573	.6906	.0153	.7778	.19024	.98174	.19378	.1606	.0186	.2566	.20734	.97827	.21195	.7181	.0222	.8229	2
59	.15615	.98773	.15809	.3257	.0124	.4042	.17336	.98486	.17603	.6809	.0154	.7683	.19052	.98168	.19408	.1525	.0186	.2487	.20763	.97821	.21225	.7114	.0223	.8163	1
60	.15643	.98769	.15838	6.3137	.0125	6.3924	.17365	.98481	.17633	5.6713	.0154	5.7588	.19081	.98163	.19438	5.1445	.0187	5.2408	.20791	.97815	.21256	4.7046	.0223	4.8097	0
′	cos	sin	cot	tan	cosec	sec	cos	sin	cot	tan	cosec	sec	cos	sin	cot	tan	cosec	sec	cos	sin	cot	tan	cosec	sec	′
	81°						80°						79°						78°						

NATURAL TRIGONOMETRIC FUNCTIONS

′	12°						13°						14°						15°						′
′	sin	cos	tan	cot	sec	cosec	sin	cos	tan	cot	sec	cosec	sin	cos	tan	cot	sec	cosec	sin	cos	tan	cot	sec	cosec	′
0	.20791	.97815	.21256	4.7046	1.0223	4.8097	.22495	.97437	.23087	4.3315	1.0263	4.4454	.24192	.97029	.24933	4.0108	1.0306	4.1336	.25882	.96592	.26795	3.7320	1.0353	3.8637	60
1	.20820	.97809	.21286	.6979	.0224	.8032	.22523	.97430	.23117	.3257	.0264	.4398	.24220	.97022	.24964	.0058	.0307	.1287	.25910	.96585	.26826	.7277	.0353	.8595	59
2	.20848	.97803	.21316	.6912	.0225	.7966	.22552	.97424	.23148	.3200	.0264	.4342	.24249	.97015	.24995	.0009	.0308	.1239	.25938	.96577	.26857	.7234	.0354	.8553	58
3	.20876	.97797	.21347	.6845	.0225	.7901	.22580	.97417	.23179	.3143	.0265	.4287	.24277	.97008	.25025	3.9959	.0308	.1191	.25966	.96570	.26888	.7191	.0355	.8512	57
4	.20905	.97790	.21377	.6778	.0226	.7835	.22608	.97411	.23209	.3086	.0266	.4231	.24305	.97001	.25056	.9910	.0309	.1144	.25994	.96562	.26920	.7147	.0356	.8470	56
5	.20933	.97784	.21408	4.6712	.0226	4.7770	.22637	.97404	.23240	4.3029	.0266	4.4176	.24333	.96994	.25087	3.9861	.0310	4.1096	.26022	.96555	.26951	3.7104	.0357	3.8428	55
6	.20962	.97778	.21438	.6646	.0227	.7706	.22665	.97398	.23270	.2972	.0267	.4121	.24361	.96987	.25118	.9812	.0311	.1048	.26050	.96547	.26982	.7062	.0358	.8387	54
7	.20990	.97772	.21468	.6580	.0228	.7641	.22693	.97391	.23301	.2916	.0268	.4065	.24390	.96980	.25149	.9763	.0311	.1001	.26078	.96540	.27013	.7019	.0358	.8346	53
8	.21019	.97766	.21499	.6514	.0228	.7576	.22722	.97384	.23332	.2859	.0268	.4011	.24418	.96973	.25180	.9714	.0312	.0953	.26107	.96532	.27044	.6976	.0359	.8304	52
9	.21047	.97760	.21529	.6448	.0229	.7512	.22750	.97378	.23363	.2803	.0269	.3956	.24446	.96966	.25211	.9665	.0313	.0906	.26135	.96524	.27076	.6933	.0360	.8263	51
10	.21076	.97754	.21560	4.6382	.0230	4.7448	.22778	.97371	.23393	4.2747	.0270	4.3901	.24474	.96959	.25242	3.9616	.0314	4.0859	.26163	.96517	.27107	3.6891	.0361	3.8222	50
11	.21104	.97748	.21590	.6317	.0230	.7384	.22807	.97364	.23424	.2691	.0271	.3847	.24502	.96952	.25273	.9568	.0314	.0812	.26191	.96509	.27138	.6848	.0362	.8181	49
12	.21132	.97741	.21621	.6252	.0231	.7320	.22835	.97358	.23455	.2635	.0271	.3792	.24531	.96944	.25304	.9520	.0315	.0765	.26219	.96502	.27169	.6806	.0362	.8140	48
13	.21161	.97735	.21651	.6187	.0232	.7257	.22863	.97351	.23485	.2579	.0272	.3738	.24559	.96937	.25335	.9471	.0316	.0718	.26247	.96494	.27201	.6764	.0363	.8100	47
14	.21189	.97729	.21682	.6122	.0232	.7193	.22892	.97344	.23516	.2524	.0273	.3684	.24587	.96930	.25366	.9423	.0317	.0672	.26275	.96486	.27232	.6722	.0364	.8059	46
15	.21218	.97723	.21712	4.6057	.0233	4.7130	.22920	.97338	.23547	4.2468	.0273	4.3630	.24615	.96923	.25397	3.9375	.0317	4.0625	.26303	.96479	.27263	3.6679	.0365	3.8018	45
16	.21246	.97717	.21742	.5993	.0234	.7067	.22948	.97331	.23577	.2413	.0274	.3576	.24643	.96916	.25428	.9327	.0318	.0579	.26331	.96471	.27294	.6637	.0366	.7978	44
17	.21275	.97711	.21773	.5928	.0234	.7004	.22977	.97324	.23608	.2358	.0275	.3522	.24672	.96909	.25459	.9279	.0319	.0532	.26359	.96463	.27326	.6596	.0367	.7937	43
18	.21303	.97704	.21803	.5864	.0235	.6942	.23005	.97318	.23639	.2303	.0276	.3469	.24700	.96901	.25490	.9231	.0320	.0486	.26387	.96456	.27357	.6554	.0367	.7897	42
19	.21331	.97698	.21834	.5800	.0235	.6879	.23033	.97311	.23670	.2248	.0276	.3415	.24728	.96894	.25521	.9184	.0320	.0440	.26415	.96448	.27388	.6512	.0368	.7857	41
20	.21360	.97692	.21864	4.5736	.0236	4.6817	.23061	.97304	.23700	4.2193	.0277	4.3362	.24756	.96887	.25552	3.9136	.0321	4.0394	.26443	.96440	.27419	3.6470	.0369	3.7816	40
21	.21388	.97686	.21895	.5673	.0237	.6754	.23090	.97298	.23731	.2139	.0278	.3309	.24784	.96880	.25583	.9089	.0322	.0348	.26471	.96433	.27451	.6429	.0370	.7776	39
22	.21417	.97680	.21925	.5609	.0237	.6692	.23118	.97291	.23762	.2084	.0278	.3256	.24813	.96873	.25614	.9042	.0323	.0302	.26499	.96425	.27482	.6387	.0371	.7736	38
23	.21445	.97673	.21956	.5546	.0238	.6631	.23146	.97284	.23793	.2030	.0279	.3203	.24841	.96865	.25645	.8994	.0323	.0256	.26527	.96417	.27513	.6346	.0371	.7697	37
24	.21473	.97667	.21986	.5483	.0239	.6569	.23175	.97277	.23823	.1976	.0280	.3150	.24869	.96858	.25676	.8947	.0324	.0211	.26556	.96409	.27544	.6305	.0372	.7657	36
25	.21502	.97661	.22017	4.5420	.0239	4.6507	.23202	.97271	.23854	4.1921	.0280	4.3098	.24897	.96851	.25707	3.8900	.0325	4.0165	.26584	.96402	.27576	3.6263	.0373	3.7617	35
26	.21530	.97655	.22047	.5357	.0240	.6446	.23231	.97264	.23885	.1867	.0281	.3045	.24925	.96844	.25738	.8853	.0326	.0120	.26612	.96394	.27607	.6222	.0374	.7577	34
27	.21559	.97648	.22078	.5294	.0241	.6385	.23260	.97257	.23916	.1814	.0282	.2993	.24953	.96836	.25769	.8807	.0327	.0074	.26640	.96386	.27638	.6181	.0375	.7538	33
28	.21587	.97642	.22108	.5232	.0241	.6324	.23288	.97250	.23946	.1760	.0283	.2941	.24982	.96829	.25800	.8760	.0327	.0029	.26668	.96378	.27670	.6140	.0376	.7498	32
29	.21615	.97636	.22139	.5169	.0242	.6263	.23316	.97244	.23977	.1706	.0283	.2838	.25010	.96822	.25831	.8713	.0328	3.9984	.26696	.96371	.27701	.6100	.0376	.7459	31
30	.21644	.97630	.22169	4.5107	.0243	4.6201	.23344	.97237	.24008	4.1653	.0284	4.2836	.25038	.96815	.25862	3.8667	.0329	3.9939	.26724	.96363	.27732	3.6059	.0377	3.7420	30
31	.21672	.97623	.22200	.5045	.0243	.6142	.23373	.97230	.24039	.1600	.0285	.2785	.25066	.96807	.25893	.8621	.0330	.9894	.26752	.96355	.27764	.6018	.0378	.7380	29
32	.21701	.97617	.22230	.4983	.0244	.6081	.23401	.97223	.24069	.1546	.0285	.2733	.25094	.96800	.25924	.8574	.0330	.9850	.26780	.96347	.27795	.5977	.0379	.7341	28
33	.21729	.97611	.22261	.4921	.0245	.6021	.23429	.97216	.24100	.1493	.0286	.2681	.25122	.96793	.25955	.8528	.0331	.9805	.26808	.96340	.27826	.5937	.0380	.7302	27
34	.21757	.97604	.22291	.4860	.0245	.5961	.23458	.97210	.24131	.1440	.0287	.2630	.25151	.96785	.25986	.8482	.0332	.97[illegible]	.26836	.96332	.27858	.5896	.0381	.7263	26
35	.21786	.97598	.22322	4.4799	.0246	4.5901	.23486	.97203	.24162	4.1388	.0288	4.2579	.25179	.96778	.26017	3.8436	.0333	3.9716	.26864	.96324	.27889	3.5856	.0382	3.7224	25
36	.21814	.97592	.22353	.4737	.0247	.5841	.23514	.97196	.24192	.1335	.0288	.2527	.25207	.96771	.26048	.8390	.0334	.9672	.26892	.96316	.27920	.5816	.0382	.7186	24
37	.21843	.97575	.22383	.4676	.0247	.5782	.23542	.97189	.24223	.1282	.0289	.2476	.25235	.96763	.26079	.8345	.0334	.9627	.26920	.96308	.27952	.5776	.0383	.7147	23
38	.21871	.97579	.22414	.4615	.0248	.5722	.23571	.97182	.24254	.1230	.0290	.2425	.25263	.96756	.26110	.8299	.0335	.9583	.26948	.96301	.27983	.5736	.0384	.7108	22
39	.21899	.97573	.22444	.4555	.0249	.5663	.23599	.97175	.24285	.1178	.0291	.2375	.25291	.96749	.26141	.8254	.0336	.9539	.26976	.96293	.28014	.5696	.0385	.7070	21
40	.21928	.97566	.22475	4.4494	.0249	4.5604	.23627	.97169	.24316	4.1126	.0291	4.2324	.25319	.96741	.26172	3.8208	.0337	3.9495	.27004	.96285	.28046	3.5656	.0386	3.7031	20
41	.21956	.97560	.22505	.4434	.0250	.5545	.23655	.97162	.24346	.1073	.0292	.2273	.25348	.96734	.26203	.8163	.0338	.9451	.27032	.96277	.28077	.5616	.0387	.6993	19
42	.21985	.97553	.22536	.4373	.0251	.5486	.23684	.97155	.24377	.1022	.0293	.2223	.25376	.96727	.26234	.8118	.0338	.9408	.27060	.96269	.28109	.5576	.0387	.6955	18
43	.22013	.97547	.22566	.4313	.0251	.5428	.23712	.97148	.24408	.0970	.0293	.2173	.25404	.96719	.26266	.8073	.0339	.9364	.27088	.96261	.28140	.5536	.0388	.6917	17
44	.22041	.97541	.22597	.4253	.0252	.5369	.23740	.97141	.24439	.0918	.0294	.2122	.25432	.96712	.26297	.8027	.0340	.9320	.27116	.96253	.28171	.5497	.0389	.6878	16
45	.22070	.97534	.22628	4.4194	.0253	4.5311	.23768	.97134	.24470	4.0867	.0295	4.2072	.25460	.96704	.26328	3.7983	.0341	3.9277	.27144	.96245	.28203	3.5457	.0390	3.6840	15
46	.22098	.97528	.22658	.4134	.0253	.5253	.23797	.97127	.24501	.0815	.0296	.2022	.25488	.96697	.26359	.7938	.0341	.9234	.27172	.96238	.28234	.5418	.0391	.6802	14
47	.22126	.97521	.22689	.4074	.0254	.5195	.23825	.97120	.24531	.0764	.0296	.1972	.25516	.96690	.26390	.7893	.0342	.9190	.27200	.96230	.28266	.5378	.0392	.6765	13
48	.22155	.97515	.22719	.4015	.0255	.5137	.23853	.97113	.24562	.0713	.0297	.1923	.25544	.96682	.26421	.7848	.0343	.9147	.27228	.96222	.28297	.5339	.0393	.6727	12
49	.22183	.97508	.22750	.3956	.0255	.5079	.23881	.97106	.24593	.0662	.0298	.1873	.25573	.96675	.26452	.7804	.0344	.9104	.27256	.96214	.28328	.5300	.0393	.6689	11
50	.22211	.97502	.22781	4.3897	.0256	4.5021	.23910	.97099	.24624	4.0611	.0299	4.1824	.25601	.96667	.26483	3.7759	.0345	3.9061	.27284	.96206	.28360	3.5261	.0394	3.6651	10
51	.22240	.97495	.22811	.3838	.0257	.4964	.23938	.97092	.24655	.0560	.0299	.1774	.25629	.96660	.26514	.7715	.0345	.9018	.27312	.96198	.28391	.5222	.0395	.6614	9
52	.22268	.97489	.22842	.3779	.0257	.4907	.23966	.97086	.24686	.0509	.0300	.1725	.25657	.96652	.26546	.7671	.0346	.8976	.27340	.96190	.28423	.5183	.0396	.6576	8
53	.22297	.97483	.22872	.3721	.0258	.4850	.23994	.97079	.24717	.0458	.0301	.1676	.25685	.96645	.26577	.7627	.0347	.8933	.27368	.96182	.28454	.5144	.0397	.6539	7
54	.22325	.97476	.22903	.3662	.0259	.4793	.24023	.97072	.24747	.0408	.0302	.1627	.25713	.96638	.26608	.7583	.0348	.8890	.27396	.96174	.28486	.5105	.0398	.6502	6
55	.22353	.97470	.22934	4.3604	.0260	4.4736	.24051	.97065	.24778	4.0358	.0302	4.1578	.25741	.96630	.26639	3.7539	.0349	3.8848	.27424	.96166	.28517	3.5066	.0399	3.6464	5
56	.22382	.97463	.22964	.3546	.0260	.4679	.24079	.97058	.24809	.0307	.0303	.1529	.25769	.96623	.26670	.7495	.0349	.8805	.27452	.96158	.28549	.5028	.0399	.6427	4
57	.22410	.97457	.22995	.3488	.0261	.4623	.24107	.97051	.24840	.0257	.0304	.1481	.25798	.96615	.26701	.7451	.0350	.8763	.27480	.96150	.28580	.4989	.0400	.6390	3
58	.22438	.97450	.23025	.3430	.0262	.4566	.24136	.97044	.24871	.0207	.0305	.1432	.25826	.96608	.26732	.7407	.0351	.8721	.27508	.96142	.28611	.4951	.0401	.6353	2
59	.22467	.97443	.23056	.3372	.0262	.4510	.24164	.97037	.24902	.0157	.0305	.1384	.25854	.96600	.26764	.7364	.0352	.8679	.27536	.96134	.28643	.4912	.0402	.6316	1
60	.22495	.97437	.23087	4.3315	.0263	4.4454	.24192	.97029	.24933	4.0108	.0306	4.1336	.25882	.96592	.26795	3.7320	.0353	3.8637	.27564	.96126	.28674	3.4874	.0403	3.6279	0
′	cos	sin	cot	tan	cosec	sec	cos	sin	cot	tan	cosec	sec	cos	sin	cot	tan	cosec	sec	cos	sin	cot	tan	cosec	sec	′
	77°						76°						75°						74°						

NATURAL TRIGONOMETRIC FUNCTIONS

	16°						17°						18°						19°						
′	sin	cos	tan	cot	sec	cosec	sin	cos	tan	cot	sec	cosec	sin	cos	tan	cot	sec	cosec	sin	cos	tan	cot	sec	cosec	′
0	.27564	.96126	.28674	3.4874	1.0403	3.6279	.29237	.95630	.30573	3.2708	1.0457	3.4203	.30902	.95106	.32492	3.0777	1.0515	3.2361	.32557	.94552	.34433	2.9042	1.0576	3.0715	60
1	.27592	.96118	.28706	.4836	.0404	.6243	.29265	.95622	.30605	.2674	.0458	.4170	.30929	.95097	.32524	.0746	.0516	.2332	.32584	.94542	.34465	.9015	.0577	.0690	59
2	.27620	.96110	.28737	.4798	.0405	.6206	.29293	.95613	.30637	.2640	.0459	.4138	.30957	.95088	.32556	.0716	.0517	.2303	.32612	.94533	.34498	.8987	.0578	.0664	58
3	.27648	.96102	.28769	.4760	.0406	.6169	.29321	.95605	.30668	.2607	.0460	.4106	.30985	.95079	.32588	.0686	.0518	.2274	.32639	.94523	.34530	.8960	.0579	.0638	57
4	.27675	.96094	.28800	.4722	.0406	.6133	.29348	.95596	.30700	.2573	.0461	.4073	.31012	.95070	.32621	.0655	.0519	.2245	.32667	.94514	.34563	.8933	.0580	.0612	56
5	.27703	.96086	.28832	3.4684	.0407	3.6096	.29376	.95588	.30732	3.2539	.0461	3.4041	.31040	.95061	.32653	3.0625	.0520	3.2216	.32694	.94504	.34595	2.8905	.0581	3.0586	55
6	.27731	.96078	.28863	.4646	.0408	.6060	.29404	.95579	.30764	.2505	.0462	.4009	.31068	.95051	.32685	.0595	.0521	.2188	.32722	.94495	.34628	.8878	.0582	.0561	54
7	.27759	.96070	.28895	.4608	.0409	.6024	.29432	.95571	.30796	.2472	.0463	.3977	.31095	.95042	.32717	.0565	.0522	.2159	.32749	.94485	.34661	.8851	.0584	.0535	53
8	.27787	.96062	.28926	.4570	.0410	.5987	.29460	.95562	.30828	.2438	.0464	.3945	.31123	.95033	.32749	.0535	.0523	.2131	.32777	.94476	.34693	.8824	.0585	.0509	52
9	.27815	.96054	.28958	.4533	.0411	.5951	.29487	.95554	.30859	.2405	.0465	.3913	.31150	.95024	.32782	.0505	.0524	.2102	.32804	.94466	.34726	.8797	.0586	.0484	51
10	.27843	.96045	.28990	3.4495	.0412	3.5915	.29515	.95545	.30891	3.2371	.0466	3.3881	.31178	.95015	.32814	3.0475	.0525	3.2074	.32832	.94457	.34758	2.8770	.0587	3.0458	50
11	.27871	.96037	.29021	.4458	.0413	.5879	.29543	.95536	.30923	.2338	.0467	.3849	.31206	.95006	.32846	.0445	.0526	.2045	.32859	.94447	.34791	.8743	.0588	.0433	49
12	.27899	.96029	.29053	.4420	.0413	.5843	.29571	.95528	.30955	.2305	.0468	.3817	.31233	.94997	.32878	.0415	.0527	.2017	.32887	.94438	.34824	.8716	.0589	.0407	48
13	.27927	.96021	.29084	.4383	.0414	.5807	.29598	.95519	.30987	.2271	.0469	.3785	.31261	.94988	.32910	.0385	.0528	.1989	.32914	.94428	.34856	.8689	.0590	.0382	47
14	.27955	.96013	.29116	.4346	.0415	.5772	.29626	.95511	.31019	.2238	.0470	.3754	.31289	.94979	.32943	.0356	.0529	.1960	.32942	.94418	.34889	.8662	.0591	.0357	46
15	.27983	.96005	.29147	3.4308	.0416	3.5736	.29654	.95502	.31051	3.2205	.0471	3.3722	.31316	.94970	.32975	3.0326	.0530	3.1932	.32969	.94409	.34921	2.8636	.0592	3.0331	45
16	.28011	.95997	.29179	.4271	.0417	.5700	.29682	.95493	.31083	.2172	.0472	.3690	.31344	.94961	.33007	.0296	.0531	.1904	.32996	.94399	.34954	.8609	.0593	.0306	44
17	.28039	.95989	.29210	.4234	.0418	.5665	.29710	.95485	.31115	.2139	.0473	.3659	.31372	.94952	.33039	.0267	.0532	.1876	.33024	.94390	.34987	.8582	.0594	.0281	43
18	.28067	.95980	.29242	.4197	.0419	.5629	.29737	.95476	.31146	.2106	.0474	.3627	.31399	.94942	.33072	.0237	.0533	.1848	.33051	.94380	.35019	.8555	.0595	.0256	42
19	.28094	.95972	.29274	.4160	.0420	.5594	.29765	.95467	.31178	.2073	.0475	.3596	.31427	.94933	.33104	.0208	.0534	.1820	.33079	.94370	.35052	.8529	.0596	.0231	41
20	.28122	.95964	.29305	3.4124	.0420	3.5559	.29793	.95459	.31210	3.2041	.0476	3.3565	.31454	.94924	.33136	3.0178	.0535	3.1792	.33106	.94361	.35085	2.8502	.0598	3.0206	40
21	.28150	.95956	.29337	.4087	.0421	.5523	.29821	.95450	.31242	.2008	.0477	.3534	.31482	.94915	.33169	.0149	.0536	.1764	.33134	.94351	.35117	.8476	.0599	.0181	39
22	.28178	.95948	.29368	.4050	.0422	.5488	.29848	.95441	.31274	.1975	.0478	.3502	.31510	.94906	.33201	.0120	.0537	.1736	.33161	.94341	.35150	.8449	.0600	.0156	38
23	.28206	.95940	.29400	.4014	.0423	.5453	.29876	.95433	.31306	.1942	.0478	.3471	.31537	.94897	.33233	.0090	.0528	.1708	.33189	.94332	.35183	.8423	.0601	.0131	37
24	.28234	.95931	.29432	.3977	.0424	.5418	.29904	.95424	.31338	.1910	.0479	.3440	.31565	.94888	.33265	.0061	.0539	.1681	.33216	.94322	.35215	.8396	.0602	.0106	36
25	.28262	.95923	.29463	3.3941	.0425	3.5383	.29932	.95415	.31370	3.1877	.0480	3.3409	.31592	.94878	.33298	3.0032	.0540	3.1653	.33243	.94313	.35248	2.8370	.0603	3.0081	35
26	.28290	.95915	.29495	.3904	.0426	.5348	.29959	.95407	.31402	.1845	.0481	.3378	.31620	.94869	.33330	.0003	.0541	.1625	.33271	.94303	.35281	.8344	.0604	.0056	34
27	.28318	.95907	.29526	.3868	.0427	.5313	.29987	.95398	.31434	.1813	.0482	.3347	.31648	.94860	.33362	2.9974	.0542	.1598	.33298	.94293	.35314	.8318	.0605	.0031	33
28	.28346	.95898	.29558	.3832	.0428	.5279	.30015	.95389	.31466	.1780	.0483	.3316	.31675	.94851	.33395	.9945	.0543	.1570	.33326	.94283	.35346	.8291	.0606	.0007	32
29	.28374	.95890	.29590	.3795	.0428	.5244	.30043	.95380	.31498	.1748	.0484	.3286	.31703	.94841	.33427	.9916	.0544	.1543	.33353	.94274	.35379	.8265	.0607	2.9982	31
30	.28401	.95882	.29621	3.3759	.0429	3.5209	.30070	.95372	.31530	3.1716	.0485	3.3255	.31730	.94832	.33459	2.9887	.0545	3.1515	.33381	.94264	.35412	2.8239	.0608	2.9957	30
31	.28429	.95874	.29653	.3723	.0430	.5175	.30098	.95363	.31562	.1684	.0486	.3224	.31758	.94823	.33492	.9858	.0546	.1488	.33408	.94254	.35445	.8213	.0609	.9933	29
32	.28457	.95865	.29685	.3687	.0431	.5140	.30126	.95354	.31594	.1652	.0487	.3194	.31786	.94814	.33524	.9829	.0547	.1461	.33435	.94245	.35477	.8187	.0611	.9908	28
33	.28485	.95857	.29716	.3651	.0432	.5106	.30154	.95345	.31626	.1620	.0488	.3163	.31813	.94805	.33557	.9800	.0548	.1433	.33463	.94235	.35510	.8161	.0612	.9884	27
34	.28513	.95849	.29748	.3616	.0433	.5072	.30181	.95337	.31658	.1588	.0489	.3133	.31841	.94795	.33589	.9772	.0549	.1406	.33490	.94225	.35543	.8135	.0613	.9859	26
35	.28541	.95840	.29780	3.3580	.0434	3.5037	.30209	.95328	.31690	3.1556	.0490	3.3102	.31868	.94786	.33621	2.9743	.0550	3.1379	.33518	.94215	.35576	2.8109	.0614	2.9835	25
36	.28569	.95832	.29811	.3544	.0435	.5003	.30237	.95319	.31722	.1524	.0491	.3072	.31896	.94777	.33654	.9714	.0551	.1352	.33545	.94206	.35608	.8083	.0615	.9810	24
37	.28597	.95824	.29843	.3509	.0436	.4969	.30265	.95310	.31754	.1492	.0492	.3042	.31923	.94767	.33686	.9686	.0552	.1325	.33572	.94196	.35641	.8057	.0616	.9786	23
38	.28624	.95816	.29875	.3473	.0437	.4935	.30292	.95301	.31786	.1460	.0493	.3011	.31951	.94758	.33718	.9657	.0553	.1298	.33600	.94186	.35674	.8032	.0617	.9762	22
39	.28652	.95807	.29906	.3438	.0438	.4901	.30320	.95293	.31818	.1429	.0494	.2981	.31978	.94749	.33751	.9629	.0554	.1271	.33627	.94176	.35707	.8006	.0618	.9738	21
40	.28680	.95799	.29938	3.3402	.0438	3.4867	.30348	.95284	.31850	3.1397	.0495	3.2951	.32006	.94740	.33783	2.9600	.0555	3.1244	.33655	.94167	.35739	2.7980	.0619	2.9713	20
41	.28708	.95791	.29970	.3367	.0439	.4833	.30375	.95275	.31882	.1366	.0496	.2921	.32034	.94730	.33816	.9572	.0556	.1217	.33682	.94157	.35772	.7954	.0620	.9689	19
42	.28736	.95782	.30001	.3332	.0440	.4799	.30403	.95266	.31914	.1334	.0497	.2891	.32061	.94721	.33848	.9544	.0557	.1190	.33709	.94147	.35805	.7929	.0622	.9665	18
43	.28764	.95774	.30033	.3296	.0441	.4766	.30431	.95257	.31946	.1303	.0498	.2861	.32089	.94712	.33880	.9515	.0558	.1163	.33737	.94137	.35838	.7903	.0623	.9641	17
44	.28792	.95765	.30065	.3261	.0442	.4732	.30459	.95248	.31978	.1271	.0499	.2831	.32116	.94702	.33913	.9487	.0559	.1137	.33764	.94127	.35871	.7878	.0624	.9617	16
45	.28820	.95757	.30096	3.3226	.0443	3.4698	.30486	.95239	.32010	3.1240	.0500	3.2801	.32144	.94693	.33945	2.9459	.0560	3.1110	.33792	.94118	.35904	2.7852	.0625	2.9593	15
46	.28847	.95749	.30128	.3191	.0444	.4665	.30514	.95231	.32042	.1209	.0501	.2772	.32171	.94684	.33978	.9431	.0561	.1083	.33819	.94108	.35936	.7827	.0626	.9569	14
47	.28875	.95740	.30160	.3156	.0445	.4632	.30542	.95222	.32074	.1177	.0502	.2742	.32199	.94674	.34010	.9403	.0562	.1057	.33846	.94098	.35969	.7801	.0627	.9545	13
48	.28903	.95732	.30192	.3121	.0446	.4598	.30569	.95213	.32106	.1146	.0503	.2712	.32226	.94665	.34043	.9375	.0563	.1030	.33874	.94088	.36002	.7776	.0628	.9521	12
49	.28931	.95723	.30223	.3087	.0447	.4565	.30597	.95204	.32138	.1115	.0504	.2683	.32254	.94655	.34075	.9347	.0565	.1004	.33901	.94078	.36035	.7751	.0629	.9497	11
50	.28959	.95715	.30255	3.3052	.0448	3.4532	.30625	.95195	.32171	3.1084	.0505	3.2653	.32282	.94646	.34108	2.9319	.0566	3.0977	.33928	.94068	.36068	2.7725	.0630	2.9474	10
51	.28987	.95707	.30287	.3017	.0448	.4498	.30653	.95186	.32203	.1053	.0506	.2624	.32309	.94637	.34140	.9291	.0567	.0951	.33956	.94058	.36101	.7700	.0632	.9450	9
52	.29014	.95698	.30319	.2983	.0449	.4465	.30680	.95177	.32235	.1022	.0507	.2594	.32337	.94627	.34173	.9263	.0568	.0925	.33983	.94049	.36134	.7675	.0633	.9426	8
53	.29042	.95690	.30350	.2948	.0450	.4432	.30708	.95168	.32267	.0991	.0508	.2565	.32364	.94618	.34205	.9235	.0569	.0898	.34011	.94039	.36167	.7650	.0634	.9402	7
54	.29070	.95681	.30382	.2914	.0451	.4399	.30736	.95159	.32299	.0960	.0509	.2535	.32392	.94608	.34238	.9208	.0570	.0872	.34038	.94029	.36199	.7625	.0635	.9379	6
55	.29098	.95673	.30414	3.2879	.0452	3.4366	.30763	.95150	.32331	3.0930	.0510	3.2506	.32419	.94599	.34270	2.9180	.0571	3.0846	.34065	.94019	.36232	2.7600	.0636	2.9355	5
56	.29126	.95664	.30446	.2845	.0453	.4334	.30791	.95141	.32363	.0899	.0511	.2477	.32447	.94590	.34303	.9152	.0572	.0820	.34093	.94009	.36265	.7575	.0637	.9332	4
57	.29154	.95656	.30478	.2811	.0454	.4301	.30819	.95132	.32395	.0868	.0512	.2448	.32474	.94580	.34335	.9125	.0573	.0793	.34120	.93999	.36298	.7550	.0638	.9308	3
58	.29181	.95647	.30509	.2777	.0455	.4268	.30846	.95124	.32428	.0838	.0513	.2419	.32502	.94571	.34368	.9097	.0574	.0767	.34147	.93989	.36331	.7525	.0639	.9285	2
59	.29209	.95639	.30541	.2742	.0456	.4236	.30874	.95115	.32460	.0807	.0514	.2390	.32529	.94561	.34400	.9069	.0575	.0741	.34175	.93979	.36364	.7500	.0641	.9261	1
60	.29237	.95630	.30573	3.2708	.0457	3.4203	.30902	.95106	.32492	3.0777	.0515	3.2361	.32557	.94552	.34433	2.9042	.0576	3.0715	.34202	.93969	.36397	2.7475	.0642	2.9238	0
′	cos	sin	cot	tan	cosec	sec	cos	sin	cot	tan	cosec	sec	cos	sin	cot	tan	cosec	sec	cos	sin	cot	tan	cosec	sec	′
	73°						72°						71°						70°						

NATURAL TRIGONOMETRIC FUNCTIONS

′	20°						21°					
′	sin	cos	tan	cot	sec	cosec	sin	cos	tan	cot	sec	cosec
0	.34202	.93969	.36397	2.7475	1.0642	2.9238	.35837	.93358	.38386	2.6051	1.0711	2.7904
1	.34229	.93959	.36430	.7450	.0643	.9215	.35864	.93348	.38420	.6028	.0713	.7883
2	.34257	.93949	.36463	.7425	.0644	.9191	.35891	.93337	.38453	.6006	.0714	.7862
3	.34284	.93939	.36496	.7400	.0645	.9168	.35918	.93327	.38486	.5983	.0715	.7841
4	.34311	.93929	.36529	.7376	.0646	.9145	.35945	.93316	.38520	.5960	.0716	.7820
5	.34339	.93919	.36562	2.7351	.0647	2.9122	.35972	.93306	.38553	2.5938	.0717	2.7799
6	.34366	.93909	.36595	.7326	.0648	.9098	.36000	.93295	.38587	.5916	.0719	.7778
7	.34393	.93899	.36628	.7302	.0650	.9075	.36027	.93285	.38620	.5893	.0720	.7757
8	.34421	.93889	.36661	.7277	.0651	.9052	.36054	.93274	.38654	.5871	.0721	.7736
9	.34448	.93879	.36694	.7252	.0652	.9029	.36081	.93264	.38687	.5848	.0722	.7715
10	.34475	.93869	.36727	2.7228	.0653	2.9006	.36108	.93253	.38720	2.5826	.0723	2.7694
11	.34502	.93859	.36760	.7204	.0654	.8983	.36135	.93243	.38754	.5804	.0725	.7674
12	.34530	.93849	.36793	.7179	.0655	.8960	.36162	.93232	.38787	.5781	.0726	.7653
13	.34557	.93839	.36826	.7155	.0656	.8937	.36189	.93222	.38821	.5759	.0727	.7632
14	.34584	.93829	.36859	.7130	.0658	.8915	.36217	.93211	.38854	.5737	.0728	.7611
15	.34612	.93819	.36892	2.7106	.0659	2.8892	.36244	.93201	.38888	2.5715	.0729	2.7591
16	.34639	.93809	.36925	.7082	.0660	.8869	.36271	.93190	.38921	.5693	.0731	.7570
17	.34666	.93799	.36958	.7058	.0661	.8846	.36298	.93180	.38955	.5671	.0732	.7550
18	.34693	.93789	.36991	.7033	.0662	.8824	.36325	.93169	.38988	.5649	.0733	.7529
19	.34721	.93779	.37024	.7009	.0663	.8801	.36352	.93158	.39022	.5627	.0734	.7509
20	.34748	.93769	.37057	2.6985	.0664	2.8778	.36379	.93148	.39055	2.5605	.0736	2.7488
21	.34775	.93758	.37090	.6961	.0666	.8756	.36406	.93137	.39089	.5583	.0737	.7468
22	.34803	.93748	.37123	.6937	.0667	.8733	.36433	.93127	.39122	.5561	.0738	.7447
23	.34830	.93738	.37156	.6913	.0668	.8711	.36460	.93116	.39156	.5539	.0739	.7427
24	.34857	.93728	.37190	.6889	.0669	.8688	.36488	.93105	.39189	.5517	.0740	.7406
25	.34884	.93718	.37223	2.6865	.0670	2.8666	.36515	.93095	.39223	2.5495	.0742	2.7386
26	.34912	.93708	.37256	.6841	.0671	.8644	.36542	.93084	.39257	.5473	.0743	.7366
27	.34939	.93698	.37289	.6817	.0673	.8621	.36569	.93074	.39290	.5451	.0744	.7346
28	.34966	.93687	.37322	.6794	.0674	.8599	.36596	.93063	.39324	.5430	.0745	.7325
29	.34993	.93677	.37355	.6770	.0675	.8577	.36623	.93052	.39357	.5408	.0747	.7305
30	.35021	.93667	.37388	2.6746	.0676	2.8554	.36650	.93042	.39391	2.5386	.0748	2.7285
31	.35048	.93657	.37422	.6722	.0677	.8532	.36677	.93031	.39425	.5365	.0749	.7265
32	.35075	.93647	.37455	.6699	.0678	.8510	.36704	.93020	.39458	.5343	.0750	.7245
33	.35102	.93637	.37488	.6675	.0679	.8488	.36731	.93010	.39492	.5322	.0751	.7225
34	.35130	.93626	.37521	.6652	.0681	.8466	.36758	.92999	.39525	.5300	.0753	.7205
35	.35157	.93616	.37554	2.6628	.0682	2.8444	.36785	.92988	.39559	2.5278	.0754	2.7185
36	.35184	.93606	.37587	.6604	.0683	.8422	.36812	.92978	.39593	.5257	.0755	.7165
37	.35211	.93596	.37621	.6581	.0684	.8400	.36839	.92967	.39626	.5236	.0756	.7145
38	.35239	.93585	.37654	.6558	.0685	.8378	.36866	.92956	.39660	.5214	.0758	.7125
39	.35266	.93575	.37687	.6534	.0686	.8356	.36893	.92945	.39694	.5193	.0759	.7105
40	.35293	.93565	.37720	2.6511	.0688	2.8334	.36921	.92935	.39727	2.5171	.0760	2.7085
41	.35320	.93555	.37754	.6487	.0689	.8312	.36948	.92924	.39761	.5150	.0761	.7065
42	.35347	.93544	.37787	.6464	.0690	.8290	.36975	.92913	.39795	.5129	.0763	.7045
43	.35375	.93534	.37820	.6441	.0691	.8269	.37002	.92902	.39828	.5108	.0764	.7026
44	.35402	.93524	.37853	.6418	.0692	.8247	.37029	.92892	.39862	.5086	.0765	.7006
45	.35429	.93513	.37887	2.6394	.0694	2.8225	.37056	.92881	.39896	2.5065	.0766	2.6986
46	.35456	.93503	.37920	.6371	.0695	.8204	.37083	.92870	.39930	.5044	.0768	.6967
47	.35483	.93493	.37953	.6348	.0696	.8182	.37110	.92859	.39963	.5023	.0769	.6947
48	.35511	.93482	.37986	.6325	.0697	.8160	.37137	.92848	.39997	.5002	.0770	.6927
49	.35538	.93472	.38020	.6302	.0698	.8139	.37164	.92838	.40031	.4981	.0771	.6908
50	.35565	.93462	.38053	2.6279	.0699	2.8117	.37191	.92827	.40065	2.4960	.0773	2.6888
51	.35592	.93451	.38086	.6256	.0701	.8096	.37218	.92816	.40098	.4939	.0774	.6869
52	.35619	.93441	.38120	.6233	.0702	.8074	.37245	.92805	.40132	.4918	.0775	.6849
53	.35647	.93431	.38153	.6210	.0703	.8053	.37272	.92794	.40166	.4897	.0776	.6830
54	.35674	.93420	.38186	.6187	.0704	.8032	.37299	.92784	.40200	.4876	.0778	.6810
55	.35701	.93410	.38220	2.6164	.0705	2.8010	.37326	.92773	.40233	2.4855	.0779	2.6791
56	.35728	.93400	.38253	.6142	.0707	.7989	.37353	.92762	.40267	.4834	.0780	.6772
57	.35755	.93389	.38286	.6119	.0708	.7968	.37380	.92751	.40301	.4813	.0781	.6752
58	.35782	.93379	.38320	.6096	.0709	.7947	.37407	.92740	.40335	.4792	.0783	.6733
59	.35810	.93368	.38353	.6073	.0710	.7925	.37434	.92729	.40369	.4772	.0784	.6714
60	.35837	.93358	.38386	2.6051	.0711	2.7904	.37461	.92718	.40403	2.4751	.0785	2.6695
′	cos	sin	cot	tan	cosec	sec	cos	sin	cot	tan	cosec	sec
	69°						68°					

′	22°						23°						′
′	sin	cos	tan	cot	sec	cosec	sin	cos	tan	cot	sec	cosec	′
0	.37461	.92718	.40403	2.4751	1.0785	2.6695	.39073	.92050	.42447	2.3558	1.0864	2.5593	60
1	.37488	.92707	.40436	.4730	.0787	.6675	.39100	.92039	.42482	.3539	.0865	.5575	59
2	.37514	.92696	.40470	.4709	.0788	.6656	.39126	.92028	.42516	.3520	.0866	.5558	58
3	.37541	.92686	.40504	.4689	.0789	.6637	.39153	.92016	.42550	.3501	.0868	.5540	57
4	.37568	.92675	.40538	.4668	.0790	.6618	.39180	.92005	.42585	.3482	.0869	.5523	56
5	.37595	.92664	.40572	2.4647	.0792	2.6599	.39207	.91993	.42619	2.3463	.0870	2.5506	55
6	.37622	.92653	.40606	.4627	.0793	.6580	.39234	.91982	.42654	.3445	.0872	.5488	54
7	.37649	.92642	.40640	.4606	.0794	.6561	.39260	.91971	.42688	.3426	.0873	.5471	53
8	.37676	.92631	.40673	.4586	.0795	.6542	.39287	.91959	.42722	.3407	.0874	.5453	52
9	.37703	.92620	.40707	.4565	.0797	.6523	.39314	.91948	.42757	.3388	.0876	.5436	51
10	.37730	.92609	.40741	2.4545	.0798	2.6504	.39341	.91936	.42791	2.3369	.0877	2.5419	50
11	.37757	.92598	.40775	.4525	.0799	.6485	.39367	.91925	.42826	.3350	.0878	.5402	49
12	.37784	.92587	.40809	.4504	.0801	.6466	.39394	.91913	.42860	.3332	.0880	.5384	48
13	.37811	.92576	.40843	.4484	.0802	.6447	.39421	.91902	.42894	.3313	.0881	.5367	47
14	.37838	.92565	.40877	.4463	.0803	.6428	.39448	.91891	.42929	.3294	.0882	.5350	46
15	.37865	.92554	.40911	2.4443	.0804	2.6410	.39474	.91879	.42963	2.3276	.0884	2.5333	45
16	.37892	.92543	.40945	.4423	.0806	.6391	.39501	.91868	.42998	.3257	.0885	.5316	44
17	.37919	.92532	.40979	.4403	.0807	.6372	.39528	.91856	.43032	.3238	.0886	.5299	43
18	.37946	.92521	.41013	.4382	.0808	.6353	.39554	.91845	.43067	.3220	.0888	.5281	42
19	.37972	.92510	.41047	.4362	.0810	.6335	.39581	.91833	.43101	.3201	.0889	.5264	41
20	.37999	.92499	.41081	2.4342	.0811	2.6316	.39608	.91822	.43136	2.3183	.0891	2.5247	40
21	.38026	.92488	.41115	.4322	.0812	.6297	.39635	.91810	.43170	.3164	.0892	.5230	39
22	.38053	.92477	.41149	.4302	.0813	.6279	.39661	.91798	.43205	.3145	.0893	.5213	38
23	.38080	.92466	.41183	.4282	.0815	.6260	.39688	.91787	.43239	.3127	.0895	.5196	37
24	.38107	.92455	.41217	.4262	.0816	.6242	.39715	.91775	.43274	.3109	.0896	.5179	36
25	.38134	.92443	.41251	2.4242	.0817	2.6223	.39741	.91764	.43308	2.3090	.0897	2.5163	35
26	.38161	.92432	.41285	.4222	.0819	.6205	.39768	.91752	.43343	.3072	.0899	.5146	34
27	.38188	.92421	.41319	.4202	.0820	.6186	.39795	.91741	.43377	.3053	.0900	.5129	33
28	.38214	.92410	.41353	.4182	.0821	.6168	.39821	.91729	.43412	.3035	.0902	.5112	32
29	.38241	.92399	.41387	.4162	.0823	.6150	.39848	.91718	.43447	.3017	.0903	.5095	31
30	.38268	.92388	.41421	2.4142	.0824	2.6131	.39875	.91706	.43481	2.2998	.0904	2.5078	30
31	.38295	.92377	.41455	.4122	.0825	.6113	.39901	.91694	.43516	.2980	.0906	.5062	29
32	.38322	.92366	.41489	.4102	.0826	.6095	.39928	.91683	.43550	.2962	.0907	.5045	28
33	.38349	.92354	.41524	.4083	.0828	.6076	.39955	.91671	.43585	.2944	.0908	.5028	27
34	.38376	.92343	.41558	.4063	.0829	.6058	.39981	.91659	.43620	.2925	.0910	.5011	26
35	.38403	.92332	.41592	2.4043	.0830	2.6040	.40008	.91648	.43654	2.2907	.0911	2.4995	25
36	.38429	.92321	.41626	.4023	.0832	.6022	.40035	.91636	.43689	.2889	.0913	.4978	24
37	.38456	.92310	.41660	.4004	.0833	.6003	.40061	.91625	.43723	.2871	.0914	.4961	23
38	.38483	.92299	.41694	.3984	.0834	.5985	.40088	.91613	.43758	.2853	.0915	.4945	22
39	.38510	.92287	.41728	.3964	.0836	.5967	.40115	.91601	.43793	.2835	.0917	.4928	21
40	.38537	.92276	.41762	2.3945	.0837	2.5949	.40141	.91590	.43827	2.2817	.0918	2.4912	20
41	.38564	.92265	.41797	.3925	.0838	.5931	.40168	.91578	.43862	.2799	.0920	.4895	19
42	.38591	.92254	.41831	.3906	.0840	.5913	.40195	.91566	.43897	.2781	.0921	.4879	18
43	.38617	.92242	.41865	.3886	.0841	.5895	.40221	.91554	.43932	.2763	.0922	.4862	17
44	.38644	.92231	.41899	.3867	.0842	.5877	.40248	.91543	.43966	.2745	.0924	.4846	16
45	.38671	.92220	.41933	2.3847	.0844	2.5958	.40205	.91531	.44001	2.2727	.0925	2.4829	15
46	.38698	.92209	.41968	.3828	.0845	.5841	.40301	.91519	.44036	.2709	.0927	.4813	14
47	.38725	.92197	.42002	.3808	.0846	.5823	.40328	.91508	.44070	.2691	.0928	.4797	13
48	.38751	.92186	.42036	.3789	.0847	.5805	.40354	.91496	.44105	.2673	.0929	.4780	12
49	.38778	.92105	.42070	.3770	.0849	.5787	.40381	.91484	.44140	.2655	.0931	.4764	11
50	.38805	.92164	.42105	2.3750	.0850	2.5770	.40408	.91472	.44175	2.2637	.0932	2.4748	10
51	.38832	.92152	.42139	.3731	.0851	.5752	.40434	.91461	.44209	.2619	.0934	.4731	9
52	.38859	.92141	.42173	.3712	.0853	.5734	.40461	.91449	.44244	.2602	.0935	.4715	8
53	.38886	.92130	.42207	.3692	.0854	.5716	.40487	.91437	.44279	.2584	.0936	.4699	7
54	.38912	.92118	.42242	.3673	.0855	.5699	.40514	.91425	.44314	.2566	.0938	.4683	6
55	.38939	.92107	.42276	2.3654	.0857	2.5681	.40541	.91414	.44349	2.2548	.0939	2.4666	5
56	.38966	.92096	.42310	.3635	.0858	.5663	.40567	.91402	.44383	.2531	.0941	.4650	4
57	.38993	.92084	.42344	.3616	.0859	.5646	.40594	.91390	.44418	.2513	.0942	.4634	3
58	.39019	.92073	.42379	.3597	.0861	.5628	.40620	.91378	.44453	.2495	.0943	.4618	2
59	.39046	.92062	.42413	.3577	.0862	.5610	.40647	.91366	.44488	.2478	.0945	.4602	1
60	.39073	.92050	.42447	2.3558	.0864	2.5593	.40674	.91354	.44523	2.2460	.0946	2.4586	0
′	cos	sin	cot	tan	cosec	sec	cos	sin	cot	tan	cosec	sec	′
	67°						66°						

NATURAL TRIGONOMETRIC FUNCTIONS

	24°						25°						26°						27°						
′	sin	cos	tan	cot	sec	cosec	sin	cos	tan	cot	sec	cosec	sin	cos	tan	cot	sec	cosec	sin	cos	tan	cot	sec	cosec	′
0	.40674	.91354	.44523	2.2460	1.0946	2.4586	.42262	.90631	.46631	2.1445	1.1034	2.3662	.43837	.89879	.48773	2.0503	1.1126	2.2812	.45399	.89101	.50952	1.9626	1.1223	2.2027	60
1	.40700	.91343	.44558	.2443	.0948	.4570	.42288	.90618	.46666	.1429	.1035	.3647	.43863	.89867	.48809	.0488	.1127	.2798	.45425	.89087	.50989	.9612	.1225	.2014	59
2	.40727	.91331	.44593	.2425	.0949	.4554	.42314	.90606	.46702	.1412	.1037	.3632	.43889	.89854	.48845	.0473	.1129	.2784	.45451	.89074	.51026	.9598	.1226	.2002	58
3	.40753	.91319	.44627	.2408	.0951	.4538	.42341	.90594	.46737	.1396	.1038	.3618	.43915	.89841	.48881	.0458	.1131	.2771	.45477	.89061	.51062	.9584	.1228	.1989	57
4	.40780	.91307	.44662	.2390	.0952	.4522	.42367	.90581	.46772	.1380	.1040	.3603	.43942	.89828	.48917	.0443	.1132	.2757	.45503	.89048	.51099	.9570	.1230	.1977	56
5	.40806	.91295	.44697	2.2373	.0953	2.4506	.42394	.90569	.46808	2.1364	.1041	2.3588	.43968	.89815	.48953	2.0427	.1134	2.2744	.45528	.89034	.51136	1.9556	.1231	2.1964	55
6	.40833	.91283	.44732	.2355	.0955	.4490	.42420	.90557	.46843	.1348	.1043	.3574	.43994	.89803	.48989	.0412	.1135	.2730	.45554	.89021	.51172	.9542	.1233	.1952	54
7	.40860	.91271	.44767	.2338	.0956	.4474	.42446	.90544	.46879	.1331	.1044	.3559	.44020	.89790	.49025	.0397	.1137	.2717	.45580	.89008	.51209	.9528	.1235	.1939	53
8	.40886	.91260	.44802	.2320	.0958	.4458	.42473	.90532	.46914	.1315	.1046	.3544	.44046	.89777	.49062	.0382	.1139	.2703	.45606	.88995	.51246	.9514	.1237	.1927	52
9	.40913	.91248	.44837	.2303	.0959	.4442	.42499	.90520	.46950	.1299	.1047	.3530	.44072	.89764	.49098	.0367	.1140	.2690	.45632	.88981	.51283	.9500	.1238	.1914	51
10	.40939	.91236	.44872	2.2286	.0961	2.4426	.42525	.90507	.46985	2.1283	.1049	2.3515	.44098	.89751	.49134	2.0352	.1142	2.2676	.45658	.88968	.51319	1.9486	.1240	2.1902	50
11	.40966	.91224	.44907	.2268	.0962	.4418	.42552	.90495	.47021	.1267	.1050	.3501	.44124	.89739	.49170	.0338	.1143	.2663	.45684	.88955	.51356	.9472	.1242	.1889	49
12	.40992	.91212	.44942	.2251	.0963	.4395	.42578	.90483	.47056	.1254	.1052	.3486	.44150	.89726	.49206	.0323	.1145	.2650	.45710	.88942	.51393	.9458	.1243	.1877	48
13	.41019	.91200	.44977	.2234	.0965	.4379	.42604	.90470	.47092	.1235	.1053	.3472	.44177	.89713	.49242	.0308	.1147	.2636	.45736	.88928	.51430	.9444	.1245	.1865	47
14	.41045	.91188	.45012	.2216	.0966	.4363	.42630	.90458	.47127	.1219	.1055	.3457	.44203	.89700	.49278	.0293	.1148	.2623	.45761	.88915	.51466	.9430	.1247	.1852	46
15	.41072	.91176	.45047	2.2199	.0968	2.4347	.42657	.90445	.47163	2.1203	.1056	2.3443	.44229	.89687	.49314	2.0278	.1150	2.2610	.45787	.88902	.51503	1.9416	.1248	2.1840	45
16	.41098	.91164	.45082	.2182	.0969	.4332	.42683	.90433	.47199	.1187	.1058	.3428	.44255	.89674	.49351	.0263	.1151	.2596	.45813	.88888	.51540	.9402	.1250	.1828	44
17	.41125	.91152	.45117	.2165	.0971	.4316	.42709	.90421	.47234	.1171	.1059	.3414	.44281	.89661	.49387	.0248	.1153	.2583	.45839	.88875	.51577	.9388	.1252	.1815	43
18	.41151	.91140	.45152	.2147	.0972	.4300	.42736	.90408	.47270	.1155	.1061	.3399	.44307	.89649	.49423	.0233	.1155	.2570	.45865	.88862	.51614	.9375	.1253	.1803	42
19	.41178	.91128	.45187	.2130	.0973	.4285	.42762	.90396	.47305	.1139	.1062	.3385	.44333	.89636	.49459	.0219	.1156	.2556	.45891	.88848	.51651	.9361	.1255	.1791	41
20	.41204	.91116	.45222	2.2113	.0975	2.4269	.42788	.90383	.47341	2.1123	.1064	2.3371	.44359	.89623	.49495	2.0204	.1158	2.2543	.45917	.88835	.51687	1.9347	.1257	2.1778	40
21	.41231	.91104	.45257	.2096	.0976	.4254	.42815	.90371	.47376	.1107	.1065	.3356	.44385	.89610	.49532	.0189	.1159	.2530	.45942	.88822	.51724	.9333	.1258	.1766	39
22	.41257	.91092	.45292	.2079	.0978	.4238	.42841	.90358	.47412	.1092	.1067	.3342	.44411	.89597	.49568	.0174	.1161	.2517	.45968	.88808	.51761	.9319	.1260	.1754	38
23	.41284	.91080	.45327	.2062	.0979	.4222	.42867	.90346	.47448	.1076	.1068	.3328	.44437	.89584	.49604	.0159	.1163	.2503	.45994	.88795	.51798	.9306	.1262	.1742	37
24	.41310	.91068	.45362	.2045	.0981	.4207	.42893	.90333	.47483	.1060	.1070	.3313	.44463	.89571	.49640	.0145	.1164	.2490	.46020	.88781	.51835	.9292	.1264	.1730	36
25	.41337	.91056	.45397	2.2028	.0982	2.4191	42920	.90321	.47519	2.1044	.1072	2.3299	.44489	.89558	.49677	2.0130	.1166	2.2477	.46046	.88768	.51872	1.9278	.1265	2.1717	35
26	.41363	.91044	.45432	.2011	.0984	.4176	.42946	.90308	.47555	.1028	.1073	.3285	.44516	.89545	.49713	.0115	.1167	.2464	.46072	.88755	.51909	.9264	.1267	.1705	34
27	.41390	.91032	.45467	.1994	.0985	.4160	.42972	.90296	.47590	.1013	.1075	.3271	.44542	.89532	.49749	.0101	.1169	.2451	.46097	.88741	.51946	.9251	.1269	.1693	33
28	.41416	.91020	.45502	.1977	.0986	.4145	.42998	.90283	.47626	.0997	.1076	.3256	.44568	.89519	.49785	.0086	.1171	.2438	.46123	.88728	.51983	.9237	.1270	.1681	32
29	.41443	.91008	.45537	.1960	.0988	.4130	.43025	.90271	.47662	.0981	.1078	.3242	.44594	.89506	.49822	.0071	.1172	.2425	.46149	.88714	.52020	.9223	.1272	.1669	31
30	.41469	.90996	.45573	2.1943	.0989	2.4114	.43051	.90258	.47697	2.0965	.1079	2.3228	.44620	.89493	.49858	2.0057	.1174	2.2411	.46175	.88701	.52057	1.9210	.1274	2.1657	30
31	.41496	.90984	.45608	.1926	.0991	.4099	.43077	.90246	.47733	.0950	.1081	.3214	.44646	.89480	.49894	.0042	.1176	.2398	.46201	.88688	.52094	.9196	.1275	.1645	29
32	.41522	.90972	.45643	.1909	.0992	.4083	.43104	.90233	.47769	.0934	.1082	.3200	.44672	.89467	.49931	.0028	.1177	.2385	.46226	.88674	.52131	.9182	.1277	.1633	28
33	.41549	.90960	.45678	.1892	.0994	.4068	.43130	.90221	.47805	.0918	.1084	.3186	.44698	.89454	.49967	.0013	.1179	.2372	.46252	.88661	.52168	.9169	.1279	.1620	27
34	.41575	.90948	.45713	.1875	.0995	.4053	.43156	.90208	.47840	.0903	.1085	.3172	.44724	.89441	.50003	1.9998	.1180	.2359	.46278	.88647	.52205	.9155	.1281	.1608	26
35	.41602	.90936	.45748	2.1859	.0997	2.4037	.43182	.90196	.47876	2.0887	.1087	2.3158	.44750	.89428	.50040	1.9984	.1182	2.2346	.46304	.88634	.52242	1.9142	.1282	2.1596	25
36	.41628	.90924	.45783	.1842	.0998	.4022	.43208	.90183	.47912	.0872	.1088	.3143	.44776	.89415	.50076	.9969	.1184	.2333	.46330	.88620	.52279	.9128	.1284	.1584	24
37	.41654	.90911	.45819	.1825	.1000	.4007	.43235	.90171	.47948	.0856	.1090	.3129	.44802	.89402	.50113	.9955	.1185	.2320	.46355	.88607	.52316	.9115	.1286	.1572	23
38	.41681	.90899	.45854	.1808	.1001	.3992	.43261	.90158	.47983	.0840	.1092	.3115	.44828	.89389	.50149	.9940	.1187	.2307	.46381	.88593	.52353	.9101	.1287	.1560	22
39	.41707	.90887	.45889	.1792	.1003	.3976	.43287	.90145	.48019	.0825	.1093	.3101	.44854	.89376	.50185	.9926	.1189	.2294	.46407	.88580	.52390	.9088	.1289	.1548	21
40	.41734	.90875	.45924	2.1775	.1004	2.3961	.43313	.90133	.48055	2.0809	.1095	2.3087	.44880	.89363	.50222	1.9912	.1190	2.2282	.46433	.88566	.52427	1.9074	.1291	2.1536	20
41	.41760	.90863	.45960	.1758	.1005	.3946	.43340	.90120	.48091	.0794	.1096	.3073	.44906	.89350	.50258	.9897	.1192	.2269	.46458	.88553	.52464	.9061	.1293	.1525	19
42	.41787	.90851	.45995	.1741	.1007	.3931	.43366	.90108	.48127	.0778	.1098	.3059	.44932	.89337	.50295	.9883	.1193	.2256	.46484	.88539	.52501	.9047	.1294	.1513	18
43	.41813	.90839	.46030	.1725	.1008	.3916	.43392	.90095	.48162	.0763	.1099	.3046	.44958	.89324	.50331	.9868	.1195	.2243	.46510	.88526	.52538	.9034	.1296	.1501	17
44	.41839	.90826	.46065	.1708	.1010	.3901	.43418	.90082	.48198	.0747	.1101	.3032	.44984	.89311	.50368	.9854	.1197	.2230	.46536	.88512	.52575	.9020	.1298	.1489	16
45	.41866	.90814	.46101	2.1692	.1011	2.3886	.43444	.90070	.48234	2.0732	.1102	2.3018	.45010	.89298	.50404	1.9840	.1198	2.2217	.46561	.88499	.52612	1.9007	.1299	2.1477	15
46	.41892	.90812	.46136	.1675	.1013	.3871	.43471	.90057	.48270	.0717	.1104	.3004	.45036	.89285	.50441	.9825	.1200	.2204	.46587	.88485	.52650	.8993	.1301	.1465	14
47	.41919	.90790	.46171	.1658	.1014	.3856	.43497	.90044	.48306	.0701	.1106	.2990	.45062	.89272	.50477	.9811	.1202	.2192	.46613	.88472	.52687	.8980	.1303	.1453	13
48	.41945	.90778	.46206	.1642	.1016	.3841	.43523	.90032	.48342	.0686	.1107	.2976	.45088	.89258	.50514	.9797	.1203	.2179	.46639	.88458	.52724	.8967	.1305	.1441	12
49	.41972	.90765	.46242	.1625	.1017	.3826	.43549	.90019	.48378	.0671	.1109	.2962	.45114	.89245	.50550	.9782	.1205	.2166	.46664	.88444	.52761	.8953	.1306	.1430	11
50	.41998	.90753	.46277	2.1609	.1019	2.3811	.43575	.90006	.48414	2.0655	.1110	2.2949	.45140	.89232	.50587	1.9768	.1207	2.2153	.46690	.88431	.52798	1.8940	.1308	2.1418	10
51	.42024	.90741	.46312	.1592	.1020	.3796	.43602	.89994	.48449	.0640	.1112	.2935	.45166	.89219	.50623	.9754	.1208	.2141	.46716	.88417	.52836	.8927	.1310	.1406	9
52	.42051	.90729	.46348	.1576	.1022	.3781	.43628	.89981	.48485	.0625	.1113	.2921	.45191	.89206	.50660	.9739	.1210	.2128	.46741	.88404	.52873	.8913	.1312	.1394	8
53	.42077	.90717	.46383	.1559	.1023	.3766	.43654	.89968	.48521	.0609	.1115	.2907	.45217	.89193	.50696	.9725	.1212	.2115	.46767	.88390	.52910	.8900	.1313	.1382	7
54	.42103	.90704	.46418	.1543	.1025	.3751	.43680	.89956	.48557	.0594	.1116	.2894	.45243	.89180	.50733	.9711	.1213	.2103	.46793	.88376	.52947	.8887	.1315	.1371	6
55	.42130	.90692	.46454	2.1527	.1026	2.3736	.43706	.89943	.48593	2.0579	.1118	2.2880	.45269	.89166	.50769	1.9697	.1215	2.2090	.46819	.88363	.52984	1.8873	.1317	2.1359	5
56	.42156	.90680	.46489	.1510	.1028	.3721	.43732	.89930	.48629	.0564	.1120	.2866	.45295	.89153	.50806	.9683	.1217	.2077	.46844	.88349	.53022	.8860	.1319	.1347	4
57	.42183	.90668	.46524	.1494	.1029	.3706	.43759	.89918	.48665	.0548	.1121	.2853	.45321	.89140	.50843	.9668	.1218	.2065	.46870	.88336	.53059	.8847	.1320	.1335	3
58	.42209	.90655	.46560	.1478	.1031	.3691	.43785	.89905	.48701	.0533	.1123	.2839	.45347	.89127	.50879	.9654	.1220	.2052	.46896	.88322	.53096	.8834	.1322	.1324	2
59	.42235	.90643	.46595	.1461	.1032	.3677	.43811	.89892	.48737	.0518	.1124	.2825	.45373	.89114	.50916	.9640	.1222	.2039	.46921	.88308	.53134	.8820	.1324	.1312	1
60	.42262	.90631	.46631	2.1445	.1034	2.3662	.43837	.89879	.48773	2.0503	.1126	2.2812	.45399	.89101	.50952	1.9626	.1223	2.2027	.46947	.88295	.53171	1.8807	.1326	2.1300	0
′	cos	sin	cot	tan	cosec	sec	cos	sin	cot	tan	cosec	sec	cos	sin	cot	tan	cosec	sec	cos	sin	cot	tan	cosec	sec	′
	65°						64°						63°						62°						

NATURAL TRIGONOMETRIC FUNCTIONS

	28°						29°						30°						31°						
′	sin	cos	tan	cot	sec	cosec	sin	cos	tan	cot	sec	cosec	sin	cos	tan	cot	sec	cosec	sin	cos	tan	cot	sec	cosec	′
0	.46947	.88295	.53171	1.8807	1.1326	2.1300	.48481	.87462	.55431	1.8040	1.1433	2.0627	.50000	.86603	.57735	1.7320	1.1547	2.0000	.51504	.85717	.60086	1.6643	1.1666	1.9416	60
1	.46973	.88281	.53208	.8794	.1327	.1289	.48506	.87448	.55469	.8028	.1435	.0616	.50025	.86588	.57774	.7309	.1549	1.9990	.51529	.85702	.60126	.6632	.1668	.9407	59
2	.46998	.88267	.53245	.8781	.1329	.1277	.48532	.87434	.55507	.8016	.1437	.0605	.50050	.86573	.57813	.7297	.1551	.9980	.51554	.85687	.60165	.6621	.1670	.9397	58
3	.47024	.88254	.53283	.8768	.1331	.1266	.48557	.87420	.55545	.8003	.1439	.0594	.50075	.86559	.57851	.7286	.1553	.9970	.51578	.85672	.60205	.6610	.1672	.9388	57
4	.47050	.88240	.53320	.8754	.1333	.1254	.48583	.87405	.55583	.7991	.1441	.0583	.50101	.86544	.57890	.7274	.1555	.9960	.51603	.85657	.60244	.6599	.1674	.9378	56
5	.47075	.88226	.53358	1.8741	.1334	2.1242	.48608	.87391	.55621	1.7979	.1443	2.0573	.50126	.86530	.57929	1.7262	.1557	1.9950	.51628	.85642	.60284	1.6588	.1676	1.9369	55
6	.47101	.88213	.53395	.8728	.1336	.1231	.48633	.87377	.55659	.7966	.1445	.0562	.50151	.86515	.57968	.7251	.1559	.9940	.51653	.85627	.60324	.6577	.1678	.9360	54
7	.47127	.88199	.53432	.8715	.1338	.1219	.48659	.87363	.55697	.7954	.1446	.0551	.50176	.86500	.58007	.7239	.1561	.9930	.51678	.85612	.60363	.6566	.1681	.9350	53
8	.47152	.88185	.53470	.8702	.1340	.1208	.48684	.87349	.55735	.7942	.1448	.0540	.50201	.86486	.58046	.7228	.1562	.9920	.51703	.85597	.60403	.6555	.1683	.9341	52
9	.47178	.88171	.53507	.8689	.1341	.1196	.48710	.87335	.55774	.7930	.1450	.0530	.50226	.86471	.58085	.7216	.1564	.9910	.51728	.85582	.60443	.6544	.1685	.9332	51
10	.47204	.88158	.53545	1.8676	.1343	2.1185	.48735	.87320	.55812	1.7917	.1452	2.0519	.50252	.86457	.58123	1.7205	.1566	1.9900	.51753	.85566	.60483	1.6534	.1687	1.9322	50
11	.47229	.88144	.53582	.8663	.1345	.1173	.48760	.87306	.55850	.7905	.1454	.0508	.50277	.86442	.58162	.7193	.1568	.9890	.51778	.85551	.60522	.6523	.1689	.9313	49
12	.47255	.88130	.53619	.8650	.1347	.1162	.48786	.87292	.55888	.7893	.1456	.0498	.50302	.86427	.58201	.7182	.1570	.9880	.51803	.85536	.60562	.6512	.1691	.9304	48
13	.47281	.88117	.53657	.8637	.1349	.1150	.48811	.87278	.55926	.7881	.1458	.0487	.50327	.86413	.58240	.7170	.1572	.9870	.51827	.85521	.60602	.6501	.1693	.9295	47
14	.47306	.88103	.53694	.8624	.1350	.1139	.48837	.87264	.55964	.7868	.1459	.0476	.50352	.86398	.58279	.7159	.1574	.9860	.51852	.85506	.60642	.6490	.1695	.9285	46
15	.47332	.88089	.53732	1.8611	.1352	2.1127	.48862	.87250	.56003	1.7856	.1461	2.0466	.50377	.86383	.58318	1.7147	.1576	1.9850	.51877	.85491	.60681	1.6479	.1697	1.9276	45
16	.47357	.88075	.53769	.8598	.1354	.1116	.48887	.87235	.56041	.7844	.1463	.0455	.50402	.86369	.58357	.7136	.1578	.9840	.51902	.85476	.60721	.6469	.1699	.9267	44
17	.47383	.88061	.53807	.8585	.1356	.1104	.48913	.87221	.56079	.7832	.1465	.0444	.50428	.86354	.58396	.7124	.1580	.9830	.51927	.85461	.60761	.6458	.1701	.9258	43
18	.47409	.88048	.53844	.8572	.1357	.1093	.48938	.87207	.56117	.7820	.1467	.0434	.50453	.86339	.58435	.7113	.1582	.9820	.51952	.85446	.60801	.6447	.1703	.9248	42
19	.47434	.88034	.53882	.8559	.1359	.1082	.48964	.87193	.56156	.7808	.1469	.0423	.50478	.86325	.58474	.7101	.1584	.9811	.51977	.85431	.60841	.6436	.1705	.9239	41
20	.47460	.88020	.53919	1.8546	.1361	2.1070	.48989	.87178	.56194	1.7795	.1471	2.0413	.50503	.86310	.58513	1.7090	.1586	1.9801	.52002	.85416	.60881	1.6425	.1707	1.9230	40
21	.47486	.88006	.53957	.8533	.1363	.1059	.49014	.87164	.56232	.7783	.1473	.0402	.50528	.86295	.58552	.7079	.1588	.9791	.52026	.85400	.60920	.6415	.1709	.9221	39
22	.47511	.87992	.53995	.8520	.1365	.1048	.49040	.87150	.56270	.7771	.1474	.0392	.50553	.86281	.58591	.7067	.1590	.9781	.52051	.85385	.60960	.6404	.1712	.9212	38
23	.47537	.87979	.54032	.8507	.1366	.1036	.49065	.87136	.56309	.7759	.1476	.0381	.50578	.86266	.58630	.7056	.1592	.9771	.52076	.85370	.61000	.6393	.1714	.9203	37
24	.47562	.87965	.54070	.8495	.1368	.1025	.49090	.87121	.56347	.7747	.1478	.0370	.50603	.86251	.58670	.7044	.1594	.9761	.52101	.85355	.61040	.6383	.1716	.9193	36
25	.47588	.87951	.54107	1.8482	.1370	2.1014	.49116	.87107	.56385	1.7735	.1480	2.0360	.50628	.86237	.58709	1.7033	.1596	1.9752	.52126	.85340	.61080	1.6372	.1718	1.9184	35
26	.47613	.87937	.54145	.8469	.1372	.1002	.49141	.87093	.56424	.7723	.1482	.0349	.50653	.86222	.58748	.7022	.1598	.9742	.52151	.85325	.61120	.6361	.1720	.9175	34
27	.47639	.87923	.54183	.8456	.1373	.0991	.49166	.87078	.56462	.7711	.1484	.0339	.50679	.86207	.58787	.7010	.1600	.9732	.52175	.85309	.61160	.6350	.1722	.9166	33
28	.47665	.87909	.54220	.8443	.1375	.0980	.49192	.87064	.56500	.7699	.1486	.0329	.50704	.86192	.58826	.6999	.1602	.9722	.52200	.85294	.61200	.6340	.1724	.9157	32
29	.47690	.87895	.54258	.8430	.1377	.0969	.49217	.87050	.56539	.7687	.1488	.0318	.50729	.86178	.58865	.6988	.1604	.9713	.52225	.85279	.61240	.6329	.1726	.9148	31
30	.47716	.87882	.54295	1.8418	.1379	2.0957	.49242	.87035	.56577	1.7675	.1489	2.0308	.50754	.86163	.58904	1.6977	.1606	1.9703	.52250	.85264	.61280	1.6318	.1728	1.9139	30
31	.47741	.87868	.54333	.8405	.1381	.0946	.49268	.87021	.56616	.7663	.1491	.0297	.50779	.86148	.58944	.6965	.1608	.9693	.52275	.85249	.61320	.6308	.1730	.9130	29
32	.47767	.87854	.54371	.8392	.1382	.0935	.49293	.87007	.56654	.7651	.1493	.0287	.50804	.86133	.58983	.6954	.1610	.9683	.52299	.85234	.61360	.6297	.1732	.9121	28
33	.47792	.87840	.54409	.8379	.1384	.0924	.49318	.86992	.56692	.7639	.1495	.0276	.50829	.86118	.59022	.6943	.1612	.9674	.52324	.85218	.61400	.6286	.1734	.9112	27
34	.47818	.87826	.54446	.8367	.1386	.0912	.49343	.86978	.56731	.7627	.1497	.0266	.50854	.86104	.59061	.6931	.1614	.9664	.52349	.85203	.61440	.6276	.1737	.9102	26
35	.47844	.87812	.54484	1.8354	.1388	2.0901	.49369	.86964	.56769	1.7615	.1499	2.0256	.50879	.86089	.59100	1.6920	.1616	1.9654	.52374	.85188	.61480	1.6265	.1739	1.9093	25
36	.47869	.87798	.54522	.8341	.1390	.0890	.49394	.86949	.56808	.7603	.1501	.0245	.50904	.86074	.59140	.6909	.1618	.9645	.52398	.85173	.61520	.6255	.1741	.9084	24
37	.47895	.87784	.54559	.8329	.1391	.0879	.49419	.86935	.56846	.7591	.1503	.0235	.50929	.86059	.59179	.6898	.1620	.9635	.52423	.85157	.61560	.6244	.1743	.9075	23
38	.47920	.87770	.54597	.8316	.1393	.0868	.49445	.86921	.56885	.7579	.1505	.0224	.50954	.86044	.59218	.6887	.1622	.9625	.52448	.85142	.61601	.6233	.1745	.9066	22
39	.47946	.87756	.54635	.8303	.1395	.0857	.49470	.86906	.56923	.7567	.1507	.0214	.50979	.86030	.59258	.6875	.1624	.9616	.52473	.85127	.61641	.6223	.1747	.9057	21
40	.47971	.87742	.54673	1.8291	.1397	2.0846	.49495	.86892	.56962	1.7555	.1508	2.0204	.51004	.86015	.59297	1.6864	.1626	1.9606	.52498	.85112	.61681	1.6212	.1749	1.9048	20
41	.47997	.87728	.54711	.8278	.1399	.0835	.49521	.86877	.57000	.7544	.1510	.0194	.51029	.86000	.59336	.6853	.1628	.9596	.52522	.85096	.61721	.6202	.1751	.9039	19
42	.48022	.87715	.54748	.8265	.1401	.0824	.49546	.86863	.57039	.7532	.1512	.0183	.51054	.85985	.59376	.6842	.1630	.9587	.52547	.85081	.61761	.6191	.1753	.9030	18
43	.48048	.87701	.54786	.8253	.1402	.0812	.49571	.86849	.57077	.7520	.1514	.0173	.51079	.85970	.59415	.6831	.1632	.9577	.52572	.85066	.61801	.6181	.1756	.9021	17
44	.48073	.87687	.54824	.8240	.1404	.0801	.49596	.86834	.57116	.7508	.1516	.0163	.51104	.85955	.59454	.6820	.1634	.9568	.52597	.85050	.61842	.6170	.1758	.9013	16
45	.48099	.87673	.54862	1.8227	.1406	2.0790	.49622	.86820	.57155	1.7496	.1518	2.0152	.51129	.85941	.59494	1.6808	.1636	1.9558	.52621	.85735	.61882	1.6160	.1760	1.9004	15
46	.48124	.87659	.54900	.8215	.1408	.0779	.49647	.86805	.57193	.7484	.1520	.0142	.51154	.85926	.59533	.6797	.1638	.9549	.52646	.85020	.61922	.6149	.1762	.8995	14
47	.48150	.87645	.54937	.8202	.1410	.0768	.49672	.86791	.57232	.7473	.1522	.0132	.51179	.85911	.59572	.6786	.1640	.9539	.52671	.85004	.61962	.6139	.1764	.8986	13
48	.48175	.87631	.54975	.8190	.1411	.0757	.49697	.86776	.57270	.7461	.1524	.0122	.51204	.85896	.59612	.6775	.1642	.9530	.52695	.84989	.62003	.6128	.1766	.8977	12
49	.48201	.87617	.55013	.8177	.1413	.0746	.49723	.86762	.57309	.7449	.1526	.0111	.51229	.85881	.59651	.6764	.1644	.9520	.52720	.84974	.62043	.6118	.1768	.8968	11
50	.48226	.87603	.55051	1.8165	.1415	2.0735	.49748	.86748	.57348	1.7437	.1528	2.0101	.51254	.85866	.59691	1.6753	.1646	1.9510	.52745	.84959	.62083	1.6107	.1770	1.8959	10
51	.48252	.87588	.55089	.8152	.1417	.0725	.49773	.86733	.57386	.7426	.1530	.0091	.51279	.85851	.59730	.6742	.1648	.9501	.52770	.84943	.62123	.6097	.1772	.8950	9
52	.48277	.87574	.55127	.8140	.1419	.0714	.49798	.86719	.57425	.7414	.1531	.0081	.51304	.85836	.59770	.6731	.1650	.9491	.52794	.84928	.62164	.6086	.1775	.8941	8
53	.48303	.87560	.55165	.8127	.1421	.0703	.49823	.86704	.57464	.7402	.1533	.0071	.51329	.85821	.59809	.6720	.1652	.9482	.52819	.84912	.62204	.6076	.1777	.8932	7
54	.48328	.87546	.55203	.8115	.1422	.0692	.49849	.86690	.57502	.7390	.1535	.0061	.51354	.85806	.59849	.6709	.1654	.9473	.52844	.84897	.62244	.6066	.1779	.8924	6
55	.48354	.87532	.55241	1.8102	.1424	2.0681	.49874	.86675	.57541	1.7379	.1537	2.0050	.51379	.85791	.59888	1.6698	.1656	1.9463	.52868	.84882	.62285	1.6055	.1781	1.8915	5
56	.48379	.87518	.55279	.8090	.1426	.0670	.49899	.86661	.57580	.7367	.1539	.0040	.51404	.85777	.59928	.6687	.1658	.9454	.52893	.84866	.62325	.6045	.1783	.8906	4
57	.48405	.87504	.55317	.8078	.1428	.0659	.49924	.86646	.57619	.7355	.1541	.0030	.51429	.85762	.59967	.6676	.1660	.9444	.52918	.84851	.62366	.6034	.1785	.8897	3
58	.48430	.87490	.55355	.8065	.1430	.0648	.49950	.86632	.57657	.7344	.1543	.0020	.51454	.85747	.60007	.6665	.1662	.9435	.52942	.84836	.62406	.6024	.1787	.8888	2
59	.48455	.87476	.55393	.8053	.1432	.0637	.49975	.86617	.57696	.7332	.1545	.0010	.51479	.85732	.60046	.6654	.1664	.9425	.52967	.84820	.62446	.6014	.1790	.8879	1
60	.48481	.87462	.55431	1.8040	.1433	2.0627	.50000	.86603	.57735	1.7320	.1547	2.0000	.51504	.85717	.60086	1.6643	.1666	1.9416	.52992	.84805	.62487	1.6003	.1792	1.8871	0
′	cos	sin	cot	tan	cosec	sec	cos	sin	cot	tan	cosec	sec	cos	sin	cot	tan	cosec	sec	cos	sin	cot	tan	cosec	sec	′
	61°						60°						59°						58°						

NATURAL TRIGONOMETRIC FUNCTIONS

′	32°						33°					
′	sin	cos	tan	cot	sec	cosec	sin	cos	tan	cot	sec	cosec
0	.52992	.84805	.62487	1.6003	1.1792	1.8871	.54464	.83867	.64941	1.5399	1.1924	1.8361
1	.53016	.84789	.62527	.5993	.1794	.8862	.54488	.83851	.64982	.5389	.1926	.8352
2	.53041	.84774	.62568	.5983	.1796	.8853	.54513	.83835	.65023	.5379	.1928	.8344
3	.53066	.84758	.62608	.5972	.1798	.8844	.54537	.83819	.65065	.5369	.1930	.8336
4	.53090	.84743	.62649	.5962	.1800	.8836	.54561	.83804	.65106	.5359	.1933	.8328
5	.53115	.84728	.62689	1.5952	.1802	1.8827	.54586	.83788	.65148	1.5350	.1935	1.8320
6	.53140	.84712	.62730	.5941	.1805	.8818	.54610	.83772	.65189	.5340	.1937	.8311
7	.53164	.84697	.62770	.5931	.1807	.8809	.54634	.83756	.65231	.5330	.1939	.8303
8	.53189	.84681	.62811	.5921	.1809	.8801	.54659	.83740	.65272	.5320	.1942	.8295
9	.53214	.84666	.62851	.5910	.1811	.8792	.54683	.83724	.65314	.5311	.1944	.8287
10	.53238	.84650	.62892	1.5900	.1813	1.8783	.54708	.83708	.65355	1.5301	.1946	1.8279
11	.53263	.84635	.62933	.5890	.1815	.8775	.54732	.83692	.65397	.5291	.1948	.8271
12	.53288	.84619	.62973	.5880	.1818	.8766	.54756	.83676	.65438	.5282	.1951	.8263
13	.53312	.84604	.63014	.5869	.1820	.8757	.54781	.83660	.65480	.5272	.1953	.8255
14	.53337	.84588	.63055	.5859	.1822	.8749	.54805	.83644	.65521	.5262	.1955	.8246
15	.53361	.84573	.63095	1.5849	.1824	1.8740	.54829	.83629	.65563	1.5252	.1958	1.8238
16	.53386	.84557	.63136	.5839	.1826	.8731	.54854	.83613	.65604	.5234	.1960	.8230
17	.53411	.84542	.63177	.5829	.1828	.8723	.54878	.83597	.65646	.5233	.1962	.8222
18	.53435	.84526	.63217	.5818	.1831	.8714	.54902	.83581	.65688	.5223	.1964	.8214
19	.53460	.84511	.63258	.5808	.1833	.8706	.54926	.83565	.65729	.5214	.1967	.8206
20	.53484	.84495	.63299	1.5798	.1835	1.8697	.54951	.83549	.65771	1.5204	.1969	1.8198
21	.53509	.84479	.63339	.5788	.1837	.8688	.54975	.83533	.65813	.5195	.1971	.8190
22	.53533	.84464	.63380	.5778	.1839	.8680	.54999	.83517	.65854	.5185	.1974	.8182
23	.53558	.84448	.63421	.5768	.1841	.8671	.55024	.83501	.65896	.5175	.1976	.8174
24	.53583	.84433	.63462	.5757	.1844	.8663	.55048	.83485	.65938	.5166	.1978	.8166
25	.53607	.84417	.63503	1.5747	.1846	1.8654	.55002	.83469	.65980	1.5156	.1980	1.8158
26	.53632	.84402	.63543	.5737	.1848	.8646	.55090	.83453	.66021	.5147	.1983	.8150
27	.53656	.84386	.63584	.5727	.1850	.8637	.55121	.83437	.66063	.5137	.1985	.8142
28	.53681	.84370	.63625	.5717	.1852	.8629	.55145	.83421	.66105	.5127	.1987	.8134
29	.53705	.84355	.63666	.5707	.1855	.8620	.55169	.83405	.66147	.5118	.1990	.8126
30	.53730	.84339	.63707	1.5697	.1857	1.8611	.55194	.83388	.56188	1.5108	.1992	1.8118
31	.53754	.84323	.63748	.5687	.1859	.8603	.55218	.83372	.66230	.5099	.1994	.8110
32	.53779	.84308	.63789	.5677	.1861	.8595	.55242	.83356	.66272	.5089	.1997	.8102
33	.53803	.84292	.63830	.5667	.1863	.8586	.55266	.83340	.66314	.5080	.1999	.8094
34	.53828	.84276	.63871	.5657	.1866	.8578	.55291	.83324	.66356	.5070	.2001	.8086
35	.53852	.84261	.63912	1.5646	.1868	1.8569	.55315	.83308	.66398	1.5061	.2004	1.8078
36	.53877	.84245	.63953	.5636	.1870	.8561	.55339	.83292	.66440	.5051	.2006	.8070
37	.53901	.84229	.63994	.5626	.1872	.8552	.55363	.83276	.66482	.5042	.2008	.8062
38	.53926	.84214	.64035	.5616	.1874	.8544	.55388	.83260	.66524	.5032	.2010	.8054
39	.53950	.84198	.64076	.5606	.1877	.8535	.55412	.83244	.66566	.5023	.2013	.8047
40	.53975	.84182	.64117	1.5596	.1879	1.8527	.55436	.83228	.66608	1.5013	.2015	1.8039
41	.53999	.84167	.64158	.5586	.1881	.8519	.55460	.83211	.66650	.5004	.2017	.8031
42	.54024	.84151	.64199	.5577	.1883	.8510	.55484	.83195	.66692	.4994	.2020	.8023
43	.54048	.84135	.64240	.5567	.1886	.8502	.55509	.83179	.66734	.4985	.2022	.8015
44	.54073	.84120	.64281	.5557	.1888	.8493	.55533	.83163	.66776	.4975	.2024	.8007
45	.54097	.84104	.64322	1.5547	.1890	1.8485	.55557	.83147	.66818	1.4966	.2027	1.7999
46	.54122	.84088	.64363	.5537	.1892	.8477	.55581	.83131	.66860	.4957	.2029	.7992
47	.54146	.84072	.64404	.5527	.1894	.8468	.55605	.83115	.66902	.4947	.2031	.7984
48	.54171	.84057	.64446	.5517	.1897	.8460	.55629	.83098	.66944	.4938	.2034	.7976
49	.54195	.84041	.64487	.5507	.1899	.8452	.55654	.83082	.66986	.4928	.2036	.7968
50	.54220	.84025	.64528	1.5497	.1901	1.8443	.55678	.83066	.67028	1.4919	.2039	1.7960
51	.54244	.84009	.64569	.5487	.1903	.8435	.55702	.83050	.67071	.4910	.2041	.7953
52	.54268	.83993	.64610	.5477	.1906	.8427	.55726	.83034	.67113	.4900	.2043	.7945
53	.54293	.83978	.64652	.5467	.1908	.8418	.55750	.83017	.67155	.4891	.2046	.7937
54	.54317	.83962	.64693	.5458	.1910	.8410	.55774	.83001	.67197	.4881	.2048	.7929
55	.54342	.83946	.64734	1.5448	.1912	1.8402	.55799	.82985	.67239	1.4872	.2050	1.7921
56	.54366	.83930	.64775	.5438	.1915	.8394	.55823	.82969	.67282	.4863	.2053	.7914
57	.54391	.83914	.64817	.5428	.1917	.8385	.55847	.82952	.67324	.4853	.2055	.7906
58	.54415	.83899	.64858	.5418	.1919	.8377	.55871	.82936	.67366	.4844	.2057	.7898
59	.54439	.83883	.64899	.5408	.1921	.8369	.55895	.82920	.67408	.4835	.2060	.7891
60	.54464	.83867	.64941	1.5399	.1922	1.8361	.55919	.82904	.67451	1.4826	.2062	1.7883
′	cos	sin	cot	tan	cosec	sec	cos	sin	cot	tan	cosec	sec
	57°						56°					

34°						35°						′
sin	cos	tan	cot	sec	cosec	sin	cos	tan	cot	sec	cosec	′
.55919	.82904	.67451	1.4826	1.2062	1.7883	.57358	.81915	.70021	1.4281	1.2208	1.7434	60
.55943	.82887	.67493	.4816	.2064	.7875	.57381	.81898	.70064	.4273	.2210	.7427	59
.55967	.82871	.67535	.4807	.2067	.7867	.57405	.81882	.70107	.4264	.2213	.7420	58
.55992	.82855	.67578	.4798	.2069	.7860	.57429	.81865	.70151	.4255	.2215	.7413	57
.56016	.82839	.67620	.4788	.2072	.7852	.57453	.81848	.70194	.4246	.2218	.7405	56
.56040	.82822	.67663	1.4779	.2074	1.7844	.57477	.81832	.70238	1.4237	.2220	1.7398	55
.56064	.82806	.67705	.4770	.2076	.7837	.57500	.81815	.70281	.4228	.2223	.7391	54
.56088	.82790	.67747	.4761	.2079	.7829	.57524	.81798	.70325	.4220	.2225	.7384	53
.56112	.82773	.67790	.4751	.2081	.7821	.57548	.81781	.70368	.4211	.2228	.7377	52
.56136	.82757	.67832	.4742	.2083	.7814	.57572	.81765	.70412	.4202	.2230	.7369	51
.56160	.82741	.67875	1.4733	.2086	1.7806	.57596	.81748	.70455	1.4193	.2233	1.7362	50
.56184	.82724	.67917	.4724	.2088	.7798	.57619	.81731	.70499	.4185	.2235	.7355	49
.56208	.82708	.67960	.4714	.2091	.7791	.57643	.81714	.70542	.4176	.2238	.7348	48
.56232	.82692	.68002	.4705	.2093	.7783	.57667	.81698	.70586	.4167	.2240	.7341	47
.56256	.82675	.68045	.4696	.2095	.7776	.57691	.81681	.70629	.4158	.2243	.7334	46
.56280	.82659	.68087	1.4687	.2098	1.7768	.57714	.81664	.70673	1.4150	.2245	1.7327	45
.56304	.82643	.68130	.4678	.2100	.7760	.57738	.81647	.70717	.4141	.2248	.7319	44
.56328	.82626	.68173	.4669	.2103	.7753	.57762	.81630	.70760	.4132	.2250	.7312	43
.56353	.82610	.68215	.4659	.2105	.7745	.57786	.81614	.70804	.4123	.2253	.7305	42
.56377	.82593	.68258	.4650	.2107	.7738	.57809	.81597	.70848	.4115	.2255	.7298	41
.56401	.82577	.68301	1.4641	.2110	1.7730	.57833	.81580	.70891	1.4106	.2258	1.7291	40
.56425	.82561	.68343	.4632	.2112	.7723	.57857	.81563	.70935	.4097	.2260	.7284	39
.56449	.82544	.68386	.4623	.2115	.7715	.57881	.81546	.70979	.4089	.2263	.7277	38
.56473	.82528	.68429	.4614	.2117	.7708	.57904	.81530	.71022	.4080	.2265	.7270	37
.56497	.82511	.68471	.4605	.2119	.7700	.57928	.81513	.71066	.4071	.2268	.7263	36
.56521	.82495	.68514	1.4595	.2122	1.7693	.57952	.81496	.71110	1.4063	.2270	1.7256	35
.56545	.82478	.68557	.4586	.2124	.7685	.57975	.81479	.71154	.4054	.2273	.7249	34
.56569	.82462	.68600	.4577	.2127	.7678	.57999	.81462	.71198	.4045	.2276	.7242	33
.56593	.82445	.68642	.4568	.2129	.7670	.58023	.81445	.71241	.4037	.2278	.7234	32
.56617	.82429	.68685	.4559	.2132	.7663	.58047	.81428	.71285	.4028	.2281	.7227	31
.56641	.82413	.68728	1.4550	.2134	1.7655	.58070	.81411	.71329	1.4019	.2283	1.7220	30
.56664	.82396	.68771	.4541	.2136	.7648	.58094	.81395	.71373	.4011	.2286	.7213	29
.56688	.82380	.68814	.4532	.2139	.7640	.58118	.81378	.71417	.4002	.2288	.7206	28
.56712	.82363	.68857	.4523	.2141	.7633	.58141	.81361	.71461	.3994	.2291	.7199	27
.56736	.82347	.68899	.4514	.2144	.7625	.58165	.81344	.71505	.3985	.2293	.7192	26
.56760	.82330	.68942	1.4505	.2146	1.7618	.58189	.81327	.71549	1.3976	.2296	1.7185	25
.56784	.82314	.68985	.4496	.2149	.7610	.58212	.81310	.71593	.3968	.2298	.7178	24
.56808	.82297	.69028	.4487	.2151	.7603	.58236	.81293	.71637	.3959	.2301	.7171	23
.56832	.82280	.69071	.4478	.2153	.7596	.58259	.81276	.71681	.3951	.2304	.7164	22
.56856	.82264	.69114	.4469	.2156	.7588	.58283	.81259	.71725	.3942	.2306	.7157	21
.56880	.82247	.69157	1.4460	.2158	1.7581	.58307	.81242	.71769	1.3933	.2309	1.7151	20
.56904	.82231	.69200	.4451	.2161	.7573	.58330	.81225	.71813	.3925	.2311	.7144	19
.56928	.82214	.69243	.4442	.2163	.7566	.58354	.81208	.71857	.3916	.2314	.7137	18
.56952	.82198	.69286	.4433	.2166	.7559	.58378	.81191	.71901	.3908	.2316	.7130	17
.56976	.82181	.69329	.4424	.2168	.7551	.58401	.81174	.71945	.3899	.2319	.7123	16
.57000	.82165	.69372	1.4415	.2171	1.7544	.58425	.81157	.71990	1.3891	.2322	1.7116	15
.57023	.82148	.69415	.4406	.2173	.7537	.58448	.81140	.72034	.3882	.2324	.7109	14
.57047	.82131	.69459	.4397	.2175	.7529	.58472	.81123	.72078	.3874	.2327	.7102	13
.57071	.82115	.69502	.4388	.2178	.7522	.58496	.81106	.72122	.3865	.2329	.7095	12
.57095	.82098	.69545	.4379	.2180	.7514	.58519	.81089	.72166	.3857	.2332	.7088	11
.57119	.82082	.69588	1.4370	.2183	1.7507	.58543	.81072	.72211	1.3848	.2335	1.7081	10
.57143	.82065	.69631	.4361	.2185	.7500	.58566	.81055	.72255	.3840	.2337	.7075	9
.57167	.82048	.69674	.4352	.2188	.7493	.58590	.81038	.72299	.3831	.2340	.7068	8
.57191	.82032	.69718	.4343	.2190	.7485	.58614	.81021	.72344	.3823	.2342	.7061	7
.57214	.82015	.69761	.4335	.2193	.7478	.58637	.81004	.72388	.3814	.2345	.7054	6
.57238	.81998	.69804	1.4326	.2195	1.7471	.58661	.80987	.72432	1.3806	.2348	1.7047	5
.57262	.81982	.69847	.4317	.2198	.7463	.58684	.80970	.72477	.3797	.2350	.7040	4
.57286	.81965	.69891	.4308	.2200	.7456	.58708	.80953	.72521	.3789	.2353	.7033	3
.57310	.81948	.69934	.4299	.2203	.7449	.58731	.80936	.72565	.3781	.2355	.7027	2
.57334	.81932	.69977	.4290	.2205	.7442	.58755	.80919	.72610	.3772	.2358	.7020	1
.57358	.81915	.70021	1.4281	.2208	1.7434	.58778	.80902	.72654	1.3764	.2361	1.7013	0
cos	sin	cot	tan	cosec	sec	cos	sin	cot	tan	cosec	sec	′
55°						54°						

′	36°					
′	sin	cos	tan	cot	sec	cosec
0	.58778	.80902	.72654	1.3764	1.2361	1.7013
1	.58802	.80885	.72699	.3755	.2363	.7006
2	.58825	.80867	.72743	.3747	.2366	.6999
3	.58849	.80850	.72788	.3738	.2368	.6993
4	.58873	.80833	.72832	.3730	.2371	.6986
5	.58896	.80816	.72877	1.3722	.2374	1.6979
6	.58920	.80799	.72921	.3713	.2376	.6972
7	.58943	.80782	.72966	.3705	.2379	.6965
8	.58967	.80765	.73010	.3697	.2382	.6959
9	.58990	.80747	.73055	.3688	.2384	.6952
10	.59014	.80730	.73100	1.3680	.2387	1.6945
11	.59037	.80713	.73144	.3672	.2389	.6938
12	.59060	.80696	.73189	.3663	.2392	.6932
13	.59084	.80679	.73234	.3655	.2395	.6925
14	.59107	.80662	.73278	.3647	.2397	.6918
15	.59131	.80644	.73323	1.3638	.2400	1.6912
16	.59154	.80627	.73368	.3630	.2403	.6905
17	.59178	.80610	.73412	.3622	.2405	.6898
18	.59201	.80593	.73457	.3613	.2408	.6891
19	.59225	.80576	.73502	.3605	.2411	.6885
20	.59248	.80558	.73547	1.3597	.2413	1.6878
21	.59272	.80541	.73592	.3588	.2416	.6871
22	.59295	.80524	.73637	.3580	.2419	.6865
23	.59318	.80507	.73681	.3572	.2421	.6858
24	.59342	.80489	.73726	.3564	.2424	.6851
25	.59365	.80472	.73771	1.3555	.2427	1.6845
26	.59389	.80455	.73816	.3547	.2429	.6838
27	.59412	.80437	.73861	.3539	.2432	.6831
28	.59435	.80420	.73906	.3531	.2435	.6825
29	.59459	.80403	.73951	.3522	.2437	.6818
30	.59482	.80386	.73996	1.3514	.2440	1.6812
31	.59506	.80368	.74041	.3506	.2443	.6805
32	.59529	.80351	.74086	.3498	.2445	.6798
33	.59552	.80334	.74131	.3489	.2448	.6792
34	.59576	.80316	.74176	.3481	.2451	.6785
35	.59599	.80299	.74221	1.3473	.2453	1.6779
36	.59622	.80282	.74266	.3465	.2456	.6772
37	.59646	.80264	.74312	.3457	.2459	.6766
38	.59669	.80247	.74357	.3449	.2461	.6759
39	.59692	.80230	.74402	.3440	.2464	.6752
40	.59716	.80212	.74447	1.3432	.2467	1.6746
41	.59739	.80195	.74492	.3424	.2470	.6739
42	.59762	.80177	.74538	.3416	.2472	.6733
43	.59786	.80160	.74583	.3408	.2475	.6726
44	.59809	.80143	.74628	.3400	.2478	.6720
45	.59832	.80125	.74673	1.3392	.2480	1.6713
46	.59856	.80108	.74019	.3383	.2483	.6707
47	.59879	.80090	.74064	.3375	.2486	.6700
48	.59902	.80073	.74809	.3367	.2488	.6694
49	.59926	.80056	.74855	.3359	.2491	.6687
50	.59949	.80038	.74900	1.3351	.2494	1.6681
51	.59972	.80021	.74946	.3343	.2497	.6674
52	.59995	.80003	.74991	.3335	.2499	.6668
53	.60019	.79986	.75037	.3327	.2502	.6661
54	.60042	.79968	.75082	.3319	.2505	.6655
55	.60065	.79951	.75128	1.3311	.2508	1.6648
56	.60088	.79933	.75173	.3303	.2510	.6642
57	.60112	.79916	.75219	.3294	.2513	.6636
58	.60135	.79898	.75264	.3286	.2516	.6629
59	.60158	.79881	.75310	.3278	.2519	.6623
60	.60181	.79863	.75355	1.3270	.2521	1.6616
′	cos	sin	cot	tan	cosec	sec
	53°					

′	37°					
′	sin	cos	tan	cot	sec	cosec
0	.60181	.79863	.75355	1.3270	1.2521	1.6616
1	.60205	.79846	.75401	.3262	.2524	.6610
2	.60228	.79828	.75447	.3254	.2527	.6603
3	.60251	.79811	.75492	.3246	.2530	.6597
4	.60274	.79793	.75538	.3238	.2532	.6591
5	.60298	.79776	.75584	1.3230	.2535	1.6584
6	.60320	.79758	.75629	.3222	.2538	.6578
7	.60344	.79741	.75675	.3214	.2541	.6572
8	.60367	.79723	.75721	.3206	.2543	.6565
9	.60390	.79706	.75767	.3198	.2546	.6559
10	.60413	.79688	.75812	1.3190	.2549	1.6552
11	.60437	.79670	.75858	.3182	.2552	.6546
12	.60460	.79653	.75904	.3174	.2554	.6540
13	.60483	.79635	.75950	.3166	.2557	.6533
14	.60506	.79618	.75996	.3159	.2560	.6527
15	.60529	.79600	.76042	1.3151	.2563	1.6521
16	.60552	.79582	.76088	.3143	.2565	.6514
17	.60576	.79565	.76134	.3135	.2568	.6508
18	.60599	.79547	.76179	.3127	.2571	.6502
19	.60622	.79530	.76225	.3119	.2574	.6496
20	.60645	.79512	.76271	1.3111	.2577	1.6489
21	.60668	.79494	.76317	.3103	.2579	.6483
22	.60691	.79477	.76364	.3095	.2582	.6477
23	.60714	.79459	.76410	.3087	.2585	.6470
24	.60737	.79441	.76456	.3079	.2588	.6464
25	.60761	.79424	.76502	1.3071	.2591	1.6458
26	.60784	.79406	.76548	.3064	.2593	.6452
27	.60807	.79388	.76594	.3056	.2596	.6445
28	.60830	.79371	.76640	.3048	.2599	.6439
29	.60853	.79353	.76686	.3040	.2602	.6433
30	.60876	.79335	.76733	1.3032	.2605	1.6427
31	.60899	.79318	.76779	.3024	.2607	.6420
32	.60922	.79300	.76825	.3016	.2610	.6414
33	.60945	.79282	.76871	.3009	.2613	.6408
34	.60968	.79264	.76918	.3001	.2616	.6402
35	.60991	.79247	.76964	1.2993	.2619	1.6396
36	.61014	.79229	.77010	.2985	.2622	.6389
37	.61037	.79211	.77057	.2977	.2624	.6383
38	.61061	.79193	.77103	.2970	.2627	.6377
39	.61084	.79176	.77149	.2962	.2630	.6371
40	.61107	.79158	.77196	1.2954	.2633	1.6365
41	.61130	.79140	.77242	.2946	.2636	.6359
42	.61153	.79122	.77289	.2938	.2639	.6352
43	.61176	.79104	.77335	.2931	.2641	.6346
44	.61199	.79087	.77382	.2923	.2644	.6340
45	.61222	.79069	.77428	1.2915	.2647	1.6334
46	.61245	.79051	.77475	.2907	.2650	.6328
47	.61268	.79033	.77521	.2900	.2653	.6322
48	.61290	.79015	.77568	.2892	.2656	.6316
49	.61314	.78998	.77614	.2884	.2659	.6309
50	.61337	.78980	.77661	1.2876	.2661	1.6303
51	.61360	.78962	.77708	.2869	.2664	.6297
52	.61383	.78944	.77754	.2861	.2667	.6291
53	.61405	.78926	.77801	.2853	.2670	.6285
54	.61428	.78908	.77848	.2845	.2673	.6279
55	.61451	.78890	.77895	1.2838	.2676	1.6273
56	.61474	.78873	.77941	.2830	.2679	.6267
57	.61497	.78855	.77988	.2822	.2681	.6261
58	.61520	.78837	.78035	.2815	.2684	.6255
59	.61543	.78819	.78082	.2807	.2687	.6249
60	.61566	.78801	.78128	1.2799	.2690	1.6243
′	cos	sin	cot	tan	cosec	sec
	52°					

′	38°					
′	sin	cos	tan	cot	sec	cosec
0	.61566	.78801	.78128	1.2799	1.2690	1.6243
1	.61589	.78783	.78175	.2792	.2693	.6237
2	.61612	.78765	.78222	.2784	.2696	.6231
3	.61635	.78747	.78269	.2776	.2699	.6224
4	.61658	.78729	.78316	.2769	.2702	.6218
5	.61681	.78711	.78363	1.2761	.2705	1.6212
6	.61703	.78693	.78410	.2753	.2707	.6206
7	.61726	.78675	.78457	.2746	.2710	.6200
8	.61749	.78657	.78504	.2738	.2713	.6194
9	.61772	.78640	.78551	.2730	.2716	.6188
10	.61795	.78622	.78598	1.2723	.2719	1.6182
11	.61818	.78604	.78645	.2715	.2722	.6176
12	.61841	.78586	.78692	.2708	.2725	.6170
13	.61864	.78568	.78739	.2700	.2728	.6164
14	.61886	.78550	.78786	.2692	.2731	.6159
15	.61909	.78532	.78834	1.2685	.2734	1.6153
16	.61932	.78514	.78881	.2677	.2737	.6147
17	.61955	.78496	.78928	.2670	.2739	.6141
18	.61978	.78478	.78975	.2662	.2742	.6135
19	.62001	.78460	.79022	.2655	.2745	.6129
20	.62023	.78441	.79070	1.2647	.2748	1.6123
21	.62046	.78423	.79117	.2639	.2751	.6117
22	.62069	.78405	.79164	.2632	.2754	.6111
23	.62092	.78387	.79212	.2624	.2757	.6105
24	.62115	.78369	.79259	.2617	.2760	.6099
25	.62137	.78351	.79306	1.2609	.2763	1.6093
26	.62160	.78333	.79354	.2602	.2766	.6087
27	.62183	.78315	.79401	.2594	.2769	.6081
28	.62206	.78297	.79449	.2587	.2772	.6077
29	.62229	.78279	.79496	.2579	.2775	.6070
30	.62251	.78261	.79543	1.2572	.2778	1.6064
31	.62274	.78243	.79591	.2564	.2781	.6058
32	.62297	.78224	.79639	.2557	.2784	.6052
33	.62320	.78206	.79686	.2549	.2787	.6046
34	.62342	.78188	.79734	.2542	.2790	.6040
35	.62365	.78170	.79781	1.2534	.2793	1.6034
36	.62388	.78152	.79829	.2527	.2795	.6029
37	.62411	.78134	.79876	.2519	.2798	.6023
38	.62433	.78116	.79924	.2512	.2801	.6017
39	.62456	.78097	.79972	.2504	.2804	.6011
40	.62479	.78079	.80020	1.2497	.2807	1.6005
41	.62501	.78061	.80067	.2489	.2810	.6000
42	.62524	.78043	.80115	.2482	.2813	.5994
43	.62547	.78025	.80163	.2475	.2816	.5988
44	.62570	.78007	.80211	.2467	.2819	.5982
45	.62592	.77988	.80258	1.2460	.2822	1.5976
46	.62615	.77970	.80306	.2452	.2825	.5971
47	.62638	.77952	.80354	.2445	.2828	.5965
48	.62660	.77934	.80402	.2437	.2831	.5959
49	.62683	.77915	.80450	.2430	.2834	.5953
50	.62706	.77897	.80498	1.2423	.2837	1.5947
51	.62728	.77879	.80546	.2415	.2840	.5942
52	.62751	.77861	.80594	.2408	.2843	.5936
53	.62774	.77842	.80642	.2400	.2846	.5930
54	.62796	.77824	.80690	.2393	.2849	.5924
55	.62819	.77806	.80738	1.2386	.2852	1.5919
56	.62841	.77788	.80786	.2378	.2855	.5913
57	.62864	.77769	.80834	.2371	.2858	.5907
58	.62887	.77751	.80882	.2364	.2861	.5901
59	.62909	.77733	.80930	.2356	.2864	.5896
60	62932	.77715	.80978	1.2349	.2867	1.5890
′	cos	sin	cot	tan	cosec	sec
	51°					

′	39°						′
′	sin	cos	tan	cot	sec	cosec	′
0	.62932	.77715	.80978	1.2349	1.2867	1.5890	60
1	.62955	.77696	.81026	.2342	.2871	.5884	59
2	.62977	.77678	.81075	.2334	.2874	.5879	58
3	.63000	.77660	.81123	.2327	.2877	.5873	57
4	.63022	.77641	.81171	.2320	.2880	.5867	56
5	.63045	.77623	.81219	1.2312	.2883	1.5862	55
6	.63067	.77605	.81268	.2305	.2886	.5856	54
7	.63090	.77586	.81316	.2297	.2889	.5850	53
8	.63113	.77568	.81364	.2290	.2892	.5854	52
9	.63135	.77549	.81413	.2283	.2895	.5839	51
10	.63158	.77531	.81461	1.2276	.2898	1.5833	50
11	.63180	.77513	.81509	.2268	.2901	.5828	49
12	.63203	.77494	.81558	.2261	.2904	.5822	48
13	.63225	.77476	.81606	.2254	.2907	.5816	47
14	.63248	.77458	.81655	.2247	.2910	.5811	46
15	.63270	.77439	.81703	1.2239	.2913	1.5805	45
16	.63293	.77421	.81752	.2232	.2916	.5799	44
17	.63315	.77402	.81800	.2225	.2919	.5794	43
18	.63338	.77384	.81849	.2218	.2922	.5788	42
19	.63360	.77365	.81898	.2210	.2926	.5783	41
20	.63383	.77347	.81946	1.2203	.2929	1.5777	40
21	.63405	.77329	.81995	.2196	.2932	.5771	39
22	.63428	.77310	.82043	.2189	.2935	.5766	38
23	.63450	.77292	.82092	.2181	.2938	.5760	37
24	.63473	.77273	.82141	.2174	.2941	.5755	36
25	.63495	.77255	.82190	1.2167	.2944	1.5749	35
26	.63518	.77236	.82238	.2160	.2947	.5743	34
27	.63540	.77218	.82287	.2152	.2950	.5738	33
28	.63563	.77199	.82336	.2145	.2953	.5732	32
29	.63585	.77181	.82385	.2138	.2956	.5727	31
30	.63608	.77162	.82434	1.2131	.2960	1.5721	30
31	.63630	.77144	.82482	.2124	.2963	.5716	29
32	.63653	.77125	.82531	.2117	.2966	.5710	28
33	.63675	.77107	.82580	.2109	.2969	.5705	27
34	.63697	.77088	.82629	.2102	.2972	.5699	26
35	.63720	.77070	.82678	1.2095	.2975	1.5694	25
36	.63742	.77051	.82727	.2088	.2978	.5688	24
37	.63765	.77033	.82776	.2081	.2981	.5683	23
38	.63787	.77014	.82825	.2074	.2985	.5677	22
39	.63810	.76996	.82874	.2066	.2988	.5672	21
40	.63832	.76977	.82923	1.2059	.2991	1.5666	20
41	.63854	.76958	.82972	.2052	.2994	.5661	19
42	.63877	.76940	.83022	.2045	.2997	.5655	18
43	.63899	.76921	.83071	.2038	.3000	.5650	17
44	.63921	.76903	.83120	.2031	.3003	.5644	16
45	.63944	.76884	.83169	1.2024	.3006	1.5639	15
46	.63966	.76865	.83218	.2016	.3010	.5633	14
47	.63989	.76847	.83267	.2009	.3013	.5628	13
48	.64011	.76828	.83317	.2002	.3016	.5622	12
49	.64033	.76810	.83366	.1995	.3019	.5617	11
50	.64056	.76791	.83415	1.1988	.3022	1.5611	10
51	.64078	.76772	.83465	.1981	.3025	.5606	9
52	.64100	.76754	.83514	.1974	.3029	.5600	8
53	.64123	.76735	.83563	.1967	.3032	.5595	7
54	.64145	.76716	.83613	.1960	.3035	.5590	6
55	.64160	.76698	.83662	1.1953	.3038	1.5584	5
56	.64189	.76679	.83712	.1946	.3041	.5579	4
57	.64212	.76660	.83761	.1939	.3044	.5573	3
58	.64234	.76642	.83811	.1932	.3048	.5568	2
59	.64256	.76623	.83860	.1924	.3051	.5563	1
60	.64279	.76604	.83910	1.1917	.3054	1.5557	0
′	cos	sin	cot	tan	cosec	sec	′
	50°						

NATURAL TRIGONOMETRIC FUNCTIONS

	40°						41°						42°						43°						
′	sin	cos	tan	cot	sec	cosec	sin	cos	tan	cot	sec	cosec	sin	cos	tan	cot	sec	cosec	sin	cos	tan	cot	sec	cosec	′
0	.64279	.76604	.83910	1.1917	1.3054	1.5557	.65606	.75471	.86929	1.1504	1.3250	1.5242	.66913	.74314	.90040	1.1106	1.3456	1.4945	.68200	.73135	.93251	1.0724	1.3673	1.4663	60
1	.64301	.76586	.83959	.1910	.3057	.5552	.65628	.75452	.86980	.1497	.3253	.5237	.66935	.74295	.90093	.1100	.3460	.4940	.68221	.73115	.93306	.0717	.3677	.4658	59
2	.64323	.76567	.84009	.1903	.3060	.5546	.65650	.75433	.87031	.1490	.3257	.5232	.66956	.74276	.90146	.1093	.3463	.4935	.68242	.73096	.93360	.0711	.3681	.4654	58
3	.64345	.76548	.84059	.1896	.3064	.5541	.65672	.75414	.87082	.1483	.3260	.5227	.66978	.74256	.90198	.1086	.3467	.4930	.68264	.73076	.93415	.0705	.3684	.4649	57
4	.64368	.76530	.84108	.1889	.3067	.5536	.65694	.75394	.87133	.1477	.3263	.5222	.66999	.74236	.90251	.1080	.3470	.4925	.68285	.73056	.93469	.0699	.3688	.4644	56
5	.64390	.76511	.84158	1.1882	.3070	1.5530	.65716	.75375	.87184	1.1470	.3267	1.5217	.67021	.74217	.90304	1.1074	.3474	1.4921	.68306	.73036	.93524	1.0692	.3692	1.4640	55
6	.64412	.76492	.84208	.1875	.3073	.5525	.65037	.75356	.87235	.1463	.3270	.5212	.67043	.74197	.90357	.1067	.3477	.4916	.68327	.73016	.93578	.0686	.3695	.4635	54
7	.64435	.76473	.84257	.1868	.3076	.5520	.65759	.75337	.87287	.1456	.3274	.5207	.67064	.74178	.90410	.1061	.3481	.4911	.68349	.72996	.93633	.0680	.3699	.4631	53
8	.64457	.76455	.84307	.1861	.3080	.5514	.65781	.75318	.87338	.1450	.3277	.5202	.67086	.74158	.90463	.1054	.3485	.4906	.68370	.72976	.93687	.0674	.3703	.4626	52
9	.64479	.76436	.84357	.1854	.3083	.5509	.65803	.75299	.87389	.1443	.3280	.5197	.67107	.74139	.90515	.1048	.3488	.4901	.68391	.72956	.93742	.0667	.3707	.4622	51
10	.64501	.76417	.84407	1.1847	.3086	1.5503	.65285	.75280	.87441	1.1436	.3284	1.5192	.67129	.74119	.90568	1.1041	.3492	1.4897	.68412	.72937	.93797	1.0661	.3710	1.4617	50
11	.64523	.76398	.84457	.1840	.3089	.5498	.65847	.75261	.87492	.1430	.3287	.5187	.67150	.74100	.90621	.1035	.3495	.4892	.68433	.72917	.93851	.0655	.3714	.4613	49
12	.64546	.76380	.84506	.1833	.3092	.5493	.65869	.75241	.87543	.1423	.3290	.5182	.67172	.74080	.90674	.1028	.3499	.4887	.68455	.72897	.93906	.0649	.3718	.4608	48
13	.64568	.76361	.84556	.1826	.3096	.5487	.65891	.75222	.87595	.1416	.3294	.5177	.67194	.74061	.90727	.1022	.3502	.4882	.68476	.72877	.93961	.0643	.3722	.4604	47
14	.64590	.76342	.84606	.1819	.3099	.5482	.65913	.75203	.87646	.1409	.3297	.5171	.67215	.74041	.90780	.1015	.3506	.4877	.68497	.72857	.94016	.0636	.3725	.4599	46
15	.64612	.76323	.84656	1.1812	.3102	1.5477	.65934	.75184	.87698	1.1403	.3301	1.5166	.67237	.74022	.90834	1.1009	.3509	1.4873	.68518	.72837	.94071	1.0630	.3729	1.4595	45
16	.64635	.76304	.84706	.1805	.3105	.5471	.65956	.75165	.87749	.1396	.3304	.5161	.67258	.74002	.90887	.1003	.3513	.4868	.68539	.72817	.94125	.0624	.3733	.4590	44
17	.64657	.76286	.84056	.1798	.3109	.5466	.65978	.75146	.87801	.1389	.3307	.5156	.67280	.73983	.90940	.0996	.3517	.4863	.68561	.72797	.94180	.0618	.3737	.4586	43
18	.64679	.76267	.84806	.1791	.3112	.5461	.66000	.75126	.87852	.1383	.3311	.5151	.67301	.73963	.90993	.0990	.3520	.4858	.68582	.72777	.94235	.0612	.3740	.4581	42
19	.64701	.76248	.84856	.1785	.3115	.5456	.66022	.75107	.87904	.1376	.3314	.5146	.67323	.73943	.91046	.0983	.3524	.4854	.68603	.72757	.94290	.0605	.3744	.4577	41
20	.64723	.76229	.84906	1.1778	.3118	1.5450	.66044	.75088	.87955	1.1369	.3318	1.5141	.67344	.73924	.91099	1.0977	.3527	1.4849	.68624	.72737	.94345	1.0599	.3748	1.4572	40
21	.64745	.76210	.84956	.1771	.3121	.5445	.66066	.75069	.88007	.1363	.3321	.5136	.67366	.73904	.91153	.0971	.3531	.4844	.68645	.72717	.94400	.0593	.3752	.4568	39
22	.64768	.76191	.85006	.1764	.3125	.5440	.66087	.75049	.88058	.1356	.3324	.5131	.67387	.73885	.91206	.0964	.3534	.4839	.68666	.72697	.94455	.0587	.3756	.4563	38
23	.64790	.76173	.85056	.1757	.3128	.5434	.66109	.75030	.88110	.1349	.3328	.5126	.67409	.73865	.91259	.0958	.3538	.4835	.68688	.72677	.94510	.0581	.3759	.4559	37
24	.64812	.76154	.85107	.1750	.3131	.5429	.66131	.75011	.88162	.1343	.3331	.5121	.67430	.73845	.91312	.0951	.3542	.4830	.68709	.72657	.94565	.0575	.3763	.4554	36
25	.64834	.76135	.85157	1.1743	.3134	1.5424	.66153	.74992	.88213	1.1336	.3335	1.5116	.67452	.73826	.91366	1.0945	.3545	1.4825	.68730	.72637	.94620	1.0568	.3767	1.4550	35
26	.64856	.76116	.85207	.1736	.3138	.5419	.66175	.74973	.88265	.1329	.3338	.5111	.67473	.73806	.91419	.0939	.3549	.4821	.68751	.72617	.94675	.0562	.3771	.4545	34
27	.64878	.76097	.85257	.1729	.3141	.5413	.66197	.74953	.88317	.1323	.3342	.5106	.67495	.73787	.91473	.0932	.3552	.4816	.68772	.72597	.94731	.0556	.3774	.4541	33
28	.64900	.76078	.85307	.1722	.3144	.5408	.66218	.74934	.88369	.1316	.3345	.5101	.67516	.73767	.91526	.0926	.3556	.4811	.68793	.72577	.94786	.0550	.3778	.4536	32
29	.64923	.76059	.85358	.1715	.3148	.5403	.66240	.74915	.88421	.1309	.3348	.5096	.67537	.73747	.91580	.0919	.3560	.4806	.68814	.72557	.94841	.0544	.3782	.4532	31
30	.64945	.76041	.85408	1.1708	.3151	1.5398	.66262	.74895	.88472	1.1303	.3352	1.5092	.67559	.73728	.91633	1.0913	.3563	1.4802	.68835	.72537	.94896	1.0538	.3786	1.4527	30
31	.64967	.76022	.85458	.1702	.3154	.5392	.66284	.74876	.88524	.1296	.3355	.5087	.67580	.73708	.91687	.0907	.3567	.4797	.68856	.72517	.94952	.0532	.3790	.4523	29
32	.64989	.76003	.85509	.1695	.3157	.5387	.66305	.74857	.88576	.1290	.3359	.5082	.67602	.73688	.91740	.0900	.3571	.4792	.68878	.72497	.95007	.0525	.3794	.4518	28
33	.65011	.75984	.85559	.1688	.3161	.5382	.66327	.74838	.88628	.1283	.3362	.5077	.67623	.73669	.91794	.0894	.3574	.4788	.68899	.72477	.95062	.0519	.3797	.4514	27
34	.65033	.75965	.85609	.1681	.3164	.5377	.66349	.74818	.88680	.1276	.3366	.5072	.67645	.73649	.91847	.0888	.3578	.4783	.68920	.72457	.95118	.0513	.3801	.4510	26
35	.65055	.75946	.85660	1.1674	.3167	1.5371	.66371	.74799	.88732	1.1270	.3369	1.5067	.67666	.73629	.91901	1.0881	.3581	1.4778	.68941	.72437	.95173	1.0507	.3805	1.4505	25
36	.65077	.75927	.85710	.1667	.3170	.5366	.66393	.74780	.88784	.1263	.3372	.5062	.67688	.73610	.91955	.0875	.3585	.4774	.68962	.72417	.95229	.0501	.3809	.4501	24
37	.65100	.75908	.85761	.1660	.3174	.5361	.66414	.74760	.88836	.1257	.3376	.5057	.67709	.73590	.92008	.0868	.3589	.4769	.68983	.72397	.95284	.0495	.3813	.4496	23
38	.65121	.75889	.85811	.1653	.3177	.5356	.66436	.74741	.88888	.1250	.3379	.5052	.67730	.73570	.92062	.0862	.3592	.4764	.69004	.72377	.95340	.0489	.3816	.4492	22
39	.65144	.75870	.85862	.1647	.3180	.5351	.66458	.74722	.88940	.1243	.3383	.5047	.67752	.73551	.92116	.0856	.3596	.4760	.69025	.72357	.95395	.0483	.3820	.4487	21
40	.65166	.75851	.85912	1.1640	.3184	1.5345	.66479	.74702	.88992	1.1237	.3386	1.5042	.67773	.73531	.92170	1.0849	.3600	1.4755	.69046	.72337	.95451	1.0476	.3824	1.4483	20
41	.65188	.75832	.85963	.1633	.3187	.5340	.66501	.74683	.89044	.1230	.3390	.5037	.67794	.73511	.92223	.0843	.3603	.4750	.69067	.72317	.95506	.0470	.3828	.4479	19
42	.65210	.75813	.86013	.1626	.3190	.5335	.66523	.74664	.89097	.1224	.3393	.5032	.67816	.73491	.92277	.0837	.3607	.4746	.69088	.72297	.95562	.0464	.3832	.4474	18
43	.65232	.75794	.86064	.1619	.3193	.5330	.66545	.74644	.89149	.1217	.3397	.5027	.67837	.73472	.92331	.0830	.3611	.4741	.69109	.72277	.95618	.0458	.3836	.4470	17
44	.65254	.75775	.86115	.1612	.3197	.5325	.66566	.74625	.89201	.1211	.3400	.5022	.67859	.73452	.92385	.0824	.3614	.4736	.69130	.72256	.95673	.0452	.3839	.4465	16
45	.65276	.75756	.86165	1.1605	.3200	1.5319	.66588	.74606	.89253	1.1204	1.3404	1.5018	.67880	.73432	.92439	1.0818	.3618	1.4732	.69151	.72236	.95729	1.0446	.3843	1.4461	15
46	.65298	.75737	.86216	.1599	.3203	.5314	.66610	.74586	.89306	.1197	.3407	.5013	.67901	.73412	.92493	.0812	.3622	.4727	.69172	.72216	.95785	.0440	.3847	.4457	14
47	.65320	.75718	.86267	.1592	.3207	.5309	.66631	.74567	.89358	.1191	.3411	.5008	.67923	.73393	.92547	.0805	.3625	.4723	.69193	.72196	.95841	.0434	.3851	.4452	13
48	.65342	.75700	.86318	.1585	.3210	.5304	.66653	.74548	.89410	.1184	.3414	.5003	.67944	.73373	.92601	.0799	.3629	.4718	.69214	.72176	.95896	.0428	.3855	.4448	12
49	.65364	.75680	.86368	.1578	.3213	.5299	.66675	.74528	.89463	.1178	.3418	.4998	.67965	.73353	.92655	.0793	.3633	.4713	.69235	.72156	.95952	.0422	.3859	.4443	11
50	.65386	.75661	.86419	1.1571	.3217	1.5294	.66697	.74509	.89515	1.1171	.3421	1.4993	.67987	.73333	.92709	1.0786	.3636	1.4709	.69256	.72136	.96008	1.0416	.3863	1.4439	10
51	.65408	.75642	.86470	.1565	.3220	.5289	.66718	.74489	.89567	.1165	.3425	.4988	.68008	.73314	.92763	.0780	.3640	.4704	.69277	.72115	.96064	.0410	.3867	.4435	9
52	.65430	.75623	.86521	.1558	.3223	.5283	.66740	.74470	.89620	.1158	.3428	.4983	.68029	.73294	.92817	.0774	.3644	.4699	.69298	.72095	.96120	.0404	.3870	.4430	8
53	.65452	.75604	.86572	.1551	.3227	.5278	.66762	.74450	.89672	.1152	.3432	.4979	.68051	.73274	.92871	.0767	.3647	.4695	.69319	.72075	.96176	.0397	.3874	.4426	7
54	.65474	.75585	.86623	.1544	.3230	.5273	.66783	.74431	.89725	.1145	.3435	.4974	.68072	.73254	.92926	.0761	.3651	.4690	.69340	.72055	.96232	.0391	.3878	.4422	6
55	.65496	.75566	.86674	1.1537	.3233	1.5268	.66805	.74412	.89777	1.1139	.3439	1.4969	.68093	.73234	.92980	1.0755	.3655	1.4686	.69361	.72035	.96288	1.0385	.3882	1.4417	5
56	.65518	.75547	.86725	.1531	.3237	.5263	.66826	.74392	.89830	.1132	.3442	.4964	.68115	.73215	.93034	.0749	.3658	.4681	.69382	.72015	.96344	.0379	.3886	.4413	4
57	.65540	.75528	.86775	.1524	.3240	.5258	.66848	.74373	.89882	.1126	.3446	.4959	.68136	.73195	.93088	.0742	.3662	.4676	.69403	.71994	.96400	.0373	.3890	.4408	3
58	.65562	.75509	.86826	.1517	.3243	.5253	.66870	.74353	.89935	.1119	.3449	.4954	.68157	.73175	.93143	.0736	.3666	.4672	.69424	.71974	.96456	.0367	.3894	.4404	2
59	.65584	.75490	.86878	.1510	.3247	.5248	.66891	.74334	.89988	.1113	.3453	.4949	.68178	.73155	.93197	.0730	.3669	.4667	.69445	.71954	.96513	.0361	.3898	.4400	1
60	.65606	.75401	.86929	1.1504	.3250	1.5242	.66913	.74314	.90040	1.1106	.3456	1.4945	.68200	.73135	.93251	1.0724	.3673	1.4663	.69466	.71934	.96569	1.0355	.3902	1.4395	0
′	cos	sin	cot	tan	cosec	sec	cos	sin	cot	tan	cosec	sec	cos	sin	cot	tan	cosec	sec	cos	sin	cot	tan	cosec	sec	′
	49°						48°						47°						46°						

44°

′	sin	cos	tan	cot	sec	cosec	′
0	.69466	.71934	.96569	1.0355	1.3902	1.4395	60
1	.69487	.71914	.96625	.0349	.3905	.4391	59
2	.69508	.71893	.96681	.0343	.3909	.4387	58
3	.69528	.71873	.96738	.0337	.3913	.4382	57
4	.69549	.71853	.96794	.0331	.3917	.4378	56
5	.69570	.71833	.96850	1.0325	.3921	1.4374	55
6	.69591	.71813	.96907	.0319	.3925	.4370	54
7	.69612	.71792	.96963	.0313	.3929	.4365	53
8	.69633	.71772	.97020	.0307	.3933	.4361	52
9	.69654	.71752	.97076	.0301	.3937	.4357	51
10	.69675	.71732	.97133	1.0295	.3941	1.4352	50
11	.69696	.71711	.97189	.0289	.3945	.4348	49
12	.69716	.71691	.97246	.0283	.3949	.4344	48
13	.69737	.71671	.97302	.0277	.3953	.4339	47
14	.69758	.71650	.97359	.0271	.3957	.4335	46
15	.69779	.71630	.97416	1.0265	.3960	1.4331	45
16	.69800	.71610	.97472	.0259	.3964	.4327	44
17	.69821	.71589	.97529	.0253	.3968	.4322	43
18	.69841	.71569	.97586	.0247	.3972	.4318	42
19	.69862	.71549	.97643	.0241	.3976	.4314	41
20	.69883	.71529	.97700	1.0235	.3980	1.4310	40
21	.69904	.71508	.97756	.0229	.3984	.4305	39
22	.69925	.71488	.97813	.0223	.3988	.4301	38
23	.69945	.71468	.97870	.0218	.3992	.4297	37
24	.69966	.71447	.97927	.0212	.3996	.4292	36
25	.69987	.71427	.97984	1.0206	.4000	1.4288	35
26	.70008	.71406	.98041	.0200	.4004	.4284	34
27	.70029	.71386	.98098	.0194	.4008	.4280	33
28	.70049	.71366	.98155	.0188	.4012	.4276	32
29	.70070	.71345	.98212	.0182	.4016	.4271	31
30	.70091	.71325	.98270	1.0176	.4020	1.4267	30
31	.70112	.71305	.98327	.0170	.4024	.4263	29
32	.70132	.71284	.98384	.0164	.4028	.4259	28
33	.70153	.71264	.98441	.0158	.4032	.4254	27
34	.70174	.71243	.98499	.0152	.4036	.4250	26
35	.70194	.71223	.98556	1.0146	.4040	1.4246	25
36	.70215	.71203	.98613	.0141	.4044	.4242	24
37	.70236	.71182	.98671	.0135	.4048	.4238	23
38	.70257	.71162	.98728	.0129	.4052	.4233	22
39	.70277	.71141	.98786	.0123	.4056	.4229	21
40	.70298	.71121	.98843	1.0117	.4060	1.4225	20
41	.70319	.71100	.98901	.0111	.4065	.4221	19
42	.70339	.71080	.98958	.0105	.4069	.4217	18
43	.70360	.71059	.99016	.0099	.4073	.4212	17
44	.70381	.71039	.99073	.0093	.4077	.4208	16
45	.70401	.71018	.99131	1.0088	.4081	1.4204	15
46	.70422	.70998	.99189	.0082	.4085	.4200	14
47	.70443	.70977	.99246	.0076	.4089	.4196	13
48	.70463	.70957	.99304	.0070	.4093	.4192	12
49	.70484	.70936	.99362	.0064	.4097	.4188	11
50	.70505	.70916	.99420	1.0058	.4101	1.4183	10
51	.70525	.70895	.99478	.0052	.4105	.4179	9
52	.70546	.70875	.99536	.0047	.4109	.4175	8
53	.70566	.70854	.99593	.0041	.4113	.4171	7
54	.70587	.70834	.99651	.0035	.4117	.4167	6
55	.70608	.70813	.99709	1.0029	.4122	1.4163	5
56	.70628	.70793	.99767	.0023	.4126	.4159	4
57	.70649	.70772	.99826	.0017	.4130	.4154	3
58	.70669	.70752	.99884	.0012	.4134	.4150	2
59	.70690	.70731	.99942	.0006	.4138	.4146	1
60	.70711	.70711	1.00000	1.0000	.4142	1.4142	0
′	cos	sin	cot	tan	cosec	sec	′

45°

INDEX

Abbreviations, 31
Abrasion, 412
Abrasives:
  aluminum oxide, 570–571, 574
  boron carbide, 572
  characteristics of, 570
  coated, 585–586
  diamonds, 573–574
  lapping, 146, 149–151
  products made with, 574
  silicon carbide, 571–572
  zirconia-aluminum oxide, 572
  See also Grinding wheels
Allowance, 30, 83
Alloys, 164, 169, 666–668
Aluminum and aluminum alloys, 165, 666–667
Aluminum oxide, 570–571, 574
Angle plate, 117, 256
Annealing, 653–654
Apprenticeship training, 12

Babbit-bearing materials, 668
Backlash, 500
Bandsaw:
  contour. *See* Contour bandsaw
  horizontal, 226
Basic dimension, 82
Bearings, 152–155
Bench work. *See* Hand tools and bench work
Bessemer converter, 637
Binary system, 525–527
Blending, 181, 707
Block square and parallel, 477–478
Bolts, 496
Boring, 248, 288, 487–489, 491–508
Boron carbide, 572–573
Briquetting, 707
Broaches, 146, 148–149
Bronze, 668
Buffing, 289–290
Burrs, 139–140

Calipers, 45–46, 59–63, 114. *See also* Micrometers
Cams and cam milling, 465–469
Capacity Resource Planning (CRP), 565
Carbide cutting tools, cemented, 169, 181–194
  angles and clearances, 171–173, 188–189
  applications of, 182–184, 192
  cutting speeds and feeds, 188
  grades of, 184–185
  grinding of, 192–194
  machine setup for, 192
  machining with, 188
  manufacture of, 181–182
  problems with, 194
  setup for, 188–191
  suggestions for using, 188
  tool selection for, 189, 192
Carbonitriding processes, 656
Carburizing, 655–656
Careers in metalworking industry, 12–16
Cartesian coordinate system, 527–529
Cast iron, 165, 634–635
Cemented-carbide cutting tools. *See* Carbide cutting tools, cemented
Cemented-oxide (ceramic) cutting tools, 169, 200–204
Center punch, 116–117
Ceramic aluminum oxide, 574
Ceramic cutting tools, 169, 200–204
Cermet, 169–170, 205
Chip formation, 160–163, 412–413
Chucking centers, computerized numerical control (CNC), 544–547
Chucks:
  accessories for, 594–595
  drill press, 254–255
  grinding, 593–595
  lathe. *See* Lathe chucks
  magnetic, 593–595
  workpieces held in, 34–35
Clamps and clamping, 117, 256–258, 496–497
Clogging, 413
Closed-loop system, 524
Clutches, 471–472, 301
Collets, 398
Combination set, 115
Compacting, 181, 707
Comparators, 88–89, 91–93, 96
Compound sine plate, 81
Computer-age machining, 6, 519–567
  chucking and turning centers in, 543–548
  computerized numerical control. *See* Numerical control
  future of, 564–567
  history of, 520–521
  robots in, 554–559
  role of computers in, 520–522
Computer-aided design (CAD), 539–542, 565
  advantages of, 542
  components of, 540
  design data and testing for, 541–542
  designing the part for, 540–541
  future use of, 565
  testing function and components of, 542
Continuous path (contouring) control, 530
Contour bandsaw, 230–245
  accuracy and finish produced on, 238
  applications of, 234–235
  band filing on, 243–245
  blade of, 234–236
  coolants for, 233–234
  external sections, 237–238
  friction sawing on, 241–242
  high-speed sawing on, 242
  internal sections, 238–241
  job requirements for, 235
  to mount saw band, 237
Contour bandsaw (*Cont.*):
  parts of, 230–236
  power feed for, 234
Coolants, 233–234
Cooling, 218
Coordinate locating system, 502–504
Copper and copper alloys, 165–166, 667
Counterboring, 248, 285
Countersinking, 248, 285–286
Cratering, 414
Cubic boron nitride (CBN) cutting tools, 171, 208–210
Cubic boron nitride (CBN) grinding wheels, 583–585
Current, electrical, 687, 693–694
Cutting fluids, 167, 214–224
  applications of, 220–224
  characteristics of good, 216
  chemical, 217–218
  for drilling machine, 270
  effects of, 167
  functions of, 167, 218–220
  history of, 214
  oils, 216
  rancidity control, 219–220
  types of, 167, 216–218
  for various metals, 223
Cutting-off or parting tools, 226–227, 307–308
Cutting speed, 178, 188, 203, 268–270, 300, 312–313, 407–409
Cutting tools, 168–177
  carbide. *See* Carbide cutting tools, cemented
  ceramic, 169, 200–204
  cermet, 169–170, 205
  cost of, 179–180
  cutting fluids and. *See* Cutting fluids
  diamond, 197–198
  drill, 176
  economic performance of, 179–180
  finishing with hand, 146–151
  geometry of, 185–188
  hand, 135–140
  lathe. *See* Lathe cutting tools
  life of, 174, 178–180
  materials used for, 168
  polycrystalline cubic boron nitride (PCBN), 207–217
  polycrystalline diamond (PCD), 211–213
  shape of, 173–174
  thread, 141–146

Decalescence point, 650, 651
Decimal (metric) systems, 39–40, 42, 525–526
Design:
  computer-aided (CAD), 539–542
  future of, 565
Destructive testing, 662–666
Dial indicators, 89–91, 320–321, 623
Diamond cutting tools, 197–198
Diamonds, 197–198, 573–574

Die sinking, 489–490
Dielectric fluid, 694–695, 698
Digital readout boxes and systems, 295, 296, 488
Direct-reading systems, 64–65, 295, 296
Dividers, 113–114
Dividing head, indexing or, 399, 426–434
Dovetails, milling, 486–487
Drafting, terms and symbols for, 29–31
Drawings, engineering, 28–31
Drilling on drilling machine or drill press, 4, 247–290
- accessories for, 253–258
- to accurate layout, 275–276
- boring, 248, 288
- buffing, 289–290
- chucks and, 254–255
- counterboring, 248, 285
- countersinking, 248, 284–286
- cutting fluids for, 221–222, 270
- cutting speed of, 268–269
- drill drift, 255
- drill jigs, 256, 290
- drill presses, 248–252
- drill sleeves, 255
- drill socket, 255
- drill vise, 255
- feed, 270
- hints for, 272–273
- holes, 271–278
- lathe center holes, 273–274
- measuring size of drill, 273
- numerically controlled, 252
- operations, 285–290
- radial, 251–252
- reaming, 248
- round work in a V-block, 277–278
- safety in, 272
- sensitive, 249–250
- spot-facing, 248
- spotting hole location with center drill, 274–275
- standard operations of, 248
- tapping, 248, 286–288
- tool-holding devices for, 254–255
- transferring hole locations, 288–289
- twist drills, 259–267, 288
- types of, 248–252
- upright, 250–251
- work held in vise, 275
- work-holding devices, 255–257

Eccentrics cut on lathes, 375–376
Edgefinder, 500
Electric spark discharge forming, 702–703
Electrical current, 687, 693–694
Electrical discharge machining (EDM), 690–699
Electro-band machining, 244–245
Electro-chemical machining (ECM), 684–686
Electrode, 693, 698–699
Electrolyte, 685–686, 687–688
Electrolytic grinding, 686–690
Electromagnetic forming, 703–704
End mills, 405
Engineering drawings, 28–31
Explosive forming, 701–702

Factories of the future, 564–567
Files, 137–140
Filing, 135
Finishing processes, hand, 146–151
Fire prevention, 26
Fit, 30
Fixtures, 397
Flycutters, 406
Follower rest, 373
Forming processes, 700–709
Friction, 166

Gage blocks, 72–76, 125–127
Gages, 67, 82–87, 115–116
- air, 94–96
- Brown and Sharpe clearance, 623
- care of, 85
- cylindrical plus, 83–84
- depth, 67–68
- dial bore, 67
- fixed, 83
- height, 69–71, 90–91
- micrometer depth, 67–69
- plain ring, 84
- precision height, 71
- small hole, 66
- snap, 86–87
- Starett cutter clearance, 623–624
- surface, 115–116
- taper plug, 84–85
- taper ring, 85
- telescope, 66–67
- thread plug, 86
- thread ring, 86
- vernier depth, 69
- vernier height, 69–71, 123–127

Gang drills, 250–251
Gear cutting and gear cutters, 441–447, 454–464
- gear tooth measurement, 447
- metric, 444–447
- rules and formulas for, 439–441
- terminology for, 438
- types of, 436–438
- worm, 356, 470–471

Graduated micrometer collars, 314–315
Grinders and grinding:
- abrasives for. *See* Abrasives
- of angular surface, 605
- of cemented-carbide tools, 192–194
- centerless grinders, 615–617
- center-type grinders, 609–610
- of ceramic tools, 204
- circle, 622–623
- clearance in, 621–623
- common types of, 6
- cutter grinding operations and setups, 624–630
- cutting fluids for, 222–224
- cutting-off operations, 607
- of cutting tools, 618–630

Grinders and grinding (*Cont.*):
- cylindrical, 6, 608–617, 628–629
- edges of workpiece, 600–603
- electrolytic, 686–690
- flat (horizontal) surface, 600
- form, 606–607, 689
- form-relieved cutter, 628
- grinding fluids, 595–597
- grinding operations, 600–606
- hollow, 622
- internal, 613–615, 629–630
- jig, 509–517
- on lathe, 389–390
- of lathe center, 389–390
- of lathe cutting tools, 321–323
- machine preparation for, 610–613
- to mount workpiece, 599
- outfeed on, 571
- parallel internal diameter, 613
- parallel outside diameter, 610–611
- plain helical milling cutter on, 624–626
- plunge, 512–513, 613, 689
- problems in, 601, 611–613, 614–615
- process, 588
- safety and, 599–600
- staggered-tooth cutter on, 626–627
- surface finish, 6, 597
- tapered hole, 615
- tapered workpiece, 611
- traverse, 689
- types of, 587–597, 618–620
- universal cutter and tool, 618–630
- universal cylindrical, 609–610, 613
- vertical surface, 603–604
- work-holding devices for, 593–595

Grinding wheels:
- balancing, 591–592
- bond types, 576, 582
- care of, 590–593
- for cemented-carbide tools, 192–193
- cubic boron nitride (CBN), 583–585
- diamond, 582–583
- to dress a convex radius on, 606–607
- for electrolytic grinding, 687
- grade of, 576–577
- handling and storage of, 582
- inspection of, 582
- for jig grinding, 513–514
- manufacture of, 578–579
- mounting, 591, 610–611
- selecting, 578, 581–582
- shapes of, standard, 579–580
- structure of, 577–578
- truing and dressing, 592–593, 687

Grooving, 337–338
Group technology (GT), 566

Hacksaw, hand, 135–137
Hammers, 130–131
Hand tools and bench work, 129–155
- bearings, 152–155
- broaches, 146, 148–149
- cutting tools, 135–140
- filing, 137–140
- finishing processes, 146–151

Hand tools and bench work (*Cont.*):
  hacksaw, 135–137, 226
  hammers, 130–131
  holding, striking, and assembling tools, 130–134
  lapping, 146, 149–151
  pliers, 134
  polishing, 138–139
  reamers, 146–148
  sawing, 135–137
  scrapers, 140
  screwdrivers, 131–132
  tapes, 142–145
  thread-cutting tools, 141–146
  vise, 130
  wrenches, 132–133
Hardening and hardness testing, 651, 655–657, 659–662
Heat:
  generated during machining, 215
  milling cutter failure and, 412
  treating with, in powder metallurgy, 708
  in treatment of steel, 643–657
Helical gearing, 454–464
Helical milling, 448–464
Honing, 194
Hydraulic systems, 671–682

Impact tests, 665–666
Impregnation, 708
Inch system of measurement, 42
Incremental system, 531–532
Indexing or dividing head, 399, 426–434
Infiltration, 708
Inspectors and inspections, 15, 567
Instrument makers, 16
International System of Units (SI), 39–41
Interviews for jobs, 18–19
Intrimik, 65

Jig borer and jig boring, 487–489, 491–497, 498–508
  accessories for, 493–496
  backlash, to eliminate, 500
  coordinate locating system in, 502–503
  inserting shanks in spindle of, 493
  locating edge, methods of, 500–502
  making settings for, 505
  measurement and inspection of holes, 504–508
  parts of, 493
  procedure of, 504
  reamers in, 495–496
  to set up work for, 498–500
  on vertical milling machine, 487–489
  work-holding devices for, 496–497
Jig grinder and jig grinding, 509–517
  advantages of, 509
  allowances, 514–515
  depth-measuring devices on, 510–511
  diamond dressing arm on, 511
  grinding head outfeed, 510–511
Jig grinder and jig grinding (*Cont.*):
  grinding methods, 511–513
  grinding wheels for, 513–514
  parts of, 510
  sequence for, 515–517
  setting up work for, 515
  shouldered holes, 516
  tapered hole, 515–516
Job planning, 27–37
Jobs and job opportunities, 12–16, 17–19

Keyseats, 121–122, 480–482
Knurling, 336–337

Lapping, 146, 149–151
Lasers, 103–104, 710–713
Lathe centers, 273–274, 318–321
Lathe chucks, 303–306, 376–385
  to bore work in, 387–388
  cutting-off work in, 384–385
  to face work in, 382–383
  to mount and remove, 377–380
  mounting work in, 380–383
  reaming in, 388
  to rough-turn work in, 383
  to spot and drill work in, 386–387
  types of, 303–306
Lathe cutting tools:
  carbide, 183–184
  grinding, 321–323
  holding devices for, 307–311
  materials, 168–171
  nomenclature for, 171
  to set up, 324
Lathe dogs, 306–307
Lathe spindle nose tapers, 342
Lathe toolbits, 168–173, 308, 322–323
Lathe toolholders, 254–255, 307–308
Lathe toolposts, 308–311
Lathes, 293–390
  accessories for, 301–307, 373–376
  boring on, 387–388
  calculating machining time and, 315–316
  carriage driver on, 300
  center holes, 273–274
  computerized numerical control (CNC) of, 294, 295
  cutting fluids for, 221
  cutting speeds on, 300, 312–313
  depth of cut of, 313–314, 329–330
  drilling on, 176, 386–387
  eccentrics cut on, 375–376
  end and face milling on, 175–176
  engine, 4–5, 294
  facing on, 324–327
  feeds on, 300, 313
  filing on, 331–332
  finish turning on, 331
  form turning on, 338–346
  graduated micrometer collars for, 314–315
  grinding on, 389–390
  grooving on, 337–338
  hydraulic tracer, 340–341
  knurling on, 336–337
Lathes (*Cont.*):
  machining between centers, 328–329
  main purpose of, 293
  mounting work between centers, 329
  parallel turning on, 329–330
  parts of, 297–301
  plain milling on, 175
  planing on, 175
  polishing on, 332
  rough turning on, 331
  safety and, 316–318
  shear pins on, 301
  shoulder turning on, 332–335
  slip clutches on, 301
  taper turning on, 346–352
  tapping on, 389
  thread cutting on, 360–370
  turning on, 174–175
  types of, 294–295
  work-holding devices for, 302–303, 307–311
Layout tools and procedures, 111–127
  accessories for, 117–118
  basic, 119–122
  of casting having cored hole, 121
  center punches, 116–117
  combination set, 115
  dividers in, 113–114
  gage blocks in, 125–127
  hermaphrodite calipers, 114
  hole locations, 119–121
  keyseat in shaft, 121–122
  layout punch, 116–117
  layout solution, 112
  layout tables, 113
  precision, 122–127
  radii, 119–121
  scriber, 69, 113
  semiprecision, 119–122
  sine bar, 125–127
  slots, 119–121
  squares, 114
  surface gage, 115–116
  surface plates, 113
  trammels, 113–114
  types of, 111
  vernier height gage, 123–127
Lead and lead alloys, 667–668
Limits, 30, 82
Linear graduating, 433–434
Linefinder, 501
Lines used on shop drawings, 29
Lubricants and lubricating, 155, 218–219

Machinability of metals, 163–167
Machine control unit (MCU), 523, 699
Machine operator, 12–13
Machine shops:
  abbreviations used in, 31
  safety in, 22–26
  symbols used in, 30–31
Machine tools:
  history of, 2–10
  introduction to, 1

Machine tools (*Cont.*):
space-age, 6
See also Computer-age machining
Machine trade opportunities, 12–19
Machining:
cost analysis and, 180
electrical discharge (EDM), 690–699
electro-band, 244–245
electro-chemical (ECM), 684–686
principles of, 174–176
procedures for, 32–37
Machining centers, 393, 548–553
Machining time, calculating, 315–316
Machinist, 13–14
Magnesium, 668
Maintenance machinist, 13
Manufacturing systems, 560–563
Material-handling devices, 566–567
Measurement and measuring instruments, 39–109
angular, 77–81
basic, 42–46
calipers, 45–46, 59–63, 114. *See also* Micrometers
comparators, 88–89, 91–93, 96
comparison, 88–97
coordinate, 97–100
depth, 67–68
dial indicators, 89–91, 320–321, 623
direct-reading instruments, 64–65, 295, 296
English system, 39
fractional, 42–43
gage blocks, 72–76, 125–127
gages. *See* Gages
height, 69–71
inch system, 42
inside measurement, 64–67
International System of Units (SI), 39–41
Intrimik, 65
lasers, 103–104, 710–713
length, 44
measuring unit, 98–100
metric (decimal) systems, 39–40, 42, 525–526
micrometers, 46, 49–58, 80
optical flats, 101–103
precision, 51–109
protractors, 77–78
sine-bar, 78–81, 125–127
sine plate, 80, 81
sine tables, 81
squares, 47–49, 114
steel rules, 43–45
straightedge, 44–45, 49–50
surface finish, 107–109
surface plates, 50, 113
systems of, 39
transfer-type instruments, 65–66
Metal cutting:
physics of, 158–163
technology of, 156–224
terminology of, 159
Metal-cutting saws, 225–246, 404–405
Metal removal, electro-chemical process of, 686
Metals and metallurgy, 631–668
alloys, 164, 169, 666–668
annealing, 653–654
bearing, 668
destructive testing, 662–666
ferrous, manufacture of, 633–641
grain structure of, 164–166
hardening and hardness testing of, 651, 655–657, 659–662
impact tests, 665–666
machinability of, 163–167
nonferrous, 666–668
normalizing, 654
plastic flow of, 159–163
powder, 700, 704–709
spheroidizing, 654
steel. *See* Steel
tensile testing, 662–665
Metric (decimal) systems, 39–40, 525–526
Micrometers (micrometer calipers), 46, 49–58, 80
Microsine plate, 496
Milling and milling machines:
accessories and attachments for, 397–399
aligning table on universal, 419
aligning vise of, 419–420, 476
angular surface, 403
arbors on, 398, 417
backlash eliminator, 395–397
block square and parallel, 477–478
cam, 465–469
clutches, 471–472
cratering, 414
cutting fluids for, 222
cutting speeds of, 407–409
depth of cut on, 411
die sinking on, 489–490
dovetails, 405–406, 486–487
ends square, 478–480
face, 175–176, 403, 423
failure of, 411–414
feeds on, 409–411
flat surface, 422–423
gang, 425
gear cutting on, 441–447
gears on, 435–441
helical, 448–464, 624–626
hints for, 416
horizontal, 392
jig boring, 487–489
keyseats, cutting, 480–482
knee-and-column-type, 394
machining centers on, 393
manufacturing-type, 392–393
milling cutters, 400–406, 418–419, 620–630
mounting and removing arbor of, 417
parts of, 394–395
plain milling, 175, 401
rack milling, 398, 469–470
radius milling, 485–487
Milling and milling machines (*Cont.*):
rotary table, 483–485
safety for, 415–416
sawing on, 425–426
setting cutter to work surface, 421
setups for, 416
side milling, 402, 424
slitting on, 404–405, 425–426
slotting on, 398, 480–482
special-type, 393, 483–490
vernier scale in, 488–489
straddle milling, 424–425
T-slots, 405, 486
universal horizontal, 394
vertical, 473–482, 487–489
worm, 470–471
Moiré fringe pattern, 98

Nickel and nickel alloys, 668
Nitriding processes, 656–657
Normalizing, 654
Numerical control and computerized numerical control (NC/CNC), 16, 523–539
advantages and disadvantages of, 535–537
of chucking centers, 543–547
computer control for, 537–538
definition of, 523
direct (DNC), 538–539
of drilling machine, 252
input media for, 531–533
of machining center, 551
measurement fundamentals for, 524–529
performance and, 533–537
productivity and, 533–535
programmers and programming for, 16, 530–531
recent advances in, 537
theory of, 524
turning centers, 547–548
types of controls, 529–531, 537–539
types of systems, 524

Open-loop systems, 524
Overcut, 695–696

Parallels, 117
Pig iron, 633–634
Plating, 708
Pliers, 134
Polishing, 138–139, 244, 332
Polycrystalline cutting tools, 207–213
Powder metallurgy, 700, 704–709
Power feed, 234
Precision surface-sensing probe, 552–553
Prick punch, 116–117
Processing center, 553
Production in future, 566
Professions, 16
Protractors, 77–78
Pumps in hydraulic system, 674–675
Punches, 116–117, 159–160

Quality control, 567
Quenching, 651

Rack milling, 398, 469–470
Radius milling, 485–486
Reamers, 146–148, 495–496
Reaming, 221–222, 248, 278–290, 388
Recalescence point, 650–651
Repressing, 708
Résumé, job, 17–18
Robots and robotics, 554–559
Rotary table, 483–485, 496–497
Rust control, 219

Safety, 21–26. *See also specific materials and procedures*
Saws and sawing, 135–137, 228–229
  abrasive cutoff, 226
  blades, 228
  cold circular cutoff, 226–227
  horizontal bandsaw, 226
  metal-cutting, 225–246, 404–405
  on milling machine, 404–405
  power hacksaw, 226
Scale size, 30
Scrapers, 140
Screwdrivers, 131–132
Scribers, 113
Sensing devices, 566
Servomechanism, 693, 698
Shear pins, 301
SI (International System of Units), 39–41
Silicon carbide, 571–572
Sine bar, 78–81, 125–127
Sine plate, 80, 81
Sine tables, 81
Sintering, 707–708
Sizing, 708
Slip clutches, 301
Slitting, 404–405, 425–426
Slotting, 398, 480–482
Spheroidizing, 654
Spot-facing, 248
Squares, 47–49, 114
Steady rest, 372–373
Steel:
  air-hardening, 646
  alloy, 164, 641–642
  carbon, 641, 648–655
  carbonitriding processes, 656–657
  chemical composition of, 640–641
  classification of, 641, 647–648
  flame-hardening, 657
  heat treatment of, 643–657
  high-carbon, 164, 641
  high-speed, 646–647
  induction-hardening, 657
Steel (*Cont.*):
  low-carbon, 164, 166
  machine, 164, 166, 641
  manufacture of, 635–640
  nitriding processes, 656–657
  oil-hardening, 646
  properties of, 632–633
  quenching, 651
  surface-hardening, 657
  tempering, 652–653, 654
  tensile strength of, 662–665
  tool, 166, 641, 646–647
  water-hardening, 646
  See also Metals and metallurgy
Steel manufacturing, 635–639
Steel rules, 43–45
Step blocks, 256
Straightedge, 44–45, 49–50
Strap clamps, 496
Straps, 256–257
Surface finish, 105–109, 167, 597
  definitions of, 105–107
  electrical discharge machining (EDM), 696
  electrolytic grinding, 688, 689
  measurement of, 107–109
  in various machining operations, 109
Surface grinders and grinding. *See* Grinders and grinding
Surface plates, 50, 113
Symbols, drafting, 30–31

T-slot, 405, 486
Tap disintegrators, 145
Tap extractor, 144
Tapers, 341–352
Tapping and taps, 142–145, 248, 286–288, 389
Technologist, 15
Temperature, 75, 166
Tempering, 652–653, 654
Tensile testing, 662–665
Threading dies, 145–146
Threads and thread cutting, 141–146, 352–370
  Acme, 368–369
  American National Standard, 142, 355
  British Standard Whitworth (BSW), 355
  Brown and Sharpe worm, 356
  calculations for, 358–359
  to convert from inch to metric, 362–363
  external, 145–146
  forms, 354–356
Threads and thread cutting (*Cont.*):
  internal, 142–145, 369–370
  international metric, 356
  ISO, 354, 355
  ISO metric tolerances and allowances, 357–358
  lathe setup for, 360
  on left-hand thread, 363
  measurement of, 364–365
  multiple, 365–366
  to reset threading tool, 362
  square, 355, 366–368
  on tapered section, 363–364
  terminology, 353–354
  thread-chasing dial, 359, 366
  Unified, 142, 355
Tin and tin alloys, 668
Tolerance, 30, 82–83
Tool design, future of, 566–567
Tool life, 174, 178–180
Toolbits, 168–173, 308, 322–323
Toolholders, 254–255, 307–308
Toolposts, 308–311
Torque control machining, 552, 566
Trammels, 113–114
Transfer punches, 288–289
Transfer screws, 289
Trial cut method, 320
Twist drills, 259–267, 288

Universal cutter and tool grinder, 618–630

V-blocks, 117–118, 255–256, 277–278, 496, 593
Valves in hydraulic systems, 675–677
Vertical milling, 473–482, 487–490
Viscosity of oil, 680–681
Vises:
  angle, 255
  bench, 130
  contour, 255
  drill press, 255
  drilling work held in, 275
  for jig boring, 496–497
  milling machine, 398–399, 419–420
  vertical milling machine, 476

Woodruff keyseat cutters, 481–482
Work hardening of workpiece, 413–414
Work-holding devices, 255–257, 302–303, 307–311, 593–595
Worms and worm gears, 356, 470–471
Wrenches, 132–133

Zinc, 668
Zirconia-aluminum oxide, 572